Linear Systems

McGraw-Hill Series in Electrical and Computer Engineering

Senior Consulting Editor
Stephen W. Director, University of Michigan, Ann Arbor

Circuits and Systems
Communications and Signal Processing
Computer Engineering
Control Theory
Electromagnetics
Electronics and VLSI Circuits
Introductory
Power and Energy
Radar and Antennas

Previous Consulting Editors

Ronald N. Bracewell, Colin Cherry, James F. Gibbons, Willis W. Harman, Hubert Heffner, Edward W. Herold, John G. Linvill, Simon Ramo, Ronald A. Rohrer, Anthony E. Siegman, Charles Susskind, Frederick E. Terman, John G. Truxal, Ernst Weber, and John R. Whinnery

Control Theory

Senior Consulting Editor
Stephen W. Director, *University of Michigan, Ann Arbor*

Antsaklis and Michel: *Linear Systems*

Bracewell: *The Fourier Transform and Its Application*

D'Azzo and Houpis: *Linear Control System Analysis and Design: Conventional and Modern*

Emanuel and Leff: *Introduction to Feedback Control Systems*

Friedland: *Control System Design: An Introduction to State-Space Methods*

Goodman: *Introduction to Fourier Optics*

Houpis and Lemont: *Digital Control Systems: Theory, Hardware, Software*

Raven: *Automatic Control Engineering*

Rohrs, Melsa, and Schultz: *Linear Control Systems*

Saadat: *Computational Aids in Control Systems Using* MATLAB™

Schultz and Melsa: *State Functions and Linear Control Systems*

Sciaviccio and Siciliano: *Modeling and Control of Robot Manipulators*

Stadler: *Analytical Robotics and Mechatronics*

Vaccaro: *Digital Control: A State-Space Approach*

LINEAR SYSTEMS

Panos J. Antsaklis
Anthony N. Michel
University of Notre Dame

THE McGRAW-HILL COMPANIES, INC.

New York St. Louis San Francisco Auckland Bogotá Caracas
Lisbon London Madrid Mexico City Milan Montreal New Delhi
San Juan Singapore Sydney Tokyo Toronto

McGraw-Hill

A Division of The **McGraw-Hill** Companies

LINEAR SYSTEMS

This book is printed on acid-free paper.

1 2 3 4 5 6 7 8 9 DOC DOC 9 0 0 9 8 7

ISBN 0-07-041433-5

This book was set in Times Roman by Publication Services.
Louis Swaim was the production supervisor, Lynn Cox was the editor,
Deborah Chusid was the jacket designer.
Project supervision was done by Publication Services.

R. R. Donnelley & Sons, Crawfordsville, was the printer and binder.

Library of Congress Cataloging-in-Publication Data

Antsaklis, Panos J.
Linear systems/Panos J. Antsaklis, Anthony N. Michel.
p. cm. – (McGraw-Hill series in electrical and computer engineering. Control theory)
Includes index.
ISBN 0-07-041433-5
1. Linear control systems. 2. Control theory. 3. Signal processing. I. Michel, Anthony N. II. Series.
TJ220.A58 1997
629.8'32–dc21

97-3753
CIP

http://www.mhcollege.com

To Our Families

To
Melinda and our daughter Lily
and to my parents
Dr. Ioannis and Marina Antsaklis
Panos J. Antsaklis

To
Leone and our children
Mary, Kathy, John,
Tony, and Pat
Anthony N. Michel

Mechanics is the paradise of the mathematical sciences because by means of it one comes to the fruits of mathematics.

Leonardo da Vinci
1452–1519

CONTENTS

PREFACE

This text is intended primarily for first-year graduate students and advanced undergraduates in engineering who are interested in control systems, signal processing, and communication systems. It is also appropriate for students in applied mathematics, economics, and certain areas in the physical and biological sciences. Designed for a challenging, one-semester systems course, the book presents an introduction to systems theory, with an emphasis on control theory. It can also be used as supplementary material for advanced systems and control courses and as a general reference on the subject.

The prerequisites for using this book are topics covered in a typical undergraduate curriculum in engineering and the sciences: undergraduate-level differential equations, linear algebra, Laplace transforms, and the modeling of electric circuits and simple mechanical systems.

The study of linear systems is a foundation for several disciplines, including control and signal processing. It is therefore very important that the coverage of linear systems be comprehensive and give readers sufficient breadth and depth in analysis and synthesis techniques of such systems. We believe that the best preparation for this is a firm understanding of the fundamentals that govern the behavior of complex systems. Indeed, only a thorough understanding of system behavior enables one to take full advantage of the various options available in the design of the best kinds of control systems and signal processors. Therefore, the primary aim of this text is to provide an understanding of these fundamentals by emphasizing mathematical descriptions of systems and their properties.

In writing this book, our goal was to clearly present the fundamental concepts of systems theory in a self-contained text. In addition to covering the fundamental principles, we provide sufficient background in analysis and algebra, to enable readers to move on to advanced topics in the systems area. The book is designed to highlight the main results and distinguish them from supporting results and extensions. Furthermore, we present the material in a sufficiently broad context to give readers a clear picture of the dynamical behavior of linear systems and the limitations of such systems.

The theory of linear systems is a mature topic, and there are literally thousands of scholarly papers reporting research on this subject. This book emphasizes fundamental results that are widely accepted as essential to the subject. For those readers interested in further detail, the end-of-chapter material includes additional results in the exercise sections as well as pertinent references and notes.

The book covers both continuous-time and discrete-time systems, which may be time-varying or time-invariant. The material is organized in such a manner that it is possible to concentrate only on the time-invariant case, if desired. The time-invariant case is treated in separate sections, and the results are presented so that they can be developed independently of the time-varying case. This type of organization provides considerable flexibility in covering the material.

Although the text is designed to serve primarily the needs of graduate students, it should also prove valuable to researchers, and practitioners, we tried to make it easy to use, and the book should prove valuable for self-study. Many simple examples are included to clarify the material and to encourage readers to actively participate in the learning process. The exercises at the end of each chapter introduce additional supporting concepts and results and encourage readers to gain additional insight by using what was learned. The exercises also encourage readers to comment, interpret, and visualize results (e.g., responses), making use of computer programs to aid in calculations and the generation of graphical results, when appropriate.

Over the past several years, the material has been class-tested in a first-year graduate-level course on linear systems, and its development has been influenced greatly by student feedback. Although there are many ways of using this book in a course, we suggest in the following several useful guidelines. Because any course on linear systems will most likely serve students with different educational experiences from a variety of disciplines and institutions, Chapters 1 and 2 provide necessary background material and develop certain systems fundamentals. Armed with this foundation, we develop essential results on controllability and observability (Chapter 3), on state observers and state feedback (Chapter 4), and on realization of systems (Chapter 5). Chapters 6 and 7 address basic issues concerning stability (Chapter 6) and the representation of systems using polynomial matrices and matrix fractions (Chapter 7). The appendix presents supplementary material (concerning numerical aspects).

How to use this book

At the beginning of each chapter is a detailed description of the chapter's contents, along with guidelines for readers. This material should be consulted when designing a course based on this book. In the following we give a general overview of the book's contents, with suggested topics for an introductory, one-semester course in linear systems.

From Chapter 1, covering a first course in linear systems should include the following: all the material on systems (Section 1.1); the material on initial-value problems (Sections 1.3 and 1.4); the material on systems of linear first-order ordinary differential equations (Sections 1.11, 1.12, and 1.13); the material on state equation descriptions of continuous-time systems (Section 1.14) and discrete-time systems (Section 1.15); and the material on input-output descriptions of systems (Section 1.16). The mathematical background material in Sections 1.2, 1.5 and Subsections 1.10A to 1.10C is included for review and to establish some needed notation. This material should not require formal class time. In Subsection 1.10D [dealing with existence, continuation, uniqueness, and continuous dependence (on initial conditions and parameters) of solutions of initial-value problems], the coverage should emphasize the results and their implications rather than the proofs of those results.

From Chapter 2, a first course in linear systems should include essentially all the material from the following sections: Section 2.3 (dealing with systems of linear homogeneous and nonhomogeneous first-order ordinary differential equations); Section 2.4 (dealing with systems of linear first-order ordinary differential equations with constant coefficients); and Sections 2.6 and 2.7 (which address the state equation description, the input-output description, and important properties, such as asymptotic stability of continuous-time and discrete-time linear systems, respectively). Section 2.5 (concerned with linear periodic systems) may be omitted without any loss of continuity. As in Chapter 1, the mathematical background material in Section 2.2 (dealing with linear algebra and matrices) is included for review and to establish some important notation and should not require much formal class time.

For Chapter 3, it is best to consider the material in two parts. From Part 1 include Section 3.1 (where the concepts of controllability and observability are introduced); and Subsections 3.2B and 3.3B (where these concepts are developed in greater detail for continuous-time time-invariant systems). From Part 2 include Subsections 3.4A and 3.4D (where special forms of system descriptions are considered); and Subsection 3.4B (where an additional controllability and observability test is presented). Similarly, the course should include the following material from Chapter 4: Section 4.1 (where state feedback and state observers are introduced); Subsections 4.2A and 4.2B (linear state feedback and eigenvalue assignment by state feedback are treated in detail); and Subsection 4.3A (where the emphasis is on identity observers); and Section 4.4 (where observer-based controllers are developed). Also, the course should include material from Chapter 5: Section 5.2 (where realization theory is introduced); Section 5.3 (where the existence, minimality, and the order of minimal realizations are developed); and Subsections 5.4A, 5.4B, 5.4C, and 5.4E (where realization algorithms are presented).

The material outlined above constitutes the major portion of a first course in linear systems. The course is rounded out, if time permits with selected topics from Chapter 6 (stability) and Chapter 7 (polynomial matrix system descriptions and fractional representations of transfer function matrices of linear time-invariant systems). The choice of these topics, and where they are presented throughout the course, depends on the interests of the instructor and the students.

We have been using this textbook in a one-semester first-year graduate course in electrical engineering. Typically, we spend the first half of the course on Chapter 1, Chapter 2, and Part 1 of Chapter 3. The second half of the course is devoted to Part 2 of Chapter 3 and Chapters 4 and 5. Selected topics from stability theory and matrix fractional descriptions of systems from Chapters 6 and 7 are included, as needed. Detailed coverage of Chapters 1 and 2, with only selective coverage of Chapters 3, 4, and 5, would be appropriate in a course that emphasizes mathematical systems theory.

Chapter 6 can also be used as an introduction to a second-level graduate course on nonlinear systems and stability. Similarly, Chapter 7 stands alone and can be used in an advanced linear systems course or as an introduction to a multi-input/multi-output linear control course. There is enough material in Chapters 3 through 7 for courses taught at several levels in a graduate program.

Acknowledgments

We are indebted to our students for their feedback and constructive suggestions during the evolution of this book. In particular, we would like to thank B. Hu, I. Konstantopoulos, and X. Koutsoukos and also Dr. K. Wang for their help and suggestions during the final editing of the manuscript. Special thanks go to Clarice Staunton for her patience in typing the many versions of our manuscript. We are also very appreciative of our excellent working relation with the staff of McGraw-Hill, especially with Lynn Cox, the Electrical Engineering Editor. Finally, we are both indebted to many individuals who have shaped our views of systems theory, in particular, Bill Wolovich, Brown University; Boyd Pearson, Rice University; David Mayne, Imperial College of the University of London (England); Sherman Wu, Marquette University; and Wolfgang Hahn, the Technical University of Graz (Austria).

Panos J. Antsaklis
Anthony N. Michel

CHAPTER 1

Mathematical Descriptions of Systems

The dynamical behavior of systems can be understood by studying their mathematical descriptions. The flight path of an airplane subject to certain engine thrust, rudder and elevator angles, and particular wind conditions, or the behavior of an automobile on cruise control when climbing a certain hill, can be predicted using mathematical descriptions of the pertinent behavior. Mathematical equations, typically differential or difference equations, are used to describe the behavior of processes and to predict their response to certain inputs. Although computer simulation is an excellent tool for verifying predicted behavior, and thus for enhancing our understanding of processes, it is certainly not an adequate substitute for generating the information captured in a mathematical model, when such a model is available. But computer simulations do complement mathematical descriptions. To be able to study the behavior of processes using mathematical descriptions, such as differential and difference equations, one needs a good working understanding of certain important mathematical concepts and procedures. Only in this way can one seriously attempt to study the behavior of complex systems and eventually design processes that exhibit the desired complex behavior.

This chapter develops mathematical descriptions for the types of systems with which we are concerned, namely, linear continuous-time and linear discrete-time finite-dimensional systems. Since such systems are frequently the result of a linearization process of nonlinear systems, or the result of the modeling process of physical systems in which the nonlinear effects have been suppressed or neglected, the origins of these linear systems are frequently nonlinear systems. For this reason, here and in Chapter 6, when we deal with certain qualitative aspects (such as existence, uniqueness, continuation, and continuity with respect to parameters of solutions of system equations, stability of an equilibrium, and so forth), we consider linear as well as nonlinear system models, although the remainder of the book deals exclusively with linear systems.

1.1 INTRODUCTION

A systematic study of (physical) *phenomena* usually begins with a *modeling process.* Examples of models include electric circuits consisting of interconnections of resistors, inductors, capacitors, transistors, diodes, voltage or current sources, etc.; mechanical circuits consisting of interconnections of point masses, springs, viscous dampers (dashpots), applied forces, etc.; and verbal characterizations of economic and societal systems, among others. Next, appropriate *laws* or *principles* are invoked to generate *equations* that describe the models (e.g., Kirchhoff's current and voltage laws, Newton's laws, conservation laws, and so forth). When using an expression such as "we consider a *system* described by ordinary differential equations," we will have in mind a phenomenon described by an appropriate set of ordinary differential equations (not the description of the physical phenomenon itself).

A. Physical Processes, Models, and Mathematical Descriptions

A physical process (physical system) will typically give rise to several different models, depending on what questions are being asked. For instance, in the study of the voltage-current characteristics of a transistor (the physical process), one may utilize a circuit (the model) that is valid at low frequencies or a circuit (a second model) that is valid at high frequencies; alternatively, if semiconductor impurities are of interest, a third model, quite different from the preceding two, is appropriate.

Over the centuries, a great deal of progress has been made in developing mathematical descriptions of physical phenomena (using models of such phenomena). In doing so, we have invoked laws (or principles) of physics, chemistry, biology, economics, etc., to derive mathematical expressions (usually equations) that characterize the evolution (in time) of the variables of interest. The availability of such mathematical descriptions enables us to make use of the vast resources offered by the many areas of applied and pure mathematics to conduct qualitative and quantitative studies of the behavior of processes. A given model of a physical process may give rise to several different mathematical descriptions. For example, when applying Kirchhoff's voltage and current laws to the low-frequency transistor model mentioned earlier, one can derive a set of differential and algebraic equations, or a set consisting of only differential equations, or a set of integro-differential equations, and so forth. *This process of mathematical modeling, "from a physical phenomenon to a model to a mathematical description," is essential in science and engineering.* To capture phenomena of interest accurately and in tractable mathematical form is a demanding task, as can be imagined, and requires a thorough understanding of the physical process involved. For this reason, the mathematical description of complex electrical systems, such as power systems, is typically accomplished by electrical engineers, the equations of flight dynamics of an aircraft are derived by aeronautical engineers, the equations of chemical processes are arrived at by chemists and chemical engineers, and the equations that characterize the behavior of economic systems are provided by economists. In most nontrivial cases, this type of modeling process is close to an art form since a good mathematical description must be detailed enough to accurately describe the phenomena of interest and at the same time simple enough to be amenable to analysis. Depending on the applications on hand, a given mathematical description of a process may be further simplified before it is used in analysis and especially in design procedures. For example, using the finite element method, one can derive a set of first-order differential equations that describe the motion of a space antenna. Typically, such mathematical descriptions contain hundreds of differential equations. Whereas all of these equations are quite useful in simulating the motion of the antenna, a lower order model is more suitable for the control design that, for example, may aim to counteract the effects of certain disturbances. Simpler mathematical models are required mainly because of our inability to deal effectively with hundreds of variables and their interactions. In such simplified mathematical descriptions, only those variables (and their interactions) that have significant effects on the phenomena of interest are included.

A point that cannot be overemphasized is that the mathematical descriptions we will encounter characterize processes only approximately. Most often, this is the case because the complexity of physical systems defies exact mathematical formulation. In many other cases, however, it is our own choice that a mathematical description of a given process approximate the actual phenomena by only a certain desired degree of accuracy. As discussed earlier, this is done in the interest of mathematical simplicity. For example, in the description of RLC circuits, one could use nonlinear differential equations that take into consideration parasitic effects in the capacitors; however, most often it suffices to use linear ordinary differential equations with constant coefficients to describe the voltage-current relations of such circuits, since typically such a description provides an adequate approximation and since it is much easier to work with linear rather than nonlinear differential equations.

In this book it will generally be assumed that the mathematical description of a system in question is given. In other words, we assume that the modeling of the process in question has taken place and that equations describing the process are given. Our main objective will be to present a qualitative theory of an important class of systems—finite-dimensional linear systems—by studying the equations representing such systems.

B. Classification of Systems

For our purposes, a comprehensive classification of systems is not particularly illuminating. However, an enumeration of the more common classes of systems encountered in engineering and science may be quite useful, if for no other reason than to show that the classes of systems considered in this book, although very important, are quite specialized.

As pointed out earlier, the particular set of equations describing a given system will in general depend on the effects one wishes to capture. Thus, one can speak of *lumped parameter* or *finite-dimensional systems* and *distributed parameter* or *infinite-dimensional systems*; *continuous-time* and *discrete-time systems*; *linear* and *nonlinear systems*; *time-varying* and *time-invariant systems*; *deterministic* and *stochastic systems*; appropriate combinations of the above, called *hybrid systems*; and perhaps others.

The appropriate mathematical settings for finite-dimensional systems are finite-dimensional vector spaces, and for infinite-dimensional systems they are most often infinite-dimensional linear spaces. Continuous-time finite-dimensional systems are usually described by ordinary differential equations or certain kinds of integral equations, while discrete-time finite-dimensional systems are usually characterized by ordinary difference equations or discrete-time counterparts to those integral equations. Equations used to describe infinite dimensional systems include partial differential equations, Volterra integro-differential equations, functional differential equations, and so forth. Hybrid system descriptions involve two or more different types of equations. Nondeterministic systems are described by stochastic counterparts to those equations (e.g., Ito differential equations).

In a broader context, not addressed in this book, most of the systems described by the equations enumerated generate *dynamical systems*. It has become customary in the engineering literature to use the term "dynamical system" rather loosely, and it has even been applied to cases where the original definition does not exactly fit. (For a discussion of general dynamical systems, refer, e.g., to Michel and Wang [13].) We will address in this book dynamical systems determined by ordinary differential equations or ordinary difference equations, considered next.

C. Finite-Dimensional Systems

The dynamical systems we will be concerned with are *continuous-time* and *discrete-time finite-dimensional systems*—primarily *linear systems*. However, since such systems are frequently a consequence of a linearization process, it is important when dealing with fundamental qualitative issues that we have an understanding of the origins of such linear systems. In particular, when dealing with questions of existence and uniqueness of solutions of the equations describing a class of systems, and with stability properties of such systems, we may consider nonlinear models as well.

Continuous-time finite-dimensional dynamical systems that we will consider are described by equations of the form

$$\dot{x}_i = f_i(t, x_1, \ldots, x_n, u_1, \ldots, u_m), \qquad i = 1, \ldots, n, \tag{1.1a}$$

$$y_i = g_i(t, x_1, \ldots, x_n, u_1, \ldots, u_m), \qquad i = 1, \ldots, p, \tag{1.1b}$$

where u_i, $i = 1, \ldots, m$, denote *inputs* or *stimuli*; y_i, $i = 1, \ldots, p$, denote *outputs* or *responses*; x_i; $i = 1, \ldots, n$, denote *state variables*; t denotes *time*; $\dot{x}_i$ denotes the time derivative of x_i; f_i, $i = 1, \ldots, n$, are real-valued functions of $1 + n + m$ real variables; and g_i, $i = 1, \ldots, p$, are real-valued functions of $1 + n + m$ real variables. A complete description of such systems will usually also require a set of *initial conditions* $x_i(t_0) = x_{i0}$, $i = 1, \ldots, n$, where t_0 denotes *initial time*. We will elaborate later on restrictions that need to be imposed on the f_i, g_i, and u_i and on the origins of the term "state variables."

Equations (1.1a) and (1.1b) can be represented in vector form as

$$\dot{x} = f(t, x, u) \tag{1.2a}$$

$$y = g(t, x, u), \tag{1.2b}$$

where x is the *state vector* with components x_i, u is the *input vector* with components u_i, y is the *output vector* with components y_i, and f and g are vector-valued functions with components f_i and g_i, respectively. We call (1.2a) a *state equation* and (1.2b) an *output equation.*

Important special cases of (1.2a) and (1.2b) are the *linear time-varying state equation* and *output equation* given by

$$\dot{x} = A(t)x + B(t)u \tag{1.3a}$$

$$y = C(t)x + D(t)u, \tag{1.3b}$$

where A, B, C, and D are real $n \times n$, $n \times m$, $p \times n$, and $p \times m$ matrices, respectively, whose elements are time-varying. Restrictions on these matrices will be provided later.

Linear time-invariant state and output equations given by

$$\dot{x} = Ax + Bu \tag{1.4a}$$

$$y = Cx + Du \tag{1.4b}$$

constitute important special cases of (1.3a) and (1.3b), respectively.

Equations (1.3a), (1.3b) and (1.4a), (1.4b) may arise in the modeling process or they may be a consequence of *linearization* of (1.1a) and (1.1b).

Discrete-time finite-dimensional dynamical systems are described by equations of the form

$$x_i(k+1) = f_i(k, x_1(k), \ldots, x_n(k), u_1(k), \ldots, u_m(k)), \qquad i = 1, \ldots, n, \tag{1.5a}$$

$$y_i(k) = g_i(k, x_1(k), \ldots, x_n(k), u_1(k), \ldots, u_m(k)), \qquad i = 1, \ldots, p, \tag{1.5b}$$

or in vector form,

$$x(k+1) = f(k, x(k), u(k)) \tag{1.6a}$$

$$y(k) = g(k, x(k), u(k)), \tag{1.6b}$$

where k is an integer that denotes *discrete time* and all other symbols are defined as before. A complete description of such systems involves a set of *initial conditions* $x(k_0) = x_{k_0}$, where k_0 denotes *initial time.* The corresponding linear time-varying and time-invariant state and output equations are given by

$$x(k+1) = A(k)x(k) + B(k)u(k) \tag{1.7a}$$

$$y(k) = C(k)x(k) + D(k)u(k) \tag{1.7b}$$

and

$$x(k+1) = Ax(k) + Bu(k) \tag{1.8a}$$

$$y(k) = Cx(k) + Du(k), \tag{1.8b}$$

respectively, where all symbols in (1.7a), (1.7b) and in (1.8a), (1.8b) are defined as in (1.3a), (1.3b) and (1.4a), (1.4b), respectively.

This type of system characterization is called *state-space description* or *state-variable description* or *internal description* of finite-dimensional systems. Another way of describing continuous-time and discrete-time finite-dimensional dynamical systems involves operators that establish a relationship between the system inputs and outputs. Such characterization, called *input-output description,* or *external description* of a system, will be addressed later in this chapter.

D. Chapter Description

In this book we will make liberal use of certain aspects of analysis and algebra. To help the reader recall some of these facts, we will provide throughout such background material as needed. This is done, e.g., in the *second section,* where we provide some of the notation used and where we recall certain facts concerning continuous functions.

In the *third section* we present the initial-value problem for nth-order ordinary differential equations and for systems of first-order ordinary differential equations, and we give a classification of ordinary differential equations. We also show that the study of nth-order ordinary differential equations can be reduced to the study of systems of first-order ordinary differential equations.

In the *fourth section* we give several specific examples of initial-value problems determined by ordinary differential equations.

In the *fifth section* we provide mathematical background material dealing with sequences, sequences of functions, and the Weierstrass M-test.

In the *sixth section* we establish conditions under which initial-value problems for ordinary differential equations possess solutions. This is accomplished in two stages. First, we establish an existence result for ϵ-approximate solutions, of which the Euler method is a special case. Next, we state and prove a preliminary result, called the Ascoli-Arzela Lemma, that we use, together with the existence result for ϵ-approximate solutions, to establish a result for the existence of solutions of initial-value problems. These solutions need not be unique. (This result is called the Peano-Cauchy Theorem.)

In the *seventh section* we make use of Zorn's Lemma to establish a result that enables us to determine the extent (in time) of the existence of solutions of initial-value problems. This is called continuation of solutions.

In the *eighth section* we prove a result that ensures the uniqueness of solutions of initial-value problems. In doing so, we utilize a useful result, called the Gronwall Inequality, that we also prove. One of the results of this section, called Picard iteration, provides a method of constructing solutions iteratively.

In the *ninth section* we show that under reasonable conditions the solutions of initial-value problems depend continuously on initial conditions and system parameters.

To simplify our presentation, we consider in Sections 6 to 9 the case of scalar first-order ordinary differential equations. In the *tenth section* we extend all results to the case of systems of first-order ordinary differential equations. In the process of accomplishing this, we introduce additional mathematical background material concerning vector spaces, normed linear spaces, and convergence on normed linear spaces.

The results in Sections 6 to 10 pertain to differential equations that in general are nonlinear. In the *eleventh section* we address linearization of such equations and provide several specific examples.

We utilize the results of Section 10 to establish in the *twelfth section* conditions for the existence, uniqueness, continuation, and continuity with respect to initial conditions and parameters of solutions of initial-value problems determined by linear ordinary differential equations.

In the *thirteenth section* we determine the solutions of linear ordinary differential equations. To arrive at some of our results (the Peano-Baker series and the matrix exponential), we make use of the Picard iteration considered in Sections 8 and 10. In this section we introduce for the first time the notions of state and state transition matrix. We also present the variations of constants formula for solving linear nonhomogeneous ordinary differential equations and introduce the notions of homogeneous and particular solutions.

Summarizing, the purpose of Sections 3 to 13 is to provide material dealing with ordinary differential equations and initial-value problems that is essential in the study of continuous-time finite-dimensional systems. This material enables us to give the state-space equation representation of continuous-time finite-dimensional systems. This is accomplished in the *fourteenth section.* We consider nonlinear as well as linear systems that may be time-varying or time-invariant.

In the *fifteenth section* we present the state-space equation representation of finite-dimensional discrete-time systems. In doing so, we introduce systems of first-order ordinary difference equations, nth-order ordinary difference equations, initial-value problems involving such equations, solutions of equations, the transition matrix, and so forth.

Finally, in the *sixteenth section* we consider an alternative description of the systems considered herein, called the input-output representation of systems. In the process of accomplishing this, we introduce several important general properties of systems (such as causality, systems with memory, linearity, time invariance, and so forth). We emphasize linear discrete-time and linear continuous-time systems. For the former we introduce the notion of pulse response, while for the latter we introduce the concepts of impulse response and the integral representation of linear continuous-time systems. For both continuous-time and discrete-time linear systems we make a connection between the state-space representation and the input-output description of systems, and we introduce the concept of system transfer function. In the first subsection of this section we encounter Dirac delta distributions.

This chapter has been organized in such a way that proofs of results may be omitted without much loss of continuity, should time constraints be a factor. However, the concepts (including the statements of most theorems) introduced in this chapter are of fundamental importance and will be utilized throughout the remainder of this book.

E. Guidelines for the Reader

In a first reading, certain material may be omitted without loss of continuity. Such material is identified throughout the book by starring the section or subsection title.

A typical graduate course in linear systems will include the following material from this chapter:

Mathematical description and classification of systems (Section 1.1).
Initial-value problems with examples (Sections 1.3 and 1.4).
Material on vector spaces and the results concerning existence and uniqueness of solutions of systems of first-order ordinary differential equations (Sections 1.10 and 1.12).
Linearization of nonlinear systems with examples (Section 1.11).

Solutions of the linear state equations $\dot{x} = A(t)x$ and $\dot{x} = A(t)x + g(t)$ (Section 1.13).

State-variable descriptions of continuous-time and discrete-time systems (Sections 1.14 and 1.15).

Input-output description of systems (Section 1.16).

1.2 PRELIMINARIES

We will employ a consistent notation and use certain facts from the calculus, analysis, and linear algebra. We will summarize this type of material, as needed, in various sections throughout the book. This is the first such section.

A. Notation

Let V and W be *sets*. Then $V \cup W$, $V \cap W$, $V - W$, and $V \times W$ denote the *union, intersection, difference,* and *Cartesian product* of V and W, respectively. If V is a *subset* of W, we write $V \subset W$; if x is an *element* of V, we write $x \in V$; and if x is not an element of V, we write $x \notin V$. We let V', ∂V, $\bar{V}$, and *int* V denote the *complement, boundary, closure,* and *interior* of V, respectively.

Let $\varnothing$ denote the *empty set,* let R denote the *real numbers,* let $R^+ = \{x \in R : x \geq 0\}$ (i.e., R^+ denotes the set of nonnegative real numbers), let Z denote the *integers,* and let $Z^+ = \{x \in Z : x \geq 0\}$.

We will let $J \subset R$ denote open, closed, or half-open *intervals.* Thus, for $a, b \in R$, $a \leq b$, J may be of the form $J = (a, b) = \{x \in R : a < x < b\}$, or $J = [a, b] = \{x \in R : a \leq x \leq b\}$, or $J = [a, b) = \{x \in R : a \leq x < b\}$, or $J = (a, b] = \{x \in R : a < x \leq b\}$.

Let R^n denote the real n-space. If $x \in R^n$, then

$$x = \begin{bmatrix} x_1 \\ \vdots \\ x_n \end{bmatrix}$$

and $x^T = (x_1, \ldots, x_n)$ denotes the *transpose* of the vector x. Also, let $R^{m \times n}$ denote the set of $m \times n$ real matrices. If $A \in R^{m \times n}$, then

$$A = [a_{ij}] = \begin{bmatrix} a_{11} & a_{12} & \cdots & a_{1n} \\ a_{21} & a_{22} & \cdots & a_{2n} \\ & & & \\ a_{m1} & a_{m2} & \cdots & a_{mn} \end{bmatrix}$$

and $A^T = [a_{ji}] \in R^{n \times m}$ denotes the *transpose* of the matrix A.

Similarly, let C^n denote the set of n-vectors with complex components and let $C^{m \times n}$ denote the set of $m \times n$ matrices with complex elements.

Let $f : V \to W$ denote a *mapping* or *function* from a set V into a set W, and denote by $D(f)$ and $R(f)$ the *domain* and the range of f, respectively. Also, let $f^{-1} : R(f) \to D(f)$, if it exists, denote the *inverse* of f.

B. Continuous Functions

First, let $J \subset R$ denote an open interval and consider a function $f : J \to R$. Recall that f is said to be *continuous at the point* $t_0 \in J$ if $\lim_{t \to t_0} f(t) = f(t_0)$ exists, i.e., if for every $\epsilon > 0$ there exists a $\delta > 0$ such that $|f(t) - f(t_0)| < \epsilon$ whenever $|t - t_0| < \delta$ and $t \in J$. The function f is said to be *continuous on J*, or simply *continuous*, if it is continuous at each point in J.

In the above definition, δ depends on the choice of t_0 and ϵ, i.e., $\delta = \delta(\epsilon, t_0)$. If at *each* $t_0 \in J$ it is true that there is a $\delta > 0$, independent of t_0 [i.e., $\delta = \delta(\epsilon)$], such that $|f(t) - f(t_0)| < \epsilon$ whenever $|t - t_0| < \delta$ and $t \in J$, then f is said to be *uniformly continuous* (on J).

Let

$$C(J, R) \triangleq \{f : J \to R \mid f \text{ is continuous on } J\}.$$

Now suppose that J contains one or both endpoints. Then continuity is interpreted as being one-sided at these points. For example, if $J = [a, b]$, then $f \in C(J, R)$ will mean that $f \in C((a, b), R)$ and that $\lim_{t \to a^+} f(t) = f(a)$ and $\lim_{t \to b^-} f(t) = f(b)$ exist.

With k any positive integer, and with J an open interval, we will use the notation

$$C^k(J, R) \triangleq \{f : J \to R \mid \text{ the derivative } f^{(j)} \text{ exists on } J \text{ and}$$
$$f^{(j)} \in C(J, R) \text{ for } j = 0, 1, \ldots, k, \text{ where } f^{(0)} \triangleq f\}$$

and will call f in this case a *C^k-function*. Also, we will call f a *piecewise C^k-function* if $f \in C^{k-1}(J, R)$ and $f^{(k-1)}$ has continuous derivatives for all $t \in J$ with the possible exception of a finite set of points where $f^{(k)}$ may have jump discontinuities. As before, when J contains one or both endpoints, then the existence and continuity of derivatives is one-sided at these points.

For any subset D of the n-space R^n with nonempty interior, we can define $C(D, R)$ and $C^k(D, R)$ in a similar manner as before. Thus, $f \in C(D, R)$ indicates that at every point $x_0 = (x_{10}, \ldots, x_{n0})^T \in D$, $\lim_{x \to x_0} f(x) = f(x_0)$ exists, or equivalently, at every $x_0 \in D$ it is true that for every $\epsilon > 0$ there exists a $\delta = \delta(\epsilon, x_0) > 0$ such that $|f(x) - f(x_0)| < \epsilon$ whenever $|x_1 - x_{10}| + \cdots + |x_n - x_{n0}| < \delta$ and $x \in D$. Also, we define $C^k(D, R)$ as

$$C^k(D, R) \triangleq \left\{f : D \to R \;\middle|\; \frac{\partial^j f}{\partial x_1^{i_1} \cdots \partial x_n^{i_n}} \in C(D, R), \qquad i_1 + \cdots + i_n = j,\right.$$
$$\left. j = 1, \ldots, k, \text{ and } f \in C(D, R)\right\}$$

(i.e., $i_1, \ldots, i_n$ take on all possible positive integer values such that their sum is j). When D contains its boundary (or part of its boundary), then the continuity of f and the existence and continuity of partial derivatives of f, $\partial^j f / \partial x_1^{i_1} \cdots \partial x_n^{i_n}$, $i_1 + \cdots + i_n = j$, $j = 1, \ldots, k$, will have to be interpreted in the appropriate way at the boundary points.

Recall that if $K \subset R^n$, $K \neq \varnothing$, and K is *compact* (i.e., K is closed and bounded), and if $f \in C(K, R)$, then f is uniformly continuous (on K) and f attains its maximum and minimum on K.

Finally, let D be a subset of R^n with nonempty interior and let $f : D \to R^m$. Then $f = (f_1, \ldots, f_m)^T$, where $f_i : D \to R$, $i = 1, \ldots, m$. We say that $f \in C(D, R^m)$ if

$f_i \in C(D, R)$, $i = 1, \ldots, m$, and that for some positive integer k, $f \in C^k(D, R^m)$ if $f_i \in C^k(D, R)$, $i = 1, \ldots, m$.

1.3 INITIAL-VALUE PROBLEMS

In this section we make precise the meaning of several concepts that arise in the study of continuous-time finite-dimensional dynamical systems.

A. Systems of First-Order Ordinary Differential Equations

Let $D \subset R^{n+1}$ denote a *domain,* i.e., an open, nonempty, and connected subset of R^{n+1}. We call R^{n+1} the *(t, x)-space*; we denote elements of R^{n+1} by (t, x) and elements of R^n by $x = (x_1, \ldots, x_n)^T$. Next, we consider the functions $f_i \in C(D, R)$, $i = 1, \ldots, n$, and if x_i is a function of t, let $x_i^{(n)} = d^n x_i/dt^n$ denote the nth derivative of x_i with respect to t (provided that it exists). In particular, when $n = 1$, we usually write

$$x_i^{(1)} = \dot{x}_i = \frac{dx_i}{dt}.$$

We call the system of equations given by

$$\dot{x}_i = f_i(t, x_1, \ldots, x_n), \qquad i = 1, \ldots, n, \tag{E_i}$$

a *system of n first-order ordinary differential equations.* By a *solution* of the system of equations (E_i) we shall mean n continuously differentiable functions $\phi_1, \ldots, \phi_n$ defined on an interval $J = (a, b)$ [i.e., $\phi \in C^1(J, R^n)$] such that $(t, \phi_1(t), \ldots, \phi_n(t)) \in D$ for all $t \in J$ and such that

$$\dot{\phi}_i(t) = f_i(t, \phi_1(t), \ldots, \phi_n(t)), \qquad i = 1, \ldots, n,$$

for all $t \in J$.

Next, we let $(t_0, x_{10}, \ldots, x_{n0}) \in D$. Then the *initial-value problem* associated with (E_i) is given by

$$\begin{aligned} \dot{x}_i &= f_i(t, x_1, \ldots, x_n), \qquad i = 1, \ldots, n, \\ x_i(t_0) &= x_{i_0}, \qquad i = 1, \ldots, n. \end{aligned} \tag{I_i}$$

A set of functions $\{\phi_1, \ldots, \phi_n\}$ is a *solution* of the initial-value problem (I_i) if $\{\phi_1, \ldots, \phi_n\}$ is a solution of (E_i) on some interval J containing t_0 and if $(\phi_1(t_0), \ldots, \phi_n(t_0)) = (x_{10}, \ldots, x_{n0})$.

In Fig. 1.1 the solution of a hypothetical initial-value problem is depicted graphically when $n = 1$. Note that $\dot{\phi}(\tau) = f(\tau, \tilde{x}) = \tan m$, where m is the slope of the line L that is tangent to the plot of the curve $\phi(t)$ vs. t, at the point $(\tau, \tilde{x})$.

In dealing with systems of equations, we will find it convenient to utilize the vector notation $x = (x_1, \ldots, x_n)^T$, $x_0 = (x_{10}, \ldots, x_{n_0})^T$, $\phi = (\phi_1, \ldots, \phi_n)^T$, $f(t, x) = (f_1(t, x_1, \ldots, x_n), \ldots, f_n(t, x_1, \ldots, x_n))^T = (f_1(t, x), \ldots, f_n(t, x))^T$, $\dot{x} = (\dot{x}_1, \ldots, \dot{x}_n)^T$, and $\int_{t_0}^{t} f(s, \phi(s))\, ds = [\int_{t_0}^{t} f_1(s, \phi(s))\, ds, \ldots, \int_{t_0}^{t} f_n(s, \phi(s))\, ds]^T$.

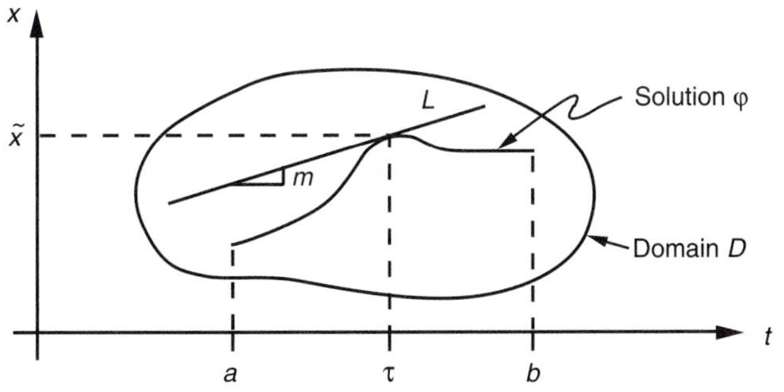

FIGURE 1.1
Solution of an initial-value problem when $n = 1$

With the above notation we can express the system of first-order ordinary differential equations (E_i) by

$$\dot{x} = f(t, x) \tag{E}$$

and the initial-value problem (I_i) by

$$\dot{x} = f(t, x), \qquad x(t_0) = x_0. \tag{I}$$

We leave it to the reader to prove that the initial-value problem (I) can be equivalently expressed by the *integral equation*

$$\phi(t) = x_0 + \int_{t_0}^{t} f(s, \phi(s))\, ds, \tag{V}$$

where ϕ denotes a solution of (I).

B. Classification of Systems of First-Order Ordinary Differential Equations

Systems of first-order ordinary differential equations have been classified in many ways. We enumerate here some of the more important cases.

If in (E), $f(t, x) \equiv f(x)$ for all $(t, x) \in D$, then

$$\dot{x} = f(x). \tag{A}$$

We call (A) an *autonomous system* of first-order ordinary differential equations.

If $(t+T, x) \in D$ whenever $(t, x) \in D$ and if $f(t, x) = f(t+T, x)$ for all $(t, x) \in D$, then (E) assumes the form

$$\dot{x} = f(t, x) = f(t + T, x). \tag{P}$$

We call such an equation a *periodic system* of first-order differential equations with *period* T. The smallest $T > 0$ for which (P) is true is called the *least period* of this system of equations.

When in (E), $f(t, x) = A(t)x$, where $A(t) = [a_{ij}(t)]$ is a real $n \times n$ matrix with elements a_{ij} that are defined and at least piecewise continuous on a t-interval J, then we have

$$\dot{x} = A(t)x \tag{LH}$$

and refer to (LH) as a *linear homogeneous system* of first-order ordinary differential equations.

If for (LH), $A(t)$ is defined for all real t, and if there is a $T > 0$ such that $A(t) = A(t + T)$ for all t, then we have

$$\dot{x} = A(t)x = A(t + T)x. \tag{LP}$$

This system is called a *linear periodic system* of first-order ordinary differential equations.

Next, if in (E), $f(t, x) = A(t)x + g(t)$, where $A(t)$ is as defined in (LH), and $g(t) = [g_1(t), \dots, g_n(t)]^T$ is a real n-vector with elements g_i that are defined and at least piecewise continuous on a t-interval J, then we have

$$\dot{x} = A(t)x + g(t). \tag{LN}$$

In this case we speak of a *linear nonhomogeneous system* of first-order ordinary differential equations.

Finally, if in (E), $f(t, x) = Ax$, where $A = [a_{ij}] \in R^{n \times n}$, then we have

$$\dot{x} = Ax. \tag{L}$$

This type of system is called a *linear, autonomous, homogeneous* system of first-order ordinary differential equations.

C. *n*th-Order Ordinary Differential Equations

Thus far we have been concerned with systems of first-order ordinary differential equations. It is also possible to characterize initial-value problems by means of nth-order ordinary differential equations. To this end we let h be a real function that is defined and continuous on a domain D of the real $(t, y, \dots, y_n)$-space [i.e., $D \subset R^{n+1}$, D is a domain, and $h \in C(D, R)$]. Then

$$y^{(n)} = h(t, y, y^{(1)}, \dots, y^{(n-1)}) \tag{E_n}$$

is an *n*th-*order ordinary differential equation.*

A *solution* of (E_n) is a function $\phi \in C^n(J, R)$ that satisfies $(t, \phi(t), \phi^{(1)}(t), \dots, \phi^{(n-1)}(t)) \in D$ for all $t \in J$ and

$$\phi^{(n)}(t) = h(t, \phi(t), \phi^{(1)}(t), \dots, \phi^{(n-1)}(t))$$

for all $t \in J$, where $J = (a, b)$ is a t-interval.

Now for a given $(t_0, x_{10}, \dots, x_{n0}) \in D$, the *initial-value problem* for (E_n) is

$$\begin{aligned} y^{(n)} &= h(t, y, y^{(1)}, \dots, y^{(n-1)}) \\ y(t_0) &= x_{10}, \dots, y^{(n-1)}(t_0) = x_{n0}. \end{aligned} \tag{I_n}$$

A function ϕ is a *solution* of (I_n) if ϕ is a solution of Eq. (E_n) on some interval containing t_0 and if $\phi(t_0) = x_{10}, \dots, \phi^{(n-1)}(t_0) = x_{n0}$.

As in the case of systems of first-order ordinary differential equations, we can point to several important special cases. Specifically, we consider equations of the form

$$y^{(n)} + a_{n-1}(t)y^{(n-1)} + \cdots + a_1(t)y^{(1)} + a_0(t)y = g(t), \tag{3.1}$$

where $a_i \in C(J, R)$, $i = 0, 1, \ldots, n-1$, and $g \in C(J, R)$. We refer to (3.1) as a *linear nonhomogeneous ordinary differential equation of order n.*

If in (3.1) we let $g(t) \equiv 0$, then

$$y^{(n)} + a_{n-1}(t)y^{(n-1)} + \cdots + a_1(t)y^{(1)} + a_0(t)y = 0. \tag{3.2}$$

We call (3.2) a *linear homogeneous ordinary differential equation of order n.*

If in (3.2) we have $a_i(t) \equiv a_i$, $i = 0, 1, \ldots, n-1$, then

$$y^{(n)} + a_{n-1}y^{(n-1)} + \cdots + a_1 y^{(1)} + a_0 y = 0, \tag{3.3}$$

and we call (3.3) a *linear, autonomous, homogeneous ordinary differential equation of order n.*

As in the case of systems of first-order ordinary differential equations, we can define *periodic* and *linear periodic ordinary differential equations of order n* in the obvious way.

It turns out that the theory of nth-order ordinary differential equations can be reduced to the theory of a system of n first-order ordinary differential equations. To demonstrate this, we let $y = x_1$, $y^{(1)} = x_2, \ldots, y^{(n-1)} = x_n$ in Eq. (I_n). We now obtain the system of first-order ordinary differential equations

$$\begin{aligned} \dot{x}_1 &= x_2 \\ \dot{x}_2 &= x_3 \\ &\vdots \\ \dot{x}_n &= h(t, x_1, \ldots, x_n) \end{aligned} \tag{3.4}$$

that is defined for all $(t, x_1, \ldots, x_n) \in D$. Assume that $\phi = (\phi_1, \ldots, \phi_n)^T$ is a solution of (3.4) on an interval J. Since $\phi_2 = \dot{\phi}_1, \phi_3 = \dot{\phi}_2, \ldots, \phi_n = \phi_1^{(n-1)}$, and since

$$\begin{aligned} h(t, \phi_1(t), \ldots, \phi_n(t)) &= h(t, \phi_1(t), \phi_1^{(1)}(t), \ldots, \phi_1^{(n-1)}(t)) \\ &= \phi_1^{(n)}(t), \end{aligned}$$

it follows that the first component ϕ_1 of the vector ϕ is a solution of Eq. (E_n) on the interval J. Conversely, if ϕ_1 is a solution of (E_n) on J, then the vector $(\phi, \phi^{(1)}, \ldots, \phi^{(n-1)})^T$ is clearly a solution of (3.4). Moreover, if $\phi_1(t_0) = x_{10}, \ldots, \phi_1^{(n-1)}(t_0) = x_{n0}$, then the vector ϕ satisfies $\phi(t_0) = x_0 = (x_{10}, \ldots, x_{n0})^T$.

1.4 EXAMPLES OF INITIAL-VALUE PROBLEMS

We now give several specific examples of initial-value problems.

EXAMPLE 4.1. The mechanical system of Fig. 1.2 consists of two point masses M_1 and M_2 that are acted upon by viscous damping forces (determined by viscous damping constants B, B_1, and B_2), spring forces (specified by the spring constants K, K_1, and K_2), and external forces f_1 and f_2. The initial displacements of M_1 and M_2 at $t_0 = 0$ are given by $y_1(0)$ and $y_2(0)$, respectively, and their initial velocities are given by $\dot{y}_1(0)$

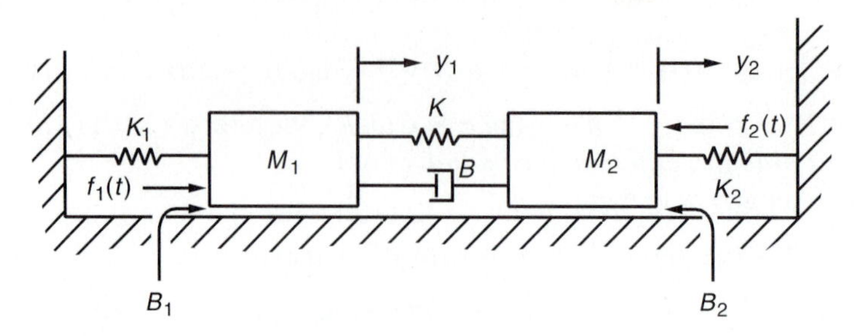

FIGURE 1.2
An example of a mechanical circuit

and $\dot{y}_2(0)$. The arrows in Fig. 1.2 indicate positive directions of displacement for M_1 and M_2.

Newton's second law yields the following coupled second-order ordinary differential equations that describe the motions of the masses in Fig. 1.2 (letting $y^{(2)} = d^2y/dt^2 = \ddot{y}$),

$$\begin{aligned} M_1\ddot{y}_1 + (B + B_1)\dot{y}_1 + (K + K_1)y_1 - B\dot{y}_2 - Ky_2 &= f_1(t) \\ M_2\ddot{y}_2 + (B + B_2)\dot{y}_2 + (K + K_2)y_2 - B_1\dot{y}_1 - Ky_1 &= -f_2(t) \end{aligned} \tag{4.1}$$

with initial data $y_1(0)$, $y_2(0)$, $\dot{y}_1(0)$, and $\dot{y}_2(0)$.

Letting $x_1 = y_1$, $x_2 = \dot{y}_1$, $x_3 = y_2$, and $x_4 = \dot{y}_2$, we can express Eq. (4.1) equivalently by the system of first-order ordinary differential equations

$$\begin{bmatrix} \dot{x}_1 \\ \dot{x}_2 \\ \dot{x}_3 \\ \dot{x}_4 \end{bmatrix} = \begin{bmatrix} 0 & 1 & 0 & 0 \\ -\dfrac{K_1 + K}{M_1} & -\dfrac{B_1 + B}{M_1} & \dfrac{K}{M_1} & \dfrac{B}{M_1} \\ 0 & 0 & 0 & 1 \\ \dfrac{K}{M_2} & \dfrac{B}{M_2} & -\dfrac{K + K_2}{M_2} & -\dfrac{B + B_2}{M_2} \end{bmatrix} \begin{bmatrix} x_1 \\ x_2 \\ x_3 \\ x_4 \end{bmatrix} + \begin{bmatrix} 0 \\ \dfrac{1}{M_1} f_1(t) \\ 0 \\ -\dfrac{1}{M_2} f_2(t) \end{bmatrix} \tag{4.2}$$

with initial data given by $x(0) = (x_1(0), x_2(0), x_3(0), x_4(0))^T$. ■

EXAMPLE 4.2. Using the node voltages v_1, v_2, and v_3 and applying Kirchhoff's current law, we can describe the behavior of the electric circuit given in Fig. 1.3 by the system of first-order ordinary differential equations

$$\begin{bmatrix} \dot{v}_1 \\ \dot{v}_2 \\ \dot{v}_3 \end{bmatrix} = \begin{bmatrix} -\dfrac{1}{C_1}\left(\dfrac{1}{R_1} + \dfrac{1}{R_2}\right) & \dfrac{1}{R_2C_1} & 0 \\ -\dfrac{1}{C_1}\left(\dfrac{1}{R_1} + \dfrac{1}{R_2}\right) & -\left(\dfrac{R_2}{L} - \dfrac{1}{R_2C_1}\right) & \dfrac{R_2}{L} \\ \dfrac{1}{R_2C_2} & -\dfrac{1}{R_2C_2} & 0 \end{bmatrix} \begin{bmatrix} v_1 \\ v_2 \\ v_3 \end{bmatrix} + \begin{bmatrix} \dfrac{v}{R_1C_1} \\ \dfrac{v}{R_1C_1} \\ 0 \end{bmatrix}. \tag{4.3}$$

To complete the description of this circuit, we specify the initial data at $t_0 = 0$, given by $v_1(0)$, $v_2(0)$, and $v_3(0)$. ■

EXAMPLE 4.3. Figure 1.4 represents a simplified model of an armature voltage-controlled dc servomotor consisting of a stationary field and a rotating armature and load. We assume that all effects of the field are negligible in the description of this

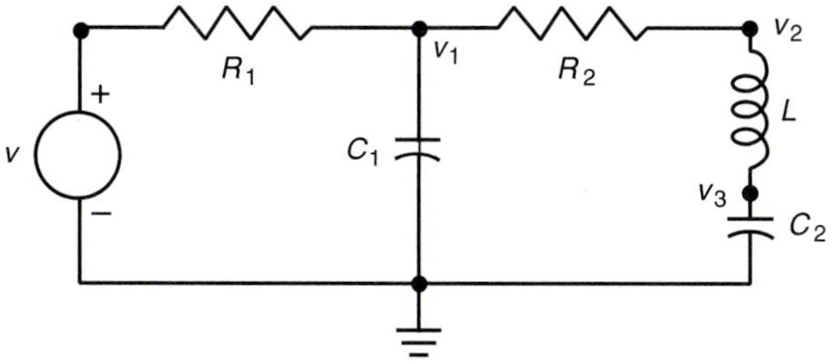

FIGURE 1.3
An example of an electric circuit

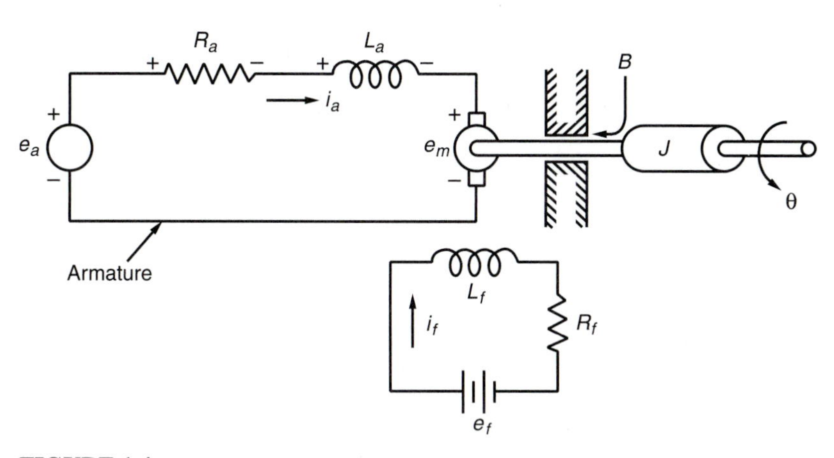

FIGURE 1.4
An example of an electromechanical system

system. The various parameters and variables in Fig. 1.4 are: e_a = externally applied armature voltage, i_a = armature current, R_a = resistance of the armature winding, L_a = armature winding inductance, e_m = back-emf voltage induced by the rotating armature winding, B = viscous damping due to bearing friction, J = moment of inertia of the armature and load, and θ = shaft position.

The back-emf voltage (with the polarity as shown) is given by

$$e_m = K_\theta \dot{\theta}, \tag{4.4}$$

where $K_\theta > 0$ is a constant and the torque T generated by the motor is given by

$$T = K_T i_a. \tag{4.5}$$

Application of Newton's second law and Kirchhoff's voltage law yields

$$J\ddot{\theta} + B\dot{\theta} = T(t) \tag{4.6}$$

and

$$L_a \frac{di_a}{dt} + R_a i_a + e_m = e_a. \tag{4.7}$$

Combining (4.4) to (4.7) and letting $x_1 = \theta$, $x_2 = \dot{\theta}$, and $x_3 = i_a$ yields the system of first-order ordinary differential equations

$$\begin{bmatrix} \dot{x}_1 \\ \dot{x}_2 \\ \dot{x}_3 \end{bmatrix} = \begin{bmatrix} 0 & 1 & 0 \\ 0 & -\dfrac{B}{J} & \dfrac{K_T}{J} \\ 0 & -\dfrac{K_\theta}{L_a} & -\dfrac{R_a}{L_a} \end{bmatrix} \begin{bmatrix} x_1 \\ x_2 \\ x_3 \end{bmatrix} + \begin{bmatrix} 0 \\ 0 \\ \dfrac{e_a}{L_a} \end{bmatrix}. \tag{4.8}$$

A suitable set of initial data for (4.8) is given by $t_0 = 0$ and $(x_1(0), x_2(0), x_3(0))^T = (\theta(0), \dot{\theta}(0), i_a(0))^T$. ■

EXAMPLE 4.4. A much studied ordinary differential equation is given by

$$\ddot{x} + f(x)\dot{x} + g(x) = 0, \tag{4.9}$$

where $f \in C^1(R, R)$ and $g \in C^1(R, R)$.

When $f(x) \geq 0$ for all $x \in R$ and $xg(x) > 0$ for all $x \neq 0$, then (4.9) is called the *Lienard Equation.* This equation can be used to represent, e.g., RLC circuits with nonlinear circuit elements.

Another important special case of (4.9) is the *van der Pol Equation* given by

$$\ddot{x} - \epsilon(1 - x^2)\dot{x} + x = 0, \tag{4.10}$$

where $\epsilon > 0$ is a parameter. This equation has been used to represent certain electronic oscillators.

If in (4.9), $f(x) \equiv 0$, we obtain

$$\ddot{x} + g(x) = 0. \tag{4.11}$$

When $xg(x) > 0$ for all $x \neq 0$, then (4.11) represents various models of so-called "mass on a nonlinear spring." In particular, if $g(x) = k(1+a^2x^2)x$, where $k > 0$ and $a^2 > 0$ are parameters, then g represents the restoring force of a *hard spring.* If $g(x) = k(1-a^2x^2)x$, where $k > 0$ and $a^2 > 0$ are parameters, then g represents the restoring force of a *soft spring.* Finally, if $g(x) = x$, then g represents the restoring force of a *linear spring.* (See Figs. 1.5 and 1.6.)

For another special case of (4.9), let $f(x) \equiv 0$ and $g(x) = k \sin x$, where $k > 0$ is a parameter. Then (4.9) assumes the form

$$\ddot{x} + k \sin x = 0. \tag{4.12}$$

This equation describes the motion of a point mass moving in a circular path about the axis of rotation normal to a constant gravitational field, as shown in Fig. 1.7. The parameter k depends on the radius l of the circular path, the gravitational acceleration g, and the mass. The symbol x denotes the angle of deflection measured from the vertical. The present model is called a *simple pendulum.*

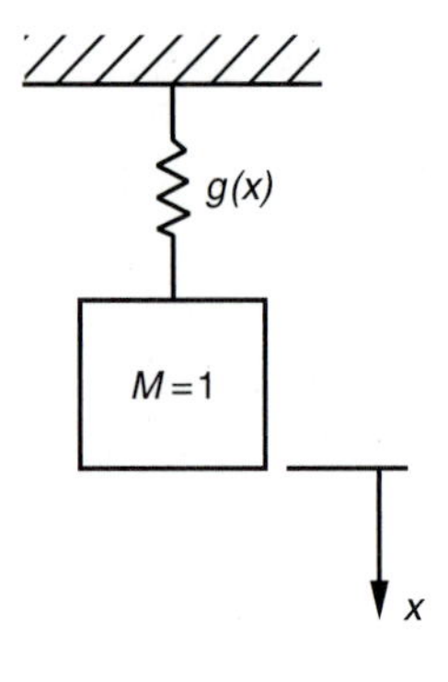

FIGURE 1.5
Mass on a nonlinear spring

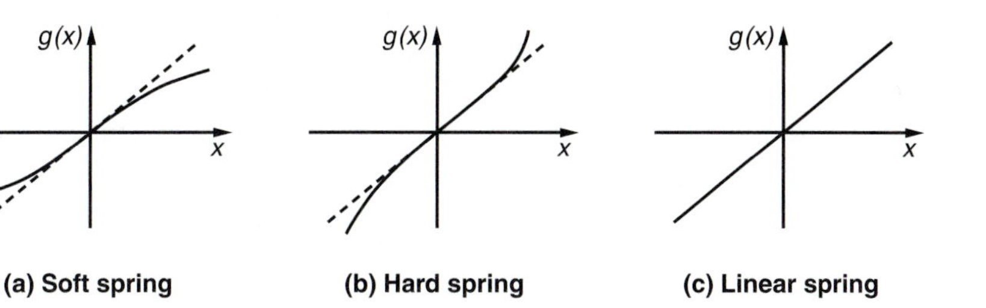

FIGURE 1.6

Letting $x_1 = x$ and $x_2 = \dot{x}$, the second-order ordinary differential equation (4.9) can be represented by the system of first-order ordinary differential equations given by

$$\begin{aligned} \dot{x}_1 &= x_2 \\ \dot{x}_2 &= -f(x_1)x_2 - g(x_1). \end{aligned} \tag{4.13}$$

The required initial data for (4.13) are given by $x_1(0)$ and $x_2(0)$.

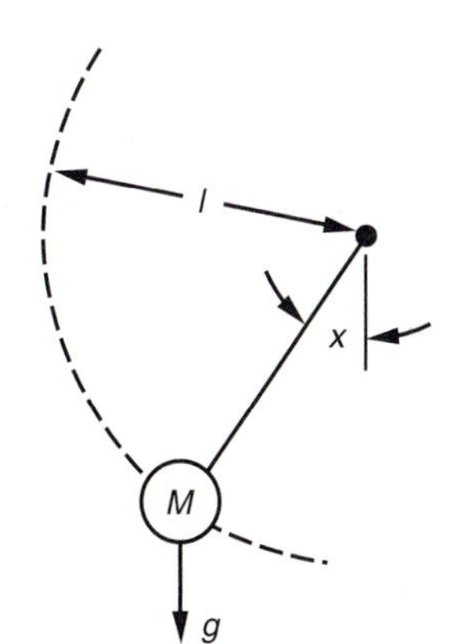

FIGURE 1.7
Model of a simple pendulum

■

*1.5 MORE MATHEMATICAL PRELIMINARIES

At this point, we need to review additional material from the calculus and analysis.

A. Sequences

Let I denote an *index set* (usually the set of positive integers). A *sequence* in R (i.e., a *real sequence*) is a mapping of I into R, say, $f(n) = x_n$. It is customary to denote such a sequence by $\{x_n\}$, rather than $\{f(n)\}$.

---

*Throughout the book, starred sections, subsections, or items may be omitted to conserve time without loss of continuity.

Let $\{x_n\}$ be a sequence in R, $n = 1, 2, 3, \ldots$, and let $n_1, n_2, \ldots, n_k, \ldots$ be a strictly increasing sequence of positive integers, i.e., $n_i > n_j$ whenever $i > j$. Then the sequence $\{x_{n_k}\}$ is called a *subsequence* of $\{x_n\}$.

Recall that a real sequence $\{x_n\}$ is said to *converge* to an element in R if for every $\epsilon > 0$ there is an integer N such that for all $n \geq N$, $|x - x_n| < \epsilon$. In general, N depends on ϵ, i.e., $N = N(\epsilon)$. We call x the *limit* of $\{x_n\}$, and we usually write this as $\lim_{n\to\infty} x_n = x$ or $x_n \to x$ as $n \to \infty$. If there is no $x \in R$ to which the sequence converges, then we say that $\{x_n\}$ *diverges.*

A real sequence $\{x_n\}$ is said to be a *Cauchy sequence* or a *fundamental sequence* if for every $\epsilon > 0$ there is an integer $N = N(\epsilon)$ such that $|x_n - x_m| < \epsilon$ whenever $m, n \geq N$.

It is easy to show that every convergent sequence is also a Cauchy sequence. One of the fundamental results in analysis shows that for R, the converse to this statement is also true: every real Cauchy sequence is a convergent sequence (i.e., it converges to an element in R). To express this property, we say that R is *complete.* Other important fundamental properties that follow from the completeness of R include the *Bolzano-Weierstrass (B-W) property* and the *Heine-Borel (H-B) property.* The *B-W property* states that every bounded sequence of real numbers contains a convergent subsequence.

To present the *H-B property,* we require the following additional concept: by a finite (or countable, or uncountable) *open covering* of a set $E \subset R$ we mean a finite (or countable, or uncountable) collection $\{G_\alpha\}$ of open sets such that $E \subset \cup_\alpha G_\alpha$. The *H-B property* states that every open covering of a compact set K contains a *finite* open subcovering of K. (Recall that a set $K \subset R$ is *compact* if it is closed and bounded.)

B. Sequences of Functions

Next, we consider *sequences of functions.* For our purposes, we let E be a nonempty subset of R, and we let $\{f_n\}$, $n = 1, 2, 3, \ldots$ denote a collection of real-valued functions defined on E (i.e., for each $n \in I$, where I denotes the positive integers, there is a mapping $f_n : E \to R$).

We say that the sequence of functions $\{f_n\}$ is *pointwise convergent* to a function f on E, if $\lim_{n\to\infty} f_n(t) = f(t)$ for all $t \in E$, i.e., if for every $\epsilon > 0$ and for every $t \in E$ there exists an integer N that may depend on ϵ and t [i.e., $N = N(\epsilon, t)$] such that $|f_n(t) - f(t)| < \epsilon$ whenever $n \geq N$.

The sequence $\{f_n\}$ is said to *converge uniformly* to a function f on E if for every $\epsilon > 0$ there is an integer N that depends only on ϵ [i.e., $N = N(\epsilon)$] such that $|f_n(t) - f(t)| < \epsilon$ whenever $n \geq N$ for all $t \in E$.

For example, the sequence $\{f_n\}$ specified by

$$f_n(t) = t^n, \qquad 0 \leq t \leq 1, \tag{5.1}$$

is *pointwise convergent* to the function

$$f(t) = \begin{cases} 0, & 0 \leq t < 1, \\ 1, & t = 1, \end{cases} \tag{5.2}$$

but it is not uniformly convergent to f. Note also that whereas for each $n = 1, 2, \ldots, f_n$ in (5.1) is continuous on $E = [0, 1]$, the limiting function f in (5.2) is not continuous on E.

As another example, we note that the sequence $\{f_n\}$ specified by $(n = 1, 2, 3, \ldots)$

$$f_n(t) = t + \frac{1}{n}, \qquad -\infty < t < \infty \tag{5.3}$$

is *pointwise convergent* and *uniformly convergent* to the function

$$f(t) = t, \qquad -\infty < t < \infty. \tag{5.4}$$

Note also that in the above example, the f_n, $n = 1, 2, 3, \ldots$ given in (5.3) as well as f given in (5.4) are continuous on R.

THEOREM 5.1. Let $f_n \in C(E, R)$, $n = 1, 2, \ldots$, and assume that the sequence $\{f_n\}$ converges uniformly to f on E. Then $f \in C(E, R)$.

Proof. Let $t_0 \in E$. We must show that $\lim_{t \to t_0} f(t) = f(t_0)$, or equivalently, we must show that for every $\epsilon > 0$, there exists a $\delta = \delta(\epsilon, t_0) > 0$ such that $|f(t) - f(t_0)| < \epsilon$ whenever $|t - t_0| < \delta$.

Since f_n converges to f uniformly on E, given $\epsilon > 0$, there exists $N = N(\epsilon)$ such that $|f_N(t) - f(t)| < \epsilon/3$ for all $t \in E$. Also, since $f_N \in C(E, R)$, there exists $\delta = \delta(\epsilon, t_0) > 0$ such that $|f_N(t) - f_N(t_0)| < \epsilon/3$ whenever $|t - t_0| < \delta$. Therefore, whenever $|t - t_0| < \delta$, we have

$$\begin{aligned} |f(t) - f(t_0)| &= |f(t) - f_N(t) + f_N(t) - f_N(t_0) + f_N(t_0) - f(t_0)| \\ &\le |f(t) - f_N(t)| + |f_N(t) - f_N(t_0)| + |f_N(t_0) - f(t_0)| \\ &< \frac{\epsilon}{3} + \frac{\epsilon}{3} + \frac{\epsilon}{3} = \epsilon. \end{aligned}$$

■

THEOREM 5.2. Let $f_n \in C(E, R)$, $n = 1, 2, \ldots$, and E be a bounded subset of R. Assume that the sequence $\{f_n\}$ converges uniformly to f on E. Then

$$\lim_{n\to\infty} \int_E f_n(t)\,dt = \int_E \lim_{n\to\infty} f_n(t)\,dt = \int_E f(t)\,dt. \tag{5.5}$$

Proof. We have

$$\int_E f_n(t)\,dt - \int_E f(t)\,dt = \int_E [f_n(t) - f(t)]\,dt.$$

Also, let $J = (a, b)$ denote a bounded interval with the property that $J \supset E$ and let $L(J) = (b - a)$.

Since f_n converges to f in t, uniformly on a bounded set E, we can choose for a given $\epsilon > 0$ an $N = N(\epsilon)$ such that $|f_n(t) - f(t)| < \epsilon/L(J)$ for all $t \in E$ whenever $n \ge N$. Therefore,

$$\begin{aligned} \left| \int_E f_n(t)\,dt - \int_E f(t)\,dt \right| &\le \int_E |f_n(t) - f(t)|\,dt \\ &\le \frac{\epsilon}{L(J)} L(J) = \epsilon \end{aligned}$$

whenever $n \ge N$. ■

As an example, consider

$$f_n(t) = \begin{cases} \dfrac{1}{n}, & 0 \le t \le n, \\ 0, & n < t, \end{cases}$$

$n = 1, 2, 3, \ldots$. The sequence $\{f_n\}$ converges uniformly to the function $f(t) = 0$ for all t. It is easily shown that

$$\lim_{n\to\infty}\int_0^\infty f_n(t)\,dt = 1$$

while
$$\int_0^\infty \lim_{n\to\infty} f_n(t)\,dt = 0.$$

Theorem 5.2 does not apply in this case, since the interval E is not bounded.

As another example, consider

$$f_n(t) = n^2 t e^{-nt}, \qquad 0 \le t \le 1,$$

$n = 1, 2, 3, \ldots$. The sequence $\{f_n\}$ converges pointwise on $[0, 1]$. It is easily shown in this case that

$$\lim_{n\to\infty}\int_0^1 f_n(t)\,dt = 1$$

while
$$\int_0^1 \lim_{n\to\infty} f_n(t)\,dt = 0.$$

Theorem 5.2 does not apply in this case, since $\{f_n\}$ is not uniformly convergent on $[0, 1]$.

The point of the above two examples is this: in the case of sequences of functions, care must be taken when interchanging limits and integration.

The next result is called the *Cauchy criterion for the uniform convergence of functions.*

THEOREM 5.3. Let $f_n : E \to R$, $n = 1, 2, 3, \ldots$. The sequence of functions $\{f_n\}$ converges uniformly on E if and only if for every $\epsilon > 0$ there exists an integer $N = N(\epsilon)$ such that $|f_n(t) - f_m(t)| < \epsilon$ for all $t \in E$ whenever $n \ge N$ and $m \ge N$.

Proof. Assume that $\{f_n\}$ converges uniformly on E to the limit function f. Then there exists an integer $N = N(\epsilon)$ such that when $n \ge N$, we have

$$|f_n(t) - f(t)| < \frac{\epsilon}{2} \qquad \text{for all } t \in E.$$

This implies that

$$\begin{aligned}|f_n(t) - f_m(t)| &= |f_n(t) - f(t) + f(t) - f_m(t)| \\ &\le |f_n(t) - f(t)| + |f(t) - f_m(t)| < \epsilon\end{aligned}$$

for all $t \in E$ whenever $n \ge N$ and $m \ge N$.

Conversely, assume that the Cauchy condition holds, i.e., for all $t \in E$,

$$|f_n(t) - f_m(t)| < \epsilon \qquad \text{when } n \ge N, m \ge N. \tag{5.6}$$

This implies that the sequence $\{f_n(t)\}$ converges, for every $t \in E$, to a limit that we call $f(t)$. (This follows since in R, every Cauchy sequence converges to an element in R.) We must show that this convergence is *uniform.* To this end, we let $\epsilon > 0$ be given and pick $N > 0$ so that (5.6) holds. Fix n and let $m \to \infty$ in (5.6). Since $f_m(t) \to f(t)$, as $m \to \infty$, this yields for all $t \in E$,

$$|f_n(t) - f(t)| \le \epsilon \qquad \text{for all } n \ge N. \qquad \blacksquare$$

C. The Weierstrass *M*-test

For an *infinite series* of real-valued functions written $\sum_{j=1}^{\infty} f_j(t)$, with each f_j defined on a set $E \subset R$, convergence is defined in terms of the sequences of *partial sums*,

$$s_n(t) = \sum_{j=1}^{n} f_j(t).$$

The series $\sum_{j=1}^{\infty} f_n(t)$ is said to *converge pointwise* to the function f if for every $t \in E$,

$$\lim_{n\to\infty} \left| f(t) - \sum_{j=1}^{n} f_j(t) \right| = 0.$$

Also, the series $\sum_{j=1}^{\infty} f_j(t)$ is said to *converge uniformly* to f on E if the sequence of partial sums $\{s_n\}$ converges uniformly to f on E.

The next result is called the *Weierstrass M-test.*

THEOREM 5.4. Let $f_n : E \to R$, $n = 1, 2, 3, \ldots$. Suppose there exist nonnegative constants M_n such that $|f_n(t)| \le M_n$ for all $t \in E$ and

$$\sum_{n=1}^{\infty} M_n < \infty.$$

Then the sum $\sum_{n=1}^{\infty} f_n(t)$ converges uniformly on E.

Proof. If $\sum_{n=1}^{\infty} M_n$ converges, then for arbitrary $\epsilon > 0$, there are $m \ge n$ sufficiently large so that

$$\left| \sum_{j=1}^{m} f_j(t) - \sum_{j=1}^{n} f_j(t) \right| = \left| \sum_{j=n}^{m} f_j(t) \right| \le \sum_{j=n}^{m} |f_j(t)|$$
$$\le \sum_{j=n}^{m} M_j \le \epsilon.$$

The uniform convergence of $\sum_{n=1}^{\infty} f_n(t)$ follows now from Theorem 5.3. ■

*1.6 EXISTENCE OF SOLUTIONS OF INITIAL-VALUE PROBLEMS

In this section we address the following question: under what conditions has the initial-value problem (I) *at least one solution* for a given set of initial data (t_0, x_0)? The significance of this question is illustrated by the following two examples.

1. For the initial-value problem,

$$\dot{x} = g(x), \qquad x(0) = 0, \qquad t \ge 0, \tag{6.1}$$

where

$$g(x) = \begin{cases} 1, & x = 0, \\ 0, & x \neq 0, \end{cases}$$

no differentiable function ϕ exists that satisfies (6.1). Hence, *no solution* (as defined in Section 1.3) exists for this initial-value problem.

2. The initial-value problem,

$$\dot{x} = x^{1/3}, \qquad x(t_0) = 0 \tag{6.2}$$

has the solution $\phi(t) = [2(t - t_0)/3]^{3/2}$ (determined by separation of variables). This solution is not unique since $\psi(t) \equiv 0$ is clearly also a solution.

To simplify our presentation, we will consider in this section and in Sections 1.7 to 1.9 one-dimensional initial-value problems [i.e., we will assume that for (I), $n = 1$]. Later, we will show how these results are modified for higher dimensional systems. Thus, we have a domain $D \subset R^2$, $f \in C(D, R)$, we are given the scalar differential equation

$$\dot{x} = f(t, x), \tag{E'}$$

we are given the initial data $(t_0, x_0) \in D$, and we *seek a solution* (or solutions) *to the one-dimensional initial-value problem*

$$\dot{x} = f(t, x), \qquad x(t_0) = x_0. \tag{I'}$$

In doing so, it suffices to find a solution of the integral equation

$$\phi(t) = x_0 + \int_{t_0}^{t} f(s, \phi(s))\, ds. \tag{V'}$$

We will solve the above problem in stages. First, we establish an existence result for a sequence of approximate solutions of (I'). Next, we show that this sequence converges to the actual solution of (I'), using a preliminary convergence result. We will establish this preliminary result first.

A. The Ascoli-Arzela Lemma

We will require the following concepts.

DEFINITION 6.1. Let $\mathcal{F}$ be a family of real-valued functions defined on a set $E \subset R$.

(i) $\mathcal{F}$ is called *uniformly bounded* if there is a nonnegative constant M such that $|f(t)| \leq M$ for all $t \in E$ and for all $f \in \mathcal{F}$.
(ii) $\mathcal{F}$ is called *equicontinuous* on E if for every $\epsilon > 0$ there is a $\delta = \delta(\epsilon) > 0$ (independent of t_1, t_2, and f) such that $|f(t_1) - f(t_2)| < \epsilon$ whenever $|t_1 - t_2| < \delta$ for all $t_1, t_2 \in E$ and for all $f \in \mathcal{F}$. ■

The next result is known as the *Ascoli-Arzela Lemma.*

THEOREM 6.1. Let E be a closed and bounded subset of R and let $\{f_m\}$ be a sequence of functions in $C(E, R)$. If $\{f_m\}$ is equicontinuous and uniformly bounded on E, then there is a subsequence $\{m_k\}$ of $\{m\}$ and a function $f \in C(E, R)$ such that $\{f_{m_k}\}$ converges to f uniformly on E.

Proof. Let $\{r_j\}$ be a dense subset of E (i.e., $\overline{\{r_j\}} = E$). (For example, let $\{r_j\}$ be the enumeration of the rational numbers contained in an interval $[a, b] \subset R$.) The sequence of real numbers $\{f_m(r_1)\}$ is bounded since $\{f_m\}$ is uniformly bounded on E. By the Bolzano-Weierstrass Theorem (see Section 1.5), $\{f_m(r_1)\}$ contains a convergent subsequence that we label $\{f_{1m}(r_1)\}$. We denote the point to which this subsequence converges by $f(r_1)$ and the sequence of functions obtained in this way by $\{f_{1m}\}$. Consider next the sequence of real numbers $\{f_{1m}(r_2)\}$, which is also bounded and contains a convergent subsequence that we label $\{f_{2m}(r_2)\}$. We denote the point to which $\{f_{2m}(r_2)\}$ converges by $f(r_2)$ and we label the sequence of functions obtained in this way by $\{f_{2m}\}$. Continuing, we obtain the subsequence $\{f_{km}\}$ of the sequence $\{f_{k-1,m}\}$ and the real number $f(r_k)$ such that $f_{km}(r_k) \to f(r_k)$ as $m \to \infty$ for $k = 1, 2, 3, \ldots$. Since the sequence $\{f_{km}\}$ is a subsequence of all the preceding sequences $\{f_{jm}\}$ for $1 \le j \le k-1$, it will converge at each point r_j with $1 \le j \le k$.

Next, we generate a subsequence by "*diagonalizing*" the preceding infinite collection of sequences. In doing so, we set $g_m = f_{mm}$ for all m. If the terms f_{km} are arranged as the elements of a semi-infinite matrix, as shown in Fig. 1.8, then the elements g_m are the diagonal elements of this matrix.

Since $\{g_m\}$ is eventually a subsequence of every sequence $\{f_{km}\}$, we have $g_m(r_k) \to f(r_k)$ as $m \to \infty$ for $k = 1, 2, 3, \ldots$. We now show that $\{g_m\}$ converges uniformly on E. Fix $\epsilon > 0$. For any rational number $r_j \in E$ there exists $M_j = M_j(\epsilon)$ such that $|g_m(r_j) - g_n(r_j)| < \epsilon$ for all $m, n \ge M_j(\epsilon)$. By the equicontinuity of $\{f_m\}$, there is a $\delta > 0$ such that $|g_n(t_1) - g_n(t_2)| < \epsilon$ for all n when $t_1, t_2 \in E$ and $|t_1 - t_2| < \delta$. Therefore, for $|t - r_j| < \delta$ and $m, n \ge M_j(\epsilon)$, we have

$$|g_m(t) - g_n(t)| \le |g_m(t) - g_m(r_j)| + |g_m(r_j) - g_n(r_j)| + |g_n(r_j) - g_n(t)| < 3\epsilon.$$

By construction, the collection of neighborhoods $B(r_j, \delta) = \{t \in R : |t - r_j| < \delta\}$ covers E (i.e., $\bigcup_j B(r_j, \delta) \supset E$). Since E is a closed and bounded subset of R (i.e., since E is compact), by the Heine-Borel Theorem there is a *finite* subcollection of the above neighborhoods, say, $\{B(r_{j1}, \delta), \ldots, B(r_{jL}, \delta)\}$ that covers E, i.e., $B(r_{j1}, \delta) \cup \cdots \cup B(r_{jL}, \delta) \supset D$ (see Section 1.5). Let $M(\epsilon) = \max\{M_{j1}(\epsilon), \ldots, M_{jL}(\epsilon)\}$. If m and n are larger than $M(\epsilon)$, and if t is any point in E, then $t \in B(r_{jl}, \delta)$ for some $l \in \{1, \ldots, L\}$. Therefore, $|g_m(t) - g_n(t)| < 3\epsilon$ for all $t \in E$ whenever $m, n > M(\epsilon)$. This shows that $\{g_m\}$ converges uniformly on E to a function f, by Theorem 5.3. Furthermore, $f \in C(E, R)$ by Theorem 5.1. ∎

$$\begin{matrix} f_{11} & f_{12} & f_{13} & f_{14} & \cdots \\ f_{21} & f_{22} & f_{23} & f_{24} & \cdots \\ f_{31} & f_{32} & f_{33} & f_{34} & \cdots \\ \vdots & \vdots & \vdots & \vdots & \end{matrix}$$

FIGURE 1.8.
Diagonalization of a collection of sequences

B. ϵ-Approximate Solutions

We will require the following concept.

DEFINITION 6.2. A real-valued function ϕ defined and continuous on a t-interval $J = (a, b)$ containing t_0 is called an ϵ-approximate solution of (I') if $\phi(t_0) = x_0$ and

(i) $(t, \phi(t)) \in D$ for all $t \in J$;
(ii) ϕ has a continuous derivative on J except possibly on a finite set I of points in J where there are jump discontinuities allowed;
(iii) $|\dot{\phi}(t) - f(t, \phi(t))| < \epsilon$ for all $t \in J - I$. ∎

Recall that if I in the above definition is not empty, ϕ is said to have a *piecewise continuous derivative on J*.

Now let

$$S = \{(t, x) \in D : |t - t_0| \le a,\ |x - x_0| \le b\} \tag{6.3}$$

be a fixed rectangle in D containing (t_0, x_0), as shown in Fig. 1.9. Since $f \in C(D, R)$, it is bounded on S and there is an $M > 0$ such that $|f(t, x)| \le M$ for all $(t, x) \in S$. We define (see Fig. 1.9)

$$c = \min\left\{a, \frac{b}{M}\right\}. \tag{6.4}$$

We now prove the following existence result.

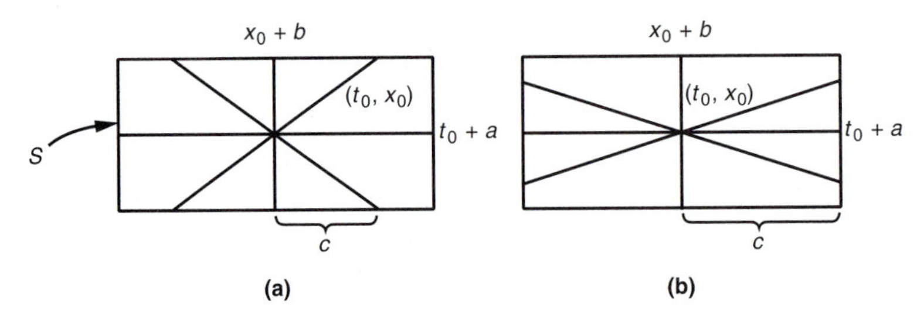

FIGURE 1.9
(a) Case $c = b/M$, (b) case $c = a$

THEOREM 6.2. If $f \in C(D, R)$ and if c is as defined in (6.4), then for any $\epsilon > 0$ there is an ϵ-approximate solution of (I') on the interval $|t - t_0| \le c$.

Proof. Given $\epsilon > 0$, we shall show that there is an ϵ-approximate solution on $[t_0, t_0 + c]$. The proof for the interval $[t_0 - c, t_0]$ is similar. The approximate solution will be made up of a finite number of straight line segments joined at their ends to achieve continuity.

Since f is continuous on S, a closed and bounded set, then f is uniformly continuous on S. Hence, there is a $\delta > 0$ such that $|f(t, x) - f(s, y)| < \epsilon$ whenever (t, x) and (s, y) are in S with $|t - s| \le \delta$ and $|x - y| \le \delta$. Now subdivide the interval $[t_0, t_0 + c]$ into m equal subintervals by a partition $t_0 < t_1 < t_2 < \cdots < t_m = t_0 + c$, where $t_{j+1} - t_j < \min\{\delta, \delta/M\}$ and where M is the bound for f given above. On the interval $t_0 \le t \le t_1$, let $\phi(t)$ be the line segment issuing from (t_0, x_0) with slope $f(t_0, x_0)$. On $t_1 \le t \le t_2$, let $\phi(t)$ be the line segment starting at $(t_1, \phi(t_1))$ with slope $f(t_1, \phi(t_1))$. Continuing in this manner, we define ϕ over $t_0 \le t \le t_m$. A typical solution is shown in Fig. 1.10. The resulting ϕ is piecewise linear and hence piecewise continuously differentiable and $\phi(t_0) = x_0$. Indeed, on $t_j \le t \le t_{j+1}$ we have

$$\phi(t) = \phi(t_j) + f(t_j, \phi(t_j))(t - t_j). \tag{6.5}$$

Since the slopes of the linear segments in (6.5) are bounded between $\pm M$, then $(t, \phi(t))$ cannot leave S before time $t_m = t_0 + c$ (see Fig. 1.10).

To see that ϕ is an ϵ-approximate solution, we use (6.5) to obtain

$$|\dot{\phi}(t) - f(t, \phi(t))| = |f(t_j, \phi(t_j)) - f(t, \phi(t))| < \epsilon.$$

This inequality is true by the choice of δ, since $|t_j - t| \le |t_j - t_{j+1}| < \delta$ and

$$|\phi(t) - \phi(t_j)| \le M|t - t_j| \le M\left(\frac{\delta}{M}\right) = \delta.$$

This completes the proof. ■

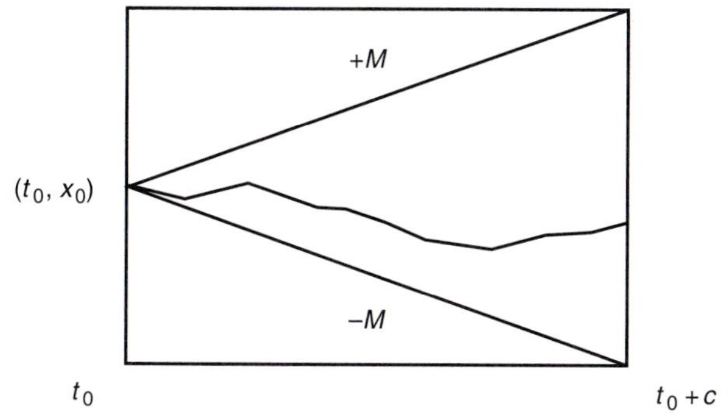

(t_0, x_0) +M −M t_0 $t_0 + c$

FIGURE 1.10
Typical ϵ-approximate solution

The approximations defined in the proof of Theorem 6.2 are called *Euler polygons,* and (6.5) with $t = t_{j+1}$ assumes the form

$$\phi(t_{j+1}) = \phi(t_j) + f(t_j, \phi(t_j))(t_{j+1} - t_j), \tag{6.6}$$

which is called *Euler's method.* This technique and more sophisticated piecewise polynomial approximations are common in determining numerical approximations to solutions of (I') via computer solutions.

C. The Cauchy-Peano Existence Theorem

We are now in a position to state and prove the main result of this section.

THEOREM 6.3. If $f \in C(D, R)$ and $(t_0, x_0) \in D$, then (I') has a solution defined on $|t - t_0| \le c$ [where c is defined in (6.4)].

Proof. Let $\{\epsilon_m\}$, $m = 1, 2, \ldots$ be a monotone decreasing sequence of positive numbers tending to zero, e.g., $\epsilon_m = 1/m$. Let c be defined in (6.4) and let ϕ_m be the ϵ_m-approximate solution given in Theorem 6.2. Then $|\phi_m(t) - \phi_m(s)| \le M|t - s|$ for all t, s in $[t_0 - c, t_0 + c]$ and for all $m \ge 1$. Therefore, $\{\phi_m\}$ is an equicontinuous sequence. Now since

$$|\phi_m(t)| \le |\phi_m(t_0)| + |\phi_m(t) - \phi_m(t_0)| \le |x_0| + Mc,$$

the sequence is also uniformly bounded. By the Ascoli-Arzela Lemma (Theorem 6.1) there is a subsequence $\{\phi_{m_k}\}$ that converges uniformly on $J = [t_0 - c, t_0 + c]$ to a continuous function ϕ.

Next, define

$$e_m(t) = \dot{\phi}_m(t) - f(t, \phi_m(t)) \tag{6.7}$$

at those points where $\dot{\phi}_m$ exists. From the proof of Theorem 6.2, e_m is piecewise continuous and $|e_m(t)| \le \epsilon_m$ on J where $\dot{\phi}_m$ exists. Integrating (6.7) and rearranging terms, we obtain

$$\phi_m(t) = x_0 + \int_{t_0}^{t} [f(s, \phi_m(s)) + e_m(s)]\, ds. \tag{6.8}$$

Now since $\{\phi_{m_k}\}$ tends to ϕ uniformly on J, and since f is uniformly continuous on the set S [defined in (6.3)], it follows that $f(t, \phi_{m_k}(t))$ tends to $f(t, \phi(t))$ uniformly on J. To see this, we note that for every $\delta' > 0$ there is an $N = N(\delta')$ such that $|\phi_{m_k}(t) - \phi(t)| < \delta'$ for all $t \in J$ whenever $k > N$, and also, for every $\epsilon > 0$ there is a $\delta = \delta(\epsilon)$ such that for all $t \in J$, $|f(t, p) - f(t, q)| < \epsilon$ for all $(t, p), (t, q) \in S$ whenever $|p - q| < \delta$. Pick N large enough so that $|\delta_{m_k}(t) - \phi(t)| < \delta$ for all $t \in J$ whenever $k > N$. Then $|f(t, \phi_{m_k}(t)) - f(t, \phi(t))| < \epsilon$ for all $t \in J$ whenever $k > N$.

Using Theorem 5.2, we now obtain

$$\begin{aligned} \lim_{k\to\infty} \int_{t_0}^{t} f(s, \phi_{m_k}(s))\, ds &= \int_{t_0}^{t} \lim_{k\to\infty} f(s, \phi_{m_k}(s))\, ds \\ &= \int_{t_0}^{t} f(s, \phi(s))\, ds. \end{aligned} \tag{6.9}$$

Also, observing that

$$\int_{t_0}^{t} e_{m_k}(s)\, ds \le \int_{t_0}^{t} |e_{m_k}(s)|\, ds = \int_{t_0}^{t} \epsilon_{m_k}\, ds \le \epsilon_{m_k} c$$

and recalling that $\lim_{k\to\infty} \epsilon_{m_k} = 0$, we obtain

$$\lim_{k\to\infty} \int_{t_0}^{t} e_{m_k}(s)\, ds = 0. \tag{6.10}$$

Letting in (6.8) $m = m_k$ and using (6.9) and (6.10), we finally obtain

$$\lim_{k\to\infty} \phi_{m_k}(t) = \phi(t) = x_0 + \int_{t_0}^{t} f(s, \phi(s))\, ds, \tag{V'}$$

which completes the proof. ■

Theorem 6.3 establishes the existence of a solution of (I') "*locally,*" i.e., only on some sufficiently short time interval. In general, this theorem cannot be changed to assert the existence of a solution for all $t \ge t_0$ or for all $t \le t_0$. As an example, consider the initial-value problem

$$\dot{x} = 1 + x^2, \qquad x(0) = x_0 = 0,$$

which has a solution given by $\phi(t) = \tan t$. This solution exists only when $-\pi/2 < t < \pi/2$.

*1.7 CONTINUATION OF SOLUTIONS

Once the existence of a solution of an initial-value problem has been established over some time interval, it is reasonable to ask whether this solution can be extended to a larger time interval. We call this process *continuation of solutions.* In this section we address this problem for the scalar initial-value problem (I'). We shall consider the continuation of solutions of an initial-value problem (I) characterized by a system of equations later (in Section 1.10).

A. Zorn's Lemma

In the proof of the main result of this section, we will require a fundamental result from analysis, called Zorn's Lemma, that we will present without proof. To state this lemma, we need to introduce the following concepts.

A *partially ordered set* $(A, \le)$ consists of a set A and a relation $\le$ on A such that for any a, b, and c in A,

1. $a \le a$,
2. $a \le b$ and $b \le c$ implies that $a \le c$,
3. $a \le b$ and $b \le a$ implies that $a = b$.

A *chain* is a subset A_0 of A such that for all a and b in A_0, either $a \le b$ or $b \le a$. An *upper bound* for a chain A_0 is an element $a_0 \in A$ such that $b \le a_0$ for all $b \in A_0$. A *maximal element* for A, if it exists, is an element a_1 of A such that for all b in A, $a_1 \le b$ implies $a_1 = b$.

THEOREM 7.1. (ZORN'S LEMMA). If each chain in a partially ordered set $(A, \le)$ has an upper bound, then A has a maximal element. ■

B. Continuable Solutions

Now let ϕ be a solution of (E') on an interval J. By a *continuation* of ϕ we mean an extension ϕ_0 of ϕ to a larger interval J_0 in such a way that the extension solves (E') on J_0; then ϕ is said to be *continued* or *extended* to the larger interval J_0. When no such continuation is possible, then ϕ is called *noncontinuable.*

To illustrate these ideas, consider the differential equation $\dot{x} = x^2$ that has a solution

$$\phi(t) = (1 - t)^{-1} \text{ on } J = (-1, 1).$$

This solution is continuable to the left to $-\infty$ and is noncontinuable to the right. As a second example, consider the differential equation $\dot{x} = x^{1/3}$ that has a solution

$$\psi(t) \equiv 0 \text{ on } J = (-1, 0).$$

This solution is continuable to the right in more than one way. For example, both $\psi_1(t) \equiv 0$ and $\psi_2(t) = (2t/3)^{3/2}$ are solutions of $\dot{x} = x^{1/3}$ for $t \ge 0$. The solution ψ can also be continued to the left using $\psi_3(t) \equiv 0$ for all $t \le -1$.

THEOREM 7.2. Let $f \in C(D, R)$ with f bounded on D. Suppose ϕ is a solution of (E') on the interval $J = (a, b)$. Then

(i) the limits

$$\lim_{t \to a^+} \phi(t) = \phi(a^+) \qquad \text{and} \qquad \lim_{t \to b^-} \phi(t) = \phi(b^-)$$

exist, and

(ii) if $(a, \phi(a^+))$ [respectively, $(b, \phi(b^-))$] is in D, then the solution ϕ can be continued to the left past the point $t = a$ (respectively, to the right past the point $t = b$).

Proof. We give the proof for the endpoint b. The proof for the endpoint a is similar. Let M be a bound for $|f(t, x)|$ on D, fix $t_0 \in J$, and let $\phi(t_0) = x_0$. Then for $t_0 < t < u < b$ the solution ϕ satisfies (V'), and thus,

$$\begin{aligned} |\phi(u) - \phi(t)| &= \left| \int_t^u f(s, \phi(s))\, ds \right| \le \int_t^u |f(s, \phi(s))|\, ds \\ &\le \int_t^u M\, ds = M(u - t). \end{aligned} \tag{7.1}$$

Given any sequence $\{t_m\} \subset (t_0, b)$ tending monotonically to b, we see from (7.1) that $\{\phi(t_m)\}$ is a Cauchy sequence, and therefore, the limit $\phi(b^-)$ exists (see Subsection 1.5A).

Next, if $(b, \phi(b^-)) \in D$, then by Theorem 6.3 there is a solution ϕ_0 of (E') that satisfies $\phi_0(b) = \phi(b^-)$. The solution ϕ_0 will be defined on some interval $b \leq t \leq b+c$ for some $c > 0$. Define $\phi_0(t) = \phi(t)$ on $a < t < b$. Then ϕ_0 is continuous on $a < t < b + c$ and satisfies

$$\phi_0(t) = x_0 + \int_{t_0}^{t} f(s, \phi_0(s))\, ds, \qquad a < t < b, \tag{7.2}$$

and

$$\phi_0(t) = \phi(b^-) + \int_{b}^{t} f(s, \phi_0(s))\, ds, \qquad b \leq t < b + c.$$

The limit of (7.2) as t tends to b is seen to be

$$\phi(b^-) = x_0 + \int_{t_0}^{b} f(s, \phi_0(s))\, ds.$$

Therefore,

$$\begin{aligned}\phi_0(t) &= x_0 + \int_{t_0}^{b} f(s, \phi_0(s))\, ds + \int_{b}^{t} f(s, \phi_0(s))\, ds \\ &= x_0 + \int_{t_0}^{t} f(s, \phi_0(s))\, ds, \qquad b \leq t < b + c.\end{aligned}$$

Hence, ϕ_0 solves (V') on $a < t < b + c$, and therefore, ϕ_0 solves (I') on $a < t < b + c$. ∎

C. Continuation of Solutions to the Boundary of D

We are now in a position to prove the following result.

THEOREM 7.3. If $f \in C(D, R)$ and if ϕ is a solution of (E') on an open interval J, then ϕ can be continued to a maximal open interval $J^* \supset J$ in such a way that $(t, \phi(t))$ tends to ∂D as $t \to \partial J^*$ when ∂D is not empty and $|t| + |\phi(t)| \to \infty$ if ∂D is empty. The extended solution ϕ^* on J^* is noncontinuable.

Proof. Let ϕ be a given solution of (E') on J. The *graph* of ϕ is the set

$$Gr(\phi) = \{(t, \phi(t)) : t \in J\}.$$

Given any two solutions ϕ_1 and ϕ_2 of (E') that extend ϕ, we define $\phi_1 \leq \phi_2$ if and only if $Gr(\phi_1) \subset Gr(\phi_2)$, i.e., if and only if ϕ_2 is an extension of ϕ_1. The relation $\leq$ determines a partial ordering on continuations of ϕ over open intervals. If $\{\phi_\alpha : \alpha \in A\}$ is any chain of such extensions, then $\cup\{Gr(\phi_\alpha) : \alpha \in A\}$ is the graph of a continuation of ϕ that we call ϕ_A. This ϕ_A is an upper bound for the chain. By Zorn's Lemma (Theorem 7.1) there is a maximal element ϕ^*. Clearly, ϕ^* is a noncontinuable extension of the original solution ϕ.

Now let J^* denote the domain of ϕ^*. By Theorem 7.2 the interval J^* must be open, for otherwise ϕ^* could not be maximal. Call $J^* = (a, b)$. If $b = \infty$, then we know that D is unbounded and $|t| + |\phi(t)| \to \infty$ as $t \to b^-$. So let us assume that $b < \infty$ and assume, for purposes of contradiction, that $(t, \phi^*(t))$ does not approach ∂D on any sequence $\{t_m\}$ that approaches b^-. Then $(t, \phi^*(t))$ must remain in a compact subset K of D when $t \in [c, b)$ for any $c \in (a, b)$. Since f must be bounded on K, then by Theorem 7.2 we can continue ϕ^*past b. But this is impossible since $\phi^*$is noncontinuable. We have shown that $(t, \phi^*(t))$ must approach ∂D on *some* sequence $\{t_m\}$ that approaches b^-.

We now claim that $(t, \phi^*(t)) \to \partial D$ as $t \to b^-$. If this is not the case, there exists a sequence $\{\tau_m\}$ that approaches b^-, and a point $(b, \xi) \in D$ such that $\phi^*(\tau_m) \to \xi$. Let ϵ be one-third the distance from (b, ξ) to ∂D. We can assume without loss of generality that $\tau_m < t_m < \tau_{m+1}$, $(\tau_m, \phi^*(\tau_m)) \in B((b, \xi), \epsilon)$, and $(t_m, \phi^*(t_m)) \notin B((b, \xi), 2\epsilon)$ for all $m \geq 1$. Let M be a bound for $|f(t, x)|$ over $B((b, \xi), 2\epsilon)$. It now follows from the properties of (E') that

$$\begin{aligned} \epsilon &\leq \{[\phi^*(t_m) - \phi^*(\tau_m)]^2 + (t_m - \tau_m)^2\}^{1/2} \leq |\phi^*(t_m) - \phi^*(\tau_m)| + (t_m - \tau_m) \\ &\leq \left| \int_{\tau_m}^{t_m} f(s, \phi^*(s))\, ds \right| + (t_m - \tau_m) \leq (M + 1)(t_m - \tau_m), \end{aligned}$$

i.e., $t_m - \tau_m \geq \epsilon/(M + 1)$ for all m. But this is impossible since $t_m \to b^-$ and $b < \infty$. We conclude that $(t, \phi^*(t)) \to \partial D$ as $t \to b^-$.

The proof for the endpoint $t = a$ is similar. ■

*1.8 UNIQUENESS OF SOLUTIONS

We now establish conditions for the uniqueness of solutions of initial-value problems determined by scalar first-order ordinary differential equations. Later, in Section 1.10, we will address the uniqueness of solutions of initial-value problems characterized by systems of first-order ordinary differential equations.

A. The Gronwall Inequality

We will require the following preliminary result on several occasions.

THEOREM 8.1. (GRONWALL INEQUALITY). Let $r, k \in C([a, b], R)$ and suppose that $r(t) \geq 0$ and $k(t) \geq 0$ for all $t \in [a, b]$. Let δ be a given nonnegative constant. If

$$r(t) \leq \delta + \int_a^t k(s)r(s)\, ds$$

for all $t \in [a, b]$, then

$$r(t) \leq \delta e^{\int_a^t k(s)\, ds}$$

for all $t \in [a, b]$.

Proof. Let $R(t) = \delta + \int_a^t k(s)r(s)\, ds$. Then $r(t) \leq R(t)$, $R(a) = \delta$, $\dot{R}(t) = k(t)r(t) \leq k(t)R(t)$, and

$$\dot{R}(t) - k(t)R(t) \leq 0 \tag{8.1}$$

for all $t \in [a, b]$. Let $K(t) = e^{-\int_a^t k(s)\, ds}$. Then

$$\dot{K}(t) = -k(t)e^{-\int_a^t k(s)\, ds} = -K(t)k(t).$$

Multiplying both sides of (8.1) by $K(t)$, we obtain

$$K(t)\dot{R}(t) - K(t)k(t)R(t) \leq 0$$

or
$$K(t)\dot{R}(t) + \dot{K}(t)R(t) \le 0$$
or
$$\frac{d}{dt}[K(t)R(t)] \le 0. \tag{8.2}$$

Integrating (8.2) from a to t, we obtain
$$K(t)R(t) - K(a)R(a) \le 0$$
or
$$K(t)R(t) - \delta \le 0$$
or
$$e^{-\int_a^t k(s)\,ds}R(t) - \delta \le 0$$
or
$$r(t) \le R(t) \le \delta e^{\int_a^t k(s)\,ds},$$

which is the desired inequality. ■

B. Unique Solutions

Before addressing the uniqueness issue, we need to introduce the notion of Lipschitz continuity.

DEFINITION 8.1. A function $f \in C(D, R)$, $D \subset R^2$, is said to satisfy a *Lipschitz condition* on D (with respect to x) with *Lipschitz constant* L if
$$|f(t, x) - f(t, y)| \le L|x - y|$$
for all (t, x), (t, y) in D. The function f is said to be *Lipschitz continuous* in x on D in this case. ■

For example, if $f \in C(D, R)$ and if $\partial f/\partial x$ exists and is continuous on D, then f is Lipschitz continuous on any compact and convex subset D_0 of D. To show this, let L_0 be a bound for $|(\partial f/\partial x)(t, x)|$ on D_0. Let (t, x) and (t, y) be in D_0. Since D_0 is convex, the straight line that connects (t, x) and (t, y) is a subset of D_0. By the Mean Value Theorem there is a point z on this line such that
$$|f(t, x) - f(t, y)| = \left|\frac{\partial f}{\partial x}(t, z)(x - y)\right| \le L_0|x - y|.$$

THEOREM 8.2. If $f \in C(D, R)$ and if f satisfies a Lipschitz condition (with respect to x) on D with Lipschitz constant L, then the initial-value problem (I') has at most one solution on any interval $|t - t_0| \le d$.

Proof. Suppose for some $d > 0$ there are two solutions ϕ_1 and ϕ_2 on $|t - t_0| \le d$. Since both solutions solve (V') on $t_0 \le t \le t_0 + d$, we have
$$\begin{aligned}|\phi_1(t) - \phi_2(t)| &\le \left|\int_{t_0}^t [f(s, \phi_1(s)) - f(s, \phi_2(s))]\,ds\right| \\ &\le \int_{t_0}^t |f(s, \phi_1(s)) - f(s, \phi_2(s))|\,ds \\ &\le \int_{t_0}^t L|\phi_1(s) - \phi_2(s)|\,ds.\end{aligned}$$

Applying the Gronwall Inequality (Theorem 8.1) with $k = L$ and $\delta = 0$, it follows that $|\phi_1(t) - \phi_2(t)| \le 0$ on the interval $t_0 \le t \le t_0 + d$. Thus, $\phi_1(t) = \phi_2(t)$ on this interval.

The proof for the interval $t_0 - d \le t \le t_0$ proceeds similarly. ■

COROLLARY 8.3. If f and $\partial f/\partial x$ are both in $C(D, R)$, then for any $(t_0, x_0) \in D$ and any J containing t_0, if a solution of (I') exists on J, it must be unique.

Proof. Let ϕ_1 and ϕ_2 be two solutions of (I') on J and define

$$b = \sup\{t \geq t_0 : \phi_1(t) = \phi_2(t)\}, \qquad a = \inf\{t \leq t_0 : \phi_1(t) = \phi_2(t)\}.$$

We claim that a and b are the endpoints of J ($\partial J = \{a, b\}$). For if b is not an endpoint of J, then by continuity we would have $\phi_1(b) = \phi_2(b)$. Since $(b, \phi_1(b)) \in D$ and D is a domain, we know that there exists $\epsilon > 0$ such that

$$D_0 = \{(t, x) : |t - b| \leq \epsilon, \qquad |x - \phi_1(b)| \leq \epsilon\} \subset D.$$

Clearly, D_0 is a compact and convex subset of D. Now from the comments following Definition 8.1 and Theorem 8.2, we have that $\phi_1(t) = \phi_2(t)$ for $t \in [b, b + \epsilon']$ for some $0 < \epsilon' < \epsilon$. This contradicts the definition of b. We conclude that b is an endpoint of J, and so is a. It follows that $\phi_1(t) \equiv \phi_2(t)$, $t \in J$, which implies the uniqueness of the solution of (I'). ∎

Using Theorems 7.3 and 8.2, we can prove the following *continuation result.*

THEOREM 8.4. Let $f \in C(J \times R, R)$ for some open *interval* $J \subset R$ and let f satisfy a Lipschitz condition on $J \times R$ (with respect to x). Then for any $(t_0, x_0) \in J \times R$, the initial value problem (I') has a unique solution that exists on the *entire* interval J.

Proof. The local existence and uniqueness of solutions $\phi(t, t_0, x_0)$ of (I') are clear from Theorem 8.2. Now if $\phi(t) \triangleq \phi(t, t_0, x_0)$ is a solution defined on $t_0 \leq t < c$, then ϕ satisfies (V'), and therefore,

$$\phi(t) - x_0 = \int_{t_0}^{t} [f(s, \phi(s)) - f(s, x_0)]\, ds + \int_{t_0}^{t} f(s, x_0)\, ds$$

and $$|\phi(t) - x_0| \leq \int_{t_0}^{t} L|\phi(s) - x_0|\, ds + \delta,$$

where $\delta = [\max_{t_0 \leq s \leq c} |f(s, x_0)|](c - t_0)$. By the Gronwall Inequality, we have

$$|\phi(t) - x_0| \leq \delta \exp[L(c - t_0)], \qquad t_0 \leq t < c.$$

Hence, $|\phi(t)|$ is bounded on $[t_0, c)$ whenever $\phi(t)$ is a solution defined on $[t_0, c)$, $c \in J$. Let $J = (a, b)$ and assume that ϕ is a noncontinuable solution of (I') that is defined on $J^* = (a', b')$. We must prove that $b' = b$. If this is not the case, we have $b' < b$. We have shown that $|\phi(t)|$ is bounded, say, $|\phi(t)| \leq M$, for $t \in [t_0, b')$. Let $D = J \times [-M - 1, M + 1]$. Applying Theorem 7.3, we have that $(t, \phi(t)) \to \partial D$ as $t \to b'$. Since $b' < b$, we must have $|\phi(t)| \to M + 1$ as $t \to b'$. But this contradicts our assumption that $|\phi(t)| \leq M$ for $t \in [t_0, b')$. Therefore, $b' = b$.

A similar argument works for $t \leq t_0$. ∎

Successive approximation of solution

If a solution ϕ of (I') is unique, then the ϵ-approximate solutions constructed in the proof of Theorem 6.2 will tend to ϕ as $\epsilon \to 0^+$, and this is the basis for justifying Euler's method, a numerical method of constructing approximations to ϕ. Now if we assume that f satisfies a Lipschitz condition, an alternative classical method of approximation is the *method of successive approximations* (also known as *Picard iterations*). Specifically, let $f \in C(D, R)$, $D \subset R^2$, and S be a rectangle in D centered at (t_0, x_0) (see Fig. 1.9), and let c and M be defined by Eq. (6.4). *Successive approximations* for (I'), or equivalently for (V'), are defined as

$$\begin{aligned}\phi_0(t) &= x_0 \\ \phi_{m+1}(t) &= x_0 + \int_{t_0}^{t} f(s, \phi_m(s))\, ds, \qquad m = 0, 1, 2, \ldots\end{aligned} \tag{8.3}$$

for $|t - t_0| \leq c$. For this sequence $\{\phi_m\}$, we have the following result.

THEOREM 8.5. If $f \in C(D, R)$ and if f is Lipschitz continuous on S (with respect to x) with constant L, then the successive approximations ϕ_m, $m = 0, 1, 2, \ldots$ given in (8.3) exist on $|t - t_0| \leq c$, are continuous there, and converge uniformly, as $m \to \infty$, to the unique solution of (I').

Proof. We give the proof for the interval $t_0 \leq t \leq t_0 + c$. The proof for the interval $t_0 - c \leq t \leq t_0$ proceeds similarly.

Using induction on the integer m, we first prove the following statements:

(i) ϕ_m exists on $[t_0, t_0 + c]$,
(ii) $\phi_m \in C^1([t_0, t_0 + c], R)$,
(iii) $|\phi_m(t) - x_0| \leq M(t - t_0)$ on $[t_0, t_0 + c]$

for all $m \geq 0$.

Each statement is clearly true when $m = 0$. Assume that each statement is true for a fixed integer $m > 0$. By (iii) and by the choice of c, it follows that $(t, \phi_m(t)) \in S \subset D$ for all $t \in [t_0, t_0 + c]$. Therefore, $f(t, \phi_m(t))$ exists and is continuous in t, while $|f(t, \phi_m(t))| \leq M$ on the time interval. This in turn means that

$$\phi_{m+1}(t) = x_0 + \int_{t_0}^{t} f(s, \phi_m(s))\, ds$$

exists, that $\phi_{m+1} \in C^1([t_0, t_0 + c], R)$, and that

$$|\phi_{m+1}(t) - x_0| = \left| \int_{t_0}^{t} f(s, \phi_m(s))\, ds \right| \leq M(t - t_0).$$

This completes the induction on m.

Next, we define $\Phi_m(t) = \phi_{m+1}(t) - \phi_m(t)$. Then

$$\begin{aligned}|\Phi_m(t)| &\leq \left| \int_{t_0}^{t} |f(s, \phi_m(s)) - f(s, \phi_{m-1}(s))|\, ds \right| \\ &\leq \int_{t_0}^{t} L|\phi_m(s) - \phi_{m-1}(s)|\, ds = L \int_{t_0}^{t} \Phi_{m-1}(s)\, ds.\end{aligned}$$

Notice in particular that

$$|\Phi_0(t)| = \left| \int_{t_0}^{t} f(s, x_0)\, ds \right| \leq M(t - t_0).$$

The above two estimates show that

$$|\Phi_1(t)| \leq L \int_{t_0}^{t} M(s - t_0)\, ds = \frac{LM(t - t_0)^2}{2!}$$

and that

$$|\Phi_2(t)| \leq L \int_{t_0}^{t} [LM(s - t_0)^2/2!]\, ds \leq \frac{L^2 M(t - t_0)^3}{3!}$$

and, by induction, that

$$|\Phi_m(t)| \leq \frac{ML^m(t-t_0)^{m+1}}{(m+1)!}.$$

Therefore, the mth term of the series

$$\phi_0(t) + \sum_{k=0}^{\infty}[\phi_{k+1}(t) - \phi_k(t)] \tag{8.4}$$

is bounded on the interval $[t_0, t_0 + c]$ by $\dfrac{M}{L}\dfrac{(Lc)^{m+1}}{(m+1)!}$. Now since

$$e^{Lc} = \sum_{k=0}^{\infty}\frac{(Lc)^k}{k!} < \infty,$$

it follows from the Weierstrass M-test (see Theorem 5.4) that the series (8.4) converges uniformly to a continuous function ϕ. This in turn means that the sequence of partial sums

$$\phi_0 + \sum_{k=0}^{m}(\phi_{k+1} - \phi_k) = \phi_0 + (\phi_1 - \phi_0) + \cdots + (\phi_{m+1} - \phi_m) = \phi_{m+1}$$

tends uniformly to ϕ as $m \to \infty$. Since the bound (iii) given above is true for all ϕ_m, it is also true in the limit, i.e.,

$$|\phi(t) - x_0| \leq M(t - t_0).$$

Therefore, $f(t, \phi(t))$ exists and is a continuous function of t. Using an identical argument as in the proof of Theorem 6.3, it now follows that

$$\begin{aligned}\phi(t) &= \lim_{m\to\infty}\phi_{m+1}(t) = x_0 + \lim_{m\to\infty}\int_{t_0}^{t} f(s, \phi_m(s))\,ds \\ &= x_0 + \int_{t_0}^{t} f(s, \phi(s))\,ds, \qquad t_0 \leq t \leq t_0 + c.\end{aligned}$$

Therefore, ϕ solves (V'). ■

We will consider the application of Theorem 8.5 to specific cases (linear systems) in Section 1.13.

*1.9 CONTINUOUS DEPENDENCE OF SOLUTIONS ON INITIAL CONDITIONS AND PARAMETERS

In practice it frequently happens that an initial-value problem may exhibit dependence on some parameter λ. An example of such a class of problems is given by

$$\begin{aligned}\dot{x} &= f(t, x, \lambda) \\ x(\tau) &= \xi,\end{aligned} \tag{$I'_{\lambda,\tau,\xi}$}$$

where $f \in C(J \times R \times D, R)$, $J \subset R$ is an open interval, and $D \subset R$.

If it is assumed that for each pair of compact subsets $J_0 \subset J$ and $D_0 \subset D$ there exists a constant $L = L_{J_0,D_0} > 0$ such that for all $(t, \lambda) \in J_0 \times D_0$, $x, y \in R$,

$$|f(t, x, \lambda) - f(t, y, \lambda)| \leq L|x - y|, \tag{9.1}$$

then by previous results we know that for every $\tau \in J$, $\lambda \in D$, and $\xi \in R$ the initial-value problem $(I'_{\lambda,\tau,\xi})$ has a unique solution $\phi(t) = \phi(t, \tau, \xi, \lambda)$ that exists for all $t \in J$. It turns out that this solution depends continuously on the initial data (τ, ξ, λ). We express this in the following result.

THEOREM 9.1. Let $f \in C(J \times R \times D, R)$, where $J \subset R$ is an open interval and $D \subset R$. Assume that for each pair of compact subsets $J_0 \subset J$ and $D_0 \subset D$, there exists a constant $L = L_{J_0,D_0} > 0$ such that for all $(t, \lambda) \in J_0 \times D_0$, $x, y \in R$, the Lipschitz condition (9.1) is true. Then the initial-value problem $(I'_{\lambda,\tau,\xi})$ has a unique solution $\phi(t, \tau, \xi, \lambda)$, where $\phi \in C(J \times J \times R \times D, R)$. Furthermore, if D is a set such that for every $\lambda_0 \in D$ there exists $\epsilon > 0$ so that $\overline{[\lambda_0 - \epsilon, \lambda_0 + \epsilon] \cap D} \subset D$, then $\phi(t, \tau, \xi, \lambda) \to \phi(t, \tau_0, \xi_0, \lambda_0)$ uniformly for $t \in J_0$ as $(\tau, \xi, \lambda) \to (\tau_0, \xi_0, \lambda_0)$, where J_0 is any compact subset of J. ■

It is because of uniform convergence that we require the restrictions on D in Theorem 9.1. However, in practice, most sets that are of interest to us satisfy these assumptions, including open and closed sets in R, intervals such as $(a, b]$ and $[a, b)$, sequences such as $\{(1/n) : n \in N\}$, $\{0\} \cup \{(1/n) : n \in N\}$, $\{m + (1/n) : m, n \in N\}$, $\{m : m \in N\} \cup \{m + (1/n) : m, n \in N\}$, and so forth.

Applying Theorem 9.1 to the initial-value problem

$$\begin{aligned} \dot{x} &= f(t, x, \lambda) \\ x(\tau) &= \xi_\lambda, \end{aligned} \qquad (I'_{\lambda,\tau})$$

where it is assumed that ξ_λ depends continuously on λ, we obtain the following result.

COROLLARY 9.2. Let $f \in C(J \times R \times D, R)$, where $J \subset R$ is an open interval and $D \subset R$. Assume that for each pair of compact subsets $J_0 \subset J$ and $D_0 \subset D$ there exists a constant $L = L_{J_0,D_0} > 0$ such that for all $(t, \lambda) \in J_0 \times D_0$, $x, y \in R$, the Lipschitz condition (9.1) is true. Then the initial-value problem $(I'_{\lambda,\tau})$ has a unique solution $\phi(t, \tau, \lambda)$, where $\phi \in C(J \times J \times D, R)$. Furthermore, if D is a set such that for every $\lambda_0 \in D$ there exists an $\epsilon > 0$ so that $\overline{[\lambda_0 - \epsilon, \lambda_0 + \epsilon] \cap D} \subset D$, then $\phi(t, \tau, \lambda) \to \phi(t, \tau_0, \lambda_0)$, uniformly for $t \in J_0$ as $(\tau, \lambda) \to (\tau_0, \lambda_0)$, where J_0 is any compact subset of J. ■

Proof of Theorem 9.1. For the solution $\phi(t, \tau, \xi, \lambda)$, we first show that $\phi \in C(J \times J \times R \times D, R)$. By $(I'_{\lambda,\tau,\xi})$ we have that

$$\phi(t, \tau, \xi, \lambda) = \xi + \int_{t_0}^{t} f(s, \phi(s, \tau, \xi, \lambda), \lambda)\, ds. \qquad (9.2)$$

We want to show that for $(t_0, \tau_0, \xi_0, \lambda_0) \in J \times J \times R \times D$,

$$\phi(t_m, \tau_m, \xi_m, \lambda_m) \to \phi(t_0, \tau_0, \xi_0, \lambda_0)$$

as $(t_m, \tau_m, \xi_m, \lambda_m) \to (t_0, \tau_0, \xi_0, \lambda_0)$, where $(t_m, \tau_m, \xi_m, \lambda_m) \in J \times J \times R \times D$ for each $m \in N$. By (9.2) we have

$$\begin{aligned} &\phi(t_m, \tau_m, \xi_m, \lambda_m) - \phi(t_0, \tau_0, \xi_0, \lambda_0) \\ &\quad = \xi_m - \xi_0 + \int_{\tau_m}^{t_m} f(s, \phi(s, \tau_m, \xi_m, \lambda_m), \lambda_m)\, ds \\ &\qquad - \int_{\tau_0}^{t_0} f(s, \phi(s, \tau_0, \xi_0, \lambda_0), \lambda_0)\, ds \\ &\quad = \xi_m - \xi_0 + \int_{\tau_m}^{t_m} (f(s, \phi(s, \tau_m, \xi_m, \lambda_m), \lambda_m) - f(s, \phi(s, \tau_0, \xi_0, \lambda_0), \lambda_m))\, ds \end{aligned}$$

$$+\int_{\tau_m}^{t_m}(f(s,\phi(s,\tau_0,\xi_0,\lambda_0),\lambda_m)-f(s,\phi(s,\tau_0,\xi_0,\lambda_0),\lambda_0))\,ds$$
$$+\int_{t_0}^{t_m} f(s,\phi(s,\tau_0,\xi_0,\lambda_0),\lambda_0)\,ds$$
$$-\int_{\tau_0}^{\tau_m} f(s,\phi(s,\tau_0,\xi_0,\lambda_0),\lambda_0)\,ds. \tag{9.3}$$

Denote

$$\langle a,b\rangle \triangleq \begin{cases}[a,b], & \text{if } a\le b,\\ [b,a], & \text{if } a>b.\end{cases}$$

Since $t_0, \tau_0 \in J$, and J is an open interval, it follows that for m sufficiently large, $\langle \tau_m, t_m\rangle$, $\langle t_0, t_m\rangle$, and $\langle \tau_0, \tau_m\rangle$ are contained in J. Also, given $(\tau_0, \xi_0, \lambda_0) \in J\times J\times D$, $\phi(s, \tau_0, \xi_0, \lambda_0)$ is a continuous function on J. Note that since $f \in C(J\times R\times D, R)$, we can assume that for m sufficiently large,

$$\max_{s\in\langle t_0,t_m\rangle} |f(s,\phi(s,\tau_0,\xi_0,\lambda_0),\lambda_0)| \le M$$

and

$$\max_{s\in\langle \tau_0,\tau_m\rangle} |f(s,\phi(s,\tau_0,\xi_0,\lambda_0),\lambda_0)| \le M$$

for some $M > 0$.

We now have

$$\begin{aligned}|\phi(t_m,\tau_m,\xi_m,\lambda_m)-\phi(t_0,\tau_0,\xi_0,\lambda_0)| &\le |\xi_m-\xi_0| + M(|t_m-t_0|+|\tau_m-\tau_0|)\\ &\quad+\left|\int_{\tau_m}^{t_m}|f(s,\phi(s,\tau_m,\xi_m,\lambda_m),\lambda_m)-f(s,\phi(s,\tau_0,\xi_0,\lambda_0),\lambda_m)|ds\right|\\ &\quad+\left|\int_{\tau_m}^{t_m}|f(s,\phi(s,\tau_0,\xi_0,\lambda_0),\lambda_m)-f(s,\phi(s,\tau_0,\xi_0,\lambda_0),\lambda_0)|ds\right|.\end{aligned}$$

Without loss of generality, we assume that $t_0 \ge \tau_0$. Then for m sufficiently large, there exists $\epsilon > 0$ such that $[\tau_0-\epsilon, t_0+\epsilon] \subset J$ and

$$\begin{aligned}|\phi(t_m,\tau_m,\xi_m,\lambda_m)-\phi(t_0,\tau_0,\xi_0,\lambda_0)| &\le |\xi_m-\xi_0| + M(|t_m-t_0|+|\tau_m-\tau_0|)\\ &\quad+\int_{\tau_0-\epsilon}^{t_0+\epsilon}|f(s,\phi(s,\tau_m,\xi_m,\lambda_m),\lambda_m)-f(s,\phi(s,\tau_0,\xi_0,\lambda_0),\lambda_m)|\,ds\\ &\quad+\int_{\tau_0-\epsilon}^{t_0+\epsilon}|f(s,\phi(s,\tau_0,\xi_0,\lambda_0),\lambda_m)-f(s,\phi(s,\tau_0,\xi_0,\lambda_0),\lambda_0)|\,ds\\ &\le |\xi_m-\xi_0| + M(|t_m-t_0|+|\tau_m-\tau_0|)\\ &\quad+\int_{\tau_0-\epsilon}^{t_0+\epsilon}|f(s,\phi(s,\tau_0,\xi_0,\lambda_0),\lambda_m)-f(s,\phi(s,\tau_0,\xi_0,\lambda_0),\lambda_0)|\,ds\\ &\quad+L\int_{\tau_0-\epsilon}^{t_0+\epsilon}|\phi(s,\tau_m,\xi_m,\lambda_m)-\phi(s,\tau_0,\xi_0,\lambda_0)|\,ds.\end{aligned}$$

By the Gronwall Inequality, we obtain that

$$\begin{aligned}|\phi(t_m,\tau_m,\xi_m,\lambda_m)-\phi(t_0,\tau_0,\xi_0,\lambda_0)| &\le (|\xi_m-\xi_0| + M(|t_m-t_0|+|\tau_m-\tau_0|)\\ &\quad+\int_{\tau_0-\epsilon}^{t_0+\epsilon}|f(s,\phi(s,\tau_0,\xi_0,\lambda_0),\lambda_m)-f(s,\phi(s,\tau_0,\xi_0,\lambda_0),\lambda_0)|\,ds)e^{L(t_0-\tau_0+2\epsilon)}.\end{aligned} \tag{9.4}$$

Since $f(s, \phi(s, \tau_0, \xi_0, \lambda_0), \lambda) \in C([\tau_0 - \epsilon, t_0 + \epsilon] \times R \times \overline{\{\lambda_m\}}, R)$ (with s, λ as variables and $\overline{\{\lambda_m\}} \subset D$ since $\lambda_m \to \lambda_0 \in D$) and since $[\tau_0 - \epsilon, t_0 + \epsilon] \times \overline{\{\lambda_m\}}$ is a compact subset of $J \times D$, we have by Theorem 5.2 and (9.4), as $m \to \infty$, that

$$\lim_{m\to\infty} |\phi(t_m, \tau_m, \xi_m, \lambda_m) - \phi(t_0, \tau_0, \xi_0, \lambda_0)| = 0.$$

Thus, $\phi \in C(J \times J \times R \times D, R)$.

Similarly, under the assumption on D, we can prove that $\phi(t, \tau, \xi, \lambda) \to \phi(t, \tau_0, \xi_0, \lambda_0)$, uniformly for $t \in J_0$ as $(\tau, \xi, \lambda) \to (\tau_0, \xi_0, \lambda_0)$, where J_0 is any compact subset of J. In place of (9.3), we have

$$\begin{aligned}
&\phi(t, \tau_0 + \Delta\tau, \xi_0 + \Delta\xi, \lambda_0 + \Delta\lambda) - \phi(t, t_0, \xi_0, \lambda_0)\\
&\quad = \Delta\xi + \int_{\tau_0+\Delta\tau}^{t} (f(s, \phi(s, \tau_0 + \Delta\tau, \xi_0 + \Delta\xi, \lambda_0 + \Delta\lambda), \lambda_0 + \Delta\lambda)\\
&\qquad - f(s, \phi(s, \tau_0, \xi_0, \lambda_0), \lambda_0 + \Delta\lambda))\, ds\\
&\qquad + \int_{\tau_0+\Delta\tau}^{t} (f(s, \phi(s, \tau_0, \xi_0, \lambda_0), \lambda_0 + \Delta\lambda) - f(s, \phi(s, \tau_0, \xi_0, \lambda_0), \lambda_0))\, ds\\
&\qquad - \int_{\tau_0}^{\tau_0+\Delta\tau} f(s, \phi(s, \tau_0, \xi_0, \lambda_0), \lambda_0)\, ds. \qquad (9.5)
\end{aligned}$$

Let $\|\Delta\| = \sqrt{(\Delta\tau)^2 + (\Delta\xi)^2 + (\Delta\lambda)^2}$, $t_{\max} \triangleq \max\{t : t \in J_0\}$, and $t_{\min} \triangleq \min\{t : t \in J_0\}$. Then there exists $\epsilon > 0$ such that when $\|\Delta\| < \epsilon$, $< \tau_0 + \Delta\tau, t > \subset < \tau_0 - \epsilon, t_{\max} > \cup < t_{\min}, \tau_0 + \epsilon > \triangleq T_{\tau_0} \subset J$, $< \tau_0, \tau_0 + \Delta\tau > \subset [\tau_0 - \epsilon, \tau_0 + \epsilon] \subset J$, and

$$\overline{[\lambda_0 - \epsilon, \lambda_0 + \epsilon] \cap D} \triangleq D_{\lambda_0} \subset D.$$

Note that both T_{τ_0} and D_{λ_0} are compact subsets of J and D, respectively. For $(t, \lambda) \in T_{\tau_0} \times D_{\lambda_0}$, let

$$M = \max_{s\in[\tau_0-\epsilon,\tau_0+\epsilon]} |f(s, \phi(s, \tau_0, \xi_0, \lambda_0), \lambda_0)|.$$

By the Lipschitz condition, we obtain

$$\begin{aligned}
&|\phi(t, \tau_0 + \Delta\tau, \xi_0 + \Delta\xi, \lambda_0 + \Delta\lambda) - \phi(t, \tau_0, \xi_0, \lambda_0)|\\
&\quad \le |\Delta\xi| + M|\Delta\tau|\\
&\qquad + \int_{s\in T_{\tau_0}} |f(s, \phi(s, \tau_0, \xi_0, \lambda_0), \lambda_0 + \Delta\lambda) - f(s, \phi(s, \tau_0, \xi_0, \lambda_0), \lambda_0)|\, ds\\
&\qquad + L\int_{s\in T_{\tau_0}} |\phi(s, \tau_0 + \Delta\tau, \xi_0 + \Delta\xi, \lambda_0 + \Delta\lambda) - \phi(s, \tau_0, \xi_0, \lambda_0)|\, ds.
\end{aligned}$$

Again, using the Gronwall Inequality, we have that

$$\begin{aligned}
&|\phi(t, \tau_0 + \Delta\tau, \xi_0 + \Delta\xi, \lambda_0 + \Delta\lambda) - \phi(t, \tau_0, \xi_0, \lambda_0)|\\
&\quad \le (|\Delta\xi| + M|\Delta\tau|\\
&\qquad + \int_{s\in T_{\tau_0}} |f(s, \phi(s, \tau_0, \xi_0, \lambda_0), \lambda_0 + \Delta\lambda) - f(s, \phi(s, \tau_0, \xi_0, \lambda_0), \lambda_0)|\, ds)\\
&\qquad \times e^{L(|t_{\max}-\tau_0|+|t_{\min}-\tau_0|+2\epsilon)}. \qquad (9.6)
\end{aligned}$$

By the first part of the proof, we already know that $f(s, \phi(s, \tau_0, \xi_0, \lambda_0), \lambda) \in C(T_{\tau_0} \times R \times D_{\lambda_0}, R)$, which implies that f is uniformly continuous on the compact set $T_{\tau_0} \times D_{\lambda_0}$. Therefore, by Theorem 5.2 and (9.6) we know that $\phi(t, \tau, \xi, \lambda) \to \phi(t, \tau_0, \xi_0, \lambda_0)$, uniformly as $(\tau, \xi, \lambda) \to (\tau_0, \xi_0, \lambda_0)$ for $t \in J_0$. ■

1.10 SYSTEMS OF FIRST-ORDER ORDINARY DIFFERENTIAL EQUATIONS

In Sections 1.6 to 1.9 we addressed the *existence* of solutions, the *continuation* of solutions, the *uniqueness* of solutions, and the *continuous dependence* of solutions on initial data and parameters for the *scalar initial-value problem for ordinary differential equations* [characterized by (E') and (I') or by (V') (resp., by $(I'_{\lambda,\tau})$]. In this section we show that these results can be extended to initial-value problems characterized by *systems of equations* [determined by (E) and (I) or by (V) (resp., by $(I_{\lambda,\tau})$)] with no essential changes in proofs. Before we can accomplish this, however, we need to introduce additional background material.

A. More Mathematical Preliminaries: Vector Spaces

We will require the notion of vector space, or linear space over a field.

DEFINITION 10.1. Let F be a set containing more than one element and let there be two operations "+" and "·" defined on F (i.e., "+" and "·" are mappings of $F \times F$ into F), called *addition* and *multiplication*, respectively. Then for each $\alpha, \beta \in F$ there is a unique element $\alpha + \beta \in F$, called the *sum* of α and β, and a unique element $\alpha\beta \triangleq \alpha \cdot \beta \in F$, called the *product* of α and β. We say that $\{F; +, \cdot\}$ is a *field* provided that the following axioms are satisfied:

(i) $\alpha + (\beta + \gamma) = (\alpha + \beta) + \gamma$ and $\alpha \cdot (\beta \cdot \gamma) = (\alpha \cdot \beta) \cdot \gamma$ for all $\alpha, \beta, \gamma \in F$ (i.e., "+" and "·" are associative operations);
(ii) $\alpha + \beta = \beta + \alpha$ and $\alpha \cdot \beta = \beta \cdot \alpha$ for all $\alpha, \beta \in F$ (i.e., "+" and "·" are commutative operations);
(iii) $\alpha \cdot (\beta + \gamma) = \alpha \cdot \beta + \alpha \cdot \gamma$ for all $\alpha, \beta, \gamma \in F$ (i.e., "·" is distributive over "+");
(iv) There exists an element $0_F \in F$ such that $0_F + \alpha = \alpha$ for all $\alpha \in F$ (i.e., 0_F is the identity element of F with respect to "+");
(v) There exists an element $1_F \in F$, $1_F \neq 0_F$, such that $1_F \cdot \alpha = \alpha$ for all $\alpha \in F$ (i.e., 1_F is the identity element of F with respect to "·");
(vi) For every $\alpha \in F$ there exists an element $-\alpha \in F$ such that $\alpha + (-\alpha) = 0_F$ (i.e., $-\alpha$ is the additive inverse of F);
(vii) For any $\alpha \neq 0_F$ there exists an $\alpha^{-1} \in F$ such that $\alpha \cdot (\alpha^{-1}) = 1_F$ (i.e., α^{-1} is the multiplicative inverse of F). ■

In the sequel, we will usually speak of a field F rather than "a field $\{F; +, \cdot\}$."

Perhaps the most widely known fields are the *field of real numbers* R and the *field of complex numbers* C. Another field we will encounter (see Chapter 2) is the *field of rational functions* (i.e., rational fractions over polynomials).

As a third example, we let $F = \{0, 1\}$ and define on F (binary) addition as $0 + 0 = 0 = 1 + 1$, $1 + 0 = 1 = 0 + 1$ and (binary) multiplication as $1 \cdot 0 = 0 \cdot 1 = 0 \cdot 0 = 0$, $1 \cdot 1 = 1$. It is easily verified that $\{F; +, \cdot\}$ is a field.

As a fourth example, let P denote the set of polynomials with real coefficients and define addition "+" and multiplication "·" on P in the usual manner. Then $\{F; +, \cdot\}$ is *not* a field since, e.g., axiom (vii) in Definition 10.1 is violated (i.e., the *multiplicative* inverse of a polynomial $p \in P$ is not necessarily a polynomial).

DEFINITION 10.2. Let V be a nonempty set, let F be a field, let "+" denote a mapping of $V \times V$ into V, and let "$\cdot$" denote a mapping of $F \times V$ into V. Let the members $x \in V$ be called *vectors,* let the elements $\alpha \in F$ be called *scalars,* let the operation "+" defined on V be called *vector addition,* and let the mapping "$\cdot$" be called *scalar multiplication* or *multiplication of vectors by scalars.* Then for each $x, y \in V$ there is a unique element, $x + y \in V$, called the *sum of x and y*, and for each $x \in V$ and $\alpha \in F$ there is a unique element, $\alpha x \triangleq \alpha \cdot x \in V$, called the *multiple of x by* α. We say that the nonempty set V and the field F, along with the two mappings of vector addition and scalar multiplication, constitute a *vector space* or a *linear space* if the following axioms are satisfied:

(i) $x + y = y + x$ for every $x, y \in V$.
(ii) $x + (y + z) = (x + y) + z$ for every $x, y, z \in V$.
(iii) There is a unique vector in V, called the *zero vector* or the *null vector* or the *origin*, that is denoted by 0_V and has the property that $0_V + x = x$ for all $x \in V$.
(iv) $\alpha(x + y) = \alpha x + \alpha y$ for all $\alpha \in F$ and for all $x, y \in V$.
(v) $(\alpha + \beta)x = \alpha x + \beta x$ for all $\alpha, \beta \in F$ and for all $x \in V$.
(vi) $(\alpha\beta)x = \alpha(\beta x)$ for all $\alpha, \beta \in F$ and for all $x \in V$.
(vii) $0_F x = 0_V$ for all $x \in V$.
(viii) $1_F x = x$ for all $x \in V$. ■

In subsequent applications, when the meaning is clear from context, we will write 0 in place of 0_F, 1 in place of 1_F, and 0 in place of 0_V. To indicate the relationship between the set of vectors V and the underlying field F, we sometimes refer to a *vector space V over the field F*, and we signify this by writing (V, F). However, usually, when the field in question is clear from context, we speak of a vector space V. If F is the field of real numbers, R, we call the space a *real vector space*. Similarly, if F is the field of complex numbers, C, we speak of a *complex vector space.*

Examples of vector spaces

EXAMPLE 10.1. Let $V = F^n$ denote the set of all ordered n-tuples of elements from a field F. Thus, if $x \in F^n$, then $x = (x_1, \dots, x_n)^T$, where $x_i \in F$, $i = 1, \dots, n$. With $x, y \in F^n$ and $\alpha \in F$, let vector addition and scalar multiplication be defined as

$$x + y = (x_1, \dots, x_n)^T + (y_1, \dots, y_n)^T \triangleq (x_1 + y_1, \dots, x_n + y_n)^T \tag{10.1}$$

and

$$\alpha x = \alpha(x_1, \dots, x_n)^T \triangleq (\alpha x_1, \dots, \alpha x_n)^T. \tag{10.2}$$

In this case the null vector is defined as $0 = (0, \dots, 0)^T$ and the vector $-x$ is defined as $-x = -(x_1, \dots, x_n)^T = (-x_1, \dots, -x_n)^T$. Then we utilize the properties of the field F, all axioms of Definition 10.2 are readily verified, and therefore, F^n is a vector space. We call this space the *space* F^n *of n-tuples of elements of F.* If in particular we let $F = R$, we have R^n, *the n-dimensional real coordinate space*. Similarly, if we let $F = C$, we have C^n, *the n-dimensional complex coordinate space.* ■

We note that the set of points in R^2, (x_1, x_2), that satisfy the linear equation

$$x_1 + x_2 + c = 0, \qquad c \neq 0,$$

with addition and multiplication defined as in Eqs. (10.1) and (10.2), is *not* a vector space (why?).

EXAMPLE 10.2. Let $V = R^\infty$ denote the set of all infinite sequences of real numbers,

$$x = \{x_1, x_2, \dots, x_k, \dots\} \triangleq \{x_i\},$$

let vector addition be defined similarly as in (10.1), and let scalar multiplication be defined as in (10.2). It is again an easy matter to show that this space is a vector space.

On some occasions we will find it convenient to modify $V = R^\infty$ to consist of the set of all real infinite sequences $\{x_i\}$, $i \in Z$. ■

EXAMPLE 10.3. Let $1 \leq p \leq \infty$ and define $V = l_p$ by

$$\begin{aligned} l_p &= \left\{ x \in R^\infty : \sum_{i=1}^{\infty} |x_i|^p < \infty \right\}, \qquad 1 \leq p < \infty, \\ l_\infty &= \{x \in R^\infty : \sup_i\{|x_i|\} < \infty\}. \end{aligned} \tag{10.3}$$

Define vector addition and scalar multiplication on l_p as in (10.1) and (10.2), respectively. It can be verified that this space, called the *l_p-space,* is a vector space. ■

In proving that l_p, $1 \leq p \leq \infty$, is indeed a vector space, in establishing some of the properties of norms defined on the l_p-spaces (see Examples 10.10 and 10.11), in defining linear transformations on l_p-spaces (see, e.g., Example 10.8), and in many other applications, we make use of the *Hölder and Minkowski Inequalities for infinite sums,* given below. (These inequalities are of course also valid for *finite sums.*) For proofs of these results, refer, e.g., to Michel and Herget [12, pp. 268–270].

Hölder's Inequality states that if $p, q \in R$ are such that $1 < p < \infty$ and $1/p + 1/q = 1$, and if $\{x_i\}$ and $\{y_i\}$ are sequences in either R or C, and if $\sum_{i=1}^{\infty} |x_i|^p < \infty$ and $\sum_{i=1}^{\infty} |y_i|^q < \infty$, then

$$\sum_{i=1}^{\infty} |x_i y_i| \leq \left(\sum_{i=1}^{\infty} |x_i|^p \right)^{1/p} \left(\sum_{i=1}^{\infty} |y_i|^q \right)^{1/q}. \tag{H_s}$$

Minkowski's Inequality states that if $p \in R$, where $1 \leq p < \infty$, and if $\{x_i\}$ and $\{y_i\}$ are sequences in either R or C, and if $\sum_{i=1}^{\infty} |x_i|^p < \infty$ and $\sum_{i=1}^{\infty} |y_i|^p < \infty$, then

$$\left(\sum_{i=1}^{\infty} |x_i \pm y_i|^p \right)^{1/p} \leq \left(\sum_{i=1}^{\infty} |x_i|^p \right)^{1/p} + \left(\sum_{i=1}^{\infty} |y_i|^p \right)^{1/p}. \tag{M_s}$$

If in particular $p = q = \frac{1}{2}$, then (H_s) reduces to *the Schwarz Inequality for sums.*

EXAMPLE 10.4. Let $V = C([a, b], R)$. We note that $x = y$ if and only if $x(t) = y(t)$ for all $t \in [a, b]$, and that the null vector is the function that is zero for all $t \in [a, b]$. Let F denote the field of real numbers, let $\alpha \in F$, and let vector addition and scalar multiplication be defined pointwise by

$$(x + y)(t) = x(t) + y(t) \qquad \text{for all } t \in [a, b] \tag{10.4}$$

and

$$(\alpha x)(t) = \alpha x(t) \qquad \text{for all } t \in [a, b]. \tag{10.5}$$

Then clearly $x + y \in V$ whenever $x, y \in V$, $\alpha x \in V$ whenever $\alpha \in F$ and $x \in V$, and all the axioms of a vector space are satisfied. We call this space the *space of real-valued continuous functions on* $[a, b]$ and we frequently denote it simply by $C[a, b]$. ■

EXAMPLE 10.5. Let $1 \leq p < \infty$ and let V denote the set of all real-valued functions x on the interval $[a, b]$ such that

$$\int_a^b |x(t)|^p \, dt < \infty. \tag{10.6}$$

Let $F = R$ and let vector addition and scalar multiplication be defined as in (10.4) and (10.5), respectively. It can be verified that this space is a vector space.

In this book we will usually assume that in (10.6), integration is in the Riemann sense. When integration in (10.6) is in the Lebesgue sense, then the vector space under discussion is called an L_p-space (or the space $L_p[a, b]$). ■

In proving that the L_p-spaces are indeed vector spaces, in establishing properties of norms defined on L_p-spaces (see, e.g., Example 10.12), in defining linear transformations on L_p-spaces (ee, e.g., Example 10.12), and in many other applications, we make use of the *Hölder and Minkowski Inequalities for integrals,* given below. (These inequalities are valid when integration is in the Riemann and the Lebesgue senses.) For proofs of these results, refer, e.g., to Michel and Herget [12, pp. 268–270].

Hölder's Inequality states that if $p, q \in R$ are such that $1 < p < \infty$ and $1/p + 1/q = 1$, if $[a, b]$ is an interval on the real line, if $f, g : [a, b] \to R$, and if $\int_a^b |f(t)|^p \, dt < \infty$ and $\int_a^b |g(t)|^q \, dt < \infty$, then

$$\int_a^b |f(t)g(t)| \, dt \leq \left(\int_a^b |f(t)|^p \, dt\right)^{1/p} \left(\int_a^b |g(t)|^q \, dt\right)^{1/q}. \qquad (H_I)$$

Minkowski's Inequality states that if $p \in R$, where $1 \leq p < \infty$, if $f, g : [a, b] \to R$, and if $\int_a^b |f(t)|^p \, dt < \infty$ and $\int_a^b |g(t)|^p \, dt < \infty$, then

$$\left(\int_a^b |f(t) \pm g(t)|^p \, dt\right)^{1/p} \leq \left(\int_a^b |f(t)|^p \, dt\right)^{1/p} + \left(\int_a^b |g(t)|^p \, dt\right)^{1/p}. \qquad (M_I)$$

If in particular $p = q = \frac{1}{2}$, then (H_I) reduces to *the Schwarz Inequality for integrals.*

EXAMPLE 10.6. Let V denote the set of all continuous real-valued functions on the interval $[a, b]$ such that

$$\sup_{a \leq t \leq b} |x(t)| < \infty. \qquad (10.7)$$

Let $F = R$ and let vector addition and scalar multiplication be defined as in (10.4) and (10.5), respectively. It can readily be verified that this space is a vector space.

In some applications it is necessary to expand the above space to the set of measurable real-valued functions on $[a, b]$ and to replace (10.7) by

$$\operatorname{ess\,sup}_{a \leq t \leq b} |x(t)| < \infty, \qquad (10.8)$$

where ess sup denotes the essential supremum, i.e.,

$$\operatorname{ess\,sup}_{a \leq t \leq b} |x(t)| = \inf \{M : m\{t : |x(t)| > M\} = 0\},$$

where m denotes the Lebesgue measure. In this case, the vector space under discussion is called the L_∞-space. ■

Next, we consider linear transformations.

DEFINITION 10.3. A mapping T of a linear space V into a linear space W, where V and W are vector spaces over the same field F, is called a *linear transformation* or a *linear operator* provided that

(L-i) $\quad T(x + y) = T(x) + T(y) \quad$ for all $x, y \in V$.

(L-ii) $\quad T(\alpha x) = \alpha T(x) \quad$ for all $x \in V$ and $\alpha \in F$. ■

In Section 1.16 we will discuss in detail the representation of linear systems by means of linear operators. This discussion will be continued in Chapter 2. In the following, we consider three specific examples of linear transformations.

EXAMPLE 10.7. Let $(V, R) = (R^n, R)$ and $(W, R) = (R^m, R)$ be vector spaces defined as in Example 10.1, let $A = [a_{ij}] \in R^{m\times n}$, and let $T : V \to W$ be defined by the equation

$$y = Ax, \qquad y \in R^m, \qquad x \in R^n.$$

It is easily verified, using the properties of matrices, that T is a linear transformation. ■

EXAMPLE 10.8. Let $(V, R) = (l_p, R)$ be the vector space defined in Example 10.3 (modified to consist of sequences $\{x_i\}$, $i \in Z$, in place of $\{x_i\}$, $i = 1, 2, \ldots$). Let $h : Z \times Z \to R$ be a function having the property that for each $x \in V$, the infinite sum

$$\sum_{k=-\infty}^{\infty} h(n, k)x(k)$$

exists and defines a function of n on Z. Let $T : V \to V$ be defined by

$$y(n) = \sum_{k=-\infty}^{\infty} h(n, k)x(k).$$

It is easily verified that T is a linear transformation.

The existence of the above sum is ensured under appropriate assumptions. For example, by using the Hölder Inequality it is readily shown that if, e.g., for fixed n, $\{h(n, k)\} \in l_2$ and $\{x(k)\} \in l_2$, then the above sum is well defined. The above sum exists also if, e.g., $\{x(k)\} \in l_\infty$ and $\{h(n, k)\} \in l_1$ for fixed n. ■

EXAMPLE 10.9. Let (V, R) denote the vector space given in Example 10.5 and let $k \in C([a, b] \times [a, b], R)$ have the property that for each $x \in V$, the Riemann integral

$$\int_a^b k(s, t)x(t)\,dt$$

exists and defines a continuous function of s on $[a, b]$. Let $T : V \to V$ be defined by

$$(Tx)(s) = y(s) = \int_a^b k(s, t)x(t)\,dt.$$

It is readily verified that T is a linear transformation of V into V. ■

B. Further Mathematical Preliminaries: Normed Linear Spaces

In the following, we require for (V, F) that F be either the field of real numbers R or the field of complex numbers C. For such linear spaces we say that a function $\|\cdot\| : V \to R^+$ is a *norm* if

(N-i) $\|x\| \geq 0$ for every *vector* $x \in V$ and $\|x\| = 0$ if and only if x is the null vector (i.e., $x = 0$);

(N-ii) For every scalar $\alpha \in F$ and for every vector $x \in V$, $\|\alpha x\| = |\alpha|\,\|x\|$, where $|\alpha|$ denotes the absolute value of α when $F = R$ and the modulus when $F = C$;

(N-iii) For every x and y in V, $\|x + y\| \leq \|x\| + \|y\|$. (This inequality is called the *triangle inequality*.)

We call a vector space on which a norm has been defined a *normed vector space* or a *normed linear space.*

EXAMPLE 10.10. On the linear space (R^n, R), we define for every $x = (x_1, \ldots, x_n)^T$,

$$\|x\|_p = \left(\sum_{i=1}^{n} |x_i|^p\right)^{1/p}, \qquad 1 \leq p < \infty, \tag{10.9}$$

and

$$\|x\|_\infty = \max\{|x_i| : 1 \leq i \leq n\}. \tag{10.10}$$

Using Minkowski's Inequality for finite sums, (M_s), it is an easy matter to show that for every p, $1 \leq p \leq \infty$, $\|\cdot\|_p$ is a norm on R^n. In addition to $\|\cdot\|_\infty$, of particular interest to us will be the cases $p = 1$ and $p = 2$, i.e.,

$$\|x\|_1 = \sum_{i=1}^{n} |x_i| \tag{10.11}$$

and

$$\|x\|_2 = \left(\sum_{i=1}^{n} |x_i|^2\right)^{1/2}. \tag{10.12}$$

The norm $\|\cdot\|_1$ is sometimes referred to as the *taxicab norm* or *Manhattan norm,* while $\|\cdot\|_2$ is called the *Euclidean norm.*

The foregoing norms are related by the inequalities

$$\|x\|_\infty \leq \|x\|_1 \leq n\|x\|_\infty \tag{10.13}$$

$$\|x\|_\infty \leq \|x\|_2 \leq \sqrt{n}\|x\|_\infty \tag{10.14}$$

$$\|x\|_2 \leq \|x\|_1 \leq \sqrt{n}\|x\|_2. \tag{10.15}$$

Also, for $p = 2$, we obtain from the Hölder Inequality for finite sums, (H_s), the *Schwarz Inequality*

$$|x^T y| = \left|\sum_{i=1}^{n} x_i y_i\right| \leq \left(\sum_{i=1}^{n} |x_i|^2\right)^{1/2} \left(\sum_{i=1}^{n} |y_i|^2\right)^{1/2} \tag{10.16}$$

for all $x, y \in R^n$. ■

The assertions made in the above example turn out to be also true for the space (C^n, C). We ask the reader to verify these relations.

EXAMPLE 10.11. On the space l_p given in Example 10.3, let

$$\|x\|_p = \left(\sum_{i=1}^{\infty} |x_i|^p\right)^{1/p}, \qquad 1 \leq p < \infty,$$

and

$$\|x\|_\infty = \sup_i |x_i|.$$

Using Minkowski's Inequality for infinite sums, (Ms), it is an easy matter to show that $\|\cdot\|_p$ is a norm for every p, $1 \leq p \leq \infty$. ■

EXAMPLE 10.12. On the space given in Example 10.5, let

$$\|x\|_p = \left(\int_a^b |x(t)|^p \, dt\right)^{1/p}, \qquad 1 \leq p < \infty.$$

Using Minkowski's Inequality for integrals, (M_I), it can readily be verified that $\|\cdot\|_p$ is a norm for every p, $1 \leq p < \infty$. Also, on the space of continuous functions given in

Example 10.6, assume that (10.7) holds. Then

$$\|x\|_\infty = \sup_{a \le t \le b} |x(t)|$$

is easily shown to define a norm. Furthermore, expression (10.8) can also be used to define a norm. ■

EXAMPLE 10.13. We can also define the norm of a matrix. To this end, consider the set of real $m \times n$ matrices, $R^{m\times n} = V$ and $F = R$. It is easily verified that $(V, F) = (R^{m\times n}, R)$ is a vector space, where vector addition is defined as matrix addition and multiplication of vectors by scalars is defined as multiplication of matrices by scalars.

For a given norm $\|\cdot\|_u$ on R^n and a given norm $\|\cdot\|_v$ on R^m, we define $\|\cdot\|_{vu} : R^{m\times n} \to R^+$ by

$$\|A\|_{vu} = \sup\{\|Ax\|_v : x \in R^n \text{ with } \|x\|_u = 1\}. \tag{10.17}$$

It is easily verified that

(M-i) $\|Ax\|_v \le \|A\|_{vu}\|x\|_u$ for any $x \in R^n$;

(M-ii) $\|A + B\|_{vu} \le \|A\|_{vu} + \|B\|_{vu}$;

(M-iii) $\|\alpha A\|_{vu} = |\alpha|\|A\|_{vu}$ for all $\alpha \in R$;

(M-iv) $\|A\|_{vu} \ge 0$ and $\|A\|_{vu} = 0$ if and only if A is the zero matrix (ie., $A = 0$);

(M-v) $\|A\|_{vu} \le \sum_{i=1}^m \sum_{j=1}^n |a_{ij}|$ for any p-vector norms defined on R^n and R^m.

Properties (M-ii) to (M-iv) clearly show that $\|\cdot\|_{vu}$ defines a norm on $R^{m\times n}$ and justifies the use of the term *matrix norm*. Since the matrix norm $\|\cdot\|_{vu}$ depends on the choice of the vector norms, $\|\cdot\|_u$, and $\|\cdot\|_v$, defined on $U \triangleq R^n$ and $V \triangleq R^m$, respectively, we say that the matrix norm $\|\cdot\|_{uv}$ is *induced* by the vector norms $\|\cdot\|_u$ and $\|\cdot\|_v$. In particular, if $\|\cdot\|_u = \|\cdot\|_p$ and $\|\cdot\|_v = \|\cdot\|_p$, then the notation $\|A\|_p$ is frequently used to denote the norm of A.

As a specific case, let $A = [a_{ij}] \in R^{m\times n}$. Then it is easily verified that

$$\|A\|_1 = \max_j \left(\sum_{i=1}^m |a_{ij}|\right)$$

$$\|A\|_2 = [\max \lambda(A^T A)]^{1/2},$$

where $\max \lambda(A^T A)$ denotes the largest eigenvalue of $A^T A$, and

$$\|A\|_\infty = \max_i \left(\sum_{j=1}^n |a_{ij}|\right).$$

When it is clear from context which vector spaces and vector norms are being used, the indicated subscripts on the matrix norms are usually not used. For example, if $A \in R^{m\times n}$ and $B \in R^{n\times k}$, it can be shown that

(M-vi) $\|AB\| \le \|A\|\,\|B\|$.

In (M-vi) we have omitted subscripts on the matrix norms to indicate inducing vector norms. ■

We conclude this subsection by noting that it is possible to define norms on $(R^{m\times n}, R)$ that need not be induced by vector norms. Furthermore, the entire discussion given in Example 10.13 holds also for norms defined on complex spaces, e.g., $(C^{m\times n}, C)$.

C. Additional Mathematical Preliminaries: Convergence

Although most of what we will present in this subsection is true in a rather general setting, we will confine ourselves to the spaces (R^n, R) or (C^n, C).

Using the concept of norm, we can define *distance* between vectors x and y in R^n [or in C^n] by $d(x, y) = \|x - y\|$. The three basic properties of distance are given next and are a consequence of the axioms of a norm:

(D-i) $\|x - y\| \geq 0$ for all vectors x, y and $\|x - y\| = 0$ if and only if $x = y$;
(D-ii) $\|x - y\| = \|y - x\|$ for all vectors x, y;
(D-iii) $\|x - z\| \leq \|x - y\| + \|y - z\|$ for all vectors x, y, z.

We can now define *spherical neighborhood* in R^n (in C^n) with *center* x_0 and *radius* $h > 0$ as

$$B(x_0, h) = \{x \in R^n : \|x - x_0\| < h\}.$$

If in particular the center of a spherical neighborhood with radius h is the origin, then we write $B(0, h) \triangleq B(h)$, i.e.,

$$B(h) = \{x \in R^n : \|x\| < h\}.$$

We shall use the notation

$$\overline{B(x_0, h)} = \{x \in R^n : \|x - x_0\| \leq h\}$$

and

$$\overline{B(h)} = \{x \in R^n : \|x\| \leq h\}.$$

The introduction of vector and matrix norms enables us to generalize the notions of convergence of sequences, continuity of functions, and the like. We will not retrace here the entire presentation given in Sections 1.2 and 1.5. Instead, to demonstrate what is involved in these generalizations, we consider a few specific cases.

A *sequence of vectors* $\{x_m\} = \{(x_{1m}, \ldots, x_{nm})^T\} \subset R^n$ is said to *converge* to a vector $x \in R^n$ (i.e., $x_m \to x$ as $m \to \infty$) if

$$\lim_{m \to \infty} \|x_m - x\| = 0,$$

or equivalently, if for every $\epsilon > 0$ there exists an integer $N = N(\epsilon)$ such that $\|x_m - x\| < \epsilon$ whenever $m \geq N$. (In this definition $\|\cdot\|$ denotes a norm on R^n.) Using the properties of norms, it is easily shown that $x_m \to x$ if and only if for each coordinate one has $x_{km} \to x_k$ as $m \to \infty$, $k = 1, \ldots, n$.

The above allows the generalization of many of the properties of R to R^n (e.g., the *Bolzano-Weierstrass* property and the *Heine-Borel* property).

As another example, consider *pointwise convergence* of a sequence of functions. We say that a sequence of functions $\{f_k\}$, $f_k : D \to R^n$, $D \subset R^m$, $k = 1, 2, \ldots$, is pointwise convergent to a function $f : D \to R^n$ if for every $\epsilon > 0$ and every $x \in D$ there is an integer $N = N(\epsilon, x)$ such that $\|f_k(x) - f(x)\| < \epsilon$ whenever $k \geq N$. (In the above definition, $\|\cdot\|$ denotes a norm on R^n.) Using the properties of norms, it is again easy to show that $f_k(x) \to f(x)$ for all $x \in D$ if and only if for each coordinate one has $f_{ik}(x) \to f_i(x)$ as $k \to \infty$, $i = 1, \ldots, n$, for all $x \in D$.

As a third example, consider continuity of a function $f : D \to R^n$, where D is an open subset of R^m. The function f is said to be *continuous at point* $x_0 \in D$ if for every $\epsilon > 0$ there is a $\delta = \delta(\epsilon, x_0) > 0$ such that

$$\|f(x) - f(x_0)\|_Y < \epsilon \qquad \text{whenever } \|x - x_0\|_X < \delta.$$

In the above definition $\|\cdot\|_Y$ is a norm defined on R^n and $\|\cdot\|_X$ is a norm defined on R^m.

Next, let $g(t) = [g_1(t), \ldots, g_n(t)]^T$ be a vector-valued function defined on some interval $J \subset R$. Assume that each component of g is differentiable and integrable on J. As pointed out earlier, differentiation and integration of g are defined componentwise, e.g.,

$$\frac{dg}{dt}(t) = \left[\frac{dg_1}{dt}(t), \ldots, \frac{dg_n}{dt}(t)\right]^T$$

and

$$\int_a^b g(t)\,dt = \left[\int_a^b g_1(t)\,dt, \ldots, \int_a^b g_n(t)\,dt\right]^T.$$

It is easily verified that for $b > a$,

$$\left\|\int_a^b g(t)\,dt\right\| \leq \int_a^b \|g(t)\|\,dt,$$

where again $\|\cdot\|$ denotes a norm on R^n.

Finally, if D is an open connected nonempty set in the (t, x)-space $R \times R^n$ and if $f : D \to R^n$, then f is said to satisfy a *Lipschitz condition* with *Lipschitz constant* L (with respect to x) if for all (t, x) and (t, y) in D,

$$\|f(t, x) - f(t, y)\| \leq L\|x - y\|.$$

This is an obvious extension of the notion of a Lipschitz condition for scalar-valued functions.

D. Solutions of Systems of First-Order Ordinary Differential Equations: Existence, Continuation, Uniqueness, and Continuous Dependence on Initial Conditions

It turns out that every result given in Sections 1.6 to 1.9 can be restated in vector form and proved, using the same methods as in the scalar case and invoking obvious modifications (such as the replacement of absolute values of scalars by the norms of vectors or the norms of matrices, and so forth). In the following we restate these results in vector form and ask the reader to prove these results.

We have a domain $D \subset R^{n+1}$, $f \in C(D, R^n)$ and we are given the system of first-order ordinary differential equations

$$\dot{x} = f(t, x). \qquad (E)$$

We are given $(t_0, x_0) \in D$ and seek a solution (or solutions) to the initial-value problem

$$\dot{x} = f(t, x), \qquad x(t_0) = x_0. \qquad (I)$$

In doing so, it suffices to find a solution of the integral equation

$$\phi(t) = x_0 + \int_{t_0}^t f(s, \phi(s))\,ds. \qquad (V)$$

As in the scalar case, this can be accomplished by the use of ϵ-approximate solutions.

DEFINITION 10.4. A function ϕ defined and continuous on a t-interval $J = (a, b)$ containing t_0 is called an *ϵ-approximate solution* of (I) if $\phi(t_0) = x_0$ and

(i) $(t, \phi(t)) \in D$ for all $t \in J$;
(ii) ϕ has a continuous derivative on J except possibly on a finite set I of points in J where there are jump discontinuities allowed;
(iii) $\|\dot{\phi}(t) - f(t, \phi(t))\| < \epsilon$ for all $t \in J - I$, where $\|\cdot\|$ denotes a norm on R^n. ■

Now let

$$S = \{(t, x) : |t - t_0| \leq a, \; |x_i - x_{i0}| \leq b_i, \qquad i = 1, \dots, n\} \subset D \quad (10.18)$$

and let $(t_0, x_0) \in S$. Since $f \in C(D, R^n)$, it is bounded on S, and hence, there are $M_i > 0$ such that $|f_i(t, x)| < M_i$ for all $(t, x) \in S$, $i = 1, \dots, n$. Define

$$\begin{aligned} c_i &= \min\left\{a, \frac{b_i}{M_i}\right\}, \qquad i = 1, \dots, n, \\ c &= \min{}_i \{c_i\}. \end{aligned} \quad (10.19)$$

THEOREM 10.1. If $f \in C(D, R^n)$ and if c is as defined in (10.19), then for any $\epsilon > 0$ there is an ϵ-approximate solution of (I) on the interval $|t - t_0| \leq c$. ■

In the proof of the next result, we require a slight generalization of the Ascoli-Arzela Lemma given in Theorem 6.1. To this end, we let $\mathcal{F}$ denote a family of real-valued functions defined on a set $G \subset R^l$. Then $\mathcal{F}$ is called *uniformly* bounded if there is a nonnegative constant M such that $|f(x)| \leq M$ for all x in G and for all f in $\mathcal{F}$. Furthermore, $\mathcal{F}$ is called *equicontinuous* on G if for any $\epsilon > 0$ there is a $\delta > 0$ (independent of x, y, and f) such that $|f(x) - f(y)| < \epsilon$ whenever $\|x - y\| < \delta$ for all x and y in G and for all $f \in \mathcal{F}$ ($\|\cdot\|$ denotes a norm on R^l). The Ascoli-Arzela Lemma now reads as follows.

THEOREM 10.2. Let G be a closed and bounded subset of R^l and let $\{f_m\}$ be a sequence of functions in $C(G, R)$. If $\{f_m\}$ is equicontinuous and uniformly bounded on G, then there is a subsequence $\{m_k\}$ and a function $f \in C(G, R)$ such that $\{f_{m_k}\}$ converges uniformly to f on G. ■

THEOREM 10.3. If $f \in C(D, R^n)$ and $(t_0, x_0) \in D$, then (I) has a solution defined on $|t - t_0| \leq c$. ■

THEOREM 10.4. Let $f \in C(D, R^n)$ with f bounded on D. Suppose that ϕ is a solution of (E) on the interval $J = (a, b)$. Then

(i) the two limits

$$\lim_{t \to a^+} \phi(t) = \phi(a^+) \qquad \text{and} \qquad \lim_{t \to b^-} \phi(t) = \phi(b^-)$$

exist;

(ii) if $(a, \phi(a^+))$ [respectively, $(b, \phi(b^-))$] is in D, the solution ϕ can be continued to the left past the point $t = a$ (resp., to the right past the point $t = b$). ■

THEOREM 10.5. If $f \in C(D, R^n)$ and if ϕ is a solution of (E) on an open interval J, then ϕ can be continued to a maximal open interval $J^* \supset J$ in such a way that $(t, \phi(t))$ tends to ∂D as $t \to \partial J^*$ when ∂D is not empty and $|t| + \|\phi(t)\| \to \infty$ if ∂D is empty. The extended solution ϕ^* on J^* is noncontinuable. ■

THEOREM 10.6. If $f \in C(D, R^n)$ and if f satisfies a Lipschitz condition on D with Lipschitz constant L (with respect to x), then the initial-value problem (I) has at most one solution on any interval $|t - t_0| \leq d$. ■

COROLLARY 10.7. If $f \in C(D, R^n)$ and $\partial f_i / \partial x_j \in C(D, R^n)$ $(i, j = 1, \ldots, n)$, then for any $(t_0, x_0) \in D$ and any J containing t_0, a solution of (I) exists on J and is unique. ■

THEOREM 10.8. Let $f \in C(J \times R^n, R^n)$ for some open interval $J \subset R$ and let f satisfy a Lipschitz condition on $J \times R^n$ (with respect to x). Then for any $(t_0, x_0) \in J \times R^n$, the initial-value problem (I) has a unique solution that exists on the *entire* interval J. ■

Next, let $f \in C(D, R^n)$, let $S \subset D$ be the set defined in (10.18), centered at (t_0, x_0), and let c be defined in (10.19). *Successive approximations* for (I), or equivalently for (V), are defined as

$$\phi_0(t) = x_0$$

$$\phi_{m+1}(t) = x_0 + \int_{t_0}^{t} f(s, \phi_m(s))\, ds, \qquad m = 0, 1, 2, \ldots \tag{10.20}$$

for $|t - t_0| \leq c$.

THEOREM 10.9. If $f \in C(D, R^n)$ and if f is Lipschitz continuous on S with constant L (with respect to x), then the successive approximations ϕ_m, $m = 0, 1, 2, \ldots$, given in (10.20) exist on $|t - t_0| \leq c$, are continuous there, and converge uniformly, as $m \to \infty$, to the unique solution of (I). ■

In the final result of this subsection, we address initial-value problems that exhibit dependence on some parameter $\lambda \in G \subset R^m$ given by

$$\begin{aligned} \dot{x} &= f(t, x, \lambda) \\ x(\tau) &= \xi_\lambda, \end{aligned} \tag{$I_{\lambda,\tau}$}$$

where $f \in C(J \times R^n \times G, R^n)$, $J \subset R$ is an open interval, and ξ_λ depends continuously on λ.

THEOREM 10.10. Let $f \in C(J \times R^n \times G, R^n)$, where $J \subset R$ is an open interval and $G \subset R^m$. Assume that for each pair of compact subsets $J_0 \subset J$ and $G_0 \subset G$ there exists a constant $L = L_{J_0, G_0} > 0$ such that for all $(t, \lambda) \in J_0 \times G_0$, $x, y \in R^n$, the Lipschitz condition

$$\|f(t, x, \lambda) - f(t, y, \lambda)\| \leq L\|x - y\|$$

is true. Then the initial-value problem $(I_{\lambda,\tau})$ has a unique solution $\phi(t, \tau, \lambda)$, where $\phi \in C(J \times J \times G, R^n)$. Furthermore, if D is a set such that for all $\lambda_0 \in D$ there exists $\epsilon > 0$ such that $[\lambda_0 - \epsilon, \lambda_0 + \epsilon] \cap D \subset D$, then $\phi(t, \tau, \lambda) \to \phi(t, \tau_0, \lambda_0)$ uniformly for $t_0 \in J_0$ as $(\tau, \lambda) \to (\tau_0, \lambda_0)$, where J_0 is any compact subset of J. ■

Theorem 10.10 is a generalization of Corollary 9.2 from the one-dimensional case $(I'_{\lambda,\tau})$ to the n-dimensional case $(I_{\lambda,\tau})$. The generalization of Theorem 9.1 for the one-dimensional case $(I'_{\lambda,\tau,\xi})$ to the n-dimensional case $(I_{\lambda,\tau,\xi})$ is of course also readily established. We leave the details to the reader.

1.11. SYSTEMS OF LINEAR FIRST-ORDER ORDINARY DIFFERENTIAL EQUATIONS

In this section we will address linear ordinary differential equations of the form

$$\dot{x} = A(t)x + g(t) \tag{LN}$$

and
$$\dot{x} = A(t)x \tag{LH}$$
and
$$\dot{x} = Ax + g(t) \tag{11.1}$$
and
$$\dot{x} = Ax, \tag{L}$$

where $x \in R^n$, $A(t) = [a_{ij}(t)] \in C(R, R^{n\times n})$, $g \in C(R, R^n)$, and $A \in R^{n\times n}$.

Linear equations of the type enumerated above may arise in a natural manner in the modeling process of physical systems (see Section 1.4 for specific examples) or in the process of linearizing equations of the form (E) or some other kind of form.

A. Linearization

We consider the system of first-order ordinary differential equations given by

$$\dot{x} = f(t, x), \tag{E}$$

where $f : R \times D \to R^n$ and $D \subset R^n$ is some domain. If $f \in C^1(R \times D, R^n)$ and if ϕ is a given solution of (E) defined for all $t \in R$, then we can *linearize* (E) about ϕ in the following manner. Define $\delta x = x - \phi(t)$ so that

$$\begin{aligned} \frac{d(\delta x)}{dt} \triangleq \delta\dot{x} &= f(t, x) - f(t, \phi(t)) \\ &= f(t, \delta x + \phi(t)) - f(t, \phi(t)) \\ &= \frac{\partial f}{\partial x}(t, \phi(t))\delta x + F(t, \delta x), \end{aligned} \tag{11.2}$$

where $(\partial f/\partial x)(t, x)$ denotes the *Jacobian matrix* of $f(t, x) = (f_1(t, x), \ldots, f_n(t, x))^T$ with respect to $x = (x_1, \ldots, x_n)^T$, i.e.,

$$\frac{\partial f}{\partial x}(t, x) = \begin{bmatrix} \dfrac{\partial f_1}{\partial x_1}(t, x) & \cdots & \dfrac{\partial f_1}{\partial x_n}(t, x) \\ \vdots & & \vdots \\ \dfrac{\partial f_n}{\partial x_1}(t, x) & \cdots & \dfrac{\partial f_n}{\partial x_n}(t, x) \end{bmatrix} \tag{11.3}$$

and
$$F(t, \delta x) \triangleq [f(t, \delta x + \phi(t)) - f(t, \phi(t))] - \frac{\partial f}{\partial x}(t, \phi(t))\,\delta x. \tag{11.4}$$

It turns out that $F(t, \delta x)$ is $o(\|\delta x\|)$ as $\|\delta x\| \to 0$ uniformly in t on compact subsets of R, i.e., for any compact subset $I \subset R$, we have

$$\lim_{\|\delta x\|\to 0} \left(\sup_{t\in I} \frac{\|F(t, \delta x)\|}{\|\delta x\|} \right) = 0.$$

To prove this, we will use the fact that for each $i = 1, \ldots, n$,

$$\begin{aligned} f_i(t, \delta x + \phi(t)) - f_i(t, \phi(t)) &= (\delta x)^T \int_0^1 \nabla f_i(t, s(\delta x) + \phi(t))\, ds \\ &= \sum_{j=1}^{n} \delta x_j \int_0^1 \frac{\partial f_i}{\partial x_j}(t, s(\delta x) + \phi(t))\, ds, \end{aligned} \tag{11.5}$$

where $\delta x_i \triangleq (\delta x)_i$ and $\nabla f_i = \left(\frac{\partial f_i}{\partial x_1}, \ldots, \frac{\partial f_i}{\partial x_n}\right)^T$. To verify (11.5), we let

$$g(s) = f_i(t, s(\delta x) + \phi(t))$$

and use the fact that

$$\begin{aligned}
g(1) - g(0) &= f_i(t, \delta x + \phi(t)) - f_i(t, \phi(t)) \\
&= \int_0^1 g'(s)\, ds = \int_0^1 df_i(t, s(\delta x) + \phi(t)) \\
&= \int_0^1 \sum_{j=1}^n \left[\frac{\partial f_i}{\partial x_j}(t, s(\delta x) + \phi(t))\, ds\right] \delta x_j \\
&= (\delta x)^T \int_0^1 \nabla f_i(t, s(\delta x) + \phi(t))\, ds.
\end{aligned}$$

Next, we note that the ith component of $F(t, \delta x)$ is given by

$$\begin{aligned}
F_i(t, \delta x) &= \sum_{j=1}^n \delta x_j \left[\int_0^1 \frac{\partial f_i}{\partial x_j}(t, s(\delta x) + \phi(t))\, ds - \frac{\partial f_i}{\partial x_j}(t, \phi(t))\right] \\
&= \sum_{j=1}^n \delta x_j \int_0^1 \left[\frac{\partial f_i}{\partial x_j}(t, s(\delta x) + \phi(t)) - \frac{\partial f_i}{\partial x_j}(t, \phi(t))\right] ds.
\end{aligned}$$

Choose $\|\delta x\| = (\sum_{i=1}^n (\delta x_i)^2)^{1/2}$ and let I be a compact interval in R. Then

$$\begin{aligned}
&\lim_{\|\delta x\| \to 0} \left(\sup_{t \in I} \frac{|F_i(t, \delta x)|}{\|\delta x\|}\right) \\
&= \lim_{\|\delta x\| \to 0} \left(\sup_{t \in I} \frac{|\sum_{j=1}^n \delta x_j \int_0^1 \left[\frac{\partial f_i}{\partial x_j}(t, s(\delta x) + \phi(t)) - \frac{\partial f_i}{\partial x_j}(t, \phi(t))\right] ds|}{\|\delta x\|}\right) \\
&\leq \lim_{\|\delta x\| \to 0} \frac{(\sum_{j=1}^n (\delta x_j)^2)^{1/2} \left(\sum_{j=1}^n (\sup_{t \in I} \int_0^1 \left[\frac{\partial f_i}{\partial x_j}(t, s(\delta x) + \phi(t)) - \frac{\partial f_i}{\partial x_j}(t, \phi(t))\right] ds)^2\right)^{1/2}}{(\sum_{j=1}^n (\delta x_j)^2)^{1/2}} \\
&= \lim_{\|\delta x\| \to 0} \left(\sum_{j=1}^n \left(\sup_{t \in I} \int_0^1 \left[\frac{\partial f_i}{\partial x_j}(t, s\delta(x) + \phi(t)) - \frac{\partial f_i}{\partial x_j}(t, \phi(t))\right] ds\right)^2\right)^{1/2} \\
&= 0,
\end{aligned}$$

where we have made use of the *Schwarz Inequality.*

To establish equality (equal to zero) in the last line of the above equation requires perhaps a bit of extra work. Since $I \subset R$ is compact and ϕ is continuous, it follows that the set $\phi(I)$ is compact. Since $\phi(I) \subset D$ and D is a domain, we have $dist\,(\phi(I), \partial D) \triangleq d > 0$. Clearly, then,

$$X_0 \triangleq \{\phi(t) + \left(\frac{d}{2\sqrt{n}}\right)(s_1, \ldots, s_n)^T : t \in I,\ -1 \leq s_i \leq 1,\ i = 1, \ldots, n\} \subset D$$

and X_0 is a compact subset of D, since $\phi(t) + [d/(2\sqrt{n})](s_1, \ldots, s_n)^T$ is a continuous vector-function of $(t, s_1, \ldots, s_n)$.

Now for $\|\delta x\| < d/2$, $0 \leq s \leq 1$, $t \in I$, we have that $s(\delta x) + \phi(t) \in X_0$. Since $(\partial f/\partial x_j)(t, x)$ is uniformly continuous on the compact set $I \times X_0$, we conclude the equality (equal to zero).

Finally, since the above argument is true for all $i = 1, \ldots, n$, it follows that $F(t, \delta x)$ is $o(\|\delta x\|)$ as $\|\delta x\| \to 0$ uniformly in t on compact subsets of R.

Letting

$$\frac{\partial f}{\partial x}(t, \phi(t)) = A(t),$$

we obtain from (11.2) the equation

$$\frac{d(\delta x)}{dt} \triangleq \delta \dot{x} = A(t)\delta x + F(t, \delta x). \tag{11.6}$$

Associated with (11.6) we have the linear differential equation

$$\dot{z} = A(t)z, \tag{LH}$$

called the *linearized equation* of (E) about the solution ϕ.

In applications, the linearization (LH) of (E), about a given solution ϕ, is frequently used as a means of approximating a nonlinear process by a linear one (in the vicinity of ϕ). In Chapter 6, where we will study the stability properties of equilibria of (E) [which are specific kinds of solutions of (E)], we will show under what conditions it makes sense to deduce qualitative properties of a nonlinear process from its linearization.

Of special interest is the case when in (E), f is independent of t, i.e.,

$$\dot{x} = f(x) \tag{A}$$

and ϕ is a constant solution of (A), say, $\phi(t) = x_0$ for all $t \in R$. Under these conditions we have

$$\frac{d(\delta x)}{dt} \triangleq \delta \dot{x} = A\delta x + F(\delta x), \tag{11.7}$$

where

$$\lim_{\|\delta x\| \to 0} \frac{\|F(\delta x)\|}{\|\delta x\|} = 0 \tag{11.8}$$

and A denotes the Jacobian $(\partial f/\partial x)(x_0)$. Again, associated with (11.7) we have the linear differential equation

$$\dot{z} = Az,$$

called the *linearized equation* of (A) about the solution $\phi(t) \equiv x_0$.

We can generalize the above to equations of the form

$$\dot{x} = f(t, x, u), \tag{E_u}$$

where $f : R \times D_1 \times D_2 \to R^n$ and $D_1 \subset R^n$, $D_2 \subset R^m$ are some domains. If $f \in C^1(R \times D_1 \times D_2, R^n)$ and if $\phi(t)$ is a given solution of (E_u) that we assume to exist for all $t \in R$ and that is determined by the initial condition x_0 and the *given specific function* $\psi \in C(R, R^m)$, i.e.,

$$\dot{\phi}(t) = f(t, \phi(t), \psi(t)), \qquad t \in R,$$

then we can linearize (E_u) in the following manner. Define $\delta x = x - \phi(t)$ and $\delta u = u - \psi(t)$. Then

$$\frac{d(\delta x)}{dt} = \delta\dot{x} = \dot{x} - \dot{\phi}(t) = f(t, x, u) - f(t, \phi(t), \psi(t))$$
$$= f(t, \delta x + \phi(t), \delta u + \psi(t)) - f(t, \phi(t), \psi(t))$$
$$= \frac{\partial f}{\partial x}(t, \phi(t), \psi(t))\delta x + \frac{\partial f}{\partial u}(t, \phi(t), \psi(t))\delta u$$
$$+ F_1(t, \delta x, u) + F_2(t, \delta u), \tag{11.9}$$

where $F_1(t, \delta x, u) = f(t, \delta x + \phi(t), u) - f(t, \phi(t), u) - \frac{\partial f}{\partial x}(t, \phi(t), \psi(t))\delta x$ is $o(\|\delta x\|)$ as $\|\delta x\| \to 0$, uniformly in t on compact subsets of R for fixed u $\left[\text{i.e., for fixed } u \text{ and for any compact subset } I \subset R, \lim_{\|\delta x\|\to 0}\left(\sup_{t\in I}\frac{\|F_1(t, \delta x, u)\|}{\|\delta x\|}\right) = 0\right]$ and where

$$F_2(t, \delta u) = f(t, \phi(t), \delta u + \psi(t)) - f(t, \phi(t), \psi(t)) - \frac{\partial f}{\partial u}(t, \phi(t), \psi(t))\delta u$$

is $o(\|\delta u\|)$ as $\|\delta u\| \to 0$, uniformly in t on compact subsets of R $\left[\text{i.e., for any compact subset } I \subset R, \lim_{\|\delta u\|\to 0}\left(\sup_{t\in I}\frac{\|F_2(t, \delta u)\|}{\|\delta u\|}\right) = 0\right]$, and where $(\partial f/\partial x)(\cdot)$ and $(\partial f/\partial u)(\cdot)$ denote the Jacobian matrix of f with respect to x and the Jacobian matrix of f with respect to u, respectively.

Letting

$$\frac{\partial f}{\partial x}(t, \phi(t), \psi(t)) = A(t) \qquad \text{and} \qquad \frac{\partial f}{\partial u}(t, \phi(t), \psi(t)) = B(t),$$

we obtain from (11.9),

$$\frac{d(\delta x)}{dt} = \delta\dot{x} = A(t)\delta x + B(t)\delta u + F_1(t, \delta x, u) + F_2(t, \delta u). \tag{11.10}$$

Associated with (11.10), we have

$$\dot{z} = A(t)z + B(t)v \tag{$\widetilde{LN}$}$$

and call $(\widetilde{LN})$ the *linearized equation* of (E_u) about the solution ϕ and the input function ψ.

As in the case of the linearization of (E) by (LH), the linearization $(\widetilde{LN})$ of system (E_u) about a given solution ϕ and a given input ψ is often used in attempting to capture the qualitative properties of a nonlinear process by a linear process (in the vicinity of ϕ and ψ). In doing so, great care must be exercised to avoid erroneous conclusions.

The motivation of linearization is of course very obvious: much more is known about linear ordinary differential equations than about nonlinear ones. For example, the explicit forms of the solutions of (L) and (11.1) are known; the structures of the solutions of (LH), (LN), and $(\widetilde{LN})$ are known; the qualitative properties of the solutions of linear equations are known; and so forth.

B. Examples

We now consider some specific cases.

EXAMPLE 11.1. We consider the *simple pendulum* discussed in Example 4.4 and described by the equation

$$\ddot{x} + k \sin x = 0, \tag{11.11}$$

where $k > 0$ is a constant. Letting $x_1 = x$ and $x_2 = \dot{x}$, (11.11) can be expressed as

$$\begin{aligned} \dot{x}_1 &= x_2 \\ \dot{x}_2 &= -k \sin x_1. \end{aligned} \tag{11.12}$$

It is easily verified that $\phi_1(t) \equiv 0$ and $\phi_2(t) \equiv 0$ is a solution of (11.12). Letting $f_1(x_1, x_2) = x_2$ and $f_2(x_1, x_2) = -k \sin x_1$, the Jacobian of $f(x_1, x_2) = (f_1(x_1, x_2), f_2(x_1, x_2))^T$ evaluated at $(x_1, x_2)^T = (0, 0)^T$ is given by

$$J(0) \triangleq A = \begin{bmatrix} 0 & 1 \\ -k\cos x_1 & 0 \end{bmatrix}_{\substack{x_1=0 \\ x_2=0}} = \begin{bmatrix} 0 & 1 \\ -k & 0 \end{bmatrix}.$$

The linearized equation of (11.12) about the solution $\phi_1(t) \equiv 0$, $\phi_2(t) \equiv 0$ is given by

$$\begin{bmatrix} \dot{z}_1 \\ \dot{z}_2 \end{bmatrix} = \begin{bmatrix} 0 & 1 \\ -k & 0 \end{bmatrix} \begin{bmatrix} z_1 \\ z_2 \end{bmatrix}.$$

■

EXAMPLE 11.2. The system of equations

$$\begin{aligned} \dot{x}_1 &= ax_1 - bx_1x_2 - cx_1^2 \\ \dot{x}_2 &= dx_2 - ex_1x_2 - fx_2^2 \end{aligned} \tag{11.13}$$

describes the growth of two competing species (e.g., two species of small fish) that prey on each other (e.g., the adult members of one species prey on the young members of the other species, and vice versa). In (11.13) a, b, c, d, e, and f are positive parameters and it is assumed that $x_1 \geq 0$ and $x_2 \geq 0$. For (11.13), $\phi_1(t) = \phi_1(t, 0, 0) \equiv 0$ and $\phi_2(t) = \phi_2(t, 0, 0) \equiv 0$, $t \geq 0$, is a solution of (11.13). A simple computation yields

$$A = \frac{\partial f}{\partial x}(0) = \begin{bmatrix} a & 0 \\ 0 & d \end{bmatrix},$$

and thus the system of equations

$$\begin{bmatrix} \dot{z}_1 \\ \dot{z}_2 \end{bmatrix} = \begin{bmatrix} a & 0 \\ 0 & d \end{bmatrix} \begin{bmatrix} z_1 \\ z_2 \end{bmatrix}$$

constitutes the linearized equation of (11.13) about the solution $\phi_1(t) = 0$, $\phi_2(t) = 0$, $t \geq 0$. ■

EXAMPLE 11.3. Consider a unit mass subjected to an inverse square law force field, as depicted in Fig. 1.11. In this figure, r denotes radius and θ denotes angle, and it is assumed that the unit mass (representing, e.g., a satellite) can thrust in the radial and in the tangential directions with thrusts u_1 and u_2, respectively. The equations that govern this system are given by

$$\begin{aligned} \ddot{r} &= r\dot{\theta}^2 - \frac{k}{r^2} + u_1 \\ \ddot{\theta} &= \frac{-2\dot{\theta}\dot{r}}{r} + \frac{1}{r}u_2. \end{aligned} \tag{11.14}$$

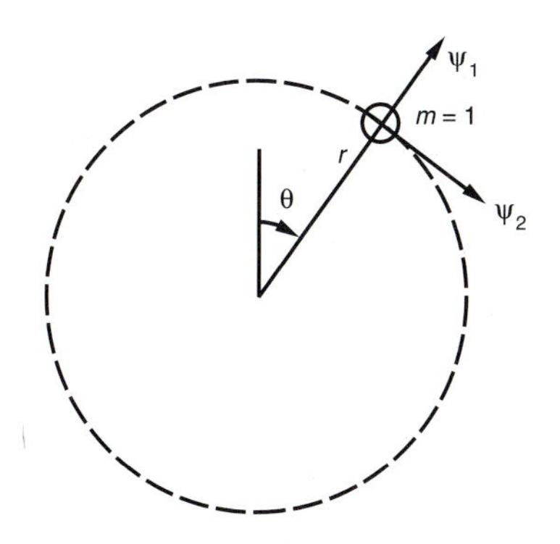

FIGURE 1.11
A unit mass subjected to an inverse square law force field

When $r(0) = r_0$, $\dot{r}(0) = 0$, $\theta(0) = \theta_0$, $\dot{\theta}(0) = \omega_0$ and $u_1(t) \equiv 0$, $u_2(t) \equiv 0$ for $t \geq 0$, it is easily verified that the system of equations (11.14) has as a solution the circular orbit given by

$$\begin{aligned} r(t) &\equiv r_0 = \text{constant} \\ \dot{\theta}(t) &= \omega_0 = \text{constant} \end{aligned} \tag{11.15}$$

for all $t \geq 0$, which implies that

$$\theta(t) = \omega_0 t + \theta_0, \tag{11.16}$$

where $\omega_0 = (k/r_0^3)^{1/2}$.

If we let $x_1 = r$, $x_2 = \dot{r}$, $x_3 = \theta$, and $x_4 = \dot{\theta}$, the equations of motion (11.14) assume the form

$$\begin{aligned} \dot{x}_1 &= x_2 \\ \dot{x}_2 &= x_1 x_4^2 - \frac{k}{x_1^2} + u_1 \\ \dot{x}_3 &= x_4 \\ \dot{x}_4 &= -\frac{2x_2x_4}{x_1} + \frac{u_2}{x_1}. \end{aligned} \tag{11.17}$$

The linearized equation of (11.17) about the solution (11.16) [with $u_1(t) \equiv 0$, $u_2(t) \equiv 0$] is given by

$$\begin{bmatrix} \dot{z}_1 \\ \dot{z}_2 \\ \dot{z}_3 \\ \dot{z}_4 \end{bmatrix} = \begin{bmatrix} 0 & 1 & 0 & 0 \\ 3\omega_0^2 & 0 & 0 & 2r_0\omega_0 \\ 0 & 0 & 0 & 1 \\ 0 & \dfrac{-2\omega_0}{r_0} & 0 & 0 \end{bmatrix} \begin{bmatrix} z_1 \\ z_2 \\ z_3 \\ z_4 \end{bmatrix} + \begin{bmatrix} 0 & 0 \\ 1 & 0 \\ 0 & 0 \\ 0 & \dfrac{1}{r_0} \end{bmatrix} \begin{bmatrix} v_1 \\ v_2 \end{bmatrix}. \qquad \blacksquare$$

EXAMPLE 11.4. In this example we consider systems described by equations of the form

$$\dot{x} + Af(x) + Bg(x) = u, \tag{11.18}$$

where $x \in R^n$, $A = [a_{ij}] \in R^{n\times n}$, $B = [b_{ij}] \in R^{n\times n}$ with $a_{ii} > 0$, $b_{ii} > 0$, $1 \leq i \leq n$, $f, g \in C^1(R^n, R^n)$, $u \in C(R^+, R^n)$, and $f(x) = 0$, $g(x) = 0$ if and only if $x = 0$.

Equation (11.18) can be used to model a great variety of physical systems. In particular, (11.18) has been used to model a large class of integrated circuits consisting of (nonlinear) transistors and diodes, (linear) capacitors and resistors, and current and voltage sources. (Figure 1.12 gives a symbolic representation of such circuits.) For such circuits, we assume that $f(x) = [f_1(x_1), \ldots, f_n(x_n)]^T$.

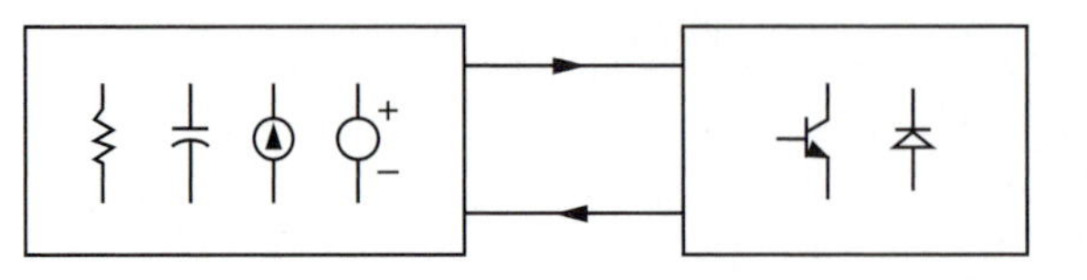

FIGURE 1.12
Integrated circuit

If $u(t) = 0$ for all $t \geq 0$, then $\phi_i(t) = 0, t \geq 0, 1 \leq i \leq n$, is a solution of (11.18). The system of equations (11.18) can be expressed equivalently as

$$\dot{x}_i = -\sum_{j=1}^{n}\left[a_{ij}\frac{f_j(x_j)}{x_j} + b_{ij}\frac{g_j(x_j)}{x_j}\right]x_j + u_i \qquad i = 1, \ldots, n. \tag{11.19}$$

The linearized equation of (11.19) about the solution $\phi_i(t) = 0$, and the input $u_i(t) = 0$, $t \geq 0$, $i = 1, \ldots, n$, is given by

$$\dot{z}_i = -\sum_{j=1}^{n}[a_{ij}f_j'(0) + b_{ij}g_j'(0)]z_j + v_i, \tag{11.20}$$

where $f_j'(0) = (df_j/dx_j)(0)$ and $g_j'(0) = (dg_j/dx_j)(0)$, $i = 1, \ldots, n$. ■

1.12 LINEAR SYSTEMS: EXISTENCE, UNIQUENESS, CONTINUATION, AND CONTINUITY WITH RESPECT TO PARAMETERS OF SOLUTIONS

In this section we address nonhomogeneous systems of first-order ordinary differential equations given by

$$\dot{x} = A(t)x + g(t), \tag{LN}$$

where $x \in R^n$, $A(t) = [a_{ij}(t)]$ is a real $n \times n$ matrix, and g is a real n-vector-valued function.

THEOREM 12.1. Suppose that $A \in C(J, R^{n\times n})$ and $g \in C(J, R^n)$, where J is some open interval. Then for any $t_0 \in J$ and any $x_0 \in R^n$, equation (LN) has a unique solution satisfying $x(t_0) = x_0$. This solution exists on the *entire* interval J and is continuous in (t, t_0, x_0).

Proof. The function $f(t, x) = A(t)x + g(t)$ is continuous in (t, x), and moreover, for any compact subinterval $J_0 \subset J$ there is an $L_0 \geq 0$ such that

$$\begin{aligned}\|f(t, x) - f(t, y)\|_1 &= \|A(t)(x - y)\|_1 \leq \|A(t)\|_1\|x - y\|_1 \\ &\leq \left(\sum_{i=1}^{n}\max_{1\leq j\leq n}|a_{ij}(t)|\right)\|x - y\|_1 \leq L_0\|x - y\|_1\end{aligned}$$

for all $(t, x), (t, y) \in J_0 \times R^n$, where L_0 is defined in the obvious way. Therefore, f satisfies a Lipschitz condition on $J_0 \times R^n$.

If $(t_0, x_0) \in J_0 \times R^n$, then the continuity of f implies the existence of solutions (Theorems 6.3 and 10.3), while the Lipschitz condition implies the uniqueness of solutions

(Theorems 8.2 and 10.6). These solutions exist for the entire interval J_0 (Theorems 8.4 and 10.8). Since this argument holds for *any* compact subinterval $J_0 \subset J$, the solutions exist and are unique for all $t \in J$. Furthermore, the solutions are continuous with respect to t_0 and x_0 (Theorems 9.1 and 10.10 modified for the case where A and g do not depend on any parameters λ). ■

For the case when in (LN) the matrix A and the vector g depend continuously on parameters λ and μ, respectively, it is possible to modify Theorem 12.1, and its proof, in the obvious way to show that the unique solutions of the system of equations

$$\dot{x} = A(t, \lambda)x + g(t, \mu) \qquad (LN_{\lambda\mu})$$

are continuous in λ and μ as well. [Assume that $A \in C(J \times R^l, R^{n\times n})$ and $g \in C(J \times R^m, R^n)$ and follow a procedure that is similar to the proof of Theorem 12.1.]

1.13 SOLUTIONS OF LINEAR STATE EQUATIONS

In this section we determine the specific form of the solutions of systems of linear first-order ordinary differential equations. We will revisit this topic in much greater detail in Chapter 2.

Homogeneous equations

We begin by considering linear homogeneous systems

$$\dot{x} = A(t)x, \qquad (LH)$$

where $A \in C(R, R^{n\times n})$. By Theorem 12.1, for every $x_0 \in R^n$, (LH) has a unique solution that exists for all $t \in R$. We will now use Theorem 10.9 to derive an expression for the solution $\phi(t, t_0, x_0)$ for (LH) for $t \in R$ with $\phi(t_0, t_0, x_0) = x_0$. In this case the successive approximations given in (10.20) assume the form

$$\begin{aligned}
\phi_0(t, t_0, x_0) &= x_0 \\
\phi_1(t, t_0, x_0) &= x_0 + \int_{t_0}^{t} A(s)x_0\, ds \\
\phi_2(t, t_0, x_0) &= x_0 + \int_{t_0}^{t} A(s)\phi_1(s, t_0, x_0)\, ds \\
\cdots\cdots\cdots & \cdots \cdots\cdots\cdots\cdots\cdots\cdots \\
\phi_m(t, t_0, x_0) &= x_0 + \int_{t_0}^{t} A(s)\phi_{m-1}(s, t_0, x_0)\, ds
\end{aligned}$$

or

$$\begin{aligned}
\phi_m(t, t_0, x_0) &= x_0 + \int_{t_0}^{t} A(s_1)x_0\, ds_1 + \int_{t_0}^{t} A(s_1)\int_{t_0}^{s_1} A(s_2)x_0\, ds_2\, ds_1 + \cdots \\
&\quad + \int_{t_0}^{t} A(s_1)\int_{t_0}^{s_1} A(s_2)\cdots\int_{t_0}^{s_{m-1}} A(s_m)x_0\, ds_m\cdots ds_1 \\
&= \Bigg[I + \int_{t_0}^{t} A(s_1)\, ds_1 + \int_{t_0}^{t} A(s_1)\int_{t_0}^{s_1} A(s_2)\, ds_2\, ds_1 + \cdots \\
&\quad + \int_{t_0}^{t} A(s_1)\int_{t_0}^{s_1} A(s_2)\cdots\int_{t_0}^{s_{m-1}} A(s_m)\, ds_m\cdots ds_1\Bigg] x_0, \qquad (13.1)
\end{aligned}$$

where I denotes the $n \times n$ identity matrix. By Theorem 10.9, the sequence $\{\phi_m\}$, $m = 0, 1, 2, \ldots$ determined by (13.1) converges uniformly, as $m \to \infty$, to the unique solution $\phi(t, t_0, x_0)$ of (LH) on compact subsets of R. We thus have

$$\phi(t, t_0, x_0) = \Phi(t, t_0)x_0, \tag{13.2}$$

where
$$\Phi(t, t_0) = \Bigg[I + \int_{t_0}^{t} A(s_1)\,ds_1 + \int_{t_0}^{t} A(s_1) \int_{t_0}^{s_1} A(s_2)\,ds_2\,ds_1$$
$$+ \int_{t_0}^{t} A(s_1) \int_{t_0}^{s_1} A(s_2) \int_{t_0}^{s_2} A(s_3)\,ds_3\,ds_2\,ds_1 + \cdots$$
$$+ \int_{t_0}^{t} A(s_1) \int_{t_0}^{s_1} A(s_2) \cdots \int_{t_0}^{s_{m-1}} A(s_m)\,ds_m\,ds_{m-1} \cdots ds_1 + \cdots \Bigg]. \tag{13.3}$$

Expression (13.3) is called the *Peano-Baker series*.

From expression (13.3) we immediately note that

$$\Phi(t, t) = I. \tag{13.4}$$

Furthermore, by differentiating expression (13.3) with respect to time and substituting into (LH), we obtain that

$$\dot{\Phi}(t, t_0) = A(t)\Phi(t, t_0). \tag{13.5}$$

From (13.2) it is clear that once the initial data are specified and once the $n \times n$ matrix $\Phi(t, t_0)$ is known, the entire behavior of system (LH) evolving in time t is known. This has motivated the *state* terminology: $x(t_0) = x_0$ is the *state of the system (LH) at time* t_0, $\phi(t, t_0, x_0)$ is the *state of the system (LH) at time t*, the solution ϕ is called the *state vector* of (LH), the components of ϕ are called the *state variables* of (LH), and the matrix $\Phi(t, t_0)$ that maps $x(t_0)$ into $\phi(t, t_0, x_0)$ is called the *state transition matrix* for (LH). Also, the vector space containing the state vectors is called the *state space* for (LH).

We can specialize the preceding discussion to linear systems of equations

$$\dot{x} = Ax. \tag{L}$$

In this case the mth term in (13.3) assumes the form

$$\int_{t_0}^{t} A(s_1) \int_{t_0}^{s_1} A(s_2) \int_{t_0}^{s_2} A(s_3) \cdots \int_{t_0}^{s_{m-1}} A(s_m)\,ds_m \cdots ds_1$$
$$= A^m \int_{t_0}^{t} \int_{t_0}^{s_1} \int_{t_0}^{s_2} \cdots \int_{t_0}^{s_{m-1}} 1\,ds_m \cdots ds_1 = \frac{A^m (t - t_0)^m}{m!},$$

and expression (13.1) for ϕ_m assumes now the form

$$\phi_m(t, t_0, x_0) = \left[I + \sum_{k=1}^{m} \frac{A^k (t - t_0)^k}{k!} \right] x_0.$$

We conclude once more from Theorem 10.9 that $\{\phi_m\}$ converges uniformly as $m \to \infty$ to the unique solution $\phi(t, t_0, x_0)$ of (L) on compact subsets of R. We have

$$\phi(t, t_0, x_0) = \left[I + \sum_{k=1}^{\infty} \frac{A^k (t - t_0)^k}{k!} \right] x_0$$
$$= \Phi(t, t_0)x_0 \triangleq \Phi(t - t_0)x_0, \tag{13.6}$$

where $\Phi(t - t_0)$ denotes the state transition matrix for (L). [Note that by writing $\Phi(t, t_0) = \Phi(t - t_0)$, we have used a slight abuse of notation.] By making the analogy with the scalar $e^a = 1 + \sum_{k=1}^{\infty}(a^k/k!)$, usage of the notation

$$e^A = I + \sum_{k=1}^{\infty} \frac{A^k}{k!} \tag{13.7}$$

should be clear. We call e^A a *matrix exponential.* In Chapter 2 we will explore several ways of determining e^A for a given A.

Nonhomogeneous equations

Next, we consider linear nonhomogeneous systems of ordinary differential equations

$$\dot{x} = A(t)x + g(t), \tag{LN}$$

where $A \in C(R, R^{n\times n})$ and $g \in C(R, R^n)$. Again, by Theorem 12.1, for every $x_0 \in R^n$, (LN) has a unique solution that exists for all $t \in R$. Instead of *deriving* the complete solution of (LN) for a given set of initial data $x(t_0) = x_0$, we will *guess* the solution and verify that it indeed satisfies (LN). To this end, let us assume that the solution is of the form

$$\phi(t, t_0, x_0) = \Phi(t, t_0)x_0 + \int_{t_0}^{t} \Phi(t, s)g(s)\, ds, \tag{13.8}$$

where $\Phi(t, t_0)$ denotes the state transition matrix for (LH).

To show that (13.8) is indeed the solution of (LN), we first let $t = t_0$. In view of (13.4) and (13.8), we have $\phi(t_0, t_0, x_0) = x_0$. Next, by differentiating both sides of (13.8) and by using (13.4), (13.5), and (13.8) we have

$$\begin{aligned}
\dot{\phi}(t, t_0, x_0) &= \dot{\Phi}(t, t_0)x_0 + \Phi(t, t)g(t) + \int_{t_0}^{t} \dot{\Phi}(t, s)g(s)\, ds \\
&= A(t)\Phi(t, t_0)x_0 + g(t) + \int_{t_0}^{t} A(t)\Phi(t, s)g(s)\, ds \\
&= A(t)[\Phi(t, t_0)x_0 + \int_{t_0}^{t} \Phi(t, s)g(s)\, ds] + g(t) \\
&= A(t)\phi(t, t_0, x_0) + g(t),
\end{aligned}$$

i.e., $\phi(t, t_0, x_0)$ given in (13.8) satisfies (LN). Therefore, $\phi(t, t_0, x_0)$ is the unique solution of (LN). Equation (13.8) is called the *variation of constants formula,* which is discussed further in Subsection 2.3C of Chapter 2. In the exercise section of Chapter 2 (refer to Exercise 2.33), we ask the reader (with hints) to *derive* the variation of constants formula (13.8), using *change of variables.*

We note that when $x_0 = 0$, (13.8) reduces to

$$\phi(t, t_0, 0) \triangleq \phi_p(t) = \int_{t_0}^{t} \Phi(t, s)g(s)\, ds, \tag{13.9}$$

and when $x_0 \neq 0$ but $g(t) = 0$ for all $t \in R$, (13.8) reduces to

$$\phi(t, t_0, x_0) \triangleq \phi_h(t) = \Phi(t, t_0)x_0. \tag{13.10}$$

Therefore, the *total solution* of (LN) may be viewed as consisting of a component that is due to the initial conditions (t_0, x_0) and another component that is due to the "*forcing term*" $g(t)$. This type of separation is in general possible only in linear systems of differential equations. We call ϕ_p a *particular solution* of the nonhomogeneous system (LN) and call ϕ_h the *homogeneous solution.*

From (13.8) it is clear that for given initial conditions $x(t_0) = x_0$ and given forcing term $g(t)$, the behavior of system (LN), summarized by ϕ, is known for all t. Thus, $\phi(t, t_0, x_0)$ specifies the *state vector* of (LH) at time t. The components ϕ_i of ϕ, $i = 1, \ldots, n$, represent the *state variables* for (LH), and the vector space that contains the state vectors is the *state space* for (LH).

Before closing this section, it should be pointed out that in applications the matrix $A(t)$ and the vector $g(t)$ in (LN) may be only *piecewise continuous* rather than continuous, as assumed above [i.e., $A(t)$ and $g(t)$ may have (at most) a finite number of discontinuities over any finite time interval]. In such cases, the derivative of x with respect to t [i.e., the right-hand side in (LN)], will be discontinuous at a finite number of instants over any finite time interval; however, the state itself, x, will still be continuous at these instants [i.e., the solutions of (LN) will still be continuous over R]. In such cases, all the results presented concerning existence, uniqueness, continuation of solutions, and so forth, as well as the explicit expressions of solutions of (LN), are either still valid or can be modified in the obvious way. For example, should $g(t)$ experience a discontinuity at, say, $t_1 > t_0$, then expression (13.8) will be modified to read

$$\phi(t, t_0, x_0) = \Phi(t, t_0)x_0 + \int_{t_0}^{t} \Phi(t, s)g(s)\,ds, \qquad t_0 \leq t < t_1, \tag{13.11}$$

$$\phi(t, t_1, x_1) = \Phi(t, t_1)x_1 + \int_{t_1}^{t} \Phi(t, s)g(s)\,ds, \qquad t \geq t_1, \tag{13.12}$$

where $x_1 = \lim_{t \to t_1^-} \phi(t, t_0, x_0)$.

1.14 STATE-SPACE DESCRIPTION OF CONTINUOUS-TIME SYSTEMS

Returning now to Section 1, let us consider once more systems described by equations of the form

$$\dot{x} = f(t, x, u) \tag{14.1a}$$

$$y = g(t, x, u), \tag{14.1b}$$

where $x \in R^n$, $y \in R^p$, $u \in R^m$, $f : R \times R^n \times R^m \to R^n$, and $g : R \times R^n \times R^m \to R^p$. Here t denotes time and u and y denote system *input* and system *output*, respectively. Equation (14.1a) is called the *state equation,* (14.1b) is called the *output equation,* and (14.1a) and (14.1b) constitute the *state-space description* of continuous-time finite-dimensional systems.

The system input may be a function of t only (i.e., $u : R \to R^m$), or as in the case of *feedback control systems,* it may be a function of t and x (i.e., $u : R \times R^n \to R^m$). In either case, for a *given* (i.e., specified) u, we let $f(t, x, u) = F(t, x)$ and rewrite (14.1a) as

$$\dot{x} = F(t, x). \tag{14.2}$$

Now according to Theorems 10.8 and 10.10, if $F \in C(R \times R^n, R^n)$ and if for any compact subinterval $J_0 \subset R$ there is a constant L_{J_0} such that $\|F(t, x) - F(t, \tilde{x})\| \leq L_{J_0}\|x - \tilde{x}\|$ for all $t \in J_0$ and for all $x, \tilde{x} \in R^n$, then the following are true:

1. For any $(t_0, x_0) \in R \times R^n$, Eq. (14.2) has a unique solution $\phi(t, t_0, x_0)$ satisfying $\phi(t_0, t_0, x_0) = x_0$ that exists for all $t \in R$.
2. The solution ϕ is continuous in t, t_0, and x_0.
3. If F depends continuously on parameters (say, $\lambda \in R^l$) and if x_0 depends continuously on λ, the solution ϕ is continuous in λ as well.

Thus, if the above conditions are satisfied, then for a given t_0, x_0, and u, Eq. (14.1a) will have a unique solution that exists for $t \in R$. Therefore, as already discussed in Section 13, $\phi(t, t_0, x_0)$ characterizes the *state* of the system at time t. Moreover, under these conditions, the system will have a unique *response* for $t \in R$, determined by Eq. (14.1b). We usually assume that $g \in C(R \times R^n \times R^m, R^p)$ or that $g \in C^1(R \times R^n \times R^m, R^p)$.

An important special case of (14.1a) and (14.1b) is systems described by linear time-varying equations of the form

$$\dot{x} = A(t)x + B(t)u \tag{14.3a}$$

$$y = C(t)x + D(t)u, \tag{14.3b}$$

where $A \in C(R, R^{n\times n})$, $B \in C(R, R^{n\times m})$, $C \in C(R, R^{p\times n})$, and $D \in C(R, R^{p\times m})$. Such equations may arise in the modeling process of a physical system, or they may be a consequence of a linearization process, as discussed in Section 11.

By applying the results of Section 12 we see that for every initial condition $x(t_0) = x_0$ and for every given input $u : R \to R^m$, system (14.3a) possesses a unique solution that exists for all $t \in R$ and that is continuous in (t, t_0, x_0). Moreover, if A and B depend continuously on parameters, say, $\lambda \in R^l$, then the solutions will be continuous in the parameters as well. Indeed, in accordance with (13.8), this solution is given by

$$\phi(t, t_0, x_0) = \Phi(t, t_0)x_0 + \int_{t_0}^{t} \Phi(t, s)B(s)u(s)\, ds, \tag{14.4}$$

where $\Phi(t, t_0)$ denotes the state transition matrix of the system of equations

$$\dot{x} = A(t)x. \tag{14.5}$$

By using (14.3b) and (14.4) we obtain the *system response* as

$$y(t) = C(t)\Phi(t, t_0)x_0 + C(t)\int_{t_0}^{t} \Phi(t, s)B(s)u(s)\, ds + D(t)u(t). \tag{14.6}$$

When in (14.3a) and (14.3b), $A(t) \equiv A$, $B(t) \equiv B$, $C(t) \equiv C$, and $D(t) \equiv D$, we have the important linear time-invariant case given by

$$\dot{x} = Ax + Bu \tag{14.7a}$$

$$y = Cx + Du. \tag{14.7b}$$

In accordance with (13.6), (13.7), (13.8), and (14.4), the solution of (14.7a) is given by

$$\phi(t, t_0, x_0) = e^{A(t-t_0)}x_0 + \int_{t_0}^{t} e^{A(t-s)}Bu(s)\, ds \tag{14.8}$$

and the response of the system is given by

$$y(t) = Ce^{A(t-t_0)}x_0 + C\int_{t_0}^{t} e^{A(t-s)}Bu(s)\,ds + Du(t). \tag{14.9}$$

Linearity

We have referred to systems described by the linear equations (14.3a) and (14.3b) [resp., (14.7a) and (14.7b)] as *linear systems.* In the following, we establish precisely the sense in which this linearity is to be understood. To this end, for (14.3a) and (14.3b) we first let y_1 and y_2 denote system outputs that correspond to system inputs given by u_1 and u_2, respectively, *under the condition that* $x_0 = 0$. By invoking (14.6), it is clear that the system output corresponding to the system input $u = \alpha_1 u_1 + \alpha_2 u_2$, where α_1 and α_2 are real scalars, is given by $y = \alpha_1 y_1 + \alpha_2 y_2$, i.e.,

$$\begin{aligned} y(t) &= C(t)\int_{t_0}^{t} \Phi(t,s)B(s)[\alpha_1 u_1(s) + \alpha_2 u_2(s)]\,ds + D(t)[\alpha_1 u_1(t) + \alpha_2 u_2(t)] \\ &= \alpha_1 C(t)\int_{t_0}^{t} \Phi(t,s)B(s)u_1(s)\,ds + \alpha_2 C(t)\int_{t_0}^{t} \Phi(t,s)B(s)u_2(s)\,ds \\ &\quad + \alpha_1 D(t)u_1(t) + \alpha_2 D(t)u_2(t) \\ &= \alpha_1 y_1(t) + \alpha_2 y_2(t). \end{aligned} \tag{14.10}$$

Next, for (14.3a) and (14.3b) we let y_1 and y_2 denote system outputs that correspond to initial conditions $x_0^{(1)}$ and $x_0^{(2)}$, respectively, *under the condition that* $u(t) = 0$ *for all* $t \in R$. Again, by invoking (14.6), it is clear that the system output corresponding to the initial condition $x_0 = \alpha_1 x_0^{(1)} + \alpha_2 x_0^{(2)}$, where α_1 and α_2 are real scalars, is given by $y = \alpha_1 y_1 + \alpha_2 y_2$, i.e.,

$$\begin{aligned} y(t) &= C(t)\Phi(t,t_0)[\alpha_1 x_0^{(1)} + \alpha_2 x_0^{(2)}] \\ &= \alpha_1 C(t)\Phi(t,t_0)x_0^{(1)} + \alpha_2 C(t)\Phi(t,t_0)x_0^{(2)} \\ &= \alpha_1 y_1(t) + \alpha_2 y_2(t). \end{aligned} \tag{14.11}$$

Equations (14.10) and (14.11) show that for systems described by the linear equations (14.3a), (14.3b) [and hence, also by (14.7a), (14.7b)], a *superposition principle* holds in terms of the input u and the corresponding output y of the system under the assumption of zero initial conditions, and in terms of the initial conditions x_0 and the corresponding output y under the assumption of zero input. It is important to note, however, that such a superposition principle will in general not hold under conditions that combine nontrivial inputs and nontrivial initial conditions. For example, with $x_0 \neq 0$ given, and with inputs u_1 and u_2 resulting in corresponding outputs y_1 and y_2 in (14.3a) and (14.3b), it does not follow that the input $\alpha_1 u_1 + \alpha_2 u_2$ will result in an output $\alpha_1 y_1 + \alpha_2 y_2$.

1.15 STATE-SPACE DESCRIPTION OF DISCRETE-TIME SYSTEMS

State-space representation

The state-space description of discrete-time finite-dimensional dynamical systems is given by equations of the form

$$x_i(k+1) = f_i(k, x_1(k), \ldots, x_n(k), u_1(k), \ldots, u_m(k)) \qquad i = 1, \ldots, n, \tag{15.1a}$$
$$y_i(k) = g_i(k, x_1(k), \ldots, x_n(k), u_1(k), \ldots, u_m(k)) \qquad i = 1, \ldots, p, \tag{15.1b}$$

for $k = k_0, k_0+1, \ldots$, where k_0 is an integer. (In the following we let Z denote the set of integers and we let Z^+ denote the set of nonnegative integers.) Letting $x(k)^T = (x_1(k), \ldots, x_n(k))$, $f(\cdot)^T = (f_1(\cdot), \ldots, f_n(\cdot))$, $u(k)^T = (u_1(k), \ldots, u_m(k))$, $y(k)^T = (y_1(k), \ldots, y_p(k))$, and $g(\cdot)^T = (g_1(\cdot), \ldots, g_m(\cdot))$, we can rewrite (15.1a) and (15.1b) more compactly as

$$x(k+1) = f(k, x(k), u(k)) \tag{15.2a}$$
$$y(k) = g(k, x(k), u(k)). \tag{15.2b}$$

Throughout this section we will assume that $f : Z \times R^n \times R^m \to R^n$ and $g : Z \times R^n \times R^m \to R^p$.

Since f is a function, for given k_0, $x(k_0) = x_0$, and for given $u(k)$, $k = k_0, k_0 + 1, \ldots$, Eq. (15.2a) possesses a unique solution $x(k)$ that exists for all $k = k_0, k_0 + 1, \ldots$. Furthermore, under these conditions, $y(k)$ is uniquely defined for $k = k_0, k_0 + 1 \ldots$.

As in the case of continuous-time finite-dimensional systems [see Eqs. (14.1a) and (14.1b)], k_0 denotes *initial time*, k denotes *time*, $u(k)$ denotes the system *input* (evaluated at time k), $y(k)$ denotes the system *output* or system *response* (evaluated at time k), $x(k)$ characterizes the *state* (evaluated at time k), $x_i(k)$, $i = 1, \ldots, n$, denote the *state variables*, (15.2a) is called the *state equation*, and (15.2b) is called the *output equation*.

A moment's reflection should make it clear that in the case of discrete-time finite-dimensional dynamical systems described by (15.2a), (15.2b), questions concerning existence, uniqueness, and continuation of solutions are not an issue, as was the case in continuous-time systems. Furthermore, continuity with respect to initial data $x(k_0) = x_0$, or with respect to system parameters, is not an issue either, provided that $f(\cdot)$ and $g(\cdot)$ have appropriate continuity properties.

In the case of continuous-time systems described by ordinary differential equations [see Eqs. (14.1a) and (14.1b)], we allow time t to evolve "forward" and "backward." Note, however, that in the case of discrete-time systems described by (15.2a) and (15.2b), we restrict the evolution of time, k, in the forward direction to ensure uniqueness of solutions. (We will revisit this issue in further detail in Chapter 2.)

Special important cases of (15.2a), (15.2b) are *linear time-varying systems* given by

$$x(k+1) = A(k)x(k) + B(k)u(k) \tag{15.3a}$$
$$y(k) = C(k)x(k) + D(k)u(k), \tag{15.3b}$$

where $A : Z \to R^{n\times n}$, $B : Z \to R^{n\times m}$, $C : Z \to R^{p\times n}$, and $D : Z \to R^{p\times m}$. When $A(k) \equiv A$, $B(k) \equiv B$, $C(k) \equiv C$, and $D(k) \equiv D$, we have *linear time-invariant systems* given by

$$x(k+1) = Ax(k) + Bu(k) \tag{15.4a}$$
$$y(k) = Cx(k) + Du(k). \tag{15.4b}$$

As in the case of continuous-time finite-dimensional dynamical systems, many of the qualitative properties of discrete-time finite-dimensional systems can be studied in terms of *initial-value problems* given by

$$x(k+1) = f(k, x(k)), \qquad x(k_0) = x_0, \tag{I_D}$$

where $x \in R^n$, $f : Z \times R^n \to R^n$, $k_0 \in Z$, and $k = k_0, k_0 + 1, \ldots$. We call the equation

$$x(k+1) = f(k, x(k)), \tag{E_D}$$

a *system of first-order ordinary difference equations.* Special important cases of (E_D) include *autonomous systems* described by

$$x(k+1) = f(x(k)), \tag{A_D}$$

periodic systems given by

$$x(k+1) = f(k, x(k)) = f(k+K, x(k)) \tag{P_D}$$

for fixed $K \in Z^+$ and for all $k \in Z$, *linear homogeneous systems* given by

$$x(k+1) = A(k)x(k), \tag{LH_D}$$

linear periodic systems characterized by

$$x(k+1) = A(k)x(k) = A(k+K)x(k) \tag{LP_D}$$

for fixed $K \in Z^+$ and for all $k \in Z$, *linear nonhomogeneous systems* given by

$$x(k+1) = A(k)x(k) + g(k), \tag{LN_D}$$

and *linear, autonomous, homogeneous systems* characterized by

$$x(k+1) = Ax(k). \tag{L_D}$$

In these equations all symbols used are defined in the obvious way by making reference to the corresponding systems of ordinary differential equations (see Subsection 1.3B).

Difference equations of order n

Thus far we have addressed systems of first-order difference equations. As in the continuous-time case, it is also possible to characterize initial-value problems by nth-order ordinary difference equations, say,

$$y(k+n) = h(k, y(k), y(k+1), \ldots, y(k+n-1)), \tag{E_{nD}}$$

where $h : Z \times R^n \to R$, $n \in Z^+$, $k = k_0, k_0 + 1, \ldots$. By specifying an *initial time* $k_0 \in Z$ and by specifying $y(k_0), y(k_0+1), \ldots, y(k_0+n-1)$, we again have an initial-value problem given by

$$\begin{aligned} y(k+n) &= h(k, y(k), y(k+1), \ldots, y(k+n-1)) \\ y(k_0) &= x_{10}, \ldots, y(k_0+n-1) = x_{n0}. \end{aligned} \tag{I_{nD}}$$

We call (E_{nD}) an n*th-order ordinary difference equation* and we note once more that in the case of initial-value problems described by such equations, there are no difficult issues involving the existence, uniqueness and continuation of solutions.

We can reduce the study of (I_{nD}) to the study of initial-value problems determined by systems of first-order ordinary difference equations. To accomplish this, we let in (I_{nD}) $y(k) = x_1(k)$, $y(k+1) = x_2(k), \ldots, y(k+n-1) = x_n(k)$. We now obtain the system of first-order ordinary difference equations

$$\begin{aligned} x_1(k+1) &= x_2(k) \\ &\cdots\cdots\cdots \\ x_{n-1}(k+1) &= x_n(k) \\ x_n(k+1) &= h(k, x_1(k), \ldots, x_n(k)). \end{aligned} \tag{15.5}$$

Equations (15.5), together with the initial data $x_0^T = (x_{10}, \ldots, x_{n0})$, are equivalent to the initial-value problem (I_{nD}) in the sense that these two problems will generate identical solutions [and in the sense that the transformation of (I_{nD}) into (15.5) can be reversed unambiguously and uniquely].

As in the case of systems of first-order ordinary difference equations, we can point to several important special cases of nth-order ordinary difference equations, including equations of the form

$$y(k+n) + a_{n-1}(k)y(k+n-1) + \cdots + a_1(k)y(k+1) + a_0(k)y(k) = g(k), \tag{15.6}$$

$$y(k+n) + a_{n-1}(k)y(k+n-1) + \cdots + a_1(k)y(k+1) + a_0(k)y(k) = 0, \tag{15.7}$$

and
$$y(k+n) + a_{n-1}y(k+n-1) + \cdots + a_1 y(k+1) + a_0 y(k) = 0. \tag{15.8}$$

We call (15.6) a *linear nonhomogeneous ordinary difference equation of order n;* we call (15.7) a *linear homogeneous ordinary difference equation of order n;* and we call (15.8) a *linear, autonomous, homogeneous ordinary difference equation of order n.* As in the case of systems of first-order ordinary difference equations, we can define *periodic* and *linear periodic ordinary difference equations of order n* in the obvious way.

Solutions of state equations

Returning now to linear homogeneous systems

$$x(k+1) = A(k)x(k), \tag{LH_D}$$

we observe that

$$\begin{aligned} x(k+2) &= A(k+1)x(k+1) = A(k+1)A(k)x(k) \\ &\cdots\cdots\cdots \\ x(n) &= A(n-1)A(n-2)\cdots A(k+1)A(k)x(k) \\ &= \prod_{j=k}^{n-1} A(j)x(k), \end{aligned}$$

i.e., the state of the system at time n is related to the state at time k by means of the $n \times n$ matrix $\prod_{j=k}^{n-1} A(j)$ (as can easily be proved by induction). This suggests that the *state transition matrix* for (LH_D) is given by

$$\Phi(n, k) = \prod_{j=k}^{n-1} A(j), \qquad n > k, \tag{15.9}$$

and that
$$\Phi(k, k) = I. \tag{15.10}$$

As in the continuous-time case, the solution to the initial-value problem

$$\begin{aligned} x(k+1) &= A(k)x(k) \\ x(k_0) &= x_{k_0}, \qquad k_0 \in Z, \end{aligned} \tag{15.11}$$

is now given by

$$x(n) = \Phi(n, k_0)x_{k_0} = \prod_{j=k_0}^{n-1} A(j)x_{k_0}, \qquad n > k_0. \tag{15.12}$$

Continuing, let us next consider initial-value problems determined by linear nonhomogeneous systems (LN_D),

$$\begin{aligned} x(k+1) &= A(k)x(k) + g(k) \\ x(k_0) &= x_{k_0}. \end{aligned} \tag{15.13}$$

Then

$$\begin{aligned} x(k_0+1) &= A(k_0)x(k_0) + g(k_0) \\ x(k_0+2) &= A(k_0+1)x(k_0+1) + g(k_0+1) \\ &= A(k_0+1)A(k_0)x(k_0) + A(k_0+1)g(k_0) + g(k_0+1) \\ x(k_0+3) &= A(k_0+2)x(k_0+2) + g(k_0+2) \\ &= A(k_0+2)A(k_0+1)A(k_0)x(k_0) + A(k_0+2)A(k_0+1)g(k_0) \\ &\quad + A(k_0+2)g(k_0+1) + g(k_0+2) \\ &= \Phi(k_0+3, k_0)x_{k_0} + \Phi(k_0+3, k_0+1)g(k_0) \\ &\quad + \Phi(k_0+3, k_0+2)g(k_0+1) + \Phi(k_0+3, k_0+3)g(k_0+2), \end{aligned}$$

and so forth. For $k \geq k_0 + 1$, we easily obtain the expression for the solution of (15.13) as

$$x(k) = \Phi(k, k_0)x_{k_0} + \sum_{j=k_0}^{k-1} \Phi(k, j+1)g(j). \tag{15.14}$$

We note that when $x_{k_0} = 0$, (15.14) reduces to

$$x_p(k) = \sum_{j=k_0}^{k-1} \Phi(k, j+1)g(j), \tag{15.15}$$

and when $x_{k_0} \neq 0$ but $g(k) \equiv 0$, (15.14) reduces to

$$x_h(k) = \Phi(k, k_0)x_{k_0}. \tag{15.16}$$

Therefore, the *total solution* of (15.13) consists of the sum of its *particular solution,* $x_p(k)$, and its *homogeneous solution,* $x_h(k)$.

System response

Finally, we observe that in view of (15.3b) and (15.14), the *system response* of the system (15.3a), (15.3b) is of the form

$$y(k) = C(k)\Phi(k, k_0)x_{k_0} + C(k)\sum_{j=k_0}^{k-1} \Phi(k, j+1)B(j)u(j) + D(k)u(k), \qquad k > k_0, \tag{15.17}$$

and

$$y(k_0) = C(k_0)x_{k_0} + D(k_0)u(k_0). \tag{15.18}$$

Discrete time systems, as discussed above, arise in several ways, including the *numerical solution* of ordinary differential equations (see, e.g., our discussion of *Euler's method* in Subsection 1.6B); the representation of *sampled-data systems* at discrete points in time (which will be discussed in further detail in Chapter 2); in the modeling process of systems that are defined only at discrete points in time (e.g., digital computer systems); and so forth.

As a specific example of a discrete-time system we consider a *second-order section digital filter* in *direct form,*

$$\begin{aligned} x_1(k+1) &= x_2(k) \\ x_2(k+1) &= ax_1(k) + bx_2(k) + u(k) \end{aligned} \tag{15.19a}$$

$$y(k) = x_1(k), \tag{15.19b}$$

$k \in Z^+$, where $x_1(k)$ and $x_2(k)$ denote the state variables, $u(k)$ denotes the input, and $y(k)$ denotes the output of the digital filter. We depict system (15.19a), (15.19b) in block diagram form in Fig. 1.13.

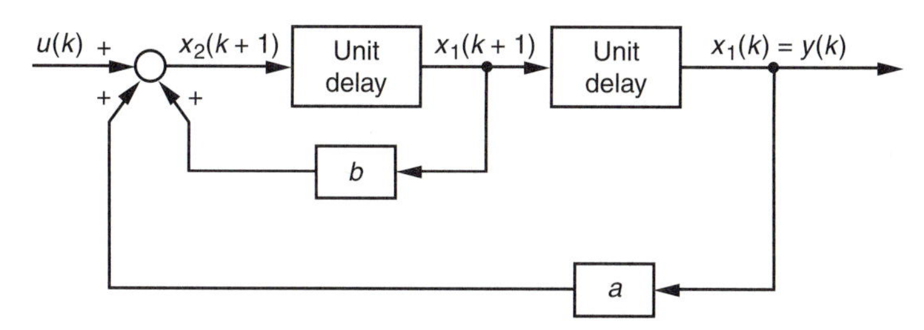

FIGURE 1.13
Second-order section digital filter in direct form

1.16 INPUT-OUTPUT DESCRIPTION OF SYSTEMS

This section consists of four subsections. First we consider rather general aspects of the input-output description of systems. Because of their simplicity, we address the characterization of linear discrete-time systems next. In the third subsection we provide a foundation for the impulse response of linear continuous-time systems. Finally, we address the external description of linear continuous-time systems.

A. External Description of Systems: General Considerations

The state-space representation of systems presupposes knowledge of the *internal structure* of the system. When this structure is unknown, it may still be possible to arrive at a system description—an *external description*—that relates system inputs to system outputs. In linear system theory, a great deal of attention is given to relating the internal description of systems (the state representation) to the external description (the input-output description).

In the present context, we view *system inputs* and *system outputs* as elements of two real vector spaces U and Y, respectively, and we view a system as being represented by an operator T that relates elements of U to elements of Y. For $u \in U$ and $y \in Y$ we will assume that $u : R \rightarrow R^m$ and $y : R \rightarrow R^p$ in the case of *continuous-time systems*, and that $u : Z \rightarrow R^m$ and $y : Z \rightarrow R^p$ in the case of *discrete-time systems*. For continuous-time systems we define vector addition (on U) and multiplication of vectors by scalars (on U) as

$$(u_1 + u_2)(t) = u_1(t) + u_2(t) \tag{16.1}$$

and

$$(\alpha u)(t) = \alpha u(t) \tag{16.2}$$

for all $u_1, u_2 \in U$, $\alpha \in R$, and $t \in R$. We similarly define vector addition and multiplication of vectors by scalars on Y. Furthermore, for discrete-time systems we define these operations on U and Y analogously. In this case the elements of U and Y are real sequences that we denote, e.g., by $u = \{u_k\}$ or $u = \{u(k)\}$. (It is easily verified that under these rather general conditions, U and Y satisfy all the axioms of a vector space, both for the continuous-time case and the discrete-time case.) In the continuous-time case as well as in the discrete-time case the system is represented by $T : U \rightarrow Y$, and we write

$$y = T(u). \tag{16.3}$$

In the subsequent development, we will impose restrictions on the vector spaces U, Y, and on the operator T, as needed. For example, if T is a linear operator, the system is called a *linear system*. In this case we have

$$\begin{aligned} y &= T(\alpha_1 u_1 + \alpha_2 u_2) \\ &= \alpha_1 T(u_1) + \alpha_2 T(u_2) \\ &= \alpha_1 y_1 + \alpha_2 y_2 \end{aligned} \tag{16.4}$$

for all $\alpha_1, \alpha_2 \in R$ and $u_1, u_2 \in U$, where $y_i = T(u_i) \in Y, i = 1, 2$, and $y \in Y$. Equation (16.4) represents the well-known *principle of superposition* of linear systems.

If in the above, $m = p = 1$, we speak of a *single-input/single-output (SISO) system*. Systems for which $m > 1$, $p > 1$, are called *multi-input/multi-output (MIMO) systems*.

We say that a system is *memoryless*, or *without memory*, if its output for each value of the independent variable (t or k) is dependent only on the input evaluated at the same value of the independent variable [e.g., $y(t_1)$ depends only on $u(t_1)$ and $y(k_1)$ depends only on $u(k_1)$]. An example of such a system is the resistor circuit shown in Fig. 1.14, where the current $i(t) = u(t)$ denotes the system input at time t and the voltage across the resistor, $v(t) = Ri(t) = y(t)$, denotes the system output at time t.

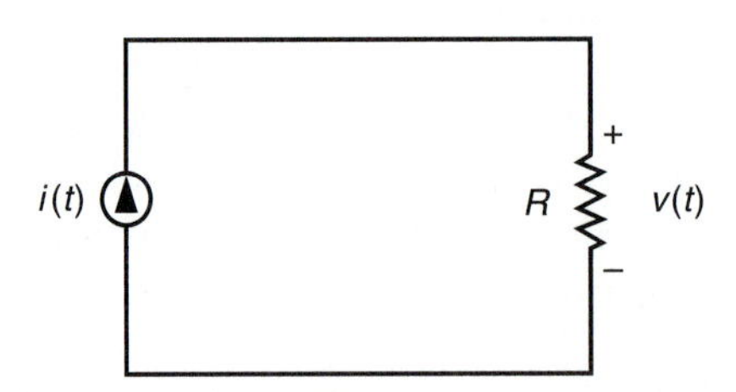

FIGURE 1.14
Resistor circuit

A system that is not memoryless is said to have memory. An example of a continuous time system *with memory* is the capacitor circuit shown in Fig. 1.15, where the current $i(t) = u(t)$ represents the system input at time t and the voltage across the capacitor,

$$y(t) = v(t) = \frac{1}{C}\int_{-\infty}^{t} i(\tau)\, d\tau,$$

denotes the system output at time t. Another example of a continuous-time system with memory is described by the scalar equation

$$y(t) = u(t-1), \qquad t \in R,$$

and an example of a discrete-time system with memory is characterized by the scalar equation

$$y(n) = \sum_{k=-\infty}^{n} x(k), \qquad n, k \in Z.$$

A system is said to be *causal* if its output at any time, say, t_1 (or k_1) depends only on values of the input evaluated for $t \leq t_1$ (for $k \leq k_1$). Thus, $y(t_1)$ depends only on $u(t), t \leq t_1$ [or $y(k_1)$ depends only on $u(k), k \leq k_1$]. Such a system is referred to as being *nonanticipative* since the system output does not anticipate future values of the input.

To make the above concept a bit more precise, we define the function $u_\tau : R \to R^m$ for $u \in U$ by

$$u_\tau(t) = \begin{cases} u(t), & t \leq \tau, \\ 0, & t > \tau, \end{cases}$$

and we similarly define the function $y_\tau : R \to R^p$ for $y \in Y$. A system that is represented by the mapping $y = T(u)$ is said to be *causal* if and only if

$$(T(u))_\tau = (T(u_\tau))_\tau \qquad \text{for all } \tau \in R, \text{ for all } u \in U.$$

Equivalently, this system is causal if and only if for $u, v \in U$ and $u_\tau = v_\tau$ it is true that

$$(T(u))_\tau = (T(v))_\tau \qquad \text{for all } \tau \in R.$$

For example, the discrete-time system described by the scalar equation

$$y(n) = u(n) - u(n+1), \qquad n \in Z$$

is *not causal*. Neither is the continuous-time system characterized by the scalar equation

$$y(t) = x(t+1), \qquad t \in R.$$

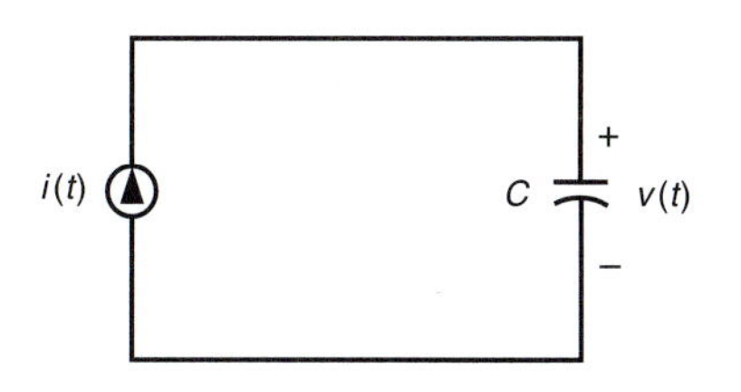

FIGURE 1.15
Capacitor circuit

It should be pointed out that systems that are not causal are by no means useless. For example, causality is *not* of fundamental importance in image-processing applications where the independent variable is not time. Even when time is the independent variable, noncausal systems may play an important role. For example, in the processing of data that have been recorded (such as speech, meteorological data, demographic data, stock market fluctuations, etc.), one is not constrained to processing the data causally. An example of this would be the smoothing of data over a time interval, say, by means of the system

$$y(n) = \frac{1}{2M+1} \sum_{k=-M}^{M} u(n-k).$$

A system is said to be *time-invariant* if a time shift in the input signal causes a corresponding time shift in the output signal. To make this concept more precise, for fixed $\alpha \in R$, we introduce the *shift operator* $Q_\alpha : U \to U$ as

$$Q_\alpha u(t) = u(t-\alpha), \qquad u \in U, t \in R.$$

A system that is represented by the mapping $y = T(u)$ is said to be *time-invariant* if and only if

$$TQ_\alpha(u) = Q_\alpha(T(u)) = Q_\alpha(y)$$

for any $\alpha \in R$ and any $u \in U$. If a system is not time-invariant, it is said to be *time-varying*.

For example, a system described by the relation

$$y(t) = \cos u(t)$$

is time-invariant. To see this, consider the inputs $u_1(t)$ and $u_2(t) = u_1(t - t_0)$. Then

$$y_1(t) = \cos u_1(t), \qquad y_2(t) = \cos u_2(t) = \cos u_1(t-t_0)$$

and

$$y_1(t-t_0) = \cos u_1(t-t_0) = y_2(t).$$

As a second example, consider a system described by the relation

$$y(n) = nu(n)$$

and consider two inputs $u_1(n)$ and $u_2(n) = u_1(n - n_0)$. Then

$$y_1(n) = nu_1(n) \qquad \text{and} \qquad y_2(n) = nu_2(n) = nu_1(n-n_0).$$

However, if we shift the output $y_1(n)$ by n_0, we obtain

$$y_1(n-n_0) = (n-n_0)u_1(n-n_0) \neq y_2(n).$$

Therefore, this system is not time-invariant.

B. Linear Discrete-Time Systems

In this subsection we investigate the representation of linear discrete-time systems. We begin our discussion by considering SISO systems.

In the following, we employ the *discrete-time impulse* (or *unit pulse* or *unit sample)*, which is defined as

$$\delta(n) = \begin{cases} 0, & n \neq 0, n \in Z, \\ 1, & n = 0. \end{cases} \tag{16.5}$$

Note that if $\{p(n)\}$ denotes the *unit step sequence,* i.e.,

$$p(n) = \begin{cases} 1, & n \geq 0, n \in Z, \\ 0, & n < 0, n \in Z, \end{cases} \tag{16.6}$$

then
$$\delta(n) = p(n) - p(n-1)$$

and
$$p(n) = \begin{cases} \sum_{k=0}^{\infty} \delta(n-k), & n \geq 0, \\ 0, & n < 0. \end{cases} \tag{16.7}$$

Furthermore, note that an arbitrary sequence $\{x(n)\}$ can be expressed as

$$x(n) = \sum_{k=-\infty}^{\infty} x(k)\delta(n-k). \tag{16.8}$$

In Example 10.8 we showed that a transformation $T : U \to Y$ determined by the equation

$$y(n) = \sum_{k=-\infty}^{\infty} h(n, k)u(k), \tag{16.9}$$

where $y \triangleq \{y(k)\} \in Y, u \triangleq \{u(k)\} \in U$, and $h : Z \times Z \to R$, is a linear transformation. We also noted in Example 10.8 that for (16.9) to make any sense, we need to impose restrictions on $\{h(n, k)\}$ and $\{u(k)\}$. For example, if for every fixed $n, \{h(n, k)\} \in l_2$ and $\{u(k)\} \in l_2 = U$, then it follows from the Hölder Inequality (resp., Schwarz Inequality), that (16.9) is well defined. There are of course other conditions that one might want to impose on (16.9) (refer to Example 10.8). For example, if for every fixed $n, \sum_{k=-\infty}^{\infty} |h(n, k)| < \infty$ (i.e., for every fixed $n, \{h(n, k)\} \in l_1$) and if $\sup_{k \in Z} |u(k)| < \infty$ (i.e., $\{u(k)\} \in l_\infty$), then (16.9) is also well defined.

We shall now elaborate on the suitability of (16.9) to represent linear discrete-time systems. To this end, we will agree once and for all that, in the ensuing discussion, all assumptions on $\{h(n, k)\}$ and $\{u(k)\}$ are satisfied that ensure that (16.9) is well defined.

We will view $y \in Y$ and $u \in U$ as system outputs and system inputs, respectively, the linear transformation $T : U \to Y$ as representing the system, and $\{h(n, k)\}$ as determining T by means of relation (16.9), in the following manner. Suppose that $h(n, k)$ denotes the response (output) of the system at time $t = n$, due to a unit pulse (input) that is applied to the system at time $t = k$. Since the system is by assumption linear, the response at time $t = n$, due to a weighted pulse, applied at $t = k$ and with weight $u(k)$, will be $h(n, k)u(k)$. Again, since the system is linear, the response of the system at time $t = n$, due to the sum of two weighted unit pulses with weights $u(k_1)$ and $u(k_2)$, applied to the system at $t = k_1$ and $t = k_2$, respectively, will be $h(n, k_1)u(k_1) + h(n, k_2)u(k_2)$. This is equivalent to saying that if an input sequence $\{u^1(k)\}$ that is zero everywhere except at $t = k_1$, where it is $u(k_1)$, and a second input sequence $\{u^2(k)\}$ that is zero everywhere except at $t = k_2$, where it equals $u(k_2)$, are applied simultaneously to the system, resulting in the input sequence $\{u^1(k)\} + \{u^2(k)\}$ that is zero everywhere except at $t = k_1$, where it is $u(k_1)$, and at $t = k_2$, where it is $u(k_2)$, then the corresponding system output at $t = n$ will equal $h(n, k_1)u(k_1) + h(n, k_2)u(k_2)$.

This process can be continued by induction to yield the *total response* of the system at time $t = n, y(n)$, due to a sequence $\{u(k)\}, k \in Z$. In doing so, one arrives precisely at the expression (16.9).

By using (16.5), we thus see that $h(n, k)$ may be viewed as representing the response of the system to $\delta(n-k)$, a unit sample (unit pulse) occurring at $n = k$. Accordingly, $h(n, k)$ is called the *unit pulse response* (or the *unit impulse response*) of a linear discrete-time system.

Next, suppose that T represents a time-invariant system. This means that if $\{h(n, 0)\}$ is the response to $\{\delta(n)\}$, then by time invariance, the response to $\{\delta(n-k)\}$ is simply $\{h(n-k, 0)\}$. By a slight abuse of notation, we let $h(n-k, 0) \triangleq h(n-k)$. Then (16.9) assumes the form

$$y(n) = \sum_{k=-\infty}^{\infty} u(k)h(n-k). \tag{16.10}$$

Expression (16.10) is called a *convolution sum* and is written more compactly as

$$y(n) = u(n) * h(n).$$

Now by a substitution of variables, we obtain for (16.10) the alternative expression

$$y(n) = \sum_{k=-\infty}^{\infty} h(k)u(n-k),$$

and therefore, we have

$$y(n) = u(n) * h(n) = h(n) * u(n),$$

i.e., the convolution operation $*$ commutes.

As a specific example, consider a linear, time-invariant, discrete-time system with unit impulse response given by

$$h(n) = \begin{Bmatrix} a^n, & n \geq 0 \\ 0, & n < 0 \end{Bmatrix} = a^n p(n), \qquad 0 < a < 1,$$

where $p(n)$ is the unit step sequence given in (16.6). It is an easy matter to show that the response of this system to an input given by

$$u(n) = p(n) - p(n-N)$$

is

$$y(n) = 0, \qquad n < 0,$$

$$y(n) = \sum_{k=0}^{n} a^{n-k} = a^n \frac{1 - 1^{-(n+1)}}{1 - a^{-1}} = \frac{1 - a^{n+1}}{1 - a}, \qquad 0 \leq n < N,$$

and

$$y(n) = \sum_{k=0}^{N-1} a^{n-k} = a^n \frac{1 - a^{-N}}{1 - a^{-1}} = \frac{a^{n-N+1} - a^{n+1}}{1 - a}, \qquad N \leq n.$$

Proceeding, with reference to (16.9), we note that $h(n, k)$ represents the system output at time n due to a δ-function input applied at time k. Now if system (16.9) is *causal,* then its output will be identically zero before an input is applied. Hence, a linear system (16.9) is causal if and only if

$$h(n, k) = 0 \qquad \text{for all } k \text{ and all } n < k.$$

Therefore, when the system (16.9) is causal, we have in fact

$$y(n) = \sum_{k=-\infty}^{n} h(n, k)u(k). \tag{16.11a}$$

We can rewrite (16.11a) as

$$y(n) = \sum_{k=-\infty}^{k_0-1} h(n,k)u(k) + \sum_{k=k_0}^{n} h(n,k)u(k)$$
$$\triangleq y(k_0-1) + \sum_{k=k_0}^{n} h(n,k)u(k). \tag{16.11b}$$

We say that the discrete-time system described by (16.9) is *at rest* at $k = k_0 \in Z$ if $u(k) = 0$ for $k \geq k_0$ implies that $y(k) = 0$ for $k \geq k_0$. Accordingly, if system (16.9) is known to be at rest at $k = k_0$, we have

$$y(n) = \sum_{k=k_0}^{\infty} h(n,k)u(k).$$

Furthermore, if system (16.9) is known to be causal and at rest at $k = k_0$, its input-output description assumes the form [in view of (16.11b)]

$$y(n) = \sum_{k=k_0}^{n} h(n,k)u(k). \tag{16.12}$$

Next, turning to linear, discrete-time, *MIMO systems,* we can generalize (16.9) to

$$y(n) = \sum_{k=-\infty}^{\infty} H(n,k)u(k), \tag{16.13}$$

where $y : Z \to R^p$, $u : Z \to R^m$, and

$$H(n,k) = \begin{bmatrix} h_{11}(n,k) & h_{12}(n,k) & \cdots & h_{1m}(n,k) \\ h_{21}(n,k) & h_{22}(n,k) & \cdots & h_{2m}(n,k) \\ \cdots & \cdots & \cdots & \cdots \\ h_{p1}(n,k) & h_{p2}(n,k) & \cdots & h_{pm}(n,k) \end{bmatrix}, \tag{16.14}$$

where $h_{ij}(n,k)$ represents the system response at time n of the ith component of y due to a discrete-time impulse δ applied at time k at the jth component of u, while the inputs at all other components of u are being held zero. The matrix H is called the *discrete-time unit impulse response matrix* of the system.

Similarly, it follows that the system (16.13) is *causal* if and only if

$$H(n,k) = 0 \qquad \text{for all } k \text{ and all } n < k,$$

and that the input-output description of linear, discrete-time, causal systems is given by

$$y(n) = \sum_{k=-\infty}^{n} H(n,k)u(k). \tag{16.15}$$

A discrete-time system described by (16.13) is said to be *at rest at* $k = k_0 \in Z$ if $u(k) = 0$ for $k \geq k_0$ implies that $y(k) = 0$ for $k \geq k_0$. Accordingly, if system (16.13) is known to be at rest at $k = k_0$, we have

$$y(n) = \sum_{k=k_0}^{\infty} H(n,k)u(k). \tag{16.16}$$

Moreover, if a linear discrete-time system that is at rest at k_0 is known to be causal, then its input-output description reduces to

$$y(n) = \sum_{k=k_0}^{n} H(n, k)u(k). \tag{16.17}$$

Finally, as in (16.10), it is easily shown that the unit impulse response $H(n, k)$ of a linear, *time-invariant*, discrete-time MIMO system depends only on the difference of n and k, i.e., by a slight abuse of notation we can write

$$H(n, k) = H(n-k, 0) \triangleq H(n-k) \tag{16.18}$$

for all n and k. Accordingly, linear, time-invariant, causal, discrete-time MIMO systems that are at rest at $k = k_0$ are described by equations of the form

$$y(n) = \sum_{k=k_0}^{n} H(n-k)u(k). \tag{16.19}$$

We conclude by supposing that the system on hand is described by the state and output equations (15.3a) and (15.3b) under the assumption that $x(k_0) = 0$, i.e., the system is at rest at $k = k_0$. Then, according to Eqs. (15.17) and (15.18), we obtain

$$H(n, k) = \begin{cases} C(n)\Phi(n, k+1)B(k), & n > k, \\ D(n), & n = k, \\ 0, & n < k. \end{cases} \tag{16.20}$$

Furthermore, for the time-invariant case we obtain

$$H(n-k) = \begin{cases} CA^{n-(k+1)}B, & n > k, \\ D, & n = k, \\ 0, & n < k. \end{cases} \tag{16.21}$$

C. The Dirac Delta Distribution

For any linear time-invariant operator P from $C(R, R)$ to itself, we say that P admits an *integral representation* if there exists an integrable function (in the Riemann or Lebesgue sense), $g_p : R \to R$, such that for any $f \in C(R, R)$,

$$(Pf)(x) = (f * g_p)(x) \triangleq \int_{-\infty}^{\infty} f(\tau)g_p(x-\tau)\,d\tau.$$

We call g_p a *kernel of the integral representation of P*.

For the identity operator I [defined by $If = f$ for any $f \in C(R, R)$] an integral representation for which g_p is a function in the usual sense does not exist (see, e.g., Z. Szmydt, *Fourier Transformation and Linear Differential Equations*, D. Reidel Publishing Company, Boston, 1977). However, there exists a sequence of functions $\{\phi_n\}$ such that for any $f \in C(R, R)$,

$$(If)(x) = f(x) = \lim_{n\to\infty}(f * \phi_n)(x). \tag{16.22}$$

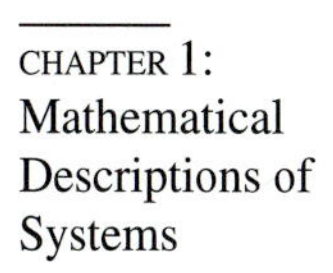

To prove (16.22) we will make use of functions $\{\phi_n\}$ given by

$$\phi_n(x) = \begin{cases} n(1 - n|x|), & \text{if } |x| \le \dfrac{1}{n}, \\ 0, & \text{if } |x| > \dfrac{1}{n}, \end{cases}$$

$n = 1, 2, 3, \ldots$. A plot of ϕ_n is depicted in Fig. 1.16.

We first establish the following useful property of $\{\phi_n\}$.

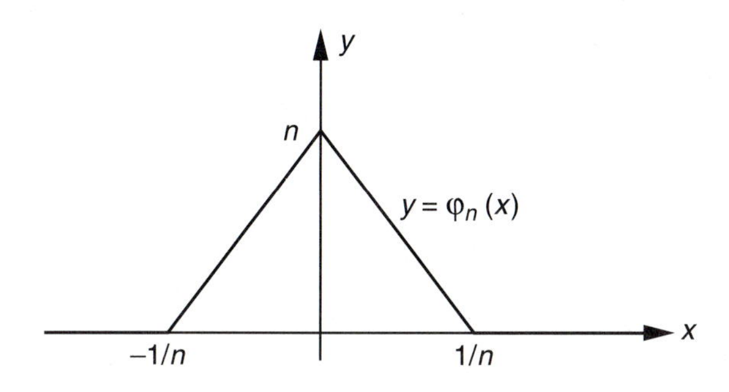

FIGURE 1.16

LEMMA 16.1. Let f be a continuous real-valued function defined on R and let ϕ_n be defined as above (Fig. 1.16). Then for any $a \in R$,

$$\lim_{n\to\infty} \int_{-\infty}^{\infty} f(\tau)\phi_n(a-\tau)\,d\tau = f(a). \tag{16.23}$$

Proof. It is easy to verify that for $n = 1, 2, \ldots$,

$$\int_{-\infty}^{\infty} \phi_n(\tau)\,d\tau = \int_{-1/n}^{1/n} \phi_n(\tau)\,d\tau = 1.$$

Then

$$\begin{aligned} \int_{-\infty}^{\infty} \phi_n(a-\tau)\,d\tau &= \int_{a-1/n}^{a+1/n} \phi_n(a-\tau)\,d\tau \\ &= \int_{-1/n}^{1/n} \phi_n(\tau)\,d\tau = 1. \end{aligned} \tag{16.24}$$

We first assume that f is a nonnegative function. By the continuity of f we may suppose that f assumes a maximum value $f(b_n)$ and a minimum value $f(c_n)$, where b_n and $c_n \in \left[a - \dfrac{1}{n}, a + \dfrac{1}{n}\right]$. Then

$$\begin{aligned} \int_{-\infty}^{\infty} \phi_n(a-\tau)f(\tau)\,d\tau &= \int_{a-1/n}^{a+1/n} \phi_n(a-\tau)f(\tau)\,d\tau \\ &\le f(b_n)\int_{a-1/n}^{a+1/n} \phi_n(a-\tau)\,d\tau = f(b_n), \end{aligned} \tag{16.25}$$

where we have used (16.24). In a similar manner we can show that

$$\int_{-\infty}^{\infty} \phi_n(a-\tau)f(\tau)\,d\tau \ge f(c_n). \tag{16.26}$$

Since b_n and $c_n \in \left[a - \dfrac{1}{n}, a + \dfrac{1}{n}\right]$, it follows that $\lim_{n\to\infty} b_n = \lim_{n\to\infty} c_n = a$. Thus, (16.25) and (16.26) together with the continuity of f imply (16.23).

To remove the assumption that f is a nonnegative function, we recall that every continuous function f can be written as the sum of two nonnegative functions, $f = f_+ - f_-$, where

$$f_+(x) = \begin{cases} f(x), & \text{if } f(x) \geq 0, \\ 0, & \text{if } f(x) < 0, \end{cases}$$

and

$$f_-(x) = \begin{cases} 0, & \text{if } f(x) \geq 0, \\ -f(x), & \text{if } f(x) < 0. \end{cases}$$

Then

$$\begin{aligned} &\lim_{n\to\infty} \int_{-\infty}^{\infty} f(\tau)\phi_n(a-\tau)\,d\tau \\ &\qquad = \lim_{n\to\infty} \int_{-\infty}^{\infty} f_+(\tau)\phi_n(a-\tau)\,d\tau - \lim_{n\to\infty} \int_{-\infty}^{\infty} f_-(\tau)\phi_n(a-\tau)\,d\tau \\ &\qquad = f_+(a) - f_-(a) = f(a). \end{aligned}$$

■

The above result, when applied to (16.22), now allows us to *define* a *generalized function* δ, (also called a *distribution*) as the kernel of a *formal* or *symbolic* integral representation of the identity operator I, i.e.,

$$f(x) = \lim_{n\to\infty} \int_{-\infty}^{\infty} f(\tau)\phi_n(x-\tau)\,d\tau \tag{16.27}$$

$$\triangleq \int_{-\infty}^{\infty} f(\tau)\delta(x-\tau)\,d\tau \tag{16.28}$$

$$= f * \delta(x). \tag{16.29}$$

It is emphasized that expression (16.28) *is not an integral at all* (in the Riemann or Lebesgue sense), *but only a symbolic representation.* To put it another way, the limit of the integral of the indicated sequence of functions in (16.27) exists [and is equal to $f(x)$], while the integral of the limit of the same indicated sequence of functions is not even defined. The *generalized function* δ is called the *unit impulse* or the *Dirac delta distribution.*

In applications we frequently encounter functions $f \in C(R^+, R)$. If we extend f to be defined on all of R by letting $f(x) = 0$ for $x < 0$, then (16.23) becomes

$$\lim_{n\to\infty} \int_0^{\infty} f(\tau)\phi_n(a-\tau)\,d\tau = f(a) \tag{16.30}$$

for any $a > 0$, where we have used the fact that in the proof of Lemma 16.1 we need f to be continuous only in a neighborhood of a. Therefore, for $f \in C(R^+, R)$, (16.27) to (16.30) yield

$$\lim_{n\to\infty} \int_0^{\infty} f(\tau)\phi_n(t-\tau)\,d\tau \triangleq \int_0^{\infty} f(\tau)\delta(t-\tau)\,d\tau = f(t) \tag{16.31}$$

for any $t > 0$. Since the ϕ_n are even functions, we have $\phi_n(t-\tau) = \phi_n(\tau-t)$, which allows for the representation $\delta(t-\tau) = \delta(\tau-t)$. We obtain from (16.31) that

$$\lim_{n\to\infty} \int_0^{\infty} f(\tau)\phi_n(\tau-t)\,d\tau \triangleq \int_0^{\infty} f(\tau)\delta(\tau-t)\,d\tau = f(t)$$

for any $t > 0$. Changing the variable $\tau' = \tau - t$, we obtain

$$\lim_{n\to\infty} \int_{-t}^{\infty} f(\tau'+t)\phi_n(\tau')\,d\tau' \triangleq \int_{-t}^{\infty} f(\tau'+t)\delta(\tau')\,d\tau' = f(t)$$

for any $t > 0$. Taking the limit $t \to 0^+$, we obtain

$$\lim_{n\to\infty} \int_{0^-}^{\infty} f(\tau' + t)\phi_n(\tau')\, d\tau' \triangleq \int_{0^-}^{\infty} f(\tau')\delta(\tau')\, d\tau' = f(0), \tag{16.32}$$

where, as in (16.24), $\int_{0^-}^{\infty} f(\tau')\delta(\tau')\, d\tau'$ is not an integral, but a symbolic representation of $\lim_{n\to\infty} \int_{0^-}^{\infty} f(\tau' + t)\phi_n(\tau')\, d\tau'$.

Now let s denote a complex variable. If in (16.31) to (16.32) we let $f(\tau) = e^{-s\tau}$, $\tau > 0$, then we obtain the *Laplace transform*

$$\lim_{n\to\infty} \int_{0^-}^{\infty} e^{-s\tau}\phi_n(\tau)\, d\tau \triangleq \int_{0^-}^{\infty} e^{-s\tau}\delta(\tau)\, d\tau = 1. \tag{16.33}$$

Symbolically we denote (16.33) by

$$\mathscr{L}(\delta) = 1, \tag{16.34}$$

and we say that the Laplace transform of the unit impulse function or the Dirac delta distribution is equal to one.

Next, we point out another important property of δ. Consider a (time-invariant) operator P and assume that P admits an integral representation with kernel g_P. If in (16.31) we let $f = g_P$, we have

$$\lim_{n\to\infty} (P\phi_n)(t) = g_P(t), \tag{16.35}$$

and we write this (symbolically) as

$$P\delta = g_P. \tag{16.36}$$

This shows that the impulse response of a linear, time-invariant, continuous-time system with integral representation is equal to the kernel of the integral representation of the system. Symbolically this is depicted in Fig. 1.17.

Next, for any linear time-varying operator P from $C(R, R)$ to itself, we say that P admits an *integral representation* if there exists an integrable function (in the Riemann or Lebesgue sense), $g_P : R \times R \to R$, such that for any $f \in C(R, R)$,

$$(Pf)(\eta) = \int_{-\infty}^{\infty} f(\tau) g_P(\eta, \tau)\, d\tau. \tag{16.37}$$

Again, we call g_P a *kernel of the integral representation of* P. It turns out that *the impulse response of a linear, time-varying, continuous-time system with integral*

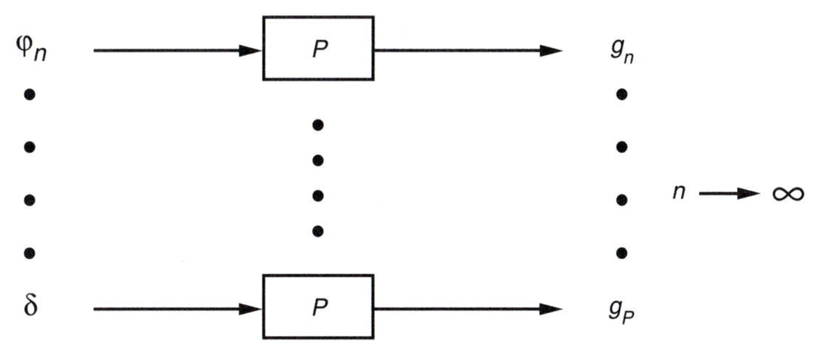

FIGURE 1.17

representation is again equal to the kernel of the integral representation of the system. To see this, we first observe that if $h \in C(R \times R, R)$, and if in Lemma 16.1 we replace $f \in C(R, R)$ by h, then all the ensuing relationships still hold, with obvious modifications. In particular, as in (16.27), we have for all $t \in R$,

$$\lim_{n\to\infty} \int_{-\infty}^{\infty} h(t, \tau)\phi_n(\eta - \tau)\, d\tau \triangleq \int_{-\infty}^{\infty} h(t, \tau)\delta(\eta - \tau)\, d\tau = h(t, \eta). \quad (16.38)$$

Also, as in (16.31), we have

$$\lim_{n\to\infty} \int_{0}^{\infty} h(t, \tau)\phi_n(\eta - \tau)\, d\tau \triangleq \int_{0}^{\infty} h(t, \tau)\delta(\eta - \tau)\, d\tau = h(t, \eta) \quad (16.39)$$

for $\eta > 0$.

Now let $h(t, \tau) = g_P(t, \tau)$. Then (16.38) yields

$$\lim_{n\to\infty} \int_{-\infty}^{\infty} g_P(t, \tau)\phi_n(\eta - \tau)\, d\tau \triangleq \int_{-\infty}^{\infty} g_P(t, \tau)\delta(\eta - \tau)\, d\tau = g_P(t, \eta), \quad (16.40)$$

which establishes our assertion. The common interpretation of (16.40) is that $g_P(t, \eta)$ represents the response of the system at time t due to an impulse applied at time η.

In establishing the results of the present subsection, we have made use of a *specific* sequence of functions $\{\phi_n\}$ given by

$$\phi_n(x) = \begin{cases} n(1 - n|x|), & \text{if } |x| \le \dfrac{1}{n}, \\ 0, & \text{if } |x| > \dfrac{1}{n}, \end{cases} \qquad n = 1, 2, \ldots .$$

We wish to emphasize that there are many other functions that could have served this purpose. An example of a commonly used sequence of functions employed in the development of the Dirac delta distribution is given by $\{\psi_n\}$, where

$$\psi_n(t) = \begin{cases} \dfrac{1}{2n}, & |t| \le n, \\ 0, & |t| > n, \end{cases}$$

$n = 1, 2, \ldots$. Another example of a sequence of functions that can be utilized for this is given by $\{\eta_n\}$, where

$$\eta_n(t) = \begin{cases} n^2 t e^{-nt}, & 0 \le t \le \infty, \\ 0, & \text{elsewhere}, \end{cases} \qquad n = 1, 2, \ldots .$$

D. Linear Continuous-Time Systems

We let P denote a linear time-varying operator from $C(R, R^m) \triangleq U$ to $C(R, R^p) = Y$ and we assume that P admits an *integral representation* given by

$$y(t) = (Pu)(t) = \int_{-\infty}^{\infty} H_P(t, \tau)u(\tau)\, d\tau, \quad (16.41)$$

where $H_P : R \times R \to R^{p\times m}$, $u \in U$, and $y \in Y$, and where H_P is assumed to be integrable. This means that each element of H_P, $h_{P_{ij}} : R \times R \to R$ is integrable (in the Riemann or Lebesgue sense).

Now let y_1 and y_2 denote the response of system (16.41) corresponding to the input u_1 and u_2, respectively, let α_1 and α_2 be real scalars, and let y denote the response of system (16.41) corresponding to the input $\alpha_1 u_1 + \alpha_2 u_2 = u$. Then

$$\begin{aligned} y = P(u) &= P(\alpha_1 u_1 + \alpha_2 u_2) = \int_{-\infty}^{\infty} H_P(t,\tau)[\alpha_1 u_1(\tau) + \alpha_2 u_2(\tau)]\, d\tau \\ &= \alpha_1 \int_{-\infty}^{\infty} H_P(t,\tau) u_1(\tau)\, d\tau + \alpha_2 \int_{-\infty}^{\infty} H_P(t,\tau) u_2(\tau)\, d\tau \\ &= \alpha_1 P(u_1) + \alpha_2 P(u_2) = \alpha_1 y_1 + \alpha_2 y_2, \end{aligned} \tag{16.42}$$

which shows that system (16.41) is indeed a *linear* system in the sense defined in (16.4).

Next, we let all components of $u(\tau)$ in (16.41) be zero, except for the jth component. Then the ith component of $y(t)$ in (16.41) assumes the form

$$y_i(t) = \int_{-\infty}^{\infty} h_{P_{ij}}(t,\tau) u_j(\tau)\, d\tau. \tag{16.43}$$

According to the results of the previous subsection [see Eq. (16.40)], $h_{P_{ij}}(t,\tau)$ denotes the response of the ith component of the output of system (16.41), measured at time t, due to an impulse applied to the jth component of the input of system (16.41), applied at time τ, while all the remaining components of the input are zero. Therefore, we call $H_P(t,\tau) = [h_{P_{ij}}(t,\tau)]$ the *impulse response matrix* of system (16.41).

Now suppose that it is known that system (16.41) is *causal*. Then its output will be identically zero before an input is applied. It follows that system (16.41) is causal if and only if

$$H_P(t,\tau) = 0 \qquad \text{for all } \tau \text{ and for all } t < \tau.$$

Therefore, when system (16.41) is causal, we have in fact that

$$y(t) = \int_{-\infty}^{t} H_P(t,\tau) u(\tau)\, d\tau. \tag{16.44}$$

We can rewrite (16.44) as

$$\begin{aligned} y(t) &= \int_{-\infty}^{t_0} H_P(t,\tau) u(\tau)\, d\tau + \int_{t_0}^{t} H_P(t,\tau) u(\tau)\, d\tau \\ &\triangleq y(t_0) + \int_{t_0}^{t} H_P(t,\tau) u(\tau)\, d\tau. \end{aligned} \tag{16.45}$$

We say that the continuous-time system (16.41) is *at rest at* $t = t_0$ if $u(t) = 0$ for $t \geq t_0$ implies that $y(t) = 0$ for $t \geq t_0$. Note that our problem formulation mandates that the system be at rest at $t_0 = -\infty$. Also, note that if a system (16.41) is known to be causal and to be at rest at $t = t_0$, then according to (16.45) we have

$$y(t) = \int_{t_0}^{t} H_P(t,\tau) u(\tau)\, d\tau. \tag{16.46}$$

Next, suppose that it is known that the system (16.41) is *time-invariant*. This means that if in (16.43) $h_{P_{ij}}(t,\tau)$ is the response y_i at time t due to an impulse applied at time τ at the jth component of the input [i.e., $u_j(\tau) = \delta(t)$], with all other input components set to zero, then a $-\tau$ time shift in the input [i.e., $u_j(t-\tau) = \delta(t-\tau)$] will result in a corresponding $-\tau$ time shift in the response, resulting in $h_{P_{ij}}(t-\tau, 0)$.

Since this argument holds for all $t, \tau \in R$ and for all $i = 1, \ldots, p$, and $j = 1, \ldots, m$, we have $H_P(t, \tau) = H_P(t - \tau, 0)$. If we define (using a slight abuse of notation) $H_P(t - \tau, 0) = H_P(t - \tau)$, then (16.41) assumes the form

$$y(t) = \int_{-\infty}^{\infty} H_P(t - \tau)u(\tau)\, d\tau. \tag{16.47}$$

Note that (16.47) is consistent with the definition of the integral representation of a linear time-invariant operator introduced in the previous subsection.

The right-hand side of (16.47) is the familiar *convolution integral* of H_P and u and is written more compactly as

$$y(t) = (H_P * u)(t). \tag{16.48}$$

We note that since $H_P(t - \tau)$ represents responses at time t due to impulse inputs applied at time τ, then $H_P(t)$ represents responses at time t due to impulse function inputs applied at $\tau = 0$. Therefore, a linear time-invariant system (16.47) is causal if and only if $H_P(t) = 0$ for all $t < 0$.

If it is known that the linear time-invariant system (16.47) is causal and is at rest at t_0, then we have

$$y(t) = \int_{t_0}^{t} H_P(t - \tau)u(\tau)\, d\tau. \tag{16.49}$$

In this case it is customary to choose, without loss of generality, $t_0 = 0$. We thus have

$$y(t) = \int_{0}^{t} H_P(t - \tau)u(\tau)\, d\tau, \qquad t \geq 0. \tag{16.50}$$

If we take the Laplace transform of both sides of (16.50), provided it exists, we obtain

$$\hat{y}(s) = \hat{H}_P(s)\hat{u}(s), \tag{16.51}$$

where $\hat{y}(s) = [\hat{y}_1(s), \ldots, \hat{y}_p(s)]^T$, $\hat{H}_P(s) = [\hat{h}_{P_{ij}}(s)]$, and $\hat{u}(s) = [\hat{u}_1(s), \ldots, \hat{u}_m(s)]^T$, where the $\hat{y}_i(s)$, $\hat{u}_j(s)$, and $\hat{h}_{P_{ij}}(s)$ denote the Laplace transforms of $y_i(t)$, $u_j(t)$, and $h_{Pij}(t)$, respectively. Consistent with Eq. (16.34), we note that $\hat{H}_P(s)$ represents the Laplace transform of the impulse response matrix $H_P(t)$. We call $\hat{H}_P(s)$ a *transfer function matrix*.

Now suppose that the input-output relation of a system is specified by the state and output equations (14.3a), (14.3b), repeated here as

$$\dot{x} = A(t)x + B(t)u \tag{16.52a}$$

$$y = C(t)x + D(t)u. \tag{16.52b}$$

If we assume that $x(t_0) = 0$ so that the system is at rest at $t_0 = 0$, we obtain for the response of this system,

$$y(t) = \int_{t_0}^{t} C(t)\Phi(t, \tau)B(\tau)u(\tau)\, d\tau + D(t)u(t) \tag{16.53}$$

$$= \int_{t_0}^{t} [C(t)\Phi(t, \tau)B(\tau) + D(t)\delta(t - \tau)]u(\tau)\, d\tau, \tag{16.54}$$

where in (16.54) we have made use of the interpretation of δ given in Subsection C. Comparing (16.54) with (16.46), we conclude that the impulse response matrix for system (16.52a), (16.52b) is given by

$$H_P(t,\tau) = \begin{cases} C(t)\Phi(t,\tau)B(\tau) + D(t)\delta(t-\tau), & t \geq \tau, \\ 0, & t < \tau. \end{cases} \tag{16.55}$$

Finally, for time-invariant systems described by the state and output equations (14.7a), (14.7b), repeated here as

$$\dot{x} = Ax + Bu \tag{16.56a}$$

$$y = Cx + Du, \tag{16.56b}$$

we obtain for the impulse response matrix the expression

$$H_P(t-\tau) = \begin{cases} Ce^{A(t-\tau)}B + D\delta(t-\tau), & t \geq \tau, \\ 0, & t < \tau, \end{cases} \tag{16.57}$$

or, as is more commonly written,

$$H_P(t) = \begin{cases} Ce^{At}B + D\delta(t), & t \geq 0, \\ 0, & t < 0. \end{cases} \tag{16.58}$$

We will pursue the topics of this section further in Chapter 2.

1.17 SUMMARY

In this chapter we addressed mathematical descriptions of systems. First, initial-value problems determined by systems of first-order nonlinear ordinary differential equations were introduced. Conditions for existence and uniqueness of solutions, and approaches to determine such solutions, were established. The main reason for considering nonlinear mathematical descriptions of systems in a book on linear systems is that the origins of most linear systems are nonlinear systems. Specifically, linear systems are frequently obtained from nonlinear systems by a linearization process, or are the result of modeling where the nonlinear effects of a physical process have been suppressed or neglected. Accordingly, the validity of linear mathematical descriptions must always be interpreted in the context of the nonlinear systems they approximate.

The linearization of nonlinear systems along a given solution (and a given input) was discussed in Section 1.11. The solutions of linear (time-varying) state equations were obtained in Section 1.13 using the method of successive approximations. The response of a linear continuous time system to an input, given initial states, was derived in Section 1.14.

The development of the theory of discrete-time systems parallels that of continuous-time systems and was addressed in Section 1.15.

State-space representations provide detailed descriptions of the internal behavior of a system, while input-output descriptions of systems emphasize external behavior and how a system interacts with its environment. Input-output descriptions of linear systems, addressed in Section 1.16, involve the convolution integral for continuous-time systems and the convolution sum for discrete-time systems.

In the next chapter, both state-space descriptions and input-output descriptions of systems are revisited, and their dynamic behavior is studied in detail.

1.18 NOTES

Standard references on linear algebra include Birkhoff and MacLane [2], Halmos [9], and Gantmacher [8]. For more recent texts on this subject, refer, e.g., to Strang [22] and Michel and Herget [12].

Excellent sources on analysis at the elementary level include Apostol [1], Rudin [17], and Taylor [24]. For treatments at an intermediate level, consult, e.g., Royden [16], Taylor [23], Naylor and Sell [15], and Michel and Herget [12].

For a classic reference on ordinary differential equations, see Coddington and Levinson [6]. Other excellent sources include Brauer and Nohel [3], Hartman [10], and Simmons [21]. Our treatment of ordinary differential equations in this chapter was greatly influenced by Coddington and Levinson [6] and Miller and Michel [14].

An original standard reference on linear systems is Zadeh and Desoer [25]. Of the many excellent texts on this subject, the reader may want to refer to Brockett [4], Kailath [11], and Chen [5]. For more recent texts on linear systems, consult, e.g., Rugh [18] and DeCarlo [7].

In Section 16 we showed that continuous-time finite-dimensional linear systems described by state equations have an input-output description given by integral equations. For a general and comprehensive treatment of the integral representation of linear systems, refer to Sandberg [19], [20].

1.19 REFERENCES

1. T. M. Apostol, *Mathematical Analysis*, Addison-Wesley, Reading, MA, 1974.
2. G. Birkhoff and S. MacLane, *A Survey of Modern Algebra*, Macmillan, New York, 1965.
3. F. Brauer and J. A. Nohel, *Qualitative Theory of Ordinary Differential Equations*, Benjamin, New York, 1969.
4. R. W. Brockett, *Finite Dimensional Linear Systems*, Wiley, New York, 1970.
5. C. T. Chen, *Linear System Theory and Design*, Holt, Rinehart and Winston, New York, 1984.
6. E. A. Coddington and N. Levinson, *Theory of Ordinary Differential Equations*, McGraw-Hill, New York, 1955.
7. R. A. DeCarlo, *Linear Systems*, Prentice-Hall, Englewood Cliffs, NJ, 1989.
8. F. R. Gantmacher, *Theory of Matrices*, Vols. I, II, Chelsea, New York, 1959.
9. P. R. Halmos, *Finite Dimensional Vector Spaces*, D. Van Nostrand, Princeton, NJ, 1958.
10. P. Hartman, *Ordinary Differential Equations*, Wiley, New York, 1964.
11. T. Kailath, *Linear Systems*, Prentice-Hall, Englewood Cliffs, NJ, 1980.
12. A. N. Michel and C. J. Herget, *Applied Algebra and Functional Analysis*, Dover, New York, 1993.
13. A. N. Michel and K. Wang, *Qualitative Theory of Dynamical Systems*, Marcel Dekker, New York, 1995.

14. R. K. Miller and A. N. Michel, *Ordinary Differential Equations*, Academic Press, New York, 1982.
15. A. W. Naylor and G. R. Sell, *Linear Operator Theory in Engineering and Science*, Holt, Rinehart and Winston, New York, 1971.
16. H. L. Royden, *Real Analysis*, Macmillan, New York, 1965.
17. W. Rudin, *Principles of Mathematical Analysis*, McGraw-Hill, New York, 1976.
18. W. J. Rugh, *Linear System Theory*, Second Edition, Prentice-Hall, Englewood Cliffs, NJ, 1996.
19. I. W. Sandberg, "Linear maps and impulse responses," *IEEE Transactions on Circuits and Systems*, Vol. 35, No. 2, pp. 201–206, 1988.
20. I. W. Sandberg, "Integral representations for linear maps," *IEEE Transactions on Circuits and Systems*, Vol. 35, No. 5, pp. 536–544, 1988.
21. G. F. Simmons, *Differential Equations*, McGraw-Hill, New York, 1972.
22. G. Strang, *Linear Algebra and Its Applications*, Academic Press, New York, 1980.
23. A. E. Taylor, *Introduction to Functional Analysis*, Wiley, New York, 1958.
24. A. E. Taylor, *Advanced Calculus*, Blaisdell, New York, 1965.
25. L. A. Zadeh and C. A. Desoer, *Linear System Theory—The State Space Approach*, McGraw-Hill, New York, 1963.

1.20 EXERCISES

1.1. (Hamiltonian dynamical systems) *Conservative dynamical systems,* also called *Hamiltonian dynamical systems,* are those systems that contain no energy-dissipating elements. Such systems with n degrees of freedom can be characterized by means of a *Hamiltonian function* $H(p, q)$, where $q^T = (q_1, \ldots, q_n)$ denotes n generalized position coordinates and $p^T = (p_1, \ldots, p_n)$ denotes n generalized momentum coordinates. We assume that $H(p, q)$ is of the form

$$H(p, q) = T(q, \dot{q}) + W(q), \tag{20.1}$$

where T denotes the kinetic energy and W denotes the potential energy of the system. These energy terms are obtained from the path-independent line integrals

$$T(q, \dot{q}) = \int_0^{\dot{q}} p(q, \xi)^T \, d\xi = \int_0^{\dot{q}} \sum_{i=1}^{n} p_i(q, \xi) \, d\xi_i \tag{20.2}$$

$$W(q) = \int_0^{q} f(\eta)^T \, d\eta = \int_0^{q} \sum_{i=1}^{n} f_i(\eta) \, d\eta_i, \tag{20.3}$$

where $f_i, i = 1, \ldots, n,$ denote generalized potential forces.

For the integral (20.2) to be path-independent, it is necessary and sufficient that

$$\frac{\partial p_i(q, \dot{q})}{\partial \dot{q}_j} = \frac{\partial p_j(q, \dot{q})}{\partial \dot{q}_i}, \qquad i, j = 1, \ldots, n. \tag{20.4}$$

A similar statement can be made about Eq. (20.3).

Conservative dynamical systems are described by the system of $2n$ ordinary differential equations

$$\begin{aligned} \dot{q}_i &= \frac{\partial H}{\partial p_i}(p, q), \qquad i = 1, \ldots, n, \\ \dot{p}_i &= -\frac{\partial H}{\partial q_i}(p, q), \qquad i = 1, \ldots, n. \end{aligned} \tag{20.5}$$

Note that if we compute the derivative of $H(p, q)$ with respect to t for (20.5) [i.e., along the solutions $q_i(t)$, $p_i(t)$, $i = 1, \ldots, n$], then we obtain, by the chain rule,

$$\begin{aligned}\frac{dH}{dt}(p(t), q(t)) &= \sum_{i=1}^{n} \frac{\partial H}{\partial p_i}(p, q)\dot{p}_i + \sum_{i=1}^{n} \frac{\partial H}{\partial q_i}(p, q)\dot{q}_i \\ &= \sum_{i=1}^{n} -\frac{\partial H}{\partial p_i}(p, q)\frac{\partial H}{\partial q_i}(p, q) + \sum_{i=1}^{n} \frac{\partial H}{\partial q_i}(p, q)\frac{\partial H}{\partial p_i}(p, q) \\ &= -\sum_{i=1}^{n} \frac{\partial H}{\partial p_i}(p, q)\frac{\partial H}{\partial q_i}(p, q) + \sum_{i=1}^{n} \frac{\partial H}{\partial p_i}(p, q)\frac{\partial H}{\partial q_i}(p, q) \equiv 0.\end{aligned}$$

In other words, in a conservative system (20.5) the Hamiltonian, i.e., the total energy, will be constant along the solutions (20.5). This constant is determined by the initial data $(p(0), q(0))$.

(a) In Fig. 1.18, M_1 and M_2 denote point masses, K_1, K_2, K denote spring constants, and x_1, x_2 denote the displacements of the masses M_1 and M_2. Use the Hamiltonian formulation of dynamical systems described above to derive a system of first-order ordinary differential equations that characterize this system. Verify your answer by using Newton's second law of motion to derive the same system of equations. Assume that $x_1(0), \dot{x}_1(0), x_2(0), \dot{x}_2(0)$ are given.

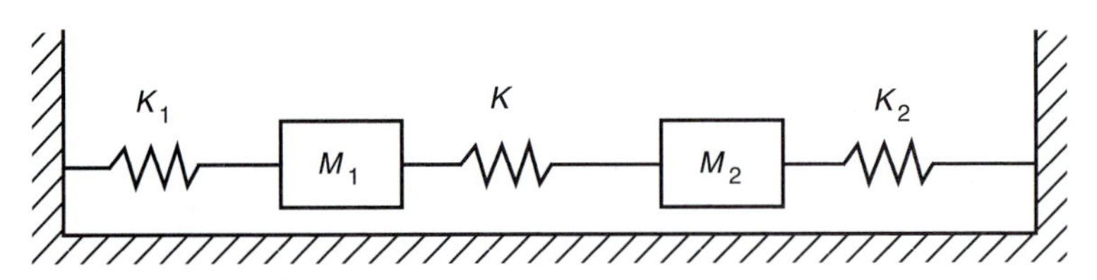

FIGURE 1.18
Example of a conservative dynamical system

(b) In Fig. 1.19, a point mass m is moving in a circular path about the axis of rotation normal to a constant gravitational field (this is called the *simple pendulum problem*). Here l is the radius of the circular path, g is the gravitational acceleration, and x denotes the angle of deflection measured from the vertical. Use the Hamiltonian formulation of dynamical systems described above to derive a system of first-order ordinary differential equations that characterize this system. Verify your answer by

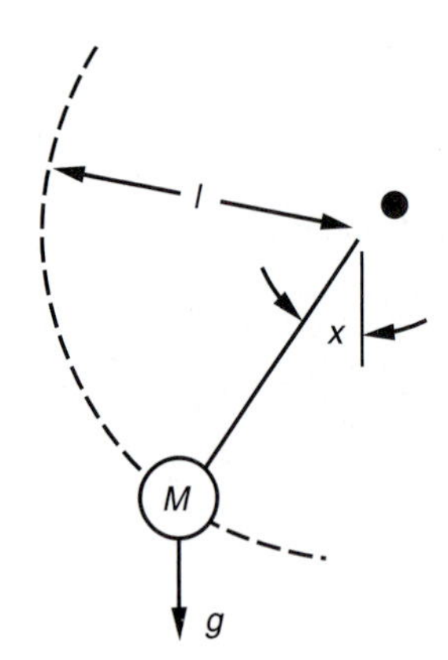

FIGURE 1.19
Simple pendulum

using Newton's second law of motion to derive the same system of equations. Assume that $x(0)$ and $\dot{x}(0)$ are given.

(c) Determine a system of first-order ordinary differential equations that characterizes the two-link pendulum depicted in Fig. 1.20. Assume that $\theta_1(0)$, $\theta_2(0)$, $\dot{\theta}_1(0)$, and $\dot{\theta}_2(0)$ are given.

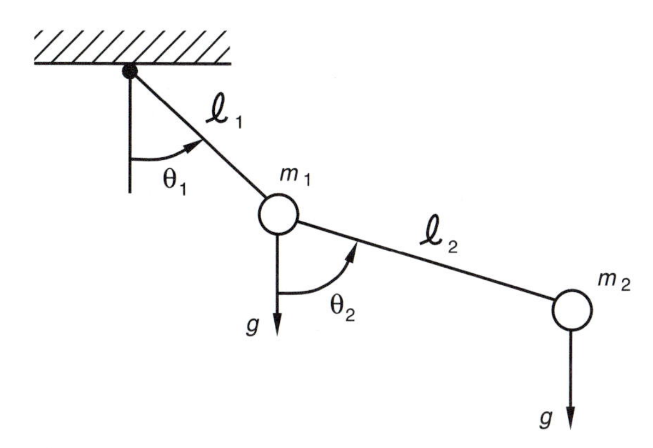

FIGURE 1.20
Two-link pendulum

1.2. (Lagrange's equation) If a dynamical system contains elements that dissipate energy, such as viscous friction elements in mechanical systems and resistors in electric circuits, then we can use *Lagrange's equation* to describe such systems. (In the following, we use some of the same notation used in Exercise 1.1.) For a system with n degrees of freedom, this equation is given by

$$\frac{d}{dt}\left(\frac{\partial L}{\partial \dot{q}_i}(q, \dot{q})\right) - \frac{\partial L}{\partial q}(q, \dot{q}) + \frac{\partial D}{\partial \dot{q}_i}(\dot{q}) = f_i, \qquad i = 1, \ldots, n, \tag{20.6}$$

where $q^T = (q_1, \ldots, q_n)$ denotes the generalized position vector. The function $L(q, \dot{q})$ is called the *Lagrangian* and is defined as

$$L(q, \dot{q}) = T(q, \dot{q}) - W(q),$$

i.e., the difference between the kinetic energy T and the potential energy W.

The function $D(\dot{q})$ denotes *Rayleigh's dissipation function,* which we shall assume to be of the form

$$D(\dot{q}) = \frac{1}{2}\sum_{i=1}^{n}\sum_{j=1}^{n} \beta_{ij}\dot{q}_i\dot{q}_j,$$

where $[\beta_{ij}]$ is a positive semidefinite matrix (i.e., $[\beta_{ij}]$ is symmetric and all its eigenvalues are nonnegative). The dissipation function D represents one-half the rate at which energy is dissipated as heat. It is produced by friction in mechanical systems and by resistance in electric circuits.

Finally, f_i in Eq. (20.6) denotes an applied force and includes all external forces associated with the q_i coordinate. The force f_i is defined as being positive when it acts to increase the value of the coordinate q_i.

(a) In Fig. 1.21, M_1 and M_2 denote point masses; K_1, K_2, K denote spring constants; y_1, y_2 denote the displacements of masses M_1 and M_2, respectively; and B_1, B_2, B denote viscous damping coefficients. Use the Lagrange formulation of dynamical

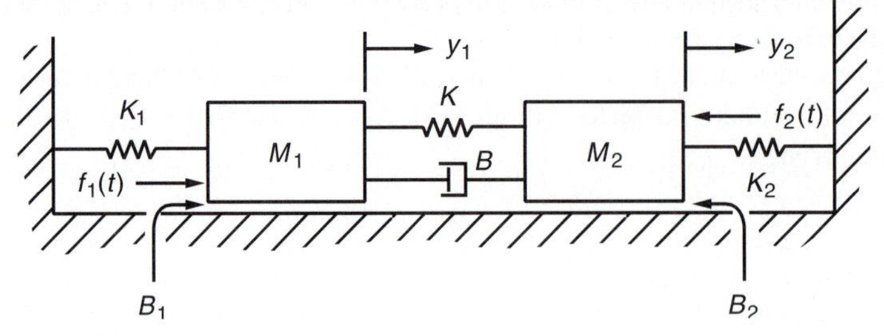

FIGURE 1.21
An example of a mechanical system with energy dissipation

systems described above to derive two second-order ordinary differential equations that characterize this system. Transform these equations into a system of first order ordinary differential equations. Verify your answer by using Newton's second law of motion to derive the same system equations. Assume that $y_1(0)$, $\dot{y}_1(0)$, $y_2(0)$, $\dot{y}_2(0)$ are given.

(b) Consider the capacitor microphone depicted in Fig. 1.22. Here we have a capacitor constructed from a fixed plate and a moving plate with mass M. The moving plate is suspended from the fixed frame by a spring that has a spring constant k and also some damping expressed by the damping constant B. Sound waves exert an external force $f(t)$ on the moving plate. The output voltage v_s, which appears across the resistor R, will reproduce electrically the sound-wave patterns that strike the moving plate.

When $f(t) \equiv 0$ there is a charge q_0 on the capacitor. This produces a force of attraction between the plates that stretches the spring by an amount x_1, and the space between the plates is x_0. When sound waves exert a force on the moving plate, there will be a resulting motion displacement x that is measured from the equilibrium position. The distance between the plates will then be $x_0 - x$, and the charge on the plates will be $q_0 + q$.

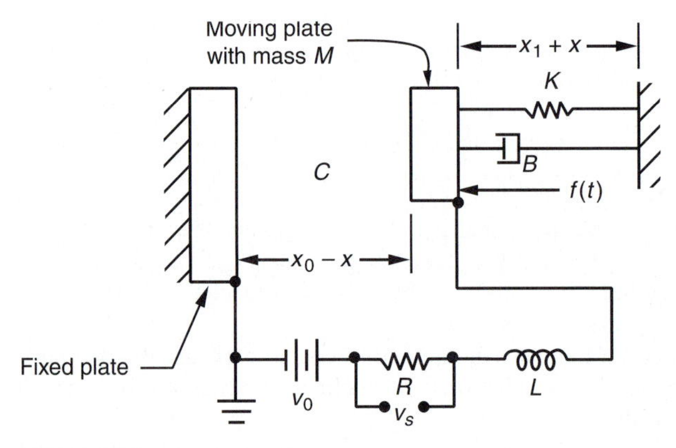

FIGURE 1.22
Capacitor microphone

When displacements are small, the expression for the capacitance is given approximately by

$$C = \frac{\epsilon A}{x_0 - x}$$

with $C_0 = \epsilon A/x_0$, where $\epsilon > 0$ is the dielectric constant for air and A is the area of the plate.

Use the Lagrange formulation of dynamical systems to derive two second-order ordinary differential equations that characterize this system. Transform these equations into a system of first-order ordinary differential equations. Verify your answer by using Newton's laws of motion and Kirchhoff's voltage/current laws. Assume that $x(0)$, $\dot{x}(0)$, $q(0)$, and $\dot{q}(0)$ are given.

(c) Use the Lagrange formulation to derive a system of first-order differential equations for the system given in Example 4.3.

1.3. Find examples of initial-value problems for which (a) no solutions exist; (b) more than one solution exists; (c) one or more solutions exist, but cannot be continued for all $t \in R$; and (d) unique solutions exist for all $t \in R$.

1.4. (Numerical solution of ordinary differential equations—Euler's method) In Subsection 1.6B it is shown that an approximation to the solution of the *scalar* initial-value problem

$$\dot{y} = f(t, y), \qquad y(t_0) = y_0 \tag{20.7}$$

is given by *Euler's method* [see Eq. (6.6)],

$$y_{k+1} = y_k + hf(t_k, y_k), \qquad k = 0, 1, 2, \ldots, \tag{20.8}$$

where $h = t_{k+1} - t_k$ is the (constant) integration step. The interpretation of this method is that the area below the solution curve [see Eq. (V) in Section 1.6] is approximated by a sequence of sums of rectangular areas. This method is also called the *forward rectangular rule* (of integration).

(a) Use Euler's method to determine the solution of the initial-value problem

$$\dot{y} = 3y, \qquad y(t_0) = 5, \qquad t_0 = 0, \qquad t_0 \leq t \leq 10.$$

(b) Use Euler's method to determine the solution of the initial-value problem

$$\ddot{y} = t(\dot{y})^2 - y^2, \qquad y(t_0) = 1, \qquad \dot{y}(t_0) = 0, \qquad t_0 = 0, \qquad t_0 \leq t \leq 10.$$

Hint: In both cases, use $h = 0.2$. For part (b), let $y = x_1, \dot{x}_1 = x_2, \dot{x}_2 = tx_2^2 - x_1^2$, and apply (20.8), appropriately adjusted to the vector case. In both cases, plot y_k vs. t_k, $k = 0, 1, 2, \ldots$.

Remark: Euler's method yields arbitrarily close approximations to the solutions of (20.7), by making h sufficiently small, *assuming infinite (computer) word length.* In practice, however, where truncation errors (quantization) and round-off errors (finite precision operations) are a reality, extremely small values of h may lead to numerical instabilities. Therefore, Euler's method is of limited value as a means of solving initial-value problems numerically.

1.5. (Numerical solution of ordinary differential equations—Runge-Kutta methods) The *Runge-Kutta* family of integration methods are among the most widely used techniques to solve initial-value problems (20.7). A simple version is given by

$$y_{i+1} = y_i + k,$$

where

$$k = \tfrac{1}{6}(k_1 + 2k_2 + 2k_3 + k_4),$$

with

$$k_1 = hf(t_i, y_i)$$
$$k_2 = hf(t_i + \tfrac{1}{2}h, y_i + \tfrac{1}{2}k_1)$$
$$k_3 = hf(t_i + \tfrac{1}{2}h, y_i + \tfrac{1}{2}k_2)$$
$$k_4 = hf(t_i + h, y_i + k_3),$$

and $t_{i+1} = t_i + h$, $y(t_0) = y_0$.

The idea of this method is to probe ahead (in time) by one-half or by a whole step h to determine the values of the derivative at several points, and then to form a weighted average.

Runge-Kutta methods can also be applied to higher order ordinary differential equations. For example, after a change of variables, suppose that a second-order differential equation has been changed to a system of two first-order differential equations, say,

$$\dot{x}_1 = f_1(t, x_1, x_2), \qquad x_1(t_0) = x_{10}$$
$$\dot{x}_2 = f_2(t, x_1, x_2), \qquad x_2(t_0) = x_{20}. \tag{20.9}$$

In solving (20.9), a simple version of the Runge-Kutta method is given by

$$y_{i+1} = y_i + \underline{k},$$

where $y_i = (x_{1i}, x_{2i})^T$ and $\underline{k} = (k, l)^T$

with $k = \frac{1}{6}(k_1 + 2k_2 + 2k_3 + k_4)$, $l = \frac{1}{6}(l_1 + 2l_2 + 2l_3 + l_4)$

and

$$k_1 = hf_1(t_i, x_{1i}, x_{2i}) \qquad l_1 = hf_2(t_i, x_{1i}, x_{2i})$$
$$k_2 = hf_1(t_i + \tfrac{1}{2}h, x_{1i} + \tfrac{1}{2}k_1, x_{2i} + \tfrac{1}{2}l_1) \qquad l_2 = hf_2(t_i + \tfrac{1}{2}h, x_{1i} + \tfrac{1}{2}k_1, x_{2i} + \tfrac{1}{2}l_1)$$
$$k_3 = hf_1(t_i + \tfrac{1}{2}h, x_{1i} + \tfrac{1}{2}k_2, x_{2i} + \tfrac{1}{2}l_2) \qquad l_3 = hf_2(t_i + \tfrac{1}{2}h, x_{1i} + \tfrac{1}{2}k_2, x_{2i} + \tfrac{1}{2}l_2)$$
$$k_4 = hf_1(t_i + h, x_{1i} + k_3, x_{2i} + l_3) \qquad l_4 = hf_2(t_i + h, x_{1i} + k_3, x_{2i} + l_3).$$

Use the Runge-Kutta method described above to obtain numerical solutions to the initial value problems given in parts (a) and (b) of Exercise 1.4. As there, plot your data.

Remark: Since Runge-Kutta methods do not use past information, they constitute attractive starting methods for more efficient numerical integration schemes (e.g., predictor-corrector methods). We note that since there are no built-in accuracy measures in the Runge-Kutta methods, significant computational efforts are frequently expended to achieve a desired accuracy.

1.6. (Numerical solution of ordinary differential equations—Predictor-Corrector methods) A common predictor-corrector technique for solving initial-value problems determined by ordinary differential equations, such as (20.7), is the *Milne* method, which we now summarize. In this method, $\dot{y}_{i-1}$ denotes the value of the first derivative at time t_{i-1}, where t_i is the time for the ith iteration step, $\dot{y}_{i-2}$ is similarly defined, and y_{i+1} represents the value of y to be determined. The details of the Milne method are:

1. $y_{i+1,p} = y_{i-3} + \dfrac{4h}{3}(2\dot{y}_{i-2} - \dot{y}_{i-1} + 2\dot{y}_i)$ *(predictor)*
2. $\dot{y}_{i+1,p} = f(t_{i+1}, y_{i+1,p})$
3. $y_{i+1,c} = y_{i-1} + \dfrac{h}{3}(\dot{y}_{i-1} + 4\dot{y}_i + \dot{y}_{i+1,p})$ *(corrector)*
4. $\dot{y}_{i+1,c} = f(t_{i+1}, y_{i+1,c})$
5. $y_{i+1,c} = y_{i-1} + \dfrac{h}{3}(\dot{y}_{i-1} + 4\dot{y}_i + \dot{y}_{i+1,c})$ *(iterating corrector)*

The first step is to obtain a predicted value of y_{i+1} and then substitute $y_{i+1,p}$ into the given differential equation to obtain a predicted value of $\dot{y}_{i+1}$, as indicated in the second equation above. This predicted value, $\dot{y}_{i+1,p}$ is then used in the second equation, the corrector equation, to obtain a corrected value of y_{i+1}. The corrected value, $y_{i+1,c}$ is next substituted into the differential equation to obtain an improved value of $\dot{y}_{i+1}$, and so on. If necessary, an iteration process involving the fourth and fifth equations continues until successive values of y_{i+1} differ by less than the value of some desirable tolerance. With y_{i+1} determined to the desired accuracy, the method steps forward one h increment.

A more complicated predictor-corrector method that is more reliable than the Milne method is the *Adams-Bashforth-Moulton* method, the essential equations of which are

$$y_{i+1,p} = y_i + \frac{h}{24}(55\dot{y}_i - 59\dot{y}_{i-1} + 37\dot{y}_{i-2} - 9\dot{y}_{i-3})$$

$$y_{i+1,c} = y_i + \frac{h}{24}(9\dot{y}_{i+1} + 19\dot{y}_i - 5\dot{y}_{i-1} + \dot{y}_{i-2}),$$

where in the corrector equation, $\dot{y}_{i+1}$ denotes the predicted value.

The application of predictor-corrector methods to systems of first-order ordinary differential equations is straightforward. For example, the application of the Milne method to the second-order system in (20.9) yields from the predictor step

$$x_{k,i+1,p} = x_{k,i-3} + \frac{4h}{3}(2\dot{x}_{k,i-2} - \dot{x}_{k,i-1} + 2\dot{x}_{k,i}), \qquad k = 1, 2.$$

Then
$$\dot{x}_{k,i+1,p} = f_k(t_{i+1}, x_{1,i+1,p}, x_{2,i+1,p}), \qquad k = 1, 2,$$

and the corrector step assumes the form

$$x_{k,i+1,c} = x_{k,i-1} + \frac{h}{3}(\dot{x}_{k,i-1} + 4\dot{x}_{k,i} + \dot{x}_{k,i+1}), \qquad k = 1, 2,$$

and
$$\dot{x}_{k,i+1,c} = f_k(t_{i+1}, x_{1,i+1,c}, x_{2,i+1,c}), \qquad k = 1, 2.$$

Use the Milne method and the Adams-Bashforth-Moulton method described above to obtain numerical solutions to the initial-value problems given in parts (a) and (b) of Exercise 1.4. To initiate the algorithm, refer to the Remark in Exercise 1.5.

Remark. Derivations and convergence properties of numerical integration schemes, such as those discussed here and in Exercises 1.4 and 1.5, can be found in many of the standard texts on numerical analysis (see, e.g., D. Kincaid and W. Cheney, *Numerical Analysis*, Brooks-Cole, Belmont, CA, 1991).

1.7. (a) Prove Theorem 6.2 for the interval $[t_0 - c, t_0]$.
(b) Prove Theorem 7.2 for the endpoint a.
(c) Prove Theorem 8.2 for the interval $[t_0 - d, t_0]$.
(d) Prove Theorem 8.4 for $t \leq t_0$.
(e) Prove Theorem 8.5 for the interval $[t_0 - c, t_0]$.
(f) Prove Theorem 10.1 through and including Theorem 10.10.

1.8. (a) Prove that the function $f(x) = x^{1/3}$ is continuous but not Lipschitz continuous.
(b) Show that the initial-value problem $\dot{x} = x^{1/3}$, $x(0) = 0$ has infinitely many different solutions. (Remember to consider $t < 0$.)

1.9. Use Theorem 8.5 to solve the initial-value problem $\dot{x} = ax + t$, $x(0) = x_0$ for $t \geq 0$. Here $a \in R$.

1.10. Consider the initial-value problem

$$\dot{x} = Ax, \qquad x(0) = x_0, \tag{20.10}$$

where $x \in R^2$ and $A \in R^{2\times 2}$. Let λ_1, λ_2 denote the eigenvalues of A, i.e., λ_1 and λ_2 are the roots of the equation $det(A - \lambda I) = 0$, where det denotes determinant, λ is a scalar, and I denotes the 2×2 identity matrix. Make specific choices of A to obtain the following cases:

1. $\lambda_1 > 0, \lambda_2 > 0$, and $\lambda_1 \neq \lambda_2$;
2. $\lambda_1 < 0, \lambda_2 < 0$, and $\lambda_1 \neq \lambda_2$;
3. $\lambda_1 = \lambda_2 > 0$;
4. $\lambda_1 = \lambda_2 < 0$;
5. $\lambda_1 > 0, \lambda_2 < 0$;
6. $\lambda_1 = \alpha + i\beta, \lambda_2 = \alpha - i\beta, i = \sqrt{-1}, \alpha > 0$;
7. $\lambda_1 = \alpha + i\beta, \lambda_2 = \alpha - i\beta, \alpha < 0$;
8. $\lambda_1 = i\beta, \lambda_2 = -i\beta$.

Using t as a parameter, plot $\phi_2(t, 0, x_0)$ vs. $\phi_1(t, 0, x_0)$ for $0 \leq t \leq t_f$ for every case enumerated above. Here $[\phi_1(t, t_0, x_0), \phi_2(t, t_0, x_0)]^T = \phi(t, t_0, x_0)$ denotes the solution of (20.10). On your plots, indicate increasing time t by means of arrows. Plots of this type are called *trajectories* for (20.10), and sufficiently many plots (using different initial conditions and sufficiently large t_f) make up a *phase portrait* for (20.10). Generate a phase portrait for each case given above.

1.11. Write two first-order ordinary differential equations for the *van der Pol equation* (4.10) by choosing $x_1 = x$ and $x_2 = \dot{x}_1$. Determine by simulation *phase portraits* (see Exercise 1.10) for this example for the cases $\epsilon = 0.05$ and $\epsilon = 10$ (refer also to Exercises 1.5 and 1.6 for numerical methods for solving differential equations). The periodic function to which the trajectories of (4.10) tend is an example of a *limit cycle*.

1.12. For (4.11) consider the *hard, linear, and soft spring models* given by

$$g(x) = k(1 + a^2x^2)x,$$
$$g(x) = kx,$$
$$g(x) = k(1 - a^2x^2)x,$$

respectively, where $k > 0$ and $a \neq 0$. Write two first-order ordinary differential equations for (4.11) by choosing $x_1 = x$ and $x_2 = \dot{x}$. Pick specific values for k and a^2. Determine by simulation *phase portraits* (see Exercise 1.10) for this example for the above three cases.

1.13. Verify that the spaces in Examples 10.1 to 10.6 satisfy the axioms of vector space.

1.14. Let $F = \{0, 1, 2, 3\}$. Determine operations "+" and "·" so that $\{F, +, \cdot\}$ is a field.

1.15. (a) Verify that the transformation given in Example 10.8 is linear.
(b) Verify that the transformation given in Example 10.9 is linear.

1.16. (a) Prove (10.13), (10.14), (10.15), and (10.16).
(b) Verify that the functions $\|\cdot\|_1, \|\cdot\|_2$, and $\|\cdot\|_\infty$ defined in (10.11), (10.12), and (10.10), respectively, each satisfy the axioms of a norm.
(c) Prove the relations (D-i), (D-ii), (D-iii) given at the beginning of Subsection 1.10C.

1.17. Let $g(t) = [g_1(t), \ldots, g_n(t)]^T$ be defined on some interval $J \subset R$ and assume that $g_i : J \to R$ is integrable over $J, i = 1, \ldots, n$. Prove that for $b > a, \|\int_a^b g(t)\,dt\| \leq \int_a^b \|g(t)\|\,dt$, where $\|\cdot\|$ denotes a norm on R^n.

1.18. (a) Show that $x^T = (0, 0)$ is a solution of the system of equations

$$\dot{x}_1 = x_1^2 + x_2^2 + x_2 \cos x_1$$
$$\dot{x}_2 = (1 + x_1)x_1 + (1 + x_2)x_2 + x_1 \sin x_2.$$

Linearize this system about the point $x^T = (0, 0)$. By means of computer simulations, compare solutions corresponding to different initial conditions in the vicinity of the origin of the above system of equations and its linearization.

(b) Linearize the (bilinear control) system

$$\ddot{x} + (3 + \dot{x}^2)\dot{x} + (1 + x + x^2)u = 0$$

about the solution $x = 0$, $\dot{x} = 0$, and the input $u(t) \equiv 0$. As in part (a), compare (by means of computer simulations) solutions of the above equation with corresponding solutions of its linearization.

(c) In the circuit given in Fig. 1.23, $v_i(t)$ is a voltage source and the nonlinear resistor obeys the relation $i_R = 1.5v_R^3$ [$v_i(t)$ is the circuit input and $v_R(t)$ is the circuit output]. Derive the differential equation for this circuit. Linearize this differential equation for the case when the circuit operates about the point $v_i = 14$.

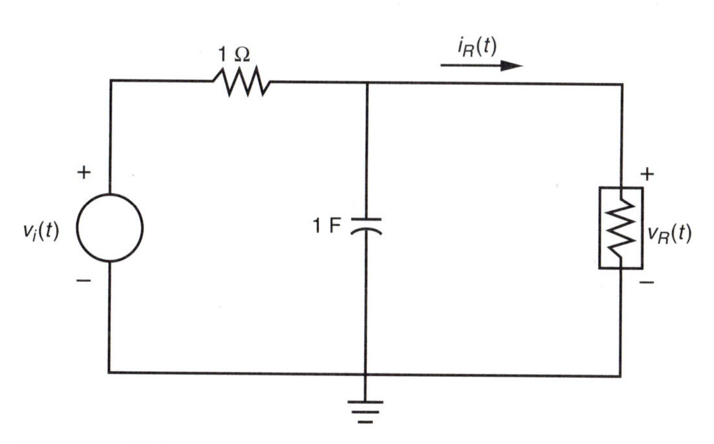

FIGURE 1.23
Nonlinear circuit

1.19. Consider a system whose state-space description is given by

$$\dot{x} = -k_1 k_2 \sqrt{x} + k_2 u(t)$$
$$y = k_1 \sqrt{x}.$$

Linearize this system about the nominal solution

$$u_0 \equiv 0, \qquad 2\sqrt{x_0(t)} = 2\sqrt{k} - k_1 k_2 t,$$

where $x_0(0) = k$.

1.20. **(Inverted pendulum)** The inverted pendulum on a moving carriage subjected to an external force $\mu(t)$ is depicted in Fig. 1.24.

The moment of inertia with respect to the center of gravity is J and the coefficient of friction of the carriage (see Fig. 1.24) is F. From Fig. 1.25 we obtain the following equations for the dynamics of this system:

$$m\frac{d^2}{dt^2}(S + L\sin\phi) \triangleq H \qquad (20.11a)$$

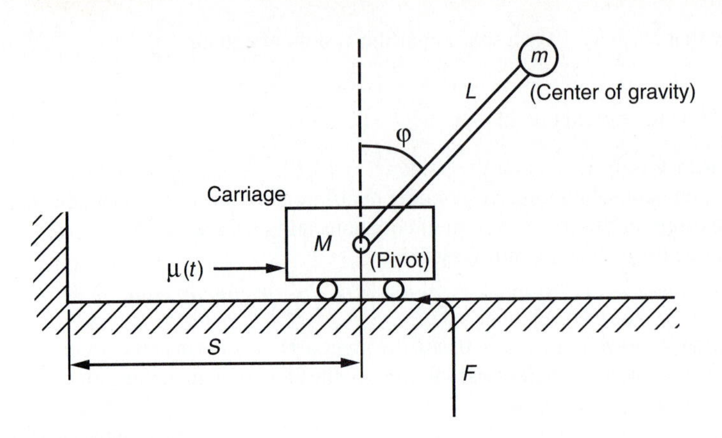

FIGURE 1.24
Inverted pendulum

$$m\frac{d^2}{dt^2}(L\cos\phi) \triangleq Y - mg \tag{20.11b}$$

$$J\frac{d^2\phi}{dt^2} = LY\sin\phi - LH\cos\phi \tag{20.11c}$$

$$M\frac{d^2S}{dt^2} = \mu(t) - H - F\frac{dS}{dt}. \tag{20.11d}$$

Assuming that $m << M$, (20.11d) reduces to

$$M\frac{d^2S}{dt^2} = \mu(t) - F\frac{dS}{dt}. \tag{20.11e}$$

Eliminating H and Y from (20.11a) to (20.11c), we obtain

$$(J + mL^2)\ddot{\phi} = mgL\sin\phi - mL\ddot{S}\cos\phi. \tag{20.11f}$$

Thus, the system of Fig. 1.24 is described by the equations

$$\begin{aligned} \ddot{\phi} - \left(\frac{g}{L'}\right)\sin\phi + \left(\frac{1}{L'}\right)\ddot{S}\cos\phi &= 0, \\ M\ddot{S} + F\dot{S} &= \mu(t), \end{aligned} \tag{20.11g}$$

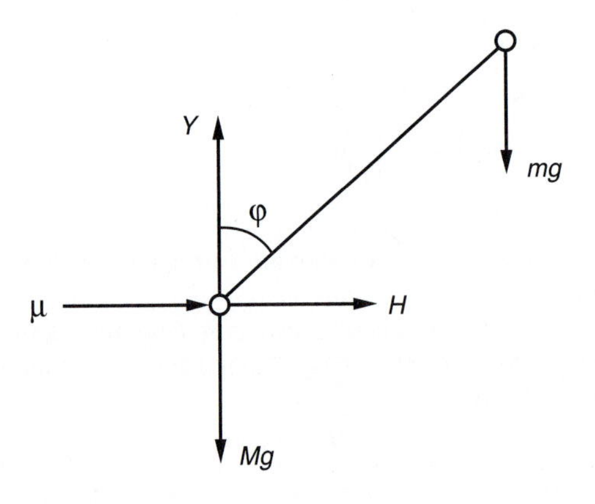

FIGURE 1.25

where $$L' = \frac{J + mL^2}{mL}$$

denotes the effective pendulum length.

Linearize system (20.11g) about $\phi = 0$.

1.21. (Magnetic ball suspension system) Figure 1.26 depicts a schematic diagram of a ball suspension control system. The steel ball is suspended in air by the electromagnetic force generated by the electromagnet. The objective of the control is to keep the steel ball suspended at a desired equilibrium position by controlling the current $i(t)$ in the magnet coil by means of the applied voltage $v(t)$, where $t \geq 0$ denotes time. The resistance and inductance of the coil are R and $L(s(t)) = L/s(t)$, respectively, where $L > 0$ is a constant and $s(t)$ denotes the distance between the center of the ball and the magnet at time t. The force produced by the magnet is $Ki^2(t)/s^2(t)$, where $K > 0$ is a proportional constant and g denotes acceleration due to gravity.

(a) Determine the differential equations governing the dynamics of this system.

(b) Let $v(t) \equiv v_{eq}$, a nominal (desired) value of $v(t)$. Determine the resulting equilibrium of this system.

(c) Linearize the equation obtained in (a) about the equilibrium solution determined in (b).

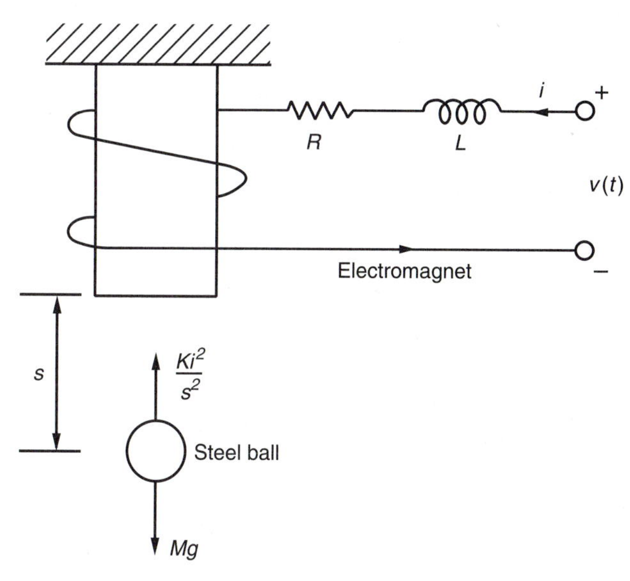

FIGURE 1.26
Magnetic ball suspension system

1.22. (a) For the mechanical system given in Exercise 1.2a, we view f_1 and f_2 as making up the system input vector, and y_1 and y_2 the system output vector. Determine a state-space description for this system.

(b) For the same mechanical system, we view $(f_1, 5f_2)^T$ as the system input and we view $8y_1 + 10y_2$ as the (scalar-valued) system output. Determine a state-space description for this system.

(c) For part (a), determine the input-output description of the system.

(d) For part (b), determine the input-output description of the system.

1.23. For the magnetic ball suspension system given in Exercise 1.21, we view v and s as the system input and output, respectively.
(a) Determine a state-space representation for this system.
(b) Using the linearized equation obtained in part (c) of Exercise 1.21, obtain the input-output description of this system.

1.24. In Example 4.3, we view e_a and θ as the system input and output, respectively.
(a) Determine a state-space representation for this system.
(b) Determine the input-output description of this system.

1.25. For the second-order section digital filter in direct form, given in Fig. 1.13, determine the input-output description, where $x_1(k)$ and $u(k)$ denote the output and input, respectively.

1.26. In the circuit of Fig. 1.27, $V_i(t)$ and $V_0(t)$ are voltages (at time t) and R_1 and R_2 are resistors. There is also an ideal diode that acts as a short circuit when V_i is positive and as an open circuit when v_i is negative. We view V_i and V_0 as the system input and output, respectively.
(a) Determine an input-output description of this system.
(b) Is this system linear? Is it time-varying or time-invariant? Is it causal? Explain your answers.

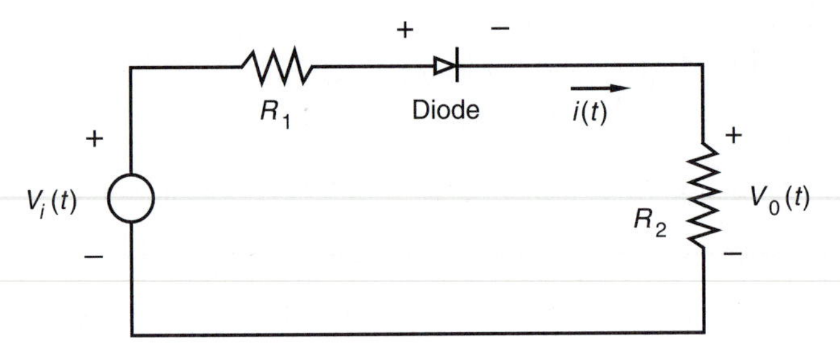

FIGURE 1.27
Diode circuit

1.27. We consider the *truncation operator* given by

$$y(t) = T_\tau(u(t))$$

as a system, where $\tau \in R$ is fixed, u and y denote system input and output, respectively, t denotes time, and $T_\tau(\cdot)$ is specified by

$$T_\tau(u(t)) = \begin{cases} u(t), & t \le \tau, \\ 0, & t > \tau. \end{cases}$$

Is this system causal? Is it linear? Is it time-invariant? What is its impulse response?

1.28. We consider the *shift operator* given by

$$y(t) = Q_\tau(u(t)) = u(t - \tau)$$

as a system, where $\tau \in R$ is fixed, u and y denote system input and system output, respectively, and t denotes time. Is this system causal? Is it linear? Is it time-invariant? What is its impulse response?

1.29. Consider the system whose input-output description is given by

$$y(t) = \min\{u_1(t), u_2(t)\},$$

where $u(t) = [u_1(t), u_2(t)]^T$ denotes the system input and $y(t)$ is the system output. Is this system linear?

1.30. Suppose it is known that a linear system has impulse response given by $h(t, \tau) = \exp(-|t - \tau|)$. Is this system causal? Is it time-invariant?

1.31. Consider a system with input-output description given by

$$y(k) = 3u(k + 1) + 1, \qquad k \in Z,$$

where y and u denote the output and input, respectively (recall that Z denotes the integers). Is this system causal? Is it linear?

1.32. Use expression (16.8),

$$x(n) = \sum_{k=-\infty}^{\infty} x(k)\delta(n - k),$$

and $\delta(n) = p(n) - p(n - 1)$ to express the system response $y(n)$ due to any input $u(k)$, as a function of the unit step response of the system [i.e., due to $u(k) = p(k)$].

1.33. **(Simple pendulum)** A system of first-order ordinary differential equations that characterize the simple pendulum considered in Exercise 1.1b is given by

$$\begin{bmatrix} \dot{x}_1 \\ \dot{x}_2 \end{bmatrix} = \begin{bmatrix} x_2 \\ -\frac{g}{l} \sin x_1 \end{bmatrix},$$

where $x_1 \triangleq \theta$ and $x_2 \triangleq \dot{\theta}$ with $x_1(0) = \theta(0)$ and $x_2(0) = \dot{\theta}(0)$ specified. A linearized model of this system about the solution $x = [0, 0]^T$ is given by

$$\begin{bmatrix} \dot{x}_1 \\ \dot{x}_2 \end{bmatrix} = \begin{bmatrix} 0 & 1 \\ -\frac{g}{l} & 0 \end{bmatrix} \begin{bmatrix} x_1 \\ x_2 \end{bmatrix}.$$

Let $g = 10$ (m/sec^2) and $l = 1$ (m).

(a) For the case when $x(0) = [\theta_0, 0]^T$ with $\theta_0 = \pi/18$, $\pi/12$, $\pi/6$, and $\pi/3$, plot the states for $t \geq 0$, for the nonlinear model.

(b) Repeat (a) for the linear model.

(c) Compare the results in (a) and (b).

CHAPTER 2

Response of Linear Systems

In system theory it is important to clearly understand how inputs and initial conditions affect the response of a system. There are many reasons for this. For example, in control theory, it is important to be able to select an input that will cause the system output to satisfy certain properties [e.g., to remain bounded (stability), to follow a given trajectory (tracking), and the like]. This is in stark contrast to the study of ordinary differential equations, where it is usually assumed that the forcing function (input) is given.

2.1 INTRODUCTION

The goal of this chapter is to study the response of linear systems in greater detail than was done in Chapter 1. To this end, solutions of linear ordinary differential equations are reexamined, this time with an emphasis on characterizing all solutions using bases (of the solution vector space) and on determining such solutions. For convenience, certain results from Chapter 1 are repeated, and time-varying and time-invariant cases are treated separately, as are continuous-time and discrete-time cases. Whereas in Chapter 1 certain fundamental issues that include input-output system descriptions, causality, linearity, and time-invariance are emphasized, we will here address in greater detail impulse (and pulse) response and transfer functions for continuous-time systems and discrete-time systems.

A. Chapter Description

As in Chapter 1, we will concern ourselves with linear, continuous-time, finite-dimensional systems represented by the state and output equations (internal description)

$$\dot{x} = A(t)x + B(t)u, \qquad y = C(t)x + D(t)u \tag{1.1}$$

and the corresponding input-output description (external description)

$$y(t) = \int_{t_0}^{t} H(t, \tau)u(\tau)\,d\tau, \tag{1.2}$$

where $H(t, \tau)$ denotes the impulse response matrix. As pointed out in Chapter 1, in the special case when $A(t) = A$, $B(t) = B$, $C(t) = C$, and $D(t) = D$, with initial time $t_0 = 0$, the external description (1.2) can be represented equivalently in terms of Laplace transform variables

$$\hat{y}(s) = \hat{H}(s)\hat{u}(s), \tag{1.3}$$

where $\hat{H}(s)$ denotes the transfer function matrix of the system. We will examine in greater detail the relationship between the internal description (1.1) and the external description (1.2) here and in subsequent chapters.

Again, as in Chapter 1, we will also concern ourselves with linear, discrete-time, finite-dimensional systems represented by the state and output equations (internal description)

$$x(k+1) = A(k)x(k) + B(k)u(k), \qquad y(k) = C(k)x(k) + D(k)u(k) \tag{1.4}$$

and the corresponding input-output description (external description)

$$y(n) = \sum_{k=k_0}^{n} H(n, k)u(k), \tag{1.5}$$

where $H(n, k)$ denotes the unit pulse response. In the time-invariant case, with $k_0 = 0$, (1.5) can be represented equivalently (in terms of z-transform variables) as

$$\hat{y}(z) = \hat{H}(z)\hat{u}(z), \tag{1.6}$$

where $\hat{H}(z)$ denotes the system transfer function matrix.

In the following, we provide a brief outline of the contents of this chapter.

In the second section we provide some mathematical background material on linear algebra and matrix theory. In the third section we further study systems of linear homogeneous and nonhomogeneous ordinary differential equations. Specifically, in this section we develop a general characterization of the solutions of such equations and we study the properties of the solutions by investigating the properties of fundamental matrices and state transition matrices. In section four we further investigate systems of linear, autonomous, homogeneous ordinary differential equations. In particular, in this section we emphasize several methods of determining the state transition matrix of such systems and we study the asymptotic behavior of the solutions of such systems. In the fifth section we study systems of linear, periodic ordinary differential equations (Floquet theory). In the sixth and seventh sections we further investigate the properties of the state representations and the input-output representations of continuous-time and discrete-time finite-dimensional systems. Specifically, in these sections we study equivalent representations of such systems, we investigate the properties of transfer function matrices, and for the discrete-time case we also address sampled-data systems and the asymptotic behavior of the system response of time-invariant systems.

B. Guidelines for the Reader

In a first reading, the background material on linear algebra, given in Section 2.2, can be reviewed rather quickly. In the study of the subsequent material of this book, selective detailed coverage of topics in Section 2.2 may also be desirable.

A typical beginning graduate course in linear systems will include Theorem 3.1 in Subsection 2.3A, which shows that the set of solutions of the linear homogeneous equation $\dot{x} = A(t)x$ forms an n-dimensional vector space. This theorem provides the basis for the definitions of the fundamental and the state transition matrix (Subsections 2.3A, 2.3B, and 2.3D). These results enable us to determine solutions of the nonhomogeneous equation $\dot{x} = A(t)x + g(t)$ in Subsection 2.3C. Background required for the above topics includes material on vector spaces, linear independence of vectors, bases for vector spaces, and linear transformations (Subsections 2.2A to 2.2E).

A typical beginning graduate course in linear systems will also address the material on time-invariant systems given in Section 2.4, including solutions of the equation $\dot{x} = Ax + g(t)$, various methods of determining the matrix exponential e^{At}, and asymptotic behavior of time-invariant systems $\dot{x} = Ax$ (including system modes and stability properties of an equilibrium). Background required for these topics includes material on equivalence and similarity, eigenvalues and eigenvectors (Subsections 2.2I and 2.2J).

In addition to the above, a beginning graduate course on linear systems will also treat the input-output description of continuous-time systems and its relation to state-space representations, including the impulse response for time-varying and time-invariant systems, the transfer function matrix for time-invariant systems, and the equivalence of state-space representations (Section 2.6).

Finally, such a course will also cover state-space and input-output descriptions of discrete-time systems (Section 2.7).

2.2 BACKGROUND MATERIAL

In this section we consider material from *linear algebra* and *matrix theory*. We assume that the reader has some background in these areas, and therefore, our presentation will constitute a *summary* rather than a development of the subject matter on hand.

This section consists of fifteen subsections. In the first three subsections we consider linear subspaces of vector spaces, linear independence of a set of vectors, and bases of vector spaces, respectively. In the next five subsections, we address general linear transformations defined on vector spaces, the representation of such transformations by matrices, some of the properties of matrices and determinants of matrices, and solutions of linear algebraic equations, respectively. In the ninth and tenth subsections, we address equivalence and similarity of matrices and eigenvalues and eigenvectors, respectively. In the eleventh subsection we digress by considering direct sums of linear subspaces. In the last four subsections we address, respectively, certain canonical forms of matrices, minimal polynomials of matrices, nilpotent operators, and the Jordan canonical form.

A. Linear Subspaces

In Section 1.10 we gave the formal definition of *vector space over a field*, say, (V, F), where V denotes the set of vectors and F denotes the set of scalars. When F (the field) is clear from context, we usually speak of a vector space V (or a *linear space* V) rather than (V, F).

A nonempty subset W of a vector space V is called a *linear subspace* (or a *linear manifold*) in V if (i) $w_1 + w_2$ is in W whenever w_1 and w_2 are in W, and (ii) αw is in W whenever $\alpha \in F$ and $w \in W$. It is an easy matter to verify that a linear subspace W satisfies all the axioms of a vector space and may as such be regarded as a linear space itself.

Two trivial examples of linear subspaces include the null vector (i.e., the set $W = \{0\}$ is a linear subspace of V) and the vector space V itself. Another example of a linear subspace is the set of all real-valued polynomials defined on the interval $[a, b]$, that is a linear subspace of the vector space consisting of all real-valued continuous functions defined on the interval $[a, b]$ (refer to Example 10.4 in Chapter 1).

As another example of a linear subspace (of R^2), we cite the set of all points on a straight line passing through the origin. On the other hand, a straight line that does not pass through the origin is *not* a linear subspace of R^2 (why?).

It is an easy matter to show that if W_1 and W_2 are linear subspaces of a vector space V, then $W_1 \cap W_2$, the intersection of W_1 and W_2, is also a linear subspace of V. A similar statement cannot be made, however, for the union of W_1 and W_2 (prove this). Note that to show that a set V is a vector space, it suffices to show that it is a linear subspace of some vector space.

B. Linear Independence

Throughout the remainder of this section, we let $\{\alpha_1, \ldots, \alpha_n\}$, $\alpha_i \in F$, denote an indexed set of scalars and we let $\{v^1, \ldots, v^n\}$, $v^i \in V$, denote an indexed set of vectors.

Now let W be a set in a linear space V (W may be a finite set or an infinite set). We say that a vector $v \in V$ is a *finite linear combination of vectors* in W if there is a finite set of elements $\{w^1, \ldots, w^n\}$ in W and a finite set of scalars $\{\alpha_1, \ldots, \alpha_n\}$ in F such that

$$v = \alpha_1 w^1 + \cdots + \alpha_n w^n.$$

Now let W be a nonempty subset of a linear space V and let $S(W)$ be the set of all finite linear combinations of the vectors from W, i.e., $w \in S(W)$ if and only if there is some set of scalars $\{\alpha_1, \ldots, \alpha_m\}$ and some finite subset $\{w^1, \ldots, w^m\}$ of W such that $w = \alpha_1 w^1 + \cdots + \alpha_m w^m$, where m may be any positive integer. Then it is easily shown that $S(W)$ is a linear subspace of V, called the *linear subspace generated* by the set W.

Now if U is a linear subspace of a vector space V and if there exists a set of vectors $W \subset V$ such that the linear space $S(W)$ generated by W is U, then we say that W *spans* U. It is easily shown that $S(W)$ is the smallest linear subspace of a vector space V containing the subset W of V. Specifically, if U is a linear subspace of V and if U contains W, then U also contains $S(W)$.

As an example, in the space (R^2, R) the set $S_1 = \{e^1\} = \{(1,0)^T\}$ spans the set consisting of all vectors of the form $(a, 0)^T$, $a \in R$, while the set $S_2 = \{e^1, e^2\}$, $e^2 = (0, 1)^T$ spans all of R^2.

We are now in a position to introduce the notion of linear dependence.

DEFINITION 2.1. Let $S = \{v^1, \ldots, v^m\}$ be a finite nonempty set in a linear space V. If there exist scalars $\alpha_1, \ldots, \alpha_m$, not all zero, such that

$$\alpha_1 v^1 + \cdots + \alpha_m v^m = 0, \tag{2.1}$$

then the set S is said to be *linearly dependent* (over F). If a set is not linearly dependent, then it is said to be *linearly independent*. In this case relation (2.1) implies that $\alpha_1 = \cdots = \alpha_m = 0$. An infinite set of vectors W in V is said to be linearly independent if every finite subset of W is linearly independent. ■

EXAMPLE 2.1. Consider the linear space (R^n, R) (see Example 10.1 in Chapter 1), and let $e^1 = (1, 0, \ldots, 0)^T$, $e^2 = (0, 1, 0, \ldots, 0)^T, \ldots, e^n = (0, \ldots, 0, 1)^T$. Clearly, $\sum_{i=1}^n \alpha_i e^i = 0$ implies that $\alpha_i = 0$, $i = 1, \ldots, n$. Therefore, the set $S = \{e^1, \ldots, e^n\}$ is a linearly independent set of vectors in R^n over the field of real numbers R. ■

EXAMPLE 2.2. Let V be the set of 2-tuples whose entries are complex-valued rational functions over the field of complex-valued rational functions. Let

$$v^1 = \begin{bmatrix} \dfrac{1}{s+1} \\[2mm] \dfrac{1}{s+2} \end{bmatrix}, \qquad v^2 = \begin{bmatrix} \dfrac{s+2}{(s+1)(s+3)} \\[2mm] \dfrac{1}{s+3} \end{bmatrix}$$

and let $\alpha_1 = -1$, $\alpha_2 = (s+3)/(s+2)$. Then $\alpha_1 v^1 + \alpha_2 v^2 = 0$, and therefore, the set $S = \{v^1, v^2\}$ is *linearly dependent over the field of rational functions.* On the other hand, since $\alpha_1 v^1 + \alpha_2 v^2 = 0$ when $\alpha_1, \alpha_2 \in R$ is true if and only if $\alpha_1 = \alpha_2 = 0$, it follows that S is linearly independent over the field of real numbers (which is a subset of the field of rational functions). This shows that *linear dependence* of a set of vectors in V depends on the field F. ■

Linear independence of functions of time

EXAMPLE 2.3. Let $V = C((a, b), R^n)$, let $F = R$, and for $x, y \in V$ and $\alpha \in F$, define addition of elements in V and multiplication of elements in V by elements in F by $(x + y)(t) = x(t) + y(t)$ for all $t \in (a, b)$ and $(\alpha x)(t) = \alpha x(t)$ for all $t \in (a, b)$. Then as in Example 10.4 of Chapter 1, we can easily show that (V, F) is a vector space. An interesting question that arises is whether for this space, linear dependence (and linear independence) of a set of vectors can be phrased in some testable form. The answer is yes. Indeed, it can readily be verified that for the present vector space (V, F), *linear dependence* of a set of vectors $S = \{\phi_1, \ldots, \phi_k\}$ in $V = C((a, b), R^n)$ over $F = R$ is equivalent to the requirement that there exist scalars $\alpha_i \in F$, $i = 1, \ldots, k$, not all zero, such that

$$\alpha_1 \phi_1(t) + \cdots + \alpha_k \phi_k(t) = 0 \qquad \text{for all } t \in (a, b).$$

Otherwise, S is *linearly independent.*

To see how the above example applies to specific cases, let $V = C((-\infty, \infty), R^2)$ and consider the vectors $\phi_1(t) = [1, t]^T$, $\phi_2(t) = [1, t^2]^T$. To show that the set $S = \{\phi_1, \phi_2\}$ is linearly independent (over $F = R$), assume for purposes of contradiction that S is linearly dependent. Then there must exist scalars α_1 and α_2, not both zero, such that $\alpha_1[1, t]^T + \alpha_2[1, t^2]^T = [0, 0]^T$ for all $t \in (-\infty, \infty)$. But in particular, for $t = 2$, the above equation is satisfied if and only if $\alpha_1 = \alpha_2 = 0$, which contradicts the assumption. Therefore, $S = \{\phi_1, \phi_2\}$ is linearly independent.

As another specific case of the above example, let $V = C((-\infty, \infty), R^2)$ and consider the set $S = \{\phi_1, \phi_2, \phi_3, \phi_4\}$, where $\phi_1(t) = [1, t]^T$, $\phi_2(t) = [1, t^2]$, $\phi_3(t) = [0, 1]^T$, and $\phi_4(t) = [e^{-t}, 0]$. The set S is clearly independent over R since $\alpha_1\phi_1(t) + \alpha_2\phi_2(t) + \alpha_3\phi_3(t) + \alpha_4\phi_4(t) = 0$ for all $t \in (-\infty, \infty)$ if and only if $\alpha_1 = \alpha_2 = \alpha_3 = \alpha_4 = 0$. ■

Next, let $S = \{v^1, \ldots, v^m\}$ be a linearly independent set in a vector space V. If $\sum_{i=1}^m \alpha_i v^i = \sum_{i=1}^m \beta_i v^i$, then it is readily shown that $\alpha_i = \beta_i$, for all $i = 1, \ldots, m$. Also, it is easily shown that the set S is linearly dependent if and only if for some index i, $1 \leq i \leq m$, we can find scalars $\gamma_1, \ldots, \gamma_{i-1}, \gamma_{i+1}, \ldots, \gamma_m$ such that $v^i = \gamma_1 v^1 + \cdots + \gamma^{i-1}v^{i-1} + \gamma^{i+1}v^{i+1} + \cdots + \gamma_m v^m$. Furthermore, it is not hard to verify that a finite nonempty set W in a linear space is linearly independent if and only if for each $v \in S(W)$, $v \neq 0$, there is a unique finite subset of W, say, $\{v^1, v^2, \ldots, v^m\}$ and a unique set of nonzero scalars $\{\delta_1, \ldots, \delta_m\}$, such that $v = \delta_1 v^1 + \cdots + \delta_m v^m$. Finally, if U is a finite set in a linear space V, then it is easily shown that U is linearly independent if and only if there is no proper subset Z of U such that $S(U) = S(Z)$. (Recall that Z is a proper subset of U if there is a $u \in U$ such that $u \notin Z$.)

C. Bases

We are now in a position to introduce another important concept.

DEFINITION 2.2. A set W in a linear space V is called a *basis* for V if
(i) W is linearly independent.
(ii) The span of W is the linear space V itself, i.e., $S(W) = V$. ■

An immediate consequence of the above definition is that if W is a linearly independent set in a vector space V, then W is a basis for $S(W)$.

To introduce the notion of dimension of a vector space, it is shown that if a linear space V is generated by a finite number of linearly independent elements, then this number of elements must be unique. The following results lead up to this.

Let $\{v^1, \ldots, v^n\}$ be a basis for a linear space V. Then it is easily shown that for each vector $v \in V$, there exist *unique* scalars $\alpha_1, \ldots, \alpha_n$ such that

$$v = \alpha_1 v^1 + \cdots + \alpha_n v^n.$$

Furthermore, if $u^1, \ldots, u^m$ is any linearly independent set of vectors in V, then $m \leq n$. Moreover, any other basis of V consists of exactly n elements. These facts allow the definitions given in the following.

If a linear space V has a basis consisting of a finite number of vectors, say, $\{v^1, \ldots, v^n\}$, then V is said to be a *finite-dimensional vector space* and the *dimension of* V is n, abbreviated $\dim V = n$. In this case we speak of an *n-dimensional vector space*. If V is not a finite-dimensional vector space, it is said to be an *infinite-dimensional vector space*.

By convention, the linear space consisting of the null vector is finite-dimensional with dimension equal to zero.

An alternative to the above definition of dimension of a (finite-dimensional) vector space is given by the following result, which is easily verified: let V be a vector space that contains n linearly independent vectors. If every set of $n + 1$ vectors in V is linearly dependent, then V is finite-dimensional and $\dim V = n$.

The preceding results enable us now to introduce the concept of coordinates of a vector. We let $\{v^1, \ldots, v^n\}$ be a basis of a vector space V and let $v \in V$ be represented by

$$v = \xi_1 v^1 + \cdots + \xi_n v^n.$$

The *unique* scalars $\xi_1, \ldots, \xi_n$ are called the *coordinates of v with respect to the basis* $\{v^1, \ldots, v^n\}$.

EXAMPLE 2.4. For the linear space (R^n, R), let $S = \{e^1, \ldots, e^n\}$, where the $e^i \in R^n$, $i = 1, \ldots, n$, were defined earlier (following Defintion 2.1). Then S is clearly a basis for (R^n, R) since it is linearly independent and since given any $v \in R^n$, there exist unique real scalars α_i, $i = 1, \ldots, n$, such that $v = \sum_{i=1}^n \alpha_i e^i = (\alpha_1, \ldots, \alpha_n)^T$, i.e., S spans R^n. It follows that with every vector $v \in R^n$, we can associate a *unique* n-tuple of scalars

$$\begin{bmatrix} \alpha_1 \\ \vdots \\ \alpha_n \end{bmatrix} \qquad \text{or} \qquad (\alpha_1, \ldots, \alpha_n)$$

relative to the basis $\{e^1, \ldots, e^n\}$, the *coordinate representation* of the vector $v \in R^n$ with respect to the basis $S = \{e^1, \ldots, e^n\}$. Henceforth, we will refer to the basis S of this example as the *natural basis for* R^n. ■

EXAMPLE 2.5. We note that the vector space of all (complex-valued) polynomials with real coefficients of degree less than n is an n-dimensional vector space over the field of real numbers. A basis for this space is given by $S = \{1, s, \ldots, s^{n-1}\}$, where s is a complex variable. Associated with a given element of this vector space, say, $p(s) = \alpha_0 + \alpha_1 s + \cdots + \alpha_{n-1} s^{n-1}$, we have the *unique* n-tuple given by $(\alpha_0, \alpha_1, \ldots, \alpha_{n-1})^T$, which constitutes the coordinate representation of $p(s)$ with respect to the basis S given above. ■

EXAMPLE 2.6. We note that the space (V, R), where $V = C([a, b], R)$, given in Example 10.4 of Chapter 1, is an infinite-dimensional vector space (why?). ■

D. Linear Transformations

In Subsection 1.10A we introduced the notion of *linear transformation* $\mathcal{T}$ from a vector space V (over the field F) into a vector space W (over the same field F). Henceforth, we will write $\mathcal{T} \in L(V, W)$ to express this. It is our objective in this subsection to identify some of the important properties of linear transformations.

Linear equations

With $\mathcal{T} \in L(V, W)$ we define the *null space of* $\mathcal{T}$ as the set

$$\mathcal{N}(\mathcal{T}) = \{v \in V : \mathcal{T} v = 0\}$$

and the *range space of* $\mathcal{T}$ as the set

$$\mathcal{R}(\mathcal{T}) = \{w \in W : w = \mathcal{T} v, v \in V\}.$$

Note that since $\mathcal{T} 0 = 0$, $\mathcal{N}(\mathcal{T})$ and $\mathcal{R}(\mathcal{T})$ are never empty. It is easily verified that $\mathcal{N}(\mathcal{T})$ is a linear subspace of V and that $\mathcal{R}(\mathcal{T})$ is a linear subspace of W. If V is finite-dimensional (of dimension n), then it is easily shown that $\dim \mathcal{R}(\mathcal{T}) \leq n$. Also, if V is finite-dimensional and if $\{w^1, \ldots, w^n\}$ is a basis for $\mathcal{R}(\mathcal{T})$ and v^i is defined by $\mathcal{T}v^i = w^i, i = 1, \ldots, n$, then it is readily proved that the vectors $v^1, \ldots, v^n$ are linearly independent.

One of the important results of linear algebra, called the *fundamental theorem of linear equations,* states that for $\mathcal{T} \in L(V, W)$ with V finite-dimensional, we have

$$\dim \mathcal{N}(\mathcal{T}) + \dim \mathcal{R}(\mathcal{T}) = \dim V.$$

For the proof of this result, refer to any of the references on linear algebra cited at the end of this chapter.

The above result gives rise to the notions of the *rank*, $\rho(\mathcal{T})$, of a linear transformation $\mathcal{T}$ of a finite-dimensional vector space V into a vector space W, which we define as the dimension of the range space $\mathcal{R}(\mathcal{T})$, and the *nullity*, $\nu(\mathcal{T})$, of $\mathcal{T}$, which we define as the dimension of the null space $\mathcal{N}(\mathcal{T})$.

With the above machinery in place, it is now easy to establish the following important results concerning *linear equations*. We let $\mathcal{T} \in L(V, W)$, where V is finite-dimensional, let $s = \dim \mathcal{N}(\mathcal{T})$, and let $\{v^1, \ldots, v^s\}$ be a basis for $\mathcal{N}(\mathcal{T})$. Then it is easily verified that (i) a vector $v \in V$ satisfies the equation $\mathcal{T}v = 0$ if and only if $v = \sum_{i=1}^{s} \alpha_i v^i$ for some set of scalars $\{\alpha_1, \ldots, \alpha_s\}$, and furthermore, for each $v \in V$ such that $\mathcal{T}v = 0$ is true, the set of scalars $\{\alpha_1, \ldots, \alpha_s\}$ is unique; (ii) if $w^0 \in W$ is a fixed vector, then $\mathcal{T}v = w^0$ holds for at least one vector $v \in V$ (called the *solution* of the equation $\mathcal{T}v = w^0$) if and only if $w^0 \in \mathcal{R}(\mathcal{T})$; and (iii) if w^0 is any fixed vector in W and if v^0 is some vector in V such that $\mathcal{T}v^0 = w^0$ (i.e., v^0 is a solution of the equation $\mathcal{T}v^0 = w^0$), then a vector $v \in V$ satisfies $\mathcal{T}v = w^0$ if and only if $v = v^0 + \sum_{i=1}^{s} \beta_i v^i$ for some set of scalars $\{\beta_1, \ldots, \beta_s\}$, and furthermore, for each $v \in V$ such that $\mathcal{T}v = w_0$, the set of scalars $\{\beta_1, \ldots, \beta_s\}$ is unique.

General properties of linear transformations

Before proceeding further, we briefly digress by recalling certain elementary properties of a function f from a set X to a set Y, written $f : X \to Y$. Letting $\mathcal{R}(f)$ denote the range of f, we can classify f in the following manner: if $\mathcal{R}(f) = Y$, then f is said to be *surjective* or a *surjection* and we say that f maps X *onto* Y; if f is such that for every $x_1, x_2 \in X$, $f(x_1) = f(x_2)$ implies that $x_1 = x_2$, then f is said to be *injective* or an *injection* or a *one-to-one mapping;* and if f is both injective and surjective, we say that f is *bijective* or a *one-to-one* and *onto mapping* or a bijection. When f is injective, its inverse $f^{-1} : \mathcal{R}(f) \to X$ exists, so that $f^{-1}(f(x)) = x$ for all $x \in X$ and $f(f^{-1}(y)) = y$ for all $y \in \mathcal{R}(f)$. Note that when f is bijective, we have $f^{-1} : Y \to X$.

Returning now to the subject on hand, we note that since a linear transformation $\mathcal{T}$ of a linear space V into a linear space W is a mapping, we distinguish in particular among linear transformations that are surjective, injective, and bijective. We will often be particularly interested in knowing when a linear transformation $\mathcal{T}$ has an inverse $\mathcal{T}^{-1}$. When this is the case, we say that $\mathcal{T}$ is *invertible*, that $\mathcal{T}^{-1}$ *exists*, or that $\mathcal{T}$ is *nonsingular*. A linear transformation that is not nonsingular is said to be *singular*.

Concerning the inverse of a linear transformation $\mathcal{T} \in L(V, W)$, it is easily shown that $\mathcal{T}^{-1}$ exists if and only if $\mathcal{T}v = 0$ implies $v = 0$, and furthermore, if $\mathcal{T}^{-1}$ exists, then $\mathcal{T}^{-1}$ is a linear transformation from $\mathcal{R}(\mathcal{T})$ onto V [i.e., $\mathcal{T}^{-1} \in L(\mathcal{R}(\mathcal{T}), V)$]. Moreover, if V is finite-dimensional, then $\mathcal{T}$ has an inverse if and only if $\mathcal{R}(\mathcal{T})$ has the same dimension as V, i.e., $\rho(\mathcal{T}) = \dim V$. Also, if both

V and W are finite-dimensional and of the same dimension, then $\mathcal{R}(\mathcal{T}) = W$ if and only if $\mathcal{T}$ has an inverse.

In the next few results, which are phrased in terms of equivalent statements, we summarize some of the important properties of linear transformations.

(*Injective linear transformations*) For $\mathcal{T} \in L(V, W)$, the following statements are equivalent: (i) $\mathcal{T}$ is injective; (ii) $\mathcal{T}$ has an inverse; (iii) $\mathcal{T}v = 0$ implies $v = 0$; (iv) for each $w \in \mathcal{R}(\mathcal{T})$, there is a unique $v \in V$ such that $\mathcal{T}v = w$; (v) if $\mathcal{T}v^1 = \mathcal{T}v^2$, then $v^1 = v^2$; (vi) if $v^1 \neq v^2$, then $\mathcal{T}v^1 \neq \mathcal{T}v^2$. If in addition, V is finite-dimensional, then the following are equivalent: (i) $\mathcal{T}$ is injective; (ii) $\rho(\mathcal{T}) = \dim V$.

(*Surjective linear transformations*) For $\mathcal{T} \in L(V, W)$, the following statements are equivalent: (i) $\mathcal{T}$ is surjective; (ii) for each $w \in W$, there is a $v \in V$ such that $\mathcal{T}v = w$. If in addition, V and W are finite-dimensional, then the following are equivalent: (i) $\mathcal{T}$ is surjective; (ii) $\dim W = \rho(\mathcal{T})$.

(*Bijective linear transformations*) For $\mathcal{T} \in L(V, W)$, the following are equivalent: (i) $\mathcal{T}$ is bijective; (ii) for every $w \in W$, there is a unique $v \in V$ such that $\mathcal{T}v = w$. If in addition V and W are finite-dimensional, then the following are equivalent: (i) $\mathcal{T}$ is bijective; (ii) $\dim V = \dim W = \rho(\mathcal{T})$.

(*Injective, surjective, and bijective linear transformations*) For $\mathcal{T} \in L(V, W)$ with V and W finite-dimensional and with $\dim V = \dim W$, the following are equivalent: (i) $\mathcal{T}$ is injective; (ii) $\mathcal{T}$ is surjective; (iii) $\mathcal{T}$ is bijective; (iv) $\mathcal{T}$ has an inverse.

Next, we examine some of the properties of $L(V, W)$, the set of all linear transformations from a vector space V into a vector space W. As before, we assume that V and W are linear spaces over the same field F.

We let $\mathcal{S}, \mathcal{T} \in L(V, W)$ and we define the *sum of* $\mathcal{S}$ *and* $\mathcal{T}$ by

$$(\mathcal{S} + \mathcal{T})v \triangleq \mathcal{S}v + \mathcal{T}v \tag{2.2}$$

for all $v \in V$. Also, with $\alpha \in F$ and $\mathcal{T} \in L(V, W)$, we define *multiplication* of $\mathcal{T}$ *by a scalar* α as

$$(\alpha\mathcal{T})v \triangleq \alpha\mathcal{T}v \tag{2.3}$$

for all $v \in V$. It is easily shown that $(\mathcal{S} + \mathcal{T}) \in L(V, W)$ and also that $\alpha\mathcal{T} \in L(V, W)$. We further note that there exists a zero element in $L(V, W)$, called the *zero transformation*, denoted by $\mathbb{O}$ and defined by

$$\mathbb{O}v = 0 \tag{2.4}$$

for all $v \in V$. Furthermore, we note that to each $\mathcal{T} \in L(V, W)$ there corresponds a unique linear transformation $-\mathcal{T} \in L(V, W)$ defined by

$$(-\mathcal{T})v = -\mathcal{T}v \tag{2.5}$$

for all $v \in V$. In this case it follows trivially that $-\mathcal{T} + \mathcal{T} = \mathbb{O}$.

With these definitions in place, it is easily proved that $L(V, W)$ is a linear space over F, called the *space of linear transformations* [with vector addition defined by (2.2) and multiplication of vectors by scalars defined by (2.3)].

To explore the properties of the space of linear transformations further, we briefly digress to recall the definition of an algebra. Specifically, a set V is called an *algebra* if it is a linear space and if in addition to each $v, w \in V$ there corresponds an element in V, denoted by $v \cdot w$ and called the *product of* v *times* w, satisfying the following axioms:

1. $v \cdot (w + u) = v \cdot w + v \cdot u$ for all $v, w, u \in V$.
2. $(v + w) \cdot u = v \cdot u + w \cdot u$ for all $v, w, u \in V$.
3. $(\alpha v) \cdot (\beta w) = (\alpha\beta)(v \cdot w)$ for all $v, w \in V$ and for all $\alpha, \beta \in F$.

If in addition to the above,

4. $(v \cdot w) \cdot u = v \cdot (w \cdot u)$ for all $v, w, u \in V$, then V is called an *associative algebra.*

If there exists an element $i \in V$ such that $i \cdot v = v \cdot i = v$ for every $v \in V$, then i is called the *identity* of the algebra. It can readily be shown that if i exists, then it is unique. Furthermore, if $v \cdot w = w \cdot v$ for all $v, w \in V$, then V is said to be a *commutative algebra.*

Returning to the subject on hand, let V, W, and U be linear spaces over F, and consider the vector spaces $L(V, W)$ and $L(W, U)$. If $\mathcal{S} \in L(W, U)$ and $\mathcal{T} \in L(V, W)$, then we define the *product* $\mathcal{S}\mathcal{T}$ as the mapping of V into U by the relation

$$(\mathcal{S}\mathcal{T})v = \mathcal{S}(\mathcal{T}v) \tag{2.6}$$

for all $v \in V$. It is easily verified that $\mathcal{S}\mathcal{T} \in L(V, U)$.

Next, let $V = W = U$. If $\mathcal{S}, \mathcal{T}, \mathcal{Q} \in L(V, V)$ and if $\alpha, \beta \in F$, then it is easily shown that

$$\mathcal{S}(\mathcal{T}\mathcal{Q}) = (\mathcal{S}\mathcal{T})\mathcal{Q} \tag{2.7}$$

$$\mathcal{S}(\mathcal{T} + \mathcal{Q}) = \mathcal{S}\mathcal{T} + \mathcal{S}\mathcal{Q} \tag{2.8}$$

$$(\mathcal{S} + \mathcal{T})\mathcal{Q} = \mathcal{S}\mathcal{Q} + \mathcal{T}\mathcal{Q} \tag{2.9}$$

and

$$(\alpha\mathcal{S})(\beta\mathcal{T}) = (\alpha\beta)\mathcal{S}\mathcal{T}. \tag{2.10}$$

We emphasize that, in general, commutativity of linear transformations does not hold, i.e., in general

$$\mathcal{S}\mathcal{T} \neq \mathcal{T}\mathcal{S}. \tag{2.11}$$

There is a special mapping from a linear space V into V, called the *identity transformation,* defined by

$$\mathcal{I}v = v \tag{2.12}$$

for all $v \in V$. We note that $\mathcal{I} \in L(V, V)$, that $\mathcal{I} \neq \mathbb{O}$ if and only if $V \neq \{0\}$, that $\mathcal{I}$ is unique, and that

$$\mathcal{T}\mathcal{I} = \mathcal{I}\mathcal{T} = \mathcal{T} \tag{2.13}$$

for all $\mathcal{T} \in L(V, V)$. Also, we can readily verify that the transformation $\alpha\mathcal{I}$, $\alpha \in F$, defined by

$$(\alpha\mathcal{I})v = \alpha\mathcal{I}v = \alpha v \tag{2.14}$$

is also a linear transformation.

Relations (2.7) to (2.14) now give rise to the following result: $L(V, V)$ is an associative algebra with identity $\mathcal{I}$. This algebra is in general not commutative.

Concerning invertible linear transformations we note that if $\mathcal{T} \in L(V, V)$ is bijective, then $\mathcal{T}^{-1} \in L(V, V)$, and furthermore,

$$\mathcal{T}^{-1}\mathcal{T} = \mathcal{T}\mathcal{T}^{-1} = \mathcal{I}, \tag{2.15}$$

where $\mathcal{I}$ denotes the identity transformation defined in (2.12).

Next, if V is a finite-dimensional vector space and $\mathcal{T} \in L(V, V)$, then we can readily show that the following are equivalent: (i) $\mathcal{T}$ is invertible; (ii) $\rho(\mathcal{T}) = \dim V$; (iii) $\mathcal{T}$ is one-to-one; (iv) $\mathcal{T}$ is onto; and (v) $\mathcal{T}v = 0$ implies that $v = 0$.

For bijective linear transformations, we can easily verify the following characterizations. Let $\mathcal{S}, \mathcal{T}, \mathcal{Q} \in L(V, V)$ and let $\mathcal{I}$ denote the identity transformation. Then (i) if $\mathcal{S}\mathcal{T} = \mathcal{Q}\mathcal{S} = \mathcal{I}$, then $\mathcal{S}$ is bijective and $\mathcal{S}^{-1} = \mathcal{T} = \mathcal{Q}$; (ii) if $\mathcal{S}$ and $\mathcal{T}$ are bijective, then $\mathcal{S}\mathcal{T}$ is bijective, and $(\mathcal{S}\mathcal{T})^{-1} = \mathcal{T}^{-1}\mathcal{S}^{-1}$; (iii) if $\mathcal{S}$ is bijective, then $(\mathcal{S}^{-1})^{-1} = \mathcal{S}$; and (iv) if $\mathcal{S}$ is bijective, then $\alpha\mathcal{S}$ is bijective and $(\alpha\mathcal{S})^{-1} = (1/\alpha)\mathcal{S}^{-1}$ for all $\alpha \in F, \alpha \neq 0$.

With the aid of the above concepts and results we can now construct certain classes of *functions of linear transformations*. Since (2.7) allows us to write the product of three or more linear transformations without the use of parentheses, we can define $\mathcal{T}^n$, where $\mathcal{T} \in L(V, V)$ and n is a positive integer, as

$$\mathcal{T}^n \triangleq \underbrace{\mathcal{T} \cdot \mathcal{T} \cdot \cdots \cdot \mathcal{T}}_{n \text{ times}}. \tag{2.16}$$

Similarly, if $\mathcal{T}^{-1}$ is the inverse of $\mathcal{T}$, then we can define $\mathcal{T}^{-m}$, where m is a positive integer, as

$$\mathcal{T}^{-m} \triangleq (\mathcal{T}^{-1})^m = \underbrace{\mathcal{T}^{-1} \cdot \mathcal{T}^{-1} \cdot \cdots \cdot \mathcal{T}^{-1}}_{m \text{ times}}. \tag{2.17}$$

Using these definitions, the usual *laws of exponents* can be verified. Thus,

$$\mathcal{T}^m \cdot \mathcal{T}^n = \mathcal{T}^{m+n} = \mathcal{T}^n \cdot \mathcal{T}^m \tag{2.18}$$

$$(\mathcal{T}^m)^n = \mathcal{T}^{mn} = \mathcal{T}^{nm} = (\mathcal{T}^n)^m \tag{2.19}$$

and

$$\mathcal{T}^m \cdot \mathcal{T}^{-n} = \mathcal{T}^{m-n}, \tag{2.20}$$

where m and n are positive integers. Consistent with the above, we also have

$$\mathcal{T}^1 = \mathcal{T} \tag{2.21}$$

and

$$\mathcal{T}^0 = \mathcal{I}. \tag{2.22}$$

We are now in a position to consider *polynomials of linear transformations*. For example, if $f(\lambda)$ is a polynomial, i.e.,

$$f(\lambda) = \alpha_0 + \alpha_1\lambda + \cdots + \alpha_n\lambda^n, \tag{2.23}$$

where $\alpha_0, \ldots, \alpha_n \in F$, then by $f(\mathcal{T})$ we mean

$$f(\mathcal{T}) = \alpha_0\mathcal{I} + \alpha_1\mathcal{T} + \cdots + \lambda_n\mathcal{T}^n. \tag{2.24}$$

The reader is cautioned that in general the above concept cannot be extended to functions of two or more linear transformations, because linear transformations in general do not commute.

E. Representation of Linear Transformations by Matrices

In the following, we let (V, F) and (W, F) be vector spaces over the *same* field and we let $\mathcal{A} : V \to W$ denote a linear mapping. We let $\{v^1, \ldots, v^n\}$ be a basis for V and we set $\bar{v}^1 = \mathcal{A}v^1, \ldots, \bar{v}^n = \mathcal{A}v^n$. Then it is an easy matter to show that if v is any vector in V and if $(\alpha_1, \ldots, \alpha_n)$ are the coordinates of v with respect to $\{v^1, \ldots, v^n\}$, then $\mathcal{A}v = \alpha_1\bar{v}^1 + \cdots + \alpha_n\bar{v}^n$. Indeed, we have $\mathcal{A}v = \mathcal{A}(\alpha_1 v^1 + \cdots + \alpha_n v^n) = \alpha_1\mathcal{A}v^1 + \cdots + \alpha_n\mathcal{A}v^n = \alpha_1\bar{v}^1 + \cdots + \alpha_n\bar{v}^n$.

Next, we let $\{\bar{v}^1, \ldots, \bar{v}^n\}$ be any set of vectors in W. Then it can be shown that there exists a unique linear transformation $\mathcal{A}$ from V into W such that $\mathcal{A}v^1 = \bar{v}^1, \ldots, \mathcal{A}v^n = \bar{v}^n$. To show this, we first observe that for each $v \in V$ we have unique scalars $\alpha_1, \ldots, \alpha_n$ such that

$$v = \alpha_1 v^1 + \cdots + \alpha_n v^n.$$

Now define a mapping $\mathcal{A} : V \to W$ as

$$\mathcal{A}(v) = \alpha_1\bar{v}^1 + \cdots + \alpha_n\bar{v}^n.$$

Clearly, $\mathcal{A}(v^i) = \bar{v}^i$, $i = 1, \ldots, n$. We first must show that $\mathcal{A}$ is linear and, then, that $\mathcal{A}$ is unique. Given $v = \alpha_1 v^1 + \cdots + \alpha_n v^n$ and $w = \beta_1 v^1 + \cdots + \beta_n v^n$, we have $\mathcal{A}(v+w) = \mathcal{A}[(\alpha_1+\beta_1)v^1 + \cdots + (\alpha_n+\beta_n)v^n] = (\alpha_1+\beta_1)\bar{v}^1 + \cdots + (\alpha_n+\beta_n)\bar{v}^n$. On the other hand, $\mathcal{A}(v) = \alpha_1\bar{v}^1 + \cdots + \alpha_n\bar{v}^n$, $\mathcal{A}(w) = \beta_1\bar{v}^1 + \cdots + \beta_n\bar{v}^n$. Thus, $\mathcal{A}(v) + \mathcal{A}(w) = (\alpha_1\bar{v}^1 + \cdots + \alpha_n\bar{v}^n) + (\beta_1\bar{v}^1 + \cdots + \beta_n\bar{v}^n) = (\alpha_1 + \beta_1)\bar{v}^1 + \cdots + (\alpha_n + \beta_n)\bar{v}^n = \mathcal{A}(v + w)$. In a similar manner, it is easily established that $\alpha\mathcal{A}(v) = \mathcal{A}(\alpha v)$ for all $\alpha \in F$ and $v \in V$. Therefore, $\mathcal{A}$ is linear. Finally, to show that $\mathcal{A}$ is unique, suppose there exists a linear transformation $\mathcal{B} : V \to W$ such that $\mathcal{B}v^i = \bar{v}^i$, $i = 1, \ldots, n$. It follows that $(\mathcal{A} - \mathcal{B})v^i = 0$, $i = 1, \ldots, n$, and, therefore, that $\mathcal{A} = \mathcal{B}$.

These results show that *a linear transformation is completely determined by knowing how it transforms the basis vectors in its domain, and that this linear transformation is uniquely determined in this way.* These results enable us to represent linear transformations defined on finite-dimensional spaces in an unambiguous way by means of matrices. We will use this fact in the following.

Let (V, F) and (W, F) denote n-dimensional and m-dimensional vector spaces, respectively and let $\{v^1, \ldots, v^n\}$ and $\{w^1, \ldots, w^m\}$ be bases for V and W, respectively. Let $\mathcal{A} : V \to W$ be a linear transformation and let $\bar{v}^i = \mathcal{A}v^i$, $i = 1, \ldots, n$. Since $\{w^1, \ldots, w^m\}$ is a basis for W, there are unique scalars $\{a_{ij}\}$, $i = 1, \ldots, m$, $j = 1, \ldots, n$, such that

$$\begin{aligned}
\mathcal{A}v^1 &= \bar{v}^1 = a_{11}w^1 + a_{21}w^2 + \cdots + a_{m1}w^m \\
\mathcal{A}v^2 &= \bar{v}^2 = a_{12}w^1 + a_{22}w^2 + \cdots + a_{m2}w^m \\
\cdots &\ \cdots \ \cdots\cdots\cdots\cdots\cdots\cdots\cdots\cdots \\
\mathcal{A}v^n &= \bar{v}^n = a_{1n}w^1 + a_{2n}w^2 + \cdots + a_{mn}w^m.
\end{aligned} \tag{2.25}$$

Next, let $v \in V$. Then v has the unique representation $v = \alpha_1 v^1 + \alpha_2 v^2 + \cdots + \alpha_n v^n$ with respect to the basis $\{v^1, \ldots, v^n\}$. In view of the result given at the beginning of this subsection, we now have

$$\mathcal{A}v = \alpha_1\bar{v}^1 + \cdots + \alpha_n\bar{v}^n. \tag{2.26}$$

Since $\mathcal{A}v \in W$, $\mathcal{A}v$ has a unique representation with respect to the basis $\{w^1, \ldots, w^m\}$, say,

$$\mathcal{A}v = \gamma_1 w^1 + \gamma_2 w^2 + \cdots + \gamma_m w^m. \tag{2.27}$$

Combining (2.25) and (2.26), we have

$$\begin{aligned}\mathcal{A}v &= \alpha_1(a_{11}w^1 + \cdots + a_{m1}w^m) \\ &\quad + \alpha_2(a_{12}w^1 + \cdots + a_{m2}w^m) \\ &\quad \cdots\cdots\cdots\cdots\cdots\cdots\cdots\cdots \\ &\quad + \alpha_n(a_{1n}w^1 + \cdots + a_{mn}w^m).\end{aligned}$$

Rearranging this expression, we have

$$\begin{aligned}\mathcal{A}v &= (a_{11}\alpha_1 + a_{12}\alpha_2 + \cdots + a_{1n}\alpha_n)w^1 \\ &\quad + (a_{21}\alpha_1 + a_{22}\alpha_2 + \cdots + a_{2n}\alpha_n)w^2 \\ &\quad \cdots\cdots\cdots\cdots\cdots\cdots\cdots\cdots\cdots\cdots \\ &\quad + (a_{m1}\alpha_1 + a_{m2}\alpha_2 + \cdots + a_{mn}\alpha_n)w^m.\end{aligned}$$

In view of the uniqueness of the representation in (2.27), we have

$$\begin{aligned}\gamma_1 &= a_{11}\alpha_1 + a_{12}\alpha_2 + \cdots + a_{1n}\alpha_n \\ \gamma_2 &= a_{21}\alpha_1 + a_{22}\alpha_2 + \cdots + a_{2n}\alpha_n \\ &\cdots\cdots\cdots\cdots\cdots\cdots\cdots\cdots\cdots \\ \gamma_m &= a_{m1}\alpha_1 + a_{m2}\alpha_2 + \cdots + a_{mn}\alpha_n,\end{aligned} \tag{2.28}$$

where $(\alpha_1, \ldots, \alpha_n)^T$ and $(\gamma_1, \ldots, \gamma_m)^T$ are coordinate representations of $v \in V$ and $\mathcal{A}v \in W$ with respect to the bases $\{v^1, \ldots, v^n\}$ of V and $\{w^1, \ldots, w^m\}$ of W, respectively. This set of equations enables us to represent the linear transformation $\mathcal{A}$ from the linear space V into the linear space W by the unique scalars $\{a_{ij}\}$, $i = 1, \ldots, m$, $j = 1, \ldots, n$. For convenience we let

$$A = [a_{ij}] = \begin{bmatrix} a_{11} & a_{12} & \cdots & a_{1n} \\ a_{21} & a_{22} & \cdots & a_{2n} \\ \cdots & \cdots & \cdots & \cdots \\ a_{m1} & a_{m2} & \cdots & a_{mn} \end{bmatrix}. \tag{2.29}$$

We see that once the bases $\{v^1, \ldots, v^n\}$, $\{w^1, \ldots, w^m\}$ are fixed, we can represent the linear transformation $\mathcal{A}$ by the array of scalars in (2.29) that are uniquely determined by (2.25). *Note that the jth column of A is the coordinate representation of the vector* $Av^j \in W$ *with respect to the basis* $\{w^1, \ldots, w^m\}$.

In view of the results given at the beginning of this subsection, the converse to the preceding statement also holds. Specifically, with the bases for V and W still fixed, the array given in (2.29) is uniquely associated with the linear transformation $\mathcal{A}$ of V into W.

The above discussion gives rise to the following important definition.

DEFINITION 2.3. The array given in (2.29) is called the *matrix A of the linear transformation $\mathcal{A}$ from a linear space V into a linear space W (over F) with respect to the basis $\{v^1, \ldots, v^n\}$ of V and the basis $\{w^1, \ldots, w^m\}$ of W.* ■

If in Definition 2.3, $V = W$, and if for both V and W the same basis $\{v^1, \ldots, v^n\}$ is used, then we simply speak of the *matrix A of the linear transformation $\mathcal{A}$ with respect to the basis $\{v^1, \ldots, v^n\}$.*

In (2.29) the scalars $(a_{i1}, a_{i2}, \ldots, a_{in})$ form the ith *row* of A and the scalars $(a_{1j}, a_{2j}, \ldots, a_{mj})^T$ form the jth *column* of A. The scalar a_{ij} refers to that element of matrix A that can be found in the ith row and jth column of A. The array in (2.29) is said to be an $m \times n$ *matrix*. If $m = n$, we speak of a *square matrix*. Consistent with the above, an $n \times 1$ matrix is called a *column vector, column matrix*, or *n-vector*, and a $1 \times n$ matrix is called a *row vector*. Finally, if $A = [a_{ij}]$ and $B = [b_{ij}]$ are two $m \times n$ matrices, then $A = B$, i.e., A and B are *equal* if and only if $a_{ij} = b_{ij}$ for all $i = 1, \ldots, m$, and for all $j = 1, \ldots, n$. Furthermore, we call $A^T = [a_{ij}]^T = [a_{ji}]$ the *transpose* of A.

The preceding discussion shows in particular that if $\mathscr{A}$ is a linear transformation of an n-dimensional vector space V into an m-dimensional vector space W,

$$w = \mathscr{A}v, \tag{2.30}$$

if $\gamma = (\gamma_1, \ldots, \gamma_m)^T$ denotes the coordinate representation of w with respect to the basis $\{w^1, \ldots, w^m\}$, if $\alpha = (\alpha_1, \ldots, \alpha_n)^T$ denotes the coordinate representation of v with respect to the basis $\{v^1, \ldots, v^n\}$, and if A denotes the matrix of $\mathscr{A}$ with respect to the bases $\{v^1, \ldots, v^n\}, \{w^1, \ldots, w^m\}$, then

$$\gamma = A\alpha, \tag{2.31}$$

or equivalently,

$$\gamma_i = \sum_{j=1}^{n} a_{ij}\alpha_j, \qquad i = 1, \ldots, m, \tag{2.32}$$

which are alternative ways to write (2.28).

Some important remarks

1. Throughout this section we use, in the interests of clarity of presentation, lowercase Greek letters to denote the coordinate representations of vectors [see (2.31)]. In the interests of simplicity, however, we will use common (Latin) lowercase letters to denote vectors throughout the remainder of this book, whether they are coordinate representations of vectors or underlying objects (elements of V).
2. We note that if, in particular, $V = R^n$, then $v \in V$ and its coordinate representation η with respect to the natural basis $\{e^1, \ldots, e^n\}$ of V will have the same form.

*F. Some Properties of Matrices

The rank of a matrix

We first consider the characterization of the rank of a linear transformation in terms of its matrix representation. To this end, let $\mathscr{A}$ be a linear transformation from a vector space V into a vector space W. It is easily shown that $\mathscr{A}$ has rank r if and only if it is possible to choose a basis $\{\bar{v}^1, \ldots, \bar{v}^n\}$ for V and a basis $\{\bar{w}^1, \ldots, \bar{w}^m\}$ for W such that the matrix $\bar{A}$ of $\mathscr{A}$ with respect to these bases is of the form

$$\bar{A} = \overbrace{\begin{bmatrix} 100 & \cdots & 000 & \cdots & 0 \\ 010 & \cdots & 000 & \cdots & 0 \\ \cdots & \cdots & \cdots & \cdots & \cdots \\ \cdots & \cdots & \cdots & \cdots & \cdots \\ 000 & \cdots & 100 & \cdots & 0 \\ 000 & \cdots & 000 & \cdots & 0 \\ \cdots & \cdots & \cdots & \cdots & \cdots \\ 000 & \cdots & 000 & \cdots & 0 \end{bmatrix}}^{r} \left.\vphantom{\begin{bmatrix} 1\\1\\1\\1\\1\\1\\1\\1 \end{bmatrix}}\right\} m = \dim W.$$

$$\underbrace{\hphantom{xxxxxxxxxxxxxxxxxxxx}}_{n=\dim V}$$

More directly, if A is the matrix representation of $\mathcal{A} \in L(V, W)$ with respect to some arbitrary bases $\{v^1, \ldots, v^n\}$ and $\{w^1, \ldots, w^m\}$, then (i) the rank of $\mathcal{A}$ is the number of vectors in the largest possible linearly independent set of columns of A; and (ii) the rank of $\mathcal{A}$ is the number of vectors in the smallest possible set of columns of A that has the property that all columns not in it can be expressed as linear combinations of the columns in it.

The above result enables us now to make the following definition: the *rank of an* $m \times n$ *matrix* A is the largest number of linearly independent columns of A.

General properties of matrices

Since matrices are representations of linear transformations on finite-dimensional vector spaces (in the sense defined in Subsection E of this section), it is reasonable to suspect that matrices inherit the properties of the transformations they represent. In the following, we address this issue.

Let V and W be n-dimensional and m-dimensional vector spaces over F, respectively, and let $\mathcal{A}$ and $\mathcal{B}$ be linear transformations of V into W. Let $A = [a_{ij}]$ and $B = [b_{ij}]$ be the matrix representation of $\mathcal{A}$ and $\mathcal{B}$, respectively, with respect to the bases $\{v^1, \ldots, v^n\}$ in V and $\{w^1, \ldots, w^m\}$ in W. Using (2.2) and Definition 2.3, the reader can readily verify that the matrix of the linear transformation $\mathcal{A} + \mathcal{B}$ (with respect to the above bases), is given by

$$A + B = [a_{ij}] + [b_{ij}] = [a_{ij} + b_{ij}] = [c_{ij}] = C. \tag{2.33}$$

Also, using (2.3) and Definition 2.3, the reader can easily show that the matrix of αA, denoted by $D \triangleq \alpha A$, is given as

$$\alpha A = \alpha[a_{ij}] = [\alpha a_{ij}] = [d_{ij}] = D. \tag{2.34}$$

From (2.33) we note that for two matrices A and B to be added, they must have the same number of rows and columns. When this is the case, we say that A and B are *comparable* matrices. Clearly, if A is an $m \times n$ matrix, then so is αA.

Next, let U be an r-dimensional vector space, let $\mathcal{A} \in L(V, W)$, and let $\mathcal{B} \in L(W, U)$. Let A be the matrix of $\mathcal{A}$ with respect to the basis $\{v^1, \ldots, v^n\}$ in V and with respect to the basis $\{w^1, \ldots, w^m\}$ in W. Let $\mathcal{B}$ be the matrix of B with respect to the basis $\{w^1, \ldots, w^m\}$ in W and with respect to the basis $\{u^1, \ldots, u^r\}$ in U. The product mapping $\mathcal{B}\mathcal{A}$ as defined by (2.6) is a linear transformation of V into U. By applying definitions, it is readily verified that the matrix of $\mathcal{B}\mathcal{A}$ with respect to the bases $\{v^1, \ldots, v^n\}$ of V and $\{u^1, \ldots, u^r\}$ of U is given by

$$C = [c_{ij}] = BA, \tag{2.35}$$

where

$$c_{ij} = \sum_{k=1}^{m} b_{ik} a_{kj} \tag{2.36}$$

for $i = 1, \ldots, r$, and $j = 1, \ldots, n$. Clearly, two matrices A and B can be multiplied to form the product BA if and only if the number of columns of B is equal to the number of rows of A. When this is true, we say that the matrices B and A are *conformal matrices*.

As mentioned earlier, the properties of general transformations established in Subsection D hold of course in the case of their matrix representations as well. We summarize some of these in the following:

1. Let A and B be $m \times n$ matrices, and let C be an $n \times r$ matrix; then

$$(A + B)C = AC + BC. \tag{2.37}$$

2. Let A be an $m \times n$ matrix, and let B and C be $n \times r$ matrices; then

$$A(B + C) = AB + AC. \tag{2.38}$$

3. Let A be an $m \times n$ matrix, let B be an $n \times r$ matrix, and let C be an $r \times s$ matrix; then

$$A(BC) = (AB)C. \tag{2.39}$$

4. Let $\alpha, \beta \in F$, and let A be an $m \times n$ matrix; then

$$(\alpha + \beta)A = \alpha A + \beta A. \tag{2.40}$$

5. Let $\alpha \in F$, and let A and B be $m \times n$ matrices; then

$$\alpha(A + B) = \alpha A + \alpha B. \tag{2.41}$$

6. Let $\alpha, \beta \in F$, let A be an $m \times n$ matrix, and let B be an $n \times r$ matrix; then

$$(\alpha A)(\beta B) = (\alpha\beta)(AB). \tag{2.42}$$

7. Let A and B be $m \times n$ matrices; then

$$A + B = B + A. \tag{2.43}$$

8. Let A, B, and C be $m \times n$ matrices; then

$$(A + B) + C = A + (B + C). \tag{2.44}$$

Next, let $\mathbb{O} \in L(V, W)$ be the zero transformation defined by (2.4). Then for any bases $\{v^1, \ldots, v^n\}$ and $\{w^1, \ldots, w^m\}$ for V and W, respectively, the zero transformation is represented by the $m \times n$ matrix

$$O = \begin{bmatrix} 0\,0 & \cdots & 0 \\ 0\,0 & \cdots & 0 \\ \cdots & \cdots & \cdots \\ 0\,0 & \cdots & 0 \end{bmatrix}, \tag{2.45}$$

called the *null matrix*. Further, let $\mathscr{I} \in L(V, V)$ be the identity transformation defined by (2.12) and let $\{v^1, \ldots, v^n\}$ be an arbitrary basis for V. Then the matrix representation of the linear transformation $\mathcal{I}$ from V into V with respect to the basis $\{v^1, \dots, v^n\}$ is given by

$$I = \begin{bmatrix} 1\,0\,0 & \cdots & 0 \\ 0\,1\,0 & \cdots & 0 \\ \cdots & \cdots & \cdots \\ 0\,0\,0 & \cdots & 1 \end{bmatrix}, \tag{2.46}$$

called the $n \times n$ *identity matrix.*

For any $m \times n$ matrix A we have that

$$A + O = O + A = A, \tag{2.47}$$

and for any $n \times n$ matrix B we have

$$BI = IB = B, \tag{2.48}$$

where I is the $n \times n$ identity matrix.

If $A = [a_{ij}]$ is a matrix representation of a linear transformation $\mathcal{A}$, then it is easily verified that the matrix $-A = (-1)A = [-a_{ij}]$ is the corresponding matrix representation of the linear transformation $-\mathcal{A}$. In this case it follows immediately that $A + (-A) = O$, where O denotes the null matrix. By convention we write that $A + (-A) = A - A$.

Next, let A and B be $n \times n$ matrices. Then we have in general that

$$AB \neq BA, \tag{2.49}$$

as was the case in (2.11).

Further, let $\mathcal{A} \in L(V, V)$, and assume that $\mathcal{A}$ is nonsingular with inverse $\mathcal{A}^{-1}$, so that $\mathcal{A}\mathcal{A}^{-1} = \mathcal{A}^{-1}\mathcal{A} = \mathcal{I}$. Now if A is the $n \times n$ matrix of $\mathcal{A}$ with respect to the basis $\{v^1, \dots, v^n\}$ in V, then there is an $n \times n$ matrix B of $\mathcal{A}^{-1}$ with respect to the basis $\{v^1, \dots, v^n\}$ in V such that

$$BA = AB = I. \tag{2.50}$$

We call B the *inverse* of A and we denote it by A^{-1}. Under the present circumstances we say that A^{-1} *exists,* or A *has an inverse,* or A *is invertible,* or A *is nonsingular.* If A^{-1} does not exist, we say that A *is singular.*

From corresponding properties given in Subsection D for arbitrary linear transformations, several properties of matrices are evident. In particular, for an $n \times n$ matrix, the following are equivalent: (i) $\mathit{rank}\, A = n$; (ii) $A\alpha = 0$ implies $\alpha = 0$; (iii) for every $\gamma_0 \in F^n$, there is a unique $\alpha_0 \in F^n$ such that $\gamma_0 = A\alpha_0$; (iv) the columns of A are linearly independent; and (v) A^{-1} exists.

In Subsection E it was shown that we can represent n linear equations by the matrix equation (2.32)

$$\gamma = A\alpha. \tag{2.51}$$

Now assume that A is nonsingular. If we premultiply both sides of this equation by A^{-1}, we obtain

$$\alpha = A^{-1}\gamma, \tag{2.52}$$

the solution to Eq. (2.51). Thus, knowledge of the inverse of A enables us to solve the system of linear equations (2.51).

Concerning inverses of matrices, the following facts are easily verified: (i) an $n \times n$ nonsingular matrix has one and only one inverse; (ii) if A and B are nonsingular

$n \times n$ matrices, then $(AB)^{-1} = B^{-1}A^{-1}$; and (iii) if A and B are $n \times n$ matrices and if AB is nonsingular, then so are A and B.

Next, we consider the principal properties of the transpose of matrices, which follow readily from definitions: (i) for any matrix A, $(A^T)^T = A$; (ii) if A and B are conformal matrices, then $(AB)^T = B^T A^T$; (iii) if A is a nonsingular matrix, then $(A^T)^{-1} = (A^{-1})^T$; (iv) if A is an $n \times n$ matrix, then A^T is nonsingular if and only if A is nonsingular; (v) if A and B are comparable matrices, then $(A+B)^T = A^T + B^T$; and (vi) if $\alpha \in F$ and A is a matrix, then $(\alpha A)^T = \alpha A^T$.

Next, we let A be an $n \times n$ matrix, and we let m be a positive integer. As in (2.16), we define the $n \times n$ matrix A^m by

$$A^m = \underbrace{A \cdot A \cdot \cdots \cdot A}_{m \text{ times}}, \tag{2.53}$$

and if A^{-1} exists, then as in (2.17) we define the $n \times n$ matrix A^{-m} as

$$A^{-m} = (A^{-1})^m = \underbrace{A^{-1} \cdot A^{-1} \cdot \cdots \cdot A^{-1}}_{m \text{ times}}. \tag{2.54}$$

As in the case of Eqs. (2.18) to (2.20), the usual laws of exponents follow from the above definitions. Specifically, if A is an $n \times n$ matrix and if r and s are positive integers, then

$$A^r \cdot A^s = A^{r+s} = A^{s+r} = A^s \cdot A^r \tag{2.55}$$

$$(A^r)^s = A^{rs} = A^{sr} = (A^s)^r, \tag{2.56}$$

and if A^{-1} exists, then

$$A^r \cdot A^{-s} = A^{r-s}. \tag{2.57}$$

Consistent with this notation, we have

$$A^1 = A \tag{2.58}$$

and

$$A^0 = I. \tag{2.59}$$

We are now once more in a position to consider functions of linear transformations, where in the present case the linear transformations are represented by matrices. For example, if $f(\lambda)$ is the polynomial in λ given in (2.23), and if A is any $n \times n$ matrix, then by $f(A)$ we mean

$$f(A) = \alpha_0 I + \alpha_1 A + \cdots + \alpha_n A^n. \tag{2.60}$$

Finally, we noted earlier that, in general, linear transformations (and in particular, matrices) do not commute [see (2.11) and (2.49)]. However, in the case of square matrices, the following facts are easily verified: let A, B, C denote $n \times n$ matrices; let O denote the $n \times n$ null matrix; and let I denote the $n \times n$ identity matrix. Then (i) O commutes with any A; (ii) A^p commutes with A^q, where p and q are positive integers; (iii) αI commutes with any A, where $\alpha \in F$; and (iv) if A commutes with B and if A commutes with C, then A commutes with $\alpha B + \beta C$, where $\alpha, \beta \in F$.

*G. Determinants of Matrices

Definition of determinant

In this section we recall and summarize some of the important properties of determinants of a matrix. To this end we let $N = \{1, 2, \ldots, n\}$ and recall that a *permutation* on N is a one-to-one mapping of N onto itself. For example, if σ denotes a permutation on N, then we can represent it symbolically as

$$\sigma = \begin{pmatrix} 1\,2 \cdots n \\ j_1\; j_2 \cdots j_n \end{pmatrix}, \tag{2.61}$$

where $j_i \in N$ for $i = 1, \ldots, n$, and $j_r \neq j_k$ whenever $r \neq k$. Henceforth, we represent σ more compactly as

$$\sigma = j_1 j_2 \cdots j_n.$$

Clearly, there are $n!$ possible permutations on N. We let $P(N)$ denote the set of all permutations on N, and we distinguish between *odd* and *even permutations*. Specifically, if there is an even number of pairs (i, k) such that $i > k$ but i precedes k in σ, then we say that σ is even. Otherwise, σ is said to be odd. Finally, we define the function *sgn* from $P(N)$ into F by

$$sgn\,(\sigma) = \begin{cases} +1, & \text{if } \sigma \text{ is even} \\ -1, & \text{if } \sigma \text{ is odd} \end{cases}$$

for all $\sigma \in P(N)$.

As a specific example, for $N = \{1, 2, 3\}$, there are six different permutations, even and odd, on N given in the following table:

σ	(j_i, j_2)	(j_i, j_3)	(j_2, j_3)	**σ is odd or even**	$sgn\,(\sigma)$
123	(1, 2)	(1, 3)	(2, 3)	Even	1
132	(1, 3)	(1, 2)	(3, 2)	Odd	−1
213	(2, 1)	(2, 3)	(1, 3)	Odd	−1
231	(2, 3)	(2, 1)	(3, 1)	Even	1
312	(3, 1)	(3, 2)	(1, 2)	Even	1
321	(3, 2)	(3, 1)	(2, 1)	Odd	−1

Now let A denote the $n \times n$ matrix given by

$$A = \begin{bmatrix} a_{11} & a_{12} & \cdots & a_{1n} \\ a_{21} & a_{22} & \cdots & a_{2n} \\ \cdots & \cdots & \cdots & \cdots \\ a_{n1} & a_{n2} & \cdots & a_{nn} \end{bmatrix}.$$

We form the product of n elements from A by taking one and only one element from each row and one and only one element from each column. We represent this product as

$$a_{1j_1} \cdot a_{2j_2} \cdot \cdots \cdot a_{nj_n},$$

where $(j_1 j_2 \cdots j_n) \in P(N)$. It is possible to find $n!$ such products, one for each $\sigma \in P(N)$. We are now in a position to define the *determinant of* A, denoted by $det\,(A)$, by the sum

$$det\,(A) = \sum_{\sigma \in P(N)} sgn\,(\sigma) \cdot a_{1j_1} \cdot a_{2j_2} \cdot \cdots \cdot a_{nj_n}, \tag{2.62}$$

where $\sigma = j_1 \cdots j_n$. We frequently denote the determinant of A by

$$det\,(A) = \begin{vmatrix} a_{11} & a_{12} & \cdots & a_{1n} \\ a_{21} & a_{22} & \cdots & a_{2n} \\ \cdots & \cdots & \cdots & \cdots \\ a_{n1} & a_{n2} & \cdots & a_{nn} \end{vmatrix} = |A|. \tag{2.63}$$

Properties of determinants

We now enumerate some of the common properties of determinants. The proofs of these follow mostly from definitions.

Let A and B be $n \times n$ matrices. Then (i) $det\,(A^T) = det\,(A)$; (ii) if all elements of a column (or row) of A are zero, then $det\,(A) = 0$; (iii) if the matrix B is the matrix obtained by multiplying every element in a column (or row) of A by a constant α, while all other columns of B are the same as those of A, then $det\,(B) = \alpha\, det\,(A)$; (iv) if B is the same as A, except that two columns (or rows) are interchanged, then $det\,(B) = -det\,(A)$; (v) if two columns (or rows) of A are identical, then $det\,(A) = 0$; and (vi) if the columns (or rows) of A are linearly dependent, then $det\,(A) = 0$.

We now introduce some additional concepts for determinants. To this end, let $A = [a_{ij}]$ be an $n \times n$ matrix. If the ith row and jth column of A are deleted, the remaining $(n-1)$ rows and $(n-1)$ columns can be used to form another matrix M_{ij} whose determinant is $det\,(M_{ij})$. We call $det\,(M_{ij})$ the *minor* of a_{ij}. If the diagonal elements of M_{ij} are diagonal elements of A, i.e., if $i = j$, then we speak of a *principal minor of* A. The *cofactor* of a_{ij} is defined as $(-1)^{i+j} det\,(M_{ij})$.

As a specific example, if A is a 3×3 matrix, then

$$det\,(A) = \begin{vmatrix} a_{11} & a_{12} & a_{13} \\ a_{21} & a_{22} & a_{23} \\ a_{31} & a_{32} & a_{33} \end{vmatrix},$$

the minor of element a_{23} is

$$det\,(M_{23}) = \begin{vmatrix} a_{11} & a_{12} \\ a_{31} & a_{32} \end{vmatrix},$$

and the cofactor of a_{23} is

$$c_{23} = (-1) \begin{vmatrix} a_{11} & a_{12} \\ a_{31} & a_{32} \end{vmatrix}.$$

Next, for an arbitrary $n \times n$ matrix A, let c_{ij} denote the cofactor of a_{ij}, $i, j = 1, \ldots, n$. It can be shown from definitions that the determinant of A is equal to the sum of the products of the elements of any column (or row) of A, each by its cofactor. Specifically,

$$det\,(A) = \sum_{i=1}^{n} a_{ij} c_{ij}, \qquad j = 1, \ldots, n, \tag{2.64}$$

or

$$det\,(A) = \sum_{j=1}^{n} a_{ij} c_{ij}, \qquad i = 1, \ldots, n. \tag{2.65}$$

For example, if A is a 2×2 matrix, we have

$$det\,(A) = \begin{vmatrix} a_{11} & a_{12} \\ a_{21} & a_{22} \end{vmatrix} = a_{11}a_{22} - a_{12}a_{21},$$

and if A is a 3×3 matrix, we have

$$det(A) = \begin{vmatrix} a_{11} & a_{12} & a_{13} \\ a_{21} & a_{22} & a_{23} \\ a_{31} & a_{32} & a_{33} \end{vmatrix} = a_{11} \begin{vmatrix} a_{22} & a_{23} \\ a_{32} & a_{33} \end{vmatrix} - a_{12} \begin{vmatrix} a_{21} & a_{23} \\ a_{31} & a_{33} \end{vmatrix} + a_{13} \begin{vmatrix} a_{21} & a_{22} \\ a_{31} & a_{32} \end{vmatrix}$$
$$= a_{11}c_{11} + a_{21}c_{21} + a_{31}c_{31}.$$

We now consider a few additional useful properties of determinants. In particular from basic definitions it can be shown that if the ith row of an $n \times n$ matrix A consists of elements of the form $a_{i1} + a'_{i1}, a_{i2} + a'_{i2}, \ldots, a_{in} + a'_{in}$, i.e., if

$$A = \begin{bmatrix} a_{11} & & a_{12} & \cdots & a_{1n} \\ a_{21} & & a_{22} & \cdots & a_{2n} \\ \cdots & \cdots & \cdots & \cdots & \cdots \\ (a_{i1} + a'_{i1}) & & (a_{i2} + a'_{i2}) & \cdots & (a_{in} + a'_{in}) \\ \cdots & \cdots & \cdots & \cdots & \cdots \\ a_{n1} & & a_{n2} & \cdots & a_{nn} \end{bmatrix},$$

then
$$det(A) = \begin{vmatrix} a_{11} & a_{12} & \cdots & a_{1n} \\ a_{21} & a_{22} & \cdots & a_{2n} \\ \cdots & \cdots & \cdots & \cdots \\ a_{i1} & a_{i2} & \cdots & a_{in} \\ \cdots & \cdots & \cdots & \cdots \\ a_{n1} & a_{n2} & \cdots & a_{nn} \end{vmatrix} + \begin{vmatrix} a_{11} & a_{12} & \cdots & a_{1n} \\ a_{21} & a_{22} & \cdots & a_{2n} \\ \cdots & \cdots & \cdots & \cdots \\ a'_{i1} & a'_{i2} & \cdots & a'_{in} \\ \cdots & \cdots & \cdots & \cdots \\ a_{n1} & a_{n2} & \cdots & a_{nn} \end{vmatrix}.$$

Next, if A and B are $n \times n$ matrices, and if B is obtained from A by adding a constant α times any column (or row) to any other column (or row) of A, then it is easily shown that $det(A) = det(B)$.

Further, if c_{ij} denotes the cofactor of a_{ij}, $i, j = 1, \ldots, n$, for an $n \times n$ matrix A, then it is easily shown that

$$\sum_{i=1}^{n} a_{ij}c_{ik} = 0 \qquad \text{for } j \neq k \tag{2.66}$$

and

$$\sum_{j=1}^{n} a_{ij}c_{kj} = 0 \qquad \text{for } i \neq k. \tag{2.67}$$

We can combine (2.64) with (2.66) and (2.65) with (2.67) to obtain the relations

$$\sum_{i=1}^{n} a_{ij}c_{i_k} = det(A)\,\delta_{jk}, \qquad j, k = 1, \ldots, n, \tag{2.68}$$

and

$$\sum_{j=1}^{n} a_{ij}c_{kj} = det(A)\,\delta_{ik}, \qquad i, k = 1, \ldots, n, \tag{2.69}$$

respectively, where δ_{mn} denotes the Kronecker delta (i.e., $\delta_{mn} = 1$ when $m = n$ and $\delta_{mn} = 0$ otherwise).

An extremely useful result concerning determinants (which can be proved by using the definition of determinant and some of the properties enumerated above) states that the determinant of the product of two matrices is equal to the product of the determinants of the matrices. Thus, if A and B are $n \times n$ matrices, then

$$det(AB) = det(A)\,det(B). \tag{2.70}$$

It is easily verified that for the $n \times n$ identity matrix I and for the $n \times n$ zero matrix O, we have $det(I) = 1$ and $det(O) = 0$.

Finally, we can readily show that an $n \times n$ matrix A is nonsingular if and only if $det(A) \neq 0$.

We conclude this discussion by considering a means of determining the inverse, A^{-1}, of a nonsingular $n \times n$ matrix. To this end, let c_{ij} be the cofactor of a_{ij}, $i, j = 1, \ldots, n$, for the matrix A, and let C be the matrix formed by the cofactors of A, i.e., $C = [c_{ij}]$. The matrix C^T is called the (*classical*) *adjoint* of A, and is denoted by $adj(A)$. It is easily verified that

$$A \cdot [adj(A)] = [adj(A)] \cdot A = [det(A)] \cdot I, \tag{2.71}$$

from which it follows that

$$A^{-1} = \frac{1}{det(A)} adj(A). \tag{2.72}$$

As a specific case, consider

$$A = \begin{bmatrix} 0 & 1 & 1 \\ 1 & 2 & 2 \\ 1 & -1 & 0 \end{bmatrix}.$$

Then $det(A) = -1$,

$$adj(A) = \begin{bmatrix} 2 & -1 & 0 \\ 2 & -1 & 1 \\ -3 & 1 & -1 \end{bmatrix},$$

and

$$A^{-1} = \begin{bmatrix} -2 & 1 & 0 \\ -2 & 1 & -1 \\ 3 & -1 & 1 \end{bmatrix}.$$

H. Solving Linear Algebraic Equations

Now consider the linear system of equations given by

$$A\alpha = \gamma, \tag{2.73}$$

where $A \in R^{m \times n}$ and $\gamma \in R^m$ are given and $\alpha \in R^n$ is to be determined. By using the results of the preceding subsections, especially Subsection 2.2D, the following important results can readily be established.

1. For a given γ, a solution α of (2.73) exists (not necessarily unique) if and only if $\gamma \in \mathfrak{R}(A)$, or equivalently, if and only if

$$\rho([A, \gamma]) = \rho(A). \tag{2.74}$$

2. A solution α of (2.73) exists for any γ if and only if

$$\rho(A) = m. \tag{2.75}$$

If (2.75) is satisfied, a solution of (2.73) can be found by using the relation

$$\alpha = A^T(AA^T)^{-1}\gamma. \tag{2.76}$$

When in (2.73), $\rho(A) = m = n$, then $A \in R^{n\times n}$ and is nonsingular and the unique solution of (2.76) is given by

$$\alpha = A^{-1}\gamma. \tag{2.77}$$

3. Every solution α of (2.73) can be expressed as a sum

$$\alpha = \alpha_p + \alpha_h, \tag{2.78}$$

where α_p is a specific solution of (2.73) and α_h satisfies $A\alpha_h = 0$. This result allows us to span the space of all solutions of (2.73). Note that there are

$$\dim \mathcal{N}(A) = n - \rho(A) \tag{2.79}$$

linearly independent solutions of the system of equations $A\beta = 0$.

EXAMPLE 2.7. Consider

$$A\alpha = \begin{bmatrix} 0 & 0 & 0 \\ 0 & 0 & 1 \\ 0 & 0 & 0 \end{bmatrix}\alpha = \gamma. \tag{2.80}$$

It is easily verified that $\{(0, 1, 0)^T\}$ is a basis for $\mathcal{R}(A)$. Since a solution of (2.80) exists if and only if $\gamma \in \mathcal{R}(A)$, γ must be of the form $\gamma = (0, k, 0)$, $k \in R$. Note that

$$\rho(A) = 1 = \rho([A, \gamma]) = rank \begin{bmatrix} 0 & 0 & 0 & \vdots & 0 \\ 0 & 0 & 1 & \vdots & k \\ 0 & 0 & 0 & \vdots & 0 \end{bmatrix},$$

as expected. To determine all solutions of (2.80), we need to determine an α_p and an α_h [see (2.78)]. In particular, $\alpha_p = (0, 0, k)^T$ will do. To determine α_h, we consider $A\beta = 0$. There are $\dim \mathcal{N}(A) = 2$ linearly independent solutions of $A\beta = 0$. In particular, $\{(1, 0, 0)^T, (0, 1, 0)^T\}$ is a basis for $\mathcal{N}(A)$. Therefore, any solution of (2.80) can be expressed as

$$\alpha = \alpha_p + \alpha_h = \begin{bmatrix} 0 \\ 0 \\ k \end{bmatrix} + \begin{bmatrix} 1 & 0 \\ 0 & 1 \\ 0 & 0 \end{bmatrix}\begin{bmatrix} c_1 \\ c_2 \end{bmatrix},$$

where c_1, c_2 are appropriately chosen real numbers. ■

We conclude by noting that an extensive discussion of determining solutions of the linear system of equations (2.73) is provided in the Appendix.

I. Equivalence and Similarity

From our previous discussion it is clear that a linear transformation $\mathcal{A}$ of a finite-dimensional vector space V into a finite-dimensional vector space W can be represented by means of different matrices, depending on the particular choice of bases in V and W. The choice of bases may in different cases result in matrices that are easy or hard to utilize. Many of the resulting "standard" forms of matrices, called *canonical forms,* arise because of practical considerations. Such canonical forms often exhibit inherent characteristics of the underlying transformation $\mathcal{A}$.

Throughout the present subsection, V and W are finite-dimensional vector spaces over the same field F, $\dim V = n$, and $\dim W = m$.

Change of bases: Vector case

Our first aim will be to consider the change of bases in the coordinate representation of vectors. Let $\{v^1, \ldots, v^n\}$ be a basis for V and let $\{\bar{v}^1, \ldots, \bar{v}^n\}$ be a set of vectors in V given by

$$\bar{v}^i = \sum_{j=1}^{n} p_{ji} v^j, \qquad i = 1, \ldots, n, \tag{2.81}$$

where $p_{ij} \in F$ for all $i, j = 1, \ldots, n$. It is easily verified that the set $\{\bar{v}^1, \ldots, \bar{v}^n\}$ forms a basis for V if and only if the $n \times n$ matrix $P = [p_{ij}]$ is nonsingular. We call *P the matrix of the basis $\{\bar{v}^1, \ldots, \bar{v}^n\}$ with respect to the basis $\{v^1, \ldots, v^n\}$*. Note that the ith column of P is the coordinate representation of $\bar{v}^i$ with respect to the basis $\{v^1, \ldots, v^n\}$.

Continuing the above discussion, let $\{v^1, \ldots, v^n\}$ and $\{\bar{v}^1, \ldots, \bar{v}^n\}$ be two bases for V, and let P be the matrix of the basis $\{\bar{v}^1, \ldots, \bar{v}^n\}$ with respect to the basis $\{v^1, \ldots, v^n\}$. Then it is easily shown that P^{-1} is the matrix of the basis $\{v^1, \ldots, v^n\}$ with respect to the basis $\{\bar{v}^1, \ldots, \bar{v}^n\}$.

Next, let the sets of vectors $\{v^1, \ldots, v^n\}$, $\{\bar{v}^1, \ldots, \bar{v}^n\}$, and $\{\tilde{v}^1, \ldots, \tilde{v}^n\}$ be bases for V. If P is the matrix of the basis $\{\bar{v}^1, \ldots, \bar{v}^n\}$ with respect to the basis $\{v^1, \ldots, v^n\}$ and if Q is the matrix of the basis $\{\tilde{v}^1, \ldots, \tilde{v}^n\}$ with respect to the basis $\{\bar{v}^1, \ldots, \bar{v}^n\}$, then it is easily verified that PQ is the matrix of the basis $\{\tilde{v}^1, \ldots, \tilde{v}^n\}$ with respect to the basis $\{v^1, \ldots, v^n\}$.

Continuing further, let $\{v^1, \ldots, v^n\}$ and $\{\bar{v}^1, \ldots, \bar{v}^n\}$ be two bases for V and let P be the matrix of the basis $\{\bar{v}^1, \ldots, \bar{v}^n\}$ with respect to the basis $\{v^1, \ldots, v^n\}$. Let $a \in V$ and let $\alpha^T = (\alpha_1, \ldots, \alpha_n)$ denote the coordinate representation of a with respect to the basis $\{v^1, \ldots, v^n\}$ (i.e., $a = \sum_{i=1}^{n} \alpha_i v^i$). Let $\bar{\alpha}^T = (\bar{\alpha}_1, \ldots, \bar{\alpha}_n)$ denote the coordinate representation of a with respect to the basis $\{\bar{v}^1, \ldots, \bar{v}^n\}$. Then it is readily verified that

$$P\bar{\alpha} = \alpha.$$

EXAMPLE 2.8. Let $V = R^3$ and $F = R$, and let $a = (1, 2, 3)^T \in R^3$ be given. Let $\{v^1, v^2, v^3\} = \{e^1, e^2, e^3\}$ denote the natural basis for R^3, i.e., $e^1 = (1, 0, 0)^T$, $e^2 = (0, 1, 0)^T$, $e^3 = (0, 0, 1)^T$. Clearly, the coordinate representation α of a with respect to the natural basis is $(1, 2, 3)^T$.

Now let $\{\bar{v}^1, \bar{v}^2, \bar{v}^3\}$ be another basis for R^3, given by $\bar{v}^1 = (1, 0, 1)^T$, $\bar{v}^2 = (0, 1, 0)^T$, $\bar{v}^3 = (0, 1, 1)^T$. From the relation

$$\begin{aligned}(1, 0, 1)^T = \bar{v}^1 &= p_{11}v^1 + p_{21}v^2 + p_{31}v^3 \\ &= p_{11}\begin{bmatrix}1\\0\\0\end{bmatrix} + p_{21}\begin{bmatrix}0\\1\\0\end{bmatrix} + p_{31}\begin{bmatrix}0\\0\\1\end{bmatrix}\end{aligned}$$

we conclude that $p_{11} = 1$, $p_{21} = 0$, and $p_{31} = 1$. Similarly, from

$$\begin{aligned}(0, 1, 0)^T = \bar{v}^2 &= p_{12}v^1 + p_{22}v^2 + p_{32}v^3 \\ &= p_{12}\begin{bmatrix}1\\0\\0\end{bmatrix} + p_{22}\begin{bmatrix}0\\1\\0\end{bmatrix} + p_{32}\begin{bmatrix}0\\0\\1\end{bmatrix}\end{aligned}$$

we conclude that $p_{12} = 0$, $p_{22} = 1$, and $p_{32} = 0$. Finally, from the relation

$$(0, 1, 1)^T = \bar{v}^3 = p_{13}\begin{bmatrix}1\\0\\0\end{bmatrix} + p_{23}\begin{bmatrix}0\\1\\0\end{bmatrix} + p_{33}\begin{bmatrix}0\\0\\1\end{bmatrix}$$

we obtain that $p_{13} = 0$, $p_{23} = 1$, and $p_{33} = 1$.

The matrix $P = [p_{ij}]$ of the basis $\{\bar{v}^1, \bar{v}^2, \bar{v}^3\}$ with respect to the basis $\{v^1, v^2, v^3\}$ is therefore determined to be

$$P = \begin{bmatrix}1 & 0 & 0\\0 & 1 & 1\\1 & 0 & 1\end{bmatrix},$$

and the coordinate representation of a with respect to the basis $\{\bar{v}^1, \bar{v}^2, \bar{v}^3\}$ is given by $\bar{\alpha} = P^{-1}\alpha$, or

$$\bar{\alpha} = \begin{bmatrix}1 & 0 & 0\\0 & 1 & 1\\1 & 0 & 1\end{bmatrix}^{-1}\begin{bmatrix}1\\2\\3\end{bmatrix} = \begin{bmatrix}1 & 0 & 0\\1 & 1 & -1\\-1 & 0 & 0\end{bmatrix}\begin{bmatrix}1\\2\\3\end{bmatrix} = \begin{bmatrix}1\\0\\2\end{bmatrix}.$$

■

Change of bases: Matrix case

Having addressed the relationship between the coordinate representations of a given vector with respect to different bases, we next consider the relationship between the matrix representations of a given linear transformation relative to different bases. To this end let $\mathcal{A} \in L(V, W)$ and let $\{v^1, \ldots, v^n\}$ and $\{w^1, \ldots, w^m\}$ be bases for V and W, respectively. Let A be the matrix of $\mathcal{A}$ with respect to the bases $\{v^1, \ldots, v^n\}$ and $\{w^1, \ldots, w^m\}$. Let $\{\bar{v}^1, \ldots, \bar{v}^n\}$ be another basis for V, and let the matrix of $\{\bar{v}^1, \ldots, \bar{v}^n\}$ with respect to $\{v^1, \ldots, v^n\}$ be P. Let $\{\bar{w}^1, \ldots, \bar{w}^m\}$ be another basis for W, and let Q be the matrix of $\{w^1, \ldots, w^m\}$ with respect to $\{\bar{w}^1, \ldots, \bar{w}^m\}$. Let $\bar{A}$ be the matrix of $\mathcal{A}$ with respect to the bases $\{\bar{v}^1, \ldots, \bar{v}^n\}$ and $\{\bar{w}^1, \ldots, \bar{w}^m\}$. Then it is readily verified that

$$\bar{A} = QAP. \tag{2.82}$$

This result is depicted schematically in Fig. 2.1.

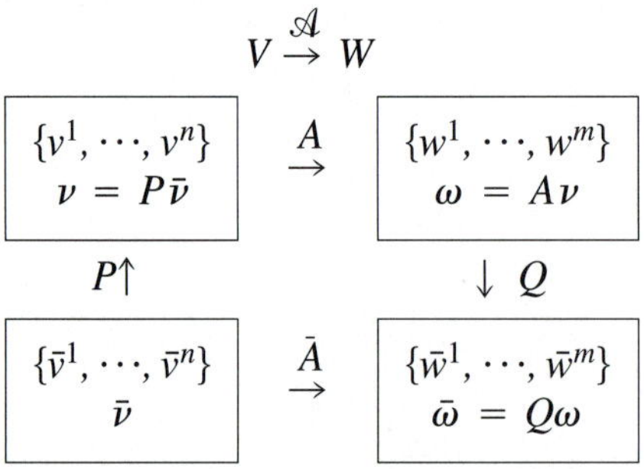

$V \overset{\mathcal{A}}{\rightarrow} W$

$\{v^1, \cdots, v^n\}$ $v = P\bar{v}$	$\overset{A}{\rightarrow}$	$\{w^1, \cdots, w^m\}$ $\omega = Av$
$P\uparrow$		$\downarrow Q$
$\{\bar{v}^1, \cdots, \bar{v}^n\}$ $\bar{v}$	$\overset{\bar{A}}{\rightarrow}$	$\{\bar{w}^1, \cdots, \bar{w}^m\}$ $\bar{\omega} = Q\omega$

FIGURE 2.1
Schematic diagram of the equivalence of two matrices

Equivalence of matrices

The preceding discussion motivates the following definition.

DEFINITION 2.4. An $m \times n$ matrix $\bar{A}$ is said to be *equivalent* to an $m \times n$ matrix A if there exists an $m \times m$ nonsingular matrix Q and an $n \times n$ nonsingular matrix P such that (2.82) is true. If $\bar{A}$ is equivalent to A, we write $\bar{A} \sim A$. ■

Thus, an $m \times n$ matrix $\bar{A}$ is equivalent to an $m \times n$ matrix A if and only if A and $\bar{A}$ can be interpreted as both being matrices of the same linear transformation $\mathscr{A}$ of a linear space V into a linear space W, but with respect to possibly different choices of bases.

It is clear that a matrix A is always equivalent to itself (i.e., $A \sim A$). Also, if a matrix A is equivalent to a matrix B, then clearly B is equivalent to A (i.e., if $A \sim B$, then $B \sim A$). Furthermore, if A is equivalent to B and B is equivalent to a matrix C, then it is evident that A is equivalent to C (i.e., if $A \sim B$ and $B \sim C$, then $A \sim C$). This shows that $\sim$ is an equivalence relation.

The reader can easily verify that every $m \times n$ matrix is equivalent to a matrix of the form

$$\begin{bmatrix} 100 & \cdots & \cdots & \cdots & 0 \\ 010 & \cdots & \cdots & \cdots & 0 \\ \cdots & \cdots & \cdots & \cdots & \cdots \\ 000 & \cdots & 100 & \cdots & 0 \\ 000 & \cdots & 000 & \cdots & 0 \\ \cdots & \cdots & \cdots & \cdots & \cdots \\ 000 & \cdots & 000 & \cdots & 0 \end{bmatrix} \left.\begin{matrix} \\ \\ \\ \\ \end{matrix}\right\} r = rank\ A$$

From this it follows that two $m \times n$ matrices A and B are equivalent if and only if they have the same rank, and furthermore, that A and A^T have the same rank.

The definition of rank of a matrix that we used in Section 2.2 is sometimes called the *column rank of a matrix*. Sometimes, an analogous definition for *row rank of a matrix* is also used. The result given in the above paragraph shows that *row rank of a matrix is equal to its column rank.*

Similarity of matrices

Next, let $V = W$, let $\mathscr{A} \in L(V, V)$, let $\{v^1, \ldots, v^n\}$ be a basis for V, and let A be the matrix of $\mathscr{A}$ with respect to $\{v^1, \ldots, v^n\}$. Let $\{\bar{v}^1, \ldots, \bar{v}^n\}$ be another basis for V whose matrix with respect to $\{v^1, \ldots, v^n\}$ is P. Let $\bar{A}$ be the matrix of $\mathscr{A}$ with respect to $\{\bar{v}^1, \ldots, \bar{v}^n\}$. Then it follows immediately from (2.82) that

$$\bar{A} = P^{-1}AP. \tag{2.83}$$

The meaning of this result is depicted schematically in Fig. 2.2.

This discussion motivates the following definition.

$$\begin{array}{ccc} V & \overset{\mathscr{A}}{\rightarrow} & V \\ \{v^1, \ldots, v^n\} & \overset{A}{\rightarrow} & \{v^1, \ldots, v^n\} \\ \uparrow P & & \downarrow P^{-1} \\ \{\bar{v}^1, \ldots, \bar{v}^n\} & \overset{\bar{A}}{\rightarrow} & \{\bar{v}^1, \ldots, \bar{v}^n\} \end{array}$$

FIGURE 2.2
Schematic diagram of the similarity of two matrices

DEFINITION 2.5. An $n \times n$ matrix $\bar{A}$ is said to be *similar* to an $n \times n$ matrix A if there exists an $n \times n$ nonsingular matrix P such that

$$\bar{A} = P^{-1}AP.$$

If $\bar{A}$ is similar to A, we write $\bar{A} \sim A$. We call P a *similarity transformation.* ■

It is easily verified that if $\bar{A}$ is similar to A [i.e., (2.83) is true], then A is similar to $\bar{A}$, i.e.,

$$A = P\bar{A}P^{-1}. \tag{2.84}$$

In view of this, there is no ambiguity in saying "two matrices are similar," and we could just as well have used (2.84) [in place of (2.83)] to define similarity of matrices.

To sum up, if two matrices A and $\bar{A}$ represent the same linear transformation $\mathcal{A} \in L(V, V)$, possibly with respect to two different bases for V, then A and $\bar{A}$ are similar matrices.

Since the similarity of two matrices is a special case of the equivalence of matrices, it follows that (i) a matrix A is similar to A; (ii) if A is similar to a matrix B, then B is similar to A; and (iii) if A is similar to B and B is similar to a matrix C, then A is similar to C. Therefore, the similarity relation of matrices is an equivalence relation.

Now let A be an $n \times n$ matrix that is similar to a matrix B. Then it is easily shown that A^k is similar to B^k, where k denotes a positive integer, i.e., $B^k = P^{-1}A^kP$. This can be extended further by letting

$$f(\lambda) = \sum_{i=0}^{m} \alpha_i \lambda^i = \alpha_0 + \alpha_1 \lambda + \cdots + \alpha_m \lambda^m \tag{2.85}$$

and by verifying that

$$f(P^{-1}AP) = P^{-1}f(A)P, \tag{2.86}$$

where $\alpha_0, \ldots, \alpha_m \in F$. This shows that if B is similar to A, then $f(B)$ is similar to $f(A)$, where in fact the same similarity transformation P is involved. Further, if $\bar{A}$ is similar to A and if $f(\lambda)$ is as given in (2.85), then $f(A) = O$ if and only if $f(\bar{A}) = O$.

Next, let $\mathcal{A} \in L(V, V)$ and let A be the matrix of $\mathcal{A}$ with respect to a basis $\{v^1, \ldots, v^n\}$ in V. Let $f(\lambda)$ denote the polynomial given in (2.85) and let A be any matrix of $\mathcal{A}$. Then it is readily verified that $f(\mathcal{A}) = \mathbb{O}$ if and only if $f(A) = O$.

We can use results such as the preceding to good advantage. For example, let $\bar{A}$ denote the *diagonal matrix* given by

$$\bar{A} = \begin{bmatrix} \lambda_1 & 0 & 0 & \cdots & 0 & 0 \\ 0 & \lambda_2 & 0 & \cdots & 0 & 0 \\ \vdots & \vdots & \vdots & & \vdots & \vdots \\ 0 & 0 & 0 & \cdots & \lambda_{n-1} & 0 \\ 0 & 0 & 0 & \cdots & 0 & \lambda_n \end{bmatrix}.$$

Then

$$(\bar{A})^k = \begin{bmatrix} \lambda_1^k & 0 & 0 & \cdots & 0 & 0 \\ 0 & \lambda_2^k & 0 & \cdots & 0 & 0 \\ \vdots & \vdots & \vdots & & \vdots & \vdots \\ 0 & 0 & 0 & \cdots & \lambda_{n-1}^k & 0 \\ 0 & 0 & 0 & \cdots & 0 & \lambda_n^k \end{bmatrix}.$$

Now let $f(\lambda)$ be given by (2.85). Then

$$f(\bar{A}) = \alpha_0 \begin{bmatrix} 1 & 0 & \cdots & \cdots & 0 \\ 0 & 1 & \cdots & \cdots & 0 \\ \vdots & \vdots & & \vdots & \vdots \\ 0 & 0 & \cdots & 1 & 0 \\ 0 & 0 & \cdots & 0 & 1 \end{bmatrix} + \alpha_1 \begin{bmatrix} \lambda_1 & 0 & \cdots & \cdots & 0 \\ 0 & \lambda_2 & \cdots & \cdots & 0 \\ \vdots & \vdots & & \vdots & \vdots \\ 0 & 0 & \cdots & \lambda_{n-1} & 0 \\ 0 & 0 & \cdots & 0 & \lambda_n \end{bmatrix} + \cdots$$

$$+ \alpha_m \begin{bmatrix} \lambda_1^m & 0 & \cdots & \cdots & 0 \\ 0 & \lambda_2^m & \cdots & \cdots & 0 \\ \vdots & \vdots & & \vdots & \vdots \\ 0 & 0 & \cdots & \lambda_{n-1}^m & 0 \\ 0 & 0 & \cdots & 0 & \lambda_n^m \end{bmatrix} = \begin{bmatrix} f(\lambda_1) & 0 & \cdots & \cdots & 0 \\ 0 & f(\lambda_2) & \cdots & \cdots & 0 \\ \vdots & \vdots & & \vdots & \vdots \\ 0 & 0 & \cdots & f(\lambda_{n-1}) & 0 \\ 0 & 0 & \cdots & 0 & f(\lambda_n) \end{bmatrix}.$$

Next, let $\mathcal{A} \in L(V, V)$, let A be the matrix of $\mathcal{A}$ with respect to a basis $\{v^1, \ldots, v^n\}$ in V, and let $\bar{A}$ be the matrix of $\mathcal{A}$ with respect to another basis $\{\bar{v}^1, \ldots, \bar{v}^n\}$ in V. Then it is easily verified that $det(A) = det(\bar{A})$. From this it follows that for any two similar matrices A and B, we have $det(A) = det(B)$.

In view of these results, there is no ambiguity in defining the *determinant of a linear transformation* $\mathcal{A}$ of a finite-dimensional vector space V into V as the determinant of any matrix A representing it, i.e., $det(\mathcal{A}) = det(A)$.

J. Eigenvalues and Eigenvectors

Definitions

Throughout this subsection, V denotes an n-dimensional vector space over a field F.

Let $\mathcal{A} \in L(V, V)$ and let us assume that there exist sets of vectors $\{v^1, \ldots, v^n\}$ and $\{\bar{v}^1, \ldots, \bar{v}^n\}$ that are bases for V such that

$$\begin{aligned} \bar{v}^1 &= \mathcal{A}v^1 = \lambda_1 v^1 \\ \bar{v}^2 &= \mathcal{A}v^2 = \lambda_2 v^2 \\ &\cdots\cdots\cdots\cdots \\ \bar{v}^n &= \mathcal{A}v^n = \lambda_n v^n, \end{aligned}$$

where $\lambda_i \in F, i = 1, \ldots, n$. If this is the case, then the matrix $\bar{A}$ of $\mathcal{A}$ with respect to the given bases is

$$\bar{A} = \begin{bmatrix} \lambda_1 & & & 0 \\ & \lambda_2 & & \\ & & \ddots & \\ 0 & & & \lambda_n \end{bmatrix}.$$

This motivates the following result that is easily verified: for $\mathcal{A} \in L(V, V)$ and $\lambda \in F$, the set of all $v \in V$ such that

$$\mathcal{A}v = \lambda v \tag{2.87}$$

is a linear subspace of V. In fact, it is the null space of the linear transformation $(\mathcal{A} - \lambda\mathcal{I})$, where $\mathcal{I}$ is the identity element of $L(V, V)$. Henceforth, we let

$$\mathcal{N}_\lambda = \{v \in V : (\mathcal{A} - \lambda\mathcal{I})v = 0\}. \tag{2.88}$$

The above gives rise to several important concepts that we introduce in the following definition.

DEFINITION 2.6. A scalar λ such that $\mathcal{N}_\lambda$ [given in (2.88)] contains more than just the zero vector is called an *eigenvalue* of $\mathcal{A}$ (i.e., if there is a $v \neq 0$ such that $\mathcal{A}v = \lambda v$, then λ is called an eigenvalue of $\mathcal{A}$). When λ is an eigenvalue of $\mathcal{A}$, then each $v \neq 0$ in $\mathcal{N}_\lambda$ is called an *eigenvector* of $\mathcal{A}$ corresponding to the eigenvalue λ. The dimension of the linear subspace $\mathcal{N}_\lambda$ is called the (geometric) *multiplicity of the eigenvalue* λ. If $\mathcal{N}_\lambda$ is of dimension one, then λ is called a *simple eigenvalue*. The set of all eigenvalues of $\mathcal{A}$ is called the *spectrum* of $\mathcal{A}$. ■

Other names for eigenvalue that are in use include *proper value, characteristic value, latent value,* or *secular value*. Similarly, other names for eigenvectors are *proper vector* or *characteristic vector*. The space $\mathcal{N}_\lambda$ is called the λth *proper subspace* of V. For matrices, we give the following corresponding definition.

DEFINITION 2.7. Let A be an $n \times n$ matrix whose elements belong to the field F. If there exists $\lambda \in F$ and a nonzero vector $\alpha \in F^n$ such that

$$A\alpha = \lambda\alpha, \tag{2.89}$$

then λ is called an *eigenvalue* of A and α is called an *eigenvector* of A corresponding to the eigenvalue λ. ■

The connection between Definitions 2.6 and 2.7 is given in the following result that the reader can verify easily: let $\mathcal{A} \in L(V, V)$ and let A be the matrix of $\mathcal{A}$ with respect to the basis $\{v^1, \dots, v^n\}$. Then λ is an eigenvalue of $\mathcal{A}$ if and only if λ is an eigenvalue of A. Also, $a \in V$ is an eigenvector of $\mathcal{A}$ corresponding to λ if and only if the coordinate representation of a with respect to the basis $\{v^1, \dots, v^n\}$, α, is an eigenvector of A corresponding to λ.

We note that if a (or α) is an eigenvector of $\mathcal{A}$ (of A), then any nonzero multiple of a (of α) is also an eigenvector of $\mathcal{A}$ (of A).

In the case of matrices, in place of (2.89), one can also consider the relationship

$$\hat{\alpha}A = \lambda\hat{\alpha}, \tag{2.90}$$

where $\hat{\alpha}$ denotes a $1 \times n$ row vector. In this context, α in (2.89) and $\hat{\alpha}$ in (2.90) are referred to as a *right eigenvector* and a *left eigenvector*, respectively. Unless explicitly stated, we will have in mind a right eigenvector when using the term eigenvector of a matrix.

Now let $\mathcal{A} \in L(V, V)$ and let A denote the matrix of $\mathcal{A}$ with respect to the basis $\{v^1, \dots, v^n\}$ in V. Then it is easily shown that $\lambda \in F$ is an eigenvalue of $\mathcal{A}$ (and hence, of A) if and only if $det\,(\mathcal{A} - \lambda\mathcal{I}) = 0$, or equivalently, if and only if $det\,(A - \lambda I) = 0$.

Characteristic polynomial

The above result enables us to determine the eigenvalues of $\mathcal{A}$ (or A) in a systematic manner. So let us examine the equation

$$det\,(\mathcal{A} - \lambda\mathcal{I}) = 0 \tag{2.91}$$

or equivalently, the equation

$$det(A - \lambda I) = 0 \tag{2.92}$$

in terms of the parameter λ. We first rewrite (2.92) as

$$det(A - \lambda I) = \begin{vmatrix} (a_{11} - \lambda) & a_{12} & \cdots & a_{1n} \\ a_{21} & (a_{22} - \lambda) & \cdots & a_{2n} \\ \vdots & \vdots & \cdots & \vdots \\ a_{n1} & a_{n2} & \cdots & (a_{nn} - \lambda) \end{vmatrix}. \tag{2.93}$$

It is clear from (2.62) that the expansion of the determinant (2.93) yields a polynomial in λ of degree n. For λ to be an eigenvalue of $\mathscr{A}$ (or A) it must satisfy (2.91) [or (2.92)] and it must belong to F. Note that in general we have no assurance that the nth-degree polynomial given by (2.92) has any roots in F. There is, however, a special class of fields for which this requirement is automatically satisfied: a field F is said to be *algebraically closed* if for every polynomial $p(\lambda)$ there is at least one $\lambda \in F$ such that

$$p(\lambda) = 0. \tag{2.94}$$

Any λ that satisfies (2.94) is said to be a *root* of the polynomial equation (2.94).

In particular the field of complex numbers is algebraically closed, whereas the field of real numbers is not (e.g., consider the equation $\lambda^2 + 1 = 0$). There are other fields besides the field of complex numbers that are algebraically closed. However, since we will not require these, we will restrict ourselves to the field of complex numbers, C, whenever the algebraic closure property is required. When considering results that are valid for a vector space over an arbitrary field, we will, as before, make use of the symbol F, or make no reference to F at all.

Summarizing the above discussion, we have the following result. Let $\mathscr{A} \in L(V, V)$ and let A be the matrix of $\mathscr{A}$ with respect to the basis $\{v^1, \ldots, v^n\}$ in V. Then (i) $det(\mathscr{A} - \lambda\mathscr{I}) = det(A - \lambda I)$ is a polynomial of degree n in the parameter λ, i.e., there exist scalars $\alpha_0, \alpha_1, \ldots, \alpha_n$, depending on $\mathscr{A}$ (and therefore on A) such that

$$det(\mathscr{A} - \lambda\mathscr{I}) = det(A - \lambda I) = \alpha_0 + \alpha_1\lambda + \alpha_2\lambda^2 + \cdots + \alpha_n\lambda^n \tag{2.95}$$

[note that $\alpha_0 = det(\mathscr{A})$ and $\alpha_n = (-1)^n$]; (ii) the eigenvalues of $\mathscr{A}$ are precisely the roots of the equation

$$det(\mathscr{A} - \lambda\mathscr{I}) = det(A - \lambda I) = \alpha_0 + \alpha_1\lambda + \alpha_2\lambda^2 + \cdots + \alpha_n\lambda^n = 0; \tag{2.96}$$

and (iii) $\mathscr{A}$ has, at most, n distinct eigenvalues.

We call (2.95) the *characteristic polynomial of* $\mathscr{A}$ (or of A) and call (2.96) *the characteristic equation of* $\mathscr{A}$ (or of A).

An important remark concerning notation

The above definition of characteristic polynomial is the one usually used in texts on linear algebra and matrix theory (refer, e.g., to some of the books on this subject cited at the end of this chapter). An alternative to the above definition is given by the expression

$$\alpha(\lambda) \triangleq det(\lambda\mathscr{I} - A) = det(\lambda I - A).$$

One of the reasons for using this convention is that this polynomial arises in a natural manner when solving systems of ordinary differential equations by operator methods [e.g., system (LH)]. Since the reader may have many occasions to consult texts on linear algebra and matrix theory, we will employ in *this section* the definition given in (2.95). Throughout the remainder of this book, however, we will follow the convention used in linear systems texts by utilizing the expression $det(\lambda I - A)$ for the characteristic polynomial. We note that because of the relationship

$$det(A - \lambda I) = (-1)^n \, det(\lambda I - A),$$

either definition can be used to develop the results considered herein. Note that $det(\lambda I - A)$ is a monic polynomial, i.e., its leading coefficient equals 1.

From the fundamental properties of polynomials over the field of complex numbers, there now follows the next important result: if V is an n-dimensional vector space over C and if $\mathscr{A} \in L(V, V)$, then it is possible to write the characteristic polynomial of $\mathscr{A}$ in the form

$$det\,(\mathscr{A} - \lambda \mathscr{I}) = (\lambda_1 - \lambda)^{m_1}(\lambda_2 - \lambda)^{m_2} \cdots (\lambda_p - \lambda)^{m_p}, \tag{2.97}$$

where λ_i, $i = 1, \ldots, p$, are the distinct roots of (2.96) (i.e., $\lambda_i \neq \lambda_j$, if $i \neq j$). In (2.97), m_i is called the *algebraic multiplicity* of the root λ_i. The m_i are positive integers, and $\sum_{i=1}^{p} m_i = n$.

The reader should make note of the distinction between the concept of algebraic multiplicity of λ_i, given above, and the (geometric) multiplicity of an eigenvalue λ_i, given earlier. In general these need not be the same, as will be seen later.

The Cayley-Hamilton Theorem and applications

We now state and prove a result that is very important in linear systems theory.

THEOREM 2.1. (CAYLEY-HAMILTON THEOREM) Every square matrix satisfies its characteristic equation. More specifically, if A is an $n \times n$ matrix and $p(\lambda) = det(A - \lambda I)$ is the characteristic polynomial of A, then $p(A) = O$.

Proof. Let the characteristic polynomial for A be $p(\lambda) = \alpha_0 + \alpha_1\lambda + \cdots + \alpha_n\lambda^n$ and let $B(\lambda) = [b_{ij}(\lambda)]$ be the classical adjoint of $(A - \lambda I)$ (refer to Subsection 2.2G). Since the $b_{ij}(\lambda)$ are cofactors of the matrix $A - \lambda I$, they are polynomials in λ of degree not more than $n - 1$. Thus, $b_{ij}(\lambda) = \beta_{ij0} + \beta_{ij1}\lambda + \cdots + \beta_{ij(n-1)}\lambda^{n-1}$. Letting $B_k = [\beta_{ijk}]$ for $k = 0, 1, \ldots, n-1$, we have $B(\lambda) = B_0 + \lambda B_1 + \cdots + \lambda^{n-1}B_{n-1}$. By (2.71), we have $(A - \lambda I)B(\lambda) = [det(A - \lambda I)]I$. Thus, $(A - \lambda I)[B_0 + \lambda B_1 + \cdots + \lambda^{n-1}B_{n-1}] = (\alpha_0 + \alpha_1\lambda + \cdots + \alpha_n\lambda^n)I$. Expanding the left-hand side of this equation and equating like powers of λ, we have $-B_{n-1} = \alpha_n I$, $AB_{n-1} - B_{n-2} = \alpha_{n-1}I, \ldots, AB_1 - B_0 = \alpha_1 I$, $AB_0 = \alpha_0 I$. Premultiplying the above matrix equations by $A^n, A^{n-1}, \ldots, A, I$, respectively, we have $-A^nB_{n-1} = \alpha_n A^n$, $A^nB_{n-1} - A^{n-1}B_{n-2} = \alpha_{n-1}A^{n-1}, \ldots, A^2B_1 - AB_0 = \alpha_1 A$, $AB_0 = \alpha_0 I$. Adding these matrix equations, we obtain $O = \alpha_0 I + \alpha_1 A + \cdots + \alpha_n A^n = p(A)$, which was to be shown. ∎

As an immediate consequence of the Cayley-Hamilton Theorem, we have the following results: let A be an $n \times n$ matrix with characteristic polynomial given by (2.96). Then (i) $A^n = (-1)^{n+1}[\alpha_0 I + \alpha_1 A + \cdots + \alpha_{n-1}A^{n-1}]$; and (ii) if $f(\lambda)$ is any polynomial in λ, then there exist $\beta_0, \beta_1, \ldots, \beta_{n-1} \in F$ such that

$$f(A) = \beta_0 I + \beta_1 A + \cdots + \beta_{n-1}A^{n-1}. \tag{2.98}$$

Part (i) follows from the Cayley-Hamilton Theorem and from the fact that $\alpha_n = (-1)^n$. To prove part (ii), let $f(\lambda)$ be any polynomial in λ and let $p(\lambda)$ denote the characteristic polynomial of A. From a result for polynomials (called the *division algorithm*), we know that there exist two unique polynomials $g(\lambda)$ and $r(\lambda)$ such that

$$f(\lambda) = p(\lambda)g(\lambda) + r(\lambda), \tag{2.99}$$

where the degree of $r(\lambda) \leq n-1$. Now since $p(A) = O$, we have that $f(A) = r(A)$ and the result follows.

Finally, we also note that if $\mathcal{A} \in L(V, V)$ and if $p(\lambda)$ denotes the characteristic polynomial of $\mathcal{A}$, then $p(\mathcal{A}) = \mathbb{O}$.

As a specific application of the Cayley-Hamilton Theorem, we evaluate the matrix A^{37}, where $A = \begin{bmatrix} 1 & 0 \\ 1 & 2 \end{bmatrix}$. Since $n = 2$, we assume, in view of (2.98), that A^{37} is of the form $A^{37} = \beta_0 I + \beta_1 A$. The characteristic polynomial of A is $p(\lambda) = (1-\lambda)(2-\lambda)$ and the eigenvalues of A are $\lambda_1 = 1$ and $\lambda_2 = 2$. In the present case $f(\lambda) = \lambda^{37}$ and $r(\lambda)$ in (2.99) is $r(\lambda) = \beta_0 + \beta_1\lambda$. To determine β_0 and β_1 we use the fact that $p(\lambda_1) = p(\lambda_2) = 0$ to conclude that $f(\lambda_1) = r(\lambda_1)$ and $f(\lambda_2) = r(\lambda_2)$. Therefore, we have that $\beta_0 + \beta_1 = 1^{37} = 1$ and $\beta_0 + 2\beta_1 = 2^{37}$. Hence, $\beta_1 = 2^{37} - 1$ and $\beta_0 = 2 - 2^{37}$. Therefore, $A^{37} = (2 - 2^{37})I + (2^{37} - 1)A$, or $A^{37} = \begin{bmatrix} 1 & 0 \\ 2^{37} - 1 & 2^{37} \end{bmatrix}$.

The Cayley-Hamilton Theorem can also be used to express matrix-valued power series (as well as other kinds of functions) as matrix polynomials of degree $n-1$. Consider in particular the matrix exponential e^{At} defined by

$$e^{At} = \sum_{k=0}^{\infty} \left(\frac{t^k}{k!}\right) A^k, \qquad t \in (-a, a). \tag{2.100}$$

In view of the Cayley-Hamilton Theorem, we can write

$$f(A) = e^{At} = \sum_{i=0}^{n-1} \alpha_i(t) A^i. \tag{2.101}$$

In the following, we present a method to determine the coefficients $\alpha_i(t)$ in (2.101) [or β_i in (2.98)].

In accordance with (2.97), let $p(\lambda) = det(A - \lambda I) = \prod_{i=1}^{p}(\lambda_i - \lambda)^{m_i}$ be the characteristic polynomial of A. Also, let $f(\lambda)$ and $g(\lambda)$ be two analytic functions. Now if

$$f^{(l)}(\lambda_i) = g^{(l)}(\lambda_i), \qquad l = 0, \ldots, m_i - 1, i = 1, \ldots, p, \tag{2.102}$$

where $f^{(l)}(\lambda_i) = (d^l f / d\lambda^l)(\lambda)|_{\lambda=\lambda_i}$, $\sum_{i=1}^{p} m_i = n$, then $f(A) = g(A)$. To see this, we note that condition (2.102) written as $(f-g)^l(\lambda_i) = 0$ implies that $f(\lambda) - g(\lambda)$ has $p(\lambda)$ as a factor, i.e., $f(\lambda) - g(\lambda) = w(\lambda)p(\lambda)$ for some analytic function $w(\lambda)$. From the Cayley-Hamilton Theorem we have that $p(A) = O$ and therefore $f(A) - g(A) = O$.

EXAMPLE 2.9. Let $A = \begin{bmatrix} -1 & 1 \\ -1 & 1 \end{bmatrix}$, and let $f(A) = e^{At}$, $f(\lambda) = e^{\lambda t}$, and $g(\lambda) = \alpha_1\lambda + \alpha_0$. The matrix A has an eigenvalue $\lambda = \lambda_1 = \lambda_2 = 0$ with multiplicity $m_1 = 2$. Conditions (2.102) are given by $f(\lambda_1) = g(\lambda_1) = 1$ and $f^{(1)}(\lambda_1) = g^{(1)}(\lambda_1)$ and imply that

$\alpha_0 = 1$ and $\alpha_1 = t$. Therefore,

$$e^{At} = f(A) = g(A) = \alpha_1 A + \alpha_0 I = \begin{bmatrix} -\alpha_1 + \alpha_0 & \alpha_1 \\ -\alpha_1 & \alpha_1 + \alpha_0 \end{bmatrix} = \begin{bmatrix} 1-t & t \\ -t & 1+t \end{bmatrix}. \quad \blacksquare$$

In the final result of this section we let $F = C$, we let A be an $n \times n$ matrix of $\mathcal{A} \in L(V, V)$, and we let $det\,(A - \lambda I) = det\,(\mathcal{A} - \lambda \mathcal{I})$ be given by (2.97). It is readily verified that (i) $det(A) = \prod_{j=1}^{p} \lambda_j^{mj}$, (ii) trace $(A) \triangleq \sum_{i=1}^{n} a_{ii} = \sum_{j=1}^{p} m_j \lambda_j$, (iii) if B is any matrix similar to A, then trace (B) = trace (A), and (iv) if $f(\lambda)$ denotes the polynomial $f(\lambda) = \gamma_0 + \gamma_1 \lambda + \cdots + \gamma_m \lambda^m$, then the roots of the characteristic polynomial of $f(A)$ are $f(\lambda_1), \ldots, f(\lambda_p)$, and $det\,[f(A) - \lambda I] = [f(\lambda_1) - \lambda]^{m_1} \cdots [f(\lambda_p) - \lambda]^{m_p}$.

K. Direct Sums of Linear Subspaces

One of the important topics in matrix theory is the development of canonical forms, including the lower (upper) triangular form, the block diagonal form, and the Jordan canonical form of matrices. Before presenting these, we need to address additional topics which are of interest and importance to us in their own right. One of these concerns direct sums of linear subspaces and linear transformations defined on such sums.

Let V be a linear space and let W and U be arbitrary subsets of V. The *sum* of sets W and U, denoted by $W + U$, is the set of all vectors in V that are of the form $w + u$, where $w \in W$ and $u \in U$. If in particular, W and U are *linear subspaces* of V, then it is easily shown that $W + U$ is also a linear subspace. If W and U are linear subspaces of V, and if $W \cap U = \{0\}$ (the singleton set consisting of the null vector of V), then we say that W and U are *disjoint*. Note that this terminology is not consistent with that used in connection with sets. If W and U are linear subspaces of V, then it is easily verified that for every $v \in U + W$, there exist *unique* elements $w \in W$ and $u \in U$ such that $v = u + w$ if and only if $U \cap W = \{0\}$.

The preceding discussion is readily extended to any number of linear subspaces of V and gives rise to the following concept. Let $V_1, \ldots, V_r$ be linear subspaces of a vector space V. The sum $V_1 + \cdots + V_r$ is said to be a *direct sum* if for each $v \in V_1 + \cdots + V_r$ there is a unique set $v^i \in V_i$, $i = 1, \ldots, r$, such that $v = v^1 + \cdots + v^r$. Henceforth, we will denote the direct sum of $V_1, \ldots, V_r$ by $V_1 \oplus \cdots \oplus V_r$.

Now let V be the direct sum of linear spaces V_1 and V_2, i.e., $V = V_1 \oplus V_2$, and let $v = v^1 + v^2$ be the unique representation of $v \in V$, where $v^1 \in V_1$ and $v^2 \in V_2$. We say that the *projection on* V_1 *along* V_2 is the transformation defined by

$$\mathcal{P}(v) = v^1. \tag{2.103}$$

We can easily verify that (i) $\mathcal{P} \in L(V, V)$, (ii) $\mathcal{R}(\mathcal{P}) = V_1$, and (ii) $\mathcal{N}(\mathcal{P}) = V_2$.

More generally, we can show that if $\mathcal{P} \in L(V, V)$, then $\mathcal{P}$ is a projection on $\mathcal{R}(\mathcal{P})$ along $\mathcal{N}(\mathcal{P})$ if and only if $\mathcal{P}\mathcal{P} = \mathcal{P}^2 = \mathcal{P}$. This gives rise to the following concept: $\mathcal{P} \in L(V, V)$ is said to be *idempotent* if $\mathcal{P}^2 = \mathcal{P}$.

We can also verify easily that $\mathcal{P}$ is a projection on a linear subspace if and only if $(\mathcal{I} - \mathcal{P})$ is a projection. If in particular $\mathcal{P}$ is the projection on V_1 along V_2, then $(\mathcal{I} - \mathcal{P})$ is the projection on V_2 along V_1.

In view of the preceding results, there is no ambiguity in simply saying a transformation $\mathcal{P}$ is a projection (rather than $\mathcal{P}$ is a projection on V_1 along V_2). We emphasize that if $\mathcal{P}$ is a projection, then

$$V = \mathcal{R}(\mathcal{P}) \oplus \mathcal{N}(\mathcal{P}). \tag{2.104}$$

This is not necessarily the case for arbitrary linear transformations $\mathcal{T} \in L(V, V)$, for in general, $\mathcal{R}(\mathcal{T})$ and $\mathcal{N}(\mathcal{T})$ need not be disjoint.

Next, let $\mathcal{T} \in L(V, V)$. A linear subspace W of V is said to be *invariant under the linear transformation* $\mathcal{T}$ if $w \in W$ implies that $\mathcal{T}w \in W$. From this definition it follows trivially that (i) V is invariant under $\mathcal{T}$, (ii) $\{0\}$ is invariant under $\mathcal{T}$, (iii) $\mathcal{R}(\mathcal{T})$ is invariant under $\mathcal{T}$, and (iv) $\mathcal{N}(\mathcal{T})$ is invariant under $\mathcal{T}$.

Next, let V be a linear space that is the direct sum of two linear subspaces W and U. If W and U are both invariant under a linear transformation $\mathcal{T}$, then $\mathcal{T}$ is said to be *reduced* by W and U. It is readily verified, using definitions, that $\mathcal{T} \in L(V, V)$ is reduced by W and U if and only if $\mathcal{P}\mathcal{T} = \mathcal{T}\mathcal{P}$, where $\mathcal{P}$ is the projection on W along U.

Next we consider briefly the matrix representation of projections. To this end, let V be an n-dimensional vector space and let $\mathcal{P} \in L(V, V)$. It is easily verified [using (2.104)] that if $\mathcal{P}$ is a projection, then there exists a basis $\{v^1, \ldots, v^n\}$ for V such that the matrix P of $\mathcal{P}$ with respect to this basis is of the form

$$P = \left[\begin{array}{cccc:ccc} 1 & 0 & \cdots & 0 & & & \\ 0 & 1 & \cdots & 0 & & & \\ \vdots & \vdots & & \vdots & & 0 & \\ 0 & 0 & \cdots & 1 & & & \\ \hdashline & & & & 0 & \cdots & 0 \\ & & 0 & & \vdots & & \vdots \\ & & & & 0 & & 0 \end{array}\right] \left.\begin{matrix} \\ \\ \\ \\ \end{matrix}\right\} r, \tag{2.105}$$

where $r = \dim \mathcal{R}(\mathcal{P})$.

We conclude with the following interesting result. Let V be a finite-dimensional vector space and let $\mathcal{A} \in L(V, V)$. If W is a p-dimensional invariant subspace of V and if $V = W \oplus U$, then there exists a basis for V such that the matrix A of $\mathcal{A}$ with respect to this basis is of the form

$$A = \left[\begin{array}{c:c} A_{11} & A_{12} \\ \hdashline 0 & A_{22} \end{array}\right], \tag{2.106}$$

where A_{11} is a $p \times p$ matrix and the remaining submatrices are of appropriate dimension.

The canonical form (2.106) will be used in Chapter 3 in developing standard forms for uncontrollable and unobservable systems.

L. Some Canonical Forms of Matrices

In this subsection we investigate under which conditions a linear transformation of a vector space into itself can be represented by (i) a diagonal matrix, (ii) a block diagonal matrix, (iii) a triangular matrix, and (iv) a companion matrix. We will also investigate when a linear transformation cannot be represented by a diagonal matrix.

Throughout this subsection, V denotes an n-dimensional vector space over a field F.

Diagonal form

We begin with the following fundamental result. Let $\lambda_1, \ldots, \lambda_p$ be the distinct eigenvalues of a linear transformation $\mathcal{A} \in L(V, V)$ and let $\bar{v}^1 \neq 0, \ldots, \bar{v}^p \neq 0$ be corresponding eigenvectors of $\mathcal{A}$. Then it is easily shown that the set $\{\bar{v}^1, \ldots, \bar{v}^p\}$ is linearly independent. We note that if in particular $\mathcal{A}$ has n distinct eigenvalues, then the corresponding n eigenvectors span the linear space and, as such, form a basis for V.

The above gives immediate rise to the next important result. Let $\mathcal{A} \in L(V, V)$ and assume that the characteristic polynomial of $\mathcal{A}$ has n distinct roots, so that

$$det\,(\mathcal{A} - \lambda\mathcal{I}) = (\lambda_1 - \lambda)(\lambda_2 - \lambda)\cdots(\lambda_n - \lambda), \tag{2.107}$$

where $\lambda_1, \ldots, \lambda_n$ are distinct eigenvalues. Then there exists a basis $\{\bar{v}^1, \ldots, \bar{v}^n\}$ of V such that $\bar{v}^i$ is an eigenvector corresponding to λ_i for $i = 1, \ldots, n$. The matrix $\bar{A}$ of $\mathcal{A}$ with respect to the basis $\{\bar{v}^1, \ldots, \bar{v}^n\}$ is

$$\bar{A} = \begin{bmatrix} \lambda_1 & & & 0 \\ & \lambda_2 & & \\ & & \ddots & \\ 0 & & & \lambda_n \end{bmatrix} \triangleq diag\,(\lambda_1, \ldots, \lambda_n). \tag{2.108}$$

In the same spirit as above, we can also easily establish the next result. Let $\mathcal{A} \in L(V, V)$, and let A be the matrix of $\mathcal{A}$ with respect to the basis $\{v^1, \ldots, v^n\}$. If the characteristic polynomial $det(\mathcal{A} - \lambda\mathcal{I}) = \alpha_0 + \alpha_1\lambda + \cdots + \alpha_n\lambda^n$ has n distinct roots, $\lambda_1, \ldots, \lambda_n$, then A is similar to the matrix $\bar{A}$ of $\mathcal{A}$ with respect to the basis $\{\bar{v}^1, \ldots, \bar{v}^n\}$, where $\bar{A}$ is given in (2.108). In this case there exists a nonsingular matrix P such that

$$\bar{A} = P^{-1}AP.$$

The matrix P is the matrix of the basis $\{\bar{v}^1, \ldots, \bar{v}^n\}$ with respect to the basis $\{v^1, \ldots, v^n\}$, and P^{-1} is the matrix of the basis $\{v^1, \ldots, v^n\}$ with respect to the basis $\{\bar{v}^1, \ldots, \bar{v}^n\}$. The matrix P can be constructed by letting its columns be eigenvectors of A corresponding to $\lambda_1, \ldots, \lambda_n$, respectively; that is,

$$P = [\pi^1, \pi^2, \ldots, \pi^n], \tag{2.109}$$

where $\pi^1, \ldots, \pi^n$ are eigenvectors of A corresponding to the eigenvalues $\lambda_1, \ldots, \lambda_n$, respectively (verify this).

The similarity transformation P given in (2.109) is called a *modal matrix*. If the conditions of the above result are satisfied and if, in particular, (2.108) holds, then we say that *the matrix A has been diagonalized*.

As a specific case, let V be a two-dimensional vector space over the real numbers, let $\mathcal{A} \in L(V, V)$, and let $\{v^1, v^2\}$ be a basis for V. Suppose the matrix A of $\mathcal{A}$ with respect to this basis is given by $A = \begin{bmatrix} -2 & 4 \\ 1 & 1 \end{bmatrix}$. The characteristic polynomial of $\mathcal{A}$ is $p(\lambda) = det\,(\mathcal{A} - \lambda\mathcal{I}) = det\,(A - \lambda I) = \lambda^2 + \lambda - 6 = (\lambda - 2)(\lambda + 3)$, and the eigenvalues of $\mathcal{A}$ are $\lambda_1 = 2$ and $\lambda_2 = -3$.

Let $\eta = (\eta_1, \eta_2)^T$ denote the coordinate representation of $v \in V$ with respect to the basis $\{v^1, v^2\}$. To find eigenvectors corresponding to λ_i, $i = 1, 2$, we solve the system of equations $(A - \lambda I)\eta = O$. An easy computation yields $\eta^1 = (1, 1)^T$ and $\eta^2 = (4, -1)$ as eigenvectors corresponding to λ_1 and λ_2, respectively. The diagonal matrix $\bar{A}$ given in (2.108) is $\bar{A} = \begin{bmatrix} \lambda_1 & 0 \\ 0 & \lambda_2 \end{bmatrix} = \begin{bmatrix} 2 & 0 \\ 0 & -3 \end{bmatrix}$. The matrix P given in (2.109) and its inverse P^{-1} are

$$P = [\eta^1, \eta^2] = \begin{bmatrix} 1 & 4 \\ 1 & -1 \end{bmatrix}, \quad P^{-1} = \begin{bmatrix} .2 & .8 \\ .2 & -.2 \end{bmatrix}.$$

As expected, we have $P^{-1}AP = \begin{bmatrix} 2 & 0 \\ 0 & -3 \end{bmatrix} = \begin{bmatrix} \lambda_1 & 0 \\ 0 & \lambda_2 \end{bmatrix}$. By (2.81), the basis $\{\bar{v}^1, \bar{v}^2\} \subset V$ with respect to which $\bar{A}$ represents $\mathcal{A}$ is given by $\bar{v}^1 = \sum_{j=1}^{2} p_{j1} v^j = v^1 + v^2$, $\bar{v}^2 = \sum_{j=1}^{2} p_{j2} v^j = 4v^1 - v^2$. If $\eta = (\eta_1, \eta_2)^T$ is the coordinate representation of v with respect to $\{v^1, v^2\}$, then $\bar{\eta} = P^{-1}\eta$ is the coordinate representation of v with respect to $\{\bar{v}^1, \bar{v}^2\}$. The vectors $\bar{v}^1, \bar{v}^2$ are of course eigenvectors of $\mathcal{A}$ corresponding to λ_1 and λ_2, respectively.

When the algebraic multiplicity of one or more of the eigenvalues of a linear transformation is greater than one, then the linear transformation is said to have *repeated eigenvalues*. In this case it is not always possible to represent the linear transformation by a diagonal matrix. However, from the preceding results of this section it should be clear that a linear transformation with repeated eigenvalues can be represented by a diagonal matrix if the number of linearly independent eigenvectors corresponding to any eigenvalue is the same as the algebraic multiplicity of the eigenvalue. We consider two specific cases to shed additional light on this.

First, we consider the matrix

$$A = \begin{bmatrix} 1 & 3 & -2 \\ 0 & 4 & -2 \\ 0 & 3 & -1 \end{bmatrix}$$

with characteristic equation $det\,(A - \lambda I) = (1 - \lambda)^2(2 - \lambda) = 0$ and eigenvalues $\lambda_1 = 1$ and $\lambda_2 = 2$. The algebraic multiplicity of λ_1 is two. Corresponding to λ_1 we can find two linearly independent eigenvectors $(1, 2, 3)^T$ and $(1, 0, 0)^T$, and corresponding to λ_2 we have an eigenvector $(1, 1, 1)^T$. Letting P denote a modal matrix, we obtain

$$P = \begin{bmatrix} 1 & 1 & 1 \\ 2 & 0 & 1 \\ 3 & 0 & 1 \end{bmatrix}, \qquad P^{-1} = \begin{bmatrix} 0 & -1 & 1 \\ 1 & -2 & 1 \\ 0 & 3 & -2 \end{bmatrix},$$

and

$$\bar{A} = P^{-1}AP = \begin{bmatrix} 1 & 0 & 0 \\ 0 & 1 & 0 \\ 0 & 0 & 1 \end{bmatrix}.$$

In this example, $\dim \mathcal{N}_{\lambda_1} = 2$, which happens to be the same as the algebraic multiplicity of λ_1. For this reason, we were able to diagonalize A.

As a second example, consider the matrix

$$A = \begin{bmatrix} 2 & 1 & -2 \\ 0 & 2 & -1 \\ 0 & 0 & 1 \end{bmatrix}$$

with characteristic equation $det\,(A - \lambda I) = (1 - \lambda)(2 - \lambda)^2 = 0$ and eigenvalues $\lambda_1 = 1$ and $\lambda_2 = 2$. The algebraic multiplicity of λ_2 is two and an eigenvector corresponding to λ_1 is $(1, 1, 1)^T$. It is easily verified that any eigenvector corresponding to λ_2 must be of the form $(\nu_1, 0, 0)$, $\nu_1 \neq 0$. We see that $\dim \mathcal{N}_{\lambda_2} = 1$, and thus, we are not able to determine a basis for the three-dimensional vector space V, which consists of eigenvectors. Consequently, we are unable to diagonalize A.

Block diagonal form

When a matrix cannot be diagonalized, we seek, for practical reasons, to represent a linear transformation by a matrix that is as nearly diagonal as possible. The next result provides the basis of representing linear transformations by such matrices, called *block diagonal* matrices. In Subsection G of this section we will consider the "simplest" type of block diagonal matrix, called the *Jordan canonical form.*

We let V be an n-dimensional vector space and we let $\mathcal{A} \in L(V, V)$. If V is the direct sum of p linear subspaces, $V_1, \ldots, V_p$, which are invariant under $\mathcal{A}$, then it can be readily shown that there exists a basis for V such that the matrix representation for $\mathcal{A}$ is in the block diagonal form given by

$$A = \begin{bmatrix} A_1 & & & 0 \\ & A_2 & & \\ & & \ddots & \\ 0 & & & A_p \end{bmatrix}. \tag{2.110}$$

Moreover, A_i is a matrix representation of $\mathcal{A}_i$, the restriction of $\mathcal{A}$ to V_i, $i = 1, \ldots, p$. Also,

$$det\,(A) = \prod_{i=1}^{p} det\,(A_i). \tag{2.111}$$

From the above it is clear that to carry out a block diagonalization of a matrix A, we need to find an appropriate set of invariant subspaces of V, and furthermore, we need to find a simple matrix representation on each of these subspaces.

As a specific case for the above, let V be an n-dimensional vector space. If $\mathcal{A} \in L(V, V)$ has n distinct eigenvalues $\lambda_1, \ldots, \lambda_n$, and if we let $\mathcal{N}_j = \{v : (\mathcal{A} - \lambda_j \mathcal{I})v = 0\}$, $j = 1, \ldots, n$, then $\mathcal{N}_j$ is an invariant linear subspace under $\mathcal{A}$ and $V = \mathcal{N}_1 \oplus \cdots \oplus \mathcal{N}_n$. For any $v \in \mathcal{N}_j$, we have $\mathcal{A}v = \lambda_j v$, and hence, $\mathcal{A}_j v = \lambda v$ for $v \in \mathcal{N}_j$. A basis for $\mathcal{N}_j$ is any nonzero $v_j \in \mathcal{N}_j$. Thus, with respect to this basis, $\mathcal{A}_j$ is represented by the matrix λ_j (in this case, simply a scalar). With respect to a basis of n linearly independent eigenvectors, $\{v^1, \ldots, v^n\}$, $\mathcal{A}$ is represented by (2.108).

Triangular form

In addition to the diagonal form and the block diagonal form, there are many other useful forms of matrices to represent linear transformations on finite-dimensional vector spaces. One of these canonical forms involves triangular matrices with one of the two forms given by

$$\begin{bmatrix} a_{11} & a_{12} & a_{13} & \dots & a_{1n} \\ 0 & a_{22} & a_{23} & \dots & a_{2n} \\ \dots & \dots & \dots & \dots & \dots \\ 0 & 0 & 0 & \dots & a_{n-1,n} \\ 0 & 0 & 0 & \dots & a_{nn} \end{bmatrix} \quad \text{or} \quad \begin{bmatrix} a_{11} & 0 & 0 & \dots & 0 \\ a_{21} & a_{22} & 0 & \dots & 0 \\ \dots & \dots & \dots & \dots & 0 \\ a_{n-1,1} & a_{n-1,2} & a_{n-1,3} & \dots & 0 \\ a_{n1} & a_{n2} & a_{n3} & \dots & a_{nn} \end{bmatrix}. \tag{2.112}$$

We call the matrix on the left an *upper triangular matrix* and the matrix on the right a *lower triangular matrix*.

Now let V be an n-dimensional vector space over C, and let $\mathcal{A} \in L(V, V)$. It can be shown that there exists a basis for V such that $\mathcal{A}$ is represented by an upper (or by a lower) triangular matrix with respect to that basis.

We note that if A is in triangular form, then

$$det(A - \lambda I) = (a_{11} - \lambda)(a_{22} - \lambda)\cdots(a_{nn} - \lambda), \tag{2.113}$$

i.e., the diagonal elements of A are in this case the eigenvalues of A.

Companion form

We conclude this subsection by considering a canonical form for real square matrices that arises frequently in systems theory, called *companion form*. As a matter of fact, a given matrix $A \in R^{n\times n}$ can be transformed via appropriate similarity transformations into four different forms of this type given by

$$\begin{bmatrix} 0 & I \\ \times & \times \end{bmatrix}, \quad \begin{bmatrix} \times & \times \\ I & 0 \end{bmatrix}, \quad \begin{bmatrix} \times & I \\ \times & 0 \end{bmatrix}, \quad \begin{bmatrix} 0 & \times \\ I & \times \end{bmatrix},$$

where the rows or columns denoted by $(\times\times)$ are made up of the coefficients $-a_0, -a_1, \dots, -a_{n-1}$, determined by $(-1)^n det(A - \lambda I) = det(\lambda I - A) = \lambda^n + a_{n-1}\lambda^{n-1} + \cdots + a_0$, where $I \in R^{(n-1)\times(n-1)}$, and where the 0 denotes an $(n-1)$-dimensional column or row vector. For example, the matrix on the left, which is perhaps the most commonly used companion form, and which henceforth we will identify as A_c, is given by

$$A_c = \begin{bmatrix} 0 & 1 & 0 & \dots & 0 \\ 0 & 0 & 1 & \dots & 0 \\ \vdots & \vdots & \vdots & \ddots & \vdots \\ 0 & 0 & 0 & \dots & 1 \\ -a_0 & -a_1 & -a_2 & \dots & -a_{n-1} \end{bmatrix}.$$

In the following, we confine our discussion to this matrix. Similar treatments apply to the other three companion forms.

Given $A \in R^{n\times n}$, it can be shown that there exists a similarity transformation P that transforms A to the companion form A_c given above if and only if there exists a vector $b \in R^n$ such that the matrix $\mathcal{C} = [b, Ab, \dots, A^{n-1}b]$ is of full rank n. Furthermore, P is given by

$$P = \begin{bmatrix} q \\ qA \\ \dots \\ qA^{n-1} \end{bmatrix}^{-1},$$

where q is the nth row of $\mathscr{C}^{-1}$, and

$$A_c = P^{-1}AP.$$

A matrix A for which such a vector b exists is called *cyclic*. It can be shown that A is cyclic if and only if the geometric multiplicity of each of its n eigenvalues is one, or equivalently, if and only if the n eigenvalues of A are exactly the roots of its minimal polynomial (refer to Subsections M and O, which follow). Finally, if λ_i is an eigenvalue of A_c (or of A), then it can be verified that $(1, \lambda_i, \ldots, \lambda_i^{n-1})^T$ is a corresponding eigenvector.

M. Minimal Polynomials

One of our goals in this section is to develop the Jordan canonical form. To accomplish this we first need to introduce and study minimal polynomials (which will be accomplished in the present subsection) and nilpotent operators (to be considered in Subsection N). Throughout this subsection, V denotes an n-dimensional vector space.

For purposes of motivation, consider the matrix

$$A = \begin{bmatrix} 1 & 3 & -2 \\ 0 & 4 & -2 \\ 0 & 3 & -1 \end{bmatrix}.$$

The characteristic polynomial of A is $p(\lambda) = (1 - \lambda)^2(2 - \lambda)$, and we know from the Cayley-Hamilton Theorem that

$$p(A) = O. \tag{2.114}$$

Now let us consider the polynomial $m(\lambda) = (1 - \lambda)(2 - \lambda) = 2 - 3\lambda + \lambda^2$. Then

$$m(A) = 2I - 3A + A^2 = O. \tag{2.115}$$

Thus, matrix A satisfies (2.115), which is of lower degree than (2.114), the characteristic equation of A.

More generally, it can be shown that for an $n \times n$ matrix A there exists a unique polynomial $m(\lambda)$ such that (i) $m(A) = O$, (ii) $m(\lambda)$ is *monic* (i.e., if m is an nth-degree polynomial in λ, then the coefficient of λ^n is unity), and (iii) if $m'(\lambda)$ is any other polynomial such that $m'(A) = O$, then the degree of $m(\lambda)$ is less or equal to the degree of $m'(\lambda)$ [i.e., $m(\lambda)$ is of the lowest degree such that $m(A) = O$]. The polynomial $m(\lambda)$ is called the *minimal polynomial of A*.

In the remainder of this section we let $p(\lambda)$ denote the characteristic polynomial of an $n \times n$ matrix A and we let $m(\lambda)$ denote the minimal polynomial of A. In what follows, we develop an explicit form for the minimal polynomial of A that makes it possible to determine it systematically.

Let $f(\lambda)$ be any polynomial such that $f(A) = O$ (e.g., the characteristic polynomial). Then it is easily shown that $m(\lambda)$ divides $f(\lambda)$ [i.e., there is a polynomial $q(\lambda)$ such that $f(\lambda) = q(\lambda)m(\lambda)$]. In particular, the minimal polynomial of A, $m(\lambda)$, divides the characteristic polynomial of A, $p(\lambda)$. Also, it can be shown that $p(\lambda)$ divides $[m(\lambda)]^n$.

Next, let $p(\lambda)$ be given by

$$p(\lambda) = (\lambda_1 - \lambda)^{m_1}(\lambda_2 - \lambda)^{m_2}\cdots(\lambda_p - \lambda)^{m_p}, \tag{2.116}$$

where $m_1, \ldots, m_p$ are the algebraic multiplicities of the distinct eigenvalues $\lambda_1, \ldots, \lambda_p$ of A, respectively. It can be shown that

$$m(\lambda) = (\lambda - \lambda_1)^{\mu_1}(\lambda - \lambda_2)^{\mu_2}\cdots(\lambda - \lambda_p)^{\mu_p}, \tag{2.117}$$

where $1 \leq \mu_i \leq m_i$, $i = 1, \ldots, p$.

The only unknowns left to determine the minimal polynomial of A are $\mu_1, \ldots, \mu_p$ in (2.117). These can be determined in several ways.

The next result is a direct consequence of (2.86). Let $\bar{A}$ be similar to A and let $\bar{m}(\lambda)$ be the minimal polynomial of $\bar{A}$. Then $\bar{m}(\lambda) = m(\lambda)$. This result, in turn, justifies the following definition. Let $\mathcal{A} \in L(V, V)$. The *minimal polynomial of* $\mathcal{A}$ is the minimal polynomial of any matrix A that represents $\mathcal{A}$.

To develop the Jordan canonical form (for linear transformations with repeated eigenvalues), we need to consider several additional preliminary results that are important in their own right.

Let $\mathcal{A} \in L(V, V)$ and let $f(\lambda)$ be any polynomial in λ. Let $\mathcal{N}_f = \{v : f(A)v = 0\}$. It can be shown that $\mathcal{N}_f$ is an invariant linear subspace of V under $\mathcal{A}$. In particular let $\lambda_1, \ldots, \lambda_p$ be the distinct eigenvalues of $\mathcal{A} \in L(V, V)$ and for $j = 1, \ldots, p$, and for any positive integer q, let

$$\mathcal{N}_j^q = \{v : (\mathcal{A} - \lambda_j \mathcal{I})^q v = 0\}. \tag{2.118}$$

In view of the above result, it follows that $\mathcal{N}_j^q$ is an invariant linear subspace of V under $\mathcal{A}$.

Next, let $\mathcal{A} \in L(V, V)$, let V_1 and V_2 be linear subspaces of V such that $V = V_1 \oplus V_2$, and let $\mathcal{A}_1$ be the restriction of $\mathcal{A}$ to V_1. Let $f(\lambda)$ be any polynomial in λ. It can be shown that if $\mathcal{A}$ is reduced by V_1 and V_2 then, for all $v^1 \in V_1$, $f(\mathcal{A}_1)v^1 = f(\mathcal{A})v^1$.

Next, let V be a vector space over $\mathcal{C}$ and let $\mathcal{A} \in L(V, V)$. Let $m(\lambda)$ be the minimal polynomial of $\mathcal{A}$ as given in (2.117). Let $g(\lambda) = (\lambda - \lambda_1)^{\mu_1}$, let $h(\lambda) = (\lambda - \lambda_2)^{\mu_2}\cdots(\lambda - \lambda_p)^{\mu_p}$ if $p \geq 2$, and let $h(\lambda) = 1$ if $p = 1$. Let $\mathcal{A}_1$ be the restriction of $\mathcal{A}$ to $\mathcal{N}_1^{\mu_1}$, i.e., $\mathcal{A}_1 v = \mathcal{A}v$ for all $v \in \mathcal{N}_1^{\mu_1}$. Let $\mathcal{M} = \{v \in V : h(\mathcal{A})v = 0\}$. Using the preceding results, we can show that (i) $V = \mathcal{N}_1^{\mu_1} \oplus \mathcal{M}$, and (ii) $(\lambda - \lambda_1)^{\mu_1}$ is the minimal polynomial of $\mathcal{A}_1$.

These make it possible to prove the next result, which is called the *primary decomposition theorem for linear transformations* and which we state next.

Let V be an n-dimensional vector space over C, let $\lambda_1, \ldots, \lambda_p$ be the distinct eigenvalues of $\mathcal{A} \in L(V, V)$, let the characteristic and minimal polynomials of $\mathcal{A}$ be

$$p(\lambda) = (\lambda_1 - \lambda)^{m_1}\cdots(\lambda_p - \lambda)^{m_p} \quad \text{and} \quad m(\lambda) = (\lambda - \lambda_1)^{\mu_1}\cdots(\lambda - \lambda_p)^{\mu_p},$$

respectively. Let

$$V_i = \{v : (\mathcal{A} - \lambda_i \mathcal{I})^{\mu_i} v = 0\}, \quad i = 1, \ldots, p. \tag{2.119}$$

Then (i) $V_i, i = 1, \ldots, p$ are invariant linear subspaces of V under $\mathcal{A}$, (ii) $V = V_1 \oplus \cdots \oplus V_p$, (iii) $(\lambda - \lambda_i)^{\mu_i}$ is the minimal polynomial of $\mathcal{A}_i$, where $\mathcal{A}_i$ is the restriction of $\mathcal{A}$ to V_i, and (iv) $\dim V_i = m_i$, $i = 1, \ldots, p$.

The above result shows that we can always present $\mathcal{A} \in L(V, V)$ by a matrix in block diagonal form, where the number of diagonal blocks is equal to the number of distinct eigenvalues of $\mathcal{A}$. We will next consider a convenient representation for each of the diagonal submatrices A_i. It may turn out that one or more of the submatrices A_i will be diagonal. The next result tells us specifically when $\mathcal{A} \in L(V, V)$ is representable by a diagonal matrix.

Let V be an n-dimensional vector space over C and let $\mathcal{A} \in L(V, V)$. Let $\lambda_1, \dots, \lambda_p$, $p \leq n$, be the distinct eigenvalues of $\mathcal{A}$. Then there exists a basis for V such that the matrix A of $\mathcal{A}$ with respect to this basis is diagonal if and only if the minimal polynomial for $\mathcal{A}$ is of the form

$$m(\lambda) = (\lambda - \lambda_1)(\lambda - \lambda_2)\cdots(\lambda - \lambda_p). \tag{2.120}$$

N. Nilpotent Operators

Let us now proceed by considering a representation for each of the $\mathcal{A}_i \in L(V_i, V_i)$ in the *primary decomposition theorem* presented in Subsection E so that the block diagonal matrix representation of $\mathcal{A} \in L(V, V)$ is as simple as possible. To accomplish this, we need to define and examine nilpotent operators.

Let $\mathcal{N} \in L(V, V)$. Then $\mathcal{N}$ is said to be *nilpotent* if there exists an integer $q > 0$ such that $\mathcal{N}^q = \mathcal{O}$. A nilpotent operator is said to be of *index* q if $\mathcal{N}^q = \mathcal{O}$ but $\mathcal{N}^{q-1} \neq \mathcal{O}$.

Recall that the primary decomposition theorem enables us to write $V = V_1 \oplus \cdots \oplus V_p$. Furthermore, the linear transformation $(\mathcal{A}_i - \lambda_i \mathcal{I})$ is nilpotent on V_i. If we let $\mathcal{N}_i = \mathcal{A}_i - \lambda_i \mathcal{I}$, then $\mathcal{A}_i = \lambda_i \mathcal{I} + \mathcal{N}_i$. Now $\lambda_i \mathcal{I}$ is clearly represented by a diagonal matrix. However, the transformation $\mathcal{N}_i$ forces the matrix representation of $\mathcal{A}_i$ to be in general nondiagonal. Therefore, the next task is to seek a simple representation of the nilpotent operator $\mathcal{N}_i$.

In the next few results, which are concerned with properties of nilpotent operators, we drop the subscript i for convenience.

Let $\mathcal{N} \in L(W, W)$, where W is an m-dimensional vector space. It can be shown that if $\mathcal{N}$ is a nilpotent linear transformation of index q and if $w \in W$ is such that $\mathcal{N}^{q-1}w \neq 0$, then the vectors $w, \mathcal{N}w, \dots, \mathcal{N}^{q-1}w$ in W are linearly independent.

Next, we examine the matrix representation of nilpotent transformations.

Let W be a q-dimensional vector space and let $\mathcal{N} \in L(W, W)$ be nilpotent of index q. Let $w^0 \in W$ be such that $\mathcal{N}^{q-1}w^0 \neq 0$. It can be shown that the matrix N of $\mathcal{N}$ with respect to the basis $\{\mathcal{N}^{q-1}w^0, \mathcal{N}^{q-2}w^0, \dots, w^0\}$ in W is given by

$$N = \begin{bmatrix} 0 & 1 & 0 & 0 & \dots & 0 & 0 \\ 0 & 0 & 1 & 0 & \dots & 0 & 0 \\ \dots & \dots & \dots & \dots & \dots & \dots & \dots \\ 0 & 0 & 0 & 0 & \dots & 0 & 1 \\ 0 & 0 & 0 & 0 & \dots & 0 & 0 \end{bmatrix}. \tag{2.121}$$

The above result characterizes the matrix representation of a nilpotent linear transformation of index q on a q-dimensional vector space. The next task is to determine the representation of a nilpotent operator of index γ on a vector space of dimension m, where $\gamma \leq m$. It is easily shown that we can dismiss the case $\gamma > m$, i.e., if $\mathcal{N} \in L(W, W)$ is nilpotent of index γ, where $\dim W = m$, then $\gamma \leq m$.

Now let W be an m-dimensional vector space, let $\mathcal{N} \in L(W, W)$, let γ be any positive integer, and let

$$\begin{aligned} W_1 &= \{w : \mathcal{N}w = 0\}, \dim W_1 = l_1 \\ W_2 &= \{w : \mathcal{N}^2 w = 0\}, \dim W_2 = l_2 \\ \cdots & \quad \cdots\cdots\cdots\cdots\cdots\cdots\cdots \\ W_\gamma &= \{w : \mathcal{N}^\gamma w = 0\}, \dim W_\gamma = l_\gamma. \end{aligned} \tag{2.122}$$

Also, for any i such that $1 < i < \gamma$, let $\{w^1, \ldots, w^m\}$ be a basis for W such that $\{w^1, \ldots, w^{l_i}\}$ is a basis for W_i. It can be shown that (i) $W_1 \subset W_2 \subset \cdots \subset W_\gamma$ and (ii) $\{w^1, \ldots, w^{l_{i-1}}, \mathcal{N}w^{l_i+1}, \ldots, \mathcal{N}w^{l_{i+1}}\}$ is a linearly independent set of vectors in W_i.

The next result, which is the principal result of this subsection, is a consequence of the above results.

Let W be an m-dimensional vector space over C and let $\mathcal{N} \in L(W, W)$ be nilpotent of index γ. Let $W_1 = \{w : \mathcal{N}w = 0\}, \ldots, W_\gamma = \{w : \mathcal{N}^\gamma w = 0\}$, and let $l_i = \dim W_i$, $i = 1, \ldots, \gamma$. Then there exists a basis in W such that the matrix N of $\mathcal{N}$ is of block diagonal form,

$$N = \begin{bmatrix} N_1 & \cdots & 0 \\ \vdots & \ddots & \vdots \\ 0 & \cdots & N_r \end{bmatrix}, \tag{2.123}$$

where
$$N_i = \begin{bmatrix} 0 & 1 & 0 & 0 & \cdots & 0 & 0 \\ 0 & 0 & 1 & 0 & \cdots & 0 & 0 \\ \cdots & \cdots & \cdots & \cdots & \cdots & \cdots & \cdots \\ 0 & 0 & 0 & 0 & \cdots & 0 & 1 \\ 0 & 0 & 0 & 0 & \cdots & 0 & 0 \end{bmatrix}, \tag{2.124}$$

$i = 1, \ldots, r$, where $r = l_1$, N_i is a $k_i \times k_i$ matrix, $1 \le k_i \le \gamma$, and k_i is determined in the following manner: there are

$$\begin{array}{lll} l_\gamma - l_{\gamma-1} & \gamma \times \gamma \text{ matrices,} & \\ 2l_i - l_{i+1} - l_{i-1} & i \times i \text{ matrices,} & i = 2, \ldots, \gamma - 1, \\ 2l_1 - l_2 & 1 \times 1 \text{ matrices.} & \end{array} \tag{2.125}$$

The basis for W consists of strings of vectors of the form $\mathcal{N}^{k_1-1}w^1, \ldots, w^1, \mathcal{N}^{k_2-1}w^2, \ldots, w^2, \ldots, \mathcal{N}^{k_r-1}w^r, \ldots, w^r$.

O. The Jordan Canonical Form

The results of the preceding three subsections can be used to prove the next result, which yields the Jordan canonical form of matrices.

Let V be an n-dimensional vector space over C and let $\mathcal{A} \in L(V, V)$. Let the characteristic polynomial of $\mathcal{A}$ be $p(\lambda) = (\lambda_1 - \lambda)^{m_1} \cdots (\lambda_p - \lambda)^{m_p}$ and let the minimal polynomial of $\mathcal{A}$ be $m(\lambda) = (\lambda - \lambda_1)^{\mu_1} \cdots (\lambda - \lambda_p)^{\mu_p}$, where $\lambda_1, \ldots, \lambda_p$ are the distinct eigenvalues of $\mathcal{A}$. Let $V_i = \{v \in V : (\mathcal{A} - \lambda_i \mathcal{I})^{\mu_i} v = 0\}$. Then (i) $V_1, \ldots, V_p$ are invariant subspaces of V under $\mathcal{A}$; (ii) $V = V_1 \oplus \cdots \oplus V_p$; (iii) $\dim V_i = m_i$, $i = 1, \ldots, p$; and (iv) there exists a basis for V such that the matrix A of $\mathcal{A}$ with respect to this basis is of the form

$$A = \begin{bmatrix} A_1 & 0 & \cdots & 0 \\ 0 & A_2 & \cdots & 0 \\ \cdots & \cdots & \cdots & \cdots \\ 0 & 0 & \cdots & A_p \end{bmatrix}, \tag{2.126}$$

where A_i is an $m_i \times m_i$ matrix of the form

$$A_i = \lambda_i I + N_i \tag{2.127}$$

and where N_i is the matrix of the nilpotent operator $(\mathscr{A}_i - \lambda_i \mathscr{I})$ of index μ_i on V_i given by (2.123) and (2.124).

Parts (i) to (iii) of the above result are restatements of the primary decomposition theorem. From this theorem we also know that $(\lambda - \lambda_i)^{\mu_i}$ is the minimal polynomial of $\mathscr{A}_i$, the restriction of $\mathscr{A}$ to V_i. Hence, if we let $\mathscr{N}_i = \mathscr{A}_i - \lambda_i \mathscr{I}$, then $\mathscr{N}_i$ is a nilpotent operator of index μ_i on V_i. We are thus able to represent $\mathscr{N}_i$ as shown in (2.124).

A little extra work shows that the representation of $\mathscr{A} \in L(V, V)$ by a matrix A of the form given in (2.126) and (2.127) is unique except for the order in which the block diagonals $A_1, \ldots, A_p$ appear in A.

The matrix A of $\mathscr{A} \in L(V, V)$ given by (2.126) and (2.127) is called the *Jordan canonical form of* $\mathscr{A}$.

An example

We conclude this section by considering a specific case.

We let $V = R^7$, we let $\{e^1, \ldots, e^7\}$ be the natural basis for V, and we let $\mathscr{A} \in L(V, V)$ be represented by the matrix

$$A = \begin{bmatrix} -1 & 0 & -1 & 1 & 1 & 3 & 0 \\ 0 & 1 & 0 & 0 & 0 & 0 & 0 \\ 2 & 1 & 2 & -1 & -1 & -6 & 0 \\ -2 & 0 & -1 & 2 & 1 & 3 & 0 \\ 0 & 0 & 0 & 0 & 1 & 0 & 0 \\ 0 & 0 & 0 & 0 & 0 & 1 & 0 \\ -1 & -1 & 0 & 1 & 2 & 4 & 1 \end{bmatrix}$$

with respect to $\{e^1, \ldots, e^7\}$. We wish to determine the matrix $\bar{A}$ that represents $\mathscr{A}$ in the Jordan canonical form.

We first determine that the characteristic polynomial of $\mathscr{A}$ is $det(A - \lambda I) = det(\mathscr{A} - \lambda \mathscr{I}) = (1 - \lambda)^7$. This indicates that $\lambda_1 = 1$ is the only distinct eigenvalue of $\mathscr{A}$, having algebraic multiplicity $m_1 = 7$. To find the minimal polynomial of $\mathscr{A}$, we let $\mathscr{N} = \mathscr{A} - \lambda_1 \mathscr{I}$, where $\mathscr{I}$ is the identity operator in $L(V, V)$. The representation of $\mathscr{N}$ with respect to the natural basis is

$$N = A - I = \begin{bmatrix} -2 & 0 & -1 & 1 & 1 & 3 & 0 \\ 0 & 0 & 0 & 0 & 0 & 0 & 0 \\ 2 & 1 & 1 & -1 & -1 & -6 & 0 \\ -2 & 0 & -1 & 1 & 1 & 3 & 0 \\ 0 & 0 & 0 & 0 & 0 & 0 & 0 \\ 0 & 0 & 0 & 0 & 0 & 0 & 0 \\ -1 & -1 & 0 & 1 & 2 & 4 & 0 \end{bmatrix}.$$

The minimal polynomial will be of the form $m(\lambda) = (\lambda - 1)^{\nu_1}$. We need to determine the smallest ν_1 such that $m(A - \lambda I) = m(N) = O$. We first obtain

$$N^2 = \begin{bmatrix} 0 & -1 & 0 & 0 & 0 & 3 & 0 \\ 0 & 0 & 0 & 0 & 0 & 0 & 0 \\ 0 & 1 & 0 & 0 & 0 & -3 & 0 \\ 0 & -1 & 0 & 0 & 0 & 3 & 0 \\ 0 & 0 & 0 & 0 & 0 & 0 & 0 \\ 0 & 0 & 0 & 0 & 0 & 0 & 0 \\ 0 & 0 & 0 & 0 & 0 & 0 & 0 \end{bmatrix}.$$

Next, we obtain $N^3 = 0$, and therefore, $\nu_1 = 3$ and $\mathcal{N}$ is a nilpotent operator of index 3. We see that $V = \mathcal{N}_1^3$ [refer to (2.118) for the notation $\mathcal{N}_1^3$]. We will now apply the results of Subsection F to obtain a representation for $\mathcal{N}$ in this space.

We let $W_1 = \{v : \mathcal{N}v = 0\}$, $W_2 = \{v : \mathcal{N}^2 v = 0\}$, and $W_3 = \{v : \mathcal{N}^3 v = 0\}$, and we observe that N has three linearly independent rows. This means that the rank of $\mathcal{N}$ is 3, and therefore, $\dim(W_1) = l_1 = 4$. Similarly, the rank of $\mathcal{N}^2$ is 1 (since N^2 has one linearly independent row), and so $\dim(W_2) = l_2 = 6$. Clearly, $\dim(W_3) = l_3 = 7$. We conclude that $\mathcal{N}$ will have a representation $\bar{N}$ of the form (2.123) with $r = 4$. Each of the $\bar{N}_i$ will be of the form (2.124). There will be $l_3 - l_2 = 1\,(3 \times 3)$ matrix, $2l_2 - l_3 - l_1 = 1\,(2 \times 2)$ matrix, and $2l_1 - l_2 = 2\,(1 \times 1)$ matrices [see (2.125)]. Hence, there is a basis for V such that $\mathcal{N}$ may be represented by the matrix

$$\bar{N} = \begin{bmatrix} 0 & 1 & 0 & 0 & 0 & 0 & 0 \\ 0 & 0 & 1 & 0 & 0 & 0 & 0 \\ 0 & 0 & 0 & 0 & 0 & 0 & 0 \\ 0 & 0 & 0 & 0 & 1 & 0 & 0 \\ 0 & 0 & 0 & 0 & 0 & 0 & 0 \\ 0 & 0 & 0 & 0 & 0 & 0 & 0 \\ 0 & 0 & 0 & 0 & 0 & 0 & 0 \end{bmatrix}.$$

The corresponding basis will consist of strings of vectors of the form $\mathcal{N}^2 v^1, \mathcal{N}v^1, v^1, \mathcal{N}v^2, v^2, v^3, v^4$.

We will represent the vectors v^1, v^2, v^3, and v^4 by η^1, η^2, η^3, and η^4, their coordinate representations, respectively, with respect to the natural basis $\{e^1, e^2, e^3, e^4, e^5, e^6, e^7\}$ in V. We begin by choosing $v^1 \in W_3$ such that $v^1 \notin W_2$, i.e., we determine a η^1 such that $N^3\eta^1 = 0$ but $N^2\eta^1 \neq 0$. The vector $\eta^1 = (0, 1, 0, 0, 0, 0, 0)^T$ will do. We see that $N\eta^1 = (0, 0, 1, 0, 0, 0, -1)^T$ and $N^2\eta^1 = (-1, 0, 1, -1, 0, 0, 0)^T$. Hence, $\mathcal{N}v^1 \in W_2$ but $\mathcal{N}v^1 \notin W_1$ and $\mathcal{N}^2 v^1 \in W_1$. We see there will be only one string of length 3, and therefore we choose next $v^2 \in W_2$ such that $v^2 \notin W_1$. Also, the pair $\{\mathcal{N}v^1, v^2\}$ must be linearly independent. The vector $\eta^2 = (1, 0, 0, 0, 0, 0, 0)^T$ will do. Next, we have $N\eta^2 = (-2, 0, 2, -2, 0, 0, -1)^T$, and therefore, $\mathcal{N}v^2 \in W_1$. We complete the basis for V by selecting two more vectors, $v^3, v^4 \in W_1$, such that $\{\mathcal{N}^2 v^1, \mathcal{N}v^2, v^3, v^4\}$ are linearly independent. The vectors $\eta^3 = (0, 0, -1, -2, 1, 0, 0)^T$ and $\eta^4 = (1, 3, 1, 0, 0, 1, 0)^T$ will do.

It now follows that the matrix $P = [N^2\eta^1, N\eta^1, \eta^1, N\eta^2, \eta^2, \eta^3, \eta^4]$ is the matrix of the new basis with respect to the natural basis. The reader can readily verify that $\bar{N} = P^{-1}NP$, where

$$P = \begin{bmatrix} -1 & 0 & 0 & -2 & 1 & 0 & 1 \\ 0 & 0 & 1 & 0 & 0 & 0 & 3 \\ 1 & 1 & 0 & 2 & 0 & -1 & 1 \\ -1 & 0 & 0 & -2 & 0 & -2 & 0 \\ 0 & 0 & 0 & 0 & 0 & 1 & 0 \\ 0 & 0 & 0 & 0 & 0 & 0 & 1 \\ 0 & -1 & 0 & -1 & 0 & 0 & 0 \end{bmatrix}, P^{-1} = \begin{bmatrix} 0 & 0 & 2 & 1 & 4 & -2 & 2 \\ 0 & 0 & 1 & 1 & 3 & -1 & 0 \\ 0 & 1 & 0 & 0 & 0 & -3 & 0 \\ 0 & 0 & -1 & -1 & -3 & 1 & -1 \\ 1 & 0 & 0 & -1 & -2 & -1 & 0 \\ 0 & 0 & 0 & 0 & 1 & 0 & 0 \\ 0 & 0 & 0 & 0 & 0 & 1 & 0 \end{bmatrix}.$$

Finally, the Jordan canonical form for A is now given by $\bar{A} = \bar{N} + I$ (recalling that the matrix representation for $\mathscr{I}$ is the same for any basis in V). Therefore, we have

$$\bar{A} = \left[\begin{array}{ccc|cc|c|c} 1 & 1 & 0 & 0 & 0 & 0 & 0 \\ 0 & 1 & 1 & 0 & 0 & 0 & 0 \\ 0 & 0 & 1 & 0 & 0 & 0 & 0 \\ \hline 0 & 0 & 0 & 1 & 1 & 0 & 0 \\ 0 & 0 & 0 & 0 & 1 & 0 & 0 \\ \hline 0 & 0 & 0 & 0 & 0 & 1 & 0 \\ \hline 0 & 0 & 0 & 0 & 0 & 0 & 1 \end{array}\right].$$

It is easily verified that $\bar{A} = P^{-1}AP$. In general it is more convenient as a check to show that $P\bar{A} = AP$.

2.3 LINEAR HOMOGENEOUS AND NONHOMOGENEOUS EQUATIONS

In this section we consider systems of linear homogeneous ordinary differential equations

$$\dot{x} = A(t)x \qquad (LH)$$

and linear nonhomogeneous ordinary differential equations

$$\dot{x} = A(t)x + g(t). \qquad (LN)$$

In Theorem 12.1 of Chapter 1 it was shown that these systems of equations, subject to initial conditions $x(t_0) = x_0$, possess unique solutions for every $(t_0, x_0) \in D$, where $D = \{(t, x) : t \in J = (a, b), x \in R^n\}$ and where it is assumed that $A \in C(J, R^{n\times n})$ and $g \in C(J, R^n)$. These solutions exist over the entire interval $J = (a, b)$ and they depend continuously on the initial conditions. Typically, we will assume that $J = (-\infty, \infty)$. We note that $\phi(t) \equiv 0$, for all $t \in J$, is a solution of (LH), with $\phi(t_0) = 0$. We call this the *trivial solution*. As in Chapter 1 (refer to Section 1.13), we recall that the preceding statements are also true when $A(t)$ and $g(t)$ are piecewise continuous on J.

In the sequel, we sometimes will encounter the case where $A(t) \equiv A$ is in Jordan canonical form that may have entries in the complex plane C. For this reason, we will allow $D = \{(t, x) : t \in J = (a, b), x \in R^n (\text{ or } x \in C^n)\}$ and $A \in C(J, R^{n\times n})$ [or $A \in C(J, C^{n\times n})$], as needed. For the case of real vectors, the field of scalars for the x-space will be the field of real numbers ($F = R$), while for the case of complex vectors, the field of scalars for the x-space will be the field of complex numbers ($F = C$). For the latter case, the theory concerning the existence and uniqueness of solutions for (LH), as presented in Chapter 1, carries over and can be modified in the obvious way.

A. The Fundamental Matrix

Solution space

We will require the following result.

THEOREM 3.1. The set of solutions of (LH) on the interval J forms an n-dimensional vector space.

Proof. Let V denote the set of all solutions of (LH) on J. Let $\alpha_1, \alpha_2 \in F$ and let $\phi_1, \phi_2 \in V$. Then $\alpha_1\phi_1 + \alpha_2\phi_2 \in V$ since $(d/dt)[\alpha_1\phi_1 + \alpha_2\phi_2] = \alpha_1(d/dt)\phi_1(t) + \alpha_2(d/dt)\phi_2(t) = \alpha_1 A(t)\phi_1(t) + \alpha_2 A(t)\phi_2(t) = A(t)[\alpha_1\phi_1(t) + \alpha_2\phi_2(t)]$ for all $t \in J$. This shows that V is a linear subspace of R^n. Hence, V is a vector space.

To complete the proof of the theorem, we must show that V is of dimension n. To accomplish this, we must find n linearly independent solutions $\phi_1, \dots, \phi_n$ that span V. To this end, we choose a set of n linearly independent vectors $x_0^1, \dots, x_0^n$ in the n-dimensional x-space (i.e., in R^n or C^n). By the existence results in Chapter 1, if $t_0 \in J$, then there exist n solutions $\phi_1, \dots, \phi_n$ of (LH) such that $\phi_1(t_0) = x_0^1, \dots, \phi_n(t_0) = x_0^n$. We first show that these solutions are linearly independent. If on the contrary, these solutions are linearly dependent, there exist scalars $\alpha_1, \dots, \alpha_n \in F$, not all zero, such that $\sum_{i=1}^n \alpha_i\phi_i(t) = 0$ for all $t \in J$. This implies in particular that $\sum_{i=1}^n \alpha_i\phi_i(t_0) = \sum_{i=1}^n \alpha_i x_0^i = 0$. But this contradicts the assumption that $\{x_0^1, \dots, x_0^n\}$ is a linearly independent set. Therefore, the solutions $\phi_1, \dots, \phi_n$ are linearly independent.

To conclude the proof, we must show that the solutions $\phi_1, \dots, \phi_n$ span V. Let ϕ be any solution of (LH) on the interval J such that $\phi(t_0) = x_0$. Then there exist unique scalars $\alpha_1, \dots, \alpha_n \in F$ such that

$$x_0 = \sum_{i=1}^n \alpha_i x_0^i,$$

since, by assumption, the vectors $x_0^1, \dots, x_0^n$ form a basis for the x-space. Now

$$\psi = \sum_{i=1}^n \alpha_i\phi_i$$

is a solution of (LH) on J such that $\psi(t_0) = x_0$. But by the uniqueness results of Chapter 1 we have that

$$\phi = \psi = \sum_{i=1}^n \alpha\phi_i.$$

Since ϕ was chosen arbitrarily, it follows that the solutions $\phi_1, \dots, \phi_n$ span V. ■

EXAMPLE 3.1. Let $A(t)$ for (LH) be given by

$$A(t) = \begin{bmatrix} -1 & e^{2t} \\ 0 & -1 \end{bmatrix}. \tag{3.1}$$

It is easily verified by direct substitution that $\phi_1(t) = (e^{-t}, 0)^T$ and $\phi_2 = (\frac{1}{2}e^t, e^{-t})^T$ are solutions of (LH) on $J = (-\infty, \infty)$ for the present case. Furthermore, it is easily shown that ϕ_1 and ϕ_2 [defined on $J = (-\infty, \infty)$] are linearly independent (using, e.g., the method in Subsection 3.1B). In view of Theorem 3.1, the solutions of (LH) with $A(t)$ specified by (3.1) form a two-dimensional vector space and $\{\phi_1, \phi_2\}$ is a basis for this solution space. Since $\{\phi_1, \phi_2\}$ spans this vector space, all solutions of (LH) with $A(t)$ specified by (3.1) are of the form $\phi(t) = \alpha_1\phi_1(t) + \alpha_2\phi_2(t) = (\alpha_1 e^{-t} + \alpha_2(\frac{1}{2})e^t, \alpha_2 e^{-t})^T$, where $\alpha_1, \alpha_2 \in R$. ■

Fundamental matrix and properties

Theorem 3.1 enables us to make the following definition.

DEFINITION 3.1. A set of n linearly independent solutions of (LH) on J, $\{\phi_1, \ldots, \phi_n\}$, is called a *fundamental set of solutions* of (LH), and the $n \times n$ matrix

$$\Phi = [\phi_1, \phi_2, \ldots, \phi_n]$$

$$= \begin{bmatrix} \phi_{11} & \phi_{12} & \ldots & \phi_{1n} \\ \phi_{21} & \phi_{22} & \ldots & \phi_{2n} \\ \vdots & \vdots & & \vdots \\ \phi_{n1} & \phi_{n2} & \ldots & \phi_{nn} \end{bmatrix}$$

is called a *fundamental matrix* of (LH). ■

We note that there are infinitely many different fundamental sets of solutions of (LH) and, hence, infinitely many different fundamental matrices for (LH). We now study some of the basic properties of fundamental matrix.

In the next result, $X = [x_{ij}]$ denotes an $n \times n$ matrix, and the derivative of X with respect to t is defined as $\dot{X} = [\dot{x}_{ij}]$. Let $A(t)$ be the $n \times n$ matrix given in (LH). We call the system of n^2 equations

$$\dot{X} = A(t)X \tag{3.2}$$

a *matrix differential equation*.

THEOREM 3.2. A fundamental matrix Φ of (LH) satisfies the matrix equation (3.2) on the interval J.

Proof. We have $\dot{\Phi} = [\dot{\phi}_1, \dot{\phi}_2, \ldots, \dot{\phi}_n] = [A(t)\phi_1, A(t)\phi_2, \ldots, A(t)\phi_n] = A(t)[\phi_1, \phi_2, \ldots, \phi_n] = A(t)\Phi$. ■

The next result is called *Abel's formula*.

THEOREM 3.3. If Φ is a solution of the matrix equation (3.2) on an interval J and τ is any point of J, then

$$det\,\Phi(t) = det\,\Phi(\tau) \exp\left[\int_\tau^t tr\,A(s)\,ds\right]$$

for every $t \in J$. [$tr\,A(s) = tr\,[a_{ij}(s)]$ denotes the trace of $A(s)$, i.e., $tr\,A(s) = \sum_{j=1}^n a_{jj}(s)$.]

Proof. Let $\Phi = [\phi_{ij}]$. Then $\dot{\phi}_{ij} = \sum_{k=1}^{n} a_{ik}(t)\phi_{kj}$. Now

$$\frac{d}{dt}[det\,\Phi(t)] = \begin{vmatrix} \dot{\phi}_{11} & \dot{\phi}_{12} & \cdots & \dot{\phi}_{1n} \\ \phi_{21} & \phi_{22} & \cdots & \phi_{2n} \\ \vdots & \vdots & & \vdots \\ \phi_{n1} & \phi_{n2} & \cdots & \phi_{nn} \end{vmatrix} + \begin{vmatrix} \phi_{11} & \phi_{12} & \cdots & \phi_{1n} \\ \dot{\phi}_{21} & \dot{\phi}_{22} & \cdots & \dot{\phi}_{2n} \\ \vdots & \vdots & & \vdots \\ \phi_{n1} & \phi_{n2} & \cdots & \phi_{nn} \end{vmatrix}$$

$$+ \cdots + \begin{vmatrix} \phi_{11} & \phi_{12} & \cdots & \phi_{1n} \\ \phi_{21} & \phi_{22} & \cdots & \phi_{2n} \\ \vdots & \vdots & & \vdots \\ \dot{\phi}_{n1} & \dot{\phi}_{n2} & \cdots & \dot{\phi}_{nn} \end{vmatrix}$$

$$= \begin{vmatrix} \sum_{k=1}^{n} a_{1k}(t)\phi_{k1} & \sum_{k=1}^{n} a_{1k}(t)\phi_{k2} & \cdots & \sum_{k=1}^{n} a_{1k}(t)\phi_{kn}(t) \\ \phi_{21} & \phi_{22} & \cdots & \phi_{2n} \\ \vdots & \vdots & & \vdots \\ \phi_{n1} & \phi_{n2} & \cdots & \phi_{nn} \end{vmatrix}$$

$$+ \begin{vmatrix} \phi_{11} & \phi_{12} & \cdots & \phi_{1n} \\ \sum_{k=1}^{n} a_{2k}(t)\phi_{k1} & \sum_{k=1}^{n} a_{2k}(t)\phi_{k2} & \cdots & \sum_{k=1}^{n} a_{2k}(t)\phi_{kn} \\ \phi_{31} & \phi_{32} & \cdots & \phi_{3n} \\ \vdots & \vdots & & \vdots \\ \phi_{n1} & \phi_{n2} & \cdots & \phi_{nn} \end{vmatrix}$$

$$+ \cdots + \begin{vmatrix} \phi_{11} & \phi_{12} & \cdots & \phi_{1n} \\ \phi_{21} & \phi_{22} & \cdots & \phi_{2n} \\ \vdots & \vdots & & \vdots \\ \phi_{n-1,1} & \phi_{n-1,2} & \cdots & \phi_{n-1,n} \\ \sum_{k=1}^{n} a_{nk}(t)\phi_{k1} & \sum_{k=1}^{n} a_{nk}(t)\phi_{k2} & \cdots & \sum_{k=1}^{n} a_{nk}(t)\phi_{kn} \end{vmatrix}.$$

The first term in the above sum of determinants is unchanged if we subtract from the first row the quantity (a_{12} times the second row) + (a_{13} times the third row) + $\cdots$ + (a_{1n} times the nth row). This yields

$$\begin{vmatrix} \dot{\phi}_{11} & \dot{\phi}_{12} & \cdots & \dot{\phi}_{1n} \\ \phi_{21} & \phi_{22} & \cdots & \phi_{2n} \\ \vdots & \vdots & & \vdots \\ \phi_{n1} & \phi_{n2} & \cdots & \phi_{nn} \end{vmatrix} = \begin{vmatrix} a_{11}(t)\phi_{11} & a_{11}(t)\phi_{12} & \cdots & a_{11}(t)\phi_{1n} \\ \phi_{21} & \phi_{22} & \cdots & \phi_{2n} \\ \vdots & \vdots & & \vdots \\ \phi_{n1} & \phi_{n2} & \cdots & \phi_{nn} \end{vmatrix}$$
$$= a_{11}(t)\,det\,\Phi(t).$$

Repeating the above procedure for the remaining terms in the above sum of determinants, we have $(d/dt)[det\,\Phi(t)] = a_{11}(t)det\,\Phi(t) + a_{22}(t)det\,\Phi(t) + \cdots + a_{nn}(t)det\,\Phi(t) = [tr\,A(t)]det\,\Phi(t)$. This implies that $det\,\Phi(t) = det\,\Phi(\tau)\exp[\int_{\tau}^{t} tr\,A(s)\,ds]$. ■

Since in Theorem 3.3 τ is arbitrary, it follows that either $det\,\Phi(t) \neq 0$ for all $t \in J$ or $det\,\Phi(t) = 0$ for each $t \in J$. The next result provides a test on whether an $n \times n$ matrix $\Phi(t)$ is a fundamental matrix of (LH).

THEOREM 3.4. A solution Φ of the matrix equation (3.2) is a fundamental matrix of (LH) if and only if its determinant is nonzero for all $t \in J$.

Proof. If $\Phi = [\phi_1, \phi_2, \ldots, \phi_n]$ is a fundamental matrix for (LH), then the columns of Φ, $\phi_1, \ldots, \phi_n$, form a linearly independent set. Now let ϕ be a nontrivial solution of (LH). Then by Theorem 3.1 there exist unique scalars $\alpha_1, \ldots, \alpha_n \in F$, not all zero, such that $\phi = \sum_{j=1}^{n} \alpha_j \phi_j = \Phi a$, where $a^T = (\alpha_1, \ldots, \alpha_n)$. Let $t = \tau \in J$. Then $\phi(\tau) = \Phi(\tau)a$, which is a system of n linear algebraic equations. By construction, this system of equations has a unique solution for any choice of $\phi(\tau)$. Therefore, $\det \Phi(\tau) \neq 0$. It now follows from Theorem 3.3 that $\det \Phi(t) \neq 0$ for any $t \in J$.

Conversely, let Φ be a solution of (3.2) and assume that $\det \Phi(t) \neq 0$ for all $t \in J$. Then the columns of Φ are linearly independent for all $t \in J$. Hence, Φ is a fundamental matrix of (LH). ■

It is emphasized that a matrix may have identically zero determinant over some interval, even though its columns are linearly independent. For example, the columns of the matrix

$$\Phi(t) = \begin{bmatrix} 1 & t & t^2 \\ 0 & 1 & t \\ 0 & 0 & 0 \end{bmatrix}$$

are linearly independent, yet $\det \Phi(t) = 0$ for all $t \in (-\infty, \infty)$. In accordance with Theorem 3.4, the above matrix cannot be a fundamental solution of the matrix equation (3.2) for any continuous matrix $A(t)$.

EXAMPLE 3.2. Using the linearly independent solutions ϕ_1, ϕ_2 given in Example 3.1, we obtain

$$\Phi(t) = \begin{bmatrix} e^{-t} & \frac{1}{2}e^{t} \\ 0 & e^{-t} \end{bmatrix}, \qquad t \in (-\infty, \infty),$$

as a fundamental matrix of (LH) with $A(t)$ given by (3.1). It is easily verified that the matrix Φ satisfies the matrix equation $\dot{\Phi} = A\Phi$. Furthermore, $\det \Phi(t) = e^{-2t} \neq 0$, $t \in (-\infty, \infty)$, and $\det \Phi(t) = \det \Phi(\tau) \exp[\int_{\tau}^{t} tr\, A(s)\, ds] = e^{-2\tau} \exp[\int_{\tau}^{t} -2\, ds] = e^{-2\tau}e^{-2(t-\tau)} = e^{-2t}$, $t \in (-\infty, \infty)$, as expected. ■

THEOREM 3.5. If Φ is a fundamental matrix of (LH) and if C is any nonsingular constant $n \times n$ matrix, then ΦC is also a fundamental matrix of (LH). Moreover, if Ψ is any other fundamental matrix of (LH), then there exists a constant $n \times n$ nonsingular matrix P such that $\Psi = \Phi P$.

Proof. For the matrix ΦC we have $(d/dt)(\Phi C) = \dot{\Phi} C = [A(t)\Phi]C = A(t)(\Phi C)$, and therefore, ΦC is a solution of the matrix equation (3.2). Furthermore, since $\det \Phi(t) \neq 0$ for $t \in J$ and $\det C \neq 0$, it follows that $\det [\Phi(t)C] = [\det \Phi(t)](\det C) \neq 0$, $t \in J$. By Theorem 3.4, ΦC is a fundamental matrix.

Next, let Ψ be any other fundamental matrix of (LH) and consider the product $\Phi^{-1}(t)\Psi$. [Notice that since $\det \Phi(t) \neq 0$ for all $t \in J$, then $\Phi^{-1}(t)$ exists for all $t \in J$.] Also, consider $\Phi\Phi^{-1} = I$, where I denotes the $n \times n$ identity matrix. Differentiating both sides, we obtain $[(d/dt)\Phi]\Phi^{-1} + \Phi[(d/dt)\Phi^{-1}] = 0$ or $(d/dt)\Phi^{-1} = -\Phi^{-1}[(d/dt)\Phi]\Phi^{-1}$. Therefore, we can compute $(d/dt)(\Phi^{-1}\Psi) = \Phi^{-1}[(d/dt)\Psi] + [(d/dt)\Phi^{-1}]\Psi = \Phi^{-1}A(t)\Psi - \{\Phi^{-1}[(d/dt)\Phi]\Phi^{-1}\}\Psi = \Phi^{-1}A(t)\Psi - (\Phi^{-1}A(t)\Phi\Phi^{-1})\Psi = \Phi^{-1}A(t)\Psi - \Phi^{-1}A(t)\Psi = 0$. Hence, $\Phi^{-1}\Psi = P$ or $\Psi = \Phi P$. ■

EXAMPLE 3.3. It is easily verified that the system of equations

$$\begin{aligned} \dot{x}_1 &= 5x_1 - 2x_2 \\ \dot{x}_2 &= 4x_1 - x_2 \end{aligned} \tag{3.3}$$

has two linearly independent solutions given by $\phi_1(t) = (e^{3t}, e^{3t})^T$, $\phi_2(t) = (e^t, 2e^t)^T$, and therefore, the matrix

$$\Phi(t) = \begin{bmatrix} e^{3t} & e^t \\ e^{3t} & 2e^t \end{bmatrix} \tag{3.4}$$

is a fundamental matrix of (3.3).

Using Theorem 3.5 we can find the particular fundamental matrix Ψ of (3.3) that satisfies the initial condition $\Psi(0) = I$ by using $\Phi(t)$ given in (3.4). We have $\Psi(0) = I = \Phi(0)C$ or $C = \Phi^{-1}(0)$, and therefore,

$$C = \begin{bmatrix} 1 & 1 \\ 1 & 2 \end{bmatrix}^{-1} = \begin{bmatrix} 2 & -1 \\ -1 & 1 \end{bmatrix}$$

and

$$\Psi(t) = \Phi C = \begin{bmatrix} (2e^{3t} - e^t) & (-e^{3t} + e^t) \\ (2e^{3t} - 2e^t) & (-e^{3t} + 2e^t) \end{bmatrix}.$$

■

B. The State Transition Matrix

In Chapter 1 we used the *Method of Successive Approximations* (Theorem 10.9) to prove that for every $(t_0, x_0) \in J \times R^n$,

$$\dot{x} = A(t)x \tag{LH}$$

possesses a unique solution of the form

$$\phi(t, t_0, x_0) = \Phi(t, t_0)x_0,$$

such that $\phi(t_0, t_0, x_0) = x_0$, which exists for all $t \in J$, where $\Phi(t, t_0)$ is the *state transition matrix* (see Section 1.13). We derived an expression for $\Phi(t, t_0)$ in series form, called the *Peano-Baker series* [see Eq. (13.3) of Chapter 1], and we showed that $\Phi(t, t_0)$ is the unique solution of the matrix differential equation

$$\frac{\partial}{\partial t}\Phi(t, t_0) = A(t)\Phi(t, t_0), \tag{3.5}$$

where

$$\Phi(t_0, t_0) = I \qquad \text{for all } t \in J. \tag{3.6}$$

We provide an alternative formulation of state transition matrix and we study some of the properties of such matrices. In the following definition, we use the natural basis $\{e_1, e_2, \ldots, e_n\}$ that was defined in Section 2.2.

DEFINITION 3.2. A fundamental matrix Φ of (LH) whose columns are determined by the linearly independent solutions $\phi_1, \ldots, \phi_n$ with

$$\phi_1(t_0) = e_1, \ldots, \phi_n(t_0) = e_n, \qquad t_0 \in J,$$

is called *the state transition matrix* Φ for (LH). Equivalently, if Ψ is *any* fundamental matrix of (LH), then the matrix Φ determined by

$$\Phi(t, t_0) \triangleq \Psi(t)\Psi^{-1}(t_0) \qquad \text{for all } t, t_0 \in J,$$

is said to be *the state transition matrix of* (LH). ■

We note that the state transition matrix of (LH) is *uniquely* determined by the matrix $A(t)$ and is *independent* of the particular choice of the fundamental matrix. To show this, let Ψ_1 and Ψ_2 be two different fundamental matrices of (LH). Then by Theorem 3.5 there exists a constant $n \times n$ nonsingular matrix P such that $\Psi_2 = \Psi_1 P$. Now by the definition of state transition matrix, we have $\Phi(t, t_0) = \Psi_2(t)[\Psi_2(t_0)]^{-1} = \Psi_1(t)PP^{-1}[\Psi_1(t_0)]^{-1} = \Psi_1(t)[\Psi_1(t_0)]^{-1}$. This shows that $\Phi(t, t_0)$ is independent of the fundamental matrix chosen.

EXAMPLE 3.4. In Examples 3.1 and 3.2 we showed that for (LH) with $A(t)$ given by (3.1),

$$\Psi(t) = \begin{bmatrix} e^{-t} & \frac{1}{2}e^{t} \\ 0 & e^{-t} \end{bmatrix}, \qquad t \in (-\infty, \infty),$$

is a fundamental matrix. For this case, the state transition matrix $\Phi(t, t_0)$ is computed as

$$\Phi(t, t_0) = \Psi(t)\Psi^{-1}(t_0) = \begin{bmatrix} e^{-t} & \frac{1}{2}e^{t} \\ 0 & e^{-t} \end{bmatrix} \begin{bmatrix} e^{-t_0} & \frac{1}{2}e^{t_0} \\ 0 & e^{-t_0} \end{bmatrix}^{-1}$$

$$= \begin{bmatrix} e^{-(t-t_0)} & \frac{1}{2}(e^{t+t_0} - e^{-t+3t_0}) \\ 0 & e^{-(t-t_0)} \end{bmatrix}.$$

The reader should verify that this matrix satisfies Eqs. (3.5) and (3.6). ■

Properties of the state transition matrix

In the following, we summarize some of the properties of state transition matrix.

THEOREM 3.6. Let $t_0 \in J$, let $\phi(t_0) = x_0$, and let $\Phi(t, t_0)$ denote the state transition matrix for (LH) for all $t \in J$. Then the following statements are true:

(i) $\Phi(t, t_0)$ is the unique solution of the matrix equation $(\partial/\partial t)\Phi(t, t_0) = A(t)\Phi(t, t_0)$ with $\Phi(t_0, t_0) = I$, the $n \times n$ identity matrix.

(ii) $\Phi(t, t_0)$ is nonsingular for all $t \in J$.

(iii) For any $t, \sigma, \tau \in J$, we have $\Phi(t, \tau) = \Phi(t, \sigma)\Phi(\sigma, \tau)$ (semigroup property).

(iv) $[\Phi(t, t_0)]^{-1} \triangleq \Phi^{-1}(t, t_0) = \Phi(t_0, t)$ for all $t, t_0 \in J$.

(v) The unique solution $\phi(t, t_0, x_0)$ of (LH), with $\phi(t_0, t_0, x_0) = x_0$ specified, is given by

$$\phi(t, t_0, x_0) = \Phi(t, t_0)x_0 \qquad \text{for all } t \in J. \tag{3.7}$$

Proof

(i) For any fundamental matrix of (LH), say, Ψ, we have, by definition, $\Phi(t, t_0) = \Psi(t)\Psi^{-1}(t_0)$, independent of the choice of Ψ. Therefore, $\partial\Phi(t, t_0)/\partial t = \dot{\Psi}(t)\Psi^{-1}(t_0) = A(t)\Psi(t)\Psi^{-1}(t_0) = A(t)\Phi(t, t_0)$. Furthermore, $\Phi(t_0, t_0) = \Psi(t_0)\Psi^{-1}(t_0) = I$.

(ii) For any fundamental matrix of (LH) we have that $det\,\Psi(t) \neq 0$ for all $t \in J$. Therefore, $det\,\Phi(t, t_0) = det\,[\Psi(t)\Psi^{-1}(t_0)] = det\,\Psi(t) det\,\Psi^{-1}(t_0) \neq 0$ for all $t, t_0 \in J$.

(iii) For any fundamental matrix Ψ of (LH) and for the state transition matrix Φ of (LH), we have $\Phi(t, \tau) = \Psi(t)\Psi^{-1}(\tau) = \Psi(t)\Psi^{-1}(\sigma)\Psi(\sigma)\Psi^{-1}(\tau) = \Phi(t, \sigma)\Phi(\sigma, \tau)$ for any $t, \sigma, \tau \in J$.

(iv) Let Ψ be any fundamental matrix of (LH) and let Φ be the state transition matrix of (LH). Then $[\Phi(t, t_0)]^{-1} = [\Psi(t)\Psi(t_0)^{-1}]^{-1} = \Psi(t_0)\Psi^{-1}(t) = \Phi(t_0, t)$ for any $t, t_0 \in J$.

(v) By the results established in Chapter 1, we know that for every $(t_0, x_0) \in D$, (LH) has a unique solution $\phi(t)$ for all $t \in J$ with $\phi(t_0) = x_0$. To verify (3.7), we note that $\dot{\phi}(t) = [\partial\Phi(t, t_0)/\partial t]x_0 = A(t)\Phi(t, t_0)x_0 = A(t)\phi(t)$. ■

EXAMPLE 3.5. Let $x(t_0) = (a_1, a_2)^T$. In view of Example 3.4, we have for (LH) given in Example 3.1,

$$\phi(t, t_0, x_0) = \begin{bmatrix} e^{-(t-t_0)} & \frac{1}{2}(e^{t+t_0} - e^{-t+3t_0}) \\ 0 & e^{-(t-t_0)} \end{bmatrix} \begin{bmatrix} a_1 \\ a_2 \end{bmatrix},$$

and therefore,

$$\phi_1(t, t_0, x_0) = e^{-(t-t_0)}a_1 + \tfrac{1}{2}(e^{t+t_0} - e^{-t+3t_0})a_2 \tag{3.8}$$

$$\phi_2(t, t_0, x_0) = e^{-(t-t_0)}a_2. \tag{3.9}$$

■

We can verify the above example by simple integration. In doing so, we first obtain $\phi_2(t, t_0, x_0)$ by integrating both sides of $\dot{x}_2 = -x_2$ and by using the initial condition $x_2(t_0) = a_2$. Next, we obtain $\phi_1(t, t_0, x_0)$ by integrating both sides of $\dot{x}_1 = -x_1 + e^{2t}\phi_2(t, t_0, x_0)$ and using the initial condition $(x_1(t_0), x_2(t_0))^T = (a_1, a_2)^T$. Note that this procedure for solving an initial-value problem determined by (LH) is valid for any triangular matrix $A(t)$.

In Chapter 1 we pointed out that the state transition matrix $\Phi(t, t_0)$ maps the solution (state) of (LH) at time t_0 to the solution (state) of (LH) at time t. Since there is no restriction on t relative to t_0 (i.e., we may have $t < t_0, t = t_0$, or $t > t_0$), we can "move forward or backward" in time. Indeed, given the solution (state) of (LH) at time t, we can solve the solution (state) of (LH) at time t_0. Thus, $x(t_0) = x_0 = [\Phi(t, t_0)]^{-1}\phi(t, t_0, x_0) = \Phi(t_0, t)\phi(t, t_0, x_0)$. This "reversibility in time" is possible because $\Phi^{-1}(t, t_0)$ always exists. [In the case of discrete-time systems described by difference equations, this reversibility in time does in general not exist (refer to Section 2.7).]

C. Nonhomogeneous Equations

In Section 1.13, we proved the following result [refer to Eqs. (13.8) to (13.10)].

THEOREM 3.7. Let $t_0 \in J$, let $(t_0, x_0) \in D$, and let $\Phi(t, t_0)$ denote the state transition matrix for (LH) for all $t \in J$. Then the unique solution $\phi(t, t_0, x_0)$ of (LN) satisfying $\phi(t_0, t_0, x_0) = x_0$ is given by

$$\phi(t, t_0, x_0) = \Phi(t, t_0)x_0 + \int_{t_0}^{t} \Phi(t, \eta)g(\eta)\,d\eta. \tag{3.10}$$

■

As pointed out in Section 1.13, when $x_0 = 0$, (3.10) reduces to

$$\phi(t, t_0, 0) \triangleq \phi_p(t) = \int_{t_0}^{t} \Phi(t, s)g(s)\,ds, \tag{3.11}$$

and when $x_0 \neq 0$, but $g(t) \equiv 0$, (3.10) reduces to

$$\phi(t, t_0, x_0) \triangleq \phi_h(t) = \Phi(t, t_0)x_0, \tag{3.12}$$

and the solution of (LN) may be viewed as consisting of a component that is due to the initial data x_0 and another component that is due to the forcing term $g(t)$. We recall that ϕ_p is called a *particular solution* of the nonhomogeneous system (LN), while ϕ_h is called the *homogeneous solution*.

There are of course other methods of solving (LN). For example, in the common approach to solving linear differential equations, all solutions $\phi(t)$ of $\dot{x} - A(t)x =$

$g(t)$ are assumed to be of the form $\phi(t) = \phi_h(t)+\phi_p(t)$, where $\phi_h(t)$ is the solution of the homogeneous equation $\dot{x} - A(t)x = 0$ and $\phi_p(t)$ is a particular solution. It is not difficult to see that $\phi_h(t) = \Phi(t, t_0)\alpha$, where $\alpha \in R^n$ is to be determined [compare to (3.12)]. Therefore, $\phi(t) = \Phi(t, t_0)\alpha + \phi_p(t)$. The vector α is determined using the initial conditions on the solution $\phi(t)$ of $\dot{x} - A(t)x = g(t)$, namely, $\phi(t_0) = x_0$. Substituting $\phi(t_0) = \Phi(t_0, t_0)\alpha + \phi_p(t_0)$, we obtain $\alpha = \phi(t_0) = x_0 - \phi_p(t_0)$. The solution to (LN) with $\phi(t_0) = x_0$ is therefore given by $\phi(t) = \phi_h(t) + \phi_p(t) = \Phi(t, t_0)[\phi(t_0) - \phi_p(t_0)] + \phi_p(t)$, or

$$\phi(t) = \Phi(t, t_0)x_0 + [\phi_p(t) - \Phi(t, t_0)\phi_p(t_0)]. \tag{3.13}$$

As expected, the "*zero-state response of* (LN)," obtained by letting $x_0 = 0$ in (3.10) and given in (3.11), is a particular solution $\phi_p(t)$ of $\dot{x} - A(t)x = g(t)$ with the property that $\phi_p(t_0) = 0$. Furthermore, the "*zero-input response of* (LN)," obtained by letting $g(t) \equiv 0$ in (3.10), is the solution $\phi_h(t)$ of $\dot{x} - A(t)x = 0$. It follows that the *variation of constants formula* (3.10) is a special form of expression (3.13), corresponding to the chosen particular solution $\phi_p(t)$; for a different choice of $\phi_p(t)$, a correspondingly different expression for the solution $\phi(t)$ is obtained.

EXAMPLE 3.6. In (LN), let $x \in R^2$, $t_0 = 0$, and

$$A(t) = \begin{bmatrix} -1 & e^{2t} \\ 0 & -1 \end{bmatrix}, \qquad g(t) = \begin{bmatrix} e^{-t} \\ 0 \end{bmatrix}, \qquad x(0) = \begin{bmatrix} 0 \\ 1 \end{bmatrix}.$$

The state transition matrix for $A(t)$ has been determined in Example 3.4 as

$$\Phi(t, t_0) = \Phi(t, 0) = \begin{bmatrix} e^{-t} & \frac{1}{2}(e^t - e^{-t}) \\ 0 & e^{-t} \end{bmatrix}.$$

Using (3.12), we obtain

$$\phi_h(t, t_0, x_0) = \Phi(t, 0)x(0) = \begin{bmatrix} \frac{1}{2}(e^t - e^{-t}) \\ e^{-t} \end{bmatrix},$$

using (3.11), we have

$$\phi_p(t, t_0, x_0) = \int_0^t \Phi(t, \eta)g(\eta)\, d\eta = \begin{bmatrix} te^{-t} \\ 0 \end{bmatrix},$$

and using (3.10), we finally obtain

$$\phi(t, t_0, x_0) = \phi_h(t, t_0, x_0) + \phi_p(t, t_0, x_0) = \begin{bmatrix} e^{-t}(t - \frac{1}{2}) + \frac{1}{2}e^t \\ e^{-t} \end{bmatrix}.$$

■

D. How to Determine $\Phi(t, t_0)$

To solve (LN) or (LH) in closed form, we require an expression for the state transition matrix $\Phi(t, t_0)$. We have seen in Example 3.5 that when $A(t)$ is triangular, we can solve (LH) [and hence, $\Phi(t, t_0)$] by sequential integration of the individual differential equations. For the general case, however, closed-form determination of $\Phi(t, t_0)$ is not possible.

In the following, we identify another important class of matrices $A(t)$ for which a closed-form expression of $\Phi(t, t_0)$ exists.

THEOREM 3.8. If for every τ, t we have

$$A(t)\left[\int_\tau^t A(\eta)\, d\eta\right] = \left[\int_\tau^t A(\eta)\, d\eta\right]A(t), \tag{3.14}$$

then
$$\Phi(t,\tau) = e^{\int_\tau^t A(\eta)d\eta} \triangleq I + \sum_{k=1}^{\infty} \frac{1}{k!}\left[\int_\tau^t A(\eta)\,d\eta\right]^k. \tag{3.15}$$

Proof. We recall that the general term of the *Peano-Baker series* [see Eq. (13.3) in Chapter 1] is given by
$$\int_\tau^t A(s_1)\int_\tau^{s_1} A(s_2)\cdots\int_\tau^{s_{m-1}} A(s_m)\,ds_m\,ds_{m-1}\cdots ds_1. \tag{3.16}$$
We wish to show that under the present assumptions, expression (3.16) is equal to the expression
$$\frac{1}{m!}\left[\int_\tau^t A(s)\,ds\right]^m. \tag{3.17}$$
To verify (3.17), we will apply the identity
$$\int_\tau^t A(r)\left[\int_\tau^r A(s)\,ds\right]^m dr = \frac{1}{m+1}\left[\int_\tau^t A(s)\,ds\right]^{m+1} \tag{3.18}$$
repeatedly to (3.16). First, however, we verify the validity of (3.18). To accomplish this, we first observe that (3.18) is true for any fixed $t = \tau$. Differentiating the left-hand side of (3.18), we obtain
$$\frac{\partial}{\partial t}\left[\int_\tau^t A(r)\left[\int_\tau^r A(s)\,ds\right]^m dr\right] = A(t)\left[\int_\tau^t A(s)\,ds\right]^m. \tag{3.19}$$
Differentiating the right-hand side of (3.18), we have that
$$\begin{aligned}
&\frac{\partial}{\partial t}\left\{\frac{1}{m+1}\left[\int_\tau^t A(s)\,ds\right]^{m+1}\right\}\\
&\quad= \frac{1}{m+1}\left\{A(t)\int_\tau^t A(s_2)\,ds_2\cdots\int_\tau^t A(s_{m+1})\,ds_{m+1}\right.\\
&\qquad+\left(\int_\tau^t A(s_1)\,ds_1\right)A(t)\int_\tau^t A(s_3)\,ds_3\cdots\int_\tau^t A(s_{m+1})\,ds_{m+1}+\cdots\\
&\qquad\left.+\int_\tau^t A(s_1)\,ds_1\cdots\left(\int_\tau^t A(s_m)\,ds_m\right)A(t)\right\} = A(t)\left[\int_\tau^t A(s)\,ds\right]^m
\end{aligned} \tag{3.20}$$
where in the last step of (3.20), the assumption (3.14) has been used repeatedly. Using (3.19) and (3.20), we obtain (3.18).

To complete the proof of the theorem, we apply (3.18) repeatedly to (3.16) to obtain (3.17), the general term of the Peano-Baker series (13.13) in Chapter 1. Indeed, we have
$$\begin{aligned}
&\int_\tau^t A(s_1)\int_\tau^{s_1} A(s_2)\cdots\int_\tau^{s_{m-2}} A(s_{m-1})\int_\tau^{s_{m-1}} A(s_m)\,ds_m\,ds_{m-1}\,ds_{m-2}\cdots ds_1\\
&\quad= \int_\tau^t A(s_1)\int_\tau^{s_1} A(s_2)\cdots\int_\tau^{s_{m-3}} A(s_{m-2})\frac{1}{2}\left[\int_\tau^{s_{m-2}} A(s)\,ds\right]^2 ds_{m-2}\cdots ds_1\\
&\quad= \frac{1}{2}\int_\tau^t A(s_1)\int_\tau^{s_1} A(s_2)\cdots\int_\tau^{s_{m-4}} A(s_{m-3})\frac{1}{3}\left[\int_\tau^{s_{m-3}} A(s)\,ds\right] ds_{m-3}\cdots ds_1\\
&\quad= \cdots = \frac{1}{m!}\left[\int_\tau^t A(s)\,ds\right]^m
\end{aligned}$$
which was to be shown. This completes the proof of the theorem. ■

For the scalar case, i.e., when $A(t) = a(t)$, relation (3.14) is always true. Also, when $A(t) = diag\,[a_{ii}(t)]$ [i.e., $A(t)$ is a diagonal matrix], relation (3.14) is true. Furthermore, for $A(t) = A$, a constant matrix, (3.14) will always hold. The reader can readily verify that $A(t)$ given in Example 3.1 also satisfies relation (3.14).

We conclude by pointing out that for $A \in C[R, R^{n\times n}]$, (3.14) is true if and only if

$$A(t)A(\tau) = A(\tau)A(t) \tag{3.21}$$

for all t and τ. We ask the reader to verify the validity of this statement.

2.4 LINEAR SYSTEMS WITH CONSTANT COEFFICIENTS

In this section we consider systems of linear, autonomous, homogeneous ordinary differential equations

$$\dot{x} = Ax \tag{L}$$

and systems of linear nonhomogeneous ordinary differential equations

$$\dot{x} = Ax + g(t), \tag{4.1}$$

where $x \in R^n$, $A \in R^{n\times n}$, and $g \in C(R, R^n)$. In the special case when $A(t) \equiv A$, system (LH) reduces to system (L) and system (LN) reduces to system (4.1). Consequently, the results of Section 2.3 are applicable to (L) as well as to (LH) and to (4.1) as well as to (LN). However, because of the special nature of (L) and (4.1), more detailed information can be determined.

A. Some Properties of e^{At}

Let $D = \{(t, x) : t \in R, x \in R^n\}$. In view of the results of Section 1.13, it follows that for every $(t_0, x_0) \in D$, the unique solution of (L) is given by

$$\begin{aligned}\phi(t, t_0, x_0) &= \left[I + \sum_{k=1}^{\infty} \frac{A^k(t-t_0)^k}{k!}\right] x_0 \\ &= \Phi(t, t_0)x_0 \triangleq \Phi(t - t_0)x_0 \triangleq e^{A(t-t_0)}x_0,\end{aligned} \tag{4.2}$$

where $\Phi(t - t_0) = e^{A(t-t_0)}$ denotes the state transition matrix for (L). [By writing $\Phi(t, t_0) = \Phi(t - t_0)$, we are using a slight abuse of notation.]

In arriving at (4.2) we invoked Theorem 10.9 of Chapter 1 in Section 1.13, to show that the sequence $\{\phi_m\}$, where

$$\phi_m(t, t_0, x_0) = \left[I + \sum_{k=1}^{m} \frac{A^k(t-t_0)^k}{k!}\right] x_0 \triangleq S_m(t - t_0)x_0, \tag{4.3}$$

converges uniformly and absolutely as $m \to \infty$ to the unique solution $\phi(t, t_0, x_0)$ of (L) given by (4.2) on compact subsets of R. In the process of arriving at this result, we also proved the following results.

THEOREM 4.1. Let A be a constant $n \times n$ matrix (which may be real or complex) and let $S_m(t)$ denote the partial sum of matrices defined by

$$S_m(t) = I + \sum_{k=1}^{m} \frac{t^k}{k!} A^k. \tag{4.4}$$

Then each element of the matrix $S_m(t)$ converges absolutely and uniformly on any finite t interval $(-a, a)$, $a > 0$, as $m \to \infty$. Furthermore, $\dot{S}_m(t) = AS_{m-1}(t) = S_{m-1}(t)A$, and thus, the limit of $S_m(t)$ as $t \to \infty$ is a C^1 function on R. Moreover, this limit commutes with A. ■

In view of the above result, the following definition makes sense (see also Section 1.13).

DEFINITION 4.1. Let A be a constant $n \times n$ matrix (which may be real or complex). We define e^{At} to be the matrix

$$e^{At} = I + \sum_{k=1}^{\infty} \frac{t^k}{k!} A^k \tag{4.5}$$

for any $-\infty < t < \infty$, and we call e^{At} a *matrix exponential*. ■

We are now in a position to provide the following characterizations of e^{At}.

THEOREM 4.2. Let $J = R$, $t_0 \in J$, and let A be a given constant matrix for (L). Then

(i) $\Phi(t) \triangleq e^{At}$ is a fundamental matrix for all $t \in J$.

(ii) The state transition matrix for (L) is given by $\Phi(t, t_0) = e^{A(t-t_0)} \triangleq \Phi(t - t_0)$, $t \in J$.

(iii) $e^{At_1} e^{At_2} = e^{A(t_1+t_2)}$ for all $t_1, t_2 \in J$.

(iv) $Ae^{At} = e^{At}A$ for all $t \in J$.

(v) $(e^{At})^{-1} = e^{-At}$ for all $t \in J$.

Proof. By (4.5) and Theorem 4.1 we have that $(d/dt)[e^{At}] = \lim_{m\to\infty} AS_m(t) = \lim_{m\to\infty} S_m(t)A = Ae^{At} = e^{At}A$. Therefore, $\Phi(t) = e^{At}$ is a solution of the matrix equation $\dot{\Phi} = A\Phi$. Next, observe that $\Phi(0) = I$. It follows from Theorem 3.3 that $det\,[e^{At}] = e^{tr(At)} \neq 0$ for all $t \in R$. Therefore, by Theorem 3.4 $\Phi(t) = e^{At}$ is a fundamental matrix for (L). We have proved parts (i) and (iv).

To prove (iii), we note that in view of Theorem 3.6(iii), we have for any $t_1, t_2 \in R$ that $\Phi(t_1, t_2) = \Phi(t_1, 0)\Phi(0, t_2)$. By Theorem 3.6(i) we see that $\Phi(t, t_0)$ solves (L) with $\Phi(t_0, t_0) = I$. It was just proved that $\Psi(t) \triangleq e^{A(t-t_0)}$ is also a solution. By uniqueness, it follows that $\Phi(t, t_0) = e^{A(t-t_0)}$. For $t = t_1, t_0 = -t_2$, we therefore obtain $e^{A(t_1+t_2)} = \Phi(t_1, -t_2) = \Phi(t_1)\Phi(-t_2)^{-1}$, and for $t = t_1, t_0 = 0$, we have $\Phi(t_1, 0) = e^{At_1} = \Phi(t_1)$. Also, for $t = 0, t_0 = -t_2$, we obtain $\Phi(0, -t_2) = e^{t_2 A} = \Phi(-t_2)^{-1}$. Therefore, $e^{A(t_1+t_2)} = e^{At_1} e^{At_2}$ for all $t_1, t_2 \in R$.

Finally, to prove (ii), we note that by (iii) we have $\Phi(t, t_0) \triangleq e^{A(t-t_0)} = I + \sum_{k=1}^{\infty} [(t - t_0)^k/k!]A^k = \Phi(t - t_0)$ is a fundamental matrix for (L) with $\Phi(t_0, t_0) = I$. Therefore, it is a state transition matrix for (L). ■

We conclude this section by stating the solution of (4.1),

$$\begin{aligned} \phi(t, t_0, x_0) &= \Phi(t - t_0)x_0 + \int_{t_0}^{t} \Phi(t - s)g(s)\,ds \\ &= e^{A(t-t_0)}x_0 + \int_{t_0}^{t} e^{A(t-s)}g(s)\,ds \\ &= e^{A(t-t_0)}x_0 + e^{At}\int_{t_0}^{t} e^{-As}g(s)\,ds, \end{aligned} \tag{4.6}$$

for all $t \in R$. In arriving at (4.6), we have used expression (13.8) of Chapter 1 and the fact that in the present case, $\Phi(t, t_0) = e^{A(t-t_0)}$.

B. How to Determine e^{At}

We begin by considering the specific case

$$A = \begin{bmatrix} 0 & \alpha \\ 0 & 0 \end{bmatrix}. \tag{4.7}$$

From (4.5) it follows immediately that

$$e^{At} = I + tA = \begin{bmatrix} 1 & \alpha t \\ 0 & 1 \end{bmatrix}. \tag{4.8}$$

As another example, we consider

$$A = \begin{bmatrix} \lambda_1 & 0 \\ 0 & \lambda_2 \end{bmatrix}, \tag{4.9}$$

where $\lambda_1, \lambda_2 \in R$. Again, from (4.5) it follows that

$$\begin{aligned} e^{At} &= \begin{bmatrix} 1 + \sum_{k=1}^{\infty} \frac{t^k}{k!}\lambda_1^k & 0 \\ 0 & 1 + \sum_{k=1}^{\infty} \frac{t^k}{k!}\lambda_2^k \end{bmatrix} \\ &= \begin{bmatrix} e^{\lambda_1 t} & 0 \\ 0 & e^{\lambda_2 t} \end{bmatrix}. \end{aligned} \tag{4.10}$$

Unfortunately, in general it is much more difficult to evaluate the matrix exponential than the preceding examples suggest. In the following, we consider several methods of evaluating e^{At}.

The infinite series method

In this case we evaluate the partial sum $S_m(t)$ (see Theorem 4.1)

$$S_m(t) = I + \sum_{k=1}^{m} \frac{t^k}{k!}A^k$$

for some fixed t, say, t_1, and for $m = 1, 2, \ldots$ until no significant changes occur in succeeding sums. This yields the matrix e^{At_1}. This method works reasonably well if the smallest and largest eigenvalues of A are not widely separated.

In the same spirit as above, we could use any of the vector differential solvers to solve $\dot{x} = Ax$, using the natural basis for R^n as n linearly independent initial conditions [i.e., using as initial conditions the vectors $e_1 = (1, 0, \ldots, 0)^T$, $e_2 = (0, 1, 0, \ldots, 0)^T, \ldots, e_n = (0, \ldots, 0, 1)^T$] and observing that in view of (4.2), the resulting solutions are the columns of e^{At} (with $t_0 = 0$).

EXAMPLE 4.1. There are cases when the definition of e^{At} (in series form) directly produces a closed-form expression. This occurs for example when $A^k = 0$ for some k. In particular, if all the eigenvalues of A are at the origin, then $A^k = 0$ for some $k \le n$. In this case, only a finite number of terms in (4.5) will be nonzero and e^{At} can be evaluated in closed form. This was precisely the case in (4.7). ■

The similarity transformation method

Let us consider the initial-value problem

$$\dot{x} = Ax, \qquad x(t_0) = x_0, \tag{4.11}$$

let P be a real $n \times n$ nonsingular matrix, and consider the transformation $x = Py$, or equivalently, $y = P^{-1}x$. Differentiating both sides with respect to t, we obtain $\dot{y} = P^{-1}\dot{x} = P^{-1}APy = Jy$, $y(t_0) = y_0 = P^{-1}x_0$. The solution of the above equation is given by

$$\psi(t, t_0, y_0) = e^{J(t-t_0)}P^{-1}x_0. \tag{4.12}$$

Using (4.12) and $x = Py$, we obtain for the solution of (4.11)

$$\phi(t, t_0, x_0) = Pe^{J(t-t_0)}P^{-1}x_0. \tag{4.13}$$

Now suppose that the similarity transformation P given above has been chosen in such a manner that

$$J = P^{-1}AP \tag{4.14}$$

is in Jordan canonical form (see Subsection 2.2O). We first consider the case when A has n linearly independent eigenvectors, say, v_i, that correspond to the eigenvalues λ_i (not necessarily distinct), $i = 1, \ldots, n$. (Necessary and sufficient conditions for this to be the case are given in Section 2.2. A sufficient condition for the eigenvectors v_i, $i = 1, \ldots, n$, to be linearly independent is that the eigenvalues of A, $\lambda_1, \ldots, \lambda_n$, be distinct.) Then P can be chosen so that $P = [v_1, \ldots, v_n]$ and the matrix $J = P^{-1}AP$ assumes the form

$$J = \begin{bmatrix} \lambda_1 & & 0 \\ & \ddots & \\ 0 & & \lambda_n \end{bmatrix}. \tag{4.15}$$

Using the power series representation

$$e^{Jt} = I + \sum_{k=1}^{\infty} \frac{t^k J^k}{k!}, \tag{4.16}$$

we immediately obtain the expression

$$e^{Jt} = \begin{bmatrix} e^{\lambda_1 t} & & 0 \\ & \ddots & \\ 0 & & e^{\lambda_n t} \end{bmatrix}. \tag{4.17}$$

Accordingly, the solution of the initial-value problem (4.11) is now given by

$$\phi(t, t_0, x_0) = P \begin{bmatrix} e^{\lambda_1(t-t_0)} & & 0 \\ & \ddots & \\ 0 & & e^{\lambda_n(t-t_0)} \end{bmatrix} P^{-1}x_0. \tag{4.18}$$

In the general case when A has repeated eigenvalues, it is no longer possible to diagonalize A (see Subsection 2.2L). However, we can generate n linearly independent vectors $v_1, \ldots, v_n$ and an $n \times n$ similarity transformation $P = [v_1, \ldots, v_n]$ that takes A into the Jordan canonical form $J = P^{-1}AP$. Here J is in the block diagonal form given by

$$J = \begin{bmatrix} J_0 & & & 0 \\ & J_1 & & \\ & & \ddots & \\ 0 & & & J_s \end{bmatrix}, \tag{4.19}$$

where J_0 is a diagonal matrix with diagonal elements $\lambda_1, \ldots, \lambda_k$ (not necessarily distinct), and each J_i, $i \geq 1$, is an $n_i \times n_i$ matrix of the form

$$J_i = \begin{bmatrix} \lambda_{k+i} & 1 & 0 & \ldots & 0 \\ 0 & \lambda_{k+i} & 1 & \ldots & 0 \\ \vdots & \vdots & \ddots & \ddots & \vdots \\ 0 & 0 & \ldots & \ddots & 1 \\ 0 & 0 & 0 & \ldots & \lambda_{k+i} \end{bmatrix}, \tag{4.20}$$

where λ_{k+i} need not be different from λ_{k+j} if $i \neq j$, and where $k+n_1+\cdots+n_s = n$.

Now since for any square block diagonal matrix

$$C = \begin{bmatrix} C_1 & & 0 \\ & \ddots & \\ 0 & & C_l \end{bmatrix}$$

with C_i, $i = 1, \ldots, l$, square, we have that

$$C^k = \begin{bmatrix} C_1^k & & 0 \\ & \ddots & \\ 0 & & C_l^k \end{bmatrix},$$

it follows from the power series representation of e^{Jt} that

$$e^{Jt} = \begin{bmatrix} e^{J_0 t} & & & 0 \\ & e^{J_1 t} & & \\ & & \ddots & \\ 0 & & & e^{J_s t} \end{bmatrix}, \tag{4.21}$$

$t \in R$. As shown earlier, we have

$$e^{J_0 t} = \begin{bmatrix} e^{\lambda_1 t} & & 0 \\ & \ddots & \\ 0 & & e^{\lambda_k t} \end{bmatrix}. \tag{4.22}$$

For J_i, $i = 1, \ldots, s$, we have

$$J_i = \lambda_{k+i} I_i + N_i, \tag{4.23}$$

where I_i denotes the $n_i \times n_i$ identity matrix and N_i is the $n_i \times n_i$ nilpotent matrix given by

$$N_i = \begin{bmatrix} 0 & 1 & \ldots & 0 \\ \vdots & \ddots & \ddots & \vdots \\ \vdots & & \ddots & 1 \\ 0 & \ldots & \ldots & 0 \end{bmatrix}. \tag{4.24}$$

Since $\lambda_{k+i}I_i$ and N_i commute, we have that

$$e^{J_i t} = e^{\lambda_{k+i}t}e^{N_i t}. \tag{4.25}$$

Repeated multiplication of N_i by itself results in $N_i^k = 0$ for all $k \geq n_i$. Therefore, the series defining e^{tN_i} terminates, resulting in

$$e^{tJ_i} = e^{\lambda_{k+i}t}\begin{bmatrix} 1 & t & \cdots & \dfrac{t^{n_i-1}}{(n_i-1)!} \\ 0 & 1 & \cdots & \dfrac{t^{n_i-2}}{(n_i-2)!} \\ \vdots & \vdots & \ddots & \vdots \\ 0 & 0 & \cdots & 1 \end{bmatrix}, \qquad i = 1, \ldots, s. \tag{4.26}$$

It now follows that the solution of (4.11) is given by

$$\phi(t, t_0, x_0) = P\begin{bmatrix} e^{J_0(t-t_0)} & 0 & \cdots & 0 \\ 0 & e^{J_1(t-t_0)} & \cdots & 0 \\ \vdots & \vdots & \ddots & \vdots \\ 0 & 0 & & e^{J_s(t-t_0)} \end{bmatrix}P^{-1}x_0. \tag{4.27}$$

EXAMPLE 4.2. In system (4.11), let $A = \begin{bmatrix} -1 & 2 \\ 0 & 1 \end{bmatrix}$. The eigenvalues of A are $\lambda_1 = -1$ and $\lambda_2 = 1$, and corresponding eigenvectors for A are given by $v_1 = (1, 0)^T$ and $v_2 = (1, 1)^T$, respectively. Then $P = [v_1, v_2] = \begin{bmatrix} 1 & 1 \\ 0 & 1 \end{bmatrix}$, $P^{-1} = \begin{bmatrix} 1 & -1 \\ 0 & 1 \end{bmatrix}$, and $J = P^{-1}AP = \begin{bmatrix} 1 & -1 \\ 0 & 1 \end{bmatrix}\begin{bmatrix} -1 & 2 \\ 0 & 1 \end{bmatrix}\begin{bmatrix} 1 & 1 \\ 0 & 1 \end{bmatrix} = \begin{bmatrix} -1 & 0 \\ 0 & 1 \end{bmatrix} = \begin{bmatrix} \lambda_1 & 0 \\ 0 & \lambda_2 \end{bmatrix}$, as expected. We obtain

$$e^{At} = Pe^{Jt}P^{-1} = \begin{bmatrix} 1 & 1 \\ 0 & 1 \end{bmatrix}\begin{bmatrix} e^{-t} & 0 \\ 0 & e^t \end{bmatrix}\begin{bmatrix} 1 & -1 \\ 0 & 1 \end{bmatrix} = \begin{bmatrix} e^{-t} & e^t - e^{-t} \\ 0 & e^t \end{bmatrix}.$$ ■

Suppose next that in (4.11) the matrix A is either in *companion form* or that it has been transformed into this form via some suitable similarity transformation P, so that $A = A_c$, where

$$A_c = \begin{bmatrix} 0 & 1 & 0 & \cdots & 0 \\ 0 & 0 & 1 & \cdots & 0 \\ \vdots & \vdots & \vdots & & \vdots \\ 0 & 0 & 0 & \cdots & 1 \\ -a_0 & -a_1 & -a_2 & \cdots & -a_{n-1} \end{bmatrix}. \tag{4.28}$$

Since in this case we have $x_{i+1} = \dot{x}_i$, $i = 1, \ldots, n-1$, it should be clear that in the calculation of e^{At} we need to determine, via some method, only the first row of e^{At}. We demonstrate this by means of a specific example.

EXAMPLE 4.3. In system (4.11), assume that $A = A_c = \begin{bmatrix} 0 & 1 \\ -2 & -3 \end{bmatrix}$, which is in companion form. To demonstrate the above observation, let us compute e^{At} by some other method, say, diagonalization. The eigenvalues of A are $\lambda_1 = -1$ and $\lambda_2 = -2$, and a set of corresponding eigenvectors is given by $v_1 = (1, -1)^T$ and $v_2 = (1, -2)^T$. We obtain

$$P = [v_1, v_2] = \begin{bmatrix} 1 & 1 \\ -1 & -2 \end{bmatrix}, P^{-1} = \begin{bmatrix} 2 & 1 \\ -1 & -1 \end{bmatrix} \text{ and } J = P^{-1}A_cP = \begin{bmatrix} -1 & 0 \\ 0 & -2 \end{bmatrix}, e^{At} =$$

$$Pe^{Jt}P^{-1} = \begin{bmatrix} 1 & 1 \\ -1 & -2 \end{bmatrix}\begin{bmatrix} e^{-t} & 0 \\ 0 & e^{-2t} \end{bmatrix}\begin{bmatrix} 2 & 1 \\ -1 & -1 \end{bmatrix} = \begin{bmatrix} (2e^{-t} - e^{-2t}) & (e^{-t} - e^{-2t}) \\ (-2e^{-t} + 2e^{-2t}) & (-e^{-t} + 2e^{-2t}) \end{bmatrix}.$$

We note that the second row of the above matrix is the derivative of the first row, as expected. ■

The Cayley-Hamilton Theorem method

If $\alpha(\lambda) = det(\lambda I - A)$ is the characteristic polynomial of an $n \times n$ matrix A, then in view of the Cayley-Hamilton Theorem, we have that $\alpha(A) = 0$, i.e., every $n \times n$ matrix satisfies its characteristic equation (refer to Subsection 2.2J). Using this result, along with the series definition of the matrix exponential e^{At}, it is easily shown that

$$e^{At} = \sum_{i=0}^{n-1} \alpha_i(t)A^i.$$

[Refer to Subsection 2.2J for the details on how to determine the terms $\alpha_i(t)$.]

The Laplace transform method

We assume that the reader is familiar with the basics of the (one-sided) Laplace transform. If $f(t) = [f_1(t), \ldots, f_n(t)]^T$, where $f_i : [0, \infty) \to R, i = 1, \ldots, n$, and if each f_i is Laplace transformable, then we define the Laplace transform of the vector f componentwise, i.e., $\hat{f}(s) = [\hat{f}_1(s), \ldots, \hat{f}_n(s)]^T$, where $\hat{f}_i(s) = \mathcal{L}[f_i(t)] \triangleq \int_0^\infty f_i(t)e^{-st}\,dt$.

We define the Laplace transform of a matrix $C(t) = [c_{ij}(t)]$ similarly. Thus, if each $c_{ij} : [0, \infty) \to R$ and if each c_{ij} is Laplace transformable, then the Laplace transform of $C(t)$ is defined as $\hat{C}(s) = \mathcal{L}[c_{ij}(t)] = [\mathcal{L}c_{ij}(t)] = [\hat{c}_{ij}(s)]$.

Laplace transforms of some of the common time signals are enumerated in Table 4.1. Also, in Table 4.2 we summarize some of the more important properties of the Laplace transform. In Table 4.1, $\delta(t)$ denotes the *Dirac delta distribution* (see Subsection 1.16C) and $p(t)$ represents the *unit step function*.

Now consider once more the initial-value problem (4.11), letting $t_0 = 0$, i.e.,

$$\dot{x} = Ax, \qquad x(0) = x_0. \tag{4.29}$$

TABLE 4.1
Laplace transforms

$f(t)(t \geq 0)$	$\hat{f}(s) = \mathcal{L}[f(t)]$
$\delta(t)$	1
$p(t)$	$1/s$
$t^k/k!$	$1/s^{k+1}$
e^{-at}	$1/(s + a)$
$t^k e^{-at}$	$k!/(s + a)^{k+1}$
$e^{-at} \sin bt$	$b/[(s + a)^2 + b^2]$
$e^{-at} \cos bt$	$(s + a)/[(s + a)^2 + b^2]$

TABLE 4.2
Laplace transform properties

Time differentiation	$df(t)/dt$	$s\hat{f}(s) - f(0)$
	$d^k f(t)/dt^k$	$s^k \hat{f}(s) - [s^{k-1} f(0) + \cdots + f^{(k-1)}(0)]$
Frequency shift	$e^{-at} f(t)$	$\hat{f}(s+a)$
Time shift	$f(t-a)s(t-a),\ a > 0$	$e^{-as}\hat{f}(s)$
Scaling	$f(t/\alpha),\ \alpha > 0$	$\alpha \hat{f}(\alpha s)$
Convolution	$\int_0^t f(\tau) g(t-\tau)\, d\tau = f(t) * g(t)$	$\hat{f}(s)\hat{g}(s)$
Initial value	$\lim_{t \to 0^+} f(t) = f(0^+)$	$\lim_{s\to\infty} s\hat{f}(s)^{\dagger}$
Final value	$\lim_{t\to\infty} f(t)$	$\lim_{s\to 0} s\hat{f}(s)^{\dagger\dagger}$

† If the limit exists.
†† If $s\hat{f}(s)$ has no singularities on the imaginary axis or in the right half s plane.

Taking the Laplace transform of both sides of $\dot{x} = Ax$, and taking into account the initial condition $x(0) = x_0$, we obtain $s\hat{x}(s) - x_0 = A\hat{x}(s)$, or $(sI - A)\hat{x}(s) = x_0$, or

$$\hat{x}(s) = (sI - A)^{-1} x_0. \tag{4.30}$$

It can be shown by analytic continuation that $(sI - A)^{-1}$ exists for all s, except at the eigenvalues of A. Taking the inverse Laplace transform of (4.30), we obtain the solution

$$\phi(t) = \mathcal{L}^{-1}[(sI - A)^{-1}]x_0 = \Phi(t, 0)x_0 = e^{At} x_0. \tag{4.31}$$

It follows from (4.29) and (4.31) that $\hat{\Phi}(s) = (sI - A)^{-1}$ and that

$$\Phi(t, 0) \triangleq \Phi(t - 0) = \Phi(t) = \mathcal{L}^{-1}[(sI - A)^{-1}] = e^{At}. \tag{4.32}$$

Finally, note that when $t_0 \neq 0$, we can immediately compute $\Phi(t, t_0) = \Phi(t - t_0) = e^{A(t-t_0)}$.

EXAMPLE 4.4. In (4.29), let $A = \begin{bmatrix} -1 & 2 \\ 0 & 1 \end{bmatrix}$. Then

$$(sI - A)^{-1} = \begin{bmatrix} s+1 & -2 \\ 0 & s-1 \end{bmatrix}^{-1} = \begin{bmatrix} \dfrac{1}{s+1} & \dfrac{2}{(s+1)(s-1)} \\ 0 & \dfrac{1}{s-1} \end{bmatrix} = \begin{bmatrix} \dfrac{1}{s+1} & \left(\dfrac{1}{s-1} - \dfrac{1}{s+1}\right) \\ 0 & \dfrac{1}{s-1} \end{bmatrix}.$$

Using Table 4.1, we obtain $\mathcal{L}^{-1}[(sI - A)^{-1}] = e^{At} = \begin{bmatrix} e^{-t} & (e^t - e^{-t}) \\ 0 & e^t \end{bmatrix}$. ■

Before concluding this subsection, we briefly consider initial-value problems determined by (4.1), i.e.,

$$\dot{x} = Ax + g(t), \qquad x(t_0) = x_0. \tag{4.33}$$

We wish to apply the Laplace transform method discussed above in solving (4.33). To this end we assume $t_0 = 0$ and we take the Laplace transform of both sides of (4.33) to obtain $s\hat{x}(s) - x_0 = A\hat{x}(s) + \hat{g}(s)$ or $(sI - A)\hat{x}(s) = x_0 + \hat{g}(s)$, or

$$\begin{aligned} \hat{x}(s) &= (sI - A)^{-1} x_0 + (sI - A)^{-1}\hat{g}(s) \\ &= \hat{\Phi}(s)x_0 + \hat{\Phi}(s)\hat{g}(s) \\ &\triangleq \hat{\phi}_h(s) + \hat{\phi}_p(s). \end{aligned} \tag{4.34}$$

Taking the inverse Laplace transform of both sides of (4.34) and using (4.6) with $t_0 = 0$, we obtain $\phi(t) = \phi_h(t) + \phi_p(t) = \mathcal{L}^{-1}[(sI - A)^{-1}]x_0 + \mathcal{L}^{-1}[(sI - A)^{-1}\hat{g}(s)] = \Phi(t)x_0 + \int_0^t \Phi(t-\eta)g(\eta)\,d\eta$, where ϕ_h denotes the homogeneous solution and ϕ_p is the particular solution, as expected.

EXAMPLE 4.5. Consider the initial-value problem given by

$$\dot{x}_1 = -x_1 + x_2$$
$$\dot{x}_2 = -2x_2 + u(t)$$

with $x_1(0) = -1$, $x_2(0) = 0$, and

$$u(t) = \begin{cases} 1 & \text{for } t > 0, \\ 0 & \text{for } t \leq 0. \end{cases}$$

It is easily verified that in this case

$$\hat{\Phi}(s) = \begin{bmatrix} \dfrac{1}{s+1} & \left(\dfrac{1}{s+1} - \dfrac{1}{s+2}\right) \\ 0 & \dfrac{1}{s+2} \end{bmatrix},$$

$$\Phi(t) = \begin{bmatrix} e^{-t} & (e^{-t} - e^{-2t}) \\ 0 & e^{-2t} \end{bmatrix},$$

$$\phi_h(t) = \begin{bmatrix} e^{-t} & (e^{-t} - e^{-2t}) \\ 0 & e^{-t} \end{bmatrix}\begin{bmatrix} -1 \\ 0 \end{bmatrix} = \begin{bmatrix} -e^{-t} \\ 0 \end{bmatrix},$$

$$\hat{\phi}_p(s) = \begin{bmatrix} \dfrac{1}{s+1} & \left(\dfrac{1}{s+1} - \dfrac{1}{s+2}\right) \\ 0 & \dfrac{1}{s+2} \end{bmatrix}\begin{bmatrix} 0 \\ \dfrac{1}{s} \end{bmatrix} = \begin{bmatrix} \dfrac{1}{2}\left(\dfrac{1}{s}\right) + \dfrac{1}{2}\left(\dfrac{1}{s+2}\right) - \dfrac{1}{s+1} \\ \dfrac{1}{2}\left(\dfrac{1}{s}\right) - \dfrac{1}{2}\left(\dfrac{1}{s+2}\right) \end{bmatrix},$$

$$\phi_p(t) = \begin{bmatrix} \frac{1}{2} + \frac{1}{2}e^{-2t} - e^{-t} \\ \frac{1}{2} - \frac{1}{2}e^{-2t} \end{bmatrix},$$

and $$\phi(t) = \phi_h(t) + \phi_p(t) = \begin{bmatrix} \frac{1}{2} - 2e^{-t} + \frac{1}{2}e^{-2t} \\ \frac{1}{2} - \frac{1}{2}e^{-2t} \end{bmatrix}.$$ ■

C. Modes and Asymptotic Behavior of Time-Invariant Systems

In this subsection we study the qualitative behavior of the solutions of linear, autonomous, homogeneous ordinary differential equations (L) by means of the modes of such systems, to be introduced shortly. Although we will not address the stability of systems in detail until Chapter 6, the results here will enable us to give some general stability characterizations for such systems.

Modes: General case

We begin by recalling that the unique solution of

$$\dot{x} = Ax, \tag{L}$$

satisfying $x(0) = x_0$, is given by

$$\phi(t, 0, x_0) = \Phi(t, 0)x(0) = \Phi(t, 0)x_0 = e^{At}x_0. \tag{4.35}$$

We also recall that $det(sI - A) = \prod_{i=1}^{\sigma}(s - \lambda_i)^{n_i}$, where $\lambda_1, \ldots, \lambda_\sigma$ denote the σ distinct eigenvalues of A, where λ_i with $i = 1, \ldots, \sigma$, is assumed to be repeated n_i times (i.e., n_i is the algebraic multiplicity of λ_i), and $\Sigma_{i=1}^{\sigma} n_i = n$.

To introduce the modes for (L), we must show that

$$e^{At} = \sum_{i=1}^{\sigma}\sum_{k=0}^{n_i-1} A_{ik}t^k e^{\lambda_i t}$$
$$= \sum_{i=1}^{\sigma}[A_{i0}e^{\lambda_i t} + A_{i1}te^{\lambda_i t} + \cdots + A_{i(n_i-1)}t^{n_i-1}e^{\lambda_i t}], \tag{4.36}$$

where
$$A_{ik} = \frac{1}{k!}\frac{1}{(n_i - 1 - k)!}\lim_{s\to\lambda_i}\{[(s - \lambda_i)^{n_i}(sI - A)^{-1}]^{(n_i-1-k)}\}. \tag{4.37}$$

In (4.37), $[\,\cdot\,]^{(l)}$ denotes the lth derivative with respect to s.

Equation (4.36) shows that e^{At} can be expressed as the sum of terms of the form $A_{ik}t^k e^{\lambda_i t}$, where $A_{ik} \in R^{n\times n}$. We call $A_{ik}t^k e^{\lambda_i t}$ *a mode of system* (L). If an eigenvalue λ_i is repeated n_i times, there are n_i modes, $A_{ik}t^k e^{\lambda_i t}$, $k = 0, 1, \ldots, n_i - 1$, in e^{At} associated with λ_i. Accordingly, the solution (4.35) of (L) is determined by the n modes of (L) corresponding to the n eigenvalues of A and by the initial condition $x(0)$. We note that by selecting $x(0)$ appropriately, modes can be combined or eliminated $[A_{ik}x(0) = 0]$, thus affecting the behavior of $\phi(t, 0, x_0)$.

To verify (4.36) we recall that $e^{At} = \mathcal{L}^{-1}[(sI - A)^{-1}]$ and we make use of the partial fraction expansion method to determine the inverse Laplace transform. As in the scalar case, it can be shown that

$$(sI - A)^{-1} = \sum_{i=1}^{\sigma}\sum_{k=0}^{n_i-1}(k!A_{ik})(s - \lambda_i)^{-(k+1)}, \tag{4.38}$$

where the $(k!A_{ik})$ are the coefficients of the partial fractions ($k!$ is for scaling). It is known that these coefficients can be evaluated for each i by multiplying both sides of (4.38) by $(s - \lambda_i)^{n_i}$, differentiating $(n_i - 1 - k)$ times with respect to s, and then evaluating the resulting expression at $s = \lambda_i$. This yields (4.37). Taking the inverse Laplace transform of (4.38) and using the fact that $\mathcal{L}[t^k e^{\lambda_i t}] = k!(s-\lambda_i)^{-(k+1)}$ (refer to Table 4.1) results in (4.36).

When all n eigenvalues λ_i of A are distinct, then $\sigma = n$, $n_i = 1$, $i = 1, \ldots, n$, and (4.36) reduces to the expression

$$e^{At} = \sum_{i=1}^{n} A_i e^{\lambda_i t}, \tag{4.39}$$

where
$$A_i = \lim_{s\to\lambda_i}[(s - \lambda_i)(sI - A)^{-1}]. \tag{4.40}$$

Expression (4.40) can also be derived directly, using a partial fraction expansion of $(sI - A)^{-1}$ given in (4.38) (verify this).

EXAMPLE 4.6. For (L) we let $A = \begin{bmatrix} 0 & 1 \\ -4 & -4 \end{bmatrix}$, for which the eigenvalue $\lambda_1 = -2$ is repeated twice, i.e., $n_1 = 2$. Applying (4.36) and (4.37), we obtain

$$e^{At} = A_{10}e^{\lambda_1 t} + A_{11}te^{\lambda_1 t} = \begin{bmatrix} 1 & 0 \\ 0 & 1 \end{bmatrix}e^{-2t} + \begin{bmatrix} 2 & 1 \\ -4 & -2 \end{bmatrix}te^{-2t}.$$ ■

EXAMPLE 4.7. For (L) we let $A = \begin{bmatrix} 0 & 1 \\ -1 & -1 \end{bmatrix}$, for which the eigenvalues are given by (the complex conjugate pair) $\lambda_1 = -\frac{1}{2} + j(\sqrt{3}/2)$, $\lambda_2 = -\frac{1}{2} - j(\sqrt{3}/2)$. Applying (4.39) and (4.40), we obtain

$$A_1 = \frac{1}{\lambda_1 - \lambda_2}\begin{bmatrix} \lambda_1 + 1 & 1 \\ -1 & \lambda_1 \end{bmatrix} = \frac{1}{j\sqrt{3}}\begin{bmatrix} \frac{1}{2} + j\frac{\sqrt{3}}{2} & 1 \\ -1 & -\frac{1}{2} + j\frac{\sqrt{3}}{2} \end{bmatrix}$$

$$A_2 = \frac{1}{\lambda_2 - \lambda_1}\begin{bmatrix} \lambda_2 + 1 & 1 \\ -1 & \lambda_2 \end{bmatrix} = \frac{1}{-j\sqrt{3}}\begin{bmatrix} \frac{1}{2} - j\frac{\sqrt{3}}{2} & 1 \\ -1 & -\frac{1}{2} - j\frac{\sqrt{3}}{2} \end{bmatrix}$$

[i.e., $A_1 = A_2^*$, where $(\cdot)^*$ denotes the complex conjugate of $(\cdot)$], and

$$\begin{aligned} e^{At} &= A_1 e^{\lambda_1 t} + A_2 e^{\lambda_2 t} = A_1 e^{\lambda_1 t} + A_1^* e^{\lambda_1^* t} \\ &= 2(Re\, A_1)(Re\, e^{\lambda_1 t}) - 2(Im\, A_1)(Im\, e^{\lambda_1 t}) \\ &= 2e^{-(1/2)t}\left[\begin{bmatrix} \frac{1}{2} & 0 \\ 0 & -\frac{1}{2} \end{bmatrix} \cos \frac{\sqrt{3}}{2}t - \begin{bmatrix} -\frac{1}{2\sqrt{3}} & -\frac{1}{\sqrt{3}} \\ \frac{1}{\sqrt{3}} & \frac{1}{2\sqrt{3}} \end{bmatrix} \sin\left(\frac{\sqrt{3}}{2}t\right)\right]. \end{aligned}$$

The last expression involves only real numbers, as expected, since A and e^{At} are real matrices. ■

EXAMPLE 4.8. For (L) we let $A = \begin{bmatrix} 1 & 0 \\ 0 & 1 \end{bmatrix}$, for which the eigenvalue $\lambda_1 = 1$ is repeated twice, i.e., $n_1 = 2$. Applying (4.36) and (4.37), we obtain

$$e^{At} = A_{10}e^{\lambda_1 t} + A_{11}te^{\lambda_1 t} = \begin{bmatrix} 1 & 0 \\ 0 & 1 \end{bmatrix} e^t + \begin{bmatrix} 0 & 0 \\ 0 & 0 \end{bmatrix} te^t = Ie^t.$$

This example shows that not all modes of the system are necessarily present in e^{At}. What is present depends in fact on the number and dimensions of the individual blocks of the Jordan canonical form of A corresponding to identical eigenvalues. To illustrate this further, we let for (L), $A = \begin{bmatrix} 1 & 1 \\ 0 & 1 \end{bmatrix}$, where the two repeated eigenvalues $\lambda_1 = 1$ belong to the same Jordan block. Then $e^{At} = \begin{bmatrix} 1 & 0 \\ 0 & 1 \end{bmatrix} e^t + \begin{bmatrix} 0 & 1 \\ 0 & 0 \end{bmatrix} te^t$. ■

Stability of an equilibrium

In Chapter 6 we will study the *qualitative properties* of linear dynamical systems, including systems described by (L). This will be accomplished by studying the *stability properties* of such systems, or more specifically, the *stability properties* of an *equilibrium* of such systems.

If $\phi(t, 0, x_e)$ denotes the solution of system (L) with $x(0) = x_e$, then x_e is said to be an *equilibrium* of (L) if $\phi(t, 0, x_e) = x_e$ for all $t \geq 0$. Clearly, $x_e = 0$ is an equilibrium of (L). In discussing the qualitative properties, it is often customary to speak, somewhat loosely, of the *stability properties of system* (L), rather than the stability properties of the equilibrium $x_e = 0$ of system (L).

We will show in Chapter 6 that the following qualitative characterizations of system (L) are actually equivalent to more fundamental qualitative characterizations of the equilibrium $x_e = 0$ of system (L):

1. The system (L) is said to be *stable* if all solutions of (L) are bounded for all $t \geq 0$ [i.e., for any solution $\phi(t, 0, x_0) = (\phi_1(t, 0, x_0), \ldots, \phi_n(t, 0, x_0))^T$ of (L), there exist constants M_i, $i = 1, \ldots, n$ (which in general will depend on the solution on hand) such that $|\phi_i(t, 0, x_0)| < M_i$ for all $t \geq 0$].
2. The system (L) is said to be *asymptotically stable* if it is stable and if all solutions of (L) tend to the origin as t tends to infinity [i.e., for any solution $\phi(t, 0, x_0) = (\phi_1(t, 0, x_0), \ldots, \phi_n(t, 0, x_0))^T$ of (L), we have $\lim_{t\to\infty} \phi_i(t, 0, x_0) = 0$, $i = 1, \ldots, n$].
3. The system (L) is said to be *unstable* if it is not stable.

By inspecting the modes of (L) given by (4.36), (4.37) and (4.39), (4.40), the following stability criteria for system (L) are now evident:

1. The system (L) is *asymptotically stable* if and only if all eigenvalues of A have negative real parts (i.e., $Re\,\lambda_j < 0$, $j = 1, \ldots, \sigma$).
2. The system (L) is *stable* if and only if $Re\,\lambda_j \leq 0$, $j = 1, \ldots, \sigma$, and for all eigenvalues with $Re\,\lambda_j = 0$ having multiplicity $n_j > 1$, it is true that

$$\lim_{s\to\lambda_j} [(s - \lambda_j)^{n_j}(sI - A)^{-1}]^{(n_j-1-k)} = 0, \qquad k = 1, \cdots, n_j - 1. \quad (4.41)$$

3. System (L) is *unstable* if and only if (2) is not true.

We note in particular that if $Re\,\lambda_j = 0$ and $n_j > 1$, then there will be modes $A_{jk}t^k$, $k = 0, \ldots, n_j - 1$, that will yield terms in (4.36) whose norm will tend to infinity as $t \to \infty$, unless their coefficients are zero. This shows why the necessary and sufficient conditions for stability of (L) include condition (4.41).

EXAMPLE 4.9. The systems in Examples 4.6 and 4.7 are asymptotically stable. A system (L) with $A = \begin{bmatrix} 0 & 1 \\ 0 & -1 \end{bmatrix}$ is stable, since the eigenvalues of A above are $\lambda_1 = 0$, $\lambda_2 = -1$. A system (L) with $A = \begin{bmatrix} -1 & 0 \\ 0 & 1 \end{bmatrix}$ is unstable since the eigenvalues of A are $\lambda_1 = 1$, $\lambda_2 = -1$. The system of Example 4.8 is also unstable. ■

Modes: Distinct eigenvalue case

When the eigenvalues λ_i of A are distinct, there is an alternative way to (4.40) of computing the matrix coefficients A_i, expressed in terms of the corresponding right and left eigenvectors of A. This method offers great insight in questions concerning the presence or absence of modes in the response of a system. Specifically, if A has n distinct eigenvalues λ_i, then

$$e^{At} = \sum_{i=1}^{n} A_i e^{\lambda_i t}, \quad (4.42)$$

where

$$A_i = v_i \tilde{v}_i, \quad (4.43)$$

where $v_i \in R^n$ and $(\tilde{v}_i)^T \in R^n$ are right and left eigenvectors of A corresponding to the eigenvalue λ_i, respectively.

To prove the above assertions, we recall that $(\lambda_i I - A)v_i = 0$ and $\tilde{v}_i(\lambda_i I - A) = 0$. If $Q \triangleq [v_1, \ldots, v_n]$, then the $\tilde{v}_i$ are the rows of

$$P = Q^{-1} = \begin{bmatrix} \tilde{v}_1 \\ \vdots \\ \tilde{v}_n \end{bmatrix}.$$

The matrix Q is of course nonsingular, since the eigenvalues λ_i, $i = 1, \ldots, n$, are by assumption distinct and since the corresponding eigenvectors are linearly independent. Notice that $Q\,diag\,[\lambda_1, \ldots, \lambda_n] = AQ$ and that $diag\,[\lambda_1, \ldots, \lambda_n]P = PA$. Also, notice that $\tilde{v}_i v_j = \delta_{ij}$, where

$$\delta_{ij} = \begin{cases} 1 & \text{when } i = j, \\ 0 & \text{when } i \neq j. \end{cases}$$

We now have $(sI - A)^{-1} = [sI - Q\,diag\,[\lambda_1, \ldots, \lambda_n]Q^{-1}]^{-1} = Q[sI - diag\,[\lambda_1, \ldots, \lambda_n]]^{-1}Q^{-1} = Q\,diag\,[(s - \lambda_1)^{-1}, \ldots, (s - \lambda_n)^{-1}]Q^{-1} = \sum_{i=1}^{n} v_i \tilde{v}_i (s - \lambda_i)^{-1}$. If we now take the inverse Laplace transform of the above expression, we obtain (4.42).

If we choose the initial value $x(0)$ for (L) to be colinear with an eigenvector v_j of A [i.e., $x(0) = \alpha v_j$ for some real $\alpha \neq 0$], then $e^{\lambda_j t}$ is the only mode that will appear in the solution ϕ of (L). This can easily be seen from our preceding discussion. In particular if $x(0) = \alpha v_j$, then (4.42) and (4.43) yield

$$\phi(t, 0, x(0)) = e^{At}x(0) = v_1\tilde{v}_1 x(0)e^{\lambda_1 t} + \cdots + v_n\tilde{v}_n x(0)e^{\lambda_n t} = \alpha v_j e^{\lambda_j t} \qquad (4.44)$$

since $\tilde{v}_i v_j = 1$ when $i = j$, and $\tilde{v}_i v_j = 0$ otherwise.

EXAMPLE 4.10. In (L) we let $A = \begin{bmatrix} -1 & 1 \\ 0 & 1 \end{bmatrix}$. The eigenvalues of A are given by $\lambda_1 = -1$, $\lambda_2 = 1$ and $Q = [v_1, v_2] = \begin{bmatrix} 1 & 1 \\ 0 & 2 \end{bmatrix}$, $Q^{-1} = \begin{bmatrix} \tilde{v}_1 \\ \tilde{v}_2 \end{bmatrix} = \begin{bmatrix} 1 & -\frac{1}{2} \\ 0 & \frac{1}{2} \end{bmatrix}$. Then $e^{At} = v_1\tilde{v}_1 e^{\lambda_1 t} + v_2\tilde{v}_2 e^{\lambda_2 t} = \begin{bmatrix} 1 & -\frac{1}{2} \\ 0 & 0 \end{bmatrix} e^{-t} + \begin{bmatrix} 0 & \frac{1}{2} \\ 0 & 1 \end{bmatrix} e^{t}$. If in particular we choose $x(0) = \alpha v_1 = (\alpha, 0)^T$, then $\phi(t, 0, x(0)) = e^{At}x(0) = \alpha(1, 0)^T e^{-t}$, which contains only the mode corresponding to the eigenvalue $\lambda_1 = -1$. Thus, for this particular choice of initial vector, the unstable behavior of the system is suppressed. ■

Remark

We conclude our discussion of modes and asymptotic behavior by briefly considering systems of linear, nonhomogeneous, ordinary differential equations (4.1) for the special case where $g(t) = Bu(t)$,

$$\dot{x} = Ax + Bu(t), \qquad (4.45)$$

where $B \in R^{n \times m}$, $u : R \to R^m$, and where it is assumed that the Laplace transform of u exists. Taking the Laplace transform of both sides of (4.45) and rearranging yields

$$\hat{x}(s) = (sI - A)^{-1}x(0) + (sI - A)^{-1}B\hat{u}(s). \qquad (4.46)$$

By taking the inverse Laplace transform of (4.46), we see that the solution ϕ is the sum of modes that correspond to the singularities or poles of $(sI - A)^{-1}x(0)$ and $(sI - A)^{-1}B\hat{u}(s)$. If in particular (L) is asymptotically stable (i.e., for $\dot{x} = Ax$, $Re\,\lambda_i <$

0, $i = 1, \ldots, n$) and if u in (4.45) is bounded (i.e., there is an M such that $|u_i(t)| < M$ for all $t \geq 0$, $i = 1, \ldots, m$), then it is easily seen that the solutions of (4.45) are bounded as well. Thus, the fact that the system (L) is asymptotically stable has repercussions on the asymptotic behavior of the solution of (4.45). Issues of this type will be addressed in greater detail in Chapter 6.

*2.5 LINEAR PERIODIC SYSTEMS

We now consider linear homogeneous systems of first-order ordinary differential equations

$$\dot{x} = A(t)x, \qquad -\infty < t < \infty, \tag{P}$$

where $A \in C(R, R^{n \times n})$ and

$$A(t) = A(t+T), \qquad -\infty < t < \infty, \tag{5.1}$$

for some $T > 0$. We call (P) a *periodic system* and T a *period* for system (P).

The principal result of this section involves the notion of the logarithm of a matrix, which we introduce in the following result.

THEOREM 5.1. For every nonsingular matrix B there exists a matrix A, called a *logarithm of B*, with the property that

$$e^A = B. \tag{5.2}$$

The matrix A is not unique.

Proof. Let $\tilde{B}$ be similar to B. Then there exists a nonsingular matrix P such that $P^{-1}BP = \tilde{B}$. Now if $e^{\tilde{A}} = \tilde{B}$, then we have $B = P\tilde{B}P^{-1} = Pe^{\tilde{A}}P^{-1} = e^{P\tilde{A}P^{-1}}$. It follows that $P\tilde{A}P^{-1}$ is also a logarithm of B. Therefore, it suffices to prove the theorem when the matrix B is in suitable canonical form.

Let $\lambda_1, \ldots, \lambda_k$ denote the distinct eigenvalues of B with respective multiplicities $n_1, \ldots, n_k$. Without loss of generality, we may assume that B is in the block diagonal form

$$B = \begin{bmatrix} B_1 & & 0 \\ & \ddots & \\ 0 & & B_k \end{bmatrix},$$

where $B_j = \lambda_j[I_{n_j} + (1/\lambda_j)N_j]$, $N_j^{n_j} = 0$, $j = 1, \ldots, k$. We note that $\lambda_j \neq 0$, $j = 1, \ldots, k$, since B is nonsingular. Using the power series expansion $\log(1+x) = \sum_{p=1}^{\infty}[(-1)^{p+1}/p]x^p$, $|x| < 1$, we formally write $A_j = \log B_j = I_{n_j}\log\lambda_j + \log[I_{n_j} + (1/\lambda_j)N_j] = I_{n_j}\log\lambda_j + \sum_{p=1}^{\infty}[(-1)^{p+1}/p](N_j/\lambda_j)^p$. Since $N_j^{n_j} = 0$ we actually have

$$A_j = I_{n_j}\log\lambda_j + \sum_{p=1}^{n_j-1}\frac{(-1)^{p+1}}{p}\left(\frac{N_j}{\lambda_j}\right)^p, \qquad j = 1, \ldots, k, \tag{5.3}$$

where we note that $\log\lambda_j$ is defined, since $\lambda_j \neq 0$. Now recall that $e^{\log(1+x)} = 1 + x$. Performing the same operations with matrices, we obtain the same terms and there is no problem with convergence, since the series (5.3) for $A_j = \log B_j$ terminates. Accordingly, we obtain $e^{A_j} = \exp(I_{n_j} \log \lambda_j) \exp\{\sum_{p=1}^{n_j-1} [(-1)^{p+1}/p](N_j/\lambda_j)^p\} = \lambda_j[I_{n_j} + (N_j/\lambda_j)] = B_j$, $j = 1, \ldots, k$. If now we let

$$A = \begin{bmatrix} A_1 & & 0 \\ & \ddots & \\ 0 & & A_k \end{bmatrix},$$

where A_j is defined in (5.3), we obtain

$$e^A = \begin{bmatrix} e^{A_1} & & 0 \\ & \ddots & \\ 0 & & e^{A_k} \end{bmatrix} = \begin{bmatrix} B_1 & & 0 \\ & \ddots & \\ 0 & & B_k \end{bmatrix} = B,$$

which is the desired result.

We conclude by noting that the matrix A is not unique, since for example, $e^{A+2\pi k_j I} = e^A e^{2\pi j} = e^A$ for all integers k (where $j = \sqrt{-1}$). ■

We are now in a position to state and prove one of the principal results of this section.

THEOREM 5.2. Assume that (5.1) is true and that $A \in C(R, R^{n\times n})$. If $\Phi(t)$ is a fundamental matrix for (P), then so is $\Phi(t+T)$, $t \in R$. Furthermore, for every Φ there exists a nonsingular matrix P that is also periodic with period T and a constant $n \times n$ matrix R, such that

$$\Phi(t) = P(t)e^{tR}.$$

Proof. Let $\Psi(t) = \Phi(t+T)$, $t \in R$. Since $\dot{\Phi}(t) = A(t)\Phi(t)$, $t \in R$, we have $\dot{\Psi}(t) = \dot{\Phi}(t+T) = A(t+T)\Phi(t+T) = A(t)\Phi(t+T)$, $t \in R$. Therefore, Ψ is also a solution of $\dot{\Psi} = A(t)\Psi$, $A(t) = A(t+T)$, $t \in R$. Furthermore, since $\Phi(t+T)$ is nonsingular for all $t \in R$, it follows that Ψ is a fundamental matrix for (P). Therefore, by Theorem 2.5, there exists a nonsingular matrix C such that $\Phi(t+T) = \Phi(t)C$, and by Theorem 5.1, there exists a constant matrix R such that $e^{TR} = C$. Therefore,

$$\Phi(t+T) = \Phi(t)e^{TR}. \tag{5.4}$$

Defining P by

$$P(t) = \Phi(t)e^{-tR} \tag{5.5}$$

and using (5.4) and (5.5), we now obtain $P(t+T) = \Phi(t+T)e^{-(t+T)R} = \Phi(t)e^{TR} \times e^{-(t+T)R} = \Phi(t)e^{-tR} = P(t)$. Therefore, $P(t)$ is nonsingular for all $t \in R$ and it is periodic. ■

The above result allows us to conclude that *the determination of a fundamental matrix Φ for system (P) over any time interval of length T leads at once to the determination of Φ over* $(-\infty, \infty)$. To show this, assume that $\Phi(t)$ is known only over the interval $[t_0, t_0, +T]$. Since $\Phi(t+T) = \Phi(t)C$, we obtain by setting $t = t_0$, $C = \Phi(t_0)^{-1}\Phi(t_0+T)$ and $R = T^{-1}\log C$. It follows that $P(t) = \Phi(t)e^{-tR}$ is now also known over $[t_0, t_0+T]$. However, $P(t)$ is periodic over $(-\infty, \infty)$. Therefore, $\Phi(t)$ is given over $(-\infty, \infty)$ by $\Phi(t) = P(t)e^{tR}$.

Next, let $\tilde{\Phi}$ be any other fundamental matrix for (P) with $A(t) = A(t+T)$. Then $\Phi = \tilde{\Phi}S$ for some constant nonsingular matrix S. Since $\Phi(t+T) = \Phi(t)e^{TR}$, we have that $\tilde{\Phi}(t+T)S = \tilde{\Phi}(t)Se^{TR}$, or

$$\tilde{\Phi}(t+T) = \tilde{\Phi}(t)(Se^{TR}S^{-1}) = \tilde{\Phi}(t)e^{T(SRS^{-1})}. \tag{5.6}$$

This shows that *every fundamental matrix* $\tilde{\Phi}$ *of* (P) *determines a matrix* $Se^{TR}S^{-1}$ *that is similar to the matrix* e^{TR}.

Conversely, let S be any constant nonsingular matrix. Then there exists a fundamental matrix of (P) such that Eq. (5.6) holds. Therefore, even though Φ does not determine R uniquely, the set of all fundamental matrices of (P), and hence of A, determines uniquely all parameters associated with e^{TR} that are invariant under a similarity transformation. In particular the set of all fundamental matrices of A determines a unique set of eigenvalues of the matrix e^{TR}, denoted by $\lambda_1, \ldots, \lambda_n$, that are called the *Floquet multipliers* associated with A. Note that none of these vanish, since $\prod_{i=1}^{n} \lambda_i = \det e^{TR} \neq 0$. The eigenvalues of R are called the *characteristic exponents*.

Next, we let Q be a constant nonsingular matrix that transforms R into its Jordan canonical form, i.e., $J = Q^{-1}RQ$, where

$$J = \begin{bmatrix} J_0 & 0 & \ldots & 0 \\ 0 & J_1 & & 0 \\ \vdots & \vdots & \ddots & \vdots \\ 0 & 0 & \ldots & J_s \end{bmatrix}.$$

Now let $\tilde{\Phi} = \Phi Q$ and let $\tilde{P} = PQ$. In view of Theorem 5.2, we have that

$$\tilde{\Phi}(t) = \tilde{P}(t)e^{tJ}, \qquad \tilde{P}(t) = \tilde{P}(t+T). \tag{5.7}$$

Let the eigenvalues of R be denoted by $\rho_1, \ldots, \rho_n$. Then

$$e^{tJ} = \begin{bmatrix} e^{tJ_0} & 0 & \ldots & 0 \\ 0 & e^{tJ_1} & \ldots & 0 \\ \vdots & \vdots & \ddots & \vdots \\ 0 & 0 & \ldots & e^{tJ_s} \end{bmatrix},$$

where

$$e^{tJ_0} = \begin{bmatrix} e^{t\rho_1} & 0 & \ldots & 0 \\ 0 & e^{t\rho_2} & \ldots & 0 \\ \vdots & \vdots & \ddots & \vdots \\ 0 & 0 & \ldots & e^{t\rho_q} \end{bmatrix}$$

and

$$e^{tJ_i} = e^{t\rho_{q+i}} \begin{bmatrix} 1 & t & \frac{t^2}{2} & \ldots & \frac{t^{r_i-1}}{(r_i-1)!} \\ 0 & 1 & t & \ldots & \frac{t^{r_i-2}}{(r_i-2)!} \\ \vdots & \vdots & \vdots & & \vdots \\ 0 & 0 & 0 & \ldots & 1 \end{bmatrix}, \qquad i = 1, \ldots, s, \qquad q + \sum_{i=1}^{s} r_i = n.$$

Now $\lambda_i = e^{T\rho_i}$. Therefore, even though the ρ_i are not uniquely determined, their real parts are. It follows from (5.7) that the columns $\tilde{\phi}_1, \ldots, \tilde{\phi}_n$ of $\tilde{\Phi}$ are linearly independent solutions of (P). Let $\tilde{p}_1, \ldots, \tilde{p}_n$ denote the periodic column vectors of $\tilde{P}$. Then

$$
\begin{aligned}
\tilde{\phi}_1(t) &= e^{t\rho_1}\tilde{p}_1(t) \\
\tilde{\phi}_2(t) &= e^{t\rho_2}\tilde{p}_2(t) \\
&\vdots \\
\tilde{\phi}_q(t) &= e^{t\rho_q}\tilde{p}_q(t) \\
\tilde{\phi}_{q+1}(t) &= e^{t\rho_{q+1}}\tilde{p}_{q+1}(t) \\
\tilde{\phi}_{q+2}(t) &= e^{t\rho_{q+1}}(t\tilde{p}_{q+1}(t) + \tilde{p}_{q+2}(t)) \\
&\vdots \\
\tilde{\phi}_{q+r_1}(t) &= e^{t\rho_{q+1}}\left[\frac{t^{r_1-1}}{(r_1-1)!}\tilde{p}_{q+1}(t) + \cdots + t\tilde{p}_{q+r_1-1}(t) + \tilde{p}_{q+r_1}(t)\right] \\
&\vdots \\
\tilde{\phi}_{n-r_s+1}(t) &= e^{t\rho_{q+s}}\tilde{p}_{n-r_s+1}(t), \\
&\vdots \\
\tilde{\phi}_n(t) &= e^{t\rho_{q+s}}\left[\frac{t^{r_s-1}}{(r_s-1)!}\tilde{p}_{n-r_s+1}(t) + \cdots + t\tilde{p}_{n-1}(t) + \tilde{p}_n(t)\right].
\end{aligned}
\tag{5.8}
$$

From (5.8) it is easy to see that when $Re\,\rho_i \triangleq \sigma_i < 0$, or equivalently, when $|\lambda_i| < 1$, there exists a $k_i > 0$ such that $|\tilde{\phi}_i(t)| \leq k_i e^{(\sigma_i/2)t} \to 0$ as $t \to \infty$. This shows that if the eigenvalues ρ_i, $i = 1, \ldots, n$, of R have negative real parts, then any norm of any solution of (P) tends to zero as $t \to +\infty$ *at an exponential rate*.

From (5.5), we can easily verify by direct computation that $AP - \dot{P} = PR$. Accordingly, for the transformation

$$x = P(t)y, \tag{5.9}$$

we obtain $\dot{x} = A(t)x = A(t)P(t)y = \dot{P}(t)y + P(t)\dot{y} = (d/dt)[P(t)y]$ or $\dot{y} = P^{-1}(t) \times [A(t)P(t) - \dot{P}(t)]y = P^{-1}(t)[P(t)R]y = Ry$. In other words, *the transformation* (5.9) *reduces the linear, homogeneous, periodic system* (P) *to the system* $\dot{y} = Ry$*, a linear homogeneous system with constant coefficients.*

We conclude this section with a specific example.

EXAMPLE 5.1. Consider the scalar system

$$\dot{x} = -(\sin t + 2)x. \tag{5.10}$$

Then $A(t) = -(\sin t + 2)$, and $A(t)$ is periodic with period $T = 2\pi$. A fundamental matrix for (5.10) is given by $\Phi(t) = \exp(\cos t - 1 - 2t)$ as can be verified by substituting into the relation $\dot{\Phi}(t) = A(t)\Phi(t)$. Letting $t = 0$ and $T = 2\pi$ in (5.4), we obtain $\Phi(2\pi) = e^{-4\pi} = \Phi(0)e^{2\pi R} = e^{2\pi R}$ or $R = -2$. The equivalence matrix $P(t)$ is now given by (5.5) as $P(t) = \exp(\cos t - 1 - 2t)e^{2t} = e^{\cos t - 1}$, which is clearly periodic with period $T = 2\pi$. The given system (5.10) is transformed by $P(t)$ into the system $\dot{y} = (e^{1-\cos t})[(-1)(\sin t + 2)e^{\cos t - 1} + \sin t e^{\cos t - 1}]y = -(e^{1-\cos t})(2e^{\cos t - 1})y = -2y = Ry$. ■

We will address some of the qualitative properties of periodic systems in further detail in Chapter 6.

2.6 STATE EQUATION AND INPUT-OUTPUT DESCRIPTION OF CONTINUOUS-TIME SYSTEMS

This section consists of three subsections. Using the material of the preceeding sections of this chapter, we first study the response of linear continuous-time systems. Next, we examine transfer functions of linear time-invariant systems, given the state equations of such systems. Finally, we explore the equivalence of internal representations of systems.

A. Response of Linear Continuous-Time Systems

Returning now to Sections 1.1 and 1.14, we consider once more systems described by linear time-varying equations of the form

$$\dot{x} = A(t)x + B(t)u \tag{6.1a}$$

$$y = C(t)x + D(t)u, \tag{6.1b}$$

where $A \in C(R, R^{n\times n})$, $B \in C(R, R^{n\times m})$, $C \in C(R, R^{p\times n})$, $D \in C(R, R^{p\times m})$, and $u : R \to R^m$ is assumed to be continuous or piecewise continuous. We recall that in (6.1a) and (6.1b), x denotes the state vector, u denotes the system input, and y denotes the system output. From Section 1.14 we recall that for given initial conditions $t_0 \in R$, $x(t_0) = x_0 \in R^n$ and for a given input u, the unique solution of (6.1a) is given by

$$\phi(t, t_0, x_0) = \Phi(t, t_0)x_0 + \int_{t_0}^{t} \Phi(t, s)B(s)u(s)\,ds \tag{6.2}$$

for $t \in R$, where Φ denotes the state transition matrix of $A(t)$. Furthermore, by substituting (6.2) into (6.1b), we obtain [as in (14.6) of Chapter 1], for all $t \in R$, the *total system response* given by

$$y(t) = C(t)\Phi(t, t_0)x_0 + C(t)\int_{t_0}^{t} \Phi(t, s)B(s)u(s)\,ds + D(t)u(t). \tag{6.3}$$

Recall that the total response (6.3) may be viewed as consisting of the sum of two components, the *zero-input response* given by the term

$$\psi(t, t_0, x_0, 0) = C(t)\Phi(t, t_0)x_0 \tag{6.4}$$

and the *zero-state response* given by the term

$$\rho(t, t_0, 0, u) = C(t)\int_{t_0}^{t} \Phi(t, s)B(s)u(s)\,ds + D(t)u(t). \tag{6.5}$$

The cause of the former is the initial condition x_0 [and can be obtained from (6.3) by letting $u(t) \equiv 0$], while for the latter the cause is the input u [and can be obtained by setting $x_0 = 0$ in (6.3)].

The zero-state response can be used to introduce the *impulse response* of the system (6.1a), (6.1b). Returning to Subsection 1.16C, we recall that by using the

Dirac delta distribution δ, we can rewrite (6.3) with $x_0 = 0$ as

$$y(t) = \int_{t_0}^{t} [C(t)\Phi(t,\tau)B(\tau) + D(t)\delta(t-\tau)]u(\tau)\,d\tau$$

$$= \int_{t_0}^{t} H(t,\tau)u(\tau)\,d\tau, \tag{6.6}$$

where $H(t,\tau)$ denotes the impulse response matrix of system (6.1a), (6.1b) given by

$$H(t,\tau) = \begin{cases} C(t)\Phi(t,\tau)B(\tau) + D(t)\delta(t-\tau), & t \geq \tau, \\ 0, & t < \tau. \end{cases} \tag{6.7}$$

When in (6.1a), (6.1b), $A(t) \equiv A$, $B(t) \equiv B$, $C(t) \equiv C$, and $D(t) \equiv D$, we obtain the time-invariant system

$$\dot{x} = Ax + Bu \tag{6.8a}$$

$$y = Cx + Du. \tag{6.8b}$$

We recall that in this case the solution of (6.8a) is given by

$$\phi(t, t_0, x_0) = e^{A(t-t_0)}x_0 + \int_{t_0}^{t} e^{A(t-s)}Bu(s)\,ds, \tag{6.9}$$

the *total response* of system (6.8a), (6.8b) is given by

$$y(t) = Ce^{A(t-t_0)}x_0 + C\int_{t_0}^{t} e^{A(t-s)}Bu(s)\,ds + Du(t), \tag{6.10}$$

and the *zero-state response* of (6.8a), (6.8b) is given by $y(t) = \int_{t_0}^{t}[Ce^{A(t-\tau)}B + D\delta(t-\tau)]u(\tau)\,d\tau = \int_{t_0}^{t} H(t,\tau)u(\tau)\,d\tau = \int_{t_0}^{t} H(t-\tau)u(\tau)\,d\tau$, where the *impulse response matrix* H of system (6.8a), (6.8b) is given by

$$H(t-\tau) = \begin{cases} Ce^{A(t-\tau)}B + D\delta(t-\tau), & t \geq \tau, \\ 0, & t < \tau, \end{cases} \tag{6.11}$$

or, as is more commonly written,

$$H(t) = \begin{cases} Ce^{At}B + D\delta(t), & t \geq 0, \\ 0, & t < 0. \end{cases} \tag{6.12}$$

At this point it may be worthwhile to consider some specific cases.

EXAMPLE 6.1. In (6.1a), (6.1b), let

$$A(t) = \begin{bmatrix} -1 & e^{2t} \\ 0 & -1 \end{bmatrix}, \qquad B(t) = \begin{bmatrix} e^{-t} \\ 0 \end{bmatrix}, \qquad C(t) = [e^t, 1], \qquad D = 0$$

and consider the case when $t_0 = 0$, $x(0) = (0, 1)^T$, u is the unit step function, and $t \geq 0$. Referring to Example 3.6, we obtain $\phi(t, t_0, x_0) = \phi_h(t, t_0, x_0) + \phi_p(t, t_0, x_0) = \begin{bmatrix} \frac{1}{2}(e^t - e^{-t}) \\ e^{-t} \end{bmatrix} + \begin{bmatrix} te^{-t} \\ 0 \end{bmatrix}$ with $t_0 = 0$ and for $t \geq 0$. The *total system response* $y(t) = C(t)x(t)$ is given by the sum of the zero-input response and the zero-state response, $y(t, t_0, x_0, u) = \psi(t, t_0, x_0, 0) + \rho(t, t_0, 0, u) = [\frac{1}{2}(e^{2t} - 1) + e^{-t}] + t$, $t \geq 0$. Note that the zero-input response ψ is due to the homogeneous part of the solution ϕ (given by ϕ_h) while the zero-state response ρ is due to the particular solution of ϕ (given by ϕ_p). ■

EXAMPLE 6.2. In (6.8a), (6.8b), let $A = \begin{bmatrix} 0 & 1 \\ 0 & 0 \end{bmatrix}$, $B = \begin{bmatrix} 0 \\ 1 \end{bmatrix}$, $C = [0, 1]$, $D = 0$ and consider the case when $t_0 = 0$, $x(0) = (1, -1)^T$, u is the unit step, and $t \geq 0$. We can easily compute the solution of (6.8a) as

$$\phi(t, t_0, x_0) = \phi_h(t, t_0, x_0) + \phi_p(t, t_0, x_0) = \begin{bmatrix} 1 - t \\ -1 \end{bmatrix} + \begin{bmatrix} \frac{1}{2}t^2 \\ t \end{bmatrix}$$

with $t_0 = 0$ and for $t \geq 0$. The total system response $y(t) = C(t)x(t)$ is given by the sum of the zero-input response and the zero-state response, $y(t, t_0, x_0, u) = \psi(t, t_0, x_0, 0) + \rho(t, t_0, 0, u) = -1 + t$, $t \geq 0$. ■

We note that when $x(0) = 0$, Example 6.1 (a time-varying system) and Example 6.2 (a time-invariant system) have identical output responses given by $y(t) = t$, $t \geq 0$, when $u(t)$ is the unit step. [Is this true for any input $u(t)$?]

EXAMPLE 6.3. Consider the time-varying system given above in Example 6.1. In this case we have $\Phi(t, \tau)B(\tau) = [e^{-t}, 0]^T$, and the impulse response has the rather unusual form

$$H(t, \tau) = \begin{cases} C(t)\Phi(t, \tau)B(\tau) = 1, & t \geq \tau, \\ 0, & t < \tau. \end{cases}$$

In other words, the response of this system to an impulse input, for zero initial conditions, is the unit step, and this is independent of the time τ at which the impulse is applied! Note that in the present case the response to a step is a ramp t, as can easily be verified from (6.6) (see also Example 6.1). Therefore, this system behaves to the outside world, for zero initial conditions, as a time-invariant system. This is interesting; however, it is not a typical situation when dealing with time-varying systems. ■

EXAMPLE 6.4. Consider the time-invariant system given above in Example 6.2. It is easily verified that in the present case

$$\Phi(t) = e^{At} = \begin{bmatrix} 1 & t \\ 0 & 1 \end{bmatrix}.$$

Then $H(t, \tau) = Ce^{A(t-\tau)}B = 1$ for $t \geq \tau$ and $H(t, \tau) = 0$ for $t < \tau$. Thus, the response of this system to an impulse input for zero initial conditions is the unit step. Comparing this with the impulse response of the system given above, in Example 6.3, we note that they are identical. In other words, the behavior of these two systems to the outside world, one a time-varying system and the other a time-invariant system, is characterized by the same response to an impulse input, when the initial conditions are zeros. Indeed, in this case, both systems behave like a time-invariant system with $H(t, \tau) = H(t - \tau, 0) = H(t, 0) = 1$. Note, however, that when the initial conditions are not zero, the responses of these two systems are quite different. ■

The preceding two examples demonstrate, as one might expect, that external descriptions of finite-dimensional linear systems are not as complete as internal descriptions of such systems. Indeed, the utility of impulse responses is found in the fact that they represent the input-output relations of a system quite well, assuming that the system is at rest. To describe other dynamic behavior, one needs in general additional information [e.g., the initial state vector (or perhaps the past history of the system input since the last time instant when the system was at rest) as well as the internal structure of the system].

Internal descriptions, such as state-space representations, constitute more complete descriptions than external descriptions. However, the latter are simpler to apply than the former. Both types of representations are useful. It is quite straightforward to obtain external descriptions of systems from internal descriptions, as was demonstrated in this section. The reverse process, however, is not quite as straightforward. The process of determining an internal system description from an external description is called *realization* and will be addressed in Chapter 5. The principal issue in system realization is to obtain minimal order internal descriptions that model a given system, avoiding the generation of unnecessary dynamics.

B. Transfer Functions

Next, if as in (16.51) in Chapter 1, we take the Laplace transform of both sides of (6.11), we obtain the input-output relation

$$\hat{y}(s) = \hat{H}(s)\hat{u}(s). \tag{6.13}$$

We recall from Section 1.16 that $\hat{H}(s)$ is called the *transfer function matrix* of system (6.8a), (6.8b). We can evaluate this matrix in a straightforward manner by first taking the Laplace transform of both sides of (6.8a) and (6.8b) to obtain

$$s\hat{x}(s) - x(0) = A\hat{x}(s) + B\hat{u}(s) \tag{6.14}$$

$$\hat{y}(s) = C\hat{x}(s) + D\hat{u}(s). \tag{6.15}$$

Using (6.14) to solve for $\hat{x}(s)$, we obtain

$$\hat{x}(s) = (sI - A)^{-1}x(0) + (sI - A)^{-1}B\hat{u}(s). \tag{6.16}$$

Substituting (6.16) into (6.15) yields

$$\hat{y}(s) = C(sI - A)^{-1}x(0) + C(sI - A)^{-1}B\hat{u}(s) + D\hat{u}(s) \tag{6.17}$$

and

$$y(t) = \mathcal{L}^{-1}\hat{y}(s) = Ce^{At}x(0) + C\int_0^t e^{A(t-s)}Bu(s)\,ds + Du(t), \tag{6.18}$$

as expected.

If in (6.17) we let $x(0) = 0$, we obtain the Laplace transform of the zero-state response given by

$$\begin{aligned}\hat{y}(s) &= [C(sI - A)^{-1}B + D]\hat{u}(s)\\ &= \hat{H}(s)\hat{u}(s),\end{aligned} \tag{6.19}$$

where $\hat{H}(s)$ denotes the transfer function of system (6.8a), (6.8b), given by

$$\hat{H}(s) = C(sI - A)^{-1}B + D. \tag{6.20}$$

Recalling that $\mathcal{L}[e^{At}] = \Phi(s) = (sI - A)^{-1}$ [refer to (4.32)], we could of course have obtained (6.20) directly by taking the Laplace transform of $H(t)$ given in (6.13).

EXAMPLE 6.5. In Example 6.2, let $t_0 = 0$ and $x(0) = 0$. Then

$$\begin{aligned}\hat{H}(s) &= C(sI - A)^{-1}B + D = [0, 1]\begin{bmatrix} s & -1 \\ 0 & s \end{bmatrix}^{-1}\begin{bmatrix} 0 \\ 1 \end{bmatrix}\\ &= [0, 1]\begin{bmatrix} \frac{1}{s} & \frac{1}{s^2} \\ 0 & \frac{1}{s} \end{bmatrix}\begin{bmatrix} 0 \\ 1 \end{bmatrix} = \frac{1}{s}\end{aligned}$$

and $H(t) = \mathcal{L}^{-1}\hat{H}(s) = 1$ for $t \geq 0$, as expected (see Example 6.2).

Next, as in Example 6.2, let $x(0) = (1, -1)^T$ and let u be the unit step. Then $\hat{y}(s) = C(sI - A)^{-1}x(0) + \hat{H}(s)\hat{u}(s) = [0, 1/s](1, -1)^T + (1/s)(1/s) = -1/s + 1/s^2$ and $y(t) = \mathcal{L}^{-1}[\hat{y}(s)] = -1 + t$ for $t \geq 0$, as expected (see Example 6.2). ■

We note that the eigenvalues of the matrix A in Example 6.5 are the roots of the equation $det\,(sI - A) = s^2 = 0$, and are given by $s_1 = 0, s_2 = 0$, while the transfer function $\hat{H}(s)$ in this example has only one pole (the zero of its denominator polynomial), located at the origin. It will be shown in Chapter 5 (on realization) that the *poles of the transfer function* $\hat{H}(s)$ *(of a SISO system)* are in general a subset of the eigenvalues of A. In Chapter 3 we will introduce and study two important system theoretic concepts, called *controllability* and *observability*. We will show in Chapter 5 that the eigenvalues of A are precisely the poles of the transfer function $\hat{H}(s) = C(sI - A)^{-1}B + D$ if and only if the system (6.8a), (6.8b) is observable and controllable. This is demonstrated in the next example.

EXAMPLE 6.6. In (6.8a), (6.8b), let $A = \begin{bmatrix} 0 & 1 \\ -1 & -2 \end{bmatrix}$, $B = \begin{bmatrix} 0 \\ 1 \end{bmatrix}$, $C = [-3, 3]$, $D = 0$. The eigenvalues of A are the roots of the equation $det\,(sI - A) = s^2 + 2s + 1 = (s+1)^2 = 0$ given by $s_1 = -1, s_2 = -1$, and the transfer function of this SISO system is given by

$$\hat{H}(s) = C(sI - A)^{-1}B + D = [-3, 3]\begin{bmatrix} s & -1 \\ 1 & s+2 \end{bmatrix}^{-1}\begin{bmatrix} 0 \\ 1 \end{bmatrix}$$

$$= 3[-1, 1]\frac{1}{(s+1)^2}\begin{bmatrix} s+2 & 1 \\ -1 & s \end{bmatrix}\begin{bmatrix} 0 \\ 1 \end{bmatrix} = \frac{3(s-1)}{(s+1)^2},$$

with poles (the zeros of the denominator polynomial) also given by $s_1 = -1, s_2 = -1$. ■

If in Example 6.6 we replace $B = [0, 1]^T$ and $D = 0$ by $B = \begin{bmatrix} 0 & -\frac{1}{2} \\ 1 & \frac{1}{2} \end{bmatrix}$ and $D = [0, 0]$, then we have a multi-input system whose transfer function is given by

$$\hat{H}(s) = \left[\frac{3(s-1)}{(s+1)^2}, \frac{3}{(s+1)}\right].$$

The concepts of poles and zeros for MIMO systems (also called multivariable systems) will be introduced in Chapter 4. The determination of the poles of such systems is not as straightforward as in the case of SISO systems. It turns out that in the present case the poles of $\hat{H}(s)$ are $s_1 = -1, s_2 = -1$, the same as the eigenvalues of A.

Before proceeding to our next topic, the equivalence of internal representations, an observation concerning the transfer function $\hat{H}(s)$ of system (6.8a), (6.8b), given by (6.20), $\hat{H}(s) = C(sI - A)^{-1}B + D$ is in order. Since the numerator matrix polynomial of $(sI - A)^{-1}$ is of degree $(n - 1)$ (refer to Subsection 2.1G), while its denominator polynomial, the characteristic polynomial $\alpha(s)$ of A, is of degree n, it is clear that

$$\lim_{s \to \infty} \hat{H}(s) = D,$$

a real-valued $m \times n$ matrix, and in particular, when the *"direct link matrix"* D in the output equation (6.8b) is zero, then

$$\lim_{s \to \infty} \hat{H}(s) = 0,$$

the $m \times n$ matrix with zeros as its entries. In the former case (when $D \neq 0$ or $D = 0$), $\hat{H}(s)$ is said to be a *proper transfer function*, while in the latter case (when $D = 0$), $\hat{H}(s)$ is said to be a *strictly proper transfer function*.

When discussing the realization of transfer functions by state-space descriptions (in Chapter 5), we will study the properties of transfer functions in greater detail. In this connection, we will also encounter systems that can be described by models corresponding to transfer functions $\hat{H}(s)$ that are *not proper*. The differential equation representation of a differentiator (or an inductor) given by $y(t) = (d/dt)u(t)$ is one such example. Indeed, in this case the system cannot be represented by Eqs. (6.8a), (6.8b) and the transfer function, given by $\hat{H}(s) = s$ is not proper. Such systems will be discussed in Chapter 7.

C. Equivalence of Internal Representations

In Subsection 2.4B it was shown that when a linear, autonomous, homogeneous system of first-order ordinary differential equations $\dot{x} = Ax$ is subjected to an appropriately chosen similarity transformation, the resulting set of equations may be considerably easier to use and may exhibit latent properties of the system of equations. It is therefore natural that we consider a similar course of action in the case of the linear systems (6.1a), (6.1b) and (6.8a), (6.8b).

We begin by considering (6.8a), (6.8b) first, letting

$$\tilde{x} = Px, \tag{6.21}$$

where P is a real, nonsingular matrix (i.e., P is a similarity transformation). Consistent with what has been said thus far, we see that such transformations bring about a *change of basis* for the state space of system (6.8a), (6.8b). Application of (6.21) to this system will result, as will be seen, in a system description of the same form as (6.8a), (6.8b), but involving different state variables. We will say that the system (6.8a), (6.8b), and the system obtained by subjecting (6.8a), (6.8b) to the transformation (6.21), constitute *equivalent internal representations* of an underlying system. We will show that equivalent internal representations (of the same system) possess identical external descriptions, as one would expect, by showing that they have identical impulse responses and transfer function matrices. In connection with this discussion, two important notions called *zero-input equivalence* and *zero-state equivalence* of a system will arise in a natural manner.

If we differentiate both sides of (6.21), and if we apply $x = P^{-1}\tilde{x}$ to (6.8a), (6.8b), we obtain the equivalent internal representation of (6.8a), (6.8b) given by

$$\dot{\tilde{x}} = \tilde{A}\tilde{x} + \tilde{B}u \tag{6.22a}$$

$$y = \tilde{C}\tilde{x} + \tilde{D}u, \tag{6.22b}$$

where
$$\tilde{A} = PAP^{-1}, \qquad \tilde{B} = PB, \qquad \tilde{C} = CP^{-1}, \qquad \tilde{D} = D \tag{6.23}$$

and where $\tilde{x}$ is given by (6.21). It is now easily verified that the system (6.8a), (6.8b) and the system (6.22a), (6.22b) have the same external representation. Recall that for (6.8a), (6.8b) and for (6.22a), (6.22b), we have for the impulse response

$$H(t,\tau) \triangleq H(t-\tau,0) = \begin{cases} Ce^{A(t-\tau)}B + D\delta(t-\tau), & t \geq \tau, \\ 0, & t < \tau, \end{cases} \tag{6.24}$$

and
$$\tilde{H}(t,\tau) \triangleq \tilde{H}(t-\tau,0) = \begin{cases} \tilde{C}e^{\tilde{A}(t-\tau)}\tilde{B} + \tilde{D}\delta(t-\tau), & t \geq \tau, \\ 0, & t < \tau. \end{cases} \tag{6.25}$$

Recalling from Subsection 2.4B [see Eq. (4.13)] that

$$e^{\tilde{A}(t-\tau)} = Pe^{A(t-\tau)}P^{-1}, \tag{6.26}$$

we obtain from (6.23) to (6.25) that $\tilde{C}e^{\tilde{A}(t-\tau)}\tilde{B}+\tilde{D}\delta(t-\tau) = CP^{-1}Pe^{A(t-\tau)}P^{-1}PB+ D\delta(t-\tau) = Ce^{A(t-\tau)}B + D\delta(t-\tau)$, which proves, in view of (6.24) and (6.25), that

$$\tilde{H}(t,\tau) = H(t,\tau), \tag{6.27}$$

and this in turn shows that

$$\hat{\tilde{H}}(s) = \hat{H}(s). \tag{6.28}$$

This last relationship can also be verified by observing that $\hat{\tilde{H}}(s) = \tilde{C}(sI - \tilde{A})^{-1} \times \tilde{B} + \tilde{D} = CP^{-1}(sI - PAP^{-1})^{-1}PB + D = CP^{-1}P(sI - A)^{-1}P^{-1}PB + D = C(sI - A)^{-1}B + D = \hat{H}(s)$.

Next, recall that in view of (6.10) we have for (6.8a), (6.8b) that

$$\begin{aligned} y(t) &= Ce^{A(t-t_0)}x_0 + \int_{t_0}^{t} H(t-\tau,0)u(\tau)\,d\tau \\ &= \psi(t,t_0,x_0,0) + \rho(t,t_0,0,u) \end{aligned} \tag{6.29}$$

and for (6.22a), (6.22b) that

$$\begin{aligned} y(t) &= \tilde{C}e^{\tilde{A}(t-t_0)}\tilde{x}_0 + \int_{t_0}^{t} \tilde{H}(t-\tau,0)u(\tau)\,d\tau \\ &= \tilde{\psi}(t,t_0,\tilde{x}_0,0) + \tilde{\rho}(t,t_0,0,u), \end{aligned} \tag{6.30}$$

where ψ and $\tilde{\psi}$ denote the zero-input response of (6.8a), (6.8b) and (6.22a), (6.22b), respectively, while ρ and $\tilde{\rho}$ denote the zero-state response of (6.8a), (6.8b) and (6.22a), (6.22b), respectively. The relations (6.29) and (6.30) give rise to the following concepts: Two state-space representations are *zero-state equivalent* if they give rise to the same impulse response (the same external description). Also, two state-space representations are *zero-input equivalent* if for any initial state vector for one representation there exists an initial state vector for the second representation such that the zero-input responses for the two representations are identical.

The following result is now clear: *if two state-space representations are equivalent, then they are both zero-state and zero-input equivalent.* They are clearly zero-state equivalent since $H(t,\tau) = \tilde{H}(t,\tau)$. Also, in view of (6.29) and (6.30), we have $\tilde{C}e^{\tilde{A}(t-t_0)}\tilde{x}_0 = (CP^{-1})[Pe^{A(t-t_0)}P^{-1}]\tilde{x}_0 = Ce^{A(t-t_0)}x_0$, where (6.26) was used. Therefore, the two state representations are also zero-input equivalent.

The converse to the above result is in general not true, since there are representations that are both zero-state and zero-input equivalent, yet not equivalent. In Chapter 5, which deals with state-space realizations of transfer functions, we will consider this topic further.

EXAMPLE 6.7. System (6.8a), (6.8b) with

$$A = \begin{bmatrix} 0 & 1 \\ -2 & -3 \end{bmatrix}, \qquad B = \begin{bmatrix} 0 \\ 1 \end{bmatrix}, \qquad C = [-1, -5], \qquad D = 1$$

has the transfer function

$$H(s) = C(sI - A)^{-1}B + D = \frac{-3s - 1}{s^2 + 3s + 2} + 1 = \frac{(s-1)^2}{(s+1)(s+2)}.$$

Using the similarity transformation

$$P = \begin{bmatrix} 1 & 1 \\ -1 & -2 \end{bmatrix}^{-1} = \begin{bmatrix} 2 & 1 \\ -1 & -1 \end{bmatrix}$$

yields the equivalent representation of the system given by

$$\tilde{A} = PAP^{-1} = \begin{bmatrix} -1 & 0 \\ 0 & -2 \end{bmatrix}, \qquad \tilde{B} = PB = \begin{bmatrix} 1 \\ -1 \end{bmatrix}, \qquad \tilde{C} = CP^{-1} = [4, 9],$$

and $\tilde{D} = D = 1$. Note that the columns of P^{-1}, given by $[1, -1]^T$ and $[1, -2]^T$, are eigenvectors of A corresponding to the eigenvalues $\lambda_1 = -1, \lambda_2 = -2$ of A, that is, P was chosen to diagonalize A. Notice that A is in companion form so that its characteristic polynomial is given by $s^2 + 3s + 2 = (s+1)(s+2)$. Notice also that the eigenvectors given above are of the form $[1, \lambda_i]^T, i = 1, 2$. The transfer function of the equivalent representation of the system is now given by

$$\hat{\tilde{H}}(s) = \tilde{C}(sI - \tilde{A})^{-1}\tilde{B} + \tilde{D} = [4, 0] \begin{bmatrix} \dfrac{1}{s+1} & 0 \\ 0 & \dfrac{1}{s+2} \end{bmatrix} \begin{bmatrix} 1 \\ -1 \end{bmatrix} + 1$$

$$= \frac{-5s - 1}{(s+1)(s+2)} + 1 = H(s).$$

Finally, it is easily verified that $e^{\tilde{A}t} = Pe^{At}P^{-1}$. ∎

From the above discussion it should be clear that systems [of the form (6.8a), (6.8b)] described by equivalent representations have identical behavior to the *outside world,* since both their zero-input and zero-state responses are the same. Their states, however, are in general not identical, but are related by the transformation $\tilde{x}(t) = Px(t)$.

In the time-invariant case considered above, the transformation P preserves the qualitative properties of the equivalent representations of a system, since in particular, the eigenvalues of A and $\tilde{A}$ are identical.

Next, we consider time-varying systems, given by

$$\dot{x} = A(t)x + B(t)u \tag{6.31a}$$

$$y = C(t)x + D(t)u, \tag{6.31b}$$

where all symbols are defined as in (6.1a), (6.1b). Let $P \in C^1(R, R^{n \times n})$ and assume that $P^{-1}(t)$ exists for all $t \in R$ and is continuous. Let

$$\tilde{x} = P(t)x. \tag{6.32}$$

Then $\dot{\tilde{x}} = \dot{P}(t)x + P(t)\dot{x} = [\dot{P}(t) + P(t)A(t)]P^{-1}(t)\tilde{x} + P(t)B(t)u \triangleq \tilde{A}(t)\tilde{x} + \tilde{B}(t)u$ and $y = C(t)P^{-1}(t)\tilde{x} + D(t)u \triangleq \tilde{C}(t)\tilde{x} + \tilde{D}(t)u$. These relations motivate the following definition: the system

$$\dot{\tilde{x}} = \tilde{A}(t)\tilde{x} + \tilde{B}(t)u \tag{6.33a}$$

$$y = \tilde{C}(t)\tilde{x} + D(t)u, \tag{6.33b}$$

where $\tilde{x} = P(t)x$, $P \in C^1(R, R^{n\times n})$, and P^{-1} is assumed to exist and be continuous for all $t \in R$, and where $\tilde{A}(t) = [P(t)A(t) + \dot{P}(t)]P^{-1}(t)$, $\tilde{B}(t) = P(t)B(t)$, $\tilde{C}(t) = C(t)P^{-1}(t)$, $\tilde{D}(t) = D(t)$, is said to be *equivalent* to the system (6.31a), (6.31b).

As in the time-invariant case, the relations between the state transition matrices $\Phi(t, t_0)$ and $\tilde{\Phi}(t, t_0)$ for the systems of equations

$$\dot{x} = A(t)x \tag{6.34}$$

and

$$\dot{\tilde{x}} = \tilde{A}(t)\tilde{x}, \tag{6.35}$$

respectively, and the relations between the impulse responses $H(t, \tau)$ and $\tilde{H}(t, \tau)$ of (6.31a), (6.31b) and (6.33a), (6.33b), respectively, are easily established. Indeed, since the solutions of (6.34) and (6.35) are given by $\phi(t, t_0, x_0) = \Phi(t, t_0)x_0$ and $\tilde{\phi}(t, t_0, x_0) = \tilde{\Phi}(t, t_0)\tilde{x}_0$, respectively, we have in view of (6.32) (assuming that P^{-1} exists for all $t \in R$), $P^{-1}(t)\tilde{\phi}(t, t_0, x_0) = \Phi(t, t_0)[P^{-1}(t_0)\tilde{x}_0]$ or $\tilde{\phi}(t, t_0, x_0) = P(t)\Phi(t, t_0)P^{-1}(t_0)\tilde{x}_0$. Since the solutions of (6.34) and (6.35) are unique, we have that

$$\tilde{\Phi}(t, \tau) = P(t)\Phi(t, \tau)P^{-1}(\tau)$$

for all $t, \tau \in R$.

Recalling that the columns of a fundamental matrix Ψ of (6.34) and a fundamental matrix $\tilde{\Psi}$ of (6.35) are linearly independent, it is not hard to show, using (6.32), that $\tilde{\Psi}(t) = P(t)\Psi(t)$ for all $t \in R$.

Next, recalling that the impulse responses of the equivalent systems (6.31a), (6.31b) and (6.33a), (6.33b) are given by

$$H(t, \tau) = \begin{cases} C(t)\Phi(t, \tau)B(\tau) + D(t)\delta(t - \tau), & t \geq \tau, \\ 0 & t < \tau, \end{cases}$$

and

$$\tilde{H}(t, \tau) = \begin{cases} \tilde{C}(t)\tilde{\Phi}(t, \tau)\tilde{B}(\tau) + \tilde{D}(t)\delta(t - \tau) & t \geq \tau, \\ 0 & t < \tau, \end{cases}$$

respectively, it is easily shown that

$$\tilde{H}(t, \tau) = H(t, \tau).$$

Indeed, we have that

$$\begin{aligned} \tilde{H}(t, \tau) &= \tilde{C}(t)\tilde{\Phi}(t, \tau)\tilde{B}(\tau) + \tilde{D}(t)\delta(t - \tau) \\ &= C(t)P^{-1}(t)P(t)\Phi(t, \tau)P^{-1}(\tau)P(\tau)B(\tau) + D(t)\delta(t - \tau) \\ &= C(t)\Phi(t, \tau)B(\tau) + D(t)\delta(t - \tau) \\ &= H(t, \tau) \end{aligned}$$

for $t \geq \tau$.

We conclude by noting that the notions of *zero-state equivalence* and *zero-input equivalence* introduced for time-invariant systems of the form (6.8a), (6.8b) carry over without changes for time-varying systems of the form (6.31a), (6.31b). Furthermore, identically to the time-invariant case, it can be shown that in the case of time-varying systems, if two state representations [such as (6.31a), (6.31b) and (6.33a), (6.33b)] are equivalent, then they are both zero-state and zero-input equivalent. The converse to this statement, however, is not true.

2.7 STATE EQUATION AND INPUT-OUTPUT DESCRIPTION OF DISCRETE-TIME SYSTEMS

In this section, which consists of five subsections, we address the state equation and input-output description of linear discrete-time systems. In the first subsection we study the response of linear time-varying systems and linear time-invariant systems described by the difference equations (1.15.3a), (1.15.3b) [or (1.1.7a), (1.1.7b)] and (1.15.4a), (1.15.4b) [or (1.1.8a), (1.1.8b)], respectively. In the second subsection we consider transfer functions for linear time-invariant systems, while in the third subsection we address the equivalence of the internal representations of time-varying and time-invariant linear discrete-time systems [described by (1.15.3a), (1.15.3b) and (1.15.4a), (1.15.4b), respectively]. Some of the most important classes of discrete-time systems include linear sampled-data systems that we develop in the fourth subsection. In the final part of this section, we address modes and asymptotic behavior of linear time-invariant discrete-time systems.

A. Response of Linear Discrete-Time Systems

We now return to Section 1.15 to consider once again systems described by linear time-varying equations of the form

$$x(k+1) = A(k)x(k) + B(k)u(k) \tag{7.1a}$$

$$y(k) = C(k)x(k) + D(k)u(k), \tag{7.1b}$$

where $A : Z \to R^{n\times n}$, $B : Z \to R^{n\times m}$, $C : Z \to R^{p\times n}$, and $D : Z \to R^{p\times m}$. When $A(k) \equiv A$, $B(k) \equiv B$, $C(k) \equiv C$, and $D(k) \equiv D$, we have systems described by linear time-invariant equations given by

$$x(k+1) = Ax(k) + Bu(k) \tag{7.2a}$$

$$y(k) = Cx(k) + Du(k). \tag{7.2b}$$

We recall that in (7.1a), (7.1b) and in (7.2a), (7.2b), x denotes the state vector, u denotes the system input, and y denotes the system output. For given initial conditions $k_0 \in Z$, $x(k_0) = x_{k_0} \in R^n$ and for a given input u, both equations (7.1a) and (7.2a) possess unique solutions $x(k)$ that are defined for all $k \geq k_0$, and thus, the response $y(k)$ for (7.1b) and for (7.2b) is also defined for all $k \geq k_0$.

Associated with (7.1a) is the linear homogeneous system of equations given by

$$x(k+1) = A(k)x(k). \tag{7.3}$$

We recall from Section 1.15 that the solution of the initial-value problem

$$x(k+1) = A(k)x(k), \qquad x(k_0) = x_{k_0} \tag{7.4}$$

is given by

$$x(k) = \Phi(k, k_0)x_{k_0} = \prod_{j=k_0}^{k-1} A(j)x_{k_0}, \qquad k > k_0, \tag{7.5}$$

where $\Phi(k, k_0)$ denotes the state transition matrix of (7.3) with

$$\Phi(k, k) = I \tag{7.6}$$

[refer to (15.9) to (15.12) in Chapter 1].

Common properties of the state transition matrix $\Phi(k, l)$, such as for example the *semigroup property* given by

$$\Phi(k, l) = \Phi(k, m)\Phi(m, l), \qquad k \geq m \geq l,$$

can quite easily be derived from (7.5), (7.6). We caution the reader, however, that not all the properties of the state transition matrix $\Phi(t, \tau)$ for continuous-time systems $\dot{x} = A(t)x$ carry over to the discrete-time case (7.3). In particular we recall that if for the continuous-time case we have $t > \tau$, then future values of the state ϕ at time t can be obtained from past values of the state ϕ at time τ, and vice versa, from the relationships $\phi(t) = \Phi(t, \tau)\phi(\tau)$ and $\phi(\tau) = \Phi^{-1}(t, \tau)\phi(t) = \Phi(\tau, t)\phi(t)$, i.e., for continuous-time systems a principle of *time reversibility exists*. This principle is in general not true for system (7.3), unless $A^{-1}(k)$ exists for all $k \in Z$. The reason for this lies in the fact that $\Phi(k, l)$ will not be nonsingular if $A(k)$ is not nonsingular for all k.

Associated with (7.2a) is the linear, autonomous, homogeneous system of equations given by

$$x(k+1) = Ax(k). \tag{7.7}$$

From (7.5) it follows that the unique solutions of the initial-value problem

$$x(k+1) = Ax(k), \qquad x(k_0) = x_{k_0} \tag{7.8}$$

are given by

$$x(k) = \Phi(k, k_0)x_{k_0} = A^{(k-k_0)}x_{k_0}, \qquad k \geq k_0. \tag{7.9}$$

EXAMPLE 7.1. In (7.3), we let $k_0 = 0$, and $A(k) = \begin{bmatrix} k & (k^2+1) \\ 0 & (\frac{1}{2})^k \end{bmatrix}$. Then $\Phi(k, 0) = A(k-1)\cdots A(0) = \begin{bmatrix} (k-1) & (k-1)^2+1 \\ 0 & (\frac{1}{2})^{k-1} \end{bmatrix} \cdots \begin{bmatrix} 0 & 1 \\ 0 & 1 \end{bmatrix}$ If, for example, $k = 3$, then $x(3) = A(2) \cdot A(1) \cdot A(0) \cdot x(0) = \begin{bmatrix} 2 & 5 \\ 0 & \frac{1}{4} \end{bmatrix} \begin{bmatrix} 1 & 2 \\ 0 & \frac{1}{2} \end{bmatrix} \begin{bmatrix} 0 & 1 \\ 0 & 1 \end{bmatrix} x(0) = \begin{bmatrix} 0 & \frac{17}{2} \\ 0 & \frac{1}{8} \end{bmatrix} x(0)$. Given $x(0)$, we can now readily determine $x(3)$. In view of the form $A(0)$, we have for any $k > 0$, $\Phi(k, 0) = \begin{bmatrix} 0 & \times \\ 0 & \times \end{bmatrix}$, that is to say, the first column of $\Phi(k, 0)$ will always be zero for any k. Clearly then, for any initial condition $x(0) = \begin{bmatrix} \alpha \\ 0 \end{bmatrix}$, $\alpha \in R$, we have $x(k) = \begin{bmatrix} 0 \\ 0 \end{bmatrix}$ for all $k > 0$. ■

EXAMPLE 7.2. In (7.5), let $A = \begin{bmatrix} 1 & 0 \\ 0 & 0 \end{bmatrix}$, $x(0) = \begin{bmatrix} 1 \\ \alpha \end{bmatrix}$, $\alpha \in R$. The initial state $x(0)$ at $k_0 = 0$ for *any* $\alpha \in R$ will map into the state $x(1) = \begin{bmatrix} 1 \\ 0 \end{bmatrix}$. Accordingly, in this case time reversibility will not apply. ■

EXAMPLE 7.3. In (7.8), let $A = \begin{bmatrix} -1 & 2 \\ 0 & 1 \end{bmatrix}$. In view of (7.9) we have that $A^{(k-k_0)} = \begin{bmatrix} (-1)^{(k-k_0)} & 1-(-1)^{(k-k_0)} \\ 0 & 1 \end{bmatrix}$, $k \geq k_0$, i.e., $A^{(k-k_0)} = A$ when $(k - k_0)$ is odd, and

$A^{(k-k_0)} = I$ when $(k - k_0)$ is even. Therefore, given $k_0 = 0$ and $x(0) = \begin{bmatrix} 2 \\ 1 \end{bmatrix}$, then $x(k) = Ax(0) = \begin{bmatrix} 0 \\ 1 \end{bmatrix}$, $k = 1, 3, 5, \ldots$, and $x(k) = Ix(0) = \begin{bmatrix} 2 \\ 1 \end{bmatrix}$, $k = 2, 4, 6, \ldots$. A plot of the states $x(k) = [x_1(k), x_2(k)]^T$ is given in Fig. 2.3. ■

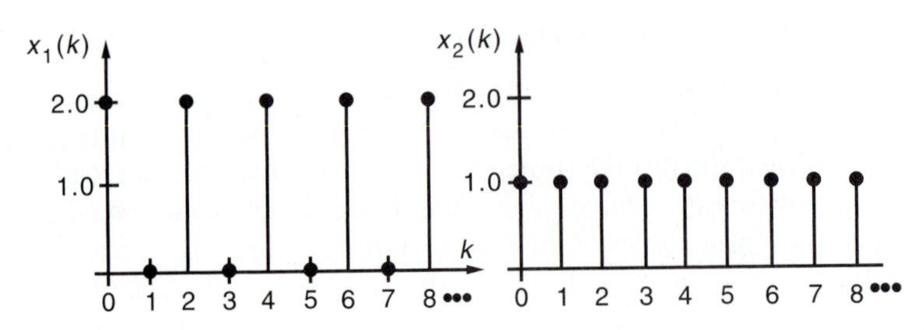

FIGURE 2.3
Plots of states for Example 7.3

Continuing, we recall that the solutions of initial-value problems determined by linear nonhomogeneous systems (15.13) of Chapter 1 are given by expression (15.14). Utilizing (15.13), the solution of (7.1a) for given $x(k_0)$ and $u(k)$ is given as

$$x(k) = \Phi(k, k_0)x(k_0) + \sum_{j=k_0}^{k-1} \Phi(k, j+1)B(j)u(j), \qquad k > k_0. \tag{7.10}$$

This expression in turn can be used to determine [as in (15.17) and (15.18) of Chapter 1] the system response for system (7.1a), (7.1b) as

$$\begin{aligned} y(k) &= C(k)\Phi(k, k_0)x(k_0) \\ &\quad + \sum_{j=k_0}^{k-1} C(k)\Phi(k, j+1)B(j)u(j) + D(k)u(k), \qquad k > k_0, \\ y(k_0) &= C(k_0)x(k_0) + D(k_0)u(k_0). \end{aligned} \tag{7.11}$$

Furthermore, for the time-invariant case (7.2a), (7.2b), we have for the system response the expression

$$\begin{aligned} y(k) &= CA^{(k-k_0)}x(k_0) + \sum_{j=k_0}^{k-1} CA^{k-(j+1)}Bu(j) + Du(k), \qquad k > k_0, \\ y(k_0) &= Cx(k_0) + Du(k_0). \end{aligned} \tag{7.12}$$

Since the system (7.2a), (7.2b) is time-invariant, we can let $k_0 = 0$ without loss of generality to obtain from (7.12) the expression

$$y(k) = CA^k x(0) + \sum_{j=0}^{k-1} CA^{k-(j+1)}Bu(j) + Du(k), \qquad k > 0. \tag{7.13}$$

As in the continuous-time case, the *total system response* (7.11) may be viewed as consisting of two components, the *zero-input response*, given by

$$\psi(k) = C(k)\Phi(k, k_0)x(k_0), \qquad k > k_0,$$

and the *zero-state response*, given by

$$
\begin{aligned}
\rho(k) &= \sum_{j=k_0}^{k-1} C(k)\Phi(k, j+1)B(j)u(j) + D(k)u(k), & k > k_0, \\
\rho(k_0) &= D(k_0)u(k_0), & k = k_0.
\end{aligned} \tag{7.14}
$$

Finally, in view of (16.20) of Chapter 1, we recall that the (discrete-time) unit impulse response matrix of system (7.1a), (7.1b) is given by

$$
H(k, l) = \begin{cases} C(k)\Phi(k, l+1)B(l), & k > l, \\ D(k), & k = l, \\ 0, & k < l, \end{cases} \tag{7.15}
$$

and the unit impulse response matrix of system (7.2a), (7.2b) is given by

$$
H(k, l) = \begin{cases} CA^{k-(l+1)}B, & k > l, \\ D, & k = l, \\ 0, & k < l, \end{cases} \tag{7.16}
$$

and in particular, when $l = 0$ (i.e., when the pulse is applied at time $l = 0$),

$$
H(k, 0) = \begin{cases} CA^{k-1}B, & k > 0, \\ D, & k = 0, \\ 0, & k < 0. \end{cases} \tag{7.17}
$$

EXAMPLE 7.4. In (7.2a), (7.2b), let

$$
A = \begin{bmatrix} 0 & 1 \\ 0 & -1 \end{bmatrix}, \qquad B = \begin{bmatrix} 0 \\ 1 \end{bmatrix}, \qquad C^T = \begin{bmatrix} 1 \\ 0 \end{bmatrix}, \qquad D = 0.
$$

We first determine A^k by using the Cayley-Hamilton Theorem (Theorem 3.1 of Chapter 2). To this end we compute the eigenvalues of A as $\lambda_1 = 0, \lambda_2 = -1$, we let $A^k = f(A)$, where $f(s) = s^k$, and we let $g(s) = \alpha_1 s + \alpha_0$. Then $f(\lambda_1) = g(\lambda_1)$, or $\alpha_0 = 0$ and $f(\lambda_2) = g(\lambda_2)$, or $(-1)^k = -\alpha_1 + \alpha_0$. Therefore, $A^k = \alpha_1 A + \alpha_0 I = -(-1)^k \begin{bmatrix} 0 & 1 \\ 0 & -1 \end{bmatrix} = \begin{bmatrix} 0 & (-1)^{k-1} \\ 0 & (-1)^k \end{bmatrix}$, $k = 1, 2, \ldots$, or $A^k = \begin{bmatrix} \delta(k) & (-1)^{k-1}p(k-1) \\ 0 & (-1)^k p(k) \end{bmatrix}$, $k = 0, 1, 2, \ldots$, where $A^0 = I$, and where $p(k)$ denotes the unit step given by

$$
p(k) = \begin{cases} 1, & k \geq 0, \\ 0, & k < 0. \end{cases}
$$

The above expression for A^k is now substituted into (7.12) to determine the response $y(k)$ for $k > 0$ for a given initial condition $x(0)$ and a given input $u(k)$, $k \geq 0$. To determine the unit impulse response, we note that $H(k, 0) = 0$ for $k < 0$ and $k = 0$. When $k > 0$, $H(k, 0) = CA^{k-1}B = (-1)^{k-2}p(k-2)$ for $k > 0$ or $H(k, 0) = 0$ for $k = 1$ and $H(k, 0) = (-1)^{k-2}$ for $k = 2, 3, \ldots$. ■

B. The Transfer Function and the z-Transform

We assume that the reader is familiar with the concept and properties of the *one-sided z-transform* of a real-valued sequence $\{f(k)\}$, given by

$$
\mathscr{Z}\{f(k)\} = \hat{f}(z) = \sum_{j=0}^{\infty} z^{-j} f(j). \tag{7.18}
$$

An important property of this transform, useful in solving difference equations, is given by the relation

$$\begin{aligned}\mathcal{Z}\{f(k+1)\} &= \sum_{j=0}^{\infty} z^{-j} f(j+1) = \sum_{j=1}^{\infty} z^{-(j-1)} f(j) \\ &= z\left[\sum_{j=0}^{\infty} z^{-j} f(j) - f(0)\right] \\ &= z[\mathcal{Z}\{f(k)\} - f(0)] = z\hat{f}(z) - zf(0). \end{aligned} \tag{7.19}$$

If we take the z-transform of both sides of Eq. (7.2a), we obtain, in view of (7.19), $z\hat{x}(z) - zx(0) = A\hat{x}(z) + B\hat{u}(z)$ or

$$\hat{x}(z) = (zI - A)^{-1} z x(0) + (zI - A)^{-1} B\hat{u}(z). \tag{7.20}$$

Next, by taking the z-transform of both sides of Eq. (7.2b), and by substituting (7.20) into the resulting expression, we obtain

$$\hat{y}(z) = C(zI - A)^{-1} z x(0) + [C(zI - A)^{-1} B + D]\hat{u}(z). \tag{7.21}$$

The time sequence $\{y(k)\}$ can be recovered from its one-sided z-transform $\hat{y}(z)$ by applying the *inverse z-transform*, denoted by $\mathcal{Z}^{-1}[\hat{y}(z)]$.

In Table 7.1 we provide the one-sided z-transforms of some of the commonly used sequences, and in Table 7.2 we enumerate some of the more frequently encountered properties of the one-sided z-transform.

The *transfer function matrix* $\hat{H}(z)$ of system (7.2a), (7.2b) relates the z-transform of the output y to the z-transform of the input u under the assumption that $x(0) = 0$. We have

$$\hat{y}(z) = \hat{H}(z)\hat{u}(z), \tag{7.22}$$

where

$$\hat{H}(z) = C(zI - A)^{-1} B + D. \tag{7.23}$$

To relate $\hat{H}(z)$ to the impulse response matrix $H(k, l)$, we notice that $\mathcal{Z}\{\delta(k - l)\} = z^{-l}$, where δ denotes the *discrete-time impulse* (or *unit pulse* or *unit sample*) defined in (16.5) of Chapter 1, i.e.,

$$\delta(k - l) = \begin{cases} 1, & k = l, \\ 0, & k \neq l. \end{cases} \tag{7.24}$$

TABLE 7.1
Some commonly used z-transforms

$\{f(k)\}, k \geq 0$	$\hat{f}(z) = \mathcal{Z}\{f(k)\}$
$\delta(k)$	1
$p(k)$	$1/(1 - z^{-1})$
k	$z^{-1}/(1 - z^{-1})^2$
k^2	$[z^{-1}(1 + z^{-1})]/(1 - z^{-1})^3$
a^k	$1/(1 - az^{-1})$
$(k + 1)a^k$	$1/(1 - az^{-1})^2$
$[(1/l!)(k + 1)\cdots(k + l)]a^k \quad l \geq 1$	$1/(1 - az^{-1})^{l+1}$
$a \cos \alpha k + b \sin \alpha k$	$[a + z^{-1}(b \sin \alpha - a \cos \alpha)]/(1 - 2z^{-1} \cos \alpha + z^{-2})$

TABLE 7.2
Some properties of z-transforms

	$\{f(k)\}\ k \geq 0$	$f(z)$
Time shift	$f(k+1)$	$z\hat{f}(z) - zf(0)$
—Advance	$f(k+l) \quad l \geq 1$	$z^l\hat{f}(z) - z\sum_{i=1}^{l} z^{l-i} f(i-1)$
Time shift	$f(k-1)$	$z^{-1}\hat{f}(z) + f(-1)$
—Delay	$f(k-l) \quad l \geq 1$	$z^{-l}\hat{f}(z) + \sum_{i=1}^{l} z^{-l+i} f(-i)$
Scaling	$a^k f(k)$	$\hat{f}(z/a)$
	$kf(k)$	$-z(d/dz)\hat{f}(z)$
Convolution	$\sum_{l=0}^{\infty} f(l)g(k-l) = f(k) * g(k)$	$\hat{f}(z)\hat{g}(z)$
Initial value	$f(l)$ with $f(k) = 0 \quad k < l$	$\lim_{z\to\infty} z^l \hat{f}(z)^{\dagger}$
Final value	$\lim_{k\to\infty} f(k)$	$\lim_{z\to 1}(1 - z^{-1})\hat{f}(z)^{\dagger\dagger}$

†If the limit exists.
†† If $(1 - z^{-1})\hat{f}(z)$ has no singularities on or outside the unit circle.

This implies that the z-transform of a unit pulse applied at time zero is $\mathcal{Z}\{\delta(k)\} = 1$. It is not difficult to see now that $\{H(k, 0)\} = \mathcal{Z}^{-1}[\hat{y}(z)]$, where $\hat{y}(z) = \hat{H}(z)\hat{u}(z)$ with $\hat{u}(z) = 1$. This shows that

$$\mathcal{Z}^{-1}[\hat{H}(z)] = \mathcal{Z}^{-1}[C(zI - A)^{-1}B + D] = \{H(k, 0)\}, \tag{7.25}$$

where the unit impulse response matrix $H(k, 0)$ is given by (7.17).

The above result can also be derived directly by taking the z-transform of $\{H(k, 0)\}$ given in (7.17) (prove this). In particular, notice that the z-transform of $\{A^{k-1}\}$, $k = 1, 2, \ldots$ is $(zI - A)^{-1}$ since

$$\begin{aligned}
\mathcal{Z}\{0, A^{k-1}\} &= \sum_{j=1}^{\infty} z^{-j} A^{j-1} = z^{-1} \sum_{j=0}^{\infty} z^{-j} A^{j} \\
&= z^{-1}(I + z^{-1}A + z^{-2}A^2 + \cdots) \\
&= z^{-1}(I - z^{-1}A)^{-1} \\
&= (zI - A)^{-1}.
\end{aligned} \tag{7.26}$$

Above, the matrix determined by the expression $(1 - \lambda)^{-1} = 1 + \lambda + \lambda^2 + \cdots$ was used. It is easily shown that the corresponding series involving A converges. Notice also that $\mathcal{Z}\{A^k\}$, $k = 0, 1, 2, \ldots$ is $z(zI - A)^{-1}$. This fact can be used to show that the inverse z-transform of (7.21) yields the time response (7.13), as expected.

We conclude this subsection with a specific example.

EXAMPLE 7.5. In system (7.2a), (7.2b), we let

$$A = \begin{bmatrix} 0 & 1 \\ 0 & -1 \end{bmatrix}, \qquad B = \begin{bmatrix} 0 \\ 1 \end{bmatrix}, \qquad C = [1, 0], \qquad D = 0.$$

To verify that $\mathcal{Z}^{-1}[z(zI - A)^{-1}] = A^k$, we compute

$$z(zI - A)^{-1} = z\begin{bmatrix} z & -1 \\ 0 & z+1 \end{bmatrix}^{-1} = z\begin{bmatrix} \dfrac{1}{z} & \dfrac{1}{z(z+1)} \\ 0 & \dfrac{1}{z+1} \end{bmatrix} = \begin{bmatrix} 1 & \dfrac{1}{z+1} \\ 0 & \dfrac{z}{z+1} \end{bmatrix}$$

and
$$\mathcal{Z}^{-1}[z(zI-A)^{-1}] = \begin{bmatrix} \delta(k) & (-1)^{k-1}p(k-1) \\ 0 & (-1)^k p(k) \end{bmatrix}$$

or
$$A^k = \begin{cases} \begin{bmatrix} 1 & 0 \\ 0 & 1 \end{bmatrix}, & \text{when } k = 0, \\ \begin{bmatrix} 0 & (-1)^{k-1} \\ 0 & (-1)^k \end{bmatrix}, & \text{when } k = 1, 2, \ldots, \end{cases}$$

as expected from Example 7.4.

Notice that

$$\begin{aligned} \mathcal{Z}^{-1}[(zI-A)^{-1}] &= \mathcal{Z}^{-1}\left[\begin{bmatrix} \dfrac{1}{z} & \dfrac{1}{z(z+1)} \\ 0 & \dfrac{1}{z+1} \end{bmatrix}\right] \\ &= \begin{bmatrix} \delta(k-1)p(k-1) & \delta(k-1)p(k-1) - (-1)^{k-1}p(k-1) \\ 0 & (-1)^{k-1}p(k-1) \end{bmatrix} \\ &= \begin{bmatrix} 0 & 0 \\ 0 & 0 \end{bmatrix} \quad \text{for } k = 0, \qquad \text{and} \qquad \begin{bmatrix} 1 & 0 \\ 0 & 1 \end{bmatrix} \quad \text{for } k = 1, \end{aligned}$$

and

$$\mathcal{Z}^{-1}[(zI-A)^{-1}] = \begin{bmatrix} 0 & -(-1)^{k-1} \\ 0 & (-1)^{k-1} \end{bmatrix} \qquad \text{for } k = 2, 3, \ldots,$$

which is equal to A^k, $k \geq 0$, delayed by one unit, i.e., it is equal to A^{k-1}, $k = 1, 2, \ldots,$ as expected.

Next, we consider the system response with $x(0) = 0$ and $u(k) = p(k)$. We have

$$\begin{aligned} y(k) &= \mathcal{Z}^{-1}[\hat{y}(z)] = \mathcal{Z}^{-1}[C(zI-A)^{-1}B \cdot \hat{u}(z)] \\ &= \mathcal{Z}^{-1}\left[\frac{1}{(z+1)(z-1)}\right] = \mathcal{Z}^{-1}\left[\frac{\frac{1}{2}}{z-1} - \frac{\frac{1}{2}}{z+1}\right] \\ &= \tfrac{1}{2}[(1)^{k-1} - (-1)^{k-1}]p(k-1) \\ &= \begin{cases} 0, & k = 0, \\ \frac{1}{2}(1 - (-1)^{k-1}), & k = 1, 2, \ldots, \end{cases} \\ &= \begin{cases} 0, & k = 0 \\ 0, & k = 1, 3, 5, \ldots, \\ 1, & k = 2, 4, 6, \ldots. \end{cases} \end{aligned}$$

Note that if $x(0) = 0$ and $u(k) = \delta(k)$, then

$$\begin{aligned} y(k) &= \mathcal{Z}^{-1}[C(zI-A)^{-1}B] = \mathcal{Z}^{-1}\left[\frac{1}{z(z+1)}\right] \\ &= \delta(k-1)p(k-1) - (-1)^{k-1}p(k-1) \\ &= \begin{cases} 0, & k = 0, 1, \\ (-1)^{k-2}, & k = 2, 3, \ldots, \end{cases} \end{aligned}$$

which is the unit impulse response of the system (refer to Example 7.4). ■

C. Equivalence of Internal Representations

Equivalent representations of linear discrete-time systems are defined in a manner analogous to the continuous-time case.

For systems (7.1a), (7.1b), we let k_0 denote initial time, we let $P(k)$ denote a real $n \times n$ matrix that is nonsingular for all $k \geq k_0$, and we consider the transformation $\tilde{x}(k) = P(k)x(k)$. Substituting the above into (7.1a), (7.1b) yields the system

$$\tilde{x}(k+1) = \tilde{A}(k)\tilde{x}(k) + \tilde{B}(k)u(k) \tag{7.27a}$$

$$y(k) = \tilde{C}(k)\tilde{x}(k) + \tilde{D}(k)u(k), \tag{7.27b}$$

where

$$\begin{aligned} \tilde{A}(k) &= P(k+1)A(k)P^{-1}(k) \\ \tilde{B}(k) &= P(k+1)B(k) \\ \tilde{C}(k) &= C(k)P^{-1} \\ \tilde{D}(k) &= D(k). \end{aligned} \tag{7.28}$$

We say that system (7.27a), (7.27b) is *equivalent* to system (7.1a), (7.1b) and we call $P(k)$ an *equivalence transformation* matrix.

If $\Phi(k, l)$ denotes the state transition matrix of (7.3) and $\tilde{\Phi}(k, l)$ denotes the state transition matrix of

$$\tilde{x}(k+1) = \tilde{A}(k)\tilde{x}(k), \tag{7.29}$$

then

$$\tilde{\Phi}(k, l) = P(k)\Phi(k, l)P^{-1}(l), \tag{7.30}$$

as can be seen by observing that $\tilde{\Phi}(k, l) = \tilde{A}(k-1)\cdots\tilde{A}(l) = [P(k)A(k-1)P^{-1}(k-1)]\cdots[P(l+1)A(l)P^{-1}(l)] = P(k)\Phi(k, l)P^{-1}(l)$.

In a similar manner as above, it can also be shown that

$$\tilde{H}(k, l) = H(k, l), \tag{7.31}$$

where $\tilde{H}(k, l)$ and $H(k, l)$ denote the unit pulse response matrices of systems (7.1a), (7.1b) and (7.27a), (7.27b), respectively. [The reader should verify (7.31).] Thus, equivalent representations of linear discrete-time system (7.1a), (7.1b) give rise to the same unit pulse response matrix. Furthermore, *zero-state equivalent representations* and *zero-input equivalent representations* are defined for system (7.1a), (7.1b) in a similar manner as in the case of linear continuous-time systems.

Turning our attention now briefly to time-invariant systems (7.2a), (7.2b), we let P denote a real nonsingular $n \times n$ matrix and we define

$$\tilde{x}(k) = Px(k). \tag{7.32}$$

Substituting (7.32) into (7.2a), (7.2b) yields the equivalent system representation

$$\tilde{x}(k+1) = \tilde{A}\tilde{x}(k) + \tilde{B}u(k) \tag{7.33a}$$

$$y(k) = \tilde{C}\tilde{x}(k) + \tilde{D}u(k), \tag{7.33b}$$

where

$$\tilde{A} = PAP^{-1}, \qquad \tilde{B} = PB, \qquad \tilde{C} = CP^{-1}, \qquad \tilde{D} = D. \tag{7.34}$$

We note that the terms in (7.34) are identical to corresponding terms obtained for the case of linear continuous-time systems.

We conclude by noting that if $\hat{H}(z)$ and $\hat{\tilde{H}}(z)$ denote the transfer functions of the unit impulse response matrices of system (7.2a), (7.2b) and system (7.33a), (7.33b), respectively, then it is easily verified that $\hat{H}(z) = \hat{\tilde{H}}(z)$.

D. Sampled-Data Systems

Discrete-time dynamical systems arise in a variety of ways in the modeling process. There are systems that are inherently defined only at discrete points in time, and there are representations of continuous-time systems at discrete points in time. Examples of the former include digital computers and devices (e.g., digital filters) where the behavior of interest of a system is adequately described by values of variables at discrete-time instants (and what happens between the discrete instants of time is quite irrelevant to the problem on hand); inventory systems where only the inventory status at the end of each day (or month) is of interest; economic systems, such as banking, where, e.g., interests are calculated and added to savings accounts at discrete time intervals only, and so forth. Examples of the latter include simulations of continuous-time processes by means of digital computers, making use of difference equations that approximate the differential equations describing the process in question; feedback control systems that employ digital controllers and give rise to sampled-data systems (as discussed further in the following); and so forth.

In providing a short discussion of sampled-data systems, we make use of the specific class of linear feedback control systems depicted in Fig. 2.4. This system may be viewed as an interconnection of a subsystem S_1, called the *plant* (the object to be controlled) and a subsystem S_2, called the *digital controller*.

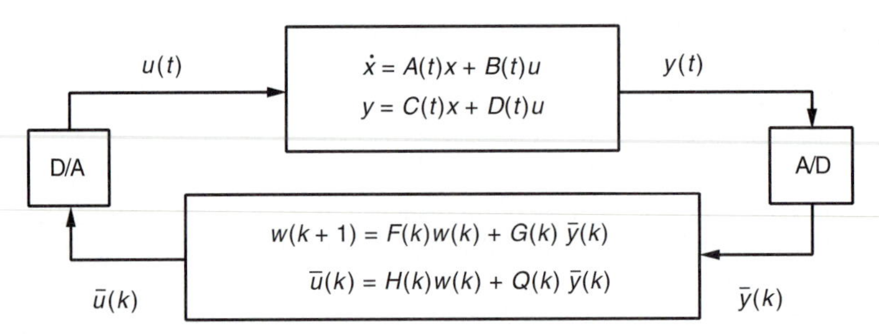

FIGURE 2.4
Digital control system

The plant is described by the equations

$$\dot{x} = A(t)x + B(t)u \tag{7.34a}$$

$$y = C(t)x + D(t)u, \tag{7.34b}$$

where all symbols in (7.34a), (7.34b) are defined as in (6.1a), (6.1b) and where we assume that $t \geq t_0 \geq 0$.

The subsystem S_2 accepts the continuous-time signal $y(t)$ as its input and it produces the piecewise continuous-time signal $u(t)$ as its output, where $t \geq t_0$. The continuous-time signal y is converted into a discrete-time signal $\{\bar{y}(k)\}$, $k \geq k_0 \geq 0$, $k, k_0 \in Z$, by means of an analog-to-digital (A/D) converter and is processed according to a control algorithm given by the difference equations

$$w(k+1) = F(k)w(k) + G(k)\bar{y}(k) \tag{7.35a}$$

$$\bar{u}(k) = H(k)w(k) + Q(k)\bar{y}(k), \tag{7.35b}$$

where the $w(k)$, $\bar{y}(k)$, $\bar{u}(k)$ are real vectors and the $F(k)$, $G(k)$, $H(k)$, and $Q(k)$ are real matrices with a consistent set of dimensions. Finally, the discrete-time signal $\{\bar{u}(k)\}$, $k \geq k_0 \geq 0$, is converted into the continuous-time signal u by means of a digital-to-analog (D/A) converter. To simplify our discussion, we assume in the following that $t_0 = t_{k_0}$.

An (ideal) *A/D converter* is a device that has as input a continuous-time signal, in our case y, and as output a sequence of real numbers, in our case $\{\bar{y}(k)\}$, $k = k_0, k_0 + 1, \ldots$, determined by the relation

$$\bar{y}(k) = y(t_k). \tag{7.36}$$

In other words, the (ideal) A/D converter is a device that *samples* an input signal, in our case $y(t)$, at times $t_0, t_1, \ldots$ producing the corresponding sequence $\{y(t_0), y(t_1), \ldots\}$.

A *D/A converter* is a device that has as input a discrete-time signal, in our case the sequence $\{\bar{u}(k)\}$, $k = k_0, k_0 + 1, \ldots$, and as output a continuous-time signal, in our case u, determined by the relation

$$u(t) = \bar{u}(k), t_k \leq t < t_{k+1}, \qquad k = k_0, k_0 + 1, \ldots. \tag{7.37}$$

In other words, the D/A converter is a device that keeps its output constant at the last value of the sequence entered. We also call such a device a *zero-order hold.*

The system of Fig. 2.4, as described above, is an example of a *sampled-data system*, since it involves truly *sampled data* (i.e., sampled signals), making use of an *ideal A/D converter*. In practice the digital controller S_2 uses *digital signals* as variables. In the scalar case, such signals are represented by real-valued sequences whose numbers belong to a subset of R consisting of a discrete set of points. (In the vector case, the previous statement applies to the components of the vector.) Specifically, in the present case, after the signal $y(t)$ has been sampled, it must be *quantized* (or *digitized*) to yield a *digital signal*, since only such signals are representable in a digital computer. If a computer uses, e.g., 8-bit words, then we can represent $2^8 = 256$ distinct levels for a variable, which determine the signal quantization. By way of a specific example, if we expect in the representation of a function a signal that varies from 9 to 25 volts, we may choose a 0.1-volt quantization step. Then 2.3 and 2.4 volts are represented by two different numbers (quantization levels); however, 2.315, 2.308, and 2.3 are all represented by the bit combination corresponding to 2.3. Quantization is an approximation, and for short wordlengths may lead to significant errors. Problems associated with *quantization effects* will not be addressed in this book.

In addition to being a sampled-data system, the system represented by Eqs. (7.34) to (7.37) constitutes a *hybrid system* as well, since it involves descriptions given by ordinary differential equations and ordinary difference equations. The analysis and synthesis of such systems can be simplified appreciably by replacing the description of subsystem S_1 (the plant) by a set of ordinary difference equations, valid only at discrete points in time t_k, $k = 0, 1, 2, \ldots$. [In terms of the blocks of Fig. 2.4, this corresponds to considering the plant S_1, together with the D/A and A/D devices, to obtain a system with input $\bar{u}(k)$ and output $\bar{y}(k)$, as shown in Fig. 2.5.] To accomplish this, we invoke the variation of constants formula in (7.34a) to obtain

$$x(t) = \Phi(t, t_k)x(t_k) + \int_{t_k}^{t} \Phi(t, \tau)B(\tau)u(\tau)\,d\tau, \tag{7.38}$$

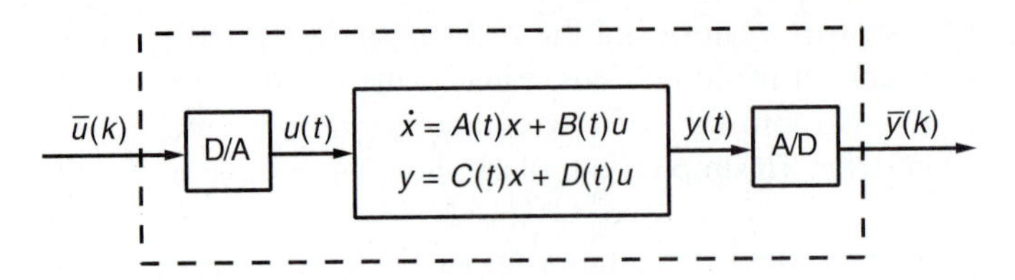

FIGURE 2.5
System described by (7.40) and (7.43)

where the notation $\phi(t, t_k, x(t_k)) = x(t)$ has been used. Since the input $u(t)$ is the output of the zero-order hold device (the D/A converter), given by (7.37), we obtain from (7.38) the expression

$$x(t_{k+1}) = \Phi(t_{t+1}, t_k)x(t_k) + \left[\int_{t_k}^{t_{k+1}} \Phi(t_{k+1}, \tau)B(\tau)\, d\tau\right] u(t_k). \tag{7.39}$$

Since $\bar{x}(k) \triangleq x(t_k)$ and $\bar{u}(k) \triangleq u(t_k)$, we obtain a discrete-time version of the state equation for the plant, given by

$$\bar{x}(k+1) = \bar{A}(k)\bar{x}(k) + \bar{B}(k)\bar{u}(k), \tag{7.40}$$

where

$$\begin{aligned} \bar{A}(k) &\triangleq \Phi(t_{k+1}, t_k) \\ \bar{B}(k) &\triangleq \int_{t_k}^{t_{k+1}} \Phi(t_{k+1}, \tau)B(\tau)\, d\tau. \end{aligned} \tag{7.41}$$

Next, we assume that the output of the plant is sampled at instants t'_k that do not necessarily coincide with the instants t_k at which the input to the plant is adjusted, and we assume that $t_k \le t'_k < t_{k+1}$. Then (7.34) and (7.38) yield

$$y(t'_k) = C(t'_k)\Phi(t'_k, t_k)x(t_k) + \left[C(t'_k)\int_{t_k}^{t'_k} \Phi(t'_k, \tau)B(\tau)\, d\tau\right] u(t_k) + D(t'_k)u(t_k). \tag{7.42}$$

Defining $\bar{y}(k) \triangleq y(t'_k)$, we obtain from (7.42),

$$\bar{y}(k) = \bar{C}(k)\bar{x}(k) + \bar{D}(k)\bar{u}(k), \tag{7.43}$$

where

$$\begin{aligned} \bar{C}(k) &\triangleq C(t'_k)\Phi(t'_k, t_k) \\ \bar{D}(k) &\triangleq C(t'_k)\int_{t_k}^{t'_k} \Phi(t'_k, \tau)B(\tau)\, d\tau + D(t'_k). \end{aligned} \tag{7.44}$$

Summarizing, (7.40) and (7.43) constitute a state-space representation, valid at discrete points in time, of the plant [given by (7.34a)] and including the A/D and D/A devices [given by (7.36) and (7.37); see Fig. 2.5]. Furthermore, the entire hybrid system of Fig. 2.4, valid at discrete points in time, can now be represented by Eqs. (7.40), (7.43), (7.35a), and (7.35b).

We now turn briefly to the case of the time-invariant plant, where $A(t) \equiv A$, $B(t) \equiv B$, $C(t) \equiv C$, and $D(t) \equiv D$, and we assume that $t_{k+1} - t_k = T$ and $t'_k - t_k = \alpha$ for all $k = 0, 1, 2, \ldots$. Then the expressions given in (7.40), (7.41), (7.43), and (7.44) assume the form

$$\bar{x}(k+1) = \bar{A}\bar{x}(k) + \bar{B}\bar{u}(k) \tag{7.45a}$$

$$\bar{y}(k) = \bar{C}\bar{x}(k) + \bar{D}\bar{u}(k), \tag{7.45b}$$

where
$$\bar{A} = e^{AT}, \bar{B} = \left(\int_0^T e^{A\tau}\,d\tau\right)B,$$
$$\bar{C} = Ce^{A\alpha}, \bar{D} = C\left(\int_0^\alpha e^{A\tau}\,d\tau\right)B + D. \tag{7.46}$$

If $t'_k = t_k$, or $\alpha = 0$, then $\bar{C} = C$ and $\bar{D} = D$.

In the preceding, T is called the *sampling period* and $1/T$ is called the *sampling rate*. Sampled-data systems are treated in great detail in texts dealing with digital control systems and with digital signal processing.

EXAMPLE 7.6. In the control system of Fig. 2.4, let
$$A = \begin{bmatrix} 0 & 1 \\ 0 & 0 \end{bmatrix}, \qquad B = \begin{bmatrix} 0 \\ 1 \end{bmatrix}, \qquad C = [1, 0], \qquad D = 0,$$
let T denote the sampling period, and assume that $\alpha = 0$. The discrete-time state-space representation of the plant, preceded by a zero-order hold (D/A converter) and followed by a sampler [an (ideal) A/D converter], both sampling at a rate of $1/T$, is given by $\bar{x}(k+1) = \bar{A}\bar{x}(k) + \bar{B}\bar{u}(k)$, $\bar{y}(k) = \bar{C}x(k)$, where
$$\bar{A} = e^{AT} = \sum_{j=1}^{\infty}\left(\frac{T^j}{j!}\right)A^j = \begin{bmatrix} 1 & 0 \\ 0 & 1 \end{bmatrix} + \begin{bmatrix} 0 & 1 \\ 0 & 0 \end{bmatrix}T = \begin{bmatrix} 1 & T \\ 0 & 1 \end{bmatrix}$$
$$\bar{B} = \left(\int_0^T e^{A\tau}\,d\tau\right)B = \left(\int_0^T \begin{bmatrix} 1 & \tau \\ 0 & 1 \end{bmatrix} d\tau\right)\begin{bmatrix} 0 \\ 1 \end{bmatrix}$$
$$= \begin{bmatrix} T & \dfrac{T^2}{2} \\ 0 & T \end{bmatrix}\begin{bmatrix} 0 \\ 1 \end{bmatrix} = \begin{bmatrix} \dfrac{T^2}{2} \\ T \end{bmatrix}$$
$$\bar{C} = C = [1, 0].$$
The transfer function (relating $\bar{y}$ to $\bar{u}$) is given by
$$\hat{H}(z) = \bar{C}(zI - \bar{A})^{-1}\bar{B}$$
$$= [1, 0]\begin{bmatrix} z-1 & -T \\ 0 & z-1 \end{bmatrix}^{-1}\begin{bmatrix} \dfrac{T^2}{2} \\ T \end{bmatrix}$$
$$= [1, 0]\begin{bmatrix} \dfrac{1}{(z-1)} & \dfrac{T}{(z-1)^2} \\ 0 & \dfrac{1}{z-1} \end{bmatrix}\begin{bmatrix} \dfrac{T^2}{2} \\ T \end{bmatrix}$$
$$= \frac{T^2}{2}\frac{(z+1)}{(z-1)^2}.$$
The transfer function of the continuous-time system (continuous-time description of the plant) is determined to be $\hat{H}(s) = C(sI - A)^{-1}B = 1/s^2$, the double integrator.

The behavior of the system between the discrete instants, t, $t_k \le t < t_{k+1}$, can be determined by using (7.38), letting $x(t_k) = x(k)$ and $u(t_k) = u(k)$. ■

An interesting observation, useful when calculating $\bar{A}$ and $\bar{B}$, is that both can be expressed in terms of a single series. In particular, $\bar{A} = e^{AT} = I + TA + (T^2/2!)A^2 + \cdots = I + TA\Psi(T)$, where $\Psi(T) = I + (T/2!)A + (T^2/3!)A^2 + \cdots = \sum_{j=0}^{\infty}(T^j \div (j+1)!)A^j$. Then $\bar{B} = (\int_0^T e^{A\tau}\,d\tau)B = (\sum_{j=0}^{\infty}(T^{j+1}/(j+1)!)A^j)B = T\Psi(T)B$. If $\Psi(T)$ is determined first, then both $\bar{A}$ and $\bar{B}$ can easily be calculated.

EXAMPLE 7.7. In Example 7.6, $\Psi(T) = I + TA = \begin{bmatrix} 1 & T \\ 0 & 1 \end{bmatrix}$. Therefore, $\bar{A} = I + TA\Psi(T) = \begin{bmatrix} 1 & T \\ 0 & 1 \end{bmatrix}$ and $\bar{B} = T\Psi(T)B = \begin{bmatrix} T^2/2 \\ T \end{bmatrix}$, as expected. ■

E. Modes and Asymptotic Behavior of Time-Invariant Systems

As in the case of continuous-time systems, we study in this subsection the qualitative behavior of the solutions of linear, autonomous, homogeneous ordinary difference equations

$$x(k+1) = Ax(k) \tag{7.47}$$

in terms of the modes of such systems, where $A \in R^{n\times n}$ and $x(k) \in R^n$ for every $k \in Z^+$. From before, the unique solution of (7.47) satisfying $x(0) = x_0$ is given by

$$\phi(k, 0, x_0) = A^k x_0. \tag{7.48}$$

Let $\lambda_1, \ldots, \lambda_\sigma$, denote the σ distinct eigenvalues of A, where λ_i with $i = 1, \ldots, \sigma$, is assumed to be repeated n_i times so that $\sum_{i=1}^{\sigma} n_i = n$. Then

$$det\,(zI - A) = \prod_{i=1}^{\sigma} (z - \lambda_i)^{n_i}. \tag{7.49}$$

To introduce the modes for (7.47), we first derive the expressions

$$\begin{aligned} A^k &= \sum_{i=1}^{\sigma} \left[A_{i0}\lambda_i^k p(k) + \sum_{l=1}^{n_i - 1} A_{il} k(k-1)\cdots(k-l+1)\lambda_i^{k-l} p(k-l) \right] \\ &= \sum_{i=1}^{\sigma} [A_{i0}\lambda_i^k p(k) + A_{i1} k\lambda_i^{k-1} p(k-1) \\ &\quad + \cdots + A_{i(n_i-1)} k(k-1)\cdots(k-n_i+2)\lambda_i^{k-(n_i-1)} p(k-n_i+1)], \end{aligned} \tag{7.50}$$

where $$A_{il} = \frac{1}{l!}\frac{1}{(n_i - 1 - l)!} \lim_{z\to\lambda_i} \{[(z-\lambda_i)^{n_i}(zI - A)^{-1}]^{(n_i-1-l)}\}. \tag{7.51}$$

In (7.51), $[\cdot]^{(q)}$ denotes the qth derivative with respect to z, and in (7.50), $p(k)$ denotes the unit step [i.e., $p(k) = 0$ for $k < 0$ and $p(k) = 1$ for $k \geq 0$]. Note that if an eigenvalue λ_i of A is zero, then (7.50) must be modified. In this case,

$$\sum_{i=0}^{n_i-1} A_{il} l!\delta(k-l) \tag{7.52}$$

are the terms in (7.50) corresponding to the zero eigenvalue.

To prove (7.50), (7.51), we proceed as in the proof of (4.36), (4.37). We recall that $\{A^k\} = \mathcal{Z}^{-1}[z(zI - A)^{-1}]$ and we use the partial fraction expansion method to determine the z-transform. In particular, as in the proof of (4.36), (4.37), we can readily verify that

$$z(zI - A)^{-1} = z\sum_{i=1}^{\sigma}\sum_{l=0}^{n_i-1} (l!A_{il})(z-\lambda_i)^{-(l+1)}, \tag{7.53}$$

where the A_{il} are given in (7.51). We now take the inverse z-transform of both sides of (7.53). We first notice that

$$\begin{aligned}
\mathscr{Z}^{-1}[z(z-\lambda_i)^{-(l+1)}] &= \mathscr{Z}^{-1}[z^{-l}z^{l+1}(z-\lambda_i)^{-(l+1)}] \\
&= \mathscr{Z}^{-1}[z^{-l}(1-\lambda_i z^{-1})^{-(l+1)}] = f(k-l)p(k-l) \\
&= \begin{cases} f(k-l) & \text{for } k \geq l, \\ 0 & \text{otherwise.} \end{cases}
\end{aligned}$$

Referring to Tables 7.1 and 7.2 we note that $f(k)p(k) = \mathscr{Z}^{-1}[(1-\lambda_i z^{-1})^{-(l+1)}] = [1/l!(k+1)\cdots(k+l)]\lambda_i^k$ for $\lambda_i \neq 0$ and $l \geq 1$. Therefore, $\mathscr{Z}^{-1}[l!z(z-\lambda_i)^{-(l+1)}] = l!f(k-l)p(k-l) = k(k-1)\cdots(k-l+1)\lambda_i^{k-l}$, $l \geq 1$. For $l = 0$, we have $\mathscr{Z}^{-1}[(1-\lambda_i z^{-1})^{-1}] = \lambda_i^k$. This shows that (7.50) is true when $\lambda_i \neq 0$. Finally, if $\lambda_i = 0$, we note that $\mathscr{Z}^{-1}[l!z^{-l}] = l!\delta(k-l)$, which implies (7.52).

Note that one can derive several alternative but equivalent expressions for (7.50) that correspond to different ways of determining the inverse z-transform of $z(zI - A)^{-1}$ or of determining A^k via some other methods.

In complete analogy with the continuous-time case, we call the terms $A_{il}k(k-1)\cdots(k-l+1)\lambda_i^{k-l}$ the *modes of the system* (7.47). There are n_i modes corresponding to the eigenvalues λ_i, $l = 0, \ldots, n_i - 1$, and the system (7.47) has a total of n modes.

It is particularly interesting to study the matrix A^k, $k = 0, 1, 2, \ldots$ using the Jordan canonical form of A, i.e., $J = P^{-1}AP$, where the similarity transformation P is constructed by using the generalized eigenvectors of A. We recall once more that $J = diag\,[J_1, \ldots, J_\sigma] \triangleq diag\,[J_i]$, where each $n_i \times n_i$ block J_i corresponds to the eigenvalue λ_i and where, in turn, $J_i = diag\,[J_{i1}, \ldots, J_{il_i}]$ with J_{ij} being smaller square blocks, the dimensions of which depend on the length of the chains of generalized eigenvectors corresponding to J_i (refer to Subsections 2.3G and 2.4B). Let J_{ij} denote a typical Jordan canonical form block. We shall investigate the matrix J_{ij}^k, since $A^k = P^{-1}J^kP = P^{-1}\,diag\,[J_{ij}^k]P$.

Let

$$J_{ij} = \begin{bmatrix} \lambda_i & 1 & 0 & \cdots & 0 \\ 0 & \lambda_i & \ddots & & \vdots \\ \vdots & \vdots & \ddots & \ddots & \vdots \\ \vdots & \vdots & & \ddots & 1 \\ 0 & 0 & \cdots & \cdots & \lambda_i \end{bmatrix} = \lambda_i I + N_i, \tag{7.54}$$

where

$$N_i = \begin{bmatrix} 0 & 1 & 0 & \cdots & 0 \\ 0 & 0 & & & 0 \\ \vdots & \vdots & \ddots & & \vdots \\ \vdots & \vdots & & \ddots & 1 \\ 0 & 0 & \cdots & \cdots & 0 \end{bmatrix}$$

and where we assume that J_{ij} is a $t \times t$ matrix. Then

$$\begin{aligned}
(J_{ij})^k &= (\lambda_i I + N_i)^k \\
&= \lambda_i^k I + k\lambda_i^{k-1}N_i + \frac{k(k-1)}{2!}\lambda_i^{k-2}N_i^2 + \cdots + k\lambda_i N_i^{k-1} + N_i^k.
\end{aligned} \tag{7.55}$$

Now since $N_i^k = 0$ for $k \geq t$, a typical $t \times t$ Jordan block J_{ij} will generate terms that involve only the scalars $\lambda_i^k, \lambda_i^{k-1}, \ldots, \lambda_i^{k-(t-1)}$. Since the largest possible block associated with the eigenvalue λ_i is of dimension $n_i \times n_i$, the expression of A^k in (7.50) should involve at most the terms λ_i^k, $\lambda_i^{k-1}, \ldots, \lambda_i^{k-(n_i-1)}$, which it does.

The above enables us to prove the following useful fact: given $A \in R^{n\times n}$, there exists an integer $k \geq 0$ such that

$$A^k = 0 \tag{7.56}$$

if and only if all the eigenvalues λ_i of A are at the origin. Furthermore, the smallest k for which (7.56) holds is equal to the dimension of the largest block J_{ij} of the Jordan canonical form of A.

The second part of the above assertion follows readily from (7.55). We ask the reader to prove the first part of the assertion.

We conclude by observing that when all n eigenvalues λ_i of A are distinct, then

$$A^k = \sum_{i=1}^{n} A_i \lambda_i^k, \; k \geq 0, \tag{7.57}$$

where

$$A_i = \lim_{z\to\lambda_i} [(z - \lambda_i)(zI - A)^{-1}]. \tag{7.58}$$

If $\lambda_i = 0$, we use $\delta(k)$, the unit pulse, in place of λ_i^k in (7.57). This result is straightforward, in view of (7.50), (7.51).

EXAMPLE 7.8. In (7.47) we let $A = \begin{bmatrix} 0 & 1 \\ -\frac{1}{4} & 1 \end{bmatrix}$. The eigenvalues of A are $\lambda_1 = \lambda_2 = \frac{1}{2}$, and therefore, $n_1 = 2$ and $\sigma = 1$. Applying (7.50), (7.51), we obtain

$$\begin{aligned} A^k &= A_{10}\lambda_1^k p(k) + A_{11}k\lambda_1^{k-1}p(k-1) \\ &= \begin{bmatrix} 1 & 0 \\ 0 & 1 \end{bmatrix}\left(\frac{1}{2}\right)^k p(k) + \begin{bmatrix} -\frac{1}{2} & 1 \\ -\frac{1}{4} & \frac{1}{2} \end{bmatrix}(k)\left(\frac{1}{2}\right)^{k-1} p(k-1). \end{aligned}$$

■

EXAMPLE 7.9. In (7.47) we let $A = \begin{bmatrix} -1 & 2 \\ 0 & 1 \end{bmatrix}$. The eigenvalues of A are $\lambda_1 = -1$, $\lambda_2 = 1$ (so that $\sigma = 2$). Applying (7.57), (7.58), we obtain

$$A^k = A_{10}\lambda_1^k + A_{20}\lambda_2^k = \begin{bmatrix} 1 & -1 \\ 0 & 0 \end{bmatrix}(-1)^k + \begin{bmatrix} 0 & 1 \\ 0 & 1 \end{bmatrix}, \qquad k \geq 0.$$

Note that this same result was obtained by an entirely different method in Example 7.3.

■

EXAMPLE 7.10. In (7.47) we let $A = \begin{bmatrix} 0 & 1 \\ 0 & -1 \end{bmatrix}$. The eigenvalues of A are $\lambda_1 = 0$, $\lambda_2 = -1$ and $\sigma = 2$. Applying (7.57), (7.58), we obtain

$$A_1 = \lim_{z\to 0} [z(zI - A)^{-1}] = \frac{1}{z+1}\begin{bmatrix} z+1 & 1 \\ 0 & z \end{bmatrix}\Bigg|_{z=0} = \begin{bmatrix} 1 & 1 \\ 0 & 0 \end{bmatrix}$$

$$A_2 = \lim_{z\to -1}\left[(z+1)\frac{1}{z(z+1)}\begin{bmatrix} z+1 & 1 \\ 0 & z \end{bmatrix}\right] = \begin{bmatrix} 0 & -1 \\ 0 & 1 \end{bmatrix}$$

and

$$A^k = A_1\delta(k) + A_2(-1)^k = \begin{bmatrix} 1 & 1 \\ 0 & 0 \end{bmatrix}\delta(k) + \begin{bmatrix} 0 & -1 \\ 0 & 1 \end{bmatrix}(-1)^k, \; k \geq 0.$$

■

As in the case of continuous-time systems described by (L), various notions of stability of an equilibrium for discrete-time systems described by linear, autonomous, homogeneous ordinary difference equations (7.47) will be studied in detail in Chapter 6. If $\phi(k, 0, x_e)$ denotes the solution of system (7.47) with $x(0) = x_e$, then x_e is said to be an *equilibrium* of (7.47) if $\phi(k, 0, x_e) = x_e$ for all $k \geq 0$. Clearly, $x_e = 0$ is an equilibrium of (7.47). In discussing the qualitative properties, it is customary to speak, somewhat informally, of the stability properties of (7.47), rather than the stability properties of the equilibrium $x_e = 0$ of system (7.47).

The concepts of *stability, asymptotic stability*, and *instability* of system (7.47) are now defined in an identical manner done in Subsection 2.4C for system (L), except that in this case continuous time $t(t \in R^+)$ is replaced by discrete time $k(k \in Z^+)$.

By inspecting the modes of system (7.47) [given by (7.50) and (7.51)], we can readily establish the following stability criteria:

1. The system (7.47) is *asymptotically stable* if and only if all eigenvalues of A are within the unit circle of the complex plane (i.e., $|\lambda_j| < 1$, $j = 1, \ldots, \sigma$).
2. The system (7.47) is *stable* if and only if $|\lambda_j| \leq 1$, $j = 1, \ldots, \sigma$, and for all eigenvalues with $|\lambda_j| = 1$ having multiplicity $n_j > 1$, it is true that

$$\lim_{z \to \lambda_j} [[z - \lambda_j)^{n_j}(zI - A)^{-1}]^{(n_j - 1 - l)}] = 0 \qquad \text{for } l = 1, \ldots, n_j - 1. \quad (7.59)$$

3. The system (7.47) is *unstable* if and only if (2) is not true.

EXAMPLE 7.11. The system given in Example 7.8 is asymptotically stable. The system given in Example 7.9 is stable. In particular, note that the solution $\phi(k, 0, x(0)) = A^k x(0)$ for Example 7.9 is bounded. ■

When the eigenvalues λ_i of A are distinct, then as in the continuous-time case [refer to (4.42), (4.43)] we can readily show that

$$A^k = \sum_{j=1}^{n} A_j \lambda_j^k, \; A_j = v_j \tilde{v}_j, \qquad k \geq 0, \tag{7.60}$$

where the v_j and $\tilde{v}_j$ are right and left eigenvectors of A corresponding to λ_j, respectively. If $\lambda_j = 0$, we use $\delta(k)$, the unit pulse, in place of λ_j^k in (7.60).

In proving (7.60), we use the same approach as in the proof of (4.42), (4.43). We have $A^k = Q\, diag\, [\lambda_1^k, \ldots, \lambda_n^k] Q^{-1}$, where the columns of Q are the n right eigenvectors and the rows of Q^{-1} are the n left eigenvectors of A.

As in the continuous-time case [system (L)], the initial condition $x(0)$ for system (7.47) can be selected to be colinear with the eigenvector v_i to eliminate from the solution of (7.47) all modes except the ones involving λ_i^k.

EXAMPLE 7.12. As in Example 7.9, we let $A = \begin{bmatrix} -1 & 2 \\ 0 & 1 \end{bmatrix}$. Corresponding to the eigenvalues $\lambda_1 = -1, \lambda_2 = 1$, we have the right and left eigenvectors $v_1 = (1, 0)^T$, $v_2 = (1, 1)^T$, $\tilde{v}_1 = (1, -1)$, and $\tilde{v}_2 = (0, 1)$. Then

$$\begin{aligned} A^k &= v_1 \tilde{v}_1 \lambda_1^k + v_2 \tilde{v}_2 \lambda_2^k \\ &= \begin{bmatrix} 1 & -1 \\ 0 & 0 \end{bmatrix} (-1)^k + \begin{bmatrix} 0 & 1 \\ 0 & 1 \end{bmatrix} (1)^k, \qquad k \geq 0. \end{aligned}$$

Choose $x(0) = \alpha(1,0)^T = \alpha v_1$ with $\alpha \neq 0$. Then

$$\phi(k, 0, x(0)) = \begin{bmatrix}\alpha\\0\end{bmatrix}(-1)^k,$$

which contains only the mode associated with $\lambda_1 = -1$. ■

We conclude the discussion of modes and asymptotic behavior by briefly considering the state equation

$$x(k+1) = Ax(k) + Bu(k), \tag{7.61}$$

where x, u, A, and B are as defined in (7.2a). Taking the $\mathcal{Z}$-transform of both sides of (7.61) and rearranging yields

$$\tilde{x}(z) = z(zI - A)^{-1}x(0) + (zI - A)^{-1}B\tilde{u}(z). \tag{7.62}$$

By taking the inverse $\mathcal{Z}$-transform of (7.62), we see that the solution ϕ of (7.61) is the sum of modes that correspond to the singularities or poles of $z(zI - A)^{-1}x(0)$ and of $(zI - A)^{-1}B\tilde{u}(z)$. If in particular, system (7.47) is asymptotically stable [i.e., for $x(k+1) = Ax(k)$, all eigenvalues λ_j of A are such that $|\lambda_j| < 1$, $j = 1, \ldots, n$] and if $u(k)$ in (7.61) is bounded [i.e., there is an M such that $|u_i(k)| < M$ for all $k \geq 0$, $i = 1, \ldots, m$], then it is easily seen that the solutions of (7.61) are bounded as well.

2.8 AN IMPORTANT COMMENT ON NOTATION

For the most part Chapters 1 and 2 are concerned with the basic (qualitative) properties of systems of first-order ordinary differential equations, such as, e.g., the system of equations given by

$$\dot{x} = Ax, \tag{8.1}$$

where $x \in R^n$ and $A \in R^{n\times n}$. In the arguments and proofs to establish various properties for such systems, we highlighted the solutions by using the ϕ-notation. Thus, the unique solution of (8.1) for a given set of initial data (t_0, x_0) was written as $\phi(t, t_0, x_0)$ with $\phi(t_0, t_0, x_0) = x_0$. A similar notation was used in the case of the equation given by

$$\dot{x} = f(t, x) \tag{8.2}$$

and the equations given by

$$\dot{x} = A(t)x + B(t)u \tag{8.3a}$$

$$y = C(t)x + D(t)u, \tag{8.3b}$$

where in (8.2) and in (8.3a), (8.3b) all symbols are defined as in (E) (see Chapter 1) and as in (6.1a), (6.1b) of this chapter, respectively.

In the study of control systems such as system (8.3a), (8.3b), the center of attention is usually the control input u and the resulting evolution of the system state in the state-space and the system output. In the development of control systems theory, the x-notation has been adopted to express the solutions of systems. Thus, the solution of (8.3a) is denoted by $x(t)$ [or $x(t, t_0, x_0)$ when t_0 and x_0 are to be emphasized] and the evolution of the system output y in (8.3b) is denoted by $y(t)$. In all subsequent chapters, except Chapter 6, we will also follow this practice, employing the usual notation utilized in the control systems literature. In Chapter 6, which is concerned with the stability properties of systems, we will use the ϕ-notation when studying the Lyapunov stability of an equilibrium [such as system (8.1)] and the x-notation when investigating the input-output properties of control systems [such as system (8.3a), (8.3b)].

2.9 SUMMARY

In this chapter the response of linear systems to specific inputs (subject to particular initial conditions) was studied in detail. State-space descriptions, as well as impulse (resp., unit pulse) response descriptions and transfer functions were used. Continuous-time, time-varying, and time-invariant systems characterized by state-space descriptions were studied first. The time-invariant case was covered in a separate section (Section 2.4) to provide flexibility in the coverage of the material. Similarly, discrete-time systems were treated in a separate section (Section 2.7). Background material on linear algebra for the present and subsequent chapters was presented in Section 2.2.

In greater detail, the solutions of the homogeneous state equation $\dot{x} = A(t)x$ were characterized first, using fundamental matrices and the state transition matrix $\Phi(t, t_0)$ in Section 2.3. The solutions of the nonhomogeneous state equations $\dot{x} = A(t)x + B(t)u$ were derived in the same section.

For time-varying systems, the state transition matrix $\Phi(t, t_0)$ can be determined in closed form only in special cases. One such case pertains to time-invariant systems $\dot{x} = Ax$, where $\Phi(t, t_0) = e^{A(t-t_0)}$. Methods of determining the matrix exponential e^{At} were addressed in Section 2.4. In addition, the asymptotic behavior and the stability of an equilibrium of linear time-invariant systems $\dot{x} = Ax$ (in terms of modes and eigenvalues) were also addressed in Section 2.4. Linear periodic systems $\dot{x} = A(t)x, A(t) = A(t + T), t \in R$, were treated in Section 2.5.

Impulse response representations (resp., transfer function representations) of linear systems, in terms of state equation and output equation parameters were discussed in Section 2.6. In addition, equivalence of state-space representations were treated in Section 2.6.

Discrete-time systems represented by state-space descriptions and by the unit pulse response descriptions we addressed in Section 2.7. Results analogous to the continuous-time case were derived. Discrete-time systems arise frequently in the description of sampled-data systems. Such systems were briefly treated in Subsection 2.7D.

2.10 NOTES

As mentioned earlier in Chapter 1, standard references on linear algebra and matrix theory include Birkhoff and McLane [2], Halmos [7], and Gantmacher [6]. For more recent texts on this subject, refer to Strang [16] and Michel and Herget [10]. Our presentation in Section 2.2 is in the spirit of the coverage given in [10].

Our treatment of basic aspects of linear ordinary differential equations in Sections 2.3, 2.4, and 2.5 follows along lines similar to the development of this subject given in Miller and Michel [11].

State-space and input-output representations of continuous-time systems and discrete-time systems, addressed in Sections 2.6 and 2.7, respectively, are covered in a variety of textbooks, including Kailath [9], Chen [4], Brockett [3], DeCarlo [5], Rugh [14], and others. For further material on sampled-data systems, refer to Aström and Wittenmark [1] and to the early works on this subject that include Jury [8] and Ragazzini and Franklin [12].

Detailed treatments of the Laplace transform and the z-transform, discussed briefly in Sections 2.4 and 2.7, respectively, can be found in numerous texts on signals and linear systems, control systems, and signal processing.

The state representation of systems received wide acceptance in systems theory beginning in the late 1950s. This was primarily due to the work of R. E. Kalman and others in filtering theory and quadratic control theory and to the work of applied mathematicians concerned with the stability theory of dynamical systems. For comments and extensive references on some of the early contributions in these areas, refer to Kailath [9] and Sontag [15]. Of course, differential equations have been used to describe the dynamical behavior of artificial systems for many years. For example, in 1868 J. C. Maxwell presented a complete treatment of the behavior of devices that regulate the steam pressure in steam engines called flyball governors (Watt governors) to explain certain phenomena.

The use of state-space representations in the systems and control area opened the way for the systematic study of systems with multi-inputs and multi-outputs. Since the 1960s an alternative description is also being used to characterize time-invariant MIMO control systems that involves usage of polynomial matrices or differential operators. Some of the original references on this approach include Rosenbrock [13] and Wolovich [17]. This method, which corresponds to system descriptions by means of higher order ordinary differential equations (rather than systems of first-order ordinary differential equations, as is the case in the state-space description) is addressed in Chapter 7.

2.11 REFERENCES

1. K. J. Aström and B. Wittenmark, *Computer-Controlled Systems. Theory and Design*, Prentice-Hall, Englewood Cliffs, NJ, 1990.
2. G. Birkhoff and S. MacLane, *A Survey of Modern Algebra*, Macmillan, New York, 1965.
3. R. W. Brockett, *Finite Dimensional Linear Systems*, Wiley, New York, 1970.
4. C. T. Chen, *Linear System Theory and Design*, Holt, Rinehart and Winston, New York, 1984.
5. R. A. DeCarlo, *Linear Systems*, Prentice-Hall, Englewood Cliffs, NJ, 1989.
6. F. R. Gantmacher, *Theory of Matrices*, Vols. I, II, Chelsea, New York, 1959.
7. P. R. Halmos, *Finite Dimensional Vector Spaces*, Van Nostrand, Princeton, NJ, 1958.
8. E. I. Jury, *Sampled-Data Control Systems*, Wiley, New York, 1958.
9. T. Kailath, *Linear Systems*, Prentice-Hall, Englewood Cliffs, NJ, 1980.

10. A. N. Michel and C. J. Herget, *Applied Algebra and Functional Analysis*, Dover, New York, 1993.
11. R. K. Miller and A. N. Michel, *Ordinary Differential Equations*, Academic Press, New York, 1982.
12. J. R. Ragazzini and G. F. Franklin, *Sampled-Data Control Systems*, McGraw-Hill, New York, 1958.
13. H. H. Rosenbrock, *State Space and Multivariable Theory*, Wiley, New York, 1970.
14. W. J. Rugh, *Linear System Theory,* Second Edition, Prentice-Hall, Englewood Cliffs, NJ, 1996.
15. E. D. Sontag, *Mathematical Control Theory. Deterministic Finite Dimensional Systems*, TAM 6, Springer-Verlag, New York, 1990.
16. G. Strang, *Linear Algebra and Its Applications*, Harcourt, Brace, Jovanovich, San Diego, 1988.
17. W. A. Wolovich, *Linear Multivariable Systems*, Springer-Verlag, New York, 1974.
18. L. A. Zadeh and C. A. Desoer, *Linear System Theory—The State Space Approach*, McGraw-Hill, New York, 1963.

2.12 EXERCISES

2.1. (a) Let $(V, F) = (R^3, R)$. Determine the representation of $v = (1, 4, 0)^T$ with respect to the basis $v^1 = (1, -1, 0)^T, v^2 = (1, 0, -1)^T$, and $v^3 = (0, 1, 0)^T$.
(b) Let $V = F^3$ and let F be the field of rational functions. Determine the representation of $\tilde{v} = (s + 2, 1/s, -2)^T$ with respect to the basis $\{v^1, v^2, v^3\}$ given in (a).

2.2. Find the relationship between the two bases $\{v^1, v^2, v^3\}$ and $\{\bar{v}^1, \bar{v}^2, \bar{v}^3\}$ (i.e., find the matrix of $\{\bar{v}^1, \bar{v}^2, \bar{v}^3\}$ with respect to $\{v^1, v^2, v^3\}$), where $v^1 = (2, 1, 0)^T, v^2 = (1, 0, -1)^T, v^3 = (1, 0, 0)^T, \bar{v}^1 = (1, 0, 0)^T, \bar{v}^2 = (0, 1, -1)$, and $\bar{v}^3 = (0, 1, 1)$. Determine the representation of the vector $e_2 = (0, 1, 0)^T$ with respect to both of the above bases.

2.3. Let $\alpha \in R$ be fixed. Show that the set of all vectors $(x, \alpha x)^T, x \in R$, determines a vector space of dimension one over $F = R$, where vector addition and multiplication of vectors by scalars is defined in the usual manner. Determine a basis for this space.

2.4. Show that the set of all real $n \times n$ matrices with the usual operation of matrix addition and the usual operation of multiplication of matrices by scalars constitutes a vector space over the reals [denoted by $(R^{n\times n}, R)$]. Determine the dimension and a basis for this space. Is the above statement still true if $R^{n\times n}$ is replaced by $R^{m\times n}$, the set of real $m \times n$ matrices? Is the above statement still true if $R^{n\times n}$ is replaced by the set of nonsingular matrices? Justify your answers.

2.5. Let $v^1 = (s^2, s)^T$ and $v^2 = (1, 1/s)^T$. Is the set $\{v^1, v^2\}$ linearly independent over the field of rational functions? Is it linearly independent over the field of real numbers?

2.6. Determine the rank of the following matrices, carefully specifying the field:

$$\text{(a)} \begin{bmatrix} j \\ 3j \\ -1 \end{bmatrix}, \quad \text{(b)} \begin{bmatrix} 1 & 4 & -5 \\ 7 & 0 & 2 \end{bmatrix}, \quad \text{(c)} \begin{bmatrix} s+4 & -2 \\ s^2-1 & 6 \\ 0 & 2s+3 \\ s & -s+4 \end{bmatrix}, \quad \text{(d)} \left(\frac{s+1}{s^2}\right),$$

where $j = \sqrt{-1}$.

2.7. Let V and W be vector spaces over the same field F and let $\mathscr{A} : V \to W$ be a linear transformation. Show that if $\{\mathscr{A}v^1, \dots, \mathscr{A}v^n\}$ is a linearly independent set, then so is the set $\{v^1, \dots, v^n\}$. Give an example to show that the converse of this statement is not true.

2.8. Let V and W be vector spaces over the same field F and let $\mathscr{A} : V \to W$ be a linear transformation. Show that $\mathscr{A}$ is a one-to-one mapping if and only if $\mathcal{N}(\mathscr{A}) = \{0\}$.

2.9. Let $\mathscr{C} \triangleq [B, AB, \dots, A^{n-1}B]$ and

$$\mathbb{O} \triangleq \begin{bmatrix} C \\ CA \\ \vdots \\ CA^{n-1} \end{bmatrix},$$

where $A \in R^{n\times n}$, $B \in R^{n\times m}$, and $C \in R^{p\times n}$.

(a) Prove that if $\eta^1 \in \mathcal{N}(\mathbb{O})$, then $A\eta^1 \in \mathcal{N}(\mathbb{O})$. ($\eta^1$ denotes the coordinate representation of a vector $\eta^1 \in R^n$ with respect to the natural basis $\{e_1, \dots, e_n\}$.)

(b) Prove that if $\eta^1 \in \mathcal{R}(\mathscr{C})$, then $A\eta^1 \in \mathcal{R}(\mathscr{C})$.

The above shows that $\mathcal{N}(\mathbb{O})$ and $\mathcal{R}(\mathscr{C})$ are invariant vector spaces under a transformation $\mathscr{A}$ that is represented by the matrix A.

2.10. Show that $\begin{bmatrix} a & b \\ c & d \end{bmatrix}^{-1} = \frac{1}{\Delta}\begin{bmatrix} d & -b \\ -c & a \end{bmatrix}$, where $\Delta = ad - bc \neq 0$.

2.11. Determine the determinant, the (classical) adjoint, and the inverse of the matrix

$$A = \begin{bmatrix} \dfrac{s^2-3}{s} & 4s+3 \\ \dfrac{1}{s^2-2} & 3 \end{bmatrix}.$$

2.12. Determine the matrix X in $\begin{bmatrix} A & B \\ O & D \end{bmatrix}^{-1} = \begin{bmatrix} A^{-1} & X \\ O & D^{-1} \end{bmatrix}$, where it is assumed that A and D are nonsingular. Also, determine the matrix $\begin{bmatrix} A & O \\ C & D \end{bmatrix}^{-1}$.

2.13. (a) Show that $det \begin{bmatrix} A & O \\ C & D \end{bmatrix} = (det\ A)(det\ D)$, where A and D are square matrices.

Hint: For D nonsingular, use the identity $\begin{bmatrix} A & O \\ C & D \end{bmatrix} = \begin{bmatrix} A & O \\ O & D \end{bmatrix}\begin{bmatrix} I & O \\ D^{-1}C & I \end{bmatrix}$.

(b) If A is nonsingular, show that

$$det \begin{bmatrix} A & B \\ C & D \end{bmatrix} = (det\ A)\ det\ (D - CA^{-1}B).$$

Hint: Note that $\begin{bmatrix} A & B \\ C & D \end{bmatrix} = \begin{bmatrix} A & O \\ O & I \end{bmatrix}\begin{bmatrix} I & A^{-1}B \\ C & D \end{bmatrix}$ and $\begin{bmatrix} I & O \\ -C & I \end{bmatrix}\begin{bmatrix} I & A^{-1}B \\ C & D \end{bmatrix} = \begin{bmatrix} I & A^{-1}B \\ O & D - CA^{-1}B \end{bmatrix}$.

(c) In part (b), derive an expression for the case when it is known only that D is nonsingular.

2.14. Show that $e^{(A_1+A_2)t} = e^{A_1 t}e^{A_2 t}$ if $A_1A_2 = A_2A_1$.

2.15. Determine the characteristic and the minimal polynomials of the matrices

$$A_1 = \begin{bmatrix} 1 & 1 & 0 & 0 \\ 0 & 1 & 1 & 0 \\ 0 & 0 & 1 & 0 \\ 0 & 0 & 0 & 1 \end{bmatrix}, \quad A_2 = \begin{bmatrix} 1 & 1 & 0 & 0 \\ 0 & 1 & 0 & 0 \\ 0 & 0 & 1 & 0 \\ 0 & 0 & 0 & 1 \end{bmatrix}, \quad A_3 = \begin{bmatrix} 1 & 1 & 0 & 0 \\ 0 & 1 & 0 & 0 \\ 0 & 0 & 1 & 1 \\ 0 & 0 & 0 & 1 \end{bmatrix}, \quad A_4 = I_4.$$

Hint: These matrices are in Jordan canonical form.

2.16. Determine the Jordan canonical form of the matrices

$$A_1 = \begin{bmatrix} 2 & 0 & 0 \\ 1 & 2 & 0 \\ 2 & 0 & 2 \end{bmatrix}, \quad A_2 = \begin{bmatrix} 2 & 0 & 0 \\ 1 & 2 & 0 \\ 0 & 1 & 2 \end{bmatrix}, \quad A_3 = \begin{bmatrix} 2 & 0 & 0 \\ 0 & 2 & 0 \\ 0 & 1 & 2 \end{bmatrix}.$$

2.17. Show that there exists a similarity transformation matrix P such that

$$PAP^{-1} = A_c = \begin{bmatrix} 0 & 1 & 0 & \cdots & 0 \\ 0 & 0 & 1 & \cdots & 0 \\ \vdots & \vdots & \vdots & & \vdots \\ 0 & 0 & 0 & \cdots & 1 \\ -\alpha_0 & -\alpha_1 & -\alpha_2 & \cdots & -\alpha_{n-1} \end{bmatrix}$$

if and only if there exists a vector $b \in R^n$ such that the rank of $[b, Ab, \ldots, A^{n-1}b]$ is n, i.e., $\rho[b, Ab, \ldots, A^{n-1}b] = n$.

2.18. Show that if λ_i is an eigenvalue of the companion matrix A_c given in Exercise 2.17, then a corresponding eigenvector is $v^i = (1, \lambda_i, \ldots, \lambda_i^{n-1})^T$.

2.19. Let λ_i be an eigenvalue of a matrix A and let v^i be a corresponding eigenvector. Let $f(\lambda) = \sum_{k=0}^{l} \alpha_k \lambda^k$ be a polynomial with real coefficients. Show that $f(\lambda_i)$ is an eigenvalue of the matrix function $f(A) = \sum_{k=0}^{l} \alpha_k A^k$. Determine an eigenvector corresponding to $f(\lambda_i)$.

2.20. For the matrices

$$A_1 = \begin{bmatrix} 1 & 2 & 0 \\ 0 & 0 & 2 \\ 0 & 0 & 1 \end{bmatrix} \quad \text{and} \quad A_2 = \begin{bmatrix} 0 & 1 & 0 & 0 \\ 0 & 0 & 1 & 0 \\ 0 & 0 & 0 & 1 \\ 0 & 0 & 0 & 0 \end{bmatrix},$$

determine the matrices A_1^{100}, A_2^{100}, $e^{A_1 t}$, and $e^{A_2 t}$, $t \in R$.

2.21. Determine some bases for the range and null spaces of the matrices

$$A_1 = [1 \quad 0 \quad 1], \quad A_2 = \begin{bmatrix} 1 & 1 \\ 0 & 0 \\ 1 & 0 \end{bmatrix}, \quad \text{and} \quad A_3 = \begin{bmatrix} 3 & 2 & 1 \\ 3 & 2 & 1 \\ 3 & 2 & 1 \end{bmatrix}.$$

2.22. Determine all solutions of the equation $A\eta = v$, where

$$A = \begin{bmatrix} 0 & 1 & 1 & 2 & -1 \\ 1 & 2 & 3 & 4 & -1 \\ 2 & 0 & 2 & 0 & 2 \end{bmatrix} \quad \text{and} \quad v = \begin{bmatrix} 0 \\ 1 \\ 2 \end{bmatrix}.$$

2.23. Let $\phi_1(t) = e^{-t}$ for $t \in [-1, 1]$ and let

$$\phi_2(t) = \begin{cases} e^t, & t \in [-1, 0], \\ e^{-t}, & t \in [0, 1]. \end{cases}$$

Show that ϕ_1 and ϕ_2 are linearly independent over the field of the real numbers on $[-1, 1]$, but not on $[0, 1]$.

Remark: This example illustrates the fact that linear independence of time functions over a time interval $[a, b]$ does not necessarily imply linear independence over a time subinterval $[a', b'] \subset [a, b]$.

2.24. Show that if two time functions $\phi_1(t), \phi_2(t)$ are linearly independent over a field F on a time interval $[a, b]$, then they are linearly independent over F on any interval that contains $[a, b]$. Give a specific example.

2.25. Prove that for $A \in C[R, R^{n\times n}]$, (3.14) is true if and only if (3.21) is true for all $t, \tau \in R$.

2.26. Determine the state transition matrix $\Phi(t, t_0)$ for (LH) with

$$A(t) = \begin{bmatrix} 0 & 0 \\ t & 0 \end{bmatrix}$$

by (a) directly solving differential equations, (b) using the Peano-Baker series, and (c) using (3.15).

2.27. Determine the state transition matrix $\Phi(t, t_0)$ for (LH) with $A(t) = \begin{bmatrix} t & 0 \\ 1 & t \end{bmatrix}$ and determine in this case the solution for (LH) when $x(1) = (1, 1)^T$.

2.28. Verify that $\phi_1(t) = (1/t^2, -1/t)^T$ and $\phi_2(t) = (2/t^3, -1/t^2)^T$ are two solutions of (LH) with

$$A(t) = \begin{bmatrix} -\dfrac{4}{t} & -\dfrac{2}{t^2} \\ 1 & 0 \end{bmatrix}.$$

(a) Determine the state transition matrix $\Phi(t, \tau)$ for this system.
(b) Determine a solution ϕ for this system that satisfies the initial conditions $x(1) = (1, 1)^T$.

2.29. Given is the system of first-order ordinary differential equations $\dot{x} = t^2 Ax$, where $A \in R^{n\times n}$ and $t \in R$. Determine the state transition matrix $\Phi(t, t_0)$. Apply your answer to the specific case when $t^2 A = \begin{bmatrix} t^2 & 0 \\ 2t^2 & -t^2 \end{bmatrix}$.

2.30. Show that the two linear systems

$$\dot{x}^{(1)} = \begin{bmatrix} 0 & 1 \\ 2 - t^2 & 2t \end{bmatrix} x^{(1)} \triangleq A_1(t)x^{(1)}$$

and

$$\dot{x}^{(2)} = \begin{bmatrix} t & 1 \\ 1 & t \end{bmatrix} x^{(2)} \triangleq A_2(t)x^{(2)}$$

are equivalent state-space representations of the differential equation

$$\ddot{y} - 2t\dot{y} - (2 - t^2)y = 0.$$

(a) For which choice is it easier to compute the state transition matrix $\Phi(t, t_0)$? For this case, compute $\Phi(t, 0)$.
(b) Determine the relation between $x^{(1)}$ and y and between $x^{(2)}$ and y.

2.31. Using the Peano-Baker series, show that when $A(t) = A$, then $\Phi(t, t_0) = e^{A(t-t_0)}$.

2.32. For (LH) with $A(t) = \begin{bmatrix} -1 & e^{2t} \\ 0 & -1 \end{bmatrix}$, determine $\lim_{t\to\infty} \phi(t, t_0, x_0)$ if $x(0) = (0, 1)^T$. This example shows that an attempt of trying to extend the concept of eigenvalue from a constant matrix A to a time-varying matrix $A(t)$, for the purpose of characterizing the asymptotic behavior of time-varying systems (LH), will in general not work.

2.33. For the system

$$\dot{x} = A(t)x + B(t)u, \tag{11.1}$$

where all symbols are as defined in (6.1a), derive the *variation of constants formula* (3.10), using the change of variables $z(t) = \Phi(t_0, t)x(t)$.

2.34. For (11.1) with $x(t_0) = x_0$, show under what conditions it is possible to determine $u(t)$ so that $\phi(t, t_0, x_0) = x_0$ for all $t \geq t_0$. Use your result to find such $u(t)$ for the particular case $\dot{x} = x + e^{-t}u$.

2.35. Show that $(\partial/\partial\tau)\Phi(t, \tau) = -\Phi(t, \tau)A(\tau)$ for all $t, \tau \in R$.

2.36. Determine the state transition matrix $\Phi(t, t_0)$ for the system of equations $\dot{x} = e^{-At}Be^{At}x$, where $A \in R^{n\times n}$ and $B \in R^{n\times n}$. Investigate the case when in particular $AB = BA$.

2.37. The *adjoint equation* of (LH) is given by

$$\dot{z} = -A(t)^T z. \tag{11.2}$$

Let $\Phi(t, t_0)$ and $\Phi_a(t, t_0)$ denote the state transition matrices of (LH) and its adjoint equation, respectively. Show that $\Phi_a(t, t_0) = [\Phi(t_0, t)]^T$.

2.38. Consider the system described by

$$\dot{x} = A(t)x + B(t)u \tag{11.3a}$$

$$y = C(t)x, \tag{11.3b}$$

where all symbols are as in (6.1a), (6.1b) with $D(t) \equiv 0$, and consider the *adjoint equation* of (11.3a), (11.3b), given by

$$\dot{z} = -A(t)^T z + C(t)^T v \tag{11.4a}$$

$$w = B(t)^T z. \tag{11.4b}$$

(a) Let $H(t, \tau)$ and $H_a(t, \tau)$ denote the impulse response matrices of (11.3a), (11.3b) and (11.4a), (11.4b), respectively. Show that at the times when the impulse responses are nonzero, they satisfy $H(t, \tau) = H_a(\tau, t)^T$.

(b) If $A(t) \equiv A$, $B(t) \equiv B$, and $C(t) \equiv C$, show that $H(s) = -H_a(-s)^T$, where $H(s)$ and $H_a(s)$ are the transfer matrices of (11.3a), (11.3b) and (11.4a), (11.4b), respectively.

2.39. Show that if for (LH),

$$A(t) = \begin{bmatrix} A_{11}(t) & A_{12}(t) \\ 0 & A_{22}(t) \end{bmatrix},$$

where $A_{11}(t)$, $A_{12}(t)$, and $A_{22}(t)$ are submatrices of appropriate dimensions, then

$$\Phi(t, t_0) = \begin{bmatrix} \Phi_{11}(t, t_0) & \Phi_{12}(t, t_0) \\ 0 & \Phi_{22}(t, t_0) \end{bmatrix},$$

where $\Phi_{ii}(t)$ satisfies the matrix equation $(\partial/\partial t)\Phi_{ii}(t, t_0) = A_{ii}(t)\Phi_{ii}(t, t_0)$ and where the matrix $\Phi_{12}(t, t_0)$ satisfies the equation $(\partial/\partial t)\Phi_{12}(t, t_0) = A_{11}(t)\Phi_{12}(t, t_0) + A_{12}(t)\Phi_{22}(t, t_0)$ with $\Phi_{12}(t_0, t_0) = O$.

Use the above result to determine the state transition matrix $\Phi(t, 0)$ for

$$A(t) = \begin{bmatrix} -1 & e^{2t} \\ 0 & -1 \end{bmatrix}.$$

2.40. Compute e^{At} for

$$A = \begin{bmatrix} 1 & 4 & 10 \\ 0 & 2 & 0 \\ 0 & 0 & 2 \end{bmatrix}.$$

2.41. Given is the matrix

$$A = \begin{bmatrix} \frac{1}{2} & -1 & 0 \\ 0 & -1 & 0 \\ 0 & 0 & -2 \end{bmatrix}.$$

(a) Determine e^{At}, using the different methods covered in this text. Discuss the advantages and disadvantages of these methods.

(b) For system (L) let A be as given. Plot the components of the solution $\phi(t, t_0, x_0)$ when $x_0 = x(0) = (1, 1, 1)^T$ and $x_0 = x(0) = (\frac{2}{3}, 1, 0)^T$. Discuss the differences in these plots, if any.

2.42. Show that for $A = \begin{bmatrix} a & b \\ -b & a \end{bmatrix}$, we have $e^{At} = e^{at}\begin{bmatrix} \cos bt & \sin bt \\ -\sin bt & \cos bt \end{bmatrix}$.

2.43. Given is the system of equations

$$\begin{bmatrix} \dot{x}_1 \\ \dot{x}_2 \end{bmatrix} = \begin{bmatrix} -1 & 0 \\ 0 & 1 \end{bmatrix}\begin{bmatrix} x_1 \\ x_2 \end{bmatrix} + \begin{bmatrix} 1 \\ 1 \end{bmatrix} u$$

with $x(0) = (1, 0)^T$ and

$$u(t) = p(t) = \begin{cases} 1, & t \geq 0, \\ 0, & \text{elsewhere.} \end{cases}$$

Plot the components of the solution of ϕ for several initial conditions $x(0) = x_0$. In particular for different initial conditions $x(0) = (a, b)^T$, investigate the changes in the asymptotic behavior of the solutions.

2.44. The system (L) with $A = \begin{bmatrix} 0 & 1 \\ -1 & 0 \end{bmatrix}$ is called the *harmonic oscillator* (refer to Chapter 1) because it has periodic solutions $\phi(t) = (\phi_1(t), \phi_2(t))^T$. Simultaneously, for the same values of t, plot $\phi_1(t)$ along the horizontal axis and $\phi_2(t)$ along the vertical axis in the x_1-x_2 plane to obtain a *trajectory* for this system for the specific initial condition $x(0) = x_0 = (x_1(0), x_2(0))^T = (1, 1)^T$. In plotting such trajectories, time t is viewed as a parameter, and arrows are used to indicate increasing time. When the horizontal axis corresponds to position and the vertical axis corresponds to velocity, the x_1-x_2 plane is called the *phase plane* and ϕ_1, ϕ_2 (resp. x_1, x_2) are called *phase variables*.

2.45. There are various ways of obtaining the coefficients $\alpha_i(t)$ given in (2.101). One of these was described in Subsection 2.2J. In the following, we present another method.

We consider the relation $(d/dt)e^{At} = Ae^{At}$ and we use (2.101) to obtain

$$\frac{d}{dt}\sum_{j=0}^{n-1} \alpha_j(t)A^j = A\sum_{j=0}^{n-2} \alpha_j(t)A^j + \alpha_{n-1}(t)[-(a_{n-1}A^{n-1} + \cdots + a_1 A + a_0 I)], \tag{11.5}$$

where the Cayley-Hamilton Theorem was used. The coefficients $\alpha_i(t)$ that satisfy this relation generate a matrix $\Phi = \sum \alpha_j(t)A^j$ that satisfies the equation $\dot{\Phi} = A\Phi$. For Φ to equal e^{At}, we also require that $\Phi(0) = \Sigma\alpha_j(0)A^j = I$ (why?).

(a) Show that the $\alpha_j(t)$ can be generated as solutions of the system of equations

$$\begin{bmatrix} \dot{\alpha}_0(t) \\ \dot{\alpha}_1(t) \\ \vdots \\ \dot{\alpha}_{n-1}(t) \end{bmatrix} = \begin{bmatrix} 0 & 0 & \cdots & 0 & -a_0 \\ 1 & 0 & \cdots & 0 & -a_1 \\ \vdots & \vdots & & \vdots & \vdots \\ 0 & 0 & \cdots & 1 & -a_{n-1} \end{bmatrix} \begin{bmatrix} \alpha_0(t) \\ \alpha_1(t) \\ \vdots \\ \alpha_{n-1}(t) \end{bmatrix} \tag{11.6}$$

with $\alpha_0(0) = 1, \alpha_j(0) = 0, j \geq 1$. Also, show that the $\alpha_j(t)$ generated via (11.6) are linearly independent.

(b) Express the solution of the equation

$$\dot{x} = Ax + Bu, \tag{11.7}$$

where all symbols are as defined in (6.8a) and $x(0) = x_0$, in terms of $\alpha_j(t)$. Also, show that for $x(0) = x_0 = 0, \phi(t, 0, 0) = \phi(t) = \sum_{j=0}^{n-1} A^j B w_j(t)$, where $w_j(t) = \int_0^t \alpha_j(t - \tau)u(\tau)\, d\tau$.

2.46. First, determine the solution ϕ of $\begin{bmatrix} \dot{x}_1 \\ \dot{x}_2 \end{bmatrix} = \begin{bmatrix} 0 & 1 \\ 1 & 0 \end{bmatrix} \begin{bmatrix} x_1 \\ x_2 \end{bmatrix}$ with $x(0) = (1, 1)^T$. Next, determine the solution ϕ of the above system for $x(0) = \alpha(1, -1)^T, \alpha \in R, \alpha \neq 0$, and discuss the properties of the two solutions.

2.47. In Subsection 2.4C it is shown that when the n eigenvalues λ_i of a real $n \times n$ matrix A are distinct, then $e^{At} = \sum_{i=1}^n A_i e^{\lambda_i t}$, where $A_i = \lim_{s \to \lambda_i} [(s - \lambda_i)(sI - A)^{-1}] = v_i \tilde{v}$ [refer to (4.39), (4.40), and (4.43)], where $v_i, \tilde{v}_i$ are the right and left eigenvectors of A, respectively, corresponding to the eigenvalue λ_i. Show that (a) $\sum_{i=1}^n A_i = I$, where I denotes the $n \times n$ identity matrix, (b) $AA_i = \lambda_i A_i$, (c) $A_i A = \lambda_i A_i$, (d) $A_i A_j = \delta_{ij} A_i$, where $\delta_{ij} = 1$ if $i = j$ and $\delta_{ij} = 0$ when $i \neq j$.

2.48. Show that two state-space representations $\{A, B, C, D\}$ and $\{\tilde{A}, \tilde{B}, \tilde{C}, \tilde{D}\}$ are zero-state equivalent if and only if $CA^kB = \tilde{C}\tilde{A}^k\tilde{B}$, $k = 0, 1, 2, \ldots$, and $D = \tilde{D}$.

2.49. Find an equivalent time-invariant representation for the system described by the scalar differential equation $\dot{x} = \sin 2tx$.

2.50. Consider the system

$$\dot{x} = Ax + Bu \tag{11.7a}$$

$$y = Cx, \tag{11.7b}$$

where all symbols are defined as in (6.8a), (6.8b) with $D = 0$. Let

$$A = \begin{bmatrix} 0 & 1 & 0 & 0 \\ 3 & 0 & 0 & 2 \\ 0 & 0 & 0 & 1 \\ 0 & -2 & 0 & 0 \end{bmatrix}, \quad B = \begin{bmatrix} 0 & 0 \\ 1 & 0 \\ 0 & 0 \\ 0 & 1 \end{bmatrix}, \quad C = [1, 0, 1, 0]. \tag{11.8}$$

(a) Find equivalent representations for system (11.7a), (11.7b), (11.8), given by

$$\dot{\tilde{x}} = \tilde{A}\tilde{x} + \tilde{B}u \tag{11.9a}$$

$$y = \tilde{C}\tilde{x}, \tag{11.9b}$$

where $\tilde{x} = Px$, when $\tilde{A}$ is in (i) the Jordan canonical (or diagonal) form, and (ii) the companion form.

(b) Determine the transfer function matrix for this system.

2.51. Consider the system (11.7a), (11.7b) with $B = 0$.

(a) Let

$$A = \begin{bmatrix} -1 & 1 & 0 \\ 0 & -1 & 0 \\ 0 & 0 & 2 \end{bmatrix} \quad \text{and} \quad C = [1,\ 1,\ 1].$$

If possible, select $x(0)$ in such a manner so that $y(t) = te^{-t}, t \geq 0$.

(b) Determine conditions under which it is possible to assign $y(t), t \geq 0$, using only the initial data $x(0)$.

2.52. Consider the system given by

$$\begin{bmatrix} \dot{x}_1 \\ \dot{x}_2 \end{bmatrix} = \begin{bmatrix} -1 & 1 \\ -\frac{1}{2} & 0 \end{bmatrix} \begin{bmatrix} x_1 \\ x_2 \end{bmatrix} + \begin{bmatrix} 0 \\ \frac{1}{2} \end{bmatrix} u, \qquad y = [1,\ 0] \begin{bmatrix} x_1 \\ x_2 \end{bmatrix}.$$

(a) Determine $x(0)$ so that for $u(t) = e^{-4t}$, $y(t) = ke^{-4t}$, where k is a real constant. Determine k for the present case. Notice that $y(t)$ does not have any transient components.

(b) Let $u(t) = e^{\alpha t}$. Determine $x(0)$ that will result in $y(t) = ke^{\alpha t}$. Determine the conditions on α for this to be true. What is k in this case?

2.53. Consider the system (11.7a), (11.7b) with

$$A = \begin{bmatrix} 0 & 0 & 1 & 0 \\ 3 & 0 & -3 & 1 \\ -1 & 1 & 4 & -1 \\ 1 & 0 & -1 & 0 \end{bmatrix}, \quad B = \begin{bmatrix} 0 & 0 \\ 1 & 0 \\ 0 & 1 \\ 0 & 0 \end{bmatrix}, \quad C = \begin{bmatrix} 1 & 0 & 0 & 0 \\ 0 & 0 & 0 & 1 \end{bmatrix}.$$

(a) For $x(0) = (1, 1, 1, 1)^T$ and for $u(t) = (1, 1)^T, t \geq 0$, determine the solution $\phi(t, 0, x(0))$ and the output $y(t)$ for this system and plot the components $\phi_i(t, 0, x(0)), i = 1, 2, 3, 4$ and $y_i(t), i = 1, 2$.

(b) Determine the transfer function matrix $H(s)$ for this system.

2.54. Consider the system

$$x(k+1) = Ax(k) + Bu(k) \qquad (11.10a)$$
$$y(k) = Cx(k), \qquad (11.10b)$$

where all symbols are defined as in (7.2a), (7.2b) with $D = 0$. Let

$$A = \begin{bmatrix} 1 & 2 \\ 0 & 1 \end{bmatrix}, \quad B = \begin{bmatrix} 2 \\ 3 \end{bmatrix}, \quad C = [1,\ 1],$$

and let $x(0) = 0$ and $u(k) = 1, k \geq 0$.

(a) Determine $\{y(k)\}, k \geq 0$, by working in the (i) time domain, and (ii) z-transform domain, using the transfer function $H(z)$.

(b) If it is known that when $u(k) = 0$, then $y(0) = y(1) = 1$, can $x(0)$ be uniquely determined? If your answer is affirmative, determine $x(0)$.

2.55. Consider $\hat{y}(z) = H(z)\hat{u}(z)$ with transfer function $H(z) = 1/(z + 0.5)$.

(a) Determine and plot the unit pulse response $\{h(k)\}$.

(b) Determine and plot the unit step response.

(c) If

$$u(k) = \begin{cases} 1, & k = 1, 2, \\ 0, & \text{elsewhere,} \end{cases}$$

determine $\{y(k)\}$ for $k = 0, 1, 2, 3$, and 4 via (i) convolution, and (ii) the z-transform. Plot your answer.

(d) For $u(k)$ given in (c), determine $y(k)$ as $k \to \infty$.

2.56. Consider the system (11.10a) with $x(0) = x_0$ and $k \geq 0$. Determine conditions under which there exists a sequence of inputs so that the state remains at x_0, i.e., so that $x(k) = x_0$ for all $k \geq 0$. How is this input sequence determined? Apply your method to the specific case

$$A = \begin{bmatrix} 2 & 0 \\ 0 & -1 \end{bmatrix}, \qquad B = \begin{bmatrix} 1 \\ 1 \end{bmatrix}, \qquad x_0 = \begin{bmatrix} -2 \\ 1 \end{bmatrix}.$$

2.57. For system (7.7) with $x(0) = x_0$ and $k \geq 0$, it is desired to have the state go to the zero state for any initial condition x_0 in at most n steps, i.e., we desire that $x(k) = 0$ for any $x_0 = x(0)$ and for all $k \geq n$.

(a) Derive conditions in terms of the eigenvalues of A under which the above is true. Determine the minimum number of steps under which the above behavior will be true.

(b) For part (a), consider the specific cases

$$A_1 = \begin{bmatrix} 0 & 1 & 0 \\ 0 & 0 & 1 \\ 0 & 0 & 0 \end{bmatrix}, \qquad A_2 = \begin{bmatrix} 0 & 1 & 0 \\ 0 & 0 & 0 \\ 0 & 0 & 0 \end{bmatrix}, \qquad A_3 = \begin{bmatrix} 0 & 0 & 0 \\ 0 & 0 & 1 \\ 0 & 0 & 0 \end{bmatrix}.$$

Hint: Use the Jordan canonical form for A. Results of this type are important in *dead-beat control*, where it is desired that a system variable attain some desired value and settle at that value in a finite number of time steps.

2.58. Consider the system representations given by

$$x(k+1) = \begin{bmatrix} -1 & 0 \\ 0 & -2 \end{bmatrix} x(k) + \begin{bmatrix} 1 \\ -1 \end{bmatrix} u(k),$$

$$y(k) = [1,\ 1]x(k) + [1,\ 0]u(k)$$

and

$$\tilde{x}(k+1) = \begin{bmatrix} 0 & 1 \\ -2 & -3 \end{bmatrix} \tilde{x}(k) + \begin{bmatrix} 0 \\ 1 \end{bmatrix} u(k),$$

$$y(k) = [1,\ 0]\tilde{x}(k) + [0,\ 1]u(k).$$

Are these representations equivalent? Are they zero-input equivalent?

2.59. For the Jordan block given by

$$J_{ij} = \begin{bmatrix} \lambda_i & 1 & 0 & \cdots & 0 \\ 0 & \lambda_i & 1 & & 0 \\ \vdots & \vdots & & \ddots & \vdots \\ 0 & 0 & \cdots & \cdots & 1 \\ 0 & 0 & \cdots & \cdots & \lambda_i \end{bmatrix},$$

where $J_{ij} \in R^{t \times t}$, show that

$$J_{ij}^k = \begin{bmatrix} \lambda_i^k & k\lambda_i^{k-1} & \frac{k(k-1)}{2!}\lambda_i^{k-2} & \cdots & \frac{k(k-1)\cdots(k-t+2)}{(t-1)!}\lambda_i^{k-(t-1)} \\ 0 & \lambda_i^k & k\lambda_i^{k-1} & \cdots & \vdots \\ 0 & 0 & \lambda_i^k & \cdots & \vdots \\ \vdots & \vdots & \vdots & & \vdots \\ 0 & 0 & 0 & \cdots & k\lambda_i^{k-1} \\ 0 & 0 & 0 & \cdots & \lambda_i^k \end{bmatrix},$$

when $k \geq t - 1$. *Hint:* Use expression (7.55).

2.60. Consider a continuous-time system described by the transfer function $H(s) = 4/(s^2 + 2s + 2)$, i.e., $\hat{y}(s) = H(s)\hat{u}(s)$.

(a) Assume that the system is at rest and assume a unit step input, i.e., $u(t) = 1$, $t \geq 0$, $u(t) = 0, t < 0$. Determine and plot $y(t)$ for $t \geq 0$.

(b) Obtain a discrete-time approximation for the above system by following these steps: (i) determine a *realization* of the form (11.7a), (11.7b) of $H(s)$ (see Exercise 2.61); (ii) assuming a sampler and a zero-order hold with sampling period T, use (7.46) to obtain a discrete-time system representation

$$\bar{x}(k+1) = \bar{A}\bar{x}(k) + \bar{B}\bar{u}(k) \tag{11.11a}$$

$$\bar{y}(k) = \bar{C}\bar{x}(k) + \bar{D}\bar{u}(k) \tag{11.11b}$$

and determine $\bar{A}$, $\bar{B}$, and $\bar{C}$ in terms of T.

(c) For the unit step input, $u(k) = 1$ for $k \geq 0$ and $u(k) = 0$ for $k < 0$, determine and plot $\bar{y}(k)$, $k \geq 0$, for different values of T, assuming the system is at rest. Compare $\bar{y}(k)$ with $y(t)$ obtained in part (a).

(d) Determine for (11.11a) and (11.11b) the transfer function $\bar{H}(z)$ in terms of T. Note that $\bar{H}(z) = \bar{C}(zI - \bar{A})^{-1}\bar{B} + \bar{D}$. It can be shown that $\bar{H}(z) = (1 - z^{-1})\mathcal{Z}\{\mathcal{L}^{-1}[H(s)/s]_{t=kT}\}$. Verify this for the given $H(s)$.

2.61. Given a proper rational transfer function matrix $H(s)$, the state-space representation $\{A, B, C, D\}$ is called a *realization of* $H(s)$ if $H(s) = C(sI - A)^{-1}B + D$. Thus, the system (6.8a), (6.8b) is a realization of $H(s)$ if its transfer function matrix is equal to $H(s)$. Realizations of $H(s)$ are studied at length in Chapter 5. When $H(s)$ is scalar, it is straightforward to derive certain realizations, and in the following, we consider one such realization.

Given a proper rational scalar transfer function $H(s)$, let $D \triangleq \lim_{s \to \infty} H(s)$ and let

$$H_{sp}(s) \triangleq H(s) - D = \frac{b_{n-1}s^{n-1} + \cdots + b_1 s + b_0}{s^n + a_{n-1}s^{n-1} + \cdots + a_1 s + a_0},$$

a strictly proper rational function.

(a) Let

$$A = \begin{bmatrix} 0 & 1 & 0 & \cdots & 0 & 0 \\ 0 & 0 & 1 & \cdots & 0 & 0 \\ \cdots & \cdots & \cdots & \cdots & \cdots & \cdots \\ 0 & 0 & 0 & \cdots & 0 & 1 \\ -a_0 & -a_1 & -a_2 & \cdots & -a_{n-2} & -a_{n-1} \end{bmatrix}, \qquad B = \begin{bmatrix} 0 \\ 0 \\ \vdots \\ 0 \\ 1 \end{bmatrix},$$

$$C = [b_0 \quad b_1 \quad \cdots \quad b_{n-1}]$$

and show that $\{A, B, C, D\}$ is indeed a realization of $H(s)$. Also, show that $\{\tilde{A} = A^T, \tilde{B} = C^T, \tilde{C} = B^T, \tilde{D} = D\}$ is a realization of $H(s)$ as well. These two state-space representations are said to be in *controller (companion) form* and in *observer (companion) form*, respectively (refer to Subsection 3.4D).

(b) In particular find realizations in controller and observer form for (i) $H(s) = 1/s^2$, (ii) $H(s) = \omega_n^2/(s^2 + 2\xi\omega_n s + \omega_n^2)$, and (iii) $H(s) = (s+1)^2/(s-1)^2$.

2.62. Given are the systems S_1 and S_2 described by the equations

$$\left.\begin{aligned} \dot{x}_1 &= A_1x_1 + B_1u_1 \\ y_1 &= C_1x_1 + D_1u_1 \end{aligned}\right\} \quad (S_1) \qquad \left.\begin{aligned} \dot{x}_2 &= A_2x_2 + B_2u_2 \\ y_2 &= C_2x_2 + D_2u_2 \end{aligned}\right\} \quad (S_2),$$

where all symbols are defined as in (6.8a), (6.8b) with an appropriate set of dimensions for all matrices and vectors.

(a) Determine state-space representations for the following *composite systems*.

(i) Systems connected in *tandem* or in *series*:

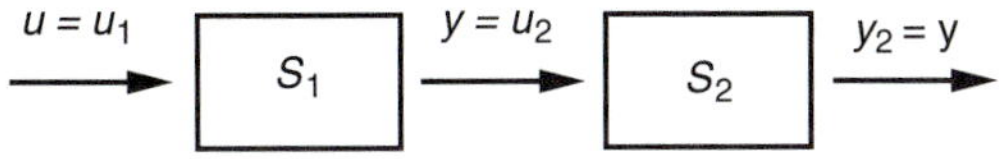

u = u1
S1
y = u2
S2
y2 = y

FIGURE 2.6
Two systems connected in series

(ii) Systems connected in *parallel*:

u1
S1
y1
+
u
y
+
S2
u2
y2

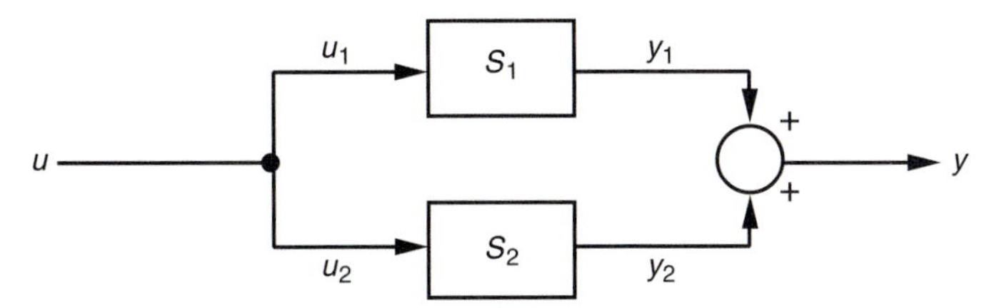

FIGURE 2.7
Two systems connected in parallel

(iii) Systems connected in a *feedback configuration*:

+
u1
S1
y1
u
y
+
S2
y2
u2

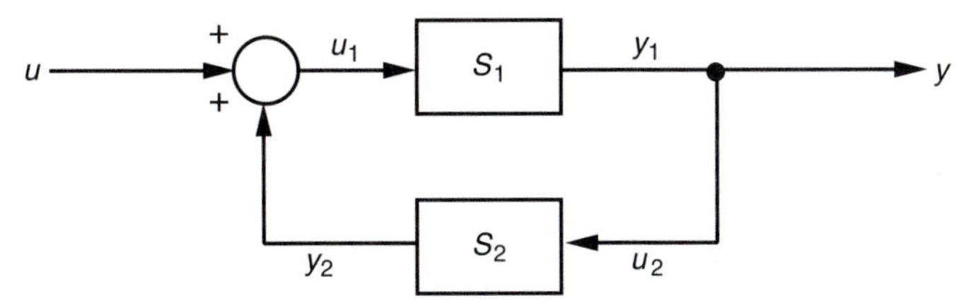

FIGURE 2.8
Feedback configuration

Hint: In each case, use $\begin{bmatrix} x_1 \\ x_2 \end{bmatrix}$ as the state of the composite system.

(b) If $H_i(s)$ is the transfer function matrix of S_i, $i = 1, 2$, determine the transfer function matrix for each of the above composite systems in terms of the $H_i(s)$, $i = 1, 2$.

2.63. Assume that $H(s)$ is a $p \times m$ proper rational transfer function matrix. Expand $H(s)$ in a Laurent series about the origin to obtain

$$H(s) = H_0 + H_1 s^{-1} + \cdots + H_k s^{-k} + \cdots = \sum_{k=0}^{\infty} H_k s^{-k}.$$

The elements of the sequence $\{H_0, H_1, \ldots, H_k, \ldots\}$ are called the *Markov parameters* of the system. These parameters provide an alternative representation of the transfer function matrix $H(s)$ (why?), and they are useful in Realization Theory (refer to Chapter 5).

(a) Show that the impulse response $H(t, 0)$ can be expressed as

$$H(t, 0) = H_0\delta(t) + \sum_{k=1}^{\infty} H_k \left[\frac{t^{k-1}}{(k-1)!}\right].$$

In the following, we assume that the system in question is described by (6.8a), (6.8b).

(b) Show that

$$H(s) = D + C(sI - A)^{-1}B = D + \sum_{k=1}^{\infty} [CA^{k-1}B]s^{-k},$$

which shows that the elements of the sequence $\{D, CB, CAB, \ldots, CA^{k-1}B, \ldots\}$ are the Markov parameters of the system, i.e., $H_0 = D$ and $H_k = CA^{k-1}B$, $k = 1, 2, \ldots$.

(c) Show that

$$H(s) = D + \frac{1}{\alpha(s)} C[R_{n-1}s^{n-1} + \cdots + R_1 s + R_0]B,$$

where $\alpha(s) = s^n + a_{n-1}s^{n-1} + \cdots + a_1 s + a_0 = det\,(sI - A)$, the characteristic polynomial of A, and $R_{n-1} = I$, $R_{n-2} = AR_{n-1} + a_{n-1}I = A + a_{n-1}I, \ldots, R_0 = A^{n-1} + a_{n-1}A^{n-2} + \cdots + a_1 I$.

Hint: Write $(sI - A)^{-1} = [1/\alpha(s)][adj\,(sI - A)] = [1/\alpha(s)][R_{n-1}s^{n-1} + \cdots + R_1 s + R_0]$, and equate the coefficients of equal powers of s in the expression

$$\alpha(s)I = (sI - A)[R_{n-1}s^{n-1} + \cdots + R_1 s + R_0].$$

2.64. Given the transfer function of a system, suggest different methods to determine its Markov parameters. Apply these methods to the specific cases given by

$$H(s) = (s^2 - 1)/(s^2 + 2s + 1) \qquad \text{and} \qquad H(s) = \begin{bmatrix} \dfrac{s}{s+1} & \dfrac{1}{s} \\ 0 & \dfrac{1}{s^2} \end{bmatrix}.$$

2.65. The *frequency response matrix* of a system described by its $p \times m$ transfer function matrix evaluated at $s = j\omega$,

$$H(\omega) \triangleq H(s)|_{s=j\omega},$$

is a very useful means of characterizing a system, since typically it can be determined experimentally, and since control system specifications are frequently expressed in terms of the frequency responses of transfer functions. When the poles of $H(s)$ have negative real parts, the system turns out to be bounded-input/bounded-output (BIBO) stable (refer to Chapter 6). Under these conditions, the frequency response $H(\omega)$ has a clear physical meaning, and this fact can be used to determine $H(\omega)$ experimentally.

(a) Consider a stable SISO system given by $\hat{y}(s) = H(s)\hat{u}(s)$. Show that if $u(t) = k\sin(\omega_0 t + \phi)$ with k constant, then $y(t)$ at steady-state (i.e., after all transients have died out) is given by

$$y_{ss}(t) = k|H(\omega_0)|\sin(\omega_0 t + \phi + \theta(\omega_0)),$$

where $|H(\omega)|$ denotes the magnitude of $H(\omega)$ and $\theta(\omega) = \arg H(\omega)$ is the argument or phase of the complex quantity $H(\omega)$.

From the above it follows that $H(\omega)$ completely characterizes the system response at steady-state (of a stable system) to a sinusoidal input. Since $u(t)$ can be expressed in terms of a series of sinusoidal terms via a Fourier series, $H(\omega)$ characterizes the steady-state response of a stable system to any bounded input $u(t)$. This physical interpretation does not apply when the system is not stable.

(b) For the $p \times m$ transfer function matrix $H(s)$, consider the frequency response matrix $H(\omega)$ and extend the discussion of part (a) above to MIMO systems to give a physical interpretation of $H(\omega)$.

2.66. Let $A \in R^{n\times n}$ and $B \in R^{n\times m}$.

(a) Is it true that $rank\,[B, AB, \dots, A^{n-1}B] = rank\,[B, AB, \dots, A^{n-1}B, A^nB]$? Justify your answer.

(b) Determine conditions under which $rank\,[B, AB, \dots, A^{n-1}B] = rank\,[AB, \dots, A^{n-1}B, A^nB]$. *Hint:* Use the *Sylvester Rank Inequality*, which relates the rank of the product of two matrices to the ranks of the individual matrices,

$$rank\,X + rank\,Y - n \le rank\,(XY) \le \min\{rank\,X, rank\,Y\},$$

where $X \in R^{n\times n}$ and $Y \in R^{n\times m}$.

2.67. (**Double integrator**) (a) Plot the response of the double integrator of Example 7.6 to a unit step input.

(b) Consider the discrete-time state-space representation of the double integrator of Example 7.6 for $T = 0.5, 1, 5$ sec and plot the unit step responses.

(c) Compare your answers in (b) with your result in (a).

2.68. (**Economic model for national income**) [D. G. Luenberger, *Introduction to Dynamic Systems*, Wiley, 1979.] A simple model describing the national income dynamics can be formulated in discrete time as follows. The national income $y(k)$ in year k in terms of consumer expenditure $c(k)$, private investment $i(k)$, and government expenditure $g(k)$ is assumed to be given by $y(k) = c(k) + i(k) + g(k)$, where the interrelations between these quantities are specified by $c(k+1) = \alpha y(k)$ and $i(k+1) = \beta[c(k-1) - c(k)]$. The constant α is called the *marginal propensity to consume,* while β is a growth coefficient. Typically, $0 < \alpha < 1$ and $\beta > 0$.

From these assumptions we obtain the difference equations $c(k+1) = \alpha c(k) + \alpha i(k) + \alpha g(k)$, $i(k+1) = (\beta\alpha - \beta)c(k) + \beta\alpha i(k) + \beta\alpha g(k)$, with discrete-time state-space representation given by

$$\begin{bmatrix} x_1(k+1) \\ x_2(k+1) \end{bmatrix} = \begin{bmatrix} \alpha & \alpha \\ \beta(\alpha - 1) & \beta\alpha \end{bmatrix}\begin{bmatrix} x_1(k) \\ x_2(k) \end{bmatrix} + \begin{bmatrix} \alpha \\ \beta\alpha \end{bmatrix}u(k)$$

$$y(k) = [1, 1]\begin{bmatrix} x_1(k) \\ x_2(k) \end{bmatrix} + u(k),$$

where $x_1(k) \triangleq c(k)$, $x_2(k) \triangleq i(k)$, and $u(k) \triangleq g(k)$. Let the parameters α, β take the values (i) $\alpha = 0.75, \beta = 1$, (ii) $\alpha = 0.75, \beta = 1.5$, and (iii) $\alpha = 1.25, \beta = 1$.

(a) Determine the eigenvalues of A for all cases and express $x(k)$ when $u = 0$ in terms of the initial conditions and the modes of the system.
(b) Plot the states for $k \geq 0$ when $u(k)$ is the unit step and $x(0) = [0, 0]^T$. Comment on your results.
(c) Plot the states for $k \geq 0$ when $u = 0$ and $x(0) = [5, 1]^T$. Comment on your results.

2.69. (**Spring mass system**) Consider the spring mass system of Example 4.1 in Chapter 1. For $M_1 = 1$ kg, $M_2 = 1$ kg, $K = 0.091$ N/m, $K_1 = 0.1$ N/m, $K_2 = 0.1$ N/m, $B = 0.0036$ N sec/m, $B_1 = 0.05$ N sec/m, and $B_2 = 0.05$ N sec/m the state-space representation of the system in (4.2) of Chapter 1 assumes the form

$$\begin{bmatrix} \dot{x}_1 \\ \dot{x}_2 \\ \dot{x}_3 \\ \dot{x}_4 \end{bmatrix} = \begin{bmatrix} 0 & 1 & 0 & 0 \\ -0.1910 & -0.0536 & 0.0910 & 0.0036 \\ 0 & 0 & 0 & 1 \\ 0.0910 & 0.0036 & -0.1910 & -0.0536 \end{bmatrix} \begin{bmatrix} x_1 \\ x_2 \\ x_3 \\ x_4 \end{bmatrix} + \begin{bmatrix} 0 & 0 \\ 1 & 0 \\ 0 & 0 \\ 0 & -1 \end{bmatrix} \begin{bmatrix} f_1 \\ f_2 \end{bmatrix},$$

where $x_1 \triangleq y_1$, $x_2 \triangleq \dot{y}_1$, $x_3 \triangleq y_2$, and $x_4 \triangleq \dot{y}_2$.
(a) Determine the eigenvalues and eigenvectors of the matrix A of the system and express $x(t)$ in terms of the modes and the initial conditions $x(0)$ of the system, assuming that $f_1 = f_2 = 0$.
(b) For $x(0) = [1, 0, -0.5, 0]^T$ and $f_1 = f_2 = 0$ plot the states for $t \geq 0$.
(c) Let $y = Cx$ with $C = \begin{bmatrix} 1 & 0 & 0 & 0 \\ 0 & 1 & 0 & 0 \end{bmatrix}$ denote the output of the system. Determine the transfer function between y and $u \triangleq [f_1, f_2]^T$.
(d) For zero initial conditions, $f_1(t) = \delta(t)$ (the unit impulse), and $f_2(t) = 0$, plot the states for $t \geq 0$ and comment on your results.
(e) It is desirable to explore what happens when the mass ratio M_2/M_1 takes on different values. For this, let $M_2 = \alpha M_1$ with $M_1 = 1$ kg and $\alpha = 0.1, 0.5, 2, 5$. All other parameter values remain the same. Repeat (a) to (d) for the different values of α and discuss your results.

2.70. (**RLC circuit**) For the circuit of Example 4.2 in Chapter 1, let $R_1 = 2\,\Omega$, $R_2 = 1\,\Omega$, $C_1 = 1$ mF, $C_2 = 1$ mF, and $L = 0.5$ H.
(a) Determine the eigenvalues of A and express $x(t)$ when $v = 0$ in terms of the initial conditions and the modes of the system.
(b) Plot the states for $t \geq 0$ when $v = 0$ and $x(0) = [5, 1, 0]^T$. Repeat for $x(0) = [0, 0, 5]^T$ and comment on your results.
(c) Compute the transfer function between $y = [v_1, v_2, v_3]^T$ and v.
(d) Plot the states when the input v is the unit step and $x(0) = [0, 0, 0]^T$. Comment on your results.

2.71. (**Armature voltage-controlled dc servomotor**) Using a consistent set of units for the armature voltage-controlled dc servomotor in Example 4.3 of Chapter 1, let $R_a = 2$, $L_a = 0.5$, $J = 1$, $B = 1$, $K_T = 2$, and $K_\theta = 1$. The state-space description of this system is given by (4.8) of Chapter 1, and here assumes the form

$$\begin{bmatrix} \dot{x}_1 \\ \dot{x}_2 \\ \dot{x}_3 \end{bmatrix} = \begin{bmatrix} 0 & 1 & 0 \\ 0 & -1 & 2 \\ 0 & -2 & -4 \end{bmatrix} \begin{bmatrix} x_1 \\ x_2 \\ x_3 \end{bmatrix} + \begin{bmatrix} 0 \\ 0 \\ 2 \end{bmatrix} e_a,$$

where $x_1 \triangleq \theta$ is the shaft position, $x_2 \triangleq \dot{\theta}$ is the angular velocity, $x_3 \triangleq i_a$ is the armature current, and the input $u = e_a$ is the armature voltage.
(a) Determine the eigenvalues and eigenvectors of A and express $x(t)$ in terms of the modes and the initial conditions of the system when $e_a = 0$.

(b) Plot the states for $t \geq 0$ when the input e_a is the unit step, and $x(0) = [0, 0, 0]^T$. Comment on your results.

2.72 (**Unit mass in an inverse square law force field**) Consider Example 11.3 of Chapter 1 where for a satellite, $r_0 = 4.218709065 \times 10^7$ m and $\omega_0 = 7.29219108 \times 10^{-5}$ rad/sec. The linearized model about the orbit $\theta(t) = \omega_0 t + \theta_0$ is given by

$$\begin{bmatrix} \dot{x}_1 \\ \dot{x}_2 \\ \dot{x}_3 \\ \dot{x}_4 \end{bmatrix} = \begin{bmatrix} 0 & 1 & 0 & 0 \\ 3\omega_0^2 & 0 & 0 & 2r_0\omega_0 \\ 0 & 0 & 0 & 1 \\ 0 & -\dfrac{2\omega_0}{r_0} & 0 & 0 \end{bmatrix} \begin{bmatrix} x_1 \\ x_2 \\ x_3 \\ x_4 \end{bmatrix} + \begin{bmatrix} 0 & 0 \\ 1 & 0 \\ 0 & 0 \\ 0 & \dfrac{1}{r_0} \end{bmatrix} \begin{bmatrix} u_1 \\ u_2 \end{bmatrix},$$

where $x_1(t) \triangleq r(t) - r_0$, $x_2(t) \triangleq \dot{r}(t)$, $x_3(t) \triangleq \theta(t)$, and $x_4(t) \triangleq \dot{\theta}(t) - \omega_0$.

(a) Determine the eigenvalues of A. Is the system asymptotically stable? Explain your answer.

(b) Plot the states for $x(0) = [100, 0, 0, 0]^T$ and zero input. Comment on your results.

(c) Plot the states for $u_1(t) = 0$, $u_2(t) = -1$ and $x(0) = 0$. If the input represents force imposed on the satellite by friction, comment on your results.

2.73. (**Magnetic ball suspension system**) Consider the magnetic ball suspension system of Exercise 1.21 in Chapter 1. It can be shown that under certain simplifying assumptions, a linearized model $\dot{x} = Ax + Bu$, $y = Cx$ of this system is given by

$$\begin{bmatrix} \dot{x}_1 \\ \dot{x}_2 \\ \dot{x}_3 \end{bmatrix} = \begin{bmatrix} -\dfrac{R}{L} & 0 & 0 \\ 0 & 0 & 1 \\ -\dfrac{2Ki_{eq}}{Ms_{eq}^2} & \dfrac{2Ki_{eq}^2}{Ms_{eq}^3} & 0 \end{bmatrix} \begin{bmatrix} x_1 \\ x_3 \\ x_3 \end{bmatrix} + \begin{bmatrix} \dfrac{1}{L} \\ 0 \\ 0 \end{bmatrix} u,$$

$$y = [0, 1, 0] \begin{bmatrix} x_1 \\ x_2 \\ x_3 \end{bmatrix}.$$

A typical set of parameters is $s_{eq} = 0.01$ m, $i_{eq} = 0.125$ A, $M = 0.01058$ kg, $K = 6.5906 \times 10^{-4}$ N m^2/A^2, $R = 31.1\ \Omega$, and $L = 0.1097$ H.

(a) Determine the eigenvalues and a set of eigenvectors of A.

(b) Compute the transfer function.

(c) Plot the states for $t \geq 0$ if the ball is slightly higher than the equilibrium position, namely, if $x(0) = [0, 0.0025, 0]^T$. Comment on your results.

2.74. (**Automobile suspension system**) [M. L. James, G. M. Smith, and J. C. Wolford, *Applied Numerical Methods for Digital Computation*, Harper & Row, 1985, p. 667.] Consider the spring mass system in Fig. 2.9, which describes part of the suspension system of an automobile. The data for this system are given as

$m_1 = \frac{1}{4} \times$ (mass of automobile) $= 375$ kg,

$m_2 =$ mass of one wheel $= 30$ kg,

$k_1 =$ spring constant $= 1500$ N/m,

$k_2 =$ linear spring constant of tire $= 6500$ N/m,

$c =$ damping constant of dashpot $= 0, 375, 750,$ and 1125 N sec/m,

$x_1 =$ displacement of automobile body from equilibrium position m,

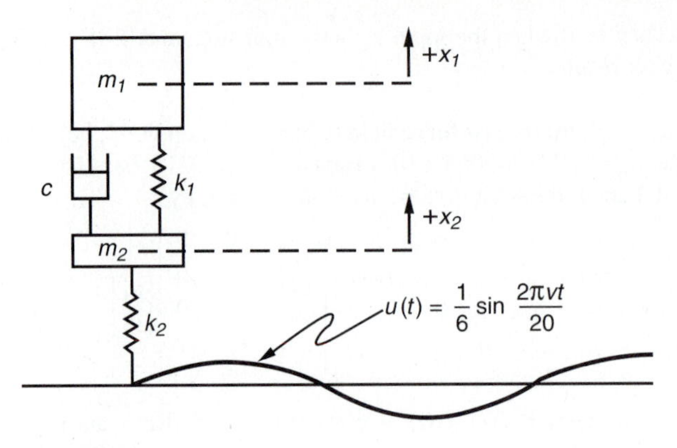

FIGURE 2.9
Model of an automobile suspension system

x_3 = displacement of wheel from equilibrium position m,

v = velocity of car = 9, 18, 27 or 36 m/sec.

A linear model $\dot{x} = Ax + Bu$ for this system is given by

$$\begin{bmatrix} \dot{x}_1 \\ \dot{x}_2 \\ \dot{x}_3 \\ \dot{x}_4 \end{bmatrix} = \begin{bmatrix} 0 & 1 & 0 & 0 \\ -\frac{k_1}{m_1} & -\frac{c}{m_1} & \frac{k_1}{m_1} & \frac{c}{m_1} \\ 0 & 0 & 0 & 1 \\ \frac{k_1}{m_1} & \frac{c}{m_2} & -\frac{k_1 + k_2}{m_2} & -\frac{c}{m_2} \end{bmatrix} \begin{bmatrix} x_1 \\ x_2 \\ x_3 \\ x_4 \end{bmatrix} + \begin{bmatrix} 0 \\ 0 \\ 0 \\ \frac{k_2}{m_2} \end{bmatrix} u(t),$$

where $u(t) = \frac{1}{6}\sin(2\pi vt/20)$ describes the profile of the roadway.

(a) Determine the eigenvalues of A for all the above cases.

(b) Plot the states for $t \geq 0$ when the input $u(t) = \frac{1}{6}\sin(2\pi vt/20)$ and $x(0) = [0, 0, 0, 0]^T$ for all the above cases. Comment on your results.

2.75. (**Building subjected to an earthquake**) [M. L. James, G. M. Smith, and J. C. Wolford, *Applied Numerical Methods for Digital Computation*, Harper & Row, 1985, p. 686.] A three-story building is modeled by a lumped mass system as shown in Fig. 2.10. For ground acceleration $\ddot{v}$, the differential equations of motion in terms of mass displacements $[q_1, q_2, q_3]$ relative to the ground are given in state-variable form $\dot{x} = Ax + Bu$ by

$$\begin{bmatrix} \dot{x}_1 \\ \dot{x}_2 \\ \dot{x}_3 \\ \dot{x}_4 \\ \dot{x}_5 \\ \dot{x}_6 \end{bmatrix} = \begin{bmatrix} 0 & 1 & 0 & 0 & 0 & 0 \\ -\frac{k_1 + k_2}{m_1} & -\frac{2c}{m_1} & \frac{k_2}{m_1} & \frac{c}{m_1} & 0 & 0 \\ 0 & 0 & 0 & 1 & 0 & 0 \\ \frac{k_2}{m_2} & \frac{c}{m_2} & -\frac{k_2 + k_3}{m_2} & -\frac{2c}{m_2} & \frac{k_2}{m_2} & \frac{c}{m_2} \\ 0 & 0 & 0 & 0 & 0 & 1 \\ 0 & 0 & \frac{k_3}{m_3} & \frac{c}{m_3} & -\frac{k_3}{m_3} & -\frac{c}{m_3} \end{bmatrix} \begin{bmatrix} x_1 \\ x_2 \\ x_3 \\ x_4 \\ x_5 \\ x_6 \end{bmatrix} + \begin{bmatrix} 0 \\ -1 \\ 0 \\ -1 \\ 0 \\ -1 \end{bmatrix} u,$$

where $x_1 \triangleq q_1$, $x_2 \triangleq \dot{q}_1$, $x_3 \triangleq q_2$, $x_4 \triangleq \dot{q}_2$, $x_5 \triangleq q_3$, $x_6 = \dot{q}_3$, and $u = \ddot{v}$. Let $k = 3.5025 \times 10^8$ N/m, $m = 1.0508 \times 10^6$ kg, and $c = 4.2030 \times 10^5$ N sec/m. Investigate the dynamic response of the structure due to the ground acceleration u for $\tau = 0.4, 0.6,$ and 0.8 sec (see Fig. 2.10). In particular:

(a) Plot the distortions

$$\begin{bmatrix} y_1 \\ y_2 \\ y_3 \end{bmatrix} = \begin{bmatrix} x_1 \\ x_3 - x_1 \\ x_5 - x_1 \end{bmatrix} = \begin{bmatrix} 1 & 0 & 0 & 0 & 0 & 0 \\ -1 & 0 & 1 & 0 & 0 & 0 \\ 0 & 0 & -1 & 0 & 1 & 0 \end{bmatrix} [x_1, x_2, x_3, x_4, x_5, x_6]^T.$$

If serious damage occurs when a distortion exceeds 0.08 m, will the given ground acceleration due to the earthquake cause serious damage to the building?

(b) Repeat (a) for different values of the damping parameter c. In particular, let $c_{new} = \alpha c_{old}$, where $\alpha = 2, 3, 10$, and repeat (a) for each value of α. Also, determine the eigenvalues of A for each α and comment on your results.

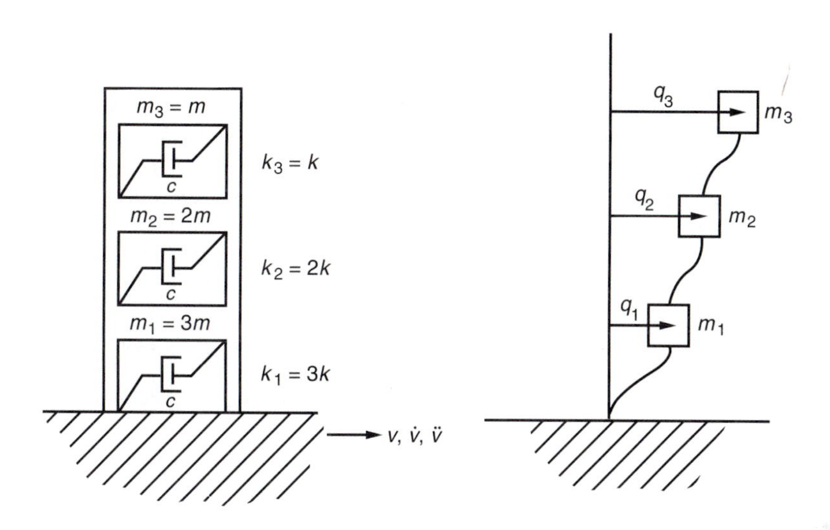

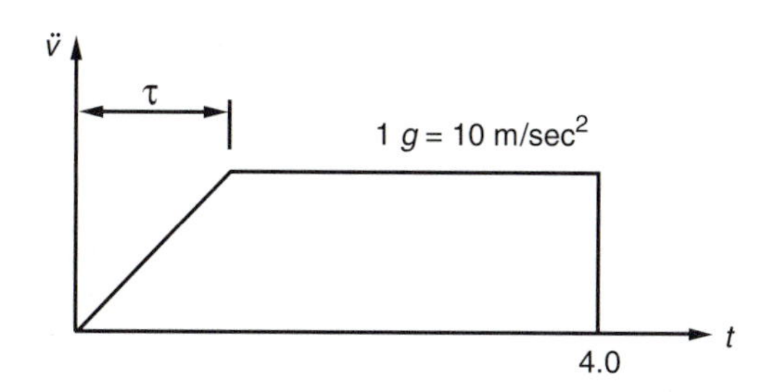

FIGURE 2.10
A model for the dynamics of a three-story structure

2.76. **(Aircraft dynamics)** [B. Friedland, *Control System Design, An Introduction to State-Space Methods,* McGraw-Hill, 1986.] For purposes of control system design, aircraft dynamics are frequently linearized about some operating condition, called a *flight regime*, where it is assumed that the aircraft velocity (Mach number) and attitude are constant. The control surfaces and engine thrust are set, or *trimmed*, to these conditions and the control system is designed to maintain these conditions, i.e., to force perturbations (deviations) from these conditions to zero.

It is customary to separate the longitudinal motion from the lateral motion, since in many cases the longitudinal and lateral dynamics are only lightly coupled. As a consequence of this the control system can be designed by considering each channel independently.

The aerodynamic variables of interest are summarized in Table 12.1 and Fig. 2.11. The aircraft body axes are denoted by x, y, and z, with the origin fixed at some reference point (typically the center of gravity of the aircraft). The positive directions of these axes are depicted in Fig. 2.11. Roll, pitch, and yaw motions constitute rotations about the x-, y-, and z-axes, respectively, using the following sign convention: looking at Fig. 2.11a we see that the pitch angle θ increases with upward rotation in the side view shown; in Fig. 2.11c, which gives the top view of the aircraft, yaw angle ψ increases in the counterclockwise direction; and looking at Fig. 2.11d, which provides the front view of the aircraft, we see that roll angle ϕ increases in the counterclockwise direction. We let $\omega_z = r, \omega_y = q$, and $\omega_x = p$ denote yaw rate, pitch rate, and roll rate, respectively. The velocity vector V is projected onto the body axes with u, v, and w being the projections onto the x-, y- and z-axes, respectively. The angle-of-attack α is the angle that the velocity vector makes with respect to the x-axis in the (positive) pitch direction, and the side-slip angle β is the angle that it makes with respect to the x-axis in the (positive) yaw direction. Note that for small angles, $\alpha \simeq w/u$ and $\beta \simeq v/u$.

The aircraft pitch motion is typically controlled by a control surface called the *elevator*, roll is controlled by a pair of *ailerons*, and yaw is controlled by a *rudder*.

Aircraft longitudinal motion. As a specific example, consider the numerical data for an actual aircraft, the AFTI-16 (a modified version of the F-16 fighter) in the landing approach configuration (speed $V = 139$ mph). The components of the state-space equation $\dot{x} = Ax + Bu$ that describe the longitudinal motion of the aircraft are given by

$$\begin{bmatrix} \dot{x}_1 \\ \dot{x}_2 \\ \dot{x}_3 \\ \dot{x}_4 \end{bmatrix} = \begin{bmatrix} -0.0507 & -3.861 & 0 & -32.2 \\ -0.00117 & -0.5164 & 1 & 0 \\ -0.000129 & 1.4168 & -0.4932 & 0 \\ 0 & 0 & 1 & 0 \end{bmatrix} \begin{bmatrix} x_1 \\ x_2 \\ x_3 \\ x_4 \end{bmatrix} + \begin{bmatrix} 0 \\ -0.0717 \\ -1.645 \\ 0 \end{bmatrix} u,$$

where the control input $u \triangleq \delta_E$ is the elevator angle and the state variables in the vector $x \triangleq [\Delta u, \alpha, q, \theta]^T$ are the change in speed, angle of attack, pitch rate, and pitch, respectively.

TABLE 12.1
Aerodynamic variables

	Lateral	Longitudinal
Rates	p: roll rate r: yaw rate β: side-slip angle	α: angle of attack q: pitch rate Δu: change in speed
Positions	ϕ: roll angle ψ: yaw angle x: forward displacement y: cross-talk displacement	θ: pitch angle z: altitude
Controls	δ_A: aileron deflection δ_R: rudder deflection	δ_E: elevator deflection

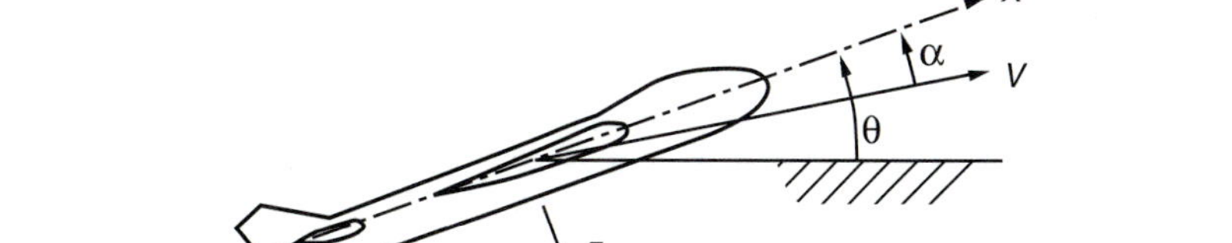

(a) Side view

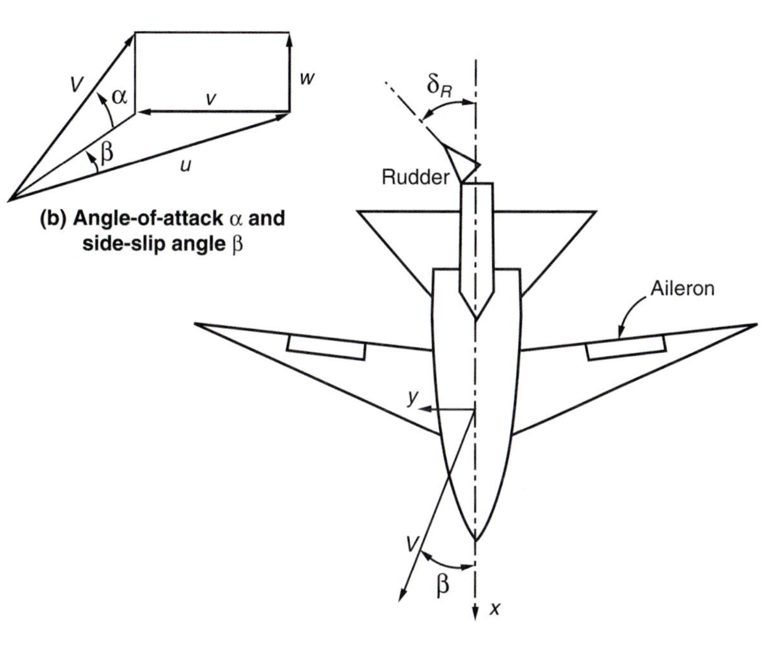

(b) Angle-of-attack α and side-slip angle β

(c) Top View

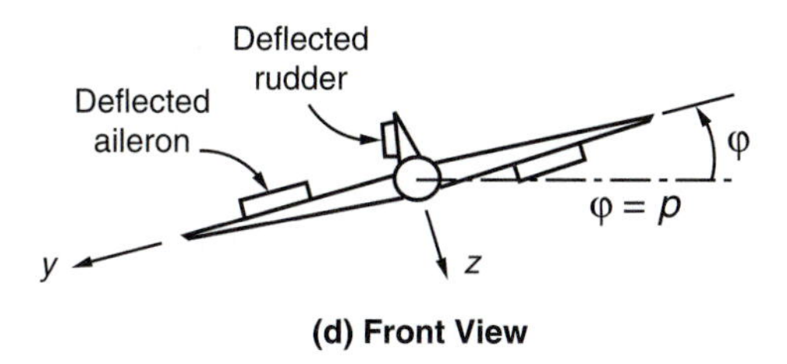

(d) Front View

FIGURE 2.11
Aircraft dynamics

The longitudinal modes of the aircraft are called *short period* and *phugoid*. The phugoid eigenvalues, which are a pair of complex conjugate eigenvalues close to the imaginary axis, cause the phugoid motion, which is a slow oscillation in altitude.

(a) For the state-space model that describes the aircraft longitudinal motion, determine the eigenvalues and eigenvectors of A. Express $x(t)$, when $u = 0$, in terms of the initial conditions and the modes of the system.

(b) Let the elevator deflection δ_E be -1 for $t \in [0, T]$ and zero afterward, where T may be taken to be the sampling period in your simulation. This corresponds to the maneuver made when the pilot pulls back on the stick to raise the nose of the airplane. (The minus sign conventionally represents pulling the stick back.) The elevator must be restored to its original position when the desired new climbing angle is reached or the plane will keep rotating. Plot the states for $x(0) = [0, 0, 0, 0]^T$ and comment on your results.

(c) Plot the states for $x(0) = [0, 0, 0, 0]^T$, using a negative unit step as the elevator input. This happens when the elevator is reset to a new position in the hope of pitching the plane up *and* climbing. Comment on your results.

(d) As a second example, consider the numerical data for a Boeing 747 jumbo jet flying near sea level at a speed of 190 mph. The state-space description $\dot{x} = Ax + Bu$ of the longitudinal motion is now given by

$$\begin{bmatrix} \dot{x}_1 \\ \dot{x}_2 \\ \dot{x}_3 \\ \dot{x}_4 \end{bmatrix} = \begin{bmatrix} -0.0188 & 11.5959 & 0 & -32.2 \\ -0.0007 & -0.5357 & 1 & 0 \\ 0.000048 & -0.4944 & -0.4935 & 0 \\ 0 & 0 & 1 & 0 \end{bmatrix} \begin{bmatrix} x_1 \\ x_2 \\ x_3 \\ x_4 \end{bmatrix} + \begin{bmatrix} 0 \\ 0 \\ -0.5632 \\ 0 \end{bmatrix} u$$

(C. E. Rohrs, J. L. Melsa, and D. G. Schultz, *Linear Control Systems*, McGraw-Hill, 1993, p. 92). Repeat (a) to (c) for the present case and discuss your answers in view of the corresponding results for the AFTI-16 fighter.

Aircraft lateral motion. As a specific example consider the lateral motion of a fighter aircraft traveling at a certain speed and altitude with state-space description $\dot{x} = Ax + Bu$ given by

$$\begin{bmatrix} \dot{x}_1 \\ \dot{x}_2 \\ \dot{x}_3 \\ \dot{x}_4 \end{bmatrix} = \begin{bmatrix} -0.746 & 0.006 & -0.999 & 0.03690 \\ -12.9 & -0.746 & 0.387 & 0 \\ 4.31 & 0.024 & -0.174 & 0 \\ 0 & 1 & 0 & 0 \end{bmatrix} \begin{bmatrix} x_1 \\ x_2 \\ x_3 \\ x_4 \end{bmatrix}$$
$$+ \begin{bmatrix} 0.0012 & 0.0092 \\ 6.05 & 0.952 \\ -0.416 & -1.76 \\ 0 & 0 \end{bmatrix} \begin{bmatrix} u_1 \\ u_2 \end{bmatrix},$$

where the control inputs $[u_1, u_2]^T \triangleq [\delta_A, \delta_R]^T$ denote the aileron and rudder deflections, respectively, and the state variables in the vector $x \triangleq [\beta, p, r, \phi]^T$ are the side-slip angle, roll rate, yaw rate, and roll angle, respectively.

(e) The eigenvalues for the aircraft lateral motion consist typically of two complex conjugate eigenvalues with relatively low damping, and two real eigenvalues. The modes caused by complex eigenvalues are called *dutch-roll*. One real eigenvalue, relatively far from the origin, defines a mode called *roll subsidence*, and a real eigenvalue near the origin defines the *spiral* mode. The spiral mode is sometimes unstable (spiral divergence). Find the modes for the aircraft lateral motion of the fighter.

(f) Plot the states when $x(0) = [0, 0, 0, 0]^T$, u_1 is the unit step and $u_2 = 0$. Repeat for $u_1 = 0$ and u_2 the unit step. Comment on your results.

2.77. (**Read/write head of a hard disk**) [MATLAB Control System Toolbox User's Guide, The MathWorks, Inc., 1993.] Using Newton's law, a simple model for the read/write head of a hard disk is described by the differential equation $J\ddot{\theta} + c\dot{\theta} + k\theta = K_T i$, where J represents the inertia of the head assembly; c denotes the viscous damping coefficient of the bearings; k is the return spring constant; K_T is the motor torque constant; $\ddot{\theta}$, $\dot{\theta}$, and θ are the angular acceleration, angular velocity, and position of the head, respectively; and i is the input current. A state-space model $\dot{x} = Ax + Bu$ of this system is given by

$$\begin{bmatrix} \dot{x}_1 \\ \dot{x}_2 \end{bmatrix} = \begin{bmatrix} 0 & 1 \\ -\frac{k}{J} & -\frac{c}{J} \end{bmatrix} \begin{bmatrix} x_1 \\ x_2 \end{bmatrix} + \begin{bmatrix} 0 \\ \frac{K_T}{J} \end{bmatrix} u,$$

where $x_1 \triangleq \theta$, $x_2 \triangleq \dot{\theta}$, and $u = i$. Let $J = 0.01$, $c = 0.004$, $k = 10$, and $K_T = 0.05$.

(a) Determine the eigenvalues of A. With $u = 0$, is the trivial solution $x = 0$ asymptotically stable? Explain.

(b) Plot the states for $t \geq 0$ when the input is the unit step and $x(0) = [0, 0]^T$.

(c) Let the plant be preceded by a zero-order hold (D/A converter) and followed by a sampler (an ideal A/D converter), both sampling at a rate of $1/T$, where $T = 5$ ms. Derive the discrete-time state-space representation of the plant. Repeat (a) and (b) for the discrete-time system and comment on your results.

CHAPTER 3

Controllability, Observability, and Special Forms

It is frequently desirable to determine an input that causes the states of a system to assume different values in finite time (e.g., to transfer the state vector from one specified vector value to another). Such is the case, for example, in satellite attitude control, where the satellite must change its orientation. This type of desirable property leads naturally to the concepts of state reachability and controllability, which will be now be studied at length.

Another desirable property of systems is the ability to determine the state from output measurements. Since it is frequently difficult or impossible to measure the state of a system directly (for example, internal temperatures and pressures in an internal combustion engine), it is extremely desirable to determine such states by observing the inputs and outputs of the system over some finite time interval. This leads to the concepts of state observability and constructibility, which will also be studied here.

The principal goals of this chapter are to introduce and study in depth the system properties of controllability and observability (and of reachability and constructibility) as well as special forms for the state-space system descriptions when a system is controllable or uncontrollable, and observable or unobservable. These special forms are very useful in the study of the relationships between state-space and input-output descriptions of a system. Note that controllability and observability play a central role when a given impulse response or a transfer function description is realized by means of a state-space description, as will be shown in Chapter 5. These special forms also provide insight into the mechanisms concerning capabilities and limitations of state controllers and state observers, as will be demonstrated in Chapter 4. The concepts of controllability and observability are central in the study of state feedback controllers (resp., output controllers) and state observers. State controllability refers to the ability to manipulate the state by applying appropriate inputs (in particular, by steering the state vector from one vector value to any other vector value in finite time). It turns out that controllability is a necessary and sufficient condition for complete eigenvalue assignment in the system matrix A by means of state feedback. State observability refers to the ability to determine the initial state vector of the system from knowledge of the input and the corresponding output over time. State observability is a necessary and sufficient condition for the arbitrary eigenvalue assignment in an asymptotic state estimator, or state observer that estimates the state of the system using input and output measurements. State feedback controllers and state observers are studied in Chapter 4.

3.1 INTRODUCTION

This chapter consists of two parts. In Part 1, consisting of Sections 3.2 and 3.3, the important concepts of state reachability (controllability) and observability (constructibility) are introduced. This is accomplished for continuous- and discrete-time systems that may be time-varying or time-invariant. In Part 2, consisting of Sections 3.4 and 3.5, special forms for state-space representations are developed for controllable or uncontrollable and observable or unobservable time-invariant (continuous- and discrete-time) systems. In addition, the Smith-McMillan form of a transfer function matrix and the poles and zeros of a system are introduced and studied.

In Subsection A of this section, the concepts of reachability and controllability and observability and constructibility are introduced, using discrete-time time-invariant systems. In this way, significant insight into the concepts is gained early, together with a clear understanding of what these properties imply for a system. Discrete-time systems are selected for this exposition because the mathematical development is simple in this case, allowing us to concentrate on explaining concepts and their implications. The continuous-time case is treated in detail in Sections 3.2 and 3.3.

A. A Brief Introduction to Reachability and Observability

Reachability and controllability are introduced first, for the case of discrete-time time-invariant systems, followed by observability and constructibility. Finally, duality is briefly discussed.

1. Reachability and controllability

The concepts of *state reachability* (or *controllability-from-the-origin*) and *controllability* (or *controllability-to-the-origin*) are introduced here and are discussed at length in Section 3.2. In the case of time-invariant systems, a state x_1 is called *reachable* if there exists an input that transfers the state of the system $x(t)$ from the zero state to x_1 in some finite time T. The definition of reachability for the discrete-time case is completely analogous.

Figure 3.1 shows that different control inputs $u_1(t)$ and $u_2(t)$ may force the state of a continuous-time system to reach the value x_1 from the origin at different finite times, following different paths. Note that reachability refers to the ability of the

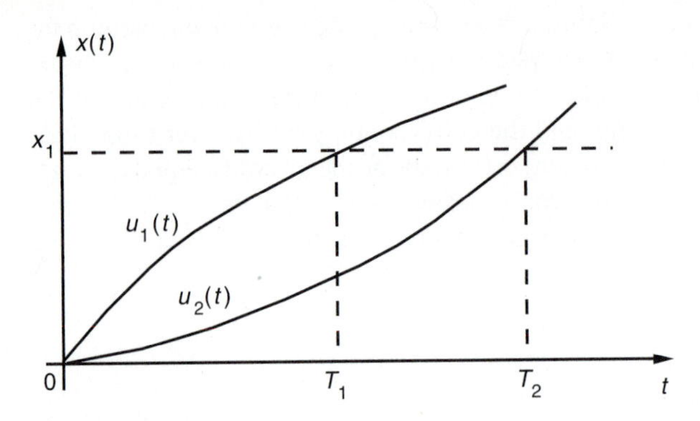

FIGURE 3.1
A reachable state x_1

system to reach x_1 from the origin in some finite time; it specifies neither the time it takes to achieve this nor the trajectory to be followed. A state x_0 is called *controllable* if there exists an input that transfers the state from x_0 to the zero state in some finite time T. See Fig. 3.2. The definition of controllability for the discrete-time case is completely analogous.

Similar to reachability, controllability refers to the ability of a system to transfer the state from x_0 to the zero state in finite time; it too specifies neither the time it takes to achieve the transfer nor the trajectory to be followed. We note that when particular types of trajectories to be followed are of interest, then one seeks particular control inputs that will achieve such transfers. This leads to various control problem formulations, including the Linear Quadratic (Optimal) Regulator (LQR). The LQR problem is discussed briefly in the next chapter.

Section 3.2 shows that reachability always implies controllability, but controllability implies reachability only when the state transition matrix Φ of the system is nonsingular. This is always true for continuous time systems, but it is true for discrete-time systems only when the matrix A of the system [or $A(k)$ for certain values of k] is nonsingular. If the system is state reachable, then there always exists an input that transfers any state x_0 to any other state x_1 in finite time.

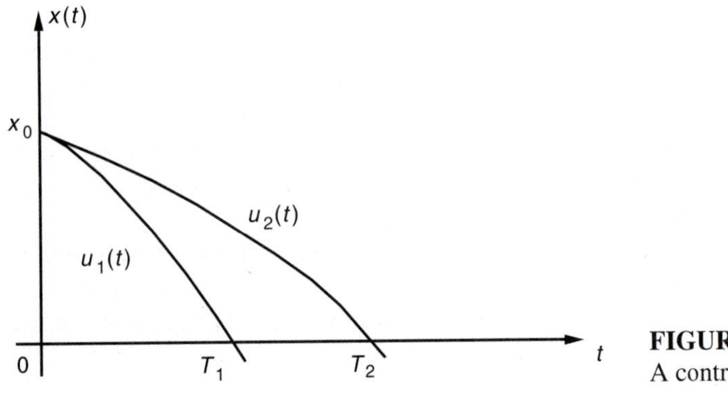

FIGURE 3.2
A controllable state x_0

In the time-invariant case, a system is said to be *reachable* (or *controllable-from-the-origin*) if and only if its *controllability matrix* $\mathscr{C}$,

$$\mathscr{C} \triangleq [B, AB, \ldots, A^{n-1}B] \in R^{n\times mn}, \tag{1.1}$$

has full row rank n, that is, *rank* $\mathscr{C} = n$. The matrices $A \in R^{n\times n}$ and $B \in R^{n\times m}$ determine either the continuous-time state equations

$$\dot{x} = Ax + Bu \tag{1.2}$$

or the discrete-time state equations

$$x(k+1) = Ax(k) + Bu(k), \tag{1.3}$$

$k \geq k_0 = 0$. Alternatively, we say that the pair (A, B) is reachable. The matrix $\mathscr{C}$ should perhaps more appropriately be called the "reachability matrix" or the "controllability-from-the-origin matrix." The term "controllability matrix," however, has been in use for some time and is expected to stay in use. Therefore, we shall call $\mathscr{C}$ the "controllability matrix," having in mind the "controllability-from-the-origin matrix."

We shall now discuss reachability and controllability for discrete-time time-invariant systems (1.3).

If the state $x(k)$ in (1.3) is expressed in terms of the initial vector $x(0)$, then (see Section 2.7)

$$x(k) = A^k x(0) + \sum_{i=0}^{k-1} A^{k-(i+1)} Bu(i) \tag{1.4}$$

for $k > 0$. It now follows that it is possible to transfer the state from some value $x(0) = x_0$ to some x_1 in n steps, that is, $x(n) = x_1$, if there exists an n-step input sequence $\{u(0), u(1), \ldots, u(n-1)\}$ which satisfies the equation

$$x_1 - A^n x_0 = \mathscr{C}_n U_n, \tag{1.5}$$

where $\mathscr{C}_n \triangleq [B, AB, \ldots, A^{n-1}B] = \mathscr{C}$ [see (1.1)] and

$$U_n \triangleq [u^T(n-1), u^T(n-2), \ldots, u^T(0)]^T. \tag{1.6}$$

From the theory of linear algebraic equations, (1.5) has a solution U_n if and only if

$$x_1 - A^n x_0 \in \mathscr{R}(\mathscr{C}), \tag{1.7}$$

where $\mathscr{R}(\mathscr{C}) =$ range $(\mathscr{C})$. Note that it is not necessary to take more than n steps in the control sequence since if this transfer cannot be accomplished in n steps, it cannot be accomplished at all. This follows from the Cayley-Hamilton Theorem, in view of which it can be shown that $\mathscr{R}(\mathscr{C}_n) = \mathscr{R}(\mathscr{C}_k)$ for $k \geq n$. Also note that $\mathscr{R}(\mathscr{C}_n)$ includes $\mathscr{R}(\mathscr{C}_k)$ for $k < n$ [i.e., $\mathscr{R}(\mathscr{C}_n) \supset \mathscr{R}(\mathscr{C}_k)$, $k < n$]. (See Exercise 3.1.)

It is now easy to see that the system (1.3) or the pair (A, B) is *reachable* (*controllable-from-the-origin*), implying that any state x_1 can be reached from the zero state ($x_0 = 0$) in finite time if and only if rank $\mathscr{C} = n$, since in this case $\mathscr{R}(\mathscr{C}) = R^n$, the entire state space. Note that $x_1 \in \mathscr{R}(\mathscr{C})$ is the condition for a particular state x_1 to be reachable from the zero state. Since $\mathscr{R}(\mathscr{C})$ contains all such states, it is called the *reachable subspace* of the system. It is also clear from (1.5) that if the system is reachable, any state x_0 can be transferred to any other state x_1 in n steps. In addition, the input that accomplishes this transfer is any solution U_n of

(1.5). Finally, depending on x_1 and x_0, this transfer may be accomplished in fewer than n steps (see Section 3.2).

EXAMPLE 1.1. Consider $x(k+1) = Ax(k) + Bu(k)$, where $A = \begin{bmatrix} 0 & 1 \\ 1 & 1 \end{bmatrix}$, $B = \begin{bmatrix} 0 \\ 1 \end{bmatrix}$.

Here the controllability (-from-the-origin) matrix $\mathscr{C}$ is $\mathscr{C} = [B, AB] = \begin{bmatrix} 0 & 1 \\ 1 & 1 \end{bmatrix}$, with rank $\mathscr{C} = 2$. Therefore the system [or the pair (A, B)] is reachable, meaning that any state x_1 can be reached from the zero state in a finite number of steps by applying at most n inputs $\{u(0), u(1), \ldots, u(n-1)\}$ (presently, $n = 2$). To see this, let $x_1 = \begin{bmatrix} a \\ b \end{bmatrix}$. Then (1.5) implies that $\begin{bmatrix} a \\ b \end{bmatrix} = \begin{bmatrix} 0 & 1 \\ 1 & 1 \end{bmatrix}\begin{bmatrix} u(1) \\ u(0) \end{bmatrix}$ or $\begin{bmatrix} u(1) \\ u(0) \end{bmatrix} = \begin{bmatrix} -1 & 1 \\ 1 & 0 \end{bmatrix}\begin{bmatrix} a \\ b \end{bmatrix} = \begin{bmatrix} b-a \\ a \end{bmatrix}$. Thus, the control $u(0) = a$, $u(1) = b - a$ will transfer the state from the origin at $k = 0$ to the state $\begin{bmatrix} a \\ b \end{bmatrix}$ at $k = 2$. To verify this, we observe that $x(1) = Ax(0) + Bu(0) = \begin{bmatrix} 0 \\ 1 \end{bmatrix} a = \begin{bmatrix} 0 \\ a \end{bmatrix}$ and $x(2) = Ax(1) + Bu(1) = \begin{bmatrix} a \\ a \end{bmatrix} + \begin{bmatrix} 0 \\ 1 \end{bmatrix}(b-a) = \begin{bmatrix} a \\ b \end{bmatrix}$.

Reachability of the system also implies that a state x_1 can be reached from any other state x_0 in at most $n = 2$ steps. To illustrate this, let $x(0) = \begin{bmatrix} 1 \\ 1 \end{bmatrix}$. Then (1.5) implies that $x_1 - A^2 x_0 = \begin{bmatrix} a \\ b \end{bmatrix} - \begin{bmatrix} 1 & 1 \\ 1 & 2 \end{bmatrix}\begin{bmatrix} 1 \\ 1 \end{bmatrix} = \begin{bmatrix} a-2 \\ b-3 \end{bmatrix} = \begin{bmatrix} 0 & 1 \\ 1 & 1 \end{bmatrix}\begin{bmatrix} u(1) \\ u(0) \end{bmatrix}$. Solving, $\begin{bmatrix} u(1) \\ u(0) \end{bmatrix} = \begin{bmatrix} b-a-1 \\ a-2 \end{bmatrix}$, which will drive the state from $\begin{bmatrix} 1 \\ 1 \end{bmatrix}$ at $k = 0$ to $\begin{bmatrix} a \\ b \end{bmatrix}$ at $k = 2$. ■

Notice that in general the solution U_n of (1.5) is not unique, i.e., there are many inputs which can accomplish the transfer from $x(0) = x_0$ to $x(n) = x_1$, each corresponding to a particular state trajectory. In control problems, particular inputs are frequently selected that, in addition to transferring the state, satisfy additional criteria, such as, e.g., minimization of an appropriate performance index (optimal control). This corresponds to selecting a particular trajectory that, e.g., may result in minimum dissipation of control energy. It is important to remember that reachability and controllability guarantee only the ability of a system to transfer an initial state to a final state by some control input action over a finite time interval. By themselves, reachability and controllability do not imply the capability of a system to follow some particular trajectory.

A system [or the pair (A, B)] is *controllable*, or *controllable-to-the-origin*, when any state x_0 can be driven to the zero state in a finite number of steps. From (1.5) we see that a system is controllable when $A^n x_0 \in \mathscr{R}(\mathscr{C})$ for any x_0. If *rank* $A = n$, a system is controllable when *rank* $\mathscr{C} = n$, i.e., when the reachability condition is satisfied. In this case the $n \times mn$ matrix

$$A^{-n}\mathscr{C} = [A^{-n}B, \ldots, A^{-1}B] \tag{1.8}$$

is of interest and the system is controllable if and only if *rank* $(A^{-n}\mathscr{C}) =$ *rank* $\mathscr{C} = n$. If, however, *rank* $A < n$, then controllability does not imply reachability (see Section 3.2).

EXAMPLE 1.2. The system in Example 1.1 is controllable (-to-the-origin). To see this, we let $x_1 = 0$ in (1.5) and write $-A^2 x_0 = -\begin{bmatrix} 1 & 1 \\ 1 & 2 \end{bmatrix}\begin{bmatrix} a \\ b \end{bmatrix} = [B, AB]\begin{bmatrix} u(1) \\ u(0) \end{bmatrix} = \begin{bmatrix} 0 & 1 \\ 1 & 1 \end{bmatrix} \times \begin{bmatrix} u(1) \\ u(0) \end{bmatrix}$, where $x_0 = \begin{bmatrix} a \\ b \end{bmatrix}$. From this we obtain $\begin{bmatrix} u(1) \\ u(0) \end{bmatrix} = -\begin{bmatrix} -1 & 1 \\ 1 & 0 \end{bmatrix}\begin{bmatrix} 1 & 1 \\ 1 & 2 \end{bmatrix}\begin{bmatrix} a \\ b \end{bmatrix} = \begin{bmatrix} 0 & -1 \\ -1 & -1 \end{bmatrix}\begin{bmatrix} a \\ b \end{bmatrix} = \begin{bmatrix} -b \\ -a-b \end{bmatrix}$, the input that will drive the state from $\begin{bmatrix} a \\ b \end{bmatrix}$ at $k = 0$ to $\begin{bmatrix} 0 \\ 0 \end{bmatrix}$ at $k = 2$. ■

EXAMPLE 1.3. The system $x(k+1) = 0$ is controllable since any state, say, $x(0) = \begin{bmatrix} a \\ b \end{bmatrix}$, can be transferred to the zero state in one step. In this system, however, the input u does not affect the state at all! This example shows that reachability is a more useful concept than controllability for discrete-time systems. ■

It should be pointed out that nothing has been said up to now about maintaining the desired system state after reaching it [refer to (1.5)]. Zeroing the input for $k \geq n$, i.e., letting $u(k) = 0$ for $k \geq n$, will not typically work, unless $Ax_1 = x_1$. In general a state starting at x_1 will remain at x_1 for all $k \geq n$ if and only if there exists an input $u(k)$, $k \geq n$, such that

$$x_1 = Ax_1 + Bu(k), \tag{1.9}$$

that is, if and only if $(I - A)x_1 \in \mathcal{R}(B)$. Clearly, there are states for which this condition may not be satisfied.

2. Observability and constructibility

In Section 3.3, definitions for state *observability* and *constructibility* are given, and appropriate tests for these concepts are derived. It is shown that observability always implies constructibility, while constructibility implies observability only when the state transition matrix Φ of the system is nonsingular. Whereas this is always true for continuous-time systems, it is true for discrete-time systems only when the matrix A of the system [or when $A(k)$ for particular values of k] is nonsingular. If a system is state observable, then its present state can be determined from knowledge of the present and future outputs and inputs. Constructibility refers to the ability to determine the present state from present and past outputs and inputs, and as such, it is of greater interest in applications.

In the time-invariant case a system [or a pair (A, C)] is observable if and only if its *observability matrix* $\mathcal{O}$, where

$$\mathcal{O} \triangleq \begin{bmatrix} C \\ CA \\ \vdots \\ CA^{n-1} \end{bmatrix} \in R^{pn \times n}, \tag{1.10}$$

has full column rank, i.e., *rank* $\mathcal{O} = n$. The matrices $A \in R^{n \times n}$ and $C \in R^{p \times n}$ are given by the system description

$$\dot{x} = Ax + Bu, \qquad y = Cx + Du \tag{1.11}$$

in the continuous-time case, and by the system description

$$x(k+1) = Ax(k) + Bu(k), \qquad y(k) = Cx(k) + Du(k), \tag{1.12}$$

with $k \geq k_0 = 0$, in the discrete-time case.

We shall now briefly discuss observability and constructibility for the discrete-time time-invariant case. As in the case of reachability and controllability, this discussion will provide insight into the underlying concepts and clarify what these imply for a system.

If the output in (1.12) is expressed in terms of the initial vector $x(0)$, then

$$y(k) = CA^k x(0) + \sum_{i=0}^{k-1} CA^{k-(i+1)}Bu(i) + Du(k) \tag{1.13}$$

for $k > 0$ (see Section 2.7). This implies that

$$\tilde{y}(k) = CA^k x_0 \tag{1.14}$$

for $k \geq 0$, where

$$\tilde{y}(k) \triangleq y(k) - \left[\sum_{i=0}^{k-1} CA^{k-(i+1)}Bu(i) + Du(k)\right]$$

for $k > 0$, $\tilde{y}(0) \triangleq y(0) - Du(0)$, and $x_0 = x(0)$. In (1.14), x_0 is to be determined assuming that the system parameters are given and the inputs and outputs are measured. Note that if $u(k) = 0$ for $k \geq 0$, then the problem is simplified since $\tilde{y}(k) = y(k)$ and the output is generated only by the initial condition x_0. It is clear that the ability to determine x_0 from output and input measurements depends only on the matrices A and C since the left-hand side of (1.14) is a known quantity. Now if $x(0) = x_0$ is known, then all $x(k)$, $k \geq 0$, can be determined by means of (1.4). To determine x_0, we apply (1.14) for $k = 0, \ldots, n-1$. Then

$$\tilde{Y}_{0,n-1} = \mathbb{O}_n x_0, \tag{1.15}$$

where $\mathbb{O}_n \triangleq [C^T, (CA)^T, \ldots, (CA^{n-1})^T]^T = \mathbb{O}$ [as in (1.10)] and

$$\tilde{Y}_{0,n-1} \triangleq [\tilde{y}^T(0), \ldots, \tilde{y}^T(n-1)]^T.$$

Now (1.15) always has a solution x_0, by construction. A system is observable if the solution x_0 is unique, i.e., if it is the only initial condition that, together with the given input sequence, can generate the observed output sequence. From the theory of linear systems of equations, (1.15) has a unique solution x_0 if and only if the null space of $\mathbb{O}$ consists of only the zero vector, i.e., *null* $(\mathbb{O}) = \mathcal{N}(\mathbb{O}) = \{0\}$, or equivalently, if and only if the only $x \in R^n$ that satisfies

$$\mathbb{O}x = 0 \tag{1.16}$$

is the zero vector. This is true if and only if *rank* $\mathbb{O} = n$. Thus, a system is observable if and only if *rank* $\mathbb{O} = n$. Any nonzero state vector $x \in R^n$ that satisfies (1.16) is said to be an unobservable state, and $\mathcal{N}(\mathbb{O})$ is said to be the *unobservable subspace*. Note that any such x satisfies $CA^k x = 0$ for $k = 0, 1, \ldots, n-1$. If *rank* $\mathbb{O} < n$, then all vectors x_0 that satisfy (1.15) are given by $x_0 = x_{0p} + x_{0h}$, where x_{0p} is a particular solution and x_{0h} is any vector in $\mathcal{N}(\mathbb{O})$. Any of these state vectors, together with the given inputs, could have generated the measured outputs.

To determine x_0 from (1.15) it is not necessary to use more than n values for $\tilde{y}(k)$, $k = 0, \ldots, n-1$, or to observe $y(k)$ for more than n steps in the future. This is true because, in view of the Cayley-Hamilton Theorem, it can be shown that $\mathcal{N}(\mathbb{O}_n) = \mathcal{N}(\mathbb{O}_k)$ for $k \geq n$. Note also that $\mathcal{N}(\mathbb{O}_n)$ is included in $\mathcal{N}(\mathbb{O}_k)$ ($\mathcal{N}(\mathbb{O}_n) \subset \mathcal{N}(\mathbb{O}_k)$) for $k < n$. Therefore, in general, one has to observe the output for n steps (see Exercise 3.1).

EXAMPLE 1.4. Consider the system $x(k+1) = Ax(k)$, $y(k) = Cx(k)$, where $A = \begin{bmatrix} 0 & 1 \\ 1 & 1 \end{bmatrix}$ and $C = [0, 1]$. Presently, $\mathbb{O} = \begin{bmatrix} C \\ CA \end{bmatrix} = \begin{bmatrix} 0 & 1 \\ 1 & 1 \end{bmatrix}$ with $rank\ \mathbb{O} = 2$. Therefore, the system [or the pair (A, C)] is observable. This means that $x(0)$ can uniquely be determined from $n = 2$ output measurements (in the present cases, the input is zero). In fact, in view of (1.15), $\begin{bmatrix} y(0) \\ y(1) \end{bmatrix} = \begin{bmatrix} 0 & 1 \\ 1 & 1 \end{bmatrix} \begin{bmatrix} x_1(0) \\ x_2(0) \end{bmatrix}$ or $\begin{bmatrix} x_1(0) \\ x_2(0) \end{bmatrix} = \begin{bmatrix} -1 & 1 \\ 1 & 0 \end{bmatrix} \begin{bmatrix} y(0) \\ y(1) \end{bmatrix} = \begin{bmatrix} y(1) - y(0) \\ y(0) \end{bmatrix}$. ■

EXAMPLE 1.5. Consider the system $x(k+1) = Ax(k)$, $y(k) = Cx(k)$, where $A = \begin{bmatrix} 1 & 0 \\ 1 & 1 \end{bmatrix}$ and $C = [1, 0]$. Presently, $\mathbb{O} = \begin{bmatrix} C \\ CA \end{bmatrix} = \begin{bmatrix} 1 & 0 \\ 1 & 0 \end{bmatrix}$ with $rank\ \mathbb{O} = 1$. Therefore, the system is not observable. Note that a basis for $\mathcal{N}(\mathbb{O})$ is $\left\{ \begin{bmatrix} 0 \\ 1 \end{bmatrix} \right\}$, which in view of (1.16) implies that all state vectors of the form $\begin{bmatrix} 0 \\ c \end{bmatrix}$, $c \in R$, are unobservable. Relation (1.15) implies that $\begin{bmatrix} y(0) \\ y(1) \end{bmatrix} = \begin{bmatrix} 1 & 0 \\ 1 & 0 \end{bmatrix} \begin{bmatrix} x_1(0) \\ x_2(0) \end{bmatrix}$. For a solution $x(0)$ to exist, as it must, we have that $y(0) = y(1) = a$. Thus, this system will generate an identical output for $k \geq 0$. Accordingly, all $x(0)$ that satisfy (1.15) and can generate this output are given by $\begin{bmatrix} x_1(0) \\ x_2(0) \end{bmatrix} = \begin{bmatrix} a \\ 0 \end{bmatrix} + \begin{bmatrix} 0 \\ c \end{bmatrix} = \begin{bmatrix} a \\ c \end{bmatrix}$, where $c \in R$. ■

In general, a system (1.12) [or a pair (A, C)] is constructible if the only vector x that satisfies $x = A^k \hat{x}$ with $C\hat{x} = 0$ for every $k \geq 0$ is the zero vector. When A is nonsingular, this condition can be stated more simply, namely, that the system is constructible if the only vector x that satisfies $CA^{-k}x = 0$ for every $k \geq 0$ is the zero vector. Compare this with the condition $CA^k x = 0$, $k \geq 0$, for x to be an unobservable state; or with the condition that a system is observable if the only vector x that satisfies $CA^k x = 0$ for every $k \geq 0$ is the zero vector. In view of (1.14), the above condition for a system to be constructible is the condition for the existence of a unique solution x_0 when past outputs and inputs are used. This, of course, makes sense since constructibility refers to determining the present state from knowledge of past outputs and inputs. Therefore, when A is nonsingular the system is constructible if and only if the $pn \times n$ matrix

$$\mathbb{O}A^{-n} = \begin{bmatrix} CA^{-n} \\ \vdots \\ CA^{-1} \end{bmatrix} \tag{1.17}$$

has full rank, since in this case the only x that satisfies $CA^{-k}x = 0$ for every $k \geq 0$ is $x = 0$. Note that if the system is observable, then it is also constructible;

however, if it is constructible, then it is also observable only when A is nonsingular (see Section 3.3).

EXAMPLE 1.6. Consider the (unobservable) system in Example 1.5. Since A is nonsingular, $\mathcal{O}A^{-2} = \begin{bmatrix} 1 & 0 \\ 1 & 0 \end{bmatrix}\begin{bmatrix} 1 & 0 \\ -2 & 1 \end{bmatrix} = \begin{bmatrix} 1 & 0 \\ 1 & 0 \end{bmatrix}$. Since *rank* $\mathcal{O}A^{-2} = 1 < 2$, the system [or the pair (A, C)] is not constructible. This can also be seen from the relation $CA^{-k}x = 0$, $k \geq 0$, that has nonzero solutions x, since $C = [1, 0] = CA^{-1} = CA^{-2} = \cdots = CA^{-k}$ for $k \geq 0$, which implies that any $x = \begin{bmatrix} 0 \\ c \end{bmatrix}$, $c \in R$ is a solution. ■

3. Dual systems

Consider the system described by

$$\dot{x} = Ax + Bu, \qquad y = Cx + Du, \tag{1.18}$$

where $A \in R^{n \times n}$, $B \in R^{n \times m}$, $C \in R^{p \times n}$, and $D \in R^{p \times m}$. The *dual system* of (1.18) is defined as the system

$$\dot{x}_D = A_D x_D + B_D u_D, \qquad y_D = C_D x_D + D_D u_D, \tag{1.19}$$

where $A_D = A^T$, $B_D = C^T$, $C_D = B^T$, and $D_D = D^T$.

LEMMA 1.1. System (1.18), denoted by $\{A, B, C, D\}$, is reachable (controllable) if and only if its dual $\{A_D, B_D, C_D, D_D\}$ in (1.19) is observable (constructible), and vice versa.

Proof. System $\{A, B, C, D\}$ is reachable if and only if $\mathcal{C} \triangleq [B, AB, \ldots, A^{n-1}B]$ has full rank n, and its dual is observable if and only if

$$\mathcal{O}_D \triangleq \begin{bmatrix} B^T \\ B^T A^T \\ \vdots \\ B^T (A^T)^{n-1} \end{bmatrix}$$

has full rank n. Since $\mathcal{O}_D^T = \mathcal{C}$, $\{A, B, C, D\}$ is reachable if and only if $\{A_D, B_D, C_D, D_D\}$ is observable. Similarly, $\{A, B, C, D\}$ is observable if and only if $\{A_D, B_D, C_D, D_D\}$ is reachable. Now $\{A, B, C, D\}$ is controllable if and only if its dual is constructible, and vice versa; recall from Sections 3.2 and 3.3, a continuous-time system is controllable if and only if it is reachable; and is constructible if and only if it is observable. ■

For the discrete-time time-invariant case, the dual system is again defined as $A_D = A^T$, $B_D = C^T$, $C_D = B^T$, and $D_D = D^T$. That such a system is reachable if and only if its dual is observable can be shown in exactly the same way as in the proof of Lemma 1.1. That such a system is controllable if and only if its dual is constructible when A is nonsingular is true because in this case the system is reachable if and only if it is controllable; and the same holds for observability and constructibility. The proof for the case when A is singular involves the controllable and unconstructible subspaces of a system and its dual. We omit the details. The reader is encouraged to complete this proof after studying Sections 3.2 and 3.3.

In the time-varying case, the dual system is defined in a similar manner as given, taking transposes of matrices, and in addition, reversing time. This will not be discussed further here. We merely wish to point out that the mappings from the original system to the dual system are in this case of the form $\{A(\alpha + t), B(\alpha + t), C(\alpha + t), D(\alpha + t)\} \rightarrow \{A^T(\alpha - t), C^T(\alpha - t), B^T(\alpha - t), D^T(\alpha - t)\}$, where α is a fixed real number. Thus, under this transformation we mirror the image of the graph of each function about a point α on the time axis and then take the transpose of each matrix. Note that the mapping between the state transition matrices is of the form $\Phi(\alpha, \alpha + t) \rightarrow \Phi^T(\alpha - t, \alpha)$. With this definition, it is now possible to establish results similar to Lemma 1.1 for the time-varying case. The proofs involve the Gramians defined in Sections 3.2 and 3.3.

Figure 3.3 summarizes the relationships between reachability (observability) and controllability (constructibility) for continuous- and discrete-time systems.

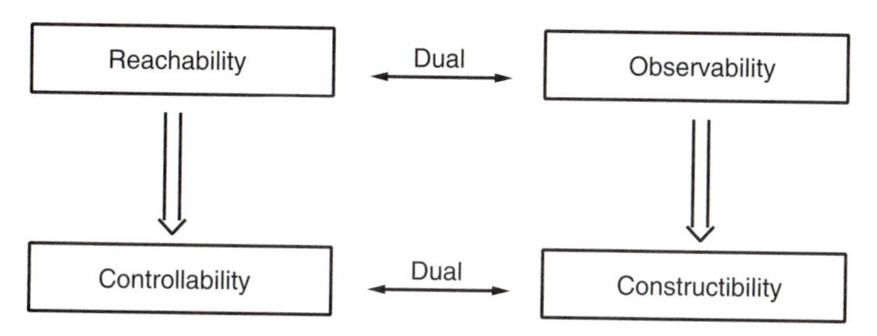

FIGURE 3.3
In continuous-time systems reachability (observability) always implies and is implied by controllability (constructibility). In discrete-time systems reachability (observability) always implies but in general is not implied by controllability (constructibility).

B. Chapter Description

This chapter consists of an introduction and two parts. In Part 1, consisting of Sections 3.2 and 3.3, reachability and controllability, and observability and constructibility, are introduced and studied. In Part 2, consisting of Sections 3.4 and 3.5, special forms for state-space representations of time-invariant systems are developed for controllable/uncontrollable, observable/unobservable (continuous- and discrete-time) systems. In addition, the poles and zeros of a system are introduced and studied.

In the introduction, Subsection 3.1A, the concepts of reachability (or controllability-from-the-origin) and observability are introduced using discrete-time time-invariant systems. The inputs that accomplish the desired transfers of a state are easily derived in terms of the controllability (-from-the-origin) matrix of a system. Conditions for reachability and observability are derived directly in terms of the controllability and observability matrices of the system. Similarly, controllability (or controllability-to-the-origin) and constructibility are also introduced. State reachability and observability are related by duality. Dual systems and the dual notions of reachability (respectively, observability) and controllability (respectively, constructibility) are also discussed.

In Section 3.2, reachability and controllability are discussed at length for both continuous- and discrete-time systems. In the continuous-time case, the inputs that accomplish the desirable state transfers are derived using the reachability (and the controllability) Gramian. Since with only minimal additional work one can treat the time-varying case as well, this is the approach pursued herein, i.e., both time-varying and time-invariant cases are studied. The time-invariant case is discussed separately and can be treated independently of the time-varying case. This adds significant flexibility to the coverage of the material in this chapter. Many criteria for reachability (controllability) are developed. It is shown that reachability implies controllability, and vice versa in the case of continuous-time systems. In discrete-time systems, although reachability implies controllability, controllability does not necessarily imply reachability. This is due to the lack of general time-reversibility in the case of difference equations, as pointed out in Chapter 2. (Note that a detailed section summary is included at the beginning of Section 3.2.)

Observability and constructibility are addressed in Section 3.3, in a manner analogous to the treatment of the dual concepts of reachability and controllability in Section 3.2. Both continuous- and discrete-time cases are considered. Observability and constructibility Gramians are used to study these properties in the case of both time-varying and time-invariant continuous-time systems. Once more, the time-invariant case is treated separately and can be studied independently of the more general time-varying case. Observability always implies constructibility in both continuous- and discrete-time systems; however, constructibility always implies observability only in the case of continuous-time systems. This is due to the lack of general time-reversibility of difference equations. (Note that a detailed section summary is included at the beginning of Section 3.3.)

In Section 3.4, similarity transformations are used to reduce the state-space representations of time-invariant systems to special forms. First, standard forms for uncontrollable and unobservable systems are developed. These lead to Kalman's Decomposition Theorem and to additional tests for controllability and observability that involve eigenvalues and eigenvectors of the system matrix A (in Subsection B) and to relations between state-space and transfer matrix descriptions (in Subsection C). Controller and observer forms for controllable and observable systems are derived next (in Subsection D). These forms are useful in state feedback control and in state observer design, discussed in Chapter 4. The Structure Theorem is introduced next. This result, which involves the controller (observer) forms and relates the state-space representations to the transfer function matrix of the system, is used in Chapter 5, where state-space realizations of transfer functions are addressed.

In Section 3.5, the poles of a system and of a transfer function matrix are introduced. There, the zeros of the system, the invariant zeros, the input and output decoupling zeros, and the transmission zeros, which are the zeros of the transfer function matrix, are also introduced. The Smith and Smith-McMillan forms of polynomial and rational matrices, respectively, are used to define poles and zeros. Utilizing zeros, one can render certain eigenvalues (system poles) and their corresponding modes unobservable from the output, using state feedback. This leads to the solution of several control problems, such as disturbance decoupling, model matching, and diagonal decoupling. The discussion of poles and zeros of a system $\{A, B, C, D\}$ and of the corresponding transfer function matrix $H(s)$ also helps to clarify the relationship between internal (state-space) descriptions and external (transfer function matrix) descriptions. This is studied in greater detail in Chapter 5.

Reachability, which is controllability-from-the-origin and controllability (-to-the-origin), together with observability and constructibility are introduced in Subsection 3.1A using discrete-time time-invariant systems. Careful study of this introductory section leads to early and significant insight into these important system properties, without requiring the mathematical sophistication needed in a careful study of these properties in the continuous-time case. Duality is also discussed in Subsection 3.1A.

In Part 1, reachability and controllability, and observability and constructibility are introduced in Sections 3.2 and 3.3, respectively, for continuous-time time-varying and time-invariant systems as well as for discrete-time systems. For convenience, detailed summaries of the results with reference to particular definitions and theorems are included at the beginning of these sections. At a first reading, one may concentrate on the time-invariant continuous-time case discussed in Subsections 3.2B and 3.3B. (Recall that an introduction to the time-invariant discrete-time case was presented in Subsection 3.1A.) The time-invariant case is developed in a self-contained manner in these sections, providing flexibility in coverage of the material. Note that in Corollary 2.12, in Subsection 3.2B, it is shown that the system is reachable if and only if the controllability matrix C has full rank. Theorem 2.13 provides an input $u(t)$ that can accomplish the transfer of the state from a vector value x_0 to another vector value x_1, provided that such transfer is possible, while Theorem 2.17 gives additional tests for reachability. A relationship between reachability and controllability is established in Theorem 2.16. In an analogous manner, in Corollary 3.8, in Subsection 3.3B, it is shown that a system is observable if and only if the observability matrix $\mathbb{O}$ has full rank. A relationship between observability and constructibility is given in Theorem 3.9, while in Theorem 3.10 additional tests for observability are presented. A useful table of all Gramians used in this chapter is provided in the summary section (Section 3.6).

In Part 2, special forms for state-space representations of continuous-time and discrete-time time-invariant systems are introduced. The standard forms for uncontrollable and unobservable representations and the Kalman Decomposition Theorem are presented in Subsection 3.4A, and useful eigenvalue/eigenvector tests for controllability and observability are developed in Subsection 3.4B. The controller and observer forms and the Structure Theorem are discussed in Subsection 3.4D. At a first reading, one could study Subsections 3.4A and 3.4B and cover Subsection 3.4D selectively, concentrating on deriving and using controller and observer forms rather than proofs and properties. Note that the controller and observer forms are used primarily in realization algorithms in Chapter 5, in a method to assign closed-loop eigenvalues via state feedback in Chapter 4, and in Chapter 7, to gain insight into the relations between state-space and polynomial matrix representations of linear time-invariant systems. Furthermore, the Structure Theorem discussed in this section introduces polynomial matrix fractional descriptions of the transfer function matrix $H(s)$. These descriptions are very useful in control problems and are discussed further in Chapter 7. In Section 3.5, the poles and zeros of a system are introduced using the Smith form of a polynomial matrix and the Smith-McMillan form of a transfer function matrix $H(s)$. The pole and zero polynomials of $H(s)$ are defined next. This gives rise to the McMillan degree of $H(s)$ and to the order of a minimal realization, discussed in Subsection 5.2C of Chapter 5. The study of poles and zeros offers significant insight into feedback control systems. At a first reading, Section 3.5 may be omitted without loss of continuity.

PART 1
CONTROLLABILITY AND OBSERVABILITY

3.2
REACHABILITY AND CONTROLLABILITY

The objective here is to study the important properties of state controllability and reachability when a system is described by a state-space representation. In Subsection 3.1A, a brief introduction to these concepts for discrete-time time-invariant systems was given, where it was shown that a system is completely reachable if and only if the controllability (-from-the-origin) matrix $\mathscr{C}$ in (1.1) has full rank n ($rank\,\mathscr{C} = n$). Furthermore, it was shown that the input sequence necessary to accomplish the transfer can be determined directly from $\mathscr{C}$ by solving a system of linear algebraic equations. In a similar manner, we would like to derive tests for reachability and controllability and determine the necessary system inputs to accomplish the state transfer for the continuous-time case. This is the main objective of this section. We note, however, that whereas the test for reachability in the time-invariant case ($rank\,\mathscr{C} = n$) can be derived by a number of methods, the appropriate sequence of system inputs to use cannot easily be determined directly from $\mathscr{C}$, as was the case for discrete-time systems. For this reason, we use an approach that utilizes ranges of maps, in particular, the range of an important $n \times n$ matrix—the reachability Gramian. The inputs that accomplish the desired state transfer can be determined directly from this matrix. However, once this is accomplished, we can develop all the results for the time-varying case as well, with hardly any additional work. This is the approach we will employ. The reader can skip the more general material, however, starting with Definition 2.1, and concentrate on the time-invariant case starting with Definition 2.9, if so desired. The contents of this section are now presented in greater detail.

Section description

In this section, the concepts of reachability and controllability are introduced and discussed in detail for linear system state-space descriptions. This is accomplished for continuous- and discrete-time systems for both time-varying and time-invariant cases.

Reachability for continuous-time systems is discussed first and the reachability Gramian $W_r(t_0, t_1)$ is defined (in Definition 2.7). It is then shown (in Corollary 2.3) that the system is reachable at t_1 if and only if $W_r(t_0, t_1)$ has full rank for some $t_0 \leq t_1$. Reachability implies that it is possible to transfer the state from a value x_0 to some value x_1, and system inputs that accomplish this transfer are given in Theorem 2.4 and Corollary 2.5. Controllability is discussed next and the controllability Gramian is defined (in Definition 2.8). It is shown (in Theorem 2.6) that a continuous-time system is reachable if and only if it is controllable. Two additional results (Theorems 2.8 and 2.9) provide further criteria for reachability and controllability. All these results are then applied to the continuous-time time-invariant case. The above material is presented in a manner that makes possible the study of time-invariant systems, independent of the time-varying case. In Lemma 2.10, a relationship between the reachability Gramian $W_r(0, T)$ and the controllability (-from-the-origin) matrix $\mathscr{C} = [B, AB, \ldots, A^{n-1}B]$ is established. It is then shown in Corollary 2.12 that a system is reachable if and only if $\mathscr{C}$ has full rank n. A system input sequence that transfers the state from x_0 to x_1 is derived in Theorem 2.13 and in Corollary 2.14. In Theorem 2.16 a relationship between reachability and controllability is established, and Theorem 2.17 provides additional tests for reachability.

For discrete-time systems, in particular, discrete-time time-invariant systems, reachability and controllability are discussed next. Here, the controllability matrix $\mathscr{C}$ plays a predominant role. It is shown that when $\mathscr{C}$ has full rank, the system is reachable (Corollary 2.19), and input sequences that transfer the state to desired values are derived (Theorem 2.20 and Corollary 2.21). In Theorem 2.22 it is shown that reachability in the case of discrete-time systems always implies controllability. In contrast to the continuous-time case, the converse to this statement is generally not true. This is due to a lack of time reversibility in difference equations. When A is nonsingular, then controllability also implies reachability. Finally, in Definitions 2.13 and 2.14 the reachability and controllability Gramians for the discrete-time case are defined for sake of completeness.

A. Continuous-Time Time-Varying Systems

We consider the state equation

$$\dot{x} = A(t)x + B(t)u, \tag{2.1}$$

where $A(t) \in R^{n\times n}$, $B(t) \in R^{n\times m}$, and $u(t) \in R^m$ are defined and (piecewise) continuous on some real open interval (a, b). The state at time t is given by

$$x(t, t_0, x_0) \triangleq x(t) = \Phi(t, t_0)x(t_0) + \int_{t_0}^{t} \Phi(t, \tau)B(\tau)u(\tau)\,d\tau, \tag{2.2}$$

where $\Phi(t, \tau)$ is the state transition matrix of the system, $t_0, t \in (a, b)$, and $x(t_0) = x_0$ denotes the initial state at initial time.

In the time-invariant case,

$$\dot{x} = Ax + Bu, \tag{2.3}$$

where $A \in R^{n\times n}$, $B \in R^{n\times m}$, (2.2) is still valid with

$$\Phi(t, \tau) = \Phi(t-\tau, 0) = \exp[(t-\tau)A] = e^{A(t-\tau)}. \tag{2.4}$$

We are interested in using the input to transfer the state from x_0 to some other value x_1 at some finite time $t_1 > t_0$, [i.e., $x(t_1) = x_1$]. Equation (2.2) assumes the form

$$x_1 = \Phi(t_1, t_0)x_0 + \int_{t_0}^{t_1} \Phi(t_1, \tau)B(\tau)u(\tau)\,d\tau, \tag{2.5}$$

and clearly, there exists $u(t)$, $t \in [t_0, t_1]$ that satisfies (2.5) if and only if such transfer of the state is possible. Rewriting (2.5) as

$$x_1 - \Phi(t_1, t_0)x_0 = \int_{t_0}^{t_1} \Phi(t_1, \tau)B(\tau)u(\tau)\, d\tau \tag{2.6}$$

and letting $\hat{x}_1 \triangleq x_1 - \Phi(t_1, t_0)x_0$, we note that the $u(t)$ that transfers the state from x_0 at t_0 to x_1 at time t_1 will also cause the state to reach $\hat{x}_1$ at t_1, starting from the origin at t_0 (i.e., $x(t_0) = 0$).

For system (2.1), we introduce the following concept.

DEFINITION 2.1. A *state* x_1 is *reachable* at time t_1 if for some finite $t_0 < t_1$ there exists an input $u(t)$, $t \in [t_0, t_1]$ that transfers the state $x(t)$ from the origin at t_0, to x_1 at time t_1 [i.e., that transfers $x(t)$ from $x(t_0) = 0$ to $x(t_1) = x_1$].

Thus, when x_1 is reachable at t_1 [with $x(t_0) = 0$], then in view of (2.5), there exists an input u such that

$$x_1 = \int_{t_0}^{t_1} \Phi(t_1, \tau)B(\tau)u(\tau)\, d\tau. \tag{2.7}$$

■

We note that the times t_1 and t_0 are important individually in the time-varying case only; in the time-invariant case, as is well known by now, $t_1 - t_0$ is the important quantity, and typically t_0 is taken to be $t_0 = 0$ with $t_1 = T$, a finite positive number.

The set of all reachable states x_1 contains the origin and constitutes a linear subspace of the state space $(X, R) = (R^n, R)$ (verify this). This gives rise to the following.

DEFINITION 2.2. The *reachable at t_1 subspace* $R_r^{t_1}$ of (2.1) is

$$R_r^{t_1} \triangleq \{\text{set of all states } x_1 \text{ reachable at } t_1\}.$$

■

When the context is clear and there is no ambiguity, we will write R_r in place of R_r^t.

DEFINITION 2.3. The *system* (2.1) is (completely state) *reachable* at t_1 if every state x_1 in the state-space is reachable at t_1 (i.e., $R_r = R^n$). In this case, we equivalently make reference to *reachable pair* $(A(t), B(t))$ at t_1. ■

A reachable state is sometimes also called *controllable-from-the-origin*. Additionally, there are also states defined to be *controllable-to-the-origin* or simply *controllable*. In particular, we have the following notion.

DEFINITION 2.4. A *state* x_0 is *controllable* at time t_0 if for some finite $t_1 > t_0$ there exists an input $u(t)$, $t \in [t_0, t_1]$ that transfers the state $x(t)$ from x_0 at t_0 to the origin at time t_1 [i.e., from $x(t_0) = x_0$ to $x(t_1) = 0$]. ■

In view of (2.5), there exists an input u such that

$$-\Phi(t_1, t_0)x_0 = \int_{t_0}^{t_1} \Phi(t_1, \tau)B(\tau)u(\tau)\, d\tau, \tag{2.8}$$

or by premultiplying by $\Phi^{-1}(t_1, t_0) = \Phi(t_0, t_1)$ (see Section 2.3),

$$-x_0 = \int_{t_0}^{t_1} \Phi(t_0, \tau)B(\tau)u(\tau)\, d\tau, \tag{2.9}$$

where the semigroup property $\Phi(t_0, t_1)\Phi(t_1, \tau) = \Phi(t_0, \tau)$ was used.

Similar to the case of reachable states, the set of all controllable states includes the origin, and is a linear subspace R_c of the state-space X (i.e., $R_c \subset X$).

DEFINITION 2.5. The *controllable at* t_0 *subspace* $R_c^{t_0}$ of (2.1) is

$$R_c^{t_0} \triangleq \{\text{set of all states } x_0 \text{ controllable at } t_0\}. \qquad \blacksquare$$

It is denoted by R_c for convenience when there is no ambiguity.

DEFINITION 2.6. The *system* (2.1) is (*completely state*) *controllable* at t_0 if every state x_0 in its state-space is controllable (i.e., if $R_c = R^n$). In this case, we equivalently make reference to *controllable pair* $(A(t), B(t))$ at t_0. $\blacksquare$

Discussion

Relation (2.7) shows that for $\dot{x} = A(t)x + B(t)u$ given in (2.1) and for given t_1, the range of the integral map

$$L \triangleq L(u, t_0, t_1) \triangleq \int_{t_0}^{t_1} \Phi(t_1, \tau)B(\tau)u(\tau)\,d\tau \tag{2.10}$$

with $u(t), t \in [t_0, t_1]$, and with t_0 varying over all finite values $t_0 \leq t_1$, is exactly the reachability subspace R_r, since a state x_1 is reachable if there exists a t_0 and u such that $x_1 \in \mathcal{R}(L)$. Notice that in view of (2.6), the input u which transfers the state from the origin at t_0 to $\hat{x}_1$ at t_1 also transfers the state from x_0 at t_0 to x_1 at t_1, where $\hat{x}_1 = x_1 - \Phi(t_1, t_0)x_0$. For fixed x_0, since $\{\hat{x}_1\}$ spans the reachability subspace $R_r^{t_1}$, this relation yields all states x_1 that can be reached from x_0 in finite time $t_1 - t_0$.

In Lemma 2.1, the range of L is shown to be equal to the range of a matrix, the reachability Gramian $W_r(t_0, t_1)$, which is rather easy to determine. In the time-invariant case, it is also shown to be equal to the range of the controllability matrix $\mathscr{C}$. Before proving these results, the relation between reachability (controllability-from-the-origin) and controllability (controllability-to-the-origin) is discussed; the exact relation is proved in Theorem 2.6.

In view of (2.7) and (2.8), a vector x is reachable (controllable-from-the-origin) at t_1 if there exists a finite t_0 and $u(t), t \in [t_0, t_1]$, so that $x \in \mathcal{R}(L)$, where L is defined in (2.10), and it is controllable (-to-the-origin) at t_0 if $\Phi(t_1, t_0)x \in \mathcal{R}(L)$. It is shown later (Theorem 2.6) that the system (2.1), or the pair (A, B), is (completely state) reachable if and only if it is controllable. This is the reason why only one term is typically used in the literature when describing these properties for continuous-time systems. For discrete-time systems, however, the situation is different. In this case, as will be shown later in this section, if the pair (A, B) is reachable, then it is also controllable, but not necessarily vice versa; that is, in the discrete-time case controllability does not necessarily imply reachability. Indeed, controllability implies reachability only when the state transition matrix $\Phi(k, k_0)$ has full rank, which is not always true in discrete-time systems. As discussed in Chapter 2, this is due to the lack of the "time reversibility" property. On the other hand, in the case of continuous-time systems, $\Phi(t, \tau)$ is always nonsingular. In such systems, reachability implies that any state x_1 can be reached from any other state x_0 in finite time $t_1 - t_0$. This property is sometimes used in the literature to define "controllability." An input that achieves this transfer is given later in Corollary 2.5. In the discrete-time systems literature, the term that is typically used is "reachability"; however, for simplicity, the term "controllability" is sometimes also used, with some sacrifice of accuracy. We will use both terms, reachability and controllability, with a warning to the reader when use of the term controllability (-from-the-origin) is made, instead of reachability.

Now suppose there exists an input u which transfers the state of the system from $x(t_0) = 0$ to $x(t_1) = x_1$, that is, (2.7) is true. The integral in (2.7) is a map $L \triangleq L(u, t_0, t_1)$ defined in (2.10) that maps an input $u(t) \in R^m$ defined over $[t_0, t_1]$ to states $x_1 \in R^n$. We are interested in the range of $L, \mathcal{R}(L)$ since it contains all the states that can be reached from the origin, $x(t_0) = 0$, at time t_1, by varying the input u. Note that L has infinite-dimensional domain, and therefore it is not easy to determine its range directly. In the following we show that $\mathcal{R}(L)$ is equal to the range of an important matrix, the reachability Gramian.

DEFINITION 2.7. The *reachability Gramian* of the system $\dot{x} = A(t)x + B(t)u$ is the $n \times n$ matrix

$$W_r(t_0, t_1) \triangleq \int_{t_0}^{t_1} \Phi(t_1, \tau)B(\tau)B^T(\tau)\Phi^T(t_1, \tau)\, d\tau, \tag{2.11}$$

where $\Phi(t, \tau)$ denotes the state transition matrix. ■

Note that W_r is symmetric and positive semidefinite for every $t_1 > t_0$; that is, $W_r = W_r^T$ and $W_r \geq 0$ (show this). Now let $t_0 < t_1$ be given. Then the following lemma can be shown.

LEMMA 2.1. $\mathcal{R}(L(u, t_0, t_1)) = \mathcal{R}(W_r(t_0, t_1))$.

Proof. We first show that $\mathcal{R}(W_r) \subset \mathcal{R}(L)$. Let $x_1 \in \mathcal{R}(W_r)$; that is, there exists $\eta_1 \in R^n$ such that $W_r\eta_1 = x_1$. Choose $u_1(\tau) = B^T(\tau)\Phi^T(t_1, \tau)\eta_1$. Then $L(u_1, t_0, t_1) = \left[\int_{t_0}^{t_1} \Phi(t_1, \tau)B(\tau)B^T(\tau)\Phi^T(t_1, \tau)\, d\tau\right]\eta_1 = W_r\eta_1 = x_1$. Therefore, $x_1 \in \mathcal{R}(L)$, and since x_1 is arbitrary, it follows that $\mathcal{R}(W_r) \subset \mathcal{R}(L)$.

We shall now show that $\mathcal{R}(L) \subset \mathcal{R}(W_r)$, which together with $\mathcal{R}(W_r) \subset \mathcal{R}(L)$ proves that $\mathcal{R}(L) = \mathcal{R}(W_r)$. Let $x_1 \in \mathcal{R}(L)$, i.e., there exists an input u_1 such that $L(u_1, t_0, t_1) = x_1$. We assume that $x_1 \notin \mathcal{R}(W_r)$ and we shall show that this leads to a contradiction. This implies that the null space of $W_r(t_0, t_1)$ is nonempty and therefore there exists a nonzero $x_2 \in R^n$ such that $W_r x_2 = 0$. Note that $x_2^T x_1 \neq 0$; that is, x_2 and x_1 are not orthogonal. This is so because W_r is symmetric (and $W_r x_2 = 0$) because this implies that the range of W_r is the orthogonal complement of its null space (prove this). Therefore, any vector x_1 such that $x_2^T x_1 = 0$ would be in the range of W_r, which is not true, and so $x_2^T x_1 \neq 0$. Now $x_2^T W_r(t_0, t_1)x_2 = 0 = \int_{t_0}^{t_1} \left[x_2^T\Phi(t_1, \tau)B(\tau)\right]\left[x_2^T\Phi(t_1, \tau)B(\tau)\right]^T d\tau = \int_{t_0}^{t_1} \|x_2^T\Phi(t_1, \tau)B(\tau)\|_2^2\, d\tau$, which shows that $x_2^T\Phi(t_1, \tau)B(\tau) = 0$ for every $\tau \in [t_0, t_1]$. This in turn implies that $x_2^T x_1 = x_2^T L(u_1) = \int_{t_0}^{t_1} [x_2^T\Phi(t_1, \tau)B(\tau)]u_1(\tau)\, d\tau = 0$, which is a contradiction since $x_2^T x_1 \neq 0$. Therefore, $x_1 \in \mathcal{R}(W_r)$, which implies that $\mathcal{R}(L) \subset \mathcal{R}(W_r)$. ■

Lemma 2.1 shows that the set of all states that can be reached at time t_1 from the origin at some finite time $t_0 < t_1$, is given by $\mathcal{R}(W_r(t_0, t_1))$, the range of the reachability Gramian.

THEOREM 2.2. Consider the system $\dot{x} = A(t)x + B(t)u$ given in (2.1). There exists an input u that transfers the state to x_1 at t_1 from the origin at some finite time $t_0 < t_1$, if and only if there exists finite time $t_0 < t_1$ so that

$$x_1 \in \mathcal{R}(W_r(t_0, t_1)).$$

Furthermore, an appropriate u that will accomplish this transfer is given by

$$u(t) = B^T(t)\Phi^T(t_1, t)\eta_1 \tag{2.12}$$

with η_1 a solution of $W_r(t_0, t_1)\eta_1 = x_1$ and $t \in [t_0, t_1]$.

Proof. In view of Lemma 2.1 and the definition of $L(u)$ in (2.10), the proof of the first part of the theorem is straightforward. To prove the second part of the theorem, note that (2.12) was used in the proof of Lemma 2.2 to accomplish the transfer to x_1. ■

COROLLARY 2.3. The system $\dot{x} = A(t)x + B(t)u$ is (completely state) reachable at t_1, or the pair $(A(t), B(t))$ is reachable at t_1, if and only if there exists finite $t_0 < t_1$ such that

$$\text{rank}\, W_r(t_0, t_1) = n. \tag{2.13}$$

Proof. In view of Theorem 2.2, all states x_1 can be reached at t_1 if and only if for some $t_0 < t_1$, $\mathcal{R}(W_r(t_0, t_1)) = R^n$, the entire state space. This is true if and only if $\text{rank}\, W_r(t_0, t_1) = n$ for some finite $t_0 < t_1$. ■

The following result is useful in accomplishing the transfer from a state x_0 to another state x_1 in some given finite time $t_1 - t_0$.

THEOREM 2.4. There exists an input u that transfers the state of the system $\dot{x} = A(t)x + B(t)u$ from x_0 at time t_0 to x_1 at time $t_1 > t_0$ if and only if

$$x_1 - \Phi(t_1, t_0)x_0 \in \mathcal{R}(W_r(t_0, t_1)). \tag{2.14}$$

Furthermore, such input is given by

$$u(t) = B^T(t)\Phi^T(t_1, t)\eta_1 \tag{2.15}$$

with η_1 a solution of

$$W_r(t_0, t_1)\eta_1 = x_1 - \Phi(t_1, t_0)x_0. \tag{2.16}$$

Proof. The proof is straightforward in view of Theorem 2.2 and the fact that there exists an input which transfers the state from x_0 at t_0 to x_1 at t_1 if and only if it transfers the state from the origin at t_0 to $\hat{x}_1 \triangleq x_1 - \Phi(t_1, t_0)x_0$ at t_1 [see (2.6)]. ■

EXAMPLE 2.1. Consider $\dot{x} = A(t)x + B(t)u$, where $A(t) = \begin{bmatrix} -1 & e^{2t} \\ 0 & -1 \end{bmatrix}$, $B(t) = \begin{bmatrix} e^{-t} \\ 0 \end{bmatrix}$. The state transition matrix was calculated in Example 3.4, Section 2.3 (of Chapter 2), to be

$$\Phi(t, \tau) = \begin{bmatrix} e^{-(t-\tau)} & \frac{1}{2}[e^{t+\tau} - e^{-t+3\tau}] \\ 0 & e^{-(t-\tau)} \end{bmatrix}.$$

Here $\Phi(t, \tau)B(\tau) = \begin{bmatrix} e^{-t} \\ 0 \end{bmatrix}$, and the reachability Gramian of the system is

$$W_r(t_0, t_1) = \int_{t_0}^{t_1} \begin{bmatrix} e^{-2t_1} & 0 \\ 0 & 0 \end{bmatrix} d\tau = \begin{bmatrix} (t_1 - t_0)e^{-2t_1} & 0 \\ 0 & 0 \end{bmatrix}.$$

It is clear that $\text{rank}\, W_r(t_0, t_1) < 2 = n$ for any $t_0 < t_1$ and therefore the system is not reachable at t_1. Note that since t_1 is arbitrary, the system is not reachable at any finite time. However, the state can be transfered from the origin to a state $x_1 \in \mathcal{R}(W_r(t_0, t_1))$. In particular, in view of Theorem 2.2, let $x_1 = \begin{bmatrix} \alpha \\ 0 \end{bmatrix}$, $\alpha \in R$, and solve $W_r(t_0, t_1)\eta_1 = x_1$ to obtain $\eta_1 = \begin{bmatrix} \frac{\alpha}{t_1 - t_0} e^{2t_1} \\ \beta \end{bmatrix}$, where $\beta \in R$ arbitrary. Then, in view of (2.12), $u(t) = [\Phi(t_1, t)B(t)]^T \eta_1 = [e^{-t_1}, 0] \begin{bmatrix} \frac{\alpha}{t_1 - t_0} e^{2t_1} \\ \beta \end{bmatrix} = \frac{\alpha}{t_1 - t_0} e^{t_1}$ will drive the state from the origin at t_0 to x_1 at t_1. To verify this, we note that $x(t_1) = \int_{t_0}^{t_1} \Phi(t_1, \tau)B(\tau)u(\tau)\,d\tau = \begin{bmatrix} e^{-t_1} \\ 0 \end{bmatrix} \frac{\alpha}{t_1 - t_0} e^{t_1}(t_1 - t_0) = \begin{bmatrix} \alpha \\ 0 \end{bmatrix}$. Notice that for the transfer to be accomplished in a short period of time, $t_1 - t_0 = \epsilon$ with ϵ small, the required control magnitude can be quite large since $u(t) = (\alpha/\epsilon)e^{t_1}$. ■

The last observation in Example 2.1 points to two important aspects that we now elaborate on.

First, we note that the faster the state of the system is required to move (the smaller the $t_1 - t_0 = \epsilon$) and the further away the desired state x_1 is (the larger the α), the larger the required control magnitude will be. This makes intuitive sense since it simply states that the more sudden and drastic the change in the state, the larger the required control force will be (think, e.g., of a simple mechanical spring system).

Second, it is clear that the property of reachability (controllability) implies the ability to change the state of the system very fast indeed, paying for this of course in terms of increased control magnitude (see Example 2.1 and Exercise 3.12). Intuitively, this is not always possible in the case of physical processes, where only limited control action is typically available. This points to some of the limitations of linear system models that do not include information about input saturation limits, nonlinear behavior, limitations of output sensors, and the like.

EXAMPLE 2.2. Consider the system described by $\dot{x} = A(t)x + B(t)u$, where $A(t) = \begin{bmatrix} -1 & e^{2t} \\ 0 & -1 \end{bmatrix}$, $B(t) = \begin{bmatrix} 0 \\ e^{-t} \end{bmatrix}$. The state transition matrix $\Phi(t, \tau)$ is given in Example 2.1. Here $\Phi(t, \tau)B(\tau) = \begin{bmatrix} \frac{1}{2}[e^{t} - e^{-t+2\tau}] \\ e^{-t} \end{bmatrix}$ and the reachability Gramian $W_r(t_0, t_1)$ is such that the system is reachable at t_1 (show this). ■

The following result demonstrates the importance of reachability in determining an input u to transfer the state from any x_0 to any x_1 in finite time.

COROLLARY 2.5. Let the system $\dot{x} = A(t)x + B(t)u$ be (completely state) reachable at time t_1, or let the pair $(A(t), B(t))$ be reachable at t_1. Then there exists an input that will transfer any state x_0 at some finite time $t_0 < t_1$, to any state x_1 at time t_1. Such input is given by

$$u(t) = B^T(t)\Phi^T(t_1, t)W_r^{-1}(t_0, t_1)[x_1 - \Phi(t_1, t_0)x_0] \tag{2.17}$$

for $t \in [t_0, t_1]$.

Proof. In view of Corollary 2.3, reachability implies that, given t_1, *rank* $W_r(t_0, t_1) = n$ for some $t_0 < t_1$ or that $\mathcal{R}(W_r(t_0, t_1)) = R^n$, the whole space, for some t_0. This implies that any vector $x_1 - \Phi(t_1, t_0)x_0 \in \mathcal{R}(W_r(t_0, t_1))$, which in view of Theorem 2.2 and (2.12) implies that the input in (2.17) is an input which will accomplish this transfer. ■

There are many different control inputs u that can accomplish the state transfer from x_0 at $t = t_0$ to x_1 at $t = t_1$. It can be shown that the input u given by

(2.17) accomplishes this transfer while expending a minimum amount of energy. In particular, among all the control inputs $u(t)$ that will transfer the state from x_0 at t_0 to x_1 at t_1, $u(t)$ in (2.17) minimizes the cost functional $\int_{t_0}^{t_1} \|u(\tau)\|^2 \, d\tau$, where $\|u(t)\| \triangleq [u^T(t)u(t)]^{1/2}$, the Euclidean norm of $u(t)$.

We shall now establish a connection between controllability and reachability of the continuous-time system $\dot{x} = A(t)x + B(t)u$.

THEOREM 2.6. If the system $\dot{x} = A(t)x + B(t)u$, or the pair $(A(t), B(t))$, is reachable at t_1, then it is controllable at some $t_0 < t_1$. Also, if it is controllable at t_0, then it is reachable at some $t_1 > t_0$.

Proof. It was shown in Corollary 2.3 that for reachability of (2.1) at t_1, we must have *rank* $W_r(t_0, t_1) = n$. A similar test for controllability can be derived in an identical manner. In particular, in view of (2.9), it is clear that the range of

$$\hat{L} \triangleq \hat{L}(u, t_0, t_1) \triangleq \int_{t_0}^{t_1} \Phi(t_0, \tau)B(\tau)u(\tau)\, d\tau \tag{2.18}$$

is of (present) interest [compare with L in (2.10) used to prove reachability]. A result similar to Lemma 2.1 can now be established using an identical approach, namely, that $\mathcal{R}(\hat{L}(u, t_0, t_1)) = \mathcal{R}(W_c(t_0, t_1))$, where $W_c(t_0, t_1)$ is the controllability Gramian, defined next. ■

DEFINITION 2.8. The *controllability Gramian* of the system $\dot{x} = A(t)x + B(t)u$ is the $n \times n$ matrix

$$W_c(t_0, t_1) \triangleq \int_{t_0}^{t_1} \Phi(t_0, \tau)B(\tau)B^T(\tau)\Phi^T(t_0, \tau)\, d\tau, \tag{2.19}$$

where $\Phi(t, \tau)$ denotes the state transition matrix.

Continuing the proof of Theorem 2.6, we note that it can be shown that the input

$$u_1(t) = -B^T(t)\Phi^T(t_0, t)\eta_1 \tag{2.20}$$

with η_1 such that $W_c(t_0, t_1)\eta_1 = x_0$ satisfies $\hat{L}(u_1, t_0, t_1) = -x_0$ or relation (2.9). Thus, $\eta_1(t)$ drives the state from x_0 at time t_0 to the origin at time $t_1 > t_0$ (compare with Theorem 2.2). As in Corollary 2.3 for the case of reachability, it can be shown in an analogous manner that the system is (completely state) controllable at t_0 if and only if there exists $t_1 > t_0$ so that

$$rank\; W_c(t_0, t_1) = n. \tag{2.21}$$

Next, we note that in view of the definitions of W_r and W_c,

$$W_r(t_0, t_1) = \Phi(t_1, t_0)W_c(t_0, t_1)\Phi^T(t_1, t_0). \tag{2.22}$$

Since $\Phi(t_1, t_0)$ is nonsingular for every t_0 and t_1, *rank* $W_r(t_0, t_1) =$ *rank* $W_c(t_0, t_1)$ for every t_0 and t_1. *Therefore, the system is reachable if and only if it is controllable.* ■

EXAMPLE 2.3. Consider the system $\dot{x} = A(t)x + B(t)u$ of Example 2.1. The controllability Gramian is given by

$$W_c(t_0, t_1) = \int_{t_0}^{t_1} \begin{bmatrix} e^{-t_0} \\ 0 \end{bmatrix} [e^{-t_0}, 0]\, d\tau = \begin{bmatrix} (t_1 - t_0)e^{-2t_0} & 0 \\ 0 & 0 \end{bmatrix}.$$

Compare this with the reachability Gramian of Example 2.1 and note that $\Phi(t_1, t_0) \times W_c(t_0, t_1)\Phi^T(t_1, t_0) = \begin{bmatrix} (t_1 - t_0)e^{-(t_1+t_0)} & 0 \\ 0 & 0 \end{bmatrix} \Phi^T(t_1, t_0) = \begin{bmatrix} e^{-2t_1}(t_1 - t_0) & 0 \\ 0 & 0 \end{bmatrix} = W_r(t_0, t_1)$, as expected [see (2.22)]. ■

Before proceeding, we note that a relation similar to (2.17) can be derived using the Gramian $W_c(t_0, t_1)$ and (2.20) in place of $W_r(t_0, t_1)$. In particular, an appropriate input that transfers the state from x_0 at t_0 to x_1 at t_1 is given by

$$u(t) = -B^T(t)\Phi^T(t_0, t)W_c^{-1}(t_0, t_1)[x_0 - \Phi(t_0, t_1)x_1]. \tag{2.23}$$

We ask the reader to show that this relation can also be derived from (2.17), using (2.22).

Additional criteria for reachability and controllability

First recall from Chapter 2 the definition of a set of linearly independent functions of time and consider in particular n complex-valued functions $f_i(t), i = 1, \dots, n$, where $f_i^T(t) \in C^m$. Recall that the set of functions $f_i, i = 1, \dots, n$, is *linearly dependent* on a time interval $[t_1, t_2]$ over the field of complex numbers C if there exist complex numbers $a_i, i = 1, \dots, n$, not all zero, such that

$$a_1 f_1(t) + \cdots + a_n f_n(t) = 0 \qquad \text{for all } t \text{ in } [t_1, t_2];$$

otherwise, the set of functions is said to be *linearly independent* on $[t_1, t_2]$ over the field of complex numbers.

It is possible to test linear independence using the Gram matrix of the functions f_i.

LEMMA 2.7. Let $F(t) \in C^{n\times m}$ be a matrix with $f_i(t) \in C^{1\times m}$ in its ith row. Define the *Gram matrix* of $f_i(t), i = 1, \dots, n$, by

$$W(t_1, t_2) \triangleq \int_{t_1}^{t_2} F(t)F^*(t)\, dt, \tag{2.24}$$

where $(\,\cdot\,)^*$ denotes the complex conjugate transpose. The set $f_i(t), i = 1, \dots, n$, is linearly independent on $[t_1, t_2]$ over the field of complex numbers if and only if the Gram matrix $W(t_1, t_2)$ is nonsingular, or equivalently, if and only if the *Gram determinant* $\det W(t_1, t_2) \neq 0$.

Proof. (*Necessity*) Assume the set $f_i, i = 1, \dots, n$, is linearly independent but $W(t_1, t_2)$ is singular. Then there exists some nonzero $\alpha \in C^{1\times n}$ so that $\alpha W(t_1, t_2) = 0$, from which $\alpha W(t_1, t_2)\alpha^* = \int_{t_1}^{t_2} (\alpha F(t))(\alpha F(t))^*\, dt = 0$. Since $(\alpha F(t))(\alpha F(t))^* \geq 0$ for all t, this implies that $\alpha F(t) = 0$ for all t in $[t_1, t_2]$, which is a contradiction. Therefore $W(t_1, t_2)$ is nonsingular.

(*Sufficiency*) Assume that $W(t_1, t_2)$ is nonsingular but the set $f_i, i = 1, \dots, n$, is linearly dependent. Then there exists some nonzero $\alpha \in C^{1\times n}$ so that $\alpha F(t) = 0$. Then $\alpha W(t_1, t_2) = \int_{t_1}^{t_2} \alpha F(t)F^*(t)\, dt = 0$, which is a contradiction. Therefore the set $f_i, i = 1, \dots, n$, is linearly independent. ■

We will use the above result to derive additional tests for reachability and controllability in this section and for observability and constructibility in the next section. In the following two theorems, we repeat some earlier results, for convenience.

THEOREM 2.8. The system $\dot{x} = A(t)x + B(t)u$ is (completely state) reachable at t_1

(i) if and only if there exists finite $t_0 < t_1$, such that

$$\operatorname{rank} W_r(t_0, t_1) = n, \tag{2.25}$$

where $W_r(t_0, t_1) \triangleq \int_{t_0}^{t_1} \Phi(t_1, \tau)B(\tau)B^T(\tau)\Phi^T(t_1, \tau)\, d\tau$, the reachability Gramian, or equivalently,

(ii) if and only if there exists finite $t_0 < t_1$, such that the n rows of

$$\Phi(t_1, t)B(t) \tag{2.26}$$

are linearly independent on $[t_0, t_1]$ over the field of complex numbers.

Proof. Part (i) was established in Corollary 2.3, while part (ii) is a direct consequence of the previous lemma and the definition of the reachability Gramian. ■

Similar results can be derived for controllability. Specifically, we have the following result.

THEOREM 2.9. The system $\dot{x} = A(t)x + B(t)u$ is (completely state) controllable at t_0
(i) if and only if there exist finite $t_1 > t_0$ such that

$$\operatorname{rank} W_c(t_0, t_1) = n, \tag{2.27}$$

where $W_c(t_0, t_1) \triangleq \int_{t_0}^{t_1} \Phi(t_0, \tau)B(\tau)B^T(\tau)\Phi^T(t_0, \tau)\,d\tau$, the controllability Gramian, or equivalently,
(ii) if and only if there exists finite $t_1 > t_0$ such that the n rows of

$$\Phi(t_0, t)B(t) \tag{2.28}$$

are linearly independent on $[t_0, t_1]$ over the field of complex numbers.

Proof. The proof is analogous to the proof for Theorem 2.8. ■

Notice that premultiplication of $\Phi(t_0, t)B(t)$ by the nonsingular matrix $\Phi(t_1, t_0)$ yields $\Phi(t_1, t)B(t)$ [refer to (2.26) in the reachability theorem; compare with (2.22)]. This can be used to prove in an alternative way the result of Theorem 2.6, that reachability (controllability-from-the-origin) implies and is implied by controllability (-to-the-origin), in the case of continuous-time systems (show this).

B. Continuous-Time Time-Invariant Systems

We shall now apply the results developed above to time-invariant systems $\dot{x} = Ax + Bu$ given in (2.3). In this case, the state transition matrix $\Phi(t, \tau)$ is explicitly known and is given by $\Phi(t, \tau) = e^{A(t-\tau)}$ in (2.4). Because of time invariance, the difference $t_1 - t_0 = T$, rather than the individual times t_0 and t_1, plays an important role. Accordingly, for the time-invariant case we can always take $t_0 = 0$ and $t_1 = T$. This practice will be adopted in the following.

The definitions of reachability, Definitions 2.1 to 2.3, and controllability, Definitions 2.4 to 2.6, are certainly also valid in the time-invariant case. We repeat them here for convenience, specializing them to the the system $\dot{x} = Ax + Bu$ given in (2.3).

DEFINITION 2.9. (i) A *state* x_1 is *reachable* if there exists an input $u(t)$, $t \in [0, T]$, that transfers the state $x(t)$ from the origin at $t = 0$ to x_1 in some finite time T.
(ii) The set of all reachable states R_r is the *reachable subspace* of the system $\dot{x} = Ax + Bu$, or of the pair (A, B).
(iii) The system $\dot{x} = Ax + Bu$, or the pair (A, B) is (*completely state*) *reachable* if every state is reachable, i.e., if $R_r = R^n$. ■

DEFINITION 2.10. (i) A *state* x_0 is *controllable* if there exists an input $u(t)$, $t \in [0, T]$, that transfers the state $x(t)$ from x_0 at $t = 0$ to the origin in some finite time T.

(ii) The set of all controllable states R_c is the *controllable subspace* of the system $\dot{x} = Ax + Bu$, or of the pair (A, B).
(iii) The system $\dot{x} = Ax + Bu$, or the pair (A, B), is *(completely state) controllable* if every state is controllable, i.e., if $R_c = R^n$. ■

DEFINITION 2.11. The $n \times n$ *reachability Gramian* of the time-invariant system $\dot{x} = Ax + Bu$ is

$$W_r(0, T) \triangleq \int_0^T e^{(T-\tau)A} BB^T e^{(T-\tau)A^T} \, d\tau. \tag{2.29}$$

■

Note that W_r is symmetric and positive semidefinite for every $T > 0$, i.e., $W_r = W_r^T$ and $W_r \geq 0$ (show this). Let the $n \times mn$ *controllability (-from-the-origin) matrix* (or more precisely, the reachability matrix) be

$$\mathscr{C} \triangleq [B, AB, \ldots, A^{n-1}B], \tag{2.30}$$

and recall that $\mathscr{C}$ was also defined in Section 3.1.

It is now shown that in the time-invariant case the range of $W_r(0, T)$, denoted by $\mathscr{R}(W_r(0, T))$, is independent of T, i.e., it is the same for any finite $T(> 0)$, and in particular, it is equal to the range of the controllability matrix $\mathscr{C}$. Thus, the reachable subspace R_r of a system is given by the range of $\mathscr{C}, \mathscr{R}(\mathscr{C})$, or the range of $W_r(0, T), \mathscr{R}(W_r(0, T))$, for some finite (and therefore for any) $T > 0$.

LEMMA 2.10. $\mathscr{R}(W_r(0, T)) = \mathscr{R}(\mathscr{C})$ for every $T > 0$.

Proof. We first show that $\mathscr{R}(W_r) \subset \mathscr{R}(\mathscr{C})$ for some $T > 0$. Let $x_1 \in \mathscr{R}(W_r)$ for some $T > 0$. In view of Lemma 2.1, $x_1 \in \mathscr{R}(L)$, i.e., there exists u_1 such that $L(u_1, 0, T) = \int_0^T \{\exp[(T-\tau)A]\}Bu_1(\tau)\, d\tau = x_1$. Using the series definition $\exp[At] = \sum_{k=0}^{\infty}(t^k/k!)A^k$, x_1 can be written as $x_1 = \sum_{k=0}^{\infty} A^k B \left[\int_0^T ((T-\tau)^k/k!)u_1(\tau)\, d\tau\right]$ or, in view of the Cayley-Hamilton Theorem, $x_1 = \sum_{k=0}^{n-1} A^k B \alpha_k(T)$, where $\alpha_k(T)$ is appropriately defined. This implies that $x_1 \in \mathscr{R}(\mathscr{C})$. Since x_1 is arbitrary, $\mathscr{R}(W_r) \subset \mathscr{R}(\mathscr{C})$.

We shall now show that $\mathscr{R}(\mathscr{C}) \subset \mathscr{R}(W_r)$. Let $x_1 \in \mathscr{R}(\mathscr{C})$, i.e., there exists $\eta_1 \in R^{nm}$ such that $\mathscr{C}\eta_1 = x_1$. Assume that $x_1 \notin \mathscr{R}(W_r)$ for some $T > 0$. We shall show that this leads to a contradiction. Indeed, this assumption implies that the null space of W_r is nonempty, and therefore there exists a nonzero $x_2 \in R^n$ such that $W_r x_2 = 0$. Note that $x_2^T x_1 \neq 0$; i.e., x_2 and x_1 are not orthogonal. As shown in the proof of Lemma 2.1, this is true because W_r, is symmetric, and $W_r x_2 = 0$, since the range of W_r is the orthogonal complement of its null space (prove this). It follows that any vector x_1 such that $x_2^T x_1 = 0$ would be in the range of W_r which is not true. Therefore, $x_2^T x_1 \neq 0$. Next, consider $x_2^T W_r(0, T) x_2 = 0 = \int_0^T [x_2^T\{\exp[(T-\tau)A]\}B][x_2^T\{\exp[(T-\tau)A]\}B]^T \, d\tau = \int_0^T \|x_2^T\{\exp[(T-\tau)A]\}B\|_2^2 \, d\tau$, which shows that $x_2^T \exp[(T-\tau)A]B = 0$ for every $\tau \in [0, T]$. Taking derivatives of both sides with respect to τ and evaluating at $\tau = T$, we obtain $x_2^T B = -x_2^T AB = \cdots = (-1)^k x_2^T A^k B = 0$ for every $k > 0$. Thus, $x_2^T A^k B = 0$ for every $k \geq 0$, and therefore, $x_2^T x_1 = x_2^T \mathscr{C} \eta_1 = 0$, which is a contradiction since $x_2^T x_1 \neq 0$. Therefore, $x_1 \in \mathscr{R}(W_r)$, which implies that $\mathscr{R}(\mathscr{C}) \subset \mathscr{R}(W_r)$. This, together with $\mathscr{R}(W_r) \subset \mathscr{R}(\mathscr{C})$, shows that $\mathscr{R}(W_r) = \mathscr{R}(\mathscr{C})$. ■

Lemma 2.10 shows that given the time-invariant system $\dot{x} = Ax + Bu$, if $x(0) = 0$, then the set of all states that can be reached in finite time; i.e., the reachability subspace R_r is given by $\mathscr{R}(\mathscr{C})$, the range of the controllability matrix, or equivalently, by $\mathscr{R}(W_r(0, T))$, the range of the reachability Gramian, where *$T > 0$ is any finite time*.

EXAMPLE 2.4. For the system $\dot{x} = Ax + Bu$ with $A = \begin{bmatrix} 0 & 1 \\ 0 & 0 \end{bmatrix}$ and $B = \begin{bmatrix} 0 \\ 1 \end{bmatrix}$, we have $e^{At} = \begin{bmatrix} 1 & t \\ 0 & 1 \end{bmatrix}$ and $e^{At}B = \begin{bmatrix} t \\ 1 \end{bmatrix}$. The reachability Gramian is $W_r(0, T) = \int_0^T \begin{bmatrix} T - \tau \\ 1 \end{bmatrix} \times [T - \tau, 1]\, d\tau = \int_0^T \begin{bmatrix} (T-\tau)^2 & T - \tau \\ T - \tau & 1 \end{bmatrix} d\tau = \begin{bmatrix} \frac{1}{3}T^3 & \frac{1}{2}T^2 \\ \frac{1}{2}T^2 & T \end{bmatrix}$. Since $det\ W_r(0, T) = \frac{1}{12}T^4 \neq 0$ for any $T > 0$, $rank\, W_r(0, T) = n$ and (A, B) is reachable. Note that $\mathscr{C} = [B, AB] = \begin{bmatrix} 0 & 1 \\ 1 & 0 \end{bmatrix}$ and that $\mathscr{R}(W_r(0, T)) = \mathscr{R}(\mathscr{C}) = R^2$, as expected (Lemma 2.10).

If $B = \begin{bmatrix} 1 \\ 0 \end{bmatrix}$, instead of $\begin{bmatrix} 0 \\ 1 \end{bmatrix}$, then $\mathscr{C} = [B, AB] = \begin{bmatrix} 1 & 0 \\ 0 & 0 \end{bmatrix}$ and (A, B) is not reachable. In this case $e^{At}B = \begin{bmatrix} 1 \\ 0 \end{bmatrix}$ and the reachability matrix is $W_r(0, T) = \int_0^T \begin{bmatrix} 1 & 0 \\ 0 & 0 \end{bmatrix} d\tau = \begin{bmatrix} T & 0 \\ 0 & 0 \end{bmatrix}$. Notice again that $\mathscr{R}(\mathscr{C}) = \mathscr{R}(W_r(0, T))$ for every $T > 0$. ■

THEOREM 2.11. Consider the system $\dot{x} = Ax + Bu$ and let $x(0) = 0$. There exists input u that transfers the state to x_1 in finite time if and only if $x_1 \in \mathscr{R}(\mathscr{C})$, or equivalently, if and only if

$$x_1 \in \mathscr{R}(W_r(0, T))$$

for some finite (and therefore for any) T. Thus, the reachable subspace $R_r = \mathscr{R}(\mathscr{C}) = \mathscr{R}(W_r(0, T))$. Furthermore, an appropriate u that will accomplish this transfer in time T is given by

$$u(t) = B^T e^{A^T(T-t)} \eta_1 \tag{2.31}$$

with η_1 such that $W_r(0, T)\eta_1 = x_1$ and $t \in [0, T]$.

Proof. Apply Theorem 2.2 to the time-invariant case and then use Lemma 2.10. ■

Note that in (2.31) no restrictions are imposed on time T, other than that T be finite. T can be as small as we wish, i.e., the transfer can be accomplished in a very short time indeed.

COROLLARY 2.12. The system $\dot{x} = Ax + Bu$, or the pair (A, B), is (completely state) reachable, if and only if

$$rank\, \mathscr{C} = n, \tag{2.32}$$

or equivalently, if and only if

$$rank\, W_r(0, T) = n \tag{2.33}$$

for some finite (and therefore for any) T.

Proof. Apply Corollary 2.3 to the time-invariant case and use Lemma 2.10. ■

THEOREM 2.13. There exists input u that transfers the state of the system $\dot{x} = Ax + Bu$ from x_0 to x_1 in some finite time T if and only if

$$x_1 - e^{AT}x_0 \in \mathscr{R}(\mathscr{C}), \tag{2.34}$$

or equivalently, if and only if

$$x_1 - e^{AT}x_0 \in \mathscr{R}(W_r(0, T)). \tag{2.35}$$

Such input is given by

$$u(t) = B^T e^{A^T(T-t)} \eta_1 \tag{2.36}$$

with $t \in [0, T]$, where η_1 is a solution of

$$W_r(0, T)\eta_1 = x_1 - e^{AT} x_0. \tag{2.37}$$

Proof. Apply Theorem 2.4 to the time-invariant case and use Lemma 2.10. ∎

The above leads to the next result, which establishes the importance of reachability in determining an input u to transfer the state from any x_0 to any x_1 in finite time.

COROLLARY 2.14. Let the system $\dot{x} = Ax + Bu$ be (completely state) reachable, or the pair (A, B) be reachable. Then there exists an input that will transfer any state x_0 to any other state x_1 in some finite time T. Such input is given by

$$u(t) = B^T e^{A^T(T-t)} W_r^{-1}(0, T)[x_1 - e^{AT} x_0] \tag{2.38}$$

for $t \in [0, T]$.

Proof. This result is the time-invariant version of Corollary 2.5. In view of Corollary 2.12, reachability implies that *rank* $W_r(0, T) = n$ for some T or that $\mathcal{R}(W_r(0, T)) = R^n$, the whole state space. This implies that any vector $x_1 - e^{AT} x_0 \in \mathcal{R}(W_r(0, T))$ that, in view of Theorem 2.13, implies that the input in (2.38) is an input which will accomplish this transfer. ∎

There are many different control inputs u that can accomplish the transfer from x_0 to x_1 in time T. It can be shown that the input u given by (2.38) accomplishes this transfer while expending a minimum amount of energy; in fact, u minimizes the cost functional $\int_0^T \|u(\tau)\|^2 d\tau$, where $\|u(t)\| \triangleq [u^T(t)u(t)]^{1/2}$ denotes the Euclidean norm of $u(t)$.

EXAMPLE 2.5. The system $\dot{x} = Ax + Bu$ with $A = \begin{bmatrix} 0 & 1 \\ 0 & 0 \end{bmatrix}$ and $B = \begin{bmatrix} 0 \\ 1 \end{bmatrix}$ is reachable (see Example 2.4). A control input $u(t)$ that will transfer any state x_0 to any other state x_1 in some finite time T is given by (see Corollary 2.14 and Example 2.4)

$$\begin{aligned} u(t) &= B^T e^{A^T(T-t)} W_r^{-1}(0, T)[x_1 - e^{AT} x_0] \\ &= [T - t, 1] \begin{bmatrix} \frac{12}{T^3} & -\frac{6}{T^2} \\ -\frac{6}{T^2} & \frac{4}{T} \end{bmatrix} \left[x_1 - \begin{bmatrix} 1 & T \\ 0 & 1 \end{bmatrix} x_0 \right]. \end{aligned}$$

∎

EXAMPLE 2.6. For the (scalar) system $\dot{x} = -ax + bu$, determine $u(t)$ that will transfer the state from $x(0) = x_0$ to the origin in T sec; i.e., $x(T) = 0$.

We shall apply Corollary 2.14. The reachability Gramian is $W_r(0, T) = \int_0^T e^{-(T-\tau)a} bbe^{-(T-\tau)a} d\tau = e^{-2aT} b^2 \int_0^T e^{2a\tau} d\tau = e^{-2aT} b^2 \frac{1}{2a}[e^{2aT} - 1] = \frac{b^2}{2a}[1 - e^{-2aT}]$. Note [see (2.41) below] that the controllability Gramian is $W_c(0, T) = \frac{b^2}{2a}[e^{2aT} - 1]$. Now in view of (2.38), we have

$$
\begin{aligned}
u(t) &= be^{-(T-t)a}\frac{2a}{b^2}\frac{1}{1-e^{-2aT}}[-e^{-aT}x_0] \\
&= -\frac{2a}{b}\frac{e^{-2aT}}{1-e^{-2aT}}e^{aT}x_0 \\
&= -\frac{2a}{b}\frac{1}{e^{2aT}-1}e^{at}x_0.
\end{aligned}
$$

To verify that this $u(t)$ accomplishes the desired transfer, we compute $x(t) = e^{At}x_0 + \int_0^t e^{A(t-\tau)}Bu(\tau)\,d\tau = e^{-at}x_0 + \int_0^t e^{-at}e^{a\tau}bu(\tau)\,d\tau = e^{-at}[x_0 + \int_0^t e^{a\tau}b \times \left(-\frac{2a}{b}\frac{1}{e^{2aT}-1}\cdot e^{a\tau}\right)d\tau = e^{-at}\left[1 - \frac{e^{2at}-1}{e^{2aT}-1}\right]x_0$. Note that $x(T) = 0$, as desired, and also that $x(0) = x_0$. The above expression shows also that for $t > T$, the state does not remain at the origin. An important point to notice here is that as $T \to 0$, the control magnitude $|u| \to \infty$. Thus, although it is (theoretically) possible to accomplish the desired transfer instantaneously, this will require infinite control magnitude. In general, the faster the transfer, the larger the control magnitude required. ■

We shall now establish the relationship between reachability and controllability for the continuous-time time-invariant systems (2.3).

Applying (2.8) to the time-invariant case, x_0 is controllable when there exists $u(t), t \in [0, T]$, so that

$$-e^{AT}x_0 = \int_0^T e^{A(T-\tau)}Bu(\tau)\,d\tau = L(u, 0, T),$$

or when $e^{AT}x_0 \in \mathcal{R}(W_r(0, T))$, or equivalently, in view of Lemma 2.1, when

$$e^{AT}x_0 \in \mathcal{R}(\mathscr{C}) \tag{2.39}$$

for some finite T. Recall that x_1 is reachable when

$$x_1 \in \mathcal{R}(\mathscr{C}). \tag{2.40}$$

We require the following technical result.

LEMMA 2.15. If $x \in \mathcal{R}(\mathscr{C})$, then $Ax \in \mathcal{R}(\mathscr{C})$; i.e., the reachable subspace $R_r = \mathcal{R}(\mathscr{C})$ is an *A-invariant* subspace.

Proof. If $x \in \mathcal{R}(\mathscr{C})$, this means that there exists a vector α such that $[B, AB, \ldots, A^{n-1}B]\alpha = x$. Then $Ax = [AB, A^2B, \ldots, A^nB]\alpha$. In view of the Cayley-Hamilton Theorem, A^n can be expressed as a linear combination of $A^{n-1}, \ldots, A, I$, which implies that $Ax = \mathscr{C}\beta$ for some appropriate vector β. Therefore, $Ax \in \mathcal{R}(\mathscr{C})$. ■

THEOREM 2.16. Consider the system $\dot{x} = Ax + Bu$.

(i) A state x is reachable if and only if it is controllable.
(ii) $R_c = R_r$.
(iii) The system (2.3), or the pair (A, B), is (completely state) reachable if and only if it is (completely state) controllable.

Proof. (i) Let x be reachable; that is, $x \in \mathcal{R}(\mathscr{C})$. Premultiply x by $e^{AT} = \sum_{k=0}^{\infty}(T^k/k!)A^k$ and notice that in view of Lemma 2.15, $Ax, A^2x, \ldots, A^kx \in \mathcal{R}(\mathscr{C})$. Therefore, $e^{AT}x \in \mathcal{R}(\mathscr{C})$ for any T, which, in view of (2.39), implies that x is also controllable. If now x is controllable, i.e., $e^{AT}x \in \mathcal{R}(\mathscr{C})$, then premultiplying by e^{-AT}, the vector $e^{-AT}(e^{AT}x) = x$ will also be in $\mathcal{R}(\mathscr{C})$. Therefore, x is also reachable. Note that the second part of (i), that controllability implies reachability, is true because the inverse $(e^{AT})^{-1} = e^{-AT}$ does exist. This is in contrast to the discrete-time case where the state transition matrix $\Phi(k, 0)$ is nonsingular if and only if A is nonsingular [nonreversibility of time in discrete-time systems (see Section 2.7)].

Parts (ii) and (iii) of the theorem follow directly from (i). ■

The reachability Gramian for the time-invariant case, $W_r(0, T)$, was defined in (2.29). Similarly, in view of Definition 2.8, we make the following definition.

DEFINITION 2.12. The *controllability Gramian* in the time-invariant case is the $n \times n$ matrix

$$W_c(0, T) \triangleq \int_0^T e^{-A\tau} BB^T e^{-A^T\tau}\, d\tau. \tag{2.41}$$

■

We note that

$$W_r(0, T) = e^{AT} W_c(0, T) e^{A^T T},$$

which can be verified directly [see also (2.22)].

As was done in the time-varying case above, we now introduce a number of additional tests for reachability and controllability of time-invariant systems. Some earlier results are also repeated here for convenience.

THEOREM 2.17. The system $\dot{x} = Ax + Bu$ is reachable (controllable-from-the-origin)

(i) if and only if

$$rank\, W_r(0, T) = n \qquad \text{for some finite } T > 0,$$

where
$$W_r(0, T) \triangleq \int_0^T e^{(T-\tau)A} BB^T e^{(T-\tau)A^T}\, d\tau, \tag{2.42}$$

the reachability Gramian, or

(ii) if and only if the n rows of

$$e^{At}B \tag{2.43}$$

are linearly independent on $[0, \infty)$ over the field of complex numbers, or alternatively, if and only if the n rows of

$$(sI - A)^{-1}B \tag{2.44}$$

are linearly independent over the field of complex numbers, or

(iii) if and only if

$$rank\, \mathscr{C} = n, \tag{2.45}$$

where $\mathscr{C} \triangleq [B, A, B, \ldots, A^{n-1}B]$, the controllability matrix, or

(iv) if and only if

$$rank\, [s_i I - A, B] = n \tag{2.46}$$

for all complex numbers s_i, or alternatively, for s_i, $i = 1, \ldots, n$, the eigenvalues of A.

Proof. Parts (i) and (iii) were proved in Corollary 2.9.

In part (ii), $rank\, W_r(0, T) = n$ implies and is implied by the linear independence of the n rows of $e^{(T-t)A}B$ on $[0, T]$ over the field of complex numbers, in view of Lemma 2.7, or by the linear independence of the n rows of $e^{\hat{t}A}B$, where $\hat{t} \triangleq T - t$, on $[0, T]$. Therefore the system is reachable if and only if the n rows of $e^{At}B$ are linearly independent on $[0, \infty)$ over the field of complex numbers. Note that the time interval can be taken to be $[0, \infty)$ since in $[0, T]$, T can be taken to be any finite positive real number. To prove the second part of (ii), recall that $\mathscr{L}(e^{At}B) = (sI - A)^{-1}B$ and that the Laplace transform is a one-to-one linear operator.

Part (iv) will be proved later in this chapter, in Corollary 4.6. ■

Results for controllability that are in the spirit of those given in Theorem 2.17 can also be established. The reader is asked to do so. This is of course not surprising since it was shown (in Theorem 2.16) that reachability implies and is implied by controllability. Therefore, the criteria developed in the theorem for reachability are typically used to test the controllability of a system.

EXAMPLE 2.7. For the system $\dot{x} = Ax + Bu$, where $A = \begin{bmatrix} 0 & 1 \\ 0 & 0 \end{bmatrix}$ and $B = \begin{bmatrix} 0 \\ 1 \end{bmatrix}$ (as in Example 2.4), we shall verify Theorem 2.17. The system is reachable since

(i) the reachability Gramian $W_r(0, T) = \begin{bmatrix} \frac{1}{3}T^3 & \frac{1}{2}T^2 \\ \frac{1}{2}T^2 & T \end{bmatrix}$ has $rank\, W_r(0, T) = 2 = n$ for any $T > 0$, or since

(ii) $e^{At}B = \begin{bmatrix} t \\ 1 \end{bmatrix}$ has rows that are linearly independent on $[0, \infty)$ over the field of complex numbers (since $a_1 \cdot t + a_2 \cdot 1 = 0$, where a_1 and a_2 are complex numbers implies that $a_1 = a_2 = 0$). Similarly, the rows of $(sI - A)^{-1}B = \begin{bmatrix} 1/s^2 \\ 1/s \end{bmatrix}$ are linearly independent over the field of complex numbers. Also, since

(iii) $rank\, \mathscr{C} = rank\, [B, AB] = rank \begin{bmatrix} 0 & 1 \\ 1 & 0 \end{bmatrix} = 2 = n$, or

(iv) $rank\, [s_iI - A, B] = rank \begin{bmatrix} s_i & -1 & 0 \\ 0 & s_i & 1 \end{bmatrix} = 2 = n$ for $s_i = 0, i = 1, 2$, the eigenvalues of A.

If $B = \begin{bmatrix} 1 \\ 0 \end{bmatrix}$ in place of $\begin{bmatrix} 0 \\ 1 \end{bmatrix}$, then

(i) $W_r(0, T) = \begin{bmatrix} T & 0 \\ 0 & 0 \end{bmatrix}$ (see Example 2.4) with $rank\, W_r(0, T) = 1 < 2 = n$, and

(ii) $e^{At}B = \begin{bmatrix} 1 \\ 0 \end{bmatrix}$ and $(sI - A)^{-1}B = \begin{bmatrix} 1/s \\ 0 \end{bmatrix}$, neither of which has rows that are linearly independent over the complex numbers. Also,

(iii) $rank\, \mathscr{C} = \begin{bmatrix} 1 & 0 \\ 0 & 0 \end{bmatrix} = 1 < 2 = n$, and

(iv) $rank\, [sI - A, B] = rank \begin{bmatrix} s_i & -1 & 1 \\ 0 & s_i & 0 \end{bmatrix} = 1 < 2 = n$ for $s_i = 0$.

Based on any of the above tests, it is concluded that the system is not reachable. ■

C. Discrete-Time Systems

The response of discrete-time systems was studied in Section 2.7 of Chapter 2. We consider systems described by equations of the form

$$x(k+1) = A(k)x(k) + B(k)u(k), \qquad k \geq k_0, \tag{2.47}$$

where $A(k) \in R^{n\times n}$, $B(k) \in R^{n\times m}$, and the input $u(k) \in R^m$ are defined for $k \geq k_0$. The state $x(k)$ for $k > k_0$ is given by

$$x(k) = \Phi(k, k_0)x(k_0) + \sum_{i=k_0}^{k-1} \Phi(k, i+1)B(i)u(i), \tag{2.48}$$

where the state transition matrix $\Phi(k, k_0)$ is given by $\Phi(k, k_0) = A(k-1) \times A(k-2)\cdots A(k_0)$ for $k > k_0$, and $\Phi(k_0, k_0) = I$.

In the time-invariant case we have

$$x(k+1) = Ax(k) + Bu(k), \qquad k \geq k_0, \tag{2.49}$$

where $A \in R^{n\times n}$ and $B \in R^{n\times m}$. The state $x(k)$ of (2.49) is given by (2.48) with

$$\Phi(k, k_0) = A^{k-k_0}, \qquad k \geq k_0. \tag{2.50}$$

Let the state at time k_0 be x_0. For the state at some time $k_1 > k_0$ to assume the value x_1, an input u must exist that satisfies

$$x_1 = \Phi(k_1, k_0)x_0 + \sum_{i=k_0}^{k_1-1} \Phi(k_1, i+1)B(i)u(i). \tag{2.51}$$

Reachability and controllability are defined for discrete-time systems in a completely analogous fashion as in the continuous-time case. The mathematical development, however, involves summations instead of integrals and is easier to deal with. The time-varying case can be developed in a manner similar to the time-invariant case. For this reason, the discrete-time time-varying case will not be developed presently. Instead, we shall concentrate on the time-invariant case. Note that some of the results given below for the discrete-time time-invariant case have already been presented in Section 3.1, Introduction.

Discrete-time time-invariant systems

For the time-invariant system $x(k+1) = Ax(k) + Bu(k)$ given in (2.49), the elapsed time $k_1 - k_0$ is of central interest, and we therefore take $k_0 = 0$ and $k_1 = K$. Recalling that $\Phi(k, 0) = A^k$, we rewrite (2.51) as

$$x_1 = A^K x_0 + \sum_{i=0}^{K-1} A^{K-(i+1)}Bu(i) \tag{2.52}$$

when $K > 0$, or

$$x_1 = A^K x_0 + \mathscr{C}_K U_K, \tag{2.53}$$

where
$$\mathscr{C}_K \triangleq [B, AB, \ldots, A^{K-1}B] \tag{2.54}$$

and
$$U_K \triangleq [u^T(K-1), u^T(K-2), \ldots, u^T(0)]^T. \tag{2.55}$$

The definitions of *reachable state* x_1, *reachable subspace* R_r, and a *system* being *(completely state) reachable*, or *the pair* (A, B) *being reachable*, are the same as in the continuous-time case (see Definition 2.9, and use integer K in place of real time T). Similarly, the definitions of *controllable state* x_0, *controllable subspace* R_c, and a *system* being (*completely state*) *controllable*, or *the pair* (A, B) *being controllable* are similar to the corresponding concepts given in Definition 2.10 for the case of continuous-time systems.

To determine the finite input sequence for discrete-time systems that will accomplish a desired state transfer, if such a sequence exists, one does not have to define matrices comparable to the reachability Gramian W_r, as in the case for continuous-time systems. In particular, we have the following result.

THEOREM 2.18. Consider the system $x(k+1) = Ax(k) + Bu(k)$ given in (2.39) and let $x(0) = 0$. There exists input u that transfers the state to x_1 in finite time if and only if

$$x_1 \in \mathcal{R}(\mathscr{C}).$$

In this case, x_1 is reachable and $R_r = \mathcal{R}(\mathscr{C})$. An appropriate input sequence $\{u(k)\}$, $k = 0, \ldots, n-1$, that accomplishes this transfer in n steps is determined by $U_n \triangleq [u^T(n-1), u^T(n-2), \ldots, u^T(0)]^T$, a solution to the equation

$$\mathscr{C}U_n = x_1. \tag{2.56}$$

Henceforth, with an abuse of language, we will refer to U_n as a control sequence when, in fact, we actually have in mind $\{u(k)\}$.

Proof. In view of (2.52), x_1 can be reached from the origin in K steps if and only if $x_1 = \mathscr{C}_K U_K$ has a solution U_K, or if and only if $x_1 \in \mathcal{R}(\mathscr{C}_K)$. Furthermore, all input sequences that accomplish this are solutions to the equation $x_1 = \mathscr{C}_K U_K$. For x_1 to be reachable we must have $x_1 \in \mathcal{R}(\mathscr{C}_K)$ for some finite K. This range, however, cannot increase beyond the range of $\mathscr{C}_n = \mathscr{C}$; i.e., $\mathcal{R}(\mathscr{C}_K) = \mathcal{R}(\mathscr{C}_n)$ for $K \geq n$. This follows from the Cayley-Hamilton Theorem, which implies that any vector x in $\mathcal{R}(\mathscr{C}_K)$, $K \geq n$, can be expressed as a linear combination of $B, AB, \ldots, A^{n-1}B$. Therefore, $x \in \mathcal{R}(\mathscr{C}_n)$. It is of course possible to have $x_1 \in \mathcal{R}(\mathscr{C}_K)$ with $K < n$ for a particular x_1; however, in this case $x_1 \in \mathcal{R}(\mathscr{C}_n)$ since $\mathscr{C}_K$ is a subset of $\mathscr{C}_n$. Thus, x_1 is reachable if and only if it is in the range of $\mathscr{C}_n = \mathscr{C}$. Clearly, any U_n that accomplishes the transfer satisfies (2.56). ■

As pointed out in the above proof, for given x_1 we may have $x_1 \in \mathcal{R}(\mathscr{C}_K)$ for some $K < n$. In this case the transfer can be accomplished in fewer than n steps, and appropriate inputs are obtained by solving the equation $\mathscr{C}_K U_K = x_1$.

COROLLARY 2.19. The system $x(k+1) = Ax(k) + Bu(k)$ given in (2.49) is (*completely state*) *reachable*, or the pair (A, B) is reachable, if and only if

$$\text{rank}\,\mathscr{C} = n. \tag{2.57}$$

Proof. Apply Theorem 2.18, noting that $\mathcal{R}(\mathscr{C}) = R_r = R^n$ if and only if rank $\mathscr{C} = n$. ■

THEOREM 2.20. There exists an input u that transfers the state of the system $x(k+1) = Ax(k) + Bu(k)$ given in (2.49) from x_0 to x_1 in some finite number of steps K, if and only if

$$x_1 - A^K x_0 \in \mathcal{R}(\mathscr{C}_K). \tag{2.58}$$

Such input sequence $U_K \triangleq [u^T(K-1), u^T(K-2), \ldots, u^T(0)]^T$ is determined by solving the equation

$$\mathscr{C}_K U_K = x_1 - A^K x_0. \tag{2.59}$$

Proof. The proof follows directly from (2.53). ■

The above theorem leads to the following result that establishes the importance of reachability in determining u to transfer the state from any x_0 to any x_1 in a finite number of steps.

COROLLARY 2.21. Let the system $x(k+1) = Ax(k)+Bu(k)$ given in (2.49) be (completely state) reachable, or the pair (A, B) be reachable. Then there exists an input sequence to transfer the state from any x_0 to any x_1 in a finite number of steps. Such input is determined by solving Eq. (2.60).

Proof. Consider (2.54). Since (A, B) is reachable, $rank\,\mathscr{C}_n = rank\,\mathscr{C} = n$ and $\mathscr{R}(\mathscr{C}) = R^n$. Then

$$\mathscr{C}U_n = x_1 - A^n x_0 \tag{2.60}$$

always has a solution $U_n = [u^T(n-1), \ldots, u^T(0)]^T$ for any x_0 and x_1. This input sequence transfers the state from x_0 to x_1 in n steps. ■

Note that, in view of Theorem 2.20, for a particular x_0 and x_1, the state transfer may be accomplished in $K < n$ steps, using (2.59).

EXAMPLE 2.8. Consider the system in Example 1.1, namely, $x(k+1) = Ax(k) + Bu(k)$, where $A = \begin{bmatrix} 0 & 1 \\ 1 & 1 \end{bmatrix}$ and $B = \begin{bmatrix} 0 \\ 1 \end{bmatrix}$. Since $rank\,\mathscr{C} = rank\,[B, AB] = rank \begin{bmatrix} 0 & 1 \\ 1 & 1 \end{bmatrix} = 2 = n$, the system is reachable, and any state x_0 can be transferred to any other state x_1 in 2 steps. Let $x_1 = \begin{bmatrix} a \\ b \end{bmatrix}$, $x_0 = \begin{bmatrix} a_0 \\ b_0 \end{bmatrix}$. Then (2.60) implies that $\begin{bmatrix} 0 & 1 \\ 1 & 1 \end{bmatrix}\begin{bmatrix} u(1) \\ u(0) \end{bmatrix} = \begin{bmatrix} a \\ b \end{bmatrix} - \begin{bmatrix} 1 & 1 \\ 1 & 2 \end{bmatrix}\begin{bmatrix} a_0 \\ b_0 \end{bmatrix}$ or $\begin{bmatrix} u(1) \\ u(0) \end{bmatrix} = \begin{bmatrix} -1 & 1 \\ 1 & 0 \end{bmatrix}\begin{bmatrix} a \\ b \end{bmatrix} - \begin{bmatrix} 0 & 1 \\ 1 & 1 \end{bmatrix}\begin{bmatrix} a_0 \\ b_0 \end{bmatrix} = \begin{bmatrix} b - 1 - b_0 \\ a - a_0 - b_0 \end{bmatrix}$. This agrees with the results obtained in Example 1.1. In view of (2.59), if x_1 and x_0 are chosen so that $x_1 - Ax_0 = \begin{bmatrix} a \\ b \end{bmatrix} - \begin{bmatrix} 0 & 1 \\ 1 & 1 \end{bmatrix}\begin{bmatrix} a_0 \\ b_0 \end{bmatrix} = \begin{bmatrix} a - b_0 \\ b - a_0 - b_0 \end{bmatrix}$ is in the $\mathscr{R}(\mathscr{C}_1) = \mathscr{R}(B) = span\left\{\begin{bmatrix} 0 \\ 1 \end{bmatrix}\right\}$, then the state transfer can be achieved in one step. For example, if $x_1 = \begin{bmatrix} 1 \\ 3 \end{bmatrix}$ and $x_0 = \begin{bmatrix} 0 \\ 1 \end{bmatrix}$, then $Bu(0) = \begin{bmatrix} 0 \\ 1 \end{bmatrix}u(0) = x_1 - Ax_0 = \begin{bmatrix} 0 \\ 2 \end{bmatrix}$ implies that the transfer from x_0 to x_1 can be accomplished in this case in $1 < 2 = n$ steps with $u(0) = 2$. ■

EXAMPLE 2.9. Consider the system $x(k+1) = Ax(k) + Bu(k)$ with $A = \begin{bmatrix} 0 & 1 \\ 0 & 0 \end{bmatrix}$ and $B = \begin{bmatrix} 0 \\ 1 \end{bmatrix}$. Since $\mathscr{C} = [B, AB] = \begin{bmatrix} 0 & 1 \\ 1 & 0 \end{bmatrix}$ has full rank, there exists an input sequence that will transfer the state from any $x(0) = x_0$ to any $x(n) = x_1$ (in n steps), given by (2.60), $U_2 = \begin{bmatrix} u(1) \\ u(0) \end{bmatrix} = \mathscr{C}^{-1}(x_1 - A^2 x_0) = \begin{bmatrix} 0 & 1 \\ 1 & 0 \end{bmatrix}(x_1 - x_0)$. Compare this with Example 2.5, where the continuous-time system had the same system parameters A and B. ■

We shall now establish the relationship between reachability and controllability for the discrete-time time-invariant systems $x(k+1) = Ax(k) + Bu(k)$ given in (2.49).

Consider (2.51). The state x_0 is controllable if it can be steered to the origin $x_1 = 0$ in a finite number of steps K. That is, x_0 is controllable if and only if

$$-A^K x_0 = \mathscr{C}_K U_K \tag{2.61}$$

for some K, or when

$$A^K x_0 \in \mathcal{R}(\mathscr{C}_K) \tag{2.62}$$

for some K. Recall that x_1 is reachable when

$$x_1 \in \mathcal{R}(\mathscr{C}). \tag{2.63}$$

THEOREM 2.22. Consider the system $x(k+1) = Ax(k) + Bu(k)$ given in (2.49).
(i) If state x is reachable, then it is controllable.
(ii) $R_r \subset R_c$.
(iii) If the system is (completely state) reachable, or the pair (A, B) is reachable, then the system is also (completely state) controllable, or the pair (A, B) is controllable.

Furthermore, if A is nonsingular, then relations (i) and (iii) become if and only if statements, since controllability also implies reachability, and relation (ii) becomes an equality, i.e., $R_c = R_r$.

Proof. (i) If x is reachable, then $x \in \mathcal{R}(\mathscr{C})$. In view of Lemma 2.14, $\mathcal{R}(\mathscr{C})$ is an A-invariant subspace and so $A^n x \in \mathcal{R}(\mathscr{C})$, which in view of (2.61) implies that x is also controllable. Since x is an arbitrary vector in R_r, this implies (ii). If $\mathcal{R}(\mathscr{C}) = R^n$, the whole state space, then $A^n x$ for any x is in $\mathcal{R}(\mathscr{C})$ and so any vector x is also controllable. Thus, reachability implies controllability. Now, if A is nonsingular, then A^{-n} exists. If x is controllable, i.e., $A^n x \in \mathcal{R}(\mathscr{C})$, then $x \in \mathcal{R}(\mathscr{C})$, i.e., x is also reachable. This can be seen by noting that A^{-n} can be written as a power series in terms of A, which in view of Lemma 2.15, implies that $A^{-n}(A^n x) = x$ is also in $\mathcal{R}(\mathscr{C})$. ■

Matrix A being nonsingular is the necessary and sufficient condition for the state transition matrix $\Phi(k, k_0)$ to be nonsingular (see Section 2.8), which in turn is the condition for *"time reversibility"* in discrete-time systems. Recall that reversibility in time may not be present in such systems since $\Phi(k, k_0)$ may be singular. In contrast to this, in continuous-time systems, $\Phi(t, t_0)$ is always nonsingular. This causes differences in behavior between continuous- and discrete-time systems and implies that in discrete-time systems controllability may not imply reachability (see Theorem 2.22). Note that, in view of Theorem 2.16, in the case of continuous-time systems, it is not only reachability which always implies controllability, but also vice versa, controllability always implies reachability.

In the following, we introduce the discrete-time reachability and controllability Gramians for system (2.49). These are defined in a manner analogous to the continuous-time case.

DEFINITION 2.13. The *reachability Gramian* is defined by

$$W_r(0, K) = \sum_{i=0}^{K-1} A^{K-(i+1)} BB^T (A^T)^{K-(i+1)}. \tag{2.64}$$

It is not difficult to verify that $W_r(0, K) = \sum_{i=0}^{K-1} A^i BB^T (A^T)^i = \mathscr{C}_K \mathscr{C}_K^T$. ■

LEMMA 2.23. $\mathcal{R}(\mathscr{C}) = \mathcal{R}(W_r(0, K))$ for every $K \geq n$.

Proof. This result can be established in a way similar to the proof of the corresponding result in the continuous-time case (Lemma 2.7). The details are left to the reader. ■

When a system is reachable, the input sequence that transfers x_0 at $k = 0$ to x_1 at $k = K$ can be determined in terms of the reachability Gramian. In particular, let

$rank\,\mathscr{C} = n = rank\, W_r(0, K)$ for $K \geq n$, and notice that the relation

$$U_K = \mathscr{C}_K^T W_r^{-1}(0, K)(x_1 - A^K x_0) \tag{2.65}$$

satisfies (2.59) since $W_r(0, K) = \mathscr{C}_K \mathscr{C}_K^T$.

DEFINITION 2.14. The *controllability Gramian* is defined as

$$W_c(0, K) = \sum_{i=0}^{K-1} A^{-(i+1)} BB^T (A^T)^{-(i+1)}. \tag{2.66}$$

■

We note that $W_c(0, K)$ is well defined only when A is nonsingular. The reachability and controllability Gramians are related by

$$W_r(0, K) = A^K W_c(0, K)(A^T)^K, \tag{2.67}$$

as can easily be verified.

When A is nonsingular, the input that will transfer the state from x_0 at $k = 0$ to $x_1 = 0$ in n steps can be determined using (2.60). In particular, one needs to solve

$$[A^{-n}\mathscr{C}]U_n = [A^{-n}B, \ldots, A^{-1}B]U_n = x_0 \tag{2.68}$$

for $U_n = [u^T(n-1), \ldots, u^T(0)]^T$. Note that x_0 is controllable if and only if $-A^n x_0 \in \mathscr{R}(\mathscr{C})$, or if and only if $x_0 \in \mathscr{R}(A^{-n}\mathscr{C})$ for A nonsingular.

Clearly, in the case of controllability (and under the assumption that A is nonsingular), the matrix $A^{-n}\mathscr{C}$ is of interest, instead of $\mathscr{C}$ [see also (1.18)]. In particular, a system is controllable if and only if $rank\,(A^{-n}\mathscr{C}) = rank\,\mathscr{C} = n$.

EXAMPLE 2.10. Consider the system $x(k+1) = Ax(k) + Bu(k)$, where $A = \begin{bmatrix} 1 & 1 \\ 0 & 1 \end{bmatrix}$ and $B = \begin{bmatrix} 1 \\ 0 \end{bmatrix}$. Since $rank\,\mathscr{C} = rank\,[B, AB] = rank \begin{bmatrix} 1 & 1 \\ 0 & 0 \end{bmatrix} = 1 < 2 = n$, this system is not (completely) reachable (controllable-from-the-origin). All reachable states are of the form $\alpha \begin{bmatrix} 1 \\ 0 \end{bmatrix}$, where $\alpha \in R$ since $\left\{\begin{bmatrix} 1 \\ 0 \end{bmatrix}\right\}$ is a basis for the $\mathscr{R}(\mathscr{C}) = R_r$, the reachability subspace. The reachability Gramian for $K = n = 2$ is $W_r(0, 2) = BB^T + (AB)(AB)^T = \begin{bmatrix} 1 & 0 \\ 0 & 0 \end{bmatrix} + \begin{bmatrix} 1 & 0 \\ 0 & 0 \end{bmatrix} = \begin{bmatrix} 2 & 0 \\ 0 & 0 \end{bmatrix}$. Note that a basis for $\mathscr{R}(W_r(0, 2))$, is $\left\{\begin{bmatrix} 1 \\ 0 \end{bmatrix}\right\}$ and $\mathscr{R}(\mathscr{C}) = \mathscr{R}(W_r(0, 2))$, which verifies Lemma 2.23.

In view of (2.62) and the Cayley-Hamilton Theorem, all controllable states x_0 satisfy $A^2 x_0 \in \mathscr{R}(\mathscr{C})$; i.e., all controllable states are of the form $\alpha \begin{bmatrix} 1 \\ 0 \end{bmatrix}$, where $\alpha \in R$. This verifies Theorem 2.22 for the case when A is nonsingular. Note that presently $R_r = R_c$.

■

EXAMPLE 2.11. Consider the system $x(k+1) = Ax(k) + Bu(k)$, where $A = \begin{bmatrix} 0 & 1 \\ 0 & 0 \end{bmatrix}$ and $B = \begin{bmatrix} 1 \\ 0 \end{bmatrix}$. Since $rank\,\mathscr{C} = rank\,[B, AB] = rank \begin{bmatrix} 1 & 0 \\ 0 & 0 \end{bmatrix} = 1 < 2 = n$, the system is not (completely) reachable. All reachable states are of the form $\alpha \begin{bmatrix} 1 \\ 0 \end{bmatrix}$, where $\alpha \in R$ since $\left\{\begin{bmatrix} 1 \\ 0 \end{bmatrix}\right\}$ is a basis for $\mathscr{R}(\mathscr{C}) = R_r$, the reachability subspace.

To determine the controllable subspace R_c, consider (2.62) for $K = n$, in view of the Cayley-Hamilton Theorem. Note that $A^{-1}\mathscr{C}$ cannot be used in the present case, since A is singular. Since $A^2 x_0 = \begin{bmatrix} 0 & 0 \\ 0 & 0 \end{bmatrix} x_0 = \begin{bmatrix} 0 \\ 0 \end{bmatrix} \in \mathscr{R}(\mathscr{C})$, any state x_0 will be a controllable state, i.e., the system is (completely) controllable and $R_c = R^n$. This verifies Theorem 2.22 and illustrates that controllability does not in general imply reachability.

Note that (2.60) can be used to determine the control sequence that will drive any state x_0 to the origin ($x_1 = 0$). In particular,

$$\mathscr{C} U_n = \begin{bmatrix} 1 & 0 \\ 0 & 0 \end{bmatrix} \begin{bmatrix} u(1) \\ u(0) \end{bmatrix} = \begin{bmatrix} 0 \\ 0 \end{bmatrix} = -A^2 x_0.$$

Therefore, $u(0) = \alpha$ and $u(1) = 0$, where $\alpha \in R$ will drive any state to the origin. To verify this, we consider $x(1) = Ax(0) + Bu(0) = \begin{bmatrix} 0 & 1 \\ 0 & 0 \end{bmatrix} \begin{bmatrix} x_{01} \\ x_{02} \end{bmatrix} + \begin{bmatrix} 1 \\ 0 \end{bmatrix} \alpha = \begin{bmatrix} x_{02} + \alpha \\ 0 \end{bmatrix}$ and $x(2) = Ax(1) + Bu(1) = \begin{bmatrix} 0 & 1 \\ 0 & 0 \end{bmatrix} \begin{bmatrix} x_{02} + \alpha \\ 0 \end{bmatrix} + \begin{bmatrix} 1 \\ 0 \end{bmatrix} 0 = \begin{bmatrix} 0 \\ 0 \end{bmatrix}$. ■

3.3 OBSERVABILITY AND CONSTRUCTIBILITY

In applications, the state of a system is frequently required but not accessible. Under such conditions, the question arises whether it is possible to determine the state by observing the response of a system to some input over some period of time. It turns out that the answer to this question is affirmative if the system is observable. *Observability* refers to the ability of determining the present state $x(t_0)$ from knowledge of future system outputs, $y(t)$, and system inputs, $u(t), t \geq t_0$. *Constructibility* refers to the ability of determining the present state $x(t_0)$ from knowledge of past system outputs, $y(t)$, and system inputs, $u(t), t \leq t_0$. Observability was briefly addressed in Section 3.1. In this section this concept is formally defined and the (present) state is explicitly determined from input and output measurements. As in Section 3.2 (dealing with reachability and controllability), the reader can concentrate on the time-invariant case, starting with Definition 3.9, if so desired, and omit the more general material (dealing with time-varying systems) that starts with Definition 3.1.

Section description

In this section, observability and constructibility are introduced and discussed in detail for given linear system state-space descriptions. This is accomplished for continuous and discrete-time systems and for both time-varying and time-invariant cases.

Observability in continuous-time systems is addressed first with introduction of the observability Gramian $W_o(t_0, t_1)$ (Definition 3.4). It is shown (Corollary 3.2) that a system is observable at t_0 if $W_o(t_0, t_1)$ has full rank for some $t_1 \geq t_0$, and furthermore, if the system is observable, how an initial state can be determined. Observability refers to the ability to determine the current state of a system from future system outputs (and inputs). Constructibility, which refers to the ability of determining the current state of a system from past system outputs (and inputs), is addressed next with the introduction of the constructibility Gramian (Definition 3.8). It is shown

(Theorem 3.3) that a continuous-time system is observable if and only if it is constructible. Next, additional tests for observability and constructibility are obtained (Theorem 3.4). All these results are then applied in the study of the continuous-time time-invariant case. The material is arranged in such a manner that the continuous-time time-invariant systems can be studied independently of the time-varying case. The relation between the observability Gramian $W_o(0, T)$ and the observability matrix $\mathbb{O} = [C^T, (CA)^T, \ldots, (CA^{n-1})^T]^T$ is established next (Lemma 3.6). It is then shown (Corollary 3.8) that a system is observable if and only if $\mathbb{O}$ has full rank n. Next, the relation between observability and constructibility is established (Theorem 3.9). Finally, additional tests for observability are derived (Theorem 3.10).

A discussion of observability and constructibility for discrete-time systems with particular emphasis on time-invariant systems is presented next. It is shown that a system is observable if the observability matrix $\mathbb{O}$ has full rank (Corollary 3.12), and for this case, an expression for the initial state x_0 is given as a function of future outputs (and inputs). A similar result involving the observability Gramian $W_o(0, K)$ in place of $\mathbb{O}$ is also established (Corollary 3.14). Next, it is shown that observability in discrete-time systems always implies constructibility. In contrast to the continuous-time case, the converse of the above statement is generally not true. This is due to the lack of time reversibility in difference equations. When A is nonsingular, then constructibility also implies observability. Finally, the constructibility Gramian $W_{cn}(0, K)$ is also introduced (Definition 3.16).

A. Continuous-Time Time-Varying Systems

We consider systems described by equations of the form

$$\dot{x} = A(t)x + B(t)u, \qquad y = C(t)x + D(t)u, \tag{3.1}$$

where $A(t) \in R^{n\times n}$, $B(t) \in R^{n\times m}$, $C(t) \in R^{p\times n}$, $D(t) \in R^{p\times m}$, and $u(t) \in R^m$ are defined and (piecewise) continuous on some real open interval (a, b). It was shown in Chapter 2 that the output $y(t)$ is given by

$$y(t) = C(t)\Phi(t, t_0)x(t_0) + \int_{t_0}^{t} C(t)\Phi(t, \tau)B(\tau)u(\tau)\, d\tau + D(t)u(t) \tag{3.2}$$

for $t_0, t \in (a, b)$, where $\Phi(t, \tau)$ denotes the state transition matrix. This can be written as

$$\tilde{y}(t) = C(t)\Phi(t, t_0)x_0, \tag{3.3}$$

where $\tilde{y}(t) \triangleq y(t) - \left[\int_{t_0}^{t} C(t)\Phi(t, \tau)B(\tau)u(\tau)\, d\tau + D(t)u(t)\right]$ and $x_0 = x(t_0)$. We will find it convenient to first give the definition of an unobservable state.

DEFINITION 3.1. A *state* x is *unobservable* at time t_0 if the zero-input response of the system is zero for every $t \geq t_0$, i.e., if

$$C(t)\Phi(t, t_0)x = 0 \qquad \text{for every } t \geq t_0. \tag{3.4}$$

■

DEFINITION 3.2. The *unobservable at* t_0 *subspace* $R_{\bar{o}}^{t_0}$ of (3.1) is

$$R_{\bar{o}}^{t_0} \triangleq \{\text{ set of all unobservable at } t_0 \text{ states } x\}.$$

■

When the context is clear and there is no ambiguity, we will write $R_{\bar{o}}$ in place of $R_{\bar{o}}^{t_0}$.

DEFINITION 3.3. The *system* (3.1) is (*completely state*) *observable* at t_0, or the pair $(A(t), C(t))$ is observable at t_0, if the only state $x \in R^n$ that is unobservable at t_0 is the zero state, $x = 0$, i.e., if $R_{\bar{o}}^{t_0} = \{0\}$. ■

We will show later that observability depends only on the pair $(A(t), C(t))$. According to Definition 3.1, a nonzero unobservable state x cannot be distinguished from the zero state if only the (future) outputs are known; that is, an unobservable state cannot be determined uniquely from knowledge of the inputs and outputs of the system. This can be seen from (3.3), where for the unobservable states x at t_0, $\tilde{y}(t) = 0$ for $t \geq t_0$. This implies that the states x, which together with the input $u(t)$ produce the output $y(t)$, cannot be distinguished from the zero state, since they both produce the same output.

In a manner analogous to the development of reachability in Section 3.2, we make the following definition.

DEFINITION 3.4. The *observability Gramian* of the system (3.1) is the $n \times n$ matrix

$$W_o(t_0, t_1) \triangleq \int_{t_0}^{t_1} \Phi^T(\tau, t_0) C^T(\tau) C(\tau) \Phi(\tau, t_0)\, d\tau. \tag{3.5}$$

■

Note that W_o is symmetric and positive semidefinite for every $t_1 > t_0$, i.e., $W_o = W_o^T$ and $W_o \geq 0$ (show this).

THEOREM 3.1. A state x is unobservable at t_0 if and only if

$$x \in \mathcal{N}(W_o(t_0, t_1)) \tag{3.6}$$

for every $t_1 \geq t_0$, where $\mathcal{N}(\cdot)$ denotes the null space of a map.

Proof. If x is unobservable, then (3.4) is satisfied. Postmultiply (3.5) by x to obtain $W_o(t_0, t_1)x = 0$ for every $t_1 \geq t_0$, i.e., $x \in \mathcal{N}(W_o(t_0, t_1))$ for every $t_1 \geq t_0$. Conversely, let x be in the null space of W_o. Then $x^T W_o(t_0, t_1) x = \int_{t_0}^{t_1} \|C(\tau)\Phi(\tau, t_0)x\|^2\, d\tau = 0$ for every $t_1 \geq t_0$. This implies that (3.4) is true, or that the state x is unobservable. ■

EXAMPLE 3.1. Consider the system $\dot{x} = A(t)x$, $y = C(t)x$, where $A(t) = \begin{bmatrix} -1 & e^{2t} \\ 0 & -1 \end{bmatrix}$ and $C(t) = [0, e^{-t}]$. The state transition matrix in this case is (see Example 2.1 in this chapter)

$$\Phi(t, \tau) = \begin{bmatrix} e^{-(t-\tau)} & \frac{1}{2}[e^{t+\tau} - e^{-t+3\tau}] \\ 0 & e^{-(t-\tau)} \end{bmatrix}.$$

Then $C(\tau)\Phi(\tau, t_0) = [0, e^{-2\tau+t_0}]$ and the observability Gramian is given by

$$W_o(t_0, t_1) = \int_{t_0}^{t_1} e^{2t_0} \begin{bmatrix} 0 \\ e^{-2\tau} \end{bmatrix} [0, e^{-2\tau}]\, d\tau$$

$$= e^{2t_0} \int_{t_0}^{t_1} \begin{bmatrix} 0 & 0 \\ 0 & e^{-4\tau} \end{bmatrix} d\tau = -\frac{1}{4} e^{2t_0} \begin{bmatrix} 0 & 0 \\ 0 & e^{-4t_1} - e^{-4t_0} \end{bmatrix}.$$

It is clear that this system is not observable, since *rank* $W_o(t_0, t_1) = 1 < 2 = n$. In view of Theorem 3.1, all unobservable states are given by $\begin{bmatrix} \alpha \\ 0 \end{bmatrix}$, where $\alpha \in R$.

Notice that $y(t) = C(t)\Phi(t, t_0)x_0 = [0, e^{-2t+t_0}]x_0 = 0$ for $x_0 = \begin{bmatrix} \alpha \\ 0 \end{bmatrix}$, that is, none of the (unobservable) states $\begin{bmatrix} \alpha \\ 0 \end{bmatrix}$ can be distinguished from the zero state. ■

Clearly, x is observable at t_0 if and only if there exists a $t_1 > t_0$ such that $W_o(t_0, t_1)x \neq 0$.

COROLLARY 3.2. The system (3.1) is (*completely state*) *observable* at t_0, or the pair $(A(t), C(t))$ is observable at t_0, if and only if there exists a finite $t_1 > t_0$ such that

$$rank\, W_o(t_0, t_1) = n. \tag{3.7}$$

If the system is observable, the state x_0 at t_0 is given by

$$x_0 = W_o^{-1}(t_0, t_1)\left[\int_{t_0}^{t_1} \Phi^T(\tau, t_0)C^T(\tau)\tilde{y}(\tau)\, d\tau\right]. \tag{3.8}$$

Proof. The system is observable if and only if the only vector x that satisfies (3.6) is the zero vector. This is true if and only if there exists (at least one) finite time t_1 for which the null space of $W_o(t_0, t_1)$ contains only the zero vector, or if and only if (3.7) is satisfied. To determine the state x_0 at t_0, given the output and input values over $[t_0, t_1]$, premultiply (3.3) by $\Phi^T(t, t_0)C^T(t)$ and integrate over $[t_0, t_1]$. Then, in view of (3.5),

$$W_o(t_0, t_1)x_0 = \int_{t_0}^{t_1} \Phi^T(\tau, t_0)C^T(\tau)\tilde{y}(\tau)\, d\tau. \tag{3.9}$$

When the system is observable, (3.9) has the unique solution (3.8). ■

It is clear that if the state at some time t_0 is found, then the state $x(t)$ for $t \geq t_0$ is easily determined, given $u(t), t \geq t_0$, via the variation of constants formula (2.2).

We mention here that alternative methods to (3.8) to determine the state of a system when the system is observable are given in the next chapter (in Section 4.3 on state estimation).

EXAMPLE 3.2 For the scalar system $\dot{x} = a(t)x$, $y = c(t)x$, where $a(t) = -1$ and $c(t) = e^t$, we have $\Phi(t, \tau) = e^{-(t-\tau)}$ and $c(\tau)\Phi(\tau, t_0) = e^{\tau}e^{-(\tau-t_0)} = e^{t_0}$. The observability Gramian in this case is $W_o(t_0, t_1) = \int_{t_0}^{t_1} e^{2t_0}\, d\tau = e^{2t_0}(t_1 - t_0)$, which implies that the system is observable at t_0 since $rank\, W_o(t_0, t_1) = 1 = n$ for any $t_1 \neq t_0$. Suppose now that the observed output is $\tilde{y}(t) = y(t) = \alpha e^{t_0}$ for $t \geq t_0$. Then x_0 can be determined using (3.8), $x_0 = [e^{-2t_0}/(t_1 - t_0)][\int_{t_0}^{t_1} e^t(\alpha e^t)\, d\tau] = \alpha$. Indeed, $y(t) = c(t)\Phi(t, t_0)\alpha = \alpha e^{t_0}$, as observed. ■

Observability utilizes future output measurements to determine the present state. In contructibility, past output measurements are used to accomplish this. Constructibility is defined below and its relation to observability is established.

DEFINITION 3.5. A *state* x is *unconstructible* at time t_1 if for every finite time $t \leq t_1$, the zero-input response of the system is zero for all t, i.e.,

$$C(t)\Phi(t, t_1)x = 0 \qquad \text{for every } t \leq t_1. \tag{3.10}$$

■

DEFINITION 3.6. The *unconstructible at* t_1 *subspace* $\mathcal{R}_{\overline{cn}}^{t_1}$ of (3.1) is

$$\mathcal{R}_{\overline{cn}}^{t_1} \triangleq \{\text{set of all states } x \text{ unconstructible at } t_1\}.$$

■

It is denoted in the following by $R_{\overline{cn}}$, for convenience, when there is no ambiguity.

DEFINITION 3.7. The system (3.1) is (*completely state*) *constructible* at t_1, or the pair $(A(t), C(t))$ is *constructible* at t_1, if the only state $x \in R^n$ that is unconstructible at t_0 is $x = 0$, i.e., if $R_{\overline{cn}}^{t_1} = \{0\}$. ■

THEOREM 3.3. If the system (3.1), or the pair $(A(t), C(t))$, is observable at t_0, then it is constructible at some $t_1 \geq t_0$; if it is constructible at t_1, then it is observable at some $t_0 \leq t_1$.

It was shown in Corollary 3.2 that the system is observable at t_0 if and only if *rank* $W_0(t_0, t_1) = n$ for some $t_1 > t_0$. Similar results can be established for constructibility. We will require the following concept.

DEFINITION 3.8. The *constructibility Gramian* of (3.1) is the $n \times n$ matrix

$$W_{cn}(t_0, t_1) \triangleq \int_{t_0}^{t_1} \Phi^T(\tau, t_1) C^T(\tau) C(\tau) \Phi(\tau, t_1)\, d\tau. \tag{3.11}$$

Proof of Theorem 3.3. A similar result as Theorem 3.1, but for unconstructible state x, can be derived. Next, using a proof similar to the proof of Corollary 3.2, it can be shown that the system is constructible at t_1 if and only if there exists finite $t_0 < t_1$ such that

$$rank\, W_{cn}(t_0, t_1) = n. \tag{3.12}$$

Note that

$$W_o(t_0, t_1) = \Phi^T(t_1, t_0) W_{cn}(t_0, t_1) \Phi(t_1, t_0), \tag{3.13}$$

which implies that *rank* $W_o(t_0, t_1) =$ *rank* $W_{cn}(t_0, t_1)$ for every t_0 and t_1. Note that $\Phi(t_1, t_0)$ is nonsingular for every t_0 and t_1. ■

EXAMPLE 3.3. (i) Consider the system $\dot{x} = A(t)x$, $y = C(t)x$ of Example 3.1. The constructibility Gramian is

$$\begin{aligned} W_{cn}(t_0, t_1) &= \int_{t_0}^{t_1} \begin{bmatrix} 0 \\ e^{-2\tau + t_1} \end{bmatrix} [0, e^{-2\tau + t_1}]\, d\tau \\ &= e^{2t_1} \int_{t_0}^{t_1} \begin{bmatrix} 0 & 0 \\ 0 & e^{-4\tau} \end{bmatrix} d\tau = -\frac{1}{4} e^{2t_1} \begin{bmatrix} 0 & 0 \\ 0 & e^{-4t_1} - e^{-4t_0} \end{bmatrix}. \end{aligned}$$

Compare this with the observability Gramian of Example 3.1 and note that

$$\Phi^T(t_1, t_0) W_{cn}(t_0, t_1)\, \Phi(t_1, t_0) = -\frac{1}{4} e^{2t_0} \begin{bmatrix} 0 & 0 \\ 0 & e^{-4t_1} - e^{-4t_0} \end{bmatrix} = W_o(t_0, t_1),$$

as expected [see (3.13)]. Presently,

$$C(t)\Phi(t, t_1)x = [0, e^{-2t + t_1}] \begin{bmatrix} x_1 \\ x_2 \end{bmatrix} = 0$$

for every $t \leq t_1$ implies, in view of Definition 3.5, that all unreconstructible states (at t_1) are of the form $\begin{bmatrix} \alpha \\ 0 \end{bmatrix}$, $\alpha \in R$. Note that they are identical to the unobservable states (see Example 3.1).

(ii) For the scalar system $\dot{x} = -x$, $y = e^t$ of Example 3.2, the constructibility Gramian is $W_{cn}(t_0, t_1) = e^{2t_1}(t_1 - t_0)$ and $\Phi^T(t_1, t_0) W_{cn}(t_0, t_1) \Phi(t_1, t_0) = e^{-(t_1 - t_0)} e^{2t_1} \times (t_1 - t_0) e^{-(t_1 - t_0)} = e^{2t_0}(t_1 - t_0) = W_o(t_0, t_1)$, as expected in view of (3.13). ■

We shall now use Lemma 2.7 in Section 3.2 to develop additional tests for observability and constructibility. These are analogous to corresponding results that we established for reachability and controllability (Theorems 2.8 and 2.9).

THEOREM 3.4. The system $\dot{x} = A(t)x+B(t)u$, $y = C(t)x+D(t)u$ is (completely state) observable at t_0

(i) if and only if there exists a finite $t_1 > t_0$ such that

$$rank\ W_o(t_0, t_1) = n, \tag{3.14}$$

where $W_o(t_0, t_1) \triangleq \int_{t_0}^{t_1} \Phi^T(\tau, t_0)C^T(\tau)C(\tau)\Phi(\tau, t_0)\,d\tau$ is the observability Gramian, or equivalently,

(ii) if and only if there exists finite $t_1 > t_0$ such that the n columns of

$$C(t)\Phi(t, t_0) \tag{3.15}$$

are linearly independent on $[t_0, t_1]$ over the field of complex numbers.

Proof. Part (i) was shown in Corollary 3.2 and part (ii) is a consequence of Lemma 2.7 (compare with the corresponding Theorem 2.8 for reachability). ■

Similar results can be derived for constructibility. In particular, we have the following result.

THEOREM 3.5. The system $\dot{x} = A(t)x + B(t)u$, $y = C(t)x + D(t)u$ is (completely state) constructible at t_1

(i) if and only if there exists finite $t_0 < t_1$ such that

$$rank\ W_{cn}(t_0, t_1) = n, \tag{3.16}$$

where $W_{cn}(t_0, t_1) \triangleq \int_{t_0}^{t_1} \Phi^T(\tau, t_1)C^T(\tau)C(\tau)\Phi(\tau, t_1)\,d\tau$, the constructibility Gramian, or equivalently,

(ii) if and only if there exists finite $t_0 < t_1$, such that the n columns of

$$C(t)\Phi(t, t_1) \tag{3.17}$$

are linearly independent on $[t_0, t_1]$ over the field of complex numbers.

Proof. The proof is analogous to the proof of the corresponding results on observability. ■

Note that postmultiplication of $C(t)\Phi(t, t_1)$ by the nonsingular matrix $\Phi(t_1, t_0)$ yields $C(t)\Phi(t, t_0)$ in (ii) of Theorem 3.4 [compare with (3.13)]. This shows again the result given in Theorem 3.3 that observability implies and is implied by constructibility, in the case of continuous-time systems.

B. Continuous-Time Time-Invariant Systems

We shall now study observability and constructibility for time-invariant systems described by equations of the form

$$\dot{x} = Ax + Bu, \qquad y = Cx + Du, \tag{3.18}$$

where $A \in R^{n\times n}$, $B \in R^{n\times m}$, $C \in R^{p\times n}$, $D \in R^{p\times m}$, and $u(t) \in R^m$ is (piecewise) continuous. As was shown in Chapter 2, the output of this system is given by

$$y(t) = Ce^{At}x(0) + \int_0^t Ce^{A(t-\tau)}Bu(\tau)\,d\tau + Du(t). \tag{3.19}$$

We recall once more that in the present case $\Phi(t, \tau) = \Phi(t - \tau, 0) = \exp[A(t - \tau)]$ and that initial time can always be taken to be $t_0 = 0$. We will find it convenient to rewrite (3.19) as

$$\tilde{y}(t) = Ce^{At}x_0, \tag{3.20}$$

where $\tilde{y}(t) \triangleq y(t) - \left[\int_0^t Ce^{A(t-\tau)}Bu(\tau)\,d\tau + Du(t)\right]$ and $x_0 = x(0)$.

DEFINITION 3.9. A *state* x is *unobservable* if the zero-input response of the system (3.18) is zero for every $t \geq 0$, i.e., if

$$Ce^{At}x = 0 \qquad \text{for every } t \geq 0. \tag{3.21}$$

The set of all unobservable states x, $R_{\bar{o}}$, is called the *unobservable subspace* of (3.18). *System* (3.18) is (*completely state*) *observable*, or the pair (A, C) is observable, if the only state $x \in R^n$ that is unobservable is $x = 0$, i.e., if $R_{\bar{o}} = \{0\}$.

Definition 3.9 states that a state is unobservable precisely when it cannot be distinguished as an initial condition at time 0 from the initial condition $x(0) = 0$. This is because in this case the output is the same as if the initial condition were the zero vector.

DEFINITION 3.10. The *observability Gramian* of system (3.18) is the $n \times n$ matrix

$$W_o(0, T) \triangleq \int_0^T e^{A^T\tau}C^TCe^{A\tau}\,d\tau. \tag{3.22}$$

■

We note that W_o is symmetric and positive semidefinite for every $T > 0$, i.e., $W_o = W_o^T$ and $W_o \geq 0$ (show this). Recall that the $pn \times n$ *observability matrix*

$$\mathbb{O} \triangleq \begin{bmatrix} C \\ CA \\ \vdots \\ CA^{n-1} \end{bmatrix} \tag{3.23}$$

was defined in Section 3.1.

We now show that the null space of $W_o(0, T)$, denoted by $\mathcal{N}(W_o(0, T))$, is independent of T, i.e., it is the same for any $T > 0$, and in particular, it is equal to the null space of the observability matrix $\mathbb{O}$. Thus, the unobservable subspace $R_{\bar{o}}$ of the system is given by the null space of $\mathbb{O}$, $\mathcal{N}(\mathbb{O})$, or the null space of $W_o(0, T)$, $\mathcal{N}(W_o(0, T))$ for some finite (and therefore for all) $T > 0$.

LEMMA 3.6. $\mathcal{N}(\mathbb{O}) = \mathcal{N}(W_o(0, T))$ for every $T > 0$.

Proof. If $x \in \mathcal{N}(\mathbb{O})$, then $\mathbb{O}x = 0$. Thus, $CA^kx = 0$ for all $0 \leq k \leq n-1$, which is also true for every $k > n-1$, in view of the Cayley-Hamilton Theorem. Then $Ce^{At}x = C[\Sigma_{k=0}^{\infty}(t^k/k!)A^i]x = 0$ for every finite t. Therefore, in view of (3.18), $W_o(0, T)x = 0$ for every $T > 0$, i.e., $x \in \mathcal{N}(W_o(0, T))$ for every $T > 0$. Now let $x \in \mathcal{N}(W_o(0, T))$ for some $T > 0$, so that $x^T W(0, T)x = \int_0^T \| Ce^{A\tau}x \|^2\, d\tau = 0$, or $Ce^{At}x = 0$ for every $t \in [0, T]$. Taking derivatives of the last equation with respect to t and evaluating at $t = 0$, we obtain $Cx = CAx = \cdots = CA^kx = 0$ for every $k > 0$. Therefore, $CA^kx = 0$ for every $k \geq 0$, i.e., $\mathbb{O}x = 0$ or $x \in \mathcal{N}(\mathbb{O})$.

■

THEOREM 3.7. A state x is unobservable if and only if

$$x \in \mathcal{N}(\mathbb{O}), \tag{3.24}$$

or equivalently, if and only if

$$x \in \mathcal{N}(W_o(0, T)) \tag{3.25}$$

for some finite (and therefore for all) $T > 0$. Thus, the unobservable subspace $R_{\bar{o}} = \mathcal{N}(\mathbb{O}) = \mathcal{N}(W_o(0, T))$ for some $T > 0$.

Proof. If x is unobservable, (3.21) is satisfied. Taking derivatives with respect to t and evaluating at $t = 0$, we obtain $Cx = CAx = \cdots = CA^k x = 0$ for $k > 0$ or $CA^k x = 0$ for every $k \geq 0$. Therefore, $\mathbb{O}x = 0$ and (3.24) is satisfied. Assume now that $\mathbb{O}x = 0$, i.e., $CA^k x = 0$ for $0 \leq k \leq n-1$, which is also true for every $k > n-1$, in view of the Cayley-Hamilton Theorem. Then $Ce^{At}x = C[\sum_{k=0}^{\infty}(t^k/k!)A^i]x = 0$ for every finite t, i.e., (3.21) is satisfied and x is unobservable. Therefore, x is unobservable if and only if (3.24) is satisfied. In view of Lemma 3.4, (3.25) follows. ■

Clearly, x is observable if and only if $\mathbb{O}x \neq 0$ or $W_o(0, T)x \neq 0$ for some $T > 0$.

COROLLARY 3.8. The system (3.19) is (completely state) observable, or the pair (A, C) is observable, if and only if

$$rank\, \mathbb{O} = n, \tag{3.26}$$

or equivalently, if and only if

$$rank\, W_o(0, T) = n \tag{3.27}$$

for some finite (and therefore for all) $T > 0$. If the system is observable, the state x_0 at $t = 0$ is given by

$$x_0 = W_o^{-1}(0, T)\left[\int_0^T e^{A^T\tau}C^T\tilde{y}(\tau)\,d\tau\right]. \tag{3.28}$$

Proof. The system is observable if and only if the only vector that satisfies (3.20) or (3.21) is the zero vector. This is true if and only if the null space is empty, i.e., if and only if (3.26) or (3.27) are true. To determine the state x_0 at $t = 0$, given the output and input values over some interval $[0, T]$, we premultiply (3.20) by $e^{A^T\tau}C^T$ and integrate over $[0, T]$ to obtain

$$W_o(0, T)x_0 = \int_0^T e^{A^T\tau}C^T\tilde{y}(\tau)\,d\tau, \tag{3.29}$$

in view of (3.22). When the system is observable, (3.29) has the unique solution (3.28). ■

Note that $T > 0$, the time span over which the input and output are observed, is arbitrary. Intuitively, one would expect in practice to have difficulties in evaluating x_0 accurately when T is small, using any numerical method. Note that for very small T, $|W_o(0, T)|$ can be very small, which can lead to numerical difficulties in solving (3.29). Compare this with the analogous case for reachability, where small T leads in general to large values in control action.

It is clear that if the state at some time t_0 is determined, then the state $x(t)$ at any subsequent time is easily determined, given $u(t)$, $t \geq t_0$, via the variation of constants formula (3.2), where $\Phi(t, \tau) = \exp[A(t-\tau)]$.

Alternative methods to (3.29) to determine the state of the system when the system is observable are provided in the next chapter, in Section 4.3.

EXAMPLE 3.4. (i) Consider the system $\dot{x} = Ax$, $y = Cx$, where $A = \begin{bmatrix} 0 & 1 \\ 0 & 0 \end{bmatrix}$ and $C = [1, 0]$. Here $e^{At} = \begin{bmatrix} 1 & t \\ 0 & 1 \end{bmatrix}$ and $Ce^{At} = [1, t]$. The observability Gramian is then

$W_o(0,T) = \int_0^T \begin{bmatrix} 1 \\ \tau \end{bmatrix} [1 \tau]\, d\tau = \int_0^T \begin{bmatrix} 1 & \tau \\ \tau & \tau^2 \end{bmatrix} d\tau = \begin{bmatrix} T & \frac{1}{2}T^2 \\ \frac{1}{2}T^2 & \frac{1}{3}T^3 \end{bmatrix}$. Notice that $det\, W_o(0,T) = \frac{1}{12}T^4 \neq 0$ for any $T > 0$, i.e., $rank\ W_o(0,T) = 2 = n$ for any $T > 0$, and therefore (Corollary 3.8), the system is observable. Alternatively, note that the observability matrix $\mathcal{O} = \begin{bmatrix} C \\ CA \end{bmatrix} = \begin{bmatrix} 1 & 0 \\ 0 & 1 \end{bmatrix}$ and $rank\, \mathcal{O} = 2 = n$. Clearly, in this case $\mathcal{N}(\mathcal{O}) = \mathcal{N}(W_o(0,T)) = \left\{ \begin{bmatrix} 0 \\ 0 \end{bmatrix} \right\}$, which verifies Lemma 3.6.

(ii) If $A = \begin{bmatrix} 0 & 1 \\ 0 & 0 \end{bmatrix}$, as before, but $C = [0, 1]$, in place of $[1, 0]$, then $Ce^{At} = [0, 1]$ and the observability Gramian is $W_o(0,T) = \int_0^T \begin{bmatrix} 0 \\ 1 \end{bmatrix} [0, 1]\, d\tau = \begin{bmatrix} 0 & 0 \\ 0 & T \end{bmatrix}$. We have $rank\ W_o(0,T) = 1 < 2 = n$ and the system is not completely observable. In view of Theorem 3.7, all unobservable states $x \in \mathcal{N}(W_o(0,T))$ and are therefore of the form $\begin{bmatrix} \alpha \\ 0 \end{bmatrix}$, $\alpha \in R$. Alternatively, the observability matrix $\mathcal{O} = \begin{bmatrix} C \\ CA \end{bmatrix} = \begin{bmatrix} 0 & 1 \\ 0 & 0 \end{bmatrix}$. Note that $\mathcal{N}(\mathcal{O}) = \mathcal{N}(W_0(0,T)) = span \left\{ \begin{bmatrix} 1 \\ 0 \end{bmatrix} \right\}$. ■

Observability utilizes future output measurements to determine the present state. In (re)constructibility, past output measurements are used. Constructibility is defined in the following, and its relation to observability is determined.

DEFINITION 3.11. A *state* x is *unconstructible* if the zero-input response of the system (3.18) is zero for all $t \leq 0$, i.e.,

$$Ce^{At}x = 0 \qquad \text{for every } t \leq 0. \tag{3.30}$$

The set of all unconstructible states x, $R_{\overline{cn}}$, is called the *unconstructible subspace* of (3.18). The system (3.18) is (completely state) (*re*)*constructible,* or the pair (A, C) is (*re*)*constructible,* if the only state $x \in R^n$ that is unconstructible is $x = 0$, i.e., $R_{\overline{cn}} = \{0\}$.

We shall now establish a relationship between observability and constructibility for the continuous-time time-invariant systems (3.18). Recall that x is unobservable if and only if

$$Ce^{At}x = 0 \qquad \text{for every } t \geq 0. \tag{3.31}$$

THEOREM 3.9. Consider the system $\dot{x} = Ax + Bu$, $y = Cx + Du$ given in (3.18).

(i) A state x is unobservable if and only if it is unconstructible.
(ii) $R_{\bar{o}} = R_{\overline{cn}}$.
(iii) The system, or the pair (A, C), is (completely state) observable if and only if it is (completely state) (re)constructible.

Proof. (i) If x is unobservable, then $Ce^{At}x = 0$ for every $t \geq 0$. Taking derivatives with respect to t and evaluating at $t = 0$, we obtain $Cx = CAx = \cdots = CA^kx = 0$ for $k > 0$ or $CA^kx = 0$ for every $k \geq 0$. This, in view of $Ce^{At}x = \sum_{k=0}^{\infty}(t^k/k!)CA^kx$, implies that $Ce^{At}x = 0$ for every $t \leq 0$, i.e., x is unconstructible. The converse is proved in a similar manner. Parts (ii) and (iii) of the theorem follow directly from (i). ■

The observability Gramian for the time-invariant case, $W_o(0,T)$, was defined in (3.22). In view of (3.11), we make the following definition.

DEFINITION 3.12. The *constructibility Gramian* of system (3.18) is the $n \times n$ matrix

$$W_{cn}(0,T) \triangleq \int_0^T e^{A^T(\tau-T)} C^T C e^{A(\tau-T)}\, d\tau. \tag{3.32}$$

■

Note that

$$W_o(0,T) = e^{A^T T} W_{cn}(0,T) e^{AT}, \tag{3.33}$$

as can be verified directly [see also (3.13)].

As in the time-varying case above, we now introduce a number of additional criteria for observability.

THEOREM 3.10. The system $\dot{x} = Ax + Bu$, $y = Cx + Du$ is observable

(i) if and only if

$$rank\ W_o(0,T) = n \tag{3.34}$$

for some finite $T > 0$, where $W_0(0,T) \triangleq \int_0^T e^{A^T\tau} C^T C e^{A\tau}\, d\tau$, the observability Gramian, or

(ii) if and only if the n columns of

$$Ce^{At} \tag{3.35}$$

are linearly independent on $[0, \infty)$ over the field of complex numbers, or alternatively, if and only if the n columns of

$$C(sI - A)^{-1} \tag{3.36}$$

are linearly independent over the field of complex numbers, or

(iii) if and only if

$$rank\ \mathbb{O} = n, \tag{3.37}$$

where $\mathbb{O} \triangleq \begin{bmatrix} C \\ CA \\ \vdots \\ CA^{n-1} \end{bmatrix}$, the observability matrix, or

(iv) if and only if

$$rank \begin{bmatrix} s_i I - A \\ C \end{bmatrix} = n \tag{3.38}$$

for all complex numbers s_i, or alternatively, for all eigenvalues of A.

Proof. The proof of this theorem is completely analogous to the (dual) results on reachability (Theorem 2.17) and is omitted. ■

Similar results as those given in Theorem 3.10 can be derived for constructibility, and the reader is encouraged to state and prove these for the cases (i) and (ii). This is of course not surprising, since it was shown (in Theorem 3.9) that observability implies and is implied by constructibility. Accordingly, the tests developed in the theorem for observability are typically also used to test for constructibility.

EXAMPLE 3.5. Consider the system $\dot{x} = Ax$, $y = Cx$, where $A = \begin{bmatrix} 0 & 1 \\ 0 & 0 \end{bmatrix}$ and $C = [1, 0]$, as in Example 3.4(i). We shall verify (i) to (iv) of Theorem 3.10 for this case.

(i) For the observability Gramian, $W_o(0, T) = \begin{bmatrix} T & \frac{1}{2}T^2 \\ \frac{1}{2}T^2 & \frac{1}{3}T^3 \end{bmatrix}$, we have $rank\, W_o(0, T) = 2 = n$ for any $T > 0$.

(ii) The columns of $Ce^{At} = [1, t]$ are linearly independent on $[0, \infty)$ over the field of complex numbers, since $a_1 \cdot 1 + a_2 \cdot t = 0$ implies that the complex numbers a_1 and a_2 must both be zero. Similarly, the columns of $C(sI - A)^{-1} = [1/s, 1/s^2]$ are linearly independent over the field of complex numbers.

(iii) $rank\, \mathcal{O} = rank \begin{bmatrix} C \\ CA \end{bmatrix} = rank \begin{bmatrix} 1 & 0 \\ 0 & 1 \end{bmatrix} = 2 = n.$

(iv) $rank \begin{bmatrix} s_i I - A \\ C \end{bmatrix} = rank \begin{bmatrix} s_i & -1 \\ 0 & s_i \\ 1 & 0 \end{bmatrix} = 2 = n$ for $s_i = 0$, $i = 1, 2$, the eigenvalues of A.

Consider again $A = \begin{bmatrix} 0 & 1 \\ 0 & 0 \end{bmatrix}$ but $C = [0, 1]$ [in place of $[1, 0]$, as in Example 3.4(ii)]. The system is not observable for the reasons given below.

(i) $W_o(0, T) = \begin{bmatrix} 0 & 0 \\ 0 & T \end{bmatrix}$ with $rank\, W_o(0, T) = 1 < 2 = n$.

(ii) $Ce^{At} = [0, 1]$ and its columns are not linearly independent. Similarly, the columns of $C(sI - A)^{-1} = [0, 1/s]$ are not linearly independent.

(iii) $rank\, \mathcal{O} = rank \begin{bmatrix} C \\ CA \end{bmatrix} = rank \begin{bmatrix} 0 & 1 \\ 0 & 0 \end{bmatrix} = 1 < 2 = n.$

(iv) $rank \begin{bmatrix} s_i I - A \\ C \end{bmatrix} = rank \begin{bmatrix} s_i & -1 \\ 0 & s_i \\ 0 & 1 \end{bmatrix} = 1 < 2 = n$ for $s_i = 0$, an eigenvalue of A. ■

C. Discrete-Time Systems

We consider systems described by equations of the form

$$x(k+1) = A(k)x(k) + B(k)u(k), \qquad y(k) = C(k)x(k) + D(k)u(k), \quad (3.39)$$

where $A(k) \in R^{n\times n}$, $B(k) \in R^{n\times m}$, $C(k) \in R^{p\times n}$, $D(k) \in R^{p\times m}$, and the input $u(k) \in R^m$ are defined for $k \geq k_0$ (see Section 2.7). The output $y(k)$ for $k > k_0$ is given by

$$y(k) = C(k)\Phi(k, k_0)x(k_0) + \sum_{i=k_0}^{k-1} C(k)\Phi(k, i+1)B(i)u(i) + D(k)u(k), \quad (3.40)$$

where the state transition matrix $\Phi(k, k_0)$ is given by $\Phi(k, k_0) = A(k-1)A(k-2) \ldots A(k_0)$ for $k > k_0$, and $\Phi(k_0, k_0) = I$.

In the time-invariant case, (3.39) assumes the form

$$x(k+1) = Ax(k) + Bu(k), \qquad y(k) = Cx(k) + Du(k), \qquad k \geq k_0, \quad (3.41)$$

where $A \in R^{n\times n}$, $C \in R^{n\times m}$, $C \in R^{p\times n}$, $D \in R^{p\times m}$, and (3.40) is still valid with the state transition matrix $\Phi(k, k_0)$ given in this case by

$$\Phi(k, k_0) = A^{k-k_0}, \qquad k \geq k_0. \quad (3.42)$$

Observability and (re)constructibility for discrete-time systems are defined as in the continuous-time case. Observability refers to the ability to uniquely determine the state from knowledge of current and future outputs and inputs, while constructibility refers to the ability to determine the state from knowledge of current and past outputs and inputs. In discrete-time systems, the time-varying case can be developed in a manner analogous to the time-invariant case and will therefore not be developed here. Instead, we shall concentrate on the time-invariant case. Note that some of the following results have already been presented in Section 3.1.

Discrete-time time-invariant systems

Consider the time-invariant system (3.41) and the expression for its output $y(k)$, given in (3.40). Without loss of generality, we take $k_0 = 0$. Then

$$y(k) = CA^k x(0) + \sum_{i=0}^{k-1} CA^{k-(i+1)} Bu(i) + Du(k) \tag{3.43}$$

for $k > 0$ and $y(0) = Cx(0) + Du(0)$. Rewrite (3.43) as

$$\tilde{y}(k) = CA^k x_0 \tag{3.44}$$

for $k \geq 0$, where $\tilde{y}(k) \triangleq y(k) - \left[\sum_{i=0}^{k-1} CA^{k-(i+1)} Bu(i) + Du(k)\right]$ for $k > 0$ and $\tilde{y}(0) \triangleq y(0) - Du(0)$, and $x_0 = x(0)$.

DEFINITION 3.13. A *state* x is *unobservable* if the zero-input response of system (3.41) is zero for all $k \geq 0$, i.e., if

$$CA^k x = 0 \qquad \text{for every } k \geq 0. \tag{3.45}$$

The set of all unobservable states x, $R_{\bar{o}}$, is called the *unobservable subspace* of (3.41). The *system* (3.41) is (*completely state*) *observable*, or the pair (A, C) is observable, if the only state $x \in R^n$ that is unobservable is $x = 0$, i.e., if $R_{\bar{o}} = \{0\}$. ■

The $pn \times n$ *observability matrix* $\mathbb{O}$ was defined in (3.23). Let $\mathcal{N}(\mathbb{O})$ denote the null space of $\mathbb{O}$.

THEOREM 3.11. A state x is unobservable if and only if

$$x \in \mathcal{N}(\mathbb{O}), \tag{3.46}$$

i.e., the unobservable subspace $R_{\bar{o}} = \mathcal{N}(\mathbb{O})$.

Proof. If $x \in \mathcal{N}(\mathbb{O})$, then $\mathbb{O}x = 0$ or $CA^k x = 0$ for $0 \leq k \leq n-1$. This statement is also true for $k > n - 1$, in view of the Cayley-Hamilton Theorem. Therefore, (3.45) is satisfied and x is unobservable. Conversely, if x is unobservable, then (3.45) is satisfied and $\mathbb{O}x = 0$. ■

Clearly, x is observable if and only if $\mathbb{O}x \neq 0$.

COROLLARY 3.12. The system (3.41) is (completely state) observable, or the pair (A, C) is observable, if and only if

$$\text{rank } \mathbb{O} = n. \tag{3.47}$$

If the system is observable, the state x_0 at $k = 0$ can be determined as the unique solution of

$$\left[Y_{0,n-1} - M_n U_{0,n-1}\right] = \mathcal{O} x_0, \tag{3.48}$$

where
$$Y_{0,n-1} \triangleq [y^T(0), y^T(1), \ldots, y^T(n-1)]^T \in R_m,$$
$$U_{0,n-1} \triangleq [u^T(0), u^T(1), \ldots, u^T(n-1)]^T \in R^{mn},$$

and M_n is the $pn \times mn$ matrix given by

$$M_n \triangleq \begin{bmatrix} D & 0 & \cdots & 0 & 0 \\ CB & D & \cdots & 0 & 0 \\ \vdots & \vdots & \ddots & \vdots & \vdots \\ CA^{n-2}B & CA^{n-3}B & \cdots & D & \\ CA^{n-1}B & CA^{n-2}B & \cdots & CB & D \end{bmatrix}.$$

Proof. The system is observable if and only if the only vector that satisfies (3.45) is the zero vector. This is true if and only if $\mathcal{N}(\mathcal{O}) = \{0\}$, or if (3.47) is true. To determine the state x_0, apply (3.43) for $k = 0, 1, \ldots, n-1$, and rearrange in a form of a system of linear equations to obtain (3.48). ∎

The matrix M_n defined above has the special structure of a *Toeplitz* matrix. Note that a matrix T is Toeplitz if its (i, j)th entry depends on the value $i - j$; that is, T is "constant along the diagonals."

Similarly to the continuous-time case, we now define the observability Gramian.

DEFINITION 3.14. The *observability Gramian* of the system (3.41) is the $n \times n$ matrix

$$W_o(0, K) \triangleq \sum_{i=0}^{K-1} (A^T)^i C^T C A^i. \tag{3.49}$$
∎

If $\mathcal{O}_k \triangleq [C^T, (CA)^T, \ldots, (CA^{k-1})^T]^T$ (with $\mathcal{O}_n = \mathcal{O}$), then

$$W_o(0, K) = \mathcal{O}_K^T \mathcal{O}_K. \tag{3.50}$$

The following result is apparent.

LEMMA 3.13. $\mathcal{N}(\mathcal{O}) = \mathcal{N}(W_o(0, K))$ for every $K \geq n$.

Proof. The proof is left as an exercise for the reader. ∎

COROLLARY 3.14. The system (3.41) is (completely state) observable, or the pair (A, C) is observable, if and only if

$$rank\, W_o(0, K) = n \tag{3.51}$$

for some (and consequently for all) $K \geq n$. If the system is observable, the state x_0 at $t = 0$ is given by

$$x_0 = W_o^{-1}(0, K) \mathcal{O}_K^T [Y_{0,K-1} - M_k U_{0,K-1}], \tag{3.52}$$

where $K \geq n$.

Proof. Statement (3.51) is a direct consequence of Corollary 3.12 and Lemma 3.13. To obtain (3.52), rewrite (3.48) in terms of K, premultiply by $\mathcal{O}_K^T$, and use relation (3.50). ∎

EXAMPLE 3.6. Consider the system in Example 1.4, $x(k+1) = Ax(k)$, $y(k) = Cx(k)$, where $A = \begin{bmatrix} 0 & 1 \\ 1 & 1 \end{bmatrix}$ and $C = [0, 1]$. The observability Gramian $W_o(0, K)$ for $K = n = 2$ is given by (3.49), $W_o(0, 2) = \sum_{i=0}^{1}(A^T)^i C^T C A^i = \begin{bmatrix} 0 \\ 1 \end{bmatrix}[0, 1] + \begin{bmatrix} 0 & 1 \\ 1 & 1 \end{bmatrix}\begin{bmatrix} 0 \\ 1 \end{bmatrix}[0, 1]\begin{bmatrix} 0 & 1 \\ 1 & 1 \end{bmatrix} = \begin{bmatrix} 0 \\ 1 \end{bmatrix}[0, 1] + \begin{bmatrix} 1 \\ 1 \end{bmatrix}[1, 1] = \begin{bmatrix} 0 & 0 \\ 0 & 1 \end{bmatrix} + \begin{bmatrix} 1 & 1 \\ 1 & 1 \end{bmatrix} = \begin{bmatrix} 1 & 1 \\ 1 & 2 \end{bmatrix}$, which is of full rank. Therefore, the system is observable. Note that $\mathcal{O} = \begin{bmatrix} 0 & 1 \\ 1 & 1 \end{bmatrix}$ is of full rank as well (see Example 1.4). Notice also that $rank\ W_o(0, K) = 2$ for every $K \geq 2$ and that [refer to (3.50)] $W_o(0, 2) = \begin{bmatrix} 1 & 1 \\ 1 & 2 \end{bmatrix} = \begin{bmatrix} 0 & 1 \\ 1 & 1 \end{bmatrix}^T \begin{bmatrix} 0 & 1 \\ 1 & 1 \end{bmatrix} = \mathcal{O}^T\mathcal{O}$. The unique vector $x(0)$ can be determined from $y(0)$ and $y(1)$ using (3.52) to obtain

$$\begin{bmatrix} x_1(0) \\ x_2(0) \end{bmatrix} = W_o^{-1}(0, 2)\mathcal{O}^T \begin{bmatrix} y(0) \\ y(1) \end{bmatrix} = \begin{bmatrix} y(1) - y(0) \\ y(0) \end{bmatrix}.$$

This is the same as the result obtained in Example 1.4, using an alternative approach. ■

Constructibility refers to the ability to determine uniquely the state $x(0)$ from knowledge of current and past outputs and inputs. This is in contrast to observability, which utilizes future outputs and inputs. The easiest way to define constructibility is by the use of (3.44), where $x(0) = x_0$ is to be determined from past data $\tilde{y}(k)$, $k \leq 0$. Note, however, that for $k \leq 0$, A^k may not exist; in fact, it exists only when A is nonsingular. To avoid making restrictive assumptions, we shall define unconstructible states in a slightly different way than anticipated. Unfortunately, this definition is not very transparent. It turns out that by using this definition, an unconstructible state can be related to an unobservable state in a manner analogous to the way a controllable state was related to a reachable state in Section 3.2 (see also the discussion of duality in Section 3.1).

DEFINITION 3.15. A *state* x is *unconstructible* if for every $k \geq 0$ there exists $\hat{x} \in R^n$ such that

$$x = A^k\hat{x}, \qquad C\hat{x} = 0. \tag{3.53}$$

The set of all unconstructible states, $R_{\overline{cn}}$, is called the *unconstructible subspace*. The system (3.41) is (*completely state*) *constructible*, or the pair (A, C) is constructible, if the only state $x \in R^n$ that is unconstructible is $x = 0$, i.e., if $R_{\overline{cn}} = \{0\}$. ■

Note that if A is nonsingular, then (3.53) simply states that x is unconstructible if $CA^{-k}x = 0$ for every $k \geq 0$ (compare this with Definition 3.13 of an unobservable state).

The results that can be derived for constructibility are simply dual to the results on controllability. They are presented briefly below, but first, a technical result must be established.

LEMMA 3.15. If $x \in \mathcal{N}(\mathcal{O})$ then $Ax \in \mathcal{N}(\mathcal{O})$, i.e., the unobservable subspace $R_{\bar{o}} = \mathcal{N}(\mathcal{O})$ is an *A-invariant subspace*.

Proof. Let $x \in \mathcal{N}(\mathcal{O})$, so that $\mathcal{O}x = 0$. Then $CA^kx = 0$ for $0 \leq k \leq n-1$. This statement is also true for $k > n-1$, in view of the Cayley-Hamilton Theorem. Therefore, $\mathcal{O}Ax = 0$, i.e., $Ax \in \mathcal{N}(\mathcal{O})$. ■

THEOREM 3.16. Consider the system $x(k+1) = Ax(k) + Bu(k)$, $y(k) = Cx(k) + Du(k)$ given in (3.41).

(i) If a state x is unconstructible, then it is unobservable.
(ii) $R_{\overline{cn}} \subset R_{\bar{o}}$.
(iii) If the system is (completely state) observable, or the pair (A, C) is observable, then the system is also (completely state) constructible, or the pair (A, C) is constructible.

If A is nonsingular, then relations (i) and (iii) are if and only if statements. In this case, constructibility also implies observability. Furthermore, in this case, (ii) becomes an equality, i.e., $R_{\overline{cn}} = R_{\bar{o}}$.

Proof. This theorem is dual to Theorem 2.22, which relates reachability and controllability in the discrete-time case. To verify (i), assume that x satisfies (3.53) and premultiply by C to obtain $Cx = CA^k\hat{x}$ for every $k \geq 0$. Note that $Cx = 0$ since for $k = 0$, $x = \hat{x}$, and $C\hat{x} = 0$. Therefore, $CA^k\hat{x} = 0$ for every $k \geq 0$, i.e., $\hat{x} \in \mathcal{N}(\mathbb{O})$. In view of Lemma 3.15, $x = A^k\hat{x} \in \mathcal{N}(\mathbb{O})$, and thus, x is unobservable. Since x is arbitrary, we have also verified (ii). When the system is observable, $R_{\bar{o}} = \{0\}$, which in view of (ii), implies that $R_{\bar{c}n} = \{0\}$ or that the system is constructible. This proves (iii). Alternatively, one could also prove this directly: assume that the system is observable but not constructible. Then there exist $x, \hat{x} \neq 0$, which satisfy (3.53). As above, this implies that $\hat{x} \in \mathcal{N}(\mathbb{O})$, which is a contradiction since the system is observable.

Consider now the case when A is nonsingular and let x be unobservable. Then, in view of Lemma 3.15, $\hat{x} \triangleq A^{-k}x$ is also in $\mathcal{N}(\mathbb{O})$, i.e., $C\hat{x} = 0$. Therefore, $x = A^k\hat{x}$ is unconstructible, in view of Definition 3.15. This implies also that $R_{\bar{o}} \subset R_{\overline{cn}}$, and therefore, $R_{\bar{o}} = R_{\overline{cn}}$, which proves that in the present case constructibility also implies observability. ∎

EXAMPLE 3.7. Consider the system in Example 1.5, $x(k+1) = Ax(k)$, $y(k) = Cx(k)$, where $A = \begin{bmatrix} 1 & 0 \\ 1 & 1 \end{bmatrix}$ and $C = [1, 0]$. As shown in Example 1.5, $rank\, \mathbb{O} = rank \begin{bmatrix} 1 & 0 \\ 1 & 0 \end{bmatrix} = 1 < 2 = n$, i.e., the system is not observable. All unobservable states are of the form $\alpha \begin{bmatrix} 0 \\ 1 \end{bmatrix}$, where $\alpha \in R$ since $\left\{\begin{bmatrix} 0 \\ 1 \end{bmatrix}\right\}$ is a basis for $\mathcal{N}(\mathbb{O}) = R_{\bar{o}}$, the unobservable subspace. The observability Gramian for $K = n = 2$ is $W_o(0, 2) = C^TC + (CA)^T(CA) = \begin{bmatrix} 1 & 0 \\ 0 & 0 \end{bmatrix} + \begin{bmatrix} 1 & 0 \\ 0 & 0 \end{bmatrix} = \begin{bmatrix} 2 & 0 \\ 0 & 0 \end{bmatrix}$. Note that a basis for $\mathcal{N}(W_o(0, 2))$ is $\left\{\begin{bmatrix} 0 \\ 1 \end{bmatrix}\right\}$ and $\mathcal{N}(\mathbb{O}) = \mathcal{N}(W_o(0, 2))$. This verifies Lemma 3.13.

In Example 1.6 it was shown that all the states x that satisfy $CA^{-k}x = 0$ for every $k \geq 0$, i.e., all the unconstructible states, are given by $\alpha \begin{bmatrix} 0 \\ 1 \end{bmatrix}$, $\alpha \in R$. This verifies Theorem 3.16(i) and (ii) for the case when A is nonsingular. ∎

EXAMPLE 3.8. Consider the system $x(k+1) = Ax(k)$, $y(k) = Cx(k)$, where $A = \begin{bmatrix} 0 & 0 \\ 1 & 0 \end{bmatrix}$ and $C = [1, 0]$. The observability matrix $\mathbb{O} = \begin{bmatrix} 1 & 0 \\ 0 & 0 \end{bmatrix}$ is of rank 1, and therefore, the system is not observable. In fact, all states of the form $\alpha \begin{bmatrix} 0 \\ 1 \end{bmatrix}$ are unobservable states since $\left\{\begin{bmatrix} 0 \\ 1 \end{bmatrix}\right\}$ is a basis for $\mathcal{N}(\mathbb{O})$.

To check constructibility, the defining relations (3.53) must be used since A is singular. $C\hat{x} = [1, 0]\hat{x} = 0$ implies $\hat{x} = \begin{bmatrix} 0 \\ \beta \end{bmatrix}$. Substituting into $x = A^k\hat{x}$, we obtain for $k = 0$, $x = \hat{x}$, and $x = 0$ for $k \geq 1$. Therefore, the only unconstructible state is $x = 0$, which implies that the system is constructible (although it is unobservable). This means that the initial state $x(0)$ can be uniquely determined from past measurements. In fact, from $x(k+1) = Ax(k)$ and $y(k) = Cx(k)$, we obtain $x(0) = \begin{bmatrix} x_1(0) \\ x_2(0) \end{bmatrix} = \begin{bmatrix} 0 & 0 \\ 1 & 0 \end{bmatrix} \begin{bmatrix} x_1(-1) \\ x_2(-1) \end{bmatrix} = \begin{bmatrix} 0 \\ x_1(-1) \end{bmatrix}$ and $y(-1) = Cx(-1) = [1, 0] \begin{bmatrix} x_1(-1) \\ x_2(-1) \end{bmatrix} = x_1(-1)$. Therefore, $x(0) = \begin{bmatrix} 0 \\ y(-1) \end{bmatrix}$. ■

DEFINITION 3.16. The *constructibility Gramian* is defined as

$$W_{cn}(0, K) = \sum_{i=0}^{K-1} (A^T)^{-(i+1)} C^T C A^{-(i+1)}. \tag{3.54}$$

■

We note that $W_{cn}(0, K)$ is well defined only when A is nonsingular. The observability and constructibility Gramians are related by

$$W_o(0, K) = (A^T)^K W_{cn}(0, K) A^K, \tag{3.55}$$

as can easily be verified.

When A is nonsingular, the state x_0 at $k = 0$ can be determined from past outputs and inputs in the following manner. We consider (3.44) and note that in this case

$$\tilde{y}(k) = CA^k x_0$$

is valid for $k \leq 0$ as well. This implies that

$$\tilde{Y}_{-1,-n} = \mathbb{O}A^{-n} x_0 = \begin{bmatrix} CA^{-n} \\ \vdots \\ CA^{-1} \end{bmatrix} x_0 \tag{3.56}$$

with $\tilde{Y}_{-1,-n} \triangleq [\tilde{y}^T(-1), \ldots, \tilde{y}^T(-n)]^T$. Equation (3.56) must be solved for x_0. Clearly, in the case of constructibility (and under the assumption that A is nonsingular), the matrix $\mathbb{O}A^{-n}$ is of interest instead of $\mathbb{O}$ [compare this with the dual results in (2.66)]. In particular, the system is constructible if and only if $rank\,(\mathbb{O}A^{-n}) = rank\,\mathbb{O} = n$.

EXAMPLE 3.9. Consider the system in Examples 1.4 and 3.6, namely, $x(k+1) = Ax(k)$, $y(k) = Cx(k)$, where $A = \begin{bmatrix} 0 & 1 \\ 1 & 1 \end{bmatrix}$ and $C = [0, 1]$. Since A is nonsingular, to check constructibility we consider $\mathbb{O}A^{-2} = \begin{bmatrix} CA^{-2} \\ CA^{-1} \end{bmatrix} = \begin{bmatrix} -1 & 1 \\ 1 & 0 \end{bmatrix}$, which has full rank. Therefore, the system is constructible (as expected), since it is observable. To determine $x(0)$, in view of (3.56), we note that $\begin{bmatrix} y(-1) \\ y(-2) \end{bmatrix} = \mathbb{O}A^{-2}x(0) = \begin{bmatrix} -1 & 1 \\ 1 & 0 \end{bmatrix} \begin{bmatrix} x_1(0) \\ x_2(0) \end{bmatrix}$, from which $\begin{bmatrix} x_1(0) \\ x_2(0) \end{bmatrix} = \begin{bmatrix} 0 & 1 \\ 1 & 1 \end{bmatrix} \begin{bmatrix} y(-1) \\ y(-2) \end{bmatrix} = \begin{bmatrix} y(-2) \\ y(-1) + y(-2) \end{bmatrix}$. It is also easy to verify (3.55), namely, $(A^T)^2 W_{cn}(0, 2) A^2 = \begin{bmatrix} 1 & 1 \\ 1 & 2 \end{bmatrix} \begin{bmatrix} 2 & -1 \\ -1 & 1 \end{bmatrix} \begin{bmatrix} 1 & 1 \\ 1 & 2 \end{bmatrix} = \begin{bmatrix} 1 & 1 \\ 1 & 2 \end{bmatrix} = W_o(0, 2)$. ■

PART 2
SPECIAL FORMS FOR TIME-INVARIANT SYSTEMS

3.4
SPECIAL FORMS

In this section, important special forms for the state-space description of time-invariant systems are presented. These forms are obtained by means of similarity transformations and are designed to reveal those features of a system that are related to the properties of controllability and observability. In Subsection A, special state-space forms that display the controllable (observable) part of a system and that separate this part from the uncontrollable (unobservable) part are presented. These forms, referred to as the standard forms for uncontrollable and unobservable systems, are very useful in establishing a number of results. In particular, these forms are used in Subsection B to derive alternative tests for controllability and observability, and in Subsection C to relate state-space and input-output descriptions. (Additionally, these forms are further used in Chapters 4 and 5.) In Subsection D, the controller and observer state-space forms are introduced. These are useful in the study of state feedback and state estimators (to be addressed in Chapter 4), and in state-space realizations (to be addressed in Chapter 5).

A. Standard Forms for Uncontrollable and Unobservable Systems

We consider time-invariant systems described by equations of the form

$$\dot{x} = Ax + Bu, \qquad y = Cx + Du, \tag{4.1}$$

where $A \in R^{n \times n}$, $B \in R^{n \times m}$, $C \in R^{p \times n}$, and $D \in R^{p \times m}$. It was shown earlier in this chapter that this system is state reachable or controllable-from-the-origin if and only if the $n \times mn$ controllability matrix

$$\mathscr{C} \triangleq [B, AB, \ldots, A^{n-1}B] \tag{4.2}$$

has full row rank n, i.e., $rank\ \mathscr{C} = n$. Recall that $\mathscr{R}(\mathscr{C}) = R_r$ is the reachable subspace, which contains all the state vectors that can be reached from the zero vector in finite time by applying an appropriate input. If the system is reachable (or controllable-from-the-origin), then it is also controllable (or controllable-to-the-origin), and vice versa (see Section 3.2).

It was also shown earlier in this chapter that system (4.1) is state observable if and only if the $pn \times n$ observability matrix

$$\mathbb{O} \triangleq \begin{bmatrix} C \\ CA \\ \vdots \\ CA^{n-1} \end{bmatrix} \tag{4.3}$$

has full column rank, i.e., $rank\ \mathbb{O} = n$. Recall that $\mathcal{N}(\mathbb{O}) = R_{\bar{o}}$ is the unobservable subspace that contains all the states that cannot be determined uniquely from input and output measurements in finite time. If the system is observable, then it is also constructible, and vice versa (see Section 3.3).

Similar results were also derived for discrete-time time-invariant systems described by equations of the form

$$x(k+1) = Ax(k) + Bu(k), \qquad y(k) = Cx(k) + Du(k) \tag{4.4}$$

in Sections 3.1, 3.2, and 3.3. Again, $rank\,\mathscr{C} = n$ and $rank\,\mathbb{O} = n$ are the necessary and sufficient conditions for state reachability and observability, respectively. Reachability always implies controllability and observability always implies constructibility, as in the continuous-time case. However, in the discrete-time case, controllability does not necessarily imply reachability and constructibility does not imply observability, unless A is nonsingular.

Next, we will address standard forms for unreachable and unobservable systems both for the continuous-time and the discrete-time time-invariant cases. These forms will be referred to as standard forms for *uncontrollable systems*, rather than unreachable systems, and standard forms for *unobservable systems*, respectively. This is to conform with the established terminology in the literature, where the term "controllable" is used instead of "reachable," perhaps because of emphasis on the continuous-time case. It should be noted, however, that by the term controllable in this section we mean controllable-from-the-origin, i.e., reachable.

1. Standard form for uncontrollable systems

If the system (4.1) [or (4.4)] is not completely controllable (-from-the-origin), then it is possible to "separate" the controllable part of the system by means of an appropriate similarity transformation. This amounts to changing the basis of the state space (see Section 2.2) so that all the vectors in the controllable (-from-the-origin) or reachable subspace R_r have certain structure. In particular, let $rank\,\mathscr{C} = n_r < n$, i.e., the pair (A, B) is not controllable. This implies that the subspace $R_r = \mathscr{R}(\mathscr{C})$ has dimension n_r. Let $\{v_1, v_2, \ldots, v_{n_r}\}$ be a basis for R_r. These n_r vectors can be, for example, any n_r linearly independent columns of $\mathscr{C}$. Define the $n \times n$ similarity transformation matrix

$$Q \triangleq [v_1, v_2, \ldots, v_{n_r}, Q_{n-n_r}], \tag{4.5}$$

where the $n \times (n - n_r)$ matrix Q_{n-n_r} contains $n - n_r$ linearly independent vectors chosen so that Q is nonsingular. There are many such choices. We are now in a position to prove the following result.

LEMMA 4.1. For (A, B) uncontrollable, there is a nonsingular matrix Q such that

$$\hat{A} = Q^{-1}AQ = \begin{bmatrix} A_1 & A_{12} \\ 0 & A_2 \end{bmatrix} \quad \text{and} \quad \hat{B} = Q^{-1}B = \begin{bmatrix} B_1 \\ 0 \end{bmatrix}, \tag{4.6}$$

where $A_1 \in R^{n_r \times n_r}$, $B_1 \in R^{n_r \times m}$, and the pair (A_1, B_1) is controllable. The pair $(\hat{A}, \hat{B})$ is in *the standard form for uncontrollable systems*.

Proof. We need to show that

$$AQ = A[v_1, \ldots, v_{n_r}, Q_{n-n_r}] = [v_1, \ldots, v_{n_r}, Q_{n-n_r}] \begin{bmatrix} A_1 & A_{12} \\ 0 & A_2 \end{bmatrix} = Q\hat{A}.$$

Since the subspace R_r is A-invariant (see Lemma 2.15 in this chapter), $Av_i \in R_r$, which can be written as a linear combination of only the n_r vectors in a basis of R_r. Thus,

A_1 in $\hat{A}$ is an $n_r \times n_r$ matrix, and the $(n - n_r) \times n_r$ matrix below it in $\hat{A}$ is a zero matrix. Similarly, we also need to show that

$$B = [v_1, \ldots, v_{n_r}, Q_{n-n_r}] \begin{bmatrix} B_1 \\ 0 \end{bmatrix} = Q\hat{B}.$$

But this is true for similar reasons: the columns of B are in the range of $\mathscr{C}$ or in R_r. ■

The $n \times nm$ controllability matrix $\hat{\mathscr{C}}$ of $(\hat{A}, \hat{B})$ is

$$\hat{\mathscr{C}} = [\hat{B}, \hat{A}\hat{B}, \ldots, \hat{A}^{n-1}\hat{B}] = \begin{bmatrix} B_1 & A_1B_1 & \cdots & A_1^{n-1}B_1 \\ 0 & 0 & \cdots & 0 \end{bmatrix}, \tag{4.7}$$

which clearly has $rank\, \hat{C} = rank\, [B_1, A_1B_1, \ldots, A_1^{n_r-1}B_1, \ldots, A_1^{n-1}B_1] = n_r$. Note that

$$\hat{\mathscr{C}} = Q^{-1}\mathscr{C}. \tag{4.8}$$

The range of $\hat{\mathscr{C}}$ is the controllable (-from-the-origin) subspace of $(\hat{A}, \hat{B})$. It contains vectors only of the form $[\alpha^T, 0]^T$, where $\alpha \in R^{n_r}$. Since dim $\mathscr{R}(\hat{\mathscr{C}}) = rank\, \hat{\mathscr{C}} = n_r$, every vector of the form $[\alpha^T, 0]^T$ is a controllable (state) vector. In other words, the similarity transformation has changed the basis of R^n in such a manner that all controllable (-from-the-origin) vectors, expressed in terms of this new basis, have this very particular structure of zeros in the last $n - n_r$ entries.

Given system (4.1) [or (4.4)], if a new state $\hat{x}(t)$ is taken to be $\hat{x}(t) = Q^{-1}x(t)$, then

$$\dot{\hat{x}} = \hat{A}\hat{x} + \hat{B}u, \qquad y = \hat{C}\hat{x} + \hat{D}u, \tag{4.9}$$

where $\hat{A} = Q^{-1}AQ, \hat{B} = Q^{-1}B, \hat{C} = CQ$, and $\hat{D} = D$ constitutes an equivalent representation (see Chapter 2). For Q as in Lemma 4.1, we obtain

$$\begin{bmatrix} \dot{\hat{x}}_1 \\ \dot{\hat{x}}_2 \end{bmatrix} = \begin{bmatrix} A_1 & A_{12} \\ 0 & A_2 \end{bmatrix} \begin{bmatrix} \hat{x}_1 \\ \hat{x}_2 \end{bmatrix} + \begin{bmatrix} B_1 \\ 0 \end{bmatrix} u, \; y = [C_1, C_2] \begin{bmatrix} \hat{x}_1 \\ \hat{x}_2 \end{bmatrix} + Du, \tag{4.10}$$

where $\hat{x} = [\hat{x}_1^T, \hat{x}_2^T]^T$ with $\hat{x}_1 \in R^{n_r}$ and where (A_1, B_1) is controllable. The matrix $\hat{C} = [C_1, C_2]$ does not have any particular structure. This representation is called a *standard form for the uncontrollable system.* The state equation can now be written as

$$\dot{\hat{x}}_1 = A_1\hat{x}_1 + B_1u + A_{12}\hat{x}_2, \dot{\hat{x}}_2 = A_2\hat{x}_2, \tag{4.11}$$

which shows that the input u does not affect the trajectory component $\hat{x}_2(t)$ at all, and therefore, $\hat{x}_2(t)$ is determined only by the value of its initial vector. The input u certainly affects $\hat{x}_1(t)$. Note also that the trajectory component $\hat{x}_1(t)$ is also influenced by $\hat{x}_2(t)$. In fact,

$$\hat{x}_1(t) = e^{A_1t}\hat{x}_1(0) + \int_0^t e^{A_1(t-\tau)}B_1u(\tau)\,d\tau + \left[\int_0^t e^{A_1(t-\tau)}A_{12}e^{A_2\tau}\,d\tau\right]\hat{x}_2(0). \tag{4.12}$$

The n_r eigenvalues of A_1 and the corresponding modes are called *controllable eigenvalues* and *controllable modes* of the pair (A, B) or of system (4.1) [or of (4.4)]. The $n - n_r$ eigenvalues of A_2 and the corresponding modes are called the *uncontrollable eigenvalues* and *uncontrollable modes*, respectively.

It is interesting to observe that in the zero-state response of the system (zero initial conditions) the uncontrollable modes are completely absent. In particular, in the solution $x(t) = e^{At}x(0) + \int_0^t e^{A(t-\tau)}Bu(\tau)\,d\tau$ of $\dot{x} = Ax + Bu$, given $x(0)$, notice that

$$e^{A(t-\tau)}B = [Qe^{\hat{A}(t-\tau)}Q^{-1}][Q\hat{B}] = Q\begin{bmatrix} e^{A_1(t-\tau)}B_1 \\ 0 \end{bmatrix}$$

(show this), where A_1 [from (4.6)] contains only the controllable eigenvalues. Therefore, the input $u(t)$ cannot directly influence the uncontrollable modes. Note, however, that the uncontrollable modes do appear in the zero-input response $e^{At}x(0)$. The same observations can be made for discrete-time systems (4.4), where the quantity A^kB is of interest (show this).

EXAMPLE 4.1. Given $A = \begin{bmatrix} 0 & -1 & 1 \\ 1 & -2 & 1 \\ 0 & 1 & -1 \end{bmatrix}$ and $B = \begin{bmatrix} 1 & 0 \\ 1 & 1 \\ 1 & 2 \end{bmatrix}$, we wish to reduce system (4.1) to the standard form (4.6). Here

$$\mathscr{C} = [B, AB, A^2B] = \left[\begin{array}{cc:cc:cc} 1 & 0 & 0 & 1 & 0 & -1 \\ 1 & 1 & 0 & 0 & 0 & 0 \\ 1 & 2 & 0 & -1 & 0 & 1 \end{array}\right]$$

and $rank\,\mathscr{C} = n_r = 2 < 3 = n$. Thus, the subspace $R_r = \mathscr{R}(\mathscr{C})$ has dimension $n_r = 2$, and a basis $\{v_1, v_2\}$ can be found by taking two linearly independent columns of $\mathscr{C}$, say, the first two, to obtain

$$Q = [v_1, v_2, Q_1] = \left[\begin{array}{cc:c} 1 & 0 & 0 \\ 1 & 1 & 0 \\ 1 & 2 & 1 \end{array}\right].$$

The third column of Q was selected so that Q is nonsingular. Note that the first two columns of Q could have been the first and third columns of $\mathscr{C}$ instead, or any other two linearly independent vectors obtained as a linear combination of the columns in $\mathscr{C}$. For the above choice for Q we have,

$$\hat{A} = Q^{-1}AQ = \begin{bmatrix} 1 & 0 & 0 \\ -1 & 1 & 0 \\ 1 & -2 & 1 \end{bmatrix}\begin{bmatrix} 0 & -1 & 1 \\ 1 & -2 & 1 \\ 0 & 1 & -1 \end{bmatrix}\begin{bmatrix} 1 & 0 & 0 \\ 1 & 1 & 0 \\ 1 & 2 & 1 \end{bmatrix}$$

$$= \begin{bmatrix} 0 & -1 & 1 \\ 1 & -1 & 0 \\ -2 & 4 & -2 \end{bmatrix}\begin{bmatrix} 1 & 0 & 0 \\ 1 & 1 & 0 \\ 1 & 2 & 1 \end{bmatrix}$$

$$= \left[\begin{array}{cc:c} 0 & 1 & 1 \\ 0 & -1 & 0 \\ \cdots & \cdots & \cdots \\ 0 & 0 & -2 \end{array}\right] = \left[\begin{array}{c:c} A_1 & A_{12} \\ \cdots & \cdots \\ 0 & A_2 \end{array}\right]$$

$$\hat{B} = Q^{-1}B = \begin{bmatrix} 1 & 0 & 0 \\ -1 & 1 & 0 \\ 1 & -2 & 1 \end{bmatrix}\begin{bmatrix} 1 & 0 \\ 1 & 1 \\ 1 & 2 \end{bmatrix} = \begin{bmatrix} 1 & 0 \\ 0 & 1 \\ \cdots & \cdots \\ 0 & 0 \end{bmatrix} = \begin{bmatrix} B_1 \\ \cdots \\ 0 \end{bmatrix},$$

where (A_1, B_1) is controllable [verify this and show that $\hat{\mathscr{C}} = Q^{-1}\mathscr{C}$, i.e., verify (4.8)].

The matrix A has three eigenvalues at $0, -1, -2$. It is clear from $(\hat{A}, \hat{B})$ that the eigenvalues $0, -1$ are controllable (in A_1), while -2 is an uncontrollable eigenvalue (in A_2). ■

2. Standard form for unobservable systems

The standard form for an unobservable system can be derived in a similar way as the standard form of uncontrollable systems. If the system (4.1) [or (4.4)] is not completely state observable, then it is possible to "separate" the unobservable part of the system by means of a similarity transformation. This amounts to changing the basis of the state space so that all the vectors in the unobservable subspace $R_{\bar{o}}$ have a certain structure.

As in the preceding discussion concerning systems or pairs (A, B) that are not completely controllable, we shall presently select a similarity transformation Q to reduce a pair (A, C), which is not completely observable, to a particular form. This can be accomplished in two ways. The simplest way is to invoke duality and work with the pair $(A_D = A^T, B_D = C^T)$, which is not controllable (refer to the discussion of dual systems in Section 3.1). If Lemma 4.1 is applied, then

$$\hat{A}_D = Q_D^{-1} A_D Q_D = \begin{bmatrix} A_{D1} & A_{D12} \\ 0 & A_{D2} \end{bmatrix}, \qquad \hat{B}_D = Q_D^{-1} B_D = \begin{bmatrix} B_{D1} \\ 0 \end{bmatrix},$$

where (A_{D1}, B_{D1}) is controllable.

Taking the dual again, we obtain the pair $(\hat{A}, \hat{C})$, which has the desired properties. In particular,

$$\begin{aligned} \hat{A} &= \hat{A}_D^T = Q_D^T A_D^T (Q_D^T)^{-1} = Q_D^T A (Q_D^T)^{-1} = \begin{bmatrix} A_{D1}^T & 0 \\ A_{D12}^T & A_{D2}^T \end{bmatrix} \\ \hat{C} &= \hat{B}_D^T = B_D^T (Q_D^T)^{-1} = C (Q_D^T)^{-1} = [B_{D1}^T, 0], \end{aligned} \tag{4.13}$$

where (A_{D1}^T, B_{D1}^T) is completely observable by duality (see Theorem 1.1).

EXAMPLE 4.2. Given $A = \begin{bmatrix} 0 & 1 & 0 \\ -1 & -2 & 1 \\ 1 & 1 & -1 \end{bmatrix}$ and $C = \begin{bmatrix} 1 & 1 & 1 \\ 0 & 1 & 2 \end{bmatrix}$, we wish to reduce system (4.1) to the standard form (4.13). To accomplish this, let $A_D = A^T$ and $B_D = C^T$. Notice that the pair (A_D, B_D) is precisely the pair (A, B) of Example 4.1. ■

A pair (A, C) can of course also be reduced directly to the standard form for unobservable systems. This is accomplished in the following.

Consider the system (4.1) [or (4.4)] and the observability matrix $\mathcal{O}$ in (4.3). Let $rank\, \mathcal{O} = n_o < n$, i.e., the pair (A, C) is not completely observable. This implies that the unobservable subspace $R_{\bar{o}} = \mathcal{N}(\mathcal{O})$ has dimension $n - n_o$. Let $\{v_1, \dots, v_{n-n_o}\}$ be a basis for $R_{\bar{o}}$ and define an $n \times n$ similarity transformation matrix Q as

$$Q \triangleq [Q_{n_o}, v_1, \dots, v_{n-n_o}], \tag{4.14}$$

where the $n \times n_o$ matrix Q_{n_o} contains n_o linearly independent vectors chosen so that Q is nonsingular. Clearly, there are many such choices.

LEMMA 4.2. For (A, C) unobservable, there is a nonsingular matrix Q such that

$$\hat{A} = Q^{-1} A Q = \begin{bmatrix} A_1 & 0 \\ A_{21} & A_2 \end{bmatrix} \quad \text{and} \quad \hat{C} = CQ = [C_1, 0], \tag{4.15}$$

where $A_1 \in R^{n_o \times n_o}$, $C_1 \in R^{p \times n_o}$, and the pair (A_1, C_1) is observable. The pair $(\hat{A}, \hat{C})$ is in the *standard form for unobservable systems.*

Proof. We need to show that

$$AQ = A[Q_{n_0}, v_1, \ldots, v_{n-n_o}] = [Q_{n_o}, v_1, \ldots, v_{n-n_o}]\begin{bmatrix} A_1 & 0 \\ A_{21} & A_2 \end{bmatrix} = Q\hat{A}.$$

Since the unobservable subspace $R_{\bar{o}}$ is A-invariant (see Lemma 3.15), $Av_i \in R_{\bar{o}}$, which can be written as a linear combination of only the $n - n_o$ vectors in a basis of $R_{\bar{o}}$. Thus, A_2 in $\hat{A}$ is an $(n - n_o) \times (n - n_o)$ matrix, and the $n_o \times (n - n_o)$ matrix above it in $\hat{A}$ is a zero matrix. Similarly, we also need to show that

$$CQ = C[Q_{n_o}, v_1, \ldots, v_{n-n_o}] = [C_1, 0] = \hat{C}.$$

This is true since $Cv_i = 0$. ■

The $pn \times n$ observability matrix $\hat{\mathcal{O}}$ of $(\hat{A}, \hat{C})$ is

$$\hat{\mathcal{O}} = \begin{bmatrix} \hat{C} \\ \hat{C}\hat{A} \\ \vdots \\ \hat{C}\hat{A}^{n-1} \end{bmatrix} = \begin{bmatrix} C_1 & 0 \\ C_1A_1 & 0 \\ \vdots & \vdots \\ C_1A_1^{n-1} & 0 \end{bmatrix}, \tag{4.16}$$

which clearly has

$$rank\, \hat{\mathcal{O}} = rank \begin{bmatrix} C_1 \\ C_1A_1 \\ \vdots \\ C_1A_1^{n_o-1} \\ \vdots \\ C_1A_1^{n-1} \end{bmatrix} = n_o.$$

Note that

$$\hat{\mathcal{O}} = \mathcal{O}Q. \tag{4.17}$$

The null space of $\hat{\mathcal{O}}$ is the unobservable subspace of $(\hat{A}, \hat{C})$. It contains vectors only of the form $[0, \alpha^T]^T$, where $\alpha \in R^{n-n_0}$. Since $\dim \mathcal{N}(\hat{\mathcal{O}}) = n - rank\, \hat{\mathcal{O}} = n - n_0$, every vector of the form $[0, \alpha^T]^T$ is an unobservable (state) vector. In other words, the similarity transformation has changed the basis of R^n in such a manner that all unobservable vectors expressed in terms of this new basis have this very particular structure—zeros in the first n_o entries.

For Q chosen as in Lemma 4.2 and (4.9), it assumes the form

$$\begin{bmatrix} \dot{\hat{x}}_1 \\ \dot{\hat{x}}_2 \end{bmatrix} = \begin{bmatrix} A_1 & 0 \\ A_{21} & A_2 \end{bmatrix}\begin{bmatrix} \hat{x}_1 \\ \hat{x}_2 \end{bmatrix} + \begin{bmatrix} B_1 \\ B_2 \end{bmatrix} u,\ y = [C_1, 0]\begin{bmatrix} \hat{x}_1 \\ \hat{x}_2 \end{bmatrix} + Du, \tag{4.18}$$

where $\hat{x} = [\hat{x}_1^T, \hat{x}_2^T]^T$ with $\hat{x}_1 \in R^{n_o}$ and where (A_1, C_1) is observable. The matrix $\hat{B} = [B_1^T, B_2^T]^T$ does not have any particular form. This representation is called a *standard form for the unobservable system.*

The n_o eigenvalues of A_1 and the corresponding modes are called *observable eigenvalues* and *observable modes* of the pair (A, C) or of the system (4.1) [or of (4.4)]. The $n - n_o$ eigenvalues of A_2 and the corresponding modes are called *unobservable eigenvalues* and *unobservable modes*, respectively.

Notice that the trajectory component $\hat{x}(t)$, which is observed via the output y, is not influenced at all by $\hat{x}_2$, the trajectory of which is determined primarily by the eigenvalues of A_2.

The unobservable modes of the system are completely absent from the output. In particular, given $\dot{x} = Ax + Bu$, $y = Cx$ with initial state $x(0)$, we have

$$y(t) = Ce^{At}x(0) + \int_0^t Ce^{A(t-\tau)}Bu(\tau)\,d\tau$$

and $Ce^{At} = [\hat{C}Q^{-1}][Qe^{\hat{A}t}Q^{-1}] = [C_1e^{A_1t}, 0]Q^{-1}$ (show this), where A_1 [from (4.15)] contains only the observable eigenvalues. Therefore, the unobservable modes cannot be seen by observing the output. The same observations can be made for discrete-time systems where the quantity CA^k is of interest (show this).

EXAMPLE 4.3. Given $A = \begin{bmatrix} 0 & 1 \\ -2 & -3 \end{bmatrix}$ and $C = [1, 1]$, we wish to reduce system (4.1) to the standard form (4.15). To accomplish this, we compute $\mathbb{O} = \begin{bmatrix} C \\ CA \end{bmatrix} = \begin{bmatrix} 1 & 1 \\ -2 & -2 \end{bmatrix}$, which has $rank\, \mathbb{O} = n_o = 1 < 2 = n$. Therefore, the unobservable subspace $R_{\bar{o}} = \mathcal{N}(\mathbb{O})$ has dimension $n - n_o = 1$. In view of (4.14),

$$Q = [Q_1, v_1] = \begin{bmatrix} 0 & \vdots & 1 \\ 1 & \vdots & -1 \end{bmatrix},$$

where $v_1 = [1, -1]^T$ is a basis for $R_{\bar{o}}$, and Q_1 was chosen so that Q is nonsingular. Then

$$\hat{A} = Q^{-1}AQ = \begin{bmatrix} 1 & 1 \\ 1 & 0 \end{bmatrix}\begin{bmatrix} 0 & 1 \\ -2 & -3 \end{bmatrix}\begin{bmatrix} 0 & 1 \\ 1 & -1 \end{bmatrix}$$

$$= \begin{bmatrix} -2 & \vdots & 0 \\ \cdots & \vdots & \cdots \\ 1 & \vdots & -1 \end{bmatrix} = \begin{bmatrix} A_1 & 0 \\ A_{21} & A_2 \end{bmatrix}$$

$$\hat{C} = CQ = [1, 1]\begin{bmatrix} 0 & 1 \\ 1 & -1 \end{bmatrix} = [1, 0] = [C_1, 0],$$

where (A_1, C_1) is observable [show this and verify that $\hat{\mathbb{O}} = \mathbb{O}Q$, i.e., verify (4.17)].

The matrix A has two eigenvalues at $-1, -2$. It is clear from $(\hat{A}, \hat{C})$ that the eigenvalue -2 is observable (in A_1), while -1 is an unobservable eigenvalue (in A_2). ■

3. Kalman's Decomposition Theorem

Lemmas 4.1 and 4.2 can be combined to obtain an equivalent representation of (4.1) where the reachable and observable parts of this system can readily be identified. To this end, we consider system (4.9) again and proceed, in the following, to construct the $n \times n$ required similarity transformation matrix Q.

As before, we let n_r denote the dimension of the controllable (-from-the-origin) subspace R_r i.e., $n_r = \dim R_r = \dim \mathcal{R}(\mathcal{C}) = rank\, \mathcal{C}$. The dimension of the unobservable subspace $R_{\bar{o}} = \mathcal{N}(\mathbb{O})$ is given by $n_{\bar{o}} = n - rank\, \mathbb{O} = n - n_o$. Let $n_{r\bar{o}}$ be the dimension of the subspace $R_{r\bar{o}} \triangleq R_r \cap R_{\bar{o}}$ that contains all the state vectors $x \in R^n$ that are controllable but unobservable. We choose

$$Q \triangleq [v_1, \ldots, v_{n_r - n_{r\bar{o}}+1}, \ldots, v_{n_r}, Q_N, \hat{v}_1, \ldots, \hat{v}_{n_{\bar{o}} - n_{r\bar{o}}}], \tag{4.19}$$

where the n_r vectors in $\{v_1, \ldots, v_{n_r}\}$ form a basis for R_r. The last $n_{r\bar{o}}$ vectors $\{v_{n_r - n_{r\bar{o}}+1}, \ldots, v_{n_r}\}$ in the basis for R_r are chosen so that they form a basis for $R_{r\bar{o}} = R_r \cap R_{\bar{o}}$. The $n_{\bar{o}} - n_{r\bar{o}} = (n - n_o - n_{r\bar{o}})$ vectors $\{\hat{v}_1, \ldots, \hat{v}_{n_{\bar{o}} - n_{r\bar{o}}}\}$ are selected so that when taken together with the $n_{r\bar{o}}$ vectors $\{v_{n_r - n_{r\bar{o}}+1}, \ldots, v_{n_r}\}$ they form a basis for $R_{\bar{o}}$, the unobservable subspace. The remaining $N = n - (n_r + n_{\bar{o}} - n_{r\bar{o}})$ columns in Q_N are simply selected so that Q is nonsingular.

The following theorem is called the *Canonical Structure Theorem* or *Kalman's Decomposition Theorem.*

THEOREM 4.3. For (A, B) uncontrollable and (A, C) unobservable, there is a nonsingular matrix Q such that

$$\hat{A} = Q^{-1}AQ = \begin{bmatrix} A_{11} & 0 & A_{13} & 0 \\ A_{21} & A_{22} & A_{23} & A_{24} \\ 0 & 0 & A_{33} & 0 \\ 0 & 0 & A_{43} & A_{44} \end{bmatrix}, \qquad \hat{B} = Q^{-1}B = \begin{bmatrix} B_1 \\ B_2 \\ 0 \\ 0 \end{bmatrix}, \tag{4.20}$$

$$\hat{C} = CQ = [C_1, 0, C_3, 0],$$

where

(i) (A_c, B_c) with

$$A_c \triangleq \begin{bmatrix} A_{11} & 0 \\ A_{21} & A_{22} \end{bmatrix} \quad \text{and} \quad B_c \triangleq \begin{bmatrix} B_1 \\ B_2 \end{bmatrix}$$

is controllable (-from-the-origin), where $A_c \in R^{n_r \times n_r}$, $B_c \in R^{n_r \times m}$,

(ii) (A_o, C_o) with

$$A_o \triangleq \begin{bmatrix} A_{11} & A_{13} \\ 0 & A_{33} \end{bmatrix} \quad \text{and} \quad C_o \triangleq [C_1, C_3]$$

is observable, where $A_o \in R^{n_o \times n_o}$ and $C_o \in R^{p \times n_o}$ and where the dimensions of the matrices A_{ij}, B_i, and C_j are

$A_{11} : (n_r - n_{r\bar{o}}) \times (n_r - n_{r\bar{o}})$, $\quad A_{22} : n_{r\bar{o}} \times n_{r\bar{o}}$,

$A_{33} : (n - (n_r + n_{\bar{o}} - n_{r\bar{o}})) \times (n - (n_r + n_{\bar{o}} - n_{r\bar{o}}))$, $\quad A_{44} : (n_{\bar{o}} - n_{r\bar{o}}) \times (n_{\bar{o}} - n_{r\bar{o}})$,

$B_1 : (n_r - n_{r\bar{o}}) \times m$, $\quad B_2 : n_{r\bar{o}} \times m$,

$C_1 : p \times (n_r - n_{r\bar{o}})$, $\quad C_3 : p \times (n - (n_r + n_{\bar{o}} - n_{r\bar{o}}))$,

(iii) the triple (A_{11}, B_1, C_1) is such that (A_{11}, B_1) is controllable (-from-the-origin) and (A_{11}, C_1) is observable.

Proof. The proof of this result is straightforward, selecting Q according to (4.19) and following the proofs of Lemmas 4.1 and 4.2. We leave the details to the reader. ■

The similarity transformation (4.19) has altered the basis of the state space in such a manner that the vectors in the controllable (-from-the-origin) subspace R_r, the vectors in the unobservable subspace $R_{\bar{o}}$, and the vectors in the subspace $R_{r\bar{o}} \cap R_{\bar{o}}$ all have specific forms. To see this, we construct the controllability matrix $\hat{\mathcal{C}} = [\hat{B}, \ldots, \hat{A}^{n-1}B]$ whose range is the controllable (-from-the-origin or reachable) subspace and the observability matrix $\hat{\mathcal{O}} = [\hat{C}^T, \ldots, (\hat{C}\hat{A}^{n-1})^T]^T$, whose null space is the unobservable subspace. Then, all controllable states are of the form $[x_1^T, x_2^T, 0, 0]^T$, all the unobservable ones have the structure $[0, x_2^T, 0, x_4^T]^T$, while states of the form $[0, x_2^T, 0, 0]^T$ characterize $R_{r\bar{o}}$, i.e., they are controllable but unobservable.

Similarly to the previous two lemmas, the eigenvalues of $\hat{A}$, or of A, are the eigenvalues of A_{11}, A_{22}, A_{33}, and A_{44}, i.e.,

$$|\lambda I - A| = |\lambda I - \hat{A}| = |\lambda I - A_{11}||\lambda I - A_{22}||\lambda I - A_{33}||\lambda I - A_{44}|. \quad (4.21)$$

If in particular we consider the representation $\{\hat{A}, \hat{B}, \hat{C}, \hat{D}\}$ given in (4.9), where Q was selected as in the canonical structure theorem given above, then

$$\begin{bmatrix} \dot{\hat{x}}_1 \\ \dot{\hat{x}}_2 \\ \dot{\hat{x}}_3 \\ \dot{\hat{x}}_4 \end{bmatrix} = \begin{bmatrix} A_{11} & 0 & A_{13} & 0 \\ A_{21} & A_{22} & A_{23} & A_{24} \\ 0 & 0 & A_{33} & 0 \\ 0 & 0 & A_{43} & A_{44} \end{bmatrix} \begin{bmatrix} \hat{x}_1 \\ \hat{x}_2 \\ \hat{x}_3 \\ \hat{x}_4 \end{bmatrix} + \begin{bmatrix} B_1 \\ B_2 \\ 0 \\ 0 \end{bmatrix} u$$

$$y = [C_1, 0, C_3, 0] \begin{bmatrix} \hat{x}_1 \\ \hat{x}_2 \\ \hat{x}_3 \\ \hat{x}_4 \end{bmatrix} + Du. \quad (4.22)$$

This shows that the trajectory components corresponding to $\hat{x}_3$ and $\hat{x}_4$ are not affected by the input u. The modes associated with the eigenvalues of A_{33} and A_{44} determine the trajectory components for $\hat{x}_3$ and $\hat{x}_4$ (compare this with the results in Lemma 4.1). Similarly to Lemma 4.2, the trajectory components for $\hat{x}_2$ and $\hat{x}_4$ cannot be observed from y, and are determined by the eigenvalues of A_{22} and A_{44}. The following is now apparent (see also Fig. 3.4):

The eigenvalues of

A_{11} are controllable and observable,
A_{22} are controllable and unobservable,
A_{33} are uncontrollable and observable,
A_{44} are uncontrollable and unobservable.

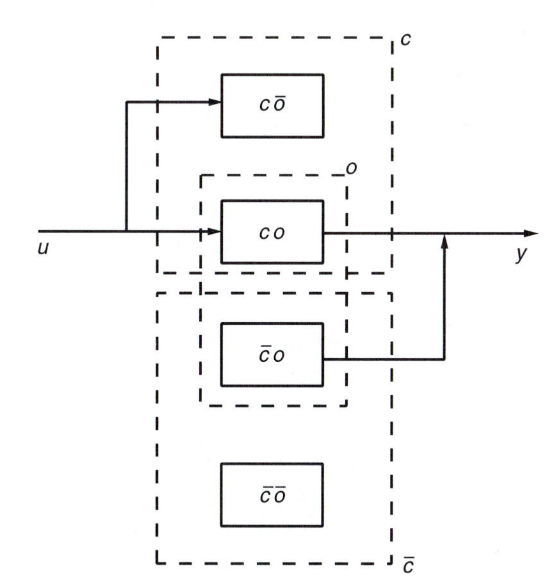

FIGURE 3.4
Canonical decomposition (c and $\bar{c}$ denote controllable and uncontrollable, respectively). The connections of the $c/\bar{c}$ and $o/\bar{o}$ parts of the system to the input and output are emphasized. Note that the impulse response (transfer function) of the system, which is an input-output description only, represents the part of the system that is both controllable and observable (see Section 3.4C).

EXAMPLE 4.4. Given $A = \begin{bmatrix} 0 & -1 & 1 \\ 1 & -2 & 1 \\ 0 & 1 & -1 \end{bmatrix}$, $B = \begin{bmatrix} 1 & 0 \\ 1 & 1 \\ 1 & 2 \end{bmatrix}$, and $C = [0, 1, 0]$, we wish to reduce system (4.1) to the canonical structure (or Kalman decomposition) form (4.20). The appropriate transformation matrix Q is given by (4.19). The matrix $\mathscr{C}$ was found in Example 4.1 and

$$\mathbb{O} = \begin{bmatrix} C \\ CA \\ CA^2 \end{bmatrix} = \begin{bmatrix} 0 & 1 & 0 \\ 1 & -2 & 1 \\ -2 & 4 & -2 \end{bmatrix}.$$

A basis for $R_{\bar{o}} = \mathcal{N}(\mathbb{O})$ is $\{(1, 0, -1)^T\}$. Note that $n_r = 2$, $n_{\bar{o}} = 1$, and $n_{r\bar{o}} = 1$. Therefore,

$$Q = [v_1, v_2, Q_N] = \begin{bmatrix} 1 & 1 & \vdots & 0 \\ 1 & 0 & \vdots & 0 \\ 1 & -1 & \vdots & 1 \end{bmatrix}$$

is an appropriate similarity matrix (check that $\det Q \neq 0$). We compute

$$\hat{A} = Q^{-1}AQ = \begin{bmatrix} 0 & 1 & 0 \\ 1 & -1 & 0 \\ 1 & -2 & 1 \end{bmatrix}\begin{bmatrix} 0 & -1 & 1 \\ 1 & -2 & 1 \\ 0 & 1 & -1 \end{bmatrix}\begin{bmatrix} 1 & 1 & 0 \\ 1 & 0 & 0 \\ 1 & -1 & 1 \end{bmatrix}$$

$$= \begin{bmatrix} 0 & \vdots & 0 & \vdots & 1 \\ \cdots & & \cdots & & \cdots \\ 0 & \vdots & -1 & \vdots & 0 \\ \cdots & & \cdots & & \cdots \\ 0 & \vdots & 0 & \vdots & -2 \end{bmatrix} = \begin{bmatrix} A_{11} & 0 & A_{13} \\ A_{21} & A_{22} & A_{23} \\ 0 & 0 & A_{33} \end{bmatrix},$$

$$\hat{B} = Q^{-1}B = \begin{bmatrix} 0 & 1 & 0 \\ 1 & -1 & 0 \\ 1 & -2 & 1 \end{bmatrix}\begin{bmatrix} 1 & 0 \\ 1 & 1 \\ 1 & 2 \end{bmatrix} = \begin{bmatrix} 1 & 1 \\ \cdots & \cdots \\ 0 & -1 \\ \cdots & \cdots \\ 0 & 0 \end{bmatrix} = \begin{bmatrix} B_1 \\ B_2 \\ 0 \end{bmatrix},$$

and $$\hat{C} = CQ = [0, 1, 0]\begin{bmatrix} 1 & 1 & 0 \\ 1 & 0 & 0 \\ 1 & -1 & 1 \end{bmatrix} = [1, 0, 0] = [C_1, 0, C_3].$$

The eigenvalue 0 (in A_{11}) is controllable and observable, the eigenvalue -1 (in A_{22}) is controllable and unobservable, and the eigenvalue -2 (in A_{33}) is uncontrollable and observable. There are no eigenvalues that are both uncontrollable and unobservable. ■

B. Eigenvalue/Eigenvector Tests for Controllability and Observability

There are tests for controllability (-from-the-origin) and observability for both continuous- and discrete-time time invariant systems that involve the eigenvalues and eigenvectors of A. Some of these criteria are called PBH tests, after the initials of the codiscoverers (Popov-Belevitch-Hautus) of these tests. These tests are useful in theoretical analysis, and in addition, they are also attractive as computational tools.

THEOREM 4.4. (i) The pair (A, B) is uncontrollable if and only if there exists a $1 \times n$ (in general) complex vector $\hat{v}_i \neq 0$ such that

$$\hat{v}_i[\lambda_i I - A, B] = 0, \tag{4.23}$$

where λ_i is some complex scalar.

(ii) The pair (A, C) is unobservable if and only if there exists an $n \times 1$ (in general) complex vector $v_i \neq 0$ such that

$$\begin{bmatrix} \lambda_i I - A \\ C \end{bmatrix} v_i = 0, \tag{4.24}$$

where λ_i is some complex scalar.

Proof. Only part (i) will be considered since (ii) can be proved using a similar argument, or directly, by duality arguments.

(*Sufficiency*) Assume that (4.23) is satisfied. In view of $\hat{v}_i A = \lambda_i \hat{v}_i$ and $\hat{v}_i B = 0$, $\hat{v}_i AB = \lambda_i \hat{v}_i B = 0$, and $\hat{v}_i A^k B = 0$, $k = 0, 1, 2, \ldots$. Therefore, $\hat{v}_i \mathscr{C} = \hat{v}_i [B, AB, \ldots, A^{n-1}B] = 0$, which shows that (A, B) is not completely controllable.

(*Necessity*) Let (A, B) be uncontrollable and assume without loss of generality the standard form for A and B given in Lemma 4.1. We will show that there exist λ_i and $\hat{v}_i$ so that (4.23) holds. Let λ_i be an uncontrollable eigenvalue and let $\hat{v}_i = [0, \alpha]$, $\alpha^T \in C^{n-n_r}$, where $\alpha(\lambda_i I - A_2) = 0$, i.e., α is a left eigenvector of A_2 corresponding to λ_i. Then $\hat{v}_i[\lambda_i I - A, B] = [0, \alpha(\lambda_i I - A_2), 0] = 0$, i.e., (4.23) is satisfied. ■

COROLLARY 4.5. (i) The pair (A, B) is controllable if and only if no left eigenvector of A is orthogonal to all the columns of B.

(ii) The pair (A, C) is observable if and only if no right eigenvector of A is orthogonal to all the rows of C.

Proof. The proof follows directly from Theorem 4.4. ■

COROLLARY 4.6. (i) λ_i is an uncontrollable eigenvalue of (A, B) if and only if there exists a $1 \times n$ (in general) complex vector $\hat{v}_i \neq 0$ that satisfies (4.23).

(ii) λ_i is an unobservable eigenvalue of (A, C) if and only if there exists an $n \times 1$ (in general) complex vector $v_i \neq 0$ that satisfies (4.24).

Proof. Only part (i) will be considered, since part (ii) can be proved using a similar argument or directly, by duality arguments.

(*Sufficiency*) Assume that (4.23) is satisfied. Now $\hat{v}_i[\lambda_i I - A, B] = 0$, in view of the sufficiency proof of Theorem 4.4, implies that $\hat{v}_i \mathscr{C} = \hat{v}_i[B, AB, \ldots, A^{n-1}B] = 0$. Therefore, (A, B) is not controllable. Without loss of generality, assume that (A, B) is in the standard form of Lemma 4.1. In this case the controllability matrix has the form (4.7) with its top n_r rows linearly independent and its lower $n - n_r$ rows being zero. Therefore, in view of $\hat{v}_i \mathscr{C} = 0$, $\hat{v}_i$ has the form $\hat{v}_i = [0, \alpha]$ for some $\alpha \in C^{n-n_r}$. Now $\hat{v}_i(\lambda_i I - A) = 0$ implies that $\alpha(\lambda_i I - A_2) = 0$, which shows that λ_i is an eigenvalue of A_2, i.e., it is an uncontrollable eigenvalue.

(*Necessity*) Let λ_i be an uncontrollable eigenvalue of (A, B). Assume without loss of generality that the pair (A, B) is in the standard form of Lemma 4.1. Then $\hat{v}_i = [0, \alpha]$, where α is such that $\alpha(\lambda_i I - A_2) = 0$ (see the proof of Theorem 4.4), satisfies $\hat{v}_i(\lambda_i I - A) = 0$. Also, $\hat{v}_i B = [0, \alpha]\begin{bmatrix} B_1 \\ 0 \end{bmatrix} = 0$. So (4.23) is satisfied. ■

EXAMPLE 4.5. Given are $A = \begin{bmatrix} 0 & -1 & 1 \\ 1 & -2 & 1 \\ 0 & 1 & -1 \end{bmatrix}$, $B = \begin{bmatrix} 1 & 0 \\ 1 & 1 \\ 1 & 2 \end{bmatrix}$, and $C = [0, 1, 0]$, as in Example 4.4. The matrix A has three eigenvalues, $\lambda_1 = 0$, $\lambda_2 = -1$, and $\lambda_3 = -2$, with corresponding right eigenvectors $v_1 = [1, 1, 1]^T$, $v_2 = [1, 0, -1]^T$, $v_3 = [1, 1, -1]^T$ and with left eigenvectors $\hat{v}_1 = [\frac{1}{2}, 0, \frac{1}{2}]$, $\hat{v}_2 = [1, -1, 0]$, and $\hat{v}_3 = [-\frac{1}{2}, 1, -\frac{1}{2}]$, respectively.

In view of Corollary 4.6, $\hat{v}_1 B = [1, 1] \neq 0$ implies that $\lambda_1 = 0$ is controllable. This is because $\hat{v}_1$ is the only nonzero vector (within a multiplication by a nonzero scalar) that satisfies $\hat{v}_1(\lambda_1 I - A) = 0$, and so $\hat{v}_1 B \neq 0$ implies that the only 1×3 vector α that satisfies $\alpha[\lambda_1 I - A, B] = 0$ is the zero vector, which in turn implies that λ_1 is controllable in view of (i) of Lemma 4.6. For similar reasons $Cv_1 = 1 \neq 0$ implies that $\lambda_1 = 0$ is observable; see (ii) of Lemma 4.6. Similarly, $\hat{v}_2 B = [0, -1] \neq 0$ implies that $\lambda_2 = -1$ is controllable, and $Cv_2 = 0$ implies that $\lambda_2 = -1$ is *unobservable*. Also, $\hat{v}_3 B = [0, 0]$ implies that $\lambda_3 = -2$ is *uncontrollable,* and $Cv_3 = 1 \neq 0$ implies that $\lambda_3 = -2$ is observable.

These results agree with the results derived in Example 4.4. ■

COROLLARY 4.7. (RANK TESTS). (ia) The pair (A, B) is controllable if and only if

$$rank\,[\lambda I - A, B] = n \tag{4.25}$$

for all complex numbers λ, or for all n eigenvalues λ_i of A.

(ib) λ_i is an uncontrollable eigenvalue of A if and only if

$$rank\,[\lambda_i I - A, B] < n. \tag{4.26}$$

(iia) The pair (A, C) is observable if and only if

$$rank \begin{bmatrix} \lambda I - A \\ C \end{bmatrix} = n \tag{4.27}$$

for all complex numbers λ, or for all n eigenvalues λ_i.

(iib) λ_i is an unobservable eigenvalue of A if and only if

$$rank \begin{bmatrix} \lambda_i I - A \\ C \end{bmatrix} < n. \tag{4.28}$$

Proof. The proofs follow in a straightforward manner from Theorem 4.4. Notice that the only values of λ that can possibly reduce the rank of $[\lambda I - A, B]$ are the eigenvalues of A. ■

EXAMPLE 4.6. If in Example 4.5 the eigenvalues $\lambda_1, \lambda_2, \lambda_3$ of A are known, but the corresponding eigenvectors are not, consider the system matrix

$$P(s) = \begin{bmatrix} sI - A & B \\ -C & 0 \end{bmatrix} = \begin{bmatrix} s & 1 & -1 & \vdots & 1 & 0 \\ -1 & s+2 & -1 & \vdots & 1 & 1 \\ 0 & -1 & s+1 & \vdots & 1 & 2 \\ \cdots & \cdots & \cdots & \vdots & \cdots & \cdots \\ 0 & -1 & 0 & & 0 & 0 \end{bmatrix}$$

and determine $rank\,[\lambda_i I - A, B]$ and $rank \begin{bmatrix} \lambda_i I - A \\ C \end{bmatrix}$. Notice that

$$rank \begin{bmatrix} sI - A \\ C \end{bmatrix}_{s=\lambda_2} = rank \begin{bmatrix} -1 & 1 & -1 \\ -1 & 1 & -1 \\ 0 & -1 & 0 \\ 0 & 1 & 0 \end{bmatrix} = 2 < 3 = n$$

$$\text{and} \quad rank\,[sI - A, B]_{s=\lambda_3} = rank \begin{bmatrix} -2 & 2 & -1 & 1 & 0 \\ -1 & 0 & -1 & 1 & 1 \\ 0 & -1 & -1 & 1 & 2 \end{bmatrix} = 2 < 3 = n.$$

In view of Corollary 4.7, $\lambda_2 = -1$ is unobservable and $\lambda_3 = -2$ is uncontrollable. Verify that these are the only uncontrollable and unobservable eigenvalues by applying the rank tests of Corollary 4.7. Compare these results with the results in Example 4.5. ■

EXAMPLE 4.7. Let $A = \begin{bmatrix} 1 & 1 \\ 0 & 1 \end{bmatrix}$, and $B = \begin{bmatrix} 1 \\ 0 \end{bmatrix}$ with $\lambda_1 = \lambda_2 = 1$ the eigenvalues of A. We would like to determine which of the eigenvalues are uncontrollable. Note that (A, B) is in the standard form for uncontrollable systems of Lemma 4.1, namely, $\left(\begin{bmatrix} A_1 & A_{12} \\ 0 & A_2 \end{bmatrix}, \begin{bmatrix} B_1 \\ 0 \end{bmatrix}\right)$. We know by inspection that the eigenvalue $\lambda_2 = 1$ of A_2, is uncontrollable. Presently, $[\lambda_i I - A, B] = \begin{bmatrix} \lambda_i - 1 & -1 & 1 \\ 0 & \lambda_i - 1 & 0 \end{bmatrix}$, and for $\hat{v}_2 = [0, 1]$, $\hat{v}_2[\lambda_2 I - A, B] = [0, 0]$, which in view of Theorem 4.4 and Corollary 4.6, implies that $\lambda_2 = 1$ is an uncontrollable eigenvalue and that $\hat{v}_2 = [0, 1]$ is the corresponding left eigenvector. Note that $\hat{v}_2$ is of the form $[0, \alpha]$, as discussed in the proof of Theorem 4.4. The other eigenvalue, $\lambda_1 = 1$, is controllable. It is the eigenvalue of $A_1 = 1$, where (A_1, B_1) is controllable. Note that the corresponding eigenvector to the controllable eigenvalue is $\hat{v}_1 = [0, 1]$, the same as $\hat{v}_2$. Therefore, when using the eigenvalue/eigenvector tests, one can detect in the present example only that at least one of the multiple eigenvalues is uncontrollable; this test is unable to detect that the other eigenvalue is controllable. This situation arises when there are multiple eigenvalues, in which case care should be taken when using the eigenvalue/eigenvector tests. When the eigenvalues are distinct, then each of them can specifically be identified as being controllable or uncontrollable by the eigenvalue/eigenvector test. For another example, try $A = \begin{bmatrix} 1 & 0 \\ 0 & 1 \end{bmatrix}$ and $B = \begin{bmatrix} 1 \\ 0 \end{bmatrix}$. ■

C. Relating State-Space and Input-Output Descriptions

The system $\dot{x} = Ax + Bu$, $y = Cx + Du$ given in (4.1) has $p \times m$ transfer function matrix

$$H(s) = C(sI - A)^{-1}B + D = \hat{C}(sI - \hat{A})^{-1}\hat{B} + \hat{D}, \tag{4.29}$$

where $\{\hat{A}, \hat{B}, \hat{C}, \hat{D}\}$ is the equivalent representation given in (4.9) (see also Sections 2.5 and 2.6). Consider now the Kalman Decomposition Theorem and the representation (4.22). We wish to investigate which of the submatrices A_{ij}, B_i, C_j determine $H(s)$ and which do not. The inverse of $sI - \hat{A}$ can be determined by repeated application of the formulas

$$\begin{bmatrix} \alpha & \beta \\ 0 & \delta \end{bmatrix}^{-1} = \begin{bmatrix} \alpha^{-1} & -\alpha^{-1}\beta\delta^{-1} \\ 0 & \delta^{-1} \end{bmatrix}$$

and

$$\begin{bmatrix} \alpha & 0 \\ \gamma & \delta \end{bmatrix}^{-1} = \begin{bmatrix} \alpha^{-1} & 0 \\ -\delta^{-1}\gamma\alpha^{-1} & \delta^{-1} \end{bmatrix}, \tag{4.30}$$

where $\alpha, \beta, \gamma, \delta$ are matrices, with α and δ square and nonsingular. It turns out (verify) that

$$H(s) = C_1(sI - A_{11})^{-1}B_1 + D, \tag{4.31}$$

that is, the only part of the system that determines the external description is $\{A_{11}, B_1, C_1, D\}$, the subsystem that is both controllable and observable [see Theorem 4.3(iii)]. Analogous results exist in the time domain. Specifically, taking the inverse Laplace transform of both sides in (4.29), the impulse response of the system for $t \geq 0$ is derived as (see Chapter 2)

$$H(t,0) = C_1 e^{A_{11}t} B_1 + D\delta(t), \tag{4.32}$$

which depends only on the controllable and observable parts of the system, as expected.

Similar results exist for discrete-time systems described by (4.4). For such systems, the transfer function matrix $H(z)$ and the pulse response $H(k,0)$ (see Chapter 2) are given by

$$H(z) = C_1(zI - A_{11})^{-1}B_1 + D \tag{4.33}$$

and

$$H(k,0) = \begin{cases} C_1 A_{11}^{k-1} B_1, & k > 0, \\ D, & k = 0. \end{cases} \tag{4.34}$$

These depend only on the part of the system that is both controllable and observable, as in the continuous-time case.

EXAMPLE 4.8. For the system $\dot{x} = Ax + Bu$, $y = Cx$, where A, B, C are as in Examples 4.4 and 4.5, we have $H(s) = C(sI - A)^{-1}B = C_1(sI - A_{11})^{-1}B_1 = (1)(1/s)[1, 1] = [1/s, 1/s]$. Notice that only the controllable and observable eigenvalue of A, $\lambda_1 = 0$ (in A_{11}), appears in the transfer function as a pole. All other eigenvalues ($\lambda_2 = -1$, $\lambda_3 = -2$) cancel out. ■

EXAMPLE 4.9. The circuit depicted in Fig. 3.5 is described by the state-space equations

$$\begin{bmatrix} \dot{x}_1(t) \\ \dot{x}_2(t) \end{bmatrix} = \begin{bmatrix} -\dfrac{1}{(R_1C)} & 0 \\ 0 & -\dfrac{R_2}{L} \end{bmatrix} \begin{bmatrix} x_1(t) \\ x_2(t) \end{bmatrix} + \begin{bmatrix} \dfrac{1}{(R_1C)} \\ \dfrac{1}{L} \end{bmatrix} v(t)$$

$$i(t) = \left[-\frac{1}{R_1}, 1\right] \begin{bmatrix} x_1(t) \\ x_2(t) \end{bmatrix} + \left(\frac{1}{R_1}\right) v(t),$$

where the voltage $v(t)$ and current $i(t)$ are the input and output variables of the system, $x_1(t)$ is the voltage across the capacitor, and $x_2(t)$ is the current through the inductor. We have $\hat{i}(s) = H(s)\hat{v}(s)$ with the transfer function given by

$$H(s) = C(sI - A)^{-1}B + D = \frac{(R_1^2C - L)s + (R_1 - R_2)}{(Ls + R_2)(R_1^2Cs + R_1)} + \frac{1}{R_1}.$$

The eigenvalues of A are $\lambda_1 = -1/(R_1C)$ and $\lambda_2 = -R_2/L$. Note that in general $rank\ [\lambda_i I\ -A, B] = rank \begin{bmatrix} \lambda_i I - A \\ C \end{bmatrix} = 2 = n$, i.e., the system is controllable and

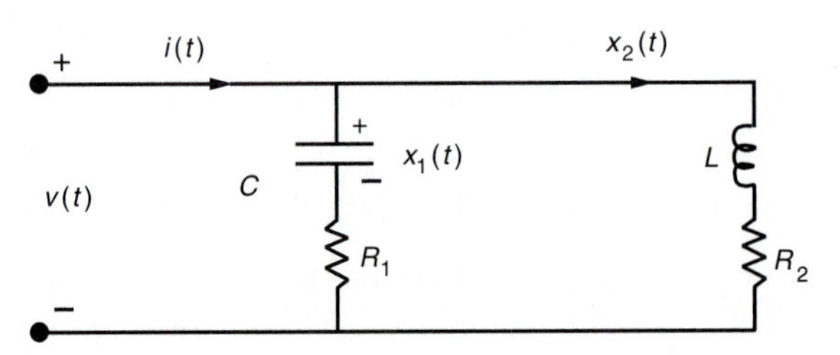

FIGURE 3.5

observable, unless the relation $R_1R_2C = L$ is satisfied. In this case $\lambda_1 = \lambda_2 = -R_2/L$ and the system matrix $P(s)$ assumes the form

$$P(s) = \begin{bmatrix} sI - A & B \\ -C & D \end{bmatrix} = \begin{bmatrix} s + \frac{R_2}{L} & 0 & \frac{R_2}{L} \\ 0 & s + \frac{R_2}{L} & \frac{1}{L} \\ \frac{1}{R_1} & -1 & \frac{1}{R_1} \end{bmatrix}.$$

In the following, assume that $R_1R_2C = L$ is satisfied.

(i) Let $R_1 \neq R_2$ and take

$$[v_1, v_2] = \begin{bmatrix} R_2 & R_1 \\ 1 & 1 \end{bmatrix}, \quad \begin{bmatrix} \hat{v}_1 \\ \hat{v}_2 \end{bmatrix} = [v_1, v_2]^{-1} = \frac{1}{R_2 - R_1} \begin{bmatrix} 1 & -R_1 \\ -1 & R_2 \end{bmatrix}$$

to be the linearly independent right and left eigenvectors corresponding to the eigenvalues $\lambda_1 = \lambda_2 = -R_2/L$. The eigenvectors could have been any two linearly independent vectors since $\lambda_i I - A = 0$. They were chosen as above because they also have the property that $\hat{v}_2 B = 0$ and $Cv_2 = 0$, which in view of Theorem 4.4, implies that $\lambda_2 = -R_2/L$ is both uncontrollable and unobservable. The eigenvalue $\lambda_1 = -R_2/L$ is both controllable and observable since it can be seen using $Q = \begin{bmatrix} R_2 & R_1 \\ 1 & 1 \end{bmatrix}$ to reduce the representation to the canonical structure form (Kalman Decomposition Theorem) (verify this). The transfer function is in this case given by

$$H(s) = \frac{(s + R_1/L)(s + R_2/L)}{R_1(s + R_2/L)(s + R_2/L)} = \frac{s + R_1/L}{R_1(s + R_2/L)},$$

that is, only the controllable and observable eigenvalue appears as a pole in $H(s)$, as expected.

(ii) Let $R_1 = R_2 = R$ and take

$$[v_1, v_2] = \begin{bmatrix} 1 & R \\ 0 & 1 \end{bmatrix}, \begin{bmatrix} \hat{v}_1 \\ \hat{v}_2 \end{bmatrix} = [v_1, v_2]^{-1} = \begin{bmatrix} 1 & -R \\ 0 & 1 \end{bmatrix}.$$

In this case $\hat{v}_1 B = 0$ and $Cv_2 = 0$. Thus, one of the eigenvalues, $\lambda_1 = -R/L$, is uncontrollable (but can be shown to be observable) and the other eigenvalue, $\lambda_2 = -R/L$, is unobservable (but can be shown to be controllable). In the present case, none of the eigenvalues appear in the transfer function. In fact,

$$H(s) = \frac{1}{R},$$

as can readily be verified. Thus, in this case the network behaves as a constant resistance network.

At this point it should be made clear that the modes that are uncontrollable and/or unobservable from certain inputs and outputs do not actually disappear; they are simply invisible from certain vantage points under certain conditions. (The voltages and currents of this network in the case of constant resistance [$H(s) = 1/R$] are studied in Exercise 3.26.) ■

EXAMPLE 4.10. Consider the system $\dot{x} = Ax + Bu$, $y = Cx$ where $A = \begin{bmatrix} 1 & 0 & 0 \\ 0 & -2 & 0 \\ 0 & 0 & -1 \end{bmatrix}$, $B = \begin{bmatrix} 1 \\ 0 \\ 1 \end{bmatrix}$, and $C = [1, 1, 0]$. Using the eigenvalue/eigenvector test it can be shown (verify this) that the three eigenvalues of A (resp., the three modes of A) are $\lambda_1 = 1$ (resp., e^t), which is controllable and observable, $\lambda_2 = -2$ (resp., e^{-2t}), which is uncontrollable and observable, and $\lambda_3 = -1$ (resp., e^{-t}), which is controllable and unobservable.

The response due to the initial condition $x(0)$ and the input $u(t)$ is

$$x(t) = e^{At}x(0) + \int_0^t e^{A(t-\tau)}Bu(\tau)\,d\tau$$

$$= \begin{bmatrix} e^t & 0 & 0 \\ 0 & e^{-2t} & 0 \\ 0 & 0 & e^{-t} \end{bmatrix} x(0) + \int_0^t \begin{bmatrix} e^{(t-\tau)} \\ 0 \\ e^{-(t-\tau)} \end{bmatrix} u(\tau)\,d\tau$$

and

$$y(t) = Ce^{At}x(0) + \int_0^t Ce^{A(t-\tau)}Bu(\tau)\,d\tau$$

$$= [e^t, e^{-2t}, 0]x(0) + \int_0^t e^{(t-\tau)}u(\tau)\,d\tau.$$

Notice that only controllable modes appear in $e^{At}B$ [resp., only controllable eigenvalues appear in $(sI - A)^{-1}B$], only observable modes appear in Ce^{At} [resp., only observable eigenvalues appear in $C(sI-A)^{-1}$], and only modes that are both controllable and observable appear in $Ce^{At}B$ [resp., only eigenvalues which are both controllable and observable appear in $C(sI - A)^{-1}B = H(s)$].

For the discrete-time case, refer to Exercise 3.17d. ■

D. Controller and Observer Forms

It has been seen several times in this book that equivalent representations of systems

$$\dot{x} = Ax + Bu, \qquad y = Cx + Du, \tag{4.35}$$

given by the equations

$$\dot{\hat{x}} = \hat{A}\hat{x} + \hat{B}u, \qquad y = \hat{C}\hat{x} + \hat{D}u, \tag{4.36}$$

where $\hat{x} = Px$, $\hat{A} = PAP^{-1}$, $\hat{B} = PB$, $\hat{C} = CP^{-1}$, and $\hat{D} = D$, may offer advantages over the original representation when P (or $Q = P^{-1}$) is chosen in an appropriate manner. This is the case when P (or Q) is such that the new basis of the state space (see Section 2.2) provides a natural setting for the properties of interest. As a specific case, refer for example to Subsection 3.4A, where Q (and the new basis) was chosen so that the controllable and uncontrollable parts of the system were separated. The same results of course apply to discrete-time systems (4.4). This subsection shows how to select Q when (A, B) is controllable [or (A, C) is observable] to obtain the controller and observer forms. These special forms are very useful, especially when studying state-feedback control (and state observers) discussed in Chapter 4 and in realizations discussed in Chapter 5. These special forms are also very useful in establishing a quick way to shift between state-space representations and another very useful class of equivalent internal representations, the polynomial matrix representations studied in Chapter 7.

Controller forms are considered first. Observer forms can of course be obtained directly in a similar manner as the controller forms, or they may be obtained by duality. This is addressed in the latter part of this section.

1. Controller forms

The controller form is a particular system representation where both matrices (A, B) have certain special structure. Since in this case A is in the companion form (see Section 2.2 in Chapter 2) the controller form is sometimes also referred to as the *controllable companion form*. Consider the system

$$\dot{x} = Ax + Bu, \qquad y = Cx + Du, \tag{4.37}$$

where $A \in R^{n\times n}$, $B \in R^{n\times m}$, $C \in R^{p\times n}$, and $D \in R^{p\times m}$ and let (A, B) be controllable (-from-the-origin). Then $rank\,\mathscr{C} = n$, where

$$\mathscr{C} = [B, AB, \ldots, A^{n-1}B]. \tag{4.38}$$

Assume that

$$rank\,B = m \leq n. \tag{4.39}$$

Under these assumptions, $rank\,\mathscr{C} = n$ and $rank\,B = m$. We will show how to obtain an equivalent pair $(\hat{A}, \hat{B})$ in controller form, first for the single-input case $(m = 1)$ and then for the multi-input case $(m > 1)$. Before this is accomplished, we discuss two special cases that do not satisfy the above assumptions that $rank\,B = m$ and that (A, B) is controllable.

1. If the m columns of B are not linearly independent $(rank\,B = r < m)$, then there exists an $m \times m$ nonsingular matrix K (or equivently, there exist elementary column operations) so that $BK = [B_r, 0]$, where the r columns of B_r are linearly independent $(rank\,B_r = r)$. Note that $\dot{x} = Ax + Bu = Ax + (BK)(K^{-1}u) = Ax + [B_r, 0]\begin{bmatrix} u_r \\ u_{m-r} \end{bmatrix} = Ax + B_r u_r$, which clearly shows that when $rank\,B = r < m$ the same input action to the system can be accomplished by only r inputs, instead of m inputs, and there is a redundancy of inputs, which in control problems clearly implies that a reconsideration of the input choices is in order. The pair (A, B_r), which is controllable when (A, B) is controllable (show this), can now be reduced to controller form, using the method developed below.
2. When (A, B) is not completely controllable, then a two-step approach can be taken. First, the controllable part is isolated (see Subsection 3.4A) and then is reduced to the controller form, using the methods of this section. In particular, consider the system $\dot{x} = Ax + Bu$ with $A \in R^{n\times n}$, $B \in R^{n\times m}$, and $rank\,B = m$. Let $rank\,[B, AB, \ldots, A^{n-1}B] = n_r < n$. Then there exists a transformation P_1 such that $P_1 A P_1^{-1} = \begin{bmatrix} A_1 & A_{12} \\ 0 & A_2 \end{bmatrix}$ and $P_1 B = \begin{bmatrix} B_1 \\ 0 \end{bmatrix}$, where $A_1 \in R^{n_r\times n_r}$, $B_1 \in R^{n_r\times m}$, and (A_1, B_1) is controllable (Subsection 3.4A). Since (A_1, B_1) is controllable, there exists a transformation P_2 such that $P_2 A_1 P_2^{-1} = A_{1c}$, and $P_2 B_1 = B_{1c}$, where A_{1c}, B_{1c} is in controller form, defined below. Combining, we obtain

$$PAP^{-1} = \begin{bmatrix} A_{1c} & P_2 A_{12} \\ 0 & A_2 \end{bmatrix}$$

and

$$PB = \begin{bmatrix} B_{1c} \\ 0 \end{bmatrix} \tag{4.40}$$

[where $A_{1c} \in R^{n_r \times n_r}$, $B_{1c} \in R^{n_r \times m}$, and (A_{1c}, B_{1c}) is controllable], which is in controller form. Note that

$$P = \begin{bmatrix} P_2 & 0 \\ 0 & I \end{bmatrix} P_1. \tag{4.41}$$

Single-input case ($m = 1$). The representation $\{A_c, B_c, C_c, D_c\}$ in controller form is given by $A_c \triangleq \hat{A} = PAP^{-1}$ and $B_c \triangleq \hat{B} = PB$ with

$$A_c = \begin{bmatrix} 0 & 1 & \cdots & 0 \\ \vdots & \vdots & \ddots & \vdots \\ 0 & 0 & \cdots & 1 \\ -\alpha_0 & -\alpha_1 & \cdots & -\alpha_{n-1} \end{bmatrix}, \qquad B_c = \begin{bmatrix} 0 \\ \vdots \\ 0 \\ 1 \end{bmatrix}, \tag{4.42}$$

where the coefficients α_i are the coefficients of the characteristic polynomial $\alpha(s)$ of A, that is,

$$\alpha(s) \triangleq det(sI - A) = s^n + \alpha_{n-1}s^{n-1} + \cdots + \alpha_1 s + \alpha_0. \tag{4.43}$$

Note that $C_c \triangleq \hat{C} = CP^{-1}$ and $D_c = D$ do not have any particular structure. The structure of (A_c, B_c) is very useful (in control problems) and the representation $\{A_c, B_c, C_c, D_c\}$ shall be referred to as the *controller form* of the system. The similarity transformation matrix P is obtained as follows. The controllability matrix $\mathscr{C} = [B, AB, \ldots, A^{n-1}B]$ is in this case an $n \times n$ nonsingular matrix and $\mathscr{C}^{-1} = \begin{bmatrix} \times \\ q \end{bmatrix}$, where q is the nth row of $\mathscr{C}^{-1}$ and $\times$ indicates the remaining entries of $\mathscr{C}^{-1}$. Then

$$P \triangleq \begin{bmatrix} q \\ qA \\ \vdots \\ qA^{n-1} \end{bmatrix}. \tag{4.44}$$

To show that $PAP^{-1} = A_c$ and $PB = B_c$ given in (4.42), note first that $qA^{i-1}B = 0 i = 1, \ldots, n-1$, and $qA^{n-1}B = 1$. This can be verified from the definition of q, which implies that $q\mathscr{C} = [0, 0, \ldots, 1]$ (verify this). Now

$$P\mathscr{C} = P[B, AB, \ldots, A^{n-1}B] = \begin{bmatrix} 0 & 0 & \cdots & \cdots & 1 \\ 0 & 0 & \cdots & 1 & \times \\ \vdots & 1 & & \vdots & \vdots \\ 1 & \times & \cdots & \times & \times \end{bmatrix} = \mathscr{C}_c, \tag{4.45}$$

which implies that $|P\mathscr{C}| = |P||\mathscr{C}| \neq 0$ or that $|P| \neq 0$. Therefore, P qualifies as a similarity transformation matrix. In view of (4.45), $PB = [0, 0, \ldots, 1]^T = B_c$. Furthermore,

$$A_c P = \begin{bmatrix} qA \\ \vdots \\ qA^{n-1} \\ qA^n \end{bmatrix} = PA, \tag{4.46}$$

where in the last row of A_cP, the relation $-\sum_{i=0}^{n-1}\alpha_i A^i = A^n$ was used [which is the Cayley-Hamilton Theorem, namely, $\alpha(A) = 0$].

EXAMPLE 4.11. Let $A = \begin{bmatrix} -1 & 0 & 0 \\ 0 & 1 & 0 \\ 0 & 0 & -2 \end{bmatrix}$ and $B = \begin{bmatrix} 1 \\ -1 \\ 1 \end{bmatrix}$. Since $n = 3$ and $|sI - A| = (s+1)(s-1)(s+2) = s^3 + 2s^2 - s - 2$, $\{A_c, B_c\}$ in controller form is given by

$$A_c = \begin{bmatrix} 0 & 1 & 0 \\ 0 & 0 & 1 \\ 2 & 1 & -2 \end{bmatrix} \quad \text{and} \quad B_c = \begin{bmatrix} 0 \\ 0 \\ 1 \end{bmatrix}.$$

The transformation matrix P that reduces (A, B) to $(A_c = PAP^{-1}, B_c = PB)$ is now derived. We have

$$\mathscr{C} = [B, AB, A^2B] = \begin{bmatrix} 1 & -1 & 1 \\ -1 & -1 & -1 \\ 1 & -2 & 4 \end{bmatrix}$$

and

$$\mathscr{C}^{-1} = \begin{bmatrix} 1 & -\frac{1}{3} & -\frac{1}{3} \\ -\frac{1}{2} & -\frac{1}{2} & 0 \\ -\frac{1}{2} & -\frac{1}{6} & \frac{1}{3} \end{bmatrix}.$$

The third (the nth) row of $\mathscr{C}^{-1}$ is $q = [-\frac{1}{2}, -\frac{1}{6}, \frac{1}{3}]$, and therefore,

$$P \triangleq \begin{bmatrix} q \\ qA \\ qA^2 \end{bmatrix} = \begin{bmatrix} -\frac{1}{2} & -\frac{1}{6} & \frac{1}{3} \\ \frac{1}{2} & -\frac{1}{6} & -\frac{2}{3} \\ -\frac{1}{2} & -\frac{1}{6} & \frac{4}{3} \end{bmatrix}.$$

It can now easily be verified that $A_c = PAP^{-1}$, or

$$A_cP = \begin{bmatrix} \frac{1}{2} & -\frac{1}{6} & -\frac{2}{3} \\ -\frac{1}{2} & -\frac{1}{6} & -\frac{2}{3} \\ \frac{1}{2} & -\frac{1}{6} & -\frac{8}{3} \end{bmatrix} = PA,$$

and that $B_c = PB$. [Verify that $P\mathscr{C} = \mathscr{C}_c$ is given by (4.45) and also compare with Exercise 3.23, which explicitly derives $\mathscr{C}_c$.] ■

An alternative form to (4.42) is

$$A_{c1} = \begin{bmatrix} -\alpha_{n-1} & \cdots & -\alpha_1 & -\alpha_0 \\ 1 & \cdots & 0 & 0 \\ \vdots & \ddots & \vdots & \vdots \\ 0 & \cdots & 1 & 0 \end{bmatrix}, \qquad B_{c1} = \begin{bmatrix} 1 \\ 0 \\ \vdots \\ 0 \end{bmatrix}, \tag{4.47}$$

which is obtained if the similarity transformation matrix is taken to be

$$P_1 \triangleq \begin{bmatrix} qA^{n-1} \\ \vdots \\ qA \\ q \end{bmatrix}, \tag{4.48}$$

i.e., by reversing the order of the rows of P in (4.44). The reader should verify this (see Exercise 3.25).

In the above, A_c is a companion matrix of the form $\begin{bmatrix} 0 & I \\ \times & \times \end{bmatrix}$ or $\begin{bmatrix} \times & \times \\ I & 0 \end{bmatrix}$ and B_c has the form $[0 \quad 0 \ldots 1]^T$ or $[1 \quad 0 \ldots 0]^T$, respectively. These representations are very useful when studying pole assignment via state feedback control law and are used in the next chapter.

A companion matrix could also be of the form $\begin{bmatrix} 0 & \times \\ I & \times \end{bmatrix}$ or $\begin{bmatrix} \times & 0 \\ \times & I \end{bmatrix}$ (see Section 2.2) with the coefficients $-[\alpha_0, \ldots, \alpha_{n-1}]^T$ in the last or the first column. It is shown here, for completeness, how to determine controller forms where A_c are such companion matrices. In particular, if

$$Q_2 = P_2^{-1} = [B, AB, \ldots, A^{n-1}B] = \mathscr{C}, \tag{4.49}$$

then
$$A_{c2} = Q_2^{-1}AQ_2 = \begin{bmatrix} 0 & \cdots & 0 & -\alpha_0 \\ 1 & \cdots & 0 & -\alpha_1 \\ \vdots & \ddots & \vdots & \vdots \\ 0 & \cdots & 1 & -\alpha_{n-1} \end{bmatrix}, \qquad B_{c2} = Q_2^{-1}B = \begin{bmatrix} 1 \\ 0 \\ \vdots \\ 0 \end{bmatrix}. \tag{4.50}$$

Also, if

$$Q_3 = P_3^{-1} = [A^{n-1}B, \ldots, B], \tag{4.51}$$

then
$$A_{c3} = Q_3^{-1}AQ_3 = \begin{bmatrix} -\alpha_{n-1} & 1 & \cdots & 0 \\ \vdots & \vdots & \ddots & \vdots \\ -\alpha_1 & 0 & \cdots & 1 \\ -\alpha_0 & 0 & \cdots & 0 \end{bmatrix}, \qquad B_{c3} = Q_3^{-1}B = \begin{bmatrix} 0 \\ \vdots \\ 0 \\ 1 \end{bmatrix}. \tag{4.52}$$

(A_c, B_c) in (4.50) and (4.52) are also in controller canonical or controllable companion form. (The reader is encouraged to verify these expressions. See also Exercise 3.25.) We also note that if the structures of A_c and B_c are specified, then P is uniquely determined (see Exercise 3.24). That is, given (A_c, B_c) in any of the above four controllable companion forms, P is readily uniquely determined in each case by $P = \hat{\mathscr{C}}\mathscr{C}^{-1}$, assuming that P also satisfies $PA^nB = A_c^nB_c$ (in view of Exercise 3.24), which it does in the above four cases. If different B_c are desirable, then an appropriate P can be found by the same formula. Note that $\hat{\mathscr{C}}$ denotes the controllability matrix of (A_c, B_c).

EXAMPLE 4.12. Let $A = \begin{bmatrix} -1 & 0 & 0 \\ 0 & 1 & 0 \\ 0 & 0 & -2 \end{bmatrix}$ and $B = \begin{bmatrix} 1 \\ -1 \\ 1 \end{bmatrix}$, as in Example 4.11. Alternative controller forms can be derived for different P. In particular, if

(i) $P = P_1 = \begin{bmatrix} qA^2 \\ qA \\ q \end{bmatrix} = \begin{bmatrix} -\frac{1}{2} & -\frac{1}{6} & \frac{4}{3} \\ \frac{1}{2} & -\frac{1}{6} & -\frac{2}{3} \\ -\frac{1}{2} & -\frac{1}{6} & \frac{1}{3} \end{bmatrix}$, as in (4.48) ($\mathscr{C}$, $\mathscr{C}^{-1}$, and q were found in Example 4.11), then

$$A_{c1} = \begin{bmatrix} -2 & 1 & 2 \\ 1 & 0 & 0 \\ 0 & 1 & 0 \end{bmatrix}, \qquad B_{c1} = \begin{bmatrix} 1 \\ 0 \\ 0 \end{bmatrix},$$

as in (4.47). Note that in the present case $A_{c1}P_1 = \begin{bmatrix} \frac{1}{2} & -\frac{1}{6} & -\frac{8}{3} \\ -\frac{1}{2} & -\frac{1}{6} & \frac{4}{3} \\ \frac{1}{2} & -\frac{1}{6} & -\frac{2}{3} \end{bmatrix} = P_1 A$, $B_{c1} = P_1 B$.

(ii) $Q_2 = \mathscr{C} = \begin{bmatrix} 1 & -1 & 1 \\ -1 & -1 & -1 \\ 1 & -2 & 4 \end{bmatrix}$, as in (4.49). Then

$$A_{c2} = \begin{bmatrix} 0 & 0 & 2 \\ 1 & 0 & 1 \\ 0 & 1 & -2 \end{bmatrix}, \qquad B_{c2} = Q_2^{-1}B = \begin{bmatrix} 1 \\ 0 \\ 0 \end{bmatrix},$$

as in (4.50).

(iii) $Q_3 = [A^2B, AB, B] = \begin{bmatrix} 1 & -1 & 1 \\ -1 & -1 & -1 \\ 4 & -2 & 1 \end{bmatrix}$, as in (4.51). Then

$$A_{c3} = \begin{bmatrix} -2 & 1 & 0 \\ 1 & 0 & 1 \\ 2 & 0 & 0 \end{bmatrix}, \qquad B_{c3} = \begin{bmatrix} 0 \\ 0 \\ 1 \end{bmatrix},$$

as in (4.52). Note that $Q_3 A_{c3} = \begin{bmatrix} -1 & 1 & -1 \\ -1 & -1 & -1 \\ -8 & 4 & -2 \end{bmatrix} = AQ_3$, $Q_3 B_{c3} = \begin{bmatrix} 1 \\ -1 \\ 1 \end{bmatrix} = B$. ■

***Multi-input case** ($m > 1$).* In this case the $n \times mn$ matrix $\mathscr{C}$ given in (4.38) is not square, and there are typically many sets of n columns of $\mathscr{C}$ that are linearly independent ($\mathit{rank}\, \mathscr{C} = n$). Depending on which columns are chosen and in what order, different controller forms (controllable companion forms) are derived. Note that in the case when $m = 1$, four different controller forms were derived, even though there was only one set of n linearly independent columns. In the present case there are many more such choices. The form that will be used most often in the following is a generalization of (A_c, B_c) given in (4.42), and this is the form that will be derived first. Other forms will be discussed as well.

Let $\hat{A} = PAP^{-1}$ and $\hat{B} = PB$, where P is constructed as follows: consider

$$\begin{aligned} \mathscr{C} &= [B, AB, \ldots, A^{n-1}B] \\ &= [b_1, \ldots, b_m, Ab_1, \ldots, Ab_m, \ldots, A^{n-1}b_1, \ldots, A^{n-1}b_m], \end{aligned} \tag{4.53}$$

where the $b_1, \ldots, b_m$ are the m columns of B. Select, starting from the left and moving to the right, the first n independent columns ($\mathit{rank}\, \mathscr{C} = n$). Reorder these columns by taking first b_1, Ab_1, A^2b_1, etc., until all columns involving b_1 have been taken; then take b_2, Ab_2, etc., and lastly, take b_m, Ab_m, etc., to obtain

$$\bar{\mathscr{C}} \triangleq [b_1, Ab_1, \ldots, A^{\mu_1-1}b_1, \ldots, b_m, \ldots, A^{\mu_m-1}b_m], \tag{4.54}$$

an $n \times n$ matrix. The integer μ_i denotes the number of columns involving b_i in the set of the first n linearly independent columns found in $\mathscr{C}$ when moving from left to right.

DEFINITION 4.1. The m integers μ_i, $i = 1, \ldots, m$, are the *controllability indices of the system,* and $\mu \triangleq \max \mu_i$ is called the *controllability index of the system.* Note that

$$\sum_{i=1}^{m} \mu_i = n \qquad \text{and} \qquad m\mu \geq n. \tag{4.55}$$

■

To illustrate the significance of μ, note that an alternative but equivalent definition for μ is that μ is the minimum integer k such that

$$rank\,[B, AB, \ldots, A^{k-1}B] = n. \tag{4.56}$$

Taking this view, one keeps adding blocks B, AB, A^2B, etc., until n independent columns appear (from left to right) for the first time. It is then not difficult to see that $\mu = \max\ \mu_i$. Alternatively, μ can be defined as the least integer such that

$$rank\,[B, AB, \ldots, A^{\mu-1}B] = rank\,[B, AB, \ldots, A^{\mu}B]. \tag{4.57}$$

Notice that in (4.54) all columns of B are always present since $rank\ B = m$. This implies also that $\mu_i \geq 1$ for all i. Notice further that if $A^k b_i$ is present, then $A^{k-1}b_i$ must also be present in (4.54). For if it were not, i.e., if it were dependent on the previous columns in $\mathscr{C}$ and had been eliminated, then $A^k b_i$ would also have been dependent (write $A^{k-1}b_i$ as a linear combination of previous columns in $\mathscr{C}$ and premultiply by A to show this). Column $A^{\mu_i}b_i$ is of course dependent on the previous ones. This relation can be expressed explicitly as

$$A^{\mu_i}b_i = \sum_{j=1}^{m} \sum_{k=1}^{\min(\mu_i, \mu_j)} \alpha_{ijk} A^{k-1} b_j + \sum_{\substack{j=1 \\ \mu_i < \mu_j}}^{i-1} \beta_{ij} A^{\mu_i} b_j \tag{4.58}$$

Note that the first sum in (4.58) indicates the dependence of $A^{\mu_i}b_i$ on the linearly independent columns in $B, AB, \ldots, A^{\mu_i-1}B$, while the second sum shows the dependence on the independent columns in $A^{\mu_i}B$ to the left of $A^{\mu_i}b_i$ (α_{ijk} and β_{ij} are appropriate reals).

Now define

$$\sigma_k \triangleq \sum_{i=1}^{k} \mu_i, \qquad k = 1, \ldots, m, \tag{4.59}$$

i.e., $\sigma_1 = \mu_1, \sigma_2 = \mu_1 + \mu_2, \ldots, \sigma_m = \mu_1 + \cdots + \mu_m = n$. Also, consider $\bar{\mathscr{C}}^{-1}$ and let q_k, where $q_k^T \in R^n$, $k = 1, \ldots, m$, denote its σ_kth row, i.e.,

$$\bar{\mathscr{C}}^{-1} = [\times, \cdots, \times, q_1^T \vdots \cdots \vdots \times, \cdots, \times, q_m^T]^T. \tag{4.60}$$

Next, define

$$P \triangleq \begin{bmatrix} q_1 \\ q_1 A \\ \vdots \\ q_1 A^{\mu_1 - 1} \\ \vdots \\ q_m \\ q_m A \\ \vdots \\ q_m A^{\mu_m - 1} \end{bmatrix}. \tag{4.61}$$

It can now be shown that $PAP^{-1} = A_c$ and $PB = B_c$ with

$$A_c = [A_{ij}], \qquad i, j = 1, \ldots, m,$$

$$A_{ii} = \begin{bmatrix} 0 & & & \\ \vdots & & I_{\mu_i - 1} & \\ 0 & & & \\ \times & \times & \cdots & \times \end{bmatrix} \in R^{\mu_i \times \mu_i},\ i = j,$$

$$A_{ij} = \begin{bmatrix} 0 & & \cdots & 0 \\ \vdots & & \vdots & \vdots \\ 0 & & \cdots & 0 \\ \times & \times & \cdots & \times \end{bmatrix} \in R^{\mu_i \times \mu_j},\ i \neq j,$$

and

$$B_c = \begin{bmatrix} B_1 \\ B_2 \\ \vdots \\ B_m \end{bmatrix}, \qquad B_i = \begin{bmatrix} 0 & \cdots & 0 & 0 & & \cdots & 0 \\ \vdots & & \vdots & \vdots & & & \vdots \\ 0 & \cdots & 0 & 1 & \times & \cdots & \times \end{bmatrix} \in R^{\mu_i \times m}, \tag{4.62}$$

where the 1 in the last row of B_i occurs at the ith column location, $i = 1, \ldots, m$, and $\times$ denotes nonfixed entries. Note that $C_c = CP^{-1}$ does not have any particular structure. The expression (4.62) is a very useful form (in control problems) and shall be referred to as the *controller form* of the system. The derivation of this result is discussed below. First, examples are given to illustrate the procedure involved, and then some alternative expressions and properties are also discussed.

EXAMPLE 4.13. Given are $A \in R^{n \times n}$ and $B \in R^{n \times m}$ with (A, B) controllable and with $rank\ B = m$. Let $n = 4$ and $m = 2$. Then there must be two controllability indices μ_1 and μ_2 such that $n = 4 = \sum_{i=1}^{2} \mu_i = \mu_1 + \mu_2$. Under these conditions, there are three possibilities:

(i) $\mu_1 = 2,\ \mu_2 = 2,$

$$A_c = \begin{bmatrix} A_{11} & A_{12} \\ A_{21} & A_{22} \end{bmatrix} = \left[\begin{array}{cc:cc} 0 & 1 & 0 & 0 \\ \times & \times & \times & \times \\ \cdots & \cdots & \cdots & \cdots \\ 0 & 0 & 0 & 1 \\ \times & \times & \times & \times \end{array}\right], \qquad B_c = \begin{bmatrix} B_1 \\ B_2 \end{bmatrix} = \begin{bmatrix} 0 & 0 \\ 1 & \times \\ \cdots & \cdots \\ 0 & 0 \\ 0 & 1 \end{bmatrix}.$$

(ii) $\mu_1 = 1,\ \mu_2 = 3,$

$$A_c = \left[\begin{array}{c:ccc} \times & \times & \times & \times \\ \cdots & \cdots & \cdots & \cdots \\ 0 & 0 & 1 & 0 \\ 0 & 0 & 0 & 1 \\ \times & \times & \times & \times \end{array}\right], \qquad B_c = \begin{bmatrix} 1 & \times \\ \cdots & \cdots \\ 0 & 0 \\ 0 & 0 \\ 0 & 1 \end{bmatrix}.$$

(iii) $\mu_1 = 3$, $\mu_2 = 1$,

$$A_c = \begin{bmatrix} 0 & 1 & 0 & \vdots & 0 \\ 0 & 0 & 1 & \vdots & 0 \\ \times & \times & \times & \vdots & \times \\ \cdots & \cdots & \cdots & \vdots & \cdots \\ \times & \times & \times & \vdots & \times \end{bmatrix}, \qquad B_c = \begin{bmatrix} 0 & 0 \\ 0 & 0 \\ 1 & \times \\ \cdots & \cdots \\ 0 & 1 \end{bmatrix}.$$

■

It is possible to write A_c, B_c in a systematic and perhaps more transparent way. In particular, notice that A_c, B_c in (4.62) can be expressed as

$$A_c = \bar{A}_c + \bar{B}_c A_m, \qquad B_c = \bar{B}_c B_m, \tag{4.63}$$

where $\bar{A}_c = block\,diag\,[\bar{A}_{11}, \bar{A}_{22}, \ldots, \bar{A}_{mm}]$ with

$$\bar{A}_{ii} = \begin{bmatrix} 0 & \\ \vdots & I_{\mu_i - 1} \\ 0 & \\ 0 & 0\cdots0 \end{bmatrix} \in R^{\mu_i \times \mu_i}, \quad \bar{B}_c = block\ diag\left(\begin{bmatrix} 0 \\ \vdots \\ 0 \\ 1 \end{bmatrix} \in R^{\mu_i \times 1}, \quad i = 1, \ldots, m \right),$$

and $A_m \in R^{m \times n}$ and $B_m \in R^{m \times m}$ are some appropriate matrices with $\sum_{i=1}^{m} \mu_i = n$. Note that the matrices $\bar{A}_c$, $\bar{B}_c$ are completely determined by the m controllability indices μ_i, $i = 1, \ldots, m$ (they are in fact in the so-called Brunovski canonical form—see the discussion following Lemma 4.8). The matrices A_m and B_m consist of the σ_1th, σ_2th, $\ldots, \sigma_m$th rows of A_c (entries denoted by $\times$) and the same rows of B_c, respectively. Note that A_m and B_m is that part of the controllable system $\dot{x} = Ax + Bu$ that can be altered by linear state feedback $u = Fx + Gv$, a fact that will be used extensively in the next chapter.

EXAMPLE 4.14. Let $A = \begin{bmatrix} 0 & 1 & 0 \\ 0 & 0 & 1 \\ 0 & 2 & -1 \end{bmatrix}$ and $B = \begin{bmatrix} 0 & 1 \\ 1 & 1 \\ 0 & 0 \end{bmatrix}$. To determine the controller form (4.62), consider

$$\begin{aligned} \mathscr{C} &= [B, AB, A^2B] \\ &= [b_1, b_2, Ab_1, Ab_2, A^2b_1, A^2b_2] = \begin{bmatrix} 0 & 1 & 1 & 1 & 0 & 0 \\ 1 & 1 & 0 & 0 & 2 & 2 \\ 0 & 0 & 2 & 2 & -2 & -2 \end{bmatrix}, \end{aligned}$$

where $rank\,\mathscr{C} = 3 = n$, i.e., (A, B) is controllable. Searching from left to right, the first three columns of $\mathscr{C}$ are selected since they are linearly independent. Then

$$\bar{\mathscr{C}} = [b_1, Ab_1, b_2] = \begin{bmatrix} 0 & 1 & 1 \\ 1 & 0 & 1 \\ 0 & 2 & 0 \end{bmatrix}$$

and the controllability indices are $\mu_1 = 2$ and $\mu_2 = 1$. Also, $\sigma_1 = \mu_1 = 2$ and $\sigma_2 = \mu_1 + \mu_2 = 3 = n$, and

$$\bar{\mathscr{C}}^{-1} = \begin{bmatrix} -1 & 1 & \frac{1}{2} \\ 0 & 0 & \frac{1}{2} \\ 1 & 0 & -\frac{1}{2} \end{bmatrix}.$$

Notice that $q_1 = [0, 0, \frac{1}{2}]$ and $q_2 = [1, 0, -\frac{1}{2}]$, the second and third rows of $\bar{\mathscr{C}}^{-1}$, respectively. In view of (4.61), $P = \begin{bmatrix} q_1 \\ q_1 A \\ q_2 \end{bmatrix} = \begin{bmatrix} 0 & 0 & \frac{1}{2} \\ 0 & 1 & -\frac{1}{2} \\ 1 & 0 & -\frac{1}{2} \end{bmatrix}$, $P^{-1} = \begin{bmatrix} 1 & 0 & 1 \\ 1 & 1 & 0 \\ 2 & 0 & 0 \end{bmatrix}$ and $A_c =$

$$PAP^{-1} = \begin{bmatrix} A_{11} & \vdots & A_{12} \\ \cdots & \vdots & \cdots \\ A_{21} & \vdots & A_{22} \end{bmatrix} = \begin{bmatrix} 0 & 1 & \vdots & 0 \\ 2 & -1 & \vdots & 0 \\ \cdots & \cdots & \vdots & \cdots \\ 1 & 0 & \vdots & 0 \end{bmatrix}, B_c = PB = \begin{bmatrix} B_1 \\ B_2 \end{bmatrix} = \begin{bmatrix} 0 & 0 \\ 1 & 1 \\ \cdots & \cdots \\ 0 & 1 \end{bmatrix}.$$

One can also verify (4.63) quite easily. We have

$$A_c = \begin{bmatrix} 0 & 1 & \vdots & 0 \\ 2 & -1 & \vdots & 0 \\ \cdots & \cdots & \vdots & \cdots \\ 1 & 0 & \vdots & 0 \end{bmatrix} = \bar{A}_c + \bar{B}_c A_m = \begin{bmatrix} 0 & 1 & \vdots & 0 \\ 0 & 0 & \vdots & 0 \\ \cdots & \cdots & \vdots & \cdots \\ 0 & 0 & \vdots & 0 \end{bmatrix} + \begin{bmatrix} 0 & 0 \\ 1 & 0 \\ \cdots & \cdots \\ 0 & 1 \end{bmatrix} \begin{bmatrix} 2 & -1 & 0 \\ 1 & 0 & 0 \end{bmatrix}$$

and

$$B_c = \begin{bmatrix} 0 & 0 \\ 1 & 1 \\ \cdots & \cdots \\ 0 & 1 \end{bmatrix} = \bar{B}_c B_m = \begin{bmatrix} 0 & 0 \\ 1 & 0 \\ \cdots & \cdots \\ 0 & 1 \end{bmatrix} \begin{bmatrix} 1 & 1 \\ 0 & 1 \end{bmatrix}.$$

It is interesting to note that in this example, the given pair (A, B) could have already been in controller form if B were different, but A the same. For example, consider the following three cases:

1. $A = \begin{bmatrix} 0 & \vdots & 1 & 0 \\ \cdots & \vdots & \cdots & \cdots \\ 0 & \vdots & 0 & 1 \\ 0 & \vdots & 2 & -1 \end{bmatrix}, \quad B = \begin{bmatrix} 1 & \times \\ \cdots & \cdots \\ 0 & 0 \\ 0 & 1 \end{bmatrix}, \quad \mu_1 = 1, \mu_2 = 2,$

2. $A = \begin{bmatrix} 0 & 1 & \vdots & 0 \\ 0 & 0 & \vdots & 1 \\ \cdots & \cdots & \vdots & \cdots \\ 0 & 2 & \vdots & -1 \end{bmatrix}, \quad B = \begin{bmatrix} 0 & 0 \\ 1 & \times \\ \cdots & \cdots \\ 0 & 1 \end{bmatrix}, \quad \mu_1 = 2, \mu_1 = 1,$

3. $A = \begin{bmatrix} 0 & 1 & 0 \\ 0 & 0 & 1 \\ 0 & 2 & -1 \end{bmatrix}, \quad B = \begin{bmatrix} 0 \\ 0 \\ 1 \end{bmatrix}, \quad \mu_1 = 3 = n.$

Note that case 3 is the single-input case (4.42). ■

Several results involving the controllability indices of (A, B) are now presented. First, it is shown that the controllability indices μ_i of a system, defined in Definition 4.1, do not change under similarity transformations [for if they were changing, then (A_c, B_c) might have different controllability indices from the original (A, B)].

In particular, let $\dot{x} = Ax + Bu$ be given, where (A, B) is controllable, and consider $\dot{\hat{x}} = \hat{A}\hat{x} + \hat{B}u$, where $\hat{A} = PAP^{-1}$, $\hat{B} = PB$, $\hat{x} = Px$, and P is nonsingular.

LEMMA 4.8. The controllability indices of (A, B) are identical to the controllability indices of $\hat{A} = PAP^{-1}$, $\hat{B} = PB$.

Proof. We have $\mathscr{C} = [B, AB, \ldots, A^{n-1}B] = P^{-1}[\hat{B}, \hat{A}\hat{B}, \ldots, \hat{A}^{n-1}\hat{B}] = P^{-1}\hat{\mathscr{C}}$. To determine the controllability indices μ_i of (A, B), the first n linearly independent columns of $\mathscr{C}$ are taken, working from left to right. It is clear that the corresponding n columns of $\hat{\mathscr{C}}$ will also be linearly independent, since they are the n columns of $\mathscr{C}$, each premultiplied by P. Furthermore, as one checks linear dependence of columns in $\hat{\mathscr{C}}$, from left to right, if a column is dependent on the previous ones, then the corresponding column in $\mathscr{C}$ will also be dependent on the previous ones in $\mathscr{C}$ (show this). Therefore, the linearly independent columns generated by this left to right search are exactly the same in $\mathscr{C}$ and $\hat{\mathscr{C}}$, and therefore, the controllability indices are identical. ■

Note that the controllability indices of (A, BG), where G is nonsingular, are equal to the controllability indices μ_i of (A, B) within reordering (see Exercise 3.20). Note also that when linear state feedback $u = Fx + Gv$ with $|G| \neq 0$ is applied to $\dot{x} = Ax + Bu$ (see Exercise 3.21), then the controllability indices of $(A + BF, BG)$ are equal to the controllability indices of (A, B), i.e., the controllability indices are invariant under linear state feedback. These results can be summarized as follows.

Given (A, B) controllable, then $(P(A + BGF)P^{-1}, PBG)$ will have the same controllability indices, within reordering, for any P, F, and G ($|P| \neq 0, |G| \neq 0$) of appropriate dimensions. In other words, *the controllability indices are invariant under similarity and input transformations P and G, and state feedback F* [*or similarity transformation P and state feedback* (F, G)]. Furthermore, it can be shown that if two pairs (A_i, B_i), $i = 1, 2$, have the same indices, then there must exist P, G, and F such that $A_1 = P(A_2 + B_2GF)P^{-1}$ and $B_1 = PB_2G$. This in fact is the *completeness property*, which together with the invariance property implies that the controllability indices $\{\mu_i\}$ constitute a *set of complete invariants* for (A, B) under operations P, G, and F. Using (4.63), it is not difficult to see that given any (A_2, B_2) with controllability indices $\{\mu_i\}$, there exist P, G, F such that $A_1 = P(A_2 + B_2GF)P^{-1}$ and $B_1 = PB_2G$ are exactly equal to $\bar{A}_c = A_1$, $\bar{B}_c = B_1$. The pair $(\bar{A}_c, \bar{B}_c)$ is unique, and since any controllable pair (A, B) with controllability indices $\{\mu_i\}$ can be reduced to $(\bar{A}_c, \bar{B}_c)$ by these operations, $(\bar{A}_c, \bar{B}_c)$ is a canonical form of such (A, B) under these operations. The pair $(\bar{A}_c, \bar{B}_c)$ is called the *Brunovsky canonical form*. It should also be mentioned here that the $\{\mu_i\}$ are the right *Kronecker indices* of the pencil $[sI - A_c, B_c]$. (Recall that $[sI - A_c, B_c] = s[I, 0] - [A_c, -B_c] = sE - A$, a matrix form called a *linear matrix pencil.*) The derivation of A_c, B_c in (4.62) is discussed next.

The exact structure of A_c and B_c depends on the selection of the n linearly independent columns in $\mathscr{C}$ and on the choice of the equivalence transformation matrix P. The n linearly independent columns in $\mathscr{C}$ were selected by a search from left to right; if a column was found dependent on previous ones, then it was dropped. The dependence relation is given in (4.58). This relation, together with the expression for P given in (4.61), are central in the derivation of A_c, B_c.

In view of $\bar{\mathscr{C}}^{-1}\bar{\mathscr{C}} = I$,

$$\begin{aligned} q_i\bar{\mathscr{C}} &= q_i[b_1, Ab_1, \ldots, A^{\mu_1-1}b_1, \ldots, A^{\mu_i-1}b_i, \ldots, b_m, \ldots, A^{\mu_m-1}b_m] \\ &= [0, \ldots, 0, 1, 0, \ldots, 0], \qquad i = 1, \ldots, m, \end{aligned} \tag{4.64}$$

where the 1 occurs at the σ_ith column. These relations can be written as

$$q_i A^{k-1} b_j = 0 \qquad k = 1, \ldots, \mu_j, \qquad i \neq j,$$
$$q_i A^{k-1} b_i = 0 \qquad k = 1, \ldots, \mu_i - 1, \qquad \text{and} \qquad q_i A^{\mu_i - 1} b_i = 1, \qquad i = j, \tag{4.65}$$

where $i = 1, \ldots, m$, and $j = 1, \ldots, m$. In view of (4.64), (4.65), and the dependence relations (4.58), it can be shown that $P\bar{\mathscr{C}}$ is nonsingular, which in view of the fact that $|\bar{\mathscr{C}}| \neq 0$ implies that $|P| \neq 0$ as well, i.e., P qualifies as a similarity transformation. We note that the proof of this result is rather involved (see Section 3.7, Notes, for appropriate references). In view of the relationship $PA = A_c P$, it is now not difficult to see that $n - m$ rows of A_c will contain fixed zeros and ones, as in (4.62). The (m) σ_1th, σ_2th, $\ldots$, σ_mth rows of A_c that are denoted by A_m in (4.63) are given by

$$A_m = \begin{bmatrix} q_1 A^{\mu_1} \\ \vdots \\ q_m A^{\mu_m} \end{bmatrix} P^{-1}, \tag{4.66}$$

where $A_c = \bar{A}_c + \bar{B}_c A_m$. Similarly, in view of the equality $B_c = PB$ and (4.63), it is not difficult to see that $n - m$ rows of B_c must be zero. The (m) remaining σ_1th, σ_2th, $\ldots$, σ_mth rows of B_c that are denoted by B_m in (4.63) are given by

$$B_m = \begin{bmatrix} q_1 A^{\mu_1 - 1} \\ \vdots \\ q_m A^{\mu_m - 1} \end{bmatrix} B, \tag{4.67}$$

where $B_c = \bar{B}_c B_m$. The matrix B_m in (4.67) is, in fact, an upper triangular matrix with ones on the diagonal. This result follows from the special form of $P\bar{\mathscr{C}}$ that was used above to show that P is nonsingular.

At this point it may perhaps be beneficial to examine a special case that is not only interesting but will also provide some indication of the type of proof required to establish these results. In particular, it will be shown that when $\mu_1 \leq \mu_2 \leq \cdots \leq \mu_m$, the upper triangular matrix B_m in (4.67) is in fact diagonal.

LEMMA 4.9. If $\mu_1 \leq \mu_2 \leq \cdots \leq \mu_m$, then B_m in (4.67) is diagonal with ones on the diagonal, i.e., $B_m = I_m$.

Proof. The matrix B_m in (4.67) is upper triangular. When, in addition, $\mu_1 \leq \mu_2 \leq \cdots \leq \mu_m$, then $q_1 A^{\mu_1 - 1} B = q_1 A^{\mu_1 - 1}[b_1, b_2, \ldots, b_m] = [1, 0, \ldots, 0]$, in view of (4.65); in particular, $q_1 A^{\mu_1 - 1} b_1 = 1$ and $q_1 A^{\mu_1 - 1} b_2 = 0$ since $q_1 A^{k-1} b_2 = 0$, $k = 1, \ldots, \mu_2$, and $\mu_1 \leq \mu_2$. Similar statements hold for the columns $b_3, b_4, \ldots, b_m$. The proof for $q_2 A^{\mu_2 - 1} B, \ldots, q_m A^{\mu_m - 1} B$ is completely analogous. ■

Note that if different relations among the controllability indices μ_i exist, then different entries of B_m (above the diagonal) will be zero.

EXAMPLE 4.15. We wish to reduce $A = \begin{bmatrix} 0 & 1 & 0 \\ 0 & 0 & 1 \\ 0 & 2 & -1 \end{bmatrix}$ and $B = \begin{bmatrix} 1 & 1 \\ 0 & 1 \\ 0 & 0 \end{bmatrix}$ to controller form. Note that A and B are almost the same as in Example 4.14; however, presently, $\mu_1 = 1 < 2 = \mu_2$, as will be seen. We have $\mathscr{C} = [B, AB, A^2B] = [b_1, b_2, Ab_1,$

$Ab_2, \ldots] = \begin{bmatrix} 1 & 1 & 0 & 1 & \\ 0 & 1 & 0 & 0 & \cdots \\ 0 & 0 & 0 & 2 & \end{bmatrix}$. Searching from left to right, the first three linearly independent columns are b_1, b_2, Ab_2 and $\bar{\mathscr{C}} = [b_1, b_2, Ab_2] = \begin{bmatrix} 1 & 1 & 1 \\ 0 & 1 & 0 \\ 0 & 0 & 2 \end{bmatrix}$, from which we conclude that $\mu_1 = 1, \mu_2 = 2, \sigma_1 = 1$, and $\sigma_2 = 3$. We compute $\bar{\mathscr{C}}^{-1} = \begin{bmatrix} 1 & -1 & -\frac{1}{2} \\ 0 & 1 & 0 \\ 0 & 0 & \frac{1}{2} \end{bmatrix}$. Note that $q_1 = [1, -1, -\frac{1}{2}]$ and $q_2 = [0, 0, \frac{1}{2}]$, the first and third rows of $\bar{\mathscr{C}}^{-1}$, respectively. Then

$$P = \begin{bmatrix} q_1 \\ q_2 \\ q_2 A \end{bmatrix} = \begin{bmatrix} 1 & -1 & -\frac{1}{2} \\ 0 & 0 & \frac{1}{2} \\ 0 & 1 & -\frac{1}{2} \end{bmatrix}, \qquad P^{-1} = \begin{bmatrix} 1 & 2 & 1 \\ 0 & 1 & 1 \\ 0 & 2 & 0 \end{bmatrix}, \qquad \text{and}$$

$$A_c = PAP^{-1} = \begin{bmatrix} A_{11} & \vdots & A_{12} \\ \cdots & \vdots & \cdots \\ A_{21} & \vdots & A_{22} \end{bmatrix} = \begin{bmatrix} 0 & \vdots & -1 & 0 \\ \cdots & \vdots & \cdots & \cdots \\ 0 & \vdots & 0 & 1 \\ 0 & \vdots & 2 & -1 \end{bmatrix},$$

$$B_c = PB = \begin{bmatrix} B_1 \\ B_2 \end{bmatrix} = \begin{bmatrix} 1 & 0 \\ \cdots & \cdots \\ 0 & 0 \\ 0 & 1 \end{bmatrix}.$$

Presently, (4.67) yields $B_m = \begin{bmatrix} q_1 \\ q_2 A \end{bmatrix} B$ and equals I_2, as expected in view of the above lemma ($\mu_1 < \mu_2$). In general, B_m is an upper triangular matrix.

For the present example, we ask the reader to also verify relations (4.65), determine A_m by (4.66), and compare these with the above results. ■

In the case of single-input systems, $m = 1$, $\mu_1 = n$, and P in (4.61) is exactly the transformation matrix given in (4.44) (verify this). In the case $m = 1$, if the order of rows in P is reversed and P_1 in (4.48) is used instead, then an alternative controller form is obtained, shown in (4.47). Similar results apply when $m > 1$: if the order of the rows of P in (4.61) is reversed within the $\mu_i \times n$ blocks, then

$$P_1 = \begin{bmatrix} q_1 A^{\mu_1 - 1} \\ \vdots \\ q_1 \\ \vdots \\ q_m A^{\mu_m - 1} \\ \vdots \\ q_m \end{bmatrix}, \tag{4.68}$$

in which case $A_{c1} = P_1 A P_1^{-1}$, $B_{c_1} = P_1$, and B are given by

$$A_{c1} = [A_{ij}], \qquad i, j = 1, \ldots, m,$$

$$A_{ii} = \begin{bmatrix} \times & \times & \cdots & \times \\ & & & 0 \\ & I_{\mu_i - 1} & & \vdots \\ & & & 0 \end{bmatrix} \in R^{\mu_i \times \mu_i}, i = j,$$

$$A_{ij} = \begin{bmatrix} \times & \cdots & \times \\ 0 & \cdots & 0 \\ \vdots & & \vdots \\ 0 & \cdots & 0 \end{bmatrix} \in R^{\mu_i \times \mu_j}, i \neq j,$$

and

$$B_{c1} = \begin{bmatrix} B_1 \\ \vdots \\ B_m \end{bmatrix}, \qquad B_i = \begin{bmatrix} 0 & \cdots & 0 & 1 & \times & \cdots & \times \\ 0 & \cdots & 0 & 0 & 0 & \cdots & 0 \\ \vdots & & \vdots & \vdots & \vdots & & \vdots \\ 0 & \cdots & 0 & 0 & 0 & \cdots & 0 \end{bmatrix} \in R^{\mu_i \times m}, \quad (4.69)$$

where the 1 on the first row of B_i occurs at the ith ($i = 1, \ldots, m$) location and $\times$ denotes nonfixed entries. These formulas are similar to (4.62) and can be derived in a completely analogous fashion.

As in the case $m = 1$ [see (4.49) to (4.52)], the columns of P^{-1} can also be selected to be n linearly independent columns of $\mathscr{C}$, in which case $\hat{A} = PAP^{-1}$ and $\hat{B} = PB$ will be in alternative controller forms. There are many ways of selecting these n linearly independent columns of the $n \times nm$ matrix $\mathscr{C}$, as was discussed before. For example, one may search $\mathscr{C}$ from left to right as above, or one may check $b_1, Ab_1, \ldots, b_2, Ab_2, \ldots$ for linear independence. The controller forms resulting in this way are generalizations of (4.50) and (4.52). These forms are not as useful to us and are omitted here.

Structure theorem—controllable version. The transfer function matrix $H(s)$ of the system $\dot{x} = Ax + Bu$, $y = Cx + Du$ is given by $H(s) = C(sI - A)^{-1}B + D$. If (A, B) is in *controller form* (4.62), then $H(s)$ can alternatively be characterized by the Structure Theorem stated in Theorem 4.10. This result is very useful in the realization of systems, addressed in Chapter 5 and in the study of state feedback in the next chapter.

Let $A = A_c = \bar{A}_c + \bar{B}_c A_m$ and $B = B_c = \bar{B}_c B_m$, as in (4.63), with $|B_m| \neq 0$ and let $C = C_c$ and $D = D_c$. Define

$$\Lambda(s) \triangleq \begin{bmatrix} s^{\mu_1} & & & \\ & s^{\mu_2} & & \\ & & \ddots & \\ & & & s^{\mu_m} \end{bmatrix}, \qquad S(s) \triangleq \textit{block diag}\left(\begin{bmatrix} 1 \\ s \\ \vdots \\ s^{\mu_i - 1} \end{bmatrix} i = 1, \ldots, m \right). \quad (4.70)$$

Note that $S(s)$ is an $n \times m$ polynomial matrix ($n = \sum_{i=1}^{m} \mu_i$), i.e., a matrix with polynomials as entries. Now define the $m \times m$ polynomial matrix $D(s)$ and the $p \times m$ polynomial matrix $N(s)$ by

$$D(s) \triangleq B_m^{-1}[\Lambda(s) - A_m S(s)], \qquad N(s) \triangleq C_c S(s) + D_c D(s). \quad (4.71)$$

The following is the controllable version of the *Structure Theorem*.

THEOREM 4.10. $H(s) = N(s)D^{-1}(s)$, where $N(s)$ and $D(s)$ are defined in (4.71).

Proof. First, note that

$$(sI - A_c)S(s) = B_c D(s). \tag{4.72}$$

To see this, we write $B_c D(s) = \bar{B}_c B_m B_m^{-1}[\Lambda(s) - A_m S(s)] = \bar{B}_c\Lambda(s) - \bar{B}_c A_m S(s)$ and $(sI - A_c)S(s) = sS(s) - (\bar{A}_c + \bar{B}_c A_m)S(s) = (sI - \bar{A}_c)S(s) - \bar{B}_c A_m S(s) = \bar{B}_c\Lambda(s) - \bar{B}_c A_m S(s)$, which proves (4.72). Now $H(s) = C_c(sI - A_c)^{-1}B_c + D_c = C_c S(s)D^{-1}(s) + D_c = [C_c S(s) + D_c D(s)]D^{-1}(s) = N(s)D^{-1}(s)$. ■

EXAMPLE 4.16. Let $A_c = \begin{bmatrix} 0 & 1 & 0 \\ 2 & -1 & 0 \\ 1 & 0 & 0 \end{bmatrix}$ and $B_c = \begin{bmatrix} 0 & 0 \\ 1 & 1 \\ 0 & 1 \end{bmatrix}$, as in Example 4.14.

Here $\mu_1 = 2$, $\mu_2 = 1$ and $A_m = \begin{bmatrix} 2 & -1 & 0 \\ 1 & 0 & 0 \end{bmatrix}$, $B_m = \begin{bmatrix} 1 & 1 \\ 0 & 1 \end{bmatrix}$. Then $\Lambda(s) = \begin{bmatrix} s^2 & 0 \\ 0 & s \end{bmatrix}$,

$S(s) = \begin{bmatrix} 1 & 0 \\ s & 0 \\ 0 & 1 \end{bmatrix}$, and

$$D(s) = B_m^{-1}[\Lambda(s) - A_m S(s)] = \begin{bmatrix} 1 & -1 \\ 0 & 1 \end{bmatrix}\left[\begin{bmatrix} s^2 & 0 \\ 0 & s \end{bmatrix} - \begin{bmatrix} -s+2 & 0 \\ 1 & 0 \end{bmatrix}\right]$$

$$= \begin{bmatrix} 1 & -1 \\ 0 & 1 \end{bmatrix}\begin{bmatrix} s^2+s-2 & 0 \\ -1 & s \end{bmatrix} = \begin{bmatrix} s^2+s-1 & -s \\ -1 & s \end{bmatrix}.$$

Now $C_c = [0, 1, 1]$, and $D_c = [0, 0]$,

$N(s) = C_c S(s) + D_c D(s) = [s, 1]$,

and $$H(s) = [s, 1]\begin{bmatrix} s^2+s-1 & -s \\ -1 & s \end{bmatrix}^{-1} = [s, 1]\begin{bmatrix} s & s \\ 1 & s^2+s-1 \end{bmatrix}\frac{1}{s(s^2+s-2)}$$

$$= \frac{1}{s(s^2+s-2)}[s^2+1, 2s^2+s-1]$$

$$= C_c(sI - A_c)^{-1}B_c + D_c.$$ ■

EXAMPLE 4.17. Let $A_c = \begin{bmatrix} 0 & 1 & 0 \\ 0 & 0 & 1 \\ 2 & 1 & -2 \end{bmatrix}$, $B_c = \begin{bmatrix} 0 \\ 0 \\ 1 \end{bmatrix}$, $C_c = [0, 1, 0]$,

and $D_c = 0$ (see Example 4.11). In the present case we have $A_m = [2, 1, -2]$, $B_m = 1$, $\Lambda(s) = s^3$, $S(s) = [1, s, s^2]^T$, and

$$D(s) = 1 \cdot [s^3 - [2, 1, -2][1, s, s^2]^T] = s^3 + 2s^2 - s - 2, \qquad N(s) = s.$$

Then

$$H(s) = N(s)D^{-1}(s) = \frac{s}{s^3 + 2s^2 - s - 2} = C_c(sI - A_c)^{-1}B_c + D_c.$$ ■

2. Observer forms

Consider the system $\dot{x} = Ax + Bu$, $y = Cx + Du$ given in (4.1) and assume that (A, C) is observable, i.e., $rank\,\mathcal{O} = n$, where

$$\mathcal{O} = \begin{bmatrix} C \\ CA \\ \vdots \\ CA^{n-1} \end{bmatrix}. \tag{4.73}$$

Also, assume that the $p \times n$ matrix C has a full row rank p, i.e.,

$$rank\, C = p \leq n. \tag{4.74}$$

Presently, it is of interest to determine a transformation matrix P so that the equivalent system representation $\{A_o, B_o, C_o, D_o\}$ with

$$A_o = PAP^{-1}, \qquad B_o = PB, \qquad C_o = CP^{-1}, \qquad D_o = D \tag{4.75}$$

will have (A_o, C_o) in an observer form (defined below). As will become clear in the following, these forms are dual to the controller forms previously discussed and can be derived by taking advantage of this fact. In particular, let $\tilde{A} \triangleq A^T$, $\tilde{B} \triangleq C^T$ [$(\tilde{A}, \tilde{B})$ is controllable] and determine a nonsingular transformation $\tilde{P}$ so that $\tilde{A}_c = \tilde{P}\tilde{A}\tilde{P}^{-1}$, $\tilde{B}_c = \tilde{P}\tilde{B}$ are in controller form given in (4.62). Then $A_o = \tilde{A}_c^T$ and $C_o = \tilde{B}_c^T$ is in observer form. In fact the equivalent representation in this case is given by (4.75), where $P = (\tilde{P}^T)^{-1}$ (show this).

It will be demonstrated in the following how to obtain observer forms directly, in a way that parallels the approach described for controller forms. This is done for the sake of completeness and to define the observability indices. Our presentation will be rather brief. The approach of using duality just given can be used in each case to verify the results.

We first note that if $rank\, C = r < p$, an approach analogous to the case when $rank\, B < m$ can be followed, as above. The fact that the rows of C are not linearly independent means that the same information can be extracted from only r outputs, and therefore, the choice for the outputs should perhaps be reconsidered. Now if (A, C) is unobservable, one may use two steps to first isolate the observable part and then reduce it to the observer form, in an analogous way to the uncontrollable case previously given.

Single-output case ($p = 1$). Let

$$P^{-1} = Q \triangleq [\tilde{q}, A\tilde{q}, \ldots, A^{n-1}\tilde{q}], \tag{4.76}$$

where $\tilde{q}$ is the nth column in $\mathcal{O}^{-1}$. Then

$$A_0 = \begin{bmatrix} 0 & \cdots & 0 & -\alpha_0 \\ 1 & \cdots & 0 & -\alpha_1 \\ \vdots & \ddots & \vdots & \vdots \\ 0 & \cdots & 1 & -\alpha_{n-1} \end{bmatrix}, \qquad C_o = [0, \ldots, 0, 1], \tag{4.77}$$

where the α_i denote the coefficients of the characteristic polynomial $\alpha(s) \triangleq det\,(sI - A) = s^n + \alpha_{n-1}s^{n-1} + \cdots + \alpha_1 s + \alpha_0$. Here $A_o = PAP^{-1} = Q^{-1}AQ$, $C_o = CP^{-1} = CQ$, and the desired result can be established by using a proof that is completely analogous to the proof in determining the (dual) controller form presented earlier in this section. Note that $B_o = PB$ does not have any particular structure. The representation $\{A_o, B_o, C_o, D_o\}$ will be referred to as the *observer form* of the system.

Reversing the order of columns in P^{-1} given in (4.76) or selecting P to be exactly $\mathcal{O}$, or to be equal to the matrix obtained after the order of the columns in $\mathcal{O}$ has been reversed, leads to alternative observer forms in a manner analogous to the controller form case. We leave it to the reader to investigate these possibilities further.

EXAMPLE 4.18. Let $A = \begin{bmatrix} -1 & 0 & 0 \\ 0 & 1 & 0 \\ 0 & 0 & -2 \end{bmatrix}$ and $C = [1, -1, 1]$. To derive the observer form (4.77), we could use duality, by defining $\tilde{A} = A^T$, $\tilde{B} = C^T$, and deriving the controller form of $\tilde{A}$, $\tilde{B}$, i.e., by following the procedure outlined above. This is left to the reader as an exercise. We note that the $\tilde{A}$, $\tilde{B}$ are exactly the matrices given in Examples 4.11 and 4.12. In the following the observer form is derived directly. In particular, we have

$$\mathcal{O} = \begin{bmatrix} C \\ CA \\ CA^2 \end{bmatrix} = \begin{bmatrix} 1 & -1 & 1 \\ -1 & -1 & -2 \\ 1 & -1 & 4 \end{bmatrix}, \qquad \mathcal{O}^{-1} = \begin{bmatrix} 1 & -\frac{1}{2} & -\frac{1}{2} \\ -\frac{1}{3} & -\frac{1}{2} & -\frac{1}{6} \\ -\frac{1}{3} & 0 & \frac{1}{3} \end{bmatrix},$$

and in view of (4.76),

$$Q = P^{-1} = [\tilde{q}, A\tilde{q}, A^2\tilde{q}] = \begin{bmatrix} -\frac{1}{2} & \frac{1}{2} & -\frac{1}{2} \\ -\frac{1}{6} & -\frac{1}{6} & -\frac{1}{6} \\ \frac{1}{3} & -\frac{2}{3} & \frac{4}{3} \end{bmatrix}.$$

Note that $\tilde{q} = \left[-\frac{1}{2}, -\frac{1}{6}, \frac{1}{3}\right]^T$, the last column of $\mathcal{O}^{-1}$. Then

$$A_o = Q^{-1}AQ = \begin{bmatrix} 0 & 0 & 2 \\ 1 & 0 & 1 \\ 0 & 1 & -2 \end{bmatrix} \quad \text{and} \quad C_o = CQ = [0, 0, 1],$$

where $|sI - A| = s^3 + 2s - s - 2 = s^3 + \alpha_2 s^2 + \alpha_1 s + \alpha_0$. Hence,

$$QA_o = \begin{bmatrix} \frac{1}{2} & -\frac{1}{2} & \frac{1}{2} \\ -\frac{1}{6} & -\frac{1}{6} & -\frac{1}{6} \\ -\frac{2}{3} & \frac{4}{3} & -\frac{8}{3} \end{bmatrix} = AQ.$$

■

Multi-output case **($p > 1$).** Consider

$$\mathcal{O} = \begin{bmatrix} C \\ CA \\ \vdots \\ CA^{n-1} \end{bmatrix} = \begin{bmatrix} c_1 \\ \vdots \\ c_p \\ c_1 A \\ \vdots \\ c_p A \\ \vdots \\ c_1 A^{n-1} \\ \vdots \\ c_p A^{n-1} \end{bmatrix}, \tag{4.78}$$

where $c_1, \ldots, c_p$ denote the p rows of C, and select the first n linearly independent rows in $\mathcal{O}$, moving from the top to bottom (rank $\mathcal{O} = n$). Next, reorder the selected rows by first taking all rows involving c_1, then c_2, etc., to obtain

$$\bar{\mathcal{O}} \triangleq \begin{bmatrix} c_1 \\ c_1 A \\ \vdots \\ c_1 A^{\nu_1 - 1} \\ \vdots \\ c_p \\ \vdots \\ c_p A^{\nu_p - 1} \end{bmatrix}, \tag{4.79}$$

an $n \times n$ matrix. The integer ν_i denotes the number of rows involving c_i in the set of the first n linearly independent rows found in $\mathcal{O}$ when moving from top to bottom.

DEFINITION 4.2. The p integers ν_i, $i = 1, \ldots, p$, are the *observability indices of the system*, and $\nu \triangleq \max \nu_i$ is called the *observability index of the system*. Note that

$$\sum_{i=1}^{p} \nu_i = n \qquad \text{and} \qquad p\nu \geq n. \tag{4.80}$$

■

When *rank* $C = p$, then $\nu_i \geq 1$. Now define

$$\tilde{\sigma}_k \triangleq \sum_{i=1}^{k} \nu_i \qquad k = 1, \ldots, p, \tag{4.81}$$

i.e., $\tilde{\sigma}_1 = \nu_1$, $\tilde{\sigma}_2 = \nu_1 + \nu_2, \ldots, \tilde{\sigma}_p = \nu_1 + \cdots + \nu_p = n$. Consider $\bar{\mathcal{O}}^{-1}$ and let $\tilde{q}_k \in R^n$, $k = 1, \ldots, p$, represent its $\tilde{\sigma}_k$th column, i.e.,

$$\bar{\mathcal{O}}^{-1} = [\times \cdots \times \tilde{q}_1 | \times \cdots \times \tilde{q}_2 | \cdots | \times \cdots \times \tilde{q}_p]. \tag{4.82}$$

Define

$$P^{-1} = Q = [\tilde{q}_1, \ldots, A^{\nu_1 - 1}\tilde{q}_1, \ldots, \tilde{q}_p, \ldots, A^{\nu_p - 1}\tilde{q}_p]. \tag{4.83}$$

Then $A_o = PAP^{-1} = Q^{-1}AQ$ and $C_o = CP^{-1} = CQ$ are given by

$$A_o = [A_{ij}],\ i, j = 1, \ldots, p,$$

$$A_{ii} = \begin{bmatrix} 0 & \cdots & 0 & \times \\ & I_{\nu_i - 1} & & \vdots \\ & & & \times \end{bmatrix} \in R^{\nu_i \times \nu_i},\ i = j, \quad A_{ij} = \begin{bmatrix} 0 & \cdots & 0 & \times \\ \vdots & & \vdots & \vdots \\ 0 & \cdots & 0 & \times \end{bmatrix} \in R^{\nu_i \times \nu_j},\ i \neq j,$$

and

$$C_o = [C_1, C_2, \ldots, C_p],\ C_i = \begin{bmatrix} 0 & \cdots & 0 & 0 \\ \vdots & & \vdots & \vdots \\ 0 & \cdots & 0 & 0 \\ 0 & \cdots & 0 & 1 \\ 0 & \cdots & 0 & \times \\ \vdots & & \vdots & \vdots \\ 0 & \cdots & 0 & \times \end{bmatrix} \in R^{p \times \nu_i}, \tag{4.84}$$

where the 1 on the last column of C_i occurs at the ith-row location ($i = 1, \ldots, p$) and $\times$ denotes nonfixed entries. Note that the matrix $B_o = PB = Q^{-1}B$ does not have any particular structure. Equation (4.84) is a very useful form (in the observer problem) and shall be referred to as the *observer form* of the system.

Analogous to (4.63), we express A_o and C_o as

$$A_o = \bar{A}_o + A_p \bar{C}_o, \qquad C_o = C_p \bar{C}_o, \tag{4.85}$$

where $\bar{A}_o = \textit{block diag}\ [A_1, A_2, \ldots, A_p]$ with $A_i = \begin{bmatrix} 0 & \cdots & 0 \\ I_{\nu_i - 1} & & \vdots \\ & & 0 \end{bmatrix} \in R^{\nu_i \times \nu_i}$, $\bar{C}_o =$ *block diag* $([0, \ldots, 0, 1]^T \in R^{\nu_i}, i = 1, \ldots, p)$, and $A_p \in R^{n \times p}$, and $C_p \in R^{p \times p}$ are appropriate matrices ($\sum_{i=1}^{p} \nu_i = n$). Note that $\bar{A}_o, \bar{C}_o$ are completely determined by the p observability indices ν_i, $i = 1, \ldots, p$, and A_p and C_p contain this information in the $\tilde{\sigma}_1$th, $\ldots, \tilde{\sigma}_p$th columns of A_o and in the same columns of C_o, respectively. Note also that A_p is that part of the observable system $\dot{x} = Ax + Bu$, $y = Cx + Du$ that can be altered by the gain of a state observer. This is discussed further in the next chapter.

Results that are in the spirit of Lemmas 4.8 and 4.9 also exist and can be established directly in a completely analogous way or by duality. The same is true for proving that Q in (4.83) is nonsingular and that A_o and C_o in (4.84) have their particular structure. Furthermore, by reversing the order of the columns of P^{-1} in (4.83) or by selecting the columns of P directly from $\mathcal{O}$, one can derive alternative observer forms. These are dual to the controller forms discussed before.

EXAMPLE 4.19. Given $A = \begin{bmatrix} 0 & 0 & 0 \\ 1 & 0 & 2 \\ 0 & 1 & -1 \end{bmatrix}$ and $C = \begin{bmatrix} 0 & 1 & 0 \\ 1 & 1 & 0 \end{bmatrix}$, we wish to reduce these to observer form. This can be accomplished using duality, i.e., by first reducing $\tilde{A} \triangleq A^T$, $\tilde{B} \triangleq C^T$ to controller form. Note that $\tilde{A}$, $\tilde{B}$ are the matrices used in Example 4.14, and therefore, the desired answer is easily obtained. Presently, we shall follow the direct algorithm described above. We have

$$\mathcal{O} = \begin{bmatrix} C \\ CA \\ CA^2 \end{bmatrix} = \begin{bmatrix} 0 & 1 & 0 \\ 1 & 1 & 0 \\ 1 & 0 & 2 \\ 1 & 0 & 2 \\ 0 & 2 & -2 \\ 0 & 2 & -2 \end{bmatrix}.$$

Searching from top to bottom, the first three linearly independent rows are c_1, c_2, $c_1 A$, and

$$\bar{\mathcal{O}} = \begin{bmatrix} c_1 \\ c_1 A \\ c_2 \end{bmatrix} = \begin{bmatrix} 0 & 1 & 0 \\ 1 & 0 & 2 \\ 1 & 1 & 0 \end{bmatrix}.$$

Note that the observability indices are $\nu_1 = 2$, $\nu_2 = 1$ and $\tilde{\sigma}_1 = 2$, $\tilde{\sigma}_2 = 3$. We compute

$$\bar{\mathcal{O}}^{-1} = \begin{bmatrix} -1 & 0 & 1 \\ 1 & 0 & 0 \\ \frac{1}{2} & \frac{1}{2} & -\frac{1}{2} \end{bmatrix} = \begin{bmatrix} \times & 0 & 1 \\ \times & 0 & 0 \\ \times & \frac{1}{2} & -\frac{1}{2} \end{bmatrix}.$$

Then, $Q = [\tilde{q}_1, A\tilde{q}_1, \tilde{q}_2] = \begin{bmatrix} 0 & 0 & 1 \\ 0 & 1 & 0 \\ \frac{1}{2} & -\frac{1}{2} & -\frac{1}{2} \end{bmatrix}$ and $Q^{-1} = \begin{bmatrix} 1 & 1 & 2 \\ 0 & 1 & 0 \\ 1 & 0 & 0 \end{bmatrix}$. Therefore,

$$A_o = Q^{-1}AQ = \begin{bmatrix} A_{11} & \vdots & A_{12} \\ \cdots & \vdots & \cdots \\ A_{21} & \vdots & A_{22} \end{bmatrix} = \begin{bmatrix} 0 & 2 & \vdots & 1 \\ 1 & -1 & \vdots & 0 \\ \cdots & \cdots & \vdots & \cdots \\ 0 & 0 & \vdots & 0 \end{bmatrix},$$

$$C_o = CQ = [C_1 \;\vdots\; C_2] = \begin{bmatrix} 0 & 1 & \vdots & 0 \\ 0 & 1 & \vdots & 1 \end{bmatrix}.$$

We can also verify (4.85), namely,

$$A_o = \begin{bmatrix} 0 & 2 & \vdots & 1 \\ 1 & -1 & \vdots & 0 \\ \cdots & \cdots & \vdots & \cdots \\ 0 & 0 & \vdots & 0 \end{bmatrix} = \bar{A}_o + A_p\bar{C}_o = \begin{bmatrix} 0 & 0 & \vdots & 0 \\ 1 & 0 & \vdots & 0 \\ \cdots & \cdots & \vdots & \cdots \\ 0 & 0 & \vdots & 0 \end{bmatrix} + \begin{bmatrix} 2 & 1 \\ -1 & 0 \\ 0 & 0 \end{bmatrix} \begin{bmatrix} 0 & 1 & 0 \\ 0 & 0 & 1 \end{bmatrix}$$

and

$$C_o = \begin{bmatrix} 0 & 1 & \vdots & 0 \\ 0 & 1 & \vdots & 1 \end{bmatrix} = C_p\bar{C}_o = \begin{bmatrix} 1 & 0 \\ 1 & 1 \end{bmatrix} \begin{bmatrix} 0 & 1 & 0 \\ 0 & 0 & 1 \end{bmatrix}.$$

■

Structure Theorem—observable version. The transfer function matrix $H(s)$ of system $\dot{x} = Ax + Bu$, $y = Cx + Du$ is given by $H(s) = C(sI - A)^{-1}B + D$. If (A, C) is in the *observer form*, given in (4.84), then $H(s)$ can alternatively be characterized by the Structure Theorem stated in Theorem 4.11. This result will be very useful in the realization of systems, addressed in Chapter 5 and also in the study of observers in the next chapter.

Let $A = A_o = \bar{A}_o + A_p\bar{C}_o$ and $C = C_o = C_p\bar{C}_o$ as in (4.85) with $|C_p| \neq 0$, let $B = B_o$ and $D = D_o$, and define

$$\tilde{\Lambda}(s) \triangleq diag\,[s^{\nu_1}, s^{\nu_2}, \ldots, s^{\nu_p}], \quad \tilde{S}(s) \triangleq block\ diag\,([1, s, \ldots, s^{\nu_i - 1}], i = 1, \ldots, p). \tag{4.86}$$

Note that $\tilde{S}(s)$ is a $p \times n$ polynomial matrix, where $n = \sum_{i=1}^{p} \nu_i$. Now define the $p \times p$ polynomial matrix $\tilde{D}(s)$ and the $p \times m$ polynomial matrix $\tilde{N}(s)$ as

$$\tilde{D}(s) \triangleq [\tilde{\Lambda}(s) - \tilde{S}(s)A_p]C_p^{-1}, \qquad \tilde{N}(s) \triangleq \tilde{S}(s)B_o + \tilde{D}(s)D_o. \tag{4.87}$$

The following result is the observable version of the *Structure Theorem*. It is the dual of Theorem 4.10 and can therefore be proved using duality arguments. The proof given is direct.

THEOREM 4.11. $H(s) = \tilde{D}^{-1}(s)\tilde{N}(s)$, where $\tilde{N}(s)$, $\tilde{D}(s)$ are defined in (4.87).

Proof. First we note that

$$\tilde{D}(s)C_o = \tilde{S}(s)(sI - A_o). \tag{4.88}$$

To see this, write $\tilde{D}(s)C_o = [\tilde{\Lambda}(s) - \tilde{S}(s)A_p]C_p^{-1}C_p\bar{C}_o = \tilde{\Lambda}(s)\bar{C}_o - \tilde{S}(s)A_p\bar{C}_o$, and also, $\tilde{S}(s)(sI - A_o) = \tilde{S}(s)s - \tilde{S}(s)(\bar{A}_o + A_p\bar{C}_o) = \tilde{S}(s)(sI - \bar{A}_o) - \tilde{S}(s)A_p\bar{C}_o = \tilde{\Lambda}(s)\bar{C}_o - \tilde{S}(s)A_p\bar{C}_o$, which proves (4.88). We now obtain $H(s) = C_o(sI - A_o)^{-1}B_o + D_o = \tilde{D}^{-1}(s)\tilde{S}(s)B_o + D_o = \tilde{D}^{-1}(s)[\tilde{S}(s)B_o + \tilde{D}(s)D_o] = \tilde{D}^{-1}(s)\tilde{N}(s)$. ■

EXAMPLE 4.20. Consider $A_o = \begin{bmatrix} 0 & 2 & 1 \\ 1 & -1 & 0 \\ 0 & 0 & 0 \end{bmatrix}$ and $C_o = \begin{bmatrix} 0 & 1 & 0 \\ 0 & 1 & 1 \end{bmatrix}$ of Example 4.19. Here $\nu_1 = 2$, $\nu_2 = 1$, $\tilde{\Lambda}(s) = \begin{bmatrix} s^2 & 0 \\ 0 & s \end{bmatrix}$, and $\tilde{S}(s) = \begin{bmatrix} 1 & s & 0 \\ 0 & 0 & 1 \end{bmatrix}$. Then $\tilde{D}(s) =$

$$[\tilde{\Lambda}(s) - \tilde{S}(s)A_p]C_p^{-1} = \left[\begin{bmatrix} s^2 & 0 \\ 0 & s \end{bmatrix} - \begin{bmatrix} 1 & s & 0 \\ 0 & 0 & 1 \end{bmatrix}\begin{bmatrix} 2 & 1 \\ -1 & 0 \\ 0 & 0 \end{bmatrix}\right] \cdot \begin{bmatrix} 1 & 0 \\ 1 & 1 \end{bmatrix}^{-1} = \left[\begin{bmatrix} s^2 & 0 \\ 0 & s \end{bmatrix} - \right.$$

$$\left.\begin{bmatrix} -s+2 & 1 \\ 0 & 0 \end{bmatrix}\right] \cdot \begin{bmatrix} 1 & 0 \\ -1 & 1 \end{bmatrix} = \begin{bmatrix} s^2+s-2 & -1 \\ 0 & s \end{bmatrix} \cdot \begin{bmatrix} 1 & 0 \\ -1 & 1 \end{bmatrix} = \begin{bmatrix} s^2+s-1 & -1 \\ -s & s \end{bmatrix}.$$

Now if $B_o = [0, 1, 1]^T$, $D_o = 0$, and $\tilde{N}(s) = \tilde{S}(s)B_o + \tilde{D}(s)D_o = [s, 1]^T$, then $H(s) = \tilde{D}^{-1}(s)\tilde{N}(s) = \{1/[s(s^2+s-2)]\}[s^2+1, 2s^2+s-1]^T = C_o(sI - A_o)^{-1}B_o + D_o$. ■

3.5 POLES AND ZEROS

In this section the poles and zeros of a time-invariant system are defined and discussed. The primary reason for considering poles and zeros of a system at this time is that poles and zeros are related to the (controllable and observable, resp., uncontrollable and unobservable) eigenvalues of A. These relationships shed light on the eigenvalue cancellation mechanisms encountered when input–output relations, such as transfer functions, are formed. These relationships also provide greater insight into the realization theory addressed later in this book. Furthermore, the relationships between uncontrollable or unobservable eigenvalues, decoupling zeros, and cancellations in the transfer function matrix to be discussed below, provide also insight into how state feedback can alter system behavior. State feedback will be studied later in Chapter 4.

In the following development, the finite *poles of a transfer function matrix* $H(s)$ [or $H(z)$] are defined first (for the definition of poles at infinity, refer to Exercise 3.36). It should be noted here that the eigenvalues of A are sometimes called *poles of the system* $\{A, B, C, D\}$. To avoid confusion, we shall use the complete term *poles of* $H(s)$, when necessary. The zeros of a system are defined using internal descriptions (state-space representations).

Smith and Smith-McMillan forms

To define the poles of $H(s)$, we shall first introduce the Smith form of a polynomial matrix $P(s)$ and the Smith-McMillan form of a rational matrix $H(s)$.

The *Smith form* $S_P(s)$ of a $p \times m$ polynomial matrix $P(s)$ (in which the entries are polynomials in s) is defined as (see also Subsection 7.2C of Chapter 7)

$$S_P(s) = \begin{bmatrix} \Lambda(s) & 0 \\ 0 & 0 \end{bmatrix} \tag{5.1}$$

with $\Lambda(s) \triangleq diag\,[\epsilon_1(s), \ldots, \epsilon_r(s)]$, where $r = rank\ P(s)$. The unique *monic* polynomials $\epsilon_i(s)$ (polynomials with leading coefficient equal to one) are the *invariant factors* of $P(s)$. It can be shown that $\epsilon_i(s)$ divides $\epsilon_{i+1}(s)$, $i = 1, \ldots, r-1$. Note that

$\epsilon_i(s)$ can be determined by

$$\epsilon_i(s) = \frac{D_i(s)}{D_{i-1}(s)}, \qquad i = 1, \ldots, r,$$

where $D_i(s)$ is the monic greatest common divisor of all the nonzero ith-order minors of $P(s)$ with $D_0(s) = 1$. The $D_i(s)$ are the *determinantal divisors* of $P(s)$. A matrix $P(s)$ can be reduced to Smith form by elementary row and column operations (see Subsections 7.2B and C in Chapter 7). This ensures that the properties of interest are preserved when $P(s)$ is reduced to its (unique) Smith form. In particular, we are interested here in the invariant factors $\epsilon_i(s)$ of $P(s)$ that can be determined directly from the determinantal divisors $D_i(s)$ without having to reduce $P(s)$ to its Smith form via elementary operations.

Consider now a $p \times m$ rational matrix $H(s)$. Let $d(s)$ be the monic least common denominator of all nonzero entries and write

$$H(s) = \frac{1}{d(s)} N(s), \tag{5.2}$$

where $N(s)$ is a polynomial matrix. Let $S_N(s) = diag\,[n_1(s), \ldots, n_r(s), 0_{p-r,m-r}]$ be the Smith form of $N(s)$, where $r = rank\ N(s) = rank\ H(s)$. Divide each $n_i(s)$ of $S_N(s)$ by $d(s)$, cancelling all common factors to obtain the *Smith-McMillan* form of $H(s)$,

$$SM_H(s) = \begin{bmatrix} \tilde{\Lambda}(s) & 0 \\ 0 & 0 \end{bmatrix}, \tag{5.3}$$

with $\tilde{\Lambda}(s) \triangleq diag\,[\epsilon_1(s)/\psi_1(s), \ldots, \epsilon_r(s)/\psi_r(s)]$, where $r = rank\ H(s)$. Note that $\epsilon_i(s)$ divides $\epsilon_{i+1}(s)$, $i = 1, 2, \ldots, r-1$, and $\psi_{i+1}(s)$ divides $\psi_i(s)$, $i = 1, 2, \ldots, r-1$.

Poles

Given a $p \times m$ rational matrix $H(s)$, its *characteristic polynomial* or *pole polynomial*, $p_H(s)$, is defined as

$$p_H(s) = \psi_1(s) \cdots \psi_r(s), \tag{5.4}$$

where the ψ_i, $i = 1, \ldots, r$, are the denominators of the Smith-McMillan form of $H(s)$. It can be shown that $p_H(s)$ is the monic least common denominator of all nonzero minors of $H(s)$.

DEFINITION 5.1. The *poles of* $H(s)$ are the roots of the pole polynomial $p_H(s)$. ■

Note that the monic least common denominator of all nonzero first-order minors (entries) of $H(s)$ is called the *minimal polynomial* of $H(s)$ and is denoted by $m_H(s)$. The $m_H(s)$ divides $p_H(s)$ and when the roots of $p_H(s)$ [poles of $H(s)$] are distinct, $m_H(s) = p_H(s)$, since the additional roots in $p_H(s)$ are repeated roots of $m_H(s)$.

It is important to note that when the minors of $H(s)$ [of order $1, 2, \ldots, \min(p, m)$] are formed by taking the determinants of all square submatrices of dimension 1×1, 2×2, etc., all cancellations of common factors between numerator and denominator polynomials should be carried out.

In the scalar case, $p = m = 1$, Definition 5.1 reduces to the well-known definition of poles of a transfer function $H(s)$, since in this case there is only one minor (of order 1), $H(s)$, and the poles are the roots of the denominator polynomial of $H(s)$. Notice that in the present case it is assumed that all the possible cancellations have taken place in the transfer function of a system. Here $p_H(s) = m_H(s)$, that is, the pole or characteristic polynomial equals the minimal polynomial of $H(s)$. Thus, $p_H(s) = m_H(s)$ are equal to the (monic) denominator of $H(s)$.

EXAMPLE 5.1. Let $H(s) = \begin{bmatrix} 1/[s(s+1)] & 1/s & 1 \\ 0 & 0 & 1/s^2 \end{bmatrix}$. The nonzero minors of order 1 are the nonzero entries. The least common denominator is $s^2(s+1) = m_H(s)$, the minimal polynomial of $H(s)$. The nonzero minors of order 2 are $1/[s^3(s+1)]$ and $1/s^3$ (taking columns 1 and 3, and 2 and 3, respectively). The least common denominator of all minors (of order 1 and 2) is $s^3(s+1) = p_H(s)$, the characteristic polynomial of $H(s)$. The poles are $\{0, 0, 0, -1\}$. Note that $m_H(s)$ is a factor of $p_H(s)$ and the additional root at $s = 0$ in $p_H(s)$ is a repeated pole.

To obtain the Smith-McMillan form of $H(s)$, write

$$H(s) = \frac{1}{s^2(s+1)} \begin{bmatrix} s & s(s+1) & s^2(s+1) \\ 0 & 0 & (s+1) \end{bmatrix} = \frac{1}{d(s)} N(s),$$

where $d(s) = s^2(s+1) = m_H(s)$ [see (5.2)]. The Smith form of $N(s)$ is

$$S_N(s) = \begin{bmatrix} 1 & 0 & 0 \\ 0 & s(s+1) & 0 \end{bmatrix}$$

since $D_0 = 1$, $D_1 = 1$, $D_2 = s(s+1)$ [the determinantal divisors of $N(s)$], and $n_1 = D_1/D_0 = 1$, $n_2 = D_2/D_1 = s(s+1)$, the invariant factors of $N(s)$. Dividing by $d(s)$, we obtain the Smith-McMillan form of $H(s)$,

$$SM_H(s) = \begin{bmatrix} \dfrac{\epsilon_1}{\psi_1} & 0 & 0 \\ 0 & \dfrac{\epsilon_2}{\psi_2} & 0 \end{bmatrix} = \begin{bmatrix} \dfrac{1}{s^2(s+1)} & 0 & 0 \\ 0 & \dfrac{1}{s} & 0 \end{bmatrix}.$$

Note that ψ_2 divides ψ_1 and ϵ_1 divides ϵ_2. Now the characteristic or pole polynomial of $H(s)$ is $p_H(s) = \psi_1\psi_2 = s^3(s+1)$ and the poles are $\{0, 0, 0, -1\}$, as expected. ■

EXAMPLE 5.2. Let $H(s) = \dfrac{1}{s+2} \begin{bmatrix} 1 & \alpha \\ 1 & 1 \end{bmatrix}$. If $\alpha \neq 1$, then the second-order minor is $|H(s)| = (1-\alpha)/(s+2)^2$. The least common denominator of this nonzero second-order minor $|H(s)|$ and of all the entries of $H(s)$ (the first-order minors) is $(s+2)^2 = p_H(s)$, i.e., the poles are at $\{-2, -2\}$. Also, $m_H(s) = s + 2$.

Now if $\alpha = 1$, then there are only first-order nonzero minors ($|H(s)| = 0$). In this case $p_H(s) = m_H(s) = s + 2$, which is quite different from the case when $\alpha \neq 1$. Presently, there is only one pole at -2. The reader should verify these results, using the Smith-McMillan form of $H(s)$. ■

In view of Subsection 3.4.C, it is clear that all the poles of $H(s)$ are roots of $|sI - A_{11}|$, that is, they are some or all of the controllable and observable eigenvalues of the system. In fact, as will be shown in Chapter 5, the poles of $H(s)$ are exactly the controllable and observable eigenvalues of the system (in A_{11}) and no factors of $|sI - A_{11}|$ in $H(s)$ cancel.

In general, for the set of poles of $H(s)$ and the eigenvalues of A, we have

$$\{\text{poles of } H(s)\} \subset \{\text{eigenvalues of } A\} \tag{5.5}$$

with equality holding when all the eigenvalues of A are controllable and observable eigenvalues of the system. Similar results hold for discrete-time systems and $H(z)$.

EXAMPLE 5.3. Consider $A = \begin{bmatrix} 0 & -1 & 1 \\ 1 & -2 & 1 \\ 0 & 1 & -1 \end{bmatrix}$, $B = \begin{bmatrix} 1 & 0 \\ 1 & 1 \\ 1 & 2 \end{bmatrix}$, and $C = [0, 1, 0]$ (refer to Example 4.6 in Section 3.5). Then the transfer function $H(s) = [1/s, 1/s]$. $H(s)$ has only one pole, $s_1 = 0(p_H(s) = s)$, and $\lambda_1 = 0$, is the only controllable and observable eigenvalue. The other two eigenvalues of A, $\lambda_2 = -1$, $\lambda_3 = -2$, that are not both controllable and observable, do not appear as poles of $H(s)$. ■

EXAMPLE 5.4. Recall the circuit in Example 4.9 in Section 3.4. If $R_1 R_2 C \neq L$, then {poles of $H(s)$} = {eigenvalues of A at $\lambda_1 = -1/(R_1 C)$ and $\lambda_2 = -R_2/L$}. In this case, both eigenvalues are controllable and observable. Now if $R_1 R_2 C = L$ with $R_1 \neq R_2$, then $H(s)$ has only one pole, $s_1 = -R_2/L$, since in this case only one eigenvalue $\lambda_1 = -R_2/L$ is controllable and observable. The other eigenvalue λ_2 at the same location $-R_2/L$ is uncontrollable and unobservable. Now if $R_1 R_2 C = L$ with $R_1 = R_2 = R$, then one of the eigenvalues becomes uncontrollable and the other (also at $-R/L$) becomes unobservable. In this case $H(s)$ has no finite poles ($H(s) = 1/R$). ■

Zeros

In a scalar transfer function $H(s)$ the roots of the denominator polynomial are the poles, and the roots of its numerator polynomial are the zeros of $H(s)$. As was discussed, the *poles of* $H(s)$ are some or all of the eigenvalues of A (the eigenvalues of A are sometimes also called *poles of the system* $\{A, B, C, D\}$). In particular, it was shown in Subsection 3.4.C that the uncontrollable and/or unobservable eigenvalues of A can never be poles of $H(s)$. In Chapter 5 it is shown that only those eigenvalues of A that are both controllable and observable appear as poles of the transfer function $H(s)$. Along similar lines, the *zeros of* $H(s)$ (to be defined later) are some or all of the characteristic values of another matrix, the system matrix $P(s)$. These characteristic values are called the *zeros of the system* $\{A, B, C, D\}$.

The *zeros of a system* for both the continuous- and discrete-time case are defined and discussed next. We consider now only finite zeros. For the case of zeros at infinity, refer to the exercises.

Let the *system matrix* (also called *Rosenbrock's system matrix*) of $\{A, B, C, D\}$ be

$$P(s) \triangleq \begin{bmatrix} sI - A & B \\ -C & D \end{bmatrix}. \tag{5.6}$$

Note that in view of the system equations $\dot{x} = Ax + Bu$, $y = Cx + Du$, we have

$$P(s) \begin{bmatrix} -\hat{x}(s) \\ \hat{u}(s) \end{bmatrix} = \begin{bmatrix} 0 \\ \hat{y}(s) \end{bmatrix},$$

where $\hat{x}(s)$ denotes the Laplace transform of $x(t)$.

Let $r = \text{rank}\, P(s)$ [note that $n \leq r \leq \min(p + n, m + n)$] and consider all those rth-order nonzero minors of $P(s)$ that are formed by taking the first n rows and n columns of $P(s)$, i.e., all rows and columns of $sI - A$, and then adding appropriate $r - n$ rows (of $[-C, D]$) and columns (of $[B^T, D^T]^T$). The *zero polynomial of the system* $\{A, B, C, D\}$, $z_P(s)$, is defined as the monic greatest common divisor of all these minors.

DEFINITION 5.2. The *zeros of the system* $\{A, B, C, D\}$ or the *system zeros* are the roots of the zero polynomial of the system, $z_P(s)$. ■

In addition, we define the *invariant zeros of the system* as the roots of the invariant polynomials of $P(s)$.

In particular, consider the $(p+n) \times (m+n)$ system matrix $P(s)$ and let

$$S_P(s) = \begin{bmatrix} \Lambda(s) & 0 \\ 0 & 0 \end{bmatrix}, \qquad \Lambda(s) = diag\,[\epsilon_1(s), \ldots, \epsilon_r(s)], \tag{5.7}$$

be its Smith form. The *invariant zero polynomial of the system* $\{A, B, C, D\}$ is defined as

$$z_P^I(s) = \epsilon_1(s)\epsilon_2(s)\cdots\epsilon_r(s) \tag{5.8}$$

and its roots are the *invariant zeros of the system*. It can be shown that the monic greatest common divisor of all the highest order nonzero minors of $P(s)$ equals $z_P^I(s)$.

In general,

$$\{\text{zeros of the system}\} \supset \{\text{invariant zeros of the system}\}.$$

When $p = m$ with $det\, P(s) \neq 0$, then the zeros of the system coincide with the invariant zeros.

Now consider the $n \times (m+n)$ matrix $[sI - A, B]$ and determine its n invariant factors $\epsilon_i(s)$ and its Smith form. The product of its invariant factors is a polynomial, the roots of which are the *input-decoupling zeros of the system* $\{A, B, C, D\}$. Note that this polynomial equals the monic greatest common divisor of all the highest order nonzero minors (of order n) of $[sI - A, B]$. Similarly, consider the $(p+n) \times n$ matrix $\begin{bmatrix} sI - A \\ -C \end{bmatrix}$ and its invariant polynomials, the roots of which define the *output-decoupling zeros of the system* $\{A, B, C, D\}$.

Using the above definitions it is not difficult to show that the input-decoupling zeros of the system are eigenvalues of A and also zeros of the system $\{A, B, C, D\}$ (show this). In addition note that if λ_i is such an input-decoupling zero, then $rank\,[\lambda_i I - A, B] < n$, and therefore, there exists a $1 \times n$ vector $\hat{v}_i \neq 0$ such that $\hat{v}_i[\lambda_i I - A, B] = 0$. This, however, implies that λ_i is an uncontrollable eigenvalue of A (and $\hat{v}_i$ is the corresponding left eigenvector), in view of Subsection 3.4B. Conversely, it can be shown that an uncontrollable eigenvalue is an input-decoupling zero. Therefore, *the input-decoupling zeros of the system* $\{A, B, C, D\}$ *are the uncontrollable eigenvalues of* A. Similarly, it can be shown that the *output-decoupling zeros of the system* $\{A, B, C, D\}$ *are the unobservable eigenvalues of* A. They are also zeros of the system, as can easily be seen from the definitions.

There are eigenvalues of A that are both uncontrollable and unobservable. These can be determined using the left and right corresponding eigenvector test or by the Canonical Structure Theorem (Kalman Decomposition Theorem) (see Subsections 3.4A and B). These uncontrollable and unobservable eigenvalues of A are zeros of the system that are both input- and output-decoupling zeros and are called *input-output decoupling zeros*. These input-output decoupling zeros can also be defined directly from $P(s)$ given in (5.6); however, care should be taken in the case of repeated zeros.

If the zeros of a system are determined and the zeros that are input- and/or output-decoupling zeros are removed, then the zeros that remain are the *zeros of* $H(s)$ and can be found directly from the transfer function $H(s)$. In particular, if the Smith-McMillan form of $H(s)$ is given by (5.3), then

$$z_H(s) = \epsilon_1(s)\epsilon_2(s)\cdots\epsilon_r(s) \tag{5.9}$$

is the *zero polynomial of* $H(s)$ and its roots are the *zeros of* $H(s)$. These are also called the *transmission zeros of the system.*

DEFINITION 5.3. The *zeros of* $H(s)$ or the *transmission zeros of the system* are the roots of the zero polynomial of $H(s)$, $z_H(s)$. ■

The relationship between the zeros of the system and the zeros of $H(s)$ can easily be determined using the identity

$$P(s) = \begin{bmatrix} sI & -A & B \\ & -C & D \end{bmatrix} = \begin{bmatrix} sI & -A & 0 \\ & -C & I \end{bmatrix} \begin{bmatrix} I & (sI - A)^{-1}B \\ 0 & H(s) \end{bmatrix}$$

for the case when $P(s)$ is square and nonsingular. Note that in the present case $|P(s)| = |sI - A||H(s)|$. It is also possible to obtain this result using the special structure of the matrices in the special form of Subsection 3.4A. In this case, the invariant zeros of the system [the roots of $|P(s)|$], which are equal here to the zeros of the system, are the zeros of $H(s)$ [the roots of $|H(s)|$] *and* those eigenvalues of A that are not both controllable and observable [the ones that do not cancel in $|sI - A||H(s)|$].

Note that the zero polynomial of $H(s)$, $z_H(s)$, equals the monic greatest common divisor of the numerators of all the highest order nonzero minors in $H(s)$ after all their denominators have been set equal to $p_H(s)$, the characteristic polynomial of $H(s)$. In the scalar case ($p = m = 1$) our definition of the zeros of $H(s)$ reduces to the well-known definition of zeros, namely, the roots of the numerator polynomial of $H(s)$.

EXAMPLE 5.5. Consider $H(s)$ of Example 5.1. From the Smith-McMillan form of $H(s)$, we obtain the zero polynomial $z_H(s) = 1$, and $H(s)$ has no (finite) zeros. Alternatively, the highest order nonzero minors are $1/[s^3(s+1)]$ and $1/s^3 = (s+1)/[s^3(s+1)]$ and the greatest common divisor of the numerators is $z_H(s) = 1$. ■

EXAMPLE 5.6. We wish to determine the zeros of $H(s) = \begin{bmatrix} \frac{s}{s+1} & 0 \\ \frac{1}{s+1} & \frac{s+1}{s^2} \end{bmatrix}$.

The first-order minors are the entries of $H(s)$, namely, $\frac{s}{s+1}, \frac{1}{s+1}, \frac{s+1}{s^2}$, and there is only one second-order minor $\frac{s}{s+1} \cdot \frac{s+1}{s^2} = \frac{1}{s}$. Then $p_H(s) = s^2(s+1)$, the least common denominator, is the characteristic polynomial. Next, write the highest (second-) order minor as $\frac{1}{s} = \frac{s(s+1)}{s^2(s+1)} = \frac{s(s+1)}{p_H(s)}$ and note that $s(s+1)$ is the zero polynomial of $H(s)$, $z_H(s)$, and the zeros of $H(s)$ are $\{0, -1\}$. It is worth noting that the poles and zeros of $H(s)$ are at the same locations. This may happen only when $H(s)$ is a matrix.

If the Smith-McMillan form of $H(s)$ is to be used, write $H(s) = \frac{1}{s^2(s+1)} \times \begin{bmatrix} s^3 & 0 \\ s^2 & (s+1)^2 \end{bmatrix} = \frac{1}{d(s)} N(s)$. The Smith form of $N(s)$ is now $\begin{bmatrix} 1 & 0 \\ 0 & s^3(s+1)^2 \end{bmatrix}$ since $D_0 = 1, D_1 = 1, D_2 = s^3(s+1)^2$ with invariant factors of $N(s)$ given by $n_1 = D_1/D_0 = 1$ and $n_2 = D_2/D_1 = s^3(s+1)^2$. Therefore, the Smith-McMillan form (5.3) of $H(s)$ is

$$SM_H(s) = \begin{bmatrix} \frac{1}{s^2(s+1)} & 0 \\ 0 & \frac{s(s+1)}{1} \end{bmatrix} = \begin{bmatrix} \frac{\epsilon_1}{\psi_1} & 0 \\ 0 & \frac{\epsilon_2}{\psi_2} \end{bmatrix}.$$

The zero polynomial is then $z_H(s) = \epsilon_1\epsilon_2 = s(s+1)$ and the zeros of $H(s)$ are $\{0, -1\}$, as expected. Also, the pole polynomial is $p_H(s) = \psi_1\psi_2 = s^2(s+1)$ and the poles are $\{0, 0, -1\}$. ■

EXAMPLE 5.7. We wish to determine the zeros of $H(s) = \begin{bmatrix} \frac{s}{s+1} & 0 \\ \frac{1}{s+1} & \frac{s+1}{s^2} \\ 0 & \frac{1}{s} \end{bmatrix}$. The second-order minors are $\frac{1}{s}, \frac{1}{s+1}, \frac{1}{s(s+1)}$ and the characteristic polynomial is $p_H(s) = s^2(s+1)$. Rewriting the highest (second-) order minors as $s(s+1)/p_H(s)$, $s^2/p_H(s)$, and $s/p_H(s)$, the greatest common divisor of the numerators is s, i.e., the zero polynomial of $H(s)$ is $z_H(s) = s$. Thus, there is only one zero of $H(s)$ located at 0. Alternatively, note that the Smith-McMillan form is

$$SM_H(s) = \begin{bmatrix} \frac{1}{s^2(s+1)} & 0 \\ 0 & \frac{s}{1} \\ 0 & 0 \end{bmatrix}.$$

■

Relations between poles, zeros, and eigenvalues of A

Consider the system $\dot{x} = Ax + Bu$, $y = Cx + Du$ and its transfer function matrix $H(s) = C(sI - A)^{-1}B + D$. Summarizing the above discussion, the following relations can be shown to be true.

1. We have the set relationship

$$\begin{aligned}\{\text{zeros of the system}\} = \{&\text{zeros of } H(s)\} \\ &\cup \{\text{input-decoupling zeros}\} \cup \{\text{output-decoupling zeros}\} \\ &- \{\text{input-output decoupling zeros}\}.\end{aligned} \tag{5.10}$$

   Note that the invariant zeros of the system contain all the zeros of $H(s)$ (transmission zeros), but not all the decoupling zeros (see Example 5.8). When $P(s)$ is square and nonsingular, the zeros of the system are exactly the invariant zeros of the system. Also, in the case when $\{A, B, C, D\}$ is controllable and observable, the zeros of the system, the invariant zeros, and the transmission zeros [zeros of $H(s)$] all coincide.

2. We have the set relationship

$$\begin{aligned}&\{\text{eigenvalues of } A \text{ (or poles of the system)}\} = \{\text{poles of } H(s)\} \\ &\quad \cup \{\text{uncontrollable eigenvalues of } A\} \cup \{\text{unobservable eigenvalues of } A\} \\ &\quad - \{\text{both uncontrollable and unobservable eigenvalues of } A\}.\end{aligned} \tag{5.11}$$

3. We have the set relationships

$$\begin{aligned}\{\text{input-decoupling zeros}\} &= \{\text{uncontrollable eigenvalues of } A\}, \\ \{\text{output-decoupling zeros}\} &= \{\text{unobservable eigenvalue of } A\}, \\ \text{and} \quad \{\text{input-output decoupling zeros}\} &= \{\text{eigenvalues of } A \text{ that are both} \\ &\qquad \text{uncontrollable and unobservable}\}.\end{aligned} \tag{5.12}$$

4. When the system $\{A, B, C, D\}$ is controllable and observable, then

$$\{\text{zeros of the system}\} = \{\text{zeros of } H(s)\}$$
$$\text{and} \quad \{\text{eigenvalues of } A \text{ (or poles of the system)}\} = \{\text{poles of } H(s)\}. \quad (5.13)$$

Note that *the eigenvalues of* A (*the poles of the system*) can be defined as the roots of the invariant factors of $sI - A$ in $P(s)$ given in (5.6).

EXAMPLE 5.8. Consider the system $\{A, B, C\}$ of Example 5.3. Let

$$P(s) = \begin{bmatrix} sI - A & B \\ -C & D \end{bmatrix} = \left[\begin{array}{ccc:cc} s & 1 & -1 & 1 & 0 \\ -1 & s+2 & -1 & 1 & 1 \\ 0 & -1 & s+1 & 1 & 2 \\ \cdots & \cdots & \cdots & \cdots & \cdots \\ 0 & -1 & 0 & 0 & 0 \end{array}\right].$$

There are two fourth-order minors that include all columns of $sI - A$ obtained by taking columns 1, 2, 3, 4 and columns 1, 2, 3, 5 of $P(s)$; they are $(s+1)(s+2)$ and $(s+1)(s+2)$ (verify this). The zero polynomial of the system is $z_P = (s+1)(s+2)$ and the zeros of the system are $\{-1, -2\}$. To determine the input-decoupling zeros, consider all the third-order minors of $[sI - A, B]$. The greatest common divisor is $s+2$ (verify this), which implies that the input-decoupling zeros are $\{-2\}$. Similarly, consider $\begin{bmatrix} sI - A \\ -C \end{bmatrix}$ and show that $s+1$ is the greatest common divisor of all the third-order minors and that the output-decoupling zeros are $\{-1\}$. The transfer function for this example was found in Example 5.3 to be $H(s) = [1/s, 1/s]$. The zero polynomial of $H(s)$ is $z_H(s) = 1$ and there are no zeros of $H(s)$. Notice that there are no input-output decoupling zeros. It is now clear that relation (5.10) holds.

The controllable (resp., uncontrollable) and the observable (resp., unobservable) eigenvalues of A (poles of the system) have been found in Examples 4.4 and 4.5 in Section 3.4. Compare these results to show that (5.12) holds. The poles of $H(s)$ are $\{0\}$. Verify that (5.11) holds.

One could work with the Smith form of the matrices of interest and the Smith-McMillan form of $H(s)$. In particular, it can be shown (do so) that the Smith form of $P(s)$ is $\begin{bmatrix} 1 & 0 & 0 & 0 & 0 \\ 0 & 1 & 0 & 0 & 0 \\ 0 & 0 & 1 & 0 & 0 \\ 0 & 0 & 0 & s+2 & 0 \end{bmatrix}$, of $[sI - A, B]$ is $\begin{bmatrix} 1 & 0 & 0 & 0 & 0 \\ 0 & 1 & 0 & 0 & 0 \\ 0 & 0 & s+2 & 0 & 0 \end{bmatrix}$, of $\begin{bmatrix} sI - A \\ -C \end{bmatrix}$ is $\begin{bmatrix} 1 & 0 & 0 \\ 0 & 1 & 0 \\ 0 & 0 & s+1 \\ 0 & 0 & 0 \end{bmatrix}$, and of $[sI - A]$ is $\begin{bmatrix} 1 & 0 & 0 \\ 0 & 1 & 0 \\ 0 & 0 & s(s+1)(s+2) \end{bmatrix}$. Also, it can be shown that the Smith-McMillan form of $H(s)$ is

$$SM_H(s) = \left[\frac{1}{s}, 0\right].$$

It is straightforward to verify the above results. Note that in the present case the invariant zero polynomial is $z_P^I(s) = s + 2$ and there is only one invariant zero at -2. ■

EXAMPLE 5.9. Consider the circuit of Example 5.4 and of Example 4.9 in this chapter and the system matrix $P(s)$ for the case when $R_1 R_2 C = L$ given by

$$P(s) = \begin{bmatrix} sI - A & B \\ -C & D \end{bmatrix} = \begin{bmatrix} \frac{s + R_2}{L} & 0 & \vdots & \frac{R_2}{L} \\ 0 & \frac{s + R_2}{L} & \vdots & \frac{1}{L} \\ \cdots & \cdots & \vdots & \cdots \\ \frac{1}{R_1} & -1 & \vdots & \frac{1}{R_1} \end{bmatrix}.$$

(i) First, let $R_1 \neq R_2$. To determine the zeros of the system, consider $|P(s)| = (1/R_1)(s + R_1/L)(s + R_2/L)$, which implies that the zeros of the system are $\{-R_1/L, -R_2/L\}$. Consider now all second-order (nonzero) minors of $[sI - A, B]$, namely, $(s + R_2/L)^2$, $(1/L)(s + R_2/L)$, and $-(R_2/L)(s + R_2/L)$, from which we see that $\{-R_2/L\}$ is the input-decoupling zero. Similarly, we also see that $\{-R_2/L\}$ is the output-decoupling zero. Therefore, $\{-R_2/L\}$ is the input-output decoupling zero. Compare this with the results in Example 5.4 to verify (5.13).

(ii) When $R_1 = R_2 = R$, then $|P(s)| = (1/R)(s + R/L)^2$, which implies that the zeros of the system are at $\{-R/L, -R/L\}$. Proceeding as in (i), it can readily be shown that $\{-R/L\}$ is the input-decoupling zero and $\{-R/L\}$ is the output-decoupling zero. To determine which are the input-output decoupling zeros, one needs additional information to the zero location. This information can be provided by the left and right eigenvectors of the two zeros at $-R/L$ to determine that there is no input-output decoupling zero in this case (see Example 4.9).

In both cases (i) and (ii), $H(s)$ has been derived in Example 4.9 of Section 3.4. Verify relation (5.10). ■

Zero directions and pole-zero cancellations

There are characteristic vectors or zero directions, associated with each invariant and decoupling zero of the system $\{A, B, C, D\}$, just as there are characteristic vectors or eigenvectors, associated with each eigenvalue of A (pole of the system).

Consider the system matrix $P(\lambda)$ given in (5.6) at $s = \lambda$. If λ is an invariant zero of the system, then there exist nonzero vectors v_z and $\hat{v}_z$ associated with λ such that

$$P(\lambda)v_z = 0, \qquad \hat{v}_z P(\lambda) = 0. \tag{5.14}$$

This is so because if λ is an invariant zero of the system, then *rank* $P(\lambda) <$ *rank* $P(s) \leq \min(n + p, n + m)$, and therefore the columns (resp., the rows) are linearly dependent. Here *rank* $P(s)$ denotes the number of linearly independent columns or rows over the field of rational functions in s [it is called *normal rank* of $P(s)$]; *rank* $P(\lambda)$ is the rank of $P(\lambda)$ over the field of complex numbers.

The vector v_z is called an *invariant zero direction or a zero direction* corresponding to λ. A physical interpretation of this is as follows. Let $v_z = \begin{bmatrix} \bar{x} \\ -\bar{u} \end{bmatrix}$, that is, let

$$\begin{bmatrix} \lambda I - A & B \\ -C & D \end{bmatrix} \begin{bmatrix} \bar{x} \\ -\bar{u} \end{bmatrix} = \begin{bmatrix} 0 \\ 0 \end{bmatrix}, \tag{5.15}$$

and assume that the system is at rest at $t = 0$. If an input of the form $u(t) = \bar{u}e^{\lambda t}, t \geq 0$, is applied, and if λ is not a pole of the system, then it will produce a state of the form $x(t) = \bar{x}e^{\lambda t}$ and an output

$$y(t) = 0, \qquad t \geq 0.$$

This is sometimes referred to in the literature as the *output-zeroing or blocking property of zeros*. This property is a generalization of the *blocking property of zeros* originally expressed in terms of the transfer function and its zeros for a SISO system.

For decoupling zeros, it is quite easy to see what the corresponding directions will be. In particular, if λ is an input-decoupling zero, then there exists a nonzero vector $\hat{v}$ such that $\hat{v}[\lambda I - A, B] = 0$, which implies that

$$\hat{v}_z P(\lambda) = [\hat{v}, 0]\begin{bmatrix} \lambda I - A & B \\ -C & D \end{bmatrix} = [0, 0]. \tag{5.16}$$

It is clear that the vector $\hat{v}$ is the left eigenvector of A corresponding to λ, which is also an eigenvalue of A. This, in fact, determines the exact relationship (location and corresponding directions) between input-decoupling zeros and uncontrollable eigenvalues.

Similar results can be derived for the output-decoupling zeros. In fact if λ is an output-decoupling zero, then there exists a nonzero vector v so that $\begin{bmatrix} \lambda I - A \\ -C \end{bmatrix} v = 0$, which implies that

$$P(\lambda)v_z = \begin{bmatrix} \lambda I - A & B \\ -C & D \end{bmatrix}\begin{bmatrix} v \\ 0 \end{bmatrix} = \begin{bmatrix} 0 \\ 0 \end{bmatrix}. \tag{5.17}$$

It is clear that the vector v is the right eigenvector of A corresponding to λ, which is also an eigenvalue of A. This provides the exact relation between output-decoupling zeros and unobservable eigenvalues.

Consider now an input-decoupling zero λ and the corresponding direction $[\hat{v}, 0]$. As was shown above λ, $\hat{v}$ are an (uncontrollable) eigenvalue and its corresponding left eigenvector, respectively. Let v be the corresponding right eigenvector to λ. If $\begin{bmatrix} \lambda I - A \\ -C \end{bmatrix} v = 0$, then λ is also an unobservable eigenvalue and an output-decoupling zero. In particular, λ is an input-output decoupling zero and also an eigenvalue that is both uncontrollable and unobservable. The vectors $[\hat{v}, 0]$ and $[v^T, 0]^T$ are the directions associated with such zero.

In general the directions v_z and $\hat{v}_z$ in (5.14) associated with the invariant zeros do not have any particular form [$\hat{v}_z = [\hat{v}, 0]$ in (5.16) for the case of an input-decoupling zero]. When the rank of $P(s)$ is full, that is,

$$rank\ P(s) = n + \min(p, m), \tag{5.18}$$

the direction that corresponds to the invariant zero at λ can be taken to be v_z [where $P(\lambda)v_z = 0$] when $\min(p, m) = m$, and $\hat{v}_z$ [where $\hat{v}_z P(\lambda) = 0$], when $\min(p, m) = p$. *Note that when* $\min(p, m) = p < m$, *there are nonzero vectors satisfying* $P(\lambda)v_z = 0$ *with λ not necessarily being an invariant zero of the system.* This situation becomes accentuated, i.e., (5.14) is satisfied for values of λ that are not necessarily zeros of the system, when rank $P(s) < n + \min(p, m)$. This phenomenon is unique to MIMO systems.

In the MIMO case it is possible for a $p \times m$ transfer function $H(s)$ to have poles and zeros at the same location (see Example 5.10). This is impossible in the scalar case, where $H(s)$ is a 1×1 matrix, since in this case common factors in the numerator and denominator will cancel in the process of forming $H(s)$. In state-space terms, this result can be expressed as follows.

LEMMA 5.1. In the SISO case, if λ is both a zero of the system and an eigenvalue of A (or pole of the system), then λ must be an input- and/or output-decoupling zero (uncontrollable and/or unobservable eigenvalue). This is not necessarily true in the MIMO case.

Proof. Assume that λ is not an input- and/or output-decoupling zero, or equivalently, that all eigenvalues of A are controllable and observable. Then all eigenvalues of A will be poles of $H(s)$ and all zeros of the system will be zeros of $H(s)$. This is not possible, however, since by definition the numerator and denominator of $H(s)$ cannot have a common factor [presently, $(s - \lambda)$]. Therefore, λ must be an input- and/or output-decoupling zero. ■

EXAMPLE 5.10. For $H(s) = \begin{bmatrix} \frac{s}{s+1} & 0 \\ \frac{1}{s+1} & \frac{s+1}{s^2} \end{bmatrix}$ the poles and zeros were determined in Example 5.6 to be {poles of $H(s)$} $= \{0, 0, -1\}$ and {zeros of $H(s)$} $= \{0, -1\}$. Note that in this case the poles and zeros are at the same locations, 0 and -1. ■

We conclude by noting that, as will be shown in Section 4.2 of Chapter 4, one can arbitrarily assign values to the controllable eigenvalues, and to a certain extent, one can alter their corresponding eigenvectors, using linear state feedback. In the scalar case, assigning a closed-loop eigenvalue at an open-loop zero location guarantees that the eigenvalue will become unobservable (linear state feedback does not alter controllability) and will not appear in the transfer function (refer to Exercises 4.6 and 4.7 in Chapter 4). In the MIMO case, the eigenvectors of the closed-loop eigenvalues must also be assigned appropriately for the eigenvalues to become unobservable and cancel out when forming the transfer function matrix (see Exercise 4.19 in Chapter 4).

3.6 SUMMARY

In this chapter the system properties of reachability (or controllability-from-the-origin) and controllability (-to-the-origin), together with the dual properties of observability and constructibility, respectively, were developed. These concepts were introduced using discrete-time time-invariant systems (Subsection 3.1A) and were further developed in Part 1 of the chapter (Sections 3.2 and 3.3).

In Section 3.2, the reachability Gramian of a continuous-time system was used to derive inputs that transfer the state of the system from one desirable vector value to another. This was accomplished for both time-varying and time-invariant systems. The time-invariant case was developed in Subsection 3.2B, so that it may be studied independently. It was shown that for continuous-time systems, reachability implies controllability, and vice-versa; however, for discrete-time systems, although reachability always implies controllability, controllability may not imply reachability (unless A has full rank). Analogous results were developed in Section 3.3 regarding observability and constructibility.

In Part 2, Section 3.4, useful special forms for (continuous-time and discrete-time) state-space descriptions of time-invariant systems were developed. The standard forms for uncontrollable (resp., unobservable) systems lead to better understanding of the relationships between state-space and transfer function descriptions of systems. The controller and observer forms provide important structural information about a system that is useful in state-space realizations of transfer function matrices and in feedback control. Polynomial matrix fractional descriptions of transfer matrices were also introduced, by the structure theorem. Finally, poles and zeros of systems were addressed, in Section 3.5. They were introduced using the Smith form of polynomial matrices and the Smith-McMillan form of transfer function matrices.

The reachability (controllability) and observability (constructibility) Gramians played an important role in this chapter. These are now summarized, for convenience.

Summary of Gramians Introduced in This Chapter

Reachability Gramians

A. $$W_r(t_0, t_1) = \int_{t_0}^{t_1} \Phi(t_1, \tau)B(\tau)B^T(\tau)\Phi(t_1, \tau)\, d\tau. \tag{2.11}$$

B. $$W_r(0, T) = \int_0^T e^{(T-\tau)A} BB^T e^{(T-\tau)A^T}\, d\tau. \tag{2.29}$$

C. $$W_r(0, K) = \sum_{i=0}^{K-1} A^{K-(i+1)}BB^T(A^T)^{K-(i+1)} = \sum_{i=0}^{K-1} A^iBB^T(A^T)^i. \tag{2.64}$$

Controllability Gramians

A. $$W_c(t_0, t_1) = \int_{t_0}^{t_1} \Phi(t_0, \tau)B(\tau)B^T(\tau)\Phi^T(t_0, \tau)\, d\tau. \tag{2.19}$$

B. $$W_c(0, T) = \int_0^T e^{-A\tau}BB^T e^{-A^T\tau}\, d\tau. \tag{2.41}$$

C. $$W_c(0, K) = \sum_{i=0}^{K-1} A^{-(i+1)}BB^T(A^T)^{-(i+1)}, \qquad |A| \neq 0. \tag{2.66}$$

Observability Gramians

A. $$W_o(t_0, t_1) = \int_{t_0}^{t_1} \Phi^T(\tau, t_0)C^T(\tau)C(\tau)\Phi(\tau, t_0)\, d\tau. \tag{3.5}$$

B. $$W_o(0, T) = \int_0^T e^{A^T\tau}C^TCe^{A\tau}\, d\tau. \tag{3.22}$$

C. $$W_o(0, K) = \sum_{i=0}^{K-1} (A^T)^iC^TCA^i. \tag{3.49}$$

Constructibility Gramians

A. $$W_{cn}(t_0, t_1) = \int_{t_0}^{t_1} \Phi^T(\tau, t_1)C^T(\tau)C(\tau)\Phi(\tau, t_1)\, d\tau. \tag{3.11}$$

B. $$W_{cn}(0, T) = \int_0^T e^{A^T(\tau-T)}C^TCe^{A(\tau-T)}\, d\tau. \tag{3.32}$$

C. $$W_{cn}(0, K) = \sum_{i=0}^{K-1} (A^T)^{-(i+1)}C^TCA^{-(i+1)}, \qquad |A| \neq 0. \tag{3.54}$$

where the Gramians in A, B, and C in each case are the Gramians of the systems

A. $\dot{x} = A(t)x + B(t)u,\ y = Cx(t) + D(t)u.$

B. $\dot{x} = Ax + Bu,\ y = Cx + Du.$

C. $x(k+1) = Ax(k) + Bu(k),\ y(k) = Cx(k) + Du(k),$ respectively.

In B and C the controllability and observability matrices are, respectively,

$$\mathscr{C} = [B, AB, \ldots, A^{n-1}B], \qquad \mathscr{O} = [C^T, (CA)^T, \ldots, (CA^{n-1})^T]^T.$$

Note that reachability is the dual concept to observability and controllability is dual to (re)constructibility.

3.7 NOTES

The concept of controllability was first encountered as a technical condition in certain optimal control problems and also in the so-called finite-settling-time design problem for discrete-time systems (see Kalman [6]). In the latter, an input must be found that returns the state x_0 to the origin as quickly as possible. Manipulating the input to assign particular values to the initial state in (analog-computer) simulations was not an issue since the individual capacitors could initially be charged independently. Also, observability was not an issue in simulations due to the particular system structures that were used (corresponding, e.g., to observer forms). The current definitions for controllability and observability and the recognition of the duality between them were worked out by Kalman in 1959–1960 (see Kalman [9] for historical comments) and were presented by Kalman in [7]. The significance of realizations that were both controllable and observable (see Chapter 5) was established later in Gilbert [3], Kalman [8], and Popov [12]. For further information regarding these historical issues, consult Kailath [5] and the original sources. Note that [5] has extensive references up to the late seventies with emphasis on the time-invariant case and a rather complete set of original references together with historical remarks for the period where the foundations of the state-space system theory were set, in the late fifties and sixties.

Special state-space forms for controllable and observable systems obtained by similarity transformations are discussed at length in Kailath [5] (refer also to the discussion on various canonical forms). Wolovich [19] discusses the algorithms for controller and observer forms and introduces the Structure Theorems. The controller form is based on results by Luenberger [11] (see also Popov [13]). A detailed derivation of the controller form can also be found in Rugh [16].

Original sources for the Canonical Structure Theorem include Kalman [8] and Gilbert [3].

The eigenvector and rank tests for controllability and observability are called PBH tests in Kailath [5]. Original sources for these include Popov [14], Belevich [1], and Hautus [4]. Consult also Rosenbrock [15], and for the case when A can be diagonalized via a similarity transformation, see Gilbert [3]. Note that in the eigenvalue/eigenvector tests presented herein the uncontrollable (unobservable) eigenvalues are also explicitly identified, which represents a modification of the above original results.

The Brunovsky canonical form is developed in Brunovsky [2].

The fact that the controllability indices appear in the work of Kronecker was recognized by Rosenbrock [15] and Kalman [10].

For an extensive introductory discussion and a formal definition of canonical forms, see Kailath [5]. Note that certain special forms exist for time-varying systems as well, but they are not considered here.

Multivariable zeros have an interesting history. For a review, see Schrader and Sain [17] and the references therein. Refer also to Vardulakis [18].

3.8 REFERENCES

1. V. Belevich, *Classical Network Theory,* Holden-Day, San Francisco, 1968.
2. P. Brunovsky, "A Classification of Linear Controllable Systems," *Kybernetika,* Vol. 3, pp. 173–187, 1970.
3. E. Gilbert, "Controllability and Observability in Multivariable Control Systems," *SIAM J. Control,* Vol. 1, pp. 128–151, 1963.
4. M. L. J. Hautus, "Controllability and Observability Conditions of Linear Automonous Systems," *Proc. Koninklijke Akademie van Wetenschappen, Serie A,* Vol. 72, pp. 443–448, 1969.
5. T. Kailath, *Linear Systems,* Prentice-Hall, Englewood Cliffs, NJ, 1980.
6. R. E. Kalman, "Optimal Nonlinear Control of Saturating Systems by Intermittent Control," IRE WESCON Rec., Sec. IV, pp. 130–135, 1957.
7. R. E. Kalman, "On the General Theory of Control Systems," in *Proc. of the First Intern. Congress on Automatic Control,* pp. 481–493, Butterworth, London, 1960.
8. R. E. Kalman, "Mathematical Descriptions of Linear Systems," *SIAM J. Control,* Vol. 1, pp. 152–192, 1963.
9. R. E. Kalman, *Lectures on Controllability and Observability,* C.I.M.E., Bologna, 1968.
10. R. E. Kalman, "Kronecker Invariants and Feedback," in *Ordinary Differential Equations,* L. Weiss, ed., pp. 459–471, Academic Press, New York, 1972.
11. D. G. Luenberger, "Canonical Forms for Linear Multivariable Systems," *IEEE Transactions on Automatic Control,* Vol. 12, pp. 290–293, 1967.
12. V. M. Popov, "On a New Problem of Stability for Control Systems," *Autom. Remote Control,* pp. 1–23, Vol. 24, No. 1, 1963.
13. V. M. Popov, "Invariant Description of Linear, Time-Invariant Controllable Systems," *SIAM Journal of Control and Optimization,* Vol. 10, No. 2, pp. 252–264, 1972.
14. V. M. Popov, *Hyperstability of Control Systems,* Springer-Verlag, Berlin, 1973.
15. H. H. Rosenbrock, *State-Space and Multivariable Theory,* Wiley, New York, 1970.
16. W. J. Rugh, *Linear System Theory,* Prentice-Hall, Englewood Cliffs, NJ, 1993.
17. C. B. Schrader and M. K. Sain, "Research on System Zeros: a Survey," *Int. Journal of Control,* Vol. 50, No. 4, pp. 1407–1433, 1989.
18. A. I. G. Vardulakis, *Linear Multivariable Control. Algebraic Analysis and Synthesis Methods,* Wiley, New York, 1991.
19. W. A. Wolovich, *Linear Multivariable Systems,* Springer-Verlag, New York, 1974.

3.9 EXERCISES

3.1. (a) Let $\mathscr{C}_k \triangleq [B, AB, \ldots, A^{k-1}B]$, where $A \in R^{n\times n}$, $B \in R^{n\times m}$. Show that

$$\mathscr{R}(\mathscr{C}_k) = \mathscr{R}(\mathscr{C}_n) \text{ for } k \geq n, \qquad \text{and} \qquad \mathscr{R}(\mathscr{C}_k) \subset \mathscr{R}(\mathscr{C}_n) \text{ for } k < n.$$

(b) Let $\mathbb{O}_k \triangleq [C^T, (CA)^T, \ldots, (CA^{k-1})^T]^T$, where $A \in R^{n\times n}$, $C \in R^{p\times n}$. Show that

$$\mathcal{N}(\mathbb{O}_k) = \mathcal{N}(\mathbb{O}_n) \text{ for } k \geq n, \qquad \text{and} \qquad \mathcal{N}(\mathbb{O}_k) \supset \mathcal{N}(\mathbb{O}_n) \text{ for } k < n.$$

3.2. Consider the state equation $\dot{x} = Ax + Bu$, where

$$A = \begin{bmatrix} 0 & 1 & 0 & 0 \\ 3w^2 & 0 & 0 & 2w \\ 0 & 0 & 0 & 1 \\ 0 & -2w & 0 & 0 \end{bmatrix}, \qquad B = \begin{bmatrix} 0 & 0 \\ 1 & 0 \\ 0 & 0 \\ 0 & 1 \end{bmatrix},$$

which was obtained by linearizing the nonlinear equations of motion of an orbiting satellite about a steady-state solution. In the state $x = [x_1, x_2, x_3, x_4]^T$, x_1 is the differential radius, while x_3 is the differential angle. In the input vector $u = [u_1, u_2]^T$, u_1 is the radial thrust and u_2 is the tangential thrust.

(a) Is this system controllable from u? If $y = \begin{bmatrix} y_1 \\ y_2 \end{bmatrix} = \begin{bmatrix} x_1 \\ x_3 \end{bmatrix}$, is the system observable from y?

(b) Can the system be controlled if the radial thruster fails? What if the tangential thruster fails?

(c) Is the system observable from y_1 only? From y_2 only?

3.3. Consider the state equation $\begin{bmatrix} \dot{x}_1 \\ \dot{x}_2 \end{bmatrix} = \begin{bmatrix} -\frac{1}{2} & 0 \\ 0 & -1 \end{bmatrix} \begin{bmatrix} x_1 \\ x_2 \end{bmatrix} + \begin{bmatrix} \frac{1}{2} \\ 1 \end{bmatrix} u.$

(a) If $x(0) = \begin{bmatrix} a \\ b \end{bmatrix}$, derive an input that will drive the state to $\begin{bmatrix} 0 \\ 0 \end{bmatrix}$ in T sec.

(b) For $x(0) = \begin{bmatrix} 5 \\ -5 \end{bmatrix}$, plot $u(t)$, $x_1(t)$, $x_2(t)$ for $T = 1, 2,$ and 5 sec. Comment on the magnitude of the input in your results.

3.4. Consider the state equation $x(k+1) = \begin{bmatrix} 1 & 1 & 0 \\ 0 & 1 & 0 \\ 0 & 0 & 1 \end{bmatrix} x(k) + \begin{bmatrix} 0 \\ 1 \\ 1 \end{bmatrix} u(k)$, $y(k) = \begin{bmatrix} 1 & 1 & 0 \\ 0 & 1 & 0 \end{bmatrix} x(k)$.

(a) Is $x^1 = \begin{bmatrix} 3 \\ 2 \\ 2 \end{bmatrix}$ reachable? If yes, what is the minimum number of steps required to transfer the state from the zero state to x^1? What inputs do you need?

(b) Determine all states that are reachable.

(c) Determine all states that are unobservable.

(d) If $\dot{x} = Ax + Bu$ is given with A, B as in (a), what is the minimum time required to transfer the state from the zero state to x^1? What is an appropriate $u(t)$?

3.5. *Output reachability (controllability)* can be defined in a manner analogous to state reachability (controllability). In particular, a system will be called output reachable if there exists an input that transfers the output from some y_0 to any y_1 in finite time.

Consider now a discrete-time time-invariant system $x(k+1) = Ax(k) + Bu(k)$, $y(k) = Cx(k) + Du(k)$ with $A \in R^{n\times n}$, $B \in R^{n\times m}$, $C \in R^{p\times n}$, and $D \in R^{p\times m}$. Recall that

$$y(k) = CA^k x(0) + \sum_{i=0}^{k-1} CA^{k-(i+1)}Bu(i) + Du(k).$$

(a) Show that the system $\{A, B, C, D\}$ is output reachable if and only if

$$rank\,[D, CB, CAB, \ldots, CA^{n-1}B] = p.$$

Note that this rank condition is also the condition for output reachability for continuous-time time-invariant systems $\dot{x} = Ax + Bu$, $y = Cx + Du$.

It should be noted that, in general, state reachability is neither necessary nor sufficient for output reachability. Notice for example that if $rank\ D = p$ then the system is output reachable.

(b) Let $D = 0$. Show that if (A, B) is (state) reachable, then $\{A, B, C, D\}$ is output reachable if and only if $rank\ C = p$.

(c) Let $A = \begin{bmatrix} 1 & 0 & 0 \\ 0 & -2 & 0 \\ 0 & 0 & -1 \end{bmatrix}$, $B = \begin{bmatrix} 1 \\ 0 \\ 1 \end{bmatrix}$, $C = [1, 1, 0]$, and $D = 0$.

(i) Is the system output reachable? Is it state reachable?

(ii) Let $x(0) = 0$. Determine an appropriate input sequence to transfer the output to $y_1 = 3$ in minimum time. Repeat for $x(0) = [1, -1, 2]^T$.

3.6. (a) Given $\dot{x} = Ax + Bu$, $y = Cx + Du$, show that this system is output reachable if and only if the rows of the $p \times m$ transfer matrix $H(s)$ are linearly independent over the field of complex numbers. In view of this result, is the system $H(s) = \begin{bmatrix} \frac{1}{s+2} \\ \frac{s}{s+1} \end{bmatrix}$ output reachable?

(b) Similarly, for discrete-time systems, the system is output reachable if and only if the rows of the transfer function matrix $H(z)$ are linearly independent over the field of complex numbers. Consider now the system of Exercise 3.5 and determine if it is output reachable.

3.7. Show that the circuit depicted in Fig. 3.6 with input u and output y is neither state reachable nor observable but is output reachable.

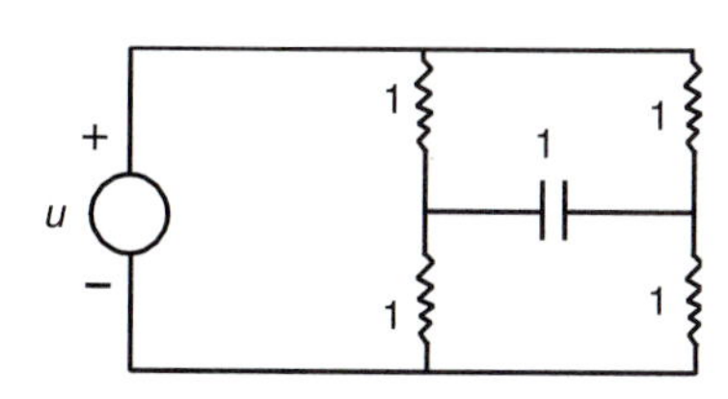

FIGURE 3.6
Circuit for Exercise 3.7

3.8. A system $\dot{x} = Ax + Bu$, $y = Cx + Du$ is called *output function controllable* if there exists an input $u(t)$, $t \in [0, \infty)$, that will cause the output $y(t)$ to follow a prescribed trajectory for $0 \leq t < \infty$, assuming that the system is at rest at $t = 0$. It is easiest to derive a test for output function controllability in terms of the $p \times m$ transfer function matrix $H(s)$, and this is the approach taken in the following. We say that the $m \times p$ rational matrix $H_R(s)$ is a *right inverse of* $H(s)$ if

$$H(s)H_R(s) = I_p.$$

(a) Show that the right inverse $H_R(s)$ exists if and only if $rank\ H(s) = p$. *Hint:* In the sufficiency proof, select $H_R = H^T(HH^T)^{-1}$, the (right) pseudoinverse of H.

(b) Show that the system is output function controllable if and only if $H(s)$ has a right inverse $H_R(s)$. *Hint:* Consider $\hat{y} = H\hat{u}$. In the necessity proof, show that if $rank\ H < p$ then the system may not be output function controllable.

Input function observability is the dual to output function controllablity. Here, the *left inverse of* $H(s)$, $H_L(s)$, is of interest and is defined by

$$H_L(s)H(s) = I_m.$$

(c) Show that the left inverse $H_L(s)$ of $H(s)$ exists if and only if $rank\, H(s) = m$. *Hint:* This is the dual result to part (a).

(d) Let $H(s) = \left[\frac{s+1}{s}, \frac{1}{s}\right]$ and characterize all inputs $u(t)$ that will cause the system (at rest at $t = 0$) to exactly follow a step, $\hat{y}(s) = 1/s$.

Part (d) points to a variety of questions that may arise when inverses are considered, including: Is $H_R(s)$ proper? Is it unique? Is it stable? What is the minimum degree possible?

3.9. Consider the system $\dot{x} = Ax + Bu$, $y = Cx$. Show that output function controllability implies output controllability (-from-the-origin, or reachability).

3.10. Given $x(k+1) = \begin{bmatrix} 1 & 1 \\ 0 & 1 \end{bmatrix} x(k) + \begin{bmatrix} 1 \\ 1 \end{bmatrix} u(k)$, $y(k) = \begin{bmatrix} 1 & 0 \\ 1 & 1 \end{bmatrix} x(k)$, and assume zero initial conditions.

(a) Is there a sequence of inputs $\{u(0), u(1), \ldots\}$ that transfers the output from $y(0) = \begin{bmatrix} 0 \\ 0 \end{bmatrix}$ to $\begin{bmatrix} 0 \\ 1 \end{bmatrix}$ in finite time? If the answer is yes, determine such a sequence.

(b) Characterize all outputs that can be reached from the zero output $\left(y(0) = \begin{bmatrix} 0 \\ 0 \end{bmatrix}\right)$ in one step.

3.11. Consider the state equation $\dot{x}(t) = \begin{bmatrix} 0 & 0 \\ 0 & 1 \end{bmatrix} x(t) + \begin{bmatrix} 1 \\ e^{-t} \end{bmatrix} u(t)$.

(a) Show that it is controllable at any $t_0 \in (-\infty, \infty)$.

(b) Suppose we are interested only in $x_2(t)$ $\left(x(t) = \begin{bmatrix} x_1(t) \\ x_2(t) \end{bmatrix}\right)$. Consider, therefore,

$$\dot{x}_2(t) = x_2(t) + e^{-t}u(t).$$

Is it possible to determine $u(t)$ so that the state $x_2(t)$ is transferred from x_{20} at $t = t_0$ $[x_2(t_0) = x_{20}]$ to the zero state at some $t = t_1$ $[x_2(t_1) = 0]$ *and* then stay there? If the answer is yes, find such a $u(t)$.

(c) In (b), let $t_0 = 0$ and study the effects of the sizes of t_1 and x_0 on the magnitude of $u(t)$.

(d) For the system in (b), determine, if possible, a $u(t)$ so that the state is transferred from x_0 at $t = t_0$ to x^1 at $t = t_1$ *and* then stay there.

3.12. Let $F(t) \in C^{n\times m}$ be a matrix with $f_i(t)$ in its ith row. Let $f_i^T \in C(R, \mathscr{C}^m)$. It was shown that the set $f_i(t)$, $i = 1, \ldots, n$, is linearly independent on $[t_1, t_2]$ over the field of complex numbers if and only if the Gram matrix $W(t_1, t_2)$ is nonsingular (see Lemma 2.7). If the $f_i(t)$, $i = 1, \ldots, n$, have continuous derivatives up to order $(n-1)$, then it can be shown that they are linearly independent if for some $t_0 \in [t_1, t_2]$,

$$rank\, [F(t_0), F^{(1)}(t_0), \ldots, F^{(n-1)}(t_0)] = n.$$

If $f_i(t)$, $i = 1, \ldots, n$, are analytic on $[t_1, t_2]$, then they are linearly independent if and only if for any fixed $t_0 \in [t_1, t_2]$,

$$rank\, [F(t_0), F^{(1)}(t_0), \ldots, F^{(n-1)}(t_0), \ldots] = n.$$

(a) The above results can be used to derive alternative tests for controllability. In particular, consider the state equation

$$\dot{x} = A(t)x + B(t)u,$$

where $A(t) \in R^{n\times n}$ and $B(t) \in R^{n\times m}$ have continuous derivatives up to order $(n-1)$. Then it can be shown that $(A(t), B(t))$ is controllable at time t_0 if there exists finite $t_1 > t_0$ such that

$$rank\,[M_0(t_1), M_1(t_1), \ldots, M_{n-1}(t_1)] = n,$$

where the $M_k(t) \in R^{n\times n}$ are defined by

$$M_{k+1}(t) = -A(t)M_k(t) + \frac{d}{dt}M_k(t), \qquad k = 0, 1, \ldots, n-1,$$

with $M_0(t) = B(t)$.

(i) Prove this result.

(ii) Show that the system $\dot{x} = A(t)x + B(t)u$ with $A(t) = \begin{bmatrix} t & 1 & 0 \\ 0 & t & 0 \\ 0 & 0 & t^2 \end{bmatrix}$, $B(t) = \begin{bmatrix} 0 \\ 1 \\ 1 \end{bmatrix}$ is controllable at any t_0.

(b) The results for linear independence of vectors $f_i(t)$, $i = 1, \ldots, n$, can also be used to derive conditions for certain specialized types of controllability. In particular, given $\dot{x} = A(t)x + B(t)u$ as in (a), a system is called *differentially controllable* at t_0, when the transfer from any $x(t_0) = x_0$ to x_1 can be accomplished in an arbitrarily small interval of time. Note that this may lead to large input magnitudes. It can be shown that when $A(t)$, $B(t)$ are analytic on $(-\infty, \infty)$, the system is differentially controllable at every $t \in (-\infty, \infty)$, if and only if for any fixed $t_0 \in (-\infty, \infty)$,

$$rank\,[M_0(t_0), M_1(t_0), \ldots, M_{n-1}(t_0), \ldots] = n.$$

The system is *instantaneously controllable*, if and only if for all $t \in (-\infty, \infty)$,

$$rank\,[M_0(t), \ldots, M_{n-1}(t)] = n.$$

In this case the transfer of the states can be achieved instantaneously at any time by using inputs that include δ-functions and their derivatives up to order of $n-1$. Note that in general instantaneous controllability implies differential controllability, which in turn, implies controllability. In the case when $A(t)$, $B(t)$ are analytic on $(-\infty, \infty)$, as above, then if $(A(t), B(t))$ is controllable at some point, it is differentially controllable at every $t \in (-\infty, \infty)$.

Show that in the time-invariant case $\dot{x} = Ax + Bu$, controllability of (A, B) always implies both differential and instantaneous controllability; i.e., if a state transfer is possible at all, it can be achieved in an arbitrarily small time interval or even instantaneously if the δ-function and its derivatives are used. For the latter case, see T. Kailath, *Linear Systems*, Prentice-Hall, 1980, for further details.

Remark: In controllability, the transfer of the state occurs in finite time, but the time interval may be very large. In differential controllability, the transfer of the state is possible in arbitrarily small intervals of time; however, this may lead to very large input magnitudes (see Example 2.1). When the system $\dot{x} = A(t)x + B(t)u$ is *uniformly controllable*, the transfer of the states can be achieved in some finite time interval using an input with magnitude not arbitrarily large. Note that uniform controllability implies controllability. Uniform controllability is useful in optimal control theory. For additional discussion on differential and uniform controllability, see C. T. Chen, *Linear System Theory and Design*, Holt, Rinehart and Winston, 1984, and the references cited therein. Note that dual results also exist for differential, instantaneous, and uniform observability.

3.13. Suppose that for the system $x(k+1) = \begin{bmatrix} 1 & 1 & 0 \\ 0 & 1 & 0 \\ 0 & 0 & 1 \end{bmatrix} x(k)$, $y(k) = \begin{bmatrix} 1 & 1 & 0 \\ 0 & 1 & 0 \end{bmatrix} x(k)$ it is known that $y(0) = y(1) = y(2) = \begin{bmatrix} 1 \\ 0 \end{bmatrix}$. Based on this information, what can be said about the initial condition $x(0)$?

3.14. (a) Consider the system $\dot{x} = Ax + Bu$, $y = Cx + Du$, where (A, C) is assumed to be observable. Express $x(t)$ as a function of $y(t)$, $u(t)$ and their derivatives. *Hint:* Write $y(t), y^{(1)}(t), \ldots, y^{(n-1)}(t)$ in terms of $x(t)$ and $u(t), u^{(1)}(t), \ldots, u^{(n-1)}(t)$ $(x(t) \in R^n)$.

(b) Given the system $\dot{x} = Ax + Bu$, $y = Cx + Du$ with (A, C) observable. Determine $x(0)$ in terms of $y(t)$, $u(t)$ and their derivatives up to order $n - 1$. Note that in general this is not a practical way of determining $x(0)$, since this method requires differentiation of signals, which is very susceptible to measurement noise.

(c) Consider the system $x(k+1) = Ax(k) + Bu(k)$, $y(k) = Cx(k) + Du(k)$, where (A, C) is observable. Express $x(k)$ as a function of $y(k), y(k+1), \ldots, y(k+n-1)$ and $u(k), u(k+1), \ldots, y(k+n-1)$. *Hint:* Express $y(k), \ldots, y(k+n-1)$ in terms of $x(k)$ and $u(k), u(k+1), \ldots, u(k+n-1)$ $[x(k) \in R^n]$. Note the relation to expression (3.48) in Section 3.3.

3.15. Write software programs to implement the algorithms of Section 3.4. In particular:

(a) Given the pair (A, B), where $A \in R^{n\times n}$, $B \in R^{n\times m}$ with

$$rank\ [B, AB, \ldots, A^{n-1}B] = n_r < n,$$

reduce this pair to the standard uncontrollable form

$$\hat{A} = PAP^{-1} = \begin{bmatrix} A_1 & A_{12} \\ 0 & A_2 \end{bmatrix}, \hat{B} = PB = \begin{bmatrix} B_1 \\ 0 \end{bmatrix},$$

where (A_1, B_1) is controllable and $A_1 \in R^{n_r\times n_r}$, $B_1 \in R^{n_r\times m}$.

(b) Given the controllable pair (A, B), where $A \in R^{n\times n}$, $B \in R^{n\times m}$ with $rank\ B = m$, reduce this pair to the controller form $A_c = PAP^{-1}$, $B_c = PB$.

3.16. Determine the uncontrollable modes of each pair (A, B) given below by

(a) reducing (A, B), using a similarity transformation,

(b) using eigenvalue/eigenvector criteria.

$$A = \begin{bmatrix} 1 & 0 & 0 \\ 0 & -1 & 0 \\ 0 & 0 & 2 \end{bmatrix}, B = \begin{bmatrix} 1 & 0 \\ 0 & 1 \\ 0 & 0 \end{bmatrix} \quad \text{and} \quad A = \begin{bmatrix} 0 & 0 & 1 & 0 \\ 0 & 0 & 1 & 0 \\ 0 & 0 & 0 & 0 \\ 0 & 0 & 0 & -1 \end{bmatrix}, B = \begin{bmatrix} 0 & 1 \\ 0 & 0 \\ 1 & 0 \\ 0 & 0 \end{bmatrix}$$

3.17. Consider the system $\dot{x} = Ax + Bu$, $y = Cx + Du$.

(a) Show that only controllable modes appear in $e^{At}B$, and therefore in the zero-state response of the state.

(b) Show that only observable modes appear in Ce^{At}, and therefore, in the zero-input response of the system.

(c) Show that only modes that are both controllable and observable appear in $Ce^{At}B$, and therefore, in the impulse response and the transfer function matrix of the system.

Consider next the system $x(k+1) = Ax(k) + Bu(k)$, $y(k) = Cx(k) + Du(k)$.

(d) Show that only controllable modes appear in A^kB, only observable modes in CA^k, and only modes that are both controllable and observable appear in CA^kB [that is, in $H(z)$].

(e) Let $A = \begin{bmatrix} 1 & 0 & 0 \\ 0 & -2 & 0 \\ 0 & 0 & -1 \end{bmatrix}$, $B = \begin{bmatrix} 1 \\ 0 \\ 1 \end{bmatrix}$, $C = [1, 1, 0]$, and $D = 0$. Verify the results obtained in (d). Compare these with Example 4.10.

3.18. Reduce the pair

$$A = \begin{bmatrix} 0 & 0 & 1 & 0 \\ 3 & 0 & -3 & 1 \\ -1 & 1 & 4 & -1 \\ 1 & 0 & -1 & 0 \end{bmatrix}, \qquad B = \begin{bmatrix} 0 & 0 \\ 1 & 0 \\ 0 & 1 \\ 0 & 0 \end{bmatrix}$$

into controller form $A_c = PAP^{-1}$, $B_c = PB$. What is the similarity transformation matrix in this case? What are the controllability indices?

3.19. Let $A = \bar{A}_c + \bar{B}_c A_m$ and $B = \bar{B}_c B_m$, where the $\bar{A}_c$, $\bar{B}_c$ are as in (4.63) with $A_m \in R^{m\times n}$, $B_m \in R^{m\times m}$, and $|B_m| \neq 0$. Show that (A, B) is reachable with controllability indices μ_i. *Hint:* Use the eigenvalue test to show that (A, B) is reachable. Use state feedback to simplify (A, B) (see Exercise 3.21) and show that the μ_i are the controllability indices.

3.20. Show that the controllability indices of the state equation $\dot{x} = Ax + BGv$, where $|G| \neq 0$ and (A, B) is reachable, with $A \in R^{n\times n}$, $B \in R^{n\times m}$, are the same as the controllability indices of $\dot{x} = Ax + Bu$, within reordering. *Hint:* Write $\bar{\mathscr{C}}_k = [BG, ABG, \ldots, A^{k-1}BG] = [B, AB, \ldots, A^{k-1}B] \cdot [block\ diag\ G] = \mathscr{C}_k \cdot [block\ diag\ G]$ and show that the number of linearly dependent columns in $A^k BG$ that occur while searching from left to right in $\bar{\mathscr{C}}_n$ is the same as the corresponding number in $\mathscr{C}_n$.

3.21. Consider the state equation $\dot{x} = Ax + Bu$, where $A \in R^{n\times n}$, $B \in R^{n\times m}$ with (A, B) reachable. Let the linear state-feedback control law be $u = Fx + Gv$, $F \in R^{m\times n}$, $G \in R^{m\times m}$ with $|G| \neq 0$. Show that

(a) $(A + BF, BG)$ is reachable.

(b) The controllability indices of $(A + BF, B)$ are identical to those of (A, B).

(c) The controllability indices of $(A + BF, BG)$ are equal to the controllability indices of (A, B) within reordering. *Hint:* Use the eigenvalue test to show (a). To show (b), use the controller forms in Section 3.4.

3.22. Show that if (A, B) is controllable (-from-the-origin), where $A \in R^{n\times n}$, and $B \in R^{n\times m}$, and $rank\ B = m$, then $rank\ A \geq n - m$.

3.23. Consider

$$A_c = \begin{bmatrix} 0 & 1 & \cdots & 0 \\ \vdots & \vdots & \ddots & \vdots \\ 0 & 0 & \cdots & 1 \\ -\alpha_0 & -\alpha_1 & \cdots & -\alpha_{n-1} \end{bmatrix}, \qquad B_c = \begin{bmatrix} 0 \\ \vdots \\ 0 \\ 1 \end{bmatrix}.$$

Show that

$$\mathscr{C} = [B_c, A_c B_c, \ldots, A_c^{n-1} B_c] = \begin{bmatrix} 0 & 0 & 0 & \cdots & 1 \\ 0 & 0 & 0 & \cdots & c_1 \\ \vdots & \vdots & \vdots & & \vdots \\ 0 & 0 & 1 & \cdots & c_{n-3} \\ 0 & 1 & c_1 & \cdots & c_{n-2} \\ 1 & c_1 & c_2 & \cdots & c_{n-1} \end{bmatrix},$$

where $c_k = -\sum_{i=0}^{k-1} \alpha_{n-i-1} c_{k-i-1}$, $k = 1, \ldots, n-1$, with $c_0 = 1$. Also, show that

$$\mathscr{C}^{-1} = \begin{bmatrix} \alpha_1 & \alpha_2 & \cdots & \alpha_{n-1} & 1 \\ \alpha_2 & \alpha_3 & \cdots & 1 & 0 \\ \vdots & \vdots & & \vdots & \vdots \\ \alpha_{n-1} & 1 & \cdots & 0 & 0 \\ 1 & 0 & \cdots & 0 & 0 \end{bmatrix}.$$

3.24. Given $A \in R^{n\times n}$, and $B \in R^{n\times m}$, let $rank\ \mathscr{C} = n$, where $\mathscr{C} = [B, AB, \ldots, A^{n-1}B]$. Consider $\hat{A} \in R^{n\times n}$, $\hat{B} \in R^{n\times m}$ with $rank\ \hat{\mathscr{C}} = n$, where $\hat{\mathscr{C}} = [\hat{B}, \hat{A}\hat{B}, \ldots, \hat{A}^{n-1}\hat{B}]$, and assume that $P \in R^{n\times n}$ with $det\ P \neq 0$ exists such that

$$P[\mathscr{C}, A^n B] = [\hat{\mathscr{C}}, \hat{A}^n \hat{B}].$$

Show that $\hat{B} = PB$ and $\hat{A} = PAP^{-1}$. *Hint:* Show that $(PA - \hat{A}P)\mathscr{C} = 0$.

3.25. Show that the matrices $A_c = PAP^{-1}$, $B_c = PB$ are
(a) given by (4.47) if P is given by (4.48),
(b) given by (4.50) if $Q(= P^{-1})$ is given by (4.49),
(c) given by (4.52) if $Q(= P^{-1})$ is given by (4.51).

3.26. In the circuit of Example 4.9, let $R_1 R_2 C = L$ and $R_1 = R_2 = R$. Determine $x(t) = [x_1(t), x_2(t)]^T$ and $i(t)$ for unit step input voltage, $v(t)$, and initial conditions $x(0) = [a, b]^T$. Comment on your results.

3.27. Consider the pair (A, b), where $A \in R^{n\times n}$, $b \in R^n$. Show that if more than one linearly independent eigenvector can be associated with a single eigenvalue, then (A, b) is uncontrollable. *Hint:* Use the eigenvector test. Let $\hat{v}_1, \hat{v}_2$ be linearly independent left eigenvectors associated with eigenvalue $\lambda_1 = \lambda_2 = \lambda$. Notice that if $\hat{v}_1 b = \alpha_1$ and $\hat{v}_2 b = \alpha_2$, then $(\alpha_1^{-1}\hat{v}_1 - \alpha_2^{-1}\hat{v}_2)b = 0$.

3.28. (a) Consider the state equation $\dot{x} = Ax + Bu$, $x(0) = x_0$, where $A = \begin{bmatrix} 0 & -1 & 1 \\ 1 & -2 & 1 \\ 0 & 1 & -1 \end{bmatrix}$ and $B = \begin{bmatrix} 1 & 0 \\ 1 & 1 \\ 1 & 2 \end{bmatrix}$. Determine $x(t)$ as a function of $u(t)$ and x_0, and verify that the uncontrollable modes do not appear in the zero-state response, but do appear in the zero-input response (see Example 4.1).
(b) Consider the state equation $x(k+1) = Ax(k) + Bu(k)$ and $x(0) = x_0$, where A and B are as in (a). Demonstrate for this case results corresponding to (a).
In (a) and (b), determine $x(t)$ and $x(k)$ for unit step inputs and $x(0) = [1, 1, 1]^T$.

3.29. (a) Consider the system $\dot{x} = Ax + Bu$, $y = Cx$ with $x(0) = x_0$, where $A = \begin{bmatrix} 0 & 1 \\ -2 & -3 \end{bmatrix}$, $B = \begin{bmatrix} 0 \\ 1 \end{bmatrix}$, and $C = [1, 1]$. Determine $y(t)$ as a function of $u(t)$ and x_0, and verify that the unobservable modes do not appear in the output (see Example 4.3).
(b) Consider the system $x(k+1) = Ax(k) + Bu(k)$, $y(k) = Cx(k)$ with $x(0) = x_0$, where A, B, and C are as in (a). Demonstrate for this case results which correspond to (a).
In (a) and (b), determine and plot $y(t)$ and $y(k)$ for unit step inputs and $x(0) = 0$.

3.30. Consider the system $x(k+1) = Ax(k) + Bu(k)$, $y(k) = Cx(k)$, where

$$A = \begin{bmatrix} 1 & 0 & 0 \\ 0 & -\frac{1}{2} & 0 \\ 0 & 0 & -\frac{1}{2} \end{bmatrix}, \qquad B = \begin{bmatrix} 1 \\ 0 \\ 1 \end{bmatrix}, \qquad C = [1, 1, 0].$$

Determine the eigenvalues that are uncontrollable and/or unobservable. Determine $x(k)$, $y(k)$ for $k \geq 0$, given $x(0)$ and $u(k)$, $k \geq 0$, and show that only controllable eigenvalues (resp., modes) appear in $A^k B$, only observable ones appear in CA^k, and only eigenvalues (resp., modes) that are both controllable and observable appear in $CA^k B$ [in $H(z)$].

3.31. For the system $\dot{x} = Ax + Bu$, $y = Cx$, consider the corresponding sampled-data system $\bar{x}(k+1) = \bar{A}\bar{x}(k) + \bar{B}\bar{u}(k)$, $\bar{y}(k) = \bar{C}\bar{x}(k)$, where

$$\bar{A} = e^{AT}, \qquad \bar{B} = \left[\int_0^T e^{A\tau}\, d\tau\right] B, \qquad \text{and} \qquad \bar{C} = C.$$

(a) Let the continuous-time system $\{A, B, C\}$ be controllable (observable) and assume it is a SISO system. Show that $\{\bar{A}, \bar{B}, \bar{C}\}$ is controllable (observable) if and only if the sampling period T is such that

$$Im\,(\lambda_i - \lambda_j) \neq \frac{2\pi k}{T}, \text{ where } k = \pm 1, \pm 2, \ldots \text{ whenever } Re\,(\lambda_i - \lambda_j) = 0,$$

where $\{\lambda_i\}$ are the eigenvalues of A. *Hint:* Use the PBH test. Also, consult Appendix D of C. T. Chen, *Linear System Theory and Design*. Holt, Rinehart, and Winston, 1984.

(b) Apply the results of (a) to the double integrator—Example 7.6 in Chapter 2—where $A = \begin{bmatrix} 0 & 1 \\ 0 & 0 \end{bmatrix}$, $B = \begin{bmatrix} 0 \\ 1 \end{bmatrix}$, $C = [1, 0]$, and also to $A = \begin{bmatrix} 0 & 1 \\ -1 & 0 \end{bmatrix}$, $B = \begin{bmatrix} 0 \\ 1 \end{bmatrix}$, $C = [1, 0]$. Determine the values of T that preserve controllability (observability).

3.32. Given is the system $\dot{x} = \begin{bmatrix} -1 & 0 & 0 \\ 0 & -1 & 0 \\ 0 & 0 & 2 \end{bmatrix} x + \begin{bmatrix} 1 & 0 \\ 0 & 1 \\ 0 & 0 \end{bmatrix} u$, $y = \begin{bmatrix} 1 & 1 & 0 \\ 1 & 0 & 0 \end{bmatrix} x$.

(a) Determine the uncontrollable and the unobservable eigenvalues (if any).
(b) What is the impulse response of this system? What is its transfer function matrix?
(c) Is the system asymptotically stable?

3.33. Consider the system $x(k+1) = \begin{bmatrix} 1 & 2 \\ 0 & 1 \end{bmatrix} x(k) + \begin{bmatrix} 2 \\ 3 \end{bmatrix} u(k)$, $y(k) = [1, 3]x(k)$. Suppose that it is known that for zero input, $y(0) = 1$ and $y(1) = 1$. Can $x(0)$ be determined? If yes, find $x(0)$ and verify your answer.

3.34. Given is the transfer function matrix $H(s) = \begin{bmatrix} \frac{s-1}{s} & 0 & \frac{s-2}{s+2} \\ 0 & \frac{s+1}{s} & 0 \end{bmatrix}$.

(a) Determine the Smith-McMillan form of $H(s)$ and its characteristic (pole) polynomial and minimal polynomial. What are the poles of $H(s)$?
(b) Determine the zero polynomial of $H(s)$. What are the zeros of $H(s)$?

3.35. Let $H(s) = \begin{bmatrix} \frac{s^2+1}{s^2} \\ \frac{s+1}{s^3} \end{bmatrix}$.

(a) Determine the Smith-McMillan form of $H(s)$ and its characteristic (pole) polynomial and minimal polynomial. What are the poles of $H(s)$?
(b) Determine the zero polynomial of $H(s)$. What are the zeros of $H(s)$?

3.36. A rational function matrix $R(s)$ may have, in addition to finite poles and zeros, *poles and zeros at infinity* ($s = \infty$). To study the poles and zeros at infinity, the bilinear transformation

$$s = \frac{b_1 w + b_0}{a_1 w + a_0}$$

with $a_1 \neq 0$, $b_1 a_0 - b_0 a_1 \neq 0$ may be used, where b_1/a_1 is not a finite pole or zero of $R(s)$. This transformation maps the point $s = b_1/a_1$ to $w = \infty$ and the point of interest, $s = \infty$, to $w = -a_0/a_1$. The rational matrix $\hat{R}(w)$ is now obtained as

$$\hat{R}(w) = R\left(\frac{b_1 w + b_0}{a_1 w + a_0}\right),$$

and the finite poles and zeros of $\hat{R}(w)$ are determined. The poles and zeros at $w = -a_0/a_1$ are the poles and zeros of $R(s)$ at $s = \infty$. Note that frequently a good choice for the bilinear transformation is $s = 1/w$, that is, $b_1 = 0$, $b_0 = 1$ and $a_1 = 1$, $a_0 = 0$.

(a) Determine the poles and zeros at infinity of

$$R_1(s) = \frac{1}{s+1}, \qquad R_2(s) = s, \qquad R_3(s) = \begin{bmatrix} 1 & 0 \\ s+1 & 1 \end{bmatrix}.$$

Note that a rational matrix may have both poles and zeros at infinity.

(b) Show that if $R(s)$ has a pole at $s = \infty$, then it is not proper ($\lim_{s\to\infty} R(s) \to \infty$).

3.37. Determine the poles and zeros at infinity of the transfer functions in Examples 5.1, 5.2, 5.6, and 5.7.

3.38. (**Spring mass system**) Consider the spring mass given in Exercise 2.69 in Chapter 2.

(a) Is the system controllable from $[f_1, f_2]^T$? If yes, reduce (A, B) to controller form.

(b) Is the system controllable from input f_1 only? Is it controllable from f_2 only? Discuss your answers.

(c) Let $y = Cx$ with $C = \begin{bmatrix} 1 & 0 & 0 & 0 \\ 0 & 1 & 0 & 0 \end{bmatrix}$. Is the system observable from y? If yes, reduce (A, C) to observer form.

3.39. (**Aircraft dynamics**) Consider the state-space description of the lateral motion of an aircraft in Exercise 2.76 in Chapter 2.

(a) Is the system controllable from $[\delta_A, \delta_R]^T$? If yes, reduce (A, B) to controller form.

(b) Is the system controllable using only the ailerons? Is the system controllable using only the rudder? Discuss your answers.

(c) Let $y = Cx$ with $C = \begin{bmatrix} 0 & 1 & 0 & 0 \\ 0 & 0 & 1 & 0 \end{bmatrix}$. Is the system observable from y? If yes, reduce (A, C) to observer form.

CHAPTER 4

State Feedback and State Observers

Feedback is a fundamental mechanism arising in nature and is present in many natural processes. Feedback is also common in manufactured systems and is essential in automatic control of dynamic processes with uncertainties in their model descriptions and their interactions with the environment. When feedback is used, the actual values of system variables are sensed, fed back, and used to control the system. Hence, a control law decision process is based not only on predictions about the system behavior derived from a process model (as in open-loop control), but also on information about the actual behavior (closed-loop feedback control). A common example of an automatic feedback control system is the cruise control system in an automobile, which maintains the speed of the automobile at a certain desired value within acceptable tolerances.

In this chapter feedback is introduced, and the problem of pole or eigenvalue assignment by means of state feedback is discussed at length in Section 4.2. It is possible to arbitrarily assign all closed-loop eigenvalues by linear static state feedback if and only if the system is completely controllable. This relation to controllability is, in fact, the motivation for introducing state feedback at this point. Feedback control is considered again in Chapter 7, where polynomial matrix descriptions are introduced.

In the study of state feedback it is assumed that it is possible to measure the values of the states using appropriate sensors. Frequently, however, it may be either impossible or impractical to obtain measurements for all states. It is therefore desirable to be able to estimate the states from measurements of input and output variables that are typically available. In addition to feedback control problems, there are many other problems where knowledge of the state vector is desirable since such knowledge contains useful information about the system. This is the case, for example, in navigation systems. State observers that asymptotically estimate the states from input and output measurements over time are also studied in this chapter.

State estimation is related to observability in an analogous way that state feedback control is related to controllability. The duality between controllability and observability makes it possible to easily solve the estimation problem once the control problem has been solved, and vice versa. In this chapter, full-order and reduced-order asymptotic estimators, also called observers, are discussed at length in Section 4.3. Finally, state feedback static controllers and state dynamic observers are combined to form dynamic output feedback controllers. Such controllers are studied in Section 4.4, using both state-space and transfer function matrix descriptions.

4.1 INTRODUCTION

A. A Brief Introduction to State Feedback Controllers and State Observers

In the following discussion, state feedback and state estimation are introduced for continuous- and discrete-time time-varying and time-invariant systems.

We consider systems described by equations of the form

$$\dot{x} = A(t)x + B(t)u, \qquad y = C(t)x + D(t)u, \tag{1.1}$$

where $t \in (a, b)$, some real open interval, and $A(t) \in R^{n \times n}$, $B(t) \in R^{n \times m}$, $C(t) \in R^{p \times n}$, $D(t) \in R^{p \times m}$, and $u(t) \in R^m$ are (piecewise) continuous in t on (a, b) (see Chapter 2).

Let the input u be determined by a *time-varying linear, state feedback control law* of the form

$$u = F(t)x + r, \tag{1.2}$$

where $F(t) \in R^{m \times n}$ is (piecewise) continuous in $t \in (a, b)$, the elements of $F(t)$ represent time-dependent gains, and $r(t) \in R^m$ is an external input (see Fig. 4.1). Substituting into (1.1), the state-space description of the *compensated or closed-loop system* is given by is given by

$$\dot{x} = [A(t) + B(t)F(t)]x + B(t)r.$$

$$y = [C(t) + D(t)F(t)]x + D(t)r. \tag{1.3}$$

We seek to select $F(t)$ so that the closed-loop system has certain desirable qualitative properties. For example, we may wish the closed-loop system to be stable. (For definitions of stability in the time-varying case, refer to Chapter 6.) Stability can be achieved under appropriate assumptions involving certain types of controllability. One way of determining such stabilizing $F(t)$ is to use results from the optimal Linear Quadratic Regulator (LQR) theory, which in fact yields the "best" $F(t)$ in some

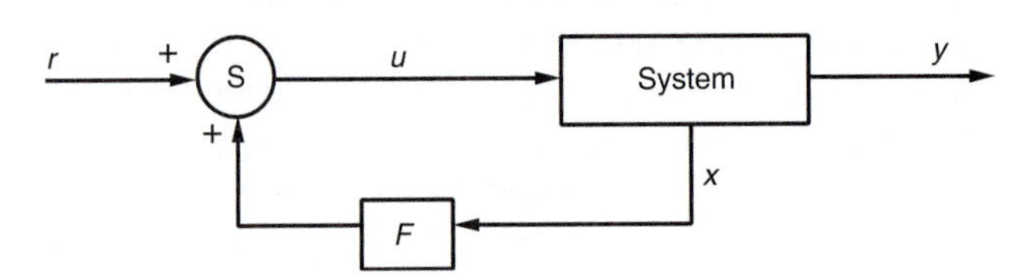

FIGURE 4.1

sense. Stabilization, or the achievement of other control objectives, for linear *time-varying* systems via LQR or other methods will not be studied in this chapter. The stabilization of linear time-invariant systems, however, is discussed at great length in Section 4.2.

In the time-invariant case we consider systems described by equations of the form

$$\dot{x} = Ax + Bu, \qquad y = Cx + Du, \tag{1.4}$$

where $A \in R^{n \times n}$, $B \in R^{n \times m}$, $C \in R^{p \times n}$, and $D \in R^{p \times m}$. Let

$$u = Fx + r \tag{1.5}$$

represent the *linear time-invariant or static state feedback control law.* The *compensated or closed-loop system* is then given by the equations

$$\dot{x} = (A + BF)x + Br, \qquad y = (C + DF)x + Dr. \tag{1.6}$$

In control design, the gain matrix F is selected so that the closed-loop system (1.6) has desired behavior. In this chapter we are particularly interested in selecting F to stabilize the system, and we will concentrate on the eigenvalue assignment problem. In particular, it is shown in the next section that the controllable eigenvalues of the system [i.e., of the pair $(A + BF, B)$] can be assigned (or shifted) to arbitrary locations, and therefore, can be assigned to locations that guarantee stable behavior.

In view of the results developed in the previous chapter, the central role played by controllability in feedback stabilization is most easily seen in the time-invariant case. Recall that controllability reflects the ability of an input to transfer the state of a system to any desirable (state) value. It was shown in Section 3.4 that given a time-invariant system $\{A, B, C, D\}$, there exists a transformation so that the controllable part of the state space can easily be recognized. In other words, there exists a basis for the state space, so that all controllable states have representations of the form $[\hat{x}_1^T, 0^T]^T$ and the uncontrollable state representations are of the form $[\hat{x}_1^T, \hat{x}_2^T]^T$, where $\dot{\hat{x}}_2 = A_2\hat{x}_2$, i.e., $\hat{x}_2$ is independent of u (see Subsection 3.4A). The eigenvalues of A_2 are in this case the uncontrollable eigenvalues that correspond to the uncontrollable modes of the system. Since u has no effect on $\hat{x}_2$, it is reasonable to conjecture that the state feedback control law (which in fact, is a particular choice for u) will not at all affect the eigenvalues of A_2, for if it did, it would also affect $\hat{x}_2$. This conjecture turns out to be true and will formally be shown in the next section. The controllable states can now arbitrarily be shifted, using u. This in turn means that the controllable eigenvalues can be arbitrarily shifted, using u, or more correctly, by appropriately selecting u in $\dot{\hat{x}}_1 = A_1\hat{x}_1 + B_1u + A_{12}\hat{x}_2$, $\hat{x}_1$ may be made to behave as if the eigenvalues of A_1 were arbitrarily shifted. This is true because if there were restrictions on the location of the eigenvalues, then the state would not be able to be shifted arbitrarily, which is a contradiction. It turns out that linear state feedback control provides such u, and therefore, it can arbitrarily shift all the eigenvalues of a system. This will formally be shown in Section 4.2. A stabilizing state feedback matrix F may also be determined as the solution to an optimal control problem, the Linear Quadratic Regulator (LQR) problem. This is briefly addressed at the end of Section 4.2.

We will also consider time-varying discrete-time systems, described by equations of the form

$$x(k+1) = A(k)x(k) + B(k)u(k), \qquad y(k) = C(k)x(k) + D(k)u(k) \quad (1.7)$$

with *linear, discrete-time, time-varying state-feedback control law* given by

$$u(k) = F(k)x(k) + r(k). \quad (1.8)$$

In this case, the closed-loop system is given by

$$\begin{aligned} x(k+1) &= [A(k) + B(k)F(k)]x(k) + B(k)r(k) \\ y(k) &= [C(k) + D(k)F(k)]x(k) + D(k)r(k). \end{aligned} \quad (1.9)$$

Also, we will consider time-invariant discrete-time systems described by

$$x(k+1) = Ax(k) + Bu(k), \qquad y(k) = Cx(k) + Du(k) \quad (1.10)$$

with *linear, discrete-time, time-invariant state-feedback control law* (time-invariant) given by

$$u(k) = Fx(k) + r(k). \quad (1.11)$$

In this case the closed-loop system assumes the form

$$\begin{aligned} x(k+1) &= [A + BF]x(k) + Br(k) \\ y(k) &= [C + DF]x(k) + Dr(k). \end{aligned} \quad (1.12)$$

As in the continuous-time case, stabilization is emphasized in this chapter, and corresponding results are developed.

In (continuous-time) state-space system representations, if the value of the state at time t_0, $x(t_0)$, is known, then the input $u(t), t \geq t_0$, uniquely determines $x(t)$ and $y(t)$ for $t \geq t_0$. Since knowledge of the initial value of the state is of such importance, methods of determining (estimating) $x(t_0)$ from input and output measurements have been devised (see Fig. 4.2).

Methods of estimating the initial value of the state, $x(0)$, in time-invariant systems are addressed in Section 4.3. In particular, full- and reduced-order state observers are designed that asymptotically estimate the state. Since controllability (-from-the-origin, or reachability) was the key property in state feedback control, the dual property of observability, studied in Section 3.3, is the key attribute in state estimation. Recall that observability refers to the ability of determining the state from measurements of the output and input over a finite time interval. (See, for instance, Corollary 3.8 in Chapter 3, where the initial state is determined using the observability Gramian of the system.) Furthermore, since the solution to the optimal

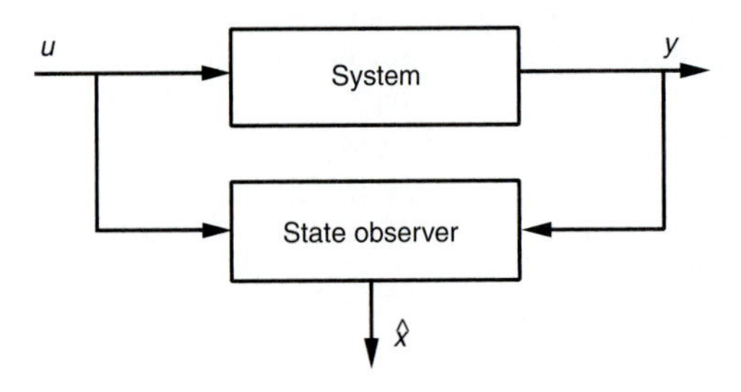

FIGURE 4.2

LQR control problem leads to optimal state-feedback control law matrices [$F(t)$ or F], a dual optimal estimation problem can be defined, the solution of which provides optimal state-observer gain matrices [$K(t)$ or K]. Corresponding discrete-time results on state estimation are also discussed. In addition, in Section 4.4, state feedback control laws and state estimation are combined to derive a dynamic-observer based controller that receives feedback information, not from the state, but from the outputs of the system to be controlled. These are typically more accessible.

B. Chapter Description

In this chapter state feedback controllers and asymptotic state estimators, also called state observers, are introduced and studied. The contents of the chapter were outlined and briefly discussed in the previous subsection (Subsection 4.1A). A detailed summary follows.

A brief introduction to state feedback controllers and state observers for continuous-time and discrete-time systems is given in Subsection 4.1A.

In Section 4.2, state feedback control is studied. Our development focuses on time-invariant systems. First, the need for feedback is demonstrated by a discussion of open- and closed-loop control laws, and their differences when uncertainties are present in the system model and its environment. The stabilization of systems is emphasized. This leads, in the time-invariant case, to the eigenvalue (or pole) assignment problem, which is studied next at length. The flexibility offered in the multi-input case in the choice of the state feedback gains to accomplish control goals, in addition to pole assignment, is stressed. A number of pole assignment methods are introduced, including the eigenvalue/eigenvector assignment method; an additional, historically important method is presented in Exercise 4.2. It is pointed out that stabilization can also be achieved by an optimal control formulation, such as the Linear Quadratic Regulator (LQR), which leads to a stabilizing state feedback control law while attaining additional control goals as well. The LQR problem is briefly discussed for both the continuous- and discrete-time cases.

In Section 4.3, full-order, full-state, and partial-state observers for continuous- and discrete-time systems are studied. Also, reduced-order and optimal observers are addressed. In the discrete-time case, current state estimators are also introduced. Optimal Linear Quadratic Gaussian (LQG) estimators are briefly discussed. The duality of the state feedback and the state observer problems is emphasized. It is pointed out that the main factor that limits the magnitude of the gains in observers is noise. This is in contrast to the limiting factor in the magnitude of the control gains, which is limitations of the control actuators and of the linear model used to describe the system.

In Section 4.4, dynamic state observers are used, together with static state feedback controllers, to derive dynamic output controllers. The *Separation Principle* is discussed and the degradation of performance in state feedback control when an observer is used to estimate the state is explained. The analysis is accomplished in both state-space and transfer function frameworks. Furthermore, additional output controller configurations are also derived in terms of state-space representations.

C. Guidelines for the Reader

In this chapter, linear state feedback controllers and linear state observers for linear time-invariant systems are studied. Subsection 4.1B describes the contents of the chapter.

At a first reading, one can cover selected topics from the material on state feedback and state observers in Sections 4.2 and 4.3, respectively. Regarding Section 4.2 on state feedback, after studying the issues associated with open- and closed-loop control in Subsection 4.2A, one could concentrate on eigenvalue assignment using linear state feedback. In particular, one could study Theorem 2.1 in Subsection 4.2B, which shows that only the controllable eigenvalues can be arbitrarily assigned via state feedback; then one could study the methodologies to assign desired eigenvalues (and in part, the corresponding eigenvectors as well). Regarding Section 4.3 on state observers, at a first reading one could concentrate only on full-order full-state observers for continuous-time systems in Subsection 4.3A. Then one could study dynamic output feedback controllers that are based on state observers in Section 4.4. The degradation of performance when an observer is used to estimate the state in a state feedback controller is also discussed in that section.

4.2 LINEAR STATE FEEDBACK

A. Continuous-Time Systems

We consider linear, time-invariant, continuous-time systems described by equations of the form

$$\dot{x} = Ax + Bu, \qquad y = Cx + Du, \tag{2.1}$$

where $A \in R^{n\times n}$, $B \in R^{n\times m}$, $C \in R^{p\times n}$, and $D \in R^{p\times m}$.

DEFINITION 2.1. The *linear, time-invariant, state feedback control law* is defined by

$$u = Fx + r, \tag{2.2}$$

where $F \in R^{m\times n}$ is a gain matrix and $r(t) \in R^m$ is an external input vector. ■

Note that $r(t)$ is an *external input*, also called a *command* or *reference input* (see Fig. 4.1). It is used to provide an input to the compensated closed-loop system and is omitted when such input is not necessary in a given discussion [$r(t) = 0$]. This is the case, e.g., when the Lyapunov stability of a system is studied. Note that the vector $r(t)$ in (2.2) has the same dimension as $u(t)$. If a different number of inputs is desired, then an input transformation map may be used to accomplish this.

The *compensated closed-loop system* of Fig. 4.1 is described by the equations

$$\dot{x} = (A + BF)x + Br$$
$$y = (C + DF)x + Dr, \tag{2.3}$$

which were determined by substituting $u = Fx + r$ into the description of the *uncompensated open-loop system* (2.1).

The *state feedback gain matrix* F affects the closed-loop system behavior. This is accomplished by altering the dynamic effects of the matrices A and C of (2.1). In fact, the main influence of F is exercised through the matrix A, resulting in the matrix $A + BF$ of the closed-loop system. The matrix F affects the eigenvalues of $A + BF$, and therefore, the modes of the closed-loop system. The effects of F can also be thought of as restricting the choices for u ($= Fx$ for $r = 0$) so that for appropriate F, certain properties, such as asymptotic Lyapunov stability, of the equilibrium $x = 0$ are obtained.

Open- and closed-loop control

The linear state feedback control law (2.2) can be expressed in terms of the initial state $x(0) = x_0$. In particular, working with Laplace transforms, we obtain $\hat{u} = F\hat{x} + \hat{r} = F[(sI - A)^{-1}x_0 + (sI - A)^{-1}B\hat{u}] + \hat{r}$, in view of $s\hat{x} - x_0 = A\hat{x} + B\hat{u}$, derived from $\dot{x} = Ax + Bu$. Collecting terms, we have $[I - F(sI - A)^{-1}B]\hat{u} = F(sI - A)^{-1}x_0 + \hat{r}$. This yields

$$\hat{u} = F[sI - (A + BF)]^{-1}x_0 + [I - F(sI - A)^{-1}B]^{-1}\hat{r}, \tag{2.4}$$

where the matrix identities $[I - F(sI - A)^{-1}B]^{-1}F(sI - A)^{-1} \equiv F(sI - A)^{-1}[I - BF(sI - A)^{-1}]^{-1} \equiv F[sI - (A + BF)]^{-1}$ have been used.

Expression (2.4) is an *open-loop* (feedforward) *control law,* expressed in the Laplace transform domain. It is phrased in terms of the initial conditions $x(0) = x_0$, and if it is applied to the open-loop system (2.1), it generates exactly the same control action $u(t)$ for $t \geq 0$ as the state feedback $u = Fx + r$ in (2.2). It can readily be verified that the descriptions of the compensated system are exactly the same when either control expressions, (2.2) or (2.4), are used (verify this). In practice, however, these two control laws hardly behave the same, as explained in the following.

First, notice that in the open-loop scheme (2.4) the initial conditions x_0 are assumed to be known exactly. It is also assumed that the plant parameters in A and B are known exactly. If there are *uncertainties* in the data, this control law may fail miserably, even when the differences are small, since it is based on incorrect information without any way of knowing that these data are not valid. In contrast to the above, the feedback law (2.2) does not require knowledge of x_0. Moreover, it receives feedback information from $x(t)$ and adjusts $u(t)$ to reflect the current system parameters, and consequently, is more robust to parameter variations. Of course the feedback control law (2.2) will also fail when the parameter variations are too large. In fact, the area of *robust control* relates feedback control law designs to bounds on the uncertainties (due to possible changes) and aims to derive the best design possible under the circumstances.

The point we wish to emphasize here is that although open- and closed-loop control laws may appear to produce identical effects, typically they do not, the reason being that the mathematical system models used are not sufficiently accurate, by necessity or design. Feedback control and closed-loop control are preferred to accommodate ever-present modeling uncertainties in the plant and the environment.

The purpose of controlling a system is to achieve certain control goals. Examples include tracking a given trajectory, regulating the state so that it returns to the origin if it is disturbed, and stabilizing a system. Stabilization is discussed at length in this chapter. Regulation and tracking, together with other control problems, are briefly discussed in Chapter 7, where output feedback compensation, in addition to state feedback compensation, is used.

At this point, a few observations are in order. First, we note that feeding back the state in synthesizing a control law is a very powerful mechanism, since the state contains all the information about the past history of a system that is needed to uniquely determine the future system behavior, given the input. We observe that the state feedback control law considered presently is linear, resulting in a closed-loop system that is also linear. Nonlinear state feedback control laws are of course also possible. Notice that when a time-invariant system is considered, the state feedback is typically static, unless there is no choice (as in certain optimal control problems), resulting in a closed-loop system that is also time-invariant. These comments justify to a certain extent the choice of linear, time-invariant, state feedback control to compensate linear time-invariant systems.

The problem of stabilizing a system by using state feedback is considered next.

Stabilization

The problem we wish to consider now is to determine a state feedback control law (2.2) having the property that the resulting compensated closed-loop system has an equilibrium $x = 0$ that is asymptotically stable (in the sense of Lyapunov) when $r = 0$. (For a discussion of asymptotic stability, refer to Chapters 2 and 6.) In particular, we wish to determine a matrix $F \in R^{m \times n}$ so that the system

$$\dot{x} = (A + BF)x, \tag{2.5}$$

where $A \in R^{n \times n}$ and $B \in R^{n \times m}$, has equilibrium $x = 0$ that is asymptotically stable. Note that (2.5) was obtained from (2.3) by letting $r = 0$.

One method of deriving such stabilizing F is by formulating the problem as an optimal control problem, e.g., as the Linear Quadratic Regulator (LQR) problem. This is discussed at the end of this section. We point out that an LQR formulation can also be used to derive stabilizing gains $F(t)$ in the time-varying case so that the equilibrium $x = 0$ of the system $\dot{x} = [A(t) + B(t)F(t)]x$ is asymptotically stable. However, this will not be pursued here.

Alternatively, in view of Chapter 2 (and Chapter 6), the equilibrium $x = 0$ of (2.5) is asymptotically stable if and only if the eigenvalues λ_i of $A + BF$ satisfy $Re\, \lambda_i < 0$, $i = 1, \ldots, n$. Therefore, the stabilization problem for the time-invariant case reduces to the problem of selecting F in such a manner that the eigenvalues of $A + BF$ are shifted into desired locations. This will be studied in the following subsection. Note that stabilization is only one of the control objectives, although a most important one, that can be achieved by shifting eigenvalues. Since the eigenvalues of a linear system determine its qualitative dynamic behavior (refer to the discussion of modes in Chapter 2), one can attain a number of control goals by shifting of eigenvalues, in addition to stability. Control system design via eigenvalue (pole) assignment is a topic that is addressed in detail in a number of control books.

B. Eigenvalue Assignment

Consider again the closed-loop system $\dot{x} = (A+BF)x$ given in (2.5). We shall show that if (A, B) is fully controllable (-from-the-origin, or reachable), all eigenvalues of $A+BF$ can be arbitrarily assigned by appropriately selecting F. In other words, "the eigenvalues of the original system can arbitrarily be changed in this case." This last statement, commonly used in the literature, is rather confusing: The eigenvalues of a given system $\dot{x} = Ax + Bu$, are *not* physically changed by the use of feedback. They are the same as they used to be before the introduction of feedback. Instead, the feedback law $u = Fx + r$, $r = 0$, generates an input $u(t)$ that, when fed back to the system, makes it behave *as if* the eigenvalues of the system were at different locations [i.e., the input $u(t)$ makes it behave as a different system, the behavior of which is, we hope, more desirable than the behavior of the original system].

THEOREM 2.1. Given $A \in R^{n\times n}$ and $B \in R^{n\times m}$, there exists $F \in R^{m\times n}$ such that the n eigenvalues of $A + BF$ can be assigned to arbitrary, real or complex conjugate, locations if and only if (A, B) is controllable (-from-the-origin, or reachable).

Proof. (*Necessity*) Suppose that the eigenvalues of $A + BF$ have been arbitrarily assigned and assume that (A, B) in (2.1) is not fully controllable. We shall show that this leads to a contradiction. Since (A, B) is not fully controllable, in view of the results in Section 3.4 in Chapter 3, there exists a similarity transformation that will separate the controllable part from the uncontrollable part in (2.5). In particular, there exists a nonsingular matrix Q such that

$$Q^{-1}(A + BF)Q = Q^{-1}AQ + (Q^{-1}B)(FQ) = \begin{bmatrix} A_1 & A_{12} \\ 0 & A_2 \end{bmatrix} + \begin{bmatrix} B_1 \\ 0 \end{bmatrix}[F_1, F_2]$$
$$= \begin{bmatrix} A_1 + B_1F_1 & A_{12} + B_1F_2 \\ 0 & A_2 \end{bmatrix}, \tag{2.6}$$

where $[F_1, F_2] \triangleq FQ$ and (A_1, B_1) is controllable. The eigenvalues of $A + BF$ are the same as the eigenvalues of $Q^{-1}(A + BF)Q$, which implies that $A + BF$ has certain fixed eigenvalues, the eigenvalues of A_2, that cannot be shifted via F. These are the uncontrollable eigenvalues of the system. Therefore, the eigenvalues of $A + BF$ have not been arbitrarily assigned, which is a contradiction. Thus, (A, B) is fully controllable.

(*Sufficiency*) Let (A, B) be fully controllable. Then by using any of the eigenvalue assignment algorithms presented later in this section, all the eigenvalues of $A + BF$ can be arbitrarily assigned. ■

LEMMA 2.2. The uncontrollable eigenvalues of (A, B) cannot be shifted via state feedback.

Proof. See the necessity part of the proof of Theorem 2.1. Note that the uncontrollable eigenvalues are the eigenvalues of A_2. ■

EXAMPLE 2.1. Consider the uncontrollable pair (A, B), where $A = \begin{bmatrix} 0 & -2 \\ 1 & -3 \end{bmatrix}$, $B = \begin{bmatrix} 1 \\ 1 \end{bmatrix}$. This pair can be transformed to a standard form for uncontrollable systems, namely, $\hat{A} = \begin{bmatrix} -2 & 1 \\ 0 & -1 \end{bmatrix}$, $\hat{B} = \begin{bmatrix} 1 \\ 0 \end{bmatrix}$, from which it can easily be seen that -1 is the uncontrollable eigenvalue, while -2 is the controllable eigenvalue (see also Example 4.3 in Chapter 3).

Now if $F = [f_1, f_2]$, then $det\,(sI - (A + BF)) = det \begin{bmatrix} s - f_1 & 2 - f_2 \\ -1 - f_1 & s + 3 - f_2 \end{bmatrix} = s^2 + s(-f_1 - f_2 + 3) + (-f_1 - f_2 + 2) = (s + 1)(s + (-f_1 - f_2 + 2))$. Clearly, the uncontrollable eigenvalue -1 cannot be shifted via state feedback. The controllable eigenvalue -2 can be shifted arbitrarily to $(f_1 + f_2 - 2)$ by $F = [f_1, f_2]$. ■

It is now quite clear that a given system (2.1) can be made asymptotically stable via the state feedback control law (2.2) only when all the uncontrollable eigenvalues of (A, B) are already in the open left part of the s-plane. This is so because state feedback can alter only the controllable eigenvalues.

DEFINITION 2.2. The pair (A, B) is called *stabilizable* if all its uncontrollable eigenvalues are stable. ■

Before presenting methods to select F for eigenvalue assignment, it is of interest to examine how the linear feedback control law $u = Fx + r$ given in (2.2) affects controllability and observability. We write

$$\begin{bmatrix} sI - (A + BF) & B \\ -(C + DF) & D \end{bmatrix} = \begin{bmatrix} sI - A & B \\ -C & D \end{bmatrix} \begin{bmatrix} I & 0 \\ -F & I \end{bmatrix} \tag{2.7}$$

and note that

$$rank\,[\lambda I - (A + BF), B] = rank\,[\lambda I - A, B]$$

for all complex λ (show this). Thus, if (A, B) is controllable, then so is $(A + BF, B)$ for any F. Further, notice that in view of

$$\begin{aligned} \mathscr{C}_F &= [B, (A + BF)B, (A + BF)^2B, \ldots, (A + BF)^{n-1}B] \\ &= [B, AB, A^2B, \ldots, A^{n-1}B] \begin{bmatrix} I & FB & F(A + BF)B & \cdot \\ 0 & I & FB & \cdot \\ & & I & \cdot \\ & & & \ddots \\ & & & I \end{bmatrix}, \end{aligned} \tag{2.8}$$

$\mathcal{R}(\mathscr{C}_F) = \mathcal{R}([B, AB, \ldots, A^{n-1}B]) = \mathcal{R}(\mathscr{C})$ (why?). This shows that F does not alter the controllability subspace of the system (see Section 3.2.). This in turn proves the following lemma.

LEMMA 2.3. The controllability subspaces of $\dot{x} = Ax + Bu$ and $\dot{x} = (A + BF)x + Br$ are the same for any F. ■

Although the controllability of the system is not altered by linear state feedback $u = Fx + r$, this is not true for the observability property. Note that the observability of the closed-loop system (2.3) depends on the matrices $(A + BF)$ and $(C + DF)$, and it is possible to select F to make certain eigenvalues unobservable from the output. In fact this mechanism is quite common and is used in several control design methods. It is also possible to make observable certain eigenvalues of the open-loop system that were unobservable; for an example see Example 2.8 below.

Several methods are now presented to select F so to arbitrarily assign the closed-loop eigenvalues.

Methods for eigenvalue assignment by state feedback

In view of Theorem 2.1, the *eigenvalue assignment problem* can now be stated as follows. Given a controllable pair (A, B), determine F to assign the n eigenvalues of $A + BF$ to arbitrary real and/or complex conjugate locations. This problem is also known as the *pole assignment problem,* where by the term "pole" is meant a "pole of the system" (or an eigenvalue of the "A" matrix). This is to be distinguished from the "poles of the transfer function" (see Section 3.5 for the appropriate definitions).

Note that all matrices A, B, and F are real, so the coefficients of the polynomial $det\,[sI - (A + BF)]$ are also real. This imposes the restriction that the complex roots of this polynomial must appear in conjugate pairs. Also, note that if (A, B) is not fully controllable, then (2.6) can be used together with the methods described a little later, to assign all the controllable eigenvalues; the uncontrollable ones will remain fixed (see Theorem 2.1 and Lemma 2.2).

It is assumed in the following that B has full column rank, i.e.,

$$rank\ B = m. \tag{2.9}$$

This means that the system $\dot{x} = Ax + Bu$ has m independent inputs. If $rank\ B = r < m$, this would imply that one could achieve the same result by manipulating only r inputs (instead of $m > r$). To assign eigenvalues in this case, one can proceed by writing

$$A + BF = A + (BM)(M^{-1}F) = A + [B_1, 0]\begin{bmatrix} F_1 \\ F_2 \end{bmatrix} = A + B_1F_1, \tag{2.10}$$

where M is chosen so that $BM = [B_1, 0]$ with $B_1 \in R^{n\times r}$ and $rank\ B_1 = r$. Then $F_1 \in R^{r\times n}$ can be determined to assign the eigenvalues of $A + B_1F_1$, using any one of the methods presented next. Note that (A, B) is controllable implies that (A, B_1) is controllable (why?). The state feedback matrix F is given in this case by

$$F = M\begin{bmatrix} F_1 \\ F_2 \end{bmatrix}, \tag{2.11}$$

where $F_2 \in R^{(m-r)\times n}$ is arbitrary.

1. Direct method. Let $F = [f_{ij}]$, $i = 1, \ldots, m$, $j = 1, \ldots, n$, and express the coefficients of the characteristic polynomial of $A + BF$ in terms of f_{ij}, i.e.,

$$det(sI - (A + BF)) = s^n + g_{n-1}(f_{ij})s^{n-1} + \cdots + g_0(f_{ij}).$$

Now if the roots of the polynomial

$$\alpha_d(s) = s^n + d_{n-1}s^{n-1} + \cdots + d_1s + d_0$$

are the n desired eigenvalues, then the f_{ij}, $i = 1, \ldots, m$, $j = 1, \ldots, n$, must be determined so that

$$g_k(f_{ij}) = d_k, \qquad k = 0, 1, \ldots, n-1. \tag{2.12}$$

In general, (2.12) constitutes a nonlinear system of algebraic equations; however, it is linear in the single-input case, $m = 1$. The main difficulty in this method is not so much in deriving a numerical solution for the nonlinear system of equations, but in carrying out the symbolic manipulations needed to determine the coefficients g_k in terms of the f_{ij} in (2.12). This difficulty usually restricts this method to the simplest cases, with $n = 2$ or 3 and $m = 1$ or 2 being typical.

EXAMPLE 2.2. For $A = \begin{bmatrix} \frac{1}{2} & 1 \\ 1 & 2 \end{bmatrix}$, $B = \begin{bmatrix} 1 \\ 1 \end{bmatrix}$, we have $det(sI - A) = s(s - 5/2)$, and therefore, the eigenvalues of A are 0 and $\frac{5}{2}$. We wish to determine F so that the eigenvalues of $A + BF$ are at $-1 \pm j$.

If $F = [f_1, f_2]$, then $det(sI - (A+BF)) = det\left(\begin{bmatrix} s-\frac{1}{2} & -1 \\ -1 & s-2 \end{bmatrix} - \begin{bmatrix} 1 \\ 1 \end{bmatrix}[f_1, f_2]\right) =$ $det\begin{bmatrix} s-\frac{1}{2}-f_1 & -1-f_2 \\ -1-f_1 & s-2-f_2 \end{bmatrix} = s^2 + s(-\frac{5}{2} - f_1 - f_2) + f_1 - \frac{1}{2}f_2$. The desired eigenvalues are the roots of the polynomial

$$\alpha_d(s) = (s - (-1 + j))(s - (-1 - j)) = s^2 + 2s + 2.$$

Equating coefficients, one obtains $-\frac{5}{2} - f_1 - f_2 = 2$, $f_1 - \frac{1}{2}f_2 = 2$, a linear system of equations. Note that it is linear because $m = 1$. In general one must solve a set of nonlinear algebraic equations. We have

$$F = [f_1, f_2] = [-\tfrac{1}{6}, -\tfrac{13}{3}]$$

as the appropriate state feedback matrix. ■

2. The use of controller forms. Given that the pair (A, B) is controllable, there exists an equivalence transformation matrix P so that the pair $(A_c = PAP^{-1}, B_c = PB)$ is in controller form (see Section 3.4). The matrices $A + BF$ and $P(A + BF)P^{-1} = PAP^{-1} + PBFP^{-1} = A_c + B_cF_c$ have the same eigenvalues, and the problem is to determine F_c so that $A_c + B_cF_c$ has desired eigenvalues. This problem is easier to solve than the original one because of the special structures of A_c and B_c. Once F_c has been determined, then the original feedback matrix F is given by

$$F = F_cP. \tag{2.13}$$

We shall now assume that (A, B) has already been reduced to (A_c, B_c) and describe methods of deriving F_c for eigenvalue assignment.

Consider first the single-input case ($m = 1$). We let

$$F_c = [f_0, \ldots, f_{n-1}]. \tag{2.14}$$

In view of Section 3.4, since A_c, B_c are in controller form, we have

$$\begin{aligned} A_{cF} &\triangleq A_c + B_cF_c \\ &= \begin{bmatrix} 0 & 1 & \cdots & 0 \\ \vdots & \vdots & & \vdots \\ 0 & 0 & \cdots & 1 \\ -\alpha_0 & -\alpha_1 & \cdots & -\alpha_{n-1} \end{bmatrix} + \begin{bmatrix} 0 \\ \vdots \\ 0 \\ 1 \end{bmatrix}[f_0, \ldots, f_{n-1}] \\ &= \begin{bmatrix} 0 & 1 & \cdots & 0 \\ \vdots & \vdots & \ddots & \vdots \\ 0 & 0 & \cdots & 1 \\ -(\alpha_0 - f_0) & -(\alpha_1 - f_1) & \cdots & -(\alpha_{n-1} - f_{n-1}) \end{bmatrix}, \end{aligned} \tag{2.15}$$

where α_i, $i = 0, \ldots, n-1$, are the coefficients of the characteristic polynomial of A_c, i.e.,

$$det(sI - A_c) = s^n + \alpha_{n-1}s^{n-1} + \cdots + \alpha_1 s + \alpha_0. \tag{2.16}$$

Notice that A_{cF} is also in companion form and its characteristic polynomial can be written directly as

$$det(sI - A_{cF}) = s^n + (\alpha_{n-1} - f_{n-1})s^{n-1} + \cdots + (\alpha_0 - f_0). \tag{2.17}$$

If the desired eigenvalues are the roots of the polynomial

$$\alpha_d(s) = s^n + d_{n-1}s^{n-1} + \cdots + d_0, \tag{2.18}$$

then by equating coefficients, $f_i, i = 0, 1, \ldots, n-1$, must satisfy the relations $d_i = \alpha_i - f_i, i = 0, 1, \ldots, n-1$, from which we obtain

$$f_i = \alpha_i - d_i, \qquad i = 0, \ldots, n-1. \tag{2.19}$$

Alternatively, note that there exists a matrix A_d in companion form, the characteristic polynomial of which is (2.18). An alternative way of deriving (2.19) is then to set $A_{cF} = A_c + B_cF_c = A_d$, from which we obtain

$$F_c = B_m^{-1}[A_{d_m} - A_m], \tag{2.20}$$

where $B_m = 1$, $A_{d_m} = [-d_0, \ldots, -d_{n-1}]$ and $A_m = [-\alpha_0, \ldots, -\alpha_{n-1}]$. Therefore, B_m, A_{d_m}, and A_m are the nth rows of B_c, A_d, and A_c, respectively (see Section 3.4). Relationship (2.20), which is an alternative formula to (2.19), has the advantage that it is in a form that can be generalized to the multi-input case.

EXAMPLE 2.3. Consider the matrices $A = \begin{bmatrix} \frac{1}{2} & 1 \\ 1 & 2 \end{bmatrix}$, $B = \begin{bmatrix} 1 \\ 1 \end{bmatrix}$ of Example 2.2. Determine F so that the eigenvalues of $A + BF$ are $-1 \pm j$, i.e., so that they are the roots of the polynomial $\alpha_d(s) = s^2 + 2s + 2$.

To reduce (A, B) into the controller form, let

$$\mathscr{C} = [B, AB] = \begin{bmatrix} 1 & \frac{3}{2} \\ 1 & 3 \end{bmatrix} \text{ and } \mathscr{C}^{-1} = \frac{2}{3}\begin{bmatrix} 3 & -\frac{3}{2} \\ -1 & 1 \end{bmatrix},$$

from which $P = \begin{bmatrix} q \\ qA \end{bmatrix} = \frac{1}{3}\begin{bmatrix} -2 & 2 \\ 1 & 2 \end{bmatrix}$ [see (4.44) in Chapter 3]. Then $P^{-1} = \begin{bmatrix} -1 & 1 \\ \frac{1}{2} & 1 \end{bmatrix}$ and

$$A_c = PAP^{-1} = \begin{bmatrix} 0 & 1 \\ 0 & \frac{5}{2} \end{bmatrix}, \qquad B_c = \begin{bmatrix} 0 \\ 1 \end{bmatrix}.$$

Thus, $A_m = [0, \frac{5}{2}]$ and $B_m = 1$. Now $A_d = \begin{bmatrix} 0 & 1 \\ -2 & -2 \end{bmatrix}$ and $A_{d_m} = [-2, -2]$ since the characteristic polynomial of A_d is $s^2 + 2s + 2 = \alpha_d(s)$. Applying (2.20), we obtain that

$$F_c = B_m^{-1}[A_{d_m} - A_m] = [-2, -\tfrac{9}{2}]$$

and $F = F_cP = [-2, -\frac{9}{2}]\begin{bmatrix} -\frac{2}{3} & \frac{2}{3} \\ \frac{1}{3} & \frac{2}{3} \end{bmatrix} = [-\frac{1}{6}, -\frac{13}{3}]$ assigns the eigenvalues of the closed-loop system at $-1 \pm j$. This is the same result as the one obtained by the direct method given in Example 2.2. If $\alpha_d(s) = s^2 + d_1s + d_0$, then $A_{d_m} = [-d_0, -d_1]$, $F_c = B_m^{-1}[A_{d_m} - A_m] = [-d_0, -d_1 - 5/2]$, and

$$F = F_cP = \tfrac{1}{3}[2d_0 - d_1 - \tfrac{5}{2}, -2d_0 - 2d_1 - 5].$$

In general the larger the difference between the coefficients of $\alpha_d(s)$ and $\alpha(s)$, $(A_{d_m} - A_m)$, the larger the gains in F. This is as expected, since larger changes require in general larger control action. ■

Note that (2.20) can also be derived using (4.63) of Section 3.4 in Chapter 3. To see this, write

$$A_{cF} = A_c + B_cF_c = (\bar{A}_c + \bar{B}_cA_m) + (\bar{B}_cB_m)F_c = \bar{A}_c + \bar{B}_c(A_m + B_mF_c),$$

where $\bar{A}_c, \bar{B}_c$ are defined in (4.63). Selecting $A_d = \bar{A}_c + \bar{B}_c A_{d_m}$ and requiring $A_{cF} = A_d$ implies

$$\bar{B}_c[A_m + B_m F_c] = \bar{B}_c A_{d_m},$$

from which $A_m + B_m F_c = A_{d_m}$, which in turn implies (2.20).

After F_c has been found, to determine F so that $A + BF$ has desired eigenvalues, one should use $F = F_c P$ given in (2.13). Note that P, which reduces (A, B) to the controller form, has a specific form in this case [see (4.44) of Section 3.4]. Combining these results, it is possible to derive a formula for the eigenvalue assigning F in terms of the original pair (A, B) and the coefficients of the desired polynomial $\alpha_d(s)$. In particular, the $1 \times n$ matrix F that assigns the n eigenvalues of $A + BF$ at the roots of $\alpha_d(s)$ is given by

$$F = -e_n^T \mathscr{C}^{-1} \alpha_d(A), \tag{2.21}$$

where $e_n = [0, \ldots, 0, 1]^T \in R^n$ and $\mathscr{C} = [B, AB, \ldots, A^{n-1}B]$ is the controllability matrix. Relation (2.21) is known as *Ackermann's formula* and its validity will be verified shortly.

Notice that the F that assigns the n eigenvalues of $A + BF$ is unique when $m = 1$. This is not difficult to see from the preceding. In particular, notice that $F_c = [f_0, \ldots, f_{n-1}]$ is uniquely determined by (2.19) and that the transformation matrix P is also unique in this case ($m = 1$) and is in fact given by (4.44) in Section 3.4.

Verification of Ackermann's Formula

Given (A_c, B_c) in controller form, it was shown that the matrix

$$F_c = [\alpha_0 - d_0, \ldots, \alpha_{n-1} - d_{n-1}] = B_m^{-1}[A_{d_m} - A_m]$$

assigns the eigenvalues of $A_c + B_c F_c$ to the desired locations, the roots of $\alpha_d(s)$ given in (2.18). The matrix F_c is related to F by $F = F_c P$, where $A_c = PAP^{-1}$, $B_c = PB$, and $P = \mathscr{C}_c \mathscr{C}^{-1}$ and where $\mathscr{C}$ and $\mathscr{C}_c$ are the controllability matrices of (A, B) and (A_c, B_c), respectively. These relations will now be combined to derive Ackermann's formula.

First, note that if (2.21) were used for a given pair (A_c, B_c) in controller form, then

$$F_c = -e_n^T \mathscr{C}_c^{-1} \alpha_d(A_c). \tag{2.22}$$

We now show that this F_c assigns the n eigenvalues at the desired locations. In doing so, we note that $\alpha_d(A_c) = A_c^n + d_{n-1}A_c^{n-1} + \cdots + d_1 A_c + d_0 I$, which in view of the Cayley-Hamilton Theorem [namely, $\alpha(A_c) = A_c^n + \alpha_{n-1}A_c^{n-1} + \cdots + \alpha_0 I = 0$] assumes the form

$$\alpha_d(A_c) = \sum_{i=0}^{n-1} (d_i - \alpha_i) A_c^i,$$

since $A_c^n = -\sum_{i=0}^{n-1} \alpha_i A_c^i$. Also note that $e_1^T \mathscr{C}_c = e_n^T$ or that $e_n^T \mathscr{C}_c^{-1} = e_1^T$. Now $-e_n^T \mathscr{C}_c^{-1} \alpha_d(A_c) = -e_1^T((d_0 - \alpha_0)I + \cdots + (d_{n-1} - \alpha_{n-1})A_c^{n-1}) = [\alpha_0 - d_0, \ldots, \alpha_{n-1} - d_{n-1}] = F_c$ of (2.20), that is, (2.22) is verified. Note that the last relation was derived using the fact that

$$e_1^T A_c = e_2^T, (e_1^T A_c) A_c = e_2^T A_c = e_3^T, \ldots, e_1^T A^{n-1} = e_n^T,$$

from which $-e_1^T[(d_0-\alpha_0)I+\cdots+(d_{n-1}-\alpha_{n-1})A_c^{n-1}] = (\alpha_0-d_0)e_1^T+(\alpha_1-d_1)e_2^T+\cdots+(\alpha_{n-1}-d_{n-1})e_n^T = [\alpha_0-d_0,\alpha_1-d_1,\ldots,\alpha_{n-1}-d_{n-1}]$. In view of (2.22), we now have

$$\begin{aligned} F &= F_cP = -e_n^T\mathscr{C}_c^{-1}\alpha_d(A_c)P \\ &= -e_n^T\mathscr{C}_c^{-1}\alpha_d(PAP^{-1})P = -e_n^T\mathscr{C}_c^{-1}P\alpha_d(A) \\ &= -e_n^T\mathscr{C}^{-1}\alpha_d(A), \end{aligned}$$

which is Ackermann's formula, (2.21).

EXAMPLE 2.4. To the system of Example 2.3 we apply (2.21) and obtain

$$\begin{aligned} F &= -e_2^T\mathscr{C}^{-1}\alpha_d(A) \\ &= -[0,1]\begin{bmatrix} 2 & -1 \\ -\frac{2}{3} & \frac{2}{3}\end{bmatrix}\left(\begin{bmatrix}\frac{1}{2} & 1 \\ 1 & 2\end{bmatrix}^2 + 2\begin{bmatrix}\frac{1}{2} & 1 \\ 1 & 2\end{bmatrix} + 2\begin{bmatrix}1 & 0 \\ 0 & 1\end{bmatrix}\right) \\ &= -[-\tfrac{2}{3},\tfrac{2}{3}]\begin{bmatrix}\frac{17}{4} & \frac{9}{2} \\ \frac{9}{2} & 11\end{bmatrix} = [-\tfrac{1}{6},-\tfrac{13}{3}], \end{aligned}$$

which is identical to the F found in Example 2.3. ■

Now consider the multi-input case ($m>1$). We proceed in a way completely analogous to the single-input case.

Assume that A_c and B_c are in the controller form, (4.62), given in Section 3.4. Notice that $A_{cF} \triangleq A_c+B_cF_c$ is also in (controller) companion form with identical block structure as A_c for any F_c. In fact, the pair (A_{cF}, B_c) has the same controllability indices μ_i, $i=1,\ldots,m$, as (A_c, B_c). This was shown in Section 3.4 of Chapter 3 and can also be seen directly, using (4.63), since

$$A_c+B_cF_c = (\bar{A}_c+\bar{B}_cA_m)+(\bar{B}_cB_m)F_c = \bar{A}_c+\bar{B}_c(A_m+B_mF_c), \quad (2.23)$$

where $\bar{A}_c$ and $\bar{B}_c$ are defined in (4.63). We can now select an $n\times n$ matrix A_d with desired characteristic polynomial

$$det(sI-A_d) = \alpha_d(s) = s^n+d_{n-1}s^{n-1}+\cdots+d_0, \quad (2.24)$$

and in companion form, having the same block structure as A_{cF} or A_c, that is, $A_d = \bar{A}_c+\bar{B}_cA_{d_m}$. Now if $A_{cF}=A_d$, then in view of (2.23), $\bar{B}_c(A_m+B_mF_c)=\bar{B}_cA_{d_m}$. From this it follows that

$$F_c = B_m^{-1}[A_{d_m}-A_m], \quad (2.25)$$

where B_m, A_{d_m}, and A_m are the m σ_jth rows of B_c, A_d, and A_c, respectively, and $\sigma_j = \sum_{i=1}^{j}\mu_i$, $j=1,\ldots,m$. Note that this is a generalization of (2.20) of the single-input case.

We shall now show how to select an $n\times n$ matrix A_d in multivariable companion form to have the desired characteristic polynomial.

One choice is

$$A_d = \begin{bmatrix} 0 & 1 & \cdots & 0 \\ \vdots & \vdots & \ddots & \vdots \\ 0 & 0 & \cdots & 1 \\ -d_0 & -d_1 & \cdots & -d_{n-1}\end{bmatrix},$$

the characteristic polynomial of which is $\alpha_d(s)$. In this case the $m \times n$ matrix A_{d_m} is given by

$$A_{d_m} = \begin{bmatrix} 0 & \cdots & 0 & 1 & \cdots & 0 & \cdots & 0 \\ \vdots & & \vdots & \vdots & & \vdots & & \vdots \\ 0 & \cdots & 0 & 0 & \cdots & 1 & \cdots & 0 \\ -d_0 & & \cdots & & & & & -d_{n-1} \end{bmatrix},$$

where the ith row, $i = 1, \ldots, m-1$, is zero everywhere except at the $\sigma_i + 1$ column location, where it is one.

Another choice is to select $A_d = [A_{ij}]$, $i, j = 1, \ldots, m$, with $A_{ij} = 0$ for $i \neq j$, i.e.,

$$A_d = \begin{bmatrix} A_{11} & 0 & \cdots & 0 \\ 0 & A_{22} & \cdots & 0 \\ \vdots & \vdots & \ddots & \vdots \\ 0 & 0 & \cdots & A_{mm} \end{bmatrix},$$

noting that $det\,(sI - A_d) = det\,(sI - A_{11})\cdots det\,(sI - A_{mm})$. Then

$$A_{ii} = \begin{bmatrix} 0 & 1 & \cdots & 0 \\ \vdots & & \ddots & \vdots \\ 0 & & & 1 \\ \times & & \cdots & \times \end{bmatrix},$$

where the last row is selected so that $det\,(sI - A_{ii})$ has desired roots. The disadvantage of this selection is that it may impose unnecessary restrictions on the number of real eigenvalues assigned. For example, if $n = 4$, $m = 2$ and the dimensions of A_{11} and A_{12}, which are equal to the controllability indices, are $d_1 = 3$ and $d_2 = 1$, then two of the eigenvalues must be real (why?).

There are of course other selections for A_d, and the reader is encouraged to come up with additional choices. A point that should be quite clear by now is that F_c (or F) is not unique in the present case, since different F_c can be derived for different A_{d_m}, all assigning the eigenvalues at the same desired locations. In the single-input case, F_c is unique, as was shown. Therefore, the following result has been shown.

LEMMA 2.4. Let (A, B) be controllable and suppose that n desired real complex conjugate eigenvalues for $A + BF$ have been selected. The state feedback matrix F that assigns all eigenvalues of $A + BF$ to desired locations is not unique in the multi-input case ($m > 1$). It is unique in the single-input case $m = 1$. ■

EXAMPLE 2.5. Consider the controllable pair (A, B), where

$$A = \begin{bmatrix} 0 & 1 & 0 \\ 0 & 0 & 1 \\ 0 & 2 & -1 \end{bmatrix} \quad \text{and} \quad B = \begin{bmatrix} 0 & 1 \\ 1 & 1 \\ 0 & 0 \end{bmatrix}.$$

It was shown in Example 4.14 of Chapter 3, that this pair can be reduced to its controller form

$$A_c = PAP^{-1} = \begin{bmatrix} 0 & 1 & 0 \\ 2 & -1 & 0 \\ 1 & 0 & 0 \end{bmatrix}, \qquad B_c = PB = \begin{bmatrix} 0 & 0 \\ 1 & 1 \\ 0 & 1 \end{bmatrix},$$

where
$$P = \begin{bmatrix} 0 & 0 & \frac{1}{2} \\ 0 & 1 & -\frac{1}{2} \\ 1 & 0 & -\frac{1}{2} \end{bmatrix}.$$

Suppose we desire to assign the eigenvalues of $A + BF$ to the locations $\{-2, -1 \pm j\}$, i.e., at the roots of the polynomial $\alpha_d(s) = (s+2)(s^2+2s+2) = s^3 + 4s^2 + 6s + 4$. A choice for A_d is

$$A_{d1} = \begin{bmatrix} 0 & 1 & 0 \\ 0 & 0 & 1 \\ -4 & -6 & -4 \end{bmatrix}, \text{ leading to } A_{d_{m1}} = \begin{bmatrix} 0 & 0 & 1 \\ -4 & -6 & -4 \end{bmatrix},$$

and
$$F_{c_1} = B_m^{-1}[A_{d_{m1}} - A_m] = \begin{bmatrix} 1 & 1 \\ 0 & 1 \end{bmatrix}^{-1} \left[\begin{bmatrix} 0 & 0 & 1 \\ -4 & -6 & -4 \end{bmatrix} - \begin{bmatrix} 2 & -1 & 0 \\ 1 & 0 & 0 \end{bmatrix} \right]$$
$$= \begin{bmatrix} 1 & -1 \\ 0 & 1 \end{bmatrix} \begin{bmatrix} -2 & 1 & 1 \\ -5 & -6 & -4 \end{bmatrix} = \begin{bmatrix} 3 & 7 & 5 \\ -5 & -6 & -4 \end{bmatrix}.$$

Alternatively,

$$A_{d_2} = \begin{bmatrix} 0 & 1 & 0 \\ -2 & -2 & 0 \\ 0 & 0 & -2 \end{bmatrix}, \text{ from which } A_{d_{m2}} = \begin{bmatrix} -2 & -2 & 0 \\ 0 & 0 & -2 \end{bmatrix}$$

and
$$F_{c_2} = B_m^{-1}[A_{d_{m2}} - A_m] = \begin{bmatrix} 1 & -1 \\ 0 & 1 \end{bmatrix} \begin{bmatrix} -4 & -1 & 0 \\ -1 & 0 & -2 \end{bmatrix}$$
$$= \begin{bmatrix} -3 & -1 & 2 \\ -1 & 0 & -2 \end{bmatrix}.$$

Both $F_1 = F_{c1}P = \begin{bmatrix} 5 & 7 & -\frac{9}{2} \\ -4 & -6 & \frac{5}{2} \end{bmatrix}$ and $F_2 = F_{c2}P = \begin{bmatrix} 2 & -1 & -2 \\ -2 & 0 & \frac{1}{2} \end{bmatrix}$ assign the eigenvalues of $A + BF$ to the locations $\{-2, -1 \pm j\}$.

The reader should plot the states of the equation $\dot{x} = (A + BF)x$ for $F = F_1$ and $F = F_2$ when $x(0) = [1,\ 1,\ 1]^T$, and should comment on the differences between the trajectories. ■

Relation (2.25) gives *all feedback matrices*, F_c (or $F = F_cP$), that assign the n eigenvalues of $A_c + B_cF_c$ (or $A + BF$) to desired locations. The freedom in selecting such F_c is expressed in terms of the different A_d, all in companion form, with $A_d = [A_{ij}]$ and A_{ij} of dimensions $\mu_i \times \mu_j$, which have the same characteristic polynomial. Deciding which one of all the possible matrices A_d to select, so that in addition to eigenvalue assignment other objectives can be achieved, is not apparent. This flexibility in selecting F can also be expressed in terms of other parameters, where both eigenvalue and eigenvector assignment are discussed, as will now be shown.

3. Assigning eigenvalues and eigenvectors. Suppose now that F was selected so that $A + BF$ has a desired eigenvalue s_j with corresponding eigenvector v_j. Then

$[s_jI - (A + BF)]v_j = 0$, which can be written as

$$[s_jI - A, B]\begin{bmatrix} v_j \\ -Fv_j \end{bmatrix} = 0. \tag{2.26}$$

To determine an F that assigns s_j as a closed-loop eigenvalue, one could first determine a basis for the right kernel (null space) of $[s_jI - A, B]$, i.e., one could determine a basis $\begin{bmatrix} M_j \\ -D_j \end{bmatrix}$ such that

$$[s_jI - A, B]\begin{bmatrix} M_j \\ -D_j \end{bmatrix} = 0. \tag{2.27}$$

Note that the dimension of this basis is $(n+m) - rank\,[s_jI - A, B] = (n+m) - n = m$, where $rank\,[s_jI - A, B] = n$ since the pair (A, B) is controllable. Since it is a basis, there exists a nonzero $m \times 1$ vector a_j so that

$$\begin{bmatrix} M_j \\ -D_j \end{bmatrix} a_j = \begin{bmatrix} v_j \\ -Fv_j \end{bmatrix}. \tag{2.28}$$

Combining the relations $-D_ja_j = -Fv_j$ and $M_ja_j = v_j$, one obtains

$$FM_ja_j = D_ja_j. \tag{2.29}$$

This is the relation that F must satisfy for s_j to be a closed-loop eigenvalue. The nonzero $m \times 1$ vector a_j can be chosen arbitrarily. Note that $M_ja_j = v_j$ is the eigenvector corresponding to s_j. Note also that a_j represents the flexibility one has in selecting the corresponding eigenvector, in addition to assigning an eigenvalue. The $n \times 1$ eigenvector v_j cannot be arbitrarily assigned; rather, the $m \times 1$ vector a_j can be (almost) arbitrarily selected. These mild conditions on a_j are stated next as a theorem.

THEOREM 2.5. The pair (s_j, v_j) is an (eigenvalue, eigenvector)-pair of $A + BF$ if and only if F satisfies (2.29) for some nonzero vector a_j such that $v_j = M_ja_j$ with $\begin{bmatrix} M_j \\ D_j \end{bmatrix}$ a basis of the null space of $[s_jI - A, B]$ as in (2.27).

Proof. Necessity has been shown. To prove sufficiency, postmultiply $s_jI - (A + BF)$ by M_ja_j and use (2.29) to obtain $(s_jI - A)M_ja_j - BD_ja_j = 0$ in view of (2.27). Thus,

$$[s_jI - (A + BF)]M_ja_j = 0,$$

which implies that s_j is an eigenvalue of $A + BF$ and $M_ja_j = v_j$ is the corresponding eigenvector. ■

If relation (2.29) is written for n desired eigenvalues s_j, where the *a_j are selected so that the corresponding eigenvectors $v_j = M_ja_j$ are linearly independent,* then

$$FV = W, \tag{2.30}$$

where $V \triangleq [M_1a_1, \ldots, M_na_n]$ and $W \triangleq [D_1a_1, \ldots, D_na_n]$ uniquely specify F as the solution to these n linearly independent equations. When s_j are distinct, it is always possible to select a_j so that V has full rank; in fact almost any set of nonzero a_j suffices. When s_j have repeated values it may still be possible under certain conditions to select a_j so that M_ja_j are linearly independent; however in general, for multiple eigenvalues, (2.30) needs to be modified, and the details for this can be found in the literature (e.g., [16]).

It can be shown that for distinct eigenvalues s_j, the n vectors $M_j a_j$, $j = 1, \ldots, n$, are linearly independent for almost any nonzero a_j. For further discussion on this, see the remarks following Example 2.6 and Subsection A.4, on Interpolation, in the Appendix. Also note that if $s_{j+1} = s_j^*$, the complex conjugate of s_j, then the corresponding eigenvector $v_{i+1} = v_j^* = M_j^* a_j^*$.

Relation (2.30) clearly shows that the F that assigns all n closed-loop eigenvalues is not unique (see also Lemma 2.4). All such F are parameterized by the vectors a_j that in turn characterize the corresponding eigenvectors. If the corresponding eigenvectors have been decided upon—of course within the set of possible eigenvectors $v_j = M_j a_j$—then F is uniquely specified. Note that in the single-input case, (2.29) becomes $FM_j = D_j$, where $v_j = M_j$. In this case, F is unique.

EXAMPLE 2.6. Consider the controllable pair (A, B) of Example 2.5 given by

$$A = \begin{bmatrix} 0 & 1 & 0 \\ 0 & 0 & 1 \\ 0 & 2 & -1 \end{bmatrix}, \qquad B = \begin{bmatrix} 0 & 1 \\ 1 & 1 \\ 0 & 0 \end{bmatrix}.$$

Again, it is desired to assign the eigenvalues of $A + BF$ at $-2, -1 \pm j$. Let $s_1 = -2$, $s_2 = -1 + j$ and $s_3 = -1 - j$. Then, in view of (2.27),

$$\begin{bmatrix} M_1 \\ -D_1 \end{bmatrix} = \begin{bmatrix} 1 & 1 \\ -1 & 0 \\ 2 & 0 \\ \cdots & \cdots \\ -1 & -2 \\ 1 & 2 \end{bmatrix}, \qquad \begin{bmatrix} M_2 \\ -D_2 \end{bmatrix} = \begin{bmatrix} 1 & 1 \\ j & 0 \\ 2 & 0 \\ \cdots & \cdots \\ 2 + j & -1 + j \\ 1 & 1 - j \end{bmatrix},$$

and $\begin{bmatrix} M_3 \\ -D_3 \end{bmatrix} = \begin{bmatrix} M_2^* \\ -D_2^* \end{bmatrix}$, the complex conjugate, since $s_3 = s_2^*$.

Each eigenvector $v_i = M_i a_i$, $i = 1, 2, 3$, is a linear combination of the columns of M_i. Note that $v_3 = v_2^*$. If we select the eigenvectors to be

$$V = [v_1, v_2, v_3] = \begin{bmatrix} 1 & 1 & 1 \\ 0 & j & -j \\ 0 & 2 & 2 \end{bmatrix},$$

i.e., $a_1 = \begin{bmatrix} 0 \\ 1 \end{bmatrix}$, $a_2 = \begin{bmatrix} 1 \\ 0 \end{bmatrix}$, and $a_3 = \begin{bmatrix} 1 \\ 0 \end{bmatrix}$, then (2.30) implies that

$$F \begin{bmatrix} 1 & 1 & 1 \\ 0 & j & -j \\ 0 & 2 & 2 \end{bmatrix} = \begin{bmatrix} 2 & -2 - j & -2 + j \\ -2 & -1 & -1 \end{bmatrix},$$

from which we have

$$\begin{aligned} F &= \frac{1}{4j} \begin{bmatrix} 2 & -2 - j & -2 + j \\ -2 & -1 & -1 \end{bmatrix} \begin{bmatrix} 4j & 0 & -2j \\ 0 & 2 & j \\ 0 & -2 & j \end{bmatrix} \\ &= \begin{bmatrix} 2 & -1 & -2 \\ -2 & 0 & \frac{1}{2} \end{bmatrix}. \end{aligned}$$

This matrix F is such that $A + BF$ has the desired eigenvalues and eigenvectors (verify this). ■

Remarks

At this point, several comments are in order.

1. In Example 2.6, if the eigenvectors were chosen to be the eigenvectors of $A+BF_1$ (instead of $A+BF_2$) of Example 2.5, then from $FV = W$ it follows that F would have been F_1 (instead of F_2).
2. When $s_i = s_{i+1}^*$, then the corresponding eigenvectors are also complex conjugates, i.e., $v_i = v_{i+1}^*$. In this case we obtain from (2.30) that

$$\begin{aligned} FV &= F[\ldots, v_{iR} + jv_{iI}, v_{iR} - jv_{iI}, \ldots] \\ &= [\ldots, w_{iR} + jw_{iI}, w_{iR} - jw_{iI}, \ldots] = W. \end{aligned}$$

Although these calculations could be performed over the complex numbers (as was done in the example), this is not necessary, since postmultiplication of $FV = W$ by

$$\begin{bmatrix} I & \vdots & & & \vdots & \\ \cdots & \vdots & \cdots & \cdots & \vdots & \cdots \\ & \vdots & \frac{1}{2} & -j\frac{1}{2} & \vdots & \\ & \vdots & \frac{1}{2} & +j\frac{1}{2} & \vdots & \\ \cdots & \vdots & \cdots & \cdots & \vdots & \cdots \\ & \vdots & & & \vdots & I \end{bmatrix}$$

shows that the above equation $FV = W$ is equivalent to

$$F[\ldots, v_{iR}, v_{iI}, \ldots] = [\ldots, w_{iR}, w_{iI}, \ldots],$$

which involves only reals.

3. The bases $\begin{bmatrix} M_j \\ -D_j \end{bmatrix}$, $j = 1, \cdots, n$, in (2.27) can be determined in an alternative way and the calculations can be simplified if the controller form of the pair (A, B) is known. In particular, note that $[sI - A, B]\begin{bmatrix} P^{-1}S(s) \\ D(s) \end{bmatrix} = 0$, where the $n \times m$ matrix $S(s)$ is given by $S(s) = \textit{block diag}\,[1, s, \ldots, s^{\mu_i - 1}]$ and the μ_i, $i = 1, \ldots, m$, are the controllability indices of (A, B). Also, the $m \times m$ matrix $D(s)$ is given by $D(s) = B_m^{-1}[\textit{diag}\,[s^{\mu_1}, \ldots, s^{\mu_m}] - A_m S(s)]$. Note that $S(s)$ and $D(s)$ were defined in the Structure Theorem (controllable version) in Chapter 3 (Subsection 3.4D). It was shown there that $(sI - A_c)S(s) = B_c D(s)$, from which it follows that $(sI - A)P^{-1}S(s) = BD(s)$, where P is a similarity transformation matrix that reduces (A, B) to the controller form $(A_c = PAP^{-1}, B_c = PB)$. Since $P^{-1}S(s)$ and $D(s)$ are right coprime polynomial matrices (see Section 7.2 of Chapter 7), we have $\textit{rank}\begin{bmatrix} P^{-1}S(s_j) \\ D(s_j) \end{bmatrix} = m$ for any s_j, and therefore, $\begin{bmatrix} P^{-1}S(s_j) \\ D(s_j) \end{bmatrix}$ qualifies as a basis for the null space of the matrix $[s_j I - A, B](P = I$ when A, B are in controller

form, i.e., $A = A_c$ and $B = B_c$.) We note that this approach is discussed further in Subsection A.4 of the Appendix.

Returning to Example 2.6, the controller form of (A, B) was found in Example 2.5 using

$$P^{-1} = \begin{bmatrix} 1 & 0 & 1 \\ 1 & 1 & 0 \\ 2 & 0 & 0 \end{bmatrix}.$$

Here $$S(s) = \begin{bmatrix} 1 & 0 \\ s & 0 \\ 0 & 1 \end{bmatrix}, \qquad D(s) = \begin{bmatrix} s^2 + s - 1 & -s \\ -1 & s \end{bmatrix},$$

and $$\begin{bmatrix} M(s) \\ -D(s) \end{bmatrix} = \begin{bmatrix} P^{-1}S(s) \\ -D(s) \end{bmatrix} = \begin{bmatrix} 1 & 1 \\ s+1 & 0 \\ 2 & 0 \\ \cdots & \cdots \\ -(s^2 + s - 1) & s \\ 1 & -s \end{bmatrix}.$$

Then

$$\begin{bmatrix} M_1 \\ -D_1 \end{bmatrix} = \begin{bmatrix} M(-2) \\ -D(-2) \end{bmatrix} = \begin{bmatrix} 1 & 1 \\ -1 & 0 \\ 2 & 0 \\ \cdots & \cdots \\ -1 & -2 \\ 1 & 2 \end{bmatrix},$$

$$\begin{bmatrix} M_2 \\ -D_2 \end{bmatrix} = \begin{bmatrix} M(-1+j) \\ -D(-1+j) \end{bmatrix} = \begin{bmatrix} 1 & 1 \\ j & 0 \\ 2 & 0 \\ \cdots & \cdots \\ 2+j & -1+j \\ 1 & 1-j \end{bmatrix},$$

and $\begin{bmatrix} M_3 \\ -D_3 \end{bmatrix} = \begin{bmatrix} M_2^* \\ -D_2^* \end{bmatrix}$, which are precisely the bases used in the example.

4. If in Example 2.6 the only requirement were that $(s_1, v_1) = (-2, (1, 0, 0)^T)$, then $F(1, 0, 0)^T = (2, -2)^T$, i.e., any $F = \begin{bmatrix} 2 & f_{12} & f_{13} \\ 2 & f_{22} & f_{23} \end{bmatrix}$ will assign the desired values to an eigenvalue of $A + BF$ and its corresponding eigenvector.
5. All possible eigenvectors v_1 and $v_2 (v_3 = v_2^*)$ in Example 2.6 are given by

$$v_1 = M_1 a_1 = \begin{bmatrix} 1 & 1 \\ -1 & 0 \\ 2 & 0 \end{bmatrix} \begin{bmatrix} a_{11} \\ a_{12} \end{bmatrix} \text{ and } v_2 = M_2 a_2 = \begin{bmatrix} 1 & 1 \\ j & 0 \\ 2 & 0 \end{bmatrix} \begin{bmatrix} a_{21} + ja_{31} \\ a_{22} + ja_{32} \end{bmatrix},$$

where the a_{ij} are such that the set $\{v_1, v_2, v_3\}$ is linearly independent (i.e., $V = [v_1, v_2, v_3]$ is nonsingular) but otherwise arbitrary. Note that in this case $(s_j$

distinct), almost any arbitrary choice for a_{ij} will satisfy the above requirement (see Appendix, Subsection A.4).

C. The Linear Quadratic Regulator (LQR): Continuous-Time Case

A linear state feedback control law that is optimal in some sense can be determined as a solution to the so-called Linear Quadratic Regulator (LQR) problem (more recently, also called the H_2 optimal control problem). The LQR problem has been studied extensively, and the interested reader should consult the extensive literature on *optimal control* for additional information on the subject. In the following, we give a brief outline of certain central results of this topic to emphasize the fact that the state feedback gain F can be determined to satisfy, in an optimal fashion, requirements other than eigenvalue assignment, discussed above. The LQR problem has been studied for the time-varying and time-invariant cases. Presently, we will concentrate on the time-invariant optimal regulator problem.

Consider the time-invariant linear system given by

$$\dot{x} = Ax + Bu, \qquad z = Mx, \tag{2.31}$$

where the vector $z(t)$ represents the variables to be regulated—to be driven to zero.

We wish to determine $u(t)$, $t \geq 0$, which minimizes the quadratic cost

$$J(u) = \int_0^\infty [z^T(t)Qz(t) + u^T(t)Ru(t)]\,dt \tag{2.32}$$

for any initial state $x(0)$. The weighting matrices Q, R are real, symmetric, and positive definite, i.e., $Q = Q^T$, $R = R^T$, and $Q > 0$, $R > 0$. This is the most common version of the LQR problem. The term $z^TQz = x^T(M^TQM)x$ is nonnegative, and minimizing its integral forces $z(t)$ to approach zero as t goes to infinity. The matrix M^TQM is in general positive semidefinite, which allows some of the states to be treated as "do not care" states. The term u^TRu with $R > 0$ is always positive for $u \neq 0$, and minimizing its integral forces $u(t)$ to remain small. The relative "size" of Q and R enforces tradeoffs between the size of the control action and the speed of response.

Assume that $(A, B, Q^{1/2}M)$ is controllable (-from-the-origin) and observable. It turns out that the solution $u^*(t)$ to this optimal control problem can be expressed in state feedback form, which is independent of the initial condition $x(0)$. In particular, the optimal control u^* is given by

$$u^*(t) = F^*x(t) = -R^{-1}B^TP_c^*x(t), \tag{2.33}$$

where P_c^* denotes the symmetric positive-definite solution of the *algebraic Riccati equation*

$$A^TP_c + P_cA - P_cBR^{-1}B^TP_c + M^TQM = 0. \tag{2.34}$$

This equation may have more than one solution, but only one that is positive-definite (see Example 2.7). It can be shown that $u^*(t) = F^*x(t)$ is a stabilizing feedback control law and that the minimum cost is given by $J_{\min} = J(u^*) = x^T(0)P_c^*x(0)$.

The assumptions that $(A, B, Q^{1/2}M)$ are controllable and observable may be relaxed somewhat. If $(A, B, Q^{1/2}M)$ is stabilizable and detectable, then the uncontrollable and unobservable eigenvalues, respectively, are stable, and P_c^* is the unique, symmetric, but now positive-semidefinite solution of the algebraic Riccati equation. The matrix F^* is still a stabilizing gain but it is understood that the uncontrollable and unobservable (but stable) eigenvalues will not be affected by F^*.

Note that if the time interval of interest in the evaluation of the cost goes from 0 to $t_1 < \infty$, instead of 0 to ∞, that is, if

$$J(u) = \int_0^{t_1} [z^T(t)Qz(t) + u^T(t)Ru(t)]\, dt, \tag{2.35}$$

then the optimal control law is time-varying and is given by

$$u^*(t) = -R^{-1}B^T P^*(t)x(t), \qquad 0 \le t \le t_1, \tag{2.36}$$

where $P^*(t)$ is the unique, symmetric, and positive-semidefinite solution of the Riccati equation, which is a matrix differential equation of the form

$$-\frac{d}{dt}P(t) = A^T P(t) + P(t)A - P(t)BR^{-1}B^T P(t) + M^T QM, \tag{2.37}$$

where $P(t_1) = 0$. It is interesting to note that if $(A, B, Q^{1/2}M)$ is stabilizable and detectable (or controllable and observable), then the solution to this problem as $t_1 \to \infty$ approaches the steady-state value P_c^* given by the algebraic Riccati equation; that is, when $t_1 \to \infty$ the optimal control policy is the time-invariant control law (2.33), which is much easier to implement than time-varying control policies.

EXAMPLE 2.7. Consider the system described by the equations $\dot{x} = Ax + Bu$, $y = Cx$, where $A = \begin{bmatrix} 0 & 1 \\ 0 & 0 \end{bmatrix}$, $B = \begin{bmatrix} 0 \\ 1 \end{bmatrix}$, $C = [1, 0]$. Then (A, B, C) is controllable and observable and $C(sI - A)^{-1}B = 1/s^2$. We wish to determine the optimal control $u^*(t)$, $t \ge 0$, which minimizes the performance index

$$J = \int_0^{\infty} (y^2(t) + \rho u^2(t))\, dt,$$

where ρ is positive and real. Then $R = \rho > 0$, $z(t) = y(t)$, $M = C$, and $Q = 1 > 0$. In the present case the algebraic Riccati equation (2.34) assumes the form

$$A^T P_c + P_c A - P_c BR^{-1}B^T P_c + M^T QM$$

$$= \begin{bmatrix} 0 & 0 \\ 1 & 0 \end{bmatrix} P_c + P_c \begin{bmatrix} 0 & 1 \\ 0 & 0 \end{bmatrix} - \frac{1}{\rho} P_c \begin{bmatrix} 0 \\ 1 \end{bmatrix} [0\ 1] P_c + \begin{bmatrix} 1 \\ 0 \end{bmatrix} [1\ 0]$$

$$= \begin{bmatrix} 0 & 0 \\ 1 & 0 \end{bmatrix} \begin{bmatrix} p_1 & p_2 \\ p_2 & p_3 \end{bmatrix} + \begin{bmatrix} p_1 & p_2 \\ p_2 & p_3 \end{bmatrix} \begin{bmatrix} 0 & 1 \\ 0 & 0 \end{bmatrix} - \frac{1}{\rho} \begin{bmatrix} p_1 & p_2 \\ p_2 & p_3 \end{bmatrix} \begin{bmatrix} 0 & 0 \\ 0 & 1 \end{bmatrix} \begin{bmatrix} p_1 & p_2 \\ p_2 & p_3 \end{bmatrix}$$

$$+ \begin{bmatrix} 1 & 0 \\ 0 & 0 \end{bmatrix} = \begin{bmatrix} 0 & 0 \\ 0 & 0 \end{bmatrix},$$

where $P_c = \begin{bmatrix} p_1 & p_2 \\ p_2 & p_3 \end{bmatrix} = P_c^T$. This implies that

$$-\frac{1}{\rho}p_2^2 + 1 = 0, \qquad p_1 - \frac{1}{\rho}p_2 p_3 = 0, \qquad 2p_2 - \frac{1}{\rho}p_3^2 = 0.$$

Now P_c is positive definite if and only if $p_1 > 0$ and $p_1 p_3 - p_2^2 > 0$. The first equation above implies that $p_2 = \pm\sqrt{\rho}$. However, the third equation, which yields $p_3^2 = 2\rho p_2$, implies that $p_2 = +\sqrt{\rho}$. Then $p_3^2 = 2\rho\sqrt{\rho}$ and $p_3 = \pm\sqrt{2\rho\sqrt{\rho}}$. The second equation yields $p_1 = (1/\rho)p_2p_3$ and implies that only $p_3 = +\sqrt{2\rho\sqrt{\rho}}$ is acceptable, since we must have $p_1 > 0$ for P_c to be positive definite. Note that $p_1 > 0$ and $p_3 - p_2^2 = 2\rho - \rho = \rho > 0$, which shows that

$$P_c^* = \begin{bmatrix} \sqrt{2\sqrt{\rho}} & \sqrt{\rho} \\ \sqrt{\rho} & \sqrt{2\rho\sqrt{\rho}} \end{bmatrix}$$

is the positive definite solution of the algebraic Riccati equation. The optimal control law is now given by

$$u^*(t) = F^*x(t) = -R^{-1}B^TP_c^*x(t) = -\frac{1}{\rho}[0, 1]P_c^*x(t).$$

The eigenvalues of the compensated system, i.e., the eigenvalues of $A+BF^*$, can now be determined for different ρ. Also, the corresponding $u^*(t)$ and $y(t)$ for given $x(0)$ can be plotted. As ρ increases, the control energy expended to drive the output to zero is forced to decrease. The reader is asked to verify this by plotting $u^*(t)$ and $y(t)$ for different values of ρ when $x(0) = [1, 1]^T$. Also, the reader is asked to plot the eigenvalues of $A + BF^*$ as a function of ρ and to comment on the results. ■

It should be pointed out that the locations of the closed-loop eigenvalues, as the weights Q and R vary, have been studied extensively. Briefly, for the single-input case and for $Q = qI$ and $R = r$ in (2.32), it can be shown that the optimal closed-loop eigenvalues are the stable zeros of $1 + (q/r)H^T(-s)H(s)$, where $H(s) = M(sI - A)^{-1}B$. As q/r varies from zero (no state weighting) to infinity (no control weighting), the optimal closed-loop eigenvalues move from the stable poles of $H^T(-s)H(s)$ to the stable zeros of $H^T(-s)H(s)$. Note that the stable poles of $H^T(-s)H(s)$ are the stable poles of $H(s)$ and the stable reflections of its unstable poles with respect to the imaginary axis in the complex plane, while its stable zeros are the stable zeros of $H(s)$ and the stable reflections of its unstable zeros.

The solution of the LQR problem relies on solving the Riccati equation. A number of numerically stable algorithms exist for solving the algebraic Riccati equation. The reader is encouraged to consult the literature for computer software packages that implement these methods. A rather straightforward method for determining P_c^* is to use the *Hamiltonian matrix* given by

$$H \triangleq \begin{bmatrix} A & -BR^{-1}B^T \\ -M^TQM & -A^T \end{bmatrix}. \tag{2.38}$$

Let $[V_1^T, V_2^T]^T$ denote the n eigenvectors of H that correspond to the n stable $[Re\,(\lambda) < 0]$ eigenvalues. Note that of the $2n$ eigenvalues of H, n are stable and are the mirror images reflected on the imaginary axis of its n unstable eigenvalues. When $(A, B, Q^{1/2}M)$ is controllable and observable, then H has no eigenvalues on the imaginary axis $[Re\,(\lambda) = 0]$. In this case the n stable eigenvalues of H are in fact the closed-loop eigenvalues of the optimally controlled system, and the solution to the algebraic Riccati equation is then given by

$$P_c^* = V_2V_1^{-1}. \tag{2.39}$$

Note that in this case the matrix V_1 consists of the n eigenvectors of $A + BF^*$, since for λ_1 a stable eigenvalue of H, and v_1 the corresponding (first) column of V_1, we have

$$
\begin{aligned}
[\lambda_1 I - (A + BF^*)]v_1 &= [\lambda_1 I - A + BR^{-1}B^T V_2 V_1^{-1}]v_1 \\
&= \left[[\lambda_1 I, 0] \begin{bmatrix} V_1 \\ V_2 \end{bmatrix} - [A, -BR^{-1}B^T] \begin{bmatrix} V_1 \\ V_2 \end{bmatrix} \right] V_1^{-1} v_1 \\
&= \begin{bmatrix} 0 & \times & \cdots & \times \\ \vdots & \vdots & & \vdots \\ 0 & \times & \cdots & \times \end{bmatrix} V_1^{-1} v_1 \\
&= \begin{bmatrix} 0 & \times & \cdots & \times \\ \vdots & \vdots & & \vdots \\ 0 & \times & \cdots & \times \end{bmatrix} \begin{bmatrix} 1 \\ 0 \\ \vdots \\ 0 \end{bmatrix} = \begin{bmatrix} 0 \\ \vdots \\ 0 \end{bmatrix},
\end{aligned}
$$

where the fact that $\begin{bmatrix} V_1 \\ V_2 \end{bmatrix}$ are eigenvectors of H was used. It is worth reflecting for a moment on the relationship between (2.39) and (2.30). The optimal control F derived by (2.39) is in the class of F derived by (2.30).

D. Input-Output Relations

It is useful to derive the input-output relations for a closed-loop system that is compensated by linear state feedback, and several are derived in this subsection. Given the uncompensated or open-loop system $\dot{x} = Ax + Bu$, $y = Cx + Du$, with initial conditions $x(0) = x_0$, we have

$$\hat{y}(s) = C(sI - A)^{-1}x_0 + H(s)\hat{u}(s), \tag{2.40}$$

where the open-loop transfer function $H(s) = C(sI - A)^{-1}B + D$. Under the feedback control law $u = Fx + r$, the compensated closed-loop system is described by the equations $\dot{x} = (A + BF)x + Br$, $y = (C + DF)x + Dr$, from which we obtain

$$\hat{y}(s) = (C + DF)[sI - (A + BF)]^{-1}x_0 + H_F(s)\hat{r}(s), \tag{2.41}$$

where the closed-loop transfer function $H_F(s)$ is given by

$$H_F(s) = (C + DF)[sI - (A + BF)]^{-1}B + D.$$

Alternative expressions for $H_F(s)$ can be derived rather easily by substituting (2.4), namely,

$$\hat{u}(s) = F[sI - (A + BF)]^{-1}x_0 + [I - F(sI - A)^{-1}B]^{-1}\hat{r}(s),$$

into (2.40). This corresponds to working with an open-loop control law that nominally produces the same results when applied to the system [see the discussion on open- and closed-loop control that follows (2.4)]. Substituting, we obtain

$$
\begin{aligned}
\hat{y}(s) = {} & [C(sI - A)^{-1} + H(s)F[sI - (A + BF)]^{-1}]x_0 \\
& + H(s)[I - F(sI - A)^{-1}B]^{-1}\hat{r}(s).
\end{aligned}
\tag{2.42}
$$

Comparing with (2.41), we see that $(C + DF)[sI - (A + BF)]^{-1} = C(sI - A)^{-1} + H(s)F[sI - (A + BF)]^{-1}$, and that

$$\begin{aligned} H_F(s) &= (C + DF)[sI - (A + BF)]^{-1}B + D \\ &= [C(sI - A)^{-1}B + D][I - F(sI - A)^{-1}B]^{-1} \\ &= H(s)[I - F(sI - A)^{-1}B]^{-1}. \end{aligned} \tag{2.43}$$

The last relation points out the fact that $\hat{y}(s) = H_F(s)\hat{r}(s)$ can be obtained from $\hat{y}(s) = H(s)\hat{u}(s)$ using the open-loop control $\hat{u}(s) = [I - F(sI - A)^{-1}B]^{-1}\hat{r}(s)$.

Relation (2.43) can easily be derived in an alternative manner, using fractional matrix descriptions for the transfer function, introduced in Section 3.4 (see the Structure Theorem). In particular, the transfer function $H(s)$ of the open-loop system $\{A, B, C, D\}$ is given by

$$H(s) = N(s)D^{-1}(s),$$

where $N(s) = CS(s) + DD(s)$ with $S(s)$ and $D(s)$ satisfying $(sI - A)S(s) = BD(s)$ (refer to the proof of the controllable version of the Structure Theorem given in Section 3.4). Notice that it has been assumed, without loss of generality, that the pair (A, B) is in controller form.

Similarly, the transfer function $H_F(s)$ of the compensated system $\{A+BF, B, C+DF, D\}$ is given by

$$H_F(s) = N_F(s)D_F^{-1}(s),$$

where $N_F(s) = (C + DF)S(s) + DD_F(s)$ with $S(s)$ and $D_F(s)$ satisfying $[sI - (A + BF)]S(s) = BD_F(s)$. This relation implies that $(sI - A)S(s) = B[D_F(s) + FS(s)]$, from which we obtain $D_F(s) + FS(s) = D(s)$. Then $N_F(s) = CS(s) + D[FS(s) + D_F(s)] = CS(s) + DD(s) = N(s)$, that is,

$$H_F(s) = N(s)D_F^{-1}(s), \tag{2.44}$$

where $D_F(s) = D(s) - FS(s)$. The full justification of the validity of these expressions will be given in Subsection 7.4B of Chapter 7, where feedback systems described in terms of polynomial matrix representations are addressed.

Note that $I - F(sI - A)^{-1}B$ in (2.43) is the transfer function of the system $\{A, B, -F, I\}$ and can be expressed as $D_F(s)D^{-1}(s)$, where $D_F(s) = -FS(s)+ID(s)$. Let $M(s) = (D_F(s)D^{-1}(s))^{-1}$. Then (2.44) assumes the form

$$H_F(s) = N(s)D_F^{-1}(s) = (N(s)D^{-1}(s))(D(s)D_F^{-1}(s)) = H(s)M(s). \tag{2.45}$$

Note that relation $H_F(s) = N(s)D_F^{-1}(s)$ also shows that the zeros of $H(s)$ [in $N(s)$, see also Subsection 7.3B] are invariant under linear state feedback; they can be changed only via cancellations with poles. Also observe that $M(s) = D(s)D_F^{-1}(s)$ is the transfer function of the system $\{A + BF, B, F, I\}$ (show this). This implies that $H_F(s)$ in (2.43) can also be written as

$$H_F(s) = H(s)[F(sI - (A + BF))^{-1}B + I],$$

a result that could also be shown directly using matrix identities.

EXAMPLE 2.8. Consider the system $\dot{x} = Ax + Bu$, $y = Cx$, where

$$A = A_c = \begin{bmatrix} 0 & 1 & 0 \\ 2 & -1 & 0 \\ 1 & 0 & 0 \end{bmatrix} \quad \text{and} \quad B = B_c = \begin{bmatrix} 0 & 0 \\ 1 & 1 \\ 0 & 1 \end{bmatrix}$$

as in Example 2.5, and let $C = C_c = [1, 1, 0]$. $H_F(s)$ will now be determined. In view of the Structure Theorem developed in Section 3.4, the transfer function is given by $H(s) = N(s)D^{-1}(s)$, where

$$N(s) = C_c S(s) = [1, 1, 0]\begin{bmatrix} 1 & 0 \\ s & 0 \\ 0 & 1 \end{bmatrix} = [s + 1, 0],$$

and $D(s) = B_m^{-1}[\Lambda(s) - A_m S(s)] = \begin{bmatrix} 1 & 1 \\ 0 & 1 \end{bmatrix}^{-1}\left[\begin{bmatrix} s^2 & 0 \\ 0 & s \end{bmatrix} - \begin{bmatrix} 2 & -1 & 0 \\ 1 & 0 & 0 \end{bmatrix}\begin{bmatrix} 1 & 0 \\ s & 0 \\ 0 & 1 \end{bmatrix}\right]$

$$= \begin{bmatrix} 1 & -1 \\ 0 & 1 \end{bmatrix}\begin{bmatrix} s^2 + s - 2 & 0 \\ -1 & s \end{bmatrix} = \begin{bmatrix} s^2 + s - 1 & -s \\ -1 & s \end{bmatrix}.$$

Then

$$H(s) = N(s)D^{-1}(s) = [s + 1, 0]\begin{bmatrix} s^2 + s - 1 & -s \\ -1 & s \end{bmatrix}^{-1}$$

$$= [s + 1, 0]\begin{bmatrix} s & s \\ 1 & s^2 + s - 1 \end{bmatrix}\frac{1}{s^3 + s^2 - 2s}$$

$$= \frac{1}{s(s^2 + s - 2)}[s(s + 1), s(s + 1)] = \frac{s + 1}{s^2 + s - 2}[1, 1].$$

If $F_c = \begin{bmatrix} 3 & 7 & 5 \\ -5 & -6 & -4 \end{bmatrix}$ (which is F_{c1} of Example 2.5), then

$$D_F(s) = D(s) - F_c S(s) = \begin{bmatrix} s^2 + s - 1 & -s \\ -1 & s \end{bmatrix} - \begin{bmatrix} 3 & 7 & 5 \\ -5 & -6 & -4 \end{bmatrix}\begin{bmatrix} 1 & 0 \\ s & 0 \\ 0 & 1 \end{bmatrix}$$

$$= \begin{bmatrix} s^2 - 6s - 4 & -s - 5 \\ 6s + 4 & s + 4 \end{bmatrix}.$$

Note that $\det D_F(s) = s^3 + 4s^2 + 6s + 4 = (s + 2)(s^2 + 2s + 2)$ with roots $-2, -1 \pm j$, as expected. Now

$$H_F(s) = N(s)D_F^{-1}(s) = [s + 1, 0]\begin{bmatrix} s + 4 & s + 5 \\ -6s - 4 & s^2 - 6s - 4 \end{bmatrix}\frac{1}{(s + 2)(s^2 + 2s + 2)}$$

$$= \frac{s + 1}{(s + 2)(s^2 + 2s + 2)}[s + 4, s + 5].$$

Note that the zeros of $H(s)$ and $H_F(s)$ are identical, located at -1. Then $H_F(s) = H(s)M(s)$, where

$$M(s) = D(s)D_F^{-1}(s) = \begin{bmatrix} s^2 + s - 1 & -s \\ -1 & s \end{bmatrix}\begin{bmatrix} s + 4 & s + 5 \\ -6s - 4 & s^2 - 6s - 4 \end{bmatrix}\frac{1}{s^3 + 4s^2 + 6s + 4}$$

$$= \begin{bmatrix} s^3 + 11s^2 + 7s - 4 & 12s^2 + 8s - 5 \\ -6s^2 - 5s - 4 & s^3 - 6s^2 - 5s - 5 \end{bmatrix}\frac{1}{s^3 + 4s^2 + 6s + 4}$$

$$= [I - F_c(sI - A_c)^{-1}B_c]^{-1}.$$

Note that the open-loop uncompensated system is unobservable, with 0 being the unobservable eigenvalue (why?), while the closed-loop system is observable, i.e., the control law changed the observability of the system.

■

E. Discrete-Time Systems

Linear state feedback control for discrete-time systems is defined in a way that is analogous to the continuous-time case. The definitions are included here for purposes of completeness.

We consider a linear, time-invariant, discrete-time system described by equations of the form

$$x(k+1) = Ax(k) + Bu(k),\ y(k) = Cx(k) + Du(k), \tag{2.46}$$

where $A \in R^{n\times n}$, $B \in R^{n\times m}$, $C \in R^{p\times n}$, $D \in R^{p\times m}$, and $k \geq k_0$, with $k \geq k_0 = 0$ being typical (see Section 2.7).

DEFINITION 2.3. The linear (discrete-time, time-invariant) state feedback control law is defined by

$$u(k) = Fx(k) + r(k), \tag{2.47}$$

where $F \in R^{m\times n}$ is a gain matrix and $r(k) \in R^m$ is the external input vector. ■

This definition is similar to Definition 2.1 for the continuous-time case. The compensated closed-loop system is now given by

$$\begin{aligned} x(k+1) &= (A+BF)x(k) + Br(k) \\ y(k) &= (C+DF)x(k) + Dr(k). \end{aligned} \tag{2.48}$$

In view of Section 2.7 of Chapter 2, the system $x(k+1) = (A+BF)x(k)$ is asymptotically stable if and only if the eigenvalues of $A+BF$ satisfy $|\lambda_i| < 1$, i.e., if they lie strictly within the unit disc of the complex plane. The stabilization problem for the time-invariant case therefore becomes a problem of shifting the eigenvalues of $A+BF$, which is precisely the problem studied before for the continuous-time case. Theorem 2.1 and Lemmas 2.2 and 2.3 apply without change, and the methods developed before for eigenvalue assignment can be used here as well. The only difference in this case is the location of the desired eigenvalues: they are assigned to be within the unit circle to achieve stability. We will not repeat here the details for these results.

Input-output relations for discrete-time systems, which are in the spirit of the results developed in the preceding subsection for continuous-time systems, can be derived in a similar fashion, this time making use of the z-transform of $x(k+1) = Ax(k) + Bu(k)$, $x(0) = x_0$ to obtain

$$\hat{x}(z) = z(zI-A)^{-1}x_0 + (zI-A)^{-1}B\hat{u}(z). \tag{2.49}$$

[Compare expression (2.49) with $\hat{x}(s) = (sI-A)^{-1}x_0 + (sI-A)^{-1}B\hat{u}(s)$.] The reader is asked to derive formulas for the discrete-time case that are analogs to expressions (2.40) to (2.45) for the continuous-time case.

F. The Linear Quadratic Regulator (LQR): Discrete-Time Case

The formulation of the LQR problem in the discrete-time case is analogous to the continuous-time LQR problem. Consider the time-invariant linear system

$$x(k+1) = Ax(k) + Bu(k),\ z(k) = Mx(k), \tag{2.50}$$

where the vector $z(t)$ represents the variables to be regulated. The LQR problem is to determine a control sequence $\{u^*(k)\}$, $k \geq 0$, which minimizes the cost function

$$J(u) = \sum_{k=0}^{\infty}[z^T(k)Qz(k) + u^T(k)Ru(k)] \tag{2.51}$$

for any initial state $x(0)$, where the weighting matrices Q and R are real symmetric and positive definite.

Assume that $(A, B, Q^{1/2}M)$ is reachable and observable. Then the solution to the LQR problem is given by the linear state feedback control law

$$u^*(k) = F^*x(k) = -[R + B^T P_c^* B]^{-1} B^T P_c^* Ax(k), \tag{2.52}$$

where P_c^* is the unique, symmetric, and positive-definite solution of the *(discrete-time) algebraic Riccati equation*, given by

$$P_c = A^T[P_c - P_c B[R + B^T P_c B]^{-1} B^T P_c]A + M^T QM. \tag{2.53}$$

The minimum value of J is $J(u^*) = J_{\min} = x^T(0)P_c^* x(0)$.

As in the continuous-time case, it can be shown that the solution P_c^* can be determined from the eigenvectors of the *Hamiltonian matrix,* which in this case is

$$H = \begin{bmatrix} A + BR^{-1}B^T A^{-T} M^T QM & -BR^{-1}B^T A^{-T} \\ -A^{-T}M^T QM & A^{-T} \end{bmatrix}, \tag{2.54}$$

where it is assumed that A^{-1} exists. Variations of the above method that relax this assumption exist and can be found in the literature. Let $[V_1^T, V_2^T]^T$ be n eigenvectors corresponding to the n stable ($|\lambda| < 1$) eigenvalues of H. Note that out of the $2n$ eigenvalues of H, n of them are stable (i.e., within the unit circle) and are the reciprocals of the remaining n unstable eigenvalues (located outside the unit circle). When $(A, B, Q^{1/2}M)$ is controllable (-from-the-origin) and observable, then H has no eigenvalues on the unit circle ($|\lambda| = 1$). In fact the n stable eigenvalues of H are in this case the closed-loop eigenvalues of the optimally controlled system.

The solution to the algebraic Riccati equation is given by

$$P_c^* = V_2 V_1^{-1}. \tag{2.55}$$

As in the continuous-time case, we note that V_1 consists of the n eigenvectors of $A + BF^*$ (show this).

EXAMPLE 2.9. We consider the system $x(k+1) = Ax(k) + Bu(k)$, $y(k) = Cx(k)$, where $A = \begin{bmatrix} 0 & 1 \\ 0 & 0 \end{bmatrix}$, $B = \begin{bmatrix} 0 \\ 1 \end{bmatrix}$, $C = [1, 0]$ and we wish to determine the optimal control sequence $\{u^*(k)\}$, $k \geq 0$, that minimizes the performance index

$$J(u) = \sum_{k=0}^{\infty}(y^2(k) + \rho u^2(k)),$$

where $\rho > 0$. In (2.51), $z(k) = y(k)$, $M = C$, $Q = 1$, and $R = \rho$. The reader is asked to determine $u^*(k)$ given in (2.52) by solving the discrete-time algebraic Riccati equation (2.53) in a manner analogous to the solution in Example 2.7 (for the continuous-time algebraic Riccati equation). ■

4.3 LINEAR STATE OBSERVERS

Since the states of a system contain a great deal of useful information, there are many applications where knowledge of the state vector over some time interval is desirable. It may be possible to measure states of a system by appropriately positioned sensors. This was in fact assumed in the previous section, where the state values were multiplied by appropriate gains and then fed back to the system in the state feedback control law. Frequently, however, it may be either impossible or simply impractical to obtain measurements for all states. In particular, some of the states may not be available for measurement at all (as in the case, for example, with temperatures and pressures in inaccessible parts of a jet engine). There are also cases where it may be impractical to obtain state measurements from otherwise available states because of economic reasons (e.g., some of the sensors may be too expensive) or because of technical reasons (e.g., the environment may be too noisy for any useful measurements). Thus, there is a need to be able to estimate the values of the state of a system from available measurements, typically outputs and inputs (see Fig. 4.3).

Given the system parameters A, B, C, D and the values of the inputs and outputs over a time interval, it is possible to estimate the state when the system is observable. This problem, a problem in state estimation, is discussed in this section. In particular, we will address at length the so-called full-order and reduced-order asymptotic estimators, which are also called full-order and reduced-order observers.

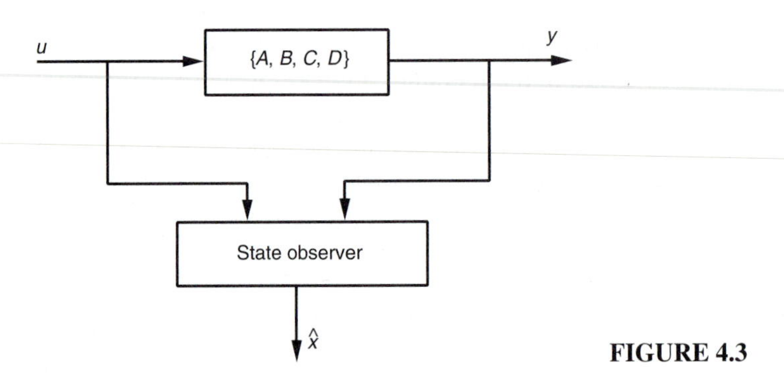

FIGURE 4.3

A. Full-Order Observers: Continuous-Time Systems

We consider systems described by equations of the form

$$\dot{x} = Ax + Bu, \qquad y = Cx + Du, \tag{3.1}$$

where $A \in R^{n \times n}$, $B \in R^{n \times m}$, $C \in R^{p \times n}$ and $D \in R^{p \times m}$.

Full-state observers: The identity observer

An estimator of the full state $x(t)$ can be constructed in the following manner. We consider the system

$$\dot{\hat{x}} = A\hat{x} + Bu + K(y - \hat{y}), \tag{3.2}$$

where $\hat{y} \triangleq C\hat{x} + Du$. Note that (3.2) can be written as

$$\dot{\hat{x}} = (A - KC)\hat{x} + [B - KD, K]\begin{bmatrix} u \\ y \end{bmatrix}, \tag{3.3}$$

which clearly reveals the role of u and y (see Fig. 4.4). The error between the actual state $x(t)$ and the estimated state $\hat{x}(t)$, $e(t) = x(t) - \hat{x}(t)$, is governed by the differential equation

$$\dot{e}(t) = \dot{x}(t) - \dot{\hat{x}}(t) = [Ax + Bu] - [A\hat{x} + Bu + KC(x - \hat{x})]$$

or

$$\dot{e}(t) = [A - KC]e(t). \tag{3.4}$$

Solving (3.4), we obtain

$$e(t) = \exp[(A - KC)t]e(0). \tag{3.5}$$

Now if the eigenvalues of $A - KC$ are in the left half-plane, then $e(t) \to 0$ as $t \to \infty$, independently of the initial condition $e(0) = x(0) - \hat{x}(0)$. This *asymptotic state estimator* is known as the *Luenberger observer.*

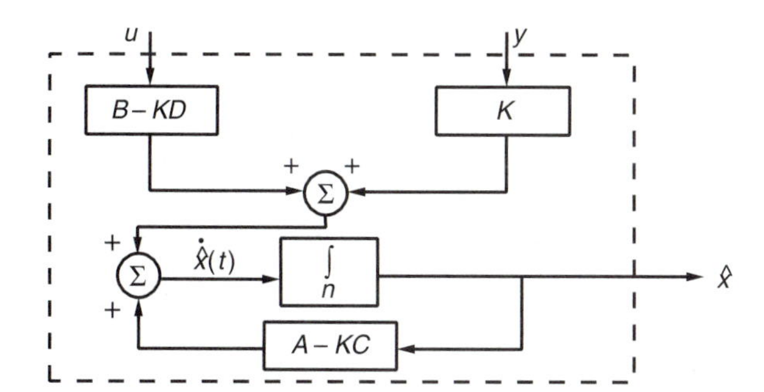

FIGURE 4.4

LEMMA 3.1. There exists $K \in R^{n \times p}$ so that the eigenvalues of $A - KC$ are assigned to arbitrary real or complex conjugate locations if and only if (A, C) is observable.

Proof. The eigenvalues of $(A - KC)^T = A^T - C^T K^T$ are arbitrarily assigned via K^T if and only if the pair (A^T, C^T) is controllable (see Theorem 2.1 of the previous section), or equivalently, if and only if the pair (A, C) is observable. ■

If (A, C) is not observable, but the unobservable eigenvalues are stable, i.e., (A, C) is *detectable,* then the error $e(t)$ will still tend to zero asymptotically. However, the unobservable eigenvalues will appear in this case as eigenvalues of $A - KC$ (show this), and they may affect the speed of the response of the estimator in an undesirable way. For example, if the unobservable eigenvalues are stable but are located close to the imaginary axis, then their corresponding modes will tend to dominate the response, most likely resulting in a state estimator that converges too slowly to the actual value of the state.

Where should the eigenvalues of $A - KC$ be located? This problem is dual to the problem of closed-loop eigenvalue placement via state feedback and is equally difficult to resolve. On one hand, the observer must estimate the state sufficiently fast, which implies that the eigenvalues should be placed sufficiently far from the imaginary axis so that the error $e(t)$ will tend to zero sufficiently fast. On the other hand, this requirement may result in a high gain K, which tends to amplify existing noise, thus reducing the accuracy of the estimate. Note that in this case, noise is the only limiting factor of how fast an estimator may be, since the gain K is realized by an algorithm and is typically implemented by means of a digital computer. Therefore, gains of any size can easily be introduced. Compare this situation with the limiting factors in the control case, which is imposed by the magnitude of the required control action (and the limits of the corresponding actuator). Typically, the faster the compensated system, the larger the required control magnitude.

One may of course balance the trade-offs between speed of response of the estimator and effects of noise by formulating an *optimal estimation problem* to derive the best K. For this, one commonly assumes certain probabilistic properties for the process. Typically, the measurement noise and the initial condition of the plant are assumed to be Gaussian random variables, and one tries to minimize a quadratic performance index. This problem is typically referred to as the Linear Quadratic Gaussian (LQG) estimation problem. This optimal estimation or filtering problem can be seen to be the dual of the quadratic optimal control problem of the previous section, a fact that will be exploited in deriving its solution. Note that the well-known *Kalman filter* is such an estimator. In the following, we shall briefly discuss the optimal estimation problem. First, however, we shall address the following related issues.

1. Is it possible to take $K = 0$ in the estimator (3.2)? Such a choice would eliminate the information contained in the term $y - \hat{y}$ from the estimator, which would now be of the form

$$\dot{\hat{x}} = A\hat{x} + Bu. \tag{3.6}$$

   In this case the estimator would operate without receiving any information on how accurate the estimate $\hat{x}$ actually is. The error $e(t) = x(t) - \hat{x}(t)$ would go to zero only when A is stable. There is no mechanism to affect the speed by which $\hat{x}(t)$ would approach $x(t)$ in this case, and this is undesirable. One could perhaps determine $x(0)$, using the methods in Section 3.3 of Chapter 3, assuming that the system is observable. Then, by setting $\hat{x}(0) = x(0)$, presumably $\hat{x}(t) = x(t)$ for all $t \geq 0$, in view of (3.6). This of course is not practical for several reasons. First, the calculated $\hat{x}(0)$ is never exactly equal to the actual $x(0)$, which implies that $e(0)$ would be nonzero. Therefore, the method would rely again on A being stable, as before, with the advantage here that $e(0)$ would be small in some sense and so $e(t) \to 0$ faster. Second, this scheme assumes that sufficient data have been collected in advance to determine (an approximation to) $x(0)$ and to initialize the estimator, which may not be possible. Third, it is assumed that this initialization process is repeated whenever the estimator is restarted, which may be impractical.
2. If derivatives of the inputs and outputs are available, then the state $x(t)$ may be determined directly (see Exercise 3.14 in Chapter 3). The estimate $\hat{x}(t)$ is in this case produced instantaneously from the values of the inputs and outputs and their derivatives. Under these circumstances, $\hat{x}(t)$ is the output of a static state estimator, as opposed to the above dynamic state estimator, which leads to a state estimate $\hat{x}(t)$ that only approaches the actual state $x(t)$ asymptotically as $t \to \infty$ $[e(t) = x(t) - \hat{x}(t) \to 0$ as $t \to \infty]$. Unfortunately, this approach is in general not viable since noise present in the measurements of $u(t)$ and $y(t)$ makes accurate calculations of the derivatives problematic, and since errors in $u(t)$, $y(t)$

and their derivatives are not smoothed by the algebraic equations of the static estimator (as opposed to the smoothing effects introduced by integration in dynamic systems). It follows that in this case the state estimates may be erroneous.

EXAMPLE 3.1. Consider the observable pair

$$A = \begin{bmatrix} 0 & 1 & 0 \\ 0 & 0 & 1 \\ 0 & 2 & -1 \end{bmatrix}, \qquad C = [1, 0, 0].$$

We wish to assign the eigenvalues of $A - KC$ in a manner that enables us to design a full-order/full-state asymptotic observer. Let the desired characteristic polynomial be $\alpha_d(s) = s^3 + d_2 s^2 + d_1 s + d_0$ and consider

$$A_D = A^T = \begin{bmatrix} 0 & 0 & 0 \\ 1 & 0 & 2 \\ 0 & 1 & -1 \end{bmatrix} \quad \text{and} \quad B_D = C^T = \begin{bmatrix} 1 \\ 0 \\ 0 \end{bmatrix}.$$

To reduce (A_D, B_D) to controller form, we consider

$$\mathscr{C} = [B_D, A_D B_D, A_D^2 B_D] = \begin{bmatrix} 1 & 0 & 0 \\ 0 & 1 & 0 \\ 0 & 0 & 1 \end{bmatrix} = \mathscr{C}^{-1}.$$

Then $$P = \begin{bmatrix} q \\ qA_D \\ qA_D^2 \end{bmatrix} = \begin{bmatrix} 0 & 0 & 1 \\ 0 & 1 & -1 \\ 1 & -1 & 3 \end{bmatrix} \quad \text{and} \quad P^{-1} = \begin{bmatrix} -2 & 1 & 1 \\ 1 & 1 & 0 \\ 1 & 0 & 0 \end{bmatrix},$$

from which we obtain

$$A_{D_c} = P A_D P^{-1} = \begin{bmatrix} 0 & 1 & 0 \\ 0 & 0 & 1 \\ 0 & 2 & -1 \end{bmatrix} \quad \text{and} \quad B_{D_c} = P B_D = \begin{bmatrix} 0 \\ 0 \\ 1 \end{bmatrix}.$$

The state feedback is then given by $F_{D_c} = B_m^{-1}[A_{d_m} - A_m] = [-d_0, -d_1 - 2, -d_2 + 1]$ and $F_D = F_{D_c} P = [-d_2 + 1, d_2 - d_1 - 3, d_1 - d_0 - 3d_2 + 5]$. Then

$$K = -F_D^T = [d_2 - 1, d_1 - d_2 + 3, d_0 - d_1 + 3d_2 - 5]^T$$

assigns the eigenvalues of $A - KC$ at the roots of $\alpha_d(s) = s^3 + d_2 s^2 + d_1 s + d_0$. Note that the same result could also have been derived using the direct method for eigenvalue assignment, using $|sI - (A - (k_0, k_1, k_2)^T C)| = \alpha_d(s)$. Also, the result could have been derived using the *observable version of Ackermann's formula,* namely,

$$K = -F_D^T = \alpha_d(A)\mathscr{O}^{-1} e_n,$$

where $F_D = -e_n^T \mathscr{C}_D^{-1} \alpha_d(A_D)$ from (2.21). Note that the given system has eigenvalues at 0, 1, -2 and is therefore unstable. The observer derived in this case will be used in the next section (Example 4.1) in combination with state feedback to stabilize the system $\dot{x} = Ax + Bu$, $y = Cx$, where

$$A = \begin{bmatrix} 0 & 1 & 0 \\ 0 & 0 & 1 \\ 0 & 2 & -1 \end{bmatrix}, \quad B = \begin{bmatrix} 0 & 1 \\ 1 & 1 \\ 0 & 0 \end{bmatrix}, \quad \text{and} \quad C = [1, 0, 0]$$

(see Example 2.5), using only output measurements. ■

EXAMPLE 3.2. Consider the system $\dot{x} = Ax$, $y = Cx$, where $A = \begin{bmatrix} 0 & -2 \\ 1 & -2 \end{bmatrix}$ and $C = [0, 1]$, and where (A, C) is in observer form. It is easy to show that $K = [d_0 - 2, d_1 - 2]^T$

assigns the eigenvalues of $A - KC$ at the roots of $s^2 + d_1 s + d_0$. To verify this, note that

$$det(sI - (A - KC)) = det\left(\begin{bmatrix} s & 0 \\ 0 & s \end{bmatrix} - \begin{bmatrix} 0 & -d_0 \\ 1 & -d_1 \end{bmatrix}\right) = s^2 + d_1 s + d_0.$$

The error $e(t) = x(t) - \hat{x}(t)$ is governed by the equation $\dot{e}(t) = (A - KC)e(t)$ given in (3.4). Noting that the eigenvalues of A are $-1 \pm j$, select different sets of eigenvalues for the observer and plot the states $x(t)$, $\hat{x}(t)$ and the error $e(t)$ for $x(0) = [2, 2]^T$ and $\hat{x}(0) = [0, 0]^T$. The further away the eigenvalues of the observer are selected from the imaginary axis (with negative real parts), the larger the gains in K will become and the faster $\hat{x}(t) \to x(t)$ (verify this). ■

Partial or linear functional state observers

The state estimator studied above is a full-state estimator or observer, i.e., $\hat{x}(t)$ is an estimate of the full-state vector $x(t)$. There are cases where only part of the state vector, or a linear combination of the states, is of interest. In control problems, for example, $F\hat{x}(t)$ is used and fed back, instead of $Fx(t)$, where F is an $m \times n$ state feedback gain matrix (see also Section 4.4). An interesting question that arises at this point is: is it possible to estimate directly a linear combination of the state, say, Tx, where $T \in R^{\tilde{n} \times n}$, $\tilde{n} \leq n$? This problem is considered next.

We consider the system

$$\dot{z} = \tilde{A}z + \tilde{B}u + Ky, \tag{3.7}$$

where $\tilde{A} \in R^{\tilde{n} \times \tilde{n}}$, $\tilde{B} \in R^{\tilde{n} \times m}$, and $K \in R^{\tilde{n} \times p}$ are to be determined. The error equation is given by

$$\begin{aligned} \dot{\tilde{e}} &= T\dot{x} - \dot{z} = TAx + TBu - \tilde{A}z - \tilde{B}u - K(Cx + Du) \\ &= \tilde{A}\tilde{e} + (TA - \tilde{A}T - KC)x + (TB - KD - \tilde{B})u. \end{aligned}$$

Now if

$$\tilde{B} = TB - KD \tag{3.8}$$

and T, $\tilde{A}$, and K satisfy

$$TA - \tilde{A}T = KC, \tag{3.9}$$

then $\dot{\tilde{e}} = \tilde{A}\tilde{e}$. If in addition, $\tilde{A}$ is stable, then $\tilde{e}(t) \to 0$ as $t \to \infty$ and $z(t)$ will approach $Tx(t)$ asymptotically. A key issue is clearly the existence (and calculation) of appropriate solutions of (3.9). This has been studied extensively in the literature (see for example O'Reilly [18]). Here we simply wish to point to a special case of (3.9), namely, the case of the identity observer ($T = I$) or of the full-state estimator: $T = I$ and $\tilde{A} = A - KC$, where for (A, C) observable, there always exists K that renders $\tilde{A}$ stable, as was shown above.

Note that in general a solution $(\tilde{A}, K)$ of (3.9) with $\tilde{A}$ stable may not exist, i.e., there may not exist an observer (3.7) of order $\tilde{n}(\tilde{A} \in R^{\tilde{n} \times \tilde{n}})$ to estimate $\tilde{n}$ linear functions of the state, Tx ($T \in R^{\tilde{n} \times n}$). It is possible, however, to decouple the order of the observer from the number of linear functions of the state to be estimated, in the following manner.

Let $w = Tx$, $w \in R^{\tilde{n}}$, and consider

$$\begin{aligned} \dot{z} &= \hat{A}z + \hat{B}u + \hat{K}y \\ \hat{w} &= T_1(y - Du) + T_2 z, \end{aligned} \tag{3.10}$$

where $z \in R^r$, $\hat{A} \in R^{r\times r}$, $\hat{B} \in R^{r\times m}$, $\hat{K} \in R^{r\times p}$, $T_1 \in R^{\tilde{n}\times p}$, and $T_2 \in R^{\tilde{n}\times r}$. Let $\hat{T} \in R^{r\times n}$ and write

$$\begin{aligned}\dot{z} - \hat{T}\dot{x} &= \hat{A}z + \hat{B}u + \hat{K}(Cx + Du) - \hat{T}(Ax + Bu)\\ &= \hat{A}z + (\hat{K}C - \hat{T}A)x + (\hat{B} - \hat{T}B + \hat{K}D)u.\end{aligned}$$

Now if $\hat{B} = \hat{T}B - \hat{K}D$ and $\hat{K}C - \hat{T}A = -\hat{A}\hat{T}$, where $\hat{A}$ is stable, then $\dot{z} - \hat{T}\dot{x} = \hat{A}(z - \hat{T}x)$ or $z - \hat{T}x = e^{\hat{A}t}(z(0) - \hat{T}x(0))$, i.e., $z(t) \to \hat{T}x(t)$. We are interested, however, in estimates of $w = Tx$. Consider therefore $\hat{w} - w = [T_1Cx + T_2(\hat{T}x + e^{\hat{A}t}(z(0) - \hat{T}x(0))] - Tx = (T_1C + T_2\hat{T} - T)x + T_2e^{\hat{A}t}(z(0) - \hat{T}x(0))$. Now if $T = T_1C + T_2\hat{T}$, then $\hat{w} - w = T_2e^{\hat{A}t}(z(0) - \hat{T}x(0))$ and $\hat{w}(t) \to w(t) = Tx(t)$, since $\hat{A}$ is stable.

Therefore, an observer (3.10) of $w = Tx$, $T \in R^{\tilde{n}\times n}$ of order r exists if there exist $\hat{T} \in R^{r\times n}$, $T_1 \in R^{\tilde{n}\times r}$, $T_2 \in R^{\tilde{n}\times r}$, and $\hat{K} \in R^{r\times p}$ such that for $\hat{A}$ stable,

$$\hat{T}A - \hat{A}\hat{T} = \hat{K}C \qquad \text{and} \qquad T_1C + T_2\hat{T} = T. \tag{3.11}$$

We thus have $\hat{B} = \hat{T}B - \hat{K}D$. We note that it can be shown that for (A, C) detectable and r sufficiently large, there will always exist a solution. An example is the case when $T = I$ and $r = n - p$. This is the case of the reduced-order observer discussed next, in Subsection B.

Another simple case of interest is when T is a row vector ($\tilde{n} = 1$). In this case it can be shown (Luenberger [14]) that an observer of order $\nu - 1$ [ν is the observability index of (A, C)] can always be constructed with its eigenvalues freely assignable, which will asymptotically estimate the linear function of the state, Tx. In general it can be shown that an observer of order $\tilde{n}(\nu - 1)$ can be constructed that will asymptotically estimate Tx. Note that the problem of determining an observer of Tx of minimum order is a very difficult problem, except in special cases.

B. Reduced-Order Observers: Continuous-Time Systems

Suppose that p states, out of the n state, can be measured directly. This information can then be used to reduce the order of the full-state estimator from n to $n-p$. Similar results are true for the estimator of a linear function of the state, but this problem will not be addressed here. To determine a full-state estimator of order $n-p$, first consider the case when $C = [I_p, 0]$. In particular, let

$$\begin{bmatrix}\dot{x}_1\\ \dot{x}_2\end{bmatrix} = \begin{bmatrix}A_{11} & A_{12}\\ A_{21} & A_{22}\end{bmatrix}\begin{bmatrix}x_1\\ x_2\end{bmatrix} + \begin{bmatrix}B_1\\ B_2\end{bmatrix}u$$

$$z = [I_p, 0]\begin{bmatrix}x_1\\ x_2\end{bmatrix}, \tag{3.12}$$

where $z = x_1$ represents the p measured states. Therefore, only $x_2(t) \in R^{n-p}$ is to be estimated. The system whose state is to be estimated is now given by

$$\begin{aligned}\dot{x}_2 &= A_{22}x_2 + [A_{21}, B_2]\begin{bmatrix}x_1\\ u\end{bmatrix}\\ &= A_{22}x_2 + \tilde{B}\tilde{u},\end{aligned} \tag{3.13}$$

where $\tilde{B} \triangleq [A_{21}, B_2]$ and $\tilde{u} \triangleq \begin{bmatrix} x_1 \\ u \end{bmatrix} = \begin{bmatrix} z \\ u \end{bmatrix}$ is a known signal. Also,

$$\tilde{y} \triangleq \dot{x}_1 - A_{11}x_1 - B_1u = A_{12}x_2, \tag{3.14}$$

where $\tilde{y}$ is known. An estimator for x_2 can now be constructed. In particular, in view of (3.2), we have that the system

$$\begin{aligned} \dot{\hat{x}}_2 &= A_{22}\hat{x}_2 + \tilde{B}\tilde{u} + \tilde{K}(\tilde{y} - A_{12}\hat{x}_2) \\ &= (A_{22} - \tilde{K}A_{12})\hat{x}_2 + (A_{21}z + B_2u) + \tilde{K}(\dot{z} - A_{11}z - B_1u) \end{aligned} \tag{3.15}$$

is an asymptotic state estimator for x_2. Note that the error e satisfies the equation

$$\dot{e} = \dot{x}_2 - \dot{\hat{x}}_2 = (A_{22} - \tilde{K}A_{12})e, \tag{3.16}$$

and if (A_{22}, A_{12}) is observable, then the eigenvalues of $A_{22} - \tilde{K}A_{12}$ can be arbitrarily assigned making use of $\tilde{K}$. It can be shown that if the pair $(A = [A_{ij}], C = [I_p, 0])$ is observable, then (A_{22}, A_{12}) is also observable (prove this using the eigenvalue observability test of Section 3.4 of Chapter 3). System (3.15) is an estimator of order $n - p$, and therefore the estimate of the entire state x is $\begin{bmatrix} z \\ \hat{x}_2 \end{bmatrix}$. To avoid using $\dot{z} = \dot{x}_1$ in $\tilde{y}$ given by (3.14), one could use $\hat{x}_2 = w + \tilde{K}z$ and obtain from (3.15) an estimator in terms of w, z, and u. In particular,

$$\dot{w} = (A_{22} - \tilde{K}A_{12})w + [(A_{22} - \tilde{K}A_{12})\tilde{K} + A_{21} - \tilde{K}A_{11}]z + [B_2 - \tilde{K}B_1]u. \tag{3.17}$$

Then w is an estimate of $\hat{x}_2 - \tilde{K}z$, and of course $w + \tilde{K}z$ is an estimate for $\hat{x}_2$ (verify this).

In the above derivation, it was assumed for simplicity that a part of the state x_1, is measured directly, i.e., $C = [I_p, 0]$. One could also derive a reduced-order estimator for the system

$$\dot{x} = Ax + Bu, \; y = Cx.$$

To see this, let rank $C = p$ and define a similarity transformation matrix $P = \begin{bmatrix} C \\ \hat{C} \end{bmatrix}$, where $\hat{C}$ is such that P is nonsingular. Then

$$\dot{\bar{x}} = \bar{A}\bar{x} + \bar{B}u, \; y = \bar{C}\bar{x} = [I_p, 0]\bar{x}, \tag{3.18}$$

where $\bar{x} = Px$, $\bar{A} = PAP^{-1}$, $\bar{B} = PB$, and $\bar{C} = CP^{-1} = [I_p, 0]$ (show this). The transformed system is now in an appropriate form for an estimator of order $n - p$ to be derived, using the procedure discussed above. The estimate of $\bar{x}$ is $\begin{bmatrix} y \\ \hat{\bar{x}}_2 \end{bmatrix}$ and the estimate of the original state x is $P^{-1}\begin{bmatrix} y \\ \hat{\bar{x}}_2 \end{bmatrix}$. In particular, $\bar{x}_2 = w + \tilde{K}y$, where w satisfies (3.17) with $z = y$, $[A_{ij}] = \bar{A} = PAP^{-1}$, and $\begin{bmatrix} B_1 \\ B_2 \end{bmatrix} = \bar{B} = PB$. The interested reader should verify this result.

EXAMPLE 3.3. Consider the system $\dot{x} = Ax + Bu$, $y = Cx$, where $A = \begin{bmatrix} 0 & -2 \\ 1 & -2 \end{bmatrix}$, $B = \begin{bmatrix} 0 \\ 1 \end{bmatrix}$, and $C = [0, 1]$. We wish to design a reduced $n - p = n - 1 = 2 - 1 =$ first-order asymptotic state estimator.

The similarity transformation matrix $P = \begin{bmatrix} C \\ \hat{C} \end{bmatrix} = \begin{bmatrix} 0 & 1 \\ 1 & 0 \end{bmatrix}$ leads to (3.18), where $\bar{x} = Px$ and $\bar{A} = PAP^{-1} = \begin{bmatrix} 0 & 1 \\ 1 & 0 \end{bmatrix}\begin{bmatrix} 0 & -2 \\ 1 & -2 \end{bmatrix}\begin{bmatrix} 0 & 1 \\ 1 & 0 \end{bmatrix} = \begin{bmatrix} -2 & 1 \\ -2 & 0 \end{bmatrix}$, $\bar{B} = PB = \begin{bmatrix} 1 \\ 0 \end{bmatrix}$, and $\bar{C} = CP^{-1} = [1, 0]$. The system $\{\bar{A}, \bar{B}, \bar{C}\}$ is now in an appropriate form for use of (3.17). We have $\bar{A} = \begin{bmatrix} A_{11} & A_{12} \\ A_{21} & A_{22} \end{bmatrix} = \begin{bmatrix} -2 & 1 \\ -2 & 0 \end{bmatrix}$, $\bar{B} = \begin{bmatrix} B_1 \\ B_2 \end{bmatrix} = \begin{bmatrix} 1 \\ 0 \end{bmatrix}$, and (3.17) assumes the form

$$\dot{w} = (-\tilde{K})w + [-\tilde{K}^2 + (-2) - \tilde{K}(-2)]y + (-\tilde{K})u,$$

a system observer of order 1.

For $\tilde{K} = -10$ we have $\dot{w} = 10w - 122y + 10u$, and $w + \tilde{K}y = w - 10y$ is an estimate for $\hat{\bar{x}}_2$. Therefore, $\begin{bmatrix} y \\ w - 10y \end{bmatrix}$ is an estimate of $\bar{x}$, and

$$P^{-1}\begin{bmatrix} y \\ w - 10y \end{bmatrix} = \begin{bmatrix} 0 & 1 \\ 1 & 0 \end{bmatrix}\begin{bmatrix} y \\ w - 10y \end{bmatrix} = \begin{bmatrix} w - 10y \\ y \end{bmatrix}$$

is an estimate of $x(t)$ for the original system. ■

C. Optimal State Estimation: Continuous-Time Systems

The gain K in the estimator (3.2) above can be determined so that it is optimal in an appropriate sense. This is discussed very briefly in the following. The interested reader should consult the extensive literature on *filtering theory* for additional information, in particular, the literature on the *Kalman-Bucy filter.*

In addressing optimal state estimation, noise with certain statistical properties is introduced in the model and an appropriate cost functional is set up that is then minimized. In the following, we shall introduce some of the key equations of the *Kalman-Bucy filter* and we will point out the duality between the optimal control and estimation problems. We concentrate on the time-invariant case, although, as in the LQR control problem discussed earlier, more general results for the time-varying case do exist.

We consider the linear time-invariant system

$$\dot{x} = Ax + Bu + \Gamma w, \; y = Cx + v, \tag{3.19}$$

where w and v represent process and measurement noise terms. Both w and v are assumed to be white, zero-mean Gaussian stochastic processes, i.e., they are uncorrelated in time and have expected values $E[w] = 0$ and $E[v] = 0$. Let

$$E[ww^T] = W, \qquad E[vv^T] = V \tag{3.20}$$

denote their covariances, where W and V are real, symmetric, and positive definite matrices, i.e., $W = W^T$, $W > 0$, and $V = V^T$, $V > 0$. Assume that the noise processes w and v are independent, i.e., $E[wv^T] = 0$. Also assume that the initial state $x(0)$ of the plant is a Gaussian random variable of known mean, $E[x(0)] = x_0$, and known covariance, $E[(x(0) - x_0)(x(0) - x_0)^T] = P_{e0}$. Assume also that $x(0)$ is independent of w and v. Note that all these are typical assumptions made in practice.

Consider now the estimator (3.2), namely,

$$\dot{\hat{x}} = A\hat{x} + Bu + K(y - C\hat{x}) = (A - KC)\hat{x} + Bu + Ky, \tag{3.21}$$

ns

and let $(A, \Gamma W^{1/2}, C)$ be controllable (-from-the-origin) and observable. It turns out that the error covariance $E[(x - \hat{x})(x - \hat{x})^T]$ is minimized when the filter gain is given by

$$K^* = P_e^* C^T V^{-1}, \tag{3.22}$$

where P_e^* denotes the symmetric, positive definite solution of the quadratic (*dual*) *algebraic Riccati equation*

$$P_e A^T + A P_e - P_e C^T V^{-1} C P_e + \Gamma W \Gamma^T = 0. \tag{3.23}$$

Note that P_e^*, which is in fact the minimum error covariance, is the positive semidefinite solution of the above Riccati equation if $(A, \Gamma W^{1/2}, C)$ is stabilizable and detectable. The optimal estimator is asymptotically stable.

The above algebraic Riccati equation is the dual to the Riccati equation given in (2.34) for optimal control and can be obtained from (2.34) making use of the substitutions

$$A \to A^T, B \to C^T, M \to \Gamma^T \quad \text{and} \quad R \to V, Q \to W. \tag{3.24}$$

Clearly, methods that are analogous to the ones developed by solving the control Riccati equation (2.34) may be applied to solve the Riccati equation (3.21) in filtering. These methods are not discussed here.

EXAMPLE 3.4 Consider the system $\dot{x} = Ax$, $y = Cx$, where $A = \begin{bmatrix} 0 & 0 \\ 1 & 0 \end{bmatrix}$, $C = [0, 1]$, and let $\Gamma = \begin{bmatrix} 1 \\ 0 \end{bmatrix}$, $V = \rho > 0$, $W = 1$. We wish to derive the optimal filter $K^* = P_e^* C^T V^{-1}$ given in (3.22). In this case, the Riccati equation (3.23) is precisely the Riccati equation of the control problem given in Example 2.7. The solution of this equation was determined to be

$$P_e^* = \begin{bmatrix} \sqrt{2\sqrt{\rho}} & \sqrt{\rho} \\ \sqrt{\rho} & \sqrt{2\rho\sqrt{\rho}} \end{bmatrix}.$$

We note that this was expected, since our example was chosen to satisfy (3.24). Therefore,

$$K^* = P_e^* \begin{bmatrix} 0 \\ 1 \end{bmatrix} \frac{1}{\rho} = \begin{bmatrix} \sqrt{\rho} \\ \sqrt{2\rho\sqrt{\rho}} \end{bmatrix} \frac{1}{\rho}.$$

■

D. Full-Order Observers: Discrete-Time Systems

We consider described by equations of the form

$$x(k+1) = Ax(k) + Bu(k), \qquad y = Cx(k) + Du(k), \tag{3.25}$$

where $A \in R^{n \times n}$, $B \in R^{n \times m}$, $C \in R^{p \times m}$, and $D \in R^{p \times m}$.

The construction of state estimators for discrete-time systems is mostly analogous to the continuous-time case, and the results that we established above for such systems are valid in here as well, subject to obvious adjustments and modifications. There are, however, some notable differences. For example, in discrete-time systems it is possible to construct a state estimator that converges to the true value of the state in finite time, instead of infinite time as in the case of asymptotic state estimators.

This is the estimator known as the *deadbeat observer*. Furthermore, in discrete-time systems it is possible to talk about current state estimators, in addition to prediction state estimators. In what follows, a brief description of the results that are analogous to the continuous-time case are given. Current estimators and deadbeat observers that are unique to the discrete-time case are discussed at greater length.

Full-state observers: The identity observer

As in the continuous-time case, following (3.2) we consider systems described by equations of the form

$$\hat{x}(k+1) = A\hat{x}(k) + Bu(k) + K[y(k) - \hat{y}(k)], \tag{3.26}$$

where $\hat{y}(k) \triangleq C\hat{x}(k) + Dx(k)$. This can also be written as

$$\hat{x}(k+1) = (A - KC)\hat{x}(k) + [B - KD, K]\begin{bmatrix} u(k) \\ y(k) \end{bmatrix}. \tag{3.27}$$

It can be shown that the error $e(k) \triangleq x(k) - \hat{x}(k)$ obeys the equation $e(k+1) = (A-KC)e(k)$. Therefore, if the eigenvalues of $A-KC$ are inside the open unit disc of the complex plane, then $e(k) \to 0$ as $k \to \infty$. There exists K so that the eigenvalues of $A-KC$ can be arbitrarily assigned if and only if the pair (A, C) is observable (see Lemma 3.1).

The discussion following Lemma 3.1 for the case when (A, C) is not completely observable, although detectable, is still valid. Also, the remarks on appropriate locations for the eigenvalues of $A - KC$ and noise being the limiting factor in state estimators are also valid in the present case. Note that the latter point should seriously be considered when deciding whether or not to use the deadbeat observer described next.

To balance the trade-offs between speed of the estimator response and noise amplification, one may formulate an *optimal estimation problem* as was done in the continuous-time case, the *Linear Quadratic Gaussian* (LQG) design being a common formulation. The Kalman filter (discrete-time case) which is based on the "current estimator" described below is such a quadratic estimator. The LQG optimal estimation problem can be seen to be the dual of the quadratic optimal control problem discussed in the previous section. As in the continuous-time case, optimal estimation in the discrete-time case will be discussed only briefly in the following. First, however, several other related issues are addressed.

Deadbeat observer. If the pair (A, C) is observable, it is possible to select K so that all the eigenvalues of $A-KC$ are at the origin. In this case $e(k) = x(k) - \hat{x}(k) = (A-KC)^k e(0) = 0$, for some $k \le n$; i.e., the error will be identically zero within at most n steps. The minimum value of k for which $(A-KC)^k = 0$ depends on the size of the largest block on the diagonal of the Jordan canonical form of $A - KC$. (Refer to the discussion on the modes of discrete-time systems in Section 2.7 of Chapter 2.)

EXAMPLE 3.5. Consider the system $x(k+1) = Ax(k)$, $y(k) = Cx(k)$, where

$$A = \left[\begin{array}{cc:c} 0 & 2 & 1 \\ 1 & -1 & 0 \\ \hdashline 0 & 0 & 0 \end{array}\right], \qquad C = \left[\begin{array}{cc:c} 0 & 1 & 0 \\ 0 & 1 & 1 \end{array}\right]$$

is in observer form. We wish to design a deadbeat observer. It is rather easy to show (compare with Example 2.5) that

$$K = \left[A_{d_m}^T - \begin{bmatrix} 2 & 1 \\ -1 & 0 \\ 0 & 0 \end{bmatrix} \right] \begin{bmatrix} -1 & 0 \\ 1 & -1 \end{bmatrix},$$

which was determined by taking the dual $A_D = A^T$, $B_D = C^T$ in controller form, using $F_D = B_m^{-1}[A_{d_m} - A_m]$ and $K = -F_D^T$.

The matrix $A_{d_m}^T$ consists of the second and third columns of a matrix $A_d = \begin{bmatrix} 0 & \times & \vdots & \times \\ 1 & \times & \vdots & \times \\ 0 & \times & \vdots & \times \end{bmatrix}$ in observer (companion) form with all its eigenvalues at 0. For $A_{d_1} = \begin{bmatrix} 0 & 0 & 0 \\ 1 & 0 & 0 \\ 0 & 1 & 0 \end{bmatrix}$, we have

$$K_1 = \left[\begin{bmatrix} 0 & 0 \\ 0 & 0 \\ 1 & 0 \end{bmatrix} - \begin{bmatrix} 2 & 1 \\ -1 & 0 \\ 0 & 0 \end{bmatrix} \right] \begin{bmatrix} -1 & 0 \\ 1 & -1 \end{bmatrix} = \begin{bmatrix} 1 & 1 \\ -1 & 0 \\ -1 & 0 \end{bmatrix},$$

and for $A_{d_2} = \begin{bmatrix} 0 & 0 & 0 \\ 1 & 0 & 0 \\ 0 & 0 & 0 \end{bmatrix}$, we obtain

$$K_2 = \left[\begin{bmatrix} 0 & 0 \\ 0 & 0 \\ 0 & 0 \end{bmatrix} - \begin{bmatrix} 2 & 1 \\ -1 & 0 \\ 0 & 0 \end{bmatrix} \right] \begin{bmatrix} -1 & 0 \\ 1 & -1 \end{bmatrix} = \begin{bmatrix} 1 & 1 \\ -1 & 0 \\ 0 & 0 \end{bmatrix}.$$

Note that $A - K_1C = A_{d_1}$, $A_{d_1}^2 = \begin{bmatrix} 0 & 0 & 0 \\ 0 & 0 & 0 \\ 1 & 0 & 0 \end{bmatrix}$ and $A_{d_1}^3 = 0$, and $A - K_2C = A_{d_2}$, and $A_{d_2}^2 = 0$. Therefore, for the observer gain K_1, the error $e(k)$ in the deadbeat observer will become zero in $n = 3$ steps since $e(3) = (A - K_1C)^3 e(0) = 0$. For the observer gain K_2, the error $e(k)$ in the deadbeat observer will become zero in $2 < n$ steps, since $e(2) = (A - K_2C)^2 e(0) = 0$. The reader should determine the Jordan canonical forms of A_{d_1} and A_{d_2} and verify that the dimension of the largest block on the diagonal is 3 and 2, respectively. ■

The comments in the discussion following Lemma 3.1 on taking $K = 0$ are valid in the discrete-time case as well. Also, the approach of determining the state instantaneously in the continuous-time case, using the derivatives of the input and output, corresponds in the discrete-time case to determining the state from current and future input and output values (see Exercise 3.14 in Chapter 3). This approach was in fact used to determine $x(0)$ when studying observability in Section 3.3. The disadvantage of this method is that it requires future measurements to calculate the current state. This issue of using future or past measurements to determine the current state is elaborated upon next.

Current estimator

The estimator (3.26) is called a *prediction estimator*. The state estimate $\hat{x}(k)$ is based on measurements up to and including $y(k-1)$. It is often of interest in applications to determine the state estimate $\hat{x}(k)$ based on measurements up to and including $y(k)$. This may seem rather odd at first; however, if the computation time required to calculate $\hat{x}(k)$ is short compared to the sample period in a sampled-data system, then it is certainly possible practically to determine the estimate $\hat{x}(k)$ before $x(k+1)$ and $y(k+1)$ are generated by the system. If this state estimate, which is based on current measurements of $y(k)$, is to be used to control the system, then the unavoidable computational delays should be taken into consideration.

Now let $\bar{x}(k)$ denote the current state estimate based on measurements up through $y(k)$. Consider the current estimator

$$\bar{x}(k) = \hat{x}(k) + K_c(y(k) - C\hat{x}(k)), \tag{3.28}$$

where

$$\hat{x}(k) = A\bar{x}(k-1) + Bu(k-1), \tag{3.29}$$

i.e., $\hat{x}(k)$ denotes the estimate based on model prediction from the previous time estimate, $\bar{x}(k-1)$. Note that in (3.28), the error is $y(k) - \hat{y}(k)$, where $\hat{y}(k) = C\hat{x}(k)$ $(D = 0)$, for simplicity.

Combining the above, we obtain

$$\hat{x}(k) = (I - K_cC)A\bar{x}(k-1) + [(I - K_cC)B, -K_c]\begin{bmatrix} u(k-1) \\ y(k) \end{bmatrix}. \tag{3.30}$$

The relation to the prediction estimator (3.26) can be seen by substituting (3.28) into (3.29) to obtain

$$\hat{x}(k+1) = A\hat{x}(k) + Bu(k) + AK_c[y(k) - C\hat{x}(k)]. \tag{3.31}$$

Comparison with the prediction estimator (3.26) (with $D = 0$) shows that if

$$K = AK_c, \tag{3.32}$$

then (3.31) is indeed the prediction estimator, and the estimate $\hat{x}(k)$ used in the current estimator (3.28) is indeed the prediction state estimate. In view of this, we expect to obtain for the error $\hat{e}(k) = x(k) - \hat{x}(k)$ the difference equation

$$\hat{e}(k+1) = (A - AK_cC)\hat{e}(k) \tag{3.33}$$

(show this). To determine the error $\bar{e}(k) = x(k) - \bar{x}(k)$ we note that $\bar{e}(k) = \hat{e}(k) - (\bar{x}(k) - \hat{x}(k))$. Equation (3.28) now implies that $\bar{x}(k) - \hat{x}(k) = K_cCe(k)$. Therefore,

$$\bar{e}(k) = (I - K_cC)\hat{e}(k). \tag{3.34}$$

This establishes the relationship between errors in current and prediction estimators.

Premultiplying (3.31) by $I - K_cC$ (assuming $|I - K_cC| \neq 0$), we obtain

$$\bar{e}(k+1) = (A - K_cCA)\bar{e}(k), \tag{3.35}$$

which is the current estimator error equation. The gain K_c is chosen so that the eigenvalues of $A - K_cCA$ are within the open unit disc of the complex plane. The pair (A, CA) must be observable for arbitrary eigenvalue assignment. Note that the two error equations (3.33) and (3.35) have identical eigenvalues (show this).

EXAMPLE 3.6. Consider the system $x(k+1) = Ax(k)$, $y(k) = Cx(k)$, where $A = \begin{bmatrix} 0 & -2 \\ 1 & -2 \end{bmatrix}$, $C = [0, 1]$, which is in observer form (see also Example 3.2). We wish to design a current estimator. In view of the error equation (3.35), we consider

$$
\begin{aligned}
det(sI - (A - K_cCA)) &= det\left(\begin{bmatrix} s & 0 \\ 0 & s \end{bmatrix} - \left(\begin{bmatrix} 0 & -2 \\ 1 & -2 \end{bmatrix} - \begin{bmatrix} k_0 \\ k_1 \end{bmatrix}[1 \ -2]\right)\right) \\
&= det\begin{bmatrix} s + k_0 & 2 - 2k_0 \\ k_1 - 1 & s + 2 - 2k_1 \end{bmatrix} \\
&= s^2 + s(2 - 2k_1 + k_0) + (2 - 2k_1) \\
&= s^2 + d_1 s + d_0 = \alpha_d(s),
\end{aligned}
$$

a desired polynomial, from which $K_c = [k_0, k_1]^T = [d_1 - d_0, \frac{1}{2}(2 - d_0)]^T$. Note that $AK_c = [d_0 - 2, d_1 - 2]^T = K$, found in Example 3.2, as noted in (3.32).

The current estimator (3.30) is now given by $\bar{x}(k) = (A - K_cCA)\bar{x}(k-1) - K_cCBu(k-1) + K_cy(k)$, or

$$
\bar{x}(k) = \begin{bmatrix} -k_0 & -2 + 2k_0 \\ 1 - k_1 & -2 + 2k_1 \end{bmatrix} \bar{x}(k-1) + \begin{bmatrix} k_0 \\ k_1 \end{bmatrix} y(k). \qquad \blacksquare
$$

Partial or linear functional state observers

The problem of estimating a linear function of the state, $Tx(k)$, $T \in R^{\tilde{n} \times n}$, where $\tilde{n} \le n$, using a prediction estimator, is completely analogous to the continuous-time case and will therefore not be discussed further here.

E. Reduced-Order Observers: Discrete-Time Systems

It is possible to estimate the full state $x(k)$ using an estimator of order $n - p$, where $p = rank\ C$. If a prediction estimator is used for that part of the state that needs to be estimated, then the problem in the discrete-time case is completely analogous to the continuous-time case, discussed before. We will omit the details.

F. Optimal State Estimation: Discrete-Time Systems

The formulation of the Kalman filtering problem in discrete-time is analogous to the continuous-time case.

Consider the linear time-invariant system given by

$$
x(k+1) = Ax(k) + Bu(k) + \Gamma w(k), \qquad y(k) = Cx(k) + v, \tag{3.36}
$$

where the process and measurement noises w, v are white, zero-mean Gaussian stochastic processes, i.e., they are uncorrelated in time with $E[w] = 0$, and $E[v] = 0$. Let the covariances be given by

$$
E[ww^T] = W, \qquad E[vv^T] = V, \tag{3.37}
$$

where $W = W^T, W > 0$ and $V = V^T, V > 0$. Assume that w, v are independent, that the initial state $x(0)$ is Gaussian of known mean ($E[x(0)] = x_0$), that $E[(x(0) - x_0)(x(0) - x_0)^T] = P_{e0}$, and that $x(0)$ is independent of w and v.

Consider now the current estimator (3.26), namely,

$$
\bar{x}(k) = \hat{x}(k) + K_c[y(k) - C\hat{x}(k)],
$$

where $\hat{x}(k) = A\bar{x}(k-1) + Bu(k-1)$, and $\hat{x}(k)$ denotes the prior estimate of the state at the time of a measurement.

It turns out that the state error covariance is minimized when the filter gain is

$$K_c^* = P_e^* C^T (C P_e^* C^T + V)^{-1}, \tag{3.38}$$

where P_e^* is the unique, symmetric, positive definite solution of the Riccati equation

$$P_e = A[P_e - P_e C^T [C P_e C^T + V]^{-1} C P_e] A^T + \Gamma W \Gamma^T. \tag{3.39}$$

It is assumed here that $(A, \Gamma W^{1/2}, C)$ is reachable and observable. This *algebraic Riccati equation* is the dual to the Riccati equation (2.53) that arose in the discrete-time LQR problem and can be obtained by substituting

$$A \to A^T, B \to C^T, M \to \Gamma^T \quad \text{and} \quad R \to V, Q \to W. \tag{3.40}$$

It is clear that, as in the case of the LQR problem, the solution of the algebraic Riccati equation can be determined using the eigenvectors of the (dual) Hamiltonian.

The filter derived above is called the *discrete-time Kalman filter*. It is based on the current estimator (3.28). Note that AK_c yields the gain K of the prediction estimator [see (3.32)].

4.4 OBSERVER-BASED DYNAMIC CONTROLLERS

State estimates, derived by the methods described in the previous section, may be used in state feedback control laws to compensate given systems. This section addresses that topic.

In Section 4.2, the linear state feedback control law was introduced. There it was implicitly assumed that the state vector $x(t)$ is available for measurement. The values of the states $x(t)$ for $t \geq t_0$ were fed back and used to generate a control input in accordance with the relation $u(t) = Fx(t) + r(t)$. There are cases, however, when it may be either impossible or impractical to measure the states directly. This has provided the motivation to develop methods for estimating the states. Some of these methods were considered in Section 4.3. A natural question that arises at this time is the following: what would happen to system performance if, in the control law $u = Fx + r$, the state estimate $\hat{x}$ were used in place of x as in Fig. 4.5? How much, if any, would the compensated system response deteriorate? What are the difficulties in designing such estimator- (observer-) based linear state feedback controllers? These questions are addressed in this section. Note that observer-based controllers of the type described in the following are widely used.

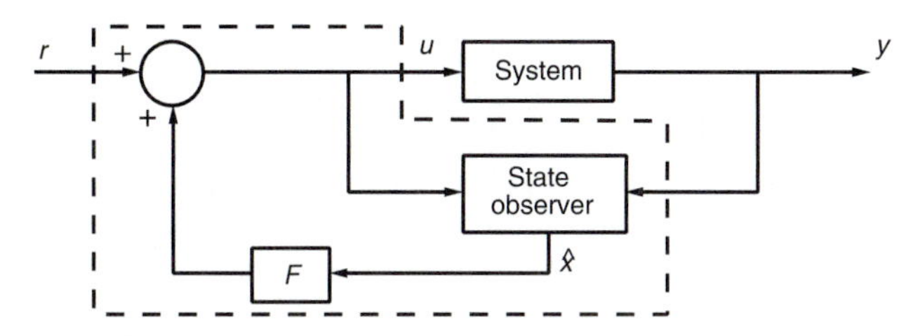

FIGURE 4.5
Observer-based controller

In the remainder of this section we will concentrate primarily on full-state/full-order observers and (static) linear state feedback, as applied to linear time-invariant systems. The analysis of partial-state and/or reduced-order observers with static or dynamic state feedback is analogous; however, it is more complex. Such control schemes are not considered at this point. Instead, they will be discussed in the exercise section (Exercise 4.8). In this section, continuous-time systems are addressed. The analysis of observer-based output controllers in the discrete-time case is completely analogous and will be omitted.

A. State-Space Analysis

We consider systems described by equations of the form

$$\dot{x} = Ax + Bu, \quad y = Cx + Du, \tag{4.1}$$

where $A \in R^{n\times n}$, $B \in R^{n\times m}$, $C \in R^{p\times n}$, and $D \in R^{p\times m}$. For such systems, we determine an estimate $\hat{x}(t) \in R^n$ of the state $x(t)$ via the (full-state/full-order) *state observer* (3.2) given by

$$\begin{aligned}\dot{\hat{x}} &= A\hat{x} + Bu + K(y - \hat{y}) \\ &= (A - KC)\hat{x} + [B - KD, K]\begin{bmatrix} u \\ y \end{bmatrix} \\ z &= \hat{x},\end{aligned} \tag{4.2}$$

where $\hat{y} = C\hat{x} + Du$. We now compensate the system by *state feedback* using the control law

$$u = F\hat{x} + r, \tag{4.3}$$

where $\hat{x}$ is the output of the state estimator and we wish to analyze the behavior of the compensated system. To this end we first eliminate y in (4.2) to obtain

$$\dot{\hat{x}} = (A - KC)\hat{x} + KCx + Bu. \tag{4.4}$$

The state equations of the compensated system are then given by

$$\begin{aligned}\dot{x} &= Ax + BF\hat{x} + Br \\ \dot{\hat{x}} &= KCx + (A - KC + BF)\hat{x} + Br,\end{aligned} \tag{4.5}$$

and the output equation assumes the form

$$y = Cx + DF\hat{x} + Dr, \tag{4.6}$$

where u was eliminated from (4.1) and (4.4), using (4.3). Rewriting in matrix form, we have

$$\begin{aligned}\begin{bmatrix} \dot{x} \\ \dot{\hat{x}} \end{bmatrix} &= \begin{bmatrix} A & BF \\ KC & A - KC + BF \end{bmatrix}\begin{bmatrix} x \\ \hat{x} \end{bmatrix} + \begin{bmatrix} B \\ B \end{bmatrix} r \\ y &= [C, DF]\begin{bmatrix} x \\ \hat{x} \end{bmatrix} + Dr,\end{aligned} \tag{4.7}$$

which is a representation of the compensated closed-loop system. Note that (4.7) constitutes a 2nth-order system. Its properties are more easily studied if an appropriate similarity transformation is used to simplify the representation. Such a transformation is given by

$$P\begin{bmatrix} x \\ \hat{x} \end{bmatrix} = \begin{bmatrix} I & 0 \\ I & -I \end{bmatrix}\begin{bmatrix} x \\ \hat{x} \end{bmatrix} = \begin{bmatrix} x \\ e \end{bmatrix}, \tag{4.8}$$

where the error $e(t) = x(t) - \hat{x}(t)$. Then the equivalent representation is

$$\begin{bmatrix} \dot{x} \\ \dot{e} \end{bmatrix} = \begin{bmatrix} A + BF & -BF \\ 0 & A - KC \end{bmatrix}\begin{bmatrix} x \\ e \end{bmatrix} + \begin{bmatrix} B \\ 0 \end{bmatrix} r$$

$$y = [C + DF, -DF]\begin{bmatrix} x \\ e \end{bmatrix} + Dr. \tag{4.9}$$

It is now quite clear that the closed-loop system is not fully controllable with respect to r (explain this in view of Subsection 3.4A). In fact, $e(t)$ does not depend on r at all. This is of course as it should be, since the error $e(t) = x(t) - \hat{x}(t)$ should converge to zero independently of the externally applied input r.

The closed-loop eigenvalues are the roots of the polynomial

$$|sI_n - (A + BF)||sI_n - (A - KC)|. \tag{4.10}$$

Recall that the roots of $|sI_n - (A + BF)|$ are the eigenvalues of $A + BF$ that can arbitrarily be assigned via F provided that the pair (A, B) is controllable. These are in fact the closed-loop eigenvalues of the system when the state x is available and the linear state feedback control law $u = Fx + r$ is used (see Section 4.2). The roots of $|sI_n - (A - KC)|$ are the eigenvalues of $(A - KC)$ that can arbitrarily be assigned via K provided that the pair (A, C) is observable. These are the eigenvalues of the estimator (4.2).

The above discussion points out that the *design of the control law* (4.3) *can be carried out independently of the design of the estimator* (4.2). This is referred to as the *Separation Property* and is generally not true for more complex systems. The separation property indicates that the linear state feedback control law may be designed as if the state x were available and the eigenvalues of $A + BF$ are assigned at appropriate locations. The feedback matrix F can also be determined by solving an optimal control problem (LQR). If state measurements are not available for feedback, a state estimator is employed. The eigenvalues of a full-state/full-order estimator are given by the eigenvalues of $A - KC$. These are typically assigned so that the error $e(t) = x(t) - \hat{x}(t)$ becomes adequately small in a short period of time. For this, the eigenvalues of $A - KC$ are (empirically) taken to be about 6 to 10 times further away from the imaginary axis (in the complex plane, for continuous-time systems) than the eigenvalues of $A + BF$. The behavior of the closed-loop system should be verified since the above is only a rule of thumb. (Refer to any good book on control for further discussion—see Section 4.6, Notes.) The estimator gain K may also be determined by solving an optimal estimation problem (the Kalman filter). In fact, under the assumption of Gaussian noise and initial conditions given earlier (see Section 4.3), F and K can be found by solving, respectively, optimal control and estimation problems with quadratic performance criteria. In particular, the deterministic LQR problem is first solved to determine the optimal control gain F^*, and then the stochastic Kalman filtering problem is solved to determine the optimal filter gain K^*. The separation property (i.e., Separation Theorem—see any optimal control textbook) guarantees that the overall (*state estimate feedback*) *Linear Quadratic Gaussian* (*LQG*) *control design* is optimal in the sense that the control law $u^*(t) = F^*\hat{x}(t)$ minimizes the quadratic performance index $E[\int_0^\infty (z^T Qz + u^T Ru)\,dt]$. As was discussed in previous sections, the gain matrices F^* and K^* are evaluated in the following manner.

Consider

$$\dot{x} = Ax + Bu + \Gamma w, \qquad y = Cx + v, z = Mx \tag{4.11}$$

with $E[ww^T] = W > 0$ and $E[vv^T] = V > 0$ and with $Q > 0$, $R > 0$ denoting the matrix weights in the performance index $E[\int_0^\infty (z^T Qx + u^T Ru)\,dt]$. Assume that both $(A, B, Q^{1/2}M)$ and $(A, \Gamma W^{1/2}, C)$ are controllable and observable. Then the optimal control law is given by

$$u^*(t) = F^*\hat{x}(t) = -R^{-1}B^T P_c^*\hat{x}(t), \tag{4.12}$$

where $P_c^* > 0$ is the solution of the algebraic Riccati equation (2.34) given by

$$A^T P_c + P_c A - P_c BR^{-1}B^T P_c + M^T QM = 0. \tag{4.13}$$

The estimate $\hat{x}$ is generated by the optimal estimator

$$\dot{\hat{x}} = A\hat{x} + Bu + K^*(y - C\hat{x}), \tag{4.14}$$

where

$$K^* = P_c^* C^T V^{-1} \tag{4.15}$$

in which $P_e^* > 0$ is the solution to the dual algebraic Riccati equation (3.21) given by

$$P_e A^T + AP_e - P_e C^T V^{-1} CP_e + \Gamma W \Gamma^T = 0. \tag{4.16}$$

Designing observer-based dynamic controllers by the LQG control design method has been quite successful, especially when the plant model is accurately known. In this approach the weight matrices Q, R and the covariance matrices W, V are used as design parameters. Unfortunately, this method does not necessarily lead to robust designs when uncertainties are present. This has led to an enhancement of this method, called the LQR/LTR (Loop Transfer Recovery) method, where the design parameters W and V are selected (iteratively) so that the robustness properties of the LQR design are recovered (refer to Section 4.6).

Finally, as was mentioned, the discrete-time case is analogous to the continuous-time case and its discussion will be omitted.

EXAMPLE 4.1. Consider the system $\dot{x} = Ax + Bu$, $y = Cx$, where

$$A = \begin{bmatrix} 0 & 1 & 0 \\ 0 & 0 & 1 \\ 0 & 2 & -1 \end{bmatrix}, \qquad B = \begin{bmatrix} 0 & 1 \\ 1 & 1 \\ 0 & 0 \end{bmatrix}, \qquad C = [1, 0, 0].$$

This is a controllable and observable but unstable system with eigenvalues of A equal to $0, -2, 1$. A linear state feedback control $u = Fx + r$ was derived in Example 2.5 to assign the eigenvalues of $A + BF$ at $-2, -1 \pm j$. An appropriate F to accomplish this was shown to be

$$F = \begin{bmatrix} 2 & -1 & -2 \\ -2 & 0 & \frac{1}{2} \end{bmatrix}.$$

If the state $x(t)$ is not available for measurement, then an estimate $\hat{x}(t)$ is used instead, i.e., the control law $u = F\hat{x} + r$ is employed. In Example 3.1, a full-order/full-state observer, given by

$$\dot{\hat{x}} = (A - KC)\hat{x} + [B, K]\begin{bmatrix} u \\ y \end{bmatrix},$$

was derived [see (3.3)] with the eigenvalues of $A - KC$ determined as the roots of the polynomial $\alpha_d(s) = s^3 + d_2s^2 + d_1s + d_0$. It was shown that the (unique) K is in this case

$$K = [d_2 - 1, d_1 - d_2 + 3, d_0 - d_1 + 3d_2 - 5]^T,$$

and the observer is given by

$$\dot{\hat{x}} = \begin{bmatrix} 1 - d_2 & 1 & 0 \\ -d_1 + d_2 - 3 & 0 & 1 \\ -d_0 + d_1 - 3d_2 + 5 & 2 & -1 \end{bmatrix}\hat{x} + \begin{bmatrix} 0 & 1 & d_2 - 1 \\ 1 & 1 & d_1 - d_2 + 3 \\ 0 & 0 & d_0 - d_1 + 3d_2 - 5 \end{bmatrix}\begin{bmatrix} u \\ y \end{bmatrix}.$$

Using the estimate $\hat{x}$ in place of the control state x in the feedback control law causes some deterioration in the behavior of the system. This deterioration can be studied experimentally. (See the next subsection for analytical results.) To this end, let the eigenvalues of the observer be at, say, $-10, -10, -10$, let $x(0) = [1, 1, 1]^T$ and $\hat{x}(0) = [0, 0, 0]^T$, plot $x(t)$, $\hat{x}(t)$, and $e(t) = x(t) - \hat{x}(t)$, and compare these with the corresponding plots of Example 2.5, where no observer was used. Repeat the above with observer eigenvalues closer to the eigenvalues of $A + BF$ (say, at $-2, -1 \pm j$) and also further away. In general the faster the observer, the faster $e(t) \to 0$, and the smaller the deterioration of response; however, in this case care should be taken if noise is present in the system. ■

B. Transfer Function Analysis

For the compensated system (4.9) [or (4.7)], the closed-loop transfer function $T(s)$ between y and r is given by

$$\tilde{y}(s) = T(s)\tilde{r}(s) = [(C + DF)[sI - (A + BF)]^{-1}B + D]\tilde{r}(s), \qquad (4.17)$$

where $\tilde{y}(s)$ and $\tilde{r}(s)$ denote the Laplace transforms of $y(t)$ and $r(t)$, respectively. The function $T(s)$ was found from (4.9), using the fact that the uncontrollable part of the system does not appear in the transfer function (see Section 3.4). Note that $T(s)$ is the transfer function of $\{A + BF, B, C + DF, D\}$, i.e., $T(s)$ is precisely the transfer function of the closed-loop system $H_F(s)$ when no state estimation is present (see Section 4.2). Therefore, the compensated system behaves to the outside world as if there were no estimator present. *Note that this statement is true only after sufficient time has elapsed from the initial time, allowing the transients to become negligible.* (Recall what the transfer function represents in a system.) Specifically, taking Laplace transforms in (4.9) and solving, we obtain

$$\begin{bmatrix} \tilde{x}(s) \\ \tilde{e}(s) \end{bmatrix} = \begin{bmatrix} [sI - (A + BF)]^{-1} & -[sI - (A + BF)]^{-1}BF[sI - (A - KC)]^{-1} \\ 0 & [sI - (A + BF)]^{-1} \end{bmatrix}\begin{bmatrix} x(0) \\ e(0) \end{bmatrix} + \begin{bmatrix} [sI - (A + BF)]^{-1}B \\ 0 \end{bmatrix}\tilde{r}(s)$$

$$\tilde{y}(s) = [C + DF, -DF]\begin{bmatrix} \tilde{x}(s) \\ \tilde{e}(s) \end{bmatrix} + D\tilde{r}(s). \qquad (4.18)$$

Therefore,

$$\begin{aligned}\tilde{y}(s) = &(C + DF)[sI - (A + BF)]^{-1}x(0)\\ &- [(C + DF)[sI - (A + BF)]^{-1}BF[sI - (A - KC)]^{-1}\\ &+ DF[sI - (A + BF)]^{-1}]e(0) + T(s)\tilde{r}(s),\end{aligned} \tag{4.19}$$

which indicates the effects of the estimator on the input-output behavior of the closed-loop system. Notice how the initial conditions for the error $e(0) = x(0) - \hat{x}(0)$ influence the response. Specifically, when $e(0) \neq 0$, its effect can be viewed as a disturbance that will become negligible at steady state. The speed by which the effect of $e(0)$ on y will diminish depends on the location of the eigenvalues of $A + BF$ and $A - KC$, as can be easily seen from relation (4.19).

Two-input controller

In the following, we will find it of interest to view the observer-based controller discussed previously as a one-vector output (u) and a two-vector input (y and r) controller. In particular, from $\dot{\hat{x}} = (A - KC)\hat{x} + (B - KD)u + Ky$ given in (4.2) and $u = F\hat{x} + r$ given in (4.3), we obtain the equations

$$\begin{aligned}\dot{\hat{x}} &= (A - KC + BF - KDF)\hat{x} + [K, B - KD]\begin{bmatrix} y \\ r \end{bmatrix}\\ u &= F\hat{x} + r.\end{aligned} \tag{4.20}$$

This is the description of the (nth order) controller shown in Fig. 4.5. The state $\hat{x}$ is of course the state of the estimator and the transfer function between u and y, r is given by

$$\begin{aligned}\tilde{u}(s) = &F[sI - (A - KC + BF - KDF)]^{-1}K\tilde{y}(s)\\ &+ [F[sI - (A - KC + BF - KDF)]^{-1}(B - KD) + I]\tilde{r}(s).\end{aligned} \tag{4.21}$$

If we are interested only in "loop properties," then r can be taken to be zero, in which case (4.21) (for $r = 0$) yields the output feedback compensator, which accomplishes the same control objectives (that are typically only "loop properties") as the original observer-based controller. This fact is used in the LQG/LTR design approach. When $r \neq 0$, (4.21) is not appropriate for the realization of the controller since the transfer function from r, which must be outside the loop, may be unstable. Note that an expression for this controller that leads to a realization of a stable closed-loop system is given by

$$\tilde{u}(s) = [F[sI - (A - KC + BF - KDF)]^{-1}[K, B - KD] + [0, I]]\begin{bmatrix} \tilde{y}(s) \\ \tilde{r}(s) \end{bmatrix} \tag{4.22}$$

(see Fig. 4.6). This was also derived from (4.20). The stability of general two-input controllers (with two degrees of freedom) is discussed at length in Section 7.4D of Chapter 7.

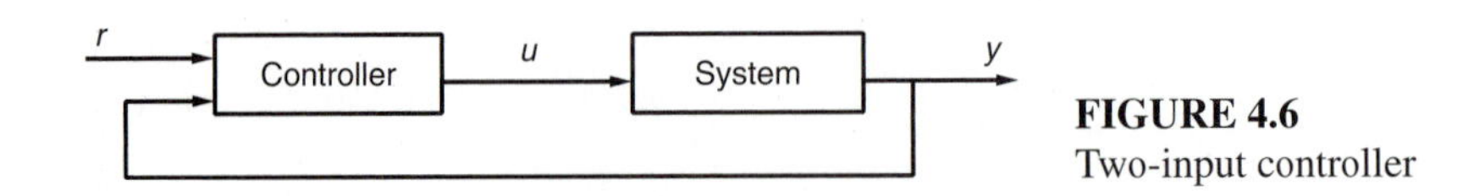

FIGURE 4.6
Two-input controller

At this point, we find it of interest to determine the relationship of the observer-based controller and the conventional one-and-two block controller configurations of Fig. 4.7. Here, the requirement is to maintain the same transfer functions between inputs y and r and output u. (For further discussion of stability and attainable response maps in systems controlled by output feedback controllers, refer to Section 7.4 of Chapter 7.) We proceed by considering once more (4.2) and (4.3) and by writing

$$\tilde{u}(s) = F[sI - (A - KC)]^{-1}(B - KD)\tilde{u}(s) + F[sI - (A - KC)]^{-1}K\tilde{y}(s) + \tilde{r}(s) = G_u\tilde{u}(s) + G_y\tilde{y}(s) + \tilde{r}(s).$$

This yields

$$\tilde{u}(s) = (I - G_u)^{-1}[G_y\tilde{y}(s) + \tilde{r}(s)] \tag{4.23}$$

(see Fig. 4.7). Notice that

$$G_y = F[sI - (A - KC)]^{-1}K, \tag{4.24}$$

i.e., the controller in the feedback path is stable (why?). The matrix $(I - G_u)^{-1}$ is not necessarily stable; however, it is inside the loop and the internal stability of the compensated system is preserved. Comparing with (4.21), we obtain

$$(I - G_u)^{-1} = F[sI - (A - KC + BF - KDF)]^{-1}(B - KD) + I. \tag{4.25}$$

Also, as expected, we have

$$(I - G_u)^{-1}G_y = F[sI - (A - KC + BF - KDF)]^{-1}K \tag{4.26}$$

(show this). These relations could have been derived directly as well by the use of matrix identities; however, such derivation is quite involved.

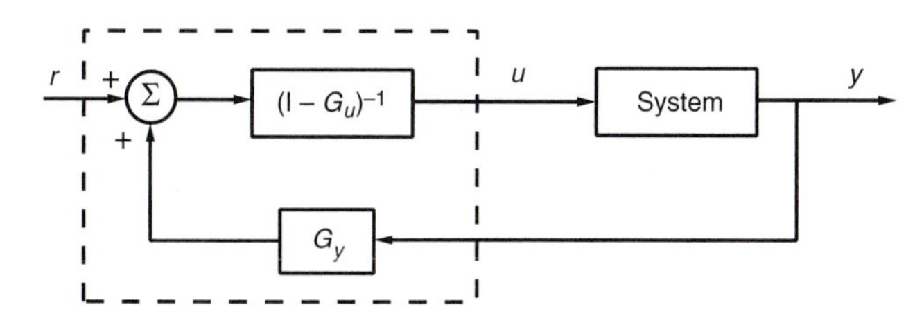

FIGURE 4.7

EXAMPLE 4.2. For the system $\dot{x} = Ax + Bu$, $y = Cx$ with $A = \begin{bmatrix} 0 & -2 \\ 1 & -2 \end{bmatrix}$, $B = \begin{bmatrix} 0 \\ 1 \end{bmatrix}$, and $C = [0, 1]$ we have $H(s) = C(sI - A)^{-1}B = s/(s^2 + 2s + 2)$. In Example 3.2, it was shown that the gain matrix $K = [d_0 - 2, d_1 - 2]^T$ assigns the eigenvalues of the asymptotic observer (of $A - KC$) at the roots of $s^2 + d_1 s + d_0$. In fact $sI - (A - KC) = \begin{bmatrix} s & d_0 \\ -1 & s + d_1 \end{bmatrix}$. It is straightforward to show that $F = [\frac{1}{2}a_0 - 1, 2 - a_1]$ will assign the eigenvalues of the closed-loop system (of $A + BF$) at the roots of $s^2 + a_1 s + a_0$. Indeed, $sI - (A + BF) = \begin{bmatrix} s & 2 \\ -\frac{1}{2}a_0 & s + a_1 \end{bmatrix}$. Now in (4.23) we have

$$G_y(s) = F(sI - (A - KC))^{-1}K$$
$$= \frac{s\,((d_0 - 2)(\frac{1}{2}a_0 - 1) + (d_1 - 2)(2 - a_1)) + ((d_0 - d_1)(a_0 - 2) + (d_0 - 2)(2 - a_1))}{s^2 + d_1 s + d_0},$$

$$G_u(s) = F(sI - (A - KC))^{-1}B = \frac{s(2 - a_1) - d_0(\frac{1}{2}a_0 - 1)}{s^2 + d_1 s + d_0},$$

$$(1 - G_u)^{-1} = \frac{s^2 + d_1 s + d_0}{s^2 + s(d_1 + a_1 - 2) + \frac{1}{2}a_0 d_0}.$$

■

4.5 SUMMARY

In this chapter, linear state feedback controllers and state observers were studied with an emphasis on time-invariant, continuous-time, and discrete-time systems (Sections 4.2 and 4.3). State feedback controllers and state observers were then combined (in Section 4.4) to develop observer-based dynamic output feedback controllers. We note that output feedback controllers are studied further in Chapter 7.

Linear state feedback was studied in Section 4.2. First, the need for feedback was explained by discussing open- and closed-loop control in the presence of uncertainties. System stabilization was then addressed which for the time-invariant case leads to the eigenvalue or pole assignment problem. It was pointed out that an optimal control formulation, such as the Linear Quadratic Regulator (LQR) problem, will also lead to stable closed-loop control systems while attaining additional control goals as well. Furthermore, the LQR problem formulation can also be used in the time-varying case. The eigenvalue assignment problem was studied at length by introducing several such methods. Analogous results for the discrete-time case were established in Subsections 4.2E and 4.2F.

In Section 4.3, full-order observers for the entire state or for a linear function of the state were presented. Reduced-order observers and optimal observers were also addressed in Subsections 4.3B and 4.3C, respectively. The duality between the state feedback controller and state observer problems was explored and emphasized. Analogous results for the discrete-time case were also described. Current state estimators were introduced and developed in Subsection 4.3D.

In Section 4.4, state observers, together with state feedback controllers were used to derive dynamic output feedback controllers and the Separation Principle was discussed. The degradation of performance in state feedback control when an observer is used to estimate the state was explained. An analysis of the closed-loop system was carried out, using both state-space and transfer function matrix descriptions.

4.6 NOTES

The fact that if a system is (state) controllable, then all its eigenvalues can arbitrarily be assigned by means of linear state feedback has been known since the 1960s. Original sources include Rissanen [19], Popov [17], and Wonham [23]. (See also remarks in Kailath [10], pp. 187, 195.)

The present approach for eigenvalue assignment via linear state feedback, using the controller form, follows the development in Wolovich [22]. Ackermann's formula first appeared in Ackermann [1].

The development of the eigenvector formulas for the feedback matrix that assign all the closed-loop eigenvalues and (in part) the corresponding eigenvectors follows

Moore [16]. The corresponding development that uses (A, B) in controller (companion) form and polynomial matrix descriptions follows Antsaklis [3]. Related results on static output feedback and on polynomial and rational matrix interpolation can be found in Antsaklis and Wolovich [4] and Antsaklis and Gao [5]. Note that the flexibility in assigning the eigenvalues via state feedback in the multi-input case can be used to assign the invariant polynomials of $sI - (A + BF)$; conditions for this are given by Rosenbrock [20].

The Linear Quadratic Regulator (LQR) problem and the Linear Quadratic Gaussian (LQG) problem have been studied extensively, particularly in the 1960s and early 1970s. Sources for these topics include the books by Anderson and Moore [2], Kwakernaak and Sivan [11], Lewis [12], and Dorato et al. [9]. Early optimal control sources include Athans and Falb [6], and Bryson and Ho [8]. A very powerful idea in optimal control is the *Principle of Optimality*, Bellman [7], which can be stated as follows: "An optimal trajectory has the property that at any intermediate point, no matter how it was reached, the remaining part of a trajectory must coincide with an optimal trajectory, computed from the intermediate point as the initial point". For historical remarks on this topic, refer, e.g., to Kailath [10], pp. 240–241.

The most influential work on state observers is the work of Luenberger. Although the asymptotic observer presented here is generally attributed to him, Luenberger's Ph.D. thesis work in 1963 was closer to the reduced-order observer presented above. Original sources on state observers include Luenberger [13], [14], and [15]. For an extensive overview of observers, refer to the book by O'Reilly [18].

When linear quadratic optimal controllers and observers are combined in control design, a procedure called LQG/LTR (Loop Transfer Recovery) is used to enhance the robustness properties of the closed loop system. For a treatment of this procedure, see Stein and Athans [21] and contemporary textbooks on multivariable control.

4.7 REFERENCES

1. J. Ackermann, "Der Entwurf linearer Regelungssysteme im Zustandsraum," *Regelungstechnik und Prozessdatenverarbeitung*, Vol. 7, pp. 297–300, 1972.
2. B. D. O. Anderson and J. B. Moore, *Optimal Control. Linear Quadratic Methods*, Prentice-Hall, Englewood Cliffs, NJ, 1990.
3. P. J. Antsaklis, "Some New Matrix Methods Applied to Multivariable System Analysis and Design," Ph.D. Dissertation, Brown University, May 1976.
4. P. J. Antsaklis and W. A. Wolovich, "Arbitrary Pole Placement Using Linear Output Feedback Compensation," *Int. J. Control*, Vol. 25, No. 6, pp. 915–925, 1977.
5. P. J. Antsaklis and Z. Gao, "Polynomial and Rational Matrix Interpolation: Theory and Control Applications," *Int. J. of Control*, Vol. 58, No. 2, 349–404, August 1993.
6. M. Athans and P. L. Falb, *Optimal Control*, McGraw-Hill, New York, 1966.
7. R. Bellman, *Dynamic Programming*, Princeton University Press, Princeton, NJ, 1957.
8. A. E. Bryson and Y. C. Ho, *Applied Optimal Control*, Holsted Press, New York, 1968.
9. P. Dorato, C. Abdallah and V. Cerone, *Linear-Quadratic Control: An Introduction*, Prentice-Hall, Englewood Cliffs, NJ, 1995.
10. T. Kailath, *Linear Systems*, Prentice-Hall, Englewood Cliffs, NJ, 1980.
11. H. Kwakernaak and R. Sivan, *Linear Optimal Control Systems*, Wiley, New York, 1972.
12. F. L. Lewis, *Optimal Control*, Wiley, New York, 1986.

13. D. G. Luenberger, "Observing the State of a Linear System," *IEEE Trans. Mil. Electron*, MIL-8, pp. 74–80, 1964.
14. D. G. Luenberger, "Observers for Multivariable Systems," *IEEE Trans. on Auto. Control*, AC-11, pp. 190–199, 1966.
15. D. G. Luenberger, "An Introduction to Observers," *IEEE Trans. on Auto. Control*, AC-16, pp. 596-603, December 1971.
16. B. C. Moore, "On the Flexibility Offered by State Feedback in Multivariable Systems Beyond Closed Loop Eigenvalue Assignment," *IEEE Trans. on Auto. Control*, AC-21, pp. 689–692, 1976; see also AC-22, pp. 140–141, 1977 for the repeated eigenvalue case.
17. V. M. Popov, "Hyperstability and Optimality of Automatic Systems with Several Control Functions," *Rev. Roum. Sci. Tech. Ser. Electrotech Energ.*, Vol. 9, pp. 629–690, 1964. See also V. M. Popov, *Hyperstability of Control Systems*, Springer-Verlag, New York, 1973.
18. J. O'Reilly, *Observers for Linear Systems*, Academic Press, New York, 1983.
19. J. Rissanen, "Control System Synthesis by Analogue Computer Based on the Generalized Linear Feedback Concept," in *Proc. of Symp. on Analog Comp. Applied to the Study of Chem. Processes*, pp. 1–13, Intern. Seminar, Brussels, 1960. Presses Académiques Européennes, Bruxelles, 1961.
20. H. H. Rosenbrock, *State-Space and Multivariable Theory*, Wiley, New York, 1970.
21. G. Stein and M. Athans, "The LQG/LTR Procedure for Multivariable Feedback Control Design," *IEEE Trans. on Automatic Control*, Vol. AC-32, pp. 105–114, February 1987.
22. W. A. Wolovich, *Linear Multivariable Systems*, Springer-Verlag, New York, 1974.
23. W. M. Wonham, "On Pole Assigment in Multi-Input Controllable Linear Systems," Vol. AC-12, pp. 660–665, 1967.

4.8 EXERCISES

4.1. Consider the system $\dot{x} = Ax + Bu$, where $A = \begin{bmatrix} -0.01 & 0 \\ 0 & -0.02 \end{bmatrix}$ and $B = \begin{bmatrix} 1 & 1 \\ -0.25 & 0.75 \end{bmatrix}$ with $u = Fx$.

(a) Verify that the three different state feedback matrices given by

$$F_1 = \begin{bmatrix} -1.1 & -3.7 \\ 0 & 0 \end{bmatrix}, \quad F_2 = \begin{bmatrix} 0 & 0 \\ -1.1 & 1.2333 \end{bmatrix}, \quad F_3 = \begin{bmatrix} -0.1 & 0 \\ 0 & -0.1 \end{bmatrix}$$

all assign the closed-loop eigenvalues at the same locations, namely, at $-0.1025 \pm j0.04944$. Note that in the first control law (F_1) only the first input is used, while in the second law (F_2) only the second input is used. For all three cases plot $x(t) = [x_1(t), x_2(t)]^T$ when $x(0) = [0, 1]^T$ and comment on your results. This example demonstrates how different the responses can be for different designs even though the eigenvalues of the compensated system are at the same locations.

(b) Use the eigenvalue/eigenvector assignment method to characterize all F that assign the closed-loop eigenvalues at $-0.1025 \pm j0.04944$. Show how to select the free parameters to obtain F_1, F_2, and F_3 above. What are the closed-loop eigenvectors in these cases?

4.2. For the system $\dot{x} = Ax + Bu$ with $A \in R^{n \times n}$ and $B \in R^{n \times m}$, where (A, B) is controllable and $m > 1$, choose $u = Fx$ as the feedback control law. It is possible to assign all eigenvalues of $A + BF$ by first reducing this problem to the case of eigenvalue assignment for single-input systems ($m = 1$). This is accomplished by first reducing the system to a *single-input controllable system*. We proceed as follows.

Let $F = g \cdot f$, where $g \in R^m$ and $f^T \in R^n$ are vectors to be selected. Let g be chosen such that (A, Bg) is controllable. Then f in

$$A + BF = A + (Bg)f$$

can be viewed as the state feedback gain vector for a single-input controllable system (A, Bg), and any of the single-input eigenvalue assignment methods can be used to select f so that the closed-loop eigenvalues are at desired locations.

The only question that remains to be addressed is whether there exists g such that (A, Bg) is controllable. It can be shown that if (A, B) is controllable and A is cyclic, then almost any $g \in R^m$ will make (A, Bg) controllable. (A matrix A is cyclic if and only if its characteristic and minimal polynomials are equal.) In the case when A is not cyclic, it can be shown that if (A, B, C) is controllable and observable, then for almost any real output feedback gain matrix H, $A + BHC$ is cyclic. So initially, by an almost arbitrary choice of H or $F = HC$, the matrix A is made cyclic, and then by employing a g, (A, Bg) is made controllable. The state feedback vector gain f is then selected so that the eigenvalues are at desired locations.

Note that $F = gf$ is always a rank one matrix, and this restriction on F reduces the applicability of the method when requirements in addition to eigenvalue assignment are to be met.

For the present approach, see W. M. Wonham, "On Pole Assignment in Multi-Input Controllable Linear Systems," *IEEE Trans. Autom. Control*, Vol. AC-12, pp. 660–665, December 1967, and F. M. Brasch and J. B. Pearson, "Pole Placement Using Dynamic Compensators," *IEEE Trans. Autom. Control*, Vol. AC-15, pp. 34–43, February 1970. For a discussion of cyclicity of matrices see Chapter 2, and also P. J. Antsaklis, "Cyclicity and Controllability in Linear Time-Invariant Systems," *IEEE Trans. Autom. Control*, Vol. AC-23, pp. 745–746, August 1978.

(a) For A, B as in Exercise 4.4, use the method described above to determine F so that the closed-loop eigenvalues are at $-1 \pm j$ and $-2 \pm j$. Comment on your choice for g.

(b) For $A = \begin{bmatrix} 0 & 1 \\ 1 & 1 \end{bmatrix}$ and $B = \begin{bmatrix} 1 & 0 \\ 0 & 1 \end{bmatrix}$, characterize all g such that the closed-loop eigenvalues are at -1.

4.3. Consider the system $x(k+1) = Ax(k) + Bu(k)$, where

$$A = \begin{bmatrix} 1 & 4 & 0 \\ 2 & -1 & 0 \\ 0 & 0 & 1 \end{bmatrix}, \qquad B = \begin{bmatrix} 0 & 0 \\ 1 & 0 \\ -1 & 1 \end{bmatrix}.$$

Determine a linear state feedback control law $u(k) = Fx(k)$ such that all the eigenvalues of $A + BF$ are located at the origin. To accomplish this, use

(a) reduction to a single-input controllable system,

(b) the controller form of (A, B),

(c) $det\,(zI - (A + BF))$ and the resulting nonlinear system of equations.

In each case, plot $x(k)$ with $x(0) = [1, 1, 1]^T$ and comment on your results. In how many steps does your compensated system go to the zero state?

4.4. For the system $\dot{x} = Ax + Bu$, where

$$A = \begin{bmatrix} 0 & 1 & 0 & 0 \\ 0 & 0 & 1 & 0 \\ 0 & 0 & 0 & 1 \\ 1 & 1 & -3 & 4 \end{bmatrix}, \qquad B = \begin{bmatrix} 1 & 0 \\ 0 & 0 \\ 0 & 0 \\ 0 & 1 \end{bmatrix},$$

determine F so that the eigenvalues of $A + BF$ are at $-1 \pm j$ and $-2 \pm j$. Use as many different methods to choose F as you can.

4.5. Consider the system $\dot{x} = Ax + Bu$, where

$$A = \begin{bmatrix} -1 & 0 & 0 & -6 & 3 & -1 \\ 1 & -2 & 1 & 0 & -1 & -1 \\ 1 & 1 & 0 & 6 & -2 & 1 \\ 1 & 0 & 0 & 0 & 0 & 0 \\ -1 & 2 & -1 & 0 & 2 & 1 \\ -2 & 0 & 0 & -2 & 0 & -1 \end{bmatrix}, \qquad B = \begin{bmatrix} 0 & 1 \\ -1 & -2 \\ 0 & -1 \\ 0 & 0 \\ 1 & 2 \\ 0 & 0 \end{bmatrix}.$$

(a) Show that -1 is an uncontrollable eigenvalue.

(b) For the control law $u = Fx$, determine F so that the controllable eigenvalues of $A + BF$ are

$$-.1, -.2, -1 \pm j, -2.$$

4.6. Consider the SISO system $\dot{x}_c = A_c x_c + B_c u$, $y = C_c x_c + D_c u$, where (A_c, B_c) is in controller form with

$$A_c = \begin{bmatrix} 0 & 1 & \cdots & 0 \\ \vdots & \vdots & \ddots & \vdots \\ 0 & 0 & \cdots & 1 \\ -\alpha_0 & -\alpha_1 & \cdots & -\alpha_{n-1} \end{bmatrix}, \qquad B_c = \begin{bmatrix} 0 \\ \vdots \\ 0 \\ 1 \end{bmatrix}, \qquad C_c = [c_0, c_1, \ldots, c_{n-1}],$$

and let $u = F_c x + r = [f_0, f_1, \ldots, f_{n-1}]x + r$ be the linear state feedback control law. Use the Structure Theorem of Chapter 3 to show that the open-loop transfer function is

$$\begin{aligned} H(s) &= C_c(sI - A_c)^{-1}B_c + D_c = \frac{c_{n-1}s^{n-1} + \cdots + c_1 s + c_0}{s^n + \alpha_{n-1}s^{n-1} + \cdots + \alpha_1 s + \alpha_0} + D_c \\ &= \frac{n(s)}{d(s)} \end{aligned}$$

and the closed-loop transfer function is

$$\begin{aligned} H_F(s) &= (C_c + D_c F_c)[sI - (A_c + B_c F_c)]^{-1}B_c + D_c \\ &= \frac{(c_{n-1} + D_c f_{n-1})s^{n-1} + \cdots + (c_1 s + D_c f_1)s + (c_0 + D_c f_0)}{s^n + (\alpha_{n-1} - f_{n-1})s^{n-1} + \cdots + (\alpha_1 - f_1)s + (\alpha_0 - f_0)} + D_c \\ &= \frac{n(s)}{d_F(s)}. \end{aligned}$$

Observe that state feedback does not change the numerator $n(s)$ of the transfer function, but it can arbitrarily assign any desired (monic) denominator polynomial $d_F(s) = d(s) - F_c[1, s, \ldots, s^{n-1}]^T$. Thus, state feedback does not (directly) alter the zeros of $H(s)$, but it can arbitrarily assign the poles of $H(s)$. Note that these results generalize to the MIMO case [see (2.44)].

4.7. Consider the system $\dot{x} = Ax + Bu$, $y = Cx$, where

$$A = \begin{bmatrix} 0 & 1 & 0 \\ 0 & 0 & 1 \\ 1 & 0 & -1 \end{bmatrix}, \qquad B = \begin{bmatrix} 0 \\ 0 \\ 1 \end{bmatrix}, \qquad C = [1, 2, 0].$$

(a) Determine an appropriate linear state feedback control law $u = Fx + Gr (G \in R)$ so that the closed-loop transfer function is equal to a given desired transfer function

$$H_m(s) = \frac{1}{s^2 + 3s + 2}.$$

We note that this is an example of *model matching*, i.e., compensating a given system so that it matches the input-output behavior of a desired model. In the present case, state feedback is used; however, output feedback is more common in model matching.

(b) Is the compensated system in (a) controllable? Is it observable? Explain your answers.

(c) Repeat (a) and (b) by assuming that the state is not available for measurement. Design an appropriate state observer, if possible.

4.8. Consider the nth order system $\dot{x} = Ax + Bu$, $y = Cx + Du$.

(a) Let $\dot{z} = \tilde{A}z + Ky + [FB - KD]u$, $\tilde{A} \in R^{m\times m}$, be an asymptotic estimator of the linear function of the state, Fx, where $F \in R^{m\times n}$ is the desired state feedback gain. Consider the control law $u = z + r$ and determine the representation and properties of the closed-loop system.

(b) Consider a reduced-order observer of order $n - p$ (as in Subsection 4.3B) and the control law $u = Fx + r$. Determine the representation and properties of the closed-loop system.

Hint: The analysis is analogous to the full-state observer case given in Section 4.4.

4.9. Design an observer for the oscillatory system $\dot{x}(t) = v(t)$, $\dot{v}(t) = -\omega_0^2 x(t)$, using measurements of the velocity v. Place both observer poles at $s = -\omega_0$.

4.10. Consider the undamped harmonic oscillator $\dot{x}_1(t) = x_2(t)$, $\dot{x}_2(t) = -\omega_0^2 x_1(t) + u(t)$. Using an observation of velocity $y = x_2$, design an observer/state feedback compensator to control the position x_1. Place the state feedback controller poles at $s = -\omega_0 \pm j\omega_0$ and both observer poles at $s = -\omega_0$. Plot $x(t)$ for $x(0) = [1, 1]^T$ and $\omega_0 = 2$.

4.11. A servomotor that drives a load is described by the equation $(d^2\theta/dt^2) + (d\theta/dt) = u$, where θ is the shaft position (output) and u is the applied voltage. Choose u so that θ and $(d\theta/dt)$ will go to zero exponentially (when their initial values are not zero). To accomplish this, proceed as follows.

(a) Derive a state-space representation of the servomotor.

(b) Determine linear state feedback, $u = Fx + r$, so that both closed-loop eigenvalues are at -1. Such F is actually optimal since it minimizes $J = \int_0^\infty [\theta^2 + (d\theta/dt)^2 + u^2]\,dt$ (show this).

(c) Since only θ and u are available for measurement, design an asymptotic state estimator (with eigenvalues at, say, -3) and use the state estimate $\hat{x}$ in the linear state feedback control law. Write the transfer function and the state-space description of the overall system and comment on stability, controllability, and observability.

(d) Plot θ and $d\theta/dt$ in (b) and (c) for $r = 0$ and initial conditions equal to $[1, 1]^T$.

(e) Repeat (c) and (d), using a reduced-order observer of order 1.

4.12. Consider the LQR problem for the system $\dot{x} = Ax + Bu$, where (A, B) is controllable and the performance index is given by

$$\tilde{J}(u) = \int_0^\infty e^{2\alpha t}[x^T(t)Qx(t) + u^T(t)Ru(t)]\,dt,$$

where $\alpha \in R$, $\alpha > 0$ and $Q \geq 0$, $R > 0$.

(a) Show that u^* that minimizes $\tilde{J}(u)$ is a fixed control law with constant gains on the states, even though the weighting matrices $\tilde{Q} = e^{2\alpha t}Q$, $\tilde{R} = e^{2\alpha t}R$ are time varying. Derive the algebraic Riccati matrix equation that characterizes this control law.

(b) The performance index given above has been used to solve the question of relative stability. In the light of your solution, how do you explain this?

Hint: Reformulate the problem in terms of the transformed variables $\tilde{x} = e^{\alpha t}x$, $\tilde{u} = e^{\alpha t}u$.

4.13. Consider the system $\dot{x} = \begin{bmatrix} 0 & 1 \\ 1 & 1 \end{bmatrix} x + \begin{bmatrix} 1 \\ 0 \end{bmatrix} u$ and the performance indices J_1, J_2 given by

$$J_1 = \int_0^\infty (x_1^2 + x_2^2 + u^2)\, dt \qquad \text{and} \qquad J_2 = \int_0^\infty (900(x_1^2 + x_2^2) + u^2)\, dt.$$

Determine the optimal control laws that minimize J_1 and J_2. In each case plot $u(t)$, $x_1(t)$, $x_2(t)$ for $x(0) = [1, 1]^T$ and comment on your results.

4.14. Consider the discrete-time system $x(k+1) = Ax(k) + Bu(k)$, $y(k) = C(k)x(k)$, where

$$A = \begin{bmatrix} 1 & T \\ 0 & 1 \end{bmatrix}, \qquad B = \begin{bmatrix} \frac{1}{2}T^2 \\ T \end{bmatrix}, \qquad C = [1, 0]$$

and where T is the sampling period. This is a sampled-data system obtained from (the double integrator) $\tilde{A} = \begin{bmatrix} 0 & 1 \\ 0 & 0 \end{bmatrix}$, $\tilde{B} = \begin{bmatrix} 0 \\ 1 \end{bmatrix}$, and $\tilde{C} = [1, 0]$ via a zero-order hold and an ideal sampler, both of period T (see Chapter 2).

Consider the cost functional $J(u) = \sum_{k=0}^{\infty}(y^2(k) + 2u^2(k))$. This represents (2.51) with $z(k) = y(k)$, $M = C$, $Q = 1$, and $R = 2$. Determine the control sequence $\{u^*(k)\}$, $k \geq 0$, that minimizes the cost functional, as a function of T. Compare your results with Examples 2.7 and 2.9.

4.15. Consider the system $\dot{x} = \begin{bmatrix} 0 & 1 \\ 1 & 0 \end{bmatrix} x + \begin{bmatrix} 0 \\ -1 \end{bmatrix} u$, $y = [1, 0]x$.

(a) Use state feedback $u = Fx$ to assign the eigenvalues of $A + BF$ at $-0.5 \pm j0.5$. Plot $x(t) = [x_1(t), x_2(t)]^T$ for the open- and closed-loop system with $x(0) = [-0.6, 0.4]^T$.

(b) Design an identity observer with eigenvalues at $-\alpha \pm j$, where $\alpha > 0$. What is the observer gain K in this case?

(c) Use the state estimate $\hat{x}$ from (b) in the linear feedback control law $u = F\hat{x}$, where F was found in (a). Derive the state-space description of the closed-loop system. If $u = F\hat{x} + r$, what is the transfer function between y and r?

(d) For $x(0) = [-0.6, 0.4]^T$ and $\hat{x}(0) = [0, 0]^T$ plot $x(t)$, $\hat{x}(t)$, $y(t)$ and $u(t)$ of the closed-loop system obtained in (c) and comment on your results. Use $\alpha = 1, 2, 5$, and 10 and comment on the effects on the system response.

Remark: This exercise illustrates the deterioration of system response when state observers are used to generate the state estimate that is used in the feedback control law.

4.16. Consider the system

$$x(k + 1) = Ax(k) + Bu(k) + Eq(k), \qquad y(k) = Cx(k),$$

where $q(k) \in R^r$ is some disturbance vector. It is desirable to completely eliminate the effects of $q(k)$ on the output $y(k)$. This can happen only when E satisfies certain conditions. Presently, it is assumed that $q(k)$ is an arbitrary $r \times 1$ vector.

(a) Express the required conditions on E in terms of the observability matrix of the system.

(b) If $A = \begin{bmatrix} 1 & 1 \\ 1 & 1 \end{bmatrix}$, $C = [1, 1]$, characterize all E that satisfy these conditions.

(c) Suppose $E \in R^{n\times 1}$, $C \in R^{1\times n}$, and $q(k)$ is a step, and let the objective be to asymptotically reduce the effects of q on the output. Note that this specification is not as strict as in (a), and in general it is more easily satisfied. Use z-transforms to derive conditions for this to happen.
Hint: Express the conditions in terms of poles and zeros of $\{A, E, C\}$.

4.17. Consider the system $\{A, B, C, D\}$ given by $\dot{x} = Ax + Bu, y = Cx + Du$, where $A \in R^{n\times n}$, $B \in R^{n\times p}$, $C \in R^{p\times n}$, $D \in R^{p\times p}$, and $\det D \neq 0$.

(a) Show that $\{\bar{A}, \bar{B}, \bar{C}, \bar{D}\} = \{A - BD^{-1}C, BD^{-1}, -D^{-1}C, D^{-1}\}$ is the *inverse system* of $\{A, B, C, D\}$, i.e.,

$$\bar{C}(sI - \bar{A})^{-1}\bar{B} + \bar{D} = H(s)^{-1},$$

where $H(s) = C(sI-A)^{-1}B+D$ is the transfer function matrix of $\{A, B, C, D\}$. Verify this result using a state-space representation of the system $H(s) = (s^2 + 1)/(s^2 + 2)$.

(b) It is possible to show (a) using results involving state feedback. In particular, let (A, B) be controllable, let $u = Fx+Gr$, and choose the linear state feedback control gain matrices (F, G) so that $H_{F,G}(s) = (C+DF)(sI-(A+BF))^{-1}BG+DG = I$. Note that this choice for F makes all eigenvalues unobservable. Use this result to prove (a). *Hint*: It can be shown using matrix identities that $H_{F,G}(s) = H(s) \times [I-F(sI-A)^{-1}C]^{-1}G = H(s)[I+F(sI-(A+BF))^{-1}B]G$ [see (2.43) to (2.45)].

Suppose now that $det\ D = 0$, but there exists a diagonal polynomial matrix

$$\tilde{X}(s) = diag\,(p_i(s)), \qquad i = 1, \ldots, p,$$

with $p_i(s)$ monic stable polynomials of degree f_i such that

$$\lim_{s\to\infty} \tilde{X}(s)H(s) = \tilde{D},$$

where $det\ \tilde{D} \neq 0$.

(c) Show that a realization of $\tilde{X}(s)H(s)$ is $\dot{x} = Ax + Bu, \tilde{y} = \tilde{C}x + \tilde{D}u$, i.e., the A, B are the same as above and $\tilde{C}, \tilde{D}$ are some new matrices.

(d) Determine a control law $u = Fx+Gr$ that when applied to the system $\{A, B, C, D\}$ yields

$$H_{F,G}(s) = \tilde{X}^{-1}(s),$$

a diagonal transfer function matrix with stable poles at the zeros of $p_i(s)$. Note that this is the problem of *diagonal decoupling via state feedback with stability* (see below and refer to Subsection 7.4D and the Notes of Chapter 7; see also Exercise 4.20). In view of the hint in (b), determine a matrix $H_R(s)$ so that $H(s)H_R(s) = \tilde{X}^{-1}(s)$.

The *diagonal decoupling problem via state feedback* is to determine a control law $u = Fx + Gr$ that when applied to the system $\{A, B, C, D\}$ yields a closed-loop transfer function $H_{F,G}(s)$ that is diagonal and nonsingular. The conditions for existence of solutions can be expressed as follows: let $X(s)$ be the diagonal matrix $X(s) = diag\,[s^{f_i}]$, where the nonnegative integers $\{f_i\}$ are so that all rows of $\lim_{s\to\infty} X(s)H(s)$ are constant and nonzero and let

$$\lim_{s\to\infty} X(s)H(s) = B^*.$$

Then the system can be diagonally decoupled via state feedback if and only if $rank\ B^* = p$. The integers $\{f_i\}$ are called the *decoupling indices* of the system. (Note that $B^* = \tilde{D}$.) It can be shown that the $p \times p$ matrix B^* can also be constructed from $\{A, B, C, D\}$ as follows: if the ith row of D is nonzero, this becomes the ith row of B^*. Otherwise, if f_i is the lowest integer, for which the ith row of $CA^{f_i-1}B$ is nonzero, then this row becomes the ith row of B^*.

(e) Consider the system $\{A, B, C\}$ of Exercise 4.20 and determine whether it can be diagonally decoupled via state feedback. Use both the state space matrices A, B, C and the transfer function $H(s)$ and verify that they result to the same matrix B^*.

4.18. Consider the system $\{A, B, C, D\}$ given by $\dot{x} = Ax + Bu$, $y = Cx + Du$ with $\det D \neq 0$, where (A, B) is controllable and (A, C) is observable. If $\{\bar{A}, \bar{B}, \bar{C}, \bar{D}\} = \{A - BD^{-1}C, BD^{-1}, -D^{-1}C, D^{-1}\}$ is its inverse system, show that the zeros of $\{A, B, C, D\}$ are the poles of $\{\bar{A}, \bar{B}, \bar{C}, \bar{D}\}$. Also, show that the poles of $\{A, B, C, D\}$ are the zeros of $\{\bar{A}, \bar{B}, \bar{C}, \bar{D}\}$.

4.19. Consider the system $\dot{x} = Ax + Bu$, $y = Cx + Du$, where

$$A = \begin{bmatrix} 0 & 0 \\ 0 & 0 \end{bmatrix}, \quad B = \begin{bmatrix} 1 & 0 \\ 0 & 1 \end{bmatrix}, \quad C = \begin{bmatrix} 1 & 0 \\ 1 & 2 \end{bmatrix}, \quad D = \begin{bmatrix} 1 & 0 \\ 0 & 1 \end{bmatrix}.$$

Let $u = Fx + r$ be a linear state feedback control law.

(a) Determine F so that the eigenvalues of $A + BF$ are $-1, -2$ and are unobservable from y. What is the closed loop transfer function $H_F(s)(\hat{y} = H_F\hat{r})$ in this case? *Hint:* Select the eigenvalues and eigenvectors of $A + BF$.

(b) Using the Structure Theorem of Chapter 3, verify that $N(s)D^{-1}(s)$

$$= \begin{bmatrix} s+1 & 0 \\ 1 & s+2 \end{bmatrix} \begin{bmatrix} s & 0 \\ 0 & s \end{bmatrix}^{-1} = \begin{bmatrix} \dfrac{s+1}{s} & 0 \\ \dfrac{1}{s} & \dfrac{s+2}{s} \end{bmatrix} = H(s) = C(sI - A)^{-1}B + D.$$

Express your results in (a) as $H(s)M(s) = (N(s)D^{-1}(s))(D(s)D_F^{-1}(s)) = N(s)D_F^{-1}(s) = H_F(s)$ (see Subsection 4.2D).

4.20. Consider the system $\dot{x} = Ax + Bu$, $y = Cx$, where

$$A = \begin{bmatrix} 0 & 1 & 0 \\ 0 & 0 & 0 \\ -1 & 0 & 1 \end{bmatrix}, \quad B = \begin{bmatrix} 0 & 0 \\ 1 & 0 \\ 0 & 1 \end{bmatrix}, \quad C = \begin{bmatrix} 1 & 1 & 0 \\ 1 & 0 & -1 \end{bmatrix}.$$

Note that

$$H(s) = N(s)D^{-1}(s) = \begin{bmatrix} s+1 & 0 \\ 1 & -1 \end{bmatrix} \begin{bmatrix} s^2 & 0 \\ 1 & s-1 \end{bmatrix}^{-1}.$$

Is it possible to determine the pair (F, G) in $u = Fx + Gr$ so that

$$H_{F,G}(s) = \begin{bmatrix} \dfrac{s+1}{(s+2)(s+3)} & 0 \\ 0 & \dfrac{1}{s+1} \end{bmatrix}?$$

If your answer is yes, determine such a pair (F, G).

Note that if it is required that $H_{F,G}(s)$ be diagonal with poles at any stable locations, then this is the problem of *diagonal decoupling via state feedback*.

Hint: Write $H(s) = \begin{bmatrix} s+1 & 0 \\ 0 & 1 \end{bmatrix} \tilde{H}(s)$ and work with $\tilde{H}(s)$ to determine (F, G) so that $\tilde{H}_{F,G}(s)$ is diagonal with poles at desired locations.

4.21. Consider the controllable and observable SISO system $\dot{x} = Ax + Bu$, $y = Cx$ with $H(s) = C(sI - A)^{-1}B$.

(a) If λ is not an eigenvalue of A, show that there exists an initial state x_0 such that the response to $u(t) = e^{\lambda t}, t \geq 0$, is $y(t) = H(\lambda)e^{\lambda t}, t \geq 0$. What happens if λ is a zero of $H(s)$?

(b) Assume that A has distinct eigenvalues. Let λ be an eigenvalue of A and show that there exists an initial state x_0 such that with "no input" ($u(t) \equiv 0$), $y(t) = ke^{\lambda t}, t \geq 0$, for some $k \in R$.

4.22. For the discrete-time case, derive expressions that correspond to formulas (2.40) to (2.45) of Subsection 4.2D.

4.23. Consider the controllable and observable SISO system $\dot{x} = Ax + bu + bw$, $y = cx$, where w is a constant unknown disturbance modelled by $\dot{w} = 0$. If an estimate of w, $\hat{w}$ is available, then we may attempt to cancel out the disturbance by selecting $u = -\hat{w}$. For this, consider w to be an additional state for the original system and determine an observer for this augmented system.

(a) Show that the augmented system is observable if and only if $s = 0$ is not a zero of the system [or of $H(s) = c(sI - A)^{-1}b$]. *Hint:* Use the eigenvalue criterion for observability.

(b) Assume that $\{A, b, c\}$ is stable (or has been stabilized) and select the gain of the observer to be $K^T = [0, k]$, where $k \in R$. Show that the compensator $u = -\hat{w}$, which asymptotically cancels out the constant disturbance, is an *integral feedback* compensator.

(c) Consider the original system $H(s) = c(sI - A)^{-1}b$ and use the results of Subsection 4.4B to design an output compensator that compensates for the above constant disturbance. For simplicity assume that $H(s)$ is stable. *Hint:* There must be a pole at $s = 0$. Note that the conditions in (a) must be satisfied.

4.24. Show that $\{A + BHC, B, C\}$ is controllable and observable for any $H \in R^{p\times m}$ if and only if $\{A, B, C\}$ is controllable and observable.

4.25. Consider the system

$$x(k+1) = \begin{bmatrix} -\frac{1}{2} & 0 & 0 \\ 0 & \frac{1}{2} & 0 \\ 0 & 0 & -\frac{1}{4} \end{bmatrix} x(k) + \begin{bmatrix} 1 & 0 \\ 0 & 1 \\ 0 & 0 \end{bmatrix} u(k), \; y(k) = \begin{bmatrix} 1 & 0 & 0 \\ 0 & 0 & 2 \end{bmatrix} x(k).$$

Is it possible to determine a linear state feedback law $u(k) = Fx(k) + r(k)$ so that the eigenvalues of $A + BF$ remain at exactly the same locations while $\{A + BF, B, C\}$ becomes observable? If the answer is affirmative, determine such F.

4.26. *Static, or constant output feedback* $u = Hy + r$, $H \in R^{m\times p}$, can be used to compensate a system $\dot{x} = Ax + Bu$, $y = Cx$, where $A \in R^{n\times n}$, $C \in R^{p\times n}$ and assign the eigenvalues of $A + BHC$ of the closed-loop system $\dot{x} = (A + BHC)x + Br$, $y = Cx$. In contrast to state feedback compensation, in general the closed-loop eigenvalues cannot be arbitrarily assigned using H even when $\{A, B, C\}$ is controllable and observable. It can be shown that one can arbitrarily assign "almost always" at least min $(m + p - 1, n)$ eigenvalues using H; note that when $p = m = 1$, only one eigenvalue can be arbitrarily assigned using H. To illustrate,

(a) Let A, B be as in Exercise 4.1, let $C = [1, 1]$, and determine H so that the eigenvalues of $A + BHC$ are at $-0.1025 \pm j0.04944$.

(b) Let A, B be as in Exercise 4.3, let $C = [1, 1]$, and determine H so that the eigenvalues of $A + BHC$ are all at zero.

(c) For an example (of a "nongenetic" case) where fewer than $p + m - 1$ eigenvalues can be arbitrarily assigned, consider

$$A = \begin{bmatrix} 0 & 1 & 0 & 0 \\ 0 & 0 & 0 & 0 \\ 0 & 0 & 1 & 0 \\ 0 & 0 & 0 & 0 \end{bmatrix}, \quad B = \begin{bmatrix} 0 & 0 \\ 1 & 0 \\ 0 & 0 \\ 0 & 1 \end{bmatrix}, \quad C = \begin{bmatrix} 1 & 0 & 0 & 0 \\ 0 & 0 & 1 & 0 \end{bmatrix}$$

and show that only $2 < p + m - 1 = 3$ eigenvalues can be assigned to $\pm\sqrt{\alpha}$, $\alpha \in R$.

Remark: When only some of the n eigenvalues are assigned, as was the case when using constant output feedback, care should be taken to guarantee that the remaining eigenvalues will also be stable, since H will typically shift all eigenvalues.

4.27. (**Inverted pendulum**) Consider the inverted pendulum described in Exercise 1.20 of Chapter 1. The linearized state-space model $\dot{x} = Ax + Bu$ (about $y = 0$) assuming the friction is zero is given by

$$\begin{bmatrix} \dot{x}_1 \\ \dot{x}_2 \\ \dot{x}_3 \\ \dot{x}_4 \end{bmatrix} = \begin{bmatrix} 0 & \frac{g}{L} + \frac{mg}{ML} & 0 & 0 \\ 1 & 0 & 0 & 0 \\ 0 & -\frac{mg}{M} & 0 & 0 \\ 0 & 0 & 1 & 0 \end{bmatrix} \begin{bmatrix} x_1 \\ x_2 \\ x_3 \\ x_4 \end{bmatrix} + \begin{bmatrix} -\frac{1}{ML} \\ 0 \\ \frac{1}{M} \\ 0 \end{bmatrix} u,$$

where $x_1 \triangleq \dot{\theta}$, $x_2 \triangleq \theta$, $x_3 = \dot{s}$, $x_4 = s$, and $u = \mu$, the force applied to the cart. Let $L = 1$ m, $m = 0.1$ kg, $M = 1$ kg, and $g = 10$ m/sec^2 to obtain

$$\begin{bmatrix} \dot{x}_1 \\ \dot{x}_2 \\ \dot{x}_3 \\ \dot{x}_4 \end{bmatrix} = \begin{bmatrix} 0 & 11 & 0 & 0 \\ 1 & 0 & 0 & 0 \\ 0 & -1 & 0 & 0 \\ 0 & 0 & 1 & 0 \end{bmatrix} \begin{bmatrix} x_1 \\ x_2 \\ x_3 \\ x_4 \end{bmatrix} + \begin{bmatrix} -1 \\ 0 \\ 1 \\ 0 \end{bmatrix} u.$$

(a) Determine the eigenvalues and eigenvectors of A.
(b) Plot the states when $x(0) = [0.01, 0, 0, 0]^T$. Repeat for zero initial conditions and u the unit step. Comment on your results.
(c) Determine the linear state feedback control law $u = Fx + r$ so that the closed-loop system eigenvalues are at $-10, -3, -1$, and -0.5.
(d) Use the LQR formulation to determine a stabilizing linear state feedback control law $u = Fx + r$. Comment on your choices for the weights.
(e) Let $x(0)^T = [0.5, 0, 0, 0]$. Repeat (b) for the closed-loop system derived in (c) and (d).

4.28. (**Armature voltage-controlled dc servomotor**) Consider the armature voltage-controlled dc servomotor of Exercise 2.71 in Chapter 2. Let $y = x_1$.
(a) Design a full-order state observer with eigenvalues at $-5, -1$, and -0.5 to derive an estimate $\hat{x}(t)$ of the state $x(t)$. For $x(0)^T = [\pi/6, 0, 0]$, $\hat{x}(0)^T = [0, 0, 0]$ plot the error $e(t) = x(t) - \hat{x}(t)$ for $t \geq 0$ and comment on your results.
(b) Assume that the system is driven by a zero-mean Gaussian, white-noise w and the measurement noise v is also zero-mean Gaussian, white noise, where w and v are uncorrelated in time with covariances $W = 10^{-3}$ and $V = 10^{-2}$, respectively. The state-space description of the system is now $\dot{x} = Ax + Bu + \Gamma w$, $y = Cx + v$, where $\Gamma = [1, 1, 0]^T$. Design an optimal observer and compare with (a).

4.29. (**Automobile suspension system**) Consider the automobile suspension system of Exercise 2.74 in Chapter 2. Assume that the state-space description of the system is $\dot{x} = Ax + Bu + \Gamma w$, $y = Cx + v$ with A, B from Exercise 2.74, $C = [0, 0, 1, 0]$, and $\Gamma = [-1, 0, 0, 0]^T$. Both process noise w and measurement noise v are assumed to be uncorrelated, zero-mean Gaussian, stochastic processes with covariances $W = 7 \times 10^{-4}$ and $V = 10^{-8}$, respectively. Let the damping constant $c = 750$ N sec/m and the velocity of the car $v = 18$ m/sec.
(a) Design an optimal LQG observer-based dynamic controller. Comment on your choice of the weights.
(b) Using the controller from (a), plot the states $x(t)$ and the control input $u(t)$ for $t \geq 0$ when $x(0) = [1, 0, 0, 0]^T$, $w = 0$, $v = 0$, and the reference input of the system is $r(t) = \frac{1}{6}\sin(2\pi vt/20)$.

4.30. (**Aircraft dynamics**) Consider the systems describing the aircraft dynamics in Exercise 2.76 in Chapter 2.

(a) For the state-space representation of the longitudinal motion of the fighter AFTI-16, design a linear state feedback control law $u = Fx + r$ so that the closed-loop system eigenvalues are at $-1.25 \pm j2.2651$ and $-0.01 \pm j0.095$.

(b) Let $y = Cx$ with $C = [0, 0, 1, 0]$. Design a full-order state observer with eigenvalues at $0, -0.421, -0.587$, and -1.

(c) Let the system be compensated via the state feedback control law $u = F\hat{x} + r$, where $\hat{x}$ is the output of the state estimator. Derive the state-space representation and the transfer function between y and r of the compensated system. Is the system fully controllable from r? Explain.

(d) Use the LQR formulation to determine a stabilizing linear state feedback control law $u = Fx + r$. Comment on your choices for the weights.

(e) Assume that process noise w and measurement noise v are present and that both are uncorrelated, zero-mean Gaussian, stochastic processes with covariances $W = 10^{-4}$ and $V = 10^{-2}$, respectively. Let $\Gamma = [0, 1, 1, 0]^T$ and design an optimal observer.

(f) Design an optimal LQG observer-based dynamic controller and determine the eigenvalues of the closed-loop system. Discuss your answer in view of the results in (c).

4.31. (**Chemical reaction process**) [C. E. Rohrs, J. M. Melsa, and D. G. Schultz, *Linear Control Systems,* McGraw-Hill, 1993, p. 70.] Consider the process depicted schematically in Fig. 4.8. A reaction tank of volume $V = 5,000$ gallons accepts a feed of reactant that contains a substance A in concentration $C_{A,0}$. The feed enters at a rate of F gallons per hour and at a temperature T_0. In the tank some of the reactant A is turned into the desired product B. The output product is removed from the tank at the same rate that the feed enters the tank. The mixture in the tank has a uniform concentration of A, C_A and a uniform temperature T.

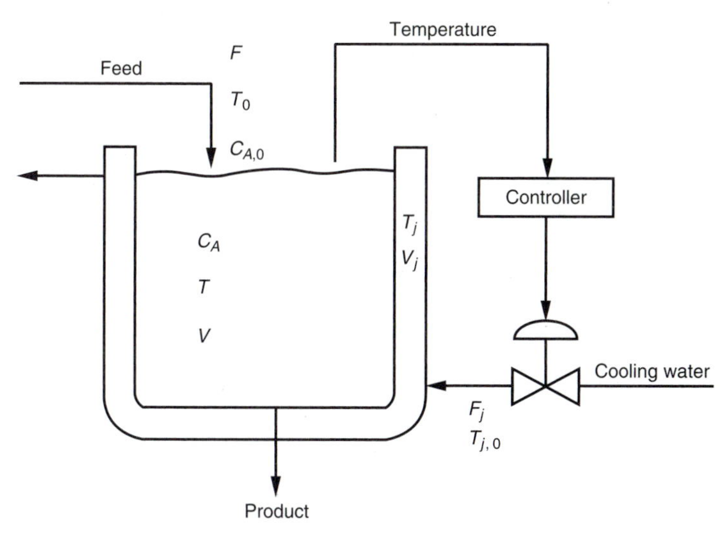

FIGURE 4.8
Chemical reaction process

The temperature of the water in the jacket is assumed uniform at T_J. The temperature of the water flowing into the jacket is $T_{J,0}$. The system is controlled by measuring the temperature T in the tank and controlling the flow of the water in the jacket F_J by activating a valve.

The equations describing the evolution of C_A, T, and T_J are

$$\dot{C}_A = \frac{F}{V}C_{A,0} - \frac{F}{V}T - k_1 C_A e^{-(k_2/T)}$$

$$\dot{T} = \frac{F}{V}T_0 - \frac{F}{V}T - k_3 k_1 C_A e^{-(k_2/T)} - k_4(T - T_J)$$

$$\dot{T}_J = \frac{F_J}{V_J}T_{J,0} - \frac{F_J}{V_J}T_J + k_5(T - T_J),$$

where $C_{A,0} = 0.5$, $T_0 = 70°\text{F}$, $T_{J,0} = 70°\text{F}$, and k_1, k_2, k_3, k_4, k_5 are appropriate constants.

A linearized state-space model $\dot{x} = Ax + Bu$, $y = Cx$ around the equilibrium point $\bar{C}_A = 0.245$, $\bar{T} = 140$, $\bar{T}_J = 93.3$ is given by

$$\begin{bmatrix} \dot{x}_1 \\ \dot{x}_2 \\ \dot{x}_3 \end{bmatrix} = \begin{bmatrix} -1.7 & -2.13 \times 10^{-4} & 0 \\ 696 & 2.9 & 2.4 \\ 0 & 6.5 & -19.5 \end{bmatrix} \begin{bmatrix} x_1 \\ x_2 \\ x_3 \end{bmatrix} + \begin{bmatrix} 0 \\ 0 \\ -0.16 \end{bmatrix} u$$

$$y = [0, 1, 0] \begin{bmatrix} x_1 \\ x_2 \\ x_3 \end{bmatrix},$$

where $x_1 \triangleq \delta C_A$, $x_2 = \delta T$, $x_3 = \delta T_J$, and $u = \delta F_J$.

(a) Determine the eigenvalues and eigenvectors of A. Is the system asymptotically stable? Explain.
(b) Let the input be the unit step indicating that the cooling flow is increased and held at the new value. Plot the states for $t \geq 0$ assuming zero initial conditions (equilibrium values).
(c) Plot the states for $t \geq 0$ for zero input but with an initial concentration of substance A slightly larger than the equilibrium value, namely, $x(0) = [0.1, 0, 0]^T$.
(d) Determine the linear state feedback control law $u = Fx + r$ so that the closed-loop system eigenvalues are at -5, -10, and -10.
(e) Use the LQR formulation to derive a stabilizing linear state feedback control law $u = Fx + r$. Comment on your choices of the weights.
(f) Repeat (b) and (c) for the closed-loop system derived in (c) and (d).

4.32. (**Economic model for national income**) Consider the economic model for national income in Exercise 2.68 in Chapter 2.
(a) In which cases, (i), (ii), or (iii), is the system reachable?
(b) For case (i), design a linear state feedback control law to place both eigenvalues at zero. This corresponds to a strategy for government spending that will return deviations in consumer expenditure and private investment to zero.

4.33. (**Read/write head of a hard disk**) Consider the discrete-time model of the read/write head of a hard disk described in Exercise 2.77 of Chapter 2.
(a) Find a linear state feedback control law to assign both eigenvalues at zero. Plot the response of the discrete-time closed-loop system to a unit step and comment on your results.
(b) Let $y(k) = \theta(k)$ be the position of the head at time k. Design an appropriate observer of the state and use it together with the control law determined in (a). Plot the response to a unit step and compare your results to the results in (a).

CHAPTER 5

Realization Theory and Algorithms

When a linear system is described by an internal description it is straightforward to derive its external description. In particular, given a state-space description for a linear system, the impulse response, and also the transfer function in the case of time-invariant systems, were readily expressed in terms of the state-space coefficient matrices in previous chapters. In this chapter the inverse problem is being addressed: given an external description of a linear system, specifically, its transfer function or its impulse response, determine an internal, state-space description for the system that generates the given transfer function. This is the problem of *system realization.* The name reflects the fact that if a (continuous-time) state-space description is known, an operational amplifier circuit can be built in a straightforward manner to realize (actually simulate) the system response.

The ability of realizing systems that exhibit desired input-output behavior is very important in applications. In the design of a system (such as a controller or a filter), the desired system behavior is frequently specified in terms of its transfer function, which is typically obtained by some desired frequency response. One must then implement a system by hardware or software that exhibits the desired input-output behavior described by the transfer function. This in effect corresponds to building a system by combining typically less complex systems in parallel, feedback, or cascade configurations. In terms of block diagrams, this corresponds to building a system by combining simpler blocks. There are of course many ways, an infinite number in fact, of realizing a given transfer function. Presently, we are interested in realizations that contain the least possible number of energy or memory storage elements, i.e., in realizations of least order (in terms of differential or difference equations). To accomplish this, the concepts of controllability and observability play a central role. Indeed, it turns out that realizations of transfer functions of least order are both controllable and observable. The theory of realizations presented in this chapter also sheds light on the behavior of systems that are built by interconnecting several other systems, as for example, in feedback control systems. In such systems, possible pole-zero cancellations between transfer functions of different subsystems can be studied, using internal descriptions and the notion of controllability and observability (see also Subsections 7.3B and 7.3C in Chapter 7).

5.1 INTRODUCTION

The goal of this chapter is to introduce the theory of realization, to establish fundamental existence and minimality results, and to develop several realization algorithms. The emphasis is on time-invariant, continuous-time, and discrete-time systems. In what follows, we first provide a glimpse of the contents of this chapter. This is followed by some guidelines for the reader.

A. Chapter Description

In Section 5.2, the problem of system realization is introduced, both for continuous-time and discrete-time systems. State-space realizations of impulse and pulse responses for time-varying and time-invariant systems and of transfer functions (in the case of time-invariant systems) are discussed.

In Section 5.3, the existence of state-space realizations is considered first, and results are presented for both time-varying and time-invariant cases. The remainder of our development of realization theory and algorithms in this chapter concentrates primarily on time-invariant, continuous-time, and discrete-time systems. Minimal or irreducible realizations are discussed. For the time-invariant case it is shown that a state-space realization is irreducible if and only if it is both controllable and observable. Also, it is shown that if two realizations are minimal, then they must be equivalent. The order of minimal realizations is considered next, and it is shown that it can be determined directly from a given transfer function matrix without first finding a realization. This is accomplished by use of the pole polynomial of the transfer function that determines its McMillan degree and also by use of the Hankel matrix of the Markov parameters of a system. In addition, it is shown that in any minimal realization, the pole polynomial of the transfer function is the characteristic polynomial of the matrix A.

In Section 5.4, a number of realization algorithms are presented. The use of duality in obtaining realizations is also highlighted. Algorithms for obtaining realizations in controller and observer form are introduced. The SISO case is treated first in the interest of clarity. Realizations with the matrix A in diagonal or block companion form are also derived. Finally, singular-value decomposition is used to obtain transfer function realizations, such as balanced realizations, in a computationally efficient manner.

B. Guidelines for the Reader

In this chapter state-space realizations of input-output descriptions, which are primarily in transfer function matrix form, are developed. In the present treatment, fundamental results are emphasized. We point out that realization theory is among the first important principal topics studied in system theory. Existence and minimality results of state-space realizations are established and several realization algorithms are developed. We note that detailed summaries of the contents of the sections are given at the beginning of Sections 5.3 and 5.4.

Existence of time-invariant state-space realizations of a transfer function matrix $H(s)$ is addressed in Subsection 5.3A, Theorem 3.3, while minimality is fully explored in two results, Theorems 3.9 and 3.10 in Subsection 5.3B. It is useful to be able to determine the order of a minimal realization directly from $H(s)$ and this is discussed in Subsection 5.3C; we note that the pole polynomial of $H(s)$ introduced in Section 3.5 of Chapter 3 is required in one of the approaches presented. In Section 5.4, several realization algorithms are developed. The use of duality in realizations is emphasized in Subsection 5.4A.

At a first reading, the reader may concentrate on minimality of realizations in Subsection 5.3B and on the order of minimal realizations in Subsection 5.3C; then study one or two realization algorithms, e.g., the one that leads to a realization with A diagonal in Subsection 5.4C, and to a realization with A, B in controller form in Subsection 5.4B. It is also important to study Subsection 5.4A on realizations using duality. If realization algorithms with good numerical properties are of primary interest, then the reader should concentrate on Subsection 5.4E, where realizations using singular-value decomposition are presented.

5.2 STATE-SPACE REALIZATIONS OF EXTERNAL DESCRIPTIONS

In this section, state-space realizations of impulse responses for time-varying and time-invariant systems and of transfer functions for time-invariant systems are introduced. Continuous-time systems are discussed first in Subsection 5.2A, followed by discrete-time systems in Subsection 5.2B.

A. Continuous-Time Systems

Before formally defining the problem of system realization, we first review some of the relations that were derived in Chapter 2.

We consider a system described by equations of the form

$$\dot{x} = A(t)x + B(t)u, \qquad y = C(t)x + D(t)u, \tag{2.1}$$

where $A(t) \in R^{n\times n}$, $B(t) \in R^{n\times m}$, $C(t) \in R^{p\times n}$, and $D(t) \in R^{p\times m}$ are continuous matrices over some open time interval (a, b). The response of this system is given by

$$y(t) = C(t)\Phi(t, t_0)x_0 + \int_{t_0}^{t} H(t, \tau)u(\tau)\, d\tau, \tag{2.2}$$

where $\Phi(t, t_0)$ is the $n \times n$ state transition matrix of $\dot{x} = A(t)x$, $x(t_0) = x_0$ (the initial condition), and $H(t, \tau)$ is the $p\times m$ impulse response matrix of this system, given by

$$H(t, \tau) = \begin{cases} C(t)\Phi(t, \tau)B(\tau) + D(t)\delta(t - \tau), & \text{for } t \geq \tau, \\ 0, & \text{for } t < \tau. \end{cases} \tag{2.3}$$

In the time-invariant case, (2.1) assumes the form

$$\dot{x} = Ax + Bu, \qquad y = Cx + Du, \tag{2.4}$$

and the system response is in this case given by

$$y(t) = Ce^{At}x_0 + \int_0^t H(t, \tau)u(\tau)\, d\tau, \tag{2.5}$$

where, without loss of generality, t_0 was taken to be zero. The impulse response is now given by the expression

$$H(t, \tau) = \begin{cases} Ce^{A(t-\tau)}B + D\delta(t - \tau), & \text{for } t \geq \tau, \\ 0, & \text{for } t < \tau. \end{cases} \tag{2.6}$$

Recall that the time invariance of system (2.4) implies that $H(t, \tau) = H(t-\tau, 0)$, and therefore, τ, which is the time at which a unit impulse input is applied to the system, can be taken to equal zero ($\tau = 0$), without loss of generality, to yield $H(t, 0)$. The transfer function matrix of the system is the (one-sided) Laplace transform of $H(t, 0)$, namely,

$$H(s) = \mathscr{L}[H(t, 0)] = C(sI - A)^{-1}B + D. \tag{2.7}$$

Let $\{A(t), B(t), C(t), D(t)\}$ denote the system description (2.1) and let $H(t, \tau)$ be a $p \times m$ matrix with real functions of arguments t and τ as entries.

DEFINITION 2.1. *A realization of* $H(t, \tau)$ is any set $\{A(t), B(t), C(t), D(t)\}$, the impulse response of which is $H(t, \tau)$. That is, $\{A(t), B(t), C(t), D(t)\}$ is a realization of $H(t, \tau)$ if (2.3) is satisfied. (See Fig. 5.1.) ■

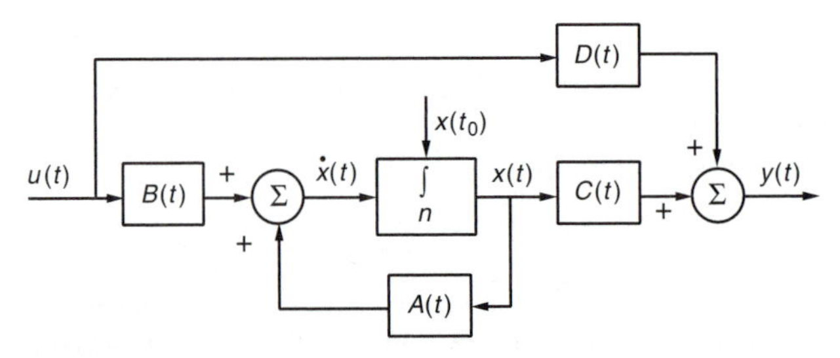

FIGURE 5.1
Block diagram realization of $\{A(t), B(t), C(t), D(t)\}$

Note that it is not necessary that any given $p \times m$ matrix $H(t, \tau)$ be the impulse response to some system of the form (2.1). The conditions on $H(t, \tau)$ under which a realization $\{A(t), B(t), C(t), D(t)\}$ exists are given in the next section.

In the time-invariant case, a realization is commonly defined in terms of the transfer function matrix. We let $\{A, B, C, D\}$ denote the system description given in (2.4) and we let $H(s)$ be a $p \times m$ matrix with entries that are functions of s.

DEFINITION 2.2. A *realization of* $H(s)$ is any set $\{A, B, C, D\}$, the transfer function matrix of which is $H(s)$, i.e., $\{A, B, C, D\}$ is a realization of $H(s)$ if (2.7) is satisfied. ■

As will be shown in the next section, given $H(s)$, a condition for a realization $\{A, B, C, D\}$ of $H(s)$ to exist is that all entries in $H(s)$ are proper, rational functions.

Alternative conditions under which a given set $\{A, B, C, D\}$ is a realization of some $H(s)$ can easily be derived. To this end, we expand $H(s)$ in a Laurent series to obtain

$$H(s) = H_0 + H_1 s^{-1} + H_2 s^{-2} + \cdots. \tag{2.8}$$

DEFINITION 2.3. The terms H_i, $i = 0, 1, 2, \ldots$ in (2.8) are the *Markov parameters* of the system. ■

The Markov parameters can be determined by the formulas

$$H_0 = \lim_{s\to\infty} H(s), \qquad H_1 = \lim_{s\to\infty} s(H(s) - H_0),$$

$$H_2 = \lim_{s\to\infty} s^2(H(s) - H_0 - H_1 s^{-1}),$$

and so forth. Recall that relations involving the Markov parameters were alluded to earlier in Exercise 2.63 of Chapter 2.

THEOREM 2.1. The set $\{A, B, C, D\}$ is a realization of $H(s)$ if and only if

$$H_0 = D \qquad \text{and} \qquad H_i = CA^{i-1}B, \qquad i = 1, 2, \ldots. \tag{2.9}$$

Proof. $H(s) = D + C(sI - A)^{-1}B = D + Cs^{-1}(I - s^{-1}A)^{-1}B = D + Cs^{-1}[\sum_{i=0}^{\infty}(s^{-1}A)^i]B = D + \sum_{i=1}^{\infty}[CA^{i-1}B]s^{-i}$, from which (2.9) is derived in view of (2.8). ■

By definition, the impulse response description of a linear system contains no information about the initial conditions, or the initial energy stored in the system. In fact, $H(t, \tau)$ is determined by assuming that the system is at rest before τ, the time when the impulse input $\delta(t - \tau)$ is applied. It is therefore apparent that different state-space realizations of $H(t, \tau)$ will yield the *same zero-state response,* while their zero-input response, which depends on initial conditions, can be quite different.

Note that if a realization of a given $H(t, \tau)$ exists, then there are *infinitely many realizations.* Given $\{A(t), B(t), C(t), D(t)\}$, a realization of $H(t, \tau)$, other realizations with the same dimension n of the state vector can readily be generated by means of equivalence transformations. Recall from Chapter 2 that equivalent representations generate the same impulse response. To illustrate, in Example 3.1 of Section 5.3, the system $\dot{x} = P(t)N(t)u(t)$ and $y(t) = M(t)P^{-1}(t)x(t)$ with $N(t) = \begin{bmatrix} 1 \\ -t \end{bmatrix}$ and $M(t) = [t, 1]$ is a realization of $H(t, \tau) = t - \tau$ for any $P(t)$ such that $P^{-1}(t)$ and $\dot{P}(t)$ exist and are continuous.

B. Discrete-Time Systems

The problem of realization in the discrete-time case is defined as in the continuous-time case: given an external description, either the unit pulse [(discrete) impulse] response $H(k, \ell)$, or in the time-invariant case, typically the transfer function matrix $H(z)$ (see Chapter 2), determine an internal state-space description, the pulse response of which is the given $H(k, \ell)$.

The realization theory in the discrete-time case essentially parallels the continuous-time case. There are of course certain notable differences because in the present case the realizations are difference equations instead of differential equations. We point to these differences in the subsequent sections.

Some of the relations derived in Section 2.7 of Chapter 2 will be recalled next. We consider systems described by equations of the form

$$x(k+1) = A(k)x(k) + B(k)u(k), \qquad y(k) = C(k)x(k) + D(k)u(k), \quad (2.10)$$

where $A(k) \in R^{n\times n}$, $B(k) \in R^{n\times m}$, $C(k) \in R^{p\times n}$, and $D(k) \in R^{p\times m}$. The response of this system is given by the expression

$$y(k) = C(k)\Phi(k, k_0)x_0 + \sum_{i=k_0}^{k-1} H(k, i)u(i), \qquad k > k_0, \quad (2.11)$$

where $\Phi(k, k_0)$ denotes the $n \times n$ state transition matrix of the system $x(k+1) = A(k)x(k)$, $x(k_0) = x_0$ is the initial condition, and $H(k, i)$ is the $p \times m$ pulse response matrix given by

$$H(k, i) = \begin{cases} C(k)\Phi(k, i+1)B(i), & k > i, \\ D(k), & k = i, \\ 0, & k < i. \end{cases} \quad (2.12)$$

The state transition matrix can readily be determined to be

$$\Phi(k, \ell) = \begin{cases} A(k-1)A(k-2)\cdots A(\ell), & k > \ell, \\ I, & k = \ell. \end{cases} \quad (2.13)$$

In the time-invariant case, (2.10) assumes the form

$$x(k+1) = Ax(k) + Bu(k), \qquad y(k) = Cx(k) + Du(k), \quad (2.14)$$

and the system response of (2.14) is given by

$$y(k) = CA^k x_0 + \sum_{i=0}^{k-1} H(k, i)u(i), \qquad k > 0, \quad (2.15)$$

where, without loss of generality, k_0 was taken to be zero. The pulse response is now given by

$$H(k, i) = \begin{cases} CA^{k-(i+1)}B, & k > i, \\ D, & k = i, \\ 0, & k < i. \end{cases} \quad (2.16)$$

Recall that since the system (2.14) is time-invariant, $H(k, i) = H(k - i, 0)$ and i, the time the pulse input is applied, can be taken to be zero, to yield $H(k, 0)$ as the external system description. The transfer function matrix for (2.14) is now the (one-sided) z-transform of $H(k, 0)$. We have,

$$H(z) = \mathcal{Z}\{H(k, 0)\} = C(zI - A)^{-1}B + D. \tag{2.17}$$

Now let $\{A(k), B(k), C(k), D(k)\}$ denote the system description (2.10) and let $H(k, i)$ be a $p \times m$ matrix with real function entries of arguments k and i defined for $k \geq i$.

DEFINITION 2.4. A *realization of* $H(k, i)$ is any set of matrices $\{A(k), B(k), C(k), D(k)\}$, the pulse response of which is $H(k, i)$. ■

Let $H(z)$ be a $p \times m$ matrix with functions of z as entries.

DEFINITION 2.5. A *realization of* $H(z)$ is any set $\{A, B, C, D\}$, the transfer function matrix of which is $H(z)$. ■

A result that is analogous to Theorem 2.1 is also valid in the discrete-time case [with $H(s)$ replaced by $H(z)$] (show this). The remarks following Theorem 2.1 concerning the zero-state response of a system and the uniqueness of realizations are also valid in the present case. Thus, all realizations of the pulse response or the transfer function will yield the same zero-state response, while their zero-input response, which depends on initial conditions, can be quite different. Also, if a realization of $H(k, i)$ or $H(z)$ exists, then there exists an infinite number of realizations (show this).

5.3 EXISTENCE AND MINIMALITY OF REALIZATIONS

In this section the existence and minimality of internal state-space realizations of a given external description are determined.

The existence of realizations is examined first. Given a $p \times m$ matrix $H(t, \tau)$, conditions under which this matrix is the impulse response of a linear system described by equations of the form $\dot{x} = A(t)x + B(t)u$, $y = C(t)x + D(t)u$ are given in Theorem 3.1. Theorem 3.2 provides the corresponding result for time-invariant systems. For such systems, $H(s)$ is typically given in place of $H(t, \tau)$, and conditions for $H(s)$ to be the transfer function matrix of a system described by equations of the form $\dot{x} = Ax + Bu$, $y = Cx + Du$ are given in Theorem 3.3. It is shown that such realizations exist if and only if $H(s)$ is a matrix of rational functions with the property that $\lim_{s\to\infty} H(s)$ is finite. The corresponding results for discrete-time systems are then developed and presented in Theorems 3.5, 3.6, and 3.7.

Realizations of least order, also called minimal or irreducible realizations, are of interest to us since they realize a system, using the least number of dynamical elements (minimum number of elements with memory). The main emphasis in this section is on time-invariant systems and realizations of transfer function matrices $H(s)$. The principal results are given in Theorems 3.9 and 3.10, where it is shown that minimal realizations are controllable (-from-the-origin) and observable and that all minimal realizations of $H(s)$ are equivalent representations. The order of any minimal realization can be determined directly without first determining a minimal realization, and this can be accomplished by using the characteristic polynomial and the degree of $H(s)$ (Theorem 3.11) or from the rank of a Hankel matrix (Theorem 3.13). All the results on minimality of realizations apply to the discrete-time case as well with no substantial changes. This is discussed at the end of the section.

A. Existence of Realizations

Continuous-time systems

Let the $p \times m$ matrix $H(t, \tau)$ with $t, \tau \in (a, b)$ be given. A realization of $H(t, \tau)$ was defined in the previous section as an internal description of the form (2.1) denoted by $\{A(t), B(t), C(t), D(t)\}$, the impulse response of which is $H(t, \tau)$.

THEOREM 3.1. $H(t, \tau)$ is realizable as the impulse response of a system described by (2.1) if and only if $H(t, \tau)$ can be decomposed into the form

$$H(t, \tau) = M(t)N(\tau) + D(t)\delta(t - \tau) \tag{3.1}$$

for $t \geq \tau$, where M, N, and D are $p \times n$, $n \times m$, and $p \times m$ matrices, respectively, with continuous real-valued entries and with n finite.

Proof. (*Sufficiency*) Assume the decomposition (3.1) is true and consider the realization $\{0, N(t), M(t), D(t)\}$, which yields the system $\dot{x} = N(t)u(t)$, $y(t) = M(t)x(t) + D(t)u(t)$. The state transition matrix is $\Phi(t, \tau) = I$, since the homogeneous equation is in this case $\dot{x} = A(t)x = 0$. Applying (2.3), we obtain as the impulse response, $\hat{H}(t, \tau) = M(t) \cdot I \cdot N(\tau) + D(t)\delta(t - \tau)$ for $t \geq \tau$, and $\hat{H}(t, \tau) = 0$ for $t < \tau$, which equals $H(t, \tau)$.

(*Necessity*) Let (2.1) be a realization of $H(t, \tau)$. Then (2.3) is true, and in view of the identity $\Phi(t, \tau) = \Phi(t, \sigma)\Phi(\sigma, \tau)$, $H(t, \tau)$ can be written for $t \geq \tau$ as $H(t, \tau) = [C(t)\Phi(t, \sigma)][\Phi(\sigma, \tau)B(\tau)] + D(t)\delta(t - \tau) = M(t)N(\tau) + D(t)\delta(t - \tau)$, where $M(t) \triangleq C(t)\Phi(t, \sigma)$ and $N(\tau) \triangleq \Phi(\sigma, \tau)B(\tau)$ (σ fixed). Therefore, the decomposition (3.1) is necessary. ∎

Since the matrices in (2.1) are taken to be continuous, we require in Theorem 3.1 that $M(t)$, $N(t)$, and $D(t)$ be continuous, although this restriction is not necessary.

EXAMPLE 3.1. Let $H(t, \tau) = t - \tau$, $t \geq \tau$. Then $H(t, \tau) = [t, 1]\begin{bmatrix} 1 \\ -\tau \end{bmatrix} = M(t)N(\tau)$. Therefore, $\dot{x} = \begin{bmatrix} 1 \\ -t \end{bmatrix} u(t)$, $y(t) = [t, 1]x(t)$ is a realization. ∎

EXAMPLE 3.2. Let $H(t, \tau) = 1/(t - \tau)$. In this case a decomposition of the form (3.1) does not exist, and therefore, $H(t, \tau)$ does not have a realization of the form (2.1). ∎

If a system is time-invariant and is described by (2.4), its impulse response is given by (2.6). The following result establishes necessary and sufficient conditions for the existence of time-invariant realizations.

THEOREM 3.2. $H(t, \tau)$ is realizable as the impulse response of a system described by (2.4) if and only if $H(t, \tau)$ can be decomposed for $t \geq \tau$ into the form

$$H(t, \tau) = M(t)N(\tau) + D(t)\delta(t - \tau), \tag{3.2}$$

where $M(t)$ and $N(t)$ are differentiable and

$$H(t, \tau) = H(t - \tau, 0). \tag{3.3}$$

The first part of this result is identical to Theorem 3.1. For time invariance, we require in addition relation (3.3) and differentiability.

Proof. (*Necessity*) Let (2.4) be a realization. Then in view of (2.6), $H(t, \tau) = H(t - \tau, 0) = Ce^{A(t-\tau)}B + D\delta(t - \tau) = (Ce^{At})(e^{-A\tau}B) + D\delta(t - \tau)$. Let $M(t) \triangleq Ce^{At}$, $N(\tau) \triangleq e^{-A\tau}B$, and note that both $M(t)$ and $N(t)$ are differentiable.

(*Sufficiency*) The proof of this part is much more involved. The complete proof can be found, for example, in Brockett [1], p. 99. In the following, we give an outline of the proof (a proof by construction). First, it can be shown that given $H(t, \tau)$ with decomposition (3.1), a realization of the form $\dot{x} = N(t)u(t)$, $y(t) = M(t)x(t) + D(t)u(t)$ can be found where n, the dimension of the state vector, is the smallest possible. Note that this system is controllable and observable. Now define $W(t_0, t_1) \triangleq \int_{t_0}^{t_1} N(\sigma)N^T(\sigma)\, d\sigma$ and $W_1(t_0, t_1) \triangleq \int_{t_0}^{t_1} [(d/d\sigma)N(\sigma)]\, N^T(\sigma)d\sigma$. Then, using $H(t, \tau) = H(t - \tau, 0)$, it can be shown that $\{A, B, C, D\} = \{-W_1(t_0, t_1)W^{-1}(t_0, t_1), N(0), M(0), D\}$ is a realization of $H(t, \tau)$. Note that $W^{-1}(t_0, t_1)$ exists because $\{0, N(t), M(t), D\}$ was taken to be of minimal dimension. Therefore, this system is controllable. ■

EXAMPLE 3.3. Consider again $H(t, \tau) = t - \tau$ given in Example 3.1, where a time-varying realization was derived. This $H(t, \tau)$ certainly satisfies the conditions of Theorem 3.2, and thus, a time-invariant realization $\dot{x} = Ax + Bu$, $y = Cx + Du$ also exists. Unfortunately, the proof of Theorem 3.2 does not provide us with a means of deriving such realizations. In general it is easier to consider the Laplace transform of $H(t, \tau)$ and use the algorithms that we will develop in Section 5.4 to show that a time-invariant realization of $H(t, \tau) = t - \tau$ is given by $\dot{x} = \begin{bmatrix} 0 & 1 \\ 0 & 0 \end{bmatrix} x + \begin{bmatrix} 0 \\ 1 \end{bmatrix} u$, $y = [1, 0]x$. (Verify this and also see Example 3.4.) ■

In the remainder of this chapter we shall concentrate primarily on time-invariant realizations of impulse responses. In fact we shall assume that the system transfer function matrix $H(s)$, which is the Laplace transform of the impulse response, is given and we shall introduce methods for deriving realizations directly from $H(s)$.

Given a $p \times m$ matrix $H(s)$, the following result establishes necessary and sufficient conditions for the existence of time-invariant realizations.

THEOREM 3.3. $H(s)$ is realizable as the transfer function matrix of a time-invariant system described by (2.4) if and only if $H(s)$ is a matrix of rational functions and satisfies

$$\lim_{s \to \infty} H(s) < \infty, \tag{3.4}$$

i.e., if and only if $H(s)$ is a *proper rational matrix*.

Proof. (*Necessity*) If the system $\dot{x} = Ax + Bu$, $y = Cx + Du$ is a realization of $H(s)$, then $C(sI - A)^{-1}B + D = H(s)$, which shows that $H(s)$ must be a rational matrix. Furthermore,

$$\lim_{s \to \infty} H(s) = D, \tag{3.5}$$

which is a real finite matrix.

(*Sufficiency*) If $H(s)$ is a proper rational matrix, then any of the algorithms discussed in the next section can be applied to derive a realization. ■

EXAMPLE 3.4. Let $H(s) = 1/s^2$, $(\hat{y}(s) = H(s)\hat{u}(s))$, which is the transfer function of the double integrator. Then, using the controller form realization algorithm of Section 5.4, a realization of $H(s)$ is given by $\dot{x} = \begin{bmatrix} 0 & 1 \\ 0 & 0 \end{bmatrix} x + \begin{bmatrix} 0 \\ 1 \end{bmatrix} u$, $y = [1, 0]x$. Notice that $\mathcal{L}^{-1}[H(s)] = \mathcal{L}^{-1}[1/s^2] = t = H(t, 0)$, the same as in Examples 3.1 and 3.3. ■

Theorem 3.3 can be used to show what types of entries are required for $H(t, 0)$ for it to be realizable as a linear time-invariant continuous-time system of the form given in (2.4). In particular, $H(t, 0) = \mathcal{L}^{-1}[H(s)]$ and the fact that all entries of $H(s)$ are proper rational functions (Theorem 3.3) implies the next result.

COROLLARY 3.4. A $p \times m$ matrix $H(t)(H(t, 0))$ is realizable as the impulse response of a system described by equations of the form (2.4) if and only if all entries of $H(t)$ are sums of terms of the form $\alpha t^k e^{\lambda t}$ and $\beta\delta(t)$, where α, β are real numbers, k is an integer ($k \geq 0$), and λ is a complex scalar.

Proof. The proof is left as an exercise for the reader. (Refer to Chapter 2, Subsection 2.4B for a review of Laplace transforms. Also, refer to the discussion of modes and asymptotic behavior of a system in that chapter.) ■

EXAMPLE 3.5. Let $H(t) = [e^t + e^{-t} + \delta(t), e^t]$. In view of the above corollary, $H(t)$ is realizable as the impulse response of a system described by (2.4). In fact, $H(s) = \mathcal{L}\{H(t)\} = \left[\dfrac{s^2 + 2s - 1}{s^2 - 1}, \dfrac{1}{s - 1}\right]$ and the system $\dot{x} = Ax + Bu$, $y = Cx + Du$ with $A = \begin{bmatrix} -1 & 2 \\ 0 & 1 \end{bmatrix}$, $B = \begin{bmatrix} 2 & 1 \\ 1 & 1 \end{bmatrix}$, $C = [1, 0]$, $D = [1, 0]$ is a realization (verify this). Comparing with Theorem 3.2, we write

$$\begin{aligned} H(t - \tau, 0) &= [e^{t-\tau} + e^{-(t-\tau)} + \delta(t - \tau), e^{t-\tau}] \\ &= [e^{-t}, e^t - e^{-t}] \begin{bmatrix} e^{\tau} + e^{-\tau} & e^{-\tau} \\ e^{-\tau} & e^{-\tau} \end{bmatrix} + [1, 0]\delta(t - \tau) \\ &= M(t)N(\tau) + D(t)\delta(t - \tau), \end{aligned}$$

which shows that the given $H(t)$ [resp. $H(t, 0)$] is indeed realizable by a system described by (2.4). Actually, in this case $M(t)$ and $N(\tau)$ were chosen so that $M(t) = Ce^{At}$ and $N(\tau) = e^{-A\tau}B$, where $e^{At} = \begin{bmatrix} e^{-t} & e^t - e^{-t} \\ 0 & e^t \end{bmatrix}$ (verify this). ■

Discrete-time systems

Results for the existence of realizations of discrete-time systems that are analogous to the continuous-time case can also be established. In the following result, which corresponds to Theorem 3.1, $H(k, i)$, $k \geq i$, denotes a $p \times m$ matrix.

THEOREM 3.5. $H(k, i)$ is realizable as the pulse response of a system described by (2.10) if and only if $H(k, i)$ can be decomposed into the form

$$H(k, i) = \begin{cases} M(k)N(i), & k > i, \\ D(k), & k = i. \end{cases} \tag{3.6}$$

Proof. (*Sufficiency*) We consider the realization $\{I, N(k), M(k), D(k)\}$, i.e., $x(k + 1) = x(k) + N(k)u(k)$ and $y(k) = M(k)x(k) + D(k)u(k)$. In this case, $\Phi(k, \ell) = I$, $k \geq \ell$. Applying (2.12), it is immediately verified that the pulse response of this system is the given $H(k, i)$.

(*Necessity*) For any realization (2.10), $C(k)\Phi(k, i + 1)B(i) = C(k)A(k - 1)\ldots A(i + 1)B(i) = C(k)A(k - 1)\ldots A(\sigma)A(\sigma - 1)\ldots A(i + 1)B(i) = (C(k)\Phi(k, \sigma)) \times (\Phi(\sigma, i + 1)B(i)) = M(k)N(i)$, $i < \sigma \leq k$, which in view of (2.10), implies (3.6). ■

The discrete-time system result corresponding to Theorem 3.2 for time-invariant systems is considered next.

THEOREM 3.6. $H(k, i)$ is realizable as the pulse response of a system described by (2.14) if and only if $H(k, i)$ can be decomposed as in (3.6), and

$$H(k, i) = H(k - i, 0). \tag{3.7}$$

Proof. (*Necessity*) This part of the proof is the same as the necessity proof of the previous theorem. Also, in view of (2.16), it is clear that $H(k, i) = H(k - i, 0)$.

(*Sufficiency*) Let $H(k, i)$ satisfy (3.6) and (3.7). The proof is by construction, considering a least-order realization $x(k+1) = x(k) + N(k)u(k)$, $y(k) = M(k)x(k) + D(k)u(k)$ and proceeding along similar lines as in the proof of Theorem 3.2. ■

EXAMPLE 3.6. Let $H(k, i) = k - i$, $k \geq i$. Here $H(k, i) = [k, 1]\begin{bmatrix} 1 \\ -i \end{bmatrix} = M(k)N(i)$, $k > i$, and $H(k, i) = 0 = D(k)$, $k = i$. In view of Theorem 3.5, there exists a realization, the pulse response of which is the given $H(k, i)$. A particular realization is given by the system equations $x(k+1) = x(k) + \begin{bmatrix} 1 \\ -k \end{bmatrix} u(k)$, $y(k) = [k, 1]x(k)$ (see the proof of Theorem 3.5). (Verify this.) This is of course a time-varying realization. However, here $H(k, i) = H(k - i, 0)$, which in view of Theorem 3.6, implies that a time-invariant realization also exists. Such a realization is $x(k+1) = Ax(k) + Bu(k)$, $y(k) = Cx(k)$, where $A = \begin{bmatrix} 0 & 1 \\ -1 & 2 \end{bmatrix}$, $B = \begin{bmatrix} 0 \\ 1 \end{bmatrix}$, $C = [0, 1]$. [Verify that in the present case $H(k, 0) = CA^{k-1}B$, $k > 0$.] This realization was determined using $H(z) = \mathcal{Z}H(k, 0) = z/(z-1)^2$ and the controller form realization algorithm in Subsection 5.4B. ■

Given a $p \times m$ matrix $H(z)$, the next theorem establishes necessary and sufficient conditions for time-invariant realizations. This result corresponds to Theorem 3.3 for the continuous-time case. Notice that the conditions in these results are identical.

THEOREM 3.7. $H(z)$ is realizable as the transfer function matrix of a time-invariant system described by (2.14) if and only if $H(z)$ is a matrix of rational functions and satisfies the condition that

$$\lim_{z \to \infty} H(z) < \infty. \tag{3.8}$$

Proof. Similar to the proof of Theorem 3.3. ■

COROLLARY 3.8. A $p \times m$ matrix $H(k)$ [resp., $H(k, 0)$], $k \geq 0$, is realizable as the pulse response of a system described by (2.14) if and only if all entries of $H(k)$ are sums of polynomial terms of the form $\alpha\lambda^k$, $\beta k(k-1)\cdots(k-\ell+1)\lambda^{k-\ell}$, and γ, where α, β, γ denote real numbers, k, ℓ are nonnegative integers, and λ is a complex scalar.

Proof. The details of the proof are left to the reader. Note that $H(k, 0) = \mathcal{Z}^{-1}[H(z)]$, where in view of Theorem 3.7, all the entries of $H(z)$ are proper rational functions. Refer to Section 2.7 of Chapter 2 for a review of z-transforms and a discussion of the modes and the asymptotic behavior of discrete-time systems. ■

EXAMPLE 3.7. Let $H(z) = z/(z-1)^2$. In view of Theorem 3.7, $H(z)$ is realizable as the transfer function of a system described by (2.14). Such a realization is given by $x(k+1) = Ax(k) + Bu(k)$, $y(k) = Cx(k) + Du(k)$, with $A = \begin{bmatrix} 0 & 1 \\ -1 & 2 \end{bmatrix}$, $B = \begin{bmatrix} 0 \\ 1 \end{bmatrix}$, $C = [0, 1]$, $D = 0$ (refer to Example 3.6). Note that in this case, $H(k, 0) = \mathcal{Z}^{-1}[H(z)] = k$, $k \geq 0$, which, in view of Corollary 3.8, implies that $H(k, 0) = k$ is realizable as the pulse response of a system described by (2.14). Such a realization is given above. ■

B. Minimality of Realizations

As was discussed in Section 5.2, realizations of an impulse response $H(t, \tau)$ can be expected to generate only the *zero-state response* of a system, since the external description $H(t, \tau)$ has, by definition, no information about the initial conditions and the zero-input response of the system.

A second important point to take note of is the fact that *if a realization of a given $H(t, \tau)$ exists, then there exist an infinite number of realizations.* It was pointed out that if (2.1), denoted by $\{A(t), B(t), C(t), D(t)\}$, is a realization of the $p \times m$ matrix $H(t, \tau)$, then realizations of the same order n, i.e., of the same dimension n of the state vector, can readily be generated by an equivalence transformation. Recall that in Subsection 2.6C of Chapter 2 it was shown that equivalent state-space (internal) representations generate identical impulse responses (external representations). There are, of course, other ways of generating alternative realizations. In particular, if (2.1) is a realization of $H(t, \tau)$, then, for example, the system

$$\begin{aligned} \dot{x} &= A(t)x + B(t)u, \qquad y = C(t)x + D(t)u \\ \dot{z} &= F(t)z + G(t)u \end{aligned} \tag{3.9}$$

is also a realization. This was accomplished by adding to (2.1) a state equation $\dot{z} = F(t)z + G(t)u$ that does not affect the system output. The dimension of F, $\dim F$, and consequently the order of the realization, $n + \dim F$, can be larger than any given finite number. In other words, *there may be no upper bound to the order of the realizations of a given $H(t, \tau)$.* There exists, however, a lower bound, and a realization of such lowest order is called a least-order minimal or irreducible realization.

DEFINITION 3.1. A realization

$$\dot{x} = A(t)x + B(t)u, \qquad y = C(t)x + D(t)u \tag{3.10}$$

of the impulse response $H(t, \tau)$ of least order n ($A(t) \in R^{n\times n}$) is called a *least order,* or *a minimal order,* or *an irreducible realization* of $H(t, \tau)$. ■

If the given impulse response $H(t, \tau)$ satisfies the conditions of Theorem 3.2, so that a time-invariant realization exists, then one usually talks about minimal or irreducible realizations of the $p \times m$ transfer function matrix $H(s) = \mathscr{L}[H(t, 0)]$, and this is the case on which we shall concentrate in the remainder of this chapter. It should be noted that results corresponding to Theorems 3.9 and 3.10 exist for time-varying systems as well. The interested reader is encouraged to consult, for instance, Brockett [1] for such results. Briefly, as will be shown for the time-invariant case, system (3.10) is a minimal realization of $H(t, \tau)$ if and only if the representation is controllable (-from-the-origin or reachable) and observable. Note that in this chapter the term controllable is used in place of reachable, to conform with accepted use in the literature. By controllability we will really mean controllability-from-the-origin or reachability. This distinction is not important in continuous-time systems, but it is important in discrete-time systems.

DEFINITION 3.2. A realization

$$\dot{x} = Ax + Bu, \qquad y = Cx + Du \tag{3.11}$$

of the transfer function matrix $H(s)$ of least order n($A \in R^{n\times n}$) is called *a least-order,* or *a minimal,* or *an irreducible realization of $H(s)$.* ■

Theorems 3.9 and 3.10 completely solve the minimal realization problem. The first of these results shows that a realization is minimal if and only if it is controllable (-from-the-origin or reachable) and observable, while the second result shows that if a minimal realization has been found, then all other minimal realizations can be obtained from the determined realization, using equivalence of representations.

Controllability (-from-the-origin, or reachability) and observability play an important role in the minimality of realizations. This is to be expected, since these properties, as was discussed at length in Chapter 3, characterize the strength of the connections between input and state, and between state and output, respectively. Therefore, it is reasonable to expect that they will play a significant role in the relation between internal and external descriptions of systems. Indeed, it was shown in Subsection 3.4C that only that part of a system that is both controllable and observable appears in $H(s)$. In other words, $H(s)$ *contains no information about the uncontrollable and/or unobservable parts of the system.* To illustrate this, consider the following specific case.

EXAMPLE 3.8. Let $H(s) = 1/(s+1)$. Four different realizations of $H(s)$ are given by

(i) $\{A = \begin{bmatrix} 0 & 1 \\ 1 & 0 \end{bmatrix}, B = \begin{bmatrix} 0 \\ 1 \end{bmatrix}, C = [-1, 1], D = 0\}.$

(ii) $\{A = \begin{bmatrix} 0 & 1 \\ 1 & 0 \end{bmatrix}, B = \begin{bmatrix} -1 \\ 1 \end{bmatrix}, C = [0, 1], D = 0\}.$

(iii) $\{A = \begin{bmatrix} 1 & 0 \\ 0 & -1 \end{bmatrix}, B = \begin{bmatrix} 0 \\ 1 \end{bmatrix}, C = [0, 1], D = 0\}.$

(iv) $\{A = -1, B = 1, C = 1, D = 0\}.$

The eigenvalue $+1$ that in (i) is unobservable, in (ii) is uncontrollable, and in (iii) is both uncontrollable and unobservable does not appear in $H(s)$ at all. Realization (iv), which is of order 1, is a minimal realization. It is controllable and observable. ■

THEOREM 3.9. An n-dimensional realization $\{A, B, C, D\}$ of $H(s)$ is minimal (irreducible, of least order) if and only if it is both controllable and observable.

Proof. (*Necessity*) Assume that $\{A, B, C, D\}$ is a minimal realization but is not both controllable and observable. Then, using Kalman's Canonical Decomposition of Subsection 3.4A, one may find another realization of lower dimension that is both controllable and observable. This contradicts the assumption that $\{A, B, C, D\}$ is a minimal realization. Therefore, it must be both controllable and observable.

(*Sufficiency*) Assume that the realization $\{A, B, C, D\}$ is controllable and observable, but there exists another realization, say, $\{\bar{A}, \bar{B}, \bar{C}, \bar{D}\}$ of order $\bar{n} < n$. Since they are both realizations of $H(s)$, or of the impulse response $H(t, 0)$, then

$$Ce^{At}B + D\delta(t) = \bar{C}e^{\bar{A}t}\bar{B} + \bar{D}\delta(t) \tag{3.12}$$

for all $t \geq 0$. Clearly, $D = \bar{D} = \lim_{s\to\infty} H(s)$. Using the power series expansion of the exponential and equating coefficients of the same power of t, we obtain

$$CA^kB = \bar{C}\bar{A}^k\bar{B}, \qquad k = 0, 1, 2, \ldots, \tag{3.13}$$

i.e., the Markov parameters of the two representations are the same (see Theorem 2.1 in Section 5.2). Let

$$\mathscr{C}_n \triangleq [B, AB, \ldots, A^{n-1}B] \in R^{n\times mn},$$

and
$$\mathcal{O}_n \triangleq \begin{bmatrix} C \\ CA \\ \vdots \\ CA^{n-1} \end{bmatrix} \in R^{pn\times n}. \tag{3.14}$$

Then the $pn \times mn$ matrix product $\mathcal{O}_n\mathcal{C}_n$ assumes the form

$$\mathcal{O}_n\mathcal{C}_n = \begin{bmatrix} CB & CAB & \cdots & CA^{n-1}B \\ CAB & CA^2B & \cdots & CA^nB \\ \vdots & \vdots & & \vdots \\ CA^{n-1}B & CA^nB & \cdots & CA^{2n}B \end{bmatrix},$$

$$= \begin{bmatrix} \bar{C}\bar{B} & \bar{C}\bar{A}\bar{B} & \cdots & \bar{C}\bar{A}^{n-1}\bar{B} \\ \bar{C}\bar{A}\bar{B} & \bar{C}\bar{A}^2\bar{B} & \cdots & \bar{C}\bar{A}^n\bar{B} \\ \vdots & \vdots & & \vdots \\ \bar{C}\bar{A}^{n-1}\bar{B} & \bar{C}\bar{A}^n\bar{B} & \cdots & \bar{C}\bar{A}^{2n}\bar{B} \end{bmatrix} = \bar{\mathcal{O}}_n\bar{\mathcal{C}}_n. \tag{3.15}$$

In view of *Sylvester's rank inequality,* which relates the rank of the product of two matrices to the rank of its factors, we have

$$rank\ \mathcal{O}_n + rank\ \mathcal{C}_n - n \leq rank\ (\bar{\mathcal{O}}_n\bar{\mathcal{C}}_n) \leq \min(rank\ \mathcal{O}_n, rank\ \mathcal{C}_n) \tag{3.16}$$

and we obtain that $rank\ \mathcal{O}_n = rank\ \mathcal{C}_n = n$, $rank\ (\bar{\mathcal{O}}_n\bar{\mathcal{C}}_n) = n$. This result, however, contradicts our assumptions, since $n = rank\ (\bar{\mathcal{O}}_n\bar{\mathcal{C}}_n) \leq \min(rank\ \bar{\mathcal{O}}_n, rank\ \bar{\mathcal{C}}_n) \leq \bar{n}$ because $\bar{n}$ is the order of $\{\bar{A}, \bar{B}, \bar{C}, \bar{D}\}$. Therefore, $n \leq \bar{n}$. Hence, $\bar{n}$ cannot be less than n and they can only be equal. Thus, $n = \bar{n}$ and $\{A, B, C, D\}$ is indeed a minimal realization. ∎

Theorem 3.9 suggests the following procedure to realize $H(s)$. First, we obtain a controllable (observable) realization of $H(s)$. Next, using a similarity transformation, we obtain an observable standard form to separate the observable from the unobservable parts (controllable from the uncontrollable parts), using the approach of Subsection 3.4A. Finally, we take the observable (controllable) part that will also be controllable (observable) as the minimal realization. We shall use this procedure in the next section.

Is the minimal realization unique? The answer to this question is of course no since we know that equivalent representations, which are of the same order, give the same transfer function matrix. The following theorem shows how to obtain *all* minimal realizations of $H(s)$.

THEOREM 3.10. Let $\{A, B, C, D\}$ and $\{\bar{A}, \bar{B}, \bar{C}, \bar{D}\}$ be realizations of $H(s)$. If $\{A, B, C, D\}$ is a minimal realization, then $\{\bar{A}, \bar{B}, \bar{C}, \bar{D}\}$ is also a minimal realization if and only if the two realizations are equivalent, i.e., if and only if $\bar{D} = D$ and there exists a nonsingular matrix P such that

$$\bar{A} = PAP^{-1}, \qquad \bar{B} = PB, \qquad \text{and} \qquad \bar{C} = CP^{-1}. \tag{3.17}$$

Furthermore, if P exists, it is given by

$$P = \mathcal{C}\bar{\mathcal{C}}^T(\bar{\mathcal{C}}\bar{\mathcal{C}}^T)^{-1} \qquad \text{or} \qquad P = (\bar{\mathcal{O}}^T\bar{\mathcal{O}})^{-1}\bar{\mathcal{O}}^T\mathcal{O}. \tag{3.18}$$

Proof. (*Sufficiency*) Let the realizations be equivalent. Since $\{A, B, C, D\}$ is minimal, it is controllable and observable and its equivalent representation $\{\bar{A}, \bar{B}, \bar{C}, \bar{D}\}$ is also controllable and observable, and therefore, minimal. Alternatively, since equivalence preserves the dimension of A, the equivalent realization $\{\bar{A}, \bar{B}, \bar{C}, \bar{D}\}$ is also minimal.

(*Necessity*) Suppose $\{\bar{A}, \bar{B}, \bar{C}, \bar{D}\}$ is also minimal. We shall show that it is equivalent to $\{A, B, C, D\}$. Since they are both realizations of $H(s)$, they satisfy $D = \bar{D}$ and

$$CA^kB = \bar{C}\bar{A}^k\bar{B}, \qquad k = 0, 1, 2\ldots, \tag{3.19}$$

as was shown in the proof of Theorem 3.9. Here, both realizations are minimal, and therefore, they are both of the same order n and are both controllable and observable.

Define $\mathscr{C} = \mathscr{C}_n$ and $\mathbb{O} = \mathbb{O}_n$, as in (3.14). Then, in view of (3.15), $\mathbb{O}\mathscr{C} = \bar{\mathbb{O}}\bar{\mathscr{C}}$ and premultiplying by $\bar{\mathbb{O}}^T$, we obtain $\bar{\mathbb{O}}^T\mathbb{O}\mathscr{C} = \bar{\mathbb{O}}^T\bar{\mathbb{O}}\bar{\mathscr{C}}$. Using Sylvester's inequality, we obtain *rank* $\bar{\mathbb{O}}^T\bar{\mathbb{O}} = n$, and therefore,

$$\bar{\mathscr{C}} = [(\bar{\mathbb{O}}^T\bar{\mathbb{O}})^{-1}\bar{\mathbb{O}}^T\mathbb{O}]\mathscr{C} = P\mathscr{C}, \tag{3.20}$$

where $P \triangleq (\bar{\mathbb{O}}^T\bar{\mathbb{O}})^{-1}\bar{\mathbb{O}}^T\mathbb{O} \in R^{n\times n}$. Note that *rank* $P = n$ since *rank* $\bar{\mathbb{O}}^T\mathbb{O}$ is also equal to n, as can be seen from *rank* $\bar{\mathbb{O}}^T\mathbb{O}\mathscr{C} = n$ and from Sylvester's inequality. Therefore, P qualifies as a similarity transformation. Similarly, $\mathbb{O}\mathscr{C} = \bar{\mathbb{O}}\bar{\mathscr{C}}$ implies that $\mathbb{O}\mathscr{C}\mathscr{C}^T = \bar{\mathbb{O}}\bar{\mathscr{C}}\mathscr{C}^T$, and

$$\mathbb{O} = \bar{\mathbb{O}}[\bar{\mathscr{C}}\mathscr{C}^T(\mathscr{C}\mathscr{C}^T)^{-1}] = \bar{\mathbb{O}}\bar{P}, \tag{3.21}$$

where $\bar{P} \triangleq \bar{\mathscr{C}}\mathscr{C}^T(\mathscr{C}\mathscr{C}^T)^{-1} \in R^{n\times n}$ with *rank* $\bar{P} = n$. Note that $P = (\bar{\mathbb{O}}^T\bar{\mathbb{O}})^{-1}\bar{\mathbb{O}}^T(\bar{\mathbb{O}}\bar{P}) = \bar{P}$. To show that P is the equivalence transformation given in (3.17), we note that $\mathbb{O}A\mathscr{C} = \bar{\mathbb{O}}\bar{A}\bar{\mathscr{C}}$ from (3.15). Premultiplying by $\bar{\mathbb{O}}^T$ and postmultiplying by $\mathscr{C}^T$, we obtain $PA = \bar{A}P$, in view of (3.20) and (3.21). To show that $PB = \bar{B}$ and $C = \bar{C}P$, we simply use the relations $P\mathscr{C} = \bar{\mathscr{C}}$ and $\mathbb{O} = \bar{\mathbb{O}}P$, respectively. ■

C. The Order of Minimal Realizations

In the next subsection algorithms to derive minimal realizations of $H(s)$ are developed. One could ask the question whether the order of a minimal realization of $H(s)$ can be determined directly, without having to actually derive a minimal realization. The answer to this question is yes, and in the following we will show how this can be accomplished. When deriving realizations, it is of advantage to know at the outset what the order of a minimal realization ought to be, since in this way erroneous results can be avoided by cross checking the validity of the results.

Determination via the characteristic or pole polynomial of $H(s)$

The *characteristic polynomial* (*or pole polynomial*), $p_H(s)$, *of a transfer function* matrix $H(s)$ was defined in Section 3.5 using the Smith-McMillan form of $H(s)$. The polynomial $p_H(s)$ is equal to the monic least common denominator of all nonzero minors of $H(s)$. *The minimal polynomial of a transfer function* matrix $H(s)$, $m_H(s)$, was defined as the monic least common denominator of all nonzero first-order minors (entries) of $H(s)$.

DEFINITION 3.3. The *McMillan degree* of $H(s)$ is the degree of $p_H(s)$. ■

The number of poles in $H(s)$, which are defined as the zeros of $p_H(s)$ in Definition 5.1 in Section 3.5, is equal to the McMillan degree of $H(s)$. The degree of $H(s)$ is in fact the order of any minimal realization of $H(s)$, as the following result shows.

THEOREM 3.11. Let $\{A, B, C, D\}$ be a minimal realization of $H(s)$. Then the characteristic polynomial of $H(s)$, $p_H(s)$, is equal to the characteristic polynomial of A, $\alpha(s) \triangleq |sI - A|$, i.e., $p_H(s) = \alpha(s)$. Therefore, the McMillan degree of $H(s)$ equals the order of any minimal realization.

Proof. The proof outlined here is based in part on results that will be established in Sections 7.2 and 7.3 of Chapter 7. (The reader may wish to postpone reading the proof until Sections 7.2 and 7.3 have been covered.) First, we note that if $H(s) = N(s)D(s)^{-1}$ with $N(s)$, $D(s)$ right coprime polynomial matrices, then the characteristic or pole polynomial of $H(s)$ is given by $p_H(s) = k\ det\ D(s)$, where the real k is such that $k\ det\ D(s)$ is monic. This can be seen, for example, by reducing $H(s)$ to its Smith-McMillan form [see (5.3) in Chapter 3]. We have,

$$U_1(s)H(s)U_2(s) = SM_H(s) = \begin{bmatrix} diag\,(\frac{\epsilon_1}{\psi_1}, \ldots, \frac{\epsilon_r}{\psi_r}) & \vdots & 0 \\ \cdots & \vdots & \cdots \\ 0 & \vdots & 0_{p-r,m-r} \end{bmatrix}$$

$$= \begin{bmatrix} diag\,(\epsilon_1, \ldots, \epsilon_r) & \vdots & 0 \\ \cdots & \vdots & \cdots \\ 0 & \vdots & 0_{p-r,m-r} \end{bmatrix} \begin{bmatrix} diag\,(\psi_1, \ldots, \psi_r) & \vdots & 0 \\ \cdots & \vdots & \cdots \\ 0 & \vdots & I_{m-r} \end{bmatrix}^{-1}$$

$$= N_s(s)D_s^{-1}(s),$$

where $r = rank\ H(s)$ and U_1, U_2 are unimodular matrices. Then $H = (U_1^{-1}N_s) \times (U_2D_s)^{-1} = ND^{-1}$, where N, D are right coprime. Note that $k\ det\ D(s) = \psi_1\psi_2\cdots\psi_r = p_H(s)$, by definition. Now this is true for any right coprime factorization since these are related by unimodular postmultiplication (see Subsection 7.3A in Chapter 7).

The controllable and observable realization $\{A, B, C, D\}$ is now equivalent to any other controllable and observable realization of the form $Dz = u$, $y = Nz$ with D, N right coprime (see Subsection 7.3A). This implies that $|sI - A| = k|D(s)|$, since such equivalence relation preserves the system eigenvalues. Note that the same result can be derived using the Structure Theorem of Chapter 3 (show this). Therefore, the pole polynomial of $H(s)$ is given by $p_H(s) = |sI - A|$. ■

It can also be shown that the minimal polynomial of $H(s)$, $m_H(s)$, is equal to the minimal polynomial of A, $\alpha_m(s)$, where $\{A, B, C, D\}$ is any controllable and observable realization of $H(s)$. This is illustrated in the following example.

EXAMPLE 3.9. Let $H(s) = \begin{bmatrix} 1/s & 2/s \\ 0 & -1/s \end{bmatrix}$. The first-order minors, the entries of $H(s)$, have denominators s, s, and s and therefore, $m_H(s) = s$. The only second-order minor is $-1/s^2$ and $p_H(s) = s^2$ with $deg\ p_H(s) = 2$. Therefore, the order of a minimal realization is 2. Such a realization is given by $\dot{x} = Ax + Bu$ and $y = Cx$ with $A = \begin{bmatrix} 0 & 0 \\ 0 & 0 \end{bmatrix}$, $B = \begin{bmatrix} 1 & 2 \\ 0 & -1 \end{bmatrix}$, $C = \begin{bmatrix} 1 & 0 \\ 0 & 1 \end{bmatrix}$. It can be verified first that this system is a realization of $H(s)$ and then we verify that it is controllable and observable, and therefore, minimal. Notice that the characteristic polynomial of A is $\alpha(s) = s^2 = p_H(s)$ and its minimal polynomial is $\alpha_m(s) = s = m_H(s)$. ■

The above example also shows that when $H(s)$ is expressed as a polynomial matrix $N(s)$ divided by a polynomial, i.e., $H(s) = (1/m_H(s))N(s)$, then the roots of m_H are not necessarily the eigenvalues of a minimal realization of $H(s)$. They are in general a subset of those eigenvalues, since the minimal polynomial always divides the characteristic polynomial. In the case when $H(s)$ is a scalar, however, the roots of $m_H = p_H$ are the eigenvalues of any minimal realization of $H(s)$, as the next example shows.

EXAMPLE 3.10. Let $H(s) = n(s)/d(s)$ be a scalar proper rational function. Applying Definition 5.1 of Section 3.5, we obtain $p_H(s) = m_H(s) = d(s)$ and the order of a minimal realization is $deg\, p_H(s) = deg\, d(s)$. Thus, given $H(s) = 1/(s^2 + 3s + 2)$, we know that a minimal realization is of second order since $deg\,(s^2 + 3s + 2) = 2$. A minimal realization in controller form is given by $A = \begin{bmatrix} 0 & 1 \\ -2 & -3 \end{bmatrix}$, $B = \begin{bmatrix} 0 \\ 1 \end{bmatrix}$, $C = [1, 0]$. Notice that the minimal polynomial of A is $\alpha_m(s) = \alpha(s) = s^2 + 3s + 2$, the characteristic polynomial of A, which is equal to $d(s) = p_H(s) = m_H(s)$, as expected. ■

The observations in Example 3.10 can be formalized as the following result.

COROLLARY 3.12. Let $H(s) = n(s)/d(s)$ be a scalar proper rational function. If $\{A, B, C, D\}$ is a minimal realization of $H(s)$, then

$$kd(s) = \alpha(s) = \alpha_m(s), \tag{3.22}$$

where $\alpha(s) = det\,(sI - A)$ and $\alpha_m(s)$ are the characteristic and minimal polynomials of A, respectively, and k is a real scalar so that $kd(s)$ is a monic polynomial.

Proof. The characteristic and minimal polynomials of $H(s)$, $p_H(s)$, and $m_H(s)$ are by definition equal to $d(s)$ in the scalar case. Applying Theorem 3.11 proves the result. ■

Determination via the Hankel matrix

There is an alternative way of determining the order of a minimal realization of $H(s)$. This is accomplished via the Hankel matrix, associated with $H(s)$.

Given $H(s)$, we express $H(s)$ as a Laurent series expansion to obtain

$$H(s) = H_0 + \hat{H}(s) = H_0 + H_1 s^{-1} + H_2 s^{-2} + H_3 s^{-3} + \cdots, \tag{3.23}$$

where $\hat{H}(s)$ is strictly proper and the real $p \times m$ matrices $H_0, H_1, \ldots$ are the Markov parameters of the system. They can be determined by the formulas

$$\begin{aligned} H_0 &= \lim_{s\to\infty} H(s), \\ H_1 &= \lim_{s\to\infty} s(H(s) - H_0), \\ H_2 &= \lim_{s\to\infty} s^2(H(s) - H_0 - H_1 s^{-1}), \end{aligned}$$

and so forth.

DEFINITION 3.4. The *Hankel matrix* $M_H(i, j)$ of order (i, j) corresponding to the (Markov parameter) sequence $H_1, H_2, \ldots$ is defined as the $ip \times jm$ matrix given by

$$M_H(i, j) \triangleq \begin{bmatrix} H_1 & H_2 & \cdots & H_j \\ H_2 & H_3 & \cdots & H_{j+1} \\ \vdots & \vdots & & \vdots \\ H_i & H_{i+1} & \cdots & H_{i+j-1} \end{bmatrix}. \tag{3.24}$$

■

THEOREM 3.13. The order of a minimal realization of $H(s)$ is the rank of $M_H(r, r)$, where r is the degree of the least common denominator of the entries of $H(s)$, i.e., $r = deg\, m_H(s)$.

Proof. Let $\{A, B, C, D\}$ be any realization of $H(s)$ of order n. The Markov parameters satisfy the relationships

$$H_i = CA^{i-1}B, \qquad i = 1, 2, \ldots, \tag{3.25}$$

(refer to Exercise 2.63 in Chapter 2). Therefore, the Hankel matrix $M_H(r, r)$ can be written as

$$M_H(r, r) = \begin{bmatrix} C \\ CA \\ \vdots \\ CA^{r-1} \end{bmatrix} [B, AB, \ldots, A^{r-1}B]. \tag{3.26}$$

Using Sylvester's Inequality, we have $rank\ M_H(r, r) \leq n$, since the common dimension in the product given by (3.26) is n. This implies that any realization of $H(s)$ is of order higher than or equal to $rank\ M_H(r, r)$. We now must show that there exists a realization of order exactly equal to $rank\ M_H(r, r)$, where r is the degree of the least common denominator of the entries of $H(s)$. Let $\{A, B, C\}$ be given by

$$A = \begin{bmatrix} 0_p & I_p & \cdots & 0_p \\ \vdots & \vdots & \ddots & \vdots \\ 0_p & 0_p & \cdots & I_p \\ -d_0 I_p & -d_1 I_p & \cdots & -d_{r-1} I_p \end{bmatrix}, \qquad B = \begin{bmatrix} H_1 \\ H_2 \\ \vdots \\ H_r \end{bmatrix}, \qquad \text{and} \qquad C = [I_p, 0, \ldots, 0],$$

where $A \in R^{pr \times pr}$, $B \in R^{pr \times m}$, and $C \in R^{p \times pr}$ with $m_H(s) = s^r + d_{r-1}s^{r-1} + \cdots + d_1 s + d_0$, the minimal polynomial of $H(s)$. We have that $\{A, B, C\}$ is an observable realization of $\hat{H}(s)$ (see Section 5.4). Furthermore, note that $|sI - A| = (m_H(s))^p$. Now

$rank\ M_H(r, r) = rank \begin{bmatrix} C \\ CA \\ \vdots \\ CA^{r-1} \end{bmatrix} [B, \ldots, A^{r-1}B] = rank\ [B, \ldots, A^{r-1}B]$. This is true because $\begin{bmatrix} C \\ CA \\ \vdots \\ CA^{r-1} \end{bmatrix} = I_{pr}$, as can easily be seen from above. Note now that $m_H(A) = 0$.

This is true since $(m_H(A))^p = 0$, in view of the Cayley-Hamilton Theorem. Therefore, $A^i, i \geq r$, can be expressed as a linear combination of $I, A, \ldots, A^{r-1}$, which implies that $rank\ M_H(r, r) = rank[B, \ldots, A^{r-1}B] = rank\ [B, \ldots, A^{pr-1}B]$. Therefore, the rank of the controllability matrix of the above realization is equal to $rank\ M_H(r, r)$. Hence, if $\{A, B, C\}$ is reduced by means of a transformation to a standard uncontrollable form $\left\{ \begin{bmatrix} A_1 & A_{12} \\ 0 & A_2 \end{bmatrix}, \begin{bmatrix} B_1 \\ 0 \end{bmatrix}, [C_1, C_2] \right\}$ (see Subsection 3.4A) with (A_1, B_1) controllable, then $\{A_1, B_1, C_1\}$ will be a controllable and observable (minimal) realization of $\hat{H}(s)$ of order equal to $rank\ M_H(r, r)$, since this is the rank of the controllablity matrix of (A, B) and the dimension of A_1. ■

EXAMPLE 3.11. Let

$$H(s) = \begin{bmatrix} \dfrac{1}{s+1} & \dfrac{2}{s+1} \\ \dfrac{-1}{(s+1)(s+2)} & \dfrac{1}{s+2} \end{bmatrix}.$$

Here the minimal polynomial is $m_H(s) = (s+1)(s+2)$, and therefore, $r = deg\ m_H(s) = 2$. The Hankel matrix $M_H(r, r)$ is then

$$M_H(r, r) = M_H(2, 2) = \begin{bmatrix} H_1 & H_2 \\ H_2 & H_3 \end{bmatrix},$$

an $rp \times rm = 4 \times 4$ matrix, and

$$H_1 = \lim_{s\to\infty} sH(s) = \lim_{s\to\infty}\begin{bmatrix} \dfrac{s}{s+1} & \dfrac{2s}{s+1} \\ \dfrac{-s}{(s+1)(s+2)} & \dfrac{s}{s+2} \end{bmatrix} = \begin{bmatrix} 1 & 2 \\ 0 & 1 \end{bmatrix}$$

and

$$H_2 = \lim_{s\to\infty} s^2(H(s) - H_1 s^{-1})$$

$$= \lim_{s\to\infty}\begin{bmatrix} \dfrac{s^2}{s+1} - s & \dfrac{2s^2}{s+1} - 2s \\ \dfrac{-s^2}{(s+1)(s+2)} & \dfrac{s^2}{s+2} - s \end{bmatrix}$$

$$= \lim_{s\to\infty}\begin{bmatrix} \dfrac{-s}{s+1} & \dfrac{-2s}{s+1} \\ \dfrac{-s^2}{(s+1)(s+2)} & \dfrac{-2s}{s+2} \end{bmatrix} = \begin{bmatrix} -1 & -2 \\ -1 & -2 \end{bmatrix}.$$

Similarly, $H_3 = \begin{bmatrix} 1 & 2 \\ 3 & 4 \end{bmatrix}$. Now

$$rank\, M_H(2,2) = rank\begin{bmatrix} 1 & 2 & -1 & -2 \\ 0 & 1 & -1 & -2 \\ -1 & -2 & 1 & 2 \\ -1 & -2 & 3 & 4 \end{bmatrix} = 3,$$

which is the order of any minimal realization, in view of Theorem 3.12. The reader should verify this result, using Theorem 3.11. ■

EXAMPLE 3.12. Consider the transfer function matrix $H(s) = \begin{bmatrix} 1/s & 2/s \\ 0 & -1/s \end{bmatrix}$, as in Example 3.9. Here $r = deg\, m_H(s) = deg\, s = 1$. Now, the Hankel matrix $M_H(r, r) = M_H(1,1) = H_1 = \lim_{s\to\infty} sH(s) = \begin{bmatrix} 1 & 2 \\ 0 & -1 \end{bmatrix}$. Its rank is 2, which is the order of a minimal realization of $H(s)$. This agrees with the results in Example 3.9. ■

D. Minimality of Realizations: Discrete-Time Systems

All definitions and theorems given thus far in this section for the continuous-time case, apply directly to the discrete-time case with no substantial changes. Thus, minimal or irreducible realizations are defined as in Definitions 3.1 and 3.2, where $H(k, i)$ and $H(z)$ should be used in place of $H(t, \tau)$ and $H(s)$, respectively. The main criteria for establishing minimality are given by results that are essentially the same as Theorems 3.9 and 3.10. The McMillan degree of $H(z)$ is as defined in Definition 3.3, while results that are essentially the same as Theorems 3.11 and 3.13 provide a means of determining the order of the minimal realizations.

The fact that the results on minimality of realizations in the discrete-time case are essentially identical to the corresponding results for the continuous-time case is not surprising since we are concentrating here on the time-invariant cases for which the transfer function matrices have the same forms: $H(s) = C(sI - A)^{-1}B + D$ and $H(z) = C(zI - A)^{-1}B + D$. Accordingly, the results on how to generate 4-tuples $\{A, B, C, D\}$ to satisfy these relations are of course the same. The *realization algorithms* developed in Section 5.4 apply directly to the discrete-time case as well, since they are algorithms for time-invariant realizations of transfer matrices. The differences in the realization theory between continuous- and discrete-time systems arise primarily in the time-varying case (compare Theorems 3.1 and 3.5, for example). However, these differences are rather insignificant.

5.4 REALIZATION ALGORITHMS

In this section, algorithms for generating time-invariant state-space realizations of external system descriptions are introduced. In particular, it is assumed that a proper rational matrix $H(s)$ of dimensions $p \times m$ is given for which a state-space realization $\{A, B, C, D\}$ of $H(s)$ given by $\dot{x} = Ax + Bu$, $y = Cx + Du$ such that $C(sI - A)^{-1}B + D = H(s)$ has been derived (see Theorem 3.3). This problem is equivalent to realizing $H(t, 0) = \mathcal{L}^{-1}[H(s)]$. A brief outline of the contents of this section follows.

Realizations of $H(s)$ can often be derived in an easier manner if duality is used, and this is demonstrated first in this section. Realizations of minimal order are both controllable and observable, as was shown in the previous section. To derive a minimal realization of $H(s)$, one typically derives a realization that is controllable (observable) and then extracts the part that is also observable (controllable), using the methods of Subsection 3.4A of Chapter 3. This involves in general a two-step procedure. However, in certain cases a minimal realization can be derived in one step, as for example, when $H(s)$ is a scalar transfer function. Algorithms for realizations in a controller/observer form are discussed first. In the interest of clarity, the SISO case is presented separately, thus providing an introduction to the general MIMO case. Realization algorithms, where A is diagonal or in block companion form, are introduced next. Finally, balanced realizations are addressed.

It is not difficult to see that the above algorithms can also be used to derive realizations described by equations of the form $x(k + 1) = Ax(k) + Bu(k)$, $y(k) = Cx(k) + Du(k)$ of transfer function matrices $H(z)$ for discrete-time time-invariant systems. Accordingly, the discrete-time case will not be treated separately in this section.

A. Realizations Using Duality

If the system described by the equations $\dot{x} = Ax + Bu$, $y = Cx + Du$ is a realization of $H(s)$, then

$$H(s) = C(sI - A)^{-1}B + D. \tag{4.1}$$

If $\tilde{H}(s) \triangleq H^T(s)$, then $\dot{\tilde{x}} = \tilde{A}\tilde{x} + \tilde{B}\tilde{u}$ and $\tilde{y} = \tilde{C}\tilde{x} + \tilde{D}\tilde{u}$, where $\tilde{A} = A^T, \tilde{B} = C^T, \tilde{C} = B^T$, and $\tilde{D} = D^T$, is a realization of $\tilde{H}(s)$ since in view of (4.1),

$$\begin{aligned}\tilde{H}(s) &= H^T(s) \\ &= B^T(sI - A^T)^{-1}C^T + D^T \\ &= \tilde{C}(sI - \tilde{A})^{-1}\tilde{B} + \tilde{D}.\end{aligned} \tag{4.2}$$

The representation $\{\tilde{A}, \tilde{B}, \tilde{C}, \tilde{D}\}$ is *the dual representation* to $\{A, B, C, D\}$, and if $\{A, B, C, D\}$ is controllable (observable), then $\{\tilde{A}, \tilde{B}, \tilde{C}, \tilde{D}\}$ is observable (controllable) (see Chapter 3). In other words, if a controllable (observable) realization $\{A, B, C, D\}$ of the $p \times m$ transfer function matrix $H(s)$ is known, then an observable (controllable) realization of the $m \times p$ transfer function matrix $\tilde{H}(s) = H^T(s)$ can be derived immediately: it is the dual representation, namely, $\{\tilde{A}, \tilde{B}, \tilde{C}, \tilde{D}\} = \{A^T, C^T, B^T, D^T\}$. This fact is used to advantage in deriving realizations in the MIMO case, since obtaining first a realization of $H^T(s)$ instead of $H(s)$ and then using duality, leads sometimes to simpler, lower order, realizations.

Duality is very useful in realizations of symmetric transfer functions, which have the property that $H(s) = H^T(s)$, as, e.g., in the case of SISO systems where $H(s)$ is a scalar. Under these conditions, if $\{A, B, C, D\}$ is a controllable (observable) realization of $H(s)$, then $\{A^T, C^T, B^T, D^T\}$ is an observable (controllable) realization of the same $H(s)$. Note that in this case,

$$H(s) = C(sI - A)^{-1}B + D = H^T(s) = B^T(sI - A^T)^{-1}C^T + D^T.$$

In realization algorithms of MIMO systems, a realization that is either controllable or observable is typically obtained first. Next, this realization is reduced to a minimal one by extracting the part of the system that is both controllable and observable, using the methods of Subsection 3.4A. Dual representations may simplify this process considerably. In the following, we summarize the process of deriving minimal realizations for the reader's convenience.

Given a proper rational $p \times m$ transfer function matrix $H(s)$, with $\lim_{s\to\infty} H(s) < \infty$ (see Theorem 3.3), we consider the strictly proper part $\hat{H}(s) = H(s) - \lim_{s\to\infty} H(s) = H(s) - D$ [noting that working with $\hat{H}(s)$ instead of $H(s)$ is optional].

1. If a realization algorithm leading to a controllable realization is used, then the following steps are taken,

   $$\hat{H}(s) \to (\tilde{H}(s) = \hat{H}^T(s)) \to \{\tilde{A}, \tilde{B}, \tilde{C}\} \to \{A = \tilde{A}^T, B = \tilde{C}^T, C = \tilde{B}^T\}, \tag{4.3a}$$

   where $\{\tilde{A}, \tilde{B}, \tilde{C}\}$ is a controllable realization of $\tilde{H}(s)$ and $\{A, B, C\}$ is an observable realization of $\hat{H}(s)$.
2. To obtain a minimal realization,

   $$\{A, B, C\} \to \left\{ \begin{bmatrix} A_1 & A_{12} \\ 0 & A_2 \end{bmatrix}, \begin{bmatrix} B_1 \\ 0 \end{bmatrix}, [C_1, C_2] \right\}, \tag{4.3b}$$

   where $\{A, B, C\}$ is an observable realization of $\hat{H}(s)$ obtained from step (1), and (A_1, B_1) is controllable (derived by using the method of Subsection 3.4A), then $\{A_1, B_1, C_1\}$ is a controllable and observable, and therefore, a minimal realization of $\hat{H}(s)$, and furthermore, $\{A_1, B_1, C_1, D\}$ is a minimal realization of $H(s)$.

B. Realizations in Controller/Observer Form

We shall first consider realizations of scalar transfer functions $H(s)$.

Single-input/single-output (SISO) systems ($p = m = 1$)
Let

$$H(s) = \frac{n(s)}{d(s)} = \frac{b_n s^n + \cdots + b_1 s + b_0}{s^n + a_{n-1}s^{n-1} + \cdots + a_1 s + a_0}, \tag{4.4}$$

where $n(s)$ and $d(s)$ are prime polynomials. This is the general form of a proper transfer function of (McMillan) degree n. Note that if the leading coefficient in the numerator $n(s)$ is zero, i.e., $b_n = 0$, then $H(s)$ is strictly proper. Also, recall that

$$y^{(n)} + a_{n-1}y^{(n-1)} + \cdots + a_1 y^{(1)} + a_0 y = b_n u^{(n)} + \cdots + b_1 u^{(1)} + b_0 u, \tag{4.5a}$$

or

$$\begin{aligned} d(q)y(t) &= (q^n + a_{n-1}q^{n-1} + \cdots + a_1 q + a_0)y(t) \\ &= (b_n q^n + \cdots b_1 q + b_0)u(t) = n(q)u(t), \end{aligned} \tag{4.5b}$$

where $q \triangleq d/dt$, the differential operator. This is the corresponding nth-order differential equation that directly gives rise to the map $\hat{y}(s) = H(s)\hat{u}(s)$ if the Laplace transform of both sides is taken, assuming that all variables and their derivatives are zero at $t = 0$.

Controller form realizations

Given $n(s)$ and $d(s)$, we proceed as follows to derive a realization in controller form.

1. Determine $C_c^T \in R^n$ and $D_c \in R$ so that

$$n(s) = C_c S(s) + D_c d(s), \tag{4.6}$$

where $S(s) \triangleq [1, s, \ldots, s^{n-1}]^T$ is an $n \times 1$ vector of polynomials. Equation (4.6) implies that

$$D_c = \lim_{s\to\infty} H(s) = b_n. \tag{4.7}$$

Then $n(s) - b_n d(s)$ is in general a polynomial of degree $n - 1$, which shows that a real vector C_c that satisfies (4.6) always exists.

If $b_n = 0$, i.e., if $H(s)$ is strictly proper, then from (4.6) we obtain $C_c = [b_0, \ldots, b_{n-1}]$, i.e., C_c consists of the coefficients of the $n - 1$ degree numerator.

If $b_n \neq 0$, then (4.6) implies that the entries of C_c are a combination of the coefficients b_i and a_i. In particular,

$$C_c = [b_0 - b_n a_0, b_1 - b_n a_1, \ldots, b_{n-1} - b_n a_{n-1}]. \tag{4.8}$$

2. A realization of $H(s)$ in controller form is given by the equations

$$\begin{aligned} \dot{x}_c &= A_c x_c + B_c u = \begin{bmatrix} 0 & 1 & \cdots & 0 \\ \vdots & \vdots & \ddots & \vdots \\ 0 & 0 & \cdots & 1 \\ -a_0 & -a_1 & \cdots & -a_{n-1} \end{bmatrix} x_c + \begin{bmatrix} 0 \\ 0 \\ \vdots \\ 1 \end{bmatrix} u \\ y &= C_c x_c + D_c u. \end{aligned} \tag{4.9}$$

The n states of the realization in (4.9) are related by

$$x_{i+1} = \dot{x}_i \quad \text{or} \quad x_{i+1} = x_1^{(i)}, \qquad i = 1, \ldots, n-1,$$

and
$$\dot{x}_n = -a_0 x_1 - \sum_{i=1}^{n-1} a_i x_{i+1} + u = -a_0 x_1 - \sum_{i=1}^{n-1} a_i x_1^{(i)} + u.$$

It can now be shown that x_1 satisfies the relationship

$$d(q)x_1(t) = u(t), \qquad y(t) = n(q)x_1(t), \tag{4.10}$$

where $q \triangleq d/dt$, the differential operator. In particular, note that $d(q)x_1(t) = u(t)$ because $\dot{x}_n = -\sum_{i=0}^{n-1} a_i x_1^{(i)} + x_1^{(n)} + u = -d(q)x_1 + u + x_1^{(n)}$, which in view of $\dot{x}_n = x_1^{(n)}$, derived from $x_n = x_1^{(n-1)}$, implies that $-d(q)x_i + u = 0$. The relation $y(t) = n(q)x_1(t)$ can easily be verified by multiplying both sides of $n(q) = C_c S(q) + D_c d(q)$ given in (4.6) by x_1.

LEMMA 4.1. The representation (4.9) is a minimal realization of $H(s)$ given in (4.4).

Proof. We must first show that (4.9) is indeed a realization, i.e., that it satisfies (4.1). This is of course true in view of the Structure Theorem in Subsection 3.4D of Chapter 3. Presently, this will be shown directly, using (4.10).

Relation $d(q)x_1(t) = u(t)$ implies that $\hat{x}_1(s) = (d(s))^{-1}\hat{u}(s)$. This yields for the state that $\hat{x}(s) = [\hat{x}_1(s), \ldots, \hat{x}_n(s)]^T = [1, s, \ldots, s^{n-1}]^T \hat{x}_1(s) = S(s)(d(s))^{-1}\hat{u}(s)$. However, we also have $\hat{x}(s) = (sI - A_c)^{-1}B_c\hat{u}(s)$. Therefore,

$$(sI - A_c)S(s) = B_c d(s). \tag{4.11}$$

Now $C_c(sI - A_c)^{-1}B_c + D_c = C_c S(s)(d(s))^{-1} + D_c = (C_c S(s) + D_c d(s))(d(s))^{-1} = n(s)/d(s) = H(s)$, i.e., (4.9) is indeed a realization.

System (4.9) is of order n, and is therefore, a minimal, controllable, and observable realization. This is because the degree of $H(s)$ is n, which in view of Theorem 3.11, is the order of any minimal realization. Controllability and observability can also be established directly by forming the controllability and observability matrices. The reader is encouraged to pursue this approach. ■

According to Definition 3.3 given in Section 5.3, the McMillan degree of a rational scalar transfer function $H(s) = n(s)/d(s)$ is n only when $n(s)$ and $d(s)$ are prime polynomials; if they are not, all cancellations must first take place before the degree can be determined. If $n(s)$ and $d(s)$ are not prime, then the above algorithm will yield a realization that is not observable. Notice that realization (4.9) is always controllable, since it is in controller form. This can also be seen directly from the expression

$$[B_c, A_c B_c, \ldots, A_c^{n-1} B_c] = \begin{bmatrix} 0 & 0 & \cdots & 1 \\ \vdots & \vdots & & \vdots \\ 0 & 1 & \cdots & \times \\ 1 & \times & \cdots & \times \end{bmatrix}, \tag{4.12}$$

which is of full rank. The realization (4.9) is observable if and only if the polynomials $d(s)$ and $n(s)$ are prime.

In Fig. 5.2 a block realization diagram of the form (4.9) for a second-order transfer function is shown. Note that the states $x_1(t)$ and $x_2(t)$ are taken to be the voltages at the outputs of the integrators. This is common when realizing transfer functions using analog computer circuits.

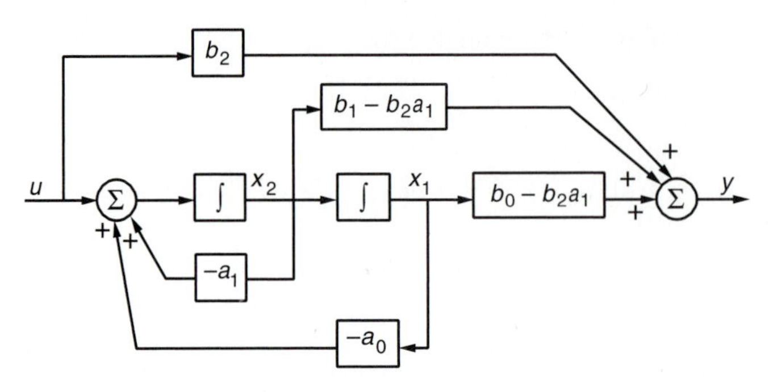

FIGURE 5.2
Block realization of $H(s)$ in controller form of the system $\begin{bmatrix} \dot{x}_1 \\ \dot{x}_2 \end{bmatrix} = \begin{bmatrix} 0 & 1 \\ -a_0 & -a_1 \end{bmatrix}\begin{bmatrix} x_1 \\ x_2 \end{bmatrix} + \begin{bmatrix} 0 \\ 1 \end{bmatrix} u$, $y = [b_0 - b_2a_0, b_1 - b_2a_1]\begin{bmatrix} x_1 \\ x_2 \end{bmatrix} + b_2u$; $H(s) = \dfrac{b_2s^2 + b_1s + b_0}{s^2a_1s + a_0}$

Observer form realizations

Given the transfer function (4.4), the nth-order realization in observer form is given by

$$\dot{x}_o = A_ox_o + B_ou$$
$$= \begin{bmatrix} 0 & \dots & 0 & -a_0 \\ 1 & & 0 & -a_1 \\ \vdots & \ddots & \vdots & \vdots \\ 0 & \dots & 1 & -a_{n-1} \end{bmatrix} x_o + \begin{bmatrix} b_0 - b_na_0 \\ b_1 - b_na_1 \\ \vdots \\ b_{n-1} - b_na_{n-1} \end{bmatrix} u,$$
$$y = C_ox_o + D_ou = [0, 0, \dots, 0, 1]x_o + b_nu. \tag{4.13}$$

This realization was derived by taking the dual of realization (4.9). Notice that $A_o = A_c^T$, $B_o = C_c^T$, $C_o = B_c^T$, and $D_o = D_c^T$.

LEMMA 4.2. The representation (4.13) is a minimal realization of $H(s)$ given in (4.4).

Proof. Note that the observer form realization $\{A_o, B_o, C_o, D_o\}$ described by (4.13) is the dual of the controller form realization $\{A_c, B_c, C_c, D_c\}$ described by (4.9), used in Lemma 4.1. ■

The realization (4.13) can also be derived directly from $H(s)$, using defining relations similar to (4.6). In particular, B_o and D_o can be determined from the expression

$$n(s) = \tilde{S}(s)B_o + d(s)D_o, \tag{4.14}$$

where $\tilde{S}(s) = [1, s, \dots, s^{n-1}]$.

It can be shown (by taking transposes) that the corresponding relation to (4.11) is now given by

$$\tilde{S}(s)(sI - A_o) = d(s)C_o \tag{4.15}$$

and that

$$d(q)z(t) = n(q)u(t),\ y(t) = z(t) \tag{4.16}$$

corresponds to (4.10).

Figure 5.3 depicts a block realization diagram of the form (4.13) for a second-order transfer function.

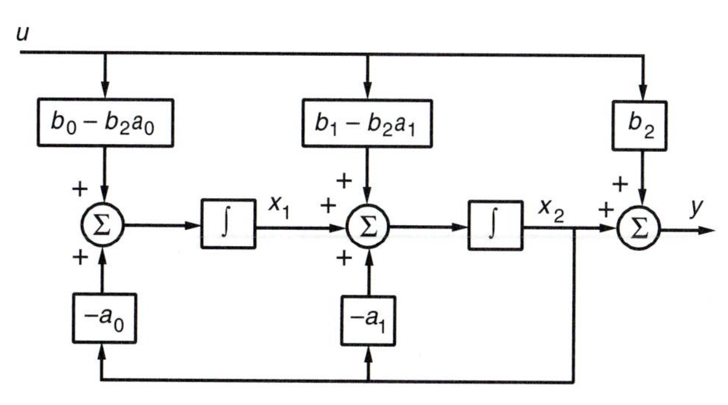

FIGURE 5.3
Block realization of $H(s)$ in observer form of the system
$\begin{bmatrix}\dot{x}_1\\ \dot{x}_2\end{bmatrix} = \begin{bmatrix}0 & -a_0\\ 1 & -a_1\end{bmatrix}\begin{bmatrix}x_1\\ x_2\end{bmatrix} + \begin{bmatrix}b_0 - b_2a_0\\ b_1 - b_2a_1\end{bmatrix}u,\ y = [0, 1]\begin{bmatrix}x_1\\ x_2\end{bmatrix} + b_2u;$
$H(s) = \dfrac{b_2s^2 + b_1s + b_0}{s^2 + a_1s + a_0}$

EXAMPLE 4.1. We wish to derive a minimal realization for the transfer function $H(s) = (s^3 + s - 1)/(s^3 + 2s^2 - s - 2)$. Consider a realization $\{A_c, B_c, C_c, D_c\}$, where (A_c, B_c) is in controller form. In view of (4.6) to (4.9), $D_c = \lim_{s\to\infty} H(s) = 1$ and $n(s) = s^3 + s - 1 = C_cS(s) + D_cd(s)$, from which we have $C_cS(s) = (s^3 + s - 1) - (s^3 + 2s^2 - s - 2) = -2s^2 + 2s + 1 = [1, 2, -2][1, s, s^2]^T$. Therefore, a realization of $H(s)$ is $\dot{x}_c = A_cx_c + B_cu$, $y = C_cx_c + D_cu$, where

$$A_c = \begin{bmatrix}0 & 1 & 0\\ 0 & 0 & 1\\ 2 & 1 & -2\end{bmatrix}, \qquad B_c = \begin{bmatrix}0\\ 0\\ 1\end{bmatrix}, \qquad C_c = [1, 2, -2], \qquad D_c = 1.$$

This is a minimal realization (verify this).

Instead of solving $n(s) = C_cS(s) + D_cd(s)$ for C_c as was done above, it is possible to derive C_c by inspection after $H(s)$ is written as

$$H(s) = \hat{H}(s) + \lim_{s\to\infty} H(s) = \hat{H}(s) + D_c, \tag{4.17}$$

where $\hat{H}(s)$ is now strictly proper. Notice that if $H(s)$ is given by (4.4), then $D_c = b_n$ and

$$\hat{H}(s) = \frac{c_{n-1}s^{n-1} + \cdots + c_1s + c_0}{s^n + a_{n-1}s^{n-1} + \cdots + a_1s + a_0}, \tag{4.18}$$

where in fact, $c_i = b_i - b_na_i$, $i = 0, \ldots, n-1$. The realization $\{A_c, B_c, C_c\}$ of $\hat{H}(s)$ has (A_c, B_c) precisely the same as before; however, C_c can now be written directly as

$$C_c = [c_0, c_1, \ldots, c_{n-1}], \tag{4.19}$$

i.e., given $H(s)$ there are three ways of determining C_c: (i) using formula (4.8), (ii) solving $C_cS(s) = n(s) - D_cd(s)$ as in (4.6), and (iii) calculating $\hat{H}(s) = H(s) - \lim_{s\to\infty} H(s)$. The reader should verify that in this example, (i) and (iii) yield the same $C_c = [1, 2, -2]$ as in method (ii).

Suppose now that it is of interest to determine a minimal realization $\{A_o, B_o, C_o, D_o\}$, where (A_o, C_o) is in observer form. This can be accomplished in ways completely analogous to the methods used to derive realizations in controller form. Alternatively, one could use duality directly and show that

$$A_o = A_c^T = \begin{bmatrix} 0 & 0 & 2 \\ 1 & 0 & 1 \\ 0 & 1 & -2 \end{bmatrix}, \qquad B_o = C_c^T = \begin{bmatrix} 1 \\ 2 \\ -2 \end{bmatrix},$$

$$C_o = B_c^T = [0, 0, 1], \qquad D_o = D_c^T = 1$$

is a minimal realization, where the pair (A_o, C_o) is in observer form. ∎

EXAMPLE 4.2. Consider now the transfer function $H(s) = (s^3 - 1)/(s^3 + 2s^2 - s - 2)$, where the numerator is $n(s) = s^3 - 1$ instead of $s^3 + s - 1$, as in Example 4.1. We wish to derive a minimal realization of $H(s)$. Using the same procedure as in the previous example, it is not difficult to derive the realization

$$A_c = \begin{bmatrix} 0 & 1 & 0 \\ 0 & 0 & 1 \\ 2 & 1 & -2 \end{bmatrix}, \qquad B_c = \begin{bmatrix} 0 \\ 0 \\ 1 \end{bmatrix}, \qquad C_c = [1, 1, -2], \qquad D_c = 1.$$

This realization is controllable, since (A_c, B_c) is in controller form (see Exercise 5.11); however, it is not observable, since $rank\ \mathbb{O} = 2 < 3 = n$, where $\mathbb{O}$ denotes the observability matrix given by

$$\mathbb{O} = \begin{bmatrix} C_c \\ C_cA_c \\ C_cA_c^2 \end{bmatrix} = \begin{bmatrix} 1 & 1 & -2 \\ -4 & -1 & 5 \\ 10 & 1 & -11 \end{bmatrix}$$

(show this). Therefore, the above is not a minimal realization. This has occurred because the numerator and denominator of $H(s)$ are not prime polynomials, i.e., $s - 1$ is a common factor. Thus, strictly speaking, the $H(s)$ given above is not a transfer function, since it is assumed that in a transfer function all cancellations of common factors have taken place. (See also the discussion following Lemma 4.1.) Correspondingly, if the algorithm for deriving an observer form would be applied to the present case, the realization $\{A_o, B_o, C_o, D_o\}$ would be an observable realization, but not a controllable one, and would therefore not be a minimal realization.

To obtain a minimal realization of the above transfer function $H(s)$, one could either extract the part of the controllable realization $\{A_c, B_c, C_c, D_c\}$ that is also observable, or simply cancel the factor $s-1$ in $H(s)$ and apply the algorithm again. The former approach of reducing a controllable realization will be illustrated when discussing the MIMO case. The latter approach is perhaps the easiest one to apply in the present case. We have

$$H(s) = \frac{s^3 - 1}{s^3 + 2s^2 - s - 2} = \frac{s^2 + s + 1}{s^2 + 3s + 2} = \frac{-2s - 1}{s^2 + 3s + 2} + 1,$$

and a minimal realization of this is then determined as

$$A_c = \begin{bmatrix} 0 & 1 \\ -2 & -3 \end{bmatrix}, \qquad B_c = \begin{bmatrix} 0 \\ 1 \end{bmatrix}, \qquad C_c = [-1, -2], \qquad D_c = 1.$$

The reader should verify this. ∎

Multi-Input-Multi-Output (MIMO) Systems (*pm* > *1*)

Let a $(p \times m)$ proper rational matrix $H(s)$ be given with $\lim_{s\to\infty} H(s) < \infty$ (see Theorem 3.3). We now present alogrithms to obtain realizations $\{A_c, B_c, C_c, D_c\}$ of $H(s)$ in controller form and realizations $\{A_o, B_o, C_o, D_o\}$ of $H(s)$ in observer form.

Minimal realizations can then be obtained by separating the observable (controllable) part of the controllable (observable) realization.

Controller form realizations

Consider a transfer function matrix $H(s) = [n_{ij}(s)/d_{ij}(s)]$, $i = 1, \ldots, p$, $j = 1, \ldots, m$, and let $\ell_j(s)$ denote the (monic) least common denominator of all entries in the jth column of $H(s)$. The $\ell_j(s)$ is the least degree polynomial divisible by all $d_{ij}(s)$, $i = 1, \ldots, p$. Then $H(s)$ can be written as

$$H(s) = N(s)D^{-1}(s), \tag{4.20}$$

a ratio of two polynomial matrices, where $N(s) \triangleq [\bar{n}_{ij}(s)]$ and $D(s) \triangleq diag\,[\ell_1(s), \ldots, \ell_m(s)]$. Note that $\bar{n}_{ij}(s)/\ell_j(s) = n_{ij}(s)/d_{ij}(s)$ for $i = 1, \ldots, p$, and all $j = 1, \ldots, m$. Let $d_j \triangleq deg\, \ell_j(s)$ and assume that $d_j \geq 1$. Define

$$\Lambda(s) \triangleq diag\,(s^{d_1}, \ldots, s^{d_m})$$

and

$$S(s) \triangleq block\ diag \left(\begin{bmatrix} 1 \\ s \\ \vdots \\ s^{d_j - 1} \end{bmatrix} \quad j = 1, \ldots, m \right), \tag{4.21}$$

and note that $S(s)$ is an $n \left(\triangleq \sum_{j=1}^{m} d_j \right) \times m$ polynomial matrix. Write

$$D(s) = D_h \Lambda(s) + D_\ell S(s) \tag{4.22}$$

and note that D_h is the highest column degree coefficient matrix of $D(s)$. Here $D(s)$ is diagonal with monic polynomial entries, and therefore, $D_h = I_m$. If, for example, $D(s) = \begin{bmatrix} 3s^2 + 1 & 2s \\ 2s & 2 \end{bmatrix}$, then the highest column degree coefficient matrix $D_h = \begin{bmatrix} 3 & 2 \\ 0 & 1 \end{bmatrix}$, and $D_\ell S(s)$ given in (4.22) accounts for the remaining lower column degree terms in $D(s)$, with D_ℓ being a matrix of coefficients.

Observe that $|D_h| \neq 0$, and define the $m \times m$ and $m \times n$ matrices

$$B_m = D_h^{-1}, \qquad A_m = -D_h^{-1} D_\ell, \tag{4.23}$$

respectively. Also, determine C_c and D_c such that

$$N(s) = C_c S(s) + D_c D(s), \tag{4.24}$$

and note that

$$D_c = \lim_{s \to \infty} H(s). \tag{4.25}$$

We have $H(s) = N(s)D^{-1}(s) = C_c S(s) D^{-1}(s) + D_c$ with $C_c S(s) D^{-1}(s)$ being strictly proper (show this). Therefore, only C_c needs to be determined from (4.24).

A controllable realization of $H(s)$ in controller form is now given by the equations

$$\dot{x}_c = A_c x_c + B_c u, \qquad y = C_c x_c + D_c u.$$

Here C_c and D_c were defined in (4.24) and (4.25), respectively,

$$A_c = \bar{A}_c + \bar{B}_c A_m, \qquad B_c = \bar{B}_c B_m, \tag{4.26}$$

where $\bar{A}_c = block\ diag\ [A_1, A_2 \ldots, A_m]$ with

$$A_j = \begin{bmatrix} 0 & & & \\ \vdots & & I_{d_{j-1}} & \\ 0 & 0 & \ldots & 0 \end{bmatrix} \in R^{d_j \times d_j},$$

$$\bar{B}_c = block\ diag \left(\begin{bmatrix} 0 \\ \vdots \\ 0 \\ 1 \end{bmatrix} \in R^{d_j},\ j = 1, \ldots, m \right),$$

and A_m, B_m were defined in (4.23). Note that if $d_j = \mu_j$, $j = 1, \ldots, m$, the controllability indices, then (4.26) is precisely the relation (4.63) given in Section 3.4.

LEMMA 4.3. The system $\{A_c, B_c, C_c, D_c\}$ is an $n(= \sum_{j=1}^{m} d_j)$-th-order controllable realization of $H(s)$ with (A_c, B_c) in controller form.

Proof. First, to show that $\{A_c, B_c, C_c, D_c\}$ is a realization of $H(s)$, we note that in view of the Structure Theorem given in Subsection 3.4D, we have $C_c(sI - A_c)^{-1}B_c + D_c = \bar{N}(s)\bar{D}(s)^{-1}$, where

$$\bar{D}(s) \triangleq B_m^{-1}[\Lambda(s) - A_m S(s)], \qquad \bar{N}(s) \triangleq C_c S(s) + D_c D(s).$$

However, $\bar{D}(s) = D(s)$ and $\bar{N}(s) = N(s)$, in view of (4.22) to (4.24). Therefore, $C_c(sI - A_c)^{-1}B_c + D_c = N(s)D^{-1}(s) = H(s)$, in view of (4.20).

It is now shown that (A_c, B_c) is controllable. We write

$$\begin{aligned} [sI - A_c, B_c] &= [sI - \bar{A}_c - \bar{B}_c A_m, \bar{B}_c B_m] \\ &= [sI - \bar{A}_c, \bar{B}_c] \begin{bmatrix} I & 0 \\ -A_m & B_m \end{bmatrix} \end{aligned} \tag{4.27}$$

and notice that $rank\ [s_j I - \bar{A}_c, \bar{B}_c] = n$ for any complex s_j. This is so because of the special form of $\bar{A}_c$, $\bar{B}_c$. (This is, in fact, the Brunovski canonical form.) Now since $|B_m| \neq 0$, Sylvester's Rank Inequality implies that $rank\ [s_j I - A_c, B_c] = n$ for any complex s_j, which in view of Subsection 3.4B, implies that (A_c, B_c) is controllable. In addition, since $B_m = I_m$, it follows that (A_c, B_c) is of the form (4.69) given in Subsection 3.4D. With $d_j = \mu_i$, the pair (A_c, B_c) is in controller form. ■

An alternative way of determining C_c is to first write $H(s)$ in the form

$$H(s) = \hat{H}(s) + \lim_{s \to \infty} H(s) = \hat{H}(s) + D_c, \tag{4.28}$$

where $\hat{H}(s) \triangleq H(s) - D_c$ is strictly proper. Now applying the above algorithm to $\hat{H}(s)$, one obtains $\hat{H}(s) = \hat{N}(s)D^{-1}(s)$, where $D(s)$ is precisely equal to the expression given in (4.20). We note, however, that $\hat{N}(s)$ is different. In fact, $\hat{N}(s) = N(s) - D_c D(s)$. The matrix C_c is now found to be of the form

$$\hat{N}(s) = C_c S(s). \tag{4.29}$$

Note that this is a generalization of the scalar case discussed in Example 4.1 [see (4.17) to (4.19)].

In the above algorithm the assumption that $d_j \geq 1$ for all $j = 1, \ldots, m$, was made. If for some j, $d_j = 0$, this would mean that the jth column of $H(s)$ will be a real $m \times 1$ vector that will be equal to the jth column of D_c [recall that $D_c =$

$\lim_{s\to\infty} H(s)$]. The strictly proper $\hat{H}(s)$ in (4.28) will then have its jth column equal to zero, and this zero column can be generated by a realization where the jth column of B_c is set to zero. Therefore, the zero column (the jth column) of $\hat{H}(s)$ is ignored in this case and the algorithm is applied to obtain a controllable realization. A zero column is then added to B_c. (See Example 4.4.)

Finally, we note that given $H(s) = N(s)D^{-1}(s)$ with $N(s)$, $D(s)$ not necessarily from (4.20), the above algorithm leads to a controllable realization if $d_j \triangleq$ column degrees of $D(s)$, provided that D_h, the highest column degree coefficient matrix is nonsingular ($|D_h| \neq 0$). This, in fact, means that $D(s)$ is column proper (see Subsection 7.2B). The resulting pair (A_c, B_c) is in controllable, companion form but not necessarily in controller form, since $B_m = D_h^{-1}$ is not necessarily upper triangular with ones on the diagonal. (See Example 4.5.)

Observer form realizations

These realizations are dual to the controller form realizations and can be obtained by duality arguments [see (4.2) and Example 4.3]. In the following, observer form realizations are obtained directly for completeness of exposition.

We consider the transfer function matrix $H(s) = [n_{ij}(s)/d_{ij}(s)]$, $i = 1, \ldots, p$, $j = 1, \ldots, m$, and let $\tilde{\ell}_i(s)$ be the (monic) least common denominator of all entries in the ith row of $H(s)$. Then $H(s)$ can be written as

$$H(s) = \tilde{D}^{-1}(s)\tilde{N}(s), \tag{4.30}$$

where $\tilde{N}(s) \triangleq [\bar{n}_{ij}(s)]$ and $\tilde{D}(s) \triangleq diag\,[\tilde{\ell}_1(s), \ldots, \tilde{\ell}_p(s)]$. Note that $\bar{n}_{ij}(s)/\tilde{\ell}_i(s) = n_{ij}(s)/d_{ij}(s)$ for $j = 1, \ldots, m$, and all $i = 1, \ldots, p$.

Let $\tilde{d}_i \triangleq deg\ \ell_i(s)$, assume that $\tilde{d}_i \geq 1$, define

$$\tilde{\Lambda}(s) \triangleq diag\,(s^{\tilde{d}_1}, \ldots, s^{\tilde{d}_p}),\ \tilde{S}(s) \triangleq block\ diag\,([1, s, \ldots, s^{\tilde{d}_i - 1}], i = 1, \ldots, p), \tag{4.31}$$

and note that $\tilde{S}(s)$ is a $p \times n$ $(\triangleq \sum_{i=1}^{p} \tilde{d}_i)$ polynomial matrix. Now, write

$$\tilde{D}(s) = \tilde{\Lambda}(s)\tilde{D}_h + \tilde{S}(s)\tilde{D}_\ell \tag{4.32}$$

and note that $\tilde{D}_h$ is the highest row degree coefficient matrix of $\tilde{D}(s)$. Note that $\tilde{D}(s)$ is diagonal, with entries monic polynomials, so that $\tilde{D}_h = I_p$, the $p \times p$ identity matrix. If, for example, $\tilde{D}(s) = \begin{bmatrix} 3s^2+1 & 2s \\ 2s & s \end{bmatrix}$, then the highest row degree coefficient matrix is $\tilde{D}_h = \begin{bmatrix} 3 & 0 \\ 2 & 1 \end{bmatrix}$ and $\tilde{S}(s)\tilde{D}_\ell$ in (4.32) accounts for the remaining lower row degree terms of $\tilde{D}(s)$, with $\tilde{D}_\ell$ a matrix of coefficients.

Observe that $|\tilde{D}_h| \neq 0$; in fact $\tilde{D}_h = I_p$. Define the $p \times p$ and $n \times p$ matrices

$$C_p = \tilde{D}_h^{-1} \qquad \text{and} \qquad A_p = -\tilde{D}_\ell \tilde{D}_h^{-1}, \tag{4.33}$$

respectively. Also, determine B_o and D_o such that

$$\tilde{N}(s) = \tilde{S}(s)B_o + \tilde{D}(s)D_o. \tag{4.34}$$

Note that

$$D_o = \lim_{s\to\infty} H(s), \tag{4.35}$$

and therefore, only B_o needs to be determined from (4.34).

An observable realization of $H(s)$ in observer form is now given by

$$\dot{x}_o = A_o x_o + B_o u, \qquad y = C_o x_o + D_o u,$$

where B_o and D_o were defined in (4.34) and (4.35), respectively, and

$$A_o = \bar{A}_o + A_p \bar{C}_o, \qquad C_o = C_p \bar{C}_o, \tag{4.36}$$

where $\bar{A}_o$ = *block diag* $[A_1, A_2, \ldots, A_p]$ with

$$A_i = \begin{bmatrix} 0 & \ldots & 0 & 0 \\ & & & 0 \\ & I_{\tilde{d}_i - 1} & & \vdots \\ & & & 0 \end{bmatrix} \in R^{\tilde{d}_i \times \tilde{d}_i},$$

$\bar{C}_o$ = *block diag* $([0, \ldots, 0, 1] \in R^{1 \times \tilde{d}_i}, i = 1, \ldots, p)$, and A_p, C_p is defined in (4.33). Note that (4.36) is exactly relation (4.85) of Section 3.4 if $\tilde{d}_i = \nu_i$, $i = 1, \ldots, p$, the observability indices.

LEMMA 4.4. The system $\{A_o, B_o, C_o, D_o\}$ is an n $(\triangleq \sum_{i=1}^{p} \tilde{d}_i)$th-order observable realization of $H(s)$ with (A_o, C_o) in observer form.

Proof. This is the dual result to Lemma 4.3. The proof is completely analogous and is omitted. ■

We conclude by noting that results dual to the results discussed after Lemma 4.3 are also valid here, i.e., results involving (i) a strictly proper $\hat{H}(s)$, (ii) an $H(s)$ with $\tilde{d}_i = 0$ for some row i, and (iii) $H(s) = \tilde{D}^{-1}(s)\tilde{N}(s)$, where $\tilde{D}(s)$, $\tilde{N}(s)$ are not necessarily determined using (4.30) (refer to the following examples). The reader is encouraged to explicitly state these results.

EXAMPLE 4.3. Let $H(s) = \left[\frac{s^2+1}{s^2}, \frac{s+1}{s^3}\right]$. We wish to derive a minimal realization for $H(s)$. To this end we consider realizations $\{A_c, B_c, C_c, D_c\}$, where (A_c, B_c) is in controller form. Here $\ell_1(s) = s^2$, $\ell_2(s) = s^3$, and $H(s)$ can therefore be written in the form (4.20) as

$$H(s) = N(s)D^{-1}(s) = [s^2 + 1, s + 1]\begin{bmatrix} s^2 & 0 \\ 0 & s^3 \end{bmatrix}^{-1}.$$

Here $d_1 = 2$, $d_2 = 3$ and $\Lambda(s) = \begin{bmatrix} s^2 & 0 \\ 0 & s^3 \end{bmatrix}$, $S(s) = \begin{bmatrix} 1 & s & 0 & 0 & 0 \\ 0 & 0 & 1 & s & s^2 \end{bmatrix}^T$. Note that $n = d_1 + d_2 = 5$, and therefore, the realization will be of order 5. Write $D(s) = D_h\Lambda(s) + D_\ell S(s)$, and note that $D_h = I_2$, $D_\ell = \begin{bmatrix} 0 & 0 & 0 & 0 & 0 \\ 0 & 0 & 0 & 0 & 0 \end{bmatrix}$. Therefore, in view of (4.23),

$$B_m = \begin{bmatrix} 1 & 0 \\ 0 & 1 \end{bmatrix} \quad \text{and} \quad A_m = -D_\ell = \begin{bmatrix} 0 & 0 & 0 & 0 & 0 \\ 0 & 0 & 0 & 0 & 0 \end{bmatrix}.$$

Here $D_c = \lim_{s \to \infty} H(s) = [1, 0]$ and (4.24) implies that $C_c S(s) = N(s) - D_c D(s) = [s^2 + 1, s + 1] - [s^2, 0] = [1, s + 1]$ from which we have $C_c = [1, 0, 1, 1, 0]$. A controllable realization in controller form is therefore given by $\dot{x} = A_c x_c + B_c u$ and $y = C_c x_c + D_c u$,

where

$$A_c = \begin{bmatrix} 0 & 1 & \vdots & 0 & 0 & 0 \\ 0 & 0 & \vdots & 0 & 0 & 0 \\ \cdots & \cdots & \vdots & \cdots & \cdots & \cdots \\ 0 & 0 & \vdots & 0 & 1 & 0 \\ 0 & 0 & \vdots & 0 & 0 & 1 \\ 0 & 0 & \vdots & 0 & 0 & 0 \end{bmatrix}, \qquad B_c = \begin{bmatrix} 0 & 0 \\ 1 & 0 \\ \cdots & \cdots \\ 0 & 0 \\ 0 & 0 \\ 0 & 1 \end{bmatrix},$$

$$C_c = [1, 0, 1, 1, 0], \qquad \text{and} \qquad D_c = [1, 0].$$

Note that the characteristic (pole) polynomial of $H(s)$ is s^3 and the McMillan degree of $H(s)$ is 3. The order of any minimal realization of $H(s)$ is therefore 3 (see Theorem 3.11). This implies that the controllable fifth-order realization derived above cannot be observable [verify that (A_c, C_c) is not observable]. To derive a minimal realization, the observable part of the system $\{A_c, B_c, C_c, D_c\}$ needs to be extracted, using the method described in Subsection 3.4A. In particular, a transformation matrix P needs to be determined so that

$$\hat{A} = PA_cP^{-1} = \begin{bmatrix} A_1 & 0 \\ A_{21} & A_2 \end{bmatrix} \qquad \text{and} \qquad \hat{C} = C_cP^{-1} = [C_1, 0],$$

where (A_1, C_1) is observable. If $\hat{B} = PB_c = \begin{bmatrix} B_1 \\ B_2 \end{bmatrix}$, then $\{A_1, B_1, C_1, D_1\}$ is a minimal realization of $H(s)$. To reduce (A_c, C_c) to such standard form for unobservable systems, we let $A_D = A^T$, $B_D = C_c^T$, and $C_D = B_c^T$ and we reduce (A_D, B_D) to a standard form for uncontrollable systems. Here the controllability matrix is

$$\mathscr{C}_D = \begin{bmatrix} 1 & 0 & 0 & 0 & 0 \\ 0 & 1 & 0 & 0 & 0 \\ 1 & 0 & 0 & 0 & 0 \\ 1 & 1 & 0 & 0 & 0 \\ 0 & 1 & 1 & 0 & 0 \end{bmatrix}.$$

Note that $rank\, \mathscr{C}_D = 3$. Now if the first three columns of $Q_D = P_D^{-1}$ are taken to be the first three linearly independent columns of $\mathscr{C}_D$, while the rest are chosen so that $|Q_D| \neq 0$ (see Subsection 3.4A), then

$$Q_D = \begin{bmatrix} 1 & 0 & 0 & 0 & 1 \\ 0 & 1 & 0 & 0 & 0 \\ 1 & 0 & 0 & 1 & 0 \\ 1 & 1 & 0 & 0 & 0 \\ 0 & 1 & 1 & 0 & 0 \end{bmatrix}$$

and

$$Q_D^{-1} = \begin{bmatrix} 0 & -1 & 0 & 1 & 0 \\ 0 & 1 & 0 & 0 & 0 \\ 0 & -1 & 0 & 0 & 1 \\ 0 & 1 & 1 & -1 & 0 \\ 1 & 1 & 0 & -1 & 0 \end{bmatrix}.$$

This implies that

$$\hat{A}_D = Q_D^{-1} A_D Q_D = \begin{bmatrix} A_{D1} & A_{D12} \\ 0 & A_{D2} \end{bmatrix} = \left[\begin{array}{ccc:cc} 0 & 0 & 0 & 1 & -1 \\ 1 & 0 & 0 & 0 & 1 \\ 0 & 1 & 0 & 0 & -1 \\ \hdashline 0 & 0 & 0 & -1 & 1 \\ 0 & 0 & 0 & -1 & 1 \end{array}\right],$$

$$\hat{B}_D = Q_D^{-1} B_D = \begin{bmatrix} B_{D1} \\ B_{D2} \end{bmatrix} = \begin{bmatrix} 1 \\ 0 \\ 0 \\ \hdashline 0 \\ 0 \end{bmatrix}, \qquad \hat{C}_D = C_D Q_D = \begin{bmatrix} 0 & 1 & 0 & 0 & 0 \\ 0 & 1 & 1 & 0 & 0 \end{bmatrix}.$$

Then

$$\hat{A} = \begin{bmatrix} A_1 & 0 \\ A_{21} & A_2 \end{bmatrix} = \hat{A}_D^T = \left[\begin{array}{ccc:cc} 0 & 1 & 0 & 0 & 0 \\ 0 & 0 & 1 & 0 & 0 \\ 0 & 0 & 0 & 0 & 0 \\ \hdashline 1 & 0 & 0 & -1 & -1 \\ -1 & 1 & -1 & 1 & 1 \end{array}\right],$$

$$\hat{B} = \hat{C}_D^T = \begin{bmatrix} 0 & 0 \\ 1 & 1 \\ 0 & 1 \\ \hdashline 0 & 0 \\ 0 & 0 \end{bmatrix}, \qquad \hat{C} = \hat{B}_D^T = [C_1, 0] = [1, 0, 0, \vdots\, 0, 0].$$

Clearly, $\hat{A} = \hat{A}_D^T$, $\hat{C} = \hat{B}_D^T$ is in standard form. Therefore, a controllable and observable realization, which is a minimal realization, is given by $\dot{x}_{co} = A_{co} x_{co} + B_{co} u$ and $y = C_{co} x_{co} + D_{co} u$, where

$$A_{co} = \begin{bmatrix} 0 & 1 & 0 \\ 0 & 0 & 1 \\ 0 & 0 & 0 \end{bmatrix}, \qquad B_{co} = \begin{bmatrix} 0 & 0 \\ 1 & 1 \\ 0 & 1 \end{bmatrix}, \qquad C_{co} = [1, 0, 0], \qquad D_{co} = [1, 0].$$

A minimal realization could also have been derived directly in the present case if a realization $\{A_o, B_o, C_o, D_o\}$ of $H(s)$, where (A_o, B_o) is in observer form, had been considered first, as is shown next. Notice that the McMillan degree of $H(s)$ is 3, and therefore, any realization of order higher than 3 will not be minimal. Here, however, the degree of the least common denominator of the (only) row is 3, and therefore, it is known in advance that the realization in observer form, which is of order three, will be minimal.

A realization $\{A_o, B_o, C_o, D_o\}$ of $H(s)$ in observer form can also be derived by considering $H^T(s)$ and deriving a realization in controller form. Presently, $\{A_o, B_o, C_o, D_o\}$

is derived directly. In particular, we write $H(s) = \tilde{D}^{-1}(s)\tilde{N}(s) = (s^3)^{-1}[s(s^2+1), s+1]$. Then $\tilde{d}_1 = 3[= deg\ \tilde{\ell}_1(s) = deg\ s^3]$, and $\tilde{\Lambda}(s) = s^3, \tilde{S}(s) = [1, s, s^2]$. Then $\tilde{D}(s) = s^3 = \tilde{\Lambda}(s)\tilde{D}_h + \tilde{S}(s)\tilde{D}_\ell$ implies that $\tilde{D}_h = 1$ and $\tilde{D}_\ell = [0, 0, 0]^T$. In view of (4.33), we have

$$C_p = 1, \qquad A_p = [0, 0, 0]^T,$$

$D_o = \lim_{s\to\infty} H(s) = [1, 0]$, and (4.34) implies that $\tilde{S}(s)B_o = \tilde{N}(s) - \tilde{D}(s)D_o = [s(s^2+1), s+1] - [s^3, 0] = [s, s+1]$, from which we have $B_o = \begin{bmatrix} 0 & 1 & 0 \\ 1 & 1 & 0 \end{bmatrix}^T$. An observable realization of $H(s)$ is the system $\dot{x} = A_o x_o + B_o u$, $y = C_o x_o + D_o u$, where

$$A_o = \begin{bmatrix} 0 & 0 & 0 \\ 1 & 0 & 0 \\ 0 & 1 & 0 \end{bmatrix}, \qquad B_o = \begin{bmatrix} 0 & 1 \\ 1 & 1 \\ 0 & 0 \end{bmatrix}, \qquad C_o = [0, 0, 1], \qquad D_o = [1, 0]$$

with (A_o, C_o) in observer form (see Lemma 4.4). This realization is minimal since it is of order 3, which is the McMillan degree of $H(s)$. (The reader should verify this.) Note how much easier it was to derive a minimal realization, using the second approach. ■

EXAMPLE 4.4. Let $H(s) = \begin{bmatrix} \frac{2}{s+1} & 1 \\ \frac{1}{s} & 0 \end{bmatrix}$. We wish to derive a minimal realization.

Here $\ell_1(s) = s(s+1)$ with $d_1 = 2$ and $\ell_2(s) = 1$ with $d_2 = 0$. In view of the discussion following Lemma 4.3, we let $D_c = \lim_{s\to\infty} H(s) = \begin{bmatrix} 0 & 1 \\ 0 & 0 \end{bmatrix}$ and $\hat{H}(s) = \begin{bmatrix} \frac{2}{s+1} & 0 \\ \frac{1}{s} & 0 \end{bmatrix}$.

We now consider the transfer function $\hat{\hat{H}}(s) = \begin{bmatrix} \frac{2}{s+1} \\ \frac{1}{s} \end{bmatrix}$ and determine a minimal realization.

Note that the McMillan degree of $\hat{\hat{H}}(s)$ is 2, and therefore, any realization of order 2 will be minimal. Minimal realizations are now derived using two alternative approaches:

1. *Via a controller form realization.* Here $\ell_1(s) = s(s+1)$, $d_1 = 2$, and $\hat{\hat{H}}(s) = \begin{bmatrix} 2s \\ s+1 \end{bmatrix}[s(s+1)]^{-1} = N(s)D^{-1}(s)$. Then $\Lambda(s) = s^2$ and $S(s) = [1, s]^T$, $D(s) = s(s+1) = 1s^2 + [0, 1][1, s]^T = D_h\Lambda(s) + D_\ell S(s)$. Therefore, $B_m = 1$ and $A_m = -[0, 1]$. Also, $C_c = \begin{bmatrix} 0 & 2 \\ 1 & 1 \end{bmatrix}$, which follows from $N(s) = \begin{bmatrix} 2s \\ s+1 \end{bmatrix} = \begin{bmatrix} 0 & 2 \\ 1 & 1 \end{bmatrix}\begin{bmatrix} 1 \\ s \end{bmatrix} = C_c S(s)$. Then a minimal realization for $H(s)$ is $A_c = \begin{bmatrix} 0 & 1 \\ 0 & -1 \end{bmatrix}$, $B_c = \begin{bmatrix} 0 \\ 1 \end{bmatrix}$, $C_c = \begin{bmatrix} 0 & 2 \\ 1 & 1 \end{bmatrix}$. Adding a zero column to B_c, a minimal realization of $H(s)$ is now derived as

$$A = \begin{bmatrix} 0 & 1 \\ 0 & -1 \end{bmatrix}, \qquad B = \begin{bmatrix} 0 & 0 \\ 1 & 0 \end{bmatrix}, \qquad C = \begin{bmatrix} 0 & 2 \\ 1 & 1 \end{bmatrix}, \qquad D = \begin{bmatrix} 0 & 1 \\ 0 & 0 \end{bmatrix}.$$

   We ask the reader to verify that by adding a zero column to B_c, controllability is preserved.

2. *Via an observer form realization.* We consider $\hat{\hat{H}}^T(s) = [2/(s+1), 1/s]$ and derive a realization in controller form. In particular, $\ell_1 = s+1$, $\ell_2 = s$, $\hat{\hat{H}}^T(s) =$

$[2, 1]\begin{bmatrix} s+1 & 0 \\ 0 & s \end{bmatrix}^{-1}$, $d_1 = d_2 = 1$, $\Lambda(s) = \begin{bmatrix} s & 0 \\ 0 & s \end{bmatrix}$, and $S(s) = \begin{bmatrix} 1 & 0 \\ 0 & 1 \end{bmatrix}$. Then $D(s) = \begin{bmatrix} s+1 & 0 \\ 0 & s \end{bmatrix} = \begin{bmatrix} 1 & 0 \\ 0 & 1 \end{bmatrix}\begin{bmatrix} s & 0 \\ 0 & s \end{bmatrix} + \begin{bmatrix} 1 & 0 \\ 0 & 0 \end{bmatrix}\begin{bmatrix} 1 & 0 \\ 0 & 1 \end{bmatrix} = D_h\Lambda(s) + D_\ell S(s)$ and $B_m = \begin{bmatrix} 1 & 0 \\ 0 & 1 \end{bmatrix}$, $A_m = \begin{bmatrix} -1 & 0 \\ 0 & 0 \end{bmatrix}$. Also, $C_c = [2, 1]$, from which we obtain $N(s) = [2, 1] = [2, 1]\begin{bmatrix} 1 & 0 \\ 0 & 1 \end{bmatrix} = C_c S(s)$. Therefore, a minimal realization $\{A, B, C\}$ of $\hat{\hat{H}}^T(s)$ is $\left\{\begin{bmatrix} -1 & 0 \\ 0 & 0 \end{bmatrix}, \begin{bmatrix} 1 & 0 \\ 0 & 1 \end{bmatrix}, [2, 1]\right\}$. The dual of this is a minimal realization of $\hat{\hat{H}}(s)$, namely, $A_o = \begin{bmatrix} -1 & 0 \\ 0 & 0 \end{bmatrix}$, $B_o = \begin{bmatrix} 2 \\ 1 \end{bmatrix}$, and $C_o = \begin{bmatrix} 1 & 0 \\ 0 & 1 \end{bmatrix}$. Therefore, a minimal realization of $H(s)$ is

$$A = \begin{bmatrix} -1 & 0 \\ 0 & 0 \end{bmatrix}, \quad B = \begin{bmatrix} 2 & 0 \\ 1 & 0 \end{bmatrix}, \quad C = \begin{bmatrix} 1 & 0 \\ 0 & 1 \end{bmatrix}, \quad D = \begin{bmatrix} 0 & 1 \\ 0 & 0 \end{bmatrix}.$$ ■

EXAMPLE 4.5. Let

$$H(s) = \begin{bmatrix} \frac{1}{s^2+1} & \frac{s-1}{s^2+1} \\ \frac{s}{s+1} & 0 \end{bmatrix} = \begin{bmatrix} 1 & s-1 \\ s & 0 \end{bmatrix}\begin{bmatrix} s+1 & 0 \\ s & s^2+1 \end{bmatrix}^{-1} = N(s)D^{-1}(s).$$

We wish to derive a minimal realization based on the given factorization $N(s)$, $D(s)$. Since the McMillan degree of $H(s)$ is 3 (show this), any realization of order 3 will be minimal. Note that in the present case $d_1 = 1$ and $d_2 = 2$, the degrees of the first and second columns of $D(s)$, respectively. Then $\Lambda(s) = \begin{bmatrix} s & 0 \\ 0 & s^2 \end{bmatrix}$, $S(s) = \begin{bmatrix} 1 & 0 & 0 \\ 0 & 1 & s \end{bmatrix}^T$, and

$D(s) = \begin{bmatrix} 1 & 0 \\ 1 & 1 \end{bmatrix}\begin{bmatrix} s & 0 \\ 0 & s^2 \end{bmatrix} + \begin{bmatrix} 1 & 0 & 0 \\ 0 & 1 & 0 \end{bmatrix}\begin{bmatrix} 1 & 0 & 0 \\ 0 & 1 & s \end{bmatrix}^T = D_h\Lambda(s) + D_\ell S(s)$. Here $|D_h| \neq 0$, and therefore, the algorithm above still applies. Note that D_h is not in upper triangular form with ones on the diagonal, and therefore, the resulting controllable realization will not be in controller form; A_c, however, will be in companion form. Let $B_m = D_h^{-1} = \begin{bmatrix} 1 & 0 \\ -1 & 1 \end{bmatrix}$, $A_m = -D_h^{-1}D_\ell = \begin{bmatrix} -1 & 0 & 0 \\ 1 & -1 & 0 \end{bmatrix}$ and

$$N(s) = C_c S(s) + D_c D(s) = \begin{bmatrix} 1 & -1 & 1 \\ -1 & 0 & 0 \end{bmatrix}\begin{bmatrix} 1 & 0 \\ 0 & 1 \\ 0 & s \end{bmatrix} + \begin{bmatrix} 0 & 0 \\ 1 & 0 \end{bmatrix}\begin{bmatrix} s+1 & 0 \\ s & s^2+1 \end{bmatrix}.$$

A minimal realization is now given by

$$A_c = \left[\begin{array}{c:cc} -1 & 0 & 0 \\ \hdashline 0 & 0 & 1 \\ 1 & -1 & 0 \end{array}\right], \quad B_c = \begin{bmatrix} 1 & 0 \\ \hdashline 0 & 0 \\ -1 & 1 \end{bmatrix},$$

$$C_c = \begin{bmatrix} 1 & -1 & 1 \\ -1 & 0 & 0 \end{bmatrix}, \quad D_c = \begin{bmatrix} 0 & 0 \\ 1 & 0 \end{bmatrix}.$$

Verify this. ■

C. Realizations with Matrix A Diagonal

When the roots of the minimal polynomial $m_H(s)$ of $H(s)$ are distinct, there is a realization algorithm due to Gilbert [2] that provides a minimal realization of $H(s)$ with A diagonal. Let

$$m_H(s) = s^r + d_{r-1}s^{r-1} + \cdots + d_1 s + d_0 \tag{4.37}$$

be the (monic) least common denominator of all nonzero entries of the $p \times m$ matrix $H(s)$ which in view of Section 3.5, is the minimal polynomial of $H(s)$. We assume that its r roots λ_i are distinct and we write

$$m_H(s) = \prod_{i=1}^{r}(s - \lambda_i). \tag{4.38}$$

Note that the pole polynomial of $H(s)$, $p_H(s)$, will have repeated roots (poles) if $p_H(s) \neq m_H(s)$ (see Section 3.5 and Example 4.6). We now consider the strictly proper matrix $\hat{H}(s) \triangleq H(s) - \lim_{s\to\infty} H(s) = H(s) - D$ and expand it into partial fractions to obtain

$$\hat{H}(s) = \frac{1}{m_H(s)}\hat{N}(s) = \sum_{i=1}^{r} \frac{1}{s - \lambda_i} R_i. \tag{4.39}$$

The $p \times m$ residue matrices R_i can be found from the relation

$$R_i = \lim_{s\to\lambda_i}(s - \lambda_i)\hat{H}(s). \tag{4.40}$$

We write

$$R_i = C_i B_i, \qquad i = 1, \ldots, r, \tag{4.41}$$

where C_i is a $p \times \rho_i$ and B_i is a $\rho_i \times m$ matrix with $\rho_i \triangleq \textit{rank}\ R_i \leq \min\,(p, m)$. Note that the above expression is always possible. Indeed, there is a systematic procedure of generating it, namely, by obtaining an LU decomposition of R_i (refer to the Appendix). Then

$$A = \begin{bmatrix} \lambda_1 I_{\rho_1} & & & \\ & \lambda_2 I_{\rho_2} & & \\ & & \ddots & \\ & & & \lambda_r I_{\rho_r} \end{bmatrix}, \qquad B = \begin{bmatrix} B_1 \\ B_2 \\ \vdots \\ B_r \end{bmatrix}, \tag{4.42}$$

$$C = [C_1, C_2, \ldots, C_r], \qquad D = \lim_{s\to\infty} H(s)$$

is a minimal realization of order $n \triangleq \sum_{i=1}^{r} \rho_i$.

LEMMA 4.5. Representation (4.42) is a minimal realization of $H(s)$.

Proof. It can be verified directly that $C(sI - A)^{-1}B + D = H(s)$, i.e., that (4.42) is a realization of $H(s)$.To verify controllability, we write

$$\mathscr{C} = [B, AB, \ldots, A^{n-1}B] = \begin{bmatrix} B_1 & & & \\ & B_2 & & \\ & & \ddots & \\ & & & B_r \end{bmatrix} \begin{bmatrix} I_m, & \lambda_1 I_m, \ldots, & \lambda_1^{n-1} I_m \\ \vdots & \vdots & \vdots \\ I_m, & \lambda_r I_m, \ldots, & \lambda_r^{n-1} I_m \end{bmatrix}.$$

The second matrix in the product is a block Vandermonde matrix of dimensions $mr \times mn$. It can be shown that this matrix has full rank mr since all λ_i are assumed to be distinct. Also note that the $(n = \Sigma \rho_i) \times mr$ matrix *block diag* $[B_i]$ has rank equal to $\sum_{i=1}^{r} \text{rank } B_i = \sum_{i=1}^{r} \rho_i = n \leq mr$. Now, in view of Sylvester's Rank Inequality, as applied to the above matrix product, we have $n + mr - mr \leq \text{rank } \mathscr{C} \leq \min(n, mr)$, from which $\text{rank } \mathscr{C} = n$. Therefore, $\{A, B, C, D\}$ is controllable. Observability is shown in a similar way. Therefore, representation (4.42) is minimal. ∎

EXAMPLE 4.6. Let

$$H(s) = \begin{bmatrix} \dfrac{1}{s} & 0 \\ \dfrac{2}{s+1} & \dfrac{1}{s(s+1)} \end{bmatrix}.$$

Here $m_H(s) = s(s+1)$ with roots $\lambda_1 = 0, \lambda_2 = -1$ distinct. We write $H(s) = (1/s)R_1 + [1/(s+1)]R_2$, where $R_1 = \lim_{s\to 0} sH(s) = \lim_{s\to 0} \begin{bmatrix} 1 & 0 \\ \dfrac{2s}{s+1} & \dfrac{1}{s+1} \end{bmatrix} = \begin{bmatrix} 1 & 0 \\ 0 & 1 \end{bmatrix}$, $R_2 = \lim_{s\to -1}(s+1)H(s) = \lim_{s\to -1} \begin{bmatrix} \dfrac{s+1}{s} & 0 \\ 2 & \dfrac{1}{s} \end{bmatrix} = \begin{bmatrix} 0 & 0 \\ 2 & -1 \end{bmatrix}$, $\rho_1 = \text{rank } R_1 = 2$, and $\rho_2 = \text{rank } R_2 = 1$, i.e., the order of a minimal realization is $n = \rho_1 + \rho_2 = 3$. We now write

$$R_1 = \begin{bmatrix} 1 & 0 \\ 0 & 1 \end{bmatrix} = \begin{bmatrix} 1 & 0 \\ 0 & 1 \end{bmatrix}\begin{bmatrix} 1 & 0 \\ 0 & 1 \end{bmatrix} = C_1 B_1,$$

$$R_2 = \begin{bmatrix} 0 & 0 \\ 2 & -1 \end{bmatrix} = \begin{bmatrix} 0 \\ 1 \end{bmatrix}[2 \; -1] = C_2 B_2.$$

Then

$$A = \begin{bmatrix} \lambda_1 I_2 & 0 \\ 0 & \lambda_2 \end{bmatrix} = \begin{bmatrix} 0 & 0 & 0 \\ 0 & 0 & 0 \\ 0 & 0 & -1 \end{bmatrix}, \qquad B = \begin{bmatrix} B_1 \\ B_2 \end{bmatrix} = \begin{bmatrix} 1 & 0 \\ 0 & 1 \\ 2 & -1 \end{bmatrix},$$

$$C = [C_1, C_2] = \begin{bmatrix} 1 & 0 & 0 \\ 0 & 1 & 1 \end{bmatrix}$$

is a minimal realization with A diagonal (show this). Note that the characteristic polynomial of $H(s)$ is $p_H(s) = s^2(s+1)$, and therefore, the McMillan degree, which is equal to the order of any minimal realization, is 3, as expected. ∎

D. Realizations with Matrix A in Block Companion Form

The realizations derived using the algorithms described below are in general either controllable or observable and of order mr or pr, where r is the degree of the minimal polynomial of $H(s)$. Most often it is also necessary to use the methods of Subsection 3.4A to reduce these realizations to the standard forms for unobservable or uncontrollable systems and in this way derive minimal realizations.

Using the numerator polynomial matrix

Consider a proper $p \times m$ matrix $H(s)$ and let

$$m_H(s) = s^r + d_{r-1}s^{r-1} + \cdots + d_1 s + d_0$$

be its minimal polynomial, as in (4.37). The polynomial $m_H(s)$ is the monic least common denominator of all entries of $H(s)$.

Let $N_b(s) \triangleq m_H(s)H(s)$ be a polynomial matrix and write

$$N_b(s) = C_{bc}S_b(s) + D_{bc}m_H(s), \tag{4.43}$$

where $S_b(s) \triangleq [I_m, sI_m, \ldots, s^{r-1}I_m]^T$. Note that $D_{bc} = \lim_{s\to\infty} H(s)$, and therefore, C_{bc} is the only unknown in (4.43). A solution C_{bc} always exists since the highest possible degree in $N_b(s) - D_{bc}m_H(s)$ is $r-1$ because $H(s)$ is proper. Expression (4.43) is analogous to (4.24). In addition, we can also write $m_H(s) = s^r - [-d_0, \ldots, -d_{r-1}][1, s, \ldots, s^{r-1}]^T$, from which we have

$$m_H(s)I_m = s^r I_m - [-d_0 I_m, \ldots, -d_{r-1}I_m]S_b(s), \tag{4.44}$$

which corresponds to $D(s)$ in (4.22).

A controllable realization in block controllable companion form is given by $\dot{x} = A_{bc}x + B_{bc}u$, $y = C_{bc}s + D_{bc}u$ with C_{bc} and D_{bc} specified in (4.43) and

$$A_{bc} = \begin{bmatrix} 0_m & I_m & \cdots & 0_m \\ \vdots & \vdots & \ddots & \vdots \\ 0_m & 0_m & \cdots & I_m \\ -d_0 I_m & -d_1 I_m & \cdots & -d_{r-1}I_m \end{bmatrix}, \qquad B_{bc} = \begin{bmatrix} 0_m \\ \vdots \\ 0_m \\ I_m \end{bmatrix}. \tag{4.45}$$

Note that if the strictly proper matrix $\hat{H}(s) \triangleq H(s) - \lim_{s\to\infty} H(s) = H(s) - D_{b_c}$ is used, then

$$\hat{N}_b(s) \triangleq m_H(s)\hat{H}(s) = R_{r-1}s^{r-1} + \cdots + R_0,$$

where R_i, $i = 0, \ldots, r-1$, are real $p \times m$ coefficient matrices. It is now not difficult to see that in this case

$$C_{b_c} = [R_0, R_1, \ldots, R_{r-1}]. \tag{4.46}$$

The realization $\{A_{bc}, B_{bc}, C_{bc}, D_{bc}\}$ in (4.45) is of order mr and is a direct generalization of the SISO realization (4.9) to the MIMO case. In general it is only controllable but not observable. It is also observable when $m = 1$, and therefore, it is minimal for $m = 1$. This is true because of Theorem 3.11 and the fact that in this case $r = deg\ m_H(s) = deg\ p_H(s)$, the McMillan degree of $H(s)$.

LEMMA 4.6. The representation (4.45) is a controllable realization of $H(s)$.

Proof. The proof is similar to the proof of Lemma 4.1. First, to show that (4.45) is a realization, we consider the $rm \times m$ matrix $X(s) = [X_1^T(s), \ldots, X_r^T(s)]^T \triangleq (sI - A_{bc})^{-1}B_{bc}$. Then $sX - A_{bc}X = B_{bc}$ and $sX_i = X_{i+1}$, $i = 1, \ldots, r-1$, or $X_{i+1} = s^i X_1$, $i = 1, \ldots, r-1$. Also, $sX_r - [-d_0 I_m, \ldots, -d_{r-1}I_m]X = I_m$, which implies, in view of $X_{i+1} = s^i X_1$, that $m_H(s)X_1 = I_m$. Thus, $X_i = (m_H(s))^{-1}s^{i-1}I_m$, $i = 1, \ldots, r$, and in view of $X(s) = (sI - A_{b_c})^{-1}B_{b_c}$,

$$(sI - A_{bc})S_b(s) = B_{bc}\delta_{m_H(s)}, \tag{4.47}$$

which is of course analogous to (4.11). (Refer also to the algorithm for the controller form realization in the MIMO case and the Structure Theorem in Subsection 3.4D, where a similar relation is valid.) Note that $|sI - A_{b_c}| = (m_H(s))^m$ (show this). To show that (A_{b_c}, B_{b_c}) is controllable, we observe that

$$[B_{bc}, A_{bc}B_{bc}, \ldots, A_{bc}^{r-1}B_{bc}, \ldots, A_{bc}^{rm-1}B_{bc}] = \begin{bmatrix} 0_m & 0_m & \ldots & I_m & \times & \ldots & \times \\ \vdots & \vdots & & \vdots & \vdots & & \vdots \\ 0_m & I_m & \ldots & & & & \\ I_m & \times & \ldots & \times & \times & \ldots & \times \end{bmatrix}, \tag{4.48}$$

which has full rank mr, since the first mr columns are linearly independent. ■

We may also easily obtain an observable realization of $H(s)$ using duality. In particular

$$A_{bo} = \begin{bmatrix} 0_p & \ldots & 0_p & -d_0 I_p \\ I_p & \ldots & 0_p & -d_1 I_p \\ \vdots & \ddots & \vdots & \vdots \\ 0_p & \ldots & I_p & -d_{r-1} I_p \end{bmatrix}, \qquad B_{bo} = \begin{bmatrix} R_0 \\ R_1 \\ \vdots \\ R_{r-1} \end{bmatrix}, \tag{4.49}$$
$$C_{bo} = [0_p, \ldots, 0_p, I_p], \qquad D_{bo} = \lim_{s\to\infty} H(s)$$

is an observable realization of order pr (use duality arguments to prove this).

In both of the above cases of controllable realization $\{A_{bc}, B_{bc}, C_{bc}, D_{bc}\}$ or observable realization $\{A_{bo}, B_{bo}, C_{bo}, D_{bo}\}$, the methods of Subsection 3.4A may be used to obtain minimal realizations [see (4.3b)].

EXAMPLE 4.7. Let

$$H(s) = \begin{bmatrix} \frac{1}{s} & 0 \\ \frac{2}{s+1} & \frac{1}{s(s+1)} \end{bmatrix}$$

as in Example 4.6. We wish to determine an observable realization with A in block companion form. To this end, duality will be used and the procedure described by (4.3a) will be followed.

In this case we have $m_H(s) = s^2 + s = s^2 + d_1 s + d_0$, from which we conclude that $r = 2$, $d_1 = 1$, and $d_0 = 0$. Note that $H(s)$ is strictly proper. Let $\tilde{H}(s) = H^T(s) = \frac{1}{s(s+1)}\begin{bmatrix} s+1 & 2s \\ 0 & 1 \end{bmatrix} = \tilde{N}_b(s)(m_H(s))^{-1}$ and write

$$\tilde{N}_b(s) = \begin{bmatrix} 1 & 2 \\ 0 & 0 \end{bmatrix} s + \begin{bmatrix} 1 & 0 \\ 0 & 1 \end{bmatrix} = R_1 s + R_0. \tag{4.50}$$

Then a controllable realization of $\tilde{H}(s)$ is given by

$$\tilde{A}_{bc} = \begin{bmatrix} 0_2 & I_2 \\ -d_0 I_2 & -d_1 I_2 \end{bmatrix} = \begin{bmatrix} 0 & 0 & \vdots & 1 & 0 \\ 0 & 0 & \vdots & 0 & 1 \\ \ldots & \ldots & \vdots & \ldots & \ldots \\ 0 & 0 & \vdots & -1 & 0 \\ 0 & 0 & \vdots & 0 & -1 \end{bmatrix}, \qquad \tilde{B}_{bc} = \begin{bmatrix} 0_2 \\ I_2 \end{bmatrix} = \begin{bmatrix} 0 & 0 \\ 0 & 0 \\ \ldots & \ldots \\ 1 & 0 \\ 0 & 1 \end{bmatrix},$$

$$\tilde{C}_{bc} = [R_0, R_1] = \begin{bmatrix} 1 & 0 & \vdots & 1 & 2 \\ 0 & 1 & \vdots & 0 & 0 \end{bmatrix},$$

and an observable realization of $H(s)$ is given by

$$A_{bo} = \tilde{A}_{bc}^T = \begin{bmatrix} 0 & 0 & \vdots & 0 & 0 \\ 0 & 0 & \vdots & 0 & 0 \\ \dots & \dots & \vdots & \dots & \dots \\ 1 & 0 & \vdots & -1 & 0 \\ 0 & 1 & \vdots & 0 & -1 \end{bmatrix}, \qquad B_{bo} = \tilde{C}_{bc}^T = \begin{bmatrix} 1 & 0 \\ 0 & 1 \\ \dots & \dots \\ 1 & 0 \\ 2 & 0 \end{bmatrix},$$

$$C_{bo} = \tilde{B}_{bc}^T = \begin{bmatrix} 0 & 0 & \vdots & 1 & 0 \\ 0 & 0 & \vdots & 0 & 1 \end{bmatrix}.$$

Note that the system $\{A_{bo}, B_{bo}, C_{bo}\}$ is not controllable. We ask the reader to use the procedure shown in (4.3b) to obtain a minimal realization. The order of any minimal realization is 3 (why?). ■

Using the Hankel Matrix

We consider a proper $p \times m$ transfer function matrix $H(s)$ and let

$$m_H(s) = s^r + d_{r-1}s^{r-1} + \cdots + d_1 s + d_0$$

be its minimal polynomial, as in (4.37). We write

$$H(s) = H_0 + H_1 s^{-1} + H_2 s^{-2} + \cdots, \tag{4.51}$$

where the H_i are the Markov parameters of the system. Let

$$A = \begin{bmatrix} 0_p & I_p & \dots & 0_p \\ \vdots & & & \vdots \\ 0_p & 0_r & \dots & I_p \\ -d_0 I_p & -d_1 I_p & \dots & -d_{r-1} I_p \end{bmatrix}, \qquad B = \begin{bmatrix} H_1 \\ H_2 \\ \vdots \\ H_r \end{bmatrix}, \tag{4.52}$$

$$C = [I_p, 0_p, \dots, 0_p], \qquad D = H_0.$$

LEMMA 4.7. The representation (4.52) is an observable realization of $H(s)$.

Proof. To show that (4.52) is a realization, we note that $\{A, B, C, D\}$ is a realization of $H(s)$ if and only if $D = H_0$ and $CA^{i-1}B = H_i, i = 1, 2, \dots$ (prove this, referring to Exercise 2.63). Here

$$\begin{bmatrix} C \\ CA \\ \vdots \\ CA^{r-1} \end{bmatrix} = \begin{bmatrix} I_p & & & \\ & I_p & & \\ & & \ddots & \\ & & & I_p \end{bmatrix}, \tag{4.53}$$

and therefore, $CA^{i-1}B = H_i, i = 1, \dots, r$. To show that this is also true for $i = r+1, \dots,$ a relationship between H_i and d_i is required. In particular, we let $\hat{H}(s) = H(s) - H_0$ and we write

$$
\begin{aligned}
m_H(s)\hat{H}(s) &= (s^r + d_{r-1}s^{r-1} + \cdots + d_1 s + d_0)(H_1 s^{-1} + H_2 s^{-2} + \cdots) \\
&= R_{r-1}s^{r-1} + \cdots + R_0
\end{aligned} \tag{4.54}
$$

Equating coefficients of equal powers of s, we obtain

$$
\begin{aligned}
s^{r-1}: \quad H_1 &= R_{r-1}, \\
s^{r-2}: \quad H_2 &= R_{r-2} - d_{r-1}H_1, \\
&\vdots \\
s^0: \quad H_r &= R_0 - d_{r-1}H_{r-1} - \cdots - d_1 H_1,
\end{aligned} \tag{4.55}
$$

and
$$H_{r+i} = -d_{r-1}H_{r+i-1} - \cdots - d_0 H_i, \qquad i = 1, 2, \ldots.$$

Using the above relations, it can be shown that $CA^{i-1}B = H_i$ for $i = 1, 2, \ldots$. Now, in view of (4.53), the observability matrix has rank pr, which is equal to the order of the system. Therefore, the system is observable. ■

We can now use duality arguments to show that

$$
A = \begin{bmatrix} 0_m & \ldots & 0_m & -d_0 I_m \\ I_m & \ldots & 0_m & -d_1 I_m \\ \vdots & \ddots & \vdots & \vdots \\ 0_m & \ldots & I_m & -d_{r-1} I_m \end{bmatrix}, \qquad B = \begin{bmatrix} I_m \\ 0_m \\ \vdots \\ 0_m \end{bmatrix}, \tag{4.56}
$$

$$
C = [H_1, H_2, \ldots, H_r], \qquad D = H_0
$$

is a controllable realization.

EXAMPLE 4.8. Let

$$
H(s) = \begin{bmatrix} \dfrac{1}{s} & 0 \\ \dfrac{2}{s+1} & \dfrac{1}{s(s+1)} \end{bmatrix}
$$

as in Example 4.7. We wish to determine an observable realization. Here $m_H(s) = s^2 + s = s^2 + d_1 s + d_0$, from which $r = 2$, $d_1 = 1$, and $d_0 = 0$. Let $H(s) = H_0 + H_1 s^{-1} + H_2 s^{-2} + \cdots$. The Markov parameters can be found directly from $H_0 = \lim_{s\to\infty} H(s) = 0$, $H_1 = \lim_{s\to\infty} s(H(s) - H_0) = \begin{bmatrix} 1 & 0 \\ 2 & 0 \end{bmatrix}$, $H_2 = \lim_{s\to\infty} s^2(H(s) - (H_0 + H_1 s^{-1})) = \begin{bmatrix} 0 & 0 \\ -2 & 1 \end{bmatrix}$, and so forth, or from (4.55) since $R_0, \ldots, R_{r-1}$ are already known from Example 4.7 (after the transpose is taken). We have $H_1 = R_1 = \begin{bmatrix} 1 & 0 \\ 2 & 0 \end{bmatrix}$, $H_2 = R_0 - d_1 H_1 = \begin{bmatrix} 1 & 0 \\ 0 & 1 \end{bmatrix} - \begin{bmatrix} 1 & 0 \\ 2 & 0 \end{bmatrix} = \begin{bmatrix} 0 & 0 \\ -2 & 1 \end{bmatrix}$, and $H_{2+i} = -d_1 H_{2+i} - d_0 H_i = -H_{1+i}$, $i = 1, 2, \ldots$; that is, $\begin{bmatrix} 0 & 0 \\ 2 & -1 \end{bmatrix} = -H_2 = H_3 = -H_4 = H_5 = \cdots$. Therefore, an observable realization (4.52) is given by

$$
A = \begin{bmatrix} 0_2 & I_2 \\ -d_0 I_2 & -d_1 I_2 \end{bmatrix} = \left[\begin{array}{cc:cc} 0 & 0 & 1 & 0 \\ 0 & 0 & 0 & 1 \\ \hdashline 0 & 0 & -1 & 0 \\ 0 & 0 & 0 & -1 \end{array}\right], \qquad B = \begin{bmatrix} H_1 \\ H_2 \end{bmatrix} = \left[\begin{array}{cc} 1 & 0 \\ 2 & 0 \\ \hdashline 0 & 0 \\ -2 & 1 \end{array}\right],
$$

$$C = [I_2, 0_2] = \begin{bmatrix} 1 & 0 & \vdots & 0 & 0 \\ 0 & 1 & \vdots & 0 & 0 \end{bmatrix}, \qquad D = \begin{bmatrix} 0 & 0 \\ 0 & 0 \end{bmatrix}.$$

This realization is not controllable, since a minimal realization is of order 3. We ask the reader to use Theorem 3.13 to show this. ■

E. Realizations Using Singular-Value Decomposition

Internally balanced realizations

Given a proper $p \times m$ matrix $H(s)$, we let r denote the degree of its minimal polynomial $m_H(s)$, we write

$$H(s) = H_0 + H_1 s^{-1} + H_2 s^{-2} + \dots$$

to obtain the Markov parameters H_i, and we define

$$T \triangleq M_H(r, r) = \begin{bmatrix} H_1 & \dots & H_r \\ \vdots & & \\ H_r & \dots & H_{2r-1} \end{bmatrix}, \qquad \hat{T} \triangleq \begin{bmatrix} H_2 & \dots & H_{r+1} \\ \vdots & & \\ H_{r+1} & \dots & H_{2r} \end{bmatrix}, \tag{4.57}$$

where $M_H(r, r)$ is the Hankel matrix (see Definition 3.4) and $T, \hat{T}$ are real matrices of dimension $rp \times rm$.

Using *singular-value decomposition* (see the Appendix), we write

$$T = K \begin{bmatrix} \Sigma & 0 \\ 0 & 0 \end{bmatrix} L \tag{4.58}$$

where $\sum = diag\ [\lambda_1, \dots, \lambda_n] \in R^{n \times n}$ with $n = rank\ T = rank\ M_H(r, r)$, which in view of Theorem 3.13 is the order of a minimal realization of $H(s)$. The λ_i with $\lambda_1 \geq \lambda_2 \geq \dots \geq \lambda_n > 0$ are the singular values of T, i.e., the nonzero eigenvalues of $T^T T$. Furthermore, $KK^T = K^T K = I_{pr}$ and $LL^T = L^T L = I_{mr}$. We write

$$T = K_1 \Sigma L_1 = \left(K_1 \Sigma^{1/2}\right)\left(\Sigma^{1/2} L_1\right) = VU, \tag{4.59}$$

where K_1 denotes the first n columns of K, L_1 denotes the first n rows of L, $K_1^T K_1 = I_n$, and $L_1 L_1^T = I_n$. Also, $V \in R^{rp \times n}$ and $U \in R^{n \times rm}$.

We let V^+ and U^+ denote pseudoinverses of V and U, respectively (see Appendix), i.e.,

$$V^+ = \Sigma^{-1/2} K_1^T \qquad \text{and} \qquad U^+ = L_1^T \Sigma^{-1/2}, \tag{4.60}$$

where $V^+ V = I_n$ and $UU^+ = I_n$. Now define

$$A = V^+ \hat{T} U^+, \qquad B = U I_{m,mr}^T, \qquad C = I_{p,pr} V, \qquad D = H_0, \tag{4.61}$$

where $I_{k,\ell} \triangleq [I_k, 0_{\ell-k}]$, $k < \ell$, i.e., $I_{k,\ell}$ is a $k \times \ell$ matrix with its first k columns determining an identity matrix and the remaining $\ell - k$ columns being equal to zero. Thus, B is defined as the first m columns of U, and C is defined as the first p rows of V. Note that $A \in R^{n \times n}$, $B \in R^{n \times m}$, $C \in R^{p \times n}$, and $D \in R^{p \times m}$.

LEMMA 4.8. The representation (4.61) is a minimal realization of $H(s)$.

Proof. It can be shown that $CA^{i-1}B = H_i$, $i = 1, 2, \ldots$ (see also the proof of Lemma 4.7). Thus, $\{A, B, C, D\}$ is a realization. We note that V and U are the observability and controllability matrices, respectively, and that both are of full rank n. Therefore, the realization is minimal. Furthermore, we notice that $V^T V = UU^T = \Sigma$. Realizations of this type are called *internally balanced realizations.* ■

The term *internally balanced* emphasizes the fact that realizations of this type are "as much controllable as they are observable," since their controllability and observability Gramians are equal and diagonal (see Exercise 5.20). Using such representations, it is possible to construct reasonable reduced-order models of systems by deleting that part of the state space that is "least controllable" and therefore "least observable" in accordance with some criterion. In fact, the realization procedure described can be used to obtain a *reduced-order model* for a given system. Specifically, if the system is to be approximated by a q-dimensional model with $q < n$, then the reduced-order model can be obtained from

$$T = K_q \, diag\,[\lambda_1, \ldots, \lambda_q] L_q, \tag{4.62}$$

where K_q denotes the first q columns of K in (4.58), and L_q denotes the first q rows of L.

5.5 SUMMARY

The theory of state-space realizations of input-output descriptions given by impulse responses, and the time-invariant case by transfer function descriptions, were studied in this chapter.

In Section 5.2 the problem of state-space realizations of input-output descriptions was defined and the existence of such realizations was addressed. In Subsection 5.3A time-varying and time-invariant continuous-time and discrete-time systems were considered. Subsequently, the focus was on time-invariant systems and transfer function matrix descriptions $H(s)$. The minimality of realizations of $H(s)$ was studied in Subsection 5.3C, culminating in two results, Theorem 3.9 and Theorem 3.10, where it was first shown that a realization is minimal if and only if it is controllable and observable, and next, that if a realization is minimal, all other minimal realizations of a given $H(s)$ can be found via similarity transformations. In Subsection 5.3C it was shown how to determine the order of minimal realizations directly from $H(s)$. Several realization algorithms were presented in Section 5.4, and the role of duality was emphasized in Section 5.4A.

5.6 NOTES

A clear understanding of the relationship between external and internal descriptions of systems is one of the principal contributions of systems theory. This topic was developed in the early sixties with original contributions by Gilbert [2] and Kalman [4].

The role of controllability and observability in minimal realizations is due to Kalman [4]. See also Kalman, Falb, and Arbib [5].

The first realization method for MIMO systems is attributed to Gilbert [2]. It was developed for systems where the matrix A can be taken to be diagonal. This method is presented in this chapter. For extensive historical comments concerning this topic, see Kailath [3].

Additional information concerning realizations for the time-varying case can be found, for example, in Brockett [1], Silverman [9], Kamen [6], Rugh [8], and the literature cited in these references.

Balanced realizations were introduced in Moore [7].

5.7 REFERENCES

1. R. W. Brockett, *Finite Dimensional Linear Systems,* Wiley, New York, 1970.
2. E. Gilbert, "Controllability and Observability in Multivariable Control Systems," *SIAM J. Control,* Vol. 1, pp. 128–151, 1963.
3. T. Kailath, *Linear Systems,* Prentice Hall, Englewood Cliffs, NJ, 1980.
4. R. E. Kalman, "Mathematical Description of Linear Systems," *SIAM J. Control,* Vol. 1, pp. 152–192, 1963.
5. R. E. Kalman, P. L. Falb, and M. A. Arbib, *Topics in Mathematical System Theory,* McGraw-Hill, New York, 1969.
6. E. W. Kamen,"New Results in Realization Theory for Linear Time-Varying Analytic Systems," *IEEE Trans. on Automatic Control,* Vol. AC-24, pp. 866–877, 1979.
7. B. C. Moore,"Principal Component Analysis in Linear Systems: Controllability, Observability and Model Reduction," *IEEE Trans. on Automatic Control,* Vol. AC-26, pp. 17–32, 1981.
8. W. J. Rugh, *Linear System Theory,* Second Edition, Prentice-Hall, England Cliffs, NJ, 1996.
9. L. M. Silverman, "Realization of Linear Dynamical Systems," *IEEE Trans. on Automatic Control,* Vol. AC-16, pp. 554–567, 1971.

5.8 EXERCISES

5.1. Given the transfer function matrix

$$H(s) = \begin{bmatrix} \frac{s-1}{s} & 0 & \frac{s-2}{s+2} \\ 0 & \frac{s+1}{s} & 0 \end{bmatrix},$$

determine the McMillan degree of $H(s)$ and find a minimal realization for $H(s)$. Verify your results.

5.2. Consider the transfer function matrix

$$H(s) = \begin{bmatrix} \frac{s-1}{s+1} & \frac{1}{s^2-1} \\ 1 & 0 \end{bmatrix}.$$

(a) Determine the pole polynomial and the McMillan degree of $H(s)$, using both the Smith-McMillan form and the Hankel matrix.

(b) Determine an observable realization of $H(s)$.
(c) Determine a minimal realization of $H(s)$.

Hint: Obtain realizations for $\left[\dfrac{s-1}{s+1}, \dfrac{1}{s^2-1}\right]$.

5.3. Consider the transfer function matrix $H(s) = \left[\dfrac{(s+1)(-s+5)}{(s-1)(s^2-9)}, \dfrac{s}{s-1}\right]^T$ and determine for $H(s)$ a minimal realization in controller form.

5.4. Consider the transfer function matrix

$$H(s) = \begin{bmatrix} \dfrac{-3s^2-6s-2}{(s+1)^3} & \dfrac{s^3-3s-1}{(s-2)(s+1)^3} & \dfrac{1}{(s-2)(s+1)^2} \\ \dfrac{s}{(s+1)^3} & \dfrac{s}{(s-2)(s+1)^3} & \dfrac{s}{(s-2)(s+1)^2} \end{bmatrix}.$$

(a) Determine the pole polynomial of $H(s)$ and the McMillan degree of $H(s)$.
(b) Determine a minimal realization of $H(s)$ in observer form.

5.5. Consider the scalar proper rational transfer function $H(s)[\hat{y}(s) = H(s)\hat{u}(s)]$, and assume that $H(s)$ can be written as $H(s) = \sum_{i=1}^{n} \dfrac{r_i}{s-\lambda_i}$. Note that this can be done when the poles λ_i, $i = 1, \ldots, n$, are distinct (see Subsection 5.4C on realizations with A in diagonal form).

(a) Show that $H(s)$ can be realized as the sum or the parallel combination of realizations of terms in the form $\dfrac{r_i}{s-\lambda_i}$; i.e., $H(s)$ is realized by the system represented in the block diagram of Fig. 5.4, where $r_i = c_i b_i$, $i = 1, \ldots, n$, with $c_i^T \in R^n$, $b_i \in R^n$, with $\dfrac{1}{q-\lambda_i}$, and $q \triangleq d/dt$ denotes the integrator circuit shown in Fig. 5.5. Note that this realization of $H(s)$ has advantages with respect to sensitivity to parameter variations.

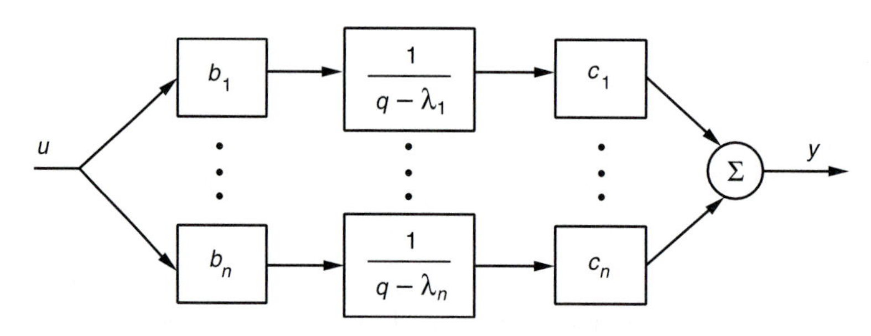

FIGURE 5.4
Block diagram of the system in Example 5.5a

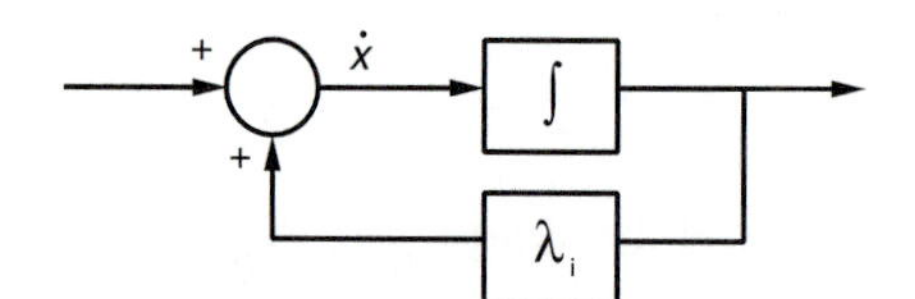

FIGURE 5.5
Block diagram of an integrator

(b) Show that when poles are repeated, as, e.g., in the transfer function $H(s) = \frac{1}{(s+1)^2} + \frac{1}{s+1} + \frac{1}{s+2}$, then a parallel realization is as shown in the block diagram given in Fig. 5.6.

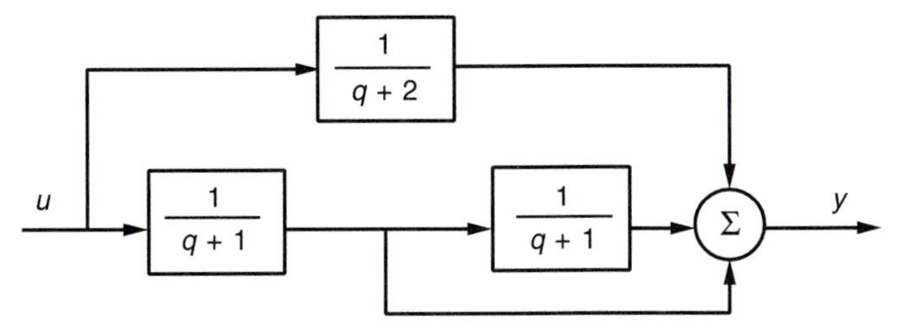

FIGURE 5.6
Block diagram of the system in Example 5.5b

(c) When there are complex conjugate poles, as, e.g., in the transfer function $H(s) = \frac{as+b}{(s+c)^2+d^2}$, then $H(s)$ can be written as $H(s) = \frac{a/(s+c)}{1+(d^2/(s+c)^2)} + \frac{(b-ac)/(s+c)^2}{1+(d^2/(s+c)^2)}$ and can be realized as indicated in the block diagram given in Fig. 5.7.

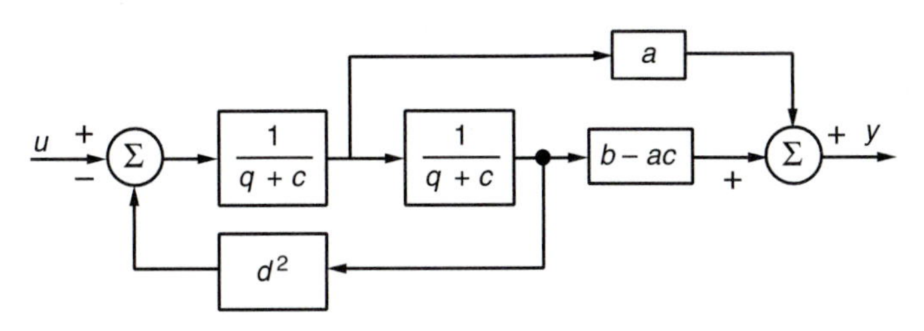

FIGURE 5.7
Block diagrams of the system in Exercise 5.5c

Note that in addition to *sum* or *parallel realizations* discussed in (a) to (c), there are also *product* or *cascade realizations* where, for example, $H(s) = \frac{s+1}{(s+2)(s+3)}$ is realized as $\frac{s+1}{s+2} \cdot \frac{1}{s+3}$ or as $\frac{1}{s+2} \cdot \frac{s+1}{s+3}$.

5.6. Consider the transfer function

$$H(s) = \begin{bmatrix} \frac{1}{s} & \frac{s+3}{s+1} \\ \frac{1}{s+3} & \frac{s}{s+1} \end{bmatrix}.$$

(a) Determine the pole polynomial of $H(s)$ and the McMillan degree of $H(s)$.
(b) Determine a minimal realization $\{A, B, C, D\}$ of $H(s)$, where A is a diagonal matrix.

5.7. Consider a system described by equations of the form $\begin{bmatrix} \dot{x}_1 \\ \dot{x}_2 \end{bmatrix} = \begin{bmatrix} A_1 & 0 \\ 0 & \bar{A}_1 \end{bmatrix} \begin{bmatrix} x_1 \\ x_2 \end{bmatrix} + \begin{bmatrix} B_1 \\ \bar{B}_1 \end{bmatrix} u$, $y = [C_1, \bar{C}_1] \begin{bmatrix} x_1 \\ x_2 \end{bmatrix}$, where A_1 is the Jordan block associated with the eigenvalue λ, and $\bar{A}_1$

is the complex conjugate of A_1 associated with $\bar{\lambda}$. Show that the similarity transformation matrix P given by $P = \begin{bmatrix} I & I \\ jI & -jI \end{bmatrix}$ can be used to reduce the representation given above to one that involves only real system parameters, namely,

$$\begin{bmatrix} \dot{\hat{x}}_1 \\ \dot{\hat{x}}_2 \end{bmatrix} = \begin{bmatrix} Re\ A_1 & Im\ A_1 \\ -Im\ A_1 & Re\ A_1 \end{bmatrix} \begin{bmatrix} \hat{x}_1 \\ \hat{x}_2 \end{bmatrix} + \begin{bmatrix} 2\ Re\ B_1 \\ -2\ Im\ B_1 \end{bmatrix} u,$$

$$y = [Re\ C_1, \quad Im\ C_1] \begin{bmatrix} \hat{x}_1 \\ \hat{x}_2 \end{bmatrix}.$$

Note that this is a way of obtaining representations involving only real coefficients from realizations with A in diagonal or in Jordan form that may contain complex numbers.

5.8. Determine an observable realization of $H(s)$ given in Exercise 5.1, where the matrix A is in block companion form, by using
(a) the numerator polynomial matrix,
(b) the Hankel matrix.

5.9. Show that if the system $\{A, B, C, D\}$ realizes $H(s) = \dfrac{s+2}{s^2+2s+1}$ and $|sI - A| = s^2 + 2s + 1$, then (A, B) must be controllable and (A, C) must be observable. *Hint:* Use Theorem 3.11.

5.10. Show that if the system $\{A, B, C, D\}$ realizes the transfer function matrix $H(s)$ and $|sI - A| = m_H(s)$, the monic least common denominator of the entries of $H(s)$, then $\{A, B, C, D\}$ is a minimal realization of $H(s)$. Can one always find a realization of $H(s)$ that satisfies this property? *Hint:* Use Theorem 3.11.

5.11. Consider a scalar proper rational transfer function $H(s) = n(s)/d(s)$, and let $\dot{x} = A_c x_c + B_c u$, $y = C_c x_c + D_c u$ be a realization of $H(s)$ in controller form (see Subsection 5.4B).
(a) Show that the realization $\{A_c, B_c, C_c, D_c\}$ is always controllable.
(b) Show that $\{A_c, B_c, C_c, D_c\}$ is observable if and only if $n(s)$ and $d(s)$ do not have any factors in common, i.e., they are prime polynomials.
(c) State the dual results to (a) and (b) involving a realization in observer form.

5.12. Consider a $p \times m$ proper rational transfer function matrix $H(s)$ and the algorithm that leads to a realization $\{A_c, B_c, C_c, D_c\}$ of $H(s) = N(s)D(s)^{-1}$ in controller form [see (4.22) and (4.24)].
(a) Show that (A_c, C_c) is observable if and only if $rank \begin{bmatrix} D(\lambda) \\ N(\lambda) \end{bmatrix} = m$ for any λ complex scalar. *Hint:* Use the eigenvalue/eigenvector tests for observability. Note that this rank condition is a necessary and sufficient condition for $N(s)$ and $D(s)$ to be right coprime (see Theorem 2.4 in Subsection 7.2D).
(b) State the dual result to (a) that involves a realization in observer form $\{A_o, B_o, C_o, D_o\}$ of $H(s) = \tilde{D}^{-1}(s)\tilde{N}(s)$.

5.13. Let $H(s) = \dfrac{n(s)}{d(s)} = \dfrac{s^2 - s + 1}{s^5 - s^4 + s^3 - s^2 + s - 1}$. Determine a realization in controller form. Is your realization minimal? Explain your answer. *Hint*: Use the results of Exercise 5.11.

5.14. For the transfer function $H(s) = \dfrac{s+1}{s^2+2}$, find
(a) an uncontrollable realization,
(b) an unobservable realization,
(c) an uncontrollable and unobservable realization,
(d) a minimal realization.

5.15. Find minimal discrete-time state-space realizations of the transfer function matrix

$$H(z) = \begin{bmatrix} \dfrac{z+2}{z+1} & \dfrac{1}{z+3} \\ \dfrac{z}{z+1} & \dfrac{z+1}{z+2} \end{bmatrix},$$

using all the realization methods described in this chapter.

5.16. Given is the system depicted in the block diagram of Fig. 5.8, where $H(s) = \dfrac{s^2+1}{(s+1)(s+2)(s+3)}$. Determine a minimal state-space representation for the closed-loop system, using two approaches. In particular:

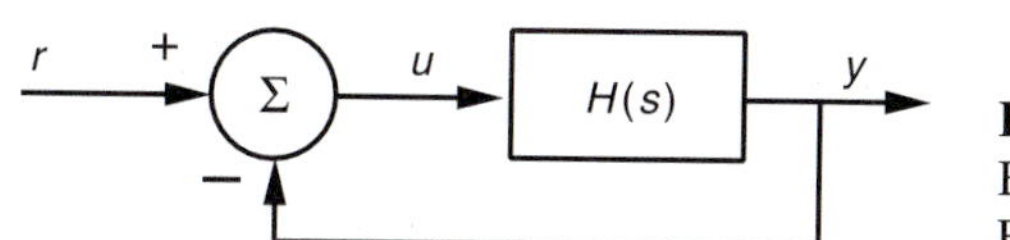

FIGURE 5.8
Block diagram of the system in Exercise 5.16

(a) First, determine a state-space realization for $H(s)$, and then, determine a minimal state-space representation for the closed-loop system.
(b) First, find the closed-loop transfer function, and then, determine a minimal state-space representation for the closed-loop system.

Compare the two approaches.

5.17. Consider the system depicted in the block diagram of Fig. 5.9, where $H(s) = \dfrac{s+1}{s(s+3)}$ and $G(s) = \dfrac{k}{s+a}$ with $k, a \in R$. Presently, $H(s)$ could be viewed as the system to be controlled and $G(s)$ could be regarded as a feedback controller.

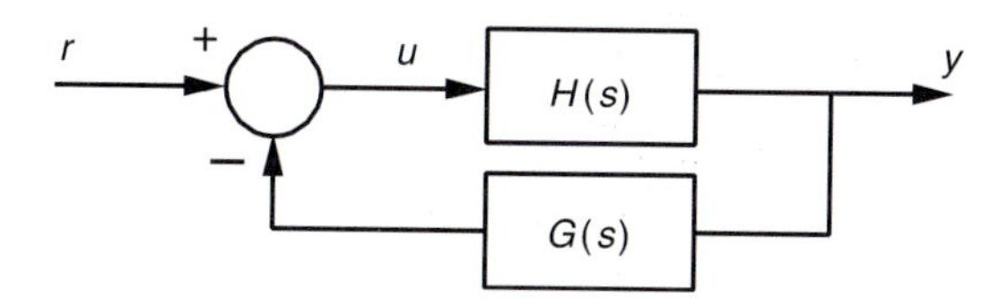

FIGURE 5.9
Block diagram of the system in Exercise 5.17

(a) Obtain a state-space representation of the closed-loop system by
  (i) first, determining realizations for $H(s)$ and $G(s)$ and then combining them;
  (ii) first, determining $H_c(s)$, the closed-loop transfer function.
(b) Are there any choices for the parameters k and a for which your closed-loop state-space representation is uncontrollable *and* unobservable? If your answer is yes, specify.

5.18. For $H(s)$ as in Exercises 5.1, 5.2, and 5.3, determine internally balanced minimal realizations, using singular-value decomposition.

5.19. Let

$$H(s) = \begin{bmatrix} \dfrac{1}{s^2+1} & \dfrac{s^2+1}{s^4+3} & \dfrac{1}{s+2} \\ \dfrac{s^3+1}{s^4} & \dfrac{s^3+1}{s^3+2s+1} & \dfrac{s^3}{s^3+1} \end{bmatrix}$$

be a transfer function matrix.

(a) Find a minimal realization for $H(s)$ in controller form.

(b) Find an internally balanced minimal realization for $H(s)$.

5.20. Consider the controllable (-from-the-origin) and observable system given by $\dot{x} = Ax + Bu$, $y = Cx + Du$ and its equivalent representation $\dot{\hat{x}} = \hat{A}\hat{x} + \hat{B}u$, $y = \hat{C}\hat{x} + \hat{D}u$, where $\hat{A} = PAP^{-1}$, $\hat{B} = PB$, $\hat{C} = CP^{-1}$, and $\hat{D} = D$. Let W_r and W_o denote the reachability and observability Gramians, respectively.

(a) Show that $\hat{W}_r = PW_rP^*$ and $\hat{W}_o = (P^{-1})^*W_oP^{-1}$, where P^* denotes the complex conjugate transpose of P. Note that $P^* = P^T$ when only real coefficients in the system equations are involved.

Using singular-value decomposition (refer to the Appendix), write

$$W_r = U_r\Sigma_r V_r^* \qquad \text{and} \qquad W_o = U_o\Sigma_o V_o^*,$$

where $U^*U = I$, $VV^* = I$, and $\Sigma = diag\,(\Sigma_1, \Sigma_2, \ldots, \Sigma_n)$ with Σ_i the singular values of W. Define

$$H = \left(\Sigma_o^{1/2}\right)^* U_o^* U_r \left(\Sigma_r^{1/2}\right),$$

and using singular-value decomposition, write

$$H = U_H \Sigma_H V_H,$$

where $U_H^* U_H = I$, $V_H V_H^* = I$. Prove the following:

(b) If $P = P_{in} \triangleq V_H(\Sigma_r^{1/2})^{-1}V_r^*$, then $\hat{W}_r = I$, $\hat{W}_o = \Sigma_H^2$.

(c) If $P = P_{out} \triangleq U_H^*(\Sigma_o^{1/2})^* V_o^*$, then $\hat{W}_r = \Sigma_H^2$, $\hat{W}_o = I$.

(d) If $P = P_{ib} = P_{in}\Sigma_H^{1/2} = \Sigma_H^{1/2} P_{out}$, then $\hat{W}_r = \hat{W}_o = \Sigma_H$. Note that the equivalent representations $\{\hat{A}, \hat{B}, \hat{C}, \hat{D}\}$ in (b), (c), and (d) are called, respectively, *input-normal, output-normal,* and *internally balanced representations.*

5.21. Show that the representation (4.61) in Section 5.4 is internally balanced, in view of Exercise 5.20(d).

5.22. Consider the transfer function $H(s) = \dfrac{s^3}{(4s^3+3s+1)}$ and determine a minimal realization for $H(s)$.

5.23. Transform the system

$$\dot{x} = \begin{bmatrix} -1 & -2 & -2 \\ 0 & -1 & 1 \\ 1 & 0 & -1 \end{bmatrix} x + \begin{bmatrix} 2 \\ 0 \\ 1 \end{bmatrix} u, \qquad y = [1, 1, 0]x$$

into $\dot{x}_c = A_c x_c + B_c u$, $y = C_c x_c$, an equivalent representation in controller form utilizing the following two methods:

(a) Use of a similarity transformation.

(b) Determination of the transfer function $H(s)$ and use of a realization algorithm.

5.24. Consider two systems S_1 and S_2 in series as depicted in the block diagram given in Fig. 5.10. Suppose that S_1 is described by $\hat{y}_1(s) = H_1(s)\hat{u}_1(s)$, where $H_1(s) = \frac{-1}{s+2} + 1$, and similarly, that S_2 is described by $\hat{y}_2(s) = H_2(s)\hat{u}_2(s)$, where $\hat{u}_2(s) = \hat{y}_1(s)$ and $H_2(s) = \frac{1}{s+1}$.

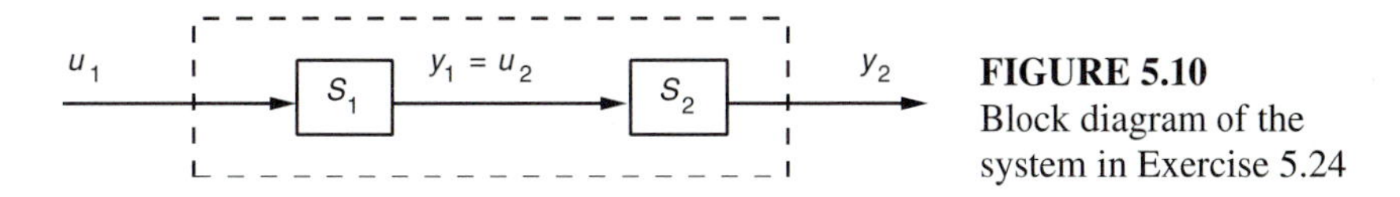

FIGURE 5.10
Block diagram of the system in Exercise 5.24

(a) Determine controllable and observable state-space descriptions for the individual subsystems S_1 and S_2.
(b) Determine a state-space representation for the *entire* system S.
(c) Is the state-space description of S in (b) controllable? Is it observable? What is the transfer function $H(s)$ of S?

5.25. Consider a system described by

$$\begin{bmatrix} \hat{y}_1(s) \\ \hat{y}_2(s) \end{bmatrix} = \begin{bmatrix} \frac{1}{(s+1)^2} & \frac{2}{s^2} \\ 0 & \frac{s+1}{s} \end{bmatrix} \begin{bmatrix} \hat{u}_1(s) \\ \hat{u}_2(s) \end{bmatrix}.$$

(a) What is the order of a controllable and observable realization of this system?
(b) If we consider such a realization, is the resulting system controllable from the input u_2? Is it observable from the output y_1? Explain your answers.

5.26. Consider the system described by $H(s) = \frac{1}{s-(1+\epsilon)}$ $(\hat{y}(s) = H(s)\hat{u}(s))$ and $C(s) = \frac{s-1}{s+2}$ $(\hat{u}(s) = C(s)\hat{r}(s))$ connected in series ($\epsilon \in R$).
(a) Derive minimal state-space realizations for $H(s)$ and $C(s)$ and determine a (second-order) state-space description for the system $\hat{y}(s) = H(s)C(s)\hat{r}(s)$.
(b) Let $\epsilon = 0$ and discuss the implications regarding the overall transfer function and your state-space representations in (a). Is the overall system now controllable, observable, asymptotically stable? Are the poles of the overall transfer function stable? [That is, is the overall system BIBO stable? (See Chapter 6.)] Plot the states and the output for some nonzero initial condition and a unit step input and comment on your results.
(c) In practice, if $H(s)$ is a given system to be controlled and $C(s)$ is a controller, it is unlikely that ϵ will be exactly equal to zero and therefore the situation in (a), rather than (b), will arise. In view of this, comment on whether open-loop stabilization can be used in practice. Carefully explain your reasoning.

CHAPTER 6

Stability

Dynamical systems, either occurring in nature or manufactured, usually function in some specified mode. The most common such modes are operating points that frequently turn out to be equilibria.

In this chapter we will concern ourselves primarily with the qualitative behavior of equilibria. Most of the time, we will be interested in the asymptotic stability of an equilibrium (operating point), which means that when the state of a given system is displaced (disturbed) from its desired operating value (equilibrium), the expectation is that the state will return to the equilibrium. For example, in the case of an automobile under cruise control, traveling at the desired constant speed of 50 mph (which determines the operating point, or equilibrium condition), perturbations due to hill climbing (hill descending), will result in decreasing (increasing) speeds. In a properly designed cruise control system, it is expected that the car will return to its desired operating speed of 50 mph.

Another qualitative characterization of dynamical systems is the expectation that bounded system inputs will result in bounded system outputs, and that small changes in inputs will result in small changes in outputs. System properties of this type are referred to as input-output stability. Such properties are important, for example, in tracking systems, where the output of the system is expected to follow a desired input. Frequently, it is possible to establish a connection between the input-output stability properties and the Lyapunov stability properties of an equilibrium. In the case of linear systems, this connection is well understood.

6.1 INTRODUCTION

In this chapter we present a brief introduction to stability theory. We are concerned primarily with linear systems and systems that are a consequence of linearizations of nonlinear systems. As in the other chapters of this book, we consider finite-dimensional continuous-time systems and finite-dimensional discrete-time systems described by systems of first-order ordinary differential equations and systems of first-order ordinary difference equations, respectively. We will consider both internal descriptions and external descriptions of systems. For the former, we present results for various types of Lyapunov stability of an equilibrium (under assumptions of no external inputs), while for the latter we develop results for input-output stability (under assumptions of zero initial conditions). We also present results that connect these two stability types. By considering both Lyapunov stability and input-output stability, we are frequently able to conduct a more complete qualitative analysis of a system than can be accomplished by applying only one type of stability analysis.

Recall that in Subsections 2.4C and 2.7E we have already encountered stability properties of an equilibrium for time-invariant, continuous-time, and discrete-time systems, respectively.

A. Chapter Description

This chapter is organized into three parts. In Part 1 (Sections 6.3 through 6.8) we address the Lyapunov stability of an equilibrium, in Part 2 (Section 6.9) we consider input-output stability, and in Part 3 (Section 6.10), we treat both Lyapunov stability and input-output stability of (time-invariant) discrete-time systems. Part 1 is preceded by Section 6.2, which contains some background material from linear algebra.

In Section 6.2 we provide additional background in linear algebra dealing with bilinear functionals and congruence, Euclidean vector spaces, and linear transformations on Euclidean vector spaces. This material constitutes a continuation of the material presented in Subsections 1.10A and 1.10B and Section 2.2, and the notation used in those sections will also be employed in the second section of this chapter.

In Section 6.3 we introduce the concept of equilibrium of dynamical systems described by systems of first-order ordinary differential equations, and in Section 6.4 we give definitions of various types of stability in the sense of Lyapunov (including stability, uniform stability, asymptotic stability, uniform asymptotic stability, exponential stability, and instability).

In Section 6.5 we establish conditions for the various Lyapunov stability and instability types enumerated in Section 6.4 for linear systems (*LH*), (*L*), and (*P*). Most of these results are phrased in terms of the properties of the state transition matrix for such systems.

In Section 6.6 we state and prove necessary and sufficient conditions for the exponential stability of the equilibrium of nth-order ordinary differential equations with constant coefficients and systems of linear first-order ordinary differential equations (*L*). These involve geometric criteria (the interlacing theorem) and algebraic criteria (the Routh-Hurwitz criterion).

In Section 6.7 we introduce the Second Method of Lyapunov, also called the Direct Method of Lyapunov, to establish necessary and sufficient conditions for various Lyapunov stability types of an equilibrium for linear systems (*L*). These results, which are phrased in terms of the system parameters [coefficients of the matrix A for system (*L*)], give rise to the Lyapunov matrix equation.

In Section 6.8 we use the Direct Method of Lyapunov in deducing the asymptotic stability and instability of an equilibrium of nonlinear autonomous systems (A) from the stability properties of their linearizations.

In Section 6.9 we establish necessary and sufficient conditions for the input-output stability (more precisely, for the uniform bounded input/bounded output stability) of continuous-time, linear, time-varying systems, and linear, time-invariant systems. These results involve the system impulse response matrix. In this section we also establish a connection between the bounded input/bounded output stability of linear systems and the exponential stability of an equilibrium of linear systems.

The stability results presented in Sections 6.3 through and including Section 6.9 pertain to continuous-time systems. In Section 6.10 we present analogous stability results for discrete time systems; however, in the interests of economy, we confine ourselves in this section to time-invariant systems. Also, to give the reader a glimpse into the qualitative theory of dynamical systems described by nonlinear equations, we present in this section Lyapunov stability results for finite-dimensional dynamical systems described by systems of nonlinear first-order ordinary difference equations.

We conclude the chapter with comments concerning some of the existing literature dealing with the present topic. As in all the other chapters, problems are provided at the end of the chapter to further clarify the subject at hand.

B. Guidelines for the Reader

In a first reading, the background material on linear algebra, given in Section 6.2, can be reviewed rather quickly, as needed.

In a first course on linear systems, the reader needs to acquire familiarity with the notion of an equilibrium (Section 6.3) and various stability concepts of an equilibrium (Section 6.4). Such a course may be confined to studying the stability properties of an equilibrium for time-invariant systems (L) (refer to Theorem 5.6 in Section 6.5). This may be followed by coverage of Section 6.7, where the principal Lyapunov stability results for time invariant systems (L) are established in terms of the properties of the Matrix Lyapunov Equation. In a first reading, Section 6.8 should also be covered in its entirety, where conditions are established that enable one to deduce the stability properties of a time-invariant nonlinear system (A) from the linearization of (A).

The reader should concentrate on the input-output stability results for linear, time-invariant, continuous-time systems given in Theorems 9.4 and 9.5 in a first course on linear systems.

Finally, in a first reading, the reader may consider some or all of the counterparts of the above results for the case of linear, time-invariant, discrete-time systems developed in Section 6.10.

6.2 MATHEMATICAL BACKGROUND MATERIAL

In this section we provide additional background material from *linear algebra* and *matrix theory.* As in previous sections where we present such material, we assume that the reader has some background in these areas, and therefore, our presentation will be in the form of a summary, rather than a development of the subject at hand.

This section consists of three subsections. In the first subsection we address bilinear functionals and congruence, in the second subsection we present material on Euclidean vector spaces, and in the third subsection we address some issues dealing with linear transformations on Euclidean spaces.

A. Bilinear Functionals and Congruence

We consider here the representation of bilinear functionals on real finite-dimensional vector spaces. Throughout this subsection, V is assumed to be an n-dimensional vector space over the field of real numbers R.

We define a *bilinear functional* on V as a mapping $f : V \times V \to R$ having the properties

$$f(\alpha v^1 + \beta v^2, w) = \alpha f(v^1, w) + \beta f(v^2, w) \tag{2.1}$$

$$f(v, \gamma w^1 + \delta w^2) = \gamma f(v, w^1) + \delta f(v, w^2) \tag{2.2}$$

for all $\alpha, \beta, \gamma, \delta \in R$ and for all v^1, v^2, w^1, w^2 in V. A direct consequence of this definition is the more general property

$$f\left(\sum_{j=1}^{n} \alpha_j v^j, \sum_{k=1}^{s} \beta_k w^k\right) = \sum_{j=1}^{r}\sum_{k=1}^{s} \alpha_j \beta_k f(v^j, w^k)$$

for all $\alpha_j, \beta_k \in R$ and $v^j, w^k \in V$, $j = 1, \ldots, r$, and $k = 1, \ldots, s$.

Now let $\{v^1, \ldots, v^n\}$ be a basis for the vector space V and let

$$f_{ij} = f(v^i, v^j), \qquad i, j = 1, \ldots, n. \tag{2.3}$$

The matrix $F = [f_{ij}]$ is called the *matrix of the bilinear functional f with respect to* $\{v^1, \ldots, v^n\}$.

The characterization of bilinear functionals on real finite-dimensional vector spaces is given in the following result, which is a direct consequence of the above definitions and the properties of bases: let f be a bilinear functional on V and let $\{v^1, \ldots, v^n\}$ be a basis for V. Let F be the matrix of the bilinear functional f with respect to the basis $\{v^1, \ldots, v^n\}$. If x and y are arbitrary vectors in V, and if ξ and η are their coordinate representations with respect to the basis $\{v^1, \ldots, v^n\}$, then

$$f(x, y) = \xi^T F \eta = \sum_{i=1}^{n}\sum_{j=1}^{n} f_{ij}\xi_i\eta_j. \tag{2.4}$$

Conversely, if we are given any $n \times n$ matrix F, we can use (2.4) to define the bilinear functional f whose matrix with respect to the given basis $\{v^1, \ldots, v^n\}$ is, in turn, F again. In general it therefore follows that on finite-dimensional vector spaces, bilinear functionals correspond in a one-to-one fashion to matrices. The particular one-to-one correspondence depends on the particular basis chosen.

A bilinear functional f on V is said to be *symmetric* if $f(x, y) = f(y, x)$ for all $x, y \in V$ and *skew symmetric* if $f(x, y) = -f(y, x)$ for all $x, y \in V$.

For symmetric and skew symmetric bilinear functionals the following results are easily proved: let $\{v^1, \ldots, v^n\}$ be a basis for V, and let F be the matrix for a bilinear functional with respect to $\{v^1, \ldots, v^n\}$. Then

1. f is symmetric if and only if $F = F^T$.
2. f is skew symmetric if and only if $F = -F^T$.
3. For every bilinear functional f, there exists a unique symmetric bilinear functional f_1 and a unique skew symmetric bilinear functional f_2 such that

$$f = f_1 + f_2. \tag{2.5}$$

We call f_1 the *symmetric part* of f and f_2 the *skew symmetric part* of f.

The above result motivates the following definitions: an $n \times n$ matrix F is said to be *symmetric* if $F = F^T$ and *skew symmetric* if $F = -F^T$.

Using definitions, the following result is easily established: let f be a bilinear functional on V and let f_1 and f_2 be the symmetric and skew symmetric parts of f, respectively. Then

$$f_1(v, w) = \tfrac{1}{2}[f(v, w) + f(w, v)] \tag{2.6}$$

and

$$f_2(v, w) = \tfrac{1}{2}[f(v, w) - f(w, v)] \tag{2.7}$$

for all $v, w \in V$.

Next, we define the *quadratic form* induced by a bilinear functional f on V as $\hat{f}(v) = f(v, v)$ for all $v \in V$. It is easily verified that in terms of matrices we have

$$\hat{f}(x) = \xi^T F \xi = \sum_{i=1}^{n} \sum_{j=1}^{n} f_{ij} \xi_i \xi_j, \tag{2.8}$$

where ξ denotes the coordinate representation of x with respect to the basis $\{v^1, \ldots, v^n\}$.

For quadratic forms we have the following result: let f and g be bilinear functionals on V. The quadratic forms induced by f and g are equal if and only if f and g have the same symmetric part. In other words, $\hat{f}(v) = \hat{g}(v)$ for all $v \in V$ if and only if

$$\tfrac{1}{2}[f(v, w) + f(w, v)] = \tfrac{1}{2}[g(v, w) + g(w, v)] \tag{2.9}$$

for all $v, w \in V$. From this result we can conclude that *when treating quadratic functionals, it suffices to work with symmetric* bilinear functionals.

It is also easily verified from definitions that a bilinear functional f on a vector space V is skew symmetric if and only if $f(v, v) = 0$ for all $v \in V$.

Next, let f be a bilinear functional on a vector space V, let $\{v^1, \ldots, v^n\}$ be a basis for V, and let F be the matrix of f with respect to this basis. Let $\{\bar{v}^1, \ldots, \bar{v}^n\}$ be another basis whose matrix with respect to $\{v^1, \ldots, v^n\}$ is P. It can readily be verified that the matrix $\bar{F}$ of f with respect to the basis $\{\bar{v}^1, \ldots, \bar{v}^n\}$ is given by

$$\bar{F} = P^T F P. \tag{2.10}$$

This result gives rise to the following concept: an $n \times n$ matrix $\bar{F}$ is said to be *congruent* to an $n \times n$ matrix F if there exists a nonsingular matrix P such that (2.10) holds. We express congruence of two matrices $F, \bar{F}$ by writing $\bar{F} \sim F$. It is easily shown that $\sim$ is reflexive, symmetric, and transitive, and as such it is an equivalence relation.

For practical reasons we are interested in determining the simplest matrix congruent to a given matrix, or what amounts to the same thing, the most convenient basis to use in expressing a given bilinear functional. If, in particular, we confine our interests to quadratic functionals, then it suffices, as observed earlier, to consider symmetric bilinear functionals. The following result, called *Sylvester's Theorem,* addresses this issue. The proof of this result is rather lengthy, and the interested reader should consult one of the references on linear algebra cited at the end of this chapter for details.

Let f be any symmetric bilinear functional on a real n-dimensional vector space V. Then there exists a basis $\{v^1, \ldots, v^n\}$ of V such that the matrix of f with respect to this basis is of the form

$$\left.\left.\left.\begin{bmatrix} 1 & & & & & & & \\ & \ddots & & & & & & \\ & & 1 & & & 0 & & \\ & & & -1 & & & & \\ & & & & \ddots & & & \\ & & 0 & & & -1 & & \\ & & & & & & 0 & \\ & & & & & & & \ddots \\ & & & & & & & & 0 \end{bmatrix}\right\}p\;\right\}r\;\right\}n. \tag{2.11}$$

The integers r and p in this matrix are uniquely determined by the bilinear form.

Sylvester's Theorem allows the following classification of symmetric bilinear functionals: the integer r in (2.11) is called the *rank* of the symmetric bilinear functional f, the integer p is called the *index* of f, and n is called the *order* of f. The integer $s = 2p - r$ (i.e., the number of $+1$'s minus the number of -1's) is called the *signature* of f.

A bilinear functional f on a vector space V is said to be *positive* (or *positive semidefinite*) if $f(v, v) \geq 0$ for all $v \in V$ and *strictly positive* (or *positive definite*) if $f(v, v) > 0$ for all $v \neq 0, v \in V$ [note that $f(v, v) = 0$ for $v = 0$]. It is readily verified that a symmetric bilinear functional is strictly positive if and only if $p = r = n$ in (2.11) and positive if and only if $p = r$.

Finally, we say that a bilinear functional f is *negative* (or *negative semidefinite*) if $-f$ is positive and *strictly negative* (or *negative definite*) if $-f$ is strictly positive. Also, a bilinear functional f is said to be *indefinite* if in (2.11) $r > p > 0$.

B. Euclidean Vector Spaces

As in the preceding subsection, we assume throughout this subsection that V is a real vector space.

A bilinear functional f defined on V is said to be an *inner product* if (i) f is symmetric and (ii) f is strictly positive. A real vector space V on which an inner product is defined is called a *real inner product space,* and a *real finite-dimensional* vector space on which an inner product is defined is called a *Euclidean space.*

Since we will always be concerned with a *given* bilinear functional on V, we will write (v, w) in place of $f(v, w)$ to denote the inner product of v and w. Accordingly, the axioms of a real inner product are given as

1. $(v, v) > 0$ for all $v \neq 0$ and $(v, v) = 0$ when $v = 0$.
2. $(v, w) = (w, v)$ for all $v, w \in V$.
3. $(\alpha v + \beta w, u) = \alpha(v, u) + \beta(w, u)$ for all $u, v, w \in V$ and all $\alpha, \beta \in R$.
4. $(u, \alpha v + \beta w) = \alpha(u, v) + \beta(u, w)$ for all $u, v, w \in V$ and all $\alpha, \beta \in R$.

In the following we enumerate several results on Euclidean spaces, all of which are easily proved:

1. The inner product $(v, w) = 0$ for all $v \in V$ if and only if $w = 0$.
2. Let $\mathcal{A} \in L(V, V)$. Then $(v, \mathcal{A}w) = 0$ for all $v, w \in V$ if and only if $\mathcal{A} = \mathcal{O}$.
3. Let $\mathcal{A}, \mathcal{B} \in L(V, V)$. If $(v, \mathcal{A}w) = (v, \mathcal{B}w)$ for all $v, w \in V$, then $\mathcal{A} = \mathcal{B}$.
4. Let $A \in R^{n \times n}$. If $\xi^T A \eta = 0$ for all $\xi, \eta \in R^n$, then $A = 0$.

We now define the function

$$\|v\| = (v, v)^{1/2} \tag{2.12}$$

for all $v \in V$. It is easily verified (using definitions and the properties of bilinear functionals) that this function satisfies the *axioms of a norm* (refer to Subsection 1.10B), i.e., for all v, w in V and for all scalars α, the following hold:

1. $\|v\| > 0$ for all $v \neq 0$ and $\|v\| = 0$ when $v = 0$.
2. $\|\alpha v\| = |\alpha| \cdot \|v\|$, where $|\alpha|$ denotes the absolute value of the scalar α.
3. $\|v + w\| \leq \|v\| + \|w\|$.

In the usual proof of 3, use is made of the *Schwarz Inequality*, which states that

$$|(v, w)| \leq \|v\| \cdot \|w\| \tag{2.13}$$

for all $v, w \in V$, and

$$|(v, w)| = \|v\| \cdot \|w\|$$

if and only if v and w are linearly dependent.

Another result for Euclidean spaces that is easily proved is the *parallelogram law,* which asserts that for all $v, w \in V$, the equality

$$\|v + w\|^2 + \|v - w\|^2 = 2\|v\|^2 + 2\|w\|^2. \tag{2.14}$$

Before proceeding, we recall that a vector space V on which a norm is defined is called a *normed linear space*. It is therefore clear that Euclidean vector spaces are *normed linear spaces* with norm defined by (2.12) (refer to Subsection 1.10B). We also recall that any norm can be used to define a *distance function*, called a *metric*, on a normed vector space by letting

$$d(u, v) = \|u - v\| \tag{2.15}$$

for all $u, v \in V$ (refer to Subsection 1.10B). Using the properties of norm, it is readily verified that for all $u, v, w \in V$, it is true that

1. $d(u, v) = d(v, u)$.
2. $d(u, v) \geq 0$ and $d(u, v) = 0$ if and only if $u = v$.
3. $d(u, v) \leq d(u, w) + d(w, v)$.

Without making use of inner products or norms, one can define a distance function having the properties enumerated above on an arbitrary set. Such a set, along with the distance function, is then called a *metric space*. It is thus clear that Euclidean vector spaces are metric spaces as well.

The concept of inner product enables us to introduce the notion of orthogonality: two vectors $v, w \in V$ are said to be *orthogonal* (to one another) if $(v, w) = 0$. This is usually written as $v \perp w$.

With the aid of the above concept we can immediately establish the famous *Pythagorean Theorem:* for $v, w \in V$, if $v \perp w$, then $\|v + w\|^2 = \|v\|^2 + \|w\|^2$.

A vector $v \in V$ is said to be a *unit vector* if $\|v\| = 1$. Let $w \neq 0$, and let $u = (1/\|w\|)w$. Then the norm of u is $\|u\| = (1/\|w\|)\|w\| = 1$, i.e., u is a unit vector. We call the process of generating a unit vector from an arbitrary nonzero vector w *normalizing the vector* w.

Now let $\{w^1, \dots, w^n\}$ be an arbitrary basis for V and let $F = [f_{ij}]$ denote the matrix of the inner product with respect to this basis, i.e., $f_{ij} = (w^i, w^j)$ for all i and j. More specifically, F denotes the matrix of the bilinear functional f that is used in determining the inner product on V with respect to the indicated basis. Let ξ and η denote the coordinate representation of vectors x and y, respectively, with respect to $\{w^1, \dots, w^n\}$. Then we have by (2.4),

$$(x, y) = \xi^T F \eta = \eta^T F \xi = \sum_{j=1}^{n} \sum_{i=1}^{n} f_{ij} \xi_j \eta_i. \tag{2.16}$$

Now by Sylvester's Theorem [see (2.11) and the discussion following that theorem], since the inner product is symmetric and strictly positive, there exists a basis $\{v^1, \dots, v^n\}$ for V such that the matrix of the inner product with respect to this basis is the $n \times n$ identity matrix I, i.e.,

$$(v^i, v^j) = \delta_{ij} = \begin{cases} 0, & \text{if } i \neq j, \\ 1, & \text{if } i = j. \end{cases} \tag{2.17}$$

This motivates the following definition: if $\{v^1, \dots, v^n\}$ is a basis for V such that $(v^i, v^j) = 0$ for all $i \neq j$, i.e., if $v^i \perp v^j$ for all $i \neq j$, then $\{v^1, \dots, v^n\}$ is called an *orthogonal basis*. If in addition, $(v^i, v^i) = 1$, i.e., $\|v^i\| = 1$ for all i, then $\{v^1, \dots, v^n\}$ is said to be an *orthonormal basis* for V [thus, $\{v^1, \dots, v^n\}$ is orthonormal if and only if $(v^i, v^j) = \delta_{ij}$].

Using the properties of inner product and the definitions of orthogonal and orthonormal bases, we can easily establish several useful results:

1. Let $\{v^1, \dots, v^n\}$ be an orthonormal basis for V. Let x and y be arbitrary vectors in V, and let the coordinate representation of x and y with respect to this basis be $\xi^T = (\xi_1, \dots, \xi_n)$ and $\eta^T = (\eta_1, \dots, \eta_n)$, respectively. Then

$$(x, y) = \xi^T \eta = \eta^T \xi = \sum_{i=1}^{n} \xi_i \eta_i \tag{2.18}$$

and

$$\|x\| = (\xi^T \xi)^{1/2} = \sqrt{\xi_1^2 + \cdots + \xi_n^2}. \tag{2.19}$$

2. Let $\{v^1, \dots, v^n\}$ be an orthonormal basis for V and let x be an arbitrary vector. The coordinates of x with respect to $\{v^1, \dots, v^n\}$ are given by the formula

$$\xi_i = (x, v^i), \qquad i = 1, \dots, n. \tag{2.20}$$

3. Let $\{v^1, \dots, v^n\}$ be an orthogonal basis for V. Then for all $x \in V$, we have

$$x = \frac{(x, v^1)}{(v^1, v^1)} v^1 + \cdots + \frac{(x, v^n)}{(v^n, v^n)} v^n. \tag{2.21}$$

4. (*Parseval's identity*) Let $\{v^1, \dots, v^n\}$ be an orthogonal basis for V. Then for any $x, y \in V$, we have

$$(x, y) = \sum_{i=1}^{n} \frac{(x, v^i)(y, v^i)}{(v^i, v^i)}. \tag{2.22}$$

5. Suppose that $x^1, \dots, x^k$ are mutually orthogonal nonzero vectors in V, i.e., $x^i \perp x^j, i \neq j$. Then $x^1, \dots, x^k$ are linearly independent.
6. A set of k nonzero mutually orthogonal vectors is a basis for V if and only if $k = \dim V = n$.
7. For V there exist not more than n mutually orthonormal vectors (called a *complete orthonormal set of vectors*).
8. (*Gram-Schmidt process*) Let $\{w^1, \dots, w^n\}$ be an arbitrary basis for V. Set

$$\begin{aligned}
u^1 &= w^1 & v^1 &= \frac{u^1}{\|u^1\|} \\
u^2 &= w^2 - (w^2, v^1)v^1 & v^2 &= \frac{u^2}{\|u^2\|} \\
&\cdots\cdots & &\cdots\cdots \\
u^n &= w^n - \sum_{j=1}^{n-1} (w^n, v^j)v^j & v^n &= \frac{u^n}{\|u^n\|}.
\end{aligned} \tag{2.23}$$

Then $\{v^1, \dots, v^n\}$ is an orthonormal basis for V.
9. If $v^1, \dots, v^k$, $k < n$, are mutually orthogonal nonzero vectors in V, then we can find a set of vectors $v^{k+1}, \dots, v^n$ such that the set $\{v^1, \dots, v^n\}$ forms a basis for V.
10. (*Bessel's inequality*) If $\{w^1, \dots, w^k\}$ is an arbitrary set of mutually orthonormal vectors in V, then

$$\sum_{i=1}^{k} |(w, w^i)|^2 \leq \|w\|^2 \tag{2.24}$$

for all $w \in V$. Moreover, the vector

$$u = w - \sum_{i=1}^{k} (w, w^i) w^i$$

is orthogonal to each w^i, $i = 1, \dots, k$.
11. Let W be a linear subspace of V, and let

$$W^{\perp} = \{v \in V : (v, w) = 0 \text{ for all } w \in W\}. \tag{2.25}$$

(i) Let $\{w^1, \dots, w^k\}$ span W. Then $v \in W^{\perp}$ if and only if $v \perp w^j$ for $j = 1, \dots, k$.
(ii) $W^{\perp}$ is a linear subspace of V.
(iii) $n = \dim V = \dim W + \dim W^{\perp}$.
(iv) $(W^{\perp})^{\perp} = W$.
(v) $V = W \oplus W^{\perp}$.

(vi) Let $u, v \in V$. If $u = u^1 + u^2$ and $v = v^1 + v^2$, where $u^1, v^1 \in W$ and $u^2, v^2 \in W^\perp$, then

$$(u, v) = (u^1, v^1) + (u^2, v^2) \tag{2.26}$$

and

$$\|u\| = \sqrt{\|u^1\|^2 + \|u^2\|^2}. \tag{2.27}$$

We conclude this subsection with the following definition: let W be a linear subspace of V. The subspace $W^\perp$ defined in (2.25) is called the *orthogonal complement* of W.

C. Linear Transformations on Euclidean Vector Spaces

In this subsection we consider some of the properties of three important classes of linear transformations defined on Euclidean spaces: orthogonal transformations, adjoint transformations, and self-adjoint transformations. Unless otherwise explicitly stated, V will denote an n-dimensional Euclidean vector space.

Let $\{v^1, \dots, v^n\}$ be an orthonormal basis for V, let $\bar{v}^i = \sum_{j=1}^n p_{ji} v^j$, $i = 1, \dots, n$, and let P denote the matrix determined by the real scalars p_{ij}. The following question arises: when is the set $\{\bar{v}^1, \dots, \bar{v}^n\}$ also an orthonormal basis for V? To determine the desired properties of P, we consider

$$(\bar{v}^i, \bar{v}^i) = \left(\sum_{k=1}^n p_{ki} v^k, \sum_{l=1}^n p_{lj} v^l \right) = \sum_{k,l} p_{ki} p_{lj} (v^k, v^l). \tag{2.28}$$

So that $(\bar{v}^i, \bar{v}^j) = 0$ for $i \neq j$ and $(\bar{v}^i, \bar{v}^j) = 1$ for $i = j$, we require that

$$(\bar{v}^i, \bar{v}^j) = \sum_{k,l=1}^n p_{ki} p_{lj} \delta_{kl} = \sum_{k=1}^n p_{ki} p_{kj} = \delta_{ij}, \tag{2.29}$$

i.e., we require that

$$P^T P = I, \tag{2.30}$$

where, as usual, I denotes the $n \times n$ identity matrix.

The above discussion is summarized in the following result: let $\{v^1, \dots, v^n\}$ be an orthonormal basis for V and let $\bar{v}^i = \sum_{j=1}^n p_{ji} e_j$, $i = 1, \dots, n$. Then $\{\bar{v}^1, \dots, \bar{v}^n\}$ is an orthonormal basis for V if and only if $P^T = P^{-1}$. This result, in turn, gives rise to the concept of orthogonal matrix. Thus, a matrix $P \in R^{n \times n}$ such that $P^T = P^{-1}$, i.e., such that $P^T P = P^{-1} P = I$, is called an *orthogonal matrix*.

It is not difficult to show that if P is an orthogonal matrix, then either $det\, P = 1$ or $det\, P = -1$. Also, if P and Q are $n \times n$ orthogonal matrices, then so is PQ.

The nomenclature used in the next definition will become clear shortly. We say that a linear transformation $\mathscr{A}$ from V into V is an *orthogonal linear transformation* if $(\mathscr{A}v, \mathscr{A}w) = (v, w)$ for all $v, w \in V$.

We now enumerate several properties of orthogonal transformations. The proofs of these statements are straightforward.

1. Let $\mathscr{A} \in L(V, V)$. Then $\mathscr{A}$ is orthogonal if and only if $\|\mathscr{A}v\| = \|v\|$ for all $v \in V$. [Note that if $\mathscr{A}$ is an orthogonal linear transformation, then $v \perp w$ for all $v, w \in V$ if and only if $\mathscr{A}v \perp \mathscr{A}w$ since $(v, w) = 0$ if and only if $(\mathscr{A}v, \mathscr{A}w) = 0$.]

2. Every orthogonal linear transformation of V into V is nonsingular.
3. Let $\{v^1, \dots, v^n\}$ be an orthonormal basis for V. Let $\mathcal{A} \in L(V, V)$, and let A be the matrix of $\mathcal{A}$ with respect to this basis. Then $\mathcal{A}$ is orthogonal if and only if A is orthogonal.
4. Let $\mathcal{A} \in L(V, V)$. If $\mathcal{A}$ is orthogonal, then $det\, \mathcal{A} = \pm 1$.
5. Let $\mathcal{A}, \mathcal{B} \in L(V, V)$. If $\mathcal{A}$ and $\mathcal{B}$ are orthogonal linear transformations, then $\mathcal{A}\mathcal{B}$ is also an orthogonal linear transformation.

The next result enables us to introduce adjoint linear transformations in a natural manner. Let $\mathcal{G} \in L(V, V)$ and define $g : V \times V \to R$ by $g(v, w) = (v, \mathcal{G}w)$ for all $v, w \in V$. Then g is a bilinear functional on V. Moreover, if $\{v^1, \dots, v^n\}$ is an orthonormal basis for V, then the matrix of g with respect to this basis, denoted by G, is the matrix of $\mathcal{G}$ with respect to $\{v^1, \dots, v^n\}$. Conversely, given an arbitrary bilinear functional g defined on V, there exists a unique linear transformation $\mathcal{G} \in L(V, V)$ such that $(v, \mathcal{G}w) = g(v, w)$ for all $v, w \in V$.

It should be noted that the correspondence between bilinear functionals and linear transformations determined by the relation $(v, \mathcal{G}w) = g(v, w)$ for all $v, w \in V$ does not depend on the particular basis chosen for V; however, it does depend on the way the inner product is chosen for V at the outset.

Now let $\mathcal{G} \in L(V, V)$, set $g(v, w) = (v, \mathcal{G}w)$, and let $h(v, w) = g(w, v) = (w, \mathcal{G}v) = (\mathcal{G}v, w)$. By the result given above, there exists a unique linear transformation, denote it by $\mathcal{G}^*$, such that $h(v, w) = (v, \mathcal{G}^*w)$ for all $v, w \in V$. We call the linear transformation $\mathcal{G}^* \in L(V, V)$ the *adjoint of* $\mathcal{G}$. We have the following results:

1. For each $\mathcal{G} \in L(V, V)$, there is a unique $\mathcal{G}^* \in L(V, V)$ such that $(v, \mathcal{G}^*w) = (\mathcal{G}v, w)$ for all $v, w \in V$.
2. Let $\{v^1, \dots, v^n\}$ be an orthonormal basis for V, and let G be the matrix of the linear transformation $\mathcal{G} \in L(V, V)$ with respect to this basis. Let G^* be the matrix of $\mathcal{G}^*$ with respect to $\{v^1, \dots, v^n\}$. Then $G^* = G^T$.

The above results allow the following equivalent definition of the adjoint linear transformation: let $\mathcal{G} \in L(V, V)$. The adjoint transformation, $\mathcal{G}^*$, is defined by the formula

$$(v, \mathcal{G}^*w) = (\mathcal{G}v, w) \tag{2.31}$$

for all $v, w \in V$.

In the following, we enumerate some of the elementary properties of the adjoint of linear transformations. The proofs of these assertions follow readily from definitions.

Let $\mathcal{A}, \mathcal{B} \in L(V, V)$, let $\mathcal{A}^*, \mathcal{B}^*$ denote their respective adjoints, and let α be a real scalar. Then

1. $(\mathcal{A}^*)^* = \mathcal{A}$.
2. $(\mathcal{A} + \mathcal{B})^* = \mathcal{A}^* + \mathcal{B}^*$.
3. $(\alpha\mathcal{A})^* = \alpha\mathcal{A}^*$.
4. $(\mathcal{A}\mathcal{B})^* = \mathcal{B}^*\mathcal{A}^*$.
5. $\mathcal{I}^* = \mathcal{I}$, where $\mathcal{I}$ denotes the identity transformation.
6. $\mathcal{O}^* = \mathcal{O}$, where $\mathcal{O}$ denotes the null transformation.
7. $\mathcal{A}$ is nonsingular if and only if $\mathcal{A}^*$ is nonsingular.
8. If $\mathcal{A}$ is nonsingular, then $(\mathcal{A}^*)^{-1} = (\mathcal{A}^{-1})^*$.

The following results (the proofs of which are straightforward) characterize orthogonal transformations in terms of their adjoints. Let $\mathcal{A} \in L(V, V)$. Then $\mathcal{A}$ is orthogonal if and only if $\mathcal{A}^* = \mathcal{A}^{-1}$. Furthermore, $\mathcal{A}$ is orthogonal if and only if $\mathcal{A}^{-1}$ is orthogonal, and $\mathcal{A}^{-1}$ is orthogonal if and only if $\mathcal{A}^*$ is orthogonal.

Using adjoints, we now introduce two additional important types of linear transformations. Let $\mathcal{A} \in L(V, V)$. Then $\mathcal{A}$ is said to be *self-adjoint* if $\mathcal{A}^* = \mathcal{A}$, and it is said to be *skew-adjoint* if $\mathcal{A}^* = -\mathcal{A}$. Some of the properties of such transformations are as follows.

1. Let $\mathcal{A} \in L(V, V)$, let $\{v^1, \ldots, v^n\}$ be an orthonormal basis for V, and let A be the matrix of $\mathcal{A}$ with respect to this basis. The following are equivalent:
   (i) $\mathcal{A}$ is self-adjoint.
   (ii) $\mathcal{A}$ is symmetric.
   (iii) $(\mathcal{A}v, w) = (v, \mathcal{A}w)$ for all $v, w \in V$.
2. Let $\mathcal{A} \in L(V, V)$, let $\{v^1, \ldots, v^n\}$ be an orthonormal basis for V, and let A be the matrix of $\mathcal{A}$ with respect to this basis. The following are equivalent:
   (i) $\mathcal{A}$ is skew-adjoint.
   (ii) $\mathcal{A}$ is skew-symmetric.
   (iii) $(\mathcal{A}v, w) = -(v, \mathcal{A}w)$ for all $v, w \in V$.

The next result follows from part (iii) of the above result. Let $\mathcal{A} \in L(V, V)$, let $\{v^1, \ldots, v^n\}$ be an orthonormal basis for V, and let A be the matrix of $\mathcal{A}$ with respect to this basis. The following are equivalent:

(i) $\mathcal{A}$ is skew-symmetric.
(ii) $(v, \mathcal{A}v) = 0$ for all $v \in V$.
(iii) $\mathcal{A}v \perp v$ for all $v \in V$.

The following result enables one to represent arbitrary linear transformations as the sum of self-adjoint and skew-adjoint transformations. Let $\mathcal{A} \in L(V, V)$. Then there exist unique $\mathcal{A}_1, \mathcal{A}_2 \in L(V, V)$ such that $\mathcal{A} = \mathcal{A}_1 + \mathcal{A}_2$, where $\mathcal{A}_1$ is self-adjoint and $\mathcal{A}_2$ is skew-adjoint. This has the direct consequence that every real $n \times n$ matrix can be written in one and only one way as the sum of a symmetric and skew-symmetric matrix.

The next result is applicable to real as well as complex vector spaces. (We will state it for complex spaces.)

Let V be a complex vector space. Then the eigenvalues of a real symmetric matrix A are all real. If all eigenvalues of A are positive (negative), then A is called *positive (negative) definite.* If all eigenvalues of A are nonnegative (nonpositive), then A is called *positive (negative) semidefinite.* If A has positive and negative eigenvalues, then A is said to be *indefinite.*

Next, let A be the matrix of a linear transformation $\mathcal{A} \in L(V, V)$ with respect to some basis. If A is symmetric, then as indicated above, all its eigenvalues are real. In this case $\mathcal{A}$ is self-adjoint and all its eigenvalues are also real; in fact, the eigenvalues of $\mathcal{A}$ and A are identical. Thus, there exist unique real scalars $\lambda_1, \ldots, \lambda_p$, $p \leq n$, such that

$$det\,(\mathcal{A} - \lambda \mathcal{I}) = det\,(A - \lambda I) = (\lambda_1 - \lambda)^{m_1}(\lambda_2 - \lambda)^{m_2} \ldots (\lambda_p - \lambda)^{m_p}. \quad (2.32)$$

We summarize the above observations in the following. Let $\mathcal{A} \in L(V, V)$. If $\mathcal{A}$ is self-adjoint, then $\mathcal{A}$ has at least one eigenvalue. Furthermore, all eigenvalues of $\mathcal{A}$

are real and there exist unique real numbers $\lambda_1, \ldots, \lambda_p$, $p \leq n$, such that Eq. (2.32) holds.

As in Eq. (3.17) of Chapter 2, we say that in (2.32) the eigenvalues λ_i, $i = 1, \ldots, p \leq n$, have *algebraic multiplicities* m_i, $i = 1, \ldots, p$, respectively.

Next, we examine some of the properties of the eigenvalues and eigenvectors of self-adjoint linear transformations. The proofs of these assertions are straightforward and follow mostly from definitions.

Let $\mathcal{A} \in L(V, V)$ be a self-adjoint transformation, and let $\lambda_1, \ldots, \lambda_p$, $p \leq n$, denote the distinct eigenvalues of $\mathcal{A}$. If v^i is an eigenvector for λ_i and if v^j is an eigenvector for λ_j, then $v^i \perp v^j$ for all $i \neq j$.

Now let $\mathcal{A} \in L(V, V)$ and let λ_i be an eigenvalue of $\mathcal{A}$. Recall that $\mathcal{N}_i$ denotes the *null space* of the linear transformation $\mathcal{A} - \lambda_i \mathcal{I}$, i.e.,

$$\mathcal{N}_i = \{v \in V : (\mathcal{A} - \lambda_i \mathcal{I})v = 0\}. \tag{2.33}$$

Recall also that $\mathcal{N}_i$ is a linear subspace of V. From the last result given above, the following result follows immediately.

Let $\mathcal{A} \in L(V, V)$ be a self-adjoint transformation and let λ_i and λ_j be eigenvalues of $\mathcal{A}$. If $\lambda_i \neq \lambda_j$, then $\mathcal{N}_i \perp \mathcal{N}_j$.

The proof of the next result is somewhat lengthy and involved. The reader should consult the references on linear algebra cited at the end of this chapter for details.

Let $\mathcal{A} \in L(V, V)$ be a self-adjoint transformation, and let $\lambda_1, \ldots, \lambda_p$, $p \leq n$, denote the distinct eigenvalues of $\mathcal{A}$. Then

$$\dim V = n = \dim \mathcal{N}_1 + \dim \mathcal{N}_2 + \ldots + \dim \mathcal{N}_p. \tag{2.34}$$

The next two results are direct consequences of the above result.

1. Let $\mathcal{A} \in L(V, V)$. If $\mathcal{A}$ is self-adjoint, then
   (i) there exists an orthonormal basis in V such that the matrix of $\mathcal{A}$ with respect to this basis is diagonal;
   (ii) for each eigenvalue λ_i of $\mathcal{A}$ we have $\dim \mathcal{N}_i$ = algebraic multiplicity of λ_i.

   We note that in the above theorem, the matrix A of the linear transformation $\mathcal{A}$ with respect to the chosen orthonormal basis in V is given by

$$A = \begin{bmatrix} \lambda_1 & & & & & & & & \\ & \ddots & & & & & & & \\ & & \lambda_1 & & & & & & \\ & & & \lambda_2 & & & 0 & & \\ & & & & \ddots & & & & \\ & & 0 & & & \lambda_2 & & & \\ & & & & & & \ddots & & \\ & & & & & & & \lambda_p & \\ & & & & & & & & \ddots \\ & & & & & & & & & \lambda_p \end{bmatrix}. \tag{2.35}$$

2. Let A be a real $n \times n$ symmetric matrix. Then there exists an orthogonal matrix P such that the matrix $\bar{A}$ defined by

$$\bar{A} = P^{-1}AP = P^T AP \tag{2.36}$$

is diagonal.

For symmetric bilinear functionals defined on Euclidean vector spaces, we have the following result. Let $f(v, w)$ be a symmetric bilinear functional on V. Then there exists an orthonormal basis for V such that the matrix of f with respect to this basis is diagonal.

For quadratic forms, the following useful result is easily proved.

Let $\hat{f}(x)$ be a quadratic form defined on V. Then there exists an orthonormal basis for V such that if $\xi^T = (\xi_1, \ldots, \xi_n)$ is the coordinate representation of x with respect to this basis, then $\hat{f}(x) = \alpha_1\xi_1^2 + \cdots + \alpha_n\xi_n^2$ for some real scalars $\alpha_1, \ldots, \alpha_n$.

The final result of this section, which we state next, is called the *Spectral Theorem for self-adjoint linear transformations.* For the proof of this result, the interested reader should consult one of the references on linear algebra cited at the end of this chapter. Before stating this theorem, we recall that a transformation $\mathcal{P} \in L(V, V)$ is a *projection* on a linear subspace of V if and only if $\mathcal{P}^2 = \mathcal{P}$ (refer to Subsection 2.2K). Also, for any projection $\mathcal{P}$, $\mathcal{V} = \mathcal{R}(\mathcal{P}) \oplus \mathcal{N}(\mathcal{P})$, where $\mathcal{R}(\mathcal{P})$ is the range of $\mathcal{P}$ and $\mathcal{N}(\mathcal{P})$ is the null space of $\mathcal{P}$ (refer to Subsection 2.2K). Furthermore, we call $\mathcal{P}$ an *orthogonal projection* if $\mathcal{R}(\mathcal{P}) \perp \mathcal{N}(\mathcal{P})$.

The Spectral Theorem for self-adjoint linear transformations: let $\mathcal{A} \in L(V, V)$ be a self-adjoint transformation, let $\lambda_1, \ldots, \lambda_p$ denote the distinct eigenvalues of $\mathcal{A}$, and let $\mathcal{N}_i$ be the null space of $\mathcal{A} - \lambda_i\mathcal{I}$. For each $i = 1, \ldots, p$, let $\mathcal{P}_i$ denote the projection on $\mathcal{N}_i$ along $\mathcal{N}_i^{\perp}$. Then

1. $\mathcal{P}_i$ is an orthogonal projection for each $i = 1, \ldots, p$.
2. $\mathcal{P}_i\mathcal{P}_j = \mathcal{O}$ for $i \neq j$, $i, j = 1, \ldots, p$.
3. $\sum_{j=1}^{p} \mathcal{P}_j = \mathcal{I}$, where $\mathcal{I} \in L(V, V)$ denotes the identity transformation.
4. $\mathcal{A} = \sum_{j=1}^{p} \lambda_j\mathcal{P}_j$.

PART 1 LYAPUNOV STABILITY

6.3 THE CONCEPT OF AN EQUILIBRIUM

In this section we concern ourselves with systems of first-order ordinary differential equations (E), i.e.,

$$\dot{x} = f(t, x), \tag{E}$$

where $x \in R^n$. When discussing global results, we shall assume that $f : R^+ \times R^n \to R^n$, while when considering local results, we may assume that $f : R^+ \times B(h) \to R^n$ for some $h > 0$. On some occasions, we may assume that $t \in R$, rather than $t \in R^+$. Unless otherwise stated, we shall assume that for every (t_0, x_0), $t_0 \in R^+$, the initial-value problem

$$\dot{x} = f(t, x), \qquad x(t_0) = x_0 \tag{I}$$

possesses a unique solution $\phi(t, t_0, x_0)$ that exists for all $t \geq t_0$ and that depends continuously on the initial data (t_0, x_0). Refer to Chapter 1 for conditions that ensure that (I) has these properties.

DEFINITION 3.1. A point $x_e \in R^n$ is called an *equilibrium point* of (E), or simply an *equilibrium* of (E) (at time $\tilde{t} \in R^+$), if

$$f(t, x_e) = 0$$

for all $t \geq \tilde{t}$. ■

We note that if x_e is an equilibrium of (E) at $\tilde{t}$, then it is also an equilibrium at all $\tau \geq \tilde{t}$. We also note that in the case of autonomous systems

$$\dot{x} = f(x), \tag{A}$$

and in the case of T-periodic systems

$$\dot{x} = f(t, x), \qquad f(t, x) = f(t + T, x), \tag{P}$$

a point $x_e \in R^n$ is an equilibrium at some time $\tilde{t}$ if and only if it is an equilibrium at all times. [Refer to Chapter 1 for the definitions of symbols in (A) and (P).] We further note that if x_e is an equilibrium at $\tilde{t}$ of (E), then the transformation $s = t - \tilde{t}$ yields

$$\frac{dx}{ds} = f(s + \tilde{t}, x),$$

and x_e is an equilibrium at $s = 0$ of this system. Accordingly, we will henceforth assume that $\tilde{t} = 0$ in Definition 3.1 and we will not mention $\tilde{t}$ again. Furthermore, we note that for any $t_0 \geq 0$,

$$\phi(t, t_0, x_e) = x_e \qquad \text{for all } t \geq t_0,$$

i.e., the equilibrium x_e is a unique solution of (E) with initial data given by $\phi(t_0, t_0, x_e) = x_e$.

We will call an equilibrium point x_e of (E) an *isolated equilibrium point* if there is an $r > 0$ such that $B(x_e, r) \subset R^n$ contains no equilibrium point of (E) other than x_e itself. [Recall that $B(x_e, r) = \{x \in R^n : \|x - x_e\| < r\}$, where $\|\cdot\|$ denotes some norm defined on R^n.] Unless stated otherwise, we will assume throughout this chapter that a given equilibrium point is an isolated equilibrium. Also, we will usually assume that in a given discussion, unless otherwise stated, the equilibrium of interest is located at the origin of R^n. This assumption can be made without loss of generality by noting that if $x_e \neq 0$ is an equilibrium point of (E), i.e., $f(t, x_e) = 0$ for all $t \geq 0$, then by letting $w = x - x_e$, we obtain the transformed system

$$\dot{w} = F(t, w) \tag{3.1}$$

with $F(t, 0) = 0$ for all $t \geq 0$, where

$$F(t, w) = f(t, w + x_e). \tag{3.2}$$

Since (3.2) establishes a one-to-one correspondence between the solutions of (E) and (3.1), we may assume henceforth that the equilibrium of interest for (E) is located at the origin. This equilibrium, $x = 0$, will be referred to as the *trivial solution* of (E).

Before concluding this section, it may be fruitful to consider some specific cases.

EXAMPLE 3.1. In Example 4.4 in Chapter 1 we considered the simple pendulum given in Fig. 1.7. Letting $x_1 = x$ and $x_2 = \dot{x}$ in Eq. (4.12) of Chapter 1, we obtain the system of equations

$$\begin{aligned} \dot{x}_1 &= x_2 \\ \dot{x}_2 &= -k \sin x_1, \end{aligned} \tag{3.3}$$

where $k > 0$ is a constant. *Physically,* the pendulum has two isolated equilibrium points: one where the mass M is located vertically at the bottom of the figure (i.e., at 6 o'clock) and the other where the mass is located vertically at the top of the figure (i.e., at 12 o'clock). The *model* of this pendulum, however, described by Eq. (3.3), has countably infinitely many isolated equilibrium points which are located in R^2 at the points $(\pi n, 0)^T$, $n = 0, \pm 1, \pm 2, \ldots$. ■

EXAMPLE 3.2. The linear homogeneous system of ordinary differential equations

$$\dot{x} = A(t)x \tag{LH}$$

has a unique equilibrium that is at the origin if $A(t_0)$ is nonsingular for all $t_0 \geq 0$. [Refer to Chapter 1 for the definitions of symbols in (LH).] ■

EXAMPLE 3.3. The linear, autonomous, homogenous system of ordinary differential equations

$$\dot{x} = Ax \tag{L}$$

has a unique equilibrium that is at the origin if and only if A is nonsingular. Otherwise, (L) has nondenumerably many equilibria. [Refer to Chapter 1 for the definitions of symbols in (L).] ■

EXAMPLE 3.4. Assume that for the autonomous system of ordinary differential equations,

$$\dot{x} = f(x), \tag{A}$$

f is continuously differentiable with respect to all of its arguments, and let

$$J(x_e) = \left.\frac{\partial f}{\partial x}(x)\right|_{x=x_e},$$

where $\partial f/\partial x$ denotes the $n \times n$ *Jacobian matrix* defined by

$$\frac{\partial f}{\partial x} = \left[\frac{\partial f_i}{\partial x_j}\right].$$

If $f(x_e) = 0$ and $J(x_e)$ is nonsingular, then x_e is an isolated equilibrium of (A). ■

EXAMPLE 3.5. The system of ordinary differential equations given by

$$\begin{aligned} \dot{x}_1 &= k + \sin(x_1 + x_2) + x_1 \\ \dot{x}_2 &= k + \sin(x_1 + x_2) - x_1, \end{aligned}$$

with $k > 1$, has no equilibrium points at all. ■

6.4 QUALITATIVE CHARACTERIZATIONS OF AN EQUILIBRIUM

In this section we consider several qualitative characterizations that are of fundamental importance in systems theory. These characterizations are concerned with various types of stability properties of an equilibrium and are referred to in the literature as *Lyapunov stability*.

Throughout this section, we consider systems of equations (E),

$$\dot{x} = f(t, x), \tag{E}$$

and we assume that (E) possesses an isolated equilibrium at the origin. We thus have $f(t, 0) = 0$ for all $t \geq 0$.

DEFINITION 4.1. The equilibrium $x = 0$ of (E) is said to be *stable* if for every $\epsilon > 0$ and any $t_0 \in R^+$ there exists a $\delta(\epsilon, t_0) > 0$ such that

$$\|\phi(t, t_0, x_0)\| < \epsilon \qquad \text{for all } t \geq t_0 \tag{4.1}$$

whenever

$$\|x_0\| < \delta(\epsilon, t_0). \tag{4.2}$$

■

In Definition 4.1, $\|\cdot\|$ denotes any one of the equivalent norms on R^n, and (as in Chapters 1 and 2) $\phi(t, t_0, x_0)$ denotes the solution of (E) with initial condition x_0 at initial time t_0. The notation $\delta(\epsilon, t_0)$ indicates that δ depends on the choice of t_0 and ϵ. If in particular it is true that δ is independent of t_0, i.e., $\delta = \delta(\epsilon)$, then the equilibrium $x = 0$ of (E) is said to be *uniformly stable.*

In words, Definition 4.1 states that by choosing the initial points in a sufficiently small spherical neighborhood, when the equilibrium $x = 0$ of (E) is stable, we can force the graph of the solution for $t \geq t_0$ to lie entirely inside a given cylinder. This is depicted in Fig. 6.1 for the case $x \in R^2$.

We note that if the equilibrium $x = 0$ of (E) satisfies condition (4.1) for a single initial condition t_0 when (4.2) is true, then it will also satisfy this condition at every initial time $t_0' > t_0$, where a different value of δ may be required. To see this, we note that the solution $\phi(t, t_0, x_0)$ determines a mapping g of $B(\delta(\epsilon, t_0))$ (at $t = t_0$) onto $g(B(\delta(\epsilon, t_0)))$ (at $t = t' \geq t_0$) that contains the origin by assigning for every $x_0 \in B(\delta(\epsilon, t_0))$ one and only one $\tilde{x}_0 \triangleq \phi(t', t_0, x_0) \in g(B(\delta(\epsilon, t_0))$. By reversing time, ϕ determines a mapping of $g(B(\delta(\epsilon, t_0))$ onto $B(\delta(\epsilon, t_0))$, which is the inverse of g, denoted by g^{-1}. Since ϕ is continuous with respect to t, t_0, and x_0, then g and g^{-1} are also continuous. Since $B(\delta(\epsilon, t_0))$ is a neighborhood (an open set), then so is $g(B(\delta(\epsilon, t_0))$ (refer, e.g., to [16], pp. 320-321). This neighborhood contains in its interior a spherical neighborhood centered at the origin and with a radius δ'. If we choose $x_0' \in B(\delta')$, then (4.1) implies that $\|\phi(t, t', x_0')\| < \epsilon$ for all $t \geq t_0'$. This argu-

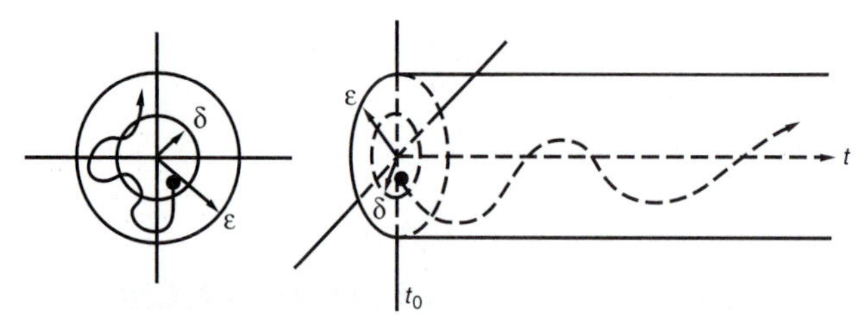

FIGURE 6.1
Stability of an equilibrium

ment shows that in Definition 4.1 we could have chosen without loss of generality the single value $t_0 = 0$ in (4.1) and (4.2).

DEFINITION 4.2. The equilibrium $x = 0$ of (E) is said to be *asymptotically stable* if

(i) it is stable,
(ii) for every $t_0 \geq 0$ there exists an $\eta(t_0) > 0$ such that $\lim_{t\to\infty} \phi(t, t_0, x_0) = 0$ whenever $\|x_0\| < \eta$. ■

The set of all $x_0 \in R^n$ such that $\phi(t, t_0, x_0) \to 0$ as $t \to \infty$ for some $t_0 \geq 0$ is called the *domain of attraction* of the equilibrium $x = 0$ of (E) (at t_0). Also, if for (E) condition (ii) is true, then the equilibrium $x = 0$ is said to be *attractive* (at t_0).

DEFINITION 4.3. The equilibrium $x = 0$ of (E) is said to be *uniformly asymptotically stable* if

(i) it is uniformly stable,
(ii) there is a $\delta_0 > 0$ such that for every $\epsilon > 0$ and for any $t_0 \in R^+$, there exists a $T(\epsilon) > 0$, independent of t_0, such that $\|\phi(t, t_0, x_0)\| < \epsilon$ for all $t \geq t_0 + T(\epsilon)$ whenever $\|x_0\| < \delta_0$. ■

Condition (ii) in Definition 4.3 can be paraphrased by saying that there exists a $\delta_0 > 0$ such that

$$\lim_{t\to\infty} \phi(t + t_0, t_0, x_0) = 0$$

uniformly in (t_0, x_0) for $t_0 \geq 0$ and for $\|x_0\| < \delta_0$. In words, this condition states that by choosing the initial points x_0 in a sufficiently small spherical neighborhood at $t = t_0$, we can force the graph of the solution to lie inside a given cylinder for all $t \geq t_0 + T(\epsilon)$. This is depicted in Fig. 6.2 for the case $x \in R^2$.

In linear systems theory, we are especially interested in the following *special case of uniform asymptotic stability*.

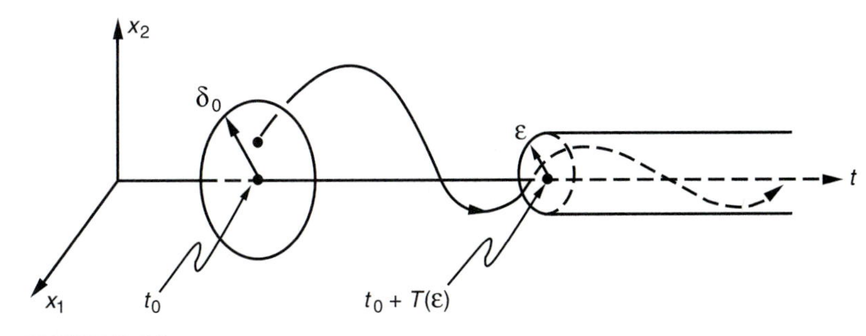

FIGURE 6.2
Attractivity of an equilibrium

DEFINITION 4.4. The equilibrium $x = 0$ of (E) is *exponentially stable* if there exists an $\alpha > 0$, and for every $\epsilon > 0$, there exists a $\delta(\epsilon) > 0$, such that

$$\|\phi(t, t_0, x_0)\| \leq \epsilon e^{-\alpha(t-t_0)} \qquad \text{for all } t \geq t_0$$

whenever $\|x_0\| < \delta(\epsilon)$ and $t_0 \geq 0$. ■

Figure 6.3 shows the behavior of a solution in the vicinity of an exponentially stable equilibrium $x = 0$.

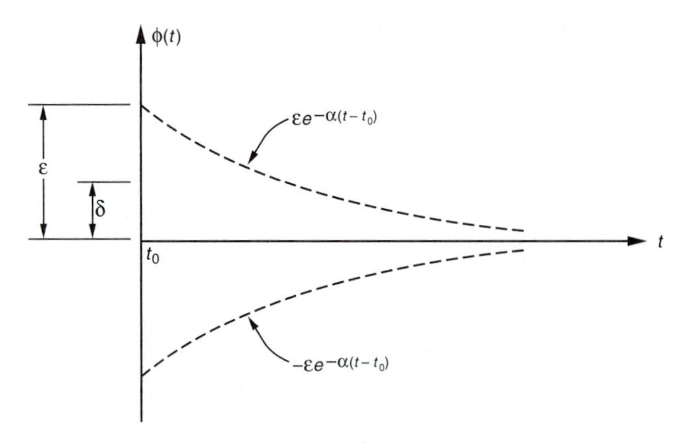

FIGURE 6.3
An exponentially stable equilibrium

DEFINITION 4.5. The equilibrium $x = 0$ of (E) is *unstable* if it is not stable. In this case, there exists a $t_0 \geq 0$, an $\epsilon > 0$, and a sequence $x_m \to 0$ of initial points and a sequence $\{t_m\}$ such that $\|\phi(t_0 + t_m, t_0, x_m)\| \geq \epsilon$ for all m, $t_m \geq 0$. ■

If $x = 0$ is an unstable equilibrium of (E), then it still can happen that all the solutions tend to zero with increasing t. This indicates that instability and attractivity of an equilibrium are compatible concepts. We note that the equilibrium $x = 0$ of (E) is necessarily unstable if every neighborhood of the origin contains initial conditions corresponding to unbounded solutions (i.e., solutions whose norm grows to infinity on a sequence $t_m \to \infty$). However, it can happen that a system (E) with unstable equilibrium $x = 0$ may have only bounded solutions.

The concepts that we have considered thus far pertain to local properties of an equilibrium. In the following, we consider global characterizations of an equilibrium.

DEFINITION 4.6. The equilibrium $x = 0$ of (E) is *asymptotically stable in the large* if it is stable and if every solution of (E) tends to zero as $t \to \infty$. ■

When the equilibrium $x = 0$ of (E) is asymptotically stable in the large, its domain of attraction is all of R^n. Note that in this case, $x = 0$ is the *only* equilibrium of (E).

DEFINITION 4.7. The equilibrium $x = 0$ of (E) is *uniformly asymptotically stable in the large* if

(i) it is uniformly stable,
(ii) for any $\alpha > 0$ and any $\epsilon > 0$, and $t_0 \in R^+$, there exists $T(\epsilon, \alpha) > 0$, independent of t_0, such that if $\|x_0\| < \alpha$, then $\|\phi(t, t_0, x_0)\| < \epsilon$ for all $t \geq t_0 + T(\epsilon, \alpha)$. ■

DEFINITION 4.8. The equilibrium $x = 0$ of (E) is *exponentially stable in the large* if there exists $\alpha > 0$ and for any $\beta > 0$, there exists $k(\beta) > 0$ such that

$$\|\phi(t, t_0, x_0)\| \leq k(\beta)\|x_0\|e^{-\alpha(t-t_0)} \qquad \text{for all } t \geq t_0$$

whenever $\|x_0\| < \beta$. ■

We conclude this section with a few specific cases.

The scalar differential equation

$$\dot{x} = 0 \tag{4.3}$$

has for any initial condition $x(0) = x_0$ the solution $\phi(t, 0, x_0) = x_0$, i.e., all solutions are equilibria of (4.3). The trivial solution is stable; in fact it is uniformly stable. However, it is not asymptotically stable.

The scalar differential equation

$$\dot{x} = ax \tag{4.4}$$

has for every $x(0) = x_0$ the solution $\phi(t, 0, x_0) = x_0 e^{at}$, and $x = 0$ is the only equilibrium of (4.4). If $a > 0$, this equilibrium is unstable, and when $a < 0$, this equilibrium is exponentially stable in the large.

The scalar differential equation

$$\dot{x} = \left(\frac{-1}{t+1}\right)x \tag{4.5}$$

has for every $x(t_0) = x_0$, $t_0 \geq 0$, a unique solution of the form

$$\phi(t, t_0, x_0) = (1 + t_0)x_0\left(\frac{1}{t+1}\right) \tag{4.6}$$

and $x = 0$ is the only equilibrium of (4.5). This equilibrium is uniformly stable and asymptotically stable in the large, but it is not uniformly asymptotically stable.

As mentioned earlier, a system

$$\dot{x} = f(t, x) \tag{E}$$

can have all solutions approaching an equilibrium, say, $x = 0$, without this equilibrium being asymptotically stable. An example of this type of behavior is given by the *nonlinear* system of equations

$$\dot{x}_1 = \frac{x_1^2(x_2 - x_1) + x_2^5}{(x_1^2 + x_2^2)[1 + (x_1^2 + x_2^2)^2]}$$

$$\dot{x}_2 = \frac{x_2^2(x_2 - 2x_1)}{(x_1^2 + x_2^2)[1 + (x_1^2 + x_2^2)^2]}.$$

For a detailed discussion of this system, refer to Hahn [7], cited at the end of this chapter.

Before proceeding any further, a few comments are in order concerning the reasons for considering equilibria and their stability properties as well as other types of stability that we will encounter. To this end we consider linear time-varying systems described by the equations

$$\dot{x} = A(t)x + B(t)u \tag{4.7a}$$

$$y = C(t)x + D(t)u \tag{4.7b}$$

and linear time-invariant systems given by

$$\dot{x} = Ax + Bu \tag{4.8a}$$

$$y = Cx + Du, \tag{4.8b}$$

where all symbols in (4.7) and (4.8) are defined as in Eqs. (6.1) and (6.8) of Chapter 2, respectively. The usual qualitative analysis of such systems involves two concepts, *internal stability* and *input-output stability.*

In the case of *internal stability,* the output equations (4.7b) and (4.8b) play no role whatsoever, the system input u is assumed to be identically zero, and the focus of the analysis is concerned with the qualitative behavior of the solutions of linear time-varying systems

$$\dot{x} = A(t)x \tag{LH}$$

or linear time-invariant systems

$$\dot{x} = Ax \tag{L}$$

near the equilibrium $x = 0$. This is accomplished by making use of the various types of Lyapunov stability concepts introduced in this section. In other words, internal stability of systems (4.7) and (4.8) concerns the Lyapunov stability of the equilibrium $x = 0$ of systems (LH) and (L), respectively.

In the case of *input-output stability,* we view systems as operators determined by (4.7) or (4.8) that relate outputs y to inputs u and the focus of the analysis is concerned with qualitative relations between system inputs and system outputs. We will address this type of stability in Section 9 of this chapter.

6.5 LYAPUNOV STABILITY OF LINEAR SYSTEMS

In this section we first study the stability properties of the equilibrium $x = 0$ of linear autonomous homogeneous systems

$$\dot{x} = Ax, \qquad t \geq 0, \tag{L}$$

and linear homogeneous systems

$$\dot{x} = A(t)x, \qquad t \geq t_0 \geq 0, \tag{LH}$$

where $A(t)$ is assumed to be continuous. Recall that $x = 0$ is always an equilibrium of (L) and (LH) and that $x = 0$ is the only equilibrium of (LH) if $A(t)$ is nonsingular for all $t \geq 0$. Recall also that the solution of (LH) for $x(t_0) = x_0$ is of the form

$$\phi(t, t_0, x_0) = \Phi(t, t_0)x_0, \qquad t \geq t_0,$$

where Φ denotes the state transition matrix of $A(t)$ and that the solution of (L) for $x(t_0) = x_0$ is given by

$$\begin{aligned} \phi(t, t_0, x_0) &= \Phi(t, t_0)x_0 = \Phi(t - t_0, 0)x_0 \\ &\triangleq \Phi(t - t_0)x_0 = e^{A(t-t_0)}x_0, \end{aligned}$$

where in the preceding equation, a slight abuse of notation has been used.

We first consider some of the basic properties of system (LH).

THEOREM 5.1. The equilibrium $x = 0$ of (LH) is *stable* if and only if the solutions of (LH) are bounded, i.e., if and only if

$$\sup_{t \geq t_0} \|\Phi(t, t_0)\| \triangleq k(t_0) < \infty,$$

where $\|\Phi(t, t_0)\|$ denotes the matrix norm induced by the vector norm used on R^n, and $k(t_0)$ denotes a constant that may depend on the choice of t_0.

Proof. Assume that the equilibrium $x = 0$ of (LH) is stable. Then for any $t_0 \geq 0$ and for $\epsilon = 1$ there is a $\delta = \delta(1, t_0) > 0$ such that $\|\phi(t, t_0, x_0)\| < 1$ for all $t \geq t_0$ and all x_0 with $\|x_0\| \leq \delta$. In this case

$$\|\phi(t, t_0, x_0)\| = \|\Phi(t, t_0)x_0\| = \left\| \frac{\Phi(t, t_0)(x_0\delta)}{\|x_0\|} \right\| \left(\frac{\|x_0\|}{\delta} \right) < \frac{\|x_0\|}{\delta}$$

for all $x_0 \neq 0$ and all $t \geq t_0$. Using the definition of matrix norm [refer to (10.17) in Chapter 1] it follows that

$$\|\Phi(t, t_0)\| \leq \delta^{-1}, \qquad t \geq t_0.$$

We have proved that if the equilibrium $x = 0$ of (LH) is stable, then the solutions of (LH) are bounded.

Conversely, suppose that all solutions $\phi(t, t_0, x_0) = \Phi(t, t_0)x_0$ are bounded. Let $\{e_1, \ldots, e_n\}$ denote the natural basis for n-space and let $\|\phi(t, t_0, e_j)\| < \beta_j$ for all $t \geq t_0$. Then for any vector $x_0 = \sum_{j=1}^{n} \alpha_j e_j$ we have that

$$\|\phi(t, t_0, x_0)\| = \left\| \sum_{j=1}^{n} \alpha_j \phi(t, t_0, e_j) \right\| \leq \sum_{j=1}^{n} |\alpha_j| \beta_j \leq (\max_j \beta_j) \sum_{j=1}^{n} |\alpha_j| \leq k\|x_0\|$$

for some constant $k > 0$ for $t \geq t_0$. For given $\epsilon > 0$, we choose $\delta = \epsilon/k$. Thus, if $\|x_0\| < \delta$, then $\|\phi(t, t_0, x_0)\| < k\|x_0\| < \epsilon$ for all $t \geq t_0$. We have proved that if the solutions of (LH) are bounded, then the equilibrium $x = 0$ of (LH) is stable. ■

THEOREM 5.2. The equilibrium $x = 0$ of (LH) is *uniformly stable* if and only if

$$\sup_{t_0 \geq 0} k(t_0) \triangleq \sup_{t_0 \geq 0}(\sup_{t \geq t_0} \|\Phi(t, t_0)\|) \triangleq k_0 < \infty.$$

■

The proof of this theorem is very similar to the proof of Theorem 5.1 and is left as an exercise.

EXAMPLE 5.1. We consider the system given by

$$\begin{bmatrix} \dot{x}_1 \\ \dot{x}_2 \end{bmatrix} = \begin{bmatrix} e^{-2t} & e^{-t} - e^{-2t} \\ 0 & e^{-t} \end{bmatrix} \begin{bmatrix} x_1 \\ x_2 \end{bmatrix} \tag{5.1}$$

with $x(0) = x_0$. We transform (5.1) by means of the relation $x = Py$, where

$$P = \begin{bmatrix} 1 & 1 \\ 0 & 1 \end{bmatrix}, \qquad P^{-1} = \begin{bmatrix} 1 & -1 \\ 0 & 1 \end{bmatrix},$$

and obtain the equivalent system

$$\begin{bmatrix} \dot{y}_1 \\ \dot{y}_2 \end{bmatrix} = \begin{bmatrix} e^{-2t} & 0 \\ 0 & e^{-t} \end{bmatrix} \begin{bmatrix} y_1 \\ y_2 \end{bmatrix} \tag{5.2}$$

with $y(0) = y_0 = P^{-1}x_0$. System (5.2) has the solution $\psi(t, 0, y_0) = \Psi(t, 0)y_0$, where

$$\Psi(t, 0) = \begin{bmatrix} e^{1/2(1-e^{-2t})} & 0 \\ 0 & e^{(1-e^{-t})} \end{bmatrix}.$$

The solution for (5.1) is obtained as $\phi(t, 0, x_0) = \Phi(t, 0)x_0$, where $\Phi(t, 0) = P\Psi(t, 0)P^{-1}$. From this we obtain for $t_0 \neq 0$, $\phi(t, t_0, x_0) = \Phi(t, t_0)x_0$, where

$$\Phi(t, t_0) = \begin{bmatrix} e^{1/2(e^{-2t_0}-e^{-2t})} & e^{(e^{-t_0}-e^{-t})} - e^{1/2(e^{-2t_0}-e^{-2t})} \\ 0 & e^{(e^{-t_0}-e^{-t})} \end{bmatrix}.$$

Now

$$\lim_{t\to\infty} \Phi(t, t_0) = \begin{bmatrix} e^{1/2e^{-2t_0}} & e^{e^{-t_0}} - e^{1/2e^{-2t_0}} \\ 0 & e^{e^{-t_0}} \end{bmatrix}. \tag{5.3}$$

We conclude that $\lim_{t_0\to\infty} \lim_{t\to\infty} \|\Phi(t, t_0)\| < \infty$, and therefore (since $\|\Phi(t, t_0)\| \le \sqrt{\sum_{i,j=1}^{n} |\phi_{ij}(t, t_0)|^2} \le \sum_{i,j=1}^{n} |\phi_{ij}(t, t_0)|$), that $\sup_{t_0\ge 0}(\sup_{t\ge t_0} \|\Phi(t, t_0)\|) < \infty$. Therefore, the equilibrium $x = 0$ of system (5.1) is *stable* by Theorem 5.1 and *uniformly stable* by Theorem 5.2. ■

THEOREM 5.3. The following statements are equivalent.

(i) The equilibrium $x = 0$ of (LH) is asymptotically stable.
(ii) The equilibrium $x = 0$ of (LH) is asymptotically stable in the large.
(iii) $\lim_{t\to\infty} \|\Phi(t, t_0)\| = 0$.

Proof. Assume that statement (i) is true. Then there is an $\eta(t_0) > 0$ such that when $\|x_0\| \le \eta(t_0)$, then $\phi(t, t_0, x_0) \to 0$ as $t \to \infty$. But then we have for any $x_0 \ne 0$ that

$$\phi(t, t_0, x_0) = \phi\left(t, t_0, \eta(t_0)\frac{x_0}{\|x_0\|}\right)\left(\frac{\|x_0\|}{\eta(t_0)}\right) \to 0$$

as $t \to \infty$. It follows that statement (ii) is true.

Next, assume that statement (ii) is true and fix $t_0 \ge 0$. For any $\epsilon > 0$ there must exist a $T(\epsilon) > 0$ such that for all $t \ge t_0 + T(\epsilon)$ we have that $\|\phi(t, t_0, x_0)\| = \|\Phi(t, t_0)x_0\| < \epsilon$. To see this, let $\{e_1, \dots, e_n\}$ be the natural basis for R^n. Thus, for some fixed constant $k > 0$, if $x_0 = (\alpha_1, \dots, \alpha_n)^T$ and if $\|x_0\| \le 1$, then $x_0 = \sum_{j=1}^{n} \alpha_j e_j$ and $\sum_{j=1}^{n} |\alpha_j| \le k$. For each j there is a $T_j(\epsilon)$ such that $\|\Phi(t, t_0)e_j\| < \epsilon/k$ and $t \ge t_0 + T_j(\epsilon)$. Define $T(\epsilon) = \max\{T_j(\epsilon) : j = 1, \dots, n\}$. For $\|x_0\| \le 1$ and $t \ge t_0 + T(\epsilon)$, we have that

$$\|\Phi(t, t_0)x_0\| = \left\| \sum_{j=1}^{n} \alpha_j \Phi(t, t_0)e_j \right\| \le \sum_{j=1}^{n} |\alpha_j| \left(\frac{\epsilon}{k}\right) \le \epsilon.$$

By the definition of the matrix norm [see (10.17) of Chapter 1], this means that $\|\Phi(t, t_0)\| \le \epsilon$ for $t \ge t_0 + T(\epsilon)$. Therefore, statement (iii) is true.

Finally, assume that statement (iii) is true. Then $\|\Phi(t, t_0)\|$ is bounded in t for all $t \ge t_0$. By Theorem 5.1, the equilibrium $x = 0$ is stable. To prove asymptotic stability, fix $t_0 \ge 0$ and $\epsilon > 0$. If $\|x_0\| < \eta(t_0) = 1$, then $\|\phi(t, t_0, x_0)\| \le \|\Phi(t, t_0)\| \, \|x_0\| \to 0$ as $t \to \infty$. Therefore, statement (i) is true. This completes the proof. ■

EXAMPLE 5.2. The equilibrium $x = 0$ of system (5.1) given in Example 5.1 is stable but it is not asymptotically stable since $\lim_{t\to\infty} \|\Phi(t, t_0)\| \ne 0$ [see Eq. (5.3)]. ■

EXAMPLE 5.3. The solution of the system

$$\dot{x} = -e^{2t}x, \qquad x(t_0) = x_0 \tag{5.4}$$

is $\phi(t, t_0, x_0) = \Phi(t, t_0)x_0$, where

$$\Phi(t, t_0) = e^{(1/2)(e^{2t_0}-e^{2t})}.$$

Since $\lim_{t\to\infty} \Phi(t, t_0) = 0$, it follows that the equilibrium $x = 0$ of system (5.4) is asymptotically stable (in the large). ■

THEOREM 5.4. The equilibrium $x = 0$ of (LH) is uniformly asymptotically stable if and only if it is exponentially stable.

Proof. The exponential stability of the equilibrium $x = 0$ implies the uniform asymptotic stability of the equilibrium $x = 0$ of systems (E) in general, and hence, for systems (LH) in particular.

Conversely, assume that the equilibrium $x = 0$ of (LH) is uniformly asymptotically stable. Then there is a $\delta > 0$ and a $T > 0$ such that if $\|x_0\| \leq \delta$, then

$$\|\Phi(t + t_0 + T, t_0)x_0\| < \frac{\delta}{2}$$

for all $t, t_0 \geq 0$. This implies that

$$\|\Phi(t + t_0 + T, t_0)\| \leq \tfrac{1}{2} \qquad \text{if } t, t_0 \geq 0. \tag{5.5}$$

From Theorem 3.6 (iii) of Chapter 2 we have that $\Phi(t, \tau) = \Phi(t, \sigma)\Phi(\sigma, \tau)$ for any t, σ, and τ. Therefore,

$$\|\Phi(t + t_0 + 2T, t_0)\| = \|\Phi(t + t_0 + 2T, t + t_0 + T)\Phi(t + t_0 + T, t_0)\| \leq \tfrac{1}{4},$$

in view of (5.5). By induction, we obtain for $t, t_0 \geq 0$ that

$$\|\Phi(t + t_0 + nT, t_0)\| \leq 2^{-n}. \tag{5.6}$$

Now let $\alpha = (\ln 2)/T$. Then (5.6) implies that for $0 \leq t < T$ we have that

$$\begin{aligned} \|\phi(t + t_0 + nT, t_0, x_0)\| &\leq 2\|x_0\|2^{-(n+1)} = 2\|x_0\|e^{-\alpha(n+1)T} \\ &\leq 2\|x_0\|e^{-\alpha(t+nT)}, \end{aligned}$$

which proves the result. ■

EXAMPLE 5.4. For system (5.4) given in Example 5.3 we have

$$\begin{aligned} \|\phi(t, t_0, x_0)\| = |\phi(t, t_0, x_0)| &= \left|x_0 e^{(1/2)e^{2t_0}} e^{-(1/2)e^{2t}}\right| \\ &\leq |x_0| e^{(1/2)e^{2t_0}} e^{-t}, \qquad t \geq t_0 \geq 0, \end{aligned}$$

since $e^{2t} > 2t$. Therefore, the equilibrium $x = 0$ of system (5.4) is uniformly asymptotically stable in the large, and exponentially stable in the large. ■

Even though the preceding results require knowledge of the state transition matrix $\Phi(t, t_0)$ of (LH), they are quite useful in the qualitative analysis of linear systems. In view of the above results, we can state the following equivalent definitions. The equilibrium $x = 0$ of (LH) is *stable* if and only if for any $t_0 \geq 0$ there exists a finite positive constant $\gamma = \gamma(t_0)$ (which in general depends on t_0) such that for any x_0, the corresponding solution satisfies the inequality

$$\|\phi(t, t_0, x_0)\| \leq \gamma(t_0)\|x_0\|, \qquad t \geq t_0.$$

Also, the equilibrium $x = 0$ of (LH) is *uniformly stable* if and only if there exists a finite positive constant γ (independent of t_0), such that for any t_0 and x_0, the corresponding solution satisfies the inequality

$$\|\phi(t, t_0, x_0)\| \leq \gamma\|x_0\|, \qquad t \geq t_0.$$

Furthermore, in view of the above results, if the equilibrium $x = 0$ of (LH) is asymptotically stable, then in fact it must be globally asymptotically stable, and if it is uniformly asymptotically stable in the large, then in fact it must be exponentially stable (in the large). In this case there exist finite constants $\gamma \geq 1$ and $\lambda > 0$ such that

$$\|\phi(t, t_0, x_0)\| \leq \gamma e^{-\lambda(t-t_0)}\|x_0\|$$

for $t \geq t_0 \geq 0$ and $x_0 \in R^n$.

The results in the next theorem, which are important in their own right, will be required later.

THEOREM 5.5. Let A be bounded on $(-\infty, \infty)$. Then any one of the following statements is equivalent to the exponential stability of the origin $x = 0$ of (LH):

(i) $\int_{t_0}^{t_1} \|\Phi(t, t_0)\|^2 \, dt \leq c_1$ for all $t_1 \geq t_0$.

(ii) $\int_{t_0}^{t_1} \|\Phi(t, t_0)\| \, dt \leq c_2$ for all $t_1 \geq t_0$.

(iii) $\int_{t_0}^{t_1} \|\Phi(t_1, \tau)\|^2 \, d\tau \leq c_3$ for all $t_1 \geq t_0$.

(iv) $\int_{t_0}^{t_1} \|\Phi(t_1, \tau)\| \, d\tau \leq c_4$ for all $t_1 \geq t_0$.

(The constants c_i are independent of t_0 or t_1, and $\|\cdot\|$ denotes a matrix norm induced by any one of the equivalent vector norms on R^n.)

Proof. If the equilibrium $x = 0$ of (LH) is exponentially stable, then there exist constants $\gamma > 0$, $\lambda > 0$ (independent of t_0) such that $\|\Phi(t, t_0)\| \leq \gamma e^{-\lambda(t-t_0)}$, $t \geq t_0$. Substituting this estimate into (i) to (iv) and evaluating the integrals yields $c_1 = c_3 = \gamma^2/(2\lambda)$ and $c_2 = c_4 = \gamma/\lambda$. Therefore, if the equilibrium $x = 0$ of (LH) is exponentially stable, then the integrals (i) to (iv) are bounded.

We now prove the converse statements by considering each case individually.

(a) Assume that the bound c_1 in (i) exists. Since A is bounded, there exists an $\alpha > 0$ such that

$$\begin{aligned} \|\dot{\Phi}(t, t_0)\| &= \|A(t)\Phi(t, t_0)\| \leq \|A(t)\| \, \|\Phi(t, t_0)\| \\ &\leq \alpha \|\Phi(t, t_0)\|, \qquad t \geq t_0. \end{aligned}$$

Therefore,

$$\begin{aligned} &\|\Phi(t_1, t_0)^T \Phi(t_1, t_0) - I\| \\ &\quad = \left\| \int_{t_0}^{t_1} [\dot{\Phi}(t, t_0)^T \Phi(t, t_0) + \Phi(t, t_0)^T \dot{\Phi}(t, t_0)] \, dt \right\| \\ &\quad \leq \int_{t_0}^{t_1} \|\dot{\Phi}(t, t_0)^T \Phi(t, t_0) + \Phi(t, t_0)^T \dot{\Phi}(t, t_0)\| \, dt \\ &\quad \leq \int_{t_0}^{t_1} \{\|[A(t)\Phi(t, t_0)]^T\| \, \|\Phi(t, t_0)\| + \|\Phi(t, t_0)^T\| \, \|A(t)\Phi(t, t_0)\|\} \, dt \\ &\quad \leq 2\alpha \int_{t_0}^{t_1} \|\Phi(t, t_0)\|^2 \, dt \leq 2\alpha c_1, \qquad t \geq t_0. \end{aligned}$$

Using the triangle inequality of norms yields

$$\begin{aligned} \|\Phi(t_1, t_0)^T \Phi(t_1, t_0)\| &= \|\Phi(t_1, t_0)^T \Phi(t_1, t_0) - I + I\| \\ &\leq \|\Phi(t_1, t_0)^T \Phi(t_1, t_0) - I\| + \|I\| \leq 2\alpha c_1 + 1, \qquad t_1 \geq t_0. \end{aligned}$$

This shows that $\|\Phi(t_1, t_0)\|$ is itself bounded for $t_1 \geq t_0$ by a constant L_1.

To determine an exponential bound for $\|\Phi(t, t_0)\|$ we note that

$$\begin{aligned} \int_{t_0}^{t} \|\Phi(t, t_0)\|^2 \, d\tau &= \int_{t_0}^{t} \|\Phi(t, \tau)\Phi(\tau, t_0)\|^2 \, d\tau \\ &\leq \int_{t_0}^{t} \|\Phi(t, \tau)\|^2 \cdot \|\Phi(\tau, t_0)\|^2 \, d\tau \leq L_1^2 \int_{t_0}^{t} \|\Phi(\tau, t_0)\|^2 \, d\tau \\ &\leq L_1^2 c_1. \end{aligned}$$

Noting that the integrand on the left side does not depend on τ, we obtain

$$(t - t_0)\|\Phi(t, t_0)\|^2 \leq L_1^2 c_1, \qquad t \geq t_0.$$

Let $t = t_0 + T$ and let $T = 4L_1^2 c_1$. Then

$$\|\Phi(t_0 + T, t_0)\| \leq \tfrac{1}{2}.$$

Repeating the above procedure, we obtain

$$\|\Phi(t_0 + 2T, t_0)\| \leq \tfrac{1}{2} \cdot \tfrac{1}{2}$$

and in general we have

$$\|\Phi(t_0 + nT, t_0)\| \leq \left(\tfrac{1}{2}\right)^n.$$

Proceeding now as in the proof of Theorem 5.4, it follows that $\|\Phi(t, t_0)\|$ is exponentially bounded.

(b) As in (a), but using $\Phi(t_1, t_0)$ instead of $\Phi(t_1, t_0)^T\Phi(t_1, t_0)$, and using the inequality $\|\dot{\Phi}(t, t_0)\| \leq \alpha\|\Phi(t, t_0)\|$, we obtain

$$\begin{aligned}
\|\Phi(t_1, t_0) - I\| &= \left\|\int_{t_0}^{t_1} \dot{\Phi}(t, t_0)\,dt\right\| \\
&\leq \int_{t_0}^{t_1} \|\dot{\Phi}(t, t_0)\|\,dt \leq \alpha \int_{t_0}^{t_1} \|\Phi(t, t_0)\|\,dt \\
&\leq \alpha c_2, \qquad t_1 \geq t_0.
\end{aligned}$$

Using the inequality property of norms, we have

$$\begin{aligned}
\|\Phi(t_1, t_0)\| &= \|\Phi(t_1, t_0) - I + I\| \leq \|\Phi(t_1, t_0) - I\| + \|I\| \\
&\leq \alpha c_2 + 1.
\end{aligned}$$

Using this estimate, we now obtain

$$\begin{aligned}
\int_{t_0}^{t_1} \|\Phi(t, t_0)\|^2\,dt &\leq (\alpha c_2 + 1)\int_{t_0}^{t_1} \|\Phi(t, t_0)\|\,dt \\
&\leq (\alpha c_2 + 1)c_2.
\end{aligned}$$

This shows that (ii) implies (i), and therefore, it follows that $\|\Phi(t, t_0)\|$ is exponentially bounded.

(c) The proof of this part follows the proof of (i) with some modifications. Since A is bounded there exists $\alpha > 0$ such that $\|A(t)\| \leq \alpha$ for all t. From Exercise 2.35 in Chapter 2, we have

$$\frac{d}{d\tau}\Phi(t, \tau) = -\Phi(t, \tau)A(\tau),$$

and therefore,

$$\left\|\frac{d}{d\tau}\Phi(t, \tau)\right\| = \|-\Phi(t, \tau)A(\tau)\| \leq \alpha\|\Phi(t, \tau)\|.$$

Then

$$\|I - \Phi(t_1, t_0)^T\Phi(t_1, t_0)\| = \left\|\int_{t_0}^{t_1} \left[\frac{d}{d\tau}\Phi(t_1, \tau)^T\right]\Phi(t_1, \tau) + \Phi(t_1, \tau)^T\left[\frac{d}{d\tau}\Phi(t_1, \tau)\right]d\tau\right\|.$$

As in (a), we now obtain

$$\|I - \Phi(t_1, t_0)^T\Phi(t_1, t_0)\| \leq 2\alpha c_3, \qquad t_1 \geq t_0.$$

Using the triangle inequality, we obtain

$$\begin{aligned}
\|\Phi(t_1, t_0)^T\Phi(t_1, t_0)\| &\leq \|\Phi(t_1, t_0)^T\Phi(t_1, t_0) - I\| + \|I\| \\
&\leq 2\alpha c_3 + 1.
\end{aligned}$$

This shows that $\|\Phi(t, t_0)\|$ is bounded for all $t > t_0$ by a constant L_3 that is independent of t and t_0.

To obtain an exponential bound we proceed similarly as in (a) to obtain

$$
\begin{aligned}
(t - t_0)\|\Phi(t_1, t_0)\|^2 &= \int_{t_0}^{t} \|\Phi(t_1, t_0)\|^2 \, d\tau \\
&= \int_{t_0}^{t} \|\Phi(t_1, \tau)\Phi(\tau, t_0)\|^2 \, d\tau \le \int_{t_0}^{t} \|\Phi(t_1, \tau)\|^2 \|\Phi(\tau, t_0)\|^2 \, d\tau \\
&\le L_3^2 \int_{t_0}^{t} \|\Phi(t_1, \tau)\|^2 \, d\tau \le L_3 c_3.
\end{aligned}
$$

Therefore, we have

$$
\|\Phi(t, t_0)\|^2 \le L_3^2 c_3 (t - t_0)^{-1}
$$

from which we obtain, letting $T = 4L_3^2 c_3$,

$$
\|\Phi(t_0 + T, t_0)\| \le \tfrac{1}{2},
$$

and more generally, we obtain

$$
\|\Phi(t_0 + nT, t_0)\| \le \left(\tfrac{1}{2}\right)^n.
$$

Again, as in part (a), we conclude that Φ is exponentially bounded.

(d) Assume that (iv) is satisfied. Then as in (b), we can show that (iv) implies (iii), which in turn implies that Φ is exponentially bounded. We obtain

$$
\|\Phi(t_1, t_0) - I\| \le \alpha c_4, \qquad t \ge t_0,
$$

where α is defined as before and c_4 is given in (iv). From this we can conclude that $\|\Phi(t_1, t_0)\|$ is bounded for all $t_1 \ge t_0$ by $\alpha c_4 + 1$. This in turn yields the inequality

$$
\begin{aligned}
\int_{t_0}^{t_1} \|\Phi(t, \tau)\|^2 \, d\tau &\le (\alpha c_4 + 1) \int_{t_0}^{t_1} \|\Phi(t_1, \tau)\| \, d\tau \\
&\le (\alpha c_4 + 1) c_4.
\end{aligned}
$$

From this it follows that (iv) implies (iii). This concludes the proof of the theorem. ■

We now turn our attention to linear, autonomous, and homogeneous systems given by

$$
\dot{x} = Ax, \qquad t \ge 0, \tag{L}
$$

referring to the discussion in Subsection 2.4A [refer to Eqs. (4.11) to (4.27) in Chapter 2] concerning the use of the Jordan canonical form to compute exp (At). We let $J = P^{-1}AP$ and define $x = Py$. Then (L) yields

$$
\dot{y} = P^{-1}APy = Jy. \tag{5.7}
$$

It is easily verified (the reader is asked to do so in the Exercises section) that the equilibrium $x = 0$ of (L) is stable (resp., asymptotically stable or unstable) if and only if $y = 0$ of (5.7) is stable (resp., asymptotically stable or unstable). In view of this, we can assume without loss of generality that the matrix A in (L) is in Jordan canonical form, given by

$$
A = diag\,[J_0, J_1, \ldots, J_s],
$$

where $\qquad J_0 = diag\,[\lambda_1, \ldots, \lambda_k] \qquad$ and $\qquad J_k = \lambda_{k+i} I_i + N_i$

for the Jordan blocks $J_1, \ldots, J_s$.

As in (4.21), (4.22), (4.26) and (4.27) of Chapter 2, we have

$$e^{At} = \begin{bmatrix} e^{J_0 t} & & & 0 \\ & e^{J_1 t} & & \\ & & \ddots & \\ 0 & & & e^{J_s t} \end{bmatrix},$$

where

$$e^{J_0 t} = diag\,[e^{\lambda_1 t}, \ldots, e^{\lambda_k t}] \tag{5.8}$$

and

$$e^{J_i t} = e^{\lambda_{k+i} t} \begin{bmatrix} 1 & t & \frac{t^2}{2} & \cdots & \frac{t^{n_i-1}}{(n_i-1)!} \\ 0 & 1 & t & \cdots & \frac{t^{n_i-2}}{(n_i-2)!} \\ \vdots & \vdots & \vdots & & \vdots \\ 0 & 0 & 0 & \cdots & 1 \end{bmatrix} \tag{5.9}$$

for $i = 1, \ldots, s$.

Now suppose that $Re\,\lambda_i \leq \beta$ for all $i = 1, \ldots, k$. Then it is clear that $\lim_{t\to\infty} (\|e^{J_0 t}\|/e^{\beta t}) < \infty$, where $\|e^{J_0 t}\|$ is the matrix norm induced by one of the equivalent vector norms defined on R^n. We write this as $\|e^{J_0 t}\| = \mathbb{O}(e^{\beta t})$. Similarly, if $\beta = Re\,\lambda_{k+i}$, then for any $\epsilon > 0$ we have that $\|e^{J_i t}\| = \mathbb{O}(t^{n_i-1}e^{\beta t}) = \mathbb{O}(e^{(\beta+\epsilon)t})$.

From the foregoing it is now clear that $\|e^{At}\| \leq K$ for some $K > 0$ if and only if all eigenvalues of A have nonpositive real parts, and the eigenvalues with zero real part occur in the Jordan form only in J_0 and not in any of the Jordan blocks J_i, $1 \leq i \leq s$. Hence, by Theorems 5.1 and 5.2, the equilibrium $x = 0$ of (L) is under these conditions stable, in fact uniformly stable.

Now suppose that all eigenvalues of A have negative real parts. From the preceding discussion it is clear that there is a constant $K > 0$ and an $\alpha > 0$ such that $\|e^{At}\| \leq Ke^{-\alpha t}$, and therefore, $\|\phi(t, t_0, x_0)\| \leq Ke^{-\alpha(t-t_0)}\|x_0\|$ for all $t \geq t_0 > 0$ and for all $x_0 \in R^n$. It follows that the equilibrium $x = 0$ is uniformly asymptotically stable in the large, in fact exponentially stable in the large. Conversely, assume that there is an eigenvalue λ_i with nonnegative real part. Then either one term in (5.8) does not tend to zero, or else a term in (5.9) is unbounded at $t \to \infty$. In either case, $e^{At}x(0)$ will not tend to zero when the initial condition $x(0) = x_0$ is properly chosen. Hence, the equilibrium $x = 0$ of (L) cannot be asymptotically stable (and hence, it cannot be exponentially stable).

Summarizing the above, we have proved the following result.

THEOREM 5.6. The equilibrium $x = 0$ of (L) is *stable*, in fact *uniformly stable*, if and only if all eigenvalues of A have nonpositive real parts, and every eigenvalue with zero real part has an associated Jordan block of order one. The equilibrium $x = 0$ of (L) is *uniformly asymptotically stable in the large*, in fact *exponentially stable in the large*, if and only if all eigenvalues of A have negative real parts. ■

A direct consequence of the above result is that the equilibrium $x = 0$ of (L) is *unstable* if and only if at least one of the eigenvalues of A has either positive real part or has zero real part that is associated with a Jordan block of order greater than one.

At this point, it may be appropriate to take note of certain conventions concerning matrices that are used in the literature. It should be noted that some of these are not entirely consistent with the terminology used in Theorem 5.6. Specifically, a real $n \times n$ matrix A is called *stable* or a *Hurwitz matrix* if all its eigenvalues have negative real parts. If at least one of the eigenvalues has positive real part, then A is called *unstable*. A matrix A, which is neither stable nor unstable, is called *critical*, and the eigenvalues with zero real parts are called *critical eigenvalues*.

We conclude our discussion concerning the stability of (L) by noting that the results given above can also be obtained by directly using the facts established in Subsection 2.4C, concerning modes and asymptotic behavior of time-invariant systems.

EXAMPLE 5.5. We consider the system (L) with

$$A = \begin{bmatrix} 0 & 1 \\ -1 & 0 \end{bmatrix}.$$

The eigenvalues of A are $\lambda_1, \lambda_2 = \pm j$. According to Theorem 5.6, the equilibrium $x = 0$ of this system is stable. This can also be verified by computing the solution of this system for a given set of initial data $x(0)^T = (x_1(0), x_2(0))$,

$$\begin{aligned} \phi_1(t, 0, x_0) &= x_1(0)\cos t + x_2(0)\sin t \\ \phi_2(t, 0, x_0) &= -x_1(0)\sin t + x_2(0)\cos t, \end{aligned}$$

$t \geq 0$, and then applying Definition 4.1. ■

EXAMPLE 5.6. We consider the system (L) with

$$A = \begin{bmatrix} 0 & 1 \\ 0 & 0 \end{bmatrix}.$$

The eigenvalues of A are $\lambda_1 = 0, \lambda_2 = 0$. According to Theorem 5.6, the equilibrium $x = 0$ of this system is unstable. This can also be verified by computing the solution of this system for a given set of initial data $x(0)^T = (x_1(0), x_2(0))$,

$$\begin{aligned} \phi_1(t, 0, x_0) &= x_1(0) + x_2(0)t \\ \phi_2(t, 0, x_0) &= x_2(0), \end{aligned}$$

$t \geq 0$, and then applying Definition 4.5. (Note that in this example, the entire x_1-axis consists of equilibria.) ■

EXAMPLE 5.7. We consider the system (L) with

$$A = \begin{bmatrix} 2.8 & 9.6 \\ 9.6 & -2.8 \end{bmatrix}.$$

The eigenvalues of A are $\lambda_1, \lambda_2 = \pm 10$. According to Theorem 5.6, the equilibrium $x = 0$ of this system is unstable. ■

EXAMPLE 5.8. We consider the system (L) with

$$A = \begin{bmatrix} -1 & 0 \\ -1 & -2 \end{bmatrix}.$$

The eigenvalues of A are $\lambda_1, \lambda_2 = -1, -2$. According to Theorem 5.6, the equilibrium $x = 0$ of this system is exponentially stable. ■

Next, we consider linear periodic systems given by

$$\dot{x} = A(t)x, \qquad A(t) = A(t + T), \tag{P}$$

where $A(t)$ is a continuous real matrix for all $t \in (-\infty, \infty)$. We recall from Section 2.5 of Chapter 2 that if $\Phi(t, t_0)$ is the state transition matrix for (P), then there exists a constant $n \times n$ matrix R and a nonsingular $n \times n$ matrix $\Psi(t, t_0)$ such that

$$\Phi(t, t_0) = \Psi(t, t_0) \exp [R(t - t_0)], \tag{5.10}$$

where
$$\Psi(t + T, t_0) = \Psi(t, t_0)$$

for all $t \in (-\infty, \infty)$. Now according to the discussion at the end of Section 2.5 of Chapter 2, the change of variables $x = \Psi(t, t_0)y$ transforms (P) to the system

$$\dot{y} = Ry, \tag{5.11}$$

where R is given in (5.10). Since $\Psi(t, t_0)$ is nonsingular, it is clear that the equilibrium $x = 0$ of (P) is uniformly stable (resp., uniformly asymptotically stable) if and only if the equilibrium $y = 0$ of system (5.11) is uniformly stable (resp., uniformly asymptotically stable). Now by applying Theorem 5.6 to system (5.11) we obtain the following result.

THEOREM 5.7. The equilibrium $x = 0$ of (P) is *uniformly stable* if and only if all eigenvalues of the matrix R [given in (5.10)] have nonpositive real parts, and every eigenvalue with zero real part has an associated Jordan block of order one. The equilibrium $x = 0$ of (P) is *uniformly asymptotically stable in the large* if and only if all eigenvalues of R have negative real parts. ■

6.6 SOME GEOMETRIC AND ALGEBRAIC STABILITY CRITERIA

In this section we concern ourselves with *nth-order linear homogeneous ordinary differential equations* of the form

$$a_0 x^{(n)} + a_1 x^{(n-1)} + \cdots + a_{n-1} x^{(1)} + a_n x = 0, \qquad a_0 \neq 0, \tag{6.1}$$

where the coefficients $a_0, \ldots, a_n$ are all real numbers. We recall from Chapter 1 that (6.1) is equivalent to the system of first-order ordinary differential equations

$$\dot{x} = Ax, \tag{6.2}$$

where in (6.2) A denotes the *companion matrix* given by

$$A = \begin{bmatrix} 0 & 1 & 0 & \cdots & 0 \\ 0 & 0 & 1 & \cdots & 0 \\ \vdots & \vdots & \vdots & & \vdots \\ -\dfrac{a_n}{a_0} & -\dfrac{a_{n-1}}{a_0} & -\dfrac{a_{n-2}}{a_0} & \cdots & -\dfrac{a_1}{a_0} \end{bmatrix}. \tag{6.3}$$

To determine whether the equilibrium $x = 0$ of (6.3) is asymptotically stable, it suffices to determine if all the eigenvalues of A have negative real parts, or what amounts to the same thing, if the roots of the polynomial

$$f(\lambda) = a_0 \lambda^n + a_1 \lambda^{n-1} + \cdots + a_{n-1} \lambda + a_n \tag{6.4}$$

all have negative real parts. To see this, we must show that the eigenvalues of A coincide with the roots of the polynomial $f(s)$. This is most easily accomplished by induction. For the first-order case $k = 1$, we have $A = -a_n/a_0$ and therefore $det(\lambda I_1 - A) = \lambda + a_n/a_0, I_1 = 1$, and so the assertion is true for $k = 1$. Next, assume that the assertion is true for $k = n - 1$. Then

$$det(\lambda I_n - A) = \begin{bmatrix} \lambda & -1 & 0 & \cdots & 0 & 0 \\ 0 & \lambda & -1 & \cdots & 0 & 0 \\ \vdots & \vdots & \vdots & & \vdots & \vdots \\ 0 & 0 & 0 & \cdots & \lambda & -1 \\ \frac{a_n}{a_0} & \frac{a_{n-1}}{a_0} & \frac{a_{n-2}}{a_0} & \cdots & \frac{a_2}{a_0} & \lambda + \frac{a_1}{a_0} \end{bmatrix}$$

$$= \lambda det(\lambda I_{n-1} - A_1) + (-1)^{n+1}\left(\frac{a_n}{a_0}\right)\begin{vmatrix} -1 & 0 & 0 & \cdots & 0 & 0 \\ \lambda & -1 & 0 & \cdots & 0 & 0 \\ 0 & \lambda & -1 & \cdots & 0 & 0 \\ \vdots & \vdots & \vdots & & \vdots & \vdots \\ 0 & 0 & 0 & \cdots & \lambda & -1 \end{vmatrix},$$

where
$$A_1 = \begin{vmatrix} 0 & 1 & 0 & \cdots & 0 \\ 0 & 0 & 1 & \cdots & 0 \\ \vdots & \vdots & \vdots & & \vdots \\ -\frac{a_{n-1}}{a_0} & -\frac{a_{n-2}}{a_0} & -\frac{a_{n-3}}{a_0} & \cdots & -\frac{a_1}{a_0} \end{vmatrix}$$

and I_n, I_{n-1} denote the $n \times n$ and $(n-1) \times (n-1)$ identity matrices. Therefore,

$$\begin{aligned} det(\lambda I_n - A) &= \lambda\, det(\lambda I_{n-1} - A_1) + \frac{a_n}{a_0} \\ &= \lambda^n + \frac{a_1}{a_0}\lambda^{n-1} + \cdots + \frac{a_{n-1}}{a_0}\lambda + \frac{a_n}{a_0} \\ &= 0 \end{aligned}$$

is equivalent to $f(\lambda) = 0$.

Analogously to matrices, we now make the following definitions. An nth-order polynomial $f(\lambda)$ with real coefficients [such as (6.4)] is called *stable* if all zeros of $f(\lambda)$ have negative real parts, it is called *unstable* if at least one of the zeros of $f(\lambda)$ has a positive real part, and it is called *critical* if $f(\lambda)$ is neither stable nor unstable. Also, a stable polynomial is called a *Hurwitz polynomial*.

In view of the above, the stability problem for nth-order differential equations with constant coefficients has now been reduced to a purely algebraic problem of determining whether the zeros of a polynomial [such as (6.4)] all have negative (resp., nonpositive) real parts. In case zeros with vanishing real parts exist, it is further necessary to determine their multiplicity.

In the following, we first present some *graphical criteria* that enable us to determine the stability of a polynomial (6.4), without determining its roots explicitly. These results, which are important in their own right, are then used to arrive at some *algebraic criteria* to determine the stability of a polynomial (6.4).

A. Some Graphical Criteria

In establishing our first result we assume that the polynomial $f(s)$ [given by (6.4)] has p zeros in the right half of the s-plane and $(n - p)$ zeros in the left half of the s-plane, and we assume that there shall be no zeros on the imaginary axis. We let C denote the counterclockwise contour formed by a semicircle $\tilde{C}$ with radius r and centered at the origin, together with its diameter on the imaginary axis, and we choose r so large that the p zeros of $f(s)$ in the right half of the s-plane lie in the interior of the circle. We now recall from a well-known result from the elementary theory of functions of a complex variable (called Cauchy's integration formula) that

$$p = \frac{1}{2\pi j}\int_C \frac{f'(s)}{f(s)}\,ds = \frac{1}{2\pi j}\delta_C \ln f(s), \tag{6.5}$$

where s is a complex variable, $j = \sqrt{-1}$, $f'(s)$ denotes the derivative of f with respect to s, $\int_C [f'(s)/f(s)]\,ds$ denotes the integral of $f'(s)/f(s)$ along the contour C, and $\delta_C \ln f(s)$ represents the increment of $\ln f(s)$ along the contour C. Let $\delta_{C'} \ln f(s)$ denote the increment of $\ln f(s)$ along the semicircular arc C' with $s = re^{j\phi}$. For s on C' we have

$$f(s) = a_0 r^n e^{jn\phi}(1 + \mathbb{O}(r^{-1}))$$

and
$$\ln f(s) = \ln a_0 + n \ln r + nj\phi + \mathbb{O}(r^{-1}).$$

Hence,

$$\begin{aligned}\delta_{C'} \ln f(s) &= nj\left(\frac{\pi}{2} + \frac{\pi}{2}\right) + \mathbb{O}(r^{-1})\\ &= n\pi j + \mathbb{O}(r^{-1}).\end{aligned}$$

Letting $r \to \infty$ and using (6.5), we conclude that

$$\begin{aligned}p &= \frac{1}{2\pi j} n\pi j + \left(\frac{1}{2\pi j}\right)\delta_I \ln f(s)\\ &= \frac{n}{2} + \frac{1}{2\pi j}\delta_I \ln f(s),\end{aligned} \tag{6.6}$$

where $\delta_I \ln f(s)$ denotes the increment of the logarithm of $f(s)$ along the imaginary axis I from $-\infty$ to $+\infty$. To determine this increment, we let $s = j\omega$ and let

$$f(s) = f(j\omega) \triangleq R(\omega)e^{j\theta(\omega)} \triangleq U(\omega) + jV(\omega), \tag{6.7}$$

and we consider R, θ as polar coordinates and U, V as coordinates in the complex plane. As the real parameter ω (the real frequency ω) ranges from $+\infty$ to $-\infty$, the points $f(j\omega)$ in (6.7) describe the *frequency response* or *frequency plot* for $f(s)$. Since $f(s)$ has real coefficents, we must have $R(\omega) = R(-\omega)$ and $\theta(\omega) = -\theta(-\omega)$. It therefore suffices to consider the part of the response curve belonging to the positive values of the parameter ω. It follows from (6.6) that

$$p = \frac{n}{2} - \frac{\theta(\infty)}{\pi} = \frac{1}{2}\left[n - \frac{\theta(\infty)}{\pi/2}\right], \tag{6.8}$$

where $\theta(\infty)$ is the limit to which the polar angle $\theta(\omega) = \tan^{-1}[V(\omega)/U(\omega)]$ of the frequency response diagram tends as ω becomes unbounded. Since $U(\omega)$ and $V(\omega)$ are polynomials of different degrees, $|V(\omega)/U(\omega)|$ will tend to either zero or infinity. In either case, in view of (6.8), $\theta(\infty)$ must be an integral multiple of $\pi/2$. Now when in particular $p = 0$, then by necessity we have that $\theta(\infty) = n(\pi/2)$. This yields the following result.

THEOREM 6.1. (LEONHARD-MIKHAILOV STABILITY CRITERION) The polynomial $f(s)$ has only zeros with negative real parts if and only if its frequency response diagram $f(j\omega)$, $0 < \omega < \infty$, passes through exactly n quadrants in the positive sense. ■

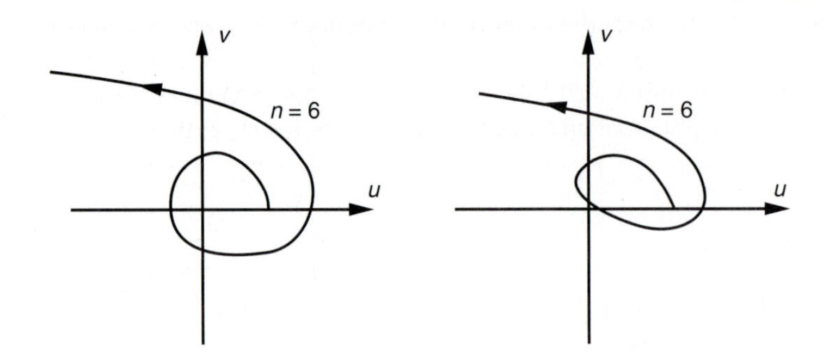

(a) **Frequency response plot for $f(s)$ stable ($n = 6$)**

(b) **Frequency response plot for $f(s)$ unstable ($n = 6$)**

FIGURE 6.4
Frequency response plot for $f(s)$ stable ($n = 6$)

In Fig. 6.4 we depict the frequency response plots of a stable and an unstable polynomial.

Next, since $f(s)$ has real coefficients, there are real polynomials f_1 and f_2 such that

$$f(j\omega) \triangleq f_1(\omega^2) + j\omega f_2(\omega^2). \tag{6.9}$$

If $n = 2k$, then $deg\, f_1(u) = k$ and $deg\, f_2(u) = k - 1$, and if $n = 2k + 1$, then $deg\, f_1(u) = k$ and $deg\, f_2(u) = k$. To the zeros u_{ik}, $i = 1, 2$ and $k = 1, 2, 3, \ldots$ of the polynomials $f_1(u)$, $f_2(u)$, respectively, correspond those values of ω^2 at which the *frequency response* diagram intersects the axes. In the case of stability, these values of ω^2 must be real and increasing, i.e., the zeros u_{ik} must be positive and alternate,

$$0 < u_{11} < u_{21} < u_{12} < u_{22} < \cdots \tag{6.10}$$

since otherwise the response curve $f(j\omega)$ will not make the proper number of turns at the appropriate locations. These considerations lead to the *interlacing* of the roots u_{1k} with the roots u_{2k}, $k = 1, 2, 3, \ldots$, which is sometimes called the *gap and position criterion*. In Fig. 6.5 we depict typical situations for stability and instability (in terms of the variables U and V).

We summarize the above in the following result.

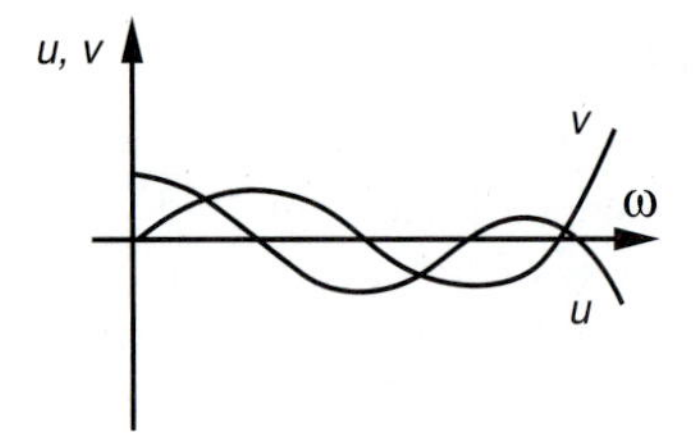

(a) Stable case

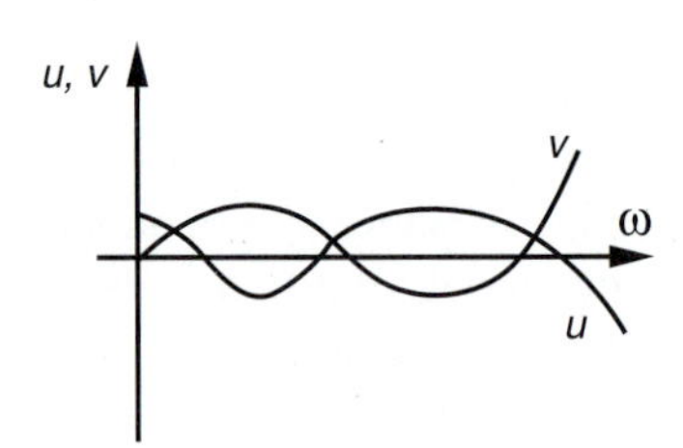

(a) Unstable case

FIGURE 6.5

THEOREM 6.2. (GAP AND POSITION STABILITY CRITERION) The polynomial $f(s)$ has only zeros with negative real parts if and only if the zeros of the polynomials defined in (6.9) are real and satisfy the inequalities (6.10). ■

The frequency response plot $f(j\omega)$ is unbounded and hence can never be displayed completely. We therefore frequently make use of the *reciprocal frequency response diagram*, i.e., the response diagram of the function $1/f(j\omega)$. Such a plot approaches zero asymptotically, and in case of stability, it rotates exactly through n quadrants in the *negative sense*. In applications of Theorem 6.1, the entire plot is not needed. It can be shown that it suffices to consider the interval $0 \leq \omega \leq \omega_0$, where

$$\omega_0 = 1 + 3 \max_i \left\{ \frac{|a_i|}{|a_0|} \right\}. \tag{6.11}$$

We will not pursue the details concerning the proof of this assertion.

B. Some Algebraic Criteria

In the next results, we develop the Routh-Hurwitz criterion, which yields necessary and sufficient conditions for $f(s)$ to be a Hurwitz polynomial. To accomplish this, we will make use of Theorem 6.2. We begin by establishing a set of necessary conditions.

THEOREM 6.3. For

$$f(s) = a_0 s^n + a_1 s^{n-1} + \cdots + a_{n-1}s + a_n \tag{6.12}$$

to be a Hurwitz polynomial it is necessary that the inequalities

$$\frac{a_1}{a_0} > 0, \frac{a_2}{a_0} > 0, \ldots, \frac{a_n}{a_0} > 0 \tag{6.13}$$

hold.

Proof. Let $s_1, \ldots, s_n$ be the zeros of (6.12), and in particular, let s'_j be the real roots and s''_k the complex roots. Then

$$\begin{aligned} f(s) &= a_0 \prod_j (s - s'_j) \prod_k (s - s''_k) \\ &= a_0 \prod_j (s - s'_j) \prod_k (s^2 - (2Re\, s''_k)s + |s''_k|^2). \end{aligned}$$

If all the numbers s'_j and $Re\, s''_k$ are negative, then we can obtain only positive coefficients for the powers of s when we multiply the product out to obtain $f(s)$. ■

Without loss of generality, we assume in the following that $a_0 > 0$. In the next result, we will require the *Routh array*:

$$\begin{array}{lllll}
 & c_{10} = a_0, & c_{20} = a_2 & c_{30} = a_4, & c_{40} = a_6, \ldots \\
 & c_{11} = a_1, & c_{21} = a_3, & c_{31} = a_5, & c_{41} = a_7, \ldots \\
r_2 = \dfrac{a_0}{a_1}, & c_{12} = a_2 - r_2 a_3, & c_{22} = a_4 - r_2 a_5, & a_{32} = a_6 - r_2 a_7, \ldots & \\
r_3 = \dfrac{c_{11}}{c_{12}}, & c_{13} = c_{21} - r_3 c_{22}, & c_{23} = c_{31} - r_3 c_{32}, & c_{33} = c_{41} - r_3 c_{42}, \ldots & \\
\cdots\cdots & \cdots\cdots\cdots & \cdots\cdots\cdots & \cdots\cdots\cdots & \cdots\cdots \\
r_j = \dfrac{c_{1,j-2}}{c_{1,j-1}}, & & c_{ij} = c_{i+1,j-2} - r_j c_{i+1,j-1}, & i = 1, 2, \ldots, & j = 2, 3, \ldots \\
 & & \cdots\cdots\cdots & & \\
 & & c_{1n} = a_n & &
\end{array}$$

Note that if $n = 2m$, we have

$$c_{m+1,0} = c_{m+1,2} = a_n, \qquad c_{m+1,1} = c_{m_1,3} = 0,$$

and if $n = 2m - 1$, we have

$$c_{m0} = a_{n-1}, \qquad c_{m1} = a_n, \qquad c_{m2} = c_{m3} = 0.$$

The above array terminates after $(n-1)$ steps in case all the numbers c_{ij} are different from zero. The last line defines c_{1n}.

In addition to the inequalities (6.13), we shall require the inequalities given by

$$c_{11} > 0, c_{12} > 0, \ldots, c_{1n} > 0. \tag{6.14}$$

THEOREM 6.4. (ROUTH-HURWITZ STABILITY CRITERION) The polynomial $f(s)$ given in (6.12) is a Hurwitz polynomial if and only if the inequalities (6.13) and (6.14) hold.

Proof. First we assume that the degree of $f(s)$ is even by letting $n = 2m$. We define the polynomials

$$h_1(s) = \tfrac{1}{2}[f(s) + f(-s)], \qquad h_2(s) = \tfrac{1}{2}[f(s) - f(-s)]. \tag{6.15}$$

Applying the Euclidean algorithm to determine the greatest common divisor of $h_1(s)$ and $h_2(s)$, we obtain

$$\begin{aligned} h_1(s) &= r_2' s h_2(s) - h_3(s) \\ h_2(s) &= r_3' s h_3(s) - h_4(s) \\ &\ldots\ldots\ldots\ldots\ldots\ldots, \end{aligned} \tag{6.16}$$

where the linear factors arising in the division have no constant term and the remainders have been written with negative signs. It is readily verified that the constants r_i' in (6.16) are related to the constants r_i in the Routh array by the expression $r_i' = (-1)^i r_i$.

Next, we define a sequence of polynomials given by

$$h_{2i-1}(s) \triangleq g_{2i-1}(s^2), \qquad h_{2i}(s) \triangleq s g_{2i}(s^2), \qquad i = 1, \ldots, m. \tag{6.17}$$

From (6.16) and (6.17) we obtain the recursion formulas given by

$$\begin{aligned} g_{2i+1}(z) &= r_{2i}' z g_{2i}(z) - g_{2i-1}(z) \\ g_{2i+2}(z) &= r_{2i+1}' g_{2i+1}(z) - g_{2i}(z). \end{aligned} \tag{6.18}$$

The first two members of this sequence are given by

$$\begin{aligned} g_1(z) &= a_0 z^m + a_2 z^{m-1} + \cdots + a_{2m} \\ g_2(z) &= a_1 z^{m-1} + a_3 z^{m-2} + \cdots + a_{2m-1}. \end{aligned} \tag{6.19}$$

We can readily verify that the above two polynomials agree with the polynomials f_1 and f_2 given in (6.9), except for sign. In fact, we have

$$f_i(u) = g_i(-u) \qquad i = 1, 2. \tag{6.20}$$

We are now in a position to construct the Routh array of a polynomial $f(s)$ by utilizing the coefficients of the sequence of polynomials $g_i(z)$. If in the process of doing so we encounter a zero row (i.e., an identically vanishing polynomial, say, g_i), then h_1 and h_2 [and thus, $f(s)$ and $f(-s)$] have a common divisor. In this case $f(s)$ possesses a divisor of the form $s^2 + \alpha$ and is not a Hurwitz polynomial.

Next, we assume that the hypotheses of this theorem are satisfied [i.e., (6.13) and (6.14) hold]. Applying definitions, we can readily verify that the numbers r_i' have alternating signs, that the signs of the leading coefficients of the polynomials g_i, $i = 1, 2, \ldots$ are given by

$$+, +, -, -, +, +, -, -, \cdots, \tag{6.21}$$

and that the degrees of these polynomials are given by

$$m, m-1, m-1, m-2, m-2, \ldots, 1, 1, 0. \tag{6.22}$$

Next, for fixed z, $-\infty < z < \infty$, we consider the sequence of numbers given by

$$g_1(z), g_2(z), \ldots, g_{2m}(z). \tag{6.23}$$

Note that the last term $g_{2m}(z)$ is a constant for all z. Let $W(z)$ denote the number of sign changes in this sequence. When $z > 0$ and is very large, then the signs of the g_i in (6.23) correspond to the signs of the leading coefficients of these polynomials. When $-z > 0$ and is very large, then the signs of g_i in (6.23) will alternate. It now follows that the difference $W(-\infty) - W(+\infty)$ is always equal to m. Thus, as z varies from $-\infty$ to $+\infty$, m sign changes in the sequence (6.23) will disappear. Such a disappearance can occur only at a zero of $g_1(z)$, since if z passes through a zero, say, z', of $g_i(z)$, $1 < i < 2m$, no disappearance in the number of sign changes occurs, because by (6.18), $sgn\, g_{i-1}(z') \neq sgn\, g_{i+1}$. We conclude that $g_1(z)$ has exactly m real zeros, and since by hypothesis all its coefficients are positive, no positive zeros can occur.

A similar argument as above shows that g_2 has exactly $(m - 1)$ negative zeros. Now, since in the sequence (6.23) the largest possible number of disappearances of sign changes occurs (as z is varied from $-\infty$ to $+\infty$), a disappearance in sign change must actually occur each time z passes through a zero of g_1. This, however, is possible only if g_2 in turn changes sign between every two zeros of g_1; otherwise there would be an additional change of signs in the sequence (6.23). It follows that the zeros of g_2 separate the zeros of g_1 (the zeros of g_2 are interlaced with the zeros of g_1). Now, in view of (6.20), the above statement concerning the zeros of g_1 and g_2 is equivalent to the inequality (6.10); thus, Theorem 6.2 applies. Therefore, condition (6.14) is sufficient for $f(s)$ to be a Hurwitz polynomial. It is also a necessary condition, since otherwise the count of the sign changes in the sequence (6.23) is too small and the hypotheses of Theorem 6.2 are not satisfied, i.e., either g_1 has too few zeros or the polynomial does not satisfy condition (6.10).

To complete the proof, we assume next that $n = 2m + 1$, i.e., n is odd. In this case we interchange the definitions of $h_1(s)$ and $h_2(s)$ given in (6.16), and we let

$$h_{2i+1}(s) \triangleq s g_{2i+1}(s^2), \qquad h_{2i}(s) \triangleq g_{2i}(s^2).$$

The degrees of the polynomials g_i formed in this manner are $m, m, m-1, m-1, \ldots, 1, 1, 0$. Following a similar procedure as before, we show that the polynomials $g_1(z)$ and $g_2(z)$ each have m negative zeros and that Theorem 6.2 applies. This concludes the proof. ■

EXAMPLE 6.1. We apply the Routh-Hurwitz criterion (Theorem 6.4) to the polynomial

$$\begin{aligned} f(s) &= (s+2)(s+1-j)(s+1+j)(s+1) \\ &= s^4 + 5s^3 + 10s^2 + 10s + 4. \end{aligned} \tag{6.24}$$

For this polynomial we form the Routh array and obtain

s^4	1	10	4
s^3	5	10	0
s^2	$\frac{1}{5}(5 \cdot 10 - 1 \cdot 10) = 8$	$\frac{1}{5}(5 \cdot 4 - 0) = 4$	0.
s^1	$\frac{1}{8}(8 \cdot 10 - 5 \cdot 4) = 7.5$	$\frac{1}{8}(8 \cdot 0 - 5 \cdot 0) = 0$	0
s^0	$\frac{1}{7.5}[(7.5) \cdot 4 - 8 \cdot 0] = 4$	0	0

The conditions of Theorem 6.4 are clearly satisfied. Hence, (6.24) is a Hurwitz polynomial. ■

EXAMPLE 6.2. We apply the Routh-Hurwitz criterion (Theorem 6.4) to the polynomial

$$\begin{aligned} f(s) &= (s+2)(s+1-j)(s+1+j)(s-1) \\ &= s^4 + 3s^3 + 2s^2 - 2s - 4. \end{aligned} \tag{6.25}$$

We note that condition (6.13) is violated, and therefore, the polynomial (6.25) is not a Hurwitz polynomial. Condition (6.14) is also not satisfied. To see this, we form the Routh array for (6.25), given as

$$\begin{array}{llll} s^4 & 1 & 2 & -4 \\ s^3 & 3 & -2 & 0 \\ s^2 & \frac{1}{3}(3\cdot 2+2) = \frac{8}{3} & \frac{1}{3}[3\cdot(-4)-0] = -4 & 0. \\ s^1 & \frac{3}{8}[\frac{8}{3}\cdot(-2)-(-4)\cdot 3] = \frac{5}{2} & 0 & 0 \\ s^0 & \frac{2}{5}[\frac{5}{2}\cdot(-4)-0] = -4 & & \end{array}$$

■

6.7 THE MATRIX LYAPUNOV EQUATION

In Section 6.6 we established a variety of stability results that require explicit knowledge of the solutions of (L) or (LH). We also derived some geometric and algebraic stability criteria for (L) when the matrix A is in *companion form* that do not require explicit knowledge of solutions, but instead, are deduced *directly* from the parameters of A.

In this section we will develop stability criteria for (L) with *arbitrary* matrix A. In doing so, we will employ *Lyapunov's Second Method* (also called *Lyapunov's Direct Method*) for the case of linear systems (L). This method utilizes auxiliary real-valued functions $v(x)$, called *Lyapunov functions*, that may be viewed as "*generalized energy functions*" or "*generalized distance functions*" (from the equilibrium $x = 0$), and the stability properties are then deduced directly from the properties of $v(x)$ and its time derivative $\dot{v}(x)$, evaluated along the solutions of (L).

A logical choice of Lyapunov function is $v(x) = x^T x = \|x\|^2$, which represents the square of the Euclidean distance of the state from the equilibrium $x = 0$ of (L). The stability properties of the equilibrium are then determined by examining the properties of $\dot{v}(x)$, the time derivative of $v(x)$ along the solutions of (L),

$$\dot{x} = Ax. \tag{L}$$

This derivative can be determined without explicitly solving for the solutions of (L) by noting that

$$\begin{aligned} \dot{v}(x) &= \dot{x}^T x + x^T \dot{x} = (Ax)^T x + x^T (Ax) \\ &= x^T (A^T + A)x. \end{aligned}$$

If the matrix A is such that $\dot{v}(x)$ is negative for all $x \neq 0$, then it is reasonable to expect that the distance of the state of (L) from $x = 0$ will decrease with increasing time, and that the state will therefore tend to the equilibrium $x = 0$ of (L) with increasing time t.

It turns out that the Lyapunov function used in the above discussion is not sufficiently flexible. In the following we will employ as a "generalized distance function" the quadratic form given by

$$v(x) = x^T P x, \qquad P = P^T, \tag{7.1}$$

where P is a real $n \times n$ matrix. The time derivative of $v(x)$ along the solutions of (L) is determined as

$$\begin{aligned}\dot{v}(x) &= \dot{x}^T P x + x^T P \dot{x} = x^T A^T P x + x^T P A x \\ &= x^T (A^T P + PA) x,\end{aligned}$$

i.e.,

$$\dot{v} = x^T C x, \tag{7.2}$$

where

$$C = A^T P + PA. \tag{7.3}$$

Note that C is real and $C^T = C$. The system of equations given in (7.3) is called the *Lyapunov Matrix Equation.*

We recall from Section 6.2 that since P is real and symmetric, all its eigenvalues are real. Also, we recall that P is said to be *positive definite* (resp., *positive semidefinite*) if all its eigenvalues are positive (resp., nonnegative), and it is called *indefinite* if P has eigenvalues of opposite sign. The definitions of *negative definite* and *negative semidefinite* (for P) are similarly defined. Furthermore, we recall that the *function* $v(x)$ given in (7.1) is said to be *positive definite, positive semidefinite, indefinite*, and so forth, if P has the corresponding definiteness properties (refer to Section 6.2).

Instead of solving for the eigenvalues of a real symmetric matrix to determine its definiteness properties, there are more efficient and direct methods of accomplishing this. We now digress to discuss some of these.

Let $G = [g_{ij}]$ be a real $n \times n$ matrix (not necessarily symmetric). Referring to Subsection 2.2G, we recall that the *minors* of G are the matrix itself and the matrix obtained by removing successively a row and a column. The *principal minors* of G are G itself and the matrices obtained by successively removing an ith row and an ith column, and the *leading principal minors* of G are G itself and the minors obtained by successively removing the last row and the last column. For example, if $G = [g_{ij}] \in R^{3\times 3}$, then the principal minors are

$$\begin{bmatrix} g_{11} & g_{12} & g_{13} \\ g_{21} & g_{22} & g_{23} \\ g_{31} & g_{32} & g_{33} \end{bmatrix}, \qquad \begin{bmatrix} g_{11} & g_{12} \\ g_{21} & g_{22} \end{bmatrix}, \qquad [g_{11}],$$

$$\begin{bmatrix} g_{11} & g_{13} \\ g_{31} & g_{33} \end{bmatrix}, \qquad \begin{bmatrix} g_{22} & g_{23} \\ g_{32} & g_{33} \end{bmatrix}, \qquad [g_{22}], \qquad [g_{33}].$$

The first three matrices above are the leading principal minors of G. On the other hand, the matrix

$$\begin{bmatrix} g_{21} & g_{22} \\ g_{31} & g_{32} \end{bmatrix}$$

is a minor but not a principal minor.

The following results, due to Sylvester, allow efficient determination of the definiteness properties of a *real, symmetric* matrix.

PROPOSITION 7.1. (i) A real symmetric matrix $P = [p_{ij}] \in R^{n\times n}$ is *positive definite* if and only if the determinants of its *leading principal minors* are positive, i.e., if and only if

$$p_{11} > 0, \qquad det\begin{bmatrix} p_{11} & p_{12} \\ p_{12} & p_{22} \end{bmatrix} > 0, \ldots, det\, P > 0.$$

(ii) A real symmetric matrix P is *positive semidefinite* if and only if the determinants of *all its principal minors* are nonnegative. ■

Still digressing, we consider next the quadratic form

$$v(w) = w^T G w, \qquad G = G^T,$$

where $G \in R^{n\times n}$. Referring to Subsection 6.2C [in particular, Eqs. (2.35) and (2.36)], there exists an orthogonal matrix Q such that the matrix P defined by

$$P = Q^{-1} G Q = Q^T G Q$$

is diagonal. Therefore, if we let $w = Qx$, then

$$v(Qx) \triangleq v(x) = x^T Q^T G Q x = x^T P x,$$

where P is in the form given in Eq. (2.35), i.e.,

$$P = \begin{bmatrix} \lambda_1 & & & & & & & & \\ & \ddots & & & & & & 0 & \\ & & \lambda_1 & & & & & & \\ & & & \lambda_2 & & & & & \\ & & & & \ddots & & & & \\ & & & & & \lambda_2 & & & \\ & & & & & & \ddots & & \\ & 0 & & & & & & \lambda_p & \\ & & & & & & & & \ddots \\ & & & & & & & & & \lambda_p \end{bmatrix}.$$

From this, we immediately obtain the following useful result.

PROPOSITION 7.2. Let $P = P^T \in R^{n\times n}$, let $\lambda_M(P)$ and $\lambda_m(P)$ denote the largest and smallest eigenvalues of P, respectively, and let $\|\cdot\|$ denote the Euclidean norm. Then

$$\lambda_m(P)\|x\|^2 \le v(x) = x^T P x \le \lambda_M(P)\|x\|^2 \tag{7.4}$$

for all $x \in R^n$. ■

Let $c_1 \triangleq \lambda_m(P)$ and $c_2 = \lambda_M(P)$. Clearly, $v(x)$ is positive definite if and only if $c_2 \ge c_1 > 0$, $v(x)$ is positive semidefinite if and only if $c_2 \ge c_1 \ge 0$, $v(x)$ is indefinite if and only if $c_2 > 0$, $c_1 < 0$, and so forth.

We are now in a position to prove several results.

THEOREM 7.1. The equilibrium $x = 0$ of (L) is *uniformly stable* if there exists a real, symmetric, and positive definite $n \times n$ matrix P such that the matrix C given in (7.3) is negative semidefinite.

Proof. Along any solution $\phi(t, t_0, x_0) \triangleq \phi(t)$ of (L) with $\phi(t_0, t_0, x_0) = \phi(t_0) = x_0$, we have

$$\phi(t)^T P \phi(t) = x_0^T P x_0 + \int_{t_0}^{t} \frac{d}{d\eta}\phi(\eta)^T P \phi(\eta)\, d\eta = x_0^T P x_0 + \int_{t_0}^{t} \phi(\eta)^T C \phi(\eta)\, d\eta$$

for all $t \geq t_0 \geq 0$. Since P is positive definite and C is negative semidefinite, we have

$$\phi(t)^T P\phi(t) - x_0^T P x_0 \leq 0$$

for all $t \geq t_0 \geq 0$, and there exist $c_2 \geq c_1 > 0$ such that

$$c_1\|\phi(t)\|^2 \leq \phi(t)^T P\phi(t) \leq x_0^T P x_0 \leq c_2\|x_0\|^2$$

for all $t \geq t_0$. It follows that

$$\|\phi(t)\| \leq \left(\frac{c_2}{c_1}\right)^{1/2} \|x_0\|$$

for all $t \geq t_0 \geq 0$ and for any $x_0 \in R^n$. Therefore, the equilibrium $x = 0$ of (L) is uniformly stable (refer to Sections 6.4 and 6.5). ■

EXAMPLE 7.1. For the system given in Example 5.5 we choose $P = I$, and we compute

$$C = A^T P + PA = A^T + A = 0.$$

According to Theorem 7.1, the equilibrium $x = 0$ of this system is stable (as expected from Example 5.5). ■

THEOREM 7.2. The equilibrium $x = 0$ of (L) is *exponentially stable in the large* if there exists a real, symmetric, and positive definite $n \times n$ matrix P such that the matrix C given in (7.3) is negative definite.

Proof. We let $\phi(t, t_0, x_0) \triangleq \phi(t)$ denote an arbitrary solution of (L) with $\phi(t_0) = x_0$. In view of the hypotheses of the theorem, there exist constants $c_2 \geq c_1 > 0$ and $c_3 \geq c_4 > 0$ such that

$$c_1\|\phi(t)\|^2 \leq v(\phi(t)) = \phi(t)^T P\phi(t) \leq c_2\|\phi(t)\|^2$$

and

$$-c_3\|\phi(t)\|^2 \leq \dot{v}(\phi(t)) = \phi(t)^T C\phi(t) \leq -c_4\|\phi(t)\|^2$$

for all $t \geq t_0 \geq 0$ and for any $x_0 \in R^n$. Then

$$\dot{v}(\phi(t)) = \frac{d}{dt}[\phi(t)^T P\phi(t)] \leq \left(-\frac{c_4}{c_2}\right)\phi(t)^T P\phi(t) = \left(-\frac{c_4}{c_2}\right)v(\phi(t))$$

for all $t \geq t_0 \geq 0$. This implies, after multiplication by the appropriate integrating factor, and integrating from t_0 to t, that

$$v(\phi(t)) = \phi(t)^T P\phi(t) \leq x_0^T P x_0 e^{-(c_4/c_2)(t-t_0)}$$

or

$$c_1\|\phi(t)\|^2 \leq \phi(t)^T P\phi(t) \leq c_2\|x_0\|^2 e^{-(c_4/c_2)(t-t_0)}$$

or

$$\|\phi(t)\| \leq \left(\frac{c_2}{c_1}\right)^{1/2} \|x_0\| e^{-(1/2)(c_4/c_2)(t-t_0)}, \qquad t \geq t_0 \geq 0.$$

This inequality holds for all $x_0 \in R^n$ and for any $t_0 \geq 0$. Therefore, the equilibrium $x = 0$ of (L) is exponentially stable in the large (refer to Sections 6.4 and 6.5). ■

In Fig. 6.6 we provide an interpretation of Theorem 7.2 for the two-dimensional case ($n = 2$). The curves C_i, called *level curves*, depict loci where $v(x)$ is constant, i.e., $C_i = \{x \in R^2 : v(x) = x^T P x = c_i\}$, $i = 0, 1, 2, 3, \ldots$. When the hypotheses of Theorem 7.2 are satisfied, trajectories determined by (L) penetrate level curves corresponding to decreasing values of c_i as t increases, tending to the origin as t becomes arbitrarily large.

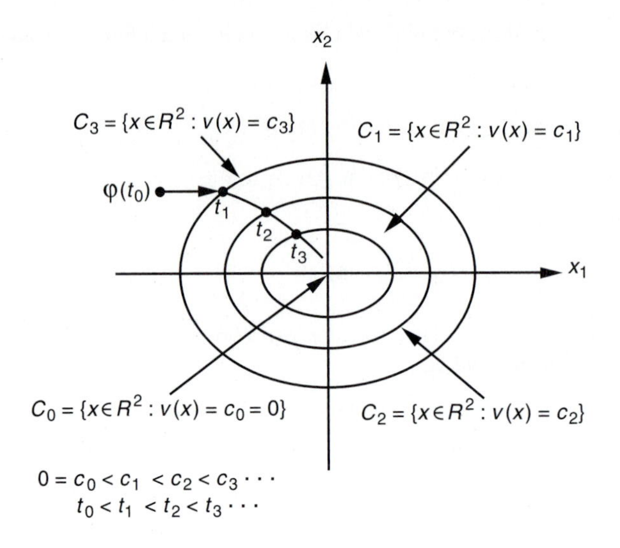

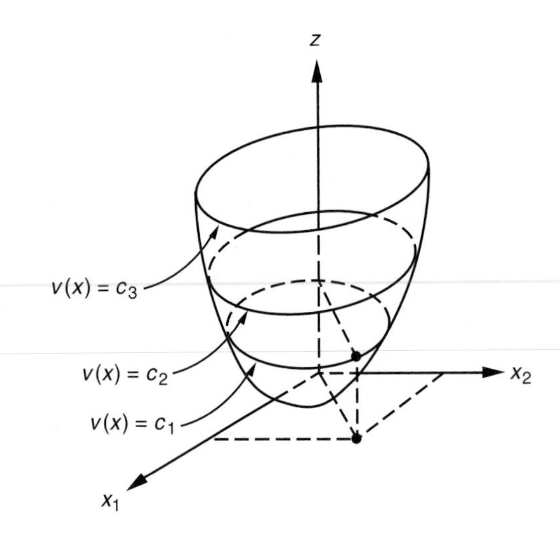

FIGURE 6.6
Asymptotic stability

EXAMPLE 7.2. For the system given in Example 5.8, we choose

$$P = \begin{bmatrix} 1 & 0 \\ 0 & 0.5 \end{bmatrix}$$

and we compute the matrix

$$C = A^T P + PA = \begin{bmatrix} -2 & 0 \\ 0 & -2 \end{bmatrix}.$$

According to Theorem 7.2, the equilibrium $x = 0$ of this system is exponentially stable in the large (as expected from Example 5.8). ■

THEOREM 7.3. The equilibrium $x = 0$ of (L) is *unstable* if there exists a real, symmetric $n \times n$ matrix P that is either negative definite or indefinite such that the matrix C given in (7.3) is negative definite.

Proof. We first assume that P is indefinite. Then P possesses eigenvalues of either sign, and every neighborhood of the origin contains points where the function

$$v(x) = x^T P x$$

is positive and negative. Consider the neighborhood

$$B(\epsilon) = \{x \in R^n : \|x\| < \epsilon\},$$

where $\|\cdot\|$ denotes the Euclidean norm, and let

$$G = \{x \in B(\epsilon) : v(x) < 0\}.$$

On the boundary of G we have either $\|x\| = \epsilon$ or $v(x) = 0$. In particular, note that the origin $x = 0$ is on the boundary of G. Now, since the matrix C is negative definite, there exist constants $c_3 > c_4 > 0$ such that

$$-c_3\|x\|^2 \leq x^T C x = \dot{v}(x) \leq -c_4\|x\|^2$$

for all $x \in R^n$. Let $\phi(t, t_0, x_0) \triangleq \phi(t)$ and let $x_0 = \phi(t_0) \in G$. Then $v(x_0) = -a < 0$. The solution $\phi(t)$ starting at x_0 must leave the set G. To see this, note that as long as $\phi(t) \in G$, $v(\phi(t)) \leq -a$ since $\dot{v}(x) < 0$ in G. Let $-c = \sup\ \{\dot{v}(x) : x \in G$ and $v(x) \leq -a\}$.

Then $c > 0$ and

$$\begin{aligned} v(\phi(t)) &= v(x_0) + \int_{t_0}^{t} \dot{v}(\phi(s))\, ds \leq -a - \int_{t_0}^{t} c\, ds \\ &= -a - (t - t_0)c, t \geq t_0. \end{aligned}$$

This inequality shows that $\phi(t)$ must escape the set G (in finite time) because $v(x)$ is bounded from below on G. But $\phi(t)$ cannot leave G through the surface determined by $v(x) = 0$ since $v(\phi(t)) \leq -a$. Hence, it must leave G through the sphere determined by $\|x\| = \epsilon$. Since the above argument holds for arbitrarily small $\epsilon > 0$, it follows that the origin $x = 0$ of (L) is unstable.

Next, we assume that P is negative definite. Then G as defined is all of $B(\epsilon)$. The proof proceeds as above. ■

The proof of Theorem 7.3 shows that for $\epsilon > 0$ sufficiently small when P is negative definite, *all* solutions $\phi(t)$ of (L) with initial conditions $x_0 \in B(\epsilon)$ will tend away from the origin. This constitutes a severe case of instability, called *complete instability*.

EXAMPLE 7.3. For the system given in Example 5.7, we choose

$$P = \begin{bmatrix} -0.28 & -0.96 \\ -0.96 & 0.28 \end{bmatrix},$$

and we compute the matrix

$$C = A^T P + PA = \begin{bmatrix} -20 & 0 \\ 0 & -20 \end{bmatrix}.$$

The eigenvalues of P are ± 1. According to Theorem 7.3, the equilibrium $x = 0$ of this system is unstable (as expected from Example 5.7). ■

In applying the results derived thus far in this section, we start by choosing (guessing) a matrix P having certain desired properties. Next, we solve for the matrix C, using Eq. (7.3). If C possesses certain desired properties (i.e., it is negative definite), we draw appropriate conclusions by applying one of the preceding theorems of this section; if not, we need to choose another matrix P. This points to the principal shortcoming of Lyapunov's Direct Method, when applied to general systems. However, in the *special case* of linear systems described by (L), it is possible to *construct* Lyapunov functions of the form $v(x) = x^T Px$ in a *systematic* manner. In doing so, one first chooses the matrix C in (7.3) (having desired properties), and then one solves (7.3) for P. Conclusions are then drawn by applying the appropriate results of this section. In applying this construction procedure, we need to know conditions under which (7.3) possesses a (unique) solution P for a given C. We will address this topic next.

We consider the quadratic form

$$v(x) = x^T Px, \qquad P = P^T, \tag{7.5}$$

and the time derivative of $v(x)$ along the solutions of (L), given by

$$\dot{v}(x) = x^T Cx, \qquad C = C^T, \tag{7.6}$$

where

$$C = A^T P + PA \tag{7.7}$$

and where all symbols are as defined in (7.1) to (7.3). Our objective is to determine the as yet unknown matrix P in such a way that $\dot{v}(x)$ becomes a preassigned negative definite quadratic form, i.e., in such a way that C is a preassigned negative definite matrix.

Equation (7.7) constitutes a system of $n(n+1)/2$ linear equations. We need to determine under what conditions we can solve for the $n(n+1)/2$ elements, p_{ik}, given C and A. To this end, we choose a similarity transformation Q such that

$$QAQ^{-1} = \bar{A}, \tag{7.8}$$

or equivalently,

$$A = Q^{-1}\bar{A}Q, \tag{7.9}$$

where $\bar{A}$ is similar to A and Q is a real $n \times n$ nonsingular matrix. From (7.9) and (7.7) we obtain

$$(\bar{A})^T(Q^{-1})^T PQ^{-1} + (Q^{-1})^T PQ^{-1}\bar{A} = (Q^{-1})^T CQ^{-1} \tag{7.10}$$

or

$$(\bar{A})^T\bar{Q} + \bar{Q}\bar{A} = \bar{C}, \qquad \bar{P} = (Q^{-1})^T PQ^{-1}, \qquad \bar{C} = (Q^{-1})^T CQ^{-1}. \tag{7.11}$$

In (7.11), P and C are subjected to a congruence transformation and $\bar{P}$ and $\bar{C}$ have the same definiteness properties as P and C, respectively. Since every real $n \times n$ matrix can be triangularized (refer to Subsection 2.2L), we can choose Q in such a fashion that $\bar{A} = [\bar{a}_{ij}]$ is *triangular*, i.e., $\bar{a}_{ij} = 0$ for $i > j$. Note that in this case the eigenvalues of A, $\lambda_1, \dots, \lambda_n$, appear in the main diagonal of $\bar{A}$. To simplify our notation, we rewrite (7.11) in the form (7.7) by dropping the bars, i.e.,

$$A^T P + PA = C, \qquad C = C^T, \tag{7.12}$$

and *we assume that* $A = [a_{ij}]$ *has been triangularized*, i.e., $a_{ij} = 0$ for $i > j$. Since the eigenvalues $\lambda_1, \dots, \lambda_n$ appear in the diagonal of A, we can rewrite (7.12) as

$$\begin{aligned} 2\lambda_1 p_{11} &= c_{11} \\ a_{21}p_{11} + (\lambda_1 + \lambda_2)p_{12} &= c_{12} \\ \cdots\cdots\cdots&\cdots\cdots\cdots \end{aligned} \tag{7.13}$$

Since this system of equations is triangular, and since its determinant is equal to

$$2^n \lambda_1 \cdots \lambda_n \prod_{i<j} (\lambda_i + \lambda_j), \tag{7.14}$$

the matrix P can be determined (uniquely) if and only if its determinant is not zero. This is true when all eigenvalues of A are nonzero and no two of them are such that $\lambda_i + \lambda_j = 0$. This condition is not affected by a similarity transformation and is therefore also valid for the original system of equations (7.7).

We summarize the above discussion in the following lemma.

LEMMA 7.1. Let $A \in R^{n \times n}$ and let $\lambda_1, \ldots, \lambda_n$ denote the (not necessarily distinct) eigenvalues of A. Then (7.12) has a unique solution for P corresponding to each $C \in R^{n \times n}$ if and only if

$$\lambda_i \neq 0, \lambda_i + \lambda_j \neq 0 \qquad \text{for all } i, j. \tag{7.15}$$

■

To construct $v(x)$, we must still check the definiteness of P. This can be done in a purely algebraic way; however, in the present case it is much easier to apply the results of this section and argue as follows:

1. If all eigenvalues of A have negative real parts [or equivalently, if the equilibrium $x = 0$ of (L) is exponentially stable in the large], and if C in (7.7) is negative definite, then $P = P^T$ must be positive definite. To prove this assertion, we choose for (L) the function v given in (7.5) with $\dot{v}$ along the solutions of (L) given by (7.6) and (7.7). For purposes of contradiction we assume that P is not positive definite. Then there exists $x_0 \neq 0$ such that $v(x_0) = x_0^T P x_0 \leq 0$. For the solution $\phi(t)$ with $\phi(t_0) = x_0$, $v(\phi(t))$ is monotone decreasing with increasing t, since $\dot{v}(\phi(t)) \leq 0$. Also, since $\dot{v}(t)|_{t=t_0} = x_0^T Q x_0 < 0$, it follows that for $t \geq t_0$, $v(\phi(t)) \leq v(x(t_0)) = v(x_0) \leq 0$. Since by assumption all the eigenvalues of A have negative real parts, we know that the equilibrium $x = 0$ of (L) is uniformly asymptotically stable. Thus, $\lim_{t \to \infty} v(\phi(t)) = 0$, which leads to a contradiction. Thus, P must be positive definite.
2. If at least one of the eigenvalues of A has positive real part and no real part of any eigenvalue of A is zero and if (7.15) is satisfied, and if C in (7.7) is negative definite, then P cannot be positive definite; otherwise we could apply Theorem 7.2 to come up with a contradiction. If in particular the real parts of all eigenvalues of A are positive, then P must be negative definite. [Note that in this case the equilibrium $x = 0$ of (L) is completely unstable.]

Now suppose that at least one of the eigenvalues of A has positive real part, and suppose that any one of the two conditions or both conditions given in (7.15) are not satisfied. Then we cannot construct $v(x)$ given in (7.5) in the manner described above (i.e., we cannot determine P in the manner described above). In this case we form a matrix $A_1 = A - \delta I$, where I denotes the $n \times n$ identity matrix and δ is chosen so that A_1 has as many eigenvalues with positive real part as A, but none of the conditions in (7.15) are violated. Then the equation

$$A_1^T P + P A_1 = C,$$

with C negative definite, can be solved for P, and P is then clearly not positive definite. The derivative of the function $v(x) = x^T P x$ is a quadratic form whose matrix is of the form

$$(A_1 + \delta I)^T P + P(A_1 + \delta I) = C + 2\delta P,$$

which is negative definite for a sufficiently small δ. The function $v(x)$ constructed in this way now satisfies the hypotheses of Theorem 7.3 for system $\dot{x} = Ax$.

Summarizing the above discussion, we have the following result.

THEOREM 7.4. If all the eigenvalues of the matrix A have negative real parts, or if at least one eigenvalue has a positive real part, then there exists a Lyapunov function of the form

$$v(x) = x^T P x, P = P^T,$$

whose derivative along the solutions of (L) is definite (i.e., it is either negative definite or positive definite). ■

EXAMPLE 7.4. We consider the system (L) with

$$A = \begin{bmatrix} 0 & 1 \\ -1 & 0 \end{bmatrix}.$$

The eigenvalues of A are $\lambda_1, \lambda_2 = \pm j$ and therefore condition (7.15) is violated. According to Lemma 7.1, the Lyapunov matrix equation

$$A^T P + PA = C$$

does not possess a unique solution for a given C. We now verify this for two specific cases.

(i) When $C = 0$, we obtain

$$\begin{bmatrix} 0 & -1 \\ 1 & 0 \end{bmatrix}\begin{bmatrix} p_{11} & p_{12} \\ p_{12} & p_{22} \end{bmatrix} + \begin{bmatrix} p_{11} & p_{12} \\ p_{12} & p_{22} \end{bmatrix}\begin{bmatrix} 0 & 1 \\ -1 & 0 \end{bmatrix} = \begin{bmatrix} -2p_{12} & p_{11} - p_{22} \\ p_{11} - p_{22} & 2p_{12} \end{bmatrix}$$
$$= \begin{bmatrix} 0 & 0 \\ 0 & 0 \end{bmatrix},$$

or $p_{12} = 0$ and $p_{11} = p_{22}$. Therefore, for any $a \in R$, the matrix $P = aI$ is a solution of the Lyapunov matrix equation. In other words, for $C = 0$, the Lyapunov matrix equation has in this example denumerably many solutions.

(ii) When $C = -2I$, we obtain

$$\begin{bmatrix} -2p_{12} & p_{11} - p_{22} \\ p_{11} - p_{22} & 2p_{12} \end{bmatrix} = \begin{bmatrix} -2 & 0 \\ 0 & -2 \end{bmatrix},$$

or $p_{11} = p_{22}$ and $p_{12} = 1$ and $p_{12} = -1$, which is impossible. Therefore, for $C = -2I$, the Lyapunov matrix equation has in this example no solutions at all. ■

It turns out that if all the eigenvalues of matrix A have negative real parts, then we can compute P in (7.7) explicitly.

THEOREM 7.5. If all eigenvalues of a real $n \times n$ matrix A have negative real parts, then for each matrix $C \in R^{n\times n}$, the unique solution of (7.7) is given by

$$P = \int_0^\infty e^{A^T t}(-C)e^{At}\, dt. \tag{7.16}$$

Proof. If all eigenvalues of A have negative real parts, then (7.15) is satisfied and therefore (7.7) has a unique solution for every $C \in R^{n\times n}$. To verify that (7.16) is indeed this solution, we first note that the right-hand side of (7.16) is well defined, since all eigenvalues of A have negative real parts. Substituting the right-hand side of (7.16) for P into (7.7), we obtain

$$\begin{aligned} A^T P + PA &= \int_0^{\infty} A^T e^{A^T t}(-C)e^{At} dt + \int_0^{\infty} e^{A^T t}(-C)e^{At} A\, dt \\ &= \int_0^{\infty} \frac{d}{dt}[e^{A^T t}(-C)e^{At}]\, dt \\ &= e^{A^T t}(-C)e^{At}\Big|_0^{\infty} = C, \end{aligned}$$

which proves the theorem. ■

6.8 LINEARIZATION

In this section we consider *nonlinear, finite-dimensional, continuous-time* dynamical systems described by equations of the form

$$\dot{w} = f(w), \tag{A}$$

where $f \in C^1(R^n, R^n)$. We assume that $w = 0$ is an equilibrium of (A). In accordance with Subsection 1.11A, we linearize system (A) about the origin to obtain

$$\dot{x} = Ax + F(x), \tag{8.1}$$

$x \in R^n$, where $F \in C(R^n, R^n)$ and where A denotes the Jacobian of $f(w)$ evaluated at $w = 0$, given by

$$A = \frac{\partial f}{\partial w}(0), \tag{8.2}$$

and where $$F(x) = o(\|x\|) \quad \text{as} \quad \|x\| \to 0. \tag{8.3}$$

Associated with (8.1) is the *linearization* of (A), given by

$$\dot{y} = Ay. \tag{L}$$

In the following, we use the results of Section 6.7 to establish criteria that allow us to deduce the stability properties of the equilibrium $w = 0$ of the nonlinear system (A) from the stability properties of the equilibrium $y = 0$ of the linear system (L).

THEOREM 8.1. Let $A \in R^{n \times n}$ be a Hurwitz matrix, let $F \in C(R^n, R^n)$, and assume that (8.3) holds. Then the equilibrium $x = 0$ of (8.1) [and hence, of (A)] is *exponentially stable*.

Proof. Theorem 7.4 applies to (L) since all the eigenvalues of A have negative real parts. In view of that theorem (and the comments following Lemma 7.1), there exists a symmetric, real, positive definite $n \times n$ matrix P such that

$$PA + A^T P = C, \tag{8.4}$$

where C is negative definite. Consider the Lyapunov function

$$v(x) = x^T P x. \tag{8.5}$$

The derivative of v with respect to t along the solutions of (8.1) is given by

$$\begin{aligned} \dot{v}(x) &= \dot{x}^T P x + x^T P \dot{x} \\ &= (Ax + F(x))^T P x + x^T P(Ax + F(x)) \\ &= x^T C x + 2x^T P F(x). \end{aligned} \tag{8.6}$$

Now choose $\gamma < 0$ such that $x^T C x \leq 3\gamma\|x\|^2$ for all $x \in R^n$. Since it is assumed that (8.3) holds, there is a $\delta > 0$ such that if $\|x\| \leq \delta$, then $\|PF(x)\| \leq -\gamma\|x\|$ for all $x \in \overline{B(\delta)} = \{x \in R^n : \|x\| \leq \delta\}$. Therefore, for all $x \in \overline{B(\delta)}$ we obtain, in view of (8.6), the estimate

$$\dot{v}(x) \leq 3\gamma\|x\|^2 - 2\gamma\|x\|^2 = \gamma\|x\|^2. \tag{8.7}$$

Now let $\alpha = \min_{\|x\|=\delta} v(x)$. Then $\alpha > 0$ (since P is positive definite). Take $\lambda \in (0, \alpha)$, and let

$$C_\lambda = \{x \in B(\delta) = \{x \in R^n : \|x\| < \delta\} : v(x) \leq \lambda\}. \tag{8.8}$$

Then $C_\lambda \subset B(\delta)$. [This can be shown by contradiction. Suppose that C_λ is not entirely inside $B(\delta)$. Then there is a point $\bar{x} \in C_\lambda$ that lies on the boundary of $B(\delta)$. At this point, $v(\bar{x}) \geq \alpha > \lambda$. We have thus arrived at a contradiction.] The set C_λ has the property that any solution of (8.1) starting in C_λ at $t = t_0$ will stay in C_λ for all $t \geq t_0 \geq 0$. To see this, we let $\phi(t, t_0, x_0) \triangleq \phi(t)$ and we recall that $\dot{v}(x) \leq \gamma\|x\|^2$, $\gamma < 0$, $x \in B(\delta) \supset C_\lambda$. Then $\dot{v}(\phi(t)) \leq 0$ implies that $v(\phi(t)) \leq v(x_0) \leq \lambda$ for all $t \geq t_0 \geq 0$. Therefore, $\phi(t) \in C_\lambda$ for all $t \geq t_0 \geq 0$.

We now proceed in a similar manner as in the proof of Theorem 7.1 to complete this proof. In doing so, we first obtain the estimate

$$\dot{v}(\phi(t)) \leq \left(\frac{\gamma}{c_2}\right) v(\phi(t)), \tag{8.9}$$

where γ is given in (8.7) and c_2 is determined by the relation

$$c_1\|x\|^2 \leq v(x) = x^T P x \leq c_2\|x\|^2. \tag{8.10}$$

Following now in an identical manner as was done in the proof of Theorem 7.1, we have

$$\|\phi(t)\| \leq \left(\frac{c_2}{c_1}\right) \|x_0\| e^{1/2(\gamma/c_2)(t-t_0)}, \qquad t \geq t_0 \geq 0, \tag{8.11}$$

whenever $x_0 \in B(r')$, where r' has been chosen sufficiently small so that $B(r') \subset C_\lambda$. This proves that the equilibrium $x = 0$ of (8.1) is exponentially stable. ■

It is important to recognize that Theorem 8.1 is a *local result* that yields sufficient conditions for the exponential stability of the equilibrium $x = 0$ of (8.1); it does not yield conditions for exponential stability in the large. The proof of Theorem 8.1, however, enables us to determine an estimate of the domain of attraction of the equilibrium $x = 0$ of (A), involving the following steps:

1. Determine an equilibrium, x_e, of (A) and transform (A) to a new system that translates x_e to the origin $x = 0$ (refer to Section 6.3).
2. Linearize (A) about the origin and determine $F(x)$, A, and the eigenvalues of A.
3. If all eigenvalues of A have negative real parts, choose a negative definite matrix C and solve the Lyapunov matrix equation

$$C = A^T P + PA.$$

4. Determine the Lyapunov function

$$v(x) = x^T P x.$$

5. Compute the derivative of v along the solutions of (8.1), given by

$$\dot{v}(x) = x^T C x + 2x^T PF(x).$$

6. Determine $\delta > 0$ such that $\dot{v}(x) < 0$ for all $x \in B(\delta) - \{0\}$.

7. Determine the largest $\lambda = \lambda_M$ such that $C_{\lambda_M} \subset B(\delta)$, where

$$C_\lambda = \{x \in R^n : v(x) < \lambda\}.$$

8. C_{λ_M} is a subset of the domain of attraction of the equilibrium $x = 0$ of (8.1), and hence, of (A).

The above procedure may be repeated for different choices of matrix C given in step (3), resulting in different matrices P_i, which in turn may result in different estimates for the domain of attraction, $C^i_{\lambda_M}$, $i \in \Lambda$, where Λ is an index set. The union of the sets $C^i_{\lambda_M} \triangleq D_i$, $D = \cup_i D_i$, is also a subset of the domain of attraction of the equilibrium $x = 0$ of (A).

THEOREM 8.2. Assume that A is a real $n \times n$ matrix that has at least one eigenvalue with positive real part. Let $F \in C(R^n, R^n)$, and assume that (8.3) holds. Then the equilibrium $x = 0$ of (8.1) [and hence, of (A)] is *unstable*.

Proof. We use Theorem 7.4 to choose a real, symmetric $n \times n$ matrix P such that the matrix $PA + A^T P = C$ is negative definite. The matrix P is not positive definite, or even positive semidefinite (refer to the comments following Lemma 7.1). Hence, the function $v(x) = x^T Px$ is negative at some points arbitrarily close to the origin. The derivative of $v(x)$ with respect to t along the solutions of (8.1) is given by (8.6). As in the proof of Theorem 8.1, we can choose a $\gamma < 0$ such that $x^T Cx \leq 3\gamma\|x\|^2$ for all $x \in R^n$, and in view of (8.3) we can choose a $\delta > 0$ such that $\|BF(x)\| \leq \gamma\|x\|$ for all $x \in B(\delta)$. Therefore, for all $x \in B(\delta)$, we obtain that

$$\dot{v}(x) \leq 3\gamma\|x\|^2 - 2\gamma\|x\|^2 = \gamma\|x\|^2.$$

Now let

$$G = \{x \in B(\delta) : v(x) < 0\}.$$

The boundary of G is made up of points where either $v(x) = 0$ or where $\|x\| = \delta$. Note in particular that the equilibrium $x = 0$ of (8.1) is in the boundary of G. Now following an identical procedure as in the proof of Theorem 7.3, we show that any solution $\phi(t)$ of (8.1) with $\phi(t_0) = x_0 \in G$ must escape G in finite time through the surface determined by $\|x\| = \delta$. Since the above argument holds for arbitrarily small $\delta > 0$, it follows that the origin $x = 0$ of (8.1) is unstable. ■

Before concluding this section, we consider a few specific cases.

EXAMPLE 8.1. The *Lienard Equation* is given by

$$\ddot{w} + g(w)\dot{w} + w = 0, \tag{8.12}$$

where $g \in C^1(R, R)$ with $g(0) > 0$. Letting $x_1 = w$ and $x_2 = \dot{w}$, we obtain

$$\begin{aligned} \dot{x}_1 &= x_2 \\ \dot{x}_2 &= -x_1 - g(x_1)x_2. \end{aligned} \tag{8.13}$$

Let $x^T = (x_1, x_2)$, $f(x)^T = (f_1(x), f_2(x))$, and let

$$J(0) = A = \begin{bmatrix} \dfrac{\partial f_1}{\partial x_1}(0) & \dfrac{\partial f_1}{\partial x_2}(0) \\ \dfrac{\partial f_2}{\partial x_1}(0) & \dfrac{\partial f_2}{\partial x_2}(0) \end{bmatrix} = \begin{bmatrix} 0 & 1 \\ -1 & -g(0) \end{bmatrix}.$$

Then

$$\dot{x} = Ax + [f(x) - Ax] = Ax + F(x),$$

where
$$F(x) = \begin{bmatrix} 0 \\ [g(0) - g(x_1)]x_2 \end{bmatrix}.$$

The origin $x = 0$ is clearly an equilibrium of (8.12) and hence of (8.13). The eigenvalues of A are given by

$$\lambda_1, \lambda_2 = \frac{-g(0) \pm \sqrt{g(0)^2 - 4}}{2},$$

and therefore, A is a Hurwitz matrix. Also, (8.3) holds. Therefore, all the conditions of Theorem 8.1 are satisfied. We conclude that the equilibrium $x = 0$ of (8.13) is *exponentially stable.* ■

EXAMPLE 8.2. We consider the system given by

$$\begin{aligned} \dot{x}_1 &= -x_1 + x_1(x_1^2 + x_2^2) \\ \dot{x}_2 &= -x_2 + x_2(x_1^2 + x_2^2). \end{aligned} \tag{8.14}$$

The origin is clearly an equilibrium of (8.14). Also, the system is already in the form (8.1) with

$$A = \begin{bmatrix} -1 & 0 \\ 0 & -1 \end{bmatrix}, \qquad F(x) = \begin{bmatrix} x_1(x_1^2 + x_2^2) \\ x_2(x_1^2 + x_2^2) \end{bmatrix},$$

and condition (8.3) is clearly satisfied. The eigenvalues of A are $\lambda_1 = -1, \lambda_2 = -1$. Therefore, all conditions of Theorem 8.1 are satisfied and we conclude that the equilibrium $x^T = (x_1, x_2) = 0$ is *exponentially stable*; however, we cannot conclude that this equilibrium is exponentially stable in the large. Accordingly, we seek to determine an estimate for the domain of attraction of this equilibrium.

We choose $C = -I$ (where $I \in R^{2\times 2}$ denotes the identity matrix) and we solve the matrix equation $A^T P + PA = C$ to obtain $P = (1/2)I$, and therefore,

$$V(x_1, x_2) = x^T P x = \tfrac{1}{2}(x_1^2 + x_2^2).$$

Along the solutions of (8.14) we obtain

$$\begin{aligned} \dot{v}(x_1, x_2) &= x^T C x + 2x^T PF(x) \\ &= -(x_1^2 + x_2^2) + (x_1^2 + x_2^2)^2. \end{aligned}$$

Clearly, $\dot{v}(x_1, x_2) < 0$ when $(x_1, x_2) \neq (0, 0)$ and $x_1^2 + x_2^2 < 1$. In the language of the proof of Theorem 8.1, we can therefore choose $\delta = 1$.

Now let

$$C_{1/2} = \{x \in R^2 : v(x_1, x_2) = \tfrac{1}{2}(x_1^2 + x_2^2) < \tfrac{1}{2}\}.$$

Then clearly, $C_{1/2} \subset B(\delta)$, $\delta = 1$, in fact $C_{1/2} = B(\delta)$. Therefore, the set $\{x \in R^2 : x_1^2 + x_2^2 < 1\}$ is a subset of the domain of attraction of the equilibrium $(x_1, x_2)^T = 0$ of system (8.14). ■

EXAMPLE 8.3. The differential equation governing the motion of a pendulum is given by

$$\ddot{\theta} + a \sin\theta = 0, \tag{8.15}$$

where $a > 0$ is a constant (refer to Chapter 1). Letting $\theta = x_1$ and $\dot{\theta} = x_2$, we obtain the system description

$$\begin{aligned} \dot{x}_1 &= x_2 \\ \dot{x}_2 &= -a \sin x_1. \end{aligned} \tag{8.16}$$

The points $x_e^{(1)} = (0, 0)^T$ and $x_e^{(2)} = (\pi, 0)^T$ are equilibria of (8.16).

(i) Linearizing (8.16) about the equilibrium $x_e^{(1)}$, we put (8.16) into the form (8.1) with

$$A = \begin{bmatrix} 0 & 1 \\ -a & 0 \end{bmatrix}.$$

The eigenvalues of A are $\lambda_1, \lambda_2 = \pm j\sqrt{a}$. Therefore, the results of this section (Theorem 8.1 and 8.2) are not applicable in the present case.

(ii) In (8.16), we let $y_1 = x_1 - \pi$ and $y_2 = x_2$. Then (8.16) assumes the form

$$\begin{aligned} \dot{y}_1 &= y_2 \\ \dot{y}_2 &= -a\sin(y_1 + \pi). \end{aligned} \tag{8.17}$$

The point $(y_1, y_2)^T = (0, 0)^T$ is clearly an equilibrium of system (8.17). Linearizing about this equilibrium, we put (8.17) into the form (8.1), where

$$A = \begin{bmatrix} 0 & 1 \\ a & 0 \end{bmatrix}, \qquad F(y_1, y_2) = \begin{bmatrix} 0 \\ -a(\sin(y_1 + \pi) + y_1) \end{bmatrix}.$$

The eigenvalues of A are $\lambda_1, \lambda_2 = a, -a$. All conditions of Theorem 8.2 are satisfied and we conclude that the equilibrium $x_e^{(2)} = (\pi, 0)^T$ of system (8.16) is *unstable*. ■

PART 2
INPUT-OUTPUT STABILITY OF CONTINUOUS-TIME SYSTEMS

6.9
INPUT-OUTPUT STABILITY

We now turn our attention to systems described by the state equations

$$\begin{aligned} \dot{x} &= A(t)x + B(t)u \\ y &= C(t)x + D(t)u, \end{aligned} \tag{9.1}$$

where $A \in C(R, R^{n\times n})$, $B \in C(R, R^{n\times m})$, $C \in C(R, R^{p\times n})$, and $D \in C(R, R^{p\times m})$ [resp., $A \in C(R^+, R^{n\times n})$, $B \in C(R^+, R^{n\times m})$, $C \in C(R^+, R^{p\times n})$, and $D \in C(R^+, R^{p\times m})$]. In the preceding sections of this chapter we investigated the *internal stability properties* of system (9.1) by studying the Lyapunov stability of the trivial solution of the associated system

$$\dot{w} = A(t)w. \tag{LH}$$

In this approach, system inputs and system outputs played no role. To account for these, we now consider the *external stability properties* of system (9.1), called *input-output stability*: every bounded input of a system should produce a bounded output. More specifically, in the present context, we say that system (9.1) is *bounded-input/bounded-output* (*BIBO*) *stable* if for all t_0 and zero initial conditions at $t = t_0$, every bounded input defined on $[t_0, \infty)$ gives rise to a bounded response on $[t_0, \infty)$.

A bounded matrix $D(t)$ does not affect the BIBO stability of (9.1), while an unbounded $D(t)$ will give rise to an unbounded response to an appropriate constant input. Accordingly, we will consider without any loss of generality the case where $D(t) \equiv 0$, i.e., throughout this section we will concern ourselves with systems described by equations of the form

$$\begin{aligned}\dot{x} &= A(t)x + B(t)u \\ y &= C(t)x.\end{aligned} \tag{9.2}$$

We will find it useful to use a more restrictive concept of input-output stability in establishing various results: we will say that the system (9.2) is *uniformly BIBO stable* if there exists a constant $k > 0$ that is independent of t_0, such that for all t_0 the conditions

$$\begin{aligned} x(t_0) &= 0 \\ \|u(t)\| &\leq 1, \qquad t \geq t_0, \end{aligned}$$

imply that $\|y(t)\| \leq k$ for all $t \geq t_0$. (The symbol $\|\cdot\|$ denotes the Euclidean norm.)

It turns out that for the class of problems considered herein, BIBO stability and uniform BIBO stability amount to the same concepts. (We will not, however, prove this assertion here.) Accordingly, we will phrase all subsequent results in terms of uniform BIBO stability, rather than BIBO stability. These results will involve the impulse response matrix of (9.2) given by

$$H(t,\tau) = \begin{cases} C(t)\Phi(t,\tau)B(\tau), & t \geq \tau, \\ 0, & t < \tau, \end{cases} \tag{9.3}$$

and the controllability and observability Gramian given, respectively, by

$$W(t_0, t_1) = \int_{t_0}^{t_1} \Phi(t_0, t)B(t)B^T(t)\Phi^T(t_0, t)\,dt \tag{9.4}$$

and

$$M(t_0, t_1) = \int_{t_0}^{t_1} \Phi(t, t_0)^T C(t)^T C(t)\Phi(t, t_0)\,dt. \tag{9.5}$$

In these results we will establish sufficient conditions for uniform BIBO stability of (9.2) and also necessary and sufficient conditions for uniform BIBO stability of (9.2). Furthermore, we will present results that make a connection between the uniform BIBO stability of (9.2) and the Lyapunov exponential stability of the equilibrium $w = 0$ of (LH). In view of the latter results, we will usually assume that $t_0 \geq 0$.

At the end of this section we will also present specialized stability results for the time-invariant systems described by equations of the form

$$\begin{aligned}\dot{x} &= Ax + Bu \\ y &= Cx,\end{aligned} \tag{9.6}$$

where $A \in R^{n\times n}$, $B \in R^{n\times m}$, and $C \in R^{p\times n}$. Associated with system (9.6) is the free system described by equations of the form

$$\dot{p} = Ap. \tag{L}$$

Recall that for system (9.6) the impulse response matrix is given by

$$\begin{aligned} H(t) &= Ce^{At}B, \qquad t \geq 0, \\ &= 0, \qquad t < 0, \end{aligned} \tag{9.7}$$

and the transfer function matrix is given by

$$\hat{H}(s) = C(sI - A)^{-1}B. \tag{9.8}$$

THEOREM 9.1. The system (9.2) is *uniformly BIBO stable* if and only if there exists a finite constant $L > 0$ such that for all t and t_0, with $t \geq t_0$,

$$\int_{t_0}^{t} \|H(t, \tau)\| \, d\tau \leq L. \tag{9.9}$$

■

The first part of the proof of Theorem 9.1 (sufficiency) is straightforward. Indeed, if $\|u(t)\| \leq 1$ for all $t \geq t_0$ and if (9.9) is true, then we have for all $t \geq t_0$ that

$$\begin{aligned} \|y(t)\| &= \left\| \int_{t_0}^{t} H(t, \tau) u(\tau) \, d\tau \right\| \\ &\leq \int_{t_0}^{t} \|H(t, \tau) u(\tau)\| \, d\tau \\ &\leq \int_{t_0}^{t} \|H(t, \tau)\| \, \|u(\tau)\| \, d\tau \\ &\leq \int_{t_0}^{t} \|H(t, \tau)\| \, d\tau \leq L. \end{aligned}$$

Therefore, system (9.2) is uniformly BIBO stable.

In proving the second part of Theorem 9.1 (necessity), we simplify matters by first considering in (9.2) the single-variable case ($n = 1$) with the input-output description given by

$$y(t) = \int_{t_0}^{t} h(t, \tau) u(\tau) \, d\tau. \tag{9.10}$$

For purposes of contradiction, we assume that the system is BIBO stable, but no finite L exists such that (9.9) is satisfied. Another way of stating this is that for *every finite* L, there exist $t_0 = t_0(L)$ and $t_1 = t_1(L)$, $t_1 > t_0$, such that

$$\int_{t_0}^{t_1} |h(t_1, \tau)| \, d\tau > L.$$

We now choose in particular the input given by

$$u(t) = \begin{cases} +1 & \text{if } h(t, \tau) > 0, \\ 0 & \text{if } h(t, \tau) = 0, \\ -1 & \text{if } h(t, \tau) < 0, \end{cases} \tag{9.11}$$

$t_0 \leq t \leq t_1$. Clearly, $|u(t)| \leq 1$ for all $t \geq t_0$. The output of the system at $t = t_1$ due to the above input, however, is

$$y(t_1) = \int_{t_0}^{t_1} h(t_1, \tau) u(\tau) \, d\tau = \int_{t_0}^{t_1} |h(t_1, \tau)| \, d\tau > L,$$

which contradicts the assumption that the system is BIBO stable.

The above can now be extended to the multivariable case. In doing so, we apply the single-variable result to every possible pair of input and output vector components, we make use of the fact that the sum of a finite number of bounded sums will be bounded, and we recall that a vector is bounded if and only if each of its components is bounded. We leave the details to the reader.

In the preceding argument we made the tacit assumption that u is continuous, or piecewise continuous. However, our particular choice of u may involve nondenumerably many switchings (discontinuities) over a given finite-time interval. In such cases, u is no longer piecewise continuous; however, it is measurable (in the

Lebesgue sense). This generalization can be handled, though in a broader mathematical setting that we do not wish to pursue here. The interested reader may want to refer, e.g., to the books by Desoer and Vidyasagar [5], Michel and Miller [17], and Vidyasagar [25] and the papers by Sandberg [21] to [23] and Zames [26], [27] for further details.

From Theorem 9.1 and from (9.7) it follows readily that a necessary and sufficient condition for the uniform BIBO stability of system (9.6) is the condition

$$\int_0^\infty \|H(t)\|\,dt < \infty. \tag{9.12}$$

COROLLARY 9.1. Assume that the equilibrium $w = 0$ of (LH) is exponentially stable and suppose there exist constants $\beta > 0$ and $\gamma > 0$ such that for all t, $\|B(t)\| \le \beta$ and $\|C(t)\| \le \gamma$. Then system (9.2) is *uniformly BIBO stable.*

Proof. Under the hypotheses of the corollary, we have

$$\begin{aligned}\left\|\int_{t_0}^t H(t,\tau)d\tau\right\| &\le \int_{t_0}^t \|H(t,\tau)\|\,d\tau\\ &= \int_{t_0}^t \|C(t)\Phi(t,\tau)B(\tau)\|\,d\tau \le \gamma\beta\int_{t_0}^t \|\Phi(t,\tau)\|\,d\tau.\end{aligned}$$

Since the equilibrium $w = 0$ of (LH) is exponentially stable, there exist $\delta > 0$, $\lambda > 0$ such that $\|\Phi(t,\tau)\| \le \delta e^{-\lambda(t-\tau)}$, $t \ge \tau$. Therefore,

$$\begin{aligned}\int_{t_0}^t \|H(t,\tau)\|\,d\tau &\le \int_{t_0}^t \gamma\beta\delta e^{-\lambda(t-\tau)}\,d\tau\\ &\le \frac{\gamma\delta\beta}{\lambda} \triangleq L\end{aligned}$$

for all τ, t with $t \ge \tau$. It now follows from Theorem 9.1 that system (9.2) is uniformly BIBO stable. ■

As indicated earlier, we seek to establish a connection between the uniform BIBO stability of (9.2) and the exponential stability of the trivial solution of (LH). We will accomplish this by means of an intermediate result for systems described by equations of the form

$$\begin{aligned}\dot{x} &= A(t)x + B(t)u\\ y &= x.\end{aligned} \tag{9.13}$$

Before stating and proving the next result, we recall that if $S \in R^{n\times n}$, $T \in R^{n\times n}$ are symmetric, then the notation $S > 0$ signifies that S is positive definite, and the notation $S > T$ indicates that the matrix $S - T$ is positive definite, i.e., $S - T > 0$. Also, if $Q(t) = Q(t)^T \in C(R, R^{n\times n})$, the condition that there is a constant $\eta > 0$ such that $Q(t) \ge \eta I$ for all $t \in R$ is equivalent to the statement that $z^T Q(t)z \ge \eta\|z\|^2$ for all $t \in R$ and all $z \in R^n$.

Next, suppose that there is a constant $\alpha > 0$ such that $\|A(t)\| \le \alpha$ for all $t \in R$, and let $\Phi(t,\tau)$ denote the state transition matrix of (LH). In the proof of the next result we will require the estimate

$$\|\Phi(t,\tau)\| \le e^{\alpha\delta}, \qquad |t-\tau| \le \delta. \tag{9.14}$$

To obtain this estimate we let $\phi(t,\tau,\xi) = \phi(t)$ denote the solution of (LH) with $\phi(\tau) = \xi$, and we compute

$$\frac{d}{dt}\phi(t)^T\phi(t) = \frac{d}{dt}\|\phi(t)\|^2 = \dot{\phi}(t)^T\phi(t) + \phi(t)^T\dot{\phi}(t)$$
$$= \phi(t)^T A(t)^T\phi(t) + \phi(t)^T A(t)\phi(t)$$

for all $t \geq \tau$. Letting $\|\phi(t)\|^2 = v(t)$, we have

$$\frac{dv}{dt} \leq 2\alpha v, \qquad v(\tau) = \|\xi\|^2,$$

which yields

$$v(t) \leq v(\tau)e^{2\alpha(t-\tau)}$$

or

$$\|\phi(t)\| \leq e^{\alpha(t-\tau)}\|\xi\| \leq e^{\alpha\delta}\|\xi\|,$$

which in turn yields (9.14). The case when $\tau > t$ is treated similarly.

THEOREM 9.2. Suppose that there exist positive constants α, β, ϵ, and δ such that for all t, $\|A(t)\| \leq \alpha$, $\|B(t)\| \leq \beta$, and $W(t_0, t_0, +\delta) \geq \epsilon I$, where I denotes the $n \times n$ identity matrix and $W(\cdot)$ denotes the controllability Gramian given in (9.4). Then the system (9.13) is uniformly BIBO stable if and only if the trivial solution of (LH) is exponentially stable.

Proof. Under the above hypotheses, it follows from Corollary 9.1 that if the trivial solution of (LH) is exponentially stable, then system (9.13) is uniformly BIBO stable.

Conversely, assume that the system (9.13) is uniformly BIBO stable and assume that the hypotheses of the theorem are satisfied with the given constants α, β, δ, and ϵ. The assumption $W(t_0, t_0 + \delta) \geq \epsilon I$ ensures that $W^{-1}(t_0, t_0 + \delta)$ exists, is bounded, and is independent of t_0. We now consider

$$I = \int_{\tau-\delta}^{\tau} [\Phi(\tau, \eta)B(\eta)B(\eta)^T\Phi(\tau, \eta)^T\, d\eta]W^{-1}(\tau - \delta, \tau). \tag{9.15}$$

Since $B(\eta)$ is bounded and since $\Phi(\tau, \eta)$ is bounded over $|\tau - \eta| \leq \delta$, there exists a constant $c > 0$ such that

$$\|B(\eta)^T\Phi(\tau, \eta)^T W^{-1}(\tau - \delta, \tau)\| \leq c. \tag{9.16}$$

Premultiplying (9.15) by $\Phi(t, \tau)$ and using the bound (9.16) and the properties of norms, we obtain

$$\|\Phi(t, \tau)\| \leq c\int_{\tau-\delta}^{\tau} \|\Phi(t, \eta)B(\eta)\|\, d\eta. \tag{9.17}$$

Since system (9.13) is uniformly BIBO stable, there exists a $k > 0$ such that

$$\int_{t-n\delta}^{t} \|\Phi(t, \eta)B(\eta)\|\, d\eta < k \tag{9.18}$$

for *all* positive integers n, where k is independent of n and t. From (9.18) it follows that

$$\int_{t-n\delta}^{t} \|\Phi(t, \eta)B(\eta)\|\, d\eta = \int_{t-\delta}^{t} \|\Phi(t, \eta)B(\eta)\|\, d\eta + \int_{t-2\delta}^{t-\delta} \|\Phi(t, \eta)B(\eta)\|\, d\eta$$
$$+ \cdots + \int_{t-n\delta}^{t-n\delta+\delta} \|\Phi(t, \eta)B(\eta)\|\, d\eta < k. \tag{9.19}$$

From (9.17) to (9.19) it follows that

$$c^{-1}\|\Phi(t, t)\| + c^{-1}\|\Phi(t, t - \delta)\| + \cdots + c^{-1}\|\Phi(t, t - n\delta + \delta)\|$$
$$\leq \int_{t-n\delta}^{t} \|\Phi(t, \eta)B(\eta)\|\, d\eta < k.$$

Since the above is true for any positive integer n, we have

$$c^{-1}(\|\Phi(t,t)\| + \|\Phi(t,t-\delta)\| + \cdots + \|\Phi(t,t-n\delta+\delta)\| + \cdots) < k$$

or

$$\|\Phi(t,t)\| + \|\Phi(t,t-\delta)\| + \cdots + \|\Phi(t,t-n\delta+\delta)\| + \cdots < ck.$$

To complete the proof, we must show that $\int_{-\infty}^{t} \|\Phi(t,\eta)\|\,d\eta$ is finite. We will accomplish this by showing that $\int_{-n\delta}^{t} \|\Phi(t,\eta)\|\,d\eta$ is finite for any positive integer. To this end we observe that for any n, t given, there exists a positive integer m such that

$$\int_{-n\delta}^{t} \|\Phi(t,\eta)\|\,d\eta \leq \int_{t-m\delta}^{t} \|\Phi(t,\eta)\|\,d\eta.$$

If we apply the Mean Value Theorem for Integrals to

$$\int_{t-m\delta}^{t} \|\Phi(t,\eta)\|\,d\eta = \int_{t-\delta}^{t} \|\Phi(t,\eta)\|\,d\eta + \int_{t-2\delta}^{t-\delta} \|\Phi(t,\eta)\|\,d\eta + \cdots + \int_{t-m\delta}^{t-(m-1)\delta} \|\Phi(t,\eta)\|\,d\eta,$$

we obtain for $\tilde{\eta}_i \in [t - i\delta, t-(i-1)\delta]$, $i = 1, \ldots, m$, that

$$\begin{aligned}\int_{-n\delta}^{t} \|\Phi(t,\eta)\|\,d\eta &\leq \int_{t-m\delta}^{t} \|\Phi(t,\eta)\|\,d\eta \\ &\leq \delta[\|\Phi(t,\tilde{\eta}_1)\| + \|\Phi(t,\tilde{\eta}_2)\| + \cdots + \|\Phi(t,\tilde{\eta}_m)\|] \\ &\leq \cdots \leq \delta e^{\alpha\delta} ck \triangleq L.\end{aligned}$$

Now invoking Theorem 5.5(iv) of this chapter, we conclude that the equilibrium $w = 0$ of (LH) is exponentially stable. ■

THEOREM 9.3. Suppose that there exist positive constants α, β, and γ such that $\|A(t)\| \leq \alpha$, $\|B(t)\| \leq \beta$, and $\|C(t)\| \leq \gamma$ for all t and assume that there exist positive constants $\epsilon_1, \epsilon_2, \delta_1$, and δ_2 such that for all t_0, $W(t_0, t_0+\delta_1) \geq \epsilon_1 I$ and $M(t_0, t_0+\delta_2) \geq \epsilon_2 I$, where $M(\cdot)$ denotes the observability Gramian given in (9.5). Then the system (9.2) is uniformly BIBO stable if and only if the equilibrium $w = 0$ of (LH) is exponentially stable.

Proof. Uniform exponential stability of the trivial solution of (LH) and the hypotheses of this theorem imply the uniform BIBO stability of system (9.2) by Corollary 9.1. To complete the proof, we show that the hypotheses of the theorem and the BIBO stability of system (9.2) imply the exponential stability of the equilibrium $w = 0$ of (LH).

To set up a contradiction, assume that uniform BIBO stability of system (9.2) does not imply exponential stability of the trivial solution of (LH). Then by Theorem 9.2, it must not imply uniform BIBO stability of system (9.13). For if there is no bound on the state, then there can be no bound on the output. To see this, let $u = 0$ on $t \leq \tau \leq t+\delta$ and obtain

$$\begin{aligned}\int_{t}^{t+\delta} \|y(\tau)\|^2\,d\tau &= x(t)^T\left[\int_{t}^{t+\delta} \Phi(t,\tau)^T C(\tau)^T C(\tau)\Phi(t,\tau)\,d\tau\right]x(t) \\ &= x(t)^T M(t,t+\delta)x(t) \leq \max_{t\leq\tau\leq t+\delta} \delta\|y(\tau)\|^2,\end{aligned}$$

and therefore,

$$\max_{t\leq\tau\leq t+\delta} \|y(\tau)\|^2 \geq \delta^{-1}\|x(t)\|^2 \lambda_{\min}[M(t,t+\delta)],$$

where $\lambda_{\min}[M(t,t+\delta)]$ denotes the smallest eigenvalue of $M(t,t+\delta)$. Thus, if the state x is not bounded for all bounded u, then y will also not be bounded. This shows that the uniform BIBO stability of (9.2) implies the uniform BIBO stability of (9.13). Applying Theorem 9.2, we conclude that the equilibrium $w = 0$ of (LH) is exponentially stable. ■

EXAMPLE 9.1. We consider the system described by the scalar equations

$$\begin{aligned} \dot{x} &= \left(-\frac{1}{t+1}\right)x + u \\ y &= x. \end{aligned} \tag{9.20}$$

This system is clearly controllable and observable. The zero-input response of this system is determined by the differential equation

$$\dot{w} = \left(-\frac{1}{t+1}\right)w, \qquad w(t_0) = x_0, \qquad t_0 \geq 0. \tag{9.21}$$

It is easily verified that the solution of (9.21) is given by

$$\phi(t, t_0, x_0) = \frac{1+t_0}{1+t}x_0 \tag{9.22}$$

[refer to Eqs. (4.5) and (4.6)]. The origin $w = 0$ is the only equilibrium of (9.21), and as seen from (9.22), this equilibrium is uniformly stable and asymptotically stable; however, it is not uniformly asymptotically stable, and hence, it is not exponentially stable.

The state transition matrix of system (9.21) is given by

$$\Phi(t, t_0) = \frac{1+t_0}{1+t}.$$

With $t_0 = 0$ and bounded input $u(t) = 1, t \geq 0$, the zero-state response of system (9.20) is

$$\begin{aligned} y(t, t_0, x_0) &= \int_0^t \Phi(t,\tau)u(\tau)\,d\tau = \int_0^t \frac{1+\tau}{1+t}\,d\tau \\ &= \frac{t + t^2/2}{1+t}. \end{aligned} \tag{9.23}$$

Summarizing, even though the zero input response of system (9.20) tends to zero as $t \to \infty$, the hypotheses of Theorem 9.3 are not satisfied since this decay is not exponential and uniform [i.e., the origin $w = 0$ of (9.21) is not exponentially stable]. In accordance with Theorem 9.3, we cannot expect the zero-state response of system (9.18) to be bounded for arbitrary bounded inputs. This is evident from expression (9.23). ■

Next, we consider in particular the time-invariant system (9.6). For this system it is easily verified that Theorem 9.3 reduces to the following appealing result that connects the uniform BIBO stability of system (9.6) and the exponential stability of the trivial solution of (L).

THEOREM 9.4. Assume that the time-invariant system (9.6) is controllable and observable. Then system (9.6) is uniformly BIBO stable if and only if the trivial solution of (L) is exponentially stable. ■

EXAMPLE 9.2. Consider the system

$$\begin{aligned} \dot{x} &= Ax + Bu \\ y &= Cx, \end{aligned}$$

where

$$A = \begin{bmatrix} 0 & 1 \\ 1 & 0 \end{bmatrix}, \qquad B = \begin{bmatrix} 0 \\ 1 \end{bmatrix}, \qquad C = [0 \quad -1].$$

The eigenvalues of A are $\lambda_1 = 1$, $\lambda_2 = -1$, and therefore the equilibrium $w = 0$ of the system $\dot{w} = Aw$ is unstable. The state transition matrix of this system is

$$\Phi(t,0) = \begin{bmatrix} \frac{1}{2}(e^t + e^{-t}) & \frac{1}{2}(e^t - e^{-t}) \\ \frac{1}{2}(e^t - e^{-t}) & \frac{1}{2}(e^t + e^{-t}) \end{bmatrix},$$

and the impulse response of the system is

$$H(t) = C\Phi(t,0)B = -e^{-t}.$$

Thus, even though the equilibrium $w = 0$ of $\dot{w} = Aw$ is unstable, the system is uniformly BIBO stable. The reason for this is that the system is not observable, as is verified by noting that

$$\begin{bmatrix} C \\ CA \end{bmatrix} = \begin{bmatrix} 1 & -1 \\ -1 & 1 \end{bmatrix},$$

which is singular. Thus, the hypotheses of Theorem 9.4 are not valid, with the consequence that the unstable mode of the system is not observable at the system output. ■

EXAMPLE 9.3. We now consider the system

$$\dot{x} = Ax + Bu$$
$$y = Cx,$$

where A and B are as in Example 9.2 and

$$C = [1 \quad 2].$$

The eigenvalues and the state transition matrix of this system are identical to those of the system given in Example 9.2. This system is both controllable and observable, and the impulse response is

$$H(t) = \tfrac{3}{4}e^{t} - \tfrac{1}{4}e^{-t}.$$

Since the system is controllable and observable and since the equilibrium $w = 0$ of the system $\dot{w} = Aw$ is unstable, it follows from Theorem 9.4 that the system cannot be uniformly BIBO stable. This can be verified directly by inspecting $H(t)$ above. ■

Next, we recall that a complex number s_p is a *pole* of $\hat{H}(s) = [\hat{h}_{ij}(s)]$ if for some pair (i, j), we have $|\hat{h}_{ij}(s_p)| = \infty$ (refer to the definition of pole in Section 3.5). If each entry of $\hat{H}(s)$ has only poles with negative real values, then, as shown in Chapter 2, each entry of $H(t) = [h_{ij}(t)]$ has a sum of exponentials with exponents with real part negative. It follows that the integral

$$\int_0^\infty \|H(t)\|\,dt$$

is finite, and any realization of $\hat{H}(s)$ will result in a system that is uniformly BIBO stable.

Now conversely, if

$$\int_0^\infty \|H(t)\|\,dt$$

is finite, then the exponential terms in any entry of $H(t)$ must have negative real parts. But then every entry of $\hat{H}(s)$ has poles whose real parts are negative.

We have proved the following result.

THEOREM 9.5. The time invariant system (9.6) is uniformly BIBO stable if and only if all poles of the transfer function $\hat{H}(s)$ given in (9.8) have only poles with negative real parts. ■

CHAPT
Stabi

PART 3 STABILITY OF DISCRETE-TIME SYSTEMS

6.10 DISCRETE-TIME SYSTEMS

In this section we address the Lyapunov stability of an equilibrium of discrete-time systems (internal stability) and the input-output stability of discrete-time systems (external stability). We could establish results for discrete-time systems that are analogous to practically all the stability results that we presented for continuous-time systems. Rather than follow such a plan, we will instead first develop Lyapunov stability results for nonlinear discrete-time systems and then apply these to obtain results for linear systems. This will broaden the reader's horizon by providing a glimpse into the qualitative analysis of dynamical systems described by nonlinear ordinary difference equations. In the Exercise section we ask the reader to imitate these results in establishing Lyapunov stability results for dynamical systems described by nonlinear ordinary differential equations. To keep our presentation simple and manageable, we will confine ourselves throughout this section to time-invariant systems. Among other issues, this approach will allow us to avoid most of the issues involving uniformity.

This section is organized into seven subsections. In the first subsection we provide essential preliminary material. In the second we establish results for the stability, instability, asymptotic stability, and global asymptotic stability of an equilibrium and boundedness of solutions of systems described by autonomous ordinary difference equations. These results are utilized in the third and fifth subsections to arrive at stability results of an equilibrium for linear time-invariant systems described by ordinary difference equations. In the fourth subsection we briefly address a result for discrete-time systems, called the Schur-Cohn criterion, which is in the same spirit as the Routh-Hurwitz criterion is for continuous-time systems. The results of the second and fifth subsections are used to develop Lyapunov stability results for linearizations of nonlinear systems described by ordinary difference equations in the sixth subsection. In the last subsection we present results for the input-output stability of time-invariant discrete-time systems.

A. Preliminaries

We concern ourselves here with finite-dimensional discrete-time systems described by difference equations of the form

$$\begin{aligned} x(k+1) &= Ax(k) + Bu(k) \\ y(k) &= Cx(k), \end{aligned} \tag{10.1}$$

where $A \in R^{n\times n}$, $B \in R^{n\times m}$, $C \in R^{p\times n}$, $k \geq k_0$, and $k, k_0 \in Z^+$. Since (10.1) is time-invariant, we will assume without loss of generality that $k_0 = 0$, and thus, $x : Z^+ \to R^n$, $y : Z^+ \to R^p$, and $u : Z^+ \to R^m$.

The internal dynamics of (10.1) under conditions of no input are described by equations of the form

$$x(k+1) = Ax(k). \tag{10.2}$$

Such equations may arise in the modeling process, or they may be the consequence of the linearization of nonlinear systems described by equations of the form

$$x(k+1) = g(x(k)), \tag{10.3}$$

where $g : R^n \to R^n$. For example, if $g \in C^1(R^n, R^n)$, then in linearizing (10.3) about, e.g., $x = 0$, we obtain

$$x(k+1) = Ax(k) + f(x(k)), \tag{10.4}$$

where $A = (\partial f/\partial x)(x)\big|_{x=0}$ and where $f : R^n \to R^n$ is $o(\|x\|)$ as a norm of x (e.g., the Euclidean norm) approaches zero. Recall that this means that given $\epsilon > 0$, there is a $\delta > 0$ such that $\|f(x)\| < \epsilon\|x\|$ for all $\|x\| < \delta$.

As in Section 6.9, we will study the *external qualitative properties* of system (10.1) by means of the *BIBO stability* of such systems. Since we are dealing with time-invariant systems, we will not have to address any issues of uniformity. Consistent with the definition of input-output stability of continuous-time systems, we will say that the system (10.1) is *BIBO stable* if there exists a constant $L > 0$ such that the conditions

$$x(0) = 0$$

$$\|u(k)\| \leq 1, \qquad k \geq 0,$$

imply that $\|y(k)\| \leq L$ for all $k \geq 0$.

We will study the *internal qualitative properties* of system (10.1) by studying the *Lyapunov stability* properties of an equilibrium of (10.2). We will accomplish this in a more general context by studying the stability properties of an equilibrium of system (10.3).

Since system (10.3) is time-invariant, we will assume without loss of generality that $k_0 = 0$. As in Chapters 1 and 2, we will denote for a given set of initial data $x(0) = x_0$ the solution of (10.3) by $\phi(k, x_0)$. When x_0 is understood or of no importance, we will frequently write $\phi(k)$ in place of $\phi(k, x_0)$. Recall that for system (10.3) [as well as systems (10.1), (10.2), and (10.4)], there are no particular difficulties concerning the existence and uniqueness of solutions, and furthermore, as long as g in (10.3) is continuous, the solutions will be continuous with respect to initial data. Recall also that in contrast to systems described by ordinary differential equations, the solutions of systems described by ordinary difference equations [such as (10.3)] exist in general only in the forward direction of time ($k \geq 0$).

We say that $x_e \in R^n$ is an *equilibrium* of system (10.3) if $\phi(k, x_e) \equiv x_e$ for all $k \geq 0$, or equivalently,

$$g(x_e) = x_e. \tag{10.5}$$

As in the continuous-time case, we will assume without loss of generality that the equilibrium of interest will be the origin, i.e., $x_e = 0$. If this is not the case, then we can always transform (similarly as in the continuous-time case) system (10.3) into a system of equations that have an equilibrium at the origin. Also, as in the case of continuous-time systems, we will generally assume that the equilibrium of (10.3) under study is an isolated equilibrium.

EXAMPLE 10.1. The system described by the equation

$$x(k+1) = x(k)[x(k) - 1]$$

has two equilibria, one at $x_{e1} = 0$ and another at $x_{e2} = 2$. ■

EXAMPLE 10.2. The system described by the equations

$$x_1(k+1) = x_2(k)$$
$$x_2(k+1) = -x_1(k)$$

has an equilibrium at $x_e^T = (0, 0)$. ■

Throughout this section we will assume that the function g in (10.3) is continuous, or if required, continuously differentiable. The various definitions of Lyapunov stability of the equilibrium $x = 0$ of system (10.3) are essentially identical to the corresponding definitions of Lyapunov stability of an equilibrium of continuous-time systems described by ordinary differential equations, replacing $t \in R^+$ by $k \in Z^+$. Since system (10.3) is time-invariant, we will not have to explicitly address the issue of uniformity in these definitions. We will concern ourselves with stability, instability, asymptotic stability, and asymptotic stability in the large of the equilibrium $x = 0$ of (10.3).

We say that the equilibrium $x = 0$ of (10.3) is *stable* if for every $\epsilon > 0$ there exists a $\delta = \delta(\epsilon) > 0$ such that $\|\phi(k, x_0)\| < \epsilon$ for all $k \geq 0$ whenever $\|x_0\| < \delta$. If the equilibrium $x = 0$ of (10.3) is not stable, it is said to be *unstable*. We say that the equilibrium $x = 0$ of (10.3) is *asymptotically stable* if (i) it is stable, and (ii) there exists an $\eta > 0$ such that if $\|x_0\| < \eta$, then $\lim_{k\to\infty} \|\phi(k, x_0)\| = 0$. If the equilibrium $x = 0$ satisfies property (ii), it is said to be *attractive,* and we call the set of all $x_0 \in R^n$ for which $x = 0$ is attractive the *domain of attraction* of this equilibrium. If $x = 0$ is asymptotically stable and if its domain of attraction is all of R^n, then it is said to be *asymptotically stable in the large* or *globally asymptotically stable*. Finally, we say that a solution of (10.3) through x_0 is *bounded* provided there is a constant M such that $\|\phi(k, x_0)\| \leq M$ for all $k \geq 0$.

In establishing results for the various stability concepts enumerated above, we will make use of auxiliary functions $v \in C(R^n, R)$, called *Lyapunov functions*. We define the *first forward difference of v along the solutions of* (10.3) as

$$Dv(x) = v(g(x)) - v(x). \tag{10.6}$$

To see that this definition makes sense, note that along *any* solution $\phi(k)$ of (10.3) we have

$$\begin{aligned} v[\phi(k+1)] - v[\phi(k)] &= v[g(\phi(k))] - v[\phi(k)] \\ &= Dv[\phi(k)] \end{aligned} \tag{10.7}$$

for all $k \geq 0$. Note that in evaluating the first forward difference of v along the solutions of (10.3), we need not know explicitly the solution $\phi(k)$ of (10.3).

We will require several characterizations of the Lyapunov functions. We say that a function $v \in C(R^n, R)$ is *positive definite* if $v(0) = 0$ and if $v(x) > 0$ for all $x \in B(\eta) - \{0\}$ for some $\eta > 0$. [Recall that $B(\eta) = \{x \in R^n : \|x\| < \eta\}$.] The function v is *negative definite* if $-v$ is positive definite. A function $v \in C(R^n, R)$ is said to be *positive semidefinite* if $v(0) = 0$ and if $v(x) \geq 0$ for all $x \in B(\eta)$, and it is said to be *negative semidefinite* if $-v$ is positive semidefinite. A positive definite function v is said to be *radially unbounded* if $v(x) > 0$ for all $x \in R^n - \{0\}$ and if $\lim_{\|x\|\to\infty} v(x) =$

∞. Finally, a function $v \in C(R^n, R)$ is said to be *indefinite* if $v(0) = 0$ and if in every neighborhood of the origin v assumes positive and negative values.

B. Lyapunov Stability of an Equilibrium

In establishing various Lyapunov stability results of the equilibrium $x = 0$ of system (10.3), we will find it useful to employ a preliminary result that is important in its own right. To present this result, we require some additional concepts given in the following.

We say that a subset $A \subset R^n$ is *positively invariant* [with respect to system (10.3)] if $g(A) \subset A$, where $g(A) = \{y \in R^n : y = g(x)$ for some $x \in A\}$. Thus, if $x \in A$, then $g(x) \in A$.

EXAMPLE 10.3. It is clear that any set consisting of an equilibrium solution of (10.3) is positively invariant. Also, the set $\{\phi(k, x_0), k \in Z^+\}$ [where ϕ is the solution of (10.3), with $\phi(k_0) = x_0$], which is called the *positive orbit* of x_0 for (10.3), is positively invariant. In particular, for arbitrary initial conditions $x(0)^T = (x_1(0), x_2(0))$, the system given in Example 10.2 has the *periodic solution* with period 4 and positive orbit given by the set

$$A = \{(x_1(0), x_2(0))^T, (x_2(0), -x_1(0))^T, (-x_1(0), -x_2(0))^T, (-x_2(0), x_1(0))^T\},$$

and in general we have $\phi(k, x_0) = \phi(k + 3, x_0)$, $k = 0, 1, 2, 3, \ldots$. The set A is clearly positively invariant. ■

Next, consider a specific solution $\phi(k, x_0)$ for system (10.3). A point $y \in R^n$ is called a *positive limit point* of $\phi(k, x_0)$ if there is a subsequence $\{k_i\}$ of the sequence $\{k\}$, $k \geq 0$, such that $\phi(k_i, x_0) \to y$. The set of all positive limit points of $\phi(k, x_0)$ is called the *positive limit set* $\omega(x_0)$ of $\phi(k, x_0)$, or simply the ω*-limit set*.

Before stating and proving our first result, which is a preliminary result, we recall that a sequence $\{x^k\} \subset R^n$ is said to approach a set $A \subset R^n$, if $d(x^k, A) = \inf\{\|x^k - y\| : y \in A\}$ approaches zero as $k \to \infty$ [$d(x, y)$ denotes the distance function defined in Subsection 1.10C].

THEOREM 10.1. If the solution $\phi(k, x_0)$ of (10.3) is bounded for all $k \in Z^+$, then $\omega(x_0)$ is a nonempty, compact, positively invariant set. Furthermore, $\phi(k, x_0) \to \omega(x_0)$ as $k \to \infty$.

Proof. The complement of $\omega(x_0)$ is open, and therefore, the set $\omega(x_0)$ is closed. Since $\phi(k, x_0)$ is bounded, $\|\phi(k, x_0)\| \leq M$ for some fixed M. Therefore, $\omega(x_0)$ is bounded and $\|y\| \leq M$ for all $y \in \omega(x_0)$. Thus, $\omega(x_0)$ is compact. Since $\phi(k, x_0)$ is bounded, the Bolzano-Weierstrass Theorem guarantees the existence of at least one limit point, and therefore, $\omega(x_0)$ is not empty (refer to Subsection 1.5A).

Next, let $y \in \omega(x_0)$ so that $\phi(k_i, x_0) \to y$ as $i \to \infty$. Since the function g in (10.3) is continuous, it follows that $\phi(k_i + 1, x_0) = g(\phi(k_i, x_0)) \to g(y)$ or $g(y) \in \omega(x_0)$. The set $\omega(x_0)$ is positively invariant [with respect to system (10.3)]. Also, $d(\phi(k, x_0), \omega)$ is bounded since both $\phi(k, x_0)$ and $\omega(x_0)$ are bounded. To set up a contradiction, assume that $d(\phi(k, x_0), \omega(x_0))$ does not converge to zero. Then there is a subsequence $\{k_i\}$ such that $\phi(k_i, x_0) \to y$ and $d(\phi(k_i, x_0), \omega(x_0)) \to a > 0$. But then $y \in \omega(x_0)$, $d(\phi(k_i, x_0), \omega(x_0)) \leq d(\phi(k_i, x_0), y) \to 0$, and therefore, $d(\phi(k_i, x_0), \omega(x_0)) \to 0$, which is a contradiction. ■

We are now in a position to state and prove several Lyapunov stability results for system (10.3).

THEOREM 10.2. The equilibrium $x = 0$ of system (10.3) is *stable* if there exists a positive definite function v such that Dv is negative semidefinite.

Proof. We take $\eta > 0$ so small that $v(x) > 0$ for all $x \in B(\eta) - \{0\}$ and $Dv(x) \leq 0$ for all $x \in B(\eta)$ [recall that $B(\eta) = \{x \in R^n : \|x\| < \eta\}$]. Let $\epsilon > 0$ be given. There is no loss of generality in choosing $0 < \epsilon < \eta$. Let $m = \min\{v(x) : \|x\| = \epsilon\}$. Then m is positive since we are taking the minimum of a positive continuous function over a compact set. Let $G = \{x \in R^n : v(x) < m/2\}$, which may consist of several disjoint connected sets called connected components of G. Let G_0 denote the connected component of G that contains the origin $x = 0$. Then both G and G_0 are open sets. Now if $x_0 \in G_0$, then $Dv(x_0) \leq 0$ and so $v(g(x_0)) \leq v(x_0) < m/2$, and therefore, $g(x_0) \in G$. Since x_0 and the origin $x = 0$ are both in the same component of G, then so are $g(0) = 0$ and $g(x_0)$. Therefore, G_0 is an open positively invariant set containing $x = 0$ and contained in $B(\epsilon)$. Since v is continuous there is a $\delta > 0$ such that $B(\delta) \subset G_0$. Therefore, if $x_0 \in B(\delta)$, then $x_0 \in G_0$ and $g(x_0) \in G_0 \subset B(\epsilon)$. This shows that the equilibrium $x = 0$ of system (10.3) is stable. ■

EXAMPLE 10.4. For the system given in Example 10.2 we choose the function $v(x_1, x_2) = x_1^2 + x_2^2$. Then

$$\begin{aligned} Dv(x_1, x_2) &= v(g(x_1, x_2)) - v(x_1, x_2) \\ &= (x_2)^2 + (-x_1)^2 - x_1^2 - x_2^2 = 0. \end{aligned}$$

Therefore, by Theorem 10.2 the equilibrium $x = 0$ of the system is stable. ■

By using a similar argument as in the proof of Theorem 10.2 we can prove the boundedness result given below. We will not present the details of the proof.

THEOREM 10.3. If v is radially unbounded and $Dv(x) \leq 0$ on the set where $\|x\| \geq M$ (M is some constant), then all solutions of system (10.3) are *bounded*. ■

EXAMPLE 10.5. For the system given in Example 10.2 we choose the radially unbounded function $v(x_1, x_2) = x_1^2 + x_2^2$. As shown in Example 10.4, $Dv(x_1, x_2) \leq 0$ for all $x \in R^2$. It follows from Theorem 10.3 that all the solutions of the system are bounded. ■

By definition, the equilibrium $x = 0$ of system (10.3) is asymptotically stable if it is stable and attractive. Theorem 10.2 provides a set of sufficient conditions for the stability of the trivial solution of system (10.3). In the next result, known as *LaSalle's Theorem* or the *Invariance Principle*, we present a method that enables us to determine the attractivity of a set. If this set consists only of the equilibrium $x = 0$, this result yields a method of determining the attractivity of the trivial solution $x = 0$ of system (10.3).

Let $v \in C(R^n, R)$ and let $v(x) = c$. In the following, we let $v^{-1}(c) = \{x \in R^n : v(x) = c\}$.

THEOREM 10.4. Let $v \in C(R^n, R)$ and let $G \subset R^n$. Assume that (i) $Dv(x) \leq 0$ for all $x \in G$, and (ii) the solution $\phi(k, x_0)$ of (10.3) is in G for all $k \geq 0$ and is bounded for all $k \geq 0$. Then there is a number c such that $\phi(k, x_0) \to M \cap v^{-1}(c)$, where M is the largest positively invariant set contained in the set $E = \{x \in R^n : Dv(x) = 0\} \cap \bar{G}$.

Proof. Since $\phi(k, x_0)$ is bounded and in G, it follows that $\omega(x_0) \neq \emptyset$, $\omega(x_0) \subset \bar{G}$, and $\phi(k, x_0)$ tends to $\omega(x_0)$. Now $v(\phi(k, x_0))$ is nonincreasing with increasing k and bounded from below. Therefore, $v(\phi(k, x_0)) \to c$. If $y \in \omega(x_0)$, there is subsequence $\{k_i\}$ of the sequence $\{k\}$ such that $\phi(k_i, x_0) \to y$ and therefore $v(\phi(k_i, x_0)) \to v(y)$ or $v(y) = c$. Therefore, $v(\omega(x_0)) = c$ [i.e., $v(y) = c$ for all $y \in \omega(x_0)$] or $\omega(x_0) \subset v^{-1}(c)$. Since $v(\omega(x_0)) = c$ and $\omega(x_0)$ is positively invariant, it follows that $Dv(\omega(x_0)) = 0$ [i.e.,

$Dv(y) = 0$ for all $y \in \omega(x_0)$]. Therefore, $\phi(k, x_0) \to \omega(x_0) \subset \{x \in R^n : Dv(x) = 0\} \cap \bar{G} \cap v^{-1}(c)$. Since $\omega(x_0)$ is positively invariant, it now follows that $\omega(x_0) \subset M$. ■

The next results, which are direct consequences of Theorems 10.2, 10.3, and 10.4, were originally established by Lyapunov.

COROLLARY 10.1. The equilibrium $x = 0$ of (10.3) is *asymptotically stable* if there exists a positive definite function v such that Dv is negative definite.

Proof. Since Dv is negative definite and v is positive definite, it follows from Theorem 10.2 that the equilibrium $x = 0$ is stable. From the proof of that theorem there is an arbitrarily small neighborhood G_0 of the origin that is positively invariant. We can make G_0 so small that $v(x) > 0$ and $\dot{v}(x) < 0$ for all $x \in G_0 - \{0\}$. So, given any $x_0 \in G_0$ it follows from the invariance principle (Theorem 10.4) that $\phi(k, x_0)$ tends to the largest invariant set in $G_0 \cap \{Dv(x) = 0\} = \{x = 0\}$ since Dv is negative definite. Therefore, the equilibrium $x = 0$ is asymptotically stable. ■

COROLLARY 10.2. The equilibrium $x = 0$ of (10.3) is *asymptotically stable in the large* (or *globally asymptotically stable*) if there exists a positive definite function v that is radially unbounded and if Dv is negative definite on R^n, i.e., $Dv(0) = 0$ and $Dv(x) < 0$ for all $x \neq 0$. ■

Proof. From Theorem 10.3 all solutions of system (10.3) are bounded. The proof now follows by modifying the proof of Corollary 10.1 in the obvious way. ■

EXAMPLE 10.6. Consider the system described by the equations

$$x_1(k+1) = \frac{a x_2(k)}{1 + x_1(k)^2}$$

$$x_2(k+1) = \frac{b x_1(k)}{1 + x_2(k)^2},$$

where it is assumed that $a^2 < 1$ and $b^2 < 1$. We choose a Lyapunov function $v(x_1, x_2) = x_1^2 + x_2^2$ that is positive definite and radially unbounded. Along the solutions of this system we compute the first forward difference as

$$\begin{aligned} Dv(x_1, x_2) &= \frac{a^2 x_2^2}{(1 + x_1^2)^2} + \frac{b^2 x_1^2}{(1 + x_2^2)^2} - (x_1^2 + x_2^2) \\ &= \left[\frac{a^2}{(1 + x_1^2)^2} - 1\right] x_2^2 + \left[\frac{b^2}{(1 + x_2^2)^2} - 1\right] x_1^2 \\ &\leq (a^2 - 1)x_2^2 + (b^2 - 1)x_1^2. \end{aligned}$$

Since by assumption $a^2 < 1$ and $b^2 < 1$, it follows that $Dv(x_1, x_2) < 0$ for all $x^T = (x_1, x_2) \neq 0$ and $Dv(x_1, x_2) = 0$ when $x = 0$. It follows from Theorem 10.3 that all solutions of the system are bounded, and from Corollary 10.2 it follows that the equilibrium $x = 0$ of the system is globally asymptotically stable. ■

EXAMPLE 10.7. We reconsider the system given in Example 10.6 under the assumption that $a^2 \leq 1$ and $b^2 \leq 1$, but $a^2 + b^2 \neq 2$. Without loss of generality we consider the case $a^2 < 1$ and $b^2 = 1$. As in Example 10.6, we again choose $v(x_1, x_2) = x_1^2 + x_2^2$, and from the computations in that example we see that

$$\begin{aligned} Dv(x_1, x_2) &\leq (a^2 - 1)x_2^2 + (b^2 - 1)x_1^2 \\ &= (a^2 - 1)x_2^2 \leq 0, \ (x_1, x_2)^T \in R^2. \end{aligned}$$

It still follows from Theorem 10.3 that all solutions of the system are bounded.

Since $Dv(x_1, x_2)$ is not negative definite, but negative semidefinite, we cannot apply Corollary 10.2 to establish the asymptotic stability of the equilibrium $x = 0$. So let us try to use Theorems 10.2 and 10.4 to accomplish this. From the former we conclude that the equilibrium is stable. Using the notation of Theorem 10.4 we note that $E = \{(x_1, 0)^T\}$, which is the x_1-axis. Now $g((x_1, 0)^T) = (0, bx_1)^T = (0, x_1)^T$ [where $g(\cdot)$ is defined in (10.3)], and therefore, the only invariant subset of (E) is the set consisting of the origin. All conditions of Theorem 10.4 are satisfied (with $G = R^2$), and we conclude that the equilibrium $x = 0$ of the system is globally asymptotically stable. ■

THEOREM 10.5. Let Dv be positive definite and assume that in every neighborhood of the origin there is $\bar{x}$ such that $v(\bar{x}) > 0$. Then the equilibrium $x = 0$ of system (10.3) is *unstable*. [Alternatively, let Dv be negative definite and assume that in every neighborhood of the origin there is $\bar{x}$ such that $v(\bar{x}) < 0$. Then the equilibrium $x = 0$ of system (10.3) is unstable.]

Proof. To set up a contradiction, assume that $x = 0$ is stable. Choose $\epsilon > 0$ so small so that $Dv(x) > 0$ for all $x \in B(\epsilon) - \{0\}$, and choose $\delta > 0$ so small so that if $x_0 \in B(\delta)$ then $\phi(k, x_0) \in B(\epsilon)$ for all $k \geq 0$. By hypothesis there is a point $x_0 \in B(\delta)$ such that $v(x_0) > 0$. Since $\phi(k, x_0)$ is bounded and remains in $B(\epsilon)$, the solution $\phi(k, x_0)$ will tend to its limit set $\{x \in R^n : Dv(x) = 0\} \cap B(\epsilon) = \{0\}$. Since $\phi(k, x_0) \to 0$, we have $v(\phi(k, x_0)) \to v(0) = 0$. But $Dv(\phi(k, x_0)) > 0$ and therefore $v(\phi(k, x_0)) \geq 0$, and thus $v(\phi(k, x_0)) \geq v(\phi(k-1, x_0)) \geq \cdots \geq v(x_0) > 0$. We have thus arrived at a contradiction that proves the theorem. ■

EXAMPLE 10.8. The system described by the equation

$$x(k + 1) = 2x(k)$$

has an equilibrium $x = 0$. The function $v(x) = x^2$ is positive definite, and along the solutions of this system we have $Dv(x) = 4x^2 - x^2 = 3x^2$, which is also positive definite. The conditions of Theorem 10.5 are satisfied and the equilibrium $x = 0$ of the system is unstable. ■

C. Linear Systems

In proving some of the results of this section, we require a result for system (10.2) that is analogous to Theorem 3.1 of Chapter 2. As in the proof of that theorem, we note that the linear combination of solutions of system (10.2) is also a solution of system (10.2), and hence, the set of solutions $\{\phi : Z^+ \times R^n \to R^n\}$ constitutes a vector space (over $F = R$ or $F = C$). The dimension of this vector space is n. To show this, we choose a set of linearly independent vectors $x_0^1, \ldots, x_0^n$ in the n-dimensional x-space (R^n or C^n) and we show, in an identical manner as in the proof of Theorem 3.1 of Chapter 2, that the set of solutions $\phi(k, x_0^i)$, $i = 1, \ldots, n$, is linearly independent and spans the set of solutions of system (10.2). (We ask the reader in the Exercise section to provide the details of the proof of the above assertions.) This yields the following result.

THEOREM 10.6. The set of solutions of system (10.2) over the time interval Z^+ forms an n-dimensional vector space. ■

Incidentally, if in particular we choose $\phi(k, e^i)$, $i = 1, \ldots, n$, where e^i, $i = 1, \ldots, n$, denotes the natural basis for R^n, and if we let $\Phi(k, k_0 = 0) \triangleq \Phi(k) = [\phi(k, e^1), \ldots, \phi(k, e^n)]$, then it is easily verified that the $n \times n$ matrix $\Phi(k)$ satisfies the matrix equation

$$\Phi(k+1) = A\Phi(k), \qquad \Phi(0) = I$$

and that $\Phi(k) = A^k$, $k \geq 0$ [i.e., $\Phi(k)$ is the state transition matrix for system (10.2)].

THEOREM 10.7. The equilibrium $x = 0$ of system (10.2) is stable if and only if the solutions of (10.2) are bounded.

Proof. Assume that the equilibrium $x = 0$ of (10.2) is stable. Then for $\epsilon = 1$ there is a $\delta > 0$ such that $\|\phi(k, x_0)\| < 1$ for all $k \geq 0$ and all $\|x_0\| \leq \delta$. In this case

$$\|\phi(k, x_0)\| = \|A^k x_0\| = \left\| \frac{A^k x_0 \delta}{\|x_0\|} \right\| \left(\frac{\|x_0\|}{\delta} \right) < \frac{\|x_0\|}{\delta}$$

for all $x_0 \neq 0$ and all $k \geq 0$. Using the definition of matrix norm [refer to (10.17) of Chapter 1] it follows that $\|A^k\| \leq \delta^{-1}$, $k \geq 0$. We have proved that if the equilibrium $x = 0$ of (10.2) is stable, then the solutions of (10.2) are bounded.

Conversely, suppose that all solutions $\phi(k, x_0) = A^k x_0$ are bounded. Let $\{e^1, \ldots, e^n\}$ denote the natural basis for n-space and let $\|\phi(k, e^j)\| < \beta_j$ for all $k \geq 0$. Then for any vector $x_0 = \sum_{j=1}^n \alpha_j e^j$ we have that

$$\|\phi(k, x_0)\| = \left\| \sum_{j=1}^n \alpha_j \phi(k, e^j) \right\| \leq \sum_{j=1}^n |\alpha_j| \beta_j \leq (\max_j \beta_j) \sum_{j=1}^n |\alpha_j|$$
$$\leq c\|x_0\|, \qquad k \geq 0,$$

for some constant c. For given $\epsilon > 0$, we choose $\delta = \epsilon/c$. Then, if $\|x_0\| < \delta$, we have $\|\phi(k, x_0)\| < c\|x_0\| < \epsilon$ for all $k \geq 0$. We have proved that if the solutions of (10.2) are bounded, then the equilibrium $x = 0$ of (10.2) is stable. ■

THEOREM 10.8. The following statements are equivalent:

(i) The equilibrium $x = 0$ of (10.2) is asymptotically stable.
(ii) The equilibrium $x = 0$ of (10.2) is asymptotically stable in the large.
(iii) $\lim_{k\to\infty} \|A^k\| = 0$.

Proof. Assume that statement (i) is true. Then there is an $\eta > 0$ such that when $\|x_0\| \leq \eta$, then $\phi(k, x_0) \to 0$ as $k \to \infty$. But then we have for *any* $x_0 \neq 0$ that

$$\phi(k, x_0) = A^k x_0 = \left[A^k \left(\frac{\eta x_0}{\|x_0\|} \right) \right] \frac{\|x_0\|}{\eta} \to 0 \text{ as } k \to \infty.$$

It follows that statement (ii) is true.

Next, assume that statement (ii) is true. Then for any $\epsilon > 0$ there must exist a $K = K(\epsilon)$ such that for all $k \geq K$ we have that $\|\phi(k, x_0)\| = \|A^k x_0\| < \epsilon$. To see this, let $\{e^1, \ldots, e^n\}$ be the natural basis for R^n. Thus, for a fixed constant $c > 0$, if $x_0 = (\alpha_1, \ldots, \alpha_n)^T$ and if $\|x_0\| \leq 1$, then $x_0 = \sum_{j=1}^n \alpha_j e^j$ and $\sum_{j=1}^n |\alpha_j| \leq c$. For each j there is a $K_j = K_j(\epsilon)$ such that $\|A^k e^j\| < \epsilon/c$ for $k \geq K_j$. Define $K = K(\epsilon) = \max\{K_j(\epsilon) : j = 1, \ldots, n\}$. For $\|x_0\| \leq 1$ and $k \geq K$ we have that

$$\|A^k x_0\| = \left\| \sum_{j=1}^n \alpha_j A^k e^j \right\| \leq \sum_{j=1}^n |\alpha_j| \left(\frac{\epsilon}{c} \right) \leq \epsilon.$$

By the definition of matrix norm [see (10.17) of Chapter 1], this means that $\|A^k\| \leq \epsilon$ for $k > K$. Therefore, statement (iii) is true.

Finally, assume that statement (iii) is true. Then $\|A^k\|$ is bounded for all $k \geq 0$. By Theorem 10.7, the equilibrium $x = 0$ is stable. To prove asymptotic stability, fix $\epsilon > 0$.

If $\|x_0\| < \eta = 1$, then $\|\phi(k, x_0)\| \leq \|A^k\| \, \|x_0\| \to 0$ as $k \to \infty$. Therefore, statement (i) is true. This completes the proof. ■

To arrive at the next result, we make reference to the results of Subsection 2.7E. Specifically, by inspecting the expressions for the modes of system (10.2) given in (7.50) and (7.51) of Chapter 2, or by utilizing the Jordan canonical form of A [refer to (7.54) and (7.55) of Chapter 2], the following result is evident.

THEOREM 10.9. (i) The equilibrium $x = 0$ of system (10.2) is *asymptotically stable* if and only if all eigenvalues of A are within the unit circle of the complex plane (i.e., if $\lambda_1, \ldots, \lambda_n$ denote the eigenvalues of A, then $|\lambda_j| < 1$, $j = 1, \ldots, n$). In this case we say that *the matrix A is Schur stable*, or simply, *the matrix A is stable*.

(ii) The equilibrium $x = 0$ of system (10.2) is *stable* if and only if $|\lambda_j| \leq 1$, $j = 1, \ldots, n$, and for each eigenvalue with $|\lambda_j| = 1$ having multiplicity $n_j > 1$, it is true that

$$\lim_{z \to \lambda_j} \left\{ \frac{d^{n_j - 1 - l}}{dz^{n_j - 1 - l}} [(z - \lambda_j)^{n_j} (zI - A)^{-1}] \right\} = 0, \qquad l = 1, \ldots, n_j - 1.$$

(iii) The equilibrium $x = 0$ of system (10.2) is *unstable* if and only if the conditions in (ii) above are not true. ■

Alternatively, it is evident that the equilibrium $x = 0$ of system (10.2) is *stable* if and only if all eigenvalues of A are within or on the unit circle of the complex plane, and every eigenvalue that is on the unit circle has an associated Jordan block of order 1.

EXAMPLE 10.9. (i) For the system in Example 10.2 we have

$$A = \begin{bmatrix} 0 & 1 \\ -1 & 0 \end{bmatrix}.$$

The eigenvalues of A are $\lambda_1, \lambda_2 = \pm\sqrt{-1}$. According to Theorem 10.9, the equilibrium $x = 0$ of the system is stable, and according to Theorem 10.7 the matrix A^k is bounded for all $k \geq 0$.

(ii) For system (10.2) let

$$A = \begin{bmatrix} 0 & -\frac{1}{2} \\ -1 & 0 \end{bmatrix}.$$

The eigenvalues of A are $\lambda_1, \lambda_2 = \pm 1/\sqrt{2}$. According to Theorem 10.9, the equilibrium $x = 0$ of the system is asymptotically stable, and according to Theorem 10.8, $\lim_{k \to \infty} A^k = 0$.

(iii) For system (10.2) let

$$A = \begin{bmatrix} 0 & -\frac{1}{2} \\ -3 & 0 \end{bmatrix}.$$

The eigenvalues of A are $\lambda_1, \lambda_2 = \pm\sqrt{3/2}$. According to Theorem 10.9, the equilibrium $x = 0$ of the system is unstable, and according to Theorem 10.7, the matrix A^k is not bounded with increasing k.

(iv) For system (10.2) let

$$A = \begin{bmatrix} 1 & 1 \\ 0 & 1 \end{bmatrix}.$$

The matrix A is a Jordan block of order 2 for the eigenvalue $\lambda = 1$. Accordingly, the equilibrium $x = 0$ of the system is unstable (refer to the remark following Theorem 10.9) and the matrix A^k is unbounded with increasing k. ■

D. The Schur-Cohn Criterion

In this section we present a method, called the *Schur-Cohn criterion*, that enables us to determine whether or not the roots of a polynomial with real coefficients given by

$$p(\lambda) = a_n\lambda^n + a_{n-1}\lambda^{n-1} + \cdots + a_0, \qquad a_n > 0, \tag{10.8}$$

lie inside of the unit circle in the complex plane by examining the polynomial coefficients, rather than solving for the roots. This method provides us with an efficient means of studying the stability of the equilibrium $x = 0$ of system (10.2) by applying it to the characteristic polynomial of the matrix A. The Schur-Cohn criterion is in the same spirit [for discrete-time systems (10.2)] as the Routh-Hurwitz criterion [for continuous-time systems (L)]. In presenting this criterion, we will require the concept of *inners* of a square matrix A, defined as the matrix itself, and all the matrices obtained by omitting successively the first and last rows and the first and last columns. In the following, we depict the inners for a matrix $A \in R^{5\times 5}$ and a matrix $A \in R^{6\times 6}$:

$$A = \begin{bmatrix} a_{11} & a_{12} & a_{13} & a_{14} & a_{15} \\ a_{21} & a_{22} & a_{23} & a_{24} & a_{25} \\ a_{31} & a_{32} & a_{33} & a_{34} & a_{35} \\ a_{41} & a_{42} & a_{43} & a_{44} & a_{45} \\ a_{51} & a_{52} & a_{53} & a_{54} & a_{55} \end{bmatrix}, \qquad A = \begin{bmatrix} a_{11} & a_{12} & a_{13} & a_{14} & a_{15} & a_{16} \\ a_{21} & a_{22} & a_{23} & a_{24} & a_{25} & a_{26} \\ a_{31} & a_{32} & a_{33} & a_{34} & a_{35} & a_{36} \\ a_{41} & a_{42} & a_{43} & a_{44} & a_{45} & a_{46} \\ a_{51} & a_{52} & a_{53} & a_{54} & a_{55} & a_{56} \\ a_{61} & a_{62} & a_{63} & a_{64} & a_{65} & a_{66} \end{bmatrix}$$

A matrix is said to be *positive innerwise* if the determinants of all of its inners are positive.

THEOREM 10.10. (SCHUR-COHN CRITERION) A necessary and sufficient condition that the polynomial (10.8) with real coefficients has all its roots inside the unit circle in the complex plane is that

(i) $p(1) > 0$ and $(-1)^n p(-1) > 0$,
(ii) the following $(n-1)\times(n-1)$ matrices are both positive innerwise:

$$\Delta_{n-1}^{\pm} = \begin{bmatrix} a_n & 0 & \cdots & \cdots & 0 \\ a_{n-1} & a_n & & & \vdots \\ \vdots & \ddots & \ddots & & \vdots \\ a_3 & \vdots & \ddots & \ddots & 0 \\ a_2 & a_3 & \cdots & a_{n-1} & a_n \end{bmatrix} \pm \begin{bmatrix} 0 & \cdots & \cdots & \cdots & 0 & a_0 \\ \vdots & & & & a_0 & a_1 \\ \vdots & & & \cdot^{\cdot^{\cdot}} & a_1 & \vdots \\ \vdots & & \cdot^{\cdot^{\cdot}} & \cdot^{\cdot^{\cdot}} & \vdots & \vdots \\ 0 & \cdot^{\cdot^{\cdot}} & \cdot^{\cdot^{\cdot}} & & \vdots & a_{n-1} \\ a_0 & a_1 & \cdots & \cdots & a_{n-1} & a_{n-2} \end{bmatrix}. \tag{10.9}$$

■

We will not present a proof of Theorem 10.10. The interested reader should consult Jury [9] cited in the reference section for a proof of this result.

EXAMPLE 10.10. For the 2×2 matrix

$$A = \begin{bmatrix} a_{11} & a_{12} \\ a_{21} & a_{22} \end{bmatrix}$$

the characteristic polynomial is given by

$$p(\lambda) = \lambda^2 + a_1\lambda + a_0.$$

We have $p(1) = 1 + a_1 + a_0$, $p(-1) = 1 - a_1 + a_0$, and $\Delta_1^{\mp} = 1 \pm a_0 > 0$. Therefore, the roots of $p(\lambda)$ lie within the unit circle of the complex plane if and only if $|a_0| < 1$ and $|a_1| < 1 + a_0$. ■

EXAMPLE 10.11. For the 3×3 matrix

$$A = \begin{bmatrix} a_{11} & a_{12} & a_{13} \\ a_{21} & a_{22} & a_{23} \\ a_{31} & a_{32} & a_{33} \end{bmatrix}$$

the characteristic polynomial is given by

$$p(\lambda) = \lambda^3 + a_2\lambda^2 + a_1\lambda + a_0.$$

We have $p(1) = 1 + a_2 + a_1 + a_0$, $-p(-1) = 1 - a_2 + a_1 - a_0$, and

$$det\,(\Delta_2^{\mp}) = \begin{vmatrix} 1 & \pm a_0 \\ a_2 \pm a_0 & 1 \pm a_1 \end{vmatrix} > 0.$$

Therefore, the roots of $p(\lambda)$ lie within the unit circle of the complex plane if and only if $|a_0 + a_2| < 1 + a_1$ and $|a_1 - a_0 a_2| < 1 - a_0^2$. ■

E. The Matrix Lyapunov Equation

In this subsection we apply the Lyapunov theorems of Subsection B to obtain another characterization of stable matrices. This gives rise to the Lyapunov matrix equation.

Returning to system (10.2) we choose as a Lyapunov function

$$v(x) = x^T B x, \qquad B = B^T, \tag{10.10}$$

and we evaluate the first forward difference of v along the solutions of (10.2) as

$$\begin{aligned} v(x(k+1)) - v(x(k)) &= x(k+1)^T B x(k+1) - x(k)^T B x(k) \\ &= x(k)^T A^T B A x(k) - x(k)^T B x(k) \\ &= x(k)^T (A^T B A - B) x(k), \end{aligned}$$

and therefore,

$$Dv(x) = x^T (A^T B A - B) x \triangleq x^T C x,$$

where

$$A^T B A - B = C, \qquad C^T = C. \tag{10.11}$$

Invocation of Theorem 10.2, Corollary 10.2, and Theorem 10.5, readily leads to the following results.

THEOREM 10.11. (i) The equilibrium $x = 0$ of system (10.2) is *stable* if there exists a real, symmetric, and positive definite matrix B such that the matrix C given in (10.11) is negative semidefinite.

(ii) The equilibrium $x = 0$ of system (10.2) is *asymptotically stable in the large* if there exists a real, symmetric, and positive definite matrix B such that the matrix C given in (10.11) is negative definite.

(iii) The equilibrium $x = 0$ of system (10.2) is *unstable* if there exists a real, symmetric matrix B that is either negative definite or indefinite such that the matrix C given in (10.11) is negative definite. ■

In applying Theorem 10.11, we start by choosing (guessing) a matrix B having certain desired properties and we then solve for the matrix C, using equation (10.11).

If C possesses certain desired properties (i.e., it is negative definite) we can draw appropriate conclusions by applying one of the results given in Theorem 10.11; if not, we need to choose another matrix B. This approach is not very satisfactory, and in the following we will derive results that will allow us (as in the case of continuous-time systems) to *construct* Lyapunov functions of the form $v(x) = x^T Bx$ in a systematic manner. In doing so, we first choose a matrix C in (10.11) that is either negative definite or positive definite, and then we solve (10.11) for B. Conclusions are then made by applying Theorem 10.11. In applying this construction procedure, we need to know conditions under which (10.11) possesses a (unique) solution B for *any* definite (i.e., positive or negative definite) matrix C. We will address this issue next.

We first show that if A is stable, i.e., if all eigenvalues of matrix A [in system (10.2)] are inside the unit circle of the complex plane, then we can compute B in (10.11) explicitly. To show this, we assume that in (10.11) C is a given matrix and that A is stable. Then

$$(A^T)^{k+1}BA^{k+1} - (A^T)^k BA^k = (A^T)^k CA^k,$$

and summing from $k = 0$ to l yields

$$A^T BA - B + (A^T)^2 BA^2 - A^T BA + \cdots - B = \sum_{k=0}^{l} (A^T)^k CA^k$$

or

$$(A^T)^{l+1} BA^{l+1} - B = \sum_{k=0}^{l} (A^T)^k CA^k.$$

Letting $l \to \infty$, we obtain

$$B = -\sum_{k=0}^{\infty} (A^T)^k CA^k. \tag{10.12}$$

It is easy to verify that (10.12) is a solution of (10.11). We have

$$-A^T \left[\sum_{k=0}^{\infty} (A^T)^k CA^k \right] A + \sum_{k=0}^{\infty} (A^T)^k CA^k = C$$

or $\quad -A^T CA + C - (A^T)^2 CA^2 + A^T CA - (A^T)^3 CA^3 + (A^T)^2 CA^2 - \cdots = C.$

Therefore (10.12) is a solution of (10.11). Furthermore, if C is negative definite, then B is positive definite.

Combining the above with Theorem 10.11(ii) we have the following result:

THEOREM 10.12. If there is a positive definite and symmetric matrix B and a negative definite and symmetric matrix C satisfying (10.11), then the matrix A is stable. Conversely, if A is stable, then, given *any* symmetric matrix C, Eq. (10.11) has a unique solution, and if C is negative definite then B is positive definite. ■

Next, we determine conditions under which the system of equations (10.11) has a (unique) solution $B = B^T \in R^{n\times n}$ for a given matrix $C = C^T \in R^{n\times n}$. To accomplish this, we consider the more general equation

$$A_1 X A_2 - X = C, \tag{10.13}$$

where $A_1 \in R^{m\times m}$, $A_2 \in R^{n\times n}$, and X and C are $m \times n$ matrices.

LEMMA 10.1. Let $A_1 \in R^{m\times m}$ and $A_2 \in R^{n\times n}$. Then Eq. (10.13) has a unique solution $X \in R^{m\times n}$ for a given $C \in R^{m\times n}$ if and only if no eigenvalue of A_1 is a reciprocal of an eigenvalue of A_2.

Proof. We need to show that the condition on A_1 and A_2 is equivalent to the condition that $A_1XA_2 = X$ implies $X = 0$. Once we have proved that $A_1XA_2 = X$ has the unique solution $X = 0$, then it can be shown that (10.13) has a unique solution for every C, since (10.13) is a linear equation.

Assume first that the condition on A_1 and A_2 is satisfied. Now $A_1XA_2 = X$ implies that $A_1^{k-j}XA_2^{k-j} = X$ and

$$A_1^jX = A_1^kXA_2^{k-j} \qquad \text{for } k \geq j \geq 0.$$

Now for a polynomial of degree k,

$$p(\lambda) = \sum_{j=0}^{k} a_j\lambda^j,$$

we define the polynomial of degree k,

$$p^*(\lambda) = \sum_{j=0}^{k} a_j\lambda^{k-j} = \lambda^k p\left(\frac{1}{\lambda}\right),$$

from which it follows that

$$p(A_1)X = A_1^kXp^*(A_2).$$

Now let $\phi_i(\lambda)$ be the characteristic polynomial of A_i, $i = 1, 2$. Since $\phi_1(\lambda)$ and $\phi_2^*(\lambda)$ are relatively prime, there are polynomials $p(\lambda)$ and $q(\lambda)$ such that

$$p(\lambda)\phi_1(\lambda) + q(\lambda)\phi_2^*(\lambda) = 1.$$

Now define $\phi(\lambda) = q(\lambda)\phi_2^*(\lambda)$ and note that $\phi^*(\lambda) = q^*(\lambda)\phi_2(\lambda)$. It follows that $\phi^*(A_2) = 0$ and $\phi(A_1) = I$. From this it follows that $A_1XA_2 = X$ implies $X = 0$.

To prove the converse, we assume that λ is an eigenvalue of A_1 and λ^{-1} is an eigenvalue of A_2 (and hence, is also an eigenvalue of A_2^T). Let $A_1x^1 = \lambda x^1$ and $A_2^Tx^2 = \lambda^{-1}x^2$, $x^1 \neq 0$, and $x^2 \neq 0$. Define $X = (x_1^2x^1, x_2^2x^1, \ldots, x_n^2x^1)$. Then $X \neq 0$ and $A_1XA_2 = X$. ■

To construct $v(x)$ by using Lemma 10.1, we must still check the definiteness of B. To accomplish this, we utilize Theorem 10.11.

1. If all eigenvalues of A [for system (10.2)] are inside the unit circle of the complex plane, then no reciprocal of an eigenvalue of A is an eigenvalue, and Lemma 10.1 gives another way of showing that Eq. (10.11) has a unique solution B for each C if A is stable. If C is negative definite, then B is positive definite. This can be shown as was done for the case of linear ordinary differential equations.
2. If at least one of the eigenvalues of A is outside the unit circle of the complex plane and if no reciprocal of an eigenvalue of A is an eigenvalue, and if C in (10.11) is negative definite, then B cannot be positive definite; otherwise we could apply Theorem 10.11(iii) to come up with a contradiction. If in particular, all eigenvalues of A are outside the unit circle of the complex plane, then B must be negative definite. [In this case the equilibrium of (10.2) is *completely unstable*.]

Now suppose that at least one of the eigenvalues of A is outside of the unit circle in the complex plane and suppose that the conditions of Lemma 10.1 are violated (i.e., an eigenvalue of A is the reciprocal of an eigenvalue of A). Then we cannot construct $v(x)$ given in (10.10) in the manner described above (i.e., we cannot determine B in the manner described above). To overcome this difficulty, we form in this case the matrix

$$A_0 = (1 + \beta)^{-1/2} A, \tag{10.14}$$

and we choose $|\beta|$ arbitrarily small and in such a manner so that A_0 has no eigenvalues on the unit circle and has the same number of eigenvalues outside of the unit circle of the complex plane as matrix A. Then A_0 satisfies the conditions of Lemma 10.1. For every given $C = C^T \in R^{n\times n}$ we can now solve the equation

$$A_0^T B A_0 - B = C \tag{10.15}$$

to obtain a unique matrix B. We use this matrix to form the Lyapunov function (10.10). Now since $A = (1+\beta)^{1/2} A_0$, the first forward difference $Dv(x)$ of the Lyapunov function $v(x) = x^T Bx$ along the solutions of (10.2) yields

$$\begin{aligned} Dv(x) &= x^T [A_0^T BA_0 - B + \beta A_0^T BA_0]x \\ &= x^T (C + \beta A_0^T BA_0)x = x^T \tilde{C} x, \end{aligned} \tag{10.16}$$

where C is given in (10.15). It now follows that if C is negative definite (positive definite), then we can choose $|\beta|$ sufficiently small so that $\tilde{C}$ will also be negative definite (positive definite).

Summarizing the above discussion, we have proved the following result.

THEOREM 10.13. If all the eigenvalues of the matrix A are within the unit circle of the complex plane, or if at least one eigenvalue is outside the unit circle of the complex plane, then there exists a Lyapunov function of the form $v(x) = x^T Bx$, $B = B^T$, whose first forward difference along the solutions of system (10.2) is definite (i.e., it is either negative definite or positive definite). ■

We conclude this subsection with some specific examples.

EXAMPLE 10.12. (i) For system (10.2) let

$$A = \begin{bmatrix} 0 & 1 \\ -1 & 0 \end{bmatrix}.$$

Let $B = I$, which is positive definite. From (10.11) we obtain

$$C = A^T A - I = \begin{bmatrix} 0 & -1 \\ 1 & 0 \end{bmatrix}\begin{bmatrix} 0 & 1 \\ -1 & 0 \end{bmatrix} - \begin{bmatrix} 1 & 0 \\ 0 & 1 \end{bmatrix} = \begin{bmatrix} 0 & 0 \\ 0 & 0 \end{bmatrix}.$$

It follows from Theorem 10.11(i) that the equilibrium $x = 0$ of this system is stable. This is the same conclusion that was made in Example 10.9.

(ii) For system (10.2) let

$$A = \begin{bmatrix} 0 & -\frac{1}{2} \\ -1 & 0 \end{bmatrix}.$$

Choose

$$B = \begin{bmatrix} \frac{8}{3} & 0 \\ 0 & \frac{5}{3} \end{bmatrix},$$

which is positive definite. From (10.11) we obtain

$$C = A^T BA - B = \begin{bmatrix} 0 & -1 \\ -\frac{1}{2} & 0 \end{bmatrix}\begin{bmatrix} \frac{8}{3} & 0 \\ 0 & \frac{5}{3} \end{bmatrix}\begin{bmatrix} 0 & -\frac{1}{2} \\ -1 & 0 \end{bmatrix} - \begin{bmatrix} \frac{8}{3} & 0 \\ 0 & \frac{5}{3} \end{bmatrix} = \begin{bmatrix} -1 & 0 \\ 0 & -1 \end{bmatrix},$$

which is negative definite. It follows from Theorem 10.11(ii) that the equilibrium $x = 0$ of this system is asymptotically stable in the large. This is the same conclusion that was made in Example 10.9(ii).

(iii) For system (10.2) let

$$A = \begin{bmatrix} 0 & -\frac{1}{2} \\ -3 & 0 \end{bmatrix}.$$

Choose

$$C = \begin{bmatrix} -1 & 0 \\ 0 & -1 \end{bmatrix},$$

which is negative definite. From (10.11) we obtain

$$C = A^T BA - B = \begin{bmatrix} 0 & -3 \\ -\frac{1}{2} & 0 \end{bmatrix}\begin{bmatrix} b_{11} & b_{12} \\ b_{12} & b_{22} \end{bmatrix}\begin{bmatrix} 0 & -\frac{1}{2} \\ -3 & 0 \end{bmatrix} - \begin{bmatrix} b_{11} & b_{12} \\ b_{12} & b_{22} \end{bmatrix}$$

or $$\begin{bmatrix} (9b_{22} - b_{11}) & \frac{1}{2}b_{12} \\ \frac{1}{2}b_{12} & (\frac{1}{4}b_{11} - b_{22}) \end{bmatrix} = \begin{bmatrix} -1 & 0 \\ 0 & -1 \end{bmatrix},$$

which yields

$$B = \begin{bmatrix} -8 & 0 \\ 0 & -1 \end{bmatrix},$$

which is also negative definite. It follows from Theorem 10.11(iii) that the equilibrium $x = 0$ of this system is unstable. This conclusion is consistent with the conclusion made in Example 10.9(iii).

(iv) For system (10.2) let

$$A = \begin{bmatrix} \frac{1}{3} & 1 \\ 0 & 3 \end{bmatrix}.$$

The eigenvalues of A are $\lambda_1 = \frac{1}{3}$ and $\lambda_2 = 3$. According to Lemma 10.1, for a given C, Eq. (10.13) does *not* have a unique solution in this case since $\lambda_1 = 1/\lambda_2$. For purposes of illustration we choose $C = -I$. Then

$$-I = A^T BA - B = \begin{bmatrix} \frac{1}{3} & 0 \\ 1 & 3 \end{bmatrix}\begin{bmatrix} b_{11} & b_{12} \\ b_{12} & b_{22} \end{bmatrix}\begin{bmatrix} \frac{1}{3} & 1 \\ 0 & 3 \end{bmatrix} - \begin{bmatrix} b_{11} & b_{12} \\ b_{12} & b_{22} \end{bmatrix}$$

or $$\begin{bmatrix} -\frac{8}{9}b_{11} & \frac{1}{3}b_{11} \\ \frac{1}{3}b_{11} & b_{11} + 6b_{12} + 8b_{22} \end{bmatrix} = \begin{bmatrix} -1 & 0 \\ 0 & -1 \end{bmatrix}$$

which shows that for $C = -I$, Eq. (10.13) does not have any solution (for B) at all. ■

F. Linearization

In this subsection we determine conditions under which the stability properties of the equilibrium $w = 0$ of the linear system

$$w(k+1) = Aw(k) \tag{10.17}$$

determine the stability properties of the equilibrium $x = 0$ of the nonlinear system

$$x(k+1) = Ax(k) + f(x(k)) \tag{10.18}$$

under the assumption that $f(x) = o(\|x\|)$ as $\|x\| \to 0$ (i.e., given $\epsilon > 0$ there exists

$\delta > 0$ such that $\|f(x(k))\| < \epsilon\|x(k)\|$ for all $k \geq 0$ and all $\|x(k)\| < \delta$). [Refer to the discussion concerning Eqs. (10.2) to (10.4) in Subsection A of this section.]

THEOREM 10.14. Assume that $f \in C(R^n, R^n)$ and that $f(x)$ is $o(\|x\|)$ as $\|x\| \to 0$. (i) If A is stable (i.e., all the eigenvalues of A are within the unit circle of the complex plane), then the equilibrium $x = 0$ of system (10.18) is *asymptotically stable*. (ii) If at least one eigenvalue of A is outside the unit circle of the complex plane, then the equilibrium $x = 0$ of system (10.18) is *unstable*.

Proof. (i) Assume that A is stable. Then for any negative definite matrix C, the equation

$$A^T BA - B = C$$

has a unique positive definite solution B. Let

$$v(x) = x^T Bx.$$

Along the solutions of (10.18) we compute the first forward difference $Dv(x)$ as

$$Dv(x) = x^T Cx + 2x^T A^T Bf(x) + v(f(x)).$$

This allows us to estimate [since $f(x)$ is $o(\|x\|)$ as $\|x\| \to 0$]

$$\begin{aligned} Dv(x) &\leq \lambda_M(C)\|x\|^2 + 2\|x\|\,\|A\|\,\|B\|\,\|f(x)\| + v(f(x)) \\ &\leq \lambda_M(C)\|x\|^2 + 2\|A\|\,\|B\|\epsilon\|x\|^2 + \lambda_M(B)\epsilon^2\|x\|^2 \\ &= [\lambda_M(C) + 2\|A\|\,\|B\|\epsilon + \lambda_M(B)\epsilon^2]\|x\|^2 = \gamma\|x\|^2 \end{aligned} \tag{10.19}$$

for all $x \in B(\delta)$ and for some $\delta > 0$, where $\lambda_M(C) < 0$, $\lambda_M(B) > 0$ denote the largest eigenvalues of C and B, respectively. We can make ϵ as small as desired by choosing δ sufficiently small, resulting in $\gamma < 0$. Therefore, $Dv(x)$ is negative definite and the equilibrium $x = 0$ of system (10.18) is asymptotically stable.

(ii) Assume that A has at least one eigenvalue outside the unit circle of the complex plane. Following the procedure given in proving Theorem 10.13 [refer to Eqs. (10.14) to (10.16)] we construct a Lyapunov function $v(x) = x^T Bx$ whose first forward difference along the solutions of (10.17) is given by $Dv(x) = x^T \tilde{C}x$ [refer to (10.16)]. We choose $\tilde{C}$ to be negative definite. Then B is indefinite, and in every neighborhood of the origin there are $\bar{x} \in R^n$ such that $v(\bar{x}) = \bar{x}^T B\bar{x} < 0$.

Next, we evaluate the first forward difference of v along the solutions of (10.18) to obtain [identically as in (10.19)],

$$Dv(x) \leq [\lambda_M(\tilde{C}) + 2\|A\|\,\|B\|\epsilon + \lambda_M(B)\epsilon^2]\|x\|^2 = \gamma\|x\|^2$$

for all $x \in B(\delta)$ for some $\delta > 0$, where $\tilde{C}$ is defined by (10.14) to (10.16). As in the proof of part (i), we can force ϵ to be as small as desired by choosing δ sufficiently small, resulting again in $\gamma < 0$. Therefore, $Dv(x)$ is negative definite. It follows from Theorem 10.5 that the equilibrium $x = 0$ of (10.18) is unstable. ■

Before concluding this subsection, we consider some specific examples.

EXAMPLE 10.13. (i) Consider the system

$$\begin{aligned} x_1(k+1) &= -\tfrac{1}{2}x_2(k) + x_1(k)^2 + x_2(k)^2 \\ x_2(k+1) &= -x_1(k) + x_1(k)^2 + x_2(k)^2. \end{aligned} \tag{10.20}$$

Using the notation of (10.18) we have

$$A = \begin{bmatrix} 0 & -\frac{1}{2} \\ -1 & 0 \end{bmatrix}, \qquad f(x_1, x_2) = \begin{bmatrix} x_1^2 + x_2^2 \\ x_1^2 + x_2^2 \end{bmatrix}.$$

The linearization of (10.20) is given by

$$w(k+1) = Aw(k). \tag{10.21}$$

From Example 10.9(ii) [and Example 10.12(ii)] it follows that the equilibrium $w = 0$ of (10.21) is asymptotically stable. Furthermore, in the present case $f(x) = o(\|x\|)$ as $\|x\| \to 0$. Therefore, in view of Theorem 10.14, the equilibrium $x = 0$ of system (10.20) is asymptotically stable.

(ii) Consider the system

$$\begin{aligned} x_1(k+1) &= -\tfrac{1}{2}x_2(k) + x_1(k)^3 + x_2(k)^2 \\ x_2(k+1) &= -3x_1(k) + x_1^4(k) - x_2(k)^5. \end{aligned} \tag{10.22}$$

Using the notation of (10.17) and (10.18), we have in the present case

$$A = \begin{bmatrix} 0 & -\frac{1}{2} \\ -3 & 0 \end{bmatrix}, \qquad f(x_1, x_2) = \begin{bmatrix} x_1^3 + x_2^2 \\ x_1^4 - x_2^5 \end{bmatrix}.$$

Since A is unstable [refer to Example 10.12(iii) and Example 10.9(iii)] and since $f(x) = o(\|x\|)$ as $\|x\| \to 0$, it follows from Theorem 10.14 that the equilibrium $x = 0$ of system (10.22) is unstable. ■

G. Input-Output Stability

We conclude this chapter by considering the input-output stability of discrete-time systems described by equations of the form

$$\begin{aligned} x(k+1) &= Ax(k) + Bu(k) \\ y(k) &= Cx(k), \end{aligned} \tag{10.23}$$

where all matrices and vectors are defined as in (10.1). Throughout this subsection we will assume that $k_0 = 0$, $x(0) = 0$, and $k \geq 0$.

As in the continuous-time case, we say that system (10.23) is *BIBO stable* if there exists a constant $c > 0$ such that the conditions

$$\begin{aligned} x(0) &= 0 \\ \|u(k)\| &\leq 1, \qquad k \geq 0, \end{aligned}$$

imply that $\|y(k)\| \leq c$ for all $k \geq 0$.

The results that we will present involve the impulse response matrix of (10.23) given by

$$H(k) = \begin{cases} CA^{k-1}B, & k > 0, \\ 0, & k \leq 0, \end{cases} \tag{10.24}$$

and the transfer function matrix given by

$$\hat{H}(z) = C(zI - A)^{-1}B. \tag{10.25}$$

Recall that

$$y(n) = \sum_{k=0}^{n} H(n-k)u(k). \tag{10.26}$$

Associated with system (10.23) is the free dynamical system described by the equation

$$p(k+1) = Ap(k). \tag{10.27}$$

THEOREM 10.15. The system (10.23) is *BIBO stable* if and only if there exists a constant $L > 0$ such that for all $n \geq 0$,

$$\sum_{k=0}^{n} \|H(k)\| \leq L. \tag{10.28}$$

■

As in the continous-time case, the first part of the proof of Theorem 10.15 (sufficiency) is straightforward. Specifically, if $\|u(k)\| \leq 1$ for all $k \geq 0$ and if (10.28) is true, then we have for all $n \geq 0$,

$$\begin{aligned} \|y(n)\| &= \|\sum_{k=0}^{n} H(n-k)u(k)\| \leq \sum_{k=0}^{n} \|H(n-k)u(k)\| \\ &\leq \sum_{k=0}^{n} \|H(n-k)\| \, \|u(k)\| \leq \sum_{k=0}^{n} \|H(n-k)\| \leq L. \end{aligned}$$

Therefore, system (10.23) is BIBO stable.

In proving the second part of Theorem 10.15 (necessity), we simplify matters by first considering in (10.23) the single-variable case ($n = 1$) with the system description given by

$$y(t) = \sum_{k=0}^{t} h(t-k)u(k), \qquad t > 0. \tag{10.29}$$

For purposes of contradiction, we assume that the system is BIBO stable, but no finite L exists such that (10.28) is satisfied. Another way of expressing the last assumption is that for *any finite* L, there exists $t = k_1(L) \triangleq k_1$ such that

$$\sum_{k=0}^{k_1} |h(k_1 - k)| > L.$$

We now choose in particular the input u given by

$$u(k) = \begin{cases} +1 & \text{if } h(t-k) > 0, \\ 0 & \text{if } h(t-k) = 0, \\ -1 & \text{if } h(t-k) < 0, \end{cases}$$

$0 \leq k \leq k_1$. Clearly, $|u(k)| \leq 1$ for all $k \geq 0$. The output of the system at $t = k_1$ due to the above input, however, is

$$y(k_1) = \sum_{k=0}^{k_1} h(k_1 - k)u(k) = \sum_{k=0}^{k_1} |h(k_1 - k)| > L,$$

which contradicts the assumption that the system is BIBO stable.

The above can now be extended to the multivariable case. In doing so we apply the single-variable result to every possible pair of input and output vector components, we make use of the fact that the sum of a finite number of bounded sums will be bounded, and we note that a vector is bounded if and only if each of its components is bounded. We leave the details to the reader.

Next, we establish a connection between the asymptotic stability of the equilibrium $p = 0$ of system (10.27) and the BIBO stability of system (10.23). First, we note that the asymptotic stability of the equilibrium $x = 0$ of system (10.27) implies the BIBO stability of system (10.23) since the sum

$$\left\| \sum_{k=1}^{\infty} CA^{k-1}B \right\| \leq \sum_{k=1}^{\infty} \|C\| \, \|A^{k-1}\| \, \|B\|$$

is finite.

The main task in proving the converse to the above statement is to show that controllability and observability of system (10.23) and the finiteness of the sum $\sum_{k=1}^{\infty} \|CA^{k-1}B\|$ imply the finiteness of the sum $\sum_{k=1}^{\infty} \|A^{k-1}\|$. So let us assume that the system (10.23) is BIBO stable. Then, since

$$\|y(k)\| = \left\| \sum_{j=0}^{k-1} CA^{k-(j+1)}Bu(j) \right\| \leq \sum_{j=0}^{k-1} \|CA^{k-(j+1)}B\|,$$

we must have that the power series

$$\sum_{k=1}^{\infty} \|CA^{k-1}B\|$$

is finite (i.e., absolutely convergent), and this implies that

$$\lim_{k \to \infty} CA^{k-1}B = 0. \tag{10.30}$$

From (10.30) we can conclude that

$$\lim_{k \to \infty} CAA^{k-1}B = \lim_{k \to \infty} CA^{k-1}AB = \lim_{k \to \infty} CA^{k}B = 0,$$

and repeating, we arrive at

$$\lim_{k \to \infty} (CA^{q})A^{k-1}(A^{r}B) = 0, \qquad q, r = 0, 1, \ldots, n-1.$$

We can write this as

$$\lim_{k \to \infty} \begin{bmatrix} C \\ CA \\ \vdots \\ CA^{n-1} \end{bmatrix} A^{k-1}[B, AB, \cdots, A^{n-1}B] = 0. \tag{10.31}$$

If we now assume that (10.23) is controllable "from-the-origin" and observable, then we can select n linearly independent columns of the controllability matrix to form an invertible $n \times n$ matrix W and n linearly independent rows of the observability matrix to form an invertible $n \times n$ matrix M. Using (10.31) we conclude that

$$\lim_{k \to \infty} MA^{k-1}W = 0, \tag{10.32}$$

which yields

$$M^{-1}\left(\lim_{k\to\infty} MA^{k-1}W\right)W^{-1} = \lim_{k\to\infty} A^{k-1} = 0. \tag{10.33}$$

From (10.33) we can conclude that the equilibrium $p = 0$ of (10.27) is asymptotically stable. To prove this assertion we assume to the contrary that $p = 0$ of (10.27) is not asymptotically stable. This implies that A has an eigenvalue λ with $|\lambda| \geq 1$. We assume the case when λ is real and leave the case when λ is complex as an exercise for the reader. Let η be an eigenvector associated with λ (which must be real). Then $A^k\eta = \lambda^k\eta$, $k \geq 0$. If $\eta = x_0$ denotes an initial condition, then the corresponding solution of (10.27) is given by $x(k) = A^k\eta = \lambda^k\eta$, $k \geq 0$, which does not go to zero as $k \to \infty$, i.e., $\lim_{k\to\infty} A^{k-1} \neq 0$, which contradicts (10.33). (*Note:* The above proof solves Exercise 6.31 for the case where the eigenvalues are real.)

We have thus arrived at the following result.

THEOREM 10.16. Assume that system (10.23) is controllable and observable. Then system (10.23) is BIBO stable if and only if the equilibrium $p = 0$ of system (10.27) is asymptotically stable. ■

Next, we recall that a complex number z_p is a *pole* of $\hat{H}(z) = [\hat{h}_{ij}(z)]$ if for some (i, j) we have $|\hat{h}_{ij}(z_p)| = \infty$ (refer to Section 3.5 for the definition of a pole). If each entry of $\hat{H}(z)$ has only poles with modulus (magnitude) less than 1, then, as shown in Chapter 2, each entry of $H(k) = [h_{ij}(k)]$ consists of a sum of converging terms. It follows that under these conditions the sum

$$\sum_{k=0}^{\infty} \|H(k)\|$$

is finite, and any realization of $\hat{H}(z)$ will result in a system that is BIBO stable.

Conversely, if

$$\sum_{k=0}^{\infty} \|H(k)\|$$

is finite, then the terms in every entry of $H(k)$ must be convergent. But then every entry of $\hat{H}(z)$ has poles whose modulus is within the unit circle of the complex plane. We have proved the final result of this section.

THEOREM 10.17. The time-invariant system (10.23) is BIBO stable if and only if the poles of the transfer function

$$\hat{H}(z) = C(zI - A)^{-1}B$$

are within the unit circle of the complex plane. ■

6.11 SUMMARY

In this chapter we first addressed the stability of an equilibrium of continuous-time finite-dimensional systems (Part 1). In doing so, we first introduced the concept of equilibrium and defined several types of stability in the sense of Lyapunov (Sections 6.3 and 6.4). Next, we established several stability conditions of an equilibrium for linear time-varying systems (LH) in terms of the state transition matrix and for linear time-invariant systems L in terms of eigenvalues (Section 6.5). In Section 6.6 we established several geometric and algebraic stability criteria for nth-order, linear, time-invariant systems [including the Leonhard-Mikhailov stability criterion (Theorem 6.1), the gap and position stability criterion (also called the interlacing stability criterion) (Theorem 6.2), and the Routh-Hurwitz criterion (Theorem 6.4)]. Next, we established various stability conditions for linear time-invariant systems that are phrased in terms of the Lyapunov Matrix Equation for system (L) (Section 6.7). In Section 6.8 we established conditions under which the asymptotic stability and the instability of an equilibrium for a nonlinear time-invariant system (A) can be deduced from the linearization of (A).

Then in Part 2 we addressed the input-output stability of time-varying and time-invariant linear, continuous-time, and finite-dimensional systems (Section 6.9). For such systems we established several conditions for bounded input bounded output stability (BIBO stability) and we related some of these to the stability properties of an equilibrium.

The chapter concluded with Part 3 (Section 6.10), where we addressed the Lyapunov stability and the input-output stability of (time-invariant) systems. For such systems, we established results that are analogous to the stability results of continuous-time systems considered in Parts 1 and 2. [Among other topics, we discussed in Subsection 6.10D the Schur-Cohn criterion (which is analogous to the Routh-Hurwitz criterion for continuous-time systems), and in Subsection 6.10E we established stability criteria involving the Lyapunov Matrix Equation for the discrete-time case.]

6.12 NOTES

The initial contributions to stability theory that took place toward the end of the last century are primarily due to physicists and mathematicians (Lyapunov [14]), while input-output stability is the brainchild of electrical engineers (Sandberg [21] to [23], Zames [26], [27]). Sources with extensive coverage of Lyapunov stability theory include, e.g., Hahn [7], Khalil [11], LaSalle [12], LaSalle and Lefschetz [13], Michel and Miller [17], Michel and Wang [18], Miller and Michel [19], and Vidyasagar [25]. Input-output stability is addressed in great detail in Desoer and Vidyasagar [5] and Vidyasagar [25]. For a survey that traces many of the important developments of stability in feedback control, refer to Michel [15].

In the context of *linear systems*, nice sources on both Lyapunov stability and input-output stability can be found in numerous texts, including Brockett [2], Chen [3], DeCarlo [4], Kailath [10], and Rugh [20]. In developing our presentation, we found the texts by Brockett [2], Hahn [7], LaSalle [12], Miller and Michel [19], and Rugh [20] especially helpful. For a proof of the Schur-Cohn criterion, and other related results, refer to the elegant book by Jury [9].

The background material summarized in the second section is developed in most standard linear algebra texts, including the classic books by Birkhoff and MacLane [1], Gantmacher [6], and Halmos [8]. For more recent references on this subject, refer, e.g., to the books by Michel and Herget [16] and Strang [24].

6.13 REFERENCES

1. G. Birkhoff and S. MacLane, *A Survey of Modern Algebra,* Macmillan, New York, 1965.
2. R. W. Brockett, *Finite Dimensional Linear Systems,* Wiley, New York, 1970.
3. C. T. Chen, *Linear System Theory and Design,* Holt, Rinehart and Winston, New York, 1984.
4. R. A. DeCarlo, *Linear Systems*, Prentice-Hall, Englewood Cliffs, NJ, 1989.
5. C. A. Desoer and M. Vidyasagar, *Feedback Systems: Input-Output Properties,* Academic Press, New York, 1975.
6. F. R. Gantmacher, *Theory of Matrices,* Vols. I, II, Chelsea, New York, 1959.
7. W. Hahn, *Stability of Motion*, Springer-Verlag, New York, 1967.
8. P. R. Halmos, *Finite Dimensional Vector Spaces,* Van Nostrand, Princeton, NJ, 1958.
9. E. I. Jury, *Inners and Stability of Dynamical Systems,* Robert E. Krieger Publisher, Malabar, FL, 1982.
10. T. Kailath, *Linear Systems,* Prentice-Hall, Englewood Cliffs, NJ, 1980.
11. H. K. Khalil, *Nonlinear Systems,* Macmillan, New York, 1992.
12. J. P. LaSalle, *The Stability and Control of Discrete Processes,* Springer-Verlag, New York, 1986.
13. J. P. LaSalle and S. Lefschetz, *Stability by Liapunov's Direct Method,* Academic Press, New York, 1961.
14. M. A. Liapounoff, "Problème générale de la stabilité de mouvement," *Ann. Fac. Sci. Toulouse,* Vol. 9, 1907, pp. 203–474. (Translation of a paper published in *Comm. Soc. Math.* Kharkow 1893, reprinted in *Ann. Math. Studies,* Vol. 17, 1949, Princeton, NJ.)
15. A. N. Michel, "Stability: the common thread in the evolution of feedback control," *IEEE Control Systems,* Vol. 16, 1996, pp. 50–60.
16. A. N. Michel and C. J. Herget, *Applied Algebra and Functional Analysis,* Dover, New York, 1993.
17. A. N. Michel and R. K. Miller, *Qualitative Analysis of Large Scale Dynamical Systems,* Academic Press, New York, 1977.
18. A. N. Michel and K. Wang, *Qualitative Theory of Dynamical Systems,* Marcel Dekker, New York, 1995.
19. R. K. Miller and A. N. Michel, *Ordinary Differential Equations,* Academic Press, New York, 1982.
20. W. J. Rugh, *Linear System Theory,* Second Edition, Prentice-Hall, Englewood Cliffs, NJ, 1996.
21. I. W. Sandberg, "On the L_2-boundedness of solutions of nonlinear functional equations," *Bell Syst. Tech. J.,* Vol. 43, 1964, pp. 1581–1599.
22. I. W. Sandberg, "A frequency-domain condition for stability of feedback systems containing a single time-varying nonlinear element," *Bell Syst. Tech. J.,* Vol. 44, 1974, pp. 1601–1608.
23. I. W. Sandberg, "Some results on the theory of physical systems governed by nonlinear functional equations," *Bell Syst. Tech. J.,* Vol. 44, 1965, pp. 821–898.
24. G. Strang, *Linear Algebra and its Applications,* Harcourt, Brace, Jovanovich, San Diego, 1988.
25. M. Vidyasagar, *Nonlinear Systems Analysis,* 2d edition, Prentice Hall, Englewood Cliffs, NJ, 1993.
26. G. Zames, "On the input-output stability of time-varying nonlinear feedback systems, Part I," *IEEE Transactions on Automatic Control,* Vol. 11, 1966, pp. 228–238.
27. G. Zames, "On the input-output stability of time-varying nonlinear feedback systems, Part II," *IEEE Transactions on Automatic Control,* Vol. 11, 1966, pp. 465–476.

6.14 EXERCISES

6.1. Determine the set of equilibrium points of a system described by the differential equations

$$\begin{aligned}\dot{x}_1 &= x_1 - x_2 + x_3\\ \dot{x}_2 &= 2x_1 + 3x_2 + x_3\\ \dot{x}_3 &= 3x_1 + 2x_2 + 2x_3.\end{aligned}$$

6.2. Determine the set of equilibria of a system described by the differential equations

$$\begin{aligned}\dot{x}_1 &= x_2\\ \dot{x}_2 &= \begin{cases} x_1 \sin\left(\dfrac{1}{x_1}\right), & \text{when } x_1 \neq 0,\\ 0, & \text{when } x_1 = 0.\end{cases}\end{aligned}$$

6.3. Determine the equilibrium points and their stability properties of a system described by the ordinary differential equation

$$\dot{x} = x(x-1) \tag{14.1}$$

by solving (14.1) and then applying the definitions of stability, uniform stability, asymptotic stability, etc.

6.4. Determine the set of equilibria and their stability properties of a system described by the ordinary differential equation

$$\dot{x} = (cost)x \tag{14.2}$$

by solving (14.2) and then applying the definitions of stability, uniform stability, asymptotic stability, etc.

6.5. Determine the set of equilibria and their stability properties of a system described by the ordinary differential equation

$$\dot{x} = (4t\sin t - 2t)x \tag{14.3}$$

by solving (14.3) and then applying the definitions of stability, uniform stability, asymptotic stability, etc.

6.6. Determine the state transition matrix $\Phi(t, t_0)$ of the system

$$\begin{bmatrix}\dot{x}_1\\ \dot{x}_2\end{bmatrix} = \begin{bmatrix}-t & 0\\ (2t-t) & -2t\end{bmatrix}\begin{bmatrix}x_1\\ x_2\end{bmatrix}.$$

Use Theorems 5.1 to 5.4 to determine the stability properties of the trivial solution of this system.

6.7. Show that the second-degree polynomial

$$f(s) = s^2 + 2as + b$$

is a Hurwitz polynomial if and only if $a > 0$ and $b > 0$ by (i) solving the equation $f(s) = 0$, and (ii) using the Routh-Hurwitz criterion (Theorem 6.4).

6.8. Determine whether the third-degree polynomial

$$f(s) = s^3 + 3s^2 + 3s + 2$$

is a Hurwitz polynomial by (i) solving the equation $f(s) = 0$, (ii) applying Theorem 6.1, (iii) applying Theorem 6.2, and (iv) applying Theorem 6.4.

6.9. Let $A \in C[R^+, R^{n\times n}]$ and $x \in R^n$ and consider

$$\dot{x} = A(t)x. \qquad (LH)$$

Show that the equilibrium $x_e = 0$ of (*LH*) is *uniformly stable* if there exists a $Q \in C^1[R^+, R^{n\times n}]$ such that $Q(t) = [Q(t)]^T$ for all t and if there exist constants $c_2 \geq c_1 > 0$ such that

$$c_1 I \leq Q(t) \leq c_2 I, \qquad t \in R, \tag{14.4}$$

and such that

$$[A(t)]^T Q(t) + Q(t)A(t) + \dot{Q}(t) \leq 0, \qquad t \in R, \tag{14.5}$$

where I is the $n \times n$ identity matrix. *Hint:* The proof of this assertion is similar to the proof of Theorem 7.1.

6.10. Show that the equilibrium $x_e = 0$ of (*LH*) is *exponentially stable* if there exists a $Q \in C^1[R^+, R^{n\times n}]$ such that $Q(t) = [Q(t)]^T$ for all t and if there exist constants $c_2 \geq c_1 > 0$ and $c_3 > 0$ such that (14.4) holds and such that

$$[A(t)]^T Q(t) + Q(t)A(t) + \dot{Q}(t) \leq -c_3 I, \qquad t \in R. \tag{14.6}$$

Hint: The proof of this assertion is similar to the proof of Theorem 7.2.

6.11. Assume that the equilibrium $x_e = 0$ of (*LH*) is *exponentially stable* and that there exists a constant $a > 0$ such that $\|A(t)\| \leq a$ for all $t \in R$. Show that the matrix given by

$$Q(t) = \int_t^\infty [\Phi(\tau, t)]^T \Phi(\tau, t)\, d\tau \tag{14.7}$$

satisfies the hypotheses of the result given in Exercise 10. *Hint:* The proof of this assertion is similar to the proof of Theorem 7.5.

6.12. For (*LH*) let $\lambda_m(t)$ and $\lambda_M(t)$ denote the smallest and largest eigenvalues of $A(t) + [A(t)]^T$ at $t \in R$, respectively. Let $\phi(t, t_0, x_0)$ denote the unique solution of (*LH*) for the initial data $x(t_0) = x_0 = \phi(t_0, t_0, x)$. Show that for any $x_0 \in R^n$ and any $t_0 \in R$, the unique solution of (*LH*) satisfies the estimate,

$$\|x_0\| e^{(1/2)\int_{t_0}^t \lambda_m(\tau)d\tau} \leq \|\phi(t, t_0, x_0)\| \leq \|x_0\| e^{(1/2)\int_{t_0}^t \lambda_M(\tau)d\tau}, \qquad t \geq t_0. \tag{14.8}$$

Hint: Let $v(t, t_0, x_0) = [\phi(t, t_0, x_0)]^T [\phi(t, t_0, x_0)] = \|\phi(t, t_0, x_0)\|^2$, evaluate $\dot{v}(t, t_0, x_0)$, and then establish (14.8).

6.13. Use Exercise 6.12 to show that the equilibrium $x_e = 0$ of (*LH*) is *uniformly stable* if there exists a constant c such that

$$\int_\sigma^t \lambda_M(\tau)\, d\tau \leq c \tag{14.9}$$

for all t, σ such that $t \geq \sigma$, where $\lambda_M(t)$ denotes the largest eigenvalue of $A(t) + [A(t)]^T$, $t \in R$. *Hint:* Use (14.8) and the definition of uniform stability.

6.14. Use Exercise 6.12 to show that the equilibrium $x_e = 0$ of (LH) is *exponentially stable* if there exist constants $\epsilon > 0$, $\alpha > 0$ such that

$$\int_\sigma^t \lambda_M(\tau)\, d\tau \leq -\alpha(t - \sigma) + \epsilon \tag{14.10}$$

for all t, σ such that $t \geq \sigma$. *Hint.* Use (14.8) and the definition of exponential stability.

6.15. Let v be a quadratic function of the form

$$v(x, t) = x^T Q(t)x, \tag{14.11}$$

where $x \in R^n$, $Q \in C^1[R, R^{n\times n}]$, $Q(t) = [Q(t)]^T$, and $Q(t) \leq kI$, $k > 0$, for all $t \in R$. Evaluate the derivative of v with respect to t, along the solutions of (LH), to obtain

$$\dot{v}_{(LH)}(x, t) = x^T[[A(t)]^T Q(t) + Q(t)A(t) + \dot{Q}(t)]x. \tag{14.12}$$

Assume that there is a quadratic form $w(x) = x^T W x \leq 0$, where $W^T = W \in R^{n\times n}$, such that

$$\dot{v}_{(LH)}(x, t) \leq w(x) \tag{14.13}$$

for all $(x, t) \in G \times R$, where G is a closed and bounded subset of R^n. Let

$$E = \{x \in G : w(x) = 0\} \tag{14.14}$$

and assume that for (LH), $\|A(t)\|$ is bounded on R. Prove that any solution of (LH) that remains in G for all $t > t_0 \geq 0$ approaches E as $t \to \infty$.

6.16. Consider the system

$$\ddot{x} + a(t)\dot{x} + x = 0,$$

which by letting $x_1 = x$ and $x_2 = \dot{x}$ can be written as

$$\dot{x}_1 = x_2$$
$$\dot{x}_2 = -a(t)x_2 - x_1.$$

Assume that $a \in C[R, R^+]$ and that there are constants c_1, c_2 such that $0 < c_1 \leq a(t) \leq c_2$ for all $t \in R$. Let $v(x) = x_1^2 + x_2^2$. First, show that all solutions of this system are bounded. Next, use the results of Exercise 6.15 to show that $\phi_2(t, t_0, x_0) \to 0$ as $t \to \infty$.

6.17. Assume that for system (LH) there exists a quadratic function of the form $v(x, t) = x^T Q(t)x$, where $Q(t) = [Q(t)]^T \in C^1[R, R^{n\times n}]$ and $Q(t) \geq cI$ for some $c > 0$, such that $\dot{v}_{(LH)}(x, t) \leq x^T W x$, where $W = W^T \in R^{n\times n}$ is negative definite. Show that if v is negative for some (x, t), then the equilibrium $x_e = 0$ of system (LH) is *unstable*. *Hint.* The proof of this assertion follows along similar lines as the proof of Theorem 7.3.

6.18. It is shown that if the equilibrium $x_e = 0$ of system (LH) is exponentially stable, then there exists a function v that satisfies the requirements of the result given in Exercise 6.10, i.e., the present result is a *converse theorem* to the result given in Exercise 6.10.

In system (LH), let A be bounded for all $t \in R$, let $L = L^T \in C^1[R, R^{n\times n}]$ and assume that L is bounded for all $t \in R$. Show that the integral

$$Q(t) = \int_t^\infty [\Phi(\sigma, t)]^T L(\sigma)\Phi(\sigma, t)\, d\sigma$$

exists for all $t \in R$. Show that the derivative of the function

$$v(x, t) = x^T Q(t)x \tag{14.15}$$

with respect to t along the solutions of (LH) is given by

$$\dot{v}_{(LH)}(x, t) = -x^T L(t)x.$$

Next, show that if $L(t) \geq c_3 I$, $c_3 > 0$, for all $t \in R$, then there exist constants $c_2 \geq c_1 > 0$ such that for all $t \in R$,

$$c_1 I \leq Q(t) \leq c_2 I, \tag{14.16}$$

where $I \in R^{n\times n}$ denotes the identity matrix.

Note that the above result constitutes a generalization to Theorem 7.5 for time-invariant systems (L).

6.19. Apply Proposition 7.1 to determine the definiteness properties of the matrix A given by

$$A = \begin{bmatrix} 1 & 2 & 1 \\ 2 & 5 & -1 \\ 1 & -1 & 10 \end{bmatrix}.$$

6.20. Use Theorem 7.3 to prove that the trivial solution of the system

$$\begin{bmatrix} \dot{x}_1 \\ \dot{x}_2 \end{bmatrix} = \begin{bmatrix} 3 & 4 \\ 2 & 1 \end{bmatrix} \begin{bmatrix} x_1 \\ x_2 \end{bmatrix}$$

is unstable.

6.21. Determine the equilibrium points of a system described by the differential equation

$$\dot{x} = -x + x^2$$

and determine the stability properties of the equilibrium points, if applicable, by using Theorem 8.1 or 8.2.

6.22. The system described by the differential equations

$$\begin{aligned} \dot{x}_1 &= x_2 + x_1(x_1^2 + x_2^2) \\ \dot{x}_2 &= -x_1 + x_2(x_1^2 + x_2^2) \end{aligned} \tag{14.17}$$

has an equilibrium at the origin $x^T = (x_1, x_2) = (0, 0)$. Show that the trivial solution of the *linearization* of system (14.17) is stable. Prove that the equilibrium $x = 0$ of system (14.17) is unstable. (This example shows that the assumptions on the matrix A in Theorems 8.1 and 8.2 are absolutely essential.)

6.23. Prove that the system given by

$$\begin{bmatrix} \dot{x}_1 \\ \dot{x}_2 \end{bmatrix} = \begin{bmatrix} -t & 0 \\ (2t - t) & -2t \end{bmatrix} \begin{bmatrix} x_1 \\ x_2 \end{bmatrix} + \begin{bmatrix} e^{-t} \\ e^{-2t} \end{bmatrix} u(t)$$

$$\begin{bmatrix} y_1 \\ y_2 \end{bmatrix} = \begin{bmatrix} \cos t & \sin t \\ \sin t & -\cos t \end{bmatrix} \begin{bmatrix} x_1 \\ x_2 \end{bmatrix}$$

is BIBO stable.

6.24. Use Theorem 9.3 to analyze the stability properties of the system given by

$$\begin{aligned} \dot{x} &= Ax + Bu \\ y &= Cx \end{aligned}$$

$$A = \begin{bmatrix} -1 & 0 \\ 1 & -1 \end{bmatrix}, \qquad B = \begin{bmatrix} 1 \\ -1 \end{bmatrix}, \qquad C = [0, 1].$$

6.25. Determine all equilibrium points for the discrete-time systems given by

(a)

$$\begin{aligned} x_1(k+1) &= x_2(k) + |x_1(k)| \\ x_2(k+1) &= -x_1(k) + |x_2(k)| \end{aligned}$$

(b)
$$x_1(k+1) = x_1(k)x_2(k) - 1$$
$$x_2(k+1) = 2x_1(k)x_2(k) + 1.$$

6.26. Consider the discrete-time system given by

$$x(k+1) = sat\,[Ax(k)] \qquad (14.18)$$

where for $\theta = (\theta_1, \ldots, \theta_n)^T \in R^n$, $sat\,\theta = [sat\,\theta_1, \ldots, sat\,\theta_n]^T$, and

$$sat\,\theta_i = \begin{cases} 1, & \theta_i > 1, \\ \theta_i, & |\theta_i| \leq 1, \\ -1, & \theta_i < 1. \end{cases}$$

(a) For $A \in R^{n\times n}$ arbitrary, use Theorem 10.1 to analyze system (14.18).
(b) Imposing various restrictions on the locations of the eigenvalues of A in the complex plane, use as many results of this chapter as you can to analyze the stability properties of the trivial solution of system (14.18).

6.27. Determine the stability properties of the trivial solution of the discrete-time system given by the equations

$$\begin{bmatrix} x_1(k+1) \\ x_2(k+1) \end{bmatrix} = \begin{bmatrix} \cos\theta & \sin\theta \\ -\sin\theta & \cos\theta \end{bmatrix} \begin{bmatrix} x_1(k) \\ x_2(k) \end{bmatrix}$$

with θ fixed.

6.28. Analyze the stability of the equilibrium $x = 0$ of the system described by the scalar-valued difference equation

$$x(k+1) = \sin[x(k)].$$

6.29 Analyze the stability of the equilibrium $x = 0$ of the system described by the difference equations

$$x_1(k+1) = x_1(k) + x_2(k)[x_1(k)^2 + x_2(k)^2]$$
$$x_2(k+1) = x_2(k) - x_1(k)[x_1(k)^2 + x_2(k)^2].$$

6.30. Determine a basis of the solution space of the system

$$\begin{bmatrix} x_1(k+1) \\ x_2(k+1) \end{bmatrix} = \begin{bmatrix} 0 & 1 \\ -6 & 5 \end{bmatrix} \begin{bmatrix} x_1(k) \\ x_2(k) \end{bmatrix}.$$

Use your answer in analyzing the stability of the trivial solution of this system.

6.31. Let $A \in R^{n\times n}$. Prove that part (iii) of Theorem 10.8 is equivalent to the statement that all eigenvalues of A have modulus less than 1, i.e.,

$$\lim_{k\to\infty} \|A^k\| = 0$$

if and only if for any eigenvalue λ of A, it is true that $|\lambda| < 1$.

6.32. Use Theorem 10.7 to show that the equilibrium $x = 0$ of the system

$$x(k+1) = \begin{bmatrix} 1 & 1 & 1 & \cdots & 1 \\ 0 & 1 & 1 & \cdots & 1 \\ \cdots & \cdots & \cdots & \cdots & \cdots \\ 0 & 0 & 0 & \cdots & 1 \end{bmatrix} x(k)$$

is unstable.

6.33. (a) Use Theorem 10.9 to determine the stability of the equilibrium $x = 0$ of the system

$$x(k+1) = \begin{bmatrix} 1 & 1 & -2 \\ 0 & 1 & 3 \\ 0 & 9 & -1 \end{bmatrix} x(k).$$

(b) Use Theorem 10.9 to determine the stability of the equilibrium $x = 0$ of the system

$$x(k+1) = \begin{bmatrix} 1 & 0 & -2 \\ 0 & 1 & 3 \\ 0 & 9 & -1 \end{bmatrix} x(k).$$

6.34. Apply the Schur-Cohn criterion (Theorem 10.10) in analyzing the stability of the trivial solution of the system given by the equations

$$\begin{bmatrix} x_1(k+1) \\ x_2(k+1) \\ x_3(k+1) \end{bmatrix} = \begin{bmatrix} -0.5 & 0 & 0.5 \\ 0.5 & 0 & 0 \\ 0 & -0.5 & 0 \end{bmatrix} \begin{bmatrix} x_1(k) \\ x_2(k) \\ x_3(k) \end{bmatrix}.$$

6.35. Apply Theorems 7.2 and 10.11 to show that if the equilibrium $x = 0$ ($x \in R^n$) of the system

$$x(k+1) = e^A x(k)$$

is asymptotically stable, then the equilibrium $x = 0$ of the system

$$\dot{x} = Ax$$

is also asymptotically stable.

6.36. Apply Theorem 10.11 to show that the trivial solution of the system given by

$$\begin{bmatrix} x_1(k+1) \\ x_2(k+1) \end{bmatrix} = \begin{bmatrix} 0 & 2 \\ 2 & 0 \end{bmatrix} \begin{bmatrix} x_1(k) \\ x_2(k) \end{bmatrix}$$

is unstable.

6.37. Determine the stability of the equilibrium $x = 0$ of the scalar-valued system given by

$$x(k+1) = \tfrac{1}{2}x(k) + \tfrac{2}{3}\sin x(k).$$

6.38. Analyze the stability properties of the discrete-time system given by

$$x(k+1) = x(k) + \tfrac{1}{2}u(k)$$
$$y(k) = \tfrac{1}{2}x(k)$$

where x, y, and u are scalar-valued variables. Is this system BIBO stable? Can Theorem 10.16 be applied in the analysis of this system?

CHAPTER 7

Polynomial Matrix Descriptions and Matrix Fractional Descriptions of Systems

In this chapter, representations of linear time-invariant systems based on polynomial matrices, called *Polynomial Matrix Description* (*PMD*) or *Differential* (*Difference*) *Operator Representation* (*DOR*) are introduced. Such representations arise naturally when differential (or difference) equations of order higher than one are used to describe the behavior of systems, and the differential (or difference) operator is introduced to represent the operation of differentiation (or of time-shift). Polynomial matrices in place of polynomials are involved since this approach is typically used to describe MIMO systems. Note that state-space system descriptions involve only first-order differential (or difference) equations, and as such, PMDs include the state-space descriptions as special cases.

A rational function matrix can be written as a ratio or fraction of two polynomial matrices or of two rational matrices. If the transfer function matrix of a system is expressed as a fraction of two polynomial or rational matrices, this leads to a *Matrix Fraction*(*al*) *Description* (*MFD*) of the system. The MFDs that involve polynomial matrices, called polynomial MFDs, can be viewed as representations of internal realizations of the transfer function matrix, i.e., as system PMDs of special form. These polynomial fractional descriptions (PMFD) help establish the relationship between internal and external system representations in a clear and transparent manner. This can be used to advantage, for example, in the study of feedback control problems, leading to clearer understanding of the phenomena that occur when systems are interconnected in feedback configurations. The MFDs that involve ratios of rational matrices, in particular, ratios of proper and stable rational matrices, offer convenient characterizations of transfer functions in feedback control problems.

MFDs that involve ratios of polynomial matrices and ratios of proper and stable rational matrices are essential in parameterizing all stabilizing feedback controllers. Appropriate selection of the parameters guarantees that a closed-loop system is not only stable but will also satisfy additional control criteria. This is precisely the approach taken in optimal control methods, such as H^∞-optimal control. Parameterizations of all stabilizing feedback controllers are studied extensively in Part 2 of this chapter. We note that extensions of MFDs are also useful in linear time-varying systems and in nonlinear systems. These extensions are not addressed here.

In addition to the importance of MFDs in characterizing all stabilizing controllers, and in H^∞-optimal control, PMFDs and PMDs have been used in other control design methodologies as well (e.g., self-tuning control). The use of PMFDs in feedback control leads in a natural way to the polynomial Diophantine matrix equation that is central in control design when PMDs are used and that directly leads to the characterization of all stabilizing controllers. The Diophantine Equation is studied at length in Part 1 of this chapter. Finally, PMDs are generalizations of state-space descriptions, and the use of PMDs to characterize the behavior of systems offers additional insight and flexibility. These issues are also explored in Part 1 of this chapter.

7.1 INTRODUCTION

In this chapter, Polynomial Matrix Descriptions (PMDs) and Matrix Fractional Descriptions (MFDs) are used to study properties such as controllability, observability, and stability, primarily of interconnected systems, and to conveniently characterize all stabilizing feedback controllers. These system descriptions are important in feedback control system analysis and design and are the key to developing control design theories such as H^∞-optimal control.

The development of the material in this chapter is concerned only with continuous-time systems; however, completely analogous results are valid for discrete-time systems and can easily be obtained by obvious modifications.

In the following, PMDs and MFDs are first introduced by an illustrative example. Next, the contents of the chapter are briefly described and some guidelines for the reader are provided.

An important comment on notation

In this chapter we will be dealing with matrices with entries that are polynomials in s or q, denoted by, e.g., $D(s)$ or $D(q)$. For simplicity of notation we frequently omit the argument s or q and we write D to denote the polynomial matrix on hand. When ambiguity may arise, or when it is important to stress the fact that the matrix in question is a polynomial matrix, the argument will be included.

A. A Brief Introduction to Polynomial and Fractional Descriptions

The PMD and the MFD of a linear time-invariant system are introduced via a simple motivating example.

EXAMPLE 1.1. In the ordinary differential equation representation of a system given by

$$\begin{aligned} \ddot{y}_1(t) + y_1(t) + y_2(t) &= \dot{u}_2(t) + u_1(t) \\ \dot{y}_1(t) + \dot{y}_2(t) + 2y_2(t) &= \dot{u}_2(t) \end{aligned} \tag{1.1}$$

$y_1(t)$, $y_2(t)$ and $u_1(t)$, $u_2(t)$ denote, respectively, outputs and inputs of interest. We assume that appropriate initial conditions for the $u_i(t)$, $y_i(t)$ and their derivatives at $t = 0$ are given.

By changing variables, one can express (1.1) by an equivalent set of first-order ordinary differential equations, in the sense that this set of equations will generate all solutions of (1.1), using appropriate initial conditions and the same inputs. To this end, let

$$x_1 = \dot{y}_1 - u_2, \qquad x_2 = y_1, \qquad x_3 = y_1 + y_2 - u_2. \tag{1.2}$$

Then (1.1) can be written as

$$\dot{x} = Ax + Bu, \qquad y = Cx + Du, \tag{1.3}$$

where $x(t) = \begin{bmatrix} x_1(t) \\ x_2(t) \\ x_3(t) \end{bmatrix}$, $u(t) = \begin{bmatrix} u_1(t) \\ u_2(t) \end{bmatrix}$, $y(t) = \begin{bmatrix} y_1(t) \\ y_2(t) \end{bmatrix}$, and

$$A = \begin{bmatrix} 0 & 0 & -1 \\ 1 & 0 & 0 \\ 0 & 2 & -2 \end{bmatrix}, \qquad B = \begin{bmatrix} 1 & -1 \\ 0 & 1 \\ 0 & -2 \end{bmatrix}, \qquad C = \begin{bmatrix} 0 & 1 & 0 \\ 0 & -1 & 1 \end{bmatrix}, \qquad D = \begin{bmatrix} 0 & 0 \\ 0 & 1 \end{bmatrix}$$

with initial conditions $x(0)$ calculated by using (1.2).

More directly, however, system (1.1) can be represented by

$$P(q)z(t) = Q(q)u(t), \qquad y(t) = R(q)z(t) + W(q)u(t), \tag{1.4}$$

where $z(t) = \begin{bmatrix} z_1(t) \\ z_2(t) \end{bmatrix}$, $u(t) = \begin{bmatrix} u_1(t) \\ u_2(t) \end{bmatrix}$, $y(t) = \begin{bmatrix} y_1(t) \\ y_2(t) \end{bmatrix}$, and

$$P(q) = \begin{bmatrix} q^2 + 1 & 1 \\ q & q + 2 \end{bmatrix}, \quad Q(q) = \begin{bmatrix} 1 & q \\ 0 & q \end{bmatrix}, \quad R(q) = \begin{bmatrix} 1 & 0 \\ 0 & 1 \end{bmatrix}, \quad W(q) = \begin{bmatrix} 0 & 0 \\ 0 & 0 \end{bmatrix}$$

with $q \triangleq d/dt$, the differential operator. The variables $z_1(t)$, $z_2(t)$ are called *partial state variables*, $z(t)$ denotes the *partial state* of the system description (1.4), and $u(t)$ and $y(t)$ denote the input and output vectors, respectively. ■

Polynomial matrix descriptions (PMD)

Representation (1.4), also denoted as $\{P(q), Q(q), R(q), W(q)\}$, is an example of a *PMD* of a system. Note that the state-space description (1.3) is a special case of (1.4). To see this, write (1.3) as

$$(qI - A)x(t) = Bu(t), \qquad y(t) = Cx(t) + Du(t). \tag{1.5}$$

Clearly, description $\{qI - A, B, C, D\}$ given in (1.5) is a special case of the general PMD $\{P(q), Q(q), R(q), W(q)\}$ with

$$P(q) = qI - A, \qquad Q(q) = B, \qquad R(q) = C, \qquad W(q) = D. \tag{1.6}$$

The above example points to the fact that a PMD of a system can be derived in a natural way from differential (or difference) equations that involve variables that are directly connected to physical quantities. By this approach, it is frequently possible to study the behavior of physical variables directly without having to transform the system to a state-space description. The latter may involve (state) variables that are quite removed from the physical phenomena they represent, thus losing physical insight when studying a given problem. The price to pay for this additional insight is that one has to deal with differential (or difference) equations of order greater than 1. This typically adds computational burdens. We note that certain special forms of PMDs, namely, the polynomial MFD, are easier to deal with than general forms. However, a change of variables may again be necessary to obtain such forms.

Consider a general PMD of a system given by

$$P(q)z(t) = Q(q)u(t), \qquad y(t) = R(q)z(t) + W(q)u(t) \tag{1.7}$$

with $P(q) \in R[q]^{l \times l}$, $Q(q) \in R[q]^{l \times m}$ and $R(q) \in R[q]^{p \times l}$, $W(q) \in R[q]^{p \times m}$, where $R[q]^{l \times l}$ denotes the set of $l \times l$ matrices with entries that are real polynomials in q. The transfer function matrix $H(s)$ of (1.7) can be determined by taking the Laplace transform of both sides of the equation assuming zero initial conditions ($z(0) = \dot{z}(0) = \cdots = 0, u(0) = \dot{u}(0) = \cdots = 0$). Then

$$H(s) = R(s)P^{-1}(s)Q(s) + W(s). \tag{1.8}$$

For the special case of state-space representations, [see (1.6)]; $H(s)$ in (1.8) assumes the well-known expression $H(s) = C(sI - A)^{-1}B + D$. For the study of the relationship between external and internal descriptions, (1.8) is not particularly convenient. Indeed, it appears that it is as difficult to investigate the relationship between $H(s)$ and PMDs as it was to study the relationship between $H(s)$ and state-space descriptions. There are, however, special cases of (1.8) that are very convenient to use in this regard. In particular, as will be shown in Section 7.3, if the system is controllable, then there exists a representation equivalent to (1.7) that is of the form

$$D_c(q)z_c(t) = u(t), \qquad y(t) = N_c(q)z_c(t), \tag{1.9}$$

where $D_c(q) \in R[q]^{m \times m}$ and $N_c(q) \in R[q]^{p \times m}$. Representation (1.9) is obtained by letting $Q(q) = I_m$ and $W(q) = 0$ in (1.7). It is common to use D and N instead of P and R, in view of

$$H(s) = N_c(s)D_c(s)^{-1}, \tag{1.10}$$

where $N_c(s)$ and $D_c(s)$ represent the matrix numerator and matrix demonimator of the transfer function, respectively. Similarly, if the system is observable, there exists a representation equivalent to (1.7) that is of the form

$$D_o(q)z_o(t) = N_o(q)u(t), \qquad y(t) = z_o(t), \tag{1.11}$$

where $D_o(q) \in R[q]^{p \times p}$ and $N_o(q) \in R[q]^{p \times m}$. Representation (1.11) is obtained by letting $R(q) = I_p$ and $W(q) = 0$ in (1.7) with $P(q) = D_o(q)$ and $Q(q) = N_o(q)$. Here,

$$H(s) = D_o^{-1}(s)N_o(s). \tag{1.12}$$

Note that (1.10) and (1.12) are generalizations to the MIMO case of the SISO system expression $H(s) = n(s)/d(s)$. In the same manner as $H(s) = n(s)/d(s)$ can be derived from the differential equation $d(q)y(t) = n(q)u(t)$, (1.12) can be derived from (1.11), usually written as $D_o(q)y(t) = N_o(q)u(t)$.

Returning now to (1.3) in the example, notice that the system is observable (state observable from the output y). Therefore, the system in this case can be represented by a description of the form $\{D_o, N_o, I_2, 0\}$. In fact (1.4) is such a description, where D_o and N_o are equal to P and Q, respectively, i.e., $D_o(q) = \begin{bmatrix} q^2+1 & 1 \\ q & q+2 \end{bmatrix}$, and $N_o(q) = \begin{bmatrix} 1 & q \\ 0 & q \end{bmatrix}$. The transfer function matrix is given by

$$
\begin{aligned}
H(s) &= C(sI - A)^{-1}B + D \\
&= \begin{bmatrix} 0 & 1 & 0 \\ 0 & -1 & 1 \end{bmatrix} \begin{bmatrix} s & 0 & 1 \\ -1 & s & 0 \\ 0 & -2 & s+2 \end{bmatrix}^{-1} \begin{bmatrix} 1 & -1 \\ 0 & 1 \\ 0 & -2 \end{bmatrix} + \begin{bmatrix} 0 & 0 \\ 0 & 1 \end{bmatrix} \\
&= D_o^{-1}(s)N_o(s) = \begin{bmatrix} s^2+1 & 1 \\ s & s+2 \end{bmatrix}^{-1} \begin{bmatrix} 1 & s \\ 0 & s \end{bmatrix} \\
&= \frac{1}{s^3+2s^2+2} \begin{bmatrix} s+2 & -1 \\ -s & s^2+1 \end{bmatrix} \begin{bmatrix} 1 & s \\ 0 & s \end{bmatrix} = \frac{1}{s^3+2s^2+2} \begin{bmatrix} s+2 & s(s+1) \\ -s & s(s^2-s+1) \end{bmatrix}.
\end{aligned}
$$

Matrix fractional descriptions (MFDs) of system transfer matrices

A given $p \times m$ proper, rational transfer function matrix $H(s)$ of a system can be represented as

$$H(s) = N_R(s)D_R^{-1}(s) = D_L^{-1}(s)N_L(s), \tag{1.13}$$

where $N_R(s) \in R[s]^{p\times m}$, $D_R(s) \in R[s]^{m\times m}$ and $N_L(s) \in R[s]^{p\times m}$, $D_L(s) \in R[s]^{p\times p}$. The pairs $\{N_R(s), D_R(s)\}$ and $\{D_L(s), N_L(s)\}$ are called *Polynomial Matrix Fractional Descriptions* (*PMFDs*) of the system transfer matrix with $\{N_R(s), D_R(s)\}$ termed a *right Fractional Description* and $\{D_L(s), N_L(s)\}$ a *left Fractional Description*. Notice that in view of (1.10), the right Polynomial Matrix Fractional Description (rPMFD) corresponds to the controllable PMD given in (1.9). That is, $\{D_R, I_m, N_R, 0\}$, or

$$D_R(q)z_R(t) = u(t), \qquad y(t) = N_R(q)z_R(t) \tag{1.14}$$

is a controllable PMD of the system with transfer function $H(s)$. The subscript c was used in (1.9) and (1.10) to emphasize the fact that N_c, D_c originated from an internal description that was controllable. In (1.13) and (1.14), the subscript R is used to emphasize that $\{N_R, D_R\}$ is a right fraction representation of the external description $H(s)$.

Similarly, in view of (1.12), the left Polynomial Matrix Fractional Description (lPMFD) corresponds to the observable PMD given in (1.11). That is, $\{D_L, N_L, I_p, 0\}$, or

$$D_L(q)z_L(t) = N_L(q)u(t), \qquad y(t) = z_L(t) \tag{1.15}$$

is an observable PMD of the system with transfer function $H(s)$. Comments analogous to the ones made above concerning controllable and right fractional descriptions (subscripts c and R) can also be made here concerning the subscripts o and L.

An MFD of a transfer function may not consist necessarily of ratios of polynomial matrices. In particular, given a $p \times m$ proper transfer function matrix $H(s)$, one can write

$$H(s) = \hat{N}_R(s)\hat{D}_R^{-1}(s) = \hat{D}_L^{-1}(s)\hat{N}_L(s), \tag{1.16}$$

where $\hat{N}_R$, $\hat{D}_R$, $\hat{D}_L$, $\hat{N}_L$ are proper and stable rational matrices. To illustrate, in the example considered above, $H(s)$ can be written as

$$\begin{aligned} H(s) &= \frac{1}{s^3 + 2s^2 + 2}\begin{bmatrix} s+2 & s(s+1) \\ -s & s(s^2 - s + 1) \end{bmatrix} \\ &= \left[\begin{bmatrix} (s+1)^2 & 0 \\ 0 & s+2 \end{bmatrix}^{-1}\begin{bmatrix} s^2+1 & 1 \\ s & s+2 \end{bmatrix}\right]^{-1}\left[\begin{bmatrix} (s+1)^2 & 0 \\ 0 & s+2 \end{bmatrix}^{-1}\begin{bmatrix} 1 & s \\ 0 & s \end{bmatrix}\right] \\ &= \begin{bmatrix} \dfrac{s^2+1}{(s+1)^2} & \dfrac{1}{(s+1)^2} \\ \dfrac{s}{s+2} & 1 \end{bmatrix}^{-1}\begin{bmatrix} \dfrac{1}{(s+1)^2} & \dfrac{s}{(s+1)^2} \\ 0 & \dfrac{s}{s+2} \end{bmatrix} = \hat{D}_L^{-1}(s)\hat{N}_L(s). \end{aligned}$$

Note that $\hat{D}_L(s)$ and $\hat{N}_L(s)$ are proper and stable rational matrices.

Such representations of proper transfer functions offer certain advantages when designing feedback control systems. They are discussed further in this chapter in Part 2, Subsection 7.4D.

B. Chapter Description

This chapter consists of two principal parts. In Part 1, the emphasis is on properties of systems described by PMDs. First, background on polynomial matrices is provided in Section 7.2, and the Diophantine Equation is studied at length in Subsection 7.2E. Equivalence of representations and system properties are addressed in Section 7.3. Properties of systems consisting of subsystems interconnected in parallel, in series (cascade), and in feedback configurations are investigated in Subsection 7.3C. In Part 2, Section 7.4, feedback control systems are studied with emphasis placed on parameterizing all stabilizing feedback controllers. Further details follow.

In Section 7.2, polynomial matrices and their properties are studied, and special forms for polynomial matrices, which are useful in subsequent developments, are introduced. In particular, polynomial matrices in column reduced, triangular, Hermite, and Smith form are defined, and algorithms to obtain such forms by pre- and postmultiplication by unimodular matrices are given in Subsections 7.2B and 7.2C. Coprimeness of polynomial matrices is related to controllability and observability of PMDs and is studied in Subsection 7.2D. The Diophantine Equation, which plays a central role in feedback control, is studied at length in Subsection 7.2E, and methods for deriving particular solutions are given.

PMDs of systems are addressed throughout Section 7.3. Controllability, observability, and stability are revisited in Subsection 7.3B. Also, PMD realizations of transfer function matrices are studied and realization algorithms are developed. The relationships among different PMDs and state-space descriptions of a system are explored in Subsection 7.3A, using equivalence of representations. The properties of systems consisting of interconnected subsystems are best explored using PMDs, and this is accomplished in Subsection 7.3C.

Feedback control systems are studied in Section 7.4 using PMDs and MFDs with emphasis on stabilizing controllers. All stabilizing controllers are parameterized using PMDs, in Subsection 7.4A, and proper and stable MFDs in Subsection 7.4C. State feedback controllers and state observers, important in the development involving MFDs, are discussed in Subsection 7.4B. The relationships among all feedback controller parameterizations discussed herein are derived and fully explained. The complete theory of parameterizing all stabilizing feedback controllers is developed in this section.

Two degrees of freedom controllers, their stability properties, and their parameterizations are explored in Subsection 7.4D. Several implementations of such controllers are introduced and their limitations are addressed. Finally, several control problems such as the model matching problem, the diagonal decoupling problem, and the static decoupling problem are formulated and briefly discussed.

C. Guidelines for the Reader

As with every chapter of this book, this chapter can be approached at different levels. If the characterization of all proper stabilizing feedback controllers using proper and stable MFDs, which arises in optimal control problems, is of primary interest, then the reader should focus on Subsection 7.4C. For better understanding of such MFDs of systems and of polynomial MFDs and their use in the study of systems, the reader at first reading should study selected topics from all sections of this chapter. In the following, the material that should be covered at first reading is described.

The reader should first study coprimeness of polynomial matrices in Subsection 7.2D with emphasis on the tests for coprimeness (Theorem 2.4). To determine a greatest common divisor of polynomial matrices, one needs the algorithms given in Subsection 7.2D. All solutions of the Diophantine matrix equation are derived in Theorem 2.15 of Subsection 7.2E with particular solutions obtained in Lemma 2.14.

The study of equivalence representations in Subsection 7.3A leads to insight concerning the relationships between different PMDs and state-space descriptions of a system. Tests for controllability, observability, and stability are given in Theorems 3.4, 3.5, and 3.6 of Subsection 7.3B. Feedback configurations of interconnected systems are studied in Subsection 7.3C. Here the closed-loop descriptions are also derived, which are then used in Section 7.4 to study the class of stabilizing feedback controllers.

All stabilizing controllers are expressed in terms of PMDs in Theorem 4.1 of Subsection 7.4A. Different parameters are introduced in Theorems 4.2 and 4.3, and Corollaries 4.5, 4.6, and 4.9. To fully understand proper and stable MFDs of systems, one needs to study state feedback controllers and state observers in terms of PMDs. This is accomplished in Subsection 7.4B. All proper stabilizing controllers are parameterized (using proper and stable MFDs) in Theorem 4.13 given in Subsection 7.4C and also in Theorem 4.16. The exact relationship between such MFDs and internal descriptions is provided by Theorem 4.20.

Two degrees of freedom controllers that offer advantages concerning attainable system responses are studied in Subsection 7.4D. Theorem 4.21 is the principal stability theorem with all stabilizing controllers being parameterized in Theorem 4.22.

Theorem 4.23 characterizes the response maps attainable via two degrees of freedom controllers under internal stability. Several control configurations are then examined. Finally, several control problems are formulated.

PART 1 ANALYSIS OF SYSTEMS

7.2 BACKGROUND MATERIAL ON POLYNOMIAL MATRICES

Let $R[s]^{p\times m}$ denote the set of $p \times m$ matrices with entries that are polynomials in s with real coefficients. If $P(s) \in R[s]^{p\times m}$, then $P(s)$ will be called a $p \times m$ *polynomial matrix*. Frequently, it will be necessary to determine the rank of $P(s)$, which is defined as the maximum number of linearly independent rows (or columns) of $P(s)$ over the field of rational functions. The rank of a polynomial matrix is discussed in Subsection A. In Subsection B, unimodular matrices are introduced and transformations of polynomial matrices to column and row reduced form are discussed. Hermite and Smith canonical forms are addressed in Subsection C, and in Subsection D the important concept of coprimeness of polynomial matrices is studied. In Subsection E the linear Diophantine Equation is examined.

A. Rank and Linear Independence

The linear independence of a set of vectors in a vector space, defined in Chapter 2, is recalled here for convenience. Let (V, F) denote a vector space V over the field F, and let $v_i \in V,\ i = 1, \ldots, k$. The set of vectors $\{v_1, \ldots, v_k\}$ is F-linearly dependent, i. e., it is linearly dependent over the field F, if there exists a set $\{a_1, \ldots, a_k\}$ of scalars in F with $a_i \neq 0$ for at least one i, such that

$$a_1v_1 + a_2v_2 + \cdots + a_kv_k = 0_V. \tag{2.1}$$

The set of vectors $\{v_1, \ldots, v_k\}$ is linearly independent over the field F if (2.1) implies that $a_i = 0$ for each $i = 1, \ldots, k$.

Linear dependence of $p \times 1$ polynomial vectors $p_i(s) \in R[s]^p$ is defined similarly. This warrants some explanation. Let $R[s]$ be the ring of polynomials with coefficients in R (see Subsection 7.2E) and let $R(s)$ be the field of rational fractions over $R[s]$ (called the field of rational functions), i.e.,

$$R(s) \triangleq \{t(s)|t(s) = \frac{n(s)}{d(s)},\ \text{with } n, d \in R[s], d \not\equiv 0\}.$$

Note that if $p(s) \in R[s]$, it can always be considered as being divided by 1, in which case $p(s)/1$ is a scalar in the field of rational fractions over $R[s]$. Thus, a polynomial vector $p_i(s) \in R[s]^p$ may be viewed as a special case of a rational vector, i.e., $p_i(s) \in R(s)^p$, whose elements are in the field of rational fractions. Now consider the polynomial vectors $p_i(s) \in R[s]^p, i = 1, \ldots, k$. The set of vectors $\{p_1(s), \ldots, p_k(s)\}$ is said to be $R(s)$-linearly dependent, (i.e., linearly dependent over the field of rational functions), if there exists a set $\{a_1(s), \ldots, a_k(s)\}$ of rational functions, (i.e., $a_i(s) \in R(s)$, $i = 1, \ldots, k$) with $a_i(s) \neq 0$ for at least one i, such that

$$a_1(s)p_1(s) + \cdots + a_k(s)p_k(s) = 0 \in R(s)^p. \tag{2.2}$$

This set of vectors is linearly independent over $R(s)$ if (2.2) implies that $a_i(s) = 0$ for each $i = 1, \ldots, k$.

EXAMPLE 2.1. Let $p_1(s) = \begin{bmatrix} s+3 \\ 0 \end{bmatrix}$, $p_2(s) = \begin{bmatrix} s+1 \\ 0 \end{bmatrix}$. Note that $a_1(s) = (s+1)/(s+3)$, $a_2(s) = -1$ satisfy (2.2) since

$$a_1(s)p_1(s) + a_2(s)p_2(s) = \frac{s+1}{s+3}\begin{bmatrix} s+3 \\ 0 \end{bmatrix} + (-1)\begin{bmatrix} s+1 \\ 0 \end{bmatrix} = \begin{bmatrix} 0 \\ 0 \end{bmatrix}.$$

Therefore, the set $\{p_1(s), p_2(s)\}$ is linearly dependent over the field of rational functions.

It is of interest to notice that $\{p_1(s), p_2(s)\}$ is linearly independent over the field of reals. In particular, if a_1, a_2 are restricted to be reals then (2.2) implies that $a_1 = a_2 = 0$ (verify this). This stresses the importance of the particular field over which linear independence is considered (refer to Section 2.2 of Chapter 2). ■

It is not difficult to see that if the set $\{p_1(s), \ldots, p_k(s)\}$ is linearly dependent, then (2.2) is also satisfied for some polynomials $a_i(s)$. To see this, simply multiply both sides of (2.2) by the least common multiple of the denominators of the rational functions $a_1(s), \ldots, a_k(s)$. This implies that linear dependence over $R(s)$ can be tested merely by searching for polynomials $a_i(s) \in R[s]$, not all zero, satisfying (2.2). To illustrate, consider:

EXAMPLE 2.2. In Example 2.1, $\{p_1(s), p_2(s)\}$ is linearly dependent over $R(s)$ since $(s+1)\begin{bmatrix} s+3 \\ 0 \end{bmatrix} + (-(s+3))\begin{bmatrix} s+1 \\ 0 \end{bmatrix} = \begin{bmatrix} 0 \\ 0 \end{bmatrix}$. ■

DEFINITION 2.1. The *normal rank of a polynomial matrix* $P(s) \in R[s]^{p\times m}$ is the maximum number of linearly independent rows (or columns) over the field of rational functions $R(s)$. ■

EXAMPLE 2.3. (i) $rank\, P_1(s) = rank \begin{bmatrix} s+1 & s+3 \\ 0 & 0 \end{bmatrix} = 1$, and (ii) $rank\, P_2(s) = rank \begin{bmatrix} s & 1 \\ s+1 & s+1 \end{bmatrix} = 2$. ■

It can be shown that the (normal) rank of $P(s)$ is also equal to the order of the largest order nonzero minor of $P(s)$.

EXAMPLE 2.4. (i) $P_1(s)$ in Example 2.3 does not have a second-order nonzero minor, since $det\, P_1(s) = 0$, and therefore its rank is less than 2. The entries are the first-order minors and since nonzero entries exist, $rank\, P_1(s) = 1$.

(ii) $P_2(s)$ in Example 2.3 has a second-order nonzero minor since $det\, P_2(s) = s^2 - 1 \neq 0$. Therefore, $rank\, P(s) = 2$. ■

Notice that if $s = 1$ or -1, then $det\, P_2(1) = det\, P_2(-1) = 0$. This is true because in this case, $P_2(1)$ [or $P_2(-1)$] has only one linearly independent column (over the field of reals R) and $rank\, P(1) = 1$. Since this loss of rank (from 2 to 1) occurred for only special values of s, $rank\, P(s)$ defined above is referred to as the normal rank of $P(s)$, instead of just the rank of $P(s)$, when there is ambiguity. Note that unless otherwise stated, the linear dependence of rows or columns of a matrix, considered for rank evaluation, is taken over the smallest field that contains the entries of the matrix.

B. Unimodular and Column (Row) Reduced Matrices

A polynomial matrix $U(s) \in R[s]^{p\times p}$ is called *unimodular* (or $R[s]$-unimodular) if there exists a $\hat{U}(s) \in R[s]^{p\times p}$ such that $U(s)\hat{U}(s) = I_p$. This is the same as saying that $U^{-1}(s) = \hat{U}(s)$ exists and is a polynomial matrix. Equivalently, $U(s)$ is unimodular if $det\, U(s) = \alpha \in R, \alpha \neq 0$.

It can be shown that every unimodular matrix is a matrix representation of a finite number of successive elementary row and column operations. The *elementary row and column operations* on any polynomial matrix $P(s) \in R[s]^{p\times m}$ consist of the

1. interchange of any two rows (or columns) of $P(s)$,
2. multiplication of any row (or column) of $P(s)$ by a nonzero real $\alpha \in R, \alpha \neq 0$ (a unit in $R[s]$),
3. addition to any row (or column) of $P(s)$ of a multiple by a nonzero polynomial $p(s)$ of another row (or column).

These elementary row (and column) operations can be performed by multiplying $P(s)$ on the left (right), [i.e., by pre-(post-) multiplying $P(s)$], by *elementary unimodular matrices.* These elementary unimodular matrices are obtained by performing the elementary operations (1) to (3) on the identity matrix I. As mentioned above, it can be shown that every unimodular matrix may be represented as the product of a finite number of elementary unimodular matrices.

EXAMPLE 2.5. The interchanging of rows 1 and 3 in a specific example is accomplished, e.g., as shown:

$$U_L(s)P(s) = \begin{bmatrix} 0 & 0 & 1 \\ 0 & 1 & 0 \\ 1 & 0 & 0 \end{bmatrix} \begin{bmatrix} 1 & 0 & s \\ s+1 & 1 & 0 \\ 0 & s+2 & 1 \end{bmatrix} = \begin{bmatrix} 0 & s+2 & 1 \\ s+1 & 1 & 0 \\ 1 & 0 & s \end{bmatrix}.$$

Also, addition to the second column of the third column multiplied by s in a specific example is accomplished, e.g., as shown:

$$P(s)U_R(s) = \begin{bmatrix} 1 & 0 & s \\ s+1 & 1 & 0 \\ 0 & s+2 & 1 \end{bmatrix} \begin{bmatrix} 1 & 0 & 0 \\ 0 & 1 & 0 \\ 0 & s & 1 \end{bmatrix} = \begin{bmatrix} 1 & s^2 & s \\ s+1 & 1 & 0 \\ 0 & 2s+2 & 1 \end{bmatrix}.$$

■

Let the *degree of a polynomial* (*row or column*) *vector* be the degree of the highest degree entry and let $deg_{r_i}(P)$ [$deg_{c_j}(P)$] denote the degree of the ith row (jth column) of $P(s)$. Also let $C_r(P)$ [$C_c(P)$] be the *highest row degree* (*column degree*) *coefficient matrix* of $P(s)$, defined as the real matrix with entries that are the coefficients of the highest degree s terms in each row (column) of $P(s)$.

We note that $P(s) \in R[s]^{p\times m}$ can be written as

$$\begin{aligned} P(s) &= diag\,(s^{d_{r1}}, s^{d_{r2}}, \ldots, s^{d_{rp}})C_r(P) + \hat{P}_r(s) \\ &= C_c(P)\, diag\,(s^{d_{c1}}, s^{d_{c2}}, \ldots, s^{d_{cm}}) + \hat{P}_c(s), \end{aligned} \quad (2.3)$$

where $d_{ri} = deg_{r_i}(P), i = 1, \ldots, p$, and $d_{cj} = deg_{c_j}(P), j = 1, \ldots, m$, with $\hat{P}_r(s)$, $\hat{P}_c(s)$ appropriate polynomial matrices.

EXAMPLE 2.6. Let $P(s) = \begin{bmatrix} s+1 & 3s^2+2 \\ s & 1 \\ s^2+3 & s^3+5 \end{bmatrix}$. The row degrees are $deg_{r1}(P) = 2$, $deg_{r2}(P) = 1$, and $deg_{r3}(P) = 3$, while the column degrees are $deg_{c1}(P) = 2$ and $deg_{c2}(P) = 3$. The highest row degree coefficient matrix of $P(s)$ is $C_r(P) = \begin{bmatrix} 0 & 3 \\ 1 & 0 \\ 0 & 1 \end{bmatrix}$ and the highest column degree coefficient matrix is $C_c(P) = \begin{bmatrix} 0 & 0 \\ 0 & 0 \\ 1 & 1 \end{bmatrix}$. We have

$$P(s) = \begin{bmatrix} s^{d_{r1}} & 0 & 0 \\ 0 & s^{d_{r2}} & 0 \\ 0 & 0 & s^{d_{r3}} \end{bmatrix} C_r + \hat{P}_r(s) = \begin{bmatrix} s^2 & 0 & 0 \\ 0 & s & 0 \\ 0 & 0 & s^3 \end{bmatrix} \begin{bmatrix} 0 & 3 \\ 1 & 0 \\ 0 & 1 \end{bmatrix} + \begin{bmatrix} s+1 & 2 \\ 0 & 1 \\ s^2+3 & 5 \end{bmatrix}$$

$$= C_c \begin{bmatrix} s^{d_{c1}} & 0 \\ 0 & s^{d_{c2}} \end{bmatrix} + \hat{P}_c(s) = \begin{bmatrix} 0 & 0 \\ 0 & 0 \\ 1 & 1 \end{bmatrix} \begin{bmatrix} s^2 & 0 \\ 0 & s^3 \end{bmatrix} + \begin{bmatrix} s+1 & 3s^2+2 \\ s & 1 \\ 3 & 5 \end{bmatrix}.$$

■

$P(s)$ is *row (column) proper, also called row (column) reduced,* if $C_r(P)$ $[C_c(P)]$ has full rank.

EXAMPLE 2.7. $P(s)$ in Example 2.6 is row proper since $rank\, C_r = 2$ but not column proper since $rank\, C_p = 1 < 2$. ■

Consider now the first two rows of $P(s)$ in Example 2.6 and note that

$$det \begin{bmatrix} s+1 & 3s^2+2 \\ s & 1 \end{bmatrix} = (-3)s^3 - s + 1$$
$$= det \begin{bmatrix} 0 & 3 \\ 1 & 0 \end{bmatrix} s^{(d_{r1}+d_{r2})} + \text{lower degree terms}.$$

This illustrates the following result that can be derived from (2.3): any highest order minor of $P(s) \in R[s]^{p\times m}$ is a polynomial of degree equal to the sum of the degrees of the rows ($p \geq m$) or of the columns ($p \leq m$), with leading coefficient equal to the corresponding minor of $C_r(P)$ or of $C_c(P)$, respectively. For the case when $P(s)$ is square, this immediately implies that

$$\begin{aligned} det\, P(s) &= det\, C_r(P) s^{\sum d_{r_i}} + \text{lower degree terms} \\ &= det\, C_c(P) s^{\sum d_{c_i}} + \text{lower degree terms}. \end{aligned} \tag{2.4}$$

Clearly then, $P(s) \in R[s]^{m\times m}$ is row (column) proper if and only if $deg\,(det\, P(s)) = \sum d_{r_i} (deg\,(det\,(P(s)) = \sum d_{c_j})$. (Show this.) Note that if $P(s) \in R[s]^{p\times m}$ is not of full rank it can be neither a column nor a row proper matrix.

Equation (2.3) leads to useful polynomial matrix representations given by

$$\begin{aligned} P(s) &= diag\,[s^{d_{r_i}}]C_r + block\ diag\,[1, s, \ldots, s^{d_{r_i}-1}]\hat{C}_r \\ &= block\ diag\,[1, s, \ldots, s^{d_{r_i}}]P_r \end{aligned} \tag{2.5a}$$

and

$$\begin{aligned} P(s) &= C_c\, diag\,[s^{d_{c_j}}] + \hat{C}_c\, block\ diag\,[[1, s, \ldots, s^{d_{c_j}-1}]^T] \\ &= P_c\, block\ diag\,[[1, s, \ldots, s^{d_{c_j}}]^T]. \end{aligned} \tag{2.5b}$$

EXAMPLE 2.8. In view of (2.5a), we have the representation

$$P(s) = \begin{bmatrix} s+1 & 3s^2+2 \\ s & 1 \\ s^2+3 & s^3+5 \end{bmatrix} = \begin{bmatrix} s^2 & 0 & 0 \\ 0 & s & 0 \\ 0 & 0 & s^3 \end{bmatrix} \begin{bmatrix} 0 & 3 \\ 1 & 0 \\ 0 & 1 \end{bmatrix} + \begin{bmatrix} 1 & s & 0 & 0 & 0 & 0 \\ 0 & 0 & 1 & 0 & 0 & 0 \\ 0 & 0 & 0 & 1 & s & s^2 \end{bmatrix} \begin{bmatrix} 1 & 2 \\ 1 & 0 \\ \cdots & \cdots \\ 0 & 1 \\ \cdots & \cdots \\ 3 & 5 \\ 0 & 0 \\ 1 & 0 \end{bmatrix}$$

$$= \begin{bmatrix} 1 & s & s^2 & \vdots & 0 & 0 & \vdots & 0 & 0 & 0 & 0 \\ 0 & 0 & 0 & \vdots & 1 & s & \vdots & 0 & 0 & 0 & 0 \\ 0 & 0 & 0 & \vdots & 0 & 0 & \vdots & 1 & s & s^2 & s^3 \end{bmatrix} \begin{bmatrix} 1 & 2 \\ 1 & 0 \\ 0 & 3 \\ \cdots & \cdots \\ 0 & 1 \\ 1 & 0 \\ \cdots & \cdots \\ 3 & 5 \\ 0 & 0 \\ 1 & 0 \\ 0 & 1 \end{bmatrix}.$$

Similarly, a representation in terms of column degrees can be obtained using (2.5b). ■

$P(s)$ can also be expressed as a matrix polynomial of the form

$$P(s) = P_k s^k + P_{k-1} s^{k-1} + \cdots + P_1 s + P_0, \tag{2.6}$$

where $P_i \in R^{p \times m}$. A square polynomial matrix is called *regular* if $rank\, P_k = m$. Note that if $P(s)$ is regular, then it is both row and column proper.

EXAMPLE 2.9. $P(s)$ of Example 2.8 can be written as $P(s) = P_3 s^3 + P_2 s^2 + P_1 s + P_0 =$

$$\begin{bmatrix} s+1 & 3s^2+2 \\ s & 1 \\ s^2+3 & s^3+5 \end{bmatrix} = \begin{bmatrix} 0 & 0 \\ 0 & 0 \\ 0 & 1 \end{bmatrix} s^3 + \begin{bmatrix} 0 & 3 \\ 0 & 1 \\ 1 & 0 \end{bmatrix} s^2 + \begin{bmatrix} 1 & 0 \\ 1 & 0 \\ 0 & 0 \end{bmatrix} s + \begin{bmatrix} 1 & 2 \\ 0 & 1 \\ 3 & 5 \end{bmatrix}.$$ ■

Reduction to a row (column) proper polynomial matrix

Given $P(s) \in R[s]^{p \times m}$ of full rank, there exists a unimodular matrix $U_L(s)$ such that $U_L(s)P(s)$ is row proper.

This is shown here by using a constructive proof. At each step of the algorithm below, the degree of a row (the highest degree row) is reduced by at least one, using elementary row operations. Since the matrix is of full rank, the algorithm will stop after a finite number of steps.

Algorithm

Let $d_{r_i} = deg_{r_i}(P)$, $i = 1, \ldots, p$.

(i) Obtain $diag\,[s^{d_{r_i}}]C_r(P)$.

(ii) Determine p monomials $p_i(s)$ such that

$$(p_1, \ldots, p_p)\, diag\,[s^{d_{r_i}}]C_r(P) = 0.$$

Take $p_k = 1$. This is accomplished by dividing all monomials by the lowest degree (nonzero) monomial, assumed here to be the kth one.

(iii) Premultiply $P(s)$ by

$$U_1(s) = \begin{bmatrix} 1 & 0 & \cdots & 0 & \cdots & 0 \\ \vdots & \vdots & & \vdots & & \vdots \\ p_1 & p_2 & \cdots & 1 & \cdots & p_p \\ \vdots & \vdots & & \vdots & & \vdots \\ 0 & 0 & \cdots & 0 & \cdots & 1 \end{bmatrix} k\text{th row.}$$

(iv) Stop if $U_1 P$ is row proper. Otherwise, set $P = U_1 P$; repeat the steps.

To determine the appropriate unimodular matrix $U_L(s)$ directly, so that $U_L(s)P(s)$ is row proper, one may decide on the necessary row operations based on $P(s)$ (as in the algorithm above), but apply these to $[P(s), I_p]$. Then $U_L(s)[P(s), I_p] = [\hat{P}(s), U_L(s)]$, where $\hat{P}(s)$ is the row proper matrix and $U_L(s)$ can be read off directly from the resulting matrix.

EXAMPLE 2.10. For $P(s) = \begin{bmatrix} s+1 & s \\ s^2 & s^2+2 \\ s & s+2 \end{bmatrix}$, the row degrees are $d_{r_1} = 1, d_{r_2} = 2,$ $d_{r_3} = 1$, and $rank\, C_r(P) = rank \begin{bmatrix} 1 & 1 \\ 1 & 1 \\ 1 & 1 \end{bmatrix} = 1 < 2$, and thus, $P(s)$ is not row proper.

The algorithm is now applied. We have

$$\text{(i) } diag\,[s^{d_{r_i}}][C_r(P)] = \begin{bmatrix} s & 0 & 0 \\ 0 & s^2 & 0 \\ 0 & 0 & s \end{bmatrix} \begin{bmatrix} 1 & 1 \\ 1 & 1 \\ 1 & 1 \end{bmatrix} = \begin{bmatrix} s & s \\ s^2 & s^2 \\ s & s \end{bmatrix},$$

$$\text{(ii), (iii) } U_1(s) = \begin{bmatrix} 1 & 0 & 0 \\ -s & 1 & 0 \\ 0 & 0 & 1 \end{bmatrix},$$

$$\text{(iv) } U_1 P = \begin{bmatrix} s+1 & s \\ -s & 2 \\ s & s+2 \end{bmatrix}, \text{ and } C_r(U_1 P) = \begin{bmatrix} 1 & 1 \\ -1 & 0 \\ 1 & 1 \end{bmatrix}, \text{ which is of full rank. Thus,}$$

$U_L P$, where $U_L = U_1$, is row proper. ■

Note that a unimodular matrix U_L such that $U_L P$ is row proper is not unique. In fact, in Example 2.10, another choice for U_L could have been $\tilde{U}_1 = \begin{bmatrix} 1 & 0 & -1 \\ 0 & 1 & 0 \\ 0 & 0 & 1 \end{bmatrix}$ since $\tilde{U}_1 P = \begin{bmatrix} 1 & -2 \\ s^2 & s^2+2 \\ s & s+2 \end{bmatrix}$ with $C_r(\tilde{U}_1 P) = \begin{bmatrix} 1 & -2 \\ 1 & 1 \\ 1 & 1 \end{bmatrix}$, which is of full rank, and therefore, $\tilde{U}_1 P$ is row proper.

It is possible to reduce a polynomial matrix to a row proper polynomial matrix using elementary column operations (in place of elementary row operations). In particular, given $P(s) \in R[s]^{p \times m}$ of full rank, there exists a unimodular matrix $U_R(s)$ such that $P(s)U_R(s)$ is row proper. Such a $U_R(s)$ is determined, for example, by the algorithm described in Subsection 7.2C, which reduces a matrix to a lower left Hermite form. Other algorithms to accomplish this can also be derived.

EXAMPLE 2.11. Consider $P(s)$ of Example 2.10 and apply the algorithm of Subsection 7.2C to reduce $P(s)$ to lower left Hermite form. Take $P(s)^T$ and determine $U_R^T(s)$ such that $U_R^T P^T$ is reduced to upper right Hermite form. Then

$$P(s)\begin{bmatrix} 1 & 0 \\ -1 & 1 \end{bmatrix}\begin{bmatrix} 1 & -s \\ 0 & 1 \end{bmatrix} = P(s)\begin{bmatrix} 1 & -s \\ -1 & s+1 \end{bmatrix}$$

$$= P(s)U_R(s) = \begin{bmatrix} 1 & 0 \\ -2 & s^2+2s+2 \\ -2 & 3s+2 \end{bmatrix},$$

which is row proper. ■

Similar results for reducing a polynomial matrix to a column proper polynomial matrix can easily be derived. Given $P(s) \in R[s]^{p\times m}$ of full rank, there exists a unimodular matrix $U_R(s)$ such that $P(s)U_R(s)$ is column proper. [Take $P(s)^T$ and apply steps (i) to (iii) of the above algorithm.] Also, there exists a unimodular matrix $U_L(s)$ such that $U_L(s)P(s)$ is column proper. [Use the algorithm to reduce $P(s)$ to upper right Hermite form described in Subsection 7.2C.] Finally, we note that if $P(s) \in R[s]^{p\times m}$ has full rank, there exist unimodular matrices U_L and U_R such that $U_L P U_R$ is both row and column proper. One such example is when U_L and U_R are chosen so that $U_L P U_R$ is in Smith form (see Subsection 7.2C).

Proper rational matrices

Recall that a rational matrix $H(s) \in R(s)^{p\times m}$ is called *proper* if

$$\lim_{s\to\infty} H(s) = D, \qquad D \in R^{p\times m}$$

and if $D = 0$, then $H(s)$ is called *strictly proper*. Frequently, $H(s)$ is expressed as

$$H(s) = N(s)D^{-1}(s),$$

where $N(s)$ and $D(s)$ are polynomial matrices ($N(s) \in R[s]^{p\times m}$ and $D(s) \in R[s]^{m\times m}$). The pair $\{N(s), D(s)\}$ can be viewed as an rPMFD of a system described by a transfer function matrix $H(s)$. It is of interest to relate the properness of the rational matrix $H(s)$ to the (column) degrees of $N(s)$ and $D(s)$. Note that when $N(s)$ and $D(s)$ are polynomials, it is easily seen that $H(s)$ is strictly proper (is proper) if and only if $deg\, N(s) < deg\, D(s)$ [if and only if $deg\, N(s) \leq deg\, D(s)$]. In the matrix case, similar necessary and sufficient conditions, given in Lemma 2.2, exist only when $D(s)$ is column reduced. Necessary conditions for properness are given in Lemma 2.1. Completely analogous results to Lemmas 2.1 and 2.2 hold for left factorizations of $H(s) = \tilde{D}^{-1}(s)\tilde{N}(s)$ as well.

LEMMA 2.1. Let $H(s)$ be a proper (or strictly proper) rational matrix and let $H(s) = N(s)D(s)^{-1}$. Then $deg_{c_j} N(s) \leq deg_{c_j} D(s)$ [or $deg_{c_j} N(s) < deg_{c_j} D(s)$] for $j = 1, \ldots, m$.

Proof. $N(s) = H(s)D(s)$ and for the jth column of $N(s)$,

$$n_{ij}(s) = \sum_{k=1}^{m} H_{ik}(s)d_{k_j}(s), \qquad i = 1, \ldots, p,$$

where $n_{ij}(s)$ denotes the ith element in the jth column of $N(s)$. Since every element $H_{ik}(s)$ of $H(s)$ is strictly proper (or proper), all entries $n_{ij}(s)$ must have degrees less than (or less than or equal to) the degree of the highest degree polynomial in the jth column of $D(s)$. ■

The converse to the above result is not always true. For example, let

$$N(s) = [1, s] \qquad \text{and} \qquad D(s) = \begin{bmatrix} s^2+1 & s \\ s+1 & 1 \end{bmatrix},$$

where $deg_{c_1} N(s) = 0 < deg_{c_1} D(s) = 2$ and $deg_{c_2} N(s) = 1 = deg_{c_2} D(s)$. Here

$$N(s)D^{-1}(s) = [1, s]\begin{bmatrix} 1 & -s \\ -s-1 & s^2+1 \end{bmatrix}\frac{1}{1-s} = \left[\frac{-s^2-s-1}{1-s}, \frac{s^3}{1-s}\right],$$

which is not proper. Notice that $D(s)$ is not column reduced. When $D(s)$ is column reduced (column proper) the conditions of the above lemma are sufficient as well, as the following result shows.

LEMMA 2.2. Let $H(s) = N(s)D^{-1}(s)$ with $D(s)$ column reduced. Then $H(s)$ is proper if and only if

$$deg_{c_j} N(s) \leq deg_{c_j} D(s), \qquad j = 1, \ldots, m.$$

$H(s)$ is strictly proper if and only if $deg_{c_j} N(s) < deg_{c_j} D(s)$, $j = 1, \ldots, m$.

Proof. Necessity was shown in the previous lemma. To show sufficiency, notice that by applying Cramer's rule for the inverse to solve $H(s)D(s) = N(s)$, we have $h_{ij}(s) = [det\, D^{ij}(s)/det\, D(s)]$, where $D^{ij}(s)$ is the matrix obtained by replacing the jth row of $D(s)$ by the ith row of $N(s)$. In view of (2.3), $D(s)$ can be written as $D(s) = C_c(D)\, diag\, [s^{d_{c_j}}] + \hat{D}_c(s)$, $j = 1, \ldots, m$, where $d_{c_j} = deg_{c_j} D(s)$, $C_c(D)$ is the highest column degree coefficient matrix of $D(s)$, and $deg_{c_j} \hat{D}_c(s) < d_{c_j}$. Similarly, $D^{ij}(s) = C_c(D^{ij})\, diag\, [s^{d_{c_j}}] + \hat{D}^{ij}(s)$, where $C_c(D^{ij})$ is the same as $C_c(D)$, except for the jth row, which may or may not be zero since each entry of the jth row of $N(s)$ is of lower than or equal degree of the corresponding entry of the jth row of $D(s)$. Since $D(s)$ is column proper, $det\, C_c(D) \neq 0$, and in view of (2.4), $deg\, det\, D(s) = \sum_{j=1}^{m} d_j$ while $deg\, det\, D^{ij}(s) \leq \sum_{j=1}^{m} d_j$. Therefore, $h_{ij}(s)$ is proper. It is strictly proper when $deg_{c_j} N(s) < deg_{c_j} D(s)$ for $j = 1, \ldots, m$. ■

EXAMPLE 2.12. $H(s) = N(s)D^{-1}(s)$, where

$$D(s) = \begin{bmatrix} s+1 & -1 \\ s & s \end{bmatrix} \qquad \text{and} \qquad N(s) = \begin{bmatrix} a_1 s + a_0 & b_1 s + b_0 \\ c_1 s + c_0 & d_1 s + d_0 \end{bmatrix}$$

is proper for any values of the parameters since $D(s)$ is column reduced and $deg_{c_j} N(s) \leq deg_{c_j} D(s)$, $j = 1, 2$. $H(s)$ is strictly proper only when $a_1 = b_1 = c_1 = d_1 = 0$. ■

Let $H(s) = \tilde{D}^{-1}(s)\tilde{N}(s)$, where $\tilde{D}(s) \in R[s]^{p\times p}$ and $\tilde{N}(s) \in R[s]^{p\times m}$. Completely analogous results to Lemmas 2.1 and 2.2 hold also for the left factorization matrices. In particular, if $H(s)$ is proper (strictly proper), then $deg_{r_i} \tilde{N}(s) \leq deg_{r_i} \tilde{D}(s)$ $[deg_{r_i} \tilde{N}(s) < deg_{r_i} \tilde{D}(s)]$ for $i = 1, \ldots, p$ (see Lemma 2.1). When $\tilde{D}(s)$ is row reduced, then the conditions are necessary and sufficient (see Lemma 2.2).

C. Hermite and Smith Forms

By elementary row and column operations, a polynomial matrix $P(s)$ can be reduced to the Hermite form or the Smith form. These special forms are studied in this subsection, together with algorithms to reduce $P(s)$ to such forms.

Hermite form

Given $P(s) \in R[s]^{p \times m}$ with $p \geq m$, there exists a unimodular matrix $U_L(s)$ such that $U_L(s)P(s)$ is an upper (right) triangular matrix of the form

$$U_L(s)P(s) = \begin{bmatrix} P_m(s) \\ 0 \end{bmatrix} = \begin{bmatrix} \times & \times & \cdots & \times \\ 0 & \times & \cdots & \times \\ \vdots & \vdots & \ddots & \vdots \\ 0 & 0 & \cdots & \times \\ \cdots & \cdots & \cdots & \cdots \\ 0 & 0 & \cdots & 0 \\ \vdots & \vdots & \ddots & \vdots \\ 0 & 0 & \cdots & 0 \end{bmatrix}, \tag{2.7}$$

where $P_m(s) \in R[s]^{m \times m}$. When $p \geq m \geq r = rank\, P(s)$, the last $p - r$ rows are identically zero. In column j, $1 \leq j \leq r$, the diagonal element is monic and of higher degree than any (nonzero) element above it. If the diagonal element is one, then all elements above it are zero. No particular form is assumed by the remaining $m - r$ columns in the top r rows. This is the *(upper triangular) column Hermite form.* Note that if $P(s)$ is of full rank, then $U_L(s)P(s)$ is column proper. By postmultiplication by a unimodular matrix $U_R(s)$, it is possible to obtain the *row Hermite form* of $P(s)$ when $p \leq m$. To accomplish this, simply determine $U_L(s)$ in such a manner that $U_L(s)P^T(s)$ is in column Hermite form, and take $(U_L(s)P^T(s))^T = P(s)U_L^T(s) = P(s)U_R(s)$, which is in row Hermite form.

The following algorithm reduces $P(s) \in R[s]^{p \times m}$, $p \geq m$, to (upper triangular) column Hermite form by elementary row operations [premultiplication by $U_L(s)$]. This algorithm can also be used to constructively prove our desired result, that there exists a unimodular matrix $U_L(s)$ such that $U_L(s)P(s)$ $(p \geq m)$ is in column Hermite form.

The algorithm is based on polynomial division. Given any two polynomials $a(s), b(s)$, $b(s) \neq 0$, there exist unique polynomials $q(s), r(s)$ such that $a(s) = q(s)b(s) + r(s)$ [$q(s)$ is the quotient and $r(s)$ is the remainder], where either $r(s) = 0$ or $deg\, r(s) < deg\, b(s)$.

By row interchange, transfer to the (1, 1) position the lowest degree element in the first column and call this element p_{11}. Every other element p_{i1} in this column can be expressed by polynomial division, as a multiple of p_{11} plus a remainder term of lower degree than p_{11}, i.e.,

$$p_{i1} = q_{i1}p_{11} + r_{i1}, \qquad \text{where } deg\, r_{i1} < deg\, p_{11}. \tag{2.8}$$

By elementary row operations, the appropriate multiple of p_{11} can be subtracted from each entry of column 1 leaving only remainders r_{i1} of lower degree than p_{11}. Repeat the above steps until all entries in the first column below (1, 1) are zero and note that the (1, 1) entry can always be taken to be a monic polynomial.

Consider next the second column and position (2, 2) while temporarily ignoring the first row. Repeat the above procedure to make all the entries below the (2, 2) entry equal to zero. If the (1, 2) entry does not have lower degree than the (2, 2) entry, use polynomial division and row operations to replace the (1, 2) entry by a polynomial of lower degree than the (2, 2) entry. If the (2, 2) entry is a nonzero constant, use row operations to make the (1, 2) entry equal to zero. Continuing this procedure with the third, fourth, and higher columns results in the desired Hermite form.

EXAMPLE 2.13.

$$P(s) = \begin{bmatrix} s(s+2) & 0 \\ 0 & (s+1)^2 \\ (s+1)(s+2) & s+1 \\ 0 & s(s+1) \end{bmatrix} \xrightarrow{U_1} \begin{bmatrix} s(s+2) & 0 \\ 0 & (s+1)^2 \\ s+2 & s+1 \\ 0 & s(s+1) \end{bmatrix} \xrightarrow{U_2} \begin{bmatrix} s+2 & s+1 \\ 0 & (s+1)^2 \\ s(s+2) & 0 \\ 0 & s(s+1) \end{bmatrix}$$

$$\xrightarrow{U_3} \begin{bmatrix} s+2 & s+1 \\ 0 & (s+1)^2 \\ 0 & -s(s+1) \\ 0 & s(s+1) \end{bmatrix} \xrightarrow{U_4} \begin{bmatrix} s+2 & s+1 \\ 0 & (s+1)^2 \\ 0 & s+1 \\ 0 & s+1 \end{bmatrix} \xrightarrow{U_5} \begin{bmatrix} s+2 & s+1 \\ 0 & s+1 \\ 0 & (s+1)^2 \\ 0 & s+1 \end{bmatrix}$$

$$\xrightarrow{U_6} \begin{bmatrix} s+2 & s+1 \\ 0 & s+1 \\ 0 & 0 \\ 0 & 0 \end{bmatrix} \xrightarrow{U_7} \begin{bmatrix} s+2 & 0 \\ 0 & s+1 \\ \cdots & \cdots \\ 0 & 0 \\ 0 & 0 \end{bmatrix} = U_L(s)P(s),$$

where $\quad U_L(s) = U_7U_6\cdots U_1$

$$= \begin{bmatrix} 1 & -1 & 0 & 0 \\ 0 & 1 & 0 & 0 \\ 0 & 0 & 1 & 0 \\ 0 & 0 & 0 & 1 \end{bmatrix} \begin{bmatrix} 1 & 0 & 0 & 0 \\ 0 & 1 & 0 & 0 \\ 0 & -(s+1) & 1 & 0 \\ 0 & -1 & 0 & 1 \end{bmatrix} \cdots \begin{bmatrix} 1 & 0 & 0 & 0 \\ 0 & 1 & 0 & 0 \\ -1 & 0 & 1 & 0 \\ 0 & 0 & 0 & 1 \end{bmatrix}$$

$$= \begin{bmatrix} -(s+2) & -1 & s+1 & 0 \\ s+1 & 1 & -s & 0 \\ -(s+1)^2 & -s & s(s+1) & 0 \\ -(s+1) & 0 & s & -1 \end{bmatrix}.$$

Notice that $P(s)$ has full rank and $U_L(s)P(s)$ here is column proper. ■

Note that to determine $U_L(s)$ directly, one may decide on the necessary operations based on $P(s)$, but apply these elementary row operations to $[P(s), I_p]$. Then $U_L(s)[P(s), I_p] = [H_p(s), U_L(s)]$, where $H_p(s)$ is the column Hermite form of $P(s)$, $(p \geq m)$. Also, note that if the algorithm is applied to $P(s) \in R[s]^{p\times m}$, where $p \leq m$, then

$$U_L(s)P(s) = [P_1(s), P_2(s)] = \left[\begin{array}{cccc:ccc} \times & \times & \cdots & \times & \times & \cdots & \times \\ 0 & \times & \cdots & \times & \times & \cdots & \times \\ \vdots & & \ddots & \vdots & \vdots & \ddots & \vdots \\ 0 & 0 & \cdots & \times & \times & \cdots & \times \end{array}\right], \tag{2.9}$$

where $P_1(s) \in R[s]^{p\times p}$.

Smith form

Given $P(s) \in R[s]^{p\times m}$ with *rank* $P(s) = r$, there exist unimodular matrices $U_L(s)$ and $U_R(s)$ such that

$$U_L(s)P(s)U_R(s) = S_P(s), \tag{2.10}$$

where $\quad S_P(s) = \begin{bmatrix} \Lambda(s) & 0 \\ 0 & 0 \end{bmatrix}, \qquad \Lambda(s) \triangleq diag\,(\epsilon_1(s), \ldots, \epsilon_r(s)).$

Each $\epsilon_i(s)$, $i = 1, \ldots, r$, is a unique monic polynomial satisfying $\epsilon_i(s) \mid \epsilon_{i+1}(s)$, $i = 1, \ldots, r-1$, where $p_2(s) | p_1(s)$ means that there exists a polynomial $p_3(s)$ such that $p_1(s) = p_2(s)p_3(s)$, that is, $\epsilon_i(s)$ divides $\epsilon_{i+1}(s)$. $S_P(s)$ is the *Smith form of* $P(s)$ and the $\epsilon_i(s)$, $i = 1, \ldots, r$, are the *invariant polynomials of* $P(s)$. It can be shown that

$$\epsilon_i(s) = \frac{D_i(s)}{D_{i-1}(s)}, \qquad i = 1, \ldots, r, \tag{2.11}$$

where $D_i(s)$ is the monic greatest common divisor of all the nonzero ith order minors of $P(s)$. Note that $D_0(s) \triangleq 1$ and $D_i(s)$ are the *determinantal divisors of* $P(s)$. The Smith form $S_P(s)$ of a matrix $P(s)$ is unique, however, $U_L(s)$, $U_R(s)$ such that $U_L(s)P(s)U_R(s) = S_P(s)$ are not unique.

EXAMPLE 2.14. $P(s) = \begin{bmatrix} s(s+2) & 0 \\ 0 & (s+1)^2 \\ (s+1)(s+2) & s+1 \\ 0 & s(s+1) \end{bmatrix}$ with *rank* $P(s) = r = 2$. Here $D_0 = 1, D_1 = 1, D_2 = (s+1)(s+2)$, and $\epsilon_1 = D_1/D_0 = 1, \epsilon_2 = D_2/D_1 = (s+1)(s+2)$. Therefore, the Smith form of $P(s)$ is

$$S_P(s) = \begin{bmatrix} \Lambda(s) \\ \cdots \\ 0 \end{bmatrix} = \begin{bmatrix} 1 & 0 \\ 0 & (s+1)(s+2) \\ \cdots & \cdots \\ 0 & 0 \\ 0 & 0 \end{bmatrix}.$$

■

The invariant factors ϵ_i of a matrix are not affected by row and column elementary operations. This follows from the fact that the determinantal divisors D_i are not affected by elementary operations (refer to the Binet-Cauchy formula in Exercise 7.3). In view of this, the following result can now be easily established: given $P_1(s), P_2(s) \in R[s]^{p \times m}$, there exist unimodular matrices $U_1(s)$, $U_2(s)$ such that

$$U_1(s)P_1(s)U_2(s) = P_2(s)$$

if and only if $P_1(s)$, $P_2(s)$ have the same Smith form.

The following algorithm reduces $P(s) \in R[s]^{p \times m}$ to its unique Smith form $S_P(s)$ and determines $U_L(s)$, $U_R(s)$ such that $U_L(s)P(s)U_R(s) = S_P(s)$. The algorithm can be used to constructively prove that the Smith form of a matrix exists.

Using row and column elementary operations, transfer the element of least degree in the matrix $P(s)$ to the (1, 1) position. By elementary row operations, make all entries in the first column below (1, 1) equal to zero (refer to the algorithm for the column Hermite form). Next, by column operations, make all entries in the first row zero except (1, 1). If nonzero entries have reappeared in the first column, repeat the above steps until all entries in the first column and row are zero except for the (1, 1) entry. [Show that at each iteration the degree of the (1, 1) element is reduced, and thus the algorithm is finite.]

If the (1, 1) element does not divide every other entry in the matrix, use polynomial division and row and column interchanges to bring a lower degree element to the (1, 1) position. Repeat the above steps until all other elements in the first column and row are zero and the (1, 1) entry divides every other entry in the matrix, that is,

$$\begin{bmatrix} \epsilon_1(s) & 0 & \cdots & 0 \\ 0 & & & \\ \vdots & & E_1(s) & \\ 0 & & & \end{bmatrix},$$

where $\epsilon_1(s)$ divides all entries of $E_1(s)$. Repeat the above steps on $E_1(s)$ and on other such terms, if necessary to obtain the Smith form of $P(s)$.

Two polynomial matrices $P_1(s), P_2(s) \in R[s]^{p\times m}$ are said to be *equivalent* if there exist unimodular matrices $U_1(s)$, $U_2(s)$ such that

$$U_1(s)P_1(s)U_2(s) = P_2(s). \tag{2.12}$$

Recall that, as was mentioned earlier, (2.12) is satisfied if and only if $P_1(s)$ and $P_2(s)$ have the same Smith form.

The Smith form of $P(s)$ is a canonical form for the relation (2.12) on $R[s]^{p\times m}$. To see this, we first recall the definition of relation and equivalence relation: a *relation* ρ in a set X is any subset of $X \times X$ and ρ is an *equivalence relation* if and only if it satisfies the following axioms:

1. $x\rho x$-reflexivity (every $x \in X$ is equivalent to itself).
2. $(x\rho y) \Rightarrow (y\rho x)$-symmetry ($x$ is equivalent to y implies that y is equivalent to x).
3. $(x\rho y)$ and $(y\rho z) \Rightarrow (x\rho z)$-transitivity ($x$ is equivalent to y and y is equivalent to z imply that x is equivalent to z).

The relation ρ described by $U_1P_1U_2 = P_2$ in (2.12), (i.e., $P_1\rho P_2$), satisfies the above axioms (verify this) and therefore it is an equivalence relation. The ρ-equivalence class or the "orbit" of a fixed $P(s) \in R[s]^{p\times m}$ is denoted by $[P(s)]_\rho$. The Smith form $S_P(s)$ of $P(s)$ is a *canonical form* for ρ on $R[s]^{p\times m}$.

D. Coprimeness and Common Divisors

Coprimeness of polynomial matrices is one of the most important concepts in the polynomial matrix representation of systems since it is directly related to controllability and observability (see Subsection 7.3B).

A polynomial $g(s)$ is a *common divisor* (cd) of polynomials $p_1(s)$, $p_2(s)$ if and only if there exist polynomials $\tilde{p}_1(s)$, $\tilde{p}_2(s)$ such that

$$p_1(s) = \tilde{p}_1(s)g(s), \qquad p_2(s) = \tilde{p}_2(s)g(s). \tag{2.13}$$

The highest degree cd of $p_1(s)$, $p_2(s)$, $g^*(s)$, is a *greatest common divisor* (gcd) of $p_1(s)$, $p_2(s)$. It is unique within multiplication by a nonzero real number. Alternatively, $g^*(s)$ is a gcd of $p_1(s)$, $p_2(s)$ if and only if *any* cd $g(s)$ of $p_1(s)$, $p_2(s)$ is a divisor of $g^*(s)$ as well, that is,

$$g^*(s) = m(s)g(s) \tag{2.14}$$

with $m(s)$ a polynomial. The polynomials $p_1(s)$, $p_2(s)$ are *coprime* (cp) if and only if a gcd $g^*(s)$ is a nonzero real.

The above can be extended to matrices. In this case, both right divisors and left divisors must be defined, since in general, two polynomial matrices do not commute. Note that one may talk about right or left divisors of polynomial matrices only when the matrices have the same number of columns or rows, respectively.

An $m \times m$ matrix $G_R(s)$ is a *common right divisor* (crd) of the $p_1 \times m$ polynomial matrix $P_1(s)$ and the $p_2 \times m$ matrix $P_2(s)$ if there exist polynomial matrices $P_{1R}(s)$, $P_{2R}(s)$ so that

$$P_1(s) = P_{1R}(s)G_R(s), \qquad P_2(s) = P_{2R}(s)G_R(s). \tag{2.15}$$

Similarly, a $p \times p$ polynomial matrix $G_L(s)$ is a *common left divisor* (cld) of the $p \times m_1$ polynomial matrix $\hat{P}_1(s)$ and the $p \times m_2$ matrix $\hat{P}_2(s)$, if there exist polynomial matrices $\hat{P}_{1L}(s)$, $\hat{P}_{2L}(s)$ so that

$$\hat{P}_1(s) = G_L(s)\hat{P}_{1L}(s), \qquad \hat{P}_2(s) = G_L(s)\hat{P}_{2L}(s). \tag{2.16}$$

Also $G_R^*(s)$ is a *greatest common right divisor* (gcrd) of $P_1(s)$ and $P_2(s)$ if and only if any crd $G_R(s)$ is an rd of $G_R^*(s)$. Similarly, $G_L^*(s)$ is a *greatest common left divisor* (gcld) of $\hat{P}_1(s)$ and $\hat{P}_2(s)$ if and only if any cld $G_L(s)$ is an ld of $G_L^*(s)$. That is,

$$G_R^*(s) = M(s)G_R(s), \qquad G_L^*(s) = G_L(s)N(s) \tag{2.17}$$

with $M(s)$ and $N(s)$ polynomial matrices and $G_R(s)$ and $G_L(s)$ any crd and cld of $P_1(s)$, $P_2(s)$, respectively.

Alternatively, it can be shown that any crd $G_R^*(s)$ [of $P_1(s)$ and $P_2(s)$ or a cld $G_L^*(s)$ of $\hat{P}_1(s)$ and $\hat{P}_2(s)$] with determinant of the highest degree possible is a gcrd (gcld) of the matrices. It is unique within a premultiplication (postmultiplication) by a unimodular matrix. Here it is assumed that $G_R(s)$ is nonsingular. Note that if *rank* $\begin{bmatrix} P_1(s) \\ P_2(s) \end{bmatrix} = m$ [a $(p_1 + p_2) \times m$ matrix], which is a typical case in polynomial matrix system descriptions, then *rank* $G_R(s) = m$, that is, $G_R(s)$ is nonsingular. Notice also that if *rank* $G_R(s) < m$, then in view of Sylvester's Rank Inequality, *rank* $\begin{bmatrix} P_1(s) \\ P_2(s) \end{bmatrix} < m$ as well.

The polynomial matrices $P_1(s)$ and $P_2(s)$ are *right coprime* (rc) if and only if a gcrd $G_R^*(s)$ is a unimodular matrix. Similarly, $\hat{P}_1(s)$ and $\hat{P}_2(s)$ are *left coprime* (lc) if and only if a gcld $G_2^*(s)$ is a unimodular matrix.

EXAMPLE 2.15. Let $P_1 = \begin{bmatrix} s(s+2) & 0 \\ 0 & (s+1)^2 \end{bmatrix}$ and $P_2 = \begin{bmatrix} (s+1)(s+2) & s+1 \\ 0 & s(s+1) \end{bmatrix}$.

Two distinct crds are $G_{R_1} = \begin{bmatrix} 1 & 0 \\ 0 & s+1 \end{bmatrix}$ and $G_{R_2} = \begin{bmatrix} s+2 & 0 \\ 0 & 1 \end{bmatrix}$ since

$$\begin{bmatrix} P_1 \\ P_2 \end{bmatrix} = \begin{bmatrix} s(s+2) & 0 \\ 0 & s+1 \\ \cdots & \cdots \\ (s+1)(s+2) & 1 \\ 0 & s \end{bmatrix} G_{R_1} = \begin{bmatrix} s & 0 \\ 0 & (s+1)^2 \\ \cdots & \cdots \\ s+2 & s+1 \\ 0 & s(s+1) \end{bmatrix} G_{R_2}.$$

A gcrd is $G_R^* = \begin{bmatrix} s+2 & 0 \\ 0 & s+1 \end{bmatrix} = \begin{bmatrix} s+2 & 0 \\ 0 & 1 \end{bmatrix} G_{R_1} = \begin{bmatrix} 1 & 0 \\ 0 & s+1 \end{bmatrix} G_{R_2}$. Now, $\begin{bmatrix} P_1 \\ P_2 \end{bmatrix} G_R^{*-1} =$

$$\begin{bmatrix} P_{1R}^* \\ P_{2R}^* \end{bmatrix} = \begin{bmatrix} s & 0 \\ 0 & s+1 \\ \cdots & \cdots \\ s+1 & 1 \\ 0 & s \end{bmatrix},$$ where P_{1R}^* and P_{2R}^* are rc. Note that a gcld of P_1 and P_2 is

$G_L^* = \begin{bmatrix} 1 & 0 \\ 0 & s+1 \end{bmatrix}$. Both G_R^* and G_L^* were determined using an algorithm to derive the Hermite form of $\begin{bmatrix} P_1 \\ P_2 \end{bmatrix}$ as will be described later in this section. ■

It can be shown that two square $p \times p$ nonsingular polynomial matrices with determinants that are prime polynomials are both rc and lc. The converse of this is not true, that is, two rc polynomial matrices do not necessarily have prime determinant polynomials. A case in point is Example 2.15, where P_{1R}^* and P_{2R}^* are rc; however, $\det P_{1R}^* = \det P_{2R}^* = s(s+1)$.

Left and right coprimeness of two polynomial matrices (provided that the matrices are compatible) are quite distinct properties. For example, two matrices can be lc but not rc, and vice versa (refer to the following example).

EXAMPLE 2.16. $P_1 = \begin{bmatrix} s(s+2) & 0 \\ 0 & s+1 \end{bmatrix}$ and $P_2 = \begin{bmatrix} (s+1)(s+2) & 1 \\ 0 & s \end{bmatrix}$ are lc but not rc since a gcrd is $G_R^* = \begin{bmatrix} s+2 & 0 \\ 0 & 1 \end{bmatrix}$ with $\det G_R^* = (s+2)$. ■

Finally, we note that all the above definitions apply also to more than two polynomial matrices. To see this, replace in all definitions P_1, P_2 by $P_1, P_2, \ldots, P_k$. This is not surprising in view of the fact that the $p_1 \times m$ matrix $P_1(s)$ and the $p_2 \times m$ matrix $P_2(s)$ consist of p_1 and p_2 rows, respectively, each of which can be viewed as a $1 \times m$ polynomial matrix; that is, instead of, e.g., the coprimeness of P_1 and P_2, one could speak of the coprimeness of the $(p_1 + p_2)$ rows of P_1 and P_2.

How to determine a greatest common right divisor (gcrd)

LEMMA 2.3. Let $P_1(s) \in R[s]^{p_1 \times m}$ and $P_2(s) \in R[s]^{p_2 \times m}$ with $p_1 + p_2 \geq m$. Let the unimodular matrix $U(s)$ be such that

$$U(s)\begin{bmatrix} P_1(s) \\ P_2(s) \end{bmatrix} = \begin{bmatrix} G_R^*(s) \\ 0 \end{bmatrix}. \tag{2.18}$$

Then $G_R^*(s)$ is a gcrd of $P_1(s), P_2(s)$.

Proof. Let

$$U = \begin{bmatrix} \bar{X} & \bar{Y} \\ -\tilde{P}_2 & \tilde{P}_1 \end{bmatrix} \tag{2.19}$$

with $\bar{X} \in R[s]^{m \times p_1}$, $\bar{Y} \in R[s]^{m \times p_2}$, $\tilde{P}_2 \in R[s]^{q \times p_1}$, and $\tilde{P}_1 \in R[s]^{q \times p_2}$, where $q \triangleq (p_1 + p_2) - m$. Note that $\bar{X}, \bar{Y}$ and $\tilde{P}_2, \tilde{P}_1$ are lc pairs. If they were not, then $\det U \neq \alpha$, a nonzero real number. Similarly, $\bar{X}, \tilde{P}_2$ and $\bar{Y}, \tilde{P}_1$ are rc pairs. Let

$$U^{-1} = \begin{bmatrix} \bar{P}_1 & -\tilde{Y} \\ \bar{P}_2 & \tilde{X} \end{bmatrix}, \tag{2.20}$$

where $\bar{P}_1 \in R[s]^{p_1 \times m}$, $\bar{P}_2 \in R[s]^{p_2 \times m}$ are rc and $\tilde{X} \in R[s]^{p_2 \times q}$, $\tilde{Y} \in R[s]^{p_1 \times q}$ are rc. Equation (2.18) implies that

$$\begin{bmatrix} P_1 \\ P_2 \end{bmatrix} = U^{-1}\begin{bmatrix} G_R^* \\ 0 \end{bmatrix} = \begin{bmatrix} \bar{P}_1 \\ \bar{P}_2 \end{bmatrix} G_R^*, \tag{2.21}$$

i.e., G_R^* is a crd of P_1, P_2. Equation (2.18) implies also that

$$\bar{X}P_1 + \bar{Y}P_2 = G_R^*. \tag{2.22}$$

This relationship shows that any crd G_R of P_1, P_2 will also be an rd of G_R^*. This can be seen directly by expressing (2.22) as $MG_R = G_R^*$, where M is a polynomial matrix. Thus, G_R^* is a crd of P_1, P_2 with the property that any crd G_R of P_1, P_2 is an rd of G_R^*. This implies that G_R^* is a gcrd of P_1, P_2. ■

EXAMPLE 2.17. Let $P_1 = \begin{bmatrix} s(s+2) & 0 \\ 0 & (s+1)^2 \end{bmatrix}$, $P_2 = \begin{bmatrix} (s+1)(s+2) & s+1 \\ 0 & s(s+1) \end{bmatrix}$. Then

$$U\begin{bmatrix} P_1 \\ P_2 \end{bmatrix} = \begin{bmatrix} \bar{X} & \bar{Y} \\ -\tilde{P}_2 & \tilde{P}_1 \end{bmatrix}\begin{bmatrix} P_1 \\ P_2 \end{bmatrix} = \begin{bmatrix} -(s+2) & -1 & \vdots & s+1 & 0 \\ s+1 & 1 & \vdots & -s & 0 \\ \cdots & \cdots & \cdots & \cdots & \cdots \\ -(s+1)^2 & -s & \vdots & s(s+1) & 0 \\ -(s+1) & 0 & \vdots & s & -1 \end{bmatrix}\begin{bmatrix} P_1 \\ P_2 \end{bmatrix}$$

$$= \begin{bmatrix} s+2 & 0 \\ 0 & s+1 \\ \cdots & \cdots \\ 0 & 0 \\ 0 & 0 \end{bmatrix} = \begin{bmatrix} G_R^* \\ \cdots \\ 0 \end{bmatrix}.$$

In view of Lemma 2.3, $G_R^* = \begin{bmatrix} s+2 & 0 \\ 0 & s+1 \end{bmatrix}$ is a gcrd (see also Example 2.15). ■

Note that to derive (2.18) and thus determine a gcrd G_R^* of P_1 and P_2, one could use the algorithm that was developed above, in Subsection 7.2C (or a variation of this algorithm) to obtain the Hermite form. Finally, note also that if the Smith form of $\begin{bmatrix} P_1 \\ P_2 \end{bmatrix}$ is known, i.e., $U_L \begin{bmatrix} P_1 \\ P_2 \end{bmatrix} U_R = S_P = \begin{bmatrix} diag\,[\epsilon_i] & 0 \\ 0 & 0 \end{bmatrix}$, then $(diag\,[\epsilon_i], 0)U_R^{-1}$ is a gcrd of P_1 and P_2 in view of Lemma 2.3. When $rank \begin{bmatrix} P_1 \\ P_2 \end{bmatrix} = m$, which is the case of interest in systems, then a gcrd of P_1 and P_2 is $diag\,[\epsilon_i]U_R^{-1}$.

Criteria for coprimeness

There are several ways of testing the coprimeness of two polynomial matrices, as shown in the following theorem.

THEOREM 2.4. Let $P_1 \in R[s]^{p_1 \times m}$ and $P_2 \in R[s]^{p_2 \times m}$ with $p_1 + p_2 \geq m$. The following statements are equivalent:

(i) P_1 and P_2 are rc.

(ii) A gcrd of P_1 and P_2 is unimodular.

(iii) There exist polynomial matrices $X \in R[s]^{m \times p_1}$ and $Y \in R[s]^{m \times p_2}$ such that

$$XP_1 + YP_2 = I_m. \tag{2.23}$$

(iv) The Smith form of $\begin{bmatrix} P_1 \\ P_2 \end{bmatrix}$ is $\begin{bmatrix} I \\ 0 \end{bmatrix}$.

(v) $rank \begin{bmatrix} P_1(s_i) \\ P_2(s_i) \end{bmatrix} = m$ for any complex number s_i.

(vi) $\begin{bmatrix} P_1 \\ P_2 \end{bmatrix}$ constitutes m columns of a unimodular matrix.

Proof. Statements (i) and (ii) are equivalent by definition. Assume now that (iii) is true. Then (2.23) implies that any crd of P_1 and P_2 must be an rd of I_m, which is of course a crd of P_1 and P_2. Therefore, in view of the definition of a gcrd, I_m is a gcrd (see also proof of Lemma 2.3) and so (ii) is true. To show that (ii) also implies (iii), determine a gcrd G_R^* as in the Lemma 2.3 and note that G_R^* is unimodular. Premultiplying (2.22) by G_R^{*-1}, we obtain (2.23).

To show (iv), recall that a gcrd of P_1 and P_2 can be determined from the Smith form of $\begin{bmatrix} P_1 \\ P_2 \end{bmatrix}$ as $G_R^* = (diag\,[\epsilon_i], 0)U_R^{-1}$ (refer to the discussion following Example 2.17). It is now clear that a gcrd will be unimodular if and only if the Smith form of $\begin{bmatrix} P_1 \\ P_2 \end{bmatrix}$ is $\begin{bmatrix} I \\ 0 \end{bmatrix}$ (i.e., (iv) and (ii) are equivalent). To show (v), consider (2.18) and note that $rank \begin{bmatrix} P_1(s_i) \\ P_2(s_i) \end{bmatrix} = rank\, G_R^*(s_i)$. The only s_i that can reduce the rank are the zeros of the determinant of G_R^*. Such s_i do not exist if and only if $det\, G_R^* = \alpha$, a nonzero real number, i.e., if and only if G_R^* is unimodular. Therefore (v) implies and is implied by (ii). Part (vi) was shown in the proof of Lemma 2.3. ■

EXAMPLE 2.18. (i) The polynomial matrices $P_1 = \begin{bmatrix} s & 0 \\ 0 & s+1 \end{bmatrix}$, $P_2 = \begin{bmatrix} s+1 & 1 \\ 0 & s \end{bmatrix}$ (see also Example 2.15) are rc in view of the following relations. To use condition (ii) of Theorem 2.4, let

$$U \begin{bmatrix} P_1 \\ P_2 \end{bmatrix} = \left[\begin{array}{cc:cc} -(s+2) & -1 & s+1 & 0 \\ s+1 & 1 & -s & 0 \\ \cdots & \cdots & \cdots & \cdots \\ -(s+1)^2 & -s & s(s+1) & 0 \\ -(s+1) & 0 & s & -1 \end{array}\right] \begin{bmatrix} P_1 \\ P_2 \end{bmatrix} = \begin{bmatrix} 1 & 0 \\ 0 & 1 \\ \cdots & \cdots \\ 0 & 0 \\ 0 & 0 \end{bmatrix} = \begin{bmatrix} G_R^* \\ \cdots \\ 0 \end{bmatrix}.$$

Then $G_R^* = I_2$, which is unimodular. Applying condition (iii), $XP_1 + YP_2 = \begin{bmatrix} -(s+2) & -1 \\ s+1 & 1 \end{bmatrix} P_1 + \begin{bmatrix} s+1 & 0 \\ -s & 0 \end{bmatrix} P_2 = I_2$.

To use (iv), note that the invariant polynomials of $\begin{bmatrix} P_1 \\ P_2 \end{bmatrix}$ are $\epsilon_1 = \epsilon_2 = 1$; the Smith form is then $\begin{bmatrix} I_2 \\ 0 \end{bmatrix}$. To use condition (v), note that presently, the only complex values s_i that may reduce the rank of $\begin{bmatrix} P_1(s_i) \\ P_2(s_i) \end{bmatrix}$ are those for which $det\, P_1(s_i)$ or $det\, P_2(s_i) = 0$, i.e., $s_1 = 0$ and $s_2 = -1$. For these values we have $rank \begin{bmatrix} P_1(s_1) \\ P_2(s_1) \end{bmatrix} = rank \begin{bmatrix} 0 & 0 \\ 0 & 1 \\ 1 & 1 \\ 0 & 0 \end{bmatrix} = 2$ and $rank \begin{bmatrix} P_1(s_2) \\ P_2(s_2) \end{bmatrix} = rank \begin{bmatrix} -1 & 0 \\ 0 & 0 \\ 0 & 1 \\ 0 & -1 \end{bmatrix} = 2$, i.e., both are of full rank. ■

Note that if the criteria for coprimeness given in Theorem 2.4 are used for a pair P_1, P_2 that is not rc, then (ii) will provide a gcrd G_R^* of P_1, P_2. The other tests provide only partial information about G_R^*. In particular, applying (iv), one obtains the Smith form of G_R^* (show this), while (v) will give information about the zeros of $det\, G_R^*(s)$.

THEOREM 2.5. Let $\hat{P}_1 \in R[s]^{p\times m_1}$ and $\hat{P}_2 \in R[s]^{p\times m_2}$ with $m_1 + m_2 \geq p$. The following statements are equivalent:

(i) $\hat{P}_1$ and $\hat{P}_2$ are lc.
(ii) A gcld of $\hat{P}_1$ and $\hat{P}_2$ is unimodular.
(iii) There exist polynomial matrices $\hat{X} \in R[s]^{m_1\times p}$ and $\hat{Y} \in R[s]^{m_2\times p}$ such that

$$\hat{P}_1\hat{X} + \hat{P}_2\hat{Y} = I_p. \tag{2.24}$$

(iv) The Smith form of $[\hat{P}_1, \hat{P}_2]$ is $[I, 0]$.
(v) *rank* $[\hat{P}_1(s_i), \hat{P}_2(s_i)] = p$ for any complex number s_i.
(vi) $[\hat{P}_1, \hat{P}_2]$ are p rows of a unimodular matrix.

Proof. The proof is completely analogous to the proof of Theorem 2.4 and is omitted. ■

E. The Diophantine Equation

The linear Diophantine Equation of interest to us is of the form

$$X(s)D(s) + Y(s)N(s) = Q(s), \tag{DIO}$$

where $D(s)$, $N(s)$, and $Q(s)$ are given polynomial matrices and $X(s)$, $Y(s)$ are to be determined. This equation will be studied in detail in this subsection. First, particular solutions of the Diophantine Equation are derived. Next, all solutions are conveniently parameterized. These parameterizations are used later in this chapter (in Section 7.4) to characterize all stabilizing linear feedback controllers of a linear time-invariant system. This subsection is concluded with some historical remarks.

In greater detail, solutions of the polynomial Diophantine Equation of low degree are derived in Lemma 2.6, using the Division Theorem for polynomials, and in Lemma 2.8, using the Sylvester Matrix of two polynomials. Corresponding results for the polynomial matrix Diophantine Equation are derived in Lemma 2.12, using the Division Theorem and in Lemma 2.14, using the Eliminant Matrix. All solutions of the Diophantine Equation are parameterized in Theorem 2.15.

The polynomial case

Given coprime polynomials $d(s)$ and $n(s)$, it was shown in Theorem 2.4 that there exist polynomials $x(s)$ and $y(s)$ such that

$$x(s)d(s) + y(s)n(s) = 1. \tag{2.25}$$

It can be shown that $x(s)$ and $y(s)$ are unique when certain restrictions are imposed on their degrees. In particular, the following result is true.

LEMMA 2.6. Let $d(s)$ and $n(s)$ be coprime polynomials. Then there exist unique polynomials $\hat{x}(s)$ and $\hat{y}(s)$ that satisfy

$$\hat{x}(s)d(s) + \hat{y}(s)n(s) = 1 \tag{2.26}$$

with

$$deg\, \hat{y}(s) < deg\, d(s) \qquad [\text{or } deg\, \hat{x}(s) < deg\, n(s)]. \tag{2.27}$$

Proof. In view of Theorem 2.4, $d(s)$ and $n(s)$ are coprime if and only if there exist polynomials $x(s)$ and $y(s)$ such that $x(s)d(s) + y(s)n(s) = 1$. The basic Division Theorem for polynomials can now be used to obtain polynomial solutions of lower degree. In particular, there exists a unique quotient polynomial $q(s)$ and a unique remainder polynomial

$r(s)$ such that

$$y(s) = q(s)d(s) + r(s) \qquad \text{with } deg\, r(s) < deg\, d(s).$$

Now define

$$\hat{y}(s) = r(s) = y(s) - q(s)d(s), \qquad \hat{x}(s) = x(s) + q(s)n(s), \tag{2.28}$$

where $deg\,\hat{y}(s) < deg\,d(s)$. Then $\hat{x}(s)d(s) + \hat{y}(s)n(s) = x(s)d(s) + y(s)n(s) + [q(s)n(s)d(s) - q(s)d(s)n(s)] = x(s)d(s) + y(s)n(s) = 1$. Also note that $deg\,[\hat{x}(s)d(s)] = deg\,[1 - \hat{y}(s)n(s)] < deg\,[d(s)n(s)]$, so that $deg\,\hat{x}(s) < deg\,n(s)$. To prove uniqueness, suppose there exists another pair $\bar{x}(s)$, $\bar{y}(s)$ that satisfies the lemma. Then

$$[\bar{x}(s) - \hat{x}(s)]d(s) + [\bar{y}(s) - \hat{y}(s)]n(s) = 0,$$

which implies that

$$[\bar{x}(s) - \hat{x}(s)] = -[\bar{y}(s) - \hat{y}(s)]\frac{n(s)}{d(s)}.$$

Since $deg\,[\bar{y}(s) - \hat{y}(s)] < deg\,d(s)$ and the left-hand side is a polynomial, this relation implies that $d(s)$ and $n(s)$ must have a common factor that contradicts the assumption that $d(s)$ and $n(s)$ are coprime. Therefore, $\hat{x}(s)$ and $\hat{y}(s)$ are unique. Note that the proof of the lemma when in (2.27) $deg\,\hat{x}(s) < deg\,n(s)$ (for the part in parentheses) is completely analogous. In this case, let $x(s)d(s) + y(s)n(s) = 1$ and divide $x(s)$ by $n(s)$. ■

Given coprime polynomials $d(s)$ and $n(s)$, one way of determining $\hat{x}(s)$ and $\hat{y}(s)$ that satisfy the above lemma is to follow the procedure outlined in its proof. In particular, $x(s)$ and $y(s)$ that satisfy (2.25) are determined first, and then polynomial division is used to reduce their degrees, if necessary.

The following example illustrates this procedure.

EXAMPLE 2.19. Let $d(s) = s(s-1)$ and $n(s) = s - 2$, which are coprime. Let $x(s)d(s) + y(s)n(s) = \frac{1}{4}s[s(s-1)] - \frac{1}{4}(s^2+s+2)[s-2] = \frac{1}{4}(s^3 - s^2) - \frac{1}{4}(s^3 - s^2 - 4) = 1$. Then $y(s) = -\frac{1}{4}(s^2 + s + 2) = q(s)d(s) + r(s) = (-\frac{1}{4})[s(s-1)] + [-\frac{1}{2}(s+1)]$, from which we obtain

$$\hat{y}(s) = r(s) = -\tfrac{1}{2}(s+1) \quad \text{and} \quad \hat{x}(s) = x(s) + q(s)n(s) = \tfrac{1}{4}s + (-\tfrac{1}{4})(s-2) = \tfrac{1}{2}.$$

Note that $\hat{x}(s)d(s) + \hat{y}(s)n(s) = \frac{1}{2}[s(s-1)] + (-\frac{1}{2}(s+1))[s-2] = 1$ and $deg\,\hat{y}(s) = 1 < deg\,d(s) = 2$ $[deg\,\hat{x}(s) = 0 < deg\,n(s) = 1]$.

The polynomials $\hat{x}(s)$ and $\hat{y}(s)$ can also be determined using the Sylvester Matrix of the polynomials $d(s)$ and $n(s)$, which in fact provides an alternative way of testing the coprimeness of $d(s)$ and $n(s)$. The Sylvester Matrix of $d(s)$ and $n(s)$ is now introduced.

Consider the polynomials,

$$\begin{aligned} d(s) &= d_n s^n + d_{n-1}s^{n-1} + \cdots + d_1 s + d_0, & d_n &\neq 0 \\ n(s) &= n_m s^m + n_{m-1}s^{m-1} + \cdots + n_1 s + n_0, & m &\leq n. \end{aligned} \tag{2.29}$$

The *Sylvester Matrix* of the polynomials $d(s)$, $n(s)$ is defined as

$$S(d,n) = \begin{bmatrix} d_n & d_{n-1} & d_{n-2} & \cdots & d_1 & d_0 & 0 & 0 & \cdots & 0 \\ 0 & d_n & d_{n-1} & \cdots & d_2 & d_1 & d_0 & 0 & \cdots & 0 \\ \vdots & \vdots & \vdots & & \vdots & \vdots & \vdots & \vdots & & \vdots \\ 0 & 0 & 0 & \cdots & 0 & d_n & d_{n-1} & d_{n-2} & \cdots & d_0 \\ n_m & n_{m-1} & n_{m-2} & \cdots & n_0 & 0 & 0 & 0 & \cdots & 0 \\ 0 & n_m & n_{m-1} & \cdots & n_1 & n_0 & 0 & 0 & \cdots & 0 \\ \vdots & \vdots & \vdots & & \vdots & \vdots & \vdots & \vdots & & \vdots \\ 0 & 0 & 0 & \cdots & 0 & 0 & n_m & n_{m-1} & & n_0 \end{bmatrix}, \tag{2.30}$$

where the block that contains the coefficients of $d(s)$ has m rows and the one that contains the coefficients of $n(s)$ has n rows. The coefficients are arranged so that there are $n+m$ columns, i.e., $S(d,n) \in R^{(n+m)\times(n+m)}$.

The determinant of $S(d,n)$ is known as the *Sylvester Resultant* or *Resultant of the polynomials* $d(s)$ and $n(s)$. Note that the matrix $S(d,n)$ is sometimes referred to as the *Eliminant Matrix* of the polynomials $d(s)$, $n(s)$. ■

THEOREM 2.7. The polynomials $d(s)$ and $n(s)$ are coprime if and only if $\det S(d,n) \neq 0$, that is, if and only if their resultant is nonzero.

Proof. *(Necessity)* Suppose that $d(s)$ and $n(s)$ are coprime but their resultant is zero. This implies that there exists a nonzero vector α, $\alpha^T \in R^{m+n}$ such that $\alpha S(d,n) = 0$ or such that

$$\alpha S(d,n)\begin{bmatrix} s^{n+m-1} \\ s^{n+m-2} \\ \vdots \\ s \\ 1 \end{bmatrix} = \alpha \begin{bmatrix} s^{m-1}d(s) \\ \vdots \\ sd(s) \\ d(s) \\ s^{n-1}n(s) \\ \vdots \\ sn(s) \\ n(s) \end{bmatrix} = [\alpha_1(s), \alpha_2(s)]\begin{bmatrix} d(s) \\ n(s) \end{bmatrix} = 0,$$

where $\alpha_1(s) = \alpha_1 \begin{bmatrix} s^{m-1} \\ \vdots \\ s \\ 1 \end{bmatrix}$ and $\alpha_2(s) = \alpha_2 \begin{bmatrix} s^{n-1} \\ \vdots \\ s \\ 1 \end{bmatrix}$ with $[\alpha_1, \alpha_2] = \alpha$. Note that both $\alpha_1(s)$ and $\alpha_2(s)$ are nonzero since otherwise either $n(s)$ or $d(s)$ would be zero and therefore $n(s)$ and $d(s)$ would not be coprime. Write $d(s)/n(s) = -\alpha_2(s)/\alpha_1(s)$. Since $\deg \alpha_2(s) < \deg d(s)$ and $\deg \alpha_1(s) < \deg n(s)$, it follows that $d(s)$ and $n(s)$ must have a common factor. Therefore, $d(s)$ and $n(s)$ are not coprime, which is a contradiction. Thus, $S(d,n)$ has full rank, or $\det S(d,n) \neq 0$.

(Sufficiency) Assume that $\det S(d,n) \neq 0$. Take $\alpha = [0,\ldots,0,1]S(d,n)^{-1}$ and let $\alpha_1(s) = \alpha_1[s^{m-1},\ldots,s,1]^T$, $\alpha_2(s) = \alpha_2[s^{n-1},\ldots,s,1]^T$, where $[\alpha_1,\alpha_2] = \alpha$. Then $\alpha_1(s)d(s) + \alpha_2(s)n(s) = \alpha S(d,n)[s^{n+m-1}, s^{n+m-2},\ldots,s,1]^T = 1$, which implies, in view of Theorem 2.4, that $d(s)$ and $n(s)$ are coprime. ■

Remarks

(i) Theorem 2.7 is attributed to Sylvester (1840).

(ii) A useful relation, used in the proof of Theorem 2.7, is

$$S(d,n)\begin{bmatrix} s^{n+m-1} \\ s^{n+m-2} \\ \vdots \\ s \\ 1 \end{bmatrix} = \begin{bmatrix} s^{m-1}d(s) \\ \vdots \\ sd(s) \\ d(s) \\ s^{n-1}n(s) \\ \vdots \\ sn(s) \\ n(s) \end{bmatrix}, \tag{2.31}$$

where $d(s)$ and $n(s)$ are given in (2.29) and $S(d,n)$ is defined in (2.30).

(iii) It is possible to arrange the coefficients in $S(d, n)$ in several different ways, recalling that elementary row and column operations leave the rank of $S(d, n)$ unchanged. For example, for $n = 3$, $m = 2$,

$$\text{(a)} \begin{bmatrix} d_3 & d_2 & d_1 & d_0 & 0 \\ 0 & d_3 & d_2 & d_1 & d_0 \\ n_2 & n_1 & n_0 & 0 & 0 \\ 0 & n_2 & n_1 & n_0 & 0 \\ 0 & 0 & n_2 & n_1 & n_0 \end{bmatrix}, \qquad \text{(b)} \begin{bmatrix} d_3 & d_2 & d_1 & d_0 & 0 \\ 0 & d_3 & d_2 & d_1 & d_0 \\ 0 & 0 & n_2 & n_1 & n_0 \\ 0 & n_2 & n_1 & n_0 & 0 \\ n_2 & n_1 & n_0 & 0 & 0 \end{bmatrix},$$

$$\text{(c)} \begin{bmatrix} d_0 & d_1 & d_2 & d_3 & 0 \\ 0 & d_0 & d_1 & d_2 & d_3 \\ n_0 & n_1 & n_2 & 0 & 0 \\ 0 & n_0 & n_1 & n_2 & 0 \\ 0 & 0 & n_0 & n_1 & n_2 \end{bmatrix}$$

are all nonsingular if and only if $d(s)$ and $n(s)$ are coprime. The matrix in (a) is the Sylvester Matrix as defined in (2.30). The matrices in (b) and (c) are variations of the Sylvester Matrix found in the literature.

(iv) By adding zero coefficients in $n(s)$, it can always be assumed that the polynomials $d(s)$ and $n(s)$ have equal degrees ($m = n$) when using the test provided by Theorem 2.7. This leads yet to another variation of the Sylvester Matrix of two polynomials of dimension $2n \times 2n$.

Using the procedure followed in the sufficiency proof of Theorem 2.7, it is straightforward to derive the unique polynomials $\hat{x}(s)$ and $\hat{y}(s)$ that satisfy (2.26) and (2.27) of Lemma 2.6. This is illustrated in the next example.

EXAMPLE 2.20. Let $d(s) = s(s-1)$ and $n(s) = s-2$, which are coprime as in Example 2.19. In this case the Sylvester Matrix or the eliminant of $d(s)$ and $n(s)$ in (2.30) is $S(d, n) = \begin{bmatrix} 1 & -1 & 0 \\ 1 & -2 & 0 \\ 0 & 1 & -2 \end{bmatrix}$ and the resultant is $det\, S(d, n) = 2 \neq 0$, as expected, since $d(s)$ and $n(s)$ are coprime polynomials. In view of the sufficiency proof of Theorem 2.7, let $\alpha = [\alpha_1, \alpha_2] = [0, 0, 1]S(d, n)^{-1} = [0, 0, 1] \begin{bmatrix} 4 & -2 & 0 \\ 2 & -2 & 0 \\ 1 & -1 & -1 \end{bmatrix} \frac{1}{2} = [\frac{1}{2}, -\frac{1}{2}, -\frac{1}{2}]$. Then $\hat{x}(s) = \alpha_1(s) = \frac{1}{2}$, $\hat{y}(s) = \alpha_2(s) = [-\frac{1}{2}, -\frac{1}{2}] \begin{bmatrix} s \\ 1 \end{bmatrix} = -\frac{1}{2}(s+1)$, which also is the answer determined in Example 2.19. ■

Using the Sylvester Matrix of coprime polynomials enables one in fact to determine solutions to the more general Diophantine Equation

$$x(s)d(s) + y(s)n(s) = q(s), \tag{2.32}$$

as the following lemma shows.

LEMMA 2.8. Let $d(s)$ and $n(s)$ be coprime polynomials with $deg\, d(s) = n$ and $deg\, n(s) = m \leq n$ given by (2.29). Let $q(s)$ be a polynomial of degree $n + m - 1$. Then there exist unique polynomials $\hat{x}(s)$ and $\hat{y}(s)$ that satisfy

$$\hat{x}(s)d(s) + \hat{y}(s)n(s) = q(s) \tag{2.33}$$

with

$$deg\, \hat{x}(s) < m \qquad \text{and} \qquad deg\, \hat{y}(s) < n. \tag{2.34}$$

Proof. The proof is constructive. Let $\hat{x}(s) = [x_{m-1}, \ldots, x_1, x_0][s^{m-1}, \ldots, s, 1]^T$, let $\hat{y}(s) = [y_{n-1}, \ldots, y_1, y_0][s^{n-1}, \ldots, s, 1]^T$, and write $\hat{x}(s)d(s) + \hat{y}(s)n(s) = [x_{m-1}, \ldots, x_0, y_{n-1}, \ldots, y_0][s^{m-1}d(s), \ldots, sd(s), d(s), s^{n-1}n(s), \ldots, sn(s), n(s)]^T = [x_{m-1}, \ldots, x_0, y_{n-1}, \ldots, y_0]S(d, n)[s^{n+m-1}, \ldots, s, 1]^T = q(s) = [q_{n+m-1}, \ldots, q_1, q_0][s^{n+m-1}, \ldots, s, 1]^T$, where relation (2.31) was used. This equality is now satisfied if and only if

$$[x_{m-1}, \ldots, x_0, y_{n-1}, \ldots, y_0]S(d, n) = [q_{n+m-1}, \ldots, q_1, q_0] \tag{2.35}$$

is satisfied. Since $S(d, n)$ is nonsingular, Eq. (2.35) has a unique solution that determines unique $\hat{x}(s)$ and $\hat{y}(s)$ that satisfy relations (2.33) and (2.34). ■

EXAMPLE 2.21. Consider $d(s) = s(s-1)$ and $n(s) = s - 2$, which are coprime as in Examples 2.19 and 2.20. Let $q(s) = q_2s^2 + q_1s + q_0$ be an arbitrary second-degree polynomial ($n + m - 1 = 2 + 1 - 1 = 2$). Relation (2.35) yields in this case $[x_0, y_1, y_0]\begin{bmatrix} 1 & -1 & 0 \\ 1 & -2 & 0 \\ 0 & 1 & -2 \end{bmatrix} = [q_2, q_1, q_0]$, from which we obtain $[x_0, y_1, y_0] = \frac{1}{2}[q_2, q_1, q_0]\begin{bmatrix} 4 & -2 & 0 \\ 2 & -2 & 0 \\ 1 & -1 & -1 \end{bmatrix} = [2q_2 + q_1 + \frac{1}{2}q_0, -q_2 - q_1 - \frac{1}{2}q_0, -\frac{1}{2}q_0]$ and

$$\hat{x}(s) = x_0 = 2q_2 + q_1 + \tfrac{1}{2}q_0$$

$$\hat{y}(s) = [y_1, y_0]\begin{bmatrix} s \\ 1 \end{bmatrix} = (-\tfrac{1}{2}q_0) + (-q_2 - q_1 - \tfrac{1}{2}q_0)s.$$

For $q(s) = s^2 + 2s + 1$, $\hat{x}(s) = \frac{9}{2}$, and $\hat{y}(s) = -\frac{1}{2} - \frac{7}{2}s$. ■

COROLLARY 2.9. Let $d(s)$ and $n(s)$ be coprime polynomials with $deg\, d(s) = n$ and $deg\, n(s) \leq n$. Let $q(s)$ be a polynomial with $deg\, q(s) \leq 2n - 1$. Then there exist unique polynomials $\hat{x}(s)$ and $\hat{y}(s)$ that satisfy the relation

$$\hat{x}(s)d(s) + \hat{y}(s)n(s) = q(s) \tag{2.36}$$

with

$$deg\, \hat{x}(s) < n \quad \text{and} \quad deg\, \hat{y}(s) < n. \tag{2.37}$$

Proof. The proof is similar to the proof of Lemma 2.8, where in place of $S(d, n) \in R^{(n+m)\times(n+m)}$, the $2n \times 2n$ Sylvester Matrix $\tilde{S}(d, n)$ of $d(s) = d_ns^n + \cdots + d_1s + d_0$, $d_n \neq 0$ and $n(s) = n_ns^n + \cdots + n_1s + n_0$ is used. Zero coefficients are added in $n(s)$ if necessary. Note that $\tilde{S}(d, n)$ is exactly $S(d, n)$ of (2.30) with $m = n$. This leads to the relation

$$[x_{n-1}, \ldots, x_0, y_{n-1}, \ldots, y_0]\tilde{S}(d, n) = [q_{2n-1}, \ldots, q_1, q_0], \tag{2.38}$$

which is the equation corresponding to (2.35). Since $d(s)$ and $n(s)$ are coprime, $\tilde{S}(d, n)$ is nonsingular and the $\hat{x}(s)$ and $\hat{y}(s)$ that satisfy (2.38) are unique. ■

EXAMPLE 2.22. As in Example 2.21, consider $d(s) = s(s-1) = s^2 - s$ and $n(s) = s - 2 = 0 \cdot s^2 + s - 2$, which are coprime. Then (2.38) implies that

$$[x_1, x_0, y_1, y_0]\begin{bmatrix} 1 & -1 & 0 & 0 \\ 0 & 1 & -1 & 0 \\ 0 & 1 & -2 & 0 \\ 0 & 0 & 1 & -2 \end{bmatrix} = [q_3, q_2, q_1, q_0],$$

where $q(s) = q_3s^3 + q_2s^2 + q_1s + q_0$ is a third-degree polynomial ($2n - 1 = 2.2 - 1 = 3$). It can easily be seen that one could equivalently solve $x_1 = q_3$ and $[x_0, y_1, y_0] \times \begin{bmatrix} 1 & -1 & 0 \\ 1 & -2 & 0 \\ 0 & 1 & -2 \end{bmatrix} = [q_2 + q_3, q_1, q_0]$ (compare with Example 2.21), from which we obtain

$$\hat{x}(s) = [x_1, x_0]\begin{bmatrix} s \\ 1 \end{bmatrix} = q_3 s + [2(q_2 + q_3) + q_1 + \tfrac{1}{2}q_0]$$

$$\hat{y}(s) = [y_1, y_0]\begin{bmatrix} s \\ 1 \end{bmatrix} = [-(q_2 + q_3) - q_1 - \tfrac{1}{2}q_0]s + (-\tfrac{1}{2}q_0).$$

Note that for $q_3 = 0$, this solution is precisely the solution of Example 2.21. ■

The polynomial matrix case

Similar results as in the preceding can be shown for polynomial matrices. First the Division Theorem for polynomial matrices is established.

THEOREM 2.10. Let $D(s) \in R[s]^{m\times m}$ be nonsingular. Then for any $N(s) \in R[s]^{p\times m}$ there exist unique polynomial matrices $Q(s)$, $R(s)$ such that

$$N(s) = Q(s)D(s) + R(s) \tag{2.39}$$

with $R(s)D(s)^{-1}$ strictly proper, or with $deg_{c_j} R(s) < deg_{c_j} D(s)$, $j = 1, \ldots, m$, when $D(s)$ is column reduced.

Proof. Let $H(s) = N(s)D^{-1}(s)$ be a rational matrix. By polynomial division in each entry, decompose this matrix uniquely into $H(s) = H_{sp}(s) + P(s)$, where $H_{sp}(s)$ is a strictly proper rational matrix and $P(s)$ is a polynomial matrix. Then $N(s) = H(s)D(s) = P(s)D(s) + R(s)$, where $R(s) = H_{sp}(s)D(s)$. Then $R(s)$ is a polynomial matrix (since it is equal to $N(s) - P(s)D(s)$), and by definition, $R(s)D^{-1}(s) = H_{sp}(s)$ is strictly proper. Note that when $D(s)$ is column reduced, then in view of Lemma 2.2, $R(s)D^{-1}(s)$ is strictly proper if and only if $deg_{c_j} R(s) < deg_{c_j} D(s)$, $j = 1, \ldots, m$. We shall now verify uniqueness of $Q(s)$ and $R(s)$ directly; it also follows from the construction. Suppose there are two pairs such that $Q(s)D(s) + R(s) = \bar{Q}(s)D(s) + \bar{R}(s)$. Then $Q(s) - \bar{Q}(s) = [\bar{R}(s) - R(s)]D^{-1}(s)$. Since the left-hand side is a polynomial matrix and the right-hand side is a strictly proper rational or zero matrix, this equality can hold only when both sides are zero, i.e., $\bar{Q}(s) = Q(s)$ and $\bar{R}(s) = R(s)$. ■

As expected, the following result also holds.

THEOREM 2.11. Let $\tilde{D}(s) \in R[s]^{p\times p}$ be nonsingular. Then for any $\tilde{N}(s) \in R[s]^{p\times m}$ there exist unique polynomial matrices $\tilde{Q}(s)$, $\tilde{R}(s)$ such that

$$\tilde{N}(s) = \tilde{D}(s)\tilde{Q}(s) + \tilde{R}(s) \tag{2.40}$$

with $\tilde{D}^{-1}(s)\tilde{R}(s)$ strictly proper, or with $deg_{r_i} \tilde{R}(s) < deg_{r_i} \tilde{D}(s)$, $i = 1, \ldots, p$, when $\tilde{D}(s)$ is row reduced.

Proof. The proof of this result is completely analogous to the proof of Theorem 2.10. ■

Given rc polynomial matrices $D(s) \in R[s]^{m\times m}$ and $N(s) \in R[s]^{p\times m}$ it was shown in Theorem 2.4 that there exist polynomial matrices $X(s)$ and $Y(s)$ such that

$$X(s)D(s) + Y(s)N(s) = I_m. \tag{2.41}$$

Let

$$H(s) = N(s)D^{-1}(s) = \tilde{D}^{-1}(s)\tilde{N}(s) \tag{2.42}$$

be proper, where $\tilde{D}(s) \in R[s]^{p\times p}$ and $\tilde{N}(s) \in R[s]^{p\times m}$ are lc with $\tilde{D}(s)$ row reduced. Let ν be the highest degree of the polynomial entries in $\tilde{D}(s)$ and denote this as

$$\nu = deg\, \tilde{D}(s). \tag{2.43}$$

We point out that it can be shown that ν is the observability index of the system $H(s) = N(s)D^{-1}(s) = \tilde{D}^{-1}(s)\tilde{N}(s)$.

LEMMA 2.12. Given $D(s) \in R[s]^{m\times m}$ and $N(s) \in R[s]^{p\times m}$ that are coprime, there exist polynomial matrices $\hat{X}(s)$ and $\hat{Y}(s)$ that satisfy

$$\hat{X}(s)D(s) + \hat{Y}(s)N(s) = I_m \tag{2.44}$$

such that

$$deg\,\hat{Y}(s) < \nu.$$

Proof. This result can also be established by using the eliminant of $N(s)$ and $D(s)$ to be introduced in Lemma 2.14. In the present proof, use is made of the Division Theorem for polynomial matrices (Theorem 2.10). Let $X(s) \in R[s]^{m\times m}$ and $Y(s) \in R[s]^{m\times p}$ satisfy (2.41). There exist unique polynomial matrices $Q(s) \in R[s]^{m\times p}$ and $R(s) \in R[s]^{m\times p}$ such that $Y(s) = Q(s)\tilde{D}(s) + R(s)$ with $R(s)\tilde{D}^{-1}(s)$ strictly proper, where $\tilde{D}(s)$ is row reduced and satisfies (2.42). Now define $\hat{Y}(s) = R(s) = Y(s) - Q(s)\tilde{D}(s)$ and $\hat{X}(s) = X(s) + Q(s)\tilde{N}(s)$. Then $\hat{X}(s)D(s) + \hat{Y}(s)N(s) = X(s)D(s) + Y(s)N(s) + [Q(s)\tilde{N}(s)D(s) - Q(s)\tilde{D}(s)N(s)] = X(s)D(s) + Y(s)N(s) = I_m$. Note also that, in view of Lemma 2.1, if $R(s)\tilde{D}^{-1}(s)$ is strictly proper, then the column degrees of $R(s)$ are less than ν, i.e., $deg\,\hat{Y}(s) < \nu$. In addition, note that from $\hat{X}(s)D(s) = I_m - \hat{Y}(s)N(s)$ it follows that $deg_{c_j}(\hat{X}(s)D(s)) = deg_{c_j}(I_m - \hat{Y}(s)N(s)) < \nu + deg_{c_j} N(s)$, $j = 1, \ldots, m$. ∎

Remarks

(i) When $\tilde{D}(s)$ is row proper, then $n = deg\,|\tilde{D}(s)| =$ sum of the row degrees of $\tilde{D}(s) \leq p\nu$, that is, $\nu \geq n/p$. If $\tilde{D}(s)$ is not row proper, then $n = deg\,|\tilde{D}(s)| <$ sum of the rows degrees of $D(s)$ and $deg\,\tilde{D}(s)$ can be large.

(ii) In the polynomial case we have that $\nu = n$ and Lemma 2.12 reduces to Lemma 2.6.

(iii) Using the Eliminant Matrix of the rc polynomial matrices $N(s)$ and $D(s)$, where $D(s)$ is column proper, it will be shown (in Lemma 2.14) that (2.44) has a solution with both $deg\,\hat{Y}(s) < \nu$ and $deg\,\hat{X}(s) < \nu$.

The Eliminant Matrix

Consider the polynomial matrices $N(s) \in R[s]^{p\times m}$ and $D(s) \in R[s]^{m\times m}$ with $D(s)$ column proper. Let $deg_{c_j} N(s) \leq deg_{c_j} D(s) = d_j$, $j = 1, \ldots, m$, and assume that $d_j \geq 1$, $j = 1, \ldots, m$. The *Eliminant Matrix* M_e of $N(s)$ and $D(s)$ is a generalization of the Sylvester Matrix, or of the Eliminant of two polynomials, introduced in (2.30), and is defined in an analogous manner (see also Remarks following Theorem 2.7). In particular, we write

$$\begin{bmatrix} N(s) \\ sN(s) \\ \vdots \\ s^{k-1}N(s) \\ D(s) \\ sD(s) \\ \vdots \\ s^{k-1}D(s) \end{bmatrix} = M_{ek}S_{ek}(s) = M_{ek}\left[block\ diag \begin{bmatrix} 1 \\ s \\ \vdots \\ s^{d_j+k-1} \end{bmatrix} \right] \tag{2.45}$$

for some integer k, where $M_{ek} \in R^{k(p+m)\times(\sum d_j+mk)}$. Note that for $k \geq (\sum d_j)/p$, M_{ek} will have as many as or more rows than columns. Let $k = \nu$ be the least integer k that minimizes $(\sum d_j + mk) - rank\, M_{ek}$, or equivalently (as can be shown), let $k = \nu$ denote the observability index of the system $\{D, I, N\}$, or of its equivalent state-space description (refer to the discussion on equivalence of representations in Subsection 7.3A).

The *Eliminant Matrix* of $N(s)$ and $D(s)$ is defined as the $\nu(p+m) \times (n + m\nu)$ matrix

$$M_e = M_{e\nu},$$

where $n = \sum d_j = deg\, det\, D(s)$, since $D(s)$ is column proper. Note that the observability index ν satisfies $\nu \geq n/p$, and therefore, M_e has as many as or more rows than columns. The following result is a generalization of Theorem 2.7 (which involved the Sylvester Matrix of two polynomials) to polynomial matrices.

THEOREM 2.13. The polynomial matrices $N(s)$ and $D(s)$ given above are rc if and only if their eliminant matrix M_e has full column rank, i.e., if and only if

$$rank\, M_e = n + m\nu. \tag{2.46}$$

Proof. The proof is omitted. For a proof of this theorem, refer to Wolovich [36]. ■

Remarks

(i) The Eliminant Matrix M_e becomes a Sylvester Matrix when applied to polynomials. In particular, for $d(s) = d_3s^3 + d_2s^2 + d_1s + d_0$ and $n(s) = n_3s^3 + n_2s^2 + n_1s + n_0$, $p = m = 1, n = d_1 = \nu = 3$, this is true where

$$\begin{bmatrix} n(s) \\ sn(s) \\ s^2n(s) \\ d(s) \\ sd(s) \\ s^2d(s) \end{bmatrix} = M_eS_e(s) = \begin{bmatrix} n_0 & n_1 & n_2 & n_3 & 0 & 0 \\ 0 & n_0 & n_1 & n_2 & n_3 & 0 \\ 0 & 0 & n_0 & n_1 & n_2 & n_3 \\ d_0 & d_1 & d_2 & d_3 & 0 & 0 \\ 0 & d_0 & d_1 & d_2 & d_3 & 0 \\ 0 & 0 & d_0 & d_1 & d_2 & d_3 \end{bmatrix} \begin{bmatrix} 1 \\ s \\ s^2 \\ s^3 \\ s^4 \\ s^5 \end{bmatrix}.$$

See also Remarks following Theorem 2.7.

(ii) It can be shown that

$$rank\, M_{ek} = rank\, M_{e\nu}, \qquad k \geq \nu,$$

i.e., the rank of M_{ek} is the maximum possible when k is equal to the observability index of the system. This is a consequence of the alternative definition for ν previously given.

The Elliminant Matrix can be used to determine solutions of the Diophantine Equation

$$X(s)D(s) + Y(s)N(s) = Q(s). \tag{2.47}$$

The Diophantine Equation is further discussed later in this subsection.

LEMMA 2.14. Consider the rc polynomial matrices $N(s) \in R[s]^{p\times m}$, $D(s) \in R[s]^{m\times m}$. Let $D(s)$ be column proper, and assume $deg_{c_j} N(s) \leq deg_{c_j} D(s) = d_j, d_j \geq 1, j = 1, \ldots, m$ [note that $H(s) = N(s)D^{-1}(s)$ is proper]. Let $Q(s) \in R[s]^{q\times m}$ with

$$deg_{c_j} Q(s) \leq d_j + \nu - 1, \tag{2.48}$$

where ν is the observability index of the system $\{D, I, N, O\}$. Then there exist $\hat{X}(s) \in R[s]^{q\times m}$ and $\hat{Y}(s) \in R[s]^{q\times p}$ that satisfy the equation

$$\hat{X}(s)D(s) + \hat{Y}(s)N(s) = Q(s) \tag{2.49}$$

with $$deg\,\hat{X}(s) \leq \nu - 1, \qquad deg\,\hat{Y}(s) \leq \nu - 1,$$

where $deg\,\hat{X}(s)$ denotes the highest polynomial degree in the entries of $\hat{X}(s)$.

Proof. The proof is by construction. Let

$$\hat{X}(s) = X_0 + X_1 s + \cdots + X_{\nu-1}s^{\nu-1} = [X_0, X_1, \ldots, X_{\nu-1}]\begin{bmatrix} I_m \\ sI_m \\ \vdots \\ s^{\nu-1}I_m \end{bmatrix} \tag{2.50}$$

and $$\hat{Y}(s) = Y_0 + Y_1 s + \cdots + Y_{\nu-1}s^{\nu-1} = [Y_0, Y_1, \ldots, Y_{\nu-1}]\begin{bmatrix} I_p \\ sI_p \\ \vdots \\ s^{\nu-1}I_p \end{bmatrix}. \tag{2.51}$$

Then $$\hat{Y}(s)N(s) + \hat{X}(s)D(s) = [Y_0, Y_1, \ldots, Y_{\nu-1}, X_0, X_1, \ldots, X_{\nu-1}]\begin{bmatrix} N(s) \\ sN(s) \\ \vdots \\ s^{\nu-1}N(s) \\ D(s) \\ sD(s) \\ \vdots \\ s^{\nu-1}D(s) \end{bmatrix} =$$

$[Y_0, \ldots, Y_{\nu-1}, X_0, \ldots, X_{\nu-1}]M_e S_e(s) = Q(s) = QS_e(s)$, in view of the definition of the eliminant matrix M_e and the assumptions on the column degrees of $Q(s)$ in (2.48). Therefore, (2.49) is satisfied with $\hat{X}(s)$ and $\hat{Y}(s)$ given in (2.50) and (2.51) if and only if

$$[Y_0, \ldots, Y_{\nu-1}, X_0, \ldots, X_{\nu-1}]M_e = Q \tag{2.52}$$

is satisfied, where $M_e \in R^{\nu(p+m)\times(n+m\nu)}$ and $Q \in R^{q\times(n+m\nu)}$ with $n = \sum_{j=1}^{m} d_j = deg\,det\,D(s)$. Since $N(s)$ and $D(s)$ are rc, it follows from Theorem 2.13 that $rank\,M_e = n + m\nu$, i.e., M_e has full column rank. This implies that a solution to (2.52) exists for arbitrary Q. Therefore, solutions $\hat{X}(s)$ and $\hat{Y}(s)$ of (2.49) with $deg\,\hat{X}(s) \leq \nu - 1$ and $deg\,\hat{Y}(s) \leq \nu - 1$ always exist. ■

Remark

When $N(s)$ and $D(s)$ are polynomials, Lemma 2.14 reduces to Corollary 2.9 (with $n(s)/d(s)$ proper). In this case, $\nu = d_1 = n = deg\,(det\,D(s))$.

EXAMPLE 2.23. The polynomial matrices $N(s) = \begin{bmatrix} s+1 & 0 \\ 1 & 1 \end{bmatrix}$, $D(s) = \begin{bmatrix} s^2 & 0 \\ 1 & -s+1 \end{bmatrix}$ are rc since $rank \begin{bmatrix} N(\lambda) \\ D(\lambda) \end{bmatrix} = 2$ for $\lambda = 0$ and $\lambda = 1$, the roots of $det\,D(s)$. $D(s)$ is column proper with $deg_{c_1} D(s) = d_1 = 2$, $deg_{c_2} D(s) = d_2 = 1$, $n = d_1 + d_2 = 3$, while $deg_{c_1} N(s) = 1 < d_1 = 2$ and $deg_{c_2} N(s) = 0 < d_2 = 1$, $p = m = 2$. To construct the Eliminant Matrix of $N(s)$ and $D(s)$, we note that $n/p = \frac{3}{2}$, and therefore, we let $k = 2$

be an initial value. Then

$$\begin{bmatrix} N(s) \\ sN(s) \\ D(s) \\ sD(s) \end{bmatrix} = M_{e2}S_{e2} = \begin{bmatrix} 1 & 1 & 0 & 0 & 0 & 0 & 0 \\ 1 & 0 & 0 & 0 & 1 & 0 & 0 \\ \cdots & \cdots & \cdots & \cdots & \cdots & \cdots & \cdots \\ 0 & 1 & 1 & 0 & 0 & 0 & 0 \\ 0 & 1 & 0 & 0 & 0 & 1 & 0 \\ \cdots & \cdots & \cdots & \cdots & \cdots & \cdots & \cdots \\ 0 & 0 & 1 & 0 & 0 & 0 & 0 \\ 1 & 0 & 0 & 0 & 1 & -1 & 0 \\ \cdots & \cdots & \cdots & \cdots & \cdots & \cdots & \cdots \\ 0 & 0 & 0 & 1 & 0 & 0 & 0 \\ 0 & 1 & 0 & 0 & 0 & 1 & -1 \end{bmatrix} \begin{bmatrix} 1 & 0 \\ s & 0 \\ s^2 & 0 \\ s^3 & 0 \\ 0 & 1 \\ 0 & s \\ 0 & s^2 \end{bmatrix}.$$

Note that $rank\, M_{e2} = 7 = n + mk$, which is full column rank. Therefore, $\nu = 2$ and the Eliminant Matrix of $N(s)$ and $D(s)$ is $M_e = M_{e2}$. The fact that $rank\, M_e = 7$ also verifies that $N(s)$ and $D(s)$ are rc.

We consider the Diophantine Equation (2.49) next.

(i) First, we let $Q(s) = [s^3 + 1, s^2]$, which satisfies (2.48), and we write $Q(s) = QS_e(s) = [1, 0, 0, 1, 0, 0, 1] \begin{bmatrix} 1 & s & s^2 & s^3 & 0 & 0 & 0 \\ 0 & 0 & 0 & 0 & 1 & s & s^2 \end{bmatrix}^T$. Then Eq. (2.52) assumes the form

$$[Y_0, Y_1, X_0, X_1]M_e = Q = [1, 0, 0, 1, 0, 0, 1],$$

which is an algebraic system of linear equations with 8 unknowns and 7 linearly independent equations that have more than one solution. One such solution is given by $[Y_0, Y_1, X_0, X_1] = [1, 0, -1, 1, 1, 0, 1, -1]$, which, in view of (2.50) and (2.51), implies that

$$\hat{X}(s) = X_0 + X_1 s = [1, 0] + [1, -1]s = [s + 1, -s]$$

$$\hat{Y}(s) = Y_0 + Y_1 s = [1, 0] + [-1, 1]s = [1 - s, s]$$

is a solution to the Diophantine Equation with $deg\, \hat{X}(s) = 1 = \nu - 1$ and $deg\, \hat{Y}(s) = 1 = \nu - 1$.

(ii) Next, we let $Q(s) = \begin{bmatrix} 1 & 0 \\ 0 & 1 \end{bmatrix}$ and we write $Q(s) = QS_e(s) =$

$$\begin{bmatrix} 1 & 0 & 0 & 0 & \vdots & 0 & 0 & 0 \\ 0 & 0 & 0 & 0 & \vdots & 1 & 0 & 0 \end{bmatrix} \begin{bmatrix} 1 & s & s^2 & s^3 & 0 & 0 & 0 \\ 0 & 0 & 0 & 0 & 1 & s & s^2 \end{bmatrix}^T.$$ Then Eq. (2.52) assumes the form

$$[Y_0, Y_1, X_0, X_1]M_e = Q = \begin{bmatrix} 1 & 0 & 0 & 0 & 0 & 0 & 0 \\ 0 & 0 & 0 & 0 & 1 & 0 & 0 \end{bmatrix}.$$

A solution of this equation is $[Y_0, Y_1, X_0, X_1] = \begin{bmatrix} 1 & 0 & -1 & 0 & 1 & 0 & 0 & 0 \\ -1 & 1 & 1 & 0 & -1 & 0 & 0 & 0 \end{bmatrix}$. This implies that

$$\hat{X}(s) = X_0 + X_1 s = \begin{bmatrix} 1 & 0 \\ -1 & 0 \end{bmatrix} + \begin{bmatrix} 0 & 0 \\ 0 & 0 \end{bmatrix} s = \begin{bmatrix} 1 & 0 \\ -1 & 0 \end{bmatrix}$$

$$\hat{Y}(s) = Y_0 + Y_1 s \begin{bmatrix} 1 & 0 \\ -1 & 1 \end{bmatrix} + \begin{bmatrix} -1 & 0 \\ 1 & 0 \end{bmatrix} s = \begin{bmatrix} 1 - s & 0 \\ -1 + s & 1 \end{bmatrix}$$

is a solution of the Diophantine Equation with $deg\, \hat{X}(s) = 0 < \nu - 1 = 1$ and $deg\, \hat{Y}(s) = 1 = \nu - 1$. ■

In the preceding development, it was shown how to derive particular solutions of the Diophantine Equation, given coprime polynomials and polynomial matrices. In the following, conditions for existence of solutions are derived and all solutions of the Diophantine Equation are conveniently parameterized.

THEOREM 2.15. Consider nonzero polynomial matrices $D(s) \in R[s]^{p_1 \times m}$ and $N(s) \in R[s]^{p_2 \times m}$ with $p_1 + p_2 \geq m$, let $G_R(s) \in R[s]^{m \times m}$ be a gcrd of $D(s)$ and $N(s)$, and let $Q(s) \in R[s]^{q \times m}$. The Diophantine Equation

$$X(s)D(s) + Y(s)N(s) = Q(s) \tag{2.53}$$

has polynomial matrix solutions $X(s) \in R[s]^{q \times p_1}$ and $Y(s) \in R[s]^{q \times p_2}$ if and only if $G_R(s)$ is an rd of $Q(s)$. If $G_R(s)$ is an rd of $Q(s)$, then (2.53) has infinitely many polynomial matrix solutions. If $X(s) = X_0(s)$ and $Y(s) = Y_0(s)$ is one such solution, then all solutions of (2.53) are given by

$$X(s) = X_0(s) - K(s)\tilde{N}_L(s) \tag{2.54a}$$

$$Y(s) = Y_0(s) + K(s)\tilde{D}_L(s), \tag{2.54b}$$

where $K(s) \in R[s]^{q \times r}$ and where $\tilde{N}_L(s) \in R[s]^{r \times p_1}$, $\tilde{D}_L \in R[s]^{r \times p_2}$, $r = (p_1 + p_2) - m$ are lc and satisfy $\tilde{N}_L(s)D(s) + \tilde{D}_L(s)N(s) = 0$, i.e., $[-\tilde{N}_L(s), \tilde{D}_L(s)]$ is a minimal basis of the left kernel of $\begin{bmatrix} D(s) \\ N(s) \end{bmatrix}$.

Proof. If there exist X and Y that satisfy (2.53), then $(XD_R + YN_R)G_R = Q$, which implies that QG_R^{-1} must be a polynomial matrix. Thus, G_R is an rd of Q that is a necessary condition for solutions of (2.53) to exist. To show that this also constitutes a sufficient condition, assume that G_R is an rd of Q, i.e., $Q = Q_R G_R$ for some polynomial matrix Q_R, and recall that there exist polynomial matrices $\hat{X} \in R[s]^{m \times p_1}$, $\hat{Y} \in R[s]^{m \times p_2}$ such that

$$\hat{X}D + \hat{Y}N = G_R. \tag{2.55}$$

The proof of this result is constructive and is based on the algorithm to determine a gcrd of D and N, using the Euclidean algorithm. (Refer to Subsection 7.2D, where it is shown that $U\begin{bmatrix} D \\ N \end{bmatrix} = \begin{bmatrix} \hat{X} & \hat{Y} \\ -\tilde{N}_L & \tilde{D}_L \end{bmatrix}\begin{bmatrix} D \\ N \end{bmatrix} = \begin{bmatrix} G_R \\ 0 \end{bmatrix}$, where U is a unimodular matrix.) Premultiplying (2.55) by Q_R, we now obtain $(Q_R\hat{X})D + (Q_R\hat{Y})N = Q_R G_R = Q$, or

$$X_0 D + Y_0 N = Q, \tag{2.56}$$

i.e., G_R being an rd of Q is also a sufficient condition for the existence of solutions of (2.53). If (2.56) is satisfied, then

$$(X_0 - K\tilde{N}_L)D + (Y_0 + K\tilde{D}_L)N = X_0 D + Y_0 N = Q,$$

and therefore, (2.53) has infinitely many solutions, among them the infinitely many given by (2.54). It remains to be shown that every solution of (2.53) can be put into the form (2.54). Suppose that X and Y satisfy (2.53). Subtracting (2.56) from (2.53), we obtain $(X - X_0)D + (Y - Y_0)N = 0$, or

$$[X - X_0, Y - Y_0]\begin{bmatrix} D \\ N \end{bmatrix} = 0.$$

Let $\tilde{N}_L \in R[s]^{r \times p_1}$ and $\tilde{D}_L \in R[s]^{r \times p_2}$, $r = (p_1 + p_2) - m$ be relatively lc and such that

$$[-\tilde{N}_L, \tilde{D}_L]\begin{bmatrix} D \\ N \end{bmatrix} = 0.$$

Then $[-\tilde{N}_L, \tilde{D}_L]$ is a minimum basis of the left kernel of $\begin{bmatrix} D \\ N \end{bmatrix}$. (A discussion of polynomial bases of vector spaces will be given shortly.) This implies that there exists some $K \in R[s]^{q\times r}$ so that

$$[X - X_0, Y - Y_0] = K[-\tilde{N}_L, \tilde{D}_L],$$

which is (2.54). Thus, every solution of (2.53) can be written in the form of (2.54). ■

Remarks

(i) If D^{-1} exists, then in the proof of Theorem 2.15, $(X-X_0)D+(Y-Y_0)N = 0$ implies that $X-X_0 = -(Y-Y_0)ND^{-1} = -(Y-Y_0)\tilde{D}_L^{-1}\tilde{N}_L$, where $\tilde{N}_L$ and $\tilde{D}_L$ are lc. Since $X - X_0$ is a polynomial matrix, this implies that $Y - Y_0 = K\tilde{D}_L$ for some K, where $X-X_0 = -K\tilde{N}_L$. Thus, when D^{-1} exists, the above results can be derived without directly using the concept of minimum basis of the left kernel of $\begin{bmatrix} D \\ N \end{bmatrix}$.

(ii) In the theory of systems and control, the Diophantine Equation is of great use, particularly for the case when D^{-1} exists and ND^{-1} is a proper rational matrix. In this case solutions of the Diophantine Equation with special properties are of interest, particularly solutions where X^{-1} exists and $X^{-1}Y$ is a proper rational matrix (see Subsections 7.4A, 7.4B, and 7.4C). Such solutions, satisfying particular properties, may also be determined via polynomial matrix interpolation (see the Appendix for theory and examples). It will be shown later in this chapter that the Diophantine Equation is central in the study of systems wherever feedback is used.

Rings, modules, and polynomial bases of rational vector spaces

Now some useful concepts from algebra are briefly reviewed. In particular, rings of polynomials, of proper rational functions, and of proper and stable rational functions are briefly discussed, together with modules and polynomial bases of rational vector spaces.

Consider the set of polynomials $R[s]$ in s with real coefficients, together with the usual operations of addition, $+$, and multiplication, $\cdot$, of polynomials. Then $(R[s], +, \cdot)$ is a *ring*, since all the axioms of a ring are satisfied. These axioms are the same as the axioms of a field (see Section 1.10 of Chapter 1), except for axiom (vii), which assumes existence of the multiplicative inverse in a field. Indeed, if $p(s) \in R[s]$ then $1/p(s) \notin R[s]$, since in general $1/p(s)$ is a rational function but not a polynomial. Another example of a ring is the set of integers, taken together with the usual operations of addition and multiplication on integers.

We note that $R[s]$ is a *Euclidean ring*, i.e., (i) for $a(s), b(s) \in R[s]$ such that $a(s)b(s) \neq 0$, $deg\,[a(s)b(s)] \geq deg\,a(s)$; and (ii) for every $a(s), b(s) \in R[s]$ with $b(s) \neq 0$, there exist $q(s), r(s) \in R[s]$ such that $a(s) = b(s)q(s) + r(s)$, where either $r(s) = 0$ or $deg\,r(s) < deg\,b(s)$ (polynomial division). If $R(s)$ is the field of rational fractions over $R[s]$, called the field of rational functions (see Subsection 7.2A), then $R[s] \subset R(s)$ and $R(s)$ is also a Euclidean ring. Note that the units of $R[s]$ are those polynomials $p(s)$ for which there exist a $p'(s) \in R[s]$ such that $p(s)p'(s) = 1$, the unity element of $R[s]$, i.e., the units of $R[s]$ are the nonzero elements of R.

Another Euclidean ring of interest is the *set of proper rational functions*, taken together with the operations of addition and multiplication. A proper rational function

$t(s) = n(s)/d(s)$ ($d(s) \neq 0$) is a *unit* of this ring if $deg\, n(s) = deg\, d(s)$, i.e., if it is a *biproper rational function*. The *set of proper and stable rational functions* (with poles in the stable region of the s-plane) is also a Euclidean ring. In this case, if $t(s)$ is a unit then $t(s)$ and $t^{-1}(s)$ are proper and stable rational functions.

Polynomial bases of rational vector spaces are discussed next. Recall that if $p_i(s) \in R[s]$, then it may always be viewed as being divided by 1, in which case $p_i(s) = p_i(s)/1 \in R(s)$, the field of rational functions. Thus, a polynomial vector $p(s) \in R[s]^p$ may be viewed as a special case of a rational vector $p(s) \in R(s)^p$.

Now let $H(s) \in R(s)^{p\times m}$, assume that $rank\, H(s) = m$, and consider the rational vector space spanned by the columns $h_j(s) \in R(s)^p$, $j = 1, \ldots, m$, of $H(s)$ and denote this space by $Y(s)$. Thus, the vectors in $Y(s)$ are generated by $H(s)\alpha(s)$, where $\alpha(s) = [\alpha_1(s), \ldots, \alpha_p(s)]^T$, $\alpha_j(s) \in R(s)$. The matrix $H(s)$ is a *basis for* $Y(s)$ and if $T(s) \in R(s)^{m\times m}$ and $rank\, T(s) = m$, then $\bar{H}(s) = H(s)T(s)$ is also a basis for $Y(s)$. Consider a right polynomial MFD of $H(s)$, $H(s) = B(s)A^{-1}(s)$, where $B(s) \in R[s]^{p\times m}$ and $A(s) \in R[s]^{m\times m}$ are not necessarily coprime. Note that $rank\, B(s) = m$ and $B(s)$ is a *polynomial basis* of the rational vector space $Y(s)$. Consider now the set M of all polynomial vectors $h(s) \in R[s]^p$ that can be written as linear combinations over the ring $R[s]$ of the columns $b_j(s) \in R[s]^p$, $j = 1, \ldots, m$, of $B(s)$. The set M is a *free* $R[s]$*-module* and $B(s)$ is a basis of M. The dimension of M is $dim\, M = rank\, B(s) = m$. If the p rows of $B(s)$ are rc, then the free $R[s]$-module M^* generated by the columns of $B(s)$ is called the *maximal module contained* in $Y(s)$ and coincides with the set of all polynomial vectors contained in the rational vector space $Y(s)$ [i.e., any polynomial vector in $Y(s)$ can be expressed as $B(s)p(s)$, for some $p(s) \in R[s]^m$]. Note that if $U(s) \in R[s]^{m\times m}$ is any unimodular matrix, then $\bar{B}(s) = B(s)U(s)$ is another basis for M, and $U(s)$ can be chosen to reduce $B(s)$ to a column proper (reduced) form, if so desired. A *minimal basis* of a rational vector space $Y(s)$ is defined as a polynomial basis $B(s)$ of $Y(s)$ with all its rows rc, i.e., if $B(s) \in R[s]^{p\times m}$ is a minimal basis of $Y(s)$, then any polynomial vector in $Y(s)$ can be expressed as $B(s)p(s)$ for some polynomial vector $p(s) \in R[s]^m$. Note that a minimal basis $B(s)$ is sometimes defined in the literature to be column proper as well.

Summarizing, given $H(s) \in R(s)^{p\times m}$ and $rank\, H(s) = m \leq p$, a minimal polynomial basis for the rational vector space spanned by the columns of $H(s)$ can be found as follows: write $H(s) = B(s)A^{-1}(s)$, where $B(s) \in R[s]^{p\times m}$ and $A(s) \in R[s]^{m\times m}$ are not necessarily coprime. Let $G_R(s) \in R[s]^{m\times m}$ be a gcrd of the p rows of $B(s)$ and write $B(s) = \hat{B}(s)G_R(s)$. Then $\hat{B}(s)$ is a minimal basis. To reduce $\hat{B}(s)$ to column proper form, consider a unimodular matrix $U(s) \in R[s]^{m\times m}$ such that $\bar{B}(s) = \hat{B}(s)U(s)$ is column proper (see Subsection 7.2B.)

Historical remarks on the Diophantine Equation

If n, d are two nonzero integers and the integer g is their gcd, then there exist integers x, y so that $xd + yn = g$ (see also Lemma 2.3). This result was known to the fourth century B.C. Greek mathematician Euclid.

The Diophantine Equation $xd + yn = q$ given above, where d, n, and q are specified and solutions x, y are to be determined, is one of the equations in more than one indeterminants introduced by Diophantus of Alexandria, who lived around A.D. 250. It is in fact called a linear Diophantine Equation or a Diophantine Equation of first-order. Diophantus, who introduced letter symbols for quantities in mathematical problems, is considered to be one of the greatest mathematicians. His treatise on algebra, considered to be the first ever, is called "ΑΡΙΘΜΗΤΙΚΑ" and consists of seven books, six of which have survived. Diophantus worked with equations involving integers. An example of a specific case of the Diophantine Equation is $3x+2y = 4$, the general solution of which is given by $x = 2-2k$, $y = -1+3k$ with k equal to any integer (see Theorem 2.15). In this case, the particular solution $x_0 = 2$, $y_0 = -1$ was used. The parameterization of all solutions in this convenient form appears to be a later development, the work of Hindu mathematicians in the fifth century A.D., apparently of Aryabhata. It is worth mentioning at this point that another class of famous Diophantine Equations is $x^n + y^n = z^n$, where x, y, z, and n are integers. When $n = 2$, then $x = 3$, $y = 4$, and $z = 5$ is a solution since $3^2 + 4^2 = 5^2$. Solutions for integer n greater than 2 do not appear to exist. In fact, the famous *Fermat's last theorem* (c. 1637) states that no solution exists for $n \geq 3$. Fermat wrote on the margin of his copy of the works of Diophantus, referring to this theorem, "I have discovered a truly remarkable proof which this margin is too small to contain." In spite of attempts over the next centuries by mathematicians like Euler, Legendre, and many others, this result has not yet been proved in its generality, although a recent proof (1994) shows great promise of being completely correct. It is, however, known to be true for n up to tens of thousands, using computer methods.

Diophantus worked with integers. However, integers and polynomials obey similar rules (they are both rings), and therefore, all results of interest on the linear Diophantine Equation, $xd + yn = q$ that were developed for integers can readily be written for polynomials, and this is what was done in Subsection 7.2E. These results for the polynomial Diophantine Equation were pointed out in the books by Gantmacher and McDuffee (refer to the bibliography at the end of the chapter). In view of this, it is not surprising that solutions of linear Diophantine Equations involving elements other than integers or polynomials, but still members of rings, can be studied and expressed in a completely analogous manner. In fact, Diophantine Equations with elements that are polynomials in z^{-1} and $(s+a)^{-1}$ were studied in the systems and control literature in the 1970s by Kucera and Pernebo, among others. Later, in 1980, Desoer et al. clearly pointed this fact out in the systems and control literature, namely, that the solutions of the linear Diophantine Equation, when the equation involves elements of rings other than integers or polynomials, are analogous to the integer and polynomial cases. They also addressed conditions on system descriptions under which such Diophantine Equations may appear.

It will be shown in Subsection 7.3C how the linear Diophantine Equation is of fundamental importance in the study of feedback systems. It should be noted here that when $q = 1$, the Diophantine Equation is sometimes referred to as the *Bezout identity* (see, e.g., Kailath [21]). For the role played by the Diophantine Equation in control, the reader should also refer to the survey paper by Kucera [24]. For further details, refer to Sections 7.6 and 7.7, Notes and References, respectively.

7.3 SYSTEMS REPRESENTED BY POLYNOMIAL MATRIX DESCRIPTIONS

We consider system representations of the form

$$P(q)z(t) = Q(q)u(t), \qquad y(t) = R(q)z(t) + W(q)u(t) \tag{3.1}$$

with $P(q) \in R[q]^{l\times l}$, $Q(q) \in R[q]^{l\times m}$, $R(q) \in R[q]^{p\times l}$, and $W(q) \in R[q]^{p\times m}$, where $R[q]^{l\times l}$ denotes the set of $l \times l$ matrices whose entries are polynomials in q with real coefficients ($q \triangleq (d/dt)$ the differential operator) and we assume that $det\, P(q) \neq 0$ and that $u(t)$ is sufficiently differentiable. We also assume that the set of homogeneous differential equations $P(q)z(t) = 0$ is "well formed," that is, for all initial values of $z(\cdot)$ and its derivatives at $t = 0$ the solution does not contain impulsive behavior at $t = 0$. This is true if and only if $P^{-1}(q)$ is a proper rational matrix, which is true for example when $P(q)$ is column or row reduced (for further details refer, e.g., to Section 3.3 of [9]). Such representations, called system *Polynomial Matrix Descriptions* (PMD), were introduced in Section 7.1.

In this section, equivalence of system representations is discussed first, in Subsection A. This notion of equivalence establishes relations not only between polynomial representations, but also with state-space representations, which are in fact a special case of PMDs (see Section 7.1). Equivalence of PMDs preserves certain system properties, including controllability, observability, and stability, that are the topics of discussion in Subsection B, together with polynomial matrix realizations of transfer function matrices. In Subsection C, Polynomial Matrix Fractional Descriptions (PMFDs) are employed to study properties of interconnected systems.

A. Equivalence of System Representations

Consider the PMDs

$$\begin{aligned} P_i(q)z_i(t) &= Q_i(q)u(t), \\ y(t) &= R_i(q)z_i(t) + W_i(q)u(t), \qquad i = 1, 2, \end{aligned} \tag{3.2}$$

where $P_i(q) \in R[q]^{l_i\times l_i}$, $Q_i(q) \in R[q]^{l_i\times m}$, $R_i(q) \in R[q]^{p\times l_i}$, and $W_i(q) \in R[q]^{p\times m}$.

DEFINITION 3.1. The representations $\{P_1, Q_1, R_1, W_1\}$ and $\{P_2, Q_2, R_2, W_2\}$ in (3.2) are called *equivalent* if there exist polynomial matrices $M \in R[q]^{l_2\times l_1}$, $N \in R[q]^{l_2\times l_1}$, $X \in R[q]^{p\times l_1}$, and $Y \in R[q]^{l_2\times m}$ such that

$$\begin{bmatrix} M & 0 \\ X & I_p \end{bmatrix}\begin{bmatrix} P_1 & Q_1 \\ -R_1 & W_1 \end{bmatrix} = \begin{bmatrix} P_2 & Q_2 \\ -R_2 & W_2 \end{bmatrix}\begin{bmatrix} N & -Y \\ 0 & I_m \end{bmatrix} \tag{3.3}$$

with (M, P_2) lc and (P_1, N) rc. ■

The equivalence of the representations $\{P_i, Q_i, R_i, W_i\}$, $i = 1, 2$, or of the corresponding *system matrices* S_1, S_2, where

$$S_i \triangleq \begin{bmatrix} P_i & Q_i \\ -R_i & W_i \end{bmatrix} \in R[s]^{(l_i+p)\times(l_i+m)} \tag{3.4}$$

is sometimes referred to as *Fuhrmann's System Equivalence (FSE)*. This equivalence relation will be denoted by ρ_F, or simply ρ. It can be shown that (3.3) with its associated conditions given in Definition 3.1 defines an equivalence relation ρ on the set of matrices S with p, m fixed, $|P| \neq 0$, and l any positive integer (see the discussion of equivalence relations at the end of Subsection 7.2C). The polynomial matrix S defined in (3.4) for system $\{P, Q, R, W\}$ is called *Rosenbrock's System Matrix* of the system $Pz = Qu$, $y = Rz + Wu$.

EXAMPLE 3.1. Consider the representations (1.3) and (1.4) in Section 7.1 and let $P_1 = qI - A, Q_1 = B, R_1 = C, W_1 = D$ given in (1.3), and let $P_2 = P, Q_2 = Q, R_2 = R, W_2 = W$ given in (1.4). If $N = C = \begin{bmatrix} 0 & 1 & 0 \\ 0 & -1 & 1 \end{bmatrix}, M = \begin{bmatrix} 1 & q & 0 \\ 0 & 0 & q \end{bmatrix}, Y = D = \begin{bmatrix} 0 & 0 \\ 0 & 1 \end{bmatrix}$, and $X = \begin{bmatrix} 0 & 0 & 0 \\ 0 & 0 & 0 \end{bmatrix}$, then (3.3) is satisfied (verify this). Note that $(M, P_2) = \begin{bmatrix} 1 & q & 0 & q^2+1 & 1 \\ 0 & 0 & q & q & q+2 \end{bmatrix}$ are lc, while (P_1, N) are rc, as can be verified by applying, for example, the rank test for coprimeness of Theorem 2.4(v) to (M, P_2) above and to $\begin{bmatrix} P_1 \\ N \end{bmatrix} = \begin{bmatrix} q & 0 & 1 \\ -1 & q & 0 \\ 0 & -2 & q+2 \\ \cdots & \cdots & \cdots \\ 0 & 1 & 0 \\ 0 & -1 & 1 \end{bmatrix}$. These representations satisfy all the conditions of Definition 3.1, and are therefore, equivalent (or Fuhrmann system equivalent). ■

There is an alternative definition of system equivalence, sometimes referred to as *strict system equivalence* or *Rosenbrock's System Equivalence* (RSE). Subsequently, ρ_R will denote this equivalence relation. It is defined as follows.

The representations given in (3.2) are called *strict system equivalent* if there exist unimodular polynomial matrices $\hat{M}, \hat{N} \in R[q]^{k\times k}$ and polynomial matrices $\hat{X} \in R[q]^{p\times k}, \hat{Y} \in R[q]^{k\times m}$ with $n_i = deg|P_i|, k \geq \max(n_1, n_2)$, such that

$$\begin{bmatrix} \hat{M} & \vdots & 0 \\ \cdots & \vdots & \cdots \\ \hat{X} & \vdots & I_p \end{bmatrix} \begin{bmatrix} I_{k-l_1} & 0 & \vdots & 0 \\ 0 & P_1 & \vdots & Q_1 \\ \cdots & \cdots & \vdots & \cdots \\ 0 & -R_1 & \vdots & W_1 \end{bmatrix}$$

$$= \begin{bmatrix} I_{k-l_2} & 0 & \vdots & 0 \\ 0 & P_2 & \vdots & Q_2 \\ \cdots & \cdots & \vdots & \cdots \\ 0 & -R_2 & \vdots & W_2 \end{bmatrix} \begin{bmatrix} \hat{N} & \vdots & -\hat{Y} \\ \cdots & \vdots & \cdots \\ 0 & \vdots & I_m \end{bmatrix}, \quad k \geq l_1, k \geq l_2 \tag{3.5}$$

Relation (3.5) is in general easier to use than (3.3), since the matrices $\hat{M}, \hat{N}, \begin{bmatrix} \hat{M} & 0 \\ \hat{X} & I_p \end{bmatrix}$, and $\begin{bmatrix} \hat{N} & -\hat{Y} \\ 0 & I_m \end{bmatrix}$ are all unimodular.

The following theorem establishes that the preceding two definitions of system equivalence [i.e., Fuhrmann's (FSE) (3.3) and Rosenbrock's (RSE) (3.5)] are, in fact equivalent. Note that it was for this reason that in Definition 3.1, which is in fact the definition of FSE, only the term "equivalent" was used.

THEOREM 3.1. Consider the representations given in (3.2). If they are Fuhrmann's System Equivalent (FSE) [satisfying (3.3)], i.e., $\{P_1, Q_1, R_1, W_1\}\rho_F\{P_2, Q_2, R_2, W_2\}$, then they are Rosenbrock's System Equivalent (RSE) [satisfying (3.5)], i.e., $\{P_1, Q_1, R_1, W_1\}\rho_R\{P_2, Q_2, R_2, W_2\}$, and vice versa.

Proof. If the representations are FSE, then $MP_1 = P_2N$ and $\begin{bmatrix} -\tilde{X} & \tilde{Y} \\ P_2 & M \end{bmatrix}\begin{bmatrix} -N & \bar{X} \\ P_1 & \bar{Y} \end{bmatrix} = \begin{bmatrix} I & 0 \\ 0 & I \end{bmatrix}$, where $[-\tilde{X}, \tilde{Y}]$ and $\begin{bmatrix} \bar{X} \\ \bar{Y} \end{bmatrix}$ were chosen so that the block matrices are unimodular. This can always be done since (P_2, M) are lc and (P_1, N) are rc. [See also the discussion on doubly coprime factorizations of a transfer function in (4.18) of Subsection 7.4A.] It can now be seen that

$$\begin{bmatrix} -\tilde{X} & \tilde{Y} & \vdots & 0 \\ P_2 & M & \vdots & 0 \\ \cdots & \cdots & \vdots & \cdots \\ -R_2 & X & \vdots & I_p \end{bmatrix}\begin{bmatrix} I_{l_2} & 0 & \vdots & 0 \\ 0 & P_1 & \vdots & Q_1 \\ \cdots & \cdots & \vdots & \cdots \\ 0 & -R_1 & \vdots & W_1 \end{bmatrix}$$
$$= \begin{bmatrix} I_{l_1} & 0 & \vdots & 0 \\ 0 & P_2 & \vdots & Q_2 \\ \cdots & \cdots & \vdots & \cdots \\ 0 & -R_2 & \vdots & W_2 \end{bmatrix}\begin{bmatrix} -\tilde{X} & \tilde{Y}P_1 & \vdots & \tilde{Y}Q_1 \\ I_{l_2} & N & \vdots & Y \\ \cdots & \cdots & \vdots & \cdots \\ 0 & 0 & \vdots & I_m \end{bmatrix}. \tag{3.6}$$

Observe that $\begin{bmatrix} I_{l_1} & \tilde{X} \\ 0 & I_{l_2} \end{bmatrix}\begin{bmatrix} -\tilde{X} & \tilde{Y}P_1 \\ I_{l_2} & N \end{bmatrix} = \begin{bmatrix} 0 & I_{l_1} \\ I_{l_2} & N \end{bmatrix}$, which implies that the second block matrix in the left-hand side is unimodular. Therefore, FSE $\Rightarrow$ RSE of the representations. To show the converse, suppose that (3.5) is satisfied. Partitioning, we can write

$$\begin{bmatrix} M_{11} & M_{12} & \vdots & 0 \\ M_{21} & M_{22} & \vdots & 0 \\ \cdots & \cdots & \vdots & \cdots \\ X_1 & X_2 & \vdots & I_p \end{bmatrix}\begin{bmatrix} I_{k-l_1} & 0 & \vdots & 0 \\ 0 & P_1 & \vdots & Q_1 \\ \cdots & \cdots & \vdots & \cdots \\ 0 & -R_1 & \vdots & W_1 \end{bmatrix}$$
$$= \begin{bmatrix} I_{k-l_2} & 0 & \vdots & 0 \\ 0 & P_2 & \vdots & Q_2 \\ \cdots & \cdots & \vdots & \cdots \\ 0 & -R_2 & \vdots & W_2 \end{bmatrix}\begin{bmatrix} N_{11} & N_{12} & \vdots & -Y_1 \\ N_{21} & N_{22} & \vdots & -Y_2 \\ \cdots & \cdots & \vdots & \cdots \\ 0 & 0 & \vdots & I_m \end{bmatrix}, \tag{3.7}$$

where $M_{22}, N_{22} \in R[q]^{l_2 \times l_1}$. Then

$$\begin{bmatrix} M_{22} & 0 \\ X_2 & I_p \end{bmatrix}\begin{bmatrix} P_1 & Q_1 \\ -R_1 & W_1 \end{bmatrix} = \begin{bmatrix} P_2 & Q_2 \\ -R_2 & W_2 \end{bmatrix}\begin{bmatrix} N_{22} & -Y_2 \\ 0 & I_m \end{bmatrix}. \tag{3.8}$$

It turns out that (M_{22}, P_2) are lc and (P_1, N_{22}) are rc. This can be shown as follows. Consider the unimodular matrices $M = \begin{bmatrix} M_{11} & M_{12} \\ P_2N_{21} & M_{22} \end{bmatrix}$ and $N = \begin{bmatrix} N_{11} & M_{12}P_1 \\ N_{21} & N_{22} \end{bmatrix}$, where the relations $M_{12}P_1 = N_{12}$, $M_{21} = P_2N_{21}$ from (3.7) were used. Now if (P_2, M_{22}) were not lc, then they would have a nonunimodular common left divisor that would have caused M not to be unimodular. Therefore, they are lc. Similarly, N is unimodular implies that (P_1, N_{22}) is rc. This shows that (3.8) is indeed the defining relation for FSE, and therefore RSE $\Rightarrow$ FSE of the representations. ■

In the following we enumerate some of the important properties of FSE and RSE.

THEOREM 3.2. Assume that the representations (3.2) are equivalent, i.e., they are FSE or RSE. Then the following are invariants of the equivalence relation $\rho = \rho_F = \rho_R$:

(i) $deg|P_i| \triangleq n, i = 1, 2$.

(ii) The nonunity invariant polynomials (in the Smith form) of P_i, $[P_i, Q_i]$, $\begin{bmatrix} P_i \\ -R_i \end{bmatrix}$, and S_i, $i = 1, 2$.

(iii) The transfer function matrix $H_i(s) = R_i(s)P_i^{-1}(s)Q_i(s) + W_i(s)$, $i = 1, 2$.

Proof. Recall from Section 7.2 that two polynomial matrices $M_1, M_2 \in R[q]^{p\times m}$ have the same Smith form if and only if there exist unimodular matrices U_1, U_2 such that $U_1M_1 = M_2U_2$. In this case M_1, M_2 are called equivalent polynomial matrices. To show (i), note from (3.5) that $\hat{M}\begin{bmatrix} I_{k-l_1} & 0 \\ 0 & P_1 \end{bmatrix} = \begin{bmatrix} I_{k-l_2} & 0 \\ 0 & P_2 \end{bmatrix}\hat{N}$, where $\hat{M}, \hat{N}$ are unimodular. Therefore, P_1 and P_2 must have the same nonunity invariant polynomials (note that P_1, P_2 may have different dimensions), which implies that $|P_1| = \alpha|P_2|$, where α is some nonzero real number (show this). Clearly then, $deg|P_1| = deg|P_2|$.

To verify (ii), note that in (3.5), $\begin{bmatrix} \hat{M} & 0 \\ \hat{X} & I_p \end{bmatrix}$ and $\begin{bmatrix} \hat{N} & -\hat{Y} \\ 0 & I_m \end{bmatrix}$ are unimodular, and therefore, the *extended system matrices*

$$S_{e_1} = \begin{bmatrix} I_{k-l_1} & 0 \\ 0 & S_1 \end{bmatrix}, \qquad S_{e_2} = \begin{bmatrix} I_{k-l_2} & 0 \\ 0 & S_2 \end{bmatrix}$$

with S_i, $i = 1, 2$, defined in (3.4), have the same Smith form. Thus, S_1 and S_2 have the same nonunity invariant polynomials. In a similar manner, (3.5) implies that

$$\hat{M}\left[\begin{array}{cc:c} I_{k-l_1} & 0 & 0 \\ 0 & P_1 & Q_1 \end{array}\right] = \left[\begin{array}{cc:c} I_{k-l_2} & 0 & 0 \\ 0 & P_2 & Q_2 \end{array}\right]\begin{bmatrix} \hat{N} & -\hat{Y} \\ 0 & I_m \end{bmatrix},$$

which shows that $[P_1, Q_1]$ and $[P_2, Q_2]$ have the same nonunity invariant polynomials. Similarly, $\begin{bmatrix} P_1 \\ -R_1 \end{bmatrix}$ and $\begin{bmatrix} P_2 \\ -R_2 \end{bmatrix}$ have the same nonunity invariant polynomials.

Finally, to show (iii), we write, in view of (3.5),

$$\begin{aligned} H_1 &= R_1P_1^{-1}Q_1 + W_1 = [0, R_1]\begin{bmatrix} I & 0 \\ 0 & P_1 \end{bmatrix}^{-1}\begin{bmatrix} 0 \\ Q_1 \end{bmatrix} + W_1 \\ &= \left[[0, R_2]\hat{N} + \hat{X}\begin{bmatrix} I & 0 \\ 0 & P_1 \end{bmatrix}\right]\begin{bmatrix} I & 0 \\ 0 & P_1 \end{bmatrix}^{-1}\begin{bmatrix} 0 \\ Q_1 \end{bmatrix} + W_1 \\ &= \left[[0, R_2]\begin{bmatrix} I & 0 \\ 0 & P_2 \end{bmatrix}^{-1}\hat{M} + \hat{X}\right]\begin{bmatrix} 0 \\ Q_1 \end{bmatrix} + W_1 \\ &= [0, R_2]\begin{bmatrix} I & 0 \\ 0 & P_2 \end{bmatrix}^{-1}\hat{M}\begin{bmatrix} 0 \\ Q_1 \end{bmatrix} + \left[\hat{X}\begin{bmatrix} 0 \\ Q_1 \end{bmatrix} + W_1\right] \\ &= [0, R_2]\begin{bmatrix} I & 0 \\ 0 & P_2 \end{bmatrix}^{-1}\left[\begin{bmatrix} 0 \\ Q_2 \end{bmatrix} - \begin{bmatrix} I & 0 \\ 0 & P_2 \end{bmatrix}\hat{Y}\right] + [W_2 + [0\; R_2]\hat{Y}] \\ &= [0, R_2]\begin{bmatrix} I & 0 \\ 0 & P_2 \end{bmatrix}^{-1}\begin{bmatrix} 0 \\ Q_2 \end{bmatrix} + W_2 = R_2P_2^{-1}Q_2 + W_2 = H_2. \end{aligned}$$

■

In the above derivations the following relations, which can directly be seen from (3.5), were used:

$$\hat{M}\begin{bmatrix} I & 0 \\ 0 & P_1 \end{bmatrix} = \begin{bmatrix} I & 0 \\ 0 & P_2 \end{bmatrix}\hat{N}, \tag{3.9}$$

$$\hat{M}\begin{bmatrix} 0 \\ Q_1 \end{bmatrix} = -\begin{bmatrix} I & 0 \\ 0 & P_2 \end{bmatrix}\hat{Y} + \begin{bmatrix} 0 \\ Q_2 \end{bmatrix}, \tag{3.10}$$

$$\hat{X}\begin{bmatrix} I & 0 \\ 0 & P_1 \end{bmatrix} + [0, -R_1] = [0, -R_2]\hat{N}, \tag{3.11}$$

$$\hat{X}\begin{bmatrix} 0 \\ Q_1 \end{bmatrix} + W_1 = -[0, -R_2]\hat{Y} + W_2. \tag{3.12}$$

Now consider the representations (3.2) and notice that they can be written as

$$S_i(q)\begin{bmatrix} z_i(t) \\ -u(t) \end{bmatrix} = \begin{bmatrix} P_i(q) & Q_i(q) \\ -R_i(q) & W_i(q) \end{bmatrix}\begin{bmatrix} z_i(t) \\ -u(t) \end{bmatrix} = \begin{bmatrix} 0 \\ -y(t) \end{bmatrix}, \tag{3.13}$$

$i = 1, 2$, where $S_i(q)$ is Rosenbrock's System Matrix given in (3.4). If these representations are equivalent, then the relation between the states z_1 and z_2 can be found in the following manner. Postmultiplying both sides of (3.3) by $[z_1^T, -u^T]^T$, we obtain $\begin{bmatrix} M & 0 \\ X & I_p \end{bmatrix}\begin{bmatrix} P_1 & Q_1 \\ -R_1 & W_1 \end{bmatrix}\begin{bmatrix} z_1 \\ -u \end{bmatrix} = \begin{bmatrix} P_2 & Q_2 \\ -R_2 & W_2 \end{bmatrix}\begin{bmatrix} N & -Y \\ 0 & I_m \end{bmatrix}\begin{bmatrix} z_1 \\ -u \end{bmatrix}$ or $\begin{bmatrix} P_2 & Q_2 \\ -R_2 & W_2 \end{bmatrix}\begin{bmatrix} Nz_1 + Yu \\ -u \end{bmatrix} = \begin{bmatrix} M & 0 \\ X & I_p \end{bmatrix}\begin{bmatrix} 0 \\ -y \end{bmatrix} = \begin{bmatrix} 0 \\ -y \end{bmatrix} = \begin{bmatrix} P_2 & Q_2 \\ -R_2 & W_2 \end{bmatrix}\begin{bmatrix} z_2 \\ -u \end{bmatrix}$. Therefore,

$$z_2(t) = N(q)z_1(t) + Y(q)u(t) \tag{3.14}$$

is a possible relation between the partial states. To obtain the inverse relation, we could express z_1 as a function of z_2 by considering (3.3) and selecting $\tilde{X}, \tilde{Y}, \bar{X}, \bar{Y}$ so that

$$\begin{bmatrix} -\tilde{X} & \tilde{Y} \\ P_2 & M \end{bmatrix}\begin{bmatrix} -N & \bar{X} \\ P_1 & \bar{Y} \end{bmatrix} = \begin{bmatrix} I & 0 \\ 0 & I \end{bmatrix} = \begin{bmatrix} -N & \bar{X} \\ P_1 & \bar{Y} \end{bmatrix}\begin{bmatrix} -\tilde{X} & \tilde{Y} \\ P_2 & M \end{bmatrix} \tag{3.15}$$

are unimodular polynomial matrices. Note that this is always possible, since $MP_1 = P_2N$ with (M, P_2) lc and (P_1, N) rc (see also the proof of Theorem 3.1). Now in view of (3.15), Eq. (3.3) implies that

$$\begin{bmatrix} P_1 & Q_1 \\ -R_1 & W_1 \end{bmatrix}\begin{bmatrix} \tilde{X} & -E \\ 0 & I_m \end{bmatrix} = \begin{bmatrix} \bar{X} & 0 \\ F & I_p \end{bmatrix}\begin{bmatrix} P_2 & Q_2 \\ -R_2 & W_2 \end{bmatrix}, \tag{3.16}$$

where $E \triangleq \tilde{Y}Q_1 - \tilde{X}Y$ and $F \triangleq R_2\bar{X} - X\bar{Y}$. This relation was of course expected, due to the symmetry property of the equivalue relation ρ_F. Postmultiplying both sides of (3.16) by $[z_2^T, -u^T]^T$ and proceeding similarly as before, we can now show that

$$z_1(t) = \tilde{X}(q)z_2(t) + E(q)u(t). \tag{3.17}$$

Equations (3.14) and (3.17) determine an invertible mapping that relates the states of the equivalent representations (3.13). It can be shown that if (3.14) and (3.17) hold for some polynomial matrices $N, Y, \tilde{X}, E$, then the representations in (3.13) are equivalent.

Polynomial Matrix Fractional Descriptions (PMFDs)

Next, we consider the right Polynomial Matrix Fractional Descriptions (rPMFDs) given by

$$D_i(q)z_i(t) = u(t), \qquad y(t) = N_i(q)z_i(t), \qquad i = 1, 2. \tag{3.18}$$

THEOREM 3.3. The rPMFDs given in (3.18) are equivalent if and only if there exists a unimodular matrix U such that

$$\begin{bmatrix} D_1 \\ N_1 \end{bmatrix} = \begin{bmatrix} D_2 \\ N_2 \end{bmatrix} U. \tag{3.19}$$

Proof. Suppose that (3.19) is true. Rewrite it as

$$\begin{bmatrix} D_1 & I_m \\ -N_1 & 0 \end{bmatrix} = \begin{bmatrix} D_2 & I_m \\ -N_2 & 0 \end{bmatrix}\begin{bmatrix} U & 0 \\ 0 & I_m \end{bmatrix} \tag{3.20}$$

and note that this is a relation of the form (3.3). Therefore, the representations $\{D_1, I, N_1\}$, $\{D_2, I, N_2\}$ are equivalent. Conversely, suppose that (3.3) is true, i.e.,

$$\begin{bmatrix} M & 0 \\ X & I_m \end{bmatrix}\begin{bmatrix} D_1 & I_m \\ -N_1 & 0 \end{bmatrix} = \begin{bmatrix} D_2 & I_m \\ -N_2 & 0 \end{bmatrix}\begin{bmatrix} N & -Y \\ 0 & I_m \end{bmatrix}$$

with (M, D_2) lc and (D_1, N) rc. Then $M = I_m - D_2Y$ and $MD_1 = D_2N$ from which it follows that $D_1 = D_2(N + YD_1) = D_2U$. Also, $XD_1 - N_1 = -N_2N$ and $X = N_2Y$ which imply that $N_1 = N_2(N + YD_1) = N_2U$. From $D_1 = D_2U$ and the fact that $deg\, D_1 = deg\, D_2$, it now follows that U is unimodular. ■

The reader should verify that in this case, the relations between the states (3.14) and (3.17) are given by $z_2 = Uz_1$ and $z_1 = U^{-1}z_2$, i.e., $N = U, Y = 0, \tilde{X} = U^{-1}$, and $E = 0$.

A similar result is true for lPMFDs given by

$$D_i(q)z_i(t) = N_i(q)u(t), \qquad y(t) = z_i(t), \qquad i = 1, 2. \tag{3.21}$$

Specifically, the lPMFDs given in (3.21) are equivalent if and only if there exists a unimodular matrix U such that

$$[D_1, N_1] = U[D_2, N_2]. \tag{3.22}$$

The proof of this result is completely analogous to the proof of Theorem 3.3 and is omitted.

State-space descriptions

We now consider the state-space representations given by

$$\dot{x}_i(t) = A_ix_i(t) + B_iu(t), \qquad y(t) = C_ix_i + D_i(q)u(t), \qquad i = 1, 2, \tag{3.23}$$

and we assume that these are related by a similarity transformation, i.e., there is a $Q \in R^{n\times n}$, $|Q| \neq 0$, such that

$$\begin{bmatrix} Q & 0 \\ 0 & I_n \end{bmatrix}\begin{bmatrix} qI_n - A_1 & B_1 \\ -C_1 & D_1(q) \end{bmatrix} = \begin{bmatrix} qI_n - A_2 & B_2 \\ -C_2 & D_2(q) \end{bmatrix}\begin{bmatrix} Q & 0 \\ 0 & I_n \end{bmatrix}. \tag{3.24}$$

Clearly, this is a relation of the form (3.3) with $M = N = Q$ and $X = Y = 0$. Therefore, $\{A_1, B_1, C_1, D_1(q)\}, \{A_2, B_2, C_2, D_2(q)\}$ are equivalent in the Fuhrmann or Rosenbrock sense. The converse is also true. In particular, if two state-space representations are equivalent in the sense of Definition 3.1, then they are related via a similarity transformation. The proof of this result is not presented here. We ask the reader to verify that in this case the relations between the states (3.14), (3.17) are given by $x_2 = Qx_1$ and $x_1 = Q^{-1}x_2$, i.e., $N = Q, Y = 0, \tilde{X} = Q^{-1}$, and $E = 0$. Note that the state-space representations considered in (3.23) are more general than the state representations of previous chapters, since the term $D(q)u(t)$ may contain derivatives of the input. In fact it can be shown that every PMD $\{P(q), Q(q), R(q), W(q)\}$ in (3.1) is equivalent to a state-space description of the form $\{A, B, C, D(q)\}$ (see, e.g., Section 2.2 of [30]). Note that a state-space description of the form $\{A, B, C, D\}$ that is equivalent to a given PMD as in (3.1) exists only when that PMD has a proper transfer function $H(s)$.

Now recall the relation between a state-space realization $\{A_c, B_c, C_c, D_c\}$ in controller form of the transfer function matrix $H(s)$ and the corresponding rPMFD $\{D_R, I, N_R\}$. Specifically, (see Subsection 3.4D of Chapter 3 and the Structure Theorem), we have

$$H(s) = N_R(s)D_R^{-1}(s)$$

with $$N_R(s) = C_cS(s) + D_cD_R(s), \qquad (sI - A_c)S(s) = B_cD_R(s), \tag{3.25}$$

where the $n \times m$ matrix $S(s) = block\, diag\, [(1, s, \ldots, s^{d_i - 1})^T]$, with $d_i, i = 1, \ldots, m$, the controllability indices of (A_c, B_c) or the dimensions of the subblocks in the controller form (A_c, B_c).

The representations $D_R(q)z(t) = u(t), y(t) = N_R(q)z(t)$ and $\dot{x}_c(t) = A_cx(t) + B_cu(t), y(t) = C_cx(t) + D_cu(t)$ are equivalent since

$$\begin{bmatrix} B_c & 0 \\ D_c & I_p \end{bmatrix} \begin{bmatrix} D_R(q) & I_m \\ -N_R(q) & 0 \end{bmatrix} = \begin{bmatrix} qI - A_c & B_c \\ -C_c & D_c \end{bmatrix} \begin{bmatrix} S(q) & 0 \\ 0 & I_m \end{bmatrix} \tag{3.26}$$

with $(B_c, qI - A_c)$ lc and $(D_R(q), S(q))$ rc (show that this is true). Verify that in this case the relation between the states given by Eq. (3.14) is $x_c(t) = S(q)z(t)$, and determine the relation between the states given by Eq. (3.17) for this case. We note that (3.25) can be written as

$$N_R(s) = C\bar{S}(s) + DD_R(s), \qquad (sI - A)\bar{S}(s) = BD_R(s), \tag{3.27}$$

where $A = Q^{-1}A_cQ, B = Q^{-1}B_c, C = C_cQ, D = D_c$ with $|Q| \neq 0$ and $\bar{S}(s) = Q^{-1}S(s)$. Equation (3.27) relates $N_R(s)$ and $D_R(s)$ to a controllable realization of $H(s)$ that is not necessarily in controller form. The matrix Q is a similarity transformation matrix.

Completely analogous results exist for a state-space realization $\{A_o, B_o, C_o, D_o\}$ in observer form and the corresponding lPMFD $\{D_L, N_L, I_p\}$.

Finally, note that the above results involving the Structure Theorem are also valid after the obvious modifications, when the state-space description is of the more general form $\{A, B, C, D(q)\}$.

B. Controllability, Observability, Stability, and Realizations

Consider now the PMD

$$P(q)z(t) = Q(q)u(t), \qquad y(t) = R(q)z(t) + W(q)u(t), \tag{3.28}$$

where $P(q) \in R[q]^{l \times l}$, $Q(q) \in R[q]^{l \times m}$, $R(q) \in R[q]^{p \times l}$, and $W(q) \in R[q]^{p \times m}$. The important system properties of controllability, observability, and asymptotic stability can be developed directly in a manner similar to that in earlier chapters, using state-space descriptions. However, instead of reintroducing these properties, we shall concentrate on establishing criteria for (3.28) to be controllable, observable, and asymptotically stable.

Assume that the PMD given in (3.28) is equivalent, in the sense of Definition 3.1 and (3.3), to some state-space representation

$$\dot{x}(t) = Ax(t) + Bu(t), \qquad y(t) = Cx(t) + Du(t), \tag{3.29}$$

where $A \in R^{n\times n}$, $B \in R^{n\times m}$, $C \in R^{p\times n}$, and $D \in R^{p\times m}$. That is, for the discussion that follows, we restrict the PMD in (3.28), at least initially, to the class of PMD that is equivalent to state-space descriptions $\{A, B, C, D\}$ that have been studied extensively throughout this book. For the more general case of (3.28) being equivalent to state-space descriptions of the form $\{A, B, C, D(q)\}$, refer to the remarks following Theorem 3.6.

Controllability

Recall from Section 3.2 of Chapter 3 that $\{A, B, C, D\}$ is controllable if and only if

1. $rank\,[s_iI - A, B] = n$ for any s_i complex number.
2. The Smith form of $[sI - A, B]$ is $[I, 0]$.

Now if $\{A, B, C, D\}$ in (3.29) is controllable, then in the equivalent description $\{P, Q, R, W\}$ given in (3.28), the Smith form of $[P, Q]$ will be $[I, 0]$ and $rank\,[P(s_i), Q(s_i)] = l$ for any complex number s_i, in view of Theorem 3.2(ii). Notice that these conditions are precisely the conditions for P, Q to be lc polynomial matrices (see Theorem 2.5 in Section 7.2). These observations give rise to the following concept and result.

DEFINITION 3.2. The representation $\{P, Q, R, W\}$ given in (3.28) is said to be *controllable* if its equivalent state-space representation $\{A, B, C, D\}$ given in (3.29) is state controllable.

THEOREM 3.4. The following statements are equivalent:

(i) $\{P, Q, R, W\}$ is controllable.
(ii) The Smith form of $[P, Q]$ is $[I, 0]$.
(iii) $rank\,[P(s_i), Q(s_i)] = l$ for any complex number s_i.
(iv) P, Q are lc.

Proof. Parts (ii) and (iii) follow from the fact that (A, B) is controllable if and only if the Smith form of $[qI - A, B]$ is $[I, 0]$, or equivalently, $rank\,[s_iI - A, B] = n$ for any complex number (see Section 3.2 of Chapter 3), and from Theorem 3.2(ii) in this section. In particular the Smith form of $[P, Q]$ in (ii) is $[I, 0]$ if and only if the Smith form of $[qI - A, B]$ is $[I, 0]$, in view of Theorem 3.2(ii). This in turn is true if and only if (A, B) is controllable, which is equivalent to (i) by definition. So (ii) is true if and only if (i) is true. Similarly, one can show that (iii) is true if and only if (i) is true. Part (iv) follows easily from (ii) or (iii), in view of Theorem 2.4 in Section 7.2. ■

Remarks

1. From the above results it follows that (A, B) is controllable if and only if $qI - A$ and B are lc polynomial matrices.

2. If P, Q are not lc, then the Smith form of $[P, Q]$ is $[\Lambda, 0]$, where $\Lambda = diag\,[\epsilon_i] \neq I$. It can easily be shown that if G_L is a gcld of $[P, Q] = G_L[\bar{P}, \bar{Q}]$, then the Smith form of G_L is Λ. The roots of the invariant polynomials ϵ_i of $[P, Q]$ or of G_L are the uncontrollable eigenvalues of the system.
3. Similar tests as the eigenvalue/eigenvector tests of Subsection 3.4B of Chapter 3 can also be developed for the representation $\{P, Q, R, W\}$.
4. The rPMFD, $\{D_R, I_m, N_R\}$, is clearly controllable since D_R and I are lc. This fact justifies the alternative notation $\{D_c, I, N_c\}$ for an rPMFD when the controllability of the representation is emphasized; here $D_c = D_R$ and $N_c = N_R$ are viewed as matrices of an internal system representation. Note that the term rPMFD stresses the relation to the transfer function $H(s) = N_R(s)D_R^{-1}(s)$, where N_R and D_R are intrepreted as the numerator and denominator of a transfer function, respectively.

Observability

Observability can be introduced in a manner completely analogous to controllability. This leads to the following concept and result.

DEFINITION 3.3. The representation $\{P, Q, R, W\}$ given in (3.28) is said to be *observable* if its equivalent state-space representation $\{A, B, C, D\}$ given in (3.29) is state observable. ■

THEOREM 3.5. The following statements are equivalent:

(i) $\{P, Q, R, W\}$ is observable.

(ii) the Smith form of $\begin{bmatrix} P \\ R \end{bmatrix}$ is $\begin{bmatrix} I \\ 0 \end{bmatrix}$.

(iii) $rank \begin{bmatrix} P(s_i) \\ R(s_i) \end{bmatrix} = l$ for any complex number s_i.

(iv) P, R are rc.

Proof. The proofs of these results are completely analogous to the proof of Theorem 3.4 and are therefore omitted. ■

Remarks

1. From the above results it follows that (A, C) is observable if and only if $qI - A$ and C are rc polynomial matrices.
2. If P, R are not rc, then the Smith form of $\begin{bmatrix} P \\ R \end{bmatrix}$ is $\begin{bmatrix} \Lambda \\ 0 \end{bmatrix}$, where $\Lambda = diag\,[\epsilon_i] \neq I$. It can easily be shown that if G_R is a gcrd of $\begin{bmatrix} P \\ R \end{bmatrix} = \begin{bmatrix} \bar{P} \\ \bar{R} \end{bmatrix} G_R$, then the Smith form of G_L is Λ. The roots of the invariant polynomials ϵ_i of $\begin{bmatrix} P \\ R \end{bmatrix}$ or of G_R are the unobservable eigenvalues of the system.
3. Similar tests as the eigenvalue/eigenvector tests of Subsection 3.4B of Chapter 3 can also be developed for the representation $\{P, Q, R, W\}$.
4. The lPMFD, $\{D_L, N_L, I_p\}$, is clearly observable since D_L and I_p are rc. This fact justifies the alternative notation $\{D_o, N_o, I_p\}$ for an lPMFD when the observability of the representation is emphasized; here $D_o = D_R$ and $N_o = N_R$ are viewed as matrices of an internal system representation. Note that the term lPMFD stresses the relation to the transfer function $H(s) = D_L^{-1}(s)N_L(s)$, where N_L and D_L are interpreted as the numerator and denominator of a transfer function, respectively.

Stability

DEFINITION 3.4. The representation $\{P, Q, R, W\}$ given in (3.28) is said to be *asymptotically stable* if for its equivalent state-space representation $\{A, B, C, D\}$ given in (3.29) the equilibrium $\dot{x} = 0$ of the free system $\dot{x} = Ax$ is is asymptotically stable. ■

THEOREM 3.6. The representation $\{P, Q, R, W\}$ is asymptotically stable if and only if $Re\ \lambda_i < 0, i = 1, \ldots, n$, where $\lambda_i, i = 1, \ldots, n$, are the roots of $det\, P(s)$; the λ_i are the eigenvalues or poles of the system.

Proof. In view of Theorem 3.2(ii), $det\,(sI - A) = \alpha\, det\, P(s)$ for some nonzero $\alpha \in R$. ■

At this point it is of interest to briefly discuss controllability, observability, and stability for the more general case, when the PMD in (3.28) is equivalent to a state-space representation of the form $\dot{x}(t) = Ax(t) + Bu(t)$, $y(t) = Cx(t) + D(q)u(t)$ instead of (3.29). First, note that the concepts of state controllability and observability in state-space descriptions of the form $\{A, B, C, D(q)\}$ are completely analogous to the $\{A, B, C, D\}$ case. Furthermore, the criteria for controllability and observability are exactly the same for the above cases, and they depend only on (A, B) and (A, C), respectively. In view of this, it can be shown that Definitions 3.2 and 3.3 and Theorems 3.4 and 3.5 for controllability and observability, respectively, are valid for the more general PMD. For similar reasons, Definition 3.4 and Theorem 3.6 on asymptotic stability are valid for the general PMD (3.28). However, care should be exercised when discussing input-output stability of a system $\{A, B, C, D(q)\}$ or of their equivalent descriptions of the form (3.28), but this topic will not be addressed here. For an extended discussion of controllability, observability, and stability of systems described by a general PMD (3.28), refer to Chapter 3 of [9] and Chapter 6 of [33].

It is of interest to note that equivalent representations not only have exactly the same eigenvalues, but also the same *invariant zeros, decoupling zeros, and transmission zeros* (see Section 3.5 of Chapter 3 for the definitions of zeros and also the discussion below). These assertions follow directly from Theorem 3.2.

Poles and zeros

In the following, we will find it useful to first recall the definitions of poles and zeros introduced in Section 3.5 of Chapter 3 and to show how these apply to PMDs. To this end, consider the representation in (3.28) with transfer function matrix $H(s) = R(s)P^{-1}(s)Q(s) + W(s)$.

Let the *Smith-McMillan form* of $H(s)$ be given by

$$SM_H(s) = \begin{bmatrix} \tilde{\Lambda}(s) & 0 \\ 0 & 0 \end{bmatrix}, \tag{3.30}$$

where $\tilde{\Lambda}(s) \triangleq diag\left(\frac{\epsilon_1(s)}{\psi_1(s)}, \ldots, \frac{\epsilon_r(s)}{\psi_r(s)}\right)$, $r = rank\, H(s)$, $\epsilon_i(s)$ divides $\epsilon_{i+1}(s)$, $i = 1, \ldots, r-1$, and $\psi_{i+1}(s)$ divides $\psi_i(s)$, $i = 1, \ldots, r-1$ [see (5.3) in Chapter 3]. Then the *poles of* $H(s)$ are the roots of the *characteristic or pole polynomial* $p_H(s)$ of $H(s)$ defined as

$$p_H(s) = \psi_1(s) \cdots \psi_r(s). \tag{3.31}$$

Note that $p_H(s)$ is the monic least common denominator of all nonzero minors of $H(s)$.

It is straightforward to show that

$$\{\text{poles of } H(s)\} \subset \{\text{roots of } det\, P(s)\}. \tag{3.32}$$

The roots of $det\, P$ are the *eigenvalues or the poles of the system* $\{P, Q, R, W\}$ and are equal to the eigenvalues of A in any equivalent state-space representation $\{A, B, C, D\}$. Relation (3.32) becomes an equality when the system is controllable and observable, since in this case the poles of H are exactly those eigenvalues of the system that are both controllable and observable.

Consider the *system matrix* or *Rosenbrock Matrix* of the representation $\{P, Q, R, W\}$,

$$S(s) = \begin{bmatrix} P(s) & Q(s) \\ -R(s) & W(s) \end{bmatrix}, \tag{3.33}$$

and an equivalent state-space representation $\{A, B, C, D\}$ given in (3.29). In view of (3.5), we have

$$\begin{bmatrix} \hat{M} & 0 \\ \hat{X} & I_p \end{bmatrix} \begin{bmatrix} I_{k-l} & 0 & \vdots & 0 \\ 0 & P & \vdots & Q \\ \cdots & \cdots & \vdots & \cdots \\ 0 & -R & \vdots & W \end{bmatrix} = \begin{bmatrix} I_{k-n} & 0 & \vdots & 0 \\ 0 & sI - A & \vdots & B \\ \cdots & \cdots & \vdots & \cdots \\ 0 & -C & \vdots & D \end{bmatrix} \begin{bmatrix} \hat{N} & -\hat{Y} \\ 0 & I_m \end{bmatrix}, \tag{3.34}$$

where $\hat{M}, \hat{N} \in R[s]^{k \times k}$ are unimodular and $k \geq \max\,(deg\,|P|, n)$ (see (3.5)).

The zeros of $\{P, Q, R, W\}$ can now be defined in a manner completely analogous to the way they were defined for state-space representations. Following the development in Section 3.5, let $k = n$ and $r = rank \begin{bmatrix} sI - A & B \\ -C & D \end{bmatrix}$, where $n \leq r \leq \min\,(p + n, m + n)$. Consider all those rth-order nonzero minors of $S_e(s) = \begin{bmatrix} I_{n-l} & 0 \\ 0 & S(s) \end{bmatrix}$ that are formed by taking the first n rows and n columns of $S_e(s)$. The *zero polynomial of the system* $\{P, Q, R, W\}$, $z_s(s)$, is defined as the monic gcd of all those minors. The roots of $z_s(s)$ are the *zeros of the system* $\{P, Q, R, W\}$. The reader is encouraged to show that this definition is consistent with the definitions given in Section 3.5.

The *invariant zeros of the system* are the roots of the invariant zero polynomial that is the product of all the invariant factors of $S(s)$. The *input-decoupling/output-decoupling* and the *input-output decoupling* zeros of $\{P, Q, R, W\}$ can be defined in a manner completely analogous to the state-space case. For example, the roots of the product of all invariant factors of $[P(s), Q(s)]$ are the input-decoupling zeros of the system; they are also the uncontrollable eigenvalues of the system. Note that the input-decoupling zeros are the roots of $det\, G_L(s)$, where $G_L(s)$ is a gcld of all the columns of $[P(s), Q(s)] = G_L(s)[\bar{P}(s), \bar{Q}(s)]$. Similar results hold for the output-decoupling zeros.

The *zeros of* $H(s)$, also called the *transmission zeros of the system*, are defined as the roots of *the zero polynomial of* $H(s)$,

$$z_H(s) = \epsilon_1(s) \ldots \epsilon_r(s), \tag{3.35}$$

where the ϵ_i are the numerator polynomials in the Smith-McMillan form of $H(s)$ given in (3.30). When $\{P, Q, R, W\}$ is controllable and observable the zeros of the system, the invariant zeros, and the transmission zeros coincide.

Consider the representation $D_R z_R = u$, $y = N_R z_R$ with $D_R \in R[s]^{m\times m}$ and $N_R \in R[s]^{p\times m}$ and notice that in this case the Rosenbrock Matrix (3.33) can be reduced via elementary column operations to the form

$$\begin{bmatrix} D_R & I \\ -N_R & 0 \end{bmatrix}\begin{bmatrix} I & 0 \\ -D_R & I \end{bmatrix}\begin{bmatrix} 0 & I \\ I & 0 \end{bmatrix} = \begin{bmatrix} 0 & I \\ -N_R & 0 \end{bmatrix}\begin{bmatrix} 0 & I \\ I & 0 \end{bmatrix} = \begin{bmatrix} I & 0 \\ 0 & -N_R \end{bmatrix}.$$

In view of the fact that the invariant factors of S do not change under elementary matrix operations, it is clear that the nonunity invariant factors of S are the nonunity invariant factors of N_R. Therefore, the *invariant zero polynomial of the system* equals the product of all invariant factors of N_R and its roots are the *invariant zeros of the system*. Note that when $rank\, N_R = p \le m$, the invariant zeros of the system are the roots of $det\, G_L$, where G_L is the gcld of all the columns of N_R, i.e., $N_R = G_L \bar{N}_R$. When N_R, D_R are rc, the system is controllable and observable. In this case it can be shown that the *zeros of* H $(= N_R D_R^{-1})$, also called the *transmission zeros of the system*, are equal to the *invariant zeros* (and to the *system zeros* of $\{D_R, I, N_R\}$) and can be determined from N_R. In fact the zero polynomial of the system, $z_s(s)$, equals $z_H(s)$, the zero polynomial of H, which equals $\epsilon_1(s)\ldots\epsilon_r(s)$, the product of the invariant factor of N_R, i.e.,

$$z_s(s) = z_H(s) = \epsilon_1(s)\ldots\epsilon_r(s). \tag{3.36}$$

The pole polynomial of $H(s)$ is

$$p_H(s) = k\, det\, D_R(s), \tag{3.37}$$

where $k \in R$.

Realization theory and algorithms

A representation $\{P, Q, R, W\}$ is a realization of a rational function $H(s)$ if the transfer function of $\{P, Q, R, W\}$ is $H(s)$, i.e., if

$$R(s)P^{-1}(s)Q(s) + W(s) = H(s). \tag{3.38}$$

The realization theory for PMDs is analogous to the theory for state-space descriptions, developed in Chapter 5. Results concerning existence and minimality can be developed in the obvious way, using the results on equivalence of representations developed in Subsection 7.3A and the above results on controllability and observability. The following theorems provide results that correspond to Theorems 3.9, 3.10, and 3.11 of Section 5.3 of Chapter 5. The reader is encouraged to give full proofs of these results.

THEOREM 3.7. A realization $\{P, Q, R, W\}$ of $H(s)$ of order $n = deg\, |P|$ is minimal (irreducible, of least order) if and only if it is both controllable and observable.

Proof. The proof is left as an exercise. ■

THEOREM 3.8. Let $\{P, Q, R, W\}$ and $\{\bar{P}, \bar{Q}, \bar{R}, \bar{W}\}$ be realizations of $H(s)$. If $\{P, Q, R, W\}$ is a minimal realization, then $\{\bar{P}, \bar{Q}, \bar{R}, \bar{W}\}$ is also a minimal realization if and only if the two realizations are equivalent.

Proof. The proof is left as an exercise. ■

THEOREM 3.9. Let $\{P, Q, R, W\}$ be a minimal realization of $H(s)$. Then the characteristic polynomial of $H(s)$, $p_H(s)$, is equal to $det\, P(s)$ within a multiplication by a nonzero real number, i.e., $p_H(s) = k\, det\, P(s)$. Therefore, the McMillan degree of $H(s)$ equals the order of any minimal realization.

Proof. To prove this result, refer to the proof of Theorem 3.11 in Chapter 5 and use the results on equivalence given in Subsection 7.3A. ■

Realization algorithms

It is rather straightforward to derive a realization of H in PMD form. In fact realizations in right (left) PMFD form were derived in Chapter 5 as a step toward determining a state-space realization in controller (observer) form (see Subsections 5.4B and 5.4D). However, these realizations, of the form $\{D_R, I_m, N_R\}$ and $\{D_L, N_L, I_p\}$, are typically not of minimal order, i.e., they are not controllable and observable. This implies that the controllable realization $\{D_R, I_m, N_R\}$, for example, is not observable, i.e., D_R, N_R are not rc. Similarly, the observable realization $\{D_L, N_L, I_p\}$ is not controllable, i.e., D_L, N_L are not lc. To obtain a minimal realization, a gcrd must be extracted from D_R, N_R, and similarly, a gcld must be extracted from D_L, N_L. This leads to the following realization algorithm that results in a minimal realization $\{D, I_m, N\}$ of H. A minimal realization of the form $\{D, N, I_p\}$ is obtained in an analogous (dual) manner.

Consider $H(s) = [n_{ij}(s)/d_{ij}(s)]$, $i = 1, \ldots, p$, $j = 1, \ldots, m$, and let $l_j(s)$ be the (monic) least common denominator of all entries in the jth column of $H(s)$. Note that $l_j(s)$ is the (monic) least degree polynomial divisible by all $d_{ij}(s)$, $i = 1, \ldots, p$. Then $H(s)$ can be written as

$$H(s) = N_R(s)D_R^{-1}(s), \tag{3.39}$$

where $N_R(s) \triangleq [\bar{n}_{ij}(s)]$ and $D_R(s) \triangleq diag(l_1(s), \ldots, l_m(s))$. Note that $\bar{n}_{ij}/l_j(s) = n_{ij}(s)/d_{ij}(s)$ for $i = 1, \ldots, p$, and all $j = 1, \ldots, m$. Now

$$D_R(q)z_R(t) = u(t), \qquad y(t) = N_R(q)z_R(t) \tag{3.40}$$

is a controllable realization of $H(s)$. If D_R, N_R are rc, it is observable as well, and therefore, minimal. If D_R and N_R are not rc, let G_R be a gcrd and let $D = D_R G_R^{-1}$ and $N = N_R G_R^{-1}$. Then

$$D(q)z(t) = u(t), \qquad y(t) = N(q)z(t) \tag{3.41}$$

is a controllable and observable, and therefore, minimal realization of $H(s)$ since D, I and D, N are lc and rc polynomial matrix pairs, respectively. Note that $ND^{-1} = (N_R G_R^{-1})(D_R G_R^{-1})^{-1} = (N_R G_R^{-1})(G_R D_R^{-1}) = N_R D_R^{-1} = H$.

There is a dual algorithm which extracts an lPMFD resulting in

$$H(s) = D_L^{-1}(s)N_L(s), \tag{3.42}$$

which corresponds to an observable realization of $H(s)$, given by

$$D_L(q)z_L(t) = N_L(q)u(t), \qquad y(t) = z_L(t). \tag{3.43}$$

The details of this procedure are completely analogous to the procedure that led to (3.39). If D_L, N_L are not lc, let G_L be a gcld and let $\tilde{D} = G_L^{-1}D_L$ and $\tilde{N} = G_L^{-1}N_L$. Then a controllable and observable, and therefore, minimal, realization of $H(s)$ is given by

$$\tilde{D}(q)\tilde{z}(t) = \tilde{N}(q)u(t), \qquad y(t) = \tilde{z}(t). \tag{3.44}$$

In Subsection 5.4B, a controllable state-space realization that is equivalent to (3.40) was obtained first using the controllable version of the Structure Theorem. Next, the unobservable part of this state-space realization was separated and a controllable, observable, and minimal realization was extracted (see Example 4.3 in Subsection 5.4B). This minimal state-space realization is of course equivalent to the realization (3.41), which is in PMFD form. It is possible to extract an equivalent minimal state-space realization directly from (3.41). However, to use the convenient Structure Theorem, typically D has to be reduced first to a column proper form, i.e., (3.41) must be reduced to

$$\bar{D}(q)\bar{z}(t) = u(t), \qquad y(t) = \bar{N}(q)\bar{z}(t), \tag{3.45}$$

where $\bar{D} = DU$ is column proper, $\bar{N} = NU$, and U is a unimodular matrix. Analogous results are valid for PMFD and realizations (3.42) through (3.44).

The following example illustrates the realization algorithms.

EXAMPLE 3.2. We wish to derive a minimal realization for $H(s) = \left[\frac{s^2+1}{s^2}, \frac{s+1}{s^3}\right]$. Note that this is the same $H(s)$ as in Example 4.3 in Section 5.4B, where minimal state-space realizations were derived. The reader is encouraged to compare those results with the realizations derived below. We shall begin with a controllable realization. In view of (3.39) $l_1 = s^2, l_2 = s^3$, and $H = N_R D_R^{-1} = [s^2+1, s+1]\begin{bmatrix} s^2 & 0 \\ 0 & s^3 \end{bmatrix}^{-1}$. Therefore, $D_R z_R = u$ and $y = N_R z_R$ constitute a controllable realization. This realization is not observable since $rank \begin{bmatrix} D_R(s) \\ N_R(s) \end{bmatrix}_{s=0} = rank \begin{bmatrix} 0 & 0 \\ 0 & 0 \\ 1 & 1 \end{bmatrix} = 1 < m = 2$, i.e., D_R and N_R are not rc.

Another way of determining that D_R and N_R are not rc would have been to observe that *deg det* $D(s) = 5 =$ order of the realization $\{D_R, I, N_R\}$. Now the McMillan degree of H, which is easily derived in the present case, is 3. Therefore, the order of any minimal realization for this example is 3. Since $\{D_R, I, N_R\}$ is of order 5 and is controllable, it cannot be observable, i.e., D_R and N_R cannot be rc.

We shall now extract a gcrd from D_R and N_R, using the procedure described in Subsection 7.2D. We have

$$\begin{bmatrix} D_R \\ N_R \end{bmatrix} = \begin{bmatrix} s^2 & 0 \\ 0 & s^3 \\ s^2+1 & s+1 \end{bmatrix} \longrightarrow \begin{bmatrix} s^2 & 0 \\ 0 & s^3 \\ 1 & s+1 \end{bmatrix} \longrightarrow \begin{bmatrix} 1 & s+1 \\ s^2 & 0 \\ 0 & s^3 \end{bmatrix} \longrightarrow$$

$$\begin{bmatrix} 1 & s+1 \\ 0 & -s^3-s^2 \\ 0 & s^3 \end{bmatrix} \longrightarrow \begin{bmatrix} 1 & s+1 \\ 0 & s^2 \\ 0 & s^3 \end{bmatrix} \longrightarrow \begin{bmatrix} 1 & s+1 \\ 0 & s^2 \\ 0 & 0 \end{bmatrix}.$$

Therefore, $G_R = \begin{bmatrix} 1 & s+1 \\ 0 & s^2 \end{bmatrix}$ is a gcrd. We now determine $D = D_R G_R^{-1}$ and $N = N_R G_R^{-1}$, using $D_R = \begin{bmatrix} s^2 & 0 \\ 0 & s^3 \end{bmatrix} = \begin{bmatrix} s^2 & -(s+1) \\ 0 & s \end{bmatrix}\begin{bmatrix} 1 & s+1 \\ 0 & s^2 \end{bmatrix} = DG_R$ and $N_R = [s^2+1, s+1] = [s^2+1, -(s+1)]\begin{bmatrix} 1 & s+1 \\ 0 & s^2 \end{bmatrix} = NG_R$, and we verify that they are rc. Then

$$\{D_R, I, N_R\} = \left\{\begin{bmatrix} q^2 & -(q+1) \\ 0 & q \end{bmatrix}, \begin{bmatrix} 1 & 0 \\ 0 & 1 \end{bmatrix}, [q^2+1, -(q+1)]\right\}$$

is a minimal realization of $H(s)$.

To determine a minimal state-space realization from $Dz = u$ and $y = Nz$ using equivalence of representations, if so desired, we notice that D is already in column proper form. Using the Structure Theorem, and the notation used in Examples 4.3

and 4.5 in Subsection 5.4B, we obtain $D(s) = \begin{bmatrix} s^2 & -(s+1) \\ 0 & s \end{bmatrix} = D_h\Lambda(s) + D_l S(s) =$ $\begin{bmatrix} 1 & -1 \\ 0 & 1 \end{bmatrix}\begin{bmatrix} s^2 & 0 \\ 0 & s \end{bmatrix} + \begin{bmatrix} 0 & 0 & -1 \\ 0 & 0 & 0 \end{bmatrix}\begin{bmatrix} 1 & 0 \\ s & 0 \\ 0 & 1 \end{bmatrix}$. Therefore, $B_m = D_h^{-1} = \begin{bmatrix} 1 & 1 \\ 0 & 1 \end{bmatrix}$ and $A_m =$ $-D_h^{-1}D_l = \begin{bmatrix} 0 & 0 & 1 \\ 0 & 0 & 0 \end{bmatrix}$. Also, $D_c = \lim_{s\to\infty} H(s) = [1, 0]$ and $N(s) = [s^2+1, -(s+1)] = C_c S(s) + D_c D(s) = [1, 0, 0]\begin{bmatrix} 1 & 0 \\ s & 0 \\ 0 & 1 \end{bmatrix} + [1, 0]\begin{bmatrix} s^2 & -(s+1) \\ 0 & s \end{bmatrix}$. Therefore, a minimal state-space realization of $H(s)$ is given by

$$\{A_c, B_c, C_c, D_c\} = \left\{\begin{bmatrix} 0 & 1 & 0 \\ 0 & 0 & 1 \\ 0 & 0 & 0 \end{bmatrix}, \begin{bmatrix} 0 & 0 \\ 1 & 1 \\ 0 & 1 \end{bmatrix}, [1, 0, 0], [1, 0]\right\}.$$

Note that this realization is precisely the minimal realization derived in Example 4.3 in Subsection 5.4B. This will not occur in general, as can be seen if for example a slightly different G_R is used. The realization $\{A_c, B_c, C_c, D_c\}$ determined here is in controller form, since D_h was an upper triangular matrix with ones on the diagonal. In general, D_h will just be nonsingular, since $D(s)$ will be column proper, and therefore, the resulting controllable realization will not be in controller form, since B_c will not be of the appropriate form. This point was also discussed in Subsection 5.4B and illustrated in Example 4.5 in that subsection.

Alternatively, given H, we shall first derive an observable realization. In view of (3.42),

$$H = D_L^{-1}N_L = (s^3)^{-1}[s(s^2+1), s+1].$$

Here $D_L(q)$ and $N_L(q)$ are lc and therefore $\tilde{D}(q)\tilde{z}(t) = \tilde{N}(q)u(t)$ and $y(t) = \tilde{z}(t)$ with $\tilde{D}(q) = D_L(q)$ and $\tilde{N}(q) = N_L(q)$ is controllable and observable, and is therefore, a minimal realization. Note that the order of this realization is *deg det* $D_L(s) = 3$, which equals the McMillan degree of $H(s)$. In Example 4.3 in Subsection 5.4B a minimal state-space realization in observer form was derived from $\tilde{D}(q)$ and $\tilde{N}(q)$. ■

C. Interconnected Systems

Interconnected systems, connected in parallel, in series, and in feedback configurations, are studied in this subsection. Polynomial matrix and transfer function matrix descriptions of interconnected systems are derived, and controllability, observability, and stability questions are addressed. It is shown that particular interconnections may introduce uncontrollable, unobservable, or unstable modes into a system. Feedback configurations, as well as series interconnections, are of particular importance in the control of systems. Feedback control systems are studied at length in the next section.

Systems connected in parallel

Consider systems S_1 and S_2 connected in parallel as shown in Fig. 7.1 and let

$$P_1(q)z_1(t) = Q_1(q)u_1(t), \qquad y_1(t) = R_1(q)z_1(t) + W_1(q)u_1(t) \qquad (3.46)$$

and

$$P_2(q)z_2(t) = Q_2(q)u_2(t), \qquad y_2(t) = R_2(q)z_2(t) + W_2(q)u_2(t) \qquad (3.47)$$

be representations (PMDs) for S_1 and S_2, respectively. Since $u(t) = u_1(t) = u_2(t)$

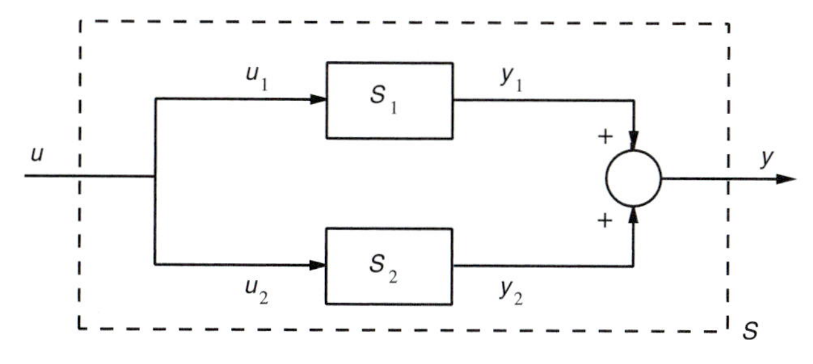

FIGURE 7.1.
Systems connected in parallel

and $y(t) = y_1(t) + y_2(t)$, the overall system description is given by

$$\begin{bmatrix} P_1(q) & 0 \\ 0 & P_2(q) \end{bmatrix} \begin{bmatrix} z_1(q) \\ z_2(q) \end{bmatrix} = \begin{bmatrix} Q_1(q) \\ Q_2(q) \end{bmatrix} u(t),$$

$$y(t) = [R_1(q), R_2(q)] \begin{bmatrix} z_1(t) \\ z_2(t) \end{bmatrix} + [W_1(q) + W_2(q)]u(t). \tag{3.48}$$

If the systems S_1 and S_2 are described by the state-space representations $\dot{x}_i = A_i x_i + B_i u_i$, $y_i = C_i x_i + D_i u_i$, $i = 1, 2$, then the overall system state-space description is given by

$$\begin{bmatrix} \dot{x}_1 \\ \dot{x}_2 \end{bmatrix} = \begin{bmatrix} A_1 & 0 \\ 0 & A_2 \end{bmatrix} \begin{bmatrix} x_1 \\ x_2 \end{bmatrix} + \begin{bmatrix} B_1 \\ B_2 \end{bmatrix} u$$

$$y = [C_1, C_2] \begin{bmatrix} x_1 \\ x_2 \end{bmatrix} + [D_1 + D_2]u. \tag{3.49}$$

If $H_1(s)$, $H_2(s)$ are the transfer function matrices of S_1 and S_2, respectively, then the overall transfer function can be found from $\hat{y}(s) = \hat{y}_1(s) + \hat{y}_2(s) = H_1(s)\hat{u}_1(s) + H_2(s)\hat{u}_2(s) = [H_1(s) + H_2(s)]\hat{u}(s)$ to be

$$H(s) = H_1(s) + H_2(s). \tag{3.50}$$

Now if both $H_1(s)$ and $H_2(s)$ are proper, then $H(s)$ is also proper.

The stability, controllability, and observability of the overall system S represented by $Pz = Qu$, $y = Rz + Wu$ in (3.48) will now be examined briefly. In view of $|P| = |P_1||P_2|$ it is clear that the overall system is internally stable if and only if each of the systems S_1 and S_2 is internally stable. Also, if both S_1 and S_2 are BIBO stable, i.e., all poles in H_1 and in H_2 have strictly negative real parts, then the overall system is also BIBO stable. The converse is also true, i.e., if the overall system S is BIBO stable, then so are S_1 and S_2. This can be seen from Fig. 7.1 since if, say, S_1 were not BIBO stable, then neither would S. Next, this result is also shown using alternative arguments.

All uncontrollable or unobservable eigenvalues of S_1 and S_2 will be uncontrollable or unobservable eigenvalues of the overall system S. To see this, consider

$$\begin{bmatrix} P_1 & 0 & Q_1 \\ 0 & P_2 & Q_2 \end{bmatrix} \tag{3.51}$$

and let G_1, G_2 be gclds of $[P_1, Q_1]$ and $[P_2, Q_2]$, respectively. Then $\begin{bmatrix} G_1 & 0 \\ 0 & G_2 \end{bmatrix}$ is an ld of the matrix (rows) in (3.51), i.e., the uncontrollable eigenvalues of S_1 and S_2 will be uncontrollable eigenvalues of the overall system. Similarly, it can be shown that the unobservable eigenvalues of S_1 and S_2 will be unobservable eigenvalues of the overall system. Note that the overall system S may have additional uncontrollable and unobservable eigenvalues, as is now shown. It is easier to show this when S_1 and S_2 are controllable and observable. Assume then that

$$D_1(q)z_1(t) = u_1(t), \qquad y_1(t) = N_1(q)z_1(t) \tag{3.52}$$

and

$$D_2(q)z_2(t) = u_2(t), \qquad y_2(t) = N_2(q)z_2(t) \tag{3.53}$$

are descriptions for S_1 and S_2, with $H_1(s) = N_1(s)D_1^{-1}(s)$ and $H_2(s) = N_2(s)D_2^{-1}(s)$. Let $D_1(q), D_2(q) \in R[q]^{m\times m}$ and $N_1(q), N_2(q) \in R[q]^{p\times m}$. The overall system is then described by

$$\begin{bmatrix} D_1(q) & 0 \\ 0 & D_2(q) \end{bmatrix}\begin{bmatrix} z_1(t) \\ z_2(t) \end{bmatrix} = \begin{bmatrix} I \\ I \end{bmatrix} u(t), \qquad y(t) = [N_1(q), N_2(q)]\begin{bmatrix} z_1(t) \\ z_2(t) \end{bmatrix}. \tag{3.54}$$

Note that $\begin{bmatrix} D_1 & 0 & I \\ 0 & D_2 & I \end{bmatrix}\begin{bmatrix} I & 0 & 0 \\ 0 & I & 0 \\ 0 & -D_2 & I \end{bmatrix} = \begin{bmatrix} D_1 & -D_2 & I \\ 0 & 0 & I \end{bmatrix}$. Now if for some complex value λ $rank\,[D_1(\lambda), -D_2(\lambda)] < m$, i.e., there are fewer than m linearly independent columns in $[D_1(\lambda), -D_2(\lambda)]$, then $rank \begin{bmatrix} D_1(\lambda) & -D_2(\lambda) & I_m \\ 0 & 0 & I_m \end{bmatrix} < 2m$, which implies that λ is an uncontrollable eigenvalue of the overall system. If G is a gcld of $[D_1, -D_2]$, then λ is a root of $det\, G$ and it is an eigenvalue that is common in both systems S_1 and S_2. Note that the uncontrollable eigenvalues in G cancel in $H = H_1 + H_2 = N_1D_1^{-1} + N_2D_2^{-1} = (N_1 + N_2D_2^{-1}D_1)D_1^{-1} = (N_1 + N_2\bar{D}_2^{-1}\bar{D}_1)D_1^{-1}$, where $D_2^{-1}D_1 = (G\bar{D}_2)^{-1}G\bar{D}_1 = \bar{D}_2^{-1}\bar{D}_1$. Completely analogous results can be shown concerning the unobservable eigenvalues by considering the representations

$$\tilde{D}_1(q)\tilde{z}_1(t) = \tilde{N}_1(q)u_1(t), \qquad y_1(t) = \tilde{z}_1(t) \tag{3.55}$$

and

$$\tilde{D}_2(q)\tilde{z}_2(t) = \tilde{N}_2(q)u_2(t), \qquad y_2(t) = \tilde{z}_2(t) \tag{3.56}$$

with $H_1(s) = \tilde{D}_1^{-1}(s)\tilde{N}_1(s)$ and $H_2(s) = \tilde{D}_2^{-1}(s)\tilde{N}_2(s)$. The unobservable eigenvalues cancel in $\tilde{D}_1(s)\tilde{D}_2^{-1}(s)$ (show this).

In view of the above it can be seen that if all the poles of H have negative real parts, then all the poles of both H_1 and H_2 also have negative real parts, since possible additional poles of H_1 and H_2 are in the same locations as some of the poles of H. Therefore, if S is BIBO stable so are both S_1 and S_2.

Systems connected in series (or in cascade or in tandem)

Consider systems S_1 and S_2 connected in series, as shown in Fig. 7.2 and let

$$P_1(q)z_1(t) = Q_1(q)u_1(t), \qquad y_1(t) = R_1(q)z_1(t) + W_1(q)u_1(t) \tag{3.57}$$

be a polynomial matrix representation for S_1 and

$$P_2(q)z_2(t) = Q_2(q)u_2(t), \qquad y_2(t) = R_2(q)z_2(t) + W_2(q)u_2(t) \tag{3.58}$$

be a representation for S_2. Here $u_2(t) = y_1(t)$. To derive the overall system description, consider $P_2z_2 = Q_2u_2 = Q_2y_1 = Q_2(R_1z_1 + W_1u_1)$ and $P_1z_1 = Q_1u_1$ and also $y_2 = R_2z_2 + W_2u_2 = R_2z_2 + W_2y_1 = R_2z_2 + W_2(R_1z_1 + W_1u_1)$ and $y_1 =$

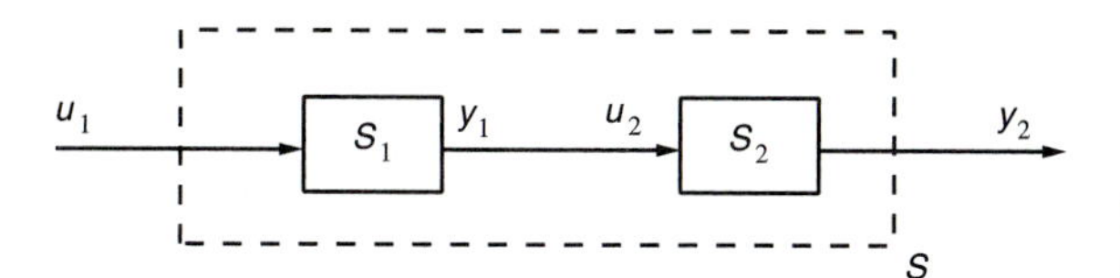

FIGURE 7.2.
Systems connected in series

$R_1 z_1 + W_1 u_1$. Then

$$\begin{bmatrix} P_1 & 0 \\ -Q_2 R_1 & P_2 \end{bmatrix} \begin{bmatrix} z_1 \\ z_2 \end{bmatrix} = \begin{bmatrix} Q_1 \\ Q_2 W_1 \end{bmatrix} u_1$$
$$y_2 = [W_2 R_1, R_2] \begin{bmatrix} z_1 \\ z_2 \end{bmatrix} + W_2 W_1 u_1. \tag{3.59}$$

If an external input r_2 is introduced as in Fig. 7.3, that is, $u_2 = y_1 + r_2$, then $P_2 z_2 = Q_2 u_2 = Q_2(y_1 + r_2) = Q_2(R_1 z_1 + W_1 u_1) + Q_2 r_2$ and $y_2 = R_2 z_2 + W_2 u_2 = R_2 z_2 + W_2(y_1 + r_2) = R_2 z_2 + W_2(R_1 z_1 + W_1 u_1) + W_2 r_2$.

In this case a more complete description of the two systems in series with inputs u_1, r_2 and outputs y_2, y_1 is given by

$$\begin{bmatrix} P_1 & 0 \\ -Q_2 R_1 & P_2 \end{bmatrix} \begin{bmatrix} z_1 \\ z_2 \end{bmatrix} = \begin{bmatrix} Q_1 & 0 \\ Q_2 W_1 & Q_2 \end{bmatrix} \begin{bmatrix} u_1 \\ r_2 \end{bmatrix}$$
$$\begin{bmatrix} y_1 \\ y_2 \end{bmatrix} = \begin{bmatrix} R_1 & 0 \\ W_2 R_1 & R_2 \end{bmatrix} \begin{bmatrix} z_1 \\ z_2 \end{bmatrix} + \begin{bmatrix} W_1 & 0 \\ W_2 W_1 & W_2 \end{bmatrix} \begin{bmatrix} u_1 \\ r_2 \end{bmatrix}. \tag{3.60}$$

If the systems S_1, S_2 are described by the state-space representations $\dot{x}_i = A_i x_i + C_i u_i$, $y_i = C_i x_i + D_i u_i$, $i = 1, 2$, then it can be shown similarly that the overall system state-space description is given by

$$\begin{bmatrix} \dot{x}_1 \\ \dot{x}_2 \end{bmatrix} = \begin{bmatrix} A_1 & 0 \\ B_2 C_1 & A_2 \end{bmatrix} \begin{bmatrix} x_1 \\ x_2 \end{bmatrix} + \begin{bmatrix} B_1 & 0 \\ B_2 D_1 & B_2 \end{bmatrix} \begin{bmatrix} u_1 \\ r_2 \end{bmatrix}$$
$$\begin{bmatrix} y_1 \\ y_2 \end{bmatrix} = \begin{bmatrix} C_1 & 0 \\ D_2 C_1 & C_2 \end{bmatrix} \begin{bmatrix} x_1 \\ x_2 \end{bmatrix} + \begin{bmatrix} D_1 & 0 \\ D_2 D_1 & D_2 \end{bmatrix} \begin{bmatrix} u_1 \\ r_2 \end{bmatrix}. \tag{3.61}$$

If $H_1(s), H_2(s)$ are the transfer function matrices of S_1 and S_2, then the overall transfer function $\hat{y}_2(s) = H(s)\hat{u}_1(s)$ is

$$H(s) = H_2(s) H_1(s). \tag{3.62}$$

It can be shown that if both H_1 and H_2 are proper, then H is also proper. Note that poles of H_1 and H_2 may cancel in the product $H_2 H_1$, and any cancellation implies

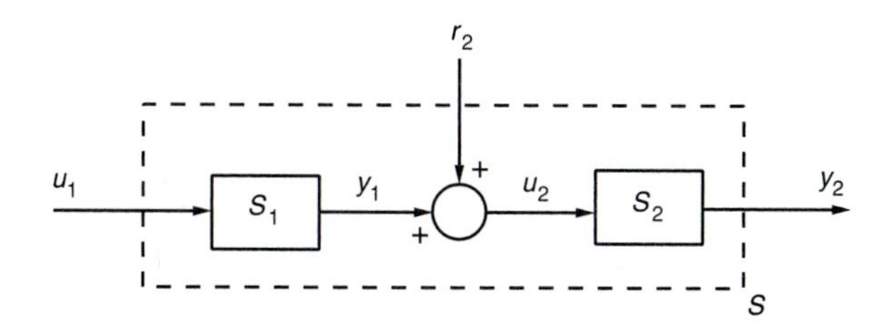

FIGURE 7.3.
Systems connected in series

that there are uncontrollable/unobservable eigenvalues in the overall system internal description. This is discussed further next.

The stability, controllability, and observability of the overall system S described by $Pz = Qu$, $y = Rz + Wu$, and given in (3.60) will now be examined. In view of $|P| = |P_1||P_2|$ it is clear that the overall system is internally stable if and only if each of the subsystems S_1 and S_2 is stable. If both S_1 and S_2 are BIBO stable, i.e., if all poles in H_1 and H_2 have strictly negative real parts, then the overall system is also BIBO stable. The converse is not necessarily true since cancellations may take place in the product H_2H_1, i.e., the unstable poles of H_1 and H_2 may cancel in $H_2H_1 = H$, thus leading to a BIBO stable system where S_1 and/or S_2 might not be BIBO stable.

Concerning controllability and observability of (3.60), we make several observations:

1. All eigenvalues of P_1 are uncontrollable from r_2 and all eigenvalues of P_2 are unobservable from y_1. This is of course as expected, since Fig. 7.3 reveals that the input r_2 does not affect system S_1 at all, and observation of the output y_1 will not reveal any information about system S_2.
2. All uncontrollable eigenvalues of S_1 and S_2 are uncontrollable from u_1. This is true because any gcld of $[P_1, Q_1]$ and any gcld of $[P_2, Q_2]$ are clds of $\begin{bmatrix} P_1 & 0 & Q_1 \\ -Q_2R_1 & P_2 & Q_2W_1 \end{bmatrix}$ (show this).
3. Finally, all unobservable eigenvalues of S_1 and S_2 are unobservable from y_2. This is true because any gcrd of $\begin{bmatrix} P_1 \\ R_1 \end{bmatrix}$ and any gcrd of $\begin{bmatrix} P_2 \\ R_2 \end{bmatrix}$ are crds of $\begin{bmatrix} P_1 & 0 \\ -Q_2R_1 & P_2 \\ W_2R_1 & R_2 \end{bmatrix}$ (show this).

It should be noted that other uncontrollable and/or unobservable eigenvalues may exist. These correspond to poles of H_1 and H_2 cancelling in the product H_2H_1 and can be found from representation (3.60) using, say, the eigenvalue tests for controllability and observability. It is of interest to determine these additional uncontrollable and unobservable eigenvalues directly from the PMFD of S_1 and S_2. To simplify the analysis, we first assume that both S_1 and S_2 are controllable and observable. Let

$$D_1(q)z_1(t) = u_1(t), \qquad y_1(t) = N_1(q)z_1(t) \tag{3.63}$$

be a description for S_1, with $H_1(s) = N_1(s)D_1^{-1}(s)$, and let

$$D_2(q)z_2(t) = u_2(t), \qquad y_2(t) = N_2(q)z_2(t) \tag{3.64}$$

be a description for S_2, with $H_2(s) = N_2(s)D_2^{-1}(s)$. Let $D_1(q) \in R[q]^{m\times m}$, $N_1(q) \in R[q]^{p\times m}$ and $D_2(q) \in R[q]^{p\times p}$, $N_2(q) \in R[q]^{r\times p}$ and recall that y_1 and u_2 have the same dimensions ($y_1 = u_2$). In this case description (3.60) becomes

$$\begin{bmatrix} D_1 & 0 \\ -N_1 & D_2 \end{bmatrix}\begin{bmatrix} z_1 \\ z_2 \end{bmatrix} = \begin{bmatrix} I & 0 \\ 0 & I \end{bmatrix}\begin{bmatrix} u_1 \\ r_2 \end{bmatrix}, \qquad \begin{bmatrix} y_1 \\ y_2 \end{bmatrix} = \begin{bmatrix} N_1 & 0 \\ 0 & N_2 \end{bmatrix}\begin{bmatrix} z_1 \\ z_2 \end{bmatrix}. \tag{3.65}$$

The uncontrollable eigenvalues from u_1 are the roots of a gcld of $\begin{bmatrix} D_1 & 0 & I \\ -N_1 & D_2 & 0 \end{bmatrix}$. Using elementary column operations corresponding to postmultiplication by a unimodular matrix, this matrix is reduced to $\begin{bmatrix} 0 & 0 & I \\ -N_1 & D_2 & 0 \end{bmatrix}$, a gcld of which is given by $\begin{bmatrix} I & 0 \\ 0 & G_L \end{bmatrix}$, where G_L is a gcld of $[-N_1, D_2]$. Thus, all the uncontrollable eigenvalues of the overall system are the roots of the determinant of a gcld of $[-N_1, D_2]$. Note that these are poles of H_2 cancelling in the product $H_2H_1 = N_2D_2^{-1}N_1D_1^{-1}$, in $D_2^{-1}N_1$, or equivalently, in H_2N_1.

The unobservable eigenvalues may be obtained similarly, using the representations

$$\tilde{D}_1(q)\tilde{z}_1(t) = \tilde{N}_1(q)u_1(t), \qquad y_1(t) = \tilde{z}_1(t) \tag{3.66}$$

and

$$\tilde{D}_2(q)\tilde{z}_2(t) = \tilde{N}_2(q)u_2(t), \qquad y_2(t) = \tilde{z}_2(t) \tag{3.67}$$

with $H_1(s) = \tilde{D}_1^{-1}(s)\tilde{N}_1(s)$ and $H_2(s) = \tilde{D}_2^{-1}(s)\tilde{N}_2(s)$. Description (3.60) becomes in this case

$$\begin{bmatrix} \tilde{D}_1 & 0 \\ -\tilde{N}_2 & \tilde{D}_2 \end{bmatrix}\begin{bmatrix} \tilde{z}_1 \\ \tilde{z}_2 \end{bmatrix} = \begin{bmatrix} \tilde{N}_1 & 0 \\ \tilde{N}_2 & \tilde{N}_2 \end{bmatrix}\begin{bmatrix} u_1 \\ r_2 \end{bmatrix}, \qquad \begin{bmatrix} y_1 \\ y_2 \end{bmatrix} = \begin{bmatrix} I & 0 \\ 0 & I \end{bmatrix}\begin{bmatrix} z_1 \\ z_2 \end{bmatrix}. \tag{3.68}$$

The unobservable eigenvalues from y_2 are the roots of a gcrd of $\begin{bmatrix} \tilde{D}_1 & 0 \\ -\tilde{N}_2 & \tilde{D}_2 \\ 0 & I \end{bmatrix}$. Using elementary row operations, corresponding to premultiplication by a unimodular matrix, this matrix is reduced to $\begin{bmatrix} \tilde{D}_1 & 0 \\ -\tilde{N}_2 & 0 \\ 0 & I \end{bmatrix}$, a gcrd of which is given by $\begin{bmatrix} G_R & 0 \\ 0 & I \end{bmatrix}$, where G_R is a gcrd of $\begin{bmatrix} \tilde{D}_1 \\ -\tilde{N}_2 \end{bmatrix}$. Thus, the unobservable eigenvalues of the overall system are the roots of the determinant of a gcrd of $\begin{bmatrix} \tilde{D}_1 \\ -\tilde{N}_2 \end{bmatrix}$. Note that these are poles of H_1 cancelling in the product $H_2H_1 = \tilde{D}_2^{-1}\tilde{N}_2\tilde{D}_1^{-1}\tilde{N}_1$, in $\tilde{N}_2\tilde{D}_1^{-1}$, or equivalently, in $\tilde{N}_2H_1$.

Systems connected in feedback configuration

Consider systems S_1 and S_2 connected in a feedback configuration as shown in Fig. 7.4A, or equivalently, as in Fig. 7.4B.
Let

$$P_1(q)z_1(t) = Q_1(q)u_1(t), \qquad y_1(t) = R_1(q)z_1(t) + W_1(q)u_1(t) \tag{3.69}$$

and

$$P_2(q)z_2(t) = Q_2(q)u_2(t), \qquad y_2(t) = R_2(q)z_2(t) + W_2(q)u_2(t) \tag{3.70}$$

be polynomial matrix representations of S_1 and S_2, respectively. Since

$$u_1(t) = y_2(t) + r_1(t), \tag{3.71}$$

$$u_2(t) = y_1(t) + r_2(t), \tag{3.72}$$

where r_1 and r_2 are external inputs, the dimensions of the vector inputs and outputs, u_1 and y_2 and also u_2 and y_1, must be the same, i.e., $u_1, y_2 \in R^m$ and $u_2, y_1 \in R^p$. To derive the overall system description, we consider $y_1 = R_1z_1 + W_1u_1 = R_1z_1 + W_1(y_2 + r_1) = R_1z_1 + W_1(R_2z_2 + W_2u_2) + W_1r_1 = R_1z_1 + W_1R_2z_2 + W_1W_2(y_1 + r_2) + W_1r_1$, from which we obtain

$$(I - W_1W_2)y_1 = R_1z_1 + W_1R_2z_2 + W_1r_1 + W_1W_2r_2. \tag{3.73}$$

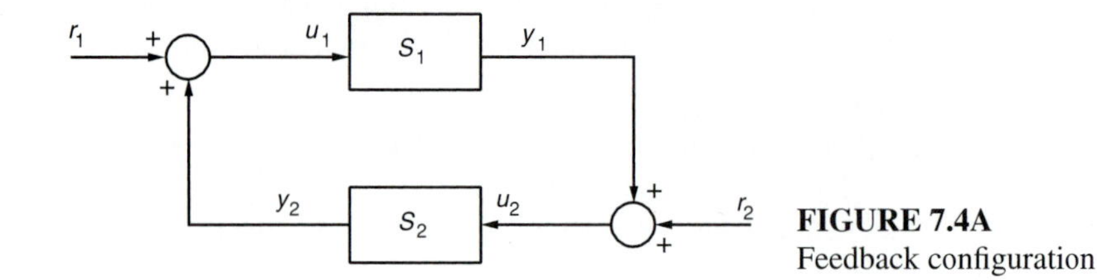

FIGURE 7.4A
Feedback configuration

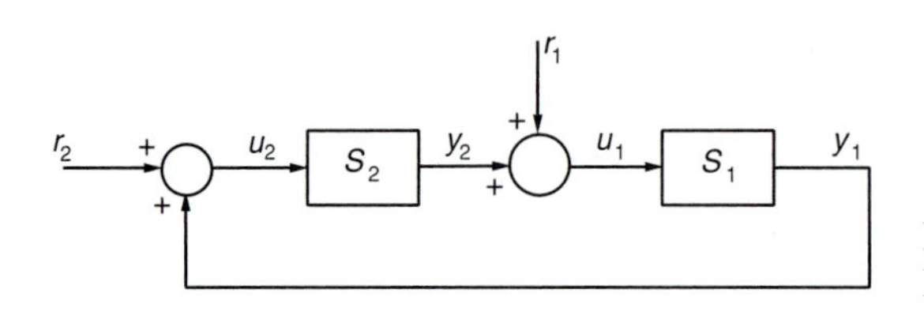

FIGURE 7.4B
Feedback configuration

Similarly, we have $y_2 = R_2z_2 + W_2u_2 = R_2z_2 + W_2(y_1 + r_2) = R_2z_2 + W_2(R_1z_1 + W_1u_1) + W_2r_2 = R_2z_2 + W_2R_1z_1 + W_2W_1(y_2 + r_1) + W_2r_2$, from which we obtain

$$(I - W_2W_1)y_2 = R_2z_2 + W_2R_1z_1 + W_2W_1r_1 + W_2r_2. \tag{3.74}$$

LEMMA 3.10. *$det(I - W_1W_2) = det(I - W_2W_1)$.*

Proof. $det(I - W_1W_2) = det\left(\begin{bmatrix} I & 0 \\ 0 & I - W_1W_2 \end{bmatrix}\right)$

$$= det\left(\begin{bmatrix} I & 0 \\ -W_1 & I \end{bmatrix}\begin{bmatrix} I & W_2 \\ W_1 & I \end{bmatrix}\begin{bmatrix} I & -W_2 \\ 0 & I \end{bmatrix}\right)$$

$$= det\left(\begin{bmatrix} I & -W_2 \\ 0 & I \end{bmatrix}\begin{bmatrix} I & W_2 \\ W_1 & I \end{bmatrix}\begin{bmatrix} I & 0 \\ -W_1 & I \end{bmatrix}\right)$$

$$= det\left(\begin{bmatrix} I - W_2W_1 & 0 \\ 0 & I \end{bmatrix}\right) = det(I - W_2W_1)$$

■

Now assume that $det(I - W_1W_2) = det(I - W_2W_1) \neq 0$. If $M_1 \triangleq (I - W_1W_2)^{-1}$ and $M_2 \triangleq (I - W_2W_1)^{-1}$, then (3.73) and (3.74) imply that $y_1 = [M_1R_1, M_1W_1R_2]\begin{bmatrix} z_1 \\ z_2 \end{bmatrix} + [M_1W_1, M_1W_1W_2]\begin{bmatrix} r_1 \\ r_2 \end{bmatrix}$ and $y_2 = [M_2W_2R_1, M_2R_2]\begin{bmatrix} z_1 \\ z_2 \end{bmatrix} + [M_2W_2W_1, M_2W_2]\begin{bmatrix} r_1 \\ r_2 \end{bmatrix}$. Now $P_1z_1 = Q_1u_1 = Q_1(y_2 + r_1)$ and $P_2z_2 = Q_2u_2 = Q_2(y_1 + r_2)$, where y_1 and y_2 are as above. Then the closed loop is described by

$$\begin{bmatrix} P_1 - Q_1M_2W_2R_1 & -Q_1M_2R_2 \\ -Q_2M_1R_1 & P_2 - Q_2M_1W_1R_2 \end{bmatrix}\begin{bmatrix} z_1 \\ z_2 \end{bmatrix} = \begin{bmatrix} Q_1M_2 & Q_1M_2W_2 \\ Q_2M_1W_1 & Q_2M_1 \end{bmatrix}\begin{bmatrix} r_1 \\ r_2 \end{bmatrix}$$

$$\begin{bmatrix} y_1 \\ y_2 \end{bmatrix} = \begin{bmatrix} M_1R_1 & M_1W_1R_2 \\ M_2W_2R_1 & M_2R_2 \end{bmatrix}\begin{bmatrix} z_1 \\ z_2 \end{bmatrix} + \begin{bmatrix} M_1W_1 & M_1W_1W_2 \\ M_2W_2W_1 & M_2W_2 \end{bmatrix}\begin{bmatrix} r_1 \\ r_2 \end{bmatrix}, \tag{3.75}$$

where the identity $I + (I - W_1W_2)^{-1}W_1W_2 = (I - W_1W_2)^{-1}$ was used.

At this point it is of interest to point out that when $W_1 = 0$, $W_2 = 0$, then the closed-loop description (3.75) is simplified to the PMD

$$\begin{bmatrix} P_1 & -Q_1R_2 \\ -Q_2R_1 & P_2 \end{bmatrix}\begin{bmatrix} z_1 \\ z_2 \end{bmatrix} = \begin{bmatrix} Q_1 & 0 \\ 0 & Q_2 \end{bmatrix}\begin{bmatrix} r_1 \\ r_2 \end{bmatrix}, \qquad \begin{bmatrix} y_1 \\ y_2 \end{bmatrix} = \begin{bmatrix} R_1 & 0 \\ 0 & R_2 \end{bmatrix}\begin{bmatrix} z_1 \\ z_2 \end{bmatrix}. \tag{3.76}$$

Note that given system descriptions (3.69) and (3.70), it is always possible to obtain equivalent representations with $W_1 = 0$ and $W_2 = 0$. Assuming this is the case, (3.76) is a general description for the closed-loop system, and the condition for the closed-loop system to be well defined is that

$$det\left(\begin{bmatrix} P_1 & -Q_1R_2 \\ -Q_2R_1 & P_2 \end{bmatrix}\right) \neq 0. \tag{3.77}$$

If this condition is not satisfied, then the closed-loop system cannot be described by the polynomial matrix representations discussed in this chapter.

If the systems S_1 and S_2 are described by the state-space representations $\dot{x}_i = A_ix_i + B_iu_i$, $y_i = C_ix_i + D_iu_i$, $i = 1, 2$, then in view of (3.75), the closed-loop system state-space description is

$$\begin{aligned} \begin{bmatrix} \dot{x}_1 \\ \dot{x}_2 \end{bmatrix} &= \begin{bmatrix} A_1 + B_1M_2D_2C_1 & B_1M_2C_2 \\ B_2M_1C_1 & A_2 + B_2M_1D_1C_2 \end{bmatrix}\begin{bmatrix} x_1 \\ x_2 \end{bmatrix} + \begin{bmatrix} B_1M_2 & B_1M_2D_2 \\ B_2M_1D_1 & B_2M_1 \end{bmatrix}\begin{bmatrix} r_1 \\ r_2 \end{bmatrix} \\ \begin{bmatrix} y_1 \\ y_2 \end{bmatrix} &= \begin{bmatrix} M_1C_1 & M_1D_1C_2 \\ M_2D_2C_1 & M_2C_2 \end{bmatrix}\begin{bmatrix} x_1 \\ x_2 \end{bmatrix} + \begin{bmatrix} M_1D_1 & M_1D_1D_2 \\ M_2D_2D_1 & M_2D_2 \end{bmatrix}\begin{bmatrix} r_1 \\ r_2 \end{bmatrix}, \end{aligned} \tag{3.78}$$

where $M_1 = (I - D_1D_2)^{-1}$ and $M_2 = (I - D_2D_1)^{-1}$. It is assumed that $det(I - D_1D_2) = det(I - D_2D_1) \neq 0$.

It is not difficult to see that in the case of state-space representations, the conditions for the closed-loop system state-space representation to be well defined is $det(I - D_1D_2) \neq 0$. When $D_1 = 0$ and $D_2 = 0$, then (3.78) simplifies to

$$\begin{aligned} \begin{bmatrix} \dot{x}_1 \\ \dot{x}_2 \end{bmatrix} &= \begin{bmatrix} A_1 & B_1C_2 \\ B_2C_1 & A_2 \end{bmatrix}\begin{bmatrix} x_1 \\ x_2 \end{bmatrix} + \begin{bmatrix} B_1 & 0 \\ 0 & B_2 \end{bmatrix}\begin{bmatrix} r_1 \\ r_2 \end{bmatrix} \\ \begin{bmatrix} y_1 \\ y_2 \end{bmatrix} &= \begin{bmatrix} C_1 & 0 \\ 0 & C_2 \end{bmatrix}\begin{bmatrix} x_1 \\ x_2 \end{bmatrix}. \end{aligned} \tag{3.79}$$

EXAMPLE 3.3. Let S_1 and S_2 be described by $z_1 = u_1$, $y_1 = qz_1$ and $qz_2 = u_2$, $y_2 = z_2$, i.e., $\{P_1, Q_1, R_1, W_1\} = \{1, 1, q, 0\}$ and $\{P_2, Q_2, R_2, W_2\} = \{q, 1, 1, 0\}$. These systems have transfer functions $H_1(s) = s$ and $H_2(s) = 1/s$, i.e., an ideal inductance S_1 is connected via feedback to an ideal capacitance S_2. Then (3.76) assumes the form

$$\begin{bmatrix} 1 & -1 \\ -q & q \end{bmatrix}\begin{bmatrix} z_1 \\ z_2 \end{bmatrix} = \begin{bmatrix} 1 & 0 \\ 0 & 1 \end{bmatrix}\begin{bmatrix} r_1 \\ r_2 \end{bmatrix}$$

$$\begin{bmatrix} y_1 \\ y_2 \end{bmatrix} = \begin{bmatrix} q & 0 \\ 0 & 1 \end{bmatrix}\begin{bmatrix} z_1 \\ z_2 \end{bmatrix}.$$

Since $det\left(\begin{bmatrix} 1 & -1 \\ -q & q \end{bmatrix}\right) = 0$, this does not constitute a well-defined closed-loop description. It is of interest to note that S_1 cannot be described by a state-space description of the form $\{A_1, B_1, C_1, D_1\}$ since its transfer function $H_1(s) = s$ is not proper. ■

EXAMPLE 3.4. Consider systems S_1 and S_2 connected in a feedback configuration with $H_1(s) = \dfrac{s+1}{s+2}$ and $H_2(s) = \dfrac{s+2}{s+1}$ and consider the realizations

$\{P_1, Q_1, R_1, W_1\} = \{q+2, q+1, 1, 0\}$ and $\{P_2, Q_2, R_2, W_2\} = \{q+1, q+2, 1, 0\}$. Then (3.76) becomes

$$\begin{bmatrix} q+2 & -(q+1) \\ -(q+2) & q+1 \end{bmatrix} \begin{bmatrix} z_1 \\ z_2 \end{bmatrix} = \begin{bmatrix} q+1 & 0 \\ 0 & q+2 \end{bmatrix} \begin{bmatrix} r_1 \\ r_2 \end{bmatrix}$$
$$\begin{bmatrix} y_1 \\ y_2 \end{bmatrix} = \begin{bmatrix} 1 & 0 \\ 0 & 1 \end{bmatrix} \begin{bmatrix} z_1 \\ z_2 \end{bmatrix}.$$

Since $det\left(\begin{bmatrix} q+2 & -(q+1) \\ -(q+2) & q+1 \end{bmatrix}\right) = 0$, the present example is not a well-defined polynomial matrix system description. It is of interest to note that here state-space realizations of H_1 and H_2 exist. Let $H_1 = \dfrac{-1}{s+2} + 1$, $H_2 = \dfrac{1}{s+1} + 1$, and consider the state-space realizations $\{A_1, B_1, C_1, D_1\} = \{-2, -1, 1, 1\}$ and $\{A_2, B_2, C_2, D_2\} = \{-1, 1, 1, 1\}$. Here $1 - D_1 D_2 = 0$, and therefore, a state-space realization of the closed-loop system does not exist, as expected. ■

EXAMPLE 3.5. Consider systems S_1 and S_2 in a feedback configuration with $H_1(s) = \dfrac{s}{s+1}$ and $H_2(s) = 1$ and consider the realizations $\{P_1, Q_1, R_1, W_1\} = \{q+1, q, 1, 0\}$ and $\{P_2, Q_2, R_2, W_2\} = \{1, 1, 1, 0\}$. Then (3.76) becomes

$$\begin{bmatrix} q+1 & -q \\ -1 & 1 \end{bmatrix} \begin{bmatrix} z_1 \\ z_2 \end{bmatrix} = \begin{bmatrix} q & 0 \\ 0 & 1 \end{bmatrix} \begin{bmatrix} r_1 \\ r_2 \end{bmatrix}$$
$$\begin{bmatrix} y_1 \\ y_2 \end{bmatrix} = \begin{bmatrix} 1 & 0 \\ 0 & 1 \end{bmatrix} \begin{bmatrix} z_1 \\ z_2 \end{bmatrix}.$$

Since $det\left(\begin{bmatrix} q+1 & -q \\ -1 & 1 \end{bmatrix}\right) = 1 \neq 0$, this is a well-defined polynomial matrix description for the closed-loop system. Note that the transfer function matrix of the closed-loop system is $H(s) = \begin{bmatrix} 1 & 0 \\ 0 & 1 \end{bmatrix} \begin{bmatrix} s+1 & -s \\ -1 & 1 \end{bmatrix}^{-1} \begin{bmatrix} s & 0 \\ 0 & 1 \end{bmatrix} = \begin{bmatrix} s & s \\ s & s+1 \end{bmatrix}$, which is not proper whereas H_1 and H_2 were both proper.

If the realizations to be considered are $\{P_1, Q_1, R_1, W_1\} = \{q+1, -1, 1, 1\}$ and $\{P_2, Q_2, R_2, W_2\} = \{1, 0, 0, 1\}$, then $1 - W_1 W_2 = 1 - 1 \cdot 1 = 0$ and no representation of the form (3.75) can be derived. This stresses the point that the conditions for the PMD of the closed-loop system to be well defined are given by (3.77), since the condition $det(I - W_1 W_2) \neq 0$ may lead to erroneous results. Note that $det(I - W_1 W_2) = 0$ does not necessarily imply that a well-defined polynomial matrix representation for the closed-loop system does not exist.

Now if state-space realizations of $H_1(s) = \dfrac{-1}{s+1} + 1$ and $H_2(s) = 1$ are considered, namely, $\{A_1, B_1, C_1, D_1\} = \{-1, 1, -1, 1\}$ and $\{A_2, B_2, C_2, D_2\} = \{0, 0, 0, 1\}$, then $1 - D_1 D_2 = 1 - 1 \cdot 1 = 0$, i.e., a state-space description of the closed-loop does not exist. This is to be expected since the closed-loop transfer function is nonproper and as such cannot be represented by a state-space realization $\{A, B, C, D\}$. ■

Next, let $H_1(s)$ and $H_2(s)$ be the transfer function matrices of S_1 and S_2, i.e., $\hat{y}_1(s) = H_1(s)\hat{u}_1(s)$ and $\hat{y}_2(s) = H_2(s)\hat{u}_2(s)$. In view of $\hat{u}_1 = \hat{y}_2 + \hat{r}_1$ and $\hat{u}_2 = \hat{y}_1 + \hat{r}_2$, we have $\hat{y}_1 = H_1\hat{u}_1 = H_1(\hat{y}_2 + \hat{r}_1) = H_1 H_2 \hat{u}_2 + H_1 \hat{r}_1 = H_1 H_2 \hat{y}_1 + H_1 H_2 \hat{r}_2 + H_1 \hat{r}_1$, or

$$(I - H_1 H_2)\hat{y}_1 = H_1 H_2 \hat{r}_2 + H_1 \hat{r}_1. \tag{3.80}$$

Also, $\hat{y}_2 = H_2\hat{u}_2 = H_2(\hat{y}_1 + \hat{r}_2) = H_2H_1\hat{u}_1 + H_2\hat{r}_2 = H_2H_1\hat{y}_2 + H_2H_1\hat{r}_1 + H_2\hat{r}_2$, or

$$(I - H_2H_1)\hat{y}_2 = H_2H_1\hat{r}_1 + H_2\hat{r}_2. \tag{3.81}$$

Assume that $det(I - H_1H_2) = det(I - H_2H_1) \neq 0$. Note that the proof of the fact that the determinants are equal is completely analogous to the proof of Lemma 3.10. Then

$$\begin{aligned}\begin{bmatrix}\hat{y}_1\\ \hat{y}_2\end{bmatrix} &= \begin{bmatrix}(I - H_1H_2)^{-1}H_1 & (I - H_1H_2)^{-1}H_1H_2\\ (I - H_2H_1)^{-1}H_2H_1 & (I - H_2H_1)^{-1}H_2\end{bmatrix}\begin{bmatrix}\hat{r}_1\\ \hat{r}_2\end{bmatrix}\\ &= \begin{bmatrix}H_{11} & H_{12}\\ H_{21} & H_{22}\end{bmatrix}\begin{bmatrix}\hat{r}_1\\ \hat{r}_2\end{bmatrix}.\end{aligned} \tag{3.82}$$

The significance of the assumption $det(I - H_1H_2) \neq 0$ can be seen as follows. Let $\tilde{D}_1\tilde{z}_1 = \tilde{N}_1u_1$, $y_1 = \tilde{z}_1$ and $D_2z_2 = u_2$, $y_2 = N_2z_2$ be representations of the systems S_1 and S_2. As will be shown, the closed-loop system description in this case is given by $(\tilde{D}_1D_2 - \tilde{N}_1N_2)z_2 = \tilde{N}_1r_1 + \tilde{D}_1r_2$ and $y_1 = D_2z_2 - r_2$ and $y_2 = N_2z_2$. Now note that $I - H_1H_2 = I - \tilde{D}_1^{-1}\tilde{N}_1N_2D_2^{-1} = \tilde{D}_1^{-1}(\tilde{D}_1D_2 - \tilde{N}_1N_2)D_2^{-1}$, which implies that $det\,(I - H_1H_2) \neq 0$ if and only if $det(\tilde{D}_1D_2 - \tilde{N}_1N_2) \neq 0$, i.e., if $det(I - H_1H_2) = 0$, then the closed-loop system cannot be described by the polynomial matrix representations discussed in this chapter. Thus, the assumption that $det(I - H_1H_2) \neq 0$ is essential for the closed-loop system to be well defined.

EXAMPLE 3.6. Consider $H_1(s) = 2$ and $H_2(s) = 1/s$ as in Example 3.3. Clearly, $1 - H_1H_2 = 0$, which implies that the closed-loop system is not well defined. This agrees with the result of Example 3.3, where it was shown that a PMD of the closed-loop system does not exist. ■

EXAMPLE 3.7. Consider $H_1(s) = \dfrac{s+1}{s+2}$, $H_2 = \dfrac{s+2}{s+1}$ as in Example 3.4. Here $1 - H_1H_2 = 0$ and the closed-loop system is not well defined, which agrees with the results in Example 3.4. ■

EXAMPLE 3.8. Consider $H_1(s) = \dfrac{s}{s+1}$, $H_2(s) = 1$ as in Example 3.5. Here $1 - H_1H_2 = \dfrac{1}{s+1} \neq 0$, and therefore, the closed-loop system is well defined. Relation (3.82) in this case assumes the form

$$\begin{bmatrix}\hat{y}_1\\ \hat{y}_2\end{bmatrix} = \begin{bmatrix}s & s\\ s & s+1\end{bmatrix}\begin{bmatrix}\hat{r}_1\\ \hat{r}_2\end{bmatrix},$$

a nonproper transfer function that is the transfer function matrix $H(s)$ derived in Example 3.5. ■

Now consider systems S_1 and S_2 with proper transfer function matrices $H_1(s)$ and $H_2(s)$, and let $\dot{x}_i = A_ix_i + B_iu_i$, $y_i = C_ix_i + D_iu_i$, $i = 1, 2$, be their state-space representations.

LEMMA 3.11. If $det(I - D_1D_2) \neq 0$, then $det(I - H_1(s)H_2(s)) \neq 0$. Furthermore, in this case $I - H_1(s)H_2(s)$ is *biproper* [i.e., both $I - H_1(s)H_2(s)$ and $(I - H_1(s)H_2(s))^{-1}$ are proper rational matrices].

Proof. It is not difficult to see that $I - H_1(s)H_2(s) = I - D_1D_2 +$ (strictly proper terms), which implies that $det(I - D_1D_2)$ is the coefficient of the highest degree s term in the numerator of $det(I - H_1(s)H_2(s))$. Then if $det(I - D_1D_2) \neq 0$, we have $det(I - H_1(s)H_2(s)) \neq 0$. Furthermore,

$$\lim_{s\to\infty}(I - H_1(s)H_2(s)) = I - D_1D_2, \tag{3.83}$$

which implies that when $det(I - D_1D_2) \neq 0$, then $I - H_1(s)H_2(s)$ is a biproper rational matrix, i.e., $I - H_1H_2$ and $(I - H_1H_2)^{-1}$ are proper. ■

In Lemma 3.11, we have used the fact that a proper rational matrix $H(s)$ has a proper inverse, e.g., $H^{-1}(s)$ proper, if and only if $det D \neq 0$, where $D = \lim_{s\to\infty} H(s)$. This result is based on Lemma 3.12 and Corollary 3.13.

LEMMA 3.12. Consider $H(s) \in R(s)^{p\times p}$, which can be expressed uniquely as $H(s) = \hat{H}(s) + Q(s)$, using polynomial division, where $\hat{H}(s) \in R(s)^{p\times p}$ is strictly proper and $Q(s) \in R[s]^{p\times p}$. Then $H^{-1}(s)$ is proper if and only if $Q^{-1}(s)$ exists and is proper.

Proof. The proof is left to the reader as an exercise. ■

COROLLARY 3.13. If $H(s) \in R(s)^{p\times p}$ is proper, then $H^{-1}(s)$ is proper if and only if $det D \neq 0$, where $\lim_{s\to\infty} H(s) = D \in R^{p\times p}$.

Proof. $D = Q(s)$ of the above lemma. ■

Note that $det(I - H_1(s)H_2(s)) \neq 0$ does not necessarily imply that $det(I - D_1D_2) \neq 0$. To see this, recall that in Examples 3.5 and 3.8, $H_1 = \frac{s}{s+1}$, $H_2 = 1$, and $1 - H_1H_2 = 1 - \frac{s}{s+1} = \frac{1}{s+1} \neq 0$. However, in any state-space realization $D_1 = 1, D_2 = 1$, and $1 - D_1D_2 = 0$. In other words, when $H_1(s)$ and $H_2(s)$ are proper, the condition $det(I - D_1D_2) = det(I - D_2D_1) \neq 0$ implies that $det(I - H_1(s)H_2(s)) = det(I - H_2(s)H_1(s)) \neq 0$, or that a closed-loop representation in state-space form exists. When $det(I - H_1(s)H_2(s)) \neq 0$, a polynomial matrix representation for the closed-loop system exists, but it is not necessarily in state-space form, i.e., $det(I - D_1D_2)$ is not necessarily nonzero.

LEMMA 3.14. Let $H_1(s), H_2(s)$ be proper with $D_1 = \lim_{s\to\infty} H_1(s)$, $D_2 = \lim_{s\to\infty} H_2(s)$ and assume that $(I - H_1(s)H_2(s))^{-1}$ exists. Then $H_{11}(s) = (I - H_1(s)H_2(s))^{-1}H_1(s)$ is proper if and only if $det(I - D_1D_2) \neq 0$.

Proof. If $det(I - D_1D_2) \neq 0$, then in view of Lemma 3.11, $I - H_1(s)H_2(s)$ is biproper and $H_{11}(s)$ is then proper. If $H_{11}(s)$ is proper, then $H_{11}(s)H_2(s)$ and $I + H_{11}(s)H_2(s)$ are proper. But $(I - H_1(s)H_2(s))^{-1} = I + (I - H_1(s)H_2(s))^{-1}H_1(s)H_2(s) = I + H_{11}(s)H_2(s)$, and therefore, $(I - H_1(s)H_2(s))^{-1}$ is proper or $I - H_1(s)H_2(s)$ is biproper. This implies, in view of Corollary 3.13, that $det(I - D_1D_2) \neq 0$. ■

Similarly, it can be shown that all the rational matrices in (3.82), namely,

$$\begin{bmatrix}\hat{y}_1\\ \hat{y}_2\end{bmatrix} = \begin{bmatrix}H_{11} & H_{12}\\ H_{21} & H_{22}\end{bmatrix}\begin{bmatrix}\hat{r}_1\\ \hat{r}_2\end{bmatrix} = \begin{bmatrix}(I - H_1H_2)^{-1}H_1 & (I - H_1H_2)^{-1}H_1H_2\\ (I - H_2H_1)H_2H_1 & (I - H_2H_1)^{-1}H_2\end{bmatrix}\begin{bmatrix}\hat{r}_1\\ \hat{r}_2\end{bmatrix}$$

are proper if and only if $det(I - D_1D_2) \neq 0$. This result can also be seen from the following system theoretic argument: The rational matrices in (3.82) exist and are proper if and only if there exists a state-space representation of the closed-loop system. This will happen if and only if $det(I - D_1D_2) \neq 0$, in view of the assumptions in (3.78).

In the following it is assumed that $det(I - H_1(s)H_2(s)) \neq 0$, that is, the closed-loop system is well defined.

Controllability and observability

Controllability and observability of closed-loop systems are studied next. In view of the representation (3.76), the following is evident:

1. If G_{1L} is a gcld of P_1, Q_1 and G_{2L} is a gcld of P_2, Q_2, then $\begin{bmatrix} G_{1L} & 0 \\ 0 & G_{2L} \end{bmatrix}$ is a cld of $\begin{bmatrix} P_1 & -Q_1R_2 \\ -Q_2R_1 & P_2 \end{bmatrix}, \begin{bmatrix} Q_1 & 0 \\ 0 & Q_2 \end{bmatrix}$. Thus, all uncontrollable eigenvalues of S_1 in G_{1L} are uncontrollable eigenvalues of the closed-loop system from r_1, or r_2, or $\begin{bmatrix} r_1 \\ r_2 \end{bmatrix}$. Also, all uncontrollable eigenvalues of S_2 in G_{2L} are uncontrollable eigenvalues of the closed-loop system from r_1 or r_2, or $\begin{bmatrix} r_1 \\ r_2 \end{bmatrix}$. There may be additional uncontrollable eigenvalues in the closed-loop system, and these are studied below.
2. If G_{1R} is a gcrd of P_1, R_1 and G_{2R} is a gcrd of P_2, R_2, then $\begin{bmatrix} G_{1R} & 0 \\ 0 & G_{2R} \end{bmatrix}$ is a crd of $\begin{bmatrix} P_1 & -Q_1R_2 \\ -Q_2R_1 & P_2 \end{bmatrix}, \begin{bmatrix} R_1 & 0 \\ 0 & R_2 \end{bmatrix}$. That is, all unobservable eigenvalues of S_1 in G_{1R} are unobservable eigenvalues of the closed-loop system from y_1, or y_2, or $\begin{bmatrix} y_1 \\ y_2 \end{bmatrix}$. Also, all unobservable eigenvalues of S_2 in G_{2R} are unobservable eigenvalues of the closed-loop system from y_1, or y_2, or $\begin{bmatrix} y_1 \\ y_2 \end{bmatrix}$. There may be additional unobservable eigenvalues in the closed-loop system, and these are studied below.

Controllability and observability can of course be studied directly from the PMDs and (3.76) as was done above. However, further insight is gained if the additional uncontrollable and unobservable eigenvalues are determined from the PMFDs of S_1 and S_2. For simplicity, assume that both S_1 and S_2 are controllable and observable and consider the following representations:

For system S_1:

(1a) $$D_1(q)z_1(t) = u_1(t), \qquad y_1(t) = N_1(q)z_1(t) \tag{3.84}$$

or (1b) $$\tilde{D}_1(q)\tilde{z}_1(t) = \tilde{N}_1(q)u_1(t), \qquad y_1(t) = \tilde{z}_1(t), \tag{3.85}$$

where $(D_1(q), N_1(q))$ are rc and $(\tilde{D}_1(q), \tilde{N}_1(q))$ are lc.

For system S_2:

(2a) $$D_2(q)z_2(t) = u_2(t), \qquad y_2(t) = N_2(q)z_2(t) \tag{3.86}$$

or (2b) $$\tilde{D}_2(q)\tilde{z}_2(t) = \tilde{N}_2(q)u_2(t), \qquad y_2(t) = \tilde{z}_2(t), \tag{3.87}$$

where $(D_2(q), N_2(q))$ are rc and $(\tilde{D}_2(q), \tilde{N}_2(q))$ are lc.

In view of the connections

$$u_1(t) = y_2(t) + r_1(t), \qquad u_2(t) = y_1(t) + r_2(t) \tag{3.88}$$

the closed-loop feedback system of Fig. 7.4 can now be characterized as follows [see also (3.76)]:

1. Using descriptions (1a) and (2a), Eqs. (3.84) and (3.86), we have

$$\begin{bmatrix} D_1 & -N_2 \\ -N_1 & D_2 \end{bmatrix}\begin{bmatrix} z_1 \\ z_2 \end{bmatrix} = \begin{bmatrix} I & 0 \\ 0 & I \end{bmatrix}\begin{bmatrix} r_1 \\ r_2 \end{bmatrix}, \qquad \begin{bmatrix} y_1 \\ y_2 \end{bmatrix} = \begin{bmatrix} N_1 & 0 \\ 0 & N_2 \end{bmatrix}\begin{bmatrix} z_1 \\ z_2 \end{bmatrix}. \tag{3.89}$$

2. Using descriptions (1b) and (2b), we have

$$\begin{bmatrix} \tilde{D}_1 & -\tilde{N}_1 \\ -\tilde{N}_2 & \tilde{D}_2 \end{bmatrix} \begin{bmatrix} \tilde{z}_1 \\ \tilde{z}_2 \end{bmatrix} = \begin{bmatrix} \tilde{N}_1 & 0 \\ 0 & \tilde{N}_2 \end{bmatrix} \begin{bmatrix} r_1 \\ r_2 \end{bmatrix}, \qquad \begin{bmatrix} y_1 \\ y_2 \end{bmatrix} = \begin{bmatrix} I & 0 \\ 0 & I \end{bmatrix} \begin{bmatrix} \tilde{z}_1 \\ \tilde{z}_2 \end{bmatrix}. \tag{3.90}$$

3. Using descriptions (1b) and (2a), we have

$$\begin{bmatrix} \tilde{D}_1 & -\tilde{N}_1 N_2 \\ -I & D_2 \end{bmatrix} \begin{bmatrix} \tilde{z}_1 \\ \tilde{z}_2 \end{bmatrix} = \begin{bmatrix} \tilde{N}_1 & 0 \\ 0 & I \end{bmatrix} \begin{bmatrix} r_1 \\ r_2 \end{bmatrix}, \qquad \begin{bmatrix} y_1 \\ y_2 \end{bmatrix} = \begin{bmatrix} I & 0 \\ 0 & N_2 \end{bmatrix} \begin{bmatrix} \tilde{z}_1 \\ \tilde{z}_2 \end{bmatrix}. \tag{3.91}$$

Also, $D_2 z_2 = u_2 = y_1 + r_2 = \tilde{D}_1^{-1}\tilde{N}_1 u_1 + r_2 = \tilde{D}_1^{-1}\tilde{N}_1(y_2 + r_1) + r_2 = \tilde{D}_1^{-1}\tilde{N}_1(N_2 z_2 + r_1) + r_2$ and $y_1 = u_2 - r_2 = D_2 z_2 - r_2$, from which we obtain

$$(\tilde{D}_1 D_2 - \tilde{N}_1 N_2) z_2 = [\tilde{N}_1, \tilde{D}_1] \begin{bmatrix} r_1 \\ r_2 \end{bmatrix}, \qquad \begin{bmatrix} y_1 \\ y_2 \end{bmatrix} = \begin{bmatrix} D_2 \\ N_2 \end{bmatrix} z_2 + \begin{bmatrix} 0 & -I \\ 0 & 0 \end{bmatrix} \begin{bmatrix} r_1 \\ r_2 \end{bmatrix}. \tag{3.92}$$

4. Using descriptions (1a) and (2b), we have

$$\begin{bmatrix} D_1 & -I \\ -\tilde{N}_2 N_1 & \tilde{D}_2 \end{bmatrix} \begin{bmatrix} z_1 \\ \tilde{z}_2 \end{bmatrix} = \begin{bmatrix} I & 0 \\ 0 & \tilde{N}_2 \end{bmatrix} \begin{bmatrix} r_1 \\ r_2 \end{bmatrix}, \qquad \begin{bmatrix} y_1 \\ y_2 \end{bmatrix} = \begin{bmatrix} N_1 & 0 \\ 0 & I \end{bmatrix} \begin{bmatrix} z_1 \\ \tilde{z}_2 \end{bmatrix}. \tag{3.93}$$

Also, $D_1 z_1 = u_1 = y_2 + r_1 = \tilde{D}_2^{-1}\tilde{N}_2 u_2 + r_1 = \tilde{D}_2^{-1}\tilde{N}_2(y_1 + r_2) + r_1 = \tilde{D}_2^{-1}\tilde{N}_2(N_1 z_1 + r_2) + r_1$ and $y_2 = u_1 - r_1 = D_1 z_1 - r_1$, from which we obtain

$$(\tilde{D}_2 D_1 - \tilde{N}_2 N_1) z_1 = [\tilde{D}_2, \tilde{N}_2] \begin{bmatrix} r_1 \\ r_2 \end{bmatrix}, \qquad \begin{bmatrix} y_1 \\ y_2 \end{bmatrix} = \begin{bmatrix} N_1 \\ D_1 \end{bmatrix} z_1 + \begin{bmatrix} 0 & 0 \\ -I & 0 \end{bmatrix} \begin{bmatrix} r_1 \\ r_2 \end{bmatrix}. \tag{3.94}$$

The preceding descriptions of the closed-loop system are of course equivalent. This can be proved directly, using the definition of equivalent representations discussed in Subsection 7.3A. Therefore, these descriptions have the same uncontrollable and unobservable modes. In the following analysis we shall carefully select different representations to study different properties in order to secure additional insight. The reader is encouraged to derive similar results, using different representations. It is stressed that in the following, both S_1 and S_2 are assumed to be controllable and observable. The uncontrollability and unobservability discussed below is due to the feedback interconnection only. As was previously shown, there will in general be additional uncontrollable and unobservable eigenvalues in a closed-loop system if S_1 and S_2 are uncontrollable and/or unobservable.

To study controllability, consider the representation (3.89) in (1). It is clear from the matrices $\begin{bmatrix} D_1 & -N_2 \\ -N_1 & D_2 \end{bmatrix}$ and $\begin{bmatrix} I & 0 \\ 0 & I \end{bmatrix}$ that the eigenvalues that are uncontrollable from r_1 will be the roots of the determinant of a gcld of $[-N_1, D_2]$, and the eigenvalues that are uncontrollable from r_2 will be the roots of a gcld of $[D_1, -N_2]$. The closed-loop system is controllable from $\begin{bmatrix} r_1 \\ r_2 \end{bmatrix}$. Clearly, all possible eigenvalues that are uncontrollable from r_1 are eigenvalues of S_2. These are the poles of $H_2 = N_2 D_2^{-1}$ that cancel in the product $H_2 N_1$ or, as can be shown, they are the poles of H_2 that cancel in $\begin{bmatrix} I \\ H_2 \end{bmatrix} H_1$. Similarly, all possible eigenvalues that are uncontrollable from r_2 are eigenvalues of S_1. These are the poles of $H_1 = N_1 D_1^{-1}$ that cancel in the product $H_1 N_2$ or, as can be shown, they are the poles of H_1 that cancel in $\begin{bmatrix} H_1 \\ I \end{bmatrix} H_2$.

To study observability, consider the representation (3.90) in (2). It is clear from the matrices $\begin{bmatrix} \tilde{D}_1 & -\tilde{N}_1 \\ -\tilde{N}_2 & \tilde{D}_2 \end{bmatrix}$ and $\begin{bmatrix} I & 0 \\ 0 & I \end{bmatrix}$ that the eigenvalues that are unobservable from y_1 will be the roots of the determinant of a gcrd of $\begin{bmatrix} -\tilde{N}_1 \\ \tilde{D}_2 \end{bmatrix}$, and the eigenvalues that are unobservable from y_2 will be the roots of the determinant of a gcrd of $\begin{bmatrix} \tilde{D}_1 \\ -\tilde{N}_2 \end{bmatrix}$. The closed-loop system is observable from $\begin{bmatrix} y_1 \\ y_2 \end{bmatrix}$. Clearly, all possible eigenvalues that are unobservable from y_1 are eigenvalues of S_2. These are the poles of $H_2 = \tilde{D}_2^{-1} \tilde{N}_2$ that cancel in the product $\tilde{N}_1 H_2$, or as can be shown, they are the poles of H_2 that cancel in $H_1[I, H_2]$. Similarly, all possible eigenvalues that are unobservable from y_2 are eigenvalues of S_1. These are the poles of $H_1 = \tilde{D}_1^{-1} \tilde{N}_1$ that cancel in the product $\tilde{N}_2 H_1$, or as can be shown, they are the poles of H_1 that cancel in $H_2[H_1, I]$.

Summary. The poles of H_2 that cancel in $H_2 N_1$ (or in $\begin{bmatrix} I \\ H_2 \end{bmatrix} H_1$) are the eigenvalues that are uncontrollable from r_1; the poles of H_2 that cancel in $\tilde{N}_1 H_2$ (or in $H_1[I, H_2]$) are the eigenvalues that are unobservable from y_1. The poles of H_1 that cancel in $H_1 N_2$ (or in $\begin{bmatrix} H_1 \\ I \end{bmatrix} H_2$) are the eigenvalues that are uncontrollable from r_2; the poles of H_1 that cancel in $\tilde{N}_2 H_1$ (or in $H_2[H_1, I]$) are the eigenvalues that are unobservable from y_2.

The above results may also be expressed in the following convenient way, which, however, should be used only when the poles of H_1 are distinct from the poles of H_2, to avoid confusion. In the product $H_2 H_1$, the poles of H_2 that cancel are the eigenvalues that are uncontrollable from r_1; the poles of H_1 that cancel are the eigenvalues that are unobservable from y_2. In the product $H_1 H_2$ the poles of H_1 that cancel are the eigenvalues that are uncontrollable from r_2; the poles of H_2 that cancel are the eigenvalues that are unobservable from y_1.

EXAMPLE 3.9. Consider systems S_1 and S_2 connected in the feedback configuration of Fig. 7.4 and let S_1 and S_2 be described by the transfer functions $H_1(s) = \frac{s+1}{s-1}$, and $H_2(s) = \frac{a_1 s + a_0}{s + b}$. For the closed loop to be well defined, we must have $1 - H_1 H_2 = 1 - \frac{s+1}{s-1} \frac{a_1 s + a_0}{s + b} = \frac{(1 - a_1)s^2 + (b - a_1 - a_0 - 1)s - (b + a_0)}{(s-1)(s+b)} \neq 0$. Note that for $a_1 = 1, a_0 = -1$, and $b = 1, H_2 = \frac{s-1}{s+1}$ and $1 - H_1 H_2 = 1 - 1 = 0$. Therefore, these values are not allowed for the parameters if the closed-loop system is to be represented by a PMD. If state-space descriptions are to be used, let $D_1 = \lim_{s\to\infty} H_1(s) = 1$ and $D_2 = \lim_{s\to\infty} H_2(s) = a_1$, from which we have $1 - D_1 D_2 = 1 - a_1 \neq 0$ for the closed-loop system to be characterized by a state-space description [and also, in view of Lemma 3.14, for the transfer functions in (3.82) to be proper]. Let us assume that $a_1 \neq 1$.

The uncontrollable and unobservable eigenvalues can be determined from a PMD such as (3.92). Alternatively, in view of the discussion just preceding this example, we conclude the following. (i) The eigenvalues that are uncontrollable from r_1 are the poles of H_2 that cancel in $H_2 N_1 = \frac{a_1 s + a_0}{s + b}(s+1)$, i.e., there is an eigenvalue that is uncontrollable from r_1 (at -1) only when $b = 1$. If this is the case, -1 is also an eigenvalue that is unobservable from y_1. (ii) The poles of H_1 that cancel in $H_1 N_2 = \frac{s+1}{s-1}(a_1 s + a_0)$ are the eigenvalues that are uncontrollable from r_2, i.e., there is an eigenvalue that is uncontrollable from r_2 (at $+1$) only when $a_0/a_1 = -1$. If this is the case, $+1$ is also an eigenvalue that is unobservable from y_2. ■

Stability

The closed-loop feedback system is internally stable if and only if all its eigenvalues have strictly negative real parts. The closed-loop eigenvalues can be determined from the closed-loop descriptions derived above. If, for example, (3.76) is considered, then the closed-loop eigenvalues are the roots of $det\left(\begin{bmatrix} P_1 & -Q_1 R_2 \\ -Q_2 R_1 & P_2 \end{bmatrix}\right)$.

Since any uncontrollable or unobservable eigenvalues of S_1 and S_2 is also an uncontrollable or unobservable eigenvalue of the closed-loop system, it is clear that for internal stability, both S_1 and S_2 should be stabilizable and detectable, i.e., any uncontrollable or unobservable eigenvalues of S_1 and S_2 should have negative real parts. This is a necessary condition for stability. We shall now concentrate on the descriptions (3.87) to (3.94) in (1) through (4). First, recall the identities

$$det\begin{bmatrix} A & D \\ C & B \end{bmatrix} = det(A)\, det(B - CA^{-1}D) = det(B)\, det(A - DB^{-1}C), \quad (3.95)$$

where in the first expression it was assumed that $det(A) \neq 0$ and in the second expression it was assumed that $det(B) \neq 0$. The proof of this result is immediate from the matrix identities $\begin{bmatrix} I & 0 \\ -CA^{-1} & I \end{bmatrix}\begin{bmatrix} A & D \\ C & B \end{bmatrix} = \begin{bmatrix} A & D \\ 0 & B - CA^{-1}D \end{bmatrix}$ and $\begin{bmatrix} I & -DB^{-1} \\ 0 & I \end{bmatrix}\begin{bmatrix} A & D \\ C & B \end{bmatrix} = \begin{bmatrix} A - DB^{-1}C & 0 \\ C & B \end{bmatrix}$.

Now consider the polynomial matrices $\begin{bmatrix} D_1 & -N_2 \\ -N_1 & D_2 \end{bmatrix}$, $\begin{bmatrix} \tilde{D}_1 & -\tilde{N}_1 \\ -\tilde{N}_2 & \tilde{D}_2 \end{bmatrix}$, $(\tilde{D}_1 D_2 - \tilde{N}_1 N_2)$, and $(\tilde{D}_2 D_1 - \tilde{N}_2 N_1)$ from the closed-loop descriptions in (1), (2), (3), and (4). Then

$$\begin{aligned} det\left(\begin{bmatrix} D_1 & -N_2 \\ -N_1 & D_2 \end{bmatrix}\right) &= det(D_1)\, det(D_2 - N_1 D_1^{-1} N_2) \\ &= det(D_1)\, det(D_2 - \tilde{D}_1^{-1}\tilde{N}_1 N_2) \\ &= det(D_1)\, det(\tilde{D}_1^{-1})\, det(\tilde{D}_1 D_2 - \tilde{N}_1 N_2) \\ &= \alpha_1\, det(\tilde{D}_1 D_2 - \tilde{N}_1 N_2), \end{aligned} \quad (3.96)$$

where α_1 is a nonzero real number. Also

$$\begin{aligned} det\left(\begin{bmatrix} D_1 & -N_2 \\ -N_1 & D_2 \end{bmatrix}\right) &= det(D_2)\, det(D_1 - N_2 D_2^{-1} N_1) \\ &= det(D_2)\, det(D_1 - \tilde{D}_2^{-1}\tilde{N}_2 N_1) \\ &= det(D_2)\, det(\tilde{D}_2^{-1})\, det(\tilde{D}_2 D_1 - \tilde{N}_2 N_1) \\ &= \alpha_2\, det(\tilde{D}_2 D_1 - \tilde{N}_2 N_1), \end{aligned} \tag{3.97}$$

where α_2 is a nonzero real number.

Similarly,

$$det\left(\begin{bmatrix} \tilde{D}_1 & -\tilde{N}_1 \\ -\tilde{N}_2 & \tilde{D}_2 \end{bmatrix}\right) = \hat{\alpha}_1\, det(\tilde{D}_2 D_1 - \tilde{N}_2 N_1), \tag{3.98}$$

where $\hat{\alpha}_1 = det(\tilde{D}_1)\, det(D_1^{-1})$ is a nonzero real number, and

$$det\left(\begin{bmatrix} \tilde{D}_1 & -\tilde{N}_1 \\ -\tilde{N}_2 & \tilde{D}_2 \end{bmatrix}\right) = \hat{\alpha}_2\, det(\tilde{D}_1 D_2 - \tilde{N}_1 N_2), \tag{3.99}$$

where $\hat{\alpha}_2 = det(\tilde{D}_2)\, det(D_2^{-1})$ is a nonzero real number. These computations verify that the equivalent representations given by (1), (2), (3), and (4) have identical eigenvalues.

The following theorem presents conditions for the internal stability of the feedback system of Fig. 7.4. These conditions are useful in a variety of circumstances. Assume that the systems S_1 and S_2 are controllable and observable and that they are described by (3.84) to (3.87) with transfer function matrices given by

$$H_1 = N_1 D_1^{-1} = \tilde{D}_1^{-1}\tilde{N}_1 \tag{3.100}$$

and

$$H_2 = N_2 D_2^{-1} = \tilde{D}_2^{-1}\tilde{N}_2 \tag{3.101}$$

where the (N_i, D_i) are rc and the $(\tilde{N}_i, \tilde{D}_i)$ are lc for $i = 1, 2$. Let $\alpha_1(s)$ and $\alpha_2(s)$ be the pole (characteristic) polynomials of $H_1(s)$ and $H_2(s)$, respectively. Note that $\alpha_i(s) = k_i\, det(D_i(s)) = \tilde{k}_i\, det(\tilde{D}_i(s))$, $i = 1, 2$, for some nonzero real numbers $k_i, \tilde{k}_i$. Consider the feedback system in Fig. 7.4.

THEOREM 3.15. The following statements are equivalent:

(i) The closed-loop feedback system in Fig. 7.4 is internally stable.
(ii) The polynomial

(a) $det\left(\begin{bmatrix} D_1 & -N_2 \\ -N_1 & D_2 \end{bmatrix}\right)$, or

(b) $det\left(\begin{bmatrix} \tilde{D}_1 & -\tilde{N}_1 \\ -\tilde{N}_2 & \tilde{D}_2 \end{bmatrix}\right)$, or

(c) $det(\tilde{D}_1 D_2 - \tilde{N}_1 N_2)$, or
(d) $det(\tilde{D}_2 D_1 - \tilde{N}_2 N_1)$
is Hurwitz; i.e., its roots have strictly negative real parts.
(iii) The polynomial

$$\alpha_1(s)\alpha_2(s)\, det(I - H_1(s)H_2(s)) = \alpha_1(s)\alpha_2(s)\, det(I - H_2(s)H_1(s)) \tag{3.102}$$

is a Hurwitz polynomial.

(iv) The poles of

$$\begin{bmatrix} \hat{u}_1 \\ \hat{u}_2 \end{bmatrix} = \begin{bmatrix} I & -H_2 \\ -H_1 & I \end{bmatrix}^{-1} \begin{bmatrix} \hat{r}_1 \\ \hat{r}_2 \end{bmatrix} = \begin{bmatrix} (I - H_2H_1)^{-1} & H_2(I - H_1H_2)^{-1} \\ H_1(I - H_2H_1)^{-1} & (I - H_1H_2)^{-1} \end{bmatrix} \begin{bmatrix} \hat{r}_1 \\ \hat{r}_2 \end{bmatrix} \tag{3.103}$$

are stable, i.e., they have negative real parts.

(v) The poles of

$$\begin{bmatrix} \hat{y}_1 \\ \hat{y}_2 \end{bmatrix} = \begin{bmatrix} -H_2 & I \\ I & -H_1 \end{bmatrix}^{-1} \begin{bmatrix} 0 & H_2 \\ H_1 & 0 \end{bmatrix} \begin{bmatrix} \hat{r}_1 \\ \hat{r}_2 \end{bmatrix} = \begin{bmatrix} (I - H_1H_2)^{-1}H_1 & (I - H_1H_2)^{-1}H_1H_2 \\ (I - H_2H_1)^{-1}H_2H_1 & (I - H_2H_1)^{-1}H_2 \end{bmatrix} \begin{bmatrix} \hat{r}_1 \\ \hat{r}_2 \end{bmatrix} \tag{3.104}$$

are stable.

Proof. It was shown previously in (3.96) to (3.98) that all the determinants in (ii) are equal within some nonzero constants, i.e., they have the same roots. These roots are the eigenvalues of the closed-loop feedback system [recall that the representations (3.89) to (3.94) are all equivalent]. The feedback system is internally stable if and only if its eigenvalues have negative real parts. This shows that (ii) is true if and only if (i) is true.

To show (iii), let $D_k = \tilde{D}_1 D_2 - \tilde{N}_1 N_2$ and write $det(I - H_1H_2) = det(\tilde{D}_1^{-1}(\tilde{D}_1 D_2 - \tilde{N}_1 N_2)D_2^{-1}) = det(\tilde{D}_1^{-1})\, det(D_2^{-1})\, det(D_k)$. Since $\alpha_1(s) = k_1\, det(\tilde{D}_1(s))$ and $\alpha_2(s) = k_2\, det(D_2(s))$, where k_1, k_2 are nonzero real numbers, it follows that, $\alpha_1(s)\alpha_2(s)\, det(I - H_1(s)H_2(s)) = k_1k_2\, det(D_k)$, which is a polynomial, the roots of which are the closed-loop eigenvalues. Note also that $det(I - H_1H_2) = det(I - H_2H_1)$. Since the roots of (3.102) are exactly the closed-loop eigenvalues, (iii) is true if and only if (i) is true.

To show (iv) we write

$$\begin{bmatrix} I & -H_2 \\ -H_1 & I \end{bmatrix} = \begin{bmatrix} \tilde{D}_2 & 0 \\ 0 & \tilde{D}_1 \end{bmatrix}^{-1} \begin{bmatrix} \tilde{D}_2 & -\tilde{N}_2 \\ -\tilde{N}_1 & \tilde{D}_1 \end{bmatrix} = \begin{bmatrix} D_1 & -N_2 \\ -N_1 & D_2 \end{bmatrix} \begin{bmatrix} D_1 & 0 \\ 0 & D_2 \end{bmatrix}^{-1} \tag{3.105}$$

and we notice that these are coprime polynomial matrix factorizations (show this). The poles of $\begin{bmatrix} I & -H_2 \\ -H_1 & I \end{bmatrix}^{-1}$ are the roots of $det\left(\begin{bmatrix} D_1 & -N_2 \\ -N_1 & D_2 \end{bmatrix}\right)$ that are precisely the closed-loop eigenvalues. Note that the poles are also equal to the roots of $det\left(\begin{bmatrix} \tilde{D}_2 & -\tilde{N}_2 \\ -\tilde{N}_1 & \tilde{D}_1 \end{bmatrix}\right) = det\left(\begin{bmatrix} \tilde{D}_1 & -\tilde{N}_1 \\ -\tilde{N}_2 & \tilde{D}_2 \end{bmatrix}\right)$ [see (ii) above] since $\begin{bmatrix} \tilde{D}_2 & -\tilde{N}_2 \\ -\tilde{N}_1 & \tilde{D}_1 \end{bmatrix} = \begin{bmatrix} 0 & I \\ I & 0 \end{bmatrix} \begin{bmatrix} \tilde{D}_1 & -\tilde{N}_1 \\ -\tilde{N}_2 & \tilde{D}_2 \end{bmatrix} \begin{bmatrix} 0 & I \\ I & 0 \end{bmatrix}$. So (iv) is true if and only if (i) is true since the poles in (3.103) are exactly the eigenvalues of the closed-loop system.

To show (v), we note that $\begin{bmatrix} \hat{y}_1 \\ \hat{y}_2 \end{bmatrix} = \begin{bmatrix} 0 & I \\ I & 0 \end{bmatrix} \left[\begin{bmatrix} \hat{u}_1 \\ \hat{u}_2 \end{bmatrix} - \begin{bmatrix} \hat{r}_1 \\ \hat{r}_2 \end{bmatrix}\right]$, which in view of the proof of (iv) above implies that

$$\begin{bmatrix} \hat{y}_1 \\ \hat{y}_2 \end{bmatrix} = \begin{bmatrix} 0 & I \\ I & 0 \end{bmatrix} \left[\begin{bmatrix} I & -H_2 \\ -H_1 & I \end{bmatrix}^{-1} \left[\begin{bmatrix} I & 0 \\ 0 & I \end{bmatrix} - \begin{bmatrix} I & -H_2 \\ -H_1 & I \end{bmatrix}\right]\right] \begin{bmatrix} \hat{r}_1 \\ \hat{r}_2 \end{bmatrix}$$

$$= \begin{bmatrix} 0 & I \\ I & 0 \end{bmatrix} \begin{bmatrix} I & -H_2 \\ -H_1 & I \end{bmatrix}^{-1} \begin{bmatrix} 0 & H_2 \\ H_1 & 0 \end{bmatrix} \begin{bmatrix} \hat{r}_1 \\ \hat{r}_2 \end{bmatrix}$$

$$= \begin{bmatrix} 0 & I \\ I & 0 \end{bmatrix} \begin{bmatrix} \tilde{D}_2 & -\tilde{N}_2 \\ -\tilde{N}_1 & \tilde{D}_1 \end{bmatrix}^{-1} \begin{bmatrix} \tilde{D}_2 & 0 \\ 0 & \tilde{D}_1 \end{bmatrix} \begin{bmatrix} \tilde{D}_2 & 0 \\ 0 & \tilde{D}_2 \end{bmatrix}^{-1} \begin{bmatrix} 0 & \tilde{N}_2 \\ \tilde{N}_1 & 0 \end{bmatrix} \begin{bmatrix} \hat{r}_1 \\ \hat{r}_2 \end{bmatrix}$$

$$= \begin{bmatrix} 0 & I \\ I & 0 \end{bmatrix} \begin{bmatrix} \tilde{D}_2 & -\tilde{N}_2 \\ -\tilde{N}_1 & \tilde{D}_1 \end{bmatrix}^{-1} \begin{bmatrix} 0 & \tilde{N}_2 \\ \tilde{N}_1 & 0 \end{bmatrix} \begin{bmatrix} \hat{r}_1 \\ \hat{r}_2 \end{bmatrix}$$

$$= \begin{bmatrix} 0 & I \\ I & 0 \end{bmatrix} \left[\begin{bmatrix} 0 & I \\ I & 0 \end{bmatrix} \begin{bmatrix} \tilde{D}_1 & -\tilde{N}_1 \\ -\tilde{N}_2 & \tilde{D}_2 \end{bmatrix} \begin{bmatrix} 0 & I \\ I & 0 \end{bmatrix} \right]^{-1} \begin{bmatrix} 0 & \tilde{N}_2 \\ \tilde{N}_1 & 0 \end{bmatrix} \begin{bmatrix} \hat{r}_1 \\ \hat{r}_2 \end{bmatrix},$$

or

$$\begin{bmatrix} \hat{y}_1 \\ \hat{y}_2 \end{bmatrix} = \begin{bmatrix} \tilde{D}_1 & -\tilde{N}_1 \\ -\tilde{N}_2 & \tilde{D}_2 \end{bmatrix}^{-1} \begin{bmatrix} \tilde{N}_1 & 0 \\ 0 & \tilde{N}_2 \end{bmatrix} \begin{bmatrix} \hat{r}_1 \\ \hat{r}_2 \end{bmatrix}, \tag{3.106}$$

which is an lc factorization. This of course is the expression for the transfer function that could have been derived directly from the internal description in (3.90). Similarly, from (3.89) it can be shown that

$$\begin{bmatrix} \hat{y}_1 \\ \hat{y}_2 \end{bmatrix} = \begin{bmatrix} N_1 & 0 \\ 0 & N_2 \end{bmatrix} \begin{bmatrix} D_1 & -N_2 \\ -N_1 & D_2 \end{bmatrix}^{-1} \begin{bmatrix} \hat{r}_1 \\ \hat{r}_2 \end{bmatrix}, \tag{3.107}$$

which is an rc factorization. Clearly, the poles of the transfer function matrix in (3.104) that relates $\begin{bmatrix} y_1 \\ y_2 \end{bmatrix}$ and $\begin{bmatrix} r_1 \\ r_2 \end{bmatrix}$ are precisely the closed-loop eigenvalues, which implies that (v) is true if and only if (i) is true. ■

Remarks

1. Parts (iv) and (v) of Theorem 3.15 can also be shown to be true by using the following brief argument. It was shown that the feedback system is controllable from $\begin{bmatrix} r_1 \\ r_2 \end{bmatrix}$ and observable from $\begin{bmatrix} y_1 \\ y_2 \end{bmatrix}$. Also, it can easily be shown that the system is observable from $\begin{bmatrix} u_1 \\ u_2 \end{bmatrix}$. Therefore, the poles of the transfer function from $\begin{bmatrix} y_1 \\ y_2 \end{bmatrix}$ or $\begin{bmatrix} u_1 \\ u_2 \end{bmatrix}$ to $\begin{bmatrix} r_1 \\ r_2 \end{bmatrix}$ are precisely the eigenvalues of the closed-loop system, which must have negative real parts for internal stability.
2. It is important to consider all four entries in the transfer function (3.104) between $\begin{bmatrix} y_1 \\ y_2 \end{bmatrix}$ and $\begin{bmatrix} r_1 \\ r_2 \end{bmatrix}$ [or in (3.103) between $\begin{bmatrix} u_1 \\ u_2 \end{bmatrix}$ and $\begin{bmatrix} r_1 \\ r_2 \end{bmatrix}$] when considering internal stability. Note that the eigenvalues that are uncontrollable from r_1 or r_2 will not appear in the first or the second column of the transfer matrix, respectively. Similarly, the eigenvalues that are unobservable from y_1 or y_2 will not appear in the first or the second row of the transfer matrix, respectively. Therefore, consideration of the poles of some of the entries may lead only to erroneous results, since possible uncontrollable or unobservable modes may be omitted from consideration, and these may lead to instabilities.
3. Write

$$(I - H_1H_2)^{-1} = D_2(\tilde{D}_1D_2 - \tilde{N}_1N_2)^{-1}\tilde{D}_1 \tag{3.108}$$

and

$$(I - H_2H_1)^{-1} = D_1(\tilde{D}_2D_1 - \tilde{N}_2N_1)^{-1}\tilde{D}_2. \tag{3.109}$$

If

$$D_k \triangleq \tilde{D}_1D_2 - \tilde{N}_1N_2 \tag{3.110}$$

and

$$\tilde{D}_k \triangleq \tilde{D}_2D_1 - \tilde{N}_2N_1, \tag{3.111}$$

then $(I - H_1H_2)^{-1} = D_2D_k^{-1}\tilde{D}_1$, $(I - H_2H_1)^{-1} = D_1\tilde{D}_k^{-1}\tilde{D}_2$, and

$$\begin{bmatrix}\hat{y}_1\\ \hat{y}_2\end{bmatrix} = \begin{bmatrix}D_2D_k^{-1}\tilde{N}_1 & N_1\tilde{D}_k^{-1}\tilde{N}_2\\ N_2D_k^{-1}\tilde{N}_1 & D_1\tilde{D}_k^{-1}\tilde{N}_2\end{bmatrix}\begin{bmatrix}\hat{r}_1\\ \hat{r}_2\end{bmatrix} = \begin{bmatrix}N_1\tilde{D}_k^{-1}\tilde{D}_2 & N_1\tilde{D}_k^{-1}\tilde{N}_2\\ N_2D_k^{-1}\tilde{N}_1 & N_2D_k^{-1}\tilde{D}_1\end{bmatrix}\begin{bmatrix}\hat{r}_1\\ \hat{r}_2\end{bmatrix}. \tag{3.112}$$

Note that the first column above can be written as $\begin{bmatrix}D_2\\ N_2\end{bmatrix}D_k^{-1}\tilde{N}_1$, and possible cancellations may only occur between D_k and $\tilde{N}_1$ since the $\begin{bmatrix}D_2\\ N_2\end{bmatrix}$ are rc. In view of the representation (3.92), these are precisely the eigenvalues of the closed-loop system that are uncontrollable from r_1. Similar results can be derived by considering the second column. These results agree with the comments made in Remark (2) above. Similarly, consider the first row $N_1\tilde{D}_k^{-1}[\tilde{D}_2, \tilde{N}_2]$. Since the $[\tilde{D}_2, \tilde{N}_2]$ are lc, possible cancellations may occur only between N_1 and $\tilde{D}_k$, which in view of representation (3.94) are precisely the eigenvalues of the closed-loop system that are unobservable from y. Analogous results can be established by considering the second row $N_2D_k^{-1}[\tilde{N}_1, \tilde{D}_1]$.

Finally, we note that

$$\begin{bmatrix}\hat{u}_1\\ \hat{u}_2\end{bmatrix} = \begin{bmatrix}D_1\tilde{D}_k^{-1}\tilde{D}_2 & N_2D_k^{-1}\tilde{D}_1\\ N_1\tilde{D}_k^{-1}\tilde{D}_2 & D_2D_k^{-1}\tilde{D}_1\end{bmatrix}\begin{bmatrix}\hat{r}_1\\ \hat{r}_2\end{bmatrix} = \begin{bmatrix}D_1\tilde{D}_k^{-1}\tilde{D}_2 & D_1\tilde{D}_k^{-1}\tilde{N}_2\\ D_2D_k^{-1}\tilde{N}_1 & D_2D_k^{-1}\tilde{D}_1\end{bmatrix}\begin{bmatrix}\hat{r}_1\\ \hat{r}_2\end{bmatrix}, \tag{3.113}$$

from which similar results concerning uncontrollable and unobservable eigenvalues can easily be derived.

4. The roots of each of the polynomials in (ii) of Theorem 3.15 are equal to the roots of the polynomials in (iii) and are equal to the poles of the transfer functions in (iv) and in (v). They are exactly the eigenvalues of the closed-loop system.
5. The open-loop characteristic polynomial of the feedback system is $\alpha_1(s)\alpha_2(s)$. The closed-loop characteristic polynomial is a monic polynomial, $\alpha_{cl}(s)$, with roots the closed-loop eigenvalues, i.e., it is equal to any of the polynomials in (ii) within a multiplication by a nonzero real number. Then, relation (3.102) implies, in view of (4), that *the determinant of the return difference matrix* $[I - H_1(s)H_2(s)]$ *is the ratio of the closed-loop characteristic polynomial over the open-loop characteristic polynomial within a multiplication by a nonzero real number.*

EXAMPLE 3.10. Consider the feedback configuration of Fig. 7.4 with $H_1 = \frac{s+1}{s-1}$, $H_2 = \frac{a_1s + a_0}{s+b}$ the transfer functions of systems S_1 and S_2, respectively. Let $a_1 \neq 1$ so that the loop is well defined in terms of state-space representations (and all transfer functions are proper). (See Example 3.9.)

All polynomials in (ii) of Theorem 3.15 are equal within a multiplication by a nonzero real number, to the closed-loop characteristic polynomial given by $\alpha_{cl}(s) = s^2 + \frac{b - a_1 - a_0 - 1}{1 - a_1}s - \frac{b + a_0}{1 - a_1}$. This polynomial must be a Hurwitz polynomial for internal stability. If $\alpha_1(s) = s - 1$ and $\alpha_2(s) = s + b$ are the pole polynomials of H_1 and H_2, then the polynomial in (iii) is given by $\alpha_1(s)\alpha_2(s)(1 - H_1(s)H_2(s)) = (1 - a_1)s^2 + (b - a_1 - a_0 - 1)s - (b + a_0) = (1 - a_1)\alpha_{cl}(s)$, which implies that the return difference $1 - H_1(s)H_2(s) = (1 - a_1)\frac{\alpha_{cl}(s)}{\alpha_1(s)\alpha_2(a)}$. Note that $(1 - H_1H_2)^{-1} = (1 - H_2H_1)^{-1} = \frac{(s-1)(s+b)}{\alpha(s)}$ with $\alpha(s) = (1 - \alpha_1)\alpha_{cl}(s)$, and the transfer function matrix in (iv) of

Theorem 3.15 is given by

$$\begin{bmatrix}\hat{u}_1\\ \hat{u}_2\end{bmatrix} = \begin{bmatrix}\dfrac{(s-1)(s+b)}{\alpha(s)} & \dfrac{(s-1)(a_1 s + a_0)}{\alpha(s)}\\ \dfrac{(s+1)(s+b)}{\alpha(s)} & \dfrac{(s-1)(s+b)}{\alpha(s)}\end{bmatrix}\begin{bmatrix}\hat{r}_1\\ \hat{r}_2\end{bmatrix}.$$

The polynomial $\alpha(s)$ has a factor $s+1$ when $b=1$. Notice that $\alpha(-1) = 2 - 2b = 0$ when $b = 1$. If this is the case ($b = 1$), then

$$\begin{bmatrix}\hat{u}_1\\ \hat{u}_2\end{bmatrix} = \begin{bmatrix}\dfrac{s-1}{\bar{\alpha}(s)} & \dfrac{(s-1)(a_1 s + a_0)}{\alpha(s)}\\ \dfrac{s+1}{\bar{\alpha}(s)} & \dfrac{s-1}{\bar{\alpha}(s)}\end{bmatrix}\begin{bmatrix}\hat{r}_1\\ \hat{r}_2\end{bmatrix},$$

where $\alpha(s) = (s+1)\bar{\alpha}(s)$. Notice that three out of four transfer functions do not contain the pole at -1 in $\bar{\alpha}(s)$. Recall that when $b = 1$, -1 is an eigenvalue that is uncontrollable from r_1, and it cancels in certain transfer functions, as expected (see Example 3.9). Similar results can be derived when $a_0/a_1 = -1$. This illustrates the necessity for considering all the transfer functions between u_1, u_2 and r_1, r_2 when studying internal stability of the feedback system. Similar results can be derived when considering the transfer functions between y_1, y_2 and r_1, r_2 in (v). ■

COROLLARY 3.16. The closed-loop feedback system in Fig. 7.4 is internally stable if and only if

(i) $D_1^{-1}[(I - H_2H_1)^{-1}, H_2(I - H_1H_2)^{-1}]$ is stable, or
(ii) $D_2^{-1}[H_1(I - H_2H_1)^{-1}, (I - H_1H_2)^{-1}]$ is stable, or
(iii) $\begin{bmatrix}(I - H_2H_1)^{-1}\\ H_1(I - H_2H_1)^{-1}\end{bmatrix}\tilde{D}_2^{-1}$ is stable, or
(iv) $\begin{bmatrix}H_2(I - H_1H_2)^{-1}\\ (I - H_1H_2)^{-1}\end{bmatrix}\tilde{D}_1^{-1}$ is stable.

Proof. The above expressions originate from rows and columns of the transfer function matrix between $\begin{bmatrix}u_1\\ u_2\end{bmatrix}$ and $\begin{bmatrix}r_1\\ r_2\end{bmatrix}$. In view of (3.113), expression (i) is equal to $D_1^{-1}D_1\tilde{D}_k^{-1}[\tilde{D}_2, \tilde{N}_2] = \tilde{D}_k^{-1}[\tilde{D}_2, \tilde{N}_2]$, the poles of which are precisely the roots of $det\,\tilde{D}_k$, the closed-loop eigenvalues, since the pair $(\tilde{D}_2, \tilde{N}_2)$ is lc. Similar arguments hold for (ii), (iii) and (iv). ■

It is of interest to point out that expression (i) of the corollary is precisely the transfer function between the partial state z_1 and the input $\begin{bmatrix}r_1\\ r_2\end{bmatrix}$, in view of (3.103) and $D_1 z_1 = u_1$. Expression (ii) corresponds to the transfer function between z_2 and $\begin{bmatrix}r_1\\ r_2\end{bmatrix}$. Notice that $\begin{bmatrix}\hat{\tilde{z}}_2\\ \hat{\tilde{z}}_1\end{bmatrix} = \begin{bmatrix}\hat{y}_2\\ \hat{y}_1\end{bmatrix} = \begin{bmatrix}(I - H_2H_1)^{-1}\\ (I - H_1H_2)^{-1}H_1\end{bmatrix}H_2\hat{r}_2 + \begin{bmatrix}(I - H_2H_1)^{-1}H_2\\ (I - H_1H_2)^{-1}\end{bmatrix}H_1\hat{r}_1$ from (3.104). Since $H_2 = \tilde{D}_2^{-1}\tilde{N}_2$ is an lc factorization, the transfer function from $\begin{bmatrix}\tilde{z}_2\\ \tilde{z}_1\end{bmatrix}$ to r_2 is stable if and only if expression (iii) is stable. A similar argument holds for (iv).

The following result can be proved in a completely analogous manner to the proof of the above corollary.

COROLLARY 3.17. The closed-loop feedback system in Fig. 7.4 is internally stable if and only if

(i) $D_1^{-1}(I - H_2H_1)^{-1}\tilde{D}_2^{-1}$ is stable, or
(ii) $D_2^{-1}(I - H_1H_2)^{-1}\tilde{D}_1^{-1}$ is stable. ■

Note that expression (i) in the corollary is stable if and only if the transfer function between z_1 and $\begin{bmatrix} u_1 \\ u_2 \end{bmatrix}$ is stable. To see this, note that $\hat{u}_1 = (I - H_2H_1)^{-1}\hat{r}_1$, from which $\hat{z}_1 = D_1^{-1}\hat{u}_1 = [D_1^{-1}(I - H_2H_1)^{-1}\tilde{D}_2^{-1}]\tilde{D}_2\hat{r}_1$. Now $\tilde{D}_2 r_1 = \tilde{D}_2 u_1 - \tilde{D}_2 y_2 = \tilde{D}_2 u_1 - \tilde{N}_2 u_2 = [\tilde{D}_2, -\tilde{N}_2]\begin{bmatrix} u_1 \\ u_2 \end{bmatrix}$, i.e., $\hat{z}_1 = [D_1^{-1}(I - H_2H_1)^{-1}\tilde{D}_2^{-1}][\tilde{D}_2, -\tilde{N}_2]\begin{bmatrix} u_1 \\ u_2 \end{bmatrix}$. Since the pair $(\tilde{D}_2, \tilde{N}_2)$ is lc, this transfer function is stable if and only if expression (i) is stable. Similarly, $\hat{u}_2 = (I - H_1H_2)^{-1}\hat{r}_2$, from which $\hat{z}_2 = [D_2^{-1}(I - H_1H_2)^{-1}\tilde{D}_1^{-1}]\tilde{D}_1\hat{r}_2 = [D_2^{-1}(I - H_1H_2)^{-1}\tilde{D}_1^{-1}][-\tilde{N}_1, \tilde{D}_1]\begin{bmatrix} u_1 \\ u_2 \end{bmatrix}$, which is stable if and only if expression (ii) is stable, since $(-\tilde{N}_1, \tilde{D}_1)$ is an lc pair of polynomial matrices.

EXAMPLE 3.11. Consider again the systems described by $H_1(s) = \dfrac{s+1}{s-1}$ and $H_2(s) = \dfrac{a_1 s + a_0}{s+b}$, as in Example 3.10. The expressions in Corollary 3.16 are given by

$$\text{(i)}\quad \frac{1}{s-1}\left[\frac{(s-1)(s+b)}{\alpha(s)}, \quad \frac{(s-1)(a_1 s + a_0)}{\alpha(s)}\right] = \frac{1}{\alpha(s)}[s+b, a_1 s + a_0],$$

$$\text{(ii)}\quad \frac{1}{s+b}\left[\frac{(s+1)(s+b)}{\alpha(s)}, \quad \frac{(s-1)(s+b)}{\alpha(s)}\right] = \frac{1}{\alpha(s)}[s+1, s-1],$$

$$\text{(iii)}\quad \begin{bmatrix} \dfrac{(s-1)(s+b)}{\alpha(s)} \\[2mm] \dfrac{(s+1)(s+b)}{\alpha(s)} \end{bmatrix} \frac{1}{s+b} = \begin{bmatrix} s-1 \\ s+1 \end{bmatrix} \frac{1}{\alpha(s)},$$

$$\text{(iv)}\quad \begin{bmatrix} \dfrac{(s-1)(a_1 s + a_0)}{\alpha(s)} \\[2mm] \dfrac{(s-1)(s+b)}{\alpha(s)} \end{bmatrix} \frac{1}{s-1} = \begin{bmatrix} a_1 s + a_0 \\ s+b \end{bmatrix} \frac{1}{\alpha(s)}.$$

Clearly, the monic pole polynomial of each of the transfer functions in (i) to (iv) is $\dfrac{\alpha(s)}{1-a_1} = s^2 + \dfrac{(b - a_1 - a_0 - 1)}{1-a_1}s - \dfrac{(b + a_0)}{1-a_1} = \alpha_{cl}(s)$, the closed-loop characteristic polynomial. This is so even when $b = 1$ or $a_0/a_1 = -1$ (see Example 3.9), in which case there are uncontrollable/unobservable eigenvalues. Notice that $s + b$ and $a_1 s + a_0$ are coprime by the definition of $H_2(s)$.

Similarly, the expressions in Corollary 3.17 are given by

$$\text{(i)}\quad \frac{1}{s-1}\,\frac{(s-1)(s+b)}{\alpha(s)}\,\frac{1}{s+b} = \frac{1}{\alpha(s)}$$

$$\text{(ii)}\quad \frac{1}{s+b}\,\frac{(s-1)(s+b)}{\alpha(s)}\,\frac{1}{s-1} = \frac{1}{\alpha(s)},$$

as expected. ■

PART 2
SYNTHESIS OF CONTROL SYSTEMS

7.4
FEEDBACK CONTROL SYSTEMS

In this section, feedback systems are studied further with an emphasis on stabilizing feedback controllers. In particular, it is shown how all the stabilizing feedback controllers can be conveniently parameterized. These parameterizations are very important in control since they are fundamental in methodologies such as the optimal H^∞-approach to control design. Our development of the subject at hand builds upon the controllability, observability, and particularly, the internal stability results introduced in Section 7.3, as well as on the Diophantine Equation results of Subsection 7.2E. First, in Subsection 7.4A all stabilizing feedback controllers are parameterized, using PMDs. A number of different parameterizations are introduced and discussed at length. State feedback and state estimation, using PMDs, are introduced in Subsection 7.4B and are then used in Subsection 7.4C, where all stabilizing feedback controllers are parameterized, using proper and stable MFDs. Two degrees of freedom feedback controllers offer additional capabilities in control design and are discussed in Subsection 7.4D. Control problems are also described in this subsection.

A. Stabilizing Feedback Controllers

Now consider systems S_1 and S_2 connected in the feedback configuration shown in Fig. 7.4, or equivalently, in Fig. 7.5. Given S_1, it is shown in this section how to parameterize all systems S_2 so that the closed-loop feedback system is internally stable. Thus, if S_1, called the *plant,* is a given system to be controlled, then S_2 is viewed as the *feedback controller* that is to be designed. Here we provide the parameterizations of all stabilizing feedback controllers.

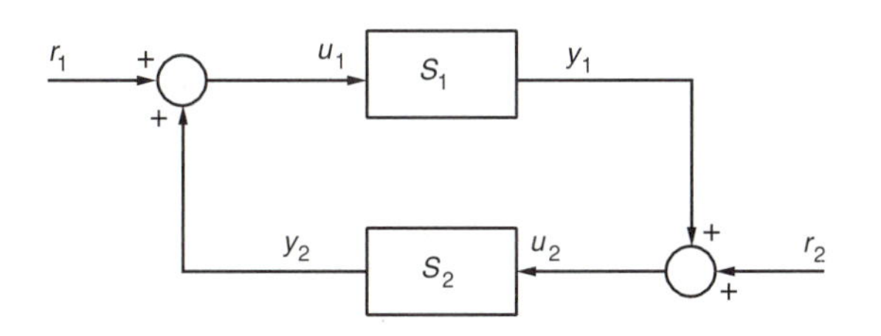

FIGURE 7.5A
Feedback configuration

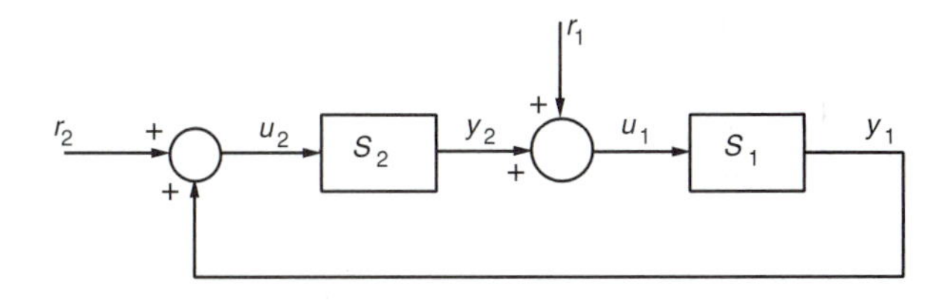

FIGURE 7.5B
Feedback configuration

Several parameters will be used to parameterize all stabilizing systems S_2. The results are based on PMDs of the systems S_1 and S_2 and of the closed-loop system, in particular, PMFDs. Parameterizations are introduced, using first the polynomial matrix parameters (1) D_k, N_k and $\tilde{D}_k$, $\tilde{N}_k$ and then the stable rational parameter (2) $K = N_k D_k^{-1} = \tilde{D}_k^{-1}\tilde{N}_k$. These parameters are very convenient in characterizing stability, but cumbersome when properness of the transfer function H_2 of S_2 is to be guaranteed. The rational proper and stable parameters (3) Q_1 and Q_2 are introduced next. These are very convenient when properness of H_2 is to be guaranteed, but cumbersome when characterizing stability, except in special cases, e.g., when H_1 is stable. The stable parameters (4) S_{12} and S_{21} that are the comparison sensitivity matrices of the feedback loop will also be used. These can be employed to conveniently parameterize all stabilizing H_2 controllers only in special cases, for example, when H_1^{-1} exists. Parameters (5) X_2 and $\tilde{X}_2$, which are closely related to Q_2, will also be introduced and discussed. The results based on X_2 in the case of SISO systems reduce to well-known classical control results. Finally, rational parameters (6) L_1, L_2 are introduced to help in providing alternative proofs for the results that involve the previously introduced parameters. These results provide a link with parameterizations of all stabilizing controllers using proper and stable factorizations. Such parameterizations are addressed in Subsection 7.4C.

The distinguishing characteristics of our approach of the parameterizations of all stabilizing feedback systems S_2 (of all stabilizing controllers) is that they are based on internal representations, which were developed at length earlier in this book. This approach builds upon previous results and provides significant insight that allows the introduction of several parameters that are useful in different circumstances. This approach also makes it possible to easily select the exact locations and number of the desired stable closed-loop eigenvalues. Utilizing the development of this section, a parameterization that uses proper and stable MFDs and involves a proper and stable parameter K' is introduced in Subsection 7.4C. The parameter K' is closely related to the stable rational parameter K used in the approach enumerated above. This type of parameterization is useful in certain control design methods such as optimal H^∞-control design.

In the following, the term "stable system S" is taken to mean that the eigenvalues of the internal description of system S have strictly negative real parts (in the continuous-time case), i.e., the system S is internally stable. Note that when the transfer functions in (3.103) and (3.104) of the feedback system S are proper, internal stability of S implies BIBO stability of the feedback system, since the poles of the various transfer functions are a subset of the closed-loop eigenvalues (see Section 7.3 and Chapter 6).

Feedback systems

The feedback system of Fig. 7.5 was studied at length in Subsection 7.3C. Recall that if

$$\begin{aligned} P_i(q)z_i(t) &= Q_i(q)u_i(t) \\ y_i(t) &= R_i(q)z_i(t), \qquad i = 1, 2, \end{aligned} \tag{4.1}$$

are polynomial matrix descriptions of S_1 and S_2, then the closed-loop system is given by (3.76), i.e.,

$$\begin{bmatrix} P_1 & -Q_1R_2 \\ -Q_2R_1 & P_2 \end{bmatrix}\begin{bmatrix} z_1 \\ z_2 \end{bmatrix} = \begin{bmatrix} Q_1 & 0 \\ 0 & Q_2 \end{bmatrix}\begin{bmatrix} r_1 \\ r_2 \end{bmatrix}$$

$$\begin{bmatrix} y_1 \\ y_2 \end{bmatrix} = \begin{bmatrix} R_1 & 0 \\ 0 & R_2 \end{bmatrix}\begin{bmatrix} z_1 \\ z_2 \end{bmatrix}, \quad (4.2)$$

where the relations

$$u_1(t) = y_2(t) + r_1(t), \qquad u_2(t) = y_1(t) + r_2(t) \quad (4.3)$$

have been used. Based on this description, it was shown that for the closed-loop eigenvalues to be stable (to have strictly negative real parts), it is necessary for S_1 and S_2 to be stabilizable and detectable. This is so because the uncontrollable and unobservable eigenvalues of both S_1 and S_2 will also be uncontrollable and unobservable eigenvalues of the closed-loop system.

Now assume that S_1 and S_2 are controllable and observable, and let system S_1 be described by

(1a) $$D_1(q)z_1(t) = u_1(t),\ y_1(t) = N_1(q)z_1(t) \quad (4.4)$$

or

(1b) $$\tilde{D}_1(q)\tilde{z}_1(t) = \tilde{N}_1(q)u_1(t),\ y_1(t) = \tilde{z}_1(t), \quad (4.5)$$

where the pair $(D_1(q), N_1(q))$ is rc and the pair $(\tilde{D}_1(q), \tilde{N}_1(q))$ is lc. Let $H_1(s) = N_1(s)D_1^{-1}(s) = \tilde{D}_1^{-1}(s)\tilde{N}_1(s)$ be the transfer function matrix of S_1. Next, let system S_2 be described by

(2a) $$D_2(q)z_2(t) = u_2(t),\ y_2(t) = N_2(q)z_2(t) \quad (4.6)$$

or

(2b) $$\tilde{D}_2(q)\tilde{z}_2(t) = \tilde{N}_2(q)u_2(t),\ y_2(t) = \tilde{z}_2(t), \quad (4.7)$$

where the pair $(D_2(q), N_2(q))$ is rc and the pair $(\tilde{D}_2(q), \tilde{N}_2(q))$ is lc. Let $H_2(s) = N_2(s)D_2^{-1}(s) = \tilde{D}_2^{-1}(s)\tilde{N}_2(s)$ be the transfer function matrix of S_2. Recall from Subsection 7.3C [see (3.89) to (3.94)] that the closed-loop system descriptions are in this case as follows:

1. Using descriptions (1a) and (2a), we have

$$\begin{bmatrix} D_1 & -N_2 \\ -N_1 & D_2 \end{bmatrix}\begin{bmatrix} z_1 \\ z_2 \end{bmatrix} = \begin{bmatrix} I & 0 \\ 0 & I \end{bmatrix}\begin{bmatrix} r_1 \\ r_2 \end{bmatrix}$$

$$\begin{bmatrix} y_1 \\ y_2 \end{bmatrix} = \begin{bmatrix} N_1 & 0 \\ 0 & N_2 \end{bmatrix}\begin{bmatrix} z_1 \\ z_2 \end{bmatrix}. \quad (4.8)$$

2. Using descriptions (1b) and (2b), we obtain

$$\begin{bmatrix} \tilde{D}_1 & -\tilde{N}_1 \\ -\tilde{N}_2 & \tilde{D}_2 \end{bmatrix}\begin{bmatrix} \tilde{z}_1 \\ \tilde{z}_2 \end{bmatrix} = \begin{bmatrix} \tilde{N}_1 & 0 \\ 0 & \tilde{N}_2 \end{bmatrix}\begin{bmatrix} r_1 \\ r_2 \end{bmatrix}$$

$$\begin{bmatrix} y_1 \\ y_2 \end{bmatrix} = \begin{bmatrix} I & 0 \\ 0 & I \end{bmatrix}\begin{bmatrix} \tilde{z}_1 \\ \tilde{z}_2 \end{bmatrix}. \quad (4.9)$$

3. Using descriptions (1b) and (2a), we have

$$(\tilde{D}_1 D_2 - \tilde{N}_1 N_2) z_2 = [\tilde{N}_1, \tilde{D}_1] \begin{bmatrix} r_1 \\ r_2 \end{bmatrix}$$

$$\begin{bmatrix} y_1 \\ y_2 \end{bmatrix} = \begin{bmatrix} D_2 \\ N_2 \end{bmatrix} z_2 + \begin{bmatrix} 0 & -I \\ 0 & 0 \end{bmatrix} \begin{bmatrix} r_1 \\ r_2 \end{bmatrix}. \tag{4.10}$$

4. Using descriptions (1a) and (2b), we obtain

$$(\tilde{D}_2 D_1 - \tilde{N}_2 N_1) z_1 = [\tilde{D}_2, \tilde{N}_2] \begin{bmatrix} r_1 \\ r_2 \end{bmatrix}$$

$$\begin{bmatrix} y_1 \\ y_2 \end{bmatrix} = \begin{bmatrix} N_1 \\ D_1 \end{bmatrix} z_1 + \begin{bmatrix} 0 & 0 \\ -I & 0 \end{bmatrix} \begin{bmatrix} r_1 \\ r_2 \end{bmatrix}. \tag{4.11}$$

These descriptions of the closed-loop system are all equivalent and the polynomials

$$det\left(\begin{bmatrix} D_1 & -N_2 \\ -N_1 & D_2 \end{bmatrix}\right), \qquad det\left(\begin{bmatrix} \tilde{D}_1 & -\tilde{N}_1 \\ -\tilde{N}_2 & \tilde{D}_2 \end{bmatrix}\right),$$

$$det\,(\tilde{D}_1 D_2 - \tilde{N}_1 N_2), \qquad \text{and} \qquad det\,(\tilde{D}_2 D_1 - \tilde{N}_2 N_1)$$

are all equal within multiplication of nonzero scalars. Their roots are the closed-loop eigenvalues (see Theorem 3.15).

Parameterizations of all stabilizing systems S_2

The six parameterizations of all stabilizing systems S_2 previously mentioned are now introduced and discussed at length.

1. *Parameters D_k, N_k and $\tilde{D}_k, \tilde{N}_k$.* Consider now the closed-loop description given in (4) above and let

$$\tilde{D}_2 D_1 - \tilde{N}_2 N_1 = \tilde{D}_k, \tag{4.12}$$

where $\tilde{D}_k$ is some desired polynomial matrix so that the roots of $det\,\tilde{D}_k$ are stable (have negative real parts). If D_1, N_1, $\tilde{D}_k$ are assumed to be given and $\tilde{D}_2$, $-\tilde{N}_2$ that satisfy (4.12) are to be determined, then this equation is seen to be a polynomial matrix linear Diophantine equation. Such equations were studied in Subsection 7.2E. In view of Theorem 2.15 of Chapter 7, Eq. (4.12) has a solution for any $\tilde{D}_k$ (of appropriate dimension) since the pair (D_1, N_1) is rc. All solutions of the Diophantine Equation (4.12) can be parameterized as follows [see (2.54)]:

$$\tilde{D}_2 = \tilde{D}_k X_1 - \tilde{N}_k \tilde{N}_1 \tag{4.13a}$$

$$-\tilde{N}_2 = \tilde{D}_k Y_1 + \tilde{N}_k \tilde{D}_1 \tag{4.13b}$$

or

$$[\tilde{D}_2, -\tilde{N}_2] = [\tilde{D}_k, \tilde{N}_k] \begin{bmatrix} X_1 & Y_1 \\ -\tilde{N}_1 & \tilde{D}_1 \end{bmatrix}, \tag{4.14}$$

where X_1, Y_1 satisfy $X_1 D_1 + Y_1 N_1 = I$, the pair $(\tilde{N}_1, \tilde{D}_1)$ is lc and is such that $-\tilde{N}_1 D_1 + \tilde{D}_1 N_1 = 0$, and $\tilde{N}_k$ is an arbitrary polynomial matrix of appropriate dimensions. Note that $[\tilde{D}_k X_1, \tilde{D}_k Y_1]$ is a particular solution of $[\tilde{D}_2, -\tilde{N}_2] \begin{bmatrix} D_1 \\ N_1 \end{bmatrix} = \tilde{D}_k$.

For the system $\tilde{D}_2\tilde{z}_2 = \tilde{N}_2 u_2$, $y_2 = \tilde{z}_2$ to be well defined, we must have $det\, \tilde{D}_2 = det\,(\tilde{D}_k X_1 - \tilde{N}_k \tilde{N}_1) \neq 0$, i.e., $\tilde{N}_k$ (and $\tilde{D}_k$) must be such that this condition is satisfied.

Similar results can be derived using the closed-loop description given in (3). Let

$$\tilde{D}_1 D_2 - \tilde{N}_1 N_2 = D_k, \tag{4.15}$$

where D_k is some desired polynomial matrix so that the roots of $det\, D_k$ are stable. In view of Theorem 2.15, all solutions of the Diophantine Equation (4.15) are given by

$$D_2 = \tilde{X}_1 D_k - N_1 N_k \tag{4.16a}$$

$$-N_2 = \tilde{Y}_1 D_k + D_1 N_k \tag{4.16b}$$

or

$$\begin{bmatrix} N_2 \\ D_2 \end{bmatrix} = \begin{bmatrix} D_1 & -\tilde{Y}_1 \\ N_1 & \tilde{X}_1 \end{bmatrix} \begin{bmatrix} -N_k \\ D_k \end{bmatrix}, \tag{4.17}$$

where $\tilde{X}_1$, $\tilde{Y}_1$ satisfy $\tilde{D}_1\tilde{X}_1 + \tilde{N}_1\tilde{Y}_1 = I$, the pair (N_1, D_1) is rc and is such that $-\tilde{D}_1 N_1 + \tilde{N}_1 D_1 = 0$, and N_k is an arbitrary polynomial matrix of appropriate dimensions. Note that $\begin{bmatrix} \tilde{X}_1 D_k \\ \tilde{Y}_1 D_k \end{bmatrix}$ is a particular solution of $[\tilde{D}_1, \tilde{N}_1]\begin{bmatrix} D_2 \\ -N_2 \end{bmatrix} = D_k$. For the system $D_2 z_2 = u_2$, $y_2 = N_2 z_2$ to be well defined, we must have $det\, D_2 = det\,(\tilde{X}_1 D_k - N_1 N_k) \neq 0$, i.e., N_k (and D_k) must be such that this condition is satisfied.

The preceding two sets of parameterizations for S_2 are of course related. A convenient way of presenting the above results in a unified way is to use the *doubly coprime factorizations* of H_1. To this end, assume that the descriptions (4.4) and (4.5) for system S_1 satisfy

$$UU^{-1} = \begin{bmatrix} X_1 & Y_1 \\ -\tilde{N}_1 & \tilde{D}_1 \end{bmatrix} \begin{bmatrix} D_1 & -\tilde{Y}_1 \\ N_1 & \tilde{X}_1 \end{bmatrix} = \begin{bmatrix} I & 0 \\ 0 & I \end{bmatrix}, \tag{4.18}$$

where U is a unimodular matrix (i.e., $det\, U$ is a nonzero real number) and the $X_1, Y_1, \tilde{X}_1, \tilde{Y}_1$ are appropriate matrices. Factorizations of $H_1 = N_1 D_1^{-1} = \tilde{D}_1^{-1}\tilde{N}_1$ that satisfy (4.18) are called *doubly coprime factorizations of* H_1. Such factorizations always exist (see, e.g., [21], [1]). In particular, in Theorem 2 of [1] it is shown that if the relations $\bar{X}_1 D_1 + \bar{Y}_1 N_1 = I$, $\tilde{D}_1\tilde{X}_1 + \tilde{N}_1\tilde{Y}_1 = I$ and $-\tilde{N}_1 D_1 + \tilde{D}_1 N_1 = 0$ are satisfied, then (4.18) is satisfied for D_1, N_1, $\tilde{D}_1$, $\tilde{N}_1$, $\tilde{X}_1$, $\tilde{Y}_1$ and $[X_1, Y_1] = [\bar{X}_1, \bar{Y}_1] + S[-\tilde{N}_1, \tilde{D}_1]$, where $S \triangleq \bar{X}_1\tilde{Y}_1 - \bar{Y}_1\tilde{X}_1$, i.e., any lc and rc factorizations of H_1 with associated matrices can be adjusted so that they are doubly coprime.

If D_1, N_1 and $\tilde{D}_1, \tilde{N}_1$ are doubly coprime factorizations and satisfy (4.18), then the solutions of the Diophantine Equations in (4.14) and (4.17) can be written as

$$[\tilde{D}_2, -\tilde{N}_2] = [\tilde{D}_k, \tilde{N}_k]U \tag{4.19}$$

and

$$\begin{bmatrix} N_2 \\ D_2 \end{bmatrix} = U^{-1}\begin{bmatrix} -N_k \\ D_k \end{bmatrix}, \tag{4.20}$$

from which it follows that

$$[\tilde{D}_k, \tilde{N}_k] = [\tilde{D}_2, -\tilde{N}_2]U^{-1} \tag{4.21}$$

and

$$\begin{bmatrix} -N_k \\ D_k \end{bmatrix} = U\begin{bmatrix} N_2 \\ D_2 \end{bmatrix}. \tag{4.22}$$

The relation between D_k, N_k and $\tilde{D}_k, \tilde{N}_k$ is now clear, namely,

$$-\tilde{D}_k N_k + \tilde{N}_k D_k = 0. \tag{4.23}$$

THEOREM 4.1. Assume that the system S_1 is controllable and observable and is described by the PMD (or PMFD) as (i) $D_1 z_1 = u_1, y_1 = N_1 z_1$ given in (4.4), or by (ii) $\tilde{D}_1 \tilde{z}_1 = \tilde{N}_1 u_1, y_1 = \tilde{z}_1$ given in (4.5). Let the pair (D_1, N_1) and the pair $(\tilde{D}_1, \tilde{N}_1)$ be doubly coprime factorizations of the transfer function matrix $H_1(s) = N_1 D_1^{-1} = \tilde{D}_1^{-1}\tilde{N}_1$ [i.e., (4.18) is satisfied]. Then all the controllable and observable systems S_2 with the property that the closed-loop feedback system eigenvalues are stable (i.e., have strictly negative real parts) are described by

$$\text{(i)} \qquad \tilde{D}_2 \tilde{z}_2 = \tilde{N}_2 u_2, \qquad y_2 = \tilde{z}_2, \tag{4.24}$$

where $\tilde{D}_2 = \tilde{D}_k X_1 - \tilde{N}_k \tilde{N}_1$ and $\tilde{N}_2 = -(\tilde{D}_k Y_1 + \tilde{N}_k \tilde{D}_1)$ with $X_1, Y_1, \tilde{N}_1, \tilde{D}_1$ given in (4.18) [see also (4.19)] and the parameters $\tilde{D}_k$ and $\tilde{N}_k$ are selected arbitrarily under the conditions that $\tilde{D}_k^{-1}$ exists and is stable, and the pair $(\tilde{D}_k, \tilde{N}_k)$ is lc and is such that $det(\tilde{D}_k X_1 - \tilde{N}_k \tilde{N}_1) \neq 0$.

Equivalently, all stabilizing S_2 can be described by

$$\text{(ii)} \qquad D_2 z_2 = u_2, \qquad y_2 = N_2 z_2, \tag{4.25}$$

where $D_2 = \tilde{X}_1 D_k - N_1 N_k$ and $N_2 = -(\tilde{Y}_1 D_k + D_1 N_k)$ with $\tilde{X}_1, \tilde{Y}_1, N_1, D_1$ given in (4.18) [see also (4.20)] and the parameters D_k and N_k are selected arbitrarily under the conditions that D_k^{-1} exists and is stable, and the pair (D_k, N_k) is rc and is such that $det(\tilde{X}_1 D_k - N_1 N_k) \neq 0$.

Furthermore, the closed-loop eigenvalues are precisely the roots of $det\,\tilde{D}_k$ or of $det\,D_k$. In addition, the transfer function matrix of S_2 is given by

$$\begin{aligned} H_2 &= -(\tilde{D}_k X_1 - \tilde{N}_k \tilde{N}_1)^{-1}(\tilde{D}_k Y_1 + \tilde{N}_k \tilde{D}_1) \\ &= -(\tilde{Y}_1 D_k + D_1 N_k)(\tilde{X}_1 D_k - N_1 N_k)^{-1}. \end{aligned} \tag{4.26}$$

Proof. The closed-loop description in case (i) is given by (4.11) in method (4) above and in case (ii) it is given by (4.10) in method (3). As was shown in Subsection 7.2E, (4.19) and (4.20) are parameterizations of all solutions of the Diophantine Equations $\tilde{D}_2 D_1 - \tilde{N}_2 N_1 = \tilde{D}_k$ [in (4.10)] and $\tilde{D}_1 D_2 - \tilde{N}_1 N_2 = D_k$ [in (4.11)], respectively. The fact that $\tilde{D}_k^{-1}$ (or D_k^{-1}) exists and is stable guarantees that the closed loop is well defined and all of its eigenvalues, which are the poles of $\tilde{D}_k^{-1}$ (or of D_k^{-1}), will be stable. The condition $det(\tilde{D}_k X_1 - \tilde{N}_k \tilde{N}_1) \neq 0$ [or $det(\tilde{X}_1 D_k - N_1 N_k) \neq 0$] guarantees that $det\,\tilde{D}_2 \neq 0$ [or $det\,D_2 \neq 0$], and therefore, the PMD for S_2 in (4.10) is well defined. Finally, note that the pair $(\tilde{D}_k, \tilde{N}_k)$ is lc if and only if the pair $(\tilde{D}_2, \tilde{N}_2)$ is lc as can be seen from $[\tilde{D}_2, -\tilde{N}_2] = [\tilde{D}_k, \tilde{N}_k]U$ given in (4.19), where U is unimodular. This then implies that the description $\{\tilde{D}_2, \tilde{N}_2, I\}$ for S_2 is both controllable and observable. Similarly, the pair (D_k, N_k) is rc guarantees that $\{D_2, I, N_2\}$ with D_2 and N_2 given in (4.20) is also a controllable and observable description for S_2. ■

2. *Parameter K.* In place of the polynomial matrix parameters $\tilde{D}_k$, $\tilde{N}_k$ or D_k, N_k, it is possible to use a single parameter, a stable rational matrix K. This is shown next in Theorem 4.2.

THEOREM 4.2. Assume that the system S_1 is controllable and observable and is described by its transfer function matrix

$$H_1 = N_1 D_1^{-1} = \tilde{D}_1^{-1}\tilde{N}_1, \tag{4.27}$$

where the pairs (N_1, D_1), $(\tilde{D}_1, \tilde{N}_1)$ are doubly coprime factorizations satisfying (4.18). Then all the controllable and observable systems S_2 with the property that the closed-loop feedback system eigenvalues are stable (i.e., they have strictly negative real parts) are described by the transfer function matrix

$$\begin{aligned} H_2 &= -(X_1 - K\tilde{N}_1)^{-1}(Y_1 + K\tilde{D}_1) \\ &= -(\tilde{Y}_1 + D_1 K)(\tilde{X}_1 - N_1 K)^{-1}, \end{aligned} \tag{4.28}$$

where the parameter K is an arbitrary rational matrix that is stable and is such that $det\,(X_1 - K\tilde{N}_1) \neq 0$ or $det\,(\tilde{X}_1 - N_1 K) \neq 0$. Furthermore, the poles of K are precisely the closed-loop eigenvalues.

Proof. This is in fact a corollary to Theorem 4.1. It is called a theorem here since it was historically one of the first results established in this area. The parameter K is called the *Youla parameter* (see Section 7.6).

In Theorem 4.1, descriptions for H_2 were given in (4.26) in terms of the parameters $\tilde{D}_k$, $\tilde{N}_k$ and D_k, N_k. Now, in view of $-\tilde{D}_k N_k + \tilde{N}_k D_k = 0$, given in (4.23), we have

$$\tilde{D}_k^{-1}\tilde{N}_k = N_k D_k^{-1} = K, \tag{4.29}$$

a stable rational matrix. Since the pair $(\tilde{D}_k, \tilde{N}_k)$ is lc and the pair (N_k, D_k) is rc, they are coprime factorizations for K. Therefore, H_2 in (4.28) can be written as the H_2 of (4.26) given in the previous theorem, from which the controllable and observable internal descriptions for S_2 in (4.24) and (4.25) can immediately be derived. Conversely, (4.28) can immediately be derived from (4.26), using (4.29). Note that the poles of K are the roots of $det\,\tilde{D}_k$ or $det\,D_k$ that are the closed-loop eigenvalues. ■

EXAMPLE 4.1. Consider $H_1 = \dfrac{s+1}{s-1}$. Here $N_1 = \tilde{N}_1 = s+1$ and $D_1 = \tilde{D}_1 = s-1$. These are doubly coprime factorizations (a trivial case) since (4.18) is satisfied. We have

$$\begin{aligned} UU^{-1} &= \begin{bmatrix} X_1 & Y_1 \\ -\tilde{N}_1 & \tilde{D}_1 \end{bmatrix}\begin{bmatrix} D_1 & -\tilde{Y}_1 \\ N_1 & \tilde{X}_1 \end{bmatrix} \\ &= \begin{bmatrix} s+\frac{1}{2} & -s+\frac{3}{2} \\ -(s+1) & s-1 \end{bmatrix}\begin{bmatrix} s-1 & -(-s+\frac{3}{2}) \\ s+1 & s+\frac{1}{2} \end{bmatrix} = \begin{bmatrix} 1 & 0 \\ 0 & 1 \end{bmatrix}. \end{aligned}$$

In view of (4.26) and (4.28), all stabilizing controllers H_2 are then given by

$$H_2 = -\frac{(-s+\frac{3}{2})d_k + (s-1)n_k}{(s+\frac{1}{2})d_k - (s+1)n_k} = -\frac{(-s+\frac{3}{2}) + (s-1)K}{(s+\frac{1}{2}) - (s+1)K},$$

where $K = n_k/d_k$, any stable rational function. ■

EXAMPLE 4.2. Consider $H_1(s) = \left[\dfrac{1}{s^2}, \dfrac{s+1}{s^2}\right] = [1, 0]\begin{bmatrix} s^2 & -(s+1) \\ 0 & 1 \end{bmatrix}^{-1} = N_1 D_1^{-1} = \dfrac{1}{s^2}[1, s+1] = \tilde{D}_1^{-1}\tilde{N}_1$, which are coprime polynomial MFDs. Relation (4.18) is given by

$$\begin{aligned} UU^{-1} &= \begin{bmatrix} X_1 & Y_1 \\ -\tilde{N}_1 & \tilde{D}_1 \end{bmatrix}\begin{bmatrix} D_1 & -\tilde{Y}_1 \\ N_1 & \tilde{X}_1 \end{bmatrix} \\ &= \begin{bmatrix} 1 & s+1 & \vdots & -s^2+1 \\ s & s^2+s+1 & \vdots & -s^3 \\ \cdots & \cdots & \vdots & \cdots \\ -1 & -(s+1) & \vdots & s^2 \end{bmatrix}\begin{bmatrix} s^2 & -(s+1) & \vdots & -(s+1) \\ 0 & 1 & \vdots & s \\ \cdots & \cdots & \vdots & \cdots \\ 1 & 0 & \vdots & 1 \end{bmatrix} \\ &= \begin{bmatrix} 1 & 0 & \vdots & 0 \\ 0 & 1 & \vdots & 0 \\ \cdots & \cdots & \vdots & \cdots \\ 0 & 0 & \vdots & 1 \end{bmatrix}. \end{aligned}$$

All stabilizing controllers may then be determined by applying (4.26) or (4.28). ■

We shall now express the transfer functions between $\begin{bmatrix} y_1 \\ y_2 \end{bmatrix}$ or $\begin{bmatrix} u_1 \\ u_2 \end{bmatrix}$ and $\begin{bmatrix} r_1 \\ r_2 \end{bmatrix}$ in terms of the parameters D_k, N_k or $\tilde{D}_k$, $\tilde{N}_k$ or K.

Recall that [see (3.103) and (3.104)]

$$\begin{bmatrix} \hat{y}_1 \\ \hat{y}_2 \end{bmatrix} = \begin{bmatrix} (I - H_1H_2)^{-1} & (I - H_1H_2)^{-1}H_1H_2 \\ (I - H_2H_1)^{-1}H_2H_1 & (I - H_2H_1)^{-1}H_2 \end{bmatrix} \begin{bmatrix} \hat{r}_1 \\ \hat{r}_2 \end{bmatrix} \tag{4.30}$$

and

$$\begin{bmatrix} \hat{u}_1 \\ \hat{u}_2 \end{bmatrix} = \begin{bmatrix} (I - H_2H_1)^{-1} & H_2(I - H_1H_2)^{-1} \\ H_1(I - H_2H_1)^{-1} & (I - H_1H_2)^{-1} \end{bmatrix} \begin{bmatrix} \hat{r}_1 \\ \hat{r}_2 \end{bmatrix}. \tag{4.31}$$

It is not difficult to see, in view of (4.26) and (4.28), that $(I - H_1H_2)^{-1} = D_2D_k^{-1}\tilde{D}_1 = (\tilde{X}_1D_k - N_1N_k)D_k^{-1}\tilde{D}_1 = (\tilde{X}_1 - N_1K)\tilde{D}_1$, and $(I - H_2H_1)^{-1} = D_1\tilde{D}_k^{-1}\tilde{D}_2 = D_1\tilde{D}_k^{-1}(\tilde{D}_kX_1 - \tilde{N}_k\tilde{N}_1) = D_1(X_1 - K\tilde{N}_1)$. Also, $(I - H_1H_2)^{-1}H_1 = D_2D_k^{-1}\tilde{N}_1 = (\tilde{X}_1D_k - N_1N_k)D_k^{-1}\tilde{N}_1 = (\tilde{X}_1 - N_1K)\tilde{N}_1 = H_1(I - H_2H_1)^{-1} = N_1\tilde{D}_k^{-1}\tilde{D}_2 = N_1\tilde{D}_k^{-1}(\tilde{D}_kX_1 - \tilde{N}_k\tilde{N}_1) = N_1(X_1 - K\tilde{N}_1)$, and $(I - H_2H_1)^{-1}H_2 = D_1\tilde{D}_k^{-1}\tilde{N}_2 = -D_1\tilde{D}_k^{-1}(\tilde{D}_kY_1 + \tilde{N}_k\tilde{D}_1) = -D_1(Y_1 + K\tilde{D}_1) = H_2(I - H_1H_2)^{-1} = N_2D_k^{-1}\tilde{D}_1 = -(\tilde{Y}_1D_k + D_1N_k)D_k^{-1}\tilde{D}_1 = -(\tilde{Y}_1 + D_1K)\tilde{D}_1$. Note that $(I - H_1H_2)^{-1}H_1H_2 = H_1(I - H_2H_1)^{-1}H_2 = N_1\tilde{D}_k^{-1}\tilde{N}_2 = -N_1\tilde{D}_k^{-1}(\tilde{D}_kY_1 + \tilde{N}_k\tilde{D}_1) = -N_1(Y_1 + K\tilde{D}_1)$. To express $(I - H_1H_2)^{-1}H_1H_2$ in terms of D_k, N_k we write it as $D_2D_k^{-1}\tilde{N}_1H_2 = -(\tilde{X}_1D_k - N_1N_k)D_k^{-1}\tilde{N}_1(\tilde{Y}_1D_k + D_1N_k)(\tilde{X}_1D_k - N_1N_k)^{-1} = -(\tilde{X}_1 - N_1K)\tilde{N}_1(\tilde{Y}_1 + D_1K)(\tilde{X}_1 - N_1K)^{-1}$, which is not as simple as the expression given in terms of $\tilde{D}_k$, $\tilde{N}_k$.

Similarly, $(I - H_2H_1)^{-1}H_2H_1 = H_2(I - H_1H_2)^{-1}H_1 = N_2D_k^{-1}\tilde{N}_1 = -(\tilde{Y}_1D_k + D_1N_k)D_k^{-1}\tilde{N}_1 = -(\tilde{Y}_1 + D_1K)\tilde{N}_1$. To express $(I - H_2H_1)^{-1}H_2H_1$ in terms of $\tilde{D}_k$, $\tilde{N}_k$, we write it as $D_1\tilde{D}_k^{-1}\tilde{N}_2H_1 = -D_1\tilde{D}_k^{-1}(\tilde{D}_kY_1 + \tilde{N}_k\tilde{D}_1)N_1D_1^{-1} = -D_1(Y_1 + K\tilde{D}_1)N_1D_1^{-1}$.

It is straightforward to express the transfer functions (4.30) and (4.31) in terms of the parameter K, in view of $K = N_kD_k^{-1} = \tilde{D}_k^{-1}\tilde{N}_k$. In particular, we have

$$\begin{bmatrix} \hat{y}_1 \\ \hat{y}_2 \end{bmatrix} = \begin{bmatrix} (\tilde{X}_1 - N_1K)\tilde{N}_1 & -(\tilde{X}_1 - N_1K)\tilde{N}_1H_2 \\ -(\tilde{Y}_1 + D_1K)\tilde{N}_1 & -(\tilde{Y}_1 + D_1K)\tilde{D}_1 \end{bmatrix} \begin{bmatrix} \hat{r}_1 \\ \hat{r}_2 \end{bmatrix} \tag{4.32}$$

$$\begin{bmatrix} \hat{u}_1 \\ \hat{u}_2 \end{bmatrix} = \begin{bmatrix} (I - H_2N_1D_1^{-1})^{-1} & -(\tilde{Y}_1 + D_1K)\tilde{D}_1 \\ (\tilde{X}_1 - N_1K)\tilde{N}_1 & (\tilde{X}_1 - N_1K)\tilde{D}_1 \end{bmatrix} \begin{bmatrix} \hat{r}_1 \\ \hat{r}_2 \end{bmatrix}. \tag{4.33}$$

These expressions were derived from the previous expressions given that involve D_k and N_k. The right factorization of H_2 in (4.28) is to be considered in these expressions. Similarly,

$$\begin{bmatrix} \hat{y}_1 \\ \hat{y}_2 \end{bmatrix} = \begin{bmatrix} N_1(X_1 - K\tilde{N}_1) & -N_1(Y_1 + K\tilde{D}_1) \\ -D_1(Y_1 + K\tilde{D}_1)N_1D_1^{-1} & -D_1(Y_1 + K\tilde{D}_1) \end{bmatrix} \begin{bmatrix} \hat{r}_1 \\ \hat{r}_2 \end{bmatrix} \tag{4.34}$$

$$\begin{bmatrix} \hat{u}_1 \\ \hat{u}_2 \end{bmatrix} = \begin{bmatrix} D_1(X_1 - K\tilde{N}_1) & -D_1(Y_1 + K\tilde{D}_1) \\ N_1(X_1 - K\tilde{N}_1) & (I - \tilde{D}_1^{-1}\tilde{N}_1H_2)^{-1} \end{bmatrix} \begin{bmatrix} \hat{r}_1 \\ \hat{r}_2 \end{bmatrix}, \tag{4.35}$$

which were derived from the previous expressions involving $\tilde{D}_k$ and $\tilde{N}_k$. The left factorization of H_2 in (4.28) is to be considered in entry (2, 2) of the $\begin{bmatrix} u_1 \\ u_2 \end{bmatrix}$ to $\begin{bmatrix} r_1 \\ r_2 \end{bmatrix}$

transfer function. These expressions for the transfer functions in terms of K can of course be combined in order to use the most convenient parameterizations.

3. *Parameters Q_1 and Q_2*. Recall from Subsection 7.3C the relations (3.103) and (3.104) given by

$$\begin{bmatrix}\hat{y}_1\\ \hat{y}_2\end{bmatrix} = \begin{bmatrix}-H_2 & I\\ I & -H_1\end{bmatrix}^{-1}\begin{bmatrix}0 & H_2\\ H_1 & 0\end{bmatrix}\begin{bmatrix}\hat{r}_1\\ \hat{r}_2\end{bmatrix}$$
$$= \begin{bmatrix}(I-H_1H_2)^{-1}H_1 & (I-H_1H_2)^{-1}H_1H_2\\ (I-H_2H_1)^{-1}H_2H_1 & (I-H_2H_1)^{-1}H_2\end{bmatrix}\begin{bmatrix}\hat{r}_1\\ \hat{r}_2\end{bmatrix} \tag{4.36}$$

and

$$\begin{bmatrix}\hat{u}_1\\ \hat{u}_2\end{bmatrix} = \begin{bmatrix}I & -H_2\\ -H_1 & I\end{bmatrix}^{-1}\begin{bmatrix}\hat{r}_1\\ \hat{r}_2\end{bmatrix}$$
$$= \begin{bmatrix}(I-H_2H_1)^{-1} & H_2(I-H_1H_2)^{-1}\\ H_1(I-H_2H_1)^{-1} & (I-H_1H_2)^{-1}\end{bmatrix}\begin{bmatrix}\hat{r}_1\\ \hat{r}_2\end{bmatrix}. \tag{4.37}$$

We shall now express the transfer function matrices in terms of some important parameters. To this end, define

$$Q_1 \triangleq H_1(I-H_2H_1)^{-1} = (I-H_1H_2)^{-1}H_1 \tag{4.38}$$
$$Q_2 \triangleq H_2(I-H_1H_2)^{-1} = (I-H_2H_1)^{-1}H_2 \tag{4.39}$$

and note that

$$H_1 = Q_1(I+H_2Q_1)^{-1} = (I+Q_1H_2)^{-1}Q_1 \tag{4.40}$$
$$H_2 = Q_2(I+H_1Q_2)^{-1} = (I+Q_2H_1)^{-1}Q_2. \tag{4.41}$$

It is not difficult to see that $Q_1H_2 = H_1Q_2$ and $Q_2H_1 = H_2Q_1$ from $Q_1H_2 = (I-H_1H_2)^{-1}H_1H_2 = H_1(I-H_2H_1)^{-1}H_2 = H_1Q_2$, and similarly for Q_2H_1. Also note that

$$(I-H_1H_2)^{-1} = I + H_1Q_2 = I + Q_1H_2 \tag{4.42}$$
$$(I-H_2H_1)^{-1} = I + H_2Q_1 = I + Q_2H_1, \tag{4.43}$$

from which by postmultiplying the first relation by H_1 and the second by H_2, we obtain $Q_1 = (I+H_1Q_2)H_1 = H_1(I+Q_2H_1)$ and $Q_2 = (I+H_2Q_1)H_2 = H_2(I+Q_1H_2)$. In view of these relations, it is now straightforward to write

$$\begin{bmatrix}\hat{y}_1\\ \hat{y}_2\end{bmatrix} = \begin{bmatrix}Q_1 & Q_1H_2\\ H_2Q_1 & (I+H_2Q_1)H_2\end{bmatrix}\begin{bmatrix}\hat{r}_1\\ \hat{r}_2\end{bmatrix} \tag{4.44}$$

$$\begin{bmatrix}\hat{u}_1\\ \hat{u}_2\end{bmatrix} = \begin{bmatrix}I+H_2Q_1 & (I+H_2Q_1)H_2\\ Q_1 & I+Q_1H_2\end{bmatrix}\begin{bmatrix}\hat{r}_1\\ \hat{r}_2\end{bmatrix}. \tag{4.45}$$

Note that (4.44) and (4.45) are useful when H_2 is given, in which case they depend only on the parameter Q_1. In this case H_1 is given by (4.40). It is not difficult to see that Q_1 is the transfer function matrix between y_1 or u_2 and r_1. Similarly,

$$\begin{bmatrix}\hat{y}_1\\ \hat{y}_2\end{bmatrix} = \begin{bmatrix}H_1(I+Q_2H_1) & H_1Q_2\\ Q_2H_1 & Q_2\end{bmatrix}\begin{bmatrix}\hat{r}_1\\ \hat{r}_2\end{bmatrix} \tag{4.46}$$

$$\begin{bmatrix}\hat{u}_1\\ \hat{u}_2\end{bmatrix} = \begin{bmatrix}I+Q_2H_1 & Q_2\\ H_1(I+Q_2H_1) & I+H_1Q_2\end{bmatrix}\begin{bmatrix}\hat{r}_1\\ \hat{r}_2\end{bmatrix}. \tag{4.47}$$

Expressions (4.46) and (4.47) are useful when H_1 is given, in which case the transfer functions depend only on the parameter Q_2. In this case H_2 is given by (4.41). Q_2 is the transfer function matrix between y_2 or u_1 and r_2.

If the given transfer function H_2 or H_1 is proper, then it is straightforward to guarantee properness of all other transfer functions by appropriately selecting the parameter Q_1 or Q_2. In particular, given that H_1 is proper, let a proper Q_2 be such that $(I + H_1Q_2)^{-1}$ exists. Then, in view of Lemma 3.14 in Subsection 7.3C, $H_2 = Q_2(I + H_1Q_2)^{-1}$ given in (4.41) is proper if and only if $det(I + (\lim_{s\to\infty} H_1)(\lim_{s\to\infty} Q_2)) \neq 0$. If this is true, then all the transfer functions in (4.46) and (4.47) will be proper. This implies that if the given H_1 is strictly proper then any proper Q_2 will guarantee that the above transfer functions exist and are proper. If H_1 is not strictly proper, i.e., $(\lim_{s\to\infty} H_1 \neq 0)$, then care should be taken in the selection of a proper Q_2; if in this case it is possible to select Q_2 strictly proper, then H_2 will be (strictly) proper for any Q_2. Similar results are valid when H_2 is given and Q_1 is selected.

In Theorem 4.3 we assume that H_1 is given and the parameter Q_2 is to be chosen to guarantee internal stability. Such a situation arises when H_1 represents the given plant and H_2 is the stabilizing controller to be designed. It should be noted that (as was shown in Subsection 7.3C), if the given system S_1 is not controllable and observable, all its uncontrollable and unobservable eigenvalues will appear as uncontrollable from $\begin{bmatrix} r_1 \\ r_2 \end{bmatrix}$ and unobservable from $\begin{bmatrix} y_1 \\ y_2 \end{bmatrix}$ or $\begin{bmatrix} u_1 \\ u_2 \end{bmatrix}$ eigenvalues in the closed-loop system. Therefore, for stability it is necessary to assume that S_1 is stabilizable and detectable. Here, working with H_1, any possible uncontrollable and unobservable parts of S_1 are ignored. Exactly analogous results exist for the parameter Q_1 when H_2 is given.

Consider the feedback system of Fig. 7.5 and assume that S_1 is controllable and observable with transfer function matrix H_1.

THEOREM 4.3. Given H_1, let

$$H_2 = Q_2(I + H_1Q_2)^{-1}, \tag{4.48}$$

where the rational matrix Q_2 is such that $(I + H_1Q_2)^{-1}$ exists. Then the eigenvalues of the closed-loop feedback system will be stable (i.e., they will have negative real parts) if and only if Q_2 is selected so that

(i) the poles of $H_1(I + Q_2H_1)$, H_1Q_2, Q_2H_1, and Q_2 are stable, or
(ii) the poles of $D_1^{-1}[I + Q_2H_1, Q_2]$ are stable, or
(iii) the poles of $\begin{bmatrix} Q_2 \\ I + H_1Q_2 \end{bmatrix} \tilde{D}_1^{-1}$ are stable.

Furthermore, the eigenvalues of the closed-loop system are precisely the poles in (i) or (ii) or (iii). In addition, if H_1 is proper, then H_2 and the transfer matrices in (4.36) and (4.37) are proper if and only if Q_2 is proper and $det(I + (\lim_{s\to\infty} H_1)(\lim_{s\to\infty} Q_2)) \neq 0$.

Proof. The expressions in (i) are precisely the entries of the transfer matrix between $\begin{bmatrix} y_1 \\ y_2 \end{bmatrix}$ and $\begin{bmatrix} r_1 \\ r_2 \end{bmatrix}$ given in (4.46), the poles of which are the closed-loop eigenvalues (see Theorem 3.15 in Subsection 7.3C). The expressions in (ii) and (iii) come from Corollary 3.16, where it was shown that the poles in these expressions are exactly the closed-loop eigenvalues. The statement about properness was proved above, just before the theorem. ■

1. It is not difficult to see that this theorem characterizes all H_2 that lead to a feedback system with stable eigenvalues. When H_1 is proper it also characterizes all proper stabilizing H_2.
2. The eigenvalues of the closed-loop system are the poles of the expressions in conditions (i) or (ii) or (iii). These will be poles of Q_2, but also stable poles of H_1, depending on the choice for Q_2.
3. It is easy to select Q_2 so that H_2 in (4.48) is proper (when H_1 is proper). Also, it is not difficult to select Q_2 so that the stability conditions are satisfied. However, to conveniently characterize all Q_2 that satisfy the stability conditions (i), (ii), or (iii) is not straightforward, unless special conditions apply. This is the case for example when H_1 is stable, as in Corollary 4.4.
4. For further discussion of the relation between properness and stability in a system described by a PMD, refer to Chapter 6 of [33] and also [9].

COROLLARY 4.4. Given H_1, assume that its poles are stable. Let H_2 be given by (4.48), where the rational matrix Q_2 is such that $(I + H_1Q_2)^{-1}$ exists. The eigenvalues of the closed-loop feedback system will be stable (i.e., will have negative real parts) if and only if Q_2 is stable. In addition, if H_1 is proper, then H_2 and all the transfer matrices in (4.36) and (4.37) are proper if and only if Q_2 is proper and $det(I + (\lim_{s\to\infty} H_1)(\lim_{s\to\infty} Q_2)) \neq 0$.

Proof. When the poles of H_1 are stable (i.e., have negative real parts), then the conditions (i) or (ii) or (iii) of Theorem 4.3 are true if and only if the poles of Q_2 are stable. ■

Corollary 4.4 greatly simplifies the conditions on Q_2. However, it is valid only when H_1 has stable poles. The result in this corollary was pointed out by Zames in [38], where it was used to introduce the H^∞-framework of optimal control system design.

4. *Parameters* S_{12} *and* S_{21}. At this point it is of interest to introduce the matrices

$$S_{12} \triangleq (I - H_1H_2)^{-1}, \qquad \text{and} \qquad S_{21} \triangleq (I - H_2H_1)^{-1}, \tag{4.49}$$

which are of importance in control system design. They are the *companion sensitivity matrices of the feedback loop.* S_{12} is also the transfer function between u_2 and r_2, while S_{21} is the transfer function between u_1 and r_1. Note that $Q_2 = H_2(I - H_1H_2)^{-1} = H_2S_{12} = S_{21}H_2$ and $Q_1 = H_1(I - H_2H_1)^{-1} = H_1S_{21} = S_{12}H_1$. Also, $H_1Q_2 = S_{12} - I$, $Q_2H_1 = S_{21} - I$ and $Q_1H_2 = S_{12} - I$, $H_2Q_1 = S_{21} - I$. Now consider Theorem 4.3 and note that

$$H_2 = Q_2S_{12}^{-1} = S_{21}^{-1}Q_2 \tag{4.50}$$

and that the conditions (ii) and (iii) can be written as $D_1^{-1}[S_{21}, Q_2]$ and $\begin{bmatrix} Q_2 \\ S_{12} \end{bmatrix} \tilde{D}_1^{-1}$, which must be stable in order to have stable closed-loop eigenvalues. This implies that a necessary condition for stable closed-loop eigenvalues is that all unstable poles of H_1 must appear as zeros of S_{21} and of S_{12}.

In general it is not possible to parameterize all stabilizing controllers in terms of only S_{12} or S_{21}, but this can be accomplished under certain assumptions on H_1. In particular, we have the following result.

COROLLARY 4.5. Assume that $det H_1 \neq 0$. Let

$$H_2 = H_1^{-1}(S_{12} - I)S_{12}^{-1} = D_1[N_1^{-1}(S_{12} - I)]S_{12}^{-1} \tag{4.51}$$

or

$$H_2 = S_{21}^{-1}(S_{21} - I)H_1^{-1} = S_{21}^{-1}[(S_{21} - I)\tilde{N}_1^{-1}]\tilde{D}_1, \tag{4.52}$$

where S_{12} or S_{21} are rational matrices such that S_{12}^{-1} and S_{21}^{-1} exist. Then the eigenvalues of the closed-loop feedback system will be stable if and only if

(i) S_{12} is such that the poles of $\begin{bmatrix} H_1^{-1}(S_{12} - I) \\ S_{12} \end{bmatrix} \tilde{D}_1^{-1} = \begin{bmatrix} D_1 N_1^{-1}(S_{12} - I)\tilde{D}_1^{-1} \\ S_{12}\tilde{D}_1^{-1} \end{bmatrix}$ are stable,

or

(ii) S_{21} is such that the poles of $D_1^{-1}[S_{21}, (S_{21} - I)H_1^{-1}] = [D_1^{-1}S_{21}, D_1^{-1}(S_{21} - I)\tilde{N}_1^{-1}\tilde{D}_1]$ are stable.

Proof. The proof of this result follows directly from Theorem 4.3. ■

It is clear from (i) of Corollary 4.5 that all the unstable zeros of H_1 must cancel with zeros of $S_{12} - I$, which is the transfer function between y_1 and r_2 [it equals $(I - H_1H_2)^{-1}H_1H_2$]. Also all the unstable poles of H_1 must cancel with zeros of S_{12}. From (ii) a similar result follows for $S_{21} - I$, which is the transfer function between y_2 and r_1 [it equals $(I - H_2H_1)^{-1}H_2H_1$] and for S_{21} [see Exercise 7.23(c) and (d)].

5. *Parameters X_2 and $\tilde{X}_2$.* Alternative parameters, closely related to Q_2, may be used in Theorem 4.3. To this end, let

$$X_2 \triangleq D_1^{-1}Q_2, \qquad \text{and} \qquad \tilde{X}_2 \triangleq Q_2\tilde{D}_1^{-1}. \tag{4.53}$$

COROLLARY 4.6. Given H_1, let

$$H_2 = D_1X_2(I + N_1X_2)^{-1} = [(I + X_2N_1)D_1^{-1}]^{-1}X_2, \tag{4.54}$$

where the rational matrix X_2 is such that $(I + N_1X_2)^{-1}$ exists. The eigenvalues of the closed-loop feedback system will be stable (i.e., will have negative real parts) if and only if X_2 is selected so that the poles of

$$[(I + X_2N_1)D_1^{-1}, X_2] \tag{4.55}$$

are stable. Furthermore, these poles are precisely the eigenvalues of the closed-loop system. In addition, if H_1 is proper, then H_2 and all the transfer matrices in (4.36) and (4.37) are proper if and only if D_1X_2 is proper and $det\,(I + (\lim_{s\to\infty} H_1)(\lim_{s\to\infty} D_1X_2)) \neq 0$.

Proof. Assume that the poles in (ii) of Theorem 4.3 are stable. Then, in view of $Q_2 = D_1X_2$, $D_1^{-1}[I + Q_2H_1, Q_2] = D_1^{-1}[I + D_1X_2N_1D_1^{-1}, D_1X_2] = [(I + X_2N_1)D_1^{-1}, X_2]$ also has stable poles. Conversely, if (4.55) has stable poles, then $D_1^{-1}[I + Q_2H_1, Q_2]$ also has (the same) stable poles, where Q_2 was selected to be $Q_2 = D_1X_2$. The remainder of the corollary follows easily in view of Theorem 4.3. ■

Note that X_2 can be seen to be the transfer function between z_1 and r_2, since Q_2 is the transfer function between $u_1 = D_1z_1$ and r_2 and $Q_2 = D_1X_2$. Also $N_1X_2 = H_1Q_2 = (I - H_1H_2)^{-1}H_1H_2$ is the transfer function matrix between y_1 and r_2. It should be pointed out that contrary to Corollary 4.4, the result in Corollary 4.6 is valid for H_1 unstable as well [7]. In a completely analogous manner, it can easily be shown that the following result is true.

COROLLARY 4.7. Given H_1, let

$$H_2 = (I + \tilde{X}_2\tilde{N}_1)^{-1}\tilde{X}_2\tilde{D}_1 = \tilde{X}_2[\tilde{D}_1^{-1}(I + \tilde{N}_1\tilde{X}_2)]^{-1}, \tag{4.56}$$

where the rational matrix $\tilde{X}_2$ is such that $(I + \tilde{X}_2\tilde{N}_1)^{-1}$ exists. The eigenvalues of the closed-loop feedback system will be stable (i.e., will have negative real parts) if and only if $\tilde{X}_2$ is selected so that the poles of

$$\begin{bmatrix} \tilde{X}_2 \\ \tilde{D}_1^{-1}(I + \tilde{N}_1\tilde{X}_2) \end{bmatrix} \tag{4.57}$$

are stable. Furthermore, these poles are precisely the eigenvalues of the closed-loop system. In addition, if H_1 is proper, then H_2 and all the transfer matrices in (4.36) and (4.37) are proper if and only if $\tilde{X}_2\tilde{D}_1$ is proper and $det\,(I + (\lim_{s\to\infty} \tilde{X}_2\tilde{D}_1)(\lim_{s\to\infty} H_1)) \neq 0$. ■

EXAMPLE 4.3. Consider $H_1 = \dfrac{s+1}{s-1}$ of Example 4.1. In view of Theorem 4.3, all stabilizing controllers H_2 are given by $H_2 = Q_2\left(1 + \dfrac{s+1}{s-1}Q_2\right)^{-1}$, where $\dfrac{1}{s-1}\left[1 + Q_2\dfrac{s+1}{s-1}, Q_2\right]$ is stable. If $Q_2 = n/d$, then $\dfrac{1}{s-1}\dfrac{n}{d}$ stable implies that $n = (s-1)\tilde{n}$, and d is stable. Also, $\dfrac{1}{s-1}\dfrac{(s-1)d + (s-1)\tilde{n}(s+1)}{d(s-1)} = \dfrac{d + \tilde{n}(s+1)}{(s-1)d}$ stable implies that $d(1) + \tilde{n}(1)2 = 0$ and d is stable. Therefore, for stability $Q_2 = n/d$ with d stable, and $n = (s-1)\tilde{n}$, $d(1) + \tilde{n}(1)2 = 0$.

In view of Corollary 4.6, all stabilizing controllers H_2 are given by $H_2 = (s-1)[1 + X_2(s+1)]^{-1}X_2$, where $\left[1 + X_2(s+1)\dfrac{1}{s-1}, X_2\right]$ is stable. If $X_2 = \tilde{n}/d$, then d is stable, and $d(1) + \tilde{n}(1)2 = 0$ for internal stability. Note that $Q_2 = n/d = D_1X_2 = (s-1)X_2 = (s-1)\tilde{n}/d$. ■

EXAMPLE 4.4. Consider $H_1(s) = \left[\dfrac{1}{s^2}, \dfrac{s+1}{s^2}\right] = \dfrac{1}{s^2}[1, s+1] = \tilde{D}_1^{-1}\tilde{N}_1$, which is an lc polynomial MFD. In view of Theorem 4.3, all stabilizing controllers H_2 are given by $H_2 = Q_2(1 + H_1Q_2)^{-1}$, where the poles of $\begin{bmatrix} Q_2 \\ 1 + H_1Q_2 \end{bmatrix}\tilde{D}_1^{-1}$ are stable. If $Q_2 = \dfrac{1}{d}\begin{bmatrix} n_1 \\ n_2 \end{bmatrix}$, then $Q_2\tilde{D}_1^{-1}$ stable implies that $n_1 = s^2\tilde{n}_1$, $n_2 = s^2\tilde{n}_2$ and d is stable. Now $(1 + H_1Q_2)\tilde{D}_1^{-1} = \dfrac{s^2 + n_1 + (s+1)n_2}{s^2 d}\cdot\dfrac{1}{s^2} = \dfrac{1 + \tilde{n}_1 + (s+1)\tilde{n}_2}{s^2 d}$ stable imposes additional conditions on $\tilde{n}_1$ and $\tilde{n}_2$. ■

EXAMPLE 4.5. Consider $H_1 = \dfrac{s-1}{s+1}$. This system is stable and therefore Corollary 4.4 applies. All stabilizing controllers H_2 are given by $H_2 = Q_2(1 + H_1Q_2)^{-1} = \dfrac{n}{d}\left(\dfrac{(s+1)d + (s-1)n}{(s+1)d}\right)^{-1} = \dfrac{n(s+1)}{(s+1)d + (s-1)n}$, where $Q_2 = n/d$ is stable. Furthermore, if Q_2 is proper with $\lim_{s\to\infty} Q_2(s) \neq -1$, then H_2 will be proper. Note that in view of Theorem 4.3, the closed-loop eigenvalues will be the poles of $\dfrac{1}{s+1} \times \left[\dfrac{(s+1) + (s-1)n}{(s+1)d}, \dfrac{n}{d}\right] = \dfrac{1}{(s+1)^2 d}[(s+1)d + (s-1)n, (s+1)n]$. ■

Relations among parameters. It is now straightforward to derive relations among the parameters Q_2 and Q_1; X_2 and $\tilde{X}_2$; $\tilde{D}_k$, $\tilde{N}_k$, and D_k, N_k; K; and also S_{12} and S_{21}. In particular, in view of (4.32) to (4.35), we have

$$
\begin{aligned}
Q_2 &= H_2(I - H_1H_2)^{-1} = (I - H_2H_1)^{-1}H_2 \\
&= N_2D_k^{-1}\tilde{D}_1 = D_1\tilde{D}_k^{-1}\tilde{N}_2 \\
&= -(\tilde{Y}_1D_k + D_1N_k)D_k^{-1}\tilde{D}_1 = -D_1\tilde{D}_k^{-1}(\tilde{D}_kY_1 + \tilde{N}_k\tilde{D}_1) \\
&= -(\tilde{Y}_1 + D_1K)\tilde{D}_1 = -D_1(Y_1 + K\tilde{D}_1).
\end{aligned} \tag{4.58}
$$

Also,

$$
\begin{aligned}
Q_1 &= H_1(I - H_2H_1)^{-1} = (I - H_1H_2)^{-1}H_1 \\
&= N_1\tilde{D}_k^{-1}\tilde{D}_2 = D_2D_k^{-1}\tilde{N}_1 \\
&= N_1\tilde{D}_k^{-1}(\tilde{D}_kX_1 - \tilde{N}_k\tilde{N}_1) = (\tilde{X}_1D_k - N_1N_k)D_k^{-1}\tilde{N}_1 \\
&= N_1(X_1 - K\tilde{N}_1) = (\tilde{X}_1 - N_1K)\tilde{N}_1.
\end{aligned} \tag{4.59}
$$

The parameters Q_2 are used when the system S_1 is assumed to be given. We have

$$
K = -D_1^{-1}(Q_2 + \tilde{Y}_1\tilde{D}_1)\tilde{D}_1^{-1} = -D_1^{-1}(Q_2 + D_1Y_1)\tilde{D}_1^{-1}. \tag{4.60}
$$

Next, we note that in view of (4.49) and (4.50), we have

$$
\begin{aligned}
S_{12} &= I + H_1Q_2 = I + N_1\tilde{D}_k^{-1}\tilde{N}_2 = I - N_1\tilde{D}_k^{-1}(\tilde{D}_kY_1 + \tilde{N}_2\tilde{D}_1) \\
&= I - N_1(Y_1 + K\tilde{D}_1)
\end{aligned} \tag{4.61}
$$

and

$$
\begin{aligned}
S_{21} &= I + Q_2H_1 = I + N_2D_k^{-1}\tilde{N}_1 = I - (\tilde{Y}_1D_k + D_1N_k)D_k^{-1}\tilde{D}_1 \\
&= I - (\tilde{Y}_1 + D_1K)\tilde{D}_1.
\end{aligned} \tag{4.62}
$$

From (4.53) it now follows that $Q_2 = D_1X_2 = \tilde{X}_2\tilde{D}_1$, from which the parameters X_2 and $\tilde{X}_2$ can be expressed as

$$
X_2 = -(Y_1 + K\tilde{D}_1) \qquad \text{and} \qquad \tilde{X}_2 = -(\tilde{Y}_1 + D_1K). \tag{4.63}
$$

The expressions of H_2 in terms of the parameters are now summarized:

$$
\begin{aligned}
H_2 &= -(\tilde{D}_kX_1 - \tilde{N}_k\tilde{N}_1)^{-1}(\tilde{D}_kY_1 + \tilde{N}_k\tilde{D}_1) = -(\tilde{Y}_1D_k + D_1N_k)(\tilde{X}_1D_k - N_1N_k)^{-1} \\
&= -(X_1 - K\tilde{N}_1)^{-1}(Y_1 + K\tilde{D}_1) = -(\tilde{Y}_1 + D_1K)(\tilde{X}_1 - N_1K)^{-1} \\
&= (I + Q_2H_1)^{-1}Q_2 = [D_1^{-1}(I + Q_2H_1)]^{-1}[D_1^{-1}Q_2] \\
&= Q_2(I + H_1Q_2)^{-1} = [Q_2\tilde{D}_1^{-1}][(I + H_1Q_2)\tilde{D}_1^{-1}]^{-1} \\
&= S_{21}^{-1}Q_2 = [D_1^{-1}S_{21}]^{-1}[D_1^{-1}Q_2] = Q_2S_{12}^{-1} = [Q_2\tilde{D}_1^{-1}][S_{12}\tilde{D}_1^{-1}]^{-1} \\
&= [(I + X_2N_1)D_1^{-1}]^{-1}X_2 = \tilde{X}_2[\tilde{D}_1^{-1}(I + \tilde{N}_1\tilde{X}_2)]^{-1}.
\end{aligned} \tag{4.64}
$$

6. *Parameters* L_1, L_2 *and* $\tilde{L}_1, \tilde{L}_2$. Expressions (4.64) that describe H_2 as a ratio of rational matrices can also be derived in an alternative way, using a theorem that we establish next. This result also provides a link between the development of this subsection and the descriptions of all stabilizing controllers using proper and stable factorizations given in Subsection 7.4C.

Consider the feedback system of Fig. 7.5 and assume that S_1 is controllable and observable. The following constitutes one of the principal results of this section.

THEOREM 4.8. Let the transfer function matrix of the system S_1 be H_1 with $H_1 = N_1D_1^{-1}(H_1 = \tilde{D}_1^{-1}\tilde{N}_1)$ being a coprime polynomial matrix factorization. If H_2 is the transfer function of S_2, then the eigenvalues of the closed-loop feedback system are stable if and only if H_2 can be written as

$$H_2 = \tilde{L}_2^{-1}\tilde{L}_1 \qquad (H_2 = L_1L_2^{-1}), \tag{4.65}$$

where the $\tilde{L}_2, \tilde{L}_1$ (L_1, L_2) are stable rational matrices with $det\,\tilde{L}_2 \neq 0$ $(det\,L_2 \neq 0)$, which satisfy

$$\tilde{L}_2D_1 - \tilde{L}_1N_1 = I \qquad (\tilde{D}_1L_2 - \tilde{N}_1L_1 = I). \tag{4.66}$$

Furthermore, if

$$[\tilde{L}_2, \tilde{L}_1] = \tilde{D}_k^{-1}[\tilde{D}_2, \tilde{N}_2] \qquad \left(\begin{bmatrix} L_2 \\ L_1 \end{bmatrix} = \begin{bmatrix} D_2 \\ N_2 \end{bmatrix} D_k^{-1}\right) \tag{4.67}$$

are coprime polynomial matrix factorizations, then the closed-loop description is given by

$$\begin{gathered} \tilde{D}_k\tilde{z} = [\tilde{D}_2, \tilde{N}_2]\begin{bmatrix} r_1 \\ r_2 \end{bmatrix}, \qquad \begin{bmatrix} y_1 \\ y_2 \end{bmatrix} = \begin{bmatrix} N_1 \\ D_1 \end{bmatrix}\tilde{z} + \begin{bmatrix} 0 & 0 \\ -I & 0 \end{bmatrix}\begin{bmatrix} r_1 \\ r_2 \end{bmatrix} \\ \left(D_kz = [\tilde{N}_1, \tilde{D}_1]\begin{bmatrix} r_1 \\ r_2 \end{bmatrix}, \qquad \begin{bmatrix} y_1 \\ y_2 \end{bmatrix} = \begin{bmatrix} D_2 \\ N_2 \end{bmatrix} z + \begin{bmatrix} 0 & -I \\ 0 & 0 \end{bmatrix}\begin{bmatrix} r_1 \\ r_2 \end{bmatrix}\right). \end{gathered} \tag{4.68}$$

Proof. The proof of the parts in parentheses will not be shown since it follows the proof given below in a completely analogous manner. Let $[\tilde{L}_1, \tilde{L}_2]$ be stable with $det\,\tilde{L}_2 \neq 0$ and satisfying (4.65) and (4.66) and write an lc polynomial matrix factorization as in (4.67). Then $\tilde{D}_2D_1 - \tilde{N}_2N_1 = \tilde{D}_k$, where $\tilde{D}_k^{-1}$ exists and is stable. Furthermore, the pair $(\tilde{D}_2, \tilde{N}_2)$ is lc [since any cld of the pair $(\tilde{D}_2, \tilde{N}_2)$ would be an ld of $\tilde{D}_k$] and $\tilde{D}_2^{-1}$ exists. Therefore, the closed-loop system with $H_2 = \tilde{L}_2^{-1}\tilde{L}_1 = \tilde{D}_2^{-1}\tilde{N}_2$ is well defined. Its internal description is given in (4.68) [see also (4.11)], and the closed-loop eigenvalues are the stable roots of $det\,\tilde{D}_k$.

Now let $H_2 = \tilde{D}_2^{-1}\tilde{N}_2$ be an lc polynomial factorization and note that the closed-loop system is given by (4.68), where $\tilde{D}_k = \tilde{D}_2D_1 - \tilde{N}_2N$. Assume that $\tilde{D}_k^{-1}$ is stable. Define $[\tilde{L}_2, \tilde{L}_1] \triangleq \tilde{D}_k^{-1}[\tilde{D}_2, \tilde{N}_2]$ and note the $\tilde{D}_k$ and $[\tilde{D}_2, \tilde{N}_2]$ are lc since the pair $(\tilde{D}_2, \tilde{N}_2)$ is lc. Then $H_2 = \tilde{L}_2^{-1}\tilde{L}_1$ where $\tilde{L}_2$ and $\tilde{L}_1$ are stable with $\tilde{L}_2D_1 - \tilde{L}_1N_1 = I$. ■

In view of the above theorem, all stabilizing H_2 are given by (4.65), where the stable rational matrices $\tilde{L}_2, \tilde{L}_1$ (L_1, L_2) span all solutions of the Diophantine Equation given in (4.66). There are different ways of parameterizing all stable solutions $\tilde{L}_2, \tilde{L}_1$ (L_2, L_1) of (4.66) with $det\,\tilde{L}_2 \neq 0$ $(det\,L_2 \neq 0)$, each one leading to a different parameterization of all stabilizing H_2. In fact, in this way one may generate the parameterizations developed above, thus providing alternative proofs for those results. This is shown in the next corollary.

COROLLARY 4.9. All stabilizing H_2 are given in the following.

(i)
$$H_2 = \tilde{L}_2^{-1}\tilde{L}_1 \qquad (H_2 = L_1L_2^{-1}), \tag{4.69}$$

where the $\tilde{L}_2, \tilde{L}_1$ (L_1, L_2) are stable with $det\,\tilde{L}_2 \neq 0$ $(det\,L_2 \neq 0)$ and $\tilde{L}_2D_1 - \tilde{L}_1N_1 = I$ $(\tilde{D}_1L_2 - \tilde{N}_1L_1 = I)$. The closed-loop eigenvalues are the poles of $[\tilde{L}_2, \tilde{L}_1]$ $\left(\begin{bmatrix} L_2 \\ L_1 \end{bmatrix}\right)$.

(ii)
$$H_2 = -(X_1 - K\tilde{N}_1)^{-1}(Y_1 + KD_1) \qquad [H_2 = -(\tilde{Y}_1 + D_1K)(\tilde{X}_1 - N_1K)^{-1}], \tag{4.70}$$

where K is any stable matrix such that $det\,(X_1 - K\tilde{N}_1) \neq 0$ $[det\,(\tilde{X}_1 - N_1K) \neq 0]$. The X_1, $Y_1, \tilde{X}_1, \tilde{Y}_1$ satisfy $UU^{-1} = \begin{bmatrix} X_1 & Y_1 \\ -\tilde{N}_1 & \tilde{D}_1 \end{bmatrix}\begin{bmatrix} D_1 & -\tilde{Y}_1 \\ N_1 & \tilde{X}_1 \end{bmatrix} = \begin{bmatrix} I & 0 \\ 0 & I \end{bmatrix}$, where U is a unimodular matrix. The closed-loop eigenvalues are the poles of

$$[X_1 - K\tilde{N}_1, Y_1 + K\tilde{D}_1] \qquad \left(\begin{bmatrix} \tilde{X}_1 - N_1 K \\ \tilde{Y}_1 + D_1 K \end{bmatrix}\right), \tag{4.71}$$

or equivalently, the poles of K.

$$\text{(iii)} \qquad H_2 = [D_1^{-1}(I + Q_2H_1)]^{-1}[D_1^{-1}Q_2] = (I + Q_2H_1)^{-1}Q_2$$
$$(H_2 = [Q_2\tilde{D}_1^{-1}][(I + H_1Q_2)\tilde{D}_1^{-1}]^{-1} = Q_2(I + Q_2H_1)), \tag{4.72}$$

where Q_2 is such that

$$[D_1^{-1}(I + Q_2H_1), D_1^{-1}Q_2] \qquad \left(\begin{bmatrix} Q_2\tilde{D}_1^{-1} \\ (I + H_1Q_2)\tilde{D}_1^{-1} \end{bmatrix}\right) \tag{4.73}$$

is stable and $det\,(I + Q_2H_1) \neq 0$ ($det\,(I + H_1Q_2) \neq 0$). The closed-loop eigenvalues are the poles of (4.73).

$$\text{(iv)} \qquad H_2 = [D_1^{-1}S_{21}]^{-1}[D_1^{-1}Q_2] \qquad (H_2 = [Q_2\tilde{D}_1^{-1}][S_{12}\tilde{D}_1^{-1}]^{-1}), \tag{4.74}$$

where S_{21} (S_{12}) and Q_2 are such that

$$[D_1^{-1}S_{21}, D_1^{-1}Q_2] \qquad \left(\begin{bmatrix} Q_2\tilde{D}_1^{-1} \\ S_{12}\tilde{D}_1^{-1} \end{bmatrix}\right) \tag{4.75}$$

is stable with $det\,S_{21} \neq 0$ ($det\,S_{12} \neq 0$) and $S_{21} - Q_2H_2 = I$ ($S_{12} - H_1Q_2 = I$). The closed-loop eigenvalues are the poles of (4.75).

$$\text{(v)} \qquad H_2 = [(I + X_2N_1)D_1^{-1}]^{-1}X_2 \qquad (H_2 = \tilde{X}_2[\tilde{D}_1^{-1}(I + \tilde{N}_1\tilde{X}_2)]^{-1}), \tag{4.76}$$

where X_2 ($\tilde{X}_2$) is such that

$$[(I + X_2N_1)D_1^{-1}, X_2] \qquad \left(\begin{bmatrix} \tilde{X}_2 \\ \tilde{D}_1^{-1}(I + \tilde{N}_1X_2) \end{bmatrix}\right) \tag{4.77}$$

is stable and $det\,(I + X_2N_1) \neq 0$ ($det\,(I + \tilde{N}_1X_2) \neq 0$). The closed-loop eigenvalues are the poles of (4.77).

Proof. The proof is straightforward in view of Theorem 4.8. Part (i) follows directly from Theorem 4.8. To show (ii), note that $[X_1 - K\tilde{N}_1]D_1 - [-(Y_1 + K\tilde{D}_1)]N_1 = I$ for any K (compare with Theorem 4.2). To show (iii), note that $[D_1^{-1}(I + Q_2H_1)]D_1 - [D_1^{-1}Q_2]N_1 = I$ for any Q_2 (compare with Theorem 4.3). To show (iv), note that $[D_1^{-1}S_{21}]D_1 - [D_1^{-1}Q_2]N_1 = I$ if and only if $S_{21} - Q_2H_1 = I$. To show (v), note that $[(I + X_2N_1)D_1^{-1}]D_1 - [X_2]N_1 = I$ for any X_2 (compare with Corollaries 4.6 and 4.7). What has not been shown yet is that all stabilizing H_2 can be expressed by, say, (4.70) in (ii). This can be accomplished in a manner analogous to the proof of the theorem. In particular, any stabilizing $H_2 = \tilde{D}_2^{-1}\tilde{N}_2$, where the pair $(\tilde{D}_2, \tilde{N}_2)$ is lc, that gives rise to a closed-loop description (4.68) implies that $\tilde{D}_k^{-1}[\tilde{D}_2, -\tilde{N}_2] = [X_1 - K\tilde{N}_1, -(Y_1 + K\tilde{D}_1)] = [I, K]\begin{bmatrix} X_1 & Y_1 \\ -\tilde{N}_1 & \tilde{D}_1 \end{bmatrix} = [I, K]U$, where U is a unimodular matrix. Therefore, from $[I, K] = \tilde{D}_k^{-1}[\tilde{D}_2, -\tilde{N}_2]U^{-1}$ it follows that a stable K can be determined uniquely. Similarly, in (iii), the relation $\tilde{D}_k^{-1}[\tilde{D}_2, \tilde{N}_2] = D_1^{-1}[I + Q_2H_1, Q_2] = D_1^{-1}[I, Q]\begin{bmatrix} I & 0 \\ H_1 & I \end{bmatrix}$ determines uniquely a stable $Q_2 = D_1\tilde{D}_k^{-1}\tilde{N}_2$. The details are left to the reader. These results were of course also shown in Theorems 4.2 and 4.3, using alternative approaches. ■

Remarks

In Theorem 4.8 and Corollary 4.9, H_1 may or may not be proper. Also, the stabilizing H_2 may or may not be proper. Thus, the above results characterize all stabilizing H_2, both proper and not proper. It is frequently desirable to restrict H_1 and H_2 to be proper rational matrices. The problem of interest then is to determine all proper stabilizing H_2, given a proper H_1. Note that this problem has already been addressed previously in this section using the parameter Q_2, and it will be studied at length in Subsection 7.4C.

We conclude by summarizing the relations among the parameters used in this subsection. Note that the relations for all parameters, except $\tilde{L}_1$, $\tilde{L}_2$ and L_2, L_1, were derived in (4.58) to (4.64). The relations to $\tilde{L}_2$, $\tilde{L}_1$ can easily be obtained in view of Corollary 4.9. These relations are summarized as:

$$\tilde{L}_2 = X_1 - K\tilde{N}_1 = D_1^{-1}(I + Q_2H_1) = D_1^{-1}S_{21} = (I + X_2N_1)D_1^{-1}$$

$$\text{and} \quad \tilde{L}_1 = -(Y_1 + K\tilde{D}_1) = D_1^{-1}Q_2 = X_2. \tag{4.78}$$

Also,

$$L_2 = \tilde{X}_1 - N_1K = (I + H_1Q_2)\tilde{D}_1^{-1} = S_{12}\tilde{D}_1^{-1} = \tilde{D}_1^{-1}(I + \tilde{N}_1\tilde{X}_2)$$

$$\text{and} \quad L_1 = -(\tilde{Y}_1 + D_1K) = Q_2\tilde{D}_1^{-1} = \tilde{X}_2. \tag{4.79}$$

EXAMPLE 4.6. Consider $H_1 = \dfrac{s+1}{s-1}$. In view of Corollary 4.9, all stabilizing controllers are given by $H_2 = \tilde{L}_2^{-1}\tilde{L}_1$, where the $\tilde{L}_2$, $\tilde{L}_1$ are stable and satisfy $\tilde{L}_2D_1 - \tilde{L}_1N_1 = I$.

Now in view of (4.78) and (4.79), $\tilde{L}_2 = (I + X_2N_1)D_1^{-1}$ and $\tilde{L}_1 = X_2$. All appropriate X_2, however, were characterized in Example 4.3 to be $X_2 = \tilde{n}/d$ with d stable and $d(1) + \tilde{n}(1)2 = 0$. Therefore, all appropriate $\tilde{L}_2$, $\tilde{L}_1$ are given by $\tilde{L}_2 = \dfrac{d + \tilde{n}(s+1)}{d}\dfrac{1}{s-1} = \dfrac{\hat{d}}{d}$, $\tilde{L}_1 = \dfrac{\tilde{n}}{d}$, where d is stable and $\tilde{n}$ is such that $d + \tilde{n}(s+1) = (s-1)\hat{d}$, i.e., $\tilde{n}(1) = -\frac{1}{2}d(1)$. ■

B. State Feedback and State Estimation

State feedback and state estimation are studied in this subsection, using PMFDs of systems. Our current development, which offers additional insight, parallels that given in Chapter 4, where state-space descriptions were used. The study of state feedback and state estimation, using PMFDs, however, is important in its own right. An additional reason for discussing state feedback and state observers at this point is to provide the necessary background needed to connect the parameterizations of all stabilizing controllers using proper and stable factorizations with the internal descriptions (PMDs or PMFDs) of systems. This is accomplished in Subsection 7.4C.

State feedback

State feedback control laws are closely related to the state-space representations of systems. Recall from Chapter 4 that given

$$\dot{x} = Ax + Bu, \qquad y = Cx + Du \tag{4.80}$$

with $A \in R^{n\times n}$, $B \in R^{n\times m}$, $C \in R^{p\times n}$, and $D \in R^{p\times m}$, the linear state feedback control law is defined by

$$u = Fx + Gr, \tag{4.81}$$

where $F \in R^{m\times n}$ and $G \in R^{m\times m}$ with G nonsingular. Frequently, $G = I$. Then the closed-loop system description is given by

$$\dot{x} = (A + BF)x + BGr, \qquad y = (C + DF)x + DGr \tag{4.82}$$

and the closed-loop eigenvalues are the zeros of $det\,[qI - (A + BF)]$. In view of the relation

$$\begin{bmatrix} qI - A - BF & BG \\ -(C + DF) & DG \end{bmatrix} = \begin{bmatrix} qI - A & B \\ -C & D \end{bmatrix} \begin{bmatrix} I & 0 \\ -F & G \end{bmatrix}, \tag{4.83}$$

it is not difficult to see that any gcld of $[qI - A, B]$ will be an ld of $qI - A - BF$ for any F. This implies that for complete eigenvalue assignment, $qI - A$ and B must be left coprime, i.e., (A, B) must be completely controllable, which is a well-known result (see Chapter 4). Also, for stability, (A, B) must be a stabilizable pair. Note that any gcld of $[qI - A, B]$ will be an ld of $[qI - A - BF, BG]$ for any F and G. Since here G is taken to be nonsingular, $[qI - A, B]$ and $[qI - A - BF, BG]$ have the same gcld, i.e., the open- and closed-loop systems have precisely the same uncontrollable eigenvalues. When G is singular, the closed-loop system may have additional uncontrollable modes (show this). Although F does not affect the uncontrollable modes of the system, it may alter its unobservable modes. For example, in an SISO controllable and observable system, it is possible to select F so that some of the closed-loop eigenvalues are at the same location as the finite zeros, and therefore, they become unobservable. This corresponds to pole/zero cancellations in the closed-loop transfer function. The closed-loop unobservable eigenvalues can also be studied by means of relation (4.83) and the gcrd of the pair $(qI - A - BF, -C - DF)$.

The effects of state feedback control laws can conveniently be studied using polynomial matrix descriptions. In particular, assume that the state-space representation (4.80) is controllable and in controller form $\{A_c, B_c, C_c, D_c\}$ (see Chapter 3). An equivalent PMFD is then given by

$$D_c(q)z_c(t) = u(t), \qquad y(t) = N_c(q)z_c(t) \tag{4.84}$$

with $D_c(q) \in R[q]^{m\times m}$ and $N_c(q) \in R[q]^{p\times m}$, where

$$\begin{bmatrix} B_c & 0 \\ D_c & I_p \end{bmatrix} \begin{bmatrix} D_c(q) & I_m \\ -N_c(q) & 0 \end{bmatrix} = \begin{bmatrix} qI - A_c & B_c \\ -C_c & D_c \end{bmatrix} \begin{bmatrix} S_c(q) & 0 \\ 0 & I_m \end{bmatrix} \tag{4.85}$$

with the pair $(B_c, qI - A_c)$ being lc and the pair $(D_c(q), S_c(q))$ being rc [see (3.26)]. The matrix $S_c(q) \triangleq block\ diag\,[(1, q, \ldots, q^{d_i-1})^T]$ is an $n \times m$ matrix with d_i, $i = 1, \ldots, m$, the controllability indices of $\{A_c, B_c, C_c, D_c\}$. Note that the above relations can be written as

$$(qI - A_c)S_c(q) = B_cD_c(q), \qquad N_c(q) = C_cS_c(q) + D_cD_c(q), \tag{4.86}$$

which were derived via the Structure Theorem in Subsection 3.4D. Note that the states are related by $x_c(t) = S_c(q)z_c(t)$. When the state feedback control law

$$u(t) = F_cx_c(t) + Gr(t) = F_cS_c(q)z_c(t) + Gr(t) = F_c(q)z_c(t) + Gr(t) \tag{4.87}$$

is applied, the closed-loop system state-space representation is $\{A_c + B_cF_c, B_cG, C_c + D_cF_c, D_cG\}$ and the polynomial matrix description is

$$D_F(q)z_c(t) = Gr(t), \qquad y(t) = N_F(q)z_c(t), \tag{4.88}$$

where $D_F(q) \triangleq D_c(q) - F_c(q) = D_c(q) - F_cS_c(q)$ and $N_F(q) = N_c(q)$. These representations are equivalent since

$$\begin{bmatrix} B_c & 0 \\ D_c & I_p \end{bmatrix}\begin{bmatrix} D_c(q) - F_cS_c(q) & G \\ -N_c(q) & 0 \end{bmatrix} = \begin{bmatrix} qI - A_c - B_cF_c & B_cG \\ -C_c - D_cF_c & D_cG \end{bmatrix}\begin{bmatrix} S_c(q) & 0 \\ 0 & I_m \end{bmatrix}, \tag{4.89}$$

where the pair $(B_c, qI - A_c - B_cF_c)$ is lc and the pair $(D_c(q) - F_cS_c(q), S_c(q))$ is rc (see Subsection 7.3A). Note that $(qI - A_c - B_cF_c)S_c(q) = B_c[D_c(q) - F_cS_c(q)]$ and $N_F(q) = (C_c + D_cF_c)S_c(q) + D_c[D_c(q) - F_cS_c(q)] = C_cS_c(q) + D_cD_c(q) = N_c(q)$, i.e., the numerator $N_c(q)$ is invariant under state feedback. Note that $N_c(q)$ contains the zeros of the system (see Subsection 7.3B).

Now assume that the state-space representation in (4.80) is controllable but not necessarily in controller form, and let $A = Q^{-1}A_cQ$, $B = Q^{-1}B_c$, $C = C_cQ$, $D = D_c$ with Q a similarity transformation matrix. Relations (4.86) then assume the form

$$(qI - A)S(q) = BD_c(q), \qquad N_c(q) = CS(q) + DD_c(q), \tag{4.90}$$

where $S(q) = Q^{-1}S_c(q)$. Then

$$D_c(q)z_c(t) = u(t), \qquad y(t) = N_c(q)z_c(t) \tag{4.91}$$

is an equivalent polynomial matrix description and now $x(t) = S(q)z_c(t)$. Note that $deg_{c_i} S(q) = deg_{c_i} S_c(q) = d_i, i = 1, \ldots, m$, which are the controllability indices of the system. The linear state feedback control law is then given by

$$u(t) = Fx(t) + Gr(t) = FS(q)z_c(t) + Gr(t) = F_c(q)z_c(t) + Gr(t). \tag{4.92}$$

We now have $(qI - A - BF)S(q) = B[D_c(q) - FS(q)]$ and $N_c(q) = (C + DF)S(q) + D[D_c(q) - FS(q)] = CS(q) + DD_c(q)$.

In view of the above, it can be seen that linear state feedback can be equivalently defined for the case of (controllable) polynomial matrix right fractional descriptions as shown in the following. Given

$$D(q)z(t) = u(t), \qquad y(t) = N(q)z(t), \tag{4.93}$$

where $D(q)$ is column reduced, define the linear state-feedback control law by

$$u(t) = F(q)z(t) + Gr(t), \tag{4.94}$$

where $deg_{c_i} F(q) < deg_{c_i} D(q)$ with $F(q) \in R[q]^{m\times m}$, $G \in R^{m\times m}$, $det\, G \neq 0$. The closed-loop system is then described by

$$D_F(q)z(t) = Gr(t), \qquad y(t) = N_F(q)z(t), \tag{4.95}$$

where $D_F(q) = D(q) - F(q)$, $N_F(q) = N(q)$. The $F(q) = F_cS_c(q)$ can be chosen to arbitrarily assign the polynomial entries of $D(q)$, up to and including the terms of degrees $d_i - 1$ in the ith column of $D(q)$, $i = 1, \ldots, m$. In fact, recall from the development in Chapter 3 (Theorem 4.10—the Structure Theorem) that $D(q) - F(q) = D(q) - F_cS_c(q) = B_m^{-1}[diag\,[q^{d_i}] - A_mS_c(q)] - F_cS_c(q) = B_m^{-1}[diag\,[q^{d_i}] -$

$(A_m + B_m F_c)S_c(q)]$. It was shown in Chapter 4 how to appropriately select F_c to arbitrarily assign all the closed-loop eigenvalues, i.e., the roots of $det(D(q) - F(q))$.

Now consider the closed-loop state-space representation (4.82). As was discussed in Chapter 4, the closed-loop transfer function can be written as

$$\begin{aligned} H_{F,G}(s) &= [(C + DF)[sI - (A + BF)]^{-1}B + D]G \\ &= [C(sI - A)^{-1}B + D][F[sI - (A + BF)]^{-1}B + I]G \\ &= H(s)H_e(s). \end{aligned} \tag{4.96}$$

Here $H(s)$ is the open-loop transfer function, and $H_e(s)$ represents the transfer function of a system that, if connected in series with the given system, will apparently produce the same overall transfer function as the feedback system. Recall that this issue was addressed at length in Chapter 4. If the PMD (4.93) is used, then the closed-loop transfer function is given by

$$\begin{aligned} H_{F,G}(s) &= N_F(s)D_F^{-1}(s)G = N(s)D_F^{-1}(s)G \\ &= [N(s)D^{-1}(s)][D(s)D_F^{-1}(s)]G = H(s)H_e(s). \end{aligned} \tag{4.97}$$

It is not difficult to verify that the system $\{D_F(q), I_m, D(q), 0\}$ is equivalent to the system $\{A+BF, B, F, I_m\}$, both of which have the same transfer function $H_e(s)$ (show this).

Relation (4.97) also implies that

$$\begin{aligned} H(s) &= N(s)D^{-1}(s) = H_{F,G}(s)H_e^{-1}(s) \\ &= [N(s)D_F^{-1}(s)G][D(s)D_F^{-1}(s)G]^{-1}. \end{aligned} \tag{4.98}$$

Note that both $H_{F,G}$ and H_e are proper and stable ($H(s)$ is proper). Furthermore, H_e^{-1} is also proper. Thus, $H_{F,G}$ and H_e are proper and stable factors in the MFD

$$H(s) = H_{F,G}(s)H_e^{-1}(s). \tag{4.99}$$

This is further discussed in the next subsection, where it is shown how all proper and stable right MFDs can be generated by means of state feedback. The left proper and stable MFDs can be generated from observers of the partial state. Observers of the state are examined next.

State observers

State observers were discussed at length in Chapter 4. Here we wish to present additional material concerning observers in terms of PMDs.

Consider the plant S and the observer S_{ob} of Fig. 7.6 and let the plant S be described by (4.93), where $D(q) \in R[q]^{m\times m}$ and $N(q) \in R[q]^{p\times m}$. As was discussed above, when $D(q)$ is column reduced, the linear state feedback control law can be defined by $u(t) = F(q)z(t) + r(t)$, where $deg_{c_i} F < deg_{c_i} D$, $i = 1, \ldots, m$. Then the

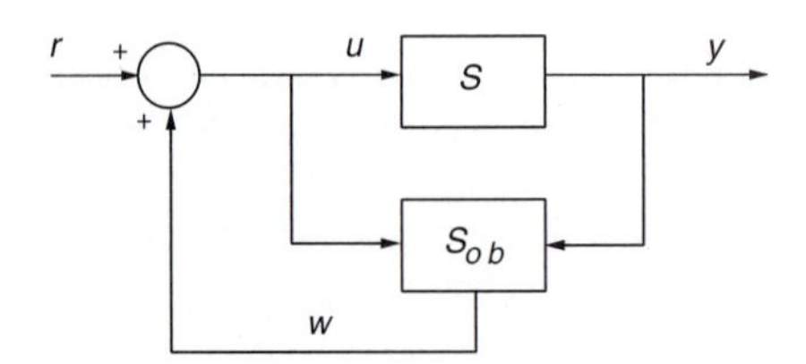

FIGURE 7.6
Plant and observer

closed-loop system is represented by

$$[D(q) - F(q)]z(t) = r(t), \qquad y(t) = N(q)z(t). \tag{4.100}$$

When the state is not readily available, then a state observer for Fz may be used. Let the observer S_{ob} be described by

$$Q(q)z_{ob}(t) = [K(q), H(q)]\begin{bmatrix} u(t) \\ y(t) \end{bmatrix}$$
$$w(t) = z_{ob}(t), \tag{4.101}$$

where $Q(q) \in R[q]^{m\times m}$, $K(q) \in R[q]^{m\times m}$, and $H(q) \in R[q]^{m\times p}$. Note that in Fig. 7.6 $u(t) = w(t) + r(t)$. Assume now that the observer polynomial matrices Q, K, and H satisfy the relation

$$K(q)D(q) + H(q)N(q) = Q(q)F(q), \tag{4.102}$$

where Q^{-1} and $Q^{-1}[K, H]$ are proper and stable ($F \in R[q]^{m\times m}$). Recall that Q^{-1} proper is needed for $Q(q)z_{ob}(t) = 0$ to be a "well-formed" set of differential equations so as to avoid impulsive behavior at $t = 0$. Note that if Q is, say, row or column proper, then Q^{-1} is proper. The rational function $Q^{-1}[K, H]$ is the transfer function of the observer. Then $w = z_{ob} = Q^{-1}[Ku + Hy] = Q^{-1}[KD + HN]z = Fz$, i.e., z_{ob} is a candidate for estimating a function $F(q)z(t)$ of the partial state $z(t)$. To show this, consider the closed-loop internal description

$$\begin{bmatrix} D(q) & -I_m \\ -Q(q)F(q) & Q(q) \end{bmatrix}\begin{bmatrix} z(t) \\ z_{ob}(t) \end{bmatrix} = \begin{bmatrix} I_m \\ 0 \end{bmatrix} r(t)$$
$$y(t) = [N(q), 0]\begin{bmatrix} z(t) \\ z_{ob}(t) \end{bmatrix} \tag{4.103}$$

derived by using $u = Dz = w + r = z_{ob} + r$ and $Qz_{ob} = Ku + Hy = (KD + HN)z = QFz$, and consider the unimodular transformation $\begin{bmatrix} I_m & 0 \\ -F(q) & I_m \end{bmatrix}\begin{bmatrix} z(t) \\ z_{ob}(t) \end{bmatrix} = \begin{bmatrix} z(t) \\ z_{ob} - F(q)z(t) \end{bmatrix} = \begin{bmatrix} z(t) \\ e(t) \end{bmatrix}$. Then the closed-loop system description becomes

$$\begin{bmatrix} D(q) - F(q) & -I_m \\ 0 & Q(q) \end{bmatrix}\begin{bmatrix} z(t) \\ e(t) \end{bmatrix} = \begin{bmatrix} I_m \\ 0 \end{bmatrix} r(t)$$
$$y(t) = [N(q), 0]\begin{bmatrix} z(t) \\ e(t) \end{bmatrix}. \tag{4.104}$$

First, note that the closed-loop eigenvalues are the roots of $det(D - F)\,det\,Q$, and therefore, the closed-loop system is stable if and only if all the roots of both $det(D - F)$ and $det\,Q$ have negative real parts. The roots of $det(D - F)$ are of course the closed-loop eigenvalues under state feedback when there is no observer, while the roots of $det\,Q$ are the observer eigenvalues that are taken to be stable. Note that in view of $Q(q)e(t) = 0$, where $Q^{-1}(q)$ is proper and stable, the error $e(t) = z_{ob}(t) - F(q)z(t) = w(t) - F(q)z(t)$ will go to zero as t goes to infinity, and therefore, the output $w(t)$ of the observer will asymptotically approach the function of the state $F(q)z(t)$. This will happen independently of $r(t)$. Note that the roots of $det\,Q$ are uncontrollable from r as can easily be seen, using, e.g., the eigenvalue test for controllability. As expected, these uncontrollable eigenvalues cancel in the closed-loop transfer function matrix given by

$$[N(s), 0]\begin{bmatrix} D(s) - F(s) & -I_m \\ 0 & Q(s) \end{bmatrix}^{-1} = N(s)[D(s) - F(s)]^{-1}. \qquad (4.105)$$

In other words, after the transients caused by initial conditions have died out (see Chapter 4), the system behaves to the outside world as though an observer were not present. The observer in (4.101) was introduced in Wolovich [36].

For an observer (4.101) to exist, K, H, and Q must satisfy the Diophantine Equation $KD + HN = QF$ given in (4.102), where Q^{-1} and $Q^{-1}[K, H]$ are proper and stable to ensure causality. Note that if $D_F = D - F$, then $F = D - D_F$ and $KD + HN = Q(D - D_F)$, or $(Q - K)D + (-H)N = QD_F$, which implies that

$$[Q^{-1}(Q - K)][DD_F^{-1}] + [-Q^{-1}H][ND_F^{-1}] = I. \qquad (4.106)$$

The pair $(Q^{-1}(Q-K), -Q^{-1}H)$ is a proper and stable solution of the equation $\hat{X}\hat{D} + \hat{Y}\hat{N} = I$, where $\hat{D} = DD_F^{-1}$ and $\hat{N} = ND_F^{-1}$ are proper and stable. This equation is important in the parameterization of all stabilizing feedback controllers when using the ring of proper and stable rational functions, discussed in the next subsection.

It is possible to implement the observer discussed above in an alternative manner. In particular, $u = w + r = Q^{-1}Ku + Q^{-1}Hy + r$ implies that $(I - Q^{-1}K)u = Q^{-1}Hy + r$, or $u = (I - Q^{-1}K)^{-1}(Q^{-1}Hy + r)$, or

$$u = (Q - K)^{-1}Q(Q^{-1}Hy + r). \qquad (4.107)$$

This corresponds to the configuration in Fig. 7.7.

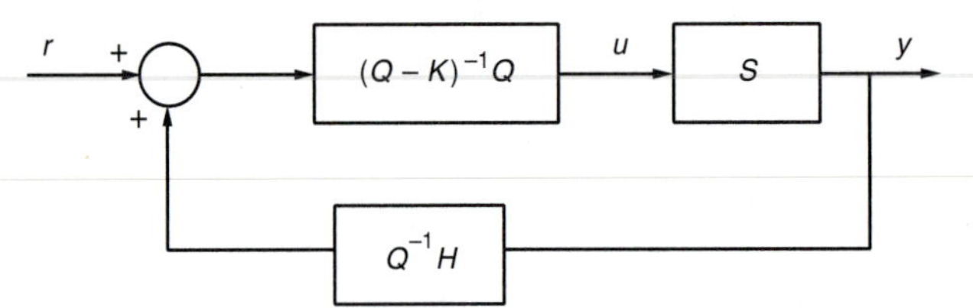

FIGURE 7.7
Observer-based controller implementation

Note that the feedback path controller $Q^{-1}H$ is always stable, while the controller in the feedforward path is biproper (i.e., it along with its inverse is proper) but not necessarily stable. If the external input r is of no interest, then take $r = 0$, in which case we have

$$u = (Q - K)^{-1}Hy. \qquad (4.108)$$

This corresponds to the configuration depicted in Fig. 7.8.

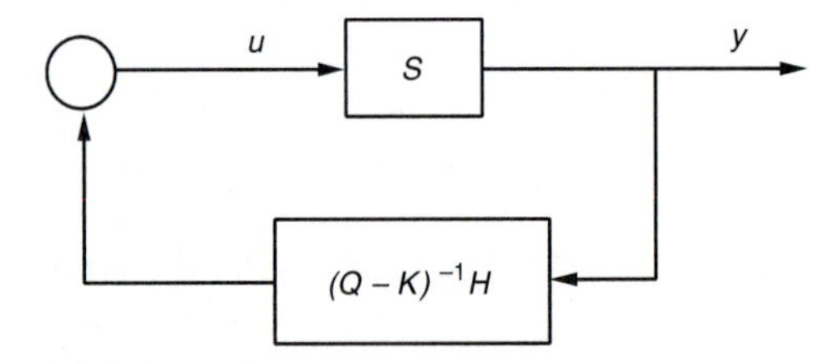

FIGURE 7.8
Observer-based controller implementation when $r = 0$

It is of interest to compare these results with the corresponding state-space results in Chapter 4. In particular, consider the state-space plant representation (4.80) and assume that it is controllable and observable. Now comparing Fig. 7.7 with Fig. 4.7 of Chapter 4, we obtain, in view of (4.25) and (4.24) of Chapter 4, the relations $(Q(s)-K(s))^{-1}Q(s) = (I-Q^{-1}(s)K(s))^{-1} = (I-G_u(s))^{-1} = F[sI-(A-KC+BF-KDF)]^{-1}(B-KD)+I$, $Q^{-1}(s)H(s) = G_y(s) = F[sI-(A-KC)]^{-1}K$ and $Q^{-1}(s)K(s) = G_u(s) = F[sI-(A-KC)]^{-1}(B-KD)$. Also, $(Q(s)-K(s))^{-1}H(s) = (I-G_u(s))^{-1}G_y(s) = F[sI-(A-KC+BF-KDF)]^{-1}K$ (see Fig. 7.8). It is therefore clear that the two degrees of freedom controller in Fig. 4.7 of Chapter 4 is a special case (of order n) of the controller in Fig. 7.7.

EXAMPLE 4.7. Consider $H(s) = \dfrac{s}{s^2+2s+2}$ given in Example 4.2 of Chapter 4. In view of the above, it is not difficult to see that the results developed in Example 4.2 can also be derived if we let $Q(s) = s^2+d_1s+d_0$, $F(s) = (2-a_1)s+(2-a_0) = [\frac{1}{2}a_0-1, 2-a_1]\begin{bmatrix}-2\\ s\end{bmatrix} = FS(s)$, $K(s) = s(2-a_1)-d_0(\frac{1}{2}a_0-1)$, and $H(s) = s((d_0-2)(\frac{1}{2}a_0-1)+(d_1-2)(2-a_1))+((d_0-d_1)(a_0-2)+(d_0-2)(s-a_1))$, which satisfy (4.102). Verify this. ■

Finally, it remains to be shown that polynomial matrices K, H, and Q, which satisfy (4.102) with Q^{-1} and $Q^{-1}[K, H]$ proper and stable, exist. Here $deg_{c_i} F < deg_{c_i} D$, where D is assumed to be column reduced and the N, D are assumed to be rc. The system S is assumed to be controllable and observable. That such K, H, and Q exist will not be shown here. This is shown in Wolovich [36], where an algorithm is given that is based on the Eliminant Matrix of D and N (see Subsection 7.2E, Theorem 2.13, and Lemma 2.14) to select an appropriate row reduced Q and to determine K and H. Note that (4.102) can also be solved by using other methodologies, such as polynomial matrix interpolation (see the Appendix).

C. Stabilizing Feedback Controllers Using Proper and Stable MFDs

Now consider systems S_1 and S_2 connected in a feedback configuration as shown in Fig. 7.5. Let system S_1 be controllable and observable and let it be described by its transfer function matrix H_1. In Subsection 7.4A, all systems S_2 that internally stabilize the closed-loop feedback system were parametrically characterized. In that development H_1 was not necessarily proper, and the stabilizing H_2 as well as the closed-loop system transfer function were not necessarily proper either. Recall that a system is said to be internally stable when all its eigenvalues, which are the roots of its characteristic polynomial, have strictly negative real parts. Polynomial matrix descriptions that can easily handle the case of nonproper transfer functions were used to derive the results in Subsection 7.4A, and the case of proper H_1 and H_2 was handled by restricting the parameters used to characterize all stabilizing controllers.

Here we concentrate exclusively on the case of proper H_1 and parametrically characterize all proper H_2 that internally stabilize the closed-loop system. For this, proper and stable MFDs of H_1 and H_2 are used. These are now described.

Consider $H(s) \in R(s)^{p\times m}$ to be proper, i.e., $\lim_{s\to\infty} H(s) < \infty$, and write the MFD as

$$H(s) = N'(s)D'(s)^{-1}, \tag{4.109}$$

where the $N'(s)$ and $D'(s)$ are proper and stable rational matrices that we denote here as $N'(s) \in RH_\infty^{p\times m}$ and $D'(s) \in RH_\infty^{m\times m}$; that is, they are matrices with elements in RH_∞, the set of all proper and stable rational functions with real coefficients. For instance, if $H(s) = \frac{s-1}{(s-2)(s+1)}$, then $H(s) = \left[\frac{s-1}{(s+2)(s+3)}\right]\left[\frac{(s-2)(s+1)}{(s+2)(s+3)}\right]^{-1} = \left[\frac{s-1}{(s+1)^2}\right]\left[\frac{s-2}{s+1}\right]^{-1}$ are examples of proper and stable MFDs.

A pair $(N', D') \in RH_\infty$ is called *right coprime* (*rc*) *in* RH_∞ if there exists a pair $(X', Y') \in RH_\infty$ such that

$$X'D' + Y'N' = I. \tag{4.110}$$

This is a *Diophantine Equation* over the ring of proper and stable rational functions. It is also called a *Bezout identity*.

Let $H = N'D'^{-1}$ and write (4.110) as $X' + Y'H = D'^{-1}$. Since the left-hand side is proper, D'^{-1} is also proper, i.e., in the MFD given by $H = N'D'^{-1}$, where the pair (N', D') is rc, D' is biproper (D' and D'^{-1} are both proper).

Note that X'^{-1}, where X' satisfies (4.110), does not necessarily exist. If, however, H is strictly proper $[\lim_{s\to\infty} H(s) = 0]$, then $\lim_{s\to\infty} X'(s) = \lim_{s\to\infty} D'(s)^{-1}$ is a nonzero real matrix, and in this case X'^{-1} exists and is proper, i.e., in this case X' is biproper.

When the Diophantine Equation (4.110) is used to characterize all stabilizing controllers, it is often desirable to have solutions (X', Y'), where X' is biproper. This is always possible. Clearly, when H is strictly proper, this is automatically true, as was shown. When H is not strictly proper, however, care should be exercised in the selection of the solutions of (4.110).

LEMMA 4.10. Let $H = N_1'D_1'^{-1} = N_2'D_2'^{-1}$ be rc factorizations. Then

$$\begin{bmatrix} D_2' \\ N_2' \end{bmatrix} = \begin{bmatrix} D_1' \\ N_1' \end{bmatrix} U', \tag{4.111}$$

where $U', U'^{-1} \in RH_\infty$.

Proof. Given the two rc factorizations, let $U' \triangleq D_1'^{-1}D_2'$ and note that $N_2' = HD_2' = N_1'D_1'^{-1}D_2' = N_1'U'$. Now $X_2'D_2' + Y_2'N_2' = (X_2'D_1' + Y_2'N_1')U' = I$, from which $U'^{-1} = X_2'D_1' + Y_2'N_1'$, i.e., $U'^{-1} \in RH_\infty$. Similarly, $X_1'D_1' + Y_1'N_1' = I$ implies that $U' \in RH_\infty$. ■

Remarks

1. A matrix U', as given above, with U' and $U'^{-1} \in RH_\infty$ is a unit in the ring RH_∞ (refer to the discussion on rings and modules in Subsection 7.2E).
2. If H is also stable, i.e., if $H \in RH_\infty$, then $H = HI^{-1}$ is an rc factorization. If now $H = N'D'^{-1}$ in any rc proper and stable MFD, then in view of Lemma 4.11, $I = D'U'$, i.e., D' and $D'^{-1} \in RH_\infty$.

EXAMPLE 4.8. Let $H = \frac{s-1}{(s-2)(s+1)} = \left(\frac{s-1}{(s+1)^2}\right)\left(\frac{s-2}{s+1}\right)^{-1} = N'D'^{-1}$. Here N' and D' are rc since $X'D' + Y'N' = \left(\frac{s-5}{s+1}\right)\left(\frac{s-2}{s+1}\right) + (9)\left(\frac{s-1}{(s+1)^2}\right) = 1$. In view of

Lemma 4.10, all $\begin{bmatrix} D_2' \\ N_2' \end{bmatrix} = \begin{bmatrix} D_1' \\ N_1' \end{bmatrix} U' = \begin{bmatrix} \frac{s-2}{s+1} \\ \frac{s-1}{(s+1)^2} \end{bmatrix} U'$, where $U', U'^{-1} \in RH_\infty$ are also rc factorizations of H. Here, $U'(s) = a(s)/b(s)$, where $a(s)$ and $b(s)$ are prime Hurwitz polynomials of the same degree n and n can be arbitrarily large, but finite. ■

The above example illustrates that when $N', D' \in RH_\infty$ are rc in RH_∞, they may have common factors that include common poles and zeros; however, these possible common poles and zeros have to be stable. Note that any common right divisor G' of an rc pair $N', D' \in RH_\infty$ must satisfy $G', G'^{-1} \in RH_\infty$. Contrast this with the case of two rc polynomial matrices.

Analogous results hold for lc proper and stable MFDs of $H = \tilde{D}'^{-1}\tilde{N}'$ with $\tilde{D}', \tilde{N}' \in RH_\infty$, where the Diophantine Equation

$$\tilde{D}'\tilde{X}' + \tilde{N}'\tilde{Y}' = I \tag{4.112}$$

is satisfied for some $\tilde{X}, \tilde{Y} \in RH_\infty$. In this case the result corresponding to Lemma 4.10 becomes $[\tilde{D}_2', \tilde{N}_2'] = \tilde{U}'[\tilde{D}_1', \tilde{N}_1']$ with $\tilde{U}', \tilde{U}'^{-1} \in RH_\infty$ when $H = \tilde{D}'^{-1}_1\tilde{N}_1 = \tilde{D}'^{-1}_2\tilde{N}_2'$ are lc MFDs.

As in the polynomial case, doubly coprime factorizations in RH_∞ of a transfer function matrix $H_1 = N_1'D'^{-1}_1 = \tilde{D}'^{-1}_1\tilde{N}_1'$, where $D_1', N_1' \in RH_\infty$ and $\tilde{D}_1', \tilde{N}_1' \in RH_\infty$ are important in obtaining parametric characterizations of all stabilizing controllers. Assume therefore that

$$U'U'^{-1} = \begin{bmatrix} X_1' & Y_1' \\ -\tilde{N}_1' & \tilde{D}_1' \end{bmatrix} \begin{bmatrix} D_1' & -\tilde{Y}_1' \\ N_1' & \tilde{X}_1' \end{bmatrix} = \begin{bmatrix} I & 0 \\ 0 & I \end{bmatrix}, \tag{4.113}$$

where U' is unimodular in RH_∞, i.e., $U', U'^{-1} \in RH_\infty$. Also, assume that $X_1', \tilde{X}_1'$ have been selected so that $\det X_1' \neq 0$ and $\det \tilde{X}_1' \neq 0$.

To see how such relations can be derived [compare with the discussion following (4.18) in Subsection 7.4A], assume that some X_o', Y_o' and $\tilde{Y}_o', \tilde{X}_o'$ have been found that satisfy $X_o'D_1' + Y_o'N_1' = I$ and $\tilde{D}_1'\tilde{X}_o' + \tilde{N}_1'\tilde{Y}_o' = I$. Note then that

$$\begin{bmatrix} X_o' & Y_o' \\ -\tilde{N}_1' & \tilde{D}_1' \end{bmatrix} \begin{bmatrix} D_1' & D'S_o' - \tilde{Y}_o' \\ N_1' & N'S_o' + \tilde{X}_o' \end{bmatrix} = \begin{bmatrix} I & 0 \\ 0 & I \end{bmatrix}, \tag{4.114}$$

where $S_o' \triangleq X_o'\tilde{Y}_o' - Y_o'\tilde{X}_o'$. Let $(X_1', -Y_1') = (X_o', Y_o')$ and $(\tilde{X}_1', -\tilde{Y}_1') = (NS_o' + \tilde{X}_o', D'S_o' - \tilde{Y}_o')$ to obtain (4.113). It can be shown that matrices are indeed unimodular.

Internal stability

Consider now the feedback system in Fig. 7.5 and let H_1 and H_2 be the transfer function matrices of S_1 and S_2, respectively, that are assumed to be controllable and observable. Internal stability of a system can be defined in a variety of equivalent ways in terms of the internal description of the system. For example in this chapter, polynomial matrix internal descriptions were used, and the system was considered as internally stable when its eigenvalues were stable, i.e., they had strictly negative real parts. In Theorem 3.15 in Subsection 7.3C, it is shown that the closed-loop feedback system is internally stable if and only if the transfer function between $\begin{bmatrix} u_1 \\ u_2 \end{bmatrix}$ and $\begin{bmatrix} r_1 \\ r_2 \end{bmatrix}$

or $\begin{bmatrix} y_1 \\ y_2 \end{bmatrix}$ and $\begin{bmatrix} r_1 \\ r_2 \end{bmatrix}$ have stable poles, i.e., if and only if the poles of $\begin{bmatrix} I & -H_2 \\ -H_1 & I \end{bmatrix}^{-1}$ or $\begin{bmatrix} -H_2 & I \\ I & -H_1 \end{bmatrix}^{-1} \begin{bmatrix} 0 & H_1 \\ H_1 & 0 \end{bmatrix}$, respectively, are stable.

In this subsection we shall regard the feedback system to be internally stable when

$$\begin{bmatrix} I & -H_2 \\ -H_1 & I \end{bmatrix}^{-1} \in RH_\infty, \tag{4.115}$$

i.e., when all the transfer function matrices in (4.115) are proper and stable. In this way, internal stability can be checked without necessarily involving internal descriptions of S_1 and S_2. This approach to stability has advantages since it can be extended to systems other than linear time-invariant systems.

THEOREM 4.11. Let $H_1 = N_1' D_1'^{-1} = \tilde{D}_1'^{-1} \tilde{N}_1'$ and $H_2 = \tilde{D}_2'^{-1} \tilde{N}_2' = N_2' D_2'^{-1}$ be doubly coprime MFDs in RH_∞. Then the closed-loop feedback system is internally stable if and only if

$$\tilde{D}_2' D_1' - \tilde{N}_2' N_1' = \tilde{U}' \tag{4.116}$$

or if and only if

$$\tilde{D}_1' D_2' - \tilde{N}_1' N_2' = U', \tag{4.117}$$

where $\tilde{U}', \tilde{U}'^{-1} \in RH_\infty$ and $U', U'^{-1} \in RH_\infty$.

Proof. Consider $H_1 = N_1' D_1'^{-1}$, $H_2 = \tilde{D}_2'^{-1} \tilde{N}_2'$ and assume that (4.116) is satisfied. We have $I = \tilde{D}_2'^{-1} \tilde{N}_2' N_1' D_1'^{-1} + \tilde{D}_2^{-1} \tilde{U}' D_1'^{-1}$, which implies that

$$\begin{bmatrix} I & -H_2 \\ -H_1 & I \end{bmatrix} = \begin{bmatrix} \tilde{D}_2' \tilde{U}' & -\tilde{D}_2'^{-1} \tilde{N}_2' \\ 0 & I \end{bmatrix} \begin{bmatrix} D_1'^{-1} & 0 \\ -N_1' D_1'^{-1} & I \end{bmatrix}$$

and

$$\begin{bmatrix} I & -H_2 \\ -H_1 & I \end{bmatrix}^{-1} = \begin{bmatrix} D_1' & 0 \\ N_1' & I \end{bmatrix} \begin{bmatrix} \tilde{U}'^{-1} \tilde{D}_2'^{-1} & \tilde{U}'^{-1} \tilde{D}_2'^{-2} \tilde{N}'_2 \\ 0 & I \end{bmatrix}, \tag{4.118}$$

which is proper and stable, since both factors in the right-hand side are proper and stable. Therefore, if (4.116) is satisfied, the closed-loop feedback system is internally stable. Similarly, it can be shown that if $H_1 = \tilde{D}_1'^{-1} \tilde{N}_1'$, $H_2 = N_2' D_2'^{-1}$ and (4.117) is satisfied, then the closed-loop feedback system is internally stable.

The converse will now be established, namely, if the feedback system is internally stable, then (4.116) is satisfied. The proof that (4.117) is also true is completely analogous. Let $H_1 = N_1' D_1'^{-1}$ be rc and $H_2 = D_2'^{-1} N_2'$ be lc MFDs in RH_∞, and let $\tilde{D}_2' D_1' - \tilde{N}_2' N_1' = \tilde{U}'$ with $\tilde{U}'$ some matrix in RH_∞. Recalling (3.103) or (4.37), we have

$$\begin{bmatrix} I & -H_2 \\ -H_1 & I \end{bmatrix}^{-1} = \begin{bmatrix} (I - H_2 H_1)^{-1} & (I - H_2 H_1)^{-1} H_2 \\ H_1 (I - H_2 H_1)^{-1} & I + H_1 (I - H_1 H_2)^{-1} H_2 \end{bmatrix}, \tag{4.119}$$

where the identities $(I - H_1 H_2)^{-1} = I + (I - H_1 H_2)^{-1} H_1 H_2$ and $(I - H_1 H_2)^{-1} H_1 H_2 = H_1 (I - H_2 H_1)^{-1} H_2$ were used. It is not difficult to see [compare also with (3.113)] that

$$\begin{aligned} \begin{bmatrix} I & -H_2 \\ -H_1 & I \end{bmatrix}^{-1} &= \begin{bmatrix} D_1' \tilde{U}'^{-1} \tilde{D}_2' & D_1' \tilde{U}'^{-1} \tilde{N}_2' \\ N_1' \tilde{U}'^{-1} \tilde{D}_2' & I + N_1' \tilde{U}'^{-1} \tilde{N}_2 \end{bmatrix} \\ &= \begin{bmatrix} 0 & 0 \\ 0 & I \end{bmatrix} + \begin{bmatrix} D_1' \\ N_1' \end{bmatrix} \tilde{U}'^{-1} [\tilde{D}'_2, \tilde{N}_2']. \end{aligned} \tag{4.120}$$

Assume now that the system is internally stable, i.e., $\begin{bmatrix} I & -H_2 \\ -H_1 & I \end{bmatrix}^{-1} \in RH_\infty$. Then, since D_1', N_1' are rc and $\tilde{D}_2', \tilde{N}_2'$ are lc, $\tilde{U}^{-1}$ is in RH_∞. To see this, premultiply $\begin{bmatrix} D_1' \\ N_1' \end{bmatrix} \tilde{U}'^{-1}[\tilde{D}_2', \tilde{N}_2']$, which is in RH_∞, by $[\tilde{D}_2', -\tilde{N}_2'] \in RH_\infty$, and postmultiply by $\begin{bmatrix} D_1' \\ -N_1' \end{bmatrix} \in RH_\infty$. These operations leave the matrix in RH_∞. Therefore, $\tilde{U}'^{-1} \in RH_\infty$. ■

COROLLARY 4.12. Let $H_1 = N_1'D_1'^{-1} = \tilde{D}_1'^{-1}\tilde{N}_1'$ be doubly coprime MFDs in RH_∞, i.e., (4.113) is satisfied. Then the closed-loop feedback system is internally stable if and only if H_2 has an lc MFD in RH_∞, $H_2 = \tilde{D}_2'^{-1}\tilde{N}_2'$, such that

$$\tilde{D}_2'D_1' - \tilde{N}_2'N_1' = I, \tag{4.121}$$

or if and only if H_2 has an rc MFD in RH_∞, $H_2 = N_2'D_2'^{-1}$, such that

$$\tilde{D}_1'D_2' - \tilde{N}_1'N_2 = I. \tag{4.122}$$

Proof. The proof is straightforward, in view of Theorem 4.11 and Lemma 4.10. ■

Parameterizations of all stabilizing controllers

Several parameterizations of all proper stabilizing controllers are now considered.

Parameter K'

THEOREM 4.13. Let $H_1 = N_1'D_1'^{-1} = \tilde{D}_1'^{-1}\tilde{N}_1'$ be doubly coprime MFDs in RH_∞ that satisfy (4.113). Then all H_2 that internally stabilize the closed-loop feedback system are given by

$$H_2 = -(X_1' - K'\tilde{N}_1')^{-1}(Y_1' + K'\tilde{D}_1') = -(\tilde{Y}_1' + D_1'K')(\tilde{X}_1' - N_1'K')^{-1}, \tag{4.123}$$

where $K' \in RH_\infty$ is such that $(X_1' - K'\tilde{N}_1')^{-1}$ [or $(\tilde{X}_1 - N_1'K')^{-1}$] exists and is proper.

Proof. It can be shown that all solutions of $\tilde{D}_2D_1' - \tilde{N}_2N_1' = I$ are given by

$$[\tilde{D}_2', -\tilde{N}_2'] = [I, K'] \begin{bmatrix} X_1' & Y_1' \\ -\tilde{N}_1' & \tilde{D}_1' \end{bmatrix}, \tag{4.124}$$

where $K' \in RH_\infty$. The proof of this result is similar to the proof of the corresponding result for the polynomial matrix Diophantine Equation in Subsection 7.2E and is omitted (see also Subsection 7.4A). Similarly, all solutions of $\tilde{D}_1'D_2' - \tilde{N}_1'N_2' = I$ are given by

$$\begin{bmatrix} N_2' \\ D_2' \end{bmatrix} = \begin{bmatrix} D_1' & -\tilde{Y}_1' \\ N_1' & \tilde{X}_1' \end{bmatrix} \begin{bmatrix} -K' \\ I \end{bmatrix}, \tag{4.125}$$

where $K' \in RH_\infty$. The result then follows directly from Corollary 4.12. ■

The above theorem is a generalization of the Youla parameterization of Theorem 4.2 over the ring of proper and stable rational functions. Generalizations of the Youla parameterization over rings other than the polynomial ring were introduced in Desoer et al. [11]; for a detailed treatment see also Vidyasagar [34].

It is interesting to note that in view of (4.113), H_2 in (4.123) can be written as follows. Assume that X_1^{-1} and $\tilde{X}_1^{-1}$ exist. Then

$$
\begin{aligned}
H_2 &= -(\tilde{Y}_1' + X_1'^{-1}(I - Y_1'N_1')K')(\tilde{X}_1 - N_1'K')^{-1} \\
&= -[\tilde{Y}_1'\tilde{X}_1'^{-1}(\tilde{X}_1' - N_1'K') + X_1'^{-1}K'](\tilde{X}_1 - N_1'K')^{-1} \\
&= -\tilde{Y}_1'\tilde{X}_1'^{-1} - X_1'^{-1}K'(\tilde{X}_1 - N_1'K')^{-1} = H_{20} + H_{2a}, \qquad (4.126)
\end{aligned}
$$

i.e., any stabilizing controller H_2 can be viewed as the sum of an initial stabilizing controller $H_{20} = -\tilde{Y}_1'\tilde{X}_1'^{-1}$ and an additional controller H_{2a} that depends on K'. When $K' = 0$, then H_{2a} is zero.

In view of (4.120), the poles of $\begin{bmatrix} I & -H_2 \\ -H_1 & I \end{bmatrix}^{-1}$, which are the closed-loop eigenvalues, are the poles of $\begin{bmatrix} D_1' \\ N_1' \end{bmatrix}\tilde{U}'^{-1}[\tilde{D}_2', \tilde{N}_2']$. Similarly, since

$$
\begin{aligned}
\begin{bmatrix} I & -H_2 \\ -H_1 & I \end{bmatrix}^{-1} &= \begin{bmatrix} I + N_2'U'^{-1}\tilde{N}_1' & N_2'U'^{-1}\tilde{D}_1' \\ D_2'U'^{-1}\tilde{N}_1 & D_2'U'^{-1}\tilde{D}_1' \end{bmatrix} \\
&= \begin{bmatrix} I & 0 \\ 0 & 0 \end{bmatrix} + \begin{bmatrix} N_2' \\ D_2' \end{bmatrix} U'^{-1}[\tilde{N}_1', \tilde{D}_1'], \qquad (4.127)
\end{aligned}
$$

where $H_1 = \tilde{D}_1'^{-1}\tilde{N}_1'$ is lc and $H_2 = N_2'D_2'^{-1}$ is rc, and both are MFDs in RH_∞ with $\tilde{D}_1'D_2' - \tilde{N}_1'N_2' = U'$ and $U', U'^{-1} \in RH_\infty$, it follows that the eigenvalues of the closed-loop feedback system are the poles of $\begin{bmatrix} N_2' \\ D_2' \end{bmatrix}U'^{-1}[\tilde{N}_1', \tilde{D}_1']$. It is now straightforward to prove the following result.

LEMMA 4.15. Given $H_1 = N_1'D_1'^{-1} = \tilde{D}_1'^{-1}\tilde{N}_1'$, doubly coprime MFDs in RH_∞, if H_2 is given by (4.123), the closed-loop eigenvalues are the poles of

$$
\begin{bmatrix} D_1' \\ N_1' \end{bmatrix}[I, K']\begin{bmatrix} X_1' & Y_1' \\ -\tilde{N}_1' & \tilde{D}_1' \end{bmatrix}\begin{bmatrix} I & 0 \\ 0 & -I \end{bmatrix} = \begin{bmatrix} D_1' \\ N_1' \end{bmatrix}[X_1' - Y_1'] - \begin{bmatrix} D_1' \\ N_1' \end{bmatrix}K'[\tilde{N}_1', \tilde{D}_1'] \qquad (4.128)
$$

or of

$$
\begin{bmatrix} D_1' & -\tilde{Y}_1' \\ N_1' & \tilde{X}_1' \end{bmatrix}\begin{bmatrix} -K' \\ I \end{bmatrix}[\tilde{N}_1', \tilde{D}_1'] = \begin{bmatrix} -\tilde{Y}_1' \\ \tilde{X}_1' \end{bmatrix}[\tilde{N}_1', \tilde{D}_1'] - \begin{bmatrix} D_1' \\ N_1' \end{bmatrix}K'[\tilde{N}_1', \tilde{D}_1']. \qquad (4.129)
$$

Proof. Note that $\tilde{D}_2'D_1' - \tilde{N}_2'N_1' = (X_1' - K'\tilde{N}_1')D_1' + (Y_1' + K'\tilde{D}_1')N_1 = X_1'D_1' + Y_1'N_1 = I = \tilde{U}'$, which in view of the above discussion, directly implies the lemma. ■

Therefore, in view of Lemma 4.15, when the parameterization for H_2 of Theorem 4.13 is used, the closed-loop eigenvalues are in general the poles of $\begin{bmatrix} D_1' \\ N_1' \end{bmatrix}$, of $\begin{bmatrix} X_1' & Y_1' \\ -\tilde{N}_1' & \tilde{D}_1' \end{bmatrix}$, and of K'.

EXAMPLE 4.9. Let $H_1 = \dfrac{1}{s-1} = \dfrac{1}{s+1}\left(\dfrac{s-1}{s+1}\right)^{-1} = N_1'D_1'^{-1} = \left(\dfrac{s-1}{s+a}\right)^{-1}\dfrac{1}{s+a} = \tilde{D}_1'^{-1}\tilde{N}_1'$ with $a > 0$, which are doubly coprime factorizations. Note that

$$
\begin{bmatrix} X_1' & Y_1' \\ -\tilde{N}_1' & \tilde{D}_1' \end{bmatrix}\begin{bmatrix} D_1' & -\tilde{Y}_1' \\ N_1' & \tilde{X}_1' \end{bmatrix} = \begin{bmatrix} \dfrac{s+3}{s+2} & \dfrac{s+5}{s+2} \\ -\dfrac{1}{s+a} & \dfrac{s-1}{s+a} \end{bmatrix}\begin{bmatrix} \dfrac{s-1}{s+1} & -\dfrac{(s+5)(s+a)}{(s+1)(s+2)} \\ \dfrac{1}{s+1} & \dfrac{(s+3)(s+a)}{(s+1)(s+2)} \end{bmatrix} = \begin{bmatrix} 1 & 0 \\ 0 & 1 \end{bmatrix}.
$$

If all stabilizing H_2 are parametrically characterized by means of (4.123), then in view of Lemma 4.15, the closed-loop eigenvalues are in general the poles of $\begin{bmatrix} D_1' \\ N_1' \end{bmatrix}$ that are at -1, the poles of $\begin{bmatrix} X_1' & Y_1' \\ -\tilde{N}_1' & \tilde{D}_1' \end{bmatrix}$ that are at -2 and $-a$, and the poles of K'. Also, H_2 in this case is given by

$$H_2 = -\frac{\dfrac{s+5}{s+2} + K'\left(\dfrac{s-1}{s+a}\right)}{\dfrac{s+3}{s+2} - K'\left(\dfrac{1}{s+a}\right)} = -\frac{\dfrac{(s+5)(s+1)}{(s+1)(s+2)} + \left(\dfrac{s-1}{s+1}\right)K'}{\dfrac{(s+3)(s+a)}{(s+1)(s+2)} - \left(\dfrac{1}{s+1}\right)K'},$$

where $K' \in RH_\infty$. ■

Note that it is possible to select $\begin{bmatrix} X_1' & Y_1' \\ -\tilde{N}_1' & \tilde{D}_1' \end{bmatrix}$ so that the poles of X_1' and Y_1' are those of $-\tilde{N}_1'$ and $\tilde{D}_1'$. In the above example, this is the case when $a = 2$. This is also the case when these quantities are expressed in terms of a state-space realization of H_1, as is shown later in this subsection.

It is always possible of course to select K' so as to minimize the number of the closed-loop eigenvalues. This corresponds to minimizing the McMillan degree (number of poles) of H_2. This follows from the fact that any proper stabilizing controller can be expressed as (4.123) for appropriate K'. If for example H_1 can be stabilized by means of a real static H_2, then a $K' \in RH_\infty$ exists for this to happen. In this case the number of closed-loop eigenvalues is equal to the number of poles of H_1. It is not easy, however, to find such K' unless for example $X_1' = I$ and Y_1' is real, in which case $K' = 0$ and $H_2 = Y_1'$.

In view of Lemma 4.15, it is recommended that the number of poles in $\begin{bmatrix} D_1' \\ N_1' \end{bmatrix}$ and in $[-\tilde{N}_1', \tilde{D}_1']$ be taken to be the minimum possible, which is the number of poles in H_1. Also, the number of poles in $[X_1', Y_1']$ should be taken to be low. This is accomplished in a systematic way in the following, using state-space descriptions.

If the desired stabilizing H_2 is known, then the appropriate K' that will modify the initial $H_{20} = X_1'^{-1} Y_1'$, to yield the desired H_2, can be calculated from (4.122) as

$$K' = -(Y_1' + X_1' H_2)(\tilde{D}_1' - \tilde{N}_1' H_2)^{-1}. \tag{4.130}$$

EXAMPLE 4.10. In the above example $H_2 = -(b+1)$, $b > 0$ characterizes all static stabilizing H_2. Then for $a = 1$, we have

$$\begin{aligned} K' &= -\left[\frac{s+5}{s+2} - \frac{s+3}{s+2}(b+1)\right]\left(\frac{s-1}{s+1} + \frac{b+1}{s+1}\right)^{-1} \\ &= -\left(\frac{-bs - 3b + 2}{s+2}\right)\left(\frac{s+b}{s+1}\right)^{-1} = \frac{(s+1)(bs+3b-2)}{(s+2)(s+b)}, \end{aligned}$$

which will yield the desired $H_2 = -(b+1)$. The closed-loop eigenvalue is in this case at $-b$, as can easily be verified. ■

Parameters Q_2, X_2'

Parameters other than K' can also be used to parametrically characterize all stabilizing H_2. These parameters were introduced in Subsection 7.4A, and in the following these results are modified to accommodate the MFDs in RH_∞.

THEOREM 4.16. Let $H_1 = N_1' D_1' = \tilde{D}_1'^{-1} \tilde{N}_1'$ be doubly coprime MFDs in RH_∞ that satisfy (4.113). Then all H_2 that internally stabilize the closed-loop system are given by

(i)
$$\begin{aligned} H_2 &= (I + Q_2 H_1)^{-1} Q_2 = [D_1'^{-1}(I + Q_2 H_1)]^{-1}[D_1'^{-1} Q_2] \\ &= [(I + X_2' N'_1) D_1'^{-1}]^{-1} X_2', \end{aligned} \tag{4.131}$$

where Q_2 is such that $D_1'^{-1}[I+Q_2H_1, Q_2] \in RH_\infty$ and $(I+Q_2H_1)^{-1}$ exists and is proper; or where X_2' is such that $[(I+X_2'N_1')^{-1}D_1'^{-1}, X_2'] \in RH_\infty$ and $(I+X_2'N_1')^{-1}$ exists and is proper. Or by

(ii)
$$\begin{aligned} H_2 &= Q_2(I+H_1Q_2)^{-1} = [Q_2\tilde{D}_1'^{-1}][(I+H_1Q_2)\tilde{D}_1'^{-1}]^{-1} \\ &= \tilde{X}_2'[\tilde{D}_1'^{-1}(I+\tilde{N}_1'\tilde{X}_2')]^{-1}, \end{aligned} \tag{4.132}$$

where Q_2 is such that $\begin{bmatrix} Q_2 \\ I+H_1Q_2 \end{bmatrix}\tilde{D}_1'^{-1} \in RH_\infty$ and $(I+H_1Q_2)^{-1}$ exists and is proper; or where $\tilde{X}_2'$ is such that $\begin{bmatrix} \tilde{X}_2' \\ \tilde{D}_1'^{-1}(I+\tilde{N}_1'\tilde{X}_2') \end{bmatrix} \in RH_\infty$ and $(I+\tilde{N}_1'\tilde{X}_2')^{-1}$ exists and is proper.

Proof. First, notice the similarity of the results in this theorem and in Theorem 4.3 and Corollary 4.6 in Subsection 7.4A. To verify (i) directly, note that if $H_2 = \tilde{D}_2'^{-1}\tilde{N}_2'$, all solutions of $\tilde{D}_2'D_1' - \tilde{N}_2'N_1' = I$ are given by

$$[\tilde{D}_2', \tilde{N}_2'] = [(I+X_2'N_1')D_1'^{-1}, X_2'] \in RH_\infty, \tag{4.133}$$

where $X_2' \in RH_\infty$ is a parameter that we set equal to $\tilde{N}_2'$. Then in view of Corollary 4.12, the result in (i) that involves X_2' follows directly. Notice that $(I+X_2'N_1')D_1'^{-1}$ is biproper, and therefore, H_2 in (4.131) is proper. Now if $Q_2 = D_1'X_2'$, then the results involving Q_2 also follow. Note that $Q_2 \in RH_\infty$. Part (ii) can be verified in an analogous manner. Here $Q_2 = \tilde{X}_2'\tilde{D}_1'$. ■

In view of Lemma 4.15 and the discussion preceding it, the next result follows readily.

LEMMA 4.17. Let $H_1 = N_1'D_1'^{-1} = \tilde{D}_1'^{-1}\tilde{N}_1'$ be doubly coprime. If H_2 is given by (4.131), then the closed-loop eigenvalues are the poles of

$$\begin{aligned} \begin{bmatrix} D_1' \\ N_1' \end{bmatrix}[D_1'^{-1}(I+Q_2H_1), D_1'^{-1}Q_2] &= \begin{bmatrix} I \\ H_1 \end{bmatrix}[I+Q_2H_1, Q_2] \\ &= \begin{bmatrix} D_1' \\ N_1' \end{bmatrix}[(I+X_2'N_1')D_1'^{-1}, X_2']. \end{aligned} \tag{4.134}$$

If H_2 is given by (4.132), then the closed-loop eigenvalues are the poles of

$$\begin{aligned} \begin{bmatrix} Q_2\tilde{D}_1'^{-1} \\ (I+H_1Q_2)\tilde{D}_1'^{-1} \end{bmatrix}[\tilde{N}_1', \tilde{D}_1'] &= \begin{bmatrix} Q_2 \\ I+H_1Q_2 \end{bmatrix}[H_1, I] \\ &= \begin{bmatrix} \tilde{X}_2' \\ \tilde{D}_1'^{-1}(I+\tilde{N}_1'\tilde{X}_2') \end{bmatrix}[\tilde{N}_1', \tilde{D}_1']. \end{aligned} \tag{4.135}$$

Proof. The proof of this result is straightforward, in view of Lemma 4.15 and the discussion preceding it. ■

An interesting case is when H_1 is stable, as the following corollary shows.

COROLLARY 4.18. Let $H_1 \in RH_\infty$. Then all H_2 that internally stabilize the closed-loop feedback system are given by

$$H_2 = (I-K'H_1)^{-1}K' = K'(I-H_1K')^{-1}, \tag{4.136}$$

where $K' \in RH_\infty$ such that $(I-K'H_1)^{-1}$ [or $(I-H_1K')^{-1}$] exists and is proper. Furthermore, the closed-loop eigenvalues are the poles of

$$\begin{bmatrix} I \\ H_1 \end{bmatrix}[I, K']\begin{bmatrix} I & 0 \\ -H_1 & -I \end{bmatrix}. \tag{4.137}$$

Proof. Note that in this case (4.113) can be written as $\begin{bmatrix} I & 0 \\ -H_1 & I \end{bmatrix}\begin{bmatrix} I & 0 \\ H_1 & I \end{bmatrix} = \begin{bmatrix} I & 0 \\ 0 & I \end{bmatrix}$ for the doubly coprime MFD $H_1 = H_1 I^{-1} = I^{-1} H_1$. Then (4.123) of Theorem 4.13 reduces to expression (4.136) for H_2. The closed-loop eigenvalues are then given by the poles of (4.137), in view of Lemma 4.15. ■

Compare this corollary with Theorem 4.16, where H_2 is expressed in terms of Q_2. In this case it is clear that $K' = -Q_2$. Note that $K' \in RH_\infty$ suffices to guarantee stability.

MFDs and internal representations

Consider $H = N'D'^{-1} = \tilde{D}'^{-1}\tilde{N}'$, a doubly coprime factorization in RH_∞, i.e., (4.113) is satisfied. It is possible to express all proper and stable matrices in (4.113) in terms of the matrices of a state-space realization of the transfer function matrix $H(s)$. In particular, we have the following result.

LEMMA 4.19. Let $\{A, B, C, D\}$ be a stabilizable and detectable realization of $H(s)$, i.e., $H(s) = C(sI - A)^{-1}B + D$, which is also denoted by $H(s) \overset{s}{=} \begin{bmatrix} A & B \\ C & D \end{bmatrix}$, and with (A, B) stabilizable and (A, C) detectable. Let F be a state feedback gain matrix such that all the eigenvalues of $A + BF$ have negative real parts, and let K be an observer gain matrix such that all the eigenvalues of $A - KC$ have negative real parts. Define

$$U' = \begin{bmatrix} X' & Y' \\ -\tilde{N}' & \tilde{D}' \end{bmatrix} \overset{s}{=} \begin{bmatrix} A - KC & \vdots & B - KD & K \\ \cdots & \vdots & \cdots & \cdots \\ -F & \vdots & I & 0 \\ -C & \vdots & -D & I \end{bmatrix} \tag{4.138}$$

and

$$\hat{U}' = \begin{bmatrix} D' & -\tilde{Y}' \\ N' & \tilde{X}' \end{bmatrix} \overset{s}{=} \begin{bmatrix} A + BF & \vdots & B & K \\ \cdots & \vdots & \cdots & \cdots \\ F & \vdots & I & 0 \\ C + DF & \vdots & D & I \end{bmatrix}. \tag{4.139}$$

Then (4.113) holds and $H = N'D'^{-1} = \tilde{D}'^{-1}\tilde{N}'$ are coprime factorizations of H.

Proof. Relation (4.113) can be shown to be true by direct computation and is left to the reader to verify. Clearly, $U', \hat{U}' \in RH_\infty$. That N', D' and $\tilde{D}', \tilde{N}'$ are coprime is a direct consequence of (4.113). That $N'D'^{-1} = \tilde{D}'^{-1}\tilde{N}' = H$ can be shown by direct computation and is left to the reader. ■

In view of Lemma 4.19, U' and $U'^{-1} \in RH_\infty$ in (4.113) can be expressed as

$$U' = \begin{bmatrix} X' & Y' \\ -\tilde{N}' & \tilde{D}' \end{bmatrix} = \begin{bmatrix} -F \\ -C \end{bmatrix}[sI - (A - KC)]^{-1}[B - KD, K] + \begin{bmatrix} I & 0 \\ -D & I \end{bmatrix} \tag{4.140}$$

and

$$U'^{-1} = \begin{bmatrix} D' & -\tilde{Y}' \\ N' & \tilde{X}' \end{bmatrix} = \begin{bmatrix} F \\ C + DF \end{bmatrix}[sI - (A + BF)]^{-1}[B, K] + \begin{bmatrix} I & 0 \\ D & I \end{bmatrix}. \tag{4.141}$$

These formulas can be used as follows. A stabilizable and detectable realization $\{A, B, C, D\}$ of $H(s)$ is first determined, and appropriate F and K are found so that $A+BF$ and $A-KC$ have eigenvalues with negative real parts. Then U' and U'^{-1} are calculated from (4.140) and (4.141). Note that appropriate state feedback gain matrices F and observer gain matrices K can be determined, using the methods discussed in Chapter 4. The matrices F and K may be determined for example by solving appropriate optimal linear quadratic control and filtering problems. All proper stabilizing controllers $H_2 = N_2' D_2'^{-1} = \tilde{D}_2'^{-1} \tilde{N}_2'$ of the plant H_1 are then characterized as in Theorem 4.13.

It can now be shown, in view of Lemma 4.19, that all stabilizing controllers are described by

$$\dot{\hat{x}} = (A + BF - K(C + DF))\hat{x} + Ky + (B - KD)r_1$$
$$u = F\hat{x} + r_1, r_2 = y - (C + DF)\hat{x} - Dr_1, r_1 = K'(q)r_2, \tag{4.142}$$

which can be rewritten as

$$\dot{\hat{x}} = A\hat{x} + Bu + K(y - (C\hat{x} + Du))$$
$$u = F\hat{x} + K'(q)(y - (C\hat{x} + Du)). \tag{4.143}$$

Thus, every stabilizing controller is a combination of an asymptotic (full-state/full-order) estimator or observer and a stabilizing state feedback, plus $K'(q)r_2$ with $r_2 = y - (C\hat{x} + Du)$, the output "error" (see Fig. 7.9).

Let

$$H = N'D'^{-1} = \tilde{D}'^{-1}\tilde{N}' \tag{4.144}$$

be coprime factorizations in RH_∞, where $N', D', \tilde{N}', \tilde{D}' \in RH_\infty$. Also, let

$$H = ND^{-1} = \tilde{D}^{-1}\tilde{N} \tag{4.145}$$

be polynomial matrix coprime factorizations. The relation of proper and stable MFDs of $H(s)$ to internal PMDs of the system is established in the next result.

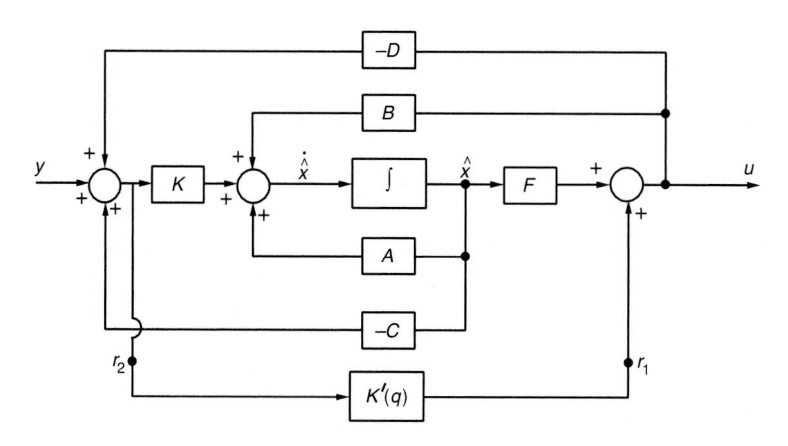

FIGURE 7.9
A state-space representation of all stabilizing controllers

THEOREM 4.20. (i) The pair $(N', D') \in RH_\infty$ defines an rc factorization of $H(s)$ as in (4.144) if and only if there exists a rational matrix Π with Π, Π^{-1} stable and $D\Pi$ biproper such that

$$\begin{bmatrix} D' \\ N' \end{bmatrix} = \begin{bmatrix} D \\ N \end{bmatrix} \Pi. \tag{4.146}$$

Furthermore, if only Π is stable and $D\Pi$ is proper, then (N', D') is a right factorization but it is not necessarily coprime.

(ii) Similarly, the pair $(\tilde{N}', \tilde{D}') \in RH_\infty$ defines an lc factorization of $H(s)$ in RH_∞ as in (4.144) if and only if there exists a rational matrix $\tilde{\Pi}$ with $\tilde{\Pi}, \tilde{\Pi}^{-1}$ stable and $\tilde{\Pi}\tilde{D}$ biproper such that

$$[\tilde{N}', \tilde{D}'] = \tilde{\Pi}[\tilde{N}, \tilde{D}]. \tag{4.147}$$

Furthermore, if only Π is stable and $\tilde{\Pi}\tilde{D}$ is proper, then $(\tilde{N}', \tilde{D}')$ is a left factorization but it is not necessarily coprime.

Proof. The proof of this result can be found in Antsaklis [4] and will not be repeated here. ∎

An interesting implication of Theorem 4.20 is the following: let $H = ND^{-1}$ be a right PMFD and let D be column proper; $Dz = u$, $y = Nz$ is a controllable PMD. Following the development in Subsection 7.4B, define the linear state feedback control law by (4.94), as $u = F(q)z + Gr$, $\det G \neq 0$, where $deg_{c_i} F(q) < \deg_{c_i} D(q)$. The closed-loop system is then $D_F z = Gr$, $y = Nz$ given in (4.95), where $D_F \triangleq D - F$. Now in view of Theorem 4.20, the relation

$$\begin{bmatrix} D' \\ N' \end{bmatrix} = \begin{bmatrix} D \\ N \end{bmatrix} D_F^{-1} G \tag{4.148}$$

defines rc proper and stable factorizations of H, when $\{D, I, N\}$ is detectable. If the pair (N, D) is rc, then $\Pi = D_F^{-1}G$. In fact, it is shown in Antsaklis [4] that all rc factorizations in RH_∞ may be obtained by means of (4.148), i.e., by using linear state feedback on controllable and detectable realizations of H. It is interesting to note that if $\{A, B, C, D\}$ is an equivalent state-space representation to $\{D, I, N\}$ with (F, G) the corresponding state feedback gain matrices, $\begin{bmatrix} D' \\ N' \end{bmatrix} = \begin{bmatrix} F \\ C + DF \end{bmatrix} [sI - (A + BF)]^{-1} BG + \begin{bmatrix} I \\ D \end{bmatrix} G$. This expression is the same as (4.141), when $G = I$. Similar results can be derived for the lc factorizations $(\tilde{N}', \tilde{D}')$ that are related to state observers.

As an application of (4.148), consider Theorem 4.8 and Corollary 4.12, where it is shown that if the given plant $H_1 = N_1 D_1^{-1} = N_1' D_1'^{-1}$ where N_1, D_1 are rc polynomial matrices and N_1', D_1' are proper and stable matrices, then H_2 is a stabilizing controller if and only if it can be written as $H_2 = \tilde{L}_2^{-1}\tilde{L}_1$, where $\tilde{L}_2 D_1 - \tilde{L}_1 N_1 = I$ (Theorem 4.8), or if and only if H_2 can be written as $H_2 = \tilde{D}_2'^{-1}\tilde{N}'_2$, where $\tilde{D}_2' D_1' - \tilde{N}_2' N_1' = I$ (Corollary 4.12). Assuming that D_1 is column proper, then in view of (4.148), the last relation can be written as $[(D_F^{-1}G)\tilde{D}_2']D_1 - [(D_F^{-1}G)\tilde{N}_2']N_1 = I$, and therefore, the relation between $\tilde{L}_2, \tilde{L}_1$ and $\tilde{D}_2', \tilde{N}_2'$ is given by

$$\begin{bmatrix} \tilde{D}_2' \\ \tilde{N}_2' \end{bmatrix} = G^{-1} D_F \begin{bmatrix} \tilde{L}_2 \\ \tilde{L}_1 \end{bmatrix}. \tag{4.149}$$

EXAMPLE 4.11. Let $H(s) = \dfrac{s-1}{(s-2)(s+1)} = ND^{-1}$, where $N = s - 1$ and $D = (s-2)(s+1)$.

(i) If $\Pi = \dfrac{1}{(s+1)^2}$, then $N' = N\Pi$, $D' = D\Pi$, and $X'D' + Y'N' = \dfrac{s-5}{s+1}\dfrac{s-2}{s+1} + 9\left[\dfrac{s-1}{(s+1)^2}\right] = 1.$

(ii) If $\Pi = \dfrac{s+2}{(s+1)^2(s+3)}$, then $X'D' + Y'N' = \dfrac{(s-5)(s+3)}{(s+1)(s+2)} \cdot \dfrac{(s-2)(s+2)}{(s+1)(s+3)} + 9\dfrac{(s+3)}{(s+2)} \cdot \dfrac{(s-1)(s+2)}{(s+1)^2(s+3)} = 1.$

Notice that in both (i) and (ii), Π, Π^{-1} are stable and $D\Pi$ is biproper as required by Theorem 4.20. ∎

The λ-approach

Given the transfer function $H(s)$, instead of obtaining proper and stable factorizations to characterize all proper stabilizing controllers via a Diophantine Equation over the ring of proper and stable rational functions, the transformation $\lambda = 1/(s+a)$, $a > 0$, may be used. Then one works with a Diophantine Equation that involves polynomial matrices in λ. This transformation maps the stable region in the s-plane into a "stable" region in the λ-plane and the point $s = \infty$ to the point $\lambda = 0$ (see Pernebo [29] for a detailed discussion). This approach corresponds to working with proper and stable factorizations of H, N', and D' with all the poles of N' and D' at $-a$ and requires only polynomial matrix manipulations [here $\Pi = (1/(s+a)^n)I$ in Theorem 4.20]. Recall that a rational $R(s)$ is proper (there are no poles at $s = \infty$) if and only if $R[(1 - \lambda a)/\lambda]$ has no pole at $\lambda = 0$. Therefore, for proper stabilizing controllers, solutions of appropriate polynomial Diophantine Equations are sought where the denominator $D_2(\lambda)$ of the stabilizing controller H_2 has no λ factors in $\det D_2(\lambda)$, i.e., $D_2^{-1}(\lambda)$ has no poles at $\lambda = 0$. Note that in this case, all the poles of the solutions of the corresponding proper and stable Diophantine Equation will also be at $-a$, and stabilizing controllers obtained by this method tend to assign multiple closed-loop eigenvalues at $-a$. As an illustration, consider $H(s) = (s-1)/[(s-2)(s+1)]$ (see the example above) and let $\lambda = 1/(s+1)$. Then $\hat{H}(\lambda) = (1-2\lambda)\lambda/(1-3\lambda)$. The polynomial Diophantine Equation in λ is solved to obtain $\hat{X}\hat{D} + \hat{Y}\hat{N} = (-6\lambda + 1)(1 - 3\lambda) + 9(1 - 2\lambda)\lambda = 1$. The corresponding (proper and stable) Diophantine Equation is obtained if we let $\lambda = 1/(s+1)$ in $\hat{X}$, $\hat{D}$, $\hat{Y}$, $\hat{N}$. Then the X', D', Y', N' of case (i) of the above example are derived. If the controller $u = Cy + r$ with $C(s) = -9(s+1)/(s-5)$ is used, all three closed-loop eigenvalues will be at -1. [$C(s) = \hat{C}(\lambda) = -\hat{X}^{-1}\hat{Y}^{-1}$ with $\lambda = 1/(s+1)$.]

D. Two Degrees of Freedom Controllers

Consider the two degrees of freedom controller S_C in the feedback configuration of Fig. 7.10. Here S_H represents the system to be controlled and is described by its transfer function matrix $H(s)$ so that

$$\hat{y}(s) = H(s)\hat{u}(s). \tag{4.150}$$

The two degrees of freedom controller S_C is described by its transfer function matrix $C(s)$ in

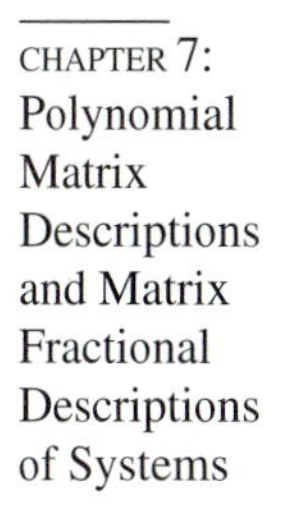

$$\hat{u}(s) = C(s)\begin{bmatrix}\hat{y}(s)\\ \hat{r}(s)\end{bmatrix} = [C_y(s), C_r(s)]\begin{bmatrix}\hat{y}(s)\\ \hat{r}(s)\end{bmatrix}. \tag{4.151}$$

Since the controller S_C generates the input u to S_H by processing independently y, the output of S_H, and r, the external input, it is called a two degrees of freedom controller.

In the following, we shall assume that H is a proper transfer function and shall determine proper controller transfer functions C that internally stabilize the feedback system in Fig. 7.10. The restriction that H and C are proper may easily be removed, if so desired. Note that in the development in Subsection 7.4C we assumed proper transfer functions in the feedback loop, while the development in Subsection 7.4A applies to nonproper transfer functions as well.

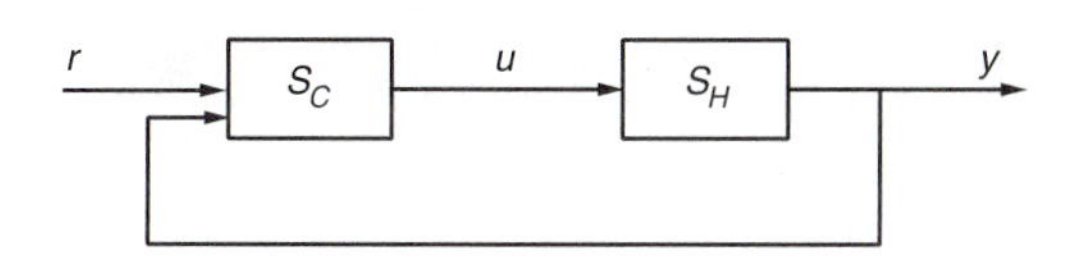

FIGURE 7.10
Two degrees of freedom controller S_C

Internal stability

THEOREM 4.21. Given is the proper transfer function H of S_H, and the proper transfer function C of S_C in (4.151), where $det(I - C_yH) \neq 0$. The closed-loop system in Fig. 7.10 is internally stable if and only if
(i) $\hat{u} = C_y\hat{y}$ internally stabilizes the system $\hat{y} = H\hat{u}$,
(ii) C_r is such that the rational matrix

$$M \triangleq (I - C_yH)^{-1}C_r \tag{4.152}$$

($u = Mr$) satisfies $D^{-1}M = X$, a stable rational matrix, where C_y satisfies (i) and $H = ND^{-1}$ is a right coprime polynomial matrix factorization.

Proof. Consider controllable and observable PMDs for S_H, given by

$$Dz = u, \qquad y = Nz, \tag{4.153}$$

and for S_C, given by

$$\tilde{D}_c\tilde{z}_c = [\tilde{N}_y, \tilde{N}_r]\begin{bmatrix}y\\ r\end{bmatrix}, \qquad u = \tilde{z}_c, \tag{4.154}$$

where the N, D are rc and the $\tilde{D}_c$, $[\tilde{N}_y, \tilde{N}_r]$ are lc polynomial matrices. The closed-loop system is then described by

$$(\tilde{D}_cD - \tilde{N}_yN)z = \tilde{N}_rr,\ y = Nz \tag{4.155}$$

and is internally stable if the roots of $det\,\tilde{D}_o$, where $\tilde{D}_o \triangleq \tilde{D}_cD - \tilde{N}_yN$, have strictly negative real parts.

(*Necessity*) Assume that the closed-loop system is internally stable, i.e., $\tilde{D}_o^{-1}$ is stable. Since $C_y = \tilde{D}_c^{-1}\tilde{N}_y$ is not necessarily an lc polynomial factorization, write $[\tilde{D}_c, \tilde{N}_y] = G_L[\tilde{D}_{C_y}, \tilde{N}_{C_y}]$, where G_L is a gcld of the pair $(\tilde{D}_c, \tilde{N}_y)$. Then $\tilde{D}_{C_y}D - \tilde{N}_{C_y}N = G_L^{-1}\tilde{D}_o = \tilde{D}_k$, where $\tilde{D}_k$ is a polynomial matrix with $\tilde{D}_k^{-1}$ stable; note also that G_L^{-1} is stable. Hence, $u = C_yy = \tilde{D}_{C_y}^{-1}\tilde{N}_{C_y}y$ internally stabilizes $y = Hu = ND^{-1}u$, i.e., part (i) of the theorem is true. To show that (ii) is true, we write $M = (I - C_yH)^{-1}C_r =$

$D\tilde{D}_k^{-1}\tilde{D}_{C_y}(\tilde{D}_c^{-1}\tilde{N}_r) = D\tilde{D}_k^{-1}G_L^{-1}\tilde{N}_r = DX$, where $X \triangleq \tilde{D}_o^{-1}\tilde{N}_r$ is a stable rational matrix. This shows that (ii) is also necessary.

(*Sufficiency*) Let C satisfy (i) and (ii) of the theorem. If $C = \tilde{D}_c^{-1}[\tilde{N}_y, \tilde{N}_r]$ is an lc polynomial MFD and G_L is a gcld of the pair $(\tilde{D}_c, \tilde{N}_y)$ then $[\tilde{D}_c, \tilde{N}_y] = G_L[\tilde{D}_{C_y}, \tilde{N}_{C_y}]$ is true for some lc matrices $\tilde{D}_{C_y}$ and $\tilde{N}_{C_y}(C_y = \tilde{D}_{C_y}^{-1}\tilde{N}_{C_y})$. Because (i) is satisfied, $\tilde{D}_{C_y}D - \tilde{N}_{C_y}N = \tilde{D}_k$, where $\tilde{D}_k^{-1}$ is stable. Premultiplying by G_L we obtain $\tilde{D}_cD - \tilde{N}_yN = G_L\tilde{D}_k$. Now if G_L^{-1} is stable, then $\tilde{D}_o^{-1}$, where $\tilde{D}_o \triangleq \tilde{D}_cD - \tilde{N}_yN = G_L\tilde{D}_k$, will be stable since $\tilde{D}_k^{-1}$ is stable. To show this, write $D^{-1}M = D^{-1}(I - C_yH)^{-1}C_r = \tilde{D}_k^{-1}\tilde{D}_{C_y}(\tilde{D}_c^{-1}\tilde{N}_r) = \tilde{D}_k^{-1}G_L^{-1}\tilde{N}_r$ and note that this is stable, in view of (ii). Observe now that the $G_L, \tilde{N}_r$ are lc; if they were not, then $C = \tilde{D}_c^{-1}[\tilde{N}_y, \tilde{N}_r]$ would not be a coprime factorization. In this case no unstable cancellations take place in $\tilde{D}_k^{-1}G_L^{-1}\tilde{N}_r$ ($\tilde{D}_k^{-1}$ is stable), and therefore, if $D^{-1}M$ is stable, then $(G_L\tilde{D}_k)^{-1} = \tilde{D}_o^{-1}$ is stable or the closed-loop system is internally stable. ∎

Remarks

(1) It is straightforward to show the same results, using proper and stable factorizations of H given by

$$H = N'D'^{-1}, \tag{4.156}$$

where the pair $(N', D') \in RH_\infty$ and (N', D') is rc, and of

$$C = \tilde{D}_c'^{-1}[\tilde{N}_y', \tilde{N}_r'], \tag{4.157}$$

where the pair $(\tilde{D}_c', [\tilde{N}_y', \tilde{N}_r']) \in RH_\infty$ and $(\tilde{D}_c', [\tilde{N}_y', \tilde{N}_r'])$ is lc. The proof is completely analogous and is left to the reader. The only change in the theorem will be in part (ii) which will now read:

(ii) C_r is such that the rational matrix $M \triangleq (I - C_yH)^{-1}C_r$ satisfies $D'^{-1}M = X' \in RH_\infty$, where C_y satisfies (i) and $H = N'D'^{-1}$ is an rc MFD in RH_∞.

(2) Theorem 4.21 separates the role of C_y, the feedback part of C, from the role of C_r, in achieving internal stability. Clearly, if only feedback action is considered, then only part (i) of the theorem is of interest; and if open-loop control is desired, then $C_y = 0$ and (i) implies that for internal stability H must be stable and $C_r = M$ must satisfy part (ii). In (ii) the parameter $M = DX$ appears naturally and in (i) the way is open to use any desired feedback parameterizations.

In view of Theorem 4.21, it is straightforward to parametrically characterize all internally stabilizing controllers C. In the theorem it is clearly stated [Part (i)] that C_y must be a stabilizing controller. Therefore, any parametric characterization of the ones developed in the previous subsections can be used for C_y. Also,
C_r is expressed in terms of $D^{-1}M = X$ (or $D'^{-1}M = X'$).

THEOREM 4.22. Given that $\hat{y} = H\hat{u}$ is proper with $H = ND^{-1} = \tilde{D}^{-1}\tilde{N}$ doubly coprime polynomial MFDs, all internally stabilizing proper controllers C in $\hat{u} = C\begin{bmatrix}\hat{y}\\ \hat{r}\end{bmatrix}$ are given by:

(i)
$$C = (I + QH)^{-1}[Q, M] = [(I + LN)D^{-1}]^{-1}[L, X], \tag{4.158}$$

where $Q = DL$ and $M = DX$ are proper with L, X and $D^{-1}(I + QH) = (I + LN)D^{-1}$ stable, so that $(I + QH)^{-1}$ exists and is proper; or

(ii)
$$C = (X_1 - K\tilde{N})^{-1}[-(X_2 + K\tilde{D}), X], \tag{4.159}$$

where K and X are stable so that $(X_1 - K\tilde{N}_1)^{-1}$ exists and C is proper. Also, X_1 and X_2 are determined from $UU^{-1} = \begin{bmatrix} X_1 & X_2 \\ -\tilde{N} & \tilde{D} \end{bmatrix}\begin{bmatrix} D & -\tilde{X}_2 \\ N & \tilde{X}_1 \end{bmatrix} = \begin{bmatrix} I & 0 \\ 0 & I \end{bmatrix}$ with U unimodular.

If $H = N'D'^{-1} = \tilde{D}'^{-1}\tilde{N}'$ are doubly coprime MFDs in RH_∞, then all stabilizing proper C are given by

(iii) $$C = (X_1' - K'\tilde{N}')^{-1}[-(X_2' + K'\tilde{D}_1'), X'], \tag{4.160}$$

where $K', X' \in RH_\infty$ so that $(X_1' - K'\tilde{N}')^{-1}$ exists and is proper. Also, $UU'^{-1} = \begin{bmatrix} X_1' & X_2' \\ -\tilde{N}' & \tilde{D}' \end{bmatrix}\begin{bmatrix} D' & -\tilde{X}_2' \\ N' & \tilde{X}_1' \end{bmatrix} = \begin{bmatrix} I & 0 \\ 0 & I \end{bmatrix}$ with $U', U'^{-1} \in RH_\infty$.

(iv) $$C = (I + QH)^{-1}[Q, M] = [(I + L'N')D'^{-1}]^{-1}[L', X'], \tag{4.161}$$

where $Q = D'L'$, $M = D'X' \in RH_\infty$ with L', X' and $D'^{-1}(I+QH) = (I+L'N')D'^{-1} \in RH_\infty$ so that $(I + QH)^{-1}$ or $(I + L'N')^{-1}$ exists and is proper.

Proof. In view of part (i) in Theorem 4.21, C_y of $C = [C_y, C_r]$ can be expressed in terms of any parameterization of all stabilizing feedback controllers developed in Subsections 7.4A and 7.4C. C_r can then be expressed as $C_r = (I - C_yH)M$ with C_y given from above and $M = DX$ with X any stable rational matrix.

If $C_y = (I + QH)^{-1}Q$ with $D^{-1}[I + QH, Q]$ stable (see Theorem 4.3) or $C_y = [(I+LN)D^{-1}]^{-1}L$ with $[(I+LN)D^{-1}, L]$ stable (see Corollary 4.6), part (i) of the theorem follows. Notice that $C_r = (I - C_yH)M = (I + QH)^{-1}M = [(I + LN)D^{-1}]^{-1}X$ since $Q = DL$ and $M = DX$. Similarly, if $C_y = -(X_1 - K\tilde{N})^{-1}(X_2 + K\tilde{D})$ with K stable (see Theorem 4.2), then $C_r = (X_1 - K\tilde{N})^{-1}D^{-1}M = (X_1 - K\tilde{N})^{-1}X$ and part (ii) of the theorem follows.

For part (iii), express C_y in terms of $K' \in RH_\infty$, as in Theorem 4.13, and use $X' \in RH_\infty$ in part (ii) of Theorem 4.20, as in remarks following (4.157). Part (iv) follows from Theorem 4.16 of Subsection 7.4C. ■

Response maps

It is straightforward to express the maps between signals of interest of Fig. 7.10 in terms of the parameters in Theorem 4.22. For instance $u = C\begin{bmatrix} y \\ r \end{bmatrix} = [C_y, C_r]\begin{bmatrix} y \\ r \end{bmatrix} = C_yHu + C_rr$, from which we have $u = (I - C_yH)^{-1}C_rr = Mr$. (In the following we will use the symbols u, y, r, etc., instead of $\hat{u}$, $\hat{y}$, $\hat{r}$, etc., for convenience.) If expressions (iv) of Theorem 4.22 are used, then

$$u = D'X'r, \qquad \text{and} \qquad y = Hu = N'D'^{-1}D'X'r = N'r, \tag{4.162}$$

in view of $(I - C_yH)^{-1} = D'(I + L'N')D'^{-1}$. Similar results can be derived using the other parameterizations in Theorem 4.22. To determine expressions for other maps of interest in control systems, consider Fig. 7.11, where d_u and d_y are assumed to be disturbances at the input and output of the plant H, respectively, and η denotes measurement noise. Then, $u = [C_y, C_r]\begin{bmatrix} y + d_y + \eta \\ r \end{bmatrix} + d_u$, from which we have $u = (I - C_yH)^{-1}[C_rr + C_yd_y + C_y\eta + d_u]$ and $y = Hu = H(I - C_yH)^{-1}[C_rr + C_yd_y + C_y\eta + d_u]$.

Then, in view of (4.161) in Theorem 4.22, we obtain

$$\begin{aligned} u &= D'X'r + D'L'd_y + D'L'\eta + D'(I + L'N')D'^{-1}d_u \\ &= Mr + Qd_y + Q\eta + S_id_u \end{aligned} \tag{4.163}$$

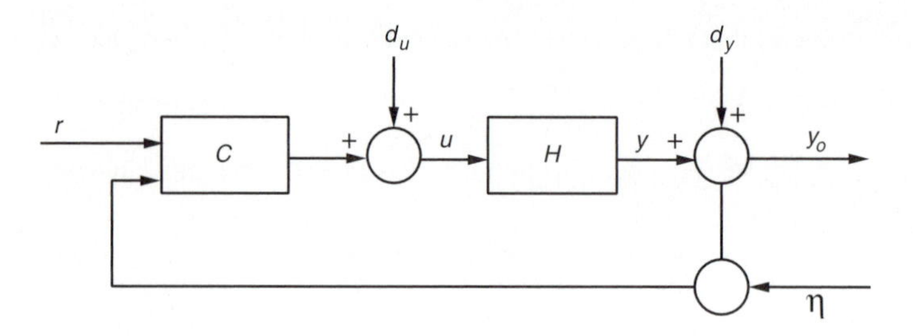

FIGURE 7.11
Two degrees of freedom control configuration

and
$$\begin{aligned} y &= N'X'r + N'L'd_y + N'L'\eta + N'(I + L'N')D'^{-1}d_u \\ &= Tr + (S_o - I)d_y + HQ\eta + HS_i d_u. \end{aligned} \tag{4.164}$$

Notice that $Q = (I - C_y H)^{-1} C_y = D'L'$ is the transfer function between u and d_y or η. Also,

$$S_i \triangleq (I - C_y H)^{-1} = D'(I + L'N')D'^{-1} = I + QH \tag{4.165}$$

is the transfer function between u and d_u. The matrix S_i is called the *input comparison sensitivity matrix*. Notice also that $y_o = y + d_y = Tr + S_o d_y + HQ\eta + HS_i d_u$; i.e.,

$$S_o = (I - HC_y)^{-1} = I + HQ \tag{4.166}$$

is the transfer function between y_o and d_y. The matrix S_o is called the *output comparison sensitivity matrix*. The sensitivity matrices S_i and S_o are important quantities in control design. Now

$$S_o - HQ = S_o - N'L' = I \tag{4.167}$$

since $HQ = H(I - C_y H)^{-1} C_y = HC_y(I - HC_y)^{-1} = -I + (I - HC_y)^{-1} = -I + S_o$, where S_o and HQ are the transfer functions from y_o to d_y and η, respectively. Equation (4.167) states that disturbance attenuation (or sensitivity reduction) and noise attenuation cannot occur over the same frequency range (show this). This is a fundamental limitation of the feedback loop and occurs also in two degrees of freedom control systems. Similarly, we note that

$$S_i - QH = I. \tag{4.168}$$

We now summarize some of the relations discussed above:

$$\begin{aligned} T &= H(I - C_y H)^{-1} C_r = HM = NX, && (y = Tr) \\ M &= (I - C_y H)^{-1} C_r = DX, && (u = Mr) \\ Q &= (I - C_y H)^{-1} C_y = DL, && (u = Qd_y) \\ S_o &= (I - HC_y)^{-1} = I + HQ, && (y_o = S_o d_y) \\ S_i &= (I - C_y H)^{-1} = I + QH, && (u = S_i d_u). \end{aligned}$$

The input-output maps attainable from r, using an internally stable two degrees of freedom configuration, can be characterized directly. In particular consider the two maps described by

$$\begin{bmatrix} y \\ u \end{bmatrix} = \begin{bmatrix} T \\ M \end{bmatrix} r, \tag{4.169}$$

i.e., the command/output map T and the command/input map M. Let $H = ND^{-1}$ be an rc polynomial MFD.

THEOREM 4.23. The stable rational function matrices T and M are realizable with internal stability by means of a two degrees of freedom control configuration [that satisfies (4.169)] if and only if there exists stable X so that

$$\begin{bmatrix} T \\ M \end{bmatrix} = \begin{bmatrix} N \\ D \end{bmatrix} X. \tag{4.170}$$

Proof. (*Necessity*) Assume that T and M in (4.169) are realizable with internal stability. Then in view of Theorem 4.20, $X \triangleq D^{-1}M$ is stable. Also, $y = Hu = (ND^{-1})(Mr) = NXr$.

(*Sufficiency*) Let (4.170) be satisfied. If X is stable then T and M are stable. Also, note that $T = HM$. We now show that in this case a controller configuration exists to implement these maps (see Fig. 7.12). Note that $u = \hat{M}r + C_y(\hat{T}r + y) = [C_y, \hat{M} + C_y\hat{T}]\begin{bmatrix} y \\ r \end{bmatrix}$, from which we obtain

$$u = (I + C_yH)^{-1}(\hat{M} + C_y\hat{T})r. \tag{4.171}$$

Now if $\hat{M} = M$ and $\hat{T} = T$, then in view of $T = HM$ this relation implies that $u = (I + C_yH)^{-1}(I + C_yH)Mr = Mr$ and $y = Hu = HMr = Tr$. Furthermore, C_y is a stabilizing feedback controller, and the system is internally stable since $\hat{T}$ and $\hat{M}$ are stable. ■

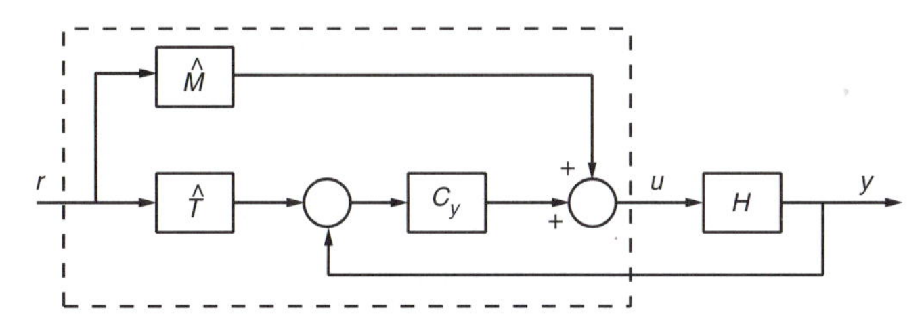

FIGURE 7.12
Feedback realization of (T, M)

Note that other internally stable controller configurations to attain these maps are possible. (The realization of both response maps T and M, instead of only T as in the case of the Model Matching Problem, makes the convenient formation in Theorem 4.24 possible. The realization of both T and M is sometimes referred to as the *Total Synthesis Problem;* see [7], [8] and the references therein.)

The results of Theorem 4.23 can be expressed in terms of $H = N'D'^{-1}$, rc MFDs in RH_∞. In particular, we have the following result.

THEOREM 4.24. $T, M \in RH_\infty$ are realizable with internal stability by means of a two degrees of freedom control configuration [that satisfies (4.169)] if and only if there exists

$X' \in RH_\infty$ so that

$$\begin{bmatrix} T \\ M \end{bmatrix} = \begin{bmatrix} N' \\ D' \end{bmatrix} X'. \tag{4.172}$$

Proof. The proof is completely analogous to the proof of Theorem 4.23 and is omitted. ∎

It is now clear that given any desirable response maps $\begin{bmatrix} y \\ u \end{bmatrix} = \begin{bmatrix} T \\ M \end{bmatrix} r$ such that $\begin{bmatrix} T \\ M \end{bmatrix} = \begin{bmatrix} N' \\ D' \end{bmatrix} X'$, where $X' \in RH_\infty$, the pair (T, M) can be realized with internal stability by using for instance a controller (4.161), $C = [(I + L'N')D'^{-1}]^{-1}[L', X']$, where $[(I + L'N')D'^{-1}, L'] \in RH_\infty$ and X' is given above, as can easily be verified. It is clear that there are many C that realize such T and M and they are all parameterized via the parameter $L' \in RH_\infty$ that, for internal stability, must satisfy the condition $(I + L'N')D'^{-1} \in RH_\infty$. Other parameterizations such as K' can also be used. In other words, the maps T, M can be realized by a variety of configurations, each with different feedback properties.

Remark

In a two degrees of freedom feedback control configuration, all admissible responses from r under condition of internal stability are characterized in terms of the parameters X (or M), while all response maps from disturbance and noise inputs that describe feedback properties of the system can be characterized in terms of parameters such as K or Q or L. This is the fundamental property of two degrees of freedom control systems: it is possible to attain the response maps from r independently from feedback properties such as response to disturbances and sensitivity to plant parameter variations.

EXAMPLE 4.12. We consider $H(s) = \dfrac{(s-1)(s+2)}{(s-2)^2}$ and wish to characterize all proper and stable transfer functions $T(s)$ that can be realized by means of some control configuration with internal stability. Let $H(s) = \dfrac{s-1}{s+2}\left[\dfrac{(s-2)^2}{(s+2)^2}\right]^{-1} = N'D'^{-1}$ be an rc MFD in RH_∞. Then in view of Theorem 4.24, all such T must satisfy $N'^{-1}T = \left(\dfrac{s+2}{s-1}\right)T = X' \in RH_\infty$. Therefore, any proper T with a zero at $+1$ can be realized via a two degrees of freedom feedback controller with internal stability.

Now if a single degree of freedom controller must be used, the class of realizable $T(s)$ under internal stability is restricted. In particular, if the unity feedback configuration $\{I; G_{ff}, I\}$ in Fig. 7.15 is used, then all proper and stable T that are realizable under internal stability are again given by $T = N'X' = \left(\dfrac{s-1}{s+2}\right)X'$, where $X' = L' \in RH_\infty$ [see (4.184)] and in addition $(I + X'N')D'^{-1} = \left[1 + X'\left(\dfrac{s-1}{s+2}\right)\right]\dfrac{(s+2)^2}{(s-2)^2} \in RH_\infty$, i.e., $X' = n_x/d_x$ is proper and stable and should also satisfy $(s+2)d_x + (s-1)n_x = (s-2)^2 p(s)$ for some polynomial $p(s)$.

This illustrates the restrictions imposed by the unity feedback controller, as opposed to a two degrees of freedom controller. Notice that these additional restrictions are imposed because the given plant has unstable poles. ∎

It is not difficult to prove the following result.

THEOREM 4.25. $T, M, S \in RH_\infty$ are realizable with internal stability by a two degrees of freedom control configuration that satisfies (4.169) and (4.167) [$S = S_o$, see Fig. 7.11 and (4.163), (4.164)] if and only if there exist $X', L' \in RH_\infty$ so that

$$\begin{bmatrix} T \\ M \\ S \end{bmatrix} = \begin{bmatrix} N' & 0 \\ D' & 0 \\ 0 & N' \end{bmatrix} \begin{bmatrix} X' \\ L' \end{bmatrix} + \begin{bmatrix} 0 \\ 0 \\ I \end{bmatrix}, \tag{4.173}$$

where $(I + L'N')D'^{-1} \in RH_\infty$. Similarly, $T, M, Q \in RH_\infty$ are realizable if and only if there exist $X', L' \in RH_\infty$ so that

$$\begin{bmatrix} T \\ M \\ Q \end{bmatrix} = \begin{bmatrix} N' & 0 \\ D' & 0 \\ 0 & D' \end{bmatrix} \begin{bmatrix} X' \\ L' \end{bmatrix}, \tag{4.174}$$

where $(I + L'N')D'^{-1} \in RH_\infty$.

Proof. The proof is straightforward in view of Theorem 4.24. Note that S or Q are selected in such a manner that the feedback loop has desirable feedback characteristics that are expressed in terms of these maps. ■

Controller implementations

The controller $C = [C_y, C_r]$ may be implemented, for example, as a system S_c as shown in Fig. 7.10 and described by (4.154), or as shown in Fig. 7.12 with $C = [C_y, M + C_y T]$, where C_y stabilizes H and T, M are desired stable maps that relate r to y and r to u, i.e., $y = Tr$ and $u = Mr$. There are also alternative ways of implementing a stabilizing controller C. In the following, the common control configuration of Fig. 7.13 is briefly discussed. It will be denoted by $\{R; G_{ff}, G_{fb}\}$.

The $\{R; G_{ff}, G_{fb}\}$ Configuration

Note that since

$$u = [C_y, C_r]\begin{bmatrix} y \\ r \end{bmatrix} = [G_{ff}G_{fb}, G_{ff}R]\begin{bmatrix} y \\ r \end{bmatrix}, \tag{4.175}$$

$\{R; G_{ff}, G_{fb}\}$ is a two degrees of freedom control configuration that is as general as the ones discussed before. To see this, let $C = [C_y, C_r] = \tilde{D}'^{-1}_c[\tilde{N}'_y, \tilde{N}'_r]$ be an lc MFD in RH_∞ and let

$$R = \tilde{N}'_r, G_{ff} = \tilde{D}'^{-1}_c, \qquad G_{fb} = \tilde{N}'_y. \tag{4.176}$$

Note that R and G_{fb} are always stable; also, G_{ff}^{-1} exists and is stable. Assume now that C was chosen so that

$$\tilde{D}'_c D' - \tilde{N}'_y N' = \tilde{U}', \tag{4.177}$$

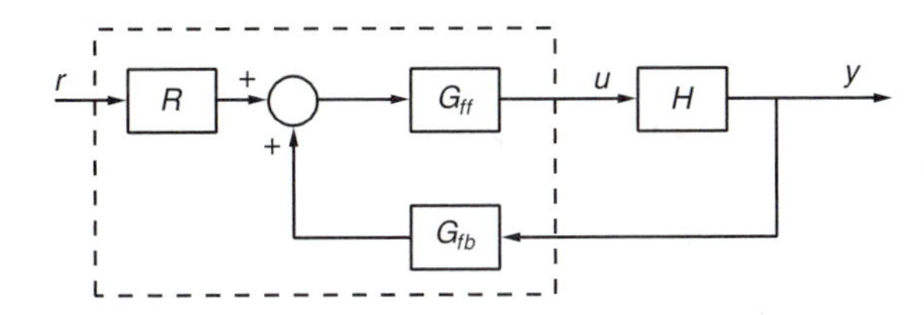

FIGURE 7.13
Two degrees of freedom controller $\{R; G_{ff}, G_{fb}\}$

where $\tilde{U}', \tilde{U}'^{-1} \in RH_\infty$. Then the system in Fig. 7.13 with R, G_{ff}, and G_{fb} given in (4.176) is internally stable, as is shown in the following.

In view of Theorem 4.21, the feedback system is stable if and only if (i) $C_y = G_{ff}G_{fb} = \tilde{D}'_c{}^{-1}\tilde{N}'_y$ internally stabilizes H, and (ii) $X' = D'^{-1}M = D'^{-1}(I - G_{ff}G_{fb}H)^{-1}G_{ff}R \in RH_\infty$. It can be shown that (4.177) implies (i). Note that any possible cancellation in the product $G_{ff}G_{fb}$, between $\tilde{D}'_c{}^{-1}$ and $\tilde{N}'_y$, will involve only stable poles; this can easily be shown, using (4.177). The cancelled stable poles will be uncontrollable or unobservable eigenvalues in the closed-loop system. In addition $X' = D'^{-1}(I - G_{ff}G_{fb}H)^{-1}G_{ff}R = D'^{-1}[D'(\tilde{D}'_c D' - \tilde{N}'_y N')^{-1}\tilde{D}'_c]\tilde{D}'^{-1}_c\tilde{N}'_r = \tilde{U}^{-1}\tilde{N}'_r \in RH_\infty$. Therefore, the control configuration $\{R; G_{ff}, G_{fb}\} = \{\tilde{N}'_r; \tilde{D}'^{-1}_c, \tilde{N}'_y\}$ with (4.177) satisfied is internally stable. In view of this result and Theorem 4.22, G_{ff}, G_{fb}, and R may also be selected as

$$R = X', G_{ff} = [(I + L'N')D'^{-1}]^{-1}, \qquad G_{fb} = L' \tag{4.178}$$

with $X', L', (I + L'N')D'^{-1} \in RH_\infty$ and $det\,(I + L'N') \neq 0$, where $X' = D'^{-1}M$ and $L' = D'^{-1}Q$. Now if X' and L' satisfy (4.173) or (4.174) of Theorem 4.25, desired command responses T, M and disturbance responses S or Q are achieved under internal stability. Note that G_{ff}, G_{fb}, and R may also be selected as

$$R = X', G_{ff} = (X'_1 - K'\tilde{N}')^{-1}, G_{fb} = -(X'_2 + K'\tilde{N}'), \tag{4.179}$$

where $X', K' \in RH_\infty$.

We shall now briefly discuss some special cases of the $\{R; G_{ff}, G_{fb}\}$ control configuration, which are quite common in practice. Note that the configurations below are simpler; however, they restrict the choices of attainable response maps and so the flexibility offered to the control designer is reduced when using these configurations.

1. The $\{I; G_{ff}, G_{fb}\}$ controller. In this case $u = [C_y, C_r]\begin{bmatrix} y \\ r \end{bmatrix} = [G_{ff}G_{fb}, G_{ff}]\begin{bmatrix} y \\ r \end{bmatrix}$, that is,

$$C_y = C_r G_{fb}. \tag{4.180}$$

In view of (4.161) given in Theorem 4.22, this implies that

$$L' = X'G_{fb}, \tag{4.181}$$

or that the choice for the parameters L' and X' is not completely independent as in the $\{R; G_{ff}, G_{fb}\}$ case. The L' and X' must of course satisfy L', X' and $(I + L'N')D'^{-1} \in RH_\infty$. In addition in this case L' and X' must be such that a proper solution G_{fb} of (4.181) exists and no unstable poles cancel in $X'G_{fb}$. Note that these poles will cancel in the product $G_{ff}G_{fb}$ and will lead to an unstable system. Since L' and X' are

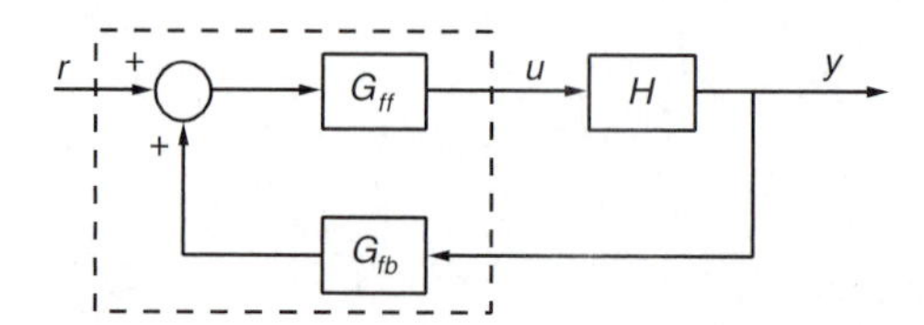

FIGURE 7.14
The $\{I; G_{ff}, G_{fb}\}$ controller

both stable, we will require that (4.181) have a solution $G_{fb} \in RH_\infty$. This implies that if for example X'^{-1} exists, then X' and L' must be such that $X'^{-1}L' \in RH_\infty$, i.e., X' and L' have the same unstable zeros and L' is "more proper" than X'. This provides some guidelines about the conditions that X' and L' must satisfy. Also,

$$G_{ff} = [(I + L'N')D'^{-1}]^{-1}X'. \tag{4.182}$$

It should be noted that the state feedback law implemented by a dynamic observer can be represented as an $\{I; G_{ff}, G_{fb}\}$ controller (see Subsection 7.4B, Fig. 7.8).

2. The $\{I; G_{ff}, I\}$ controller. A special case of (i) is the unity feedback control configuration. Here $u = [C_y, C_r]\begin{bmatrix} y \\ r \end{bmatrix} = [G_{ff}, G_{ff}]\begin{bmatrix} y \\ r \end{bmatrix}$; i.e.,

$$C_r = C_y, \tag{4.183}$$

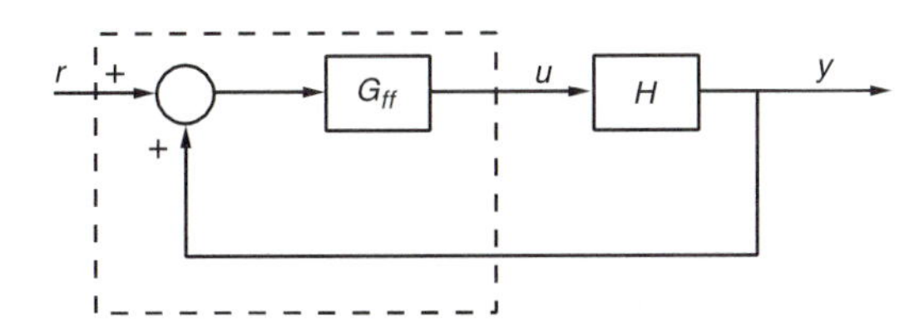

FIGURE 7.15
The $\{I; G_{ff}, I\}$ controller

which in view of (4.161) implies that

$$X' = L'. \tag{4.184}$$

In this case the responses between y or u and r (characterized by X') cannot be designed independently of feedback properties such as sensitivity (characterized by L'). This is a single degree of freedom controller and is used primarily to attain feedback control specifications.

3. The $\{R; G_{ff}, I\}$ controller. Here $u = [C_y, C_r]\begin{bmatrix} y \\ r \end{bmatrix} = [G_{ff}, G_{ff}R]\begin{bmatrix} y \\ r \end{bmatrix}$; i.e.,

$$C_r = C_y R. \tag{4.185}$$

In view of (4.161) given in Theorem 4.22, this implies that

$$X' = L'R. \tag{4.186}$$

The L' and X' must satisfy $L', X', (I + L'N')D'^{-1} \in RH_\infty$. In addition they must be such that (4.186) has a solution $R \in RH_\infty$. Note that R stable is necessary for internal

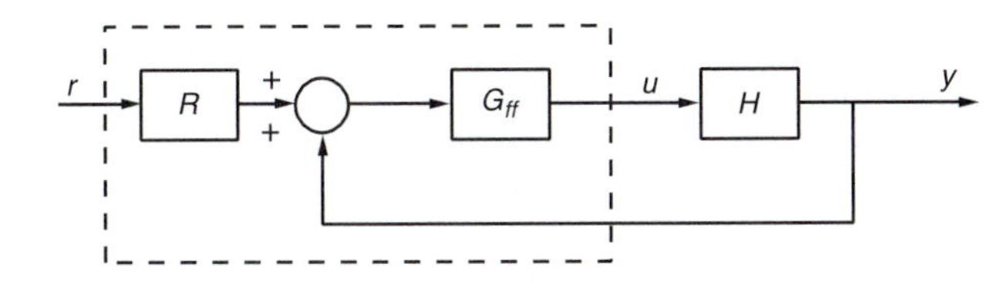

FIGURE 7.16
The $\{R; G_{ff}, I\}$ controller

stability. The reader should refer to the discussion in (1) above for the implications of such assumptions on X' and L'. Also,

$$G_{ff} = [(I + L'N')D'^{-1}]^{-1}L'. \tag{4.187}$$

4. *The* $\{R; I, G_{fb}\}$ *controller.* In this case

$$u = [C_y, C_r]\begin{bmatrix} y \\ r \end{bmatrix} = [G_{fb}, R]\begin{bmatrix} y \\ r \end{bmatrix}. \tag{4.188}$$

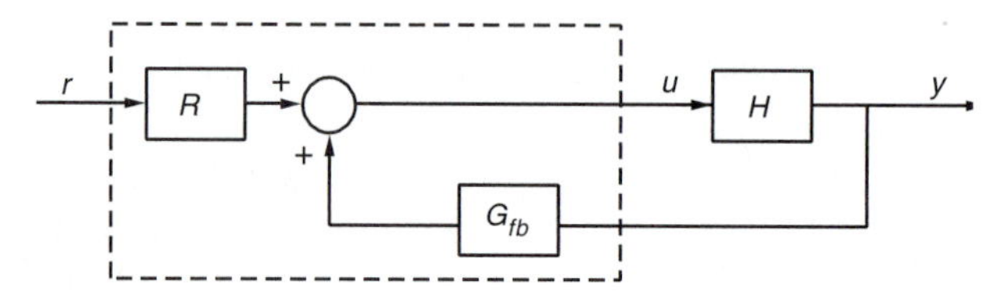

FIGURE 7.17
The $\{R; I, G_{fb}\}$ controller

For internal stability, R must be stable. In view of (4.161) given in Theorem 4.22, this implies the requirement $[(I + L'N')D'^{-1}]^{-1}X' \in RH_\infty$, in addition to $L', X', (I + L'N')D^{-1} \in RH_\infty$, which imposes significant additional restrictions on L'. Here

$$[G_{fb}, R] = [(I + L'N')D'^{-1}]^{-1}[L', X']. \tag{4.189}$$

5. *The* $\{I; I, G_{fb}\}$ *controller.* This is a special case of (4), a single degree of freedom case, where $R = I$. Here $R = I$ implies that

$$X' = (I + L'N')D'^{-1}, \tag{4.190}$$

or that X' and L' must satisfy additionally the relation

$$D'X' - L'N' = I, \tag{4.191}$$

a (skew) Diophantine Equation. This is in addition to the condition that $L', X', (I + L'N')D^{-1} \in RH_\infty$.

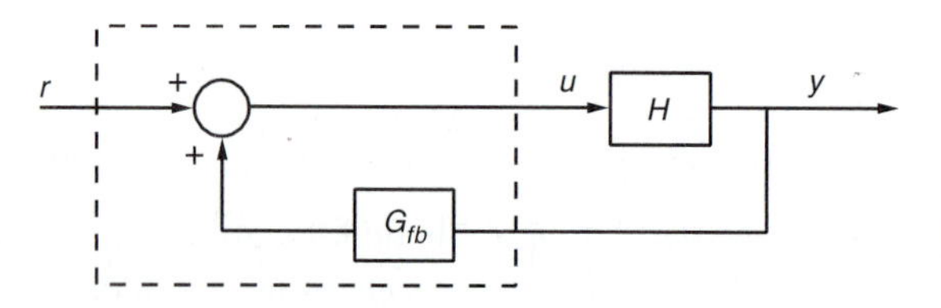

FIGURE 7.18
The $\{I; I, G_{fb}\}$ controller

Control problems

In control problems, design specifications typically include requirements for internal stability or pole placement, low sensitivity to parameter variations, disturbance attenuation, and noise reduction. Also, requirements such as model matching, diagonal decoupling, static decoupling, regulation, and tracking are included in the specifications.

Internal stability has of course been a central theme throughout the book, and in this section all stabilizing controllers were parameterized. Pole placement was studied in Chapter 4, using state feedback, and output feedback via the Diophantine Equation was addressed earlier in Chapter 7. Sensitivity and disturbance/noise reduction are treated by appropriately selecting the feedback controller C_y. Methodologies to accomplish these control goals, frequently in an optimal way, are developed in many control books. It should be noted that many important design approaches such as the H_∞-optimal control design method, are based on the parameterizations of all feedback stabilizing controllers discussed earlier. In particular an appropriate or optimal controller is selected by restricting the parameters used, so that additional control goals are accomplished optimally, while guaranteeing internal stability in the loop.

Our development of the theory of two degrees of freedom controllers can be used directly to study model matching and decoupling, and a brief outline of this approach is now given. Note that this does not, by any means, constitute a complete treatment of these important control problems, but rather, an illustration of the methodologies introduced in this section.

In the *model matching* problem, the transfer function of the plant $H(s)$ $(y = Hu)$ and a desired transfer function $T(s)$ $(y = Tr)$ are given and a transfer function $M(s)$ $(u = Mr)$ is sought so that

$$T(s) = H(s)M(s). \tag{4.192}$$

Typically, $H(s)$ is proper, and the proper and stable $T(s)$ is to be obtained from $H(s)$ using a controller under the condition of internal stability. Therefore, $M(s)$ can in general not be implemented as an open-loop controller, but rather, as a two degrees of freedom controller. In view of Theorem 4.24 (or Theorem 4.23), if $H = N'D'^{-1}$ is an rc MFD in RH_∞, then the pair (T, M) can be realized with internal stability if and only if there exists $X' \in RH_\infty$ so that $\begin{bmatrix} T \\ M \end{bmatrix} = \begin{bmatrix} N' \\ D' \end{bmatrix} X'$. Note that an M that satisfies (4.192) must first be selected (there may be an infinite number of solutions M). In the case when $det\, H(s) \neq 0$, T can be realized with internal stability by means of a two degrees of freedom control configuration if and only if $N'^{-1}T = X' \in RH_\infty$ (see Example 4.12). In this case $M = D'X'$. Now if the model matching is to be achieved by a more restricted control configuration, then additional conditions are imposed on T for this to happen, which are expressed in terms of X' (see for instance Example 4.12 for the case of the unity feedback configuration and Exercises 7.23, 7.2b.).

In the problem of *diagonal decoupling,* $T(s)$ in (4.192) is not completely specified, but is required to be diagonal, proper, and stable. In this problem the first input affects only the first output, the second input affects only the second output, and so forth. If $H(s)^{-1}$ exists, then diagonal decoupling under internal stability via a two degrees of freedom control configuration is possible if and only if

$$N'^{-1}T = N'^{-1} \begin{bmatrix} \frac{n_1}{d_1} & & \\ & \ddots & \\ & & \frac{n_m}{d_m} \end{bmatrix} = X' \in RH_\infty, \tag{4.193}$$

where $H = N'D'^{-1}$ is an rc MFD in RH_∞ and $T(s) = diag\,[n_i(s)/d_i(s)], i = 1, \ldots, m$. It is clear that if $H(s)$ has only stable zeros, then no additional restrictions are imposed on $T(s)$. Relation (4.193) implies restrictions on the zeros of $n_i(s)$ when $H(s)$ has unstable zeros.

It is straightforward to show that if diagonal decoupling is to be accomplished by means of more restricted control configurations, then additional restrictions will be imposed on $T(s)$ via X'. (See Exercise 7.25 and Exercise 4.17, 4.20 of Chapter 4 for the case of diagonal decoupling via linear state feedback.)

The problem of diagonal decoupling has a long and interesting history and a very rich literature. The original solution of the problem involved linear state feedback and state-space descriptions and is due to Falb and Wolovich [12]. For an approach involving PMDs and the transfer function matrix, see Williams and Antsaklis [35]. Other types of decoupling, such as block, dynamic, and static, are also treated there.

A problem closely related to diagonal decoupling is that of the *inverse* of $H(s)$. In this case, $T(s) = I$. There is also a very rich literature on this problem and the interested reader is encouraged to find out more about it. A starting point could be Williams and Antsaklis [35]. (See also Exercises 4.17, 4.18 in Chapter 4.)

In the problem of *static decoupling*, $T(s) \in RH_\infty$ is square and also satisfies $T(0) = \Lambda$, a real nonsingular diagonal matrix. An example of such $T(s)$ is $T(s) = \frac{1}{d(s)}\begin{bmatrix} s^2+1 & s(s^2+2) \\ s(s+2) & s^2+3s+1 \end{bmatrix}$, where $d(s)$ is a Hurwitz polynomial. Note that if $T(0) = \Lambda$, then a step change in the first input r will affect only the first output in y at steady-state, and so forth. Here $y = Tr = T(1/s)$ and $\lim_{s\to 0} sT(1/s) = T(0) = \Lambda$, which is diagonal and nonsingular. For this to happen, with internal stability when $H(s)$ is nonsingular (see Theorem 4.24), we must have $N'^{-1}T = X' \in RH_\infty$, from which can be seen that static decoupling is possible if and only if $H(s)$ does not have zeros at $s = 0$. If this is the case and if in addition $H(s)$ is stable, static decoupling can be achieved with just a precompensation by a real gain matrix G, where $G = H^{-1}(0)\Lambda$. In this case $T(s) = H(s)G = H(s)H^{-1}(0)\Lambda$, from which $T(0) = \Lambda$.

7.5 SUMMARY

In this chapter alternatives to state-space descriptions were introduced and used to further study the behavior of linear time-invariant systems and to study in depth structural properties of feedback control systems.

In Part 1, the properties of systems described by Polynomial Matrix Descriptions (PMDs) were explored in Section 7.3 and background on polynomial matrices was provided in Section 7.2. The Diophantine Equation, which plays an important role in feedback systems, was studied at length in Subsection 7.2E.

An in-depth study of the theory of parameterizations of all stabilizing controllers with emphasis on PMDs was undertaken in Part 2, Subsection 7.4A, and the parameterizations of all proper stabilizing controllers in terms of proper and stable Matrix Fraction Descriptions (MFDs) were derived in Subsection 7.4C. State feedback and state estimation using PMDs were studied in Subsection 7.4B. Finally, control systems with two degrees of freedom controllers were explored in Subsection 7.4D, with an emphasis on stability, parameterizations of all stabilizing controllers, and attainable response maps.

7.6 NOTES

Two books that are original sources on the use of polynomial matrix descriptions in systems and control are Rosenbrock [30] and Wolovich [36]. In the former, what is now called Rosenbrock's Matrix is employed and relations to state-space descriptions are emphasized. In the latter, what are now called Polynomial Matrix Fractional Descriptions (PMFDs) are emphasized, and the relation to state space is accomplished primarily by using controller forms and the Structure Theorem, which was presented in Chapter 3. Good general sources for the polynomial matrix description approach also include the books by Vardulakis [33], Kailath [21], and Chen [10].

Basic references for the material on polynomial matrices discussed in Section 7.2 are the books by MacDuffee [25] and Gantmacher [14]. These books also include material on the Diophantine Equation discussed in Subsection 7.2E. Additional sources for the properties of polynomial matrices that we found useful include Wolovich [36], Vardulakis [33], and Kailath [21].

A key concept in our development of polynomial descriptions for the study of systems is the notion of equivalence of representations, discussed in Subsection 7.3A, since it establishes not only relations between polynomial descriptions but also between polynomial and state-space representations. Original sources for this include Rosenbrock [30] and Fuhrmann [13]. See also Pernebo [28] and the comments by Rosenbrock in [31] and [32], noting that the definition of equivalence given by Wolovich [36] was shown by Pernebo in [28] to be the same as strict system equivalence. Additional material on this topic can be found in Kailath [21], Wolovich [36], and Vardulakis [33]. A good source for the study of feedback systems using PMDs and MFDs is the book by Callier and Desoer [9].

The development of the properties of interconnected systems, addressed in Subsection 7.3C, which include controllability, observability, and stability of systems in parallel, in series, and in feedback configurations is based primarily on the approach taken in Antsaklis and Sain [8], Antsaklis [2] and [3], and Gonzalez and Antsaklis [18].

Parameterizations of all stabilizing controllers are of course very important in control theory today. Historically, their development appears to have evolved in the following manner (see also the historical remarks on the Diophantine Equation in Subsection 7.2E): Youla et al. [37] introduced the K parameterization (as in Theorem 4.2) in 1976 and used it in the Wiener-Hopf design of optimal controllers. This work is considered to be the seminal contribution in this area. The proofs of the results on the parameterizations in Youla et al. [37] involve transfer functions and their characteristic polynomials. Neither the Diophantine Equation nor PMDs of the system are used (explicitly). It should be recalled that in the middle 1970s most of the control results in the literature concerning MIMO systems involved state-space descriptions and a few employed transfer function matrices. The PMD descriptions of systems presented in the books by Rosenbrock [30] and Wolovich [36] were only beginning to make some impact. A version of the linear Diophantine Equation, namely, $AX + YB = C$ polynomial in z^{-1} was used in control design by Kucera in work reported in 1974 and 1975. In that work, parameterizations of all stabilizing controllers were implicit, in the sense that the stabilizing controllers were expressed in terms of the general solution of the Diophantine Equation, which in turn can be described parametrically. Explicit parameterizations were reported in Kucera [23]. In Antsaklis

[1] the doubly coprime MFDs were used with the polynomial Diophantine Equation, working over the ring of polynomials, to derive the results by Youla in an alternative way. In Desoer et al. [11] parameterizations K' of all stabilizing controllers using coprime MFDs in rings other than polynomial rings (including the ring of proper and stable rational functions) were derived. It should also be noted that proper and stable MFDs had apparently been used earlier by Vidyasagar. In Zames [38], a parameterization Q of all stabilizing controllers, but only for stable plants, was introduced and used in H_∞-optimal control design. (Similar parameterizations were also used elsewhere, but apparently not to characterize all stabilizing controllers; for example, they were used in the design of the closed-loop transfer function in control systems and in sensitivity studies in the 50s and 60s, and also in the "internal model control" studies in chemical process control in the 80s.) A parameterization X of all stabilizing controllers (where X is closely related to the attainable response in an error feedback control system), valid for unstable plants as well, was introduced in Antsaklis and Sain [7]. Parameterizations involving proper and stable MFDs were further developed in the 80s in connection with optimal control design methodologies, such as H_∞-optimal control, and connections to state-space approaches were derived. Two degrees of freedom controllers were also studied, and the limitations of the different control configurations became better understood. By now, MFDs and PMDs have become important system representations and their study is essential, if optimal control design methodologies are to be well understood.

In Subsection 7.4A, the discussion of parameters N_k, D_k, and K (Theorems 4.1 and 4.2) follows Antsaklis [1]. The material for the parameter X_2 (Corollary 4.6) follows Antsaklis and Sain [7], where X_2 was introduced in connection with the error feedback control configuration. Q_2 was used in Zames [38] for stable systems (Corollary 4.4); however, Theorem 4.3 is valid for unstable systems as well. The discussion of the parameters S_{12} and L_1, L_2 follows Antsaklis and Sain [8], and Antsaklis [3]. Subsection 7.4B is based on Wolovich [36] and Antsaklis [2] and [3].

The results on proper and stable MFDs in Subsection 7.4C and their use in parameterizing all proper stabilizing controllers are due to Desoer et al. [11]. The development in Subsection 7.4C was based on Vidyasagar [34], Antsaklis [3], Green and Limebeer [20], and Maciejowski [26]. The development of the relations between MFDs in RH_∞ and PMDs of a system follows Antsaklis [4], and Nett et al. [27]; see also Khargonekar and Sontag [22]. The λ-approach was developed in Pernebo [29].

The material on two degrees of freedom controllers in Subsection 7.4D is based on Antsaklis [3], Antsaklis and Gonzalez [6], and Gonzalez and Antsaklis [16], [17], [18], [19]; a good source for this topic is also Vidyasagar [34]. Note that the main stability theorem (Theorem 4.21) first appeared in Antsaklis [3] and Antsaklis and Gonzalez [6]. For additional material on model matching and decoupling, consult Chen [10], Kailath [21], Falb and Wolovich [12], Williams and Antsaklis [35], and the extensive list of references therein.

7.7 REFERENCES

1. P. J. Antsaklis, "Some Relations Satisfied by Prime Polynomial Matrices and Their Role in Linear Multivariable System Theory," *IEEE Trans. on Autom. Control,* Vol. AC-24, No. 4, pp. 611–616, August 1979.

2. P. J. Antsaklis, *Notes on: Polynomial Matrix Representation of Linear Control Systems,* Pub. No. 80/17, Dept. of Elec. Engr., Imperial College, London, 1980.
3. P. J. Antsaklis, Lecture Notes of the graduate course, *Feedback Systems,* University of Notre Dame, Spring 1985.
4. P. J. Antsaklis, "Proper, Stable Transfer Matrix Factorizations and Internal System Descriptions," *IEEE Trans. on Autom. Control,* Vol. AC-31, No. 7, pp. 634–638, July 1986.
5. P. J. Antsaklis, "On the Order of the Compensator and the Closed-Loop Eigenvalues in the Fractional Approach to Design," *Intern. J. Control,* Vol. 49, No. 3, pp. 929–936, 1989.
6. P. J. Antsaklis and O. R. Gonzalez, "Stability Parameterizations and Stable Hidden Modes in Two Degrees of Freedom Control Design," *Proc. of the 25th Annual Allerton Conference on Communication, Control and Computing,* pp. 546–555, Monticello, IL, Sept. 30–Oct. 2, 1987.
7. P. J. Antsaklis and M. K. Sain, "Unity Feedback Compensation of Unstable Plants," *Proc. of the 20th IEEE Conf. on Decision and Control,* pp. 305–308, San Diego, December 1981.
8. P. J. Antsaklis and M. K. Sain, "Feedback Controller Parameterizations: Finite Hidden Modes and Causality," in *Multivariable Control: New Concepts and Tools,* S. G. Tzafestas, ed., D. Reidel Pub., Dordrecht, Holland, 1984.
9. F. M. Callier and C. A. Desoer, *Multivariable Feedback Systems,* Springer-Verlag, New York, 1982.
10. C. T. Chen, *Linear System Theory and Design,* Holt, Rinehart and Winston, New York, 1984.
11. C. A. Desoer, R. W. Liu, J. Murray, and R. Saeks, "Feedback System Design: The Fractional Approach to Analysis and Synthesis," *IEEE Trans. on Autom. Control,* Vol. AC-25, pp. 399–412, June 1980.
12. P. L. Falb and W. A. Wolovich, "Decoupling in the Design of Multivariable Control Systems," *IEEE Trans. on Autom. Control,* Vol. AC-12, pp. 651–659, 1967.
13. P. A. Fuhrmann, "On Strict System Equivalence and Similarity," *Int. J. Control,* Vol. 25, pp. 5–10, 1977.
14. F. R. Gantmacher, *The Theory of Matrices,* Vol. 1 and 2, Chelsea, New York, 1959.
15. Z. Gao and P. J. Antsaklis, "On Stable Solutions of the One and Two Sided Model Matching Problems," *IEEE Trans. on Autom. Control,* Vol. 34, No. 9, pp. 978–982, September 1989.
16. O. R. Gonzalez and P. J. Antsaklis, "Implementations of Two Degrees of Freedom Controllers," *Proc. of the 1989 American Control Conference,* pp. 269–273, Pittsburgh, PA, June 21–23, 1989.
17. O. R. Gonzalez and P. J. Antsaklis, "Sensitivity Considerations in the Control of Generalized Plants," *IEEE Trans. on Autom. Control,* Vol. 34, No. 8, pp. 885–888, August 1989.
18. O. R. Gonzalez and P. J. Antsaklis, "Hidden Modes of Two Degrees of Freedom Systems in Control Design," *IEEE Trans. on Autom. Control,* Vol. 35, No. 4, pp. 502–506, April 1990.
19. O. R. Gonzalez and P. J. Antsaklis, "Internal Models in Regulation, Stabilization and Tracking," *Int. J. of Control,* Vol. 53, No. 2, pp. 411–430, 1991.
20. M. Green and D. J. N. Limebeer, *Linear Robust Control,* Prentice Hall, Englewood Cliffs, NJ, 1995.
21. T. Kailath, *Linear Systems,* Prentice-Hall, Englewood Cliffs, NJ, 1980.
22. P. Khargonekar and E. D. Sontag, "On the Relation Between Stable Matrix Fraction Factorizations and Regulable Realizations of Linear Systems Over Rings," *IEEE Trans. on Autom. Control,* Vol. AC-27, pp. 627–638, June 1982.
23. V. Kucera, *Discrete Linear Control,* Wiley, New York, 1979.

24. V. Kucera, "Diophantine Equations in Control—A Survey," *Automatica,* Vol. 29, pp. 1361–1375, 1993.
25. C. C. MacDuffee, *The Theory of Matrices,* Chelsea, New York, 1946.
26. J. M. Maciejowski, *Multivariable Feedback Design,* Addison-Wesley, Reading, MA, 1989.
27. C. N. Nett, C. A. Jacobson, and M. J. Balas, "A Connection Between State-Space and Doubly Coprime Fractional Representations," *IEEE Trans. on Autom. Control,* Vol. AC-29, pp. 831–832, September 1984.
28. L. Pernebo, "Notes on Strict System Equivalence," *Int. J. of Control,* Vol. 25, pp. 21–38, 1977.
29. L. Pernebo, "An Algebraic Theory for the Design of Controllers for Linear Multivariable Systems," *IEEE Trans. on Autom. Control,* Vol. AC-26, pp. 171–194, February 1981.
30. H. H. Rosenbrock, *State-Space and Multivariable Theory,* Nelson, London, 1970.
31. H. H. Rosenbrock, "A Comment on Three Papers," *Int. J. of Control,* Vol. 25, pp. 1–3, 1977.
32. H. H. Rosenbrock, "The Transformation of Strict System Equivalence," *Int. J. Control,* Vol. 25, pp. 11–19, 1977.
33. A. I. G. Vardulakis, *Linear Multivariable Control,* Wiley, New York, 1991.
34. M. Vidyasagar, *Control System Synthesis. A Factorization Approach,* MIT Press, Cambridge, MA, 1985.
35. T. Williams and P. J. Antsaklis, "Decoupling," *The Control Handbook*, Chap. 50, pp. 745–804, CRC Press and IEEE Press, Boca Raton, FL, 1996.
36. W. A. Wolovich, *Linear Multivariable Systems,* Springer-Verlag, New York, 1974.
37. D. C. Youla, H. A. Jabr, and J. J. Bongiorno, Jr., "Modern Wiener-Hopf Design of Optimal Controllers—Part II: The Multivariable Case," *IEEE Trans. on Autom. Control,* Vol. AC-21, pp. 319–338, June 1976.
38. G. Zames, "Feedback and Optimal Sensitivity: Model Reference Transformations, Multiplicative Seminorms, and Approximate Inverses," *IEEE Trans. on Autom. Control,* Vol. 26, pp. 301–320, 1981.

7.8 EXERCISES

7.1. Given a polynomial matrix $P(s) \in R[s]^{p\times m}$ with *rank* $P(s) = \min(p, m)$, write a computer program to reduce $P(s)$ to a row proper (row reduced) matrix via row elementary operations and to derive the corresponding unimodular matrix $U_L(s)$. Use your computer algorithm to verify the results of Example 2.10. *Hint:* Apply the algorithm to $[P(s), I_p]$ so that in $U_L(s)[P(s), I_p] = [\hat{P}(s), U_L(s)]$, $\hat{P}(s)$ is in row proper form and $U_L(s)$ is the appropriate unimodular matrix.

7.2. Given a polynomial matrix $P(s) \in R[s]^{p\times m}$ with $p \geq m$, write a computer program to reduce $P(s)$ to column Hermite form and to derive the corresponding unimodular matrix $U_L(s)$. Use your computer algorithm to verify the results of Example 2.13. *Hint:* Apply the algorithm to $[P(s), I_p]$ so that in $U_L(s)[P(s), I_p] = [\hat{P}(s), U_L(s)]$, $\hat{P}(s)$ is in Hermite form and $U_L(s)$ is the appropriate unimodular matrix.

7.3. Consider $A \in R^{m\times n}$ and let $|A|_{\{k_1,\ldots,k_r\}}^{\{i_1,\ldots,i_r\}}$ be the $r \times r$ minor formed by selecting r rows denoted by $\{i_1, \ldots, i_r\}$, and r columns of A, denoted by $\{k_1, \ldots k_r\}$, where $r \leq \min(m, n)$. Let $B \in R^{n\times p}$ and consider the minors of the product AB. These can be determined using

$$|AB|_{\{k_1,\ldots,k_r\}}^{\{i_1,\ldots,i_r\}} = \sum_{l_j} |A|_{\{l_1,\ldots,l_r\}}^{\{i_1,\ldots,i_r\}} |B|_{\{k_1,\ldots,k_r\}}^{\{l_1,\ldots,l_r\}},$$

where $\{l_1, \ldots, l_r\}$ denotes all possible sets of r integers among the n columns of A and n rows of B, with $r \le \min(m, n, p)$. This formula for the minors of the product of matrices is known as the *Binet-Cauchy formula*.

(a) If $A \in R^{m \times n}$, $B \in R^{n \times m}$, with $m \le n$, determine an expression for $det\, AB$.

(b) Show that if A and B are both square, then $det\, AB = det\, A\, det\, B = det\, BA$.

(c) Suppose that $P_1(s)$ and $P_2(s)$, polynomial matrices of the same dimensions, are related by $P_1(s) = U(s)P_2(s)V(s)$, where $U(s)$ and $V(s)$ are also polynomial matrices. Show that the (monic) gcd of all $j \times j$ minors of $P_2(s)$, i.e., the determinantal divisor $D_j(s)$ of $P_2(s)$, divides the gcd of all $j \times j$ minors of $P_1(s)$, i.e., the corresponding determinantal divisor of $P_1(s)$.

(d) If $U(s)$ and $V(s)$ are unimodular, show that all the determinantal divisors $D_j(s)$, and therefore, all the invariant factors $\epsilon_j(s)$ of $P_1(s)$ and $P_2(s)$, are the same. That is, *the invariant factors of a matrix are not affected by row and column elementary operations.* This also shows that if $P_1(s) = U(s)P_2(s)V(s)$ with $U(s)$ and $V(s)$ unimodular, then $P_1(s)$ and $P_2(s)$ have precisely the same Smith form. It is straightforward to also show the opposite, i.e., if $P_1(s)$ and $P_2(s)$ have the same Smith form, then there exist unimodular matrices $U(s)$ and $V(s)$ such that $P_1(s) = U(s)P_2(s)V(s)$ (see Subsection 7.2C).

7.4. Given a polynomial matrix $P(s) \in R[s]^{p \times m}$ with $p \ge m$, write a computer program to determine a gcrd $G_R^*(s)$ of all the rows of $P(s)$ and to derive the corresponding unimodular matrix $U(s)$. Use your computer program to determine a gcrd and a gcld of the matrices $P_1(s)$ and $P_2(s)$ of Example 2.15. *Hint:* Determine a unimodular matrix $U(s)$ such that $U(s)P(s) = \begin{bmatrix} G_R^*(s) \\ 0 \end{bmatrix}$. Apply your algorithm to $[P(s), I_p]$ so that $U(s)[P(s), I_p] = \left[\begin{pmatrix} G_R^*(s) \\ 0 \end{pmatrix}, U(s)\right]$.

7.5. Consider the polynomial matrices $P(s) = \begin{bmatrix} s^2 + s & -s \\ -s^2 - 1 & s^2 \end{bmatrix}$, $R(s) = \begin{bmatrix} s & 0 \\ -s - 1 & 1 \end{bmatrix}$.

(a) Are they rc? If they are not, find a greatest common right divisor (gcrd).

(b) Are they lc? If they are not, find a gcld.

7.6. (a) Show that two square and nonsingular polynomial matrices, the determinants of which are coprime polynomials, are both rc and lc. *Hint:* Assume they are not, say, rc and then use the determinants of their gcrd to arrive at a contradiction.

(b) Show that the opposite is not true, i.e., two rc (lc) polynomial matrices do not necessarily have determinants that are coprime polynomials.

7.7. (a) Show that if (the nonsingular) $G_{R_1}^*(s)$ and $G_{R_2}^*(s)$ are both gcrds of P_1 and P_2, then there exists a unimodular matrix $U_R(s)$ such that $G_{R_2}^*(s) = U_R(s)G_{R_1}^*(s)$.

(b) Similarly, show that if the nonsingular $G_{L_1}^*(s)$ and $G_{L_2}^*(s)$ are both gclds of $\hat{P}_1$ and $\hat{P}_2$, then there exists a unimodular matrix $U_L(s)$ such that $G_{L_2}^*(s) = G_{L_1}^*(s)U_L(s)$.

7.8. Show that ν defined in (2.43) is indeed the observability index of the system.

7.9. Let $P(s) = P_n s^n + P_{n-1}s^{n-1} + \cdots + P_0$ be a matrix polynomial with $P_i \in R^{n \times n}$ and let $A \in R^{n \times n}$. Show that

(a) $P(s) = Q_r(s)(sI - A) + R_r$ with $R_r = P_n A^n + P_{n-1}A^{n-1} + \cdots + P_0$,

(b) $P(s) = (sI - A)Q_l(s) + R_l$ with $R_l = A^n P_n + A^{n-1}P_{n-1} + \cdots + P_0$.

Hint: $Q_r(s) = P_n s^{n-1} + (P_n A + P_{n-1})s^{n-2} + \cdots + (P_n A^{n-1} + \cdots + P_1)$.

7.10. Let $P(s)$ be a polynomial matrix of full column rank and let $y(s)$ be a given polynomial vector. Show that the equation $P(s)x(s) = y(s)$ will have a polynomial solution $x(s)$ for any $y(s)$ if and only if the columns of $P(s)$ are lc, or equivalently, if and only if $P(\lambda)$ has full column rank for any complex number λ.

7.11. Let $P(s) = P_d s^d + P_{d-1}s^{d-1} + \cdots + P_0 \in R[s]^{p\times m}$ and let $P_e(s) = block\ diag\,(I_m, \ldots, I_m, P(s))$ with d blocks on the diagonal. It can be shown that by means of elementary row and column operations, $P_e(s)$ can be transformed to a (linear) *matrix pencil* $sE - A$. In particular,

$$P_e(s) = U(s)(sE - A)V(s),$$

where E and A are real matrices given by

$$E = block\ diag\,(P_d, I_m, \ldots, I_m)$$

$$A = \begin{bmatrix} -P_{d-1} & \cdots & -P_1 & -P_0 \\ I & \cdots & 0 & 0 \\ \vdots & \ddots & \vdots & \vdots \\ 0 & \cdots & I & 0 \end{bmatrix}$$

and $U(s)$ and $V(s)$ are unimodular matrices given by

$$V(s) = \begin{bmatrix} I & sI & \cdots & s^{d-1}I \\ & \ddots & \ddots & \vdots \\ & & \ddots & sI \\ & & & I \end{bmatrix}, \qquad U(s) = \begin{bmatrix} 0 & -I & & \\ \vdots & \vdots & \ddots & \\ 0 & 0 & \cdots & -I \\ I & B_1(s) & \cdots & B_{d-1}(s) \end{bmatrix},$$

with $B_{i+1}(s) = sB_i(s) + P_{d-(i+1)}$, $i = 1, \ldots, d-2$, and $B_1(s) = sP_d + P_{d-1}$. Note that $P(s)$ and $sE - A$ have the same nonunity invariant polynomials. Let $P(s) = \begin{bmatrix} s^2 + s & -s \\ -s^2 - 1 & s^2 \end{bmatrix}$ and obtain the equivalent matrix pencil $sE - A$.

7.12. Let $\{D_R, I, N_R\}$ and $\{D_L, N_L, I\}$ be minimal realizations of $H(s)$, where $H(s) = N_R D_R^{-1} = D_L^{-1} N_L$ with (N_R, D_R) rc and (D_L, N_L) lc and related by $UU^{-1} = \begin{bmatrix} X & Y \\ -N_L & D_L \end{bmatrix}\begin{bmatrix} D_R & -\tilde{Y} \\ N_R & \tilde{X} \end{bmatrix} = \begin{bmatrix} I & 0 \\ 0 & I \end{bmatrix}$, where U is unimodular, i.e., they are doubly coprime factorizations of $H(s)$. Show that these realizations are equivalent representations.

Hint:
$$\begin{bmatrix} N_L & 0 \\ 0 & I \end{bmatrix}\begin{bmatrix} D_R & I \\ -N_R & 0 \end{bmatrix} = \begin{bmatrix} D_L & N_L \\ -I & 0 \end{bmatrix}\begin{bmatrix} N_R & 0 \\ 0 & I \end{bmatrix}$$

and also,
$$\begin{bmatrix} \tilde{Y} & 0 \\ \tilde{X} & I \end{bmatrix}\begin{bmatrix} D_L & N_L \\ -I & 0 \end{bmatrix} = \begin{bmatrix} D_R & I \\ -N_R & 0 \end{bmatrix}\begin{bmatrix} Y & -X \\ 0 & I \end{bmatrix}.$$

7.13. Consider $P(q)z(t) = Q(q)u(t)$ and $y(t) = R(q)z(t) + W(q)u(t)$, where

$$P(q) = \begin{bmatrix} q^3 - q & q^2 - 1 \\ -q - 2 & 0 \end{bmatrix}, \qquad Q(q) = \begin{bmatrix} q - 1 & -2q + 2 \\ 1 & 3q \end{bmatrix}$$

$$R(q) = \begin{bmatrix} 2q^2 + q + 2 & 2q \\ -q - 2 & 0 \end{bmatrix}, \qquad W(q) = \begin{bmatrix} -1 & 3q + 4 \\ -1 & -3q \end{bmatrix}$$

with $q \triangleq d/dt$.

(a) Is this system representation controllable? Is it observable?

(b) Find the transfer function matrix $H(s)$ $(\hat{y}(s) = H(s)\hat{u}(s))$.

(c) Determine an equivalent state-space representation $\dot{x} = Ax + Bu$, $y = Cx + Du$ and repeat (a) and (b) for this representation.

7.14. Use system theoretic arguments to show that two polynomials $d(s) = s^n + d_{n-1}s^{n-1} + \cdots + d_1 s + d_0$ and $n(s) = n_{n-1}s^{n-1} + n_{n-2}s^{n-2} + \cdots + n_1 s + n_0$ are coprime if and only if

$$rank \begin{bmatrix} C_c \\ C_c A_c \\ \vdots \\ C_c A_c^{n-1} \end{bmatrix} = n,$$

where $A_c = \begin{bmatrix} 0 & 1 & 0 & \cdots & 0 \\ 0 & 0 & 1 & \cdots & 0 \\ \vdots & \vdots & \vdots & & \vdots \\ 0 & 0 & 0 & \cdots & 1 \\ -d_0 & -d_1 & -d_2 & & -d_{n-1} \end{bmatrix}$ and $C_c = [n_0, n_1, \ldots, n_{n-1}]$.

7.15. Consider the transfer function $H(s) = \begin{bmatrix} 1 & \frac{1}{s} & \frac{s-1}{s} \\ 0 & \frac{s+1}{s^2} & 0 \end{bmatrix}$. Determine a minimal realization in

(a) Polynomial Matrix Fractional Description (PMFD) form,
(b) State-Space Description (SSD) form.

7.16. In the following, assume that a PMD, $\{P, Q, R, W\}$, realizes an invertible $m \times m$ transfer function matrix H.

(a) Show that the system matrix for H^{-1} is given by

$$S_I = \begin{bmatrix} P & Q & \vdots & 0 \\ -R & W & \vdots & -I_m \\ \cdots & \cdots & \cdots & \cdots \\ 0 & I_m & \vdots & 0 \end{bmatrix}.$$

Hint: Note that $S_I[z^T, -u^T, -y^T]^T = [0, 0, -u^T]^T$ when $Pz = Qu$, $y = Rz + Wu$.

(b) Show that the following system matrix can also be used to characterize H^{-1},

$$S_M = \begin{bmatrix} I & R & \vdots & -W \\ 0 & P & \vdots & Q \\ \cdots & \cdots & \vdots & \cdots \\ 0 & 0 & \vdots & I \end{bmatrix}.$$

(c) Let $det\, W \neq 0$. Show that $H^{-1} = \tilde{R}\tilde{P}^{-1}\tilde{Q} + \tilde{W}$, where $\tilde{W} = W^{-1}$, $\tilde{Q} = QW^{-1}$, $\tilde{R} = -W^{-1}R$, and $\tilde{P} = P + QW^{-1}R$.

7.17. Consider the system $Dz = u$, $y = Nz$, where $D = \begin{bmatrix} s^2 & 0 \\ 0 & s^3 \end{bmatrix}$ and $N = [s^2 - 1, s + 1]$.

(a) Determine an equivalent state-space representation.
(b) Is the system controllable? Is it observable? Determine all uncontrollable and/or unobservable eigenvalues, if any.
(c) Determine the invariant and the transmission zeros of the system.

7.18. Consider the series connection depicted in Fig. 7.2 and the PMFDs for S_1 and S_2, given in (3.63), (3.64) and (3.66), (3.67).

(a) Show that the system S is controllable if and only if

(i) (N_1, D_2) is lc, or

(ii) $(\tilde{D}_1 D_2, \tilde{N}_1)$ is lc, or

(iii) $(\tilde{N}_2 N_1, \tilde{D}_2)$ is lc.

(b) Determine similar conditions [as in (a)] for S to be observable.

7.19. Consider the parallel connection depicted in Fig. 7.1 and the PMFDs for S_1 and S_2 given in (3.52), (3.53) and (3.55), (3.66). Derive conditions for controllability and observability of S analogous to the ones derived for the series connection in Exercise 7.18.

7.20. Consider the double integrator $H_1 = 1/s^2$.

(a) Characterize all stabilizing controllers H_2 for H_1 using all the methods developed in Subsections 7.4A and 7.4C.

(b) Characterize all proper stabilizing controllers H_2 for H_1 of order 1.

7.21. Consider the double integrator $H_1 = 1/s^2$.

(a) Derive a minimal state-space realization for H_1 and use Lemma 4.19 to derive doubly coprime factorizations in RH_∞.

(b) Use the polynomial Diophantine Equations (4.102) and (4.106) to derive factorizations in RH_∞.

7.22. Consider $H_1 = \left[\frac{s^2+1}{s^2}, \frac{s+1}{s^3}\right]$.

(a) Derive a minimal state-space realization $\{A, B, C, D\}$ and use Lemma 4.19 and Theorem 4.13 to parameterize all stabilizing controllers H_2.

(b) Derive a stabilizing controller H_2 of order 3 by appropriately selecting K'. What are the closed-loop eigenvalues in this case? Comment on your results.
Hint: A minimal state-space realization was derived in Example 3.2. The eigenvalues of $A + BF$ and $A - KC$ are stable, but otherwise arbitrary. Note that some of these eigenvalues become closed-loop eigenvalues.

7.23. Consider the unity feedback (error feedback) control system depicted in Fig. 7.19, where H and C are the transfer function matrices of the plant and controller, respectively.

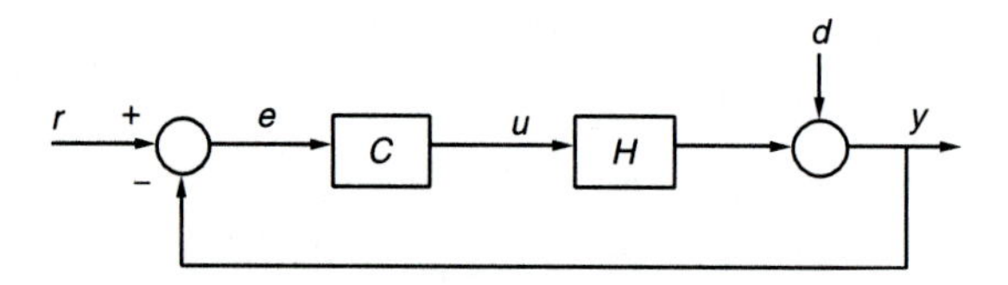

FIGURE 7.19

Assume that $(I + HC)^{-1}$ exists.

(a) Verify the relations

$$y = (I + HC)^{-1}HCr + (I + HC)^{-1}d \triangleq Tr + Sd$$

$$u = (I + CH)^{-1}Cr - (I + CH)^{-1}Cd \triangleq Mr - Md.$$

Compare these with relations (4.163) to (4.168) for the two degrees of freedom controller $u = C_y y + C_r r$. Note here that $u = -Cy + Cr$, and therefore, $C_y = -C$ and $C_r = C$. Hence, for the error feedback system of Fig. 7.19, the relations following (4.168) assume the forms

$$\begin{aligned} M &= (I + CH)^{-1}C = DX = -Q = -DL \\ T &= H(I + CH)^{-1}C = (I + HC)^{-1}HC = HM = NX \\ S_o &= (I + HC)^{-1} = I + HQ = I - HM = I - T \\ S_i &= (I + CH)^{-1} = I + QH = I - MH. \end{aligned}$$

(b) (i) Let $H = ND^{-1}$ be an rc polynomial factorization. Show that all stabilizing controllers are given by

$$C = [(I - XN)D^{-1}]^{-1}X,$$

where $[(I - XN)D^{-1}, X]$ is stable and $(I - XN)^{-1}$ exists. *Hint:* Apply Theorem 4.22 to the error feedback case.

(ii) If H is proper and $H = N'D'^{-1}$ is an rc MFD in RH_∞, show that all proper stabilizing controllers are given by

$$C = [(I - X'N')D'^{-1}]^{-1}X',$$

where $[(I - X'N')D'^{-1}, X'] \in RH_\infty$ and $(I - X'N')^{-1}$ exists and it is proper. *Hint:* Apply Theorem 4.22 to the error feedback case.

(c) Assume that H is proper and H^{-1} exists, i.e., H is square and nonsingular. Let $H = ND^{-1}$ be an rc polynomial MFD. If T is the closed-loop transfer function between y and r, show that the system will be internally stable if and only if

$$[N^{-1}(I - T)H, N^{-1}T]$$

is stable. Assume that $T \neq I$ for the loop to be well defined. Note that if T is proper, then

$$C = H^{-1}T(I - T)^{-1}$$

is proper if and only if $H^{-1}T$ is proper and $I - T$ is biproper.

(d) Assume that in (c) H and T are SISO transfer functions. Let $H = n/d$. Show that the closed-loop system will be stable if and only if

$$(1 - T)d^{-1} = Sd^{-1} \quad \text{and} \quad Tn^{-1}$$

are stable, i.e., if and only if the sensitivity matrix has as zeros all the unstable poles of the plant and the closed-loop transfer function has as zeros all the unstable zeros of the plant.

Remark: Note that this is a result known in the classical control literature (refer to J. R. Ragazzini and G. F. Franklin, *Sampled Data Control Systems*, McGraw-Hill, 1958). It is derived here by specializing the more general MIMO case results to the SISO case.

(e) Given $H(s) = \dfrac{s - 1}{(s - 2)(s + 1)}$, characterize all scalar proper transfer functions T that can be realized via the error feedback configuration shown in Fig. 7.19, under internal stability. For comparison purposes, characterize all T that can be realized via a two degrees of freedom controller and comment on your results. In both cases, comment on the location and number of the closed-loop eigenvalues. *Hint:* Use Theorems 4.22 and 4.24.

7.24. Consider $H(s) = \frac{(s-1)(s+2)}{(s-2)^2}$ and characterize all proper and stable transfer functions $T(s)$ that can be realized via linear state feedback under internal stability. *Hint:* From Subsection 7.4B, note that $T = ND_F^{-1}G$ and consider Theorem 4.23.

7.25. Consider $H = \begin{bmatrix} \frac{1}{s+1} & \frac{2}{s+3} \\ \frac{1}{s+1} & \frac{1}{s+1} \end{bmatrix}$.

(a) Derive an rc MFD in RH_∞, $H = N'D'^{-1}$.

(b) Let $T = \begin{bmatrix} \frac{n_1}{d_1} & 0 \\ 0 & \frac{n_2}{d_2} \end{bmatrix}$ and characterize all diagonal T that can be realized under internal stability via a two degrees of freedom control configuration.

(c) Repeat (b) for a unity feedback configuration $\{I; G_{ff}, I\}$ (see Fig. 7.15).

7.26. In the model matching problem, the transfer function matrices $H \in R^{p\times m}(s)$ of the plant and $T \in R^{p\times m}(s)$ of the model must be found so that $T = HM$ [see (4.192)]. M is to be realized via a feedback control configuration under internal stability. Here we are interested in the *model matching problem via linear state feedback.* For this, let $H = ND^{-1}$ an rc polynomial factorization with D column reduced. Then $Dz = u$, $y = Nz$ is a minimal realization of H. Let the state-feedback control law be defined by $u = Fz + Gr$, where $F \in R[s]^{m\times m}$, $G \in R^{m\times m}$ with $\det G \neq 0$ and $deg_{c_j} F < deg_{c_j} D$. To allow additional flexibility, let $r = Kv$, $K \in R^{m\times q}$. Note that (see Subsection 7.4C) $H_{F,GK} = ND_F^{-1}GK = (ND^{-1})(DD_F^{-1}GK) = (ND^{-1})[D(G^{-1}D_F)^{-1}K] = HM$.

In view of the above, solve the model matching problem via linear state feedback, determine F, G, and K, and comment on your results when

(a) $H = \frac{(s+1)(s+2)}{2s^2-3s+2}$, $\quad T = \frac{s+1}{s+2}$

(b) $H = \begin{bmatrix} \frac{s+1}{s} & 0 \\ \frac{1}{s} & \frac{s+2}{s} \end{bmatrix}$, $\quad T = I_2$

(c) $H = \begin{bmatrix} \frac{s+2}{s+1} & \frac{s+3}{s+2} \\ \frac{1}{s+1} & 0 \end{bmatrix}$, $\quad T = \begin{bmatrix} \frac{s+1}{s+4} \\ \frac{-2}{(s+2)(s+4)} \end{bmatrix}$.

Hint: The model matching problem via linear state feedback is not difficult to solve when $p = m$ and $\operatorname{rank} H = m$, in view of $(G^{-1}D_F)^{-1}K = D^{-1}M = D^{-1}H^{-1}T = N^{-1}T$.

APPENDIX

Numerical Considerations

A.1 INTRODUCTION

To compute the rank of the controllability matrix $[B, AB, \ldots, A^{n-1}B]$, or the eigenvalues of A, or the zeros of the system $\{A, B, C, D\}$, typically requires use of a digital computer. When this is the case, one must deal with selection of an algorithm and interpret numerical results. In doing so, two issues arise that play important roles in numerical computations using a computer, namely, the *numerical stability or instability* of the computational method used, and how *well or ill conditioned* the problem is numerically.

An example of a problem that can be ill conditioned is the problem of calculating the roots of a polynomial, given its coefficients. This is so because for certain polynomials, small variations in the values of the coefficients, introduced, say, via round-off errors, can lead to great changes in the roots of the polynomial. That is to say, the roots of a polynomial can be very sensitive to changes in its coefficients. Note that ill conditioning is a property of the problem to be solved and depends on neither the floating-point system used in the computer nor on the particular algorithm being implemented.

A computational method is numerically stable if it yields a solution that is near the true solution of a problem with slightly changed data. An example of a numerically unstable method to compute the roots of $ax^2 + 2bx + c = 0$ is the formula $(-b \pm \sqrt{(b^2 - ac)})/a$ that for certain parameters a, b, c may give erroneous results in finite arithmetic. This instability is due to subtraction of two approximately equal large numbers in the numerator when $b^2 >> ac$. Note that the roots may be calculated in a numerically stable way, using the mathematically equivalent, but numerically very different, expression $c/(-b \mp \sqrt{(b^2 - ac)})$.

We would of course always like to use numerically stable methods and we would prefer to have well-conditioned problems. In the following section, we briefly discuss the problem of solving a set of algebraic equations given by $Ax = b$. We will show that a measure of how ill conditioned a given problem is, is the size of the *condition number* (to be defined) of the matrix A. There are many algorithms to numerically solve $Ax = b$, and we will briefly discuss numerically stable ones.

Conditioning of a problem and numerical stability of a method are key issues in the area of numerical analysis. Our aim in this appendix is to make the reader aware that, depending on the problem, the numerical considerations in the calculation of a solution may be nontrivial. These issues are discussed at length in many textbooks on numerical analysis. Examples of good books in this area include, Golub and Van Loan [6] and Stewart [9], where matrix computations are emphasized. Also, see Petkov et al. [8] and Patel et al. [7] for computational methods with emphasis on system and control problems. For background on the theory of algorithms, on optimization algorithms, and their numerical properties, see Bazaran et al. [2] and Bertsekas and Tsitsiklis [3].

In Section 2 we present methods for solving linear algebraic equations. Singular values and singular-value decompositions are discussed in Section 3. In Section 4, an approach for solving polynomial matrix and rational matrix equations based on polynomial matrix interpolation is presented.

A.2 SOLVING LINEAR ALGEBRAIC EQUATIONS

Consider the set of linear algebraic equations given by

$$Ax = b, \tag{2.1}$$

where $A \in R^{m \times n}$. Its solution is important in many engineering problems. It is of interest to know the effects of small variations of A and b to the solution x of this system of equations. Note that such variations may be introduced for example by rounding errors when calculating a solution or by noisy data.

Condition number

Let $A \in R^{n \times n}$ be nonsingular. If A is known exactly and b has some uncertainty Δb, associated with it, then $A(x + \Delta x) = b + \Delta b$. It can then be shown that the variation in the solution x is bounded by

$$\frac{\|\Delta x\|}{\|x\|} \leq cond(A) \frac{\|\Delta b\|}{\|b\|}, \tag{2.2}$$

where $\|\cdot\|$ denotes any vector norm (and consistent matrix norm) and $cond(A)$ denotes the *condition number of* A, where $cond(A) \triangleq \|A\| \|A^{-1}\|$. Note that

$$cond(A) = \frac{\sigma_{\max}(A)}{\sigma_{\min}(A)}, \tag{2.3}$$

where $\sigma_{\max}(A)$, and $\sigma_{\min}(A)$ are the maximum and minimum singular values of A, respectively (see next section). From the property of matrix norms, $\|AA^{-1}\| \leq$

$\|A\|\,\|A^{-1}\|$, it follows that $cond\,(A) \geq 1$. This also follows from the expression involving singular values. If $cond\,(A)$ is small, then A is said to be *well conditioned* with respect to the problem of solving linear equations. If $cond\,(A)$ is large, then A is *ill conditioned* with respect to the problem of solving linear equations. In this case the relative uncertainty in the solution ($\|\Delta x\|/\|x\|$) can be many times the relative uncertainty in $b(\|\Delta b\|/\|b\|)$. This is of course undesirable. Similar results can be derived when variations in both b and A are considered, i.e., when b and A become $b + \Delta b$ and $A + \Delta A$. Note that the conditioning of A, and of the given problem, is independent of the algorithm used to determine a solution.

The condition number of A provides a measure of the distance of A to the set of singular (reduced rank) matrices. In particular, if $\|\Delta A\|$ is the norm of the smallest perturbation ΔA such that $A + \Delta A$ is singular, and is denoted by $d(A)$, then $dA/\|A\| = 1/cond\,(A)$. Thus, a large condition number indicates a short distance to a singularity, and it should not be surprising that this implies great sensitivity of the numerical solution x of $Ax = b$ to variations in the problem data.

The condition number of A plays a similar role in the case when A is not square. It can be determined in terms of the singular values of A defined in the next section.

Computational methods

The system of equations $Ax = b$ is easily solved if A has some special form, (e.g., if it is diagonal or triangular). Using the method of *Gaussian elimination*, any nonsingular matrix A can be reduced to an upper triangular matrix U. These operations can be represented by premultiplication of A by a sequence of lower triangular matrices. It can then be shown that A can be represented as

$$A = LU, \tag{2.4}$$

where L is a lower triangular matrix with all diagonal elements equal to 1 and U is an upper triangular matrix. The solution of $Ax = b$ is then reduced to the solution of two systems of equations with triangular matrices, $Ly = b$ and $Ux = y$. This method of solving $Ax = b$ is based on the decomposition (2.4) of A, which is called the *LU decomposition* of A.

If A is a symmetric positive definite matrix, then it may be represented as

$$A = U^T U, \tag{2.5}$$

where U is an upper triangular matrix. This is known as the *Cholesky decomposition* of a positive definite matrix. It can be obtained using a variant of Gaussian elimination. Note that this method requires half of the operations necessary for Gaussian elimination on an arbitrary nonsingular matrix A, since A is symmetric.

Now consider the system of equations $Ax = b$, where $A \in R^{m \times n}$, and let $rank\,A = n (\leq m)$. Then

$$A = Q\begin{bmatrix} R \\ O \end{bmatrix} = [Q_1, Q_2]\begin{bmatrix} R \\ O \end{bmatrix} = Q_1 R, \tag{2.6}$$

where Q is an orthogonal matrix ($Q^T = Q^{-1}$) and $R \in R^{n \times n}$ is an upper triangular matrix of full rank n. Expression (2.6) is called the *QR decomposition* of A. When $rank\,A = r$, the QR decomposition of A is expressed as

$$AP = Q\begin{bmatrix} R_1 & R_2 \\ 0 & 0 \end{bmatrix}, \tag{2.7}$$

where Q is orthogonal, $R_1 \in R^{r\times r}$ is nonsingular and upper triangular, and P is a permutation matrix that represents the moving of the columns during the reduction (in $Q^T AP$).

QR decomposition can be used to determine solutions of $Ax = b$. In particular, consider $A \in R^{m\times n}$ with $rank\, A = n(\leq m)$ and assume that a solution exists. First, determine the QR decomposition of A given in (2.6). Then $Q^T Ax = Q^T b$ or $\begin{bmatrix} R \\ 0 \end{bmatrix} x = Q^T b$ (since $Q^T = Q^{-1}$) or $Rx = c$. Solve this system of equations, where R is triangular and $c = [I_n, 0]Q^T b$. In the general case when $rank(A) = r \leq \min(n, m)$, determine the QR decomposition of A (2.7) and assume that a solution exists. The solutions are given by $x = P\begin{bmatrix} R_1^{-1}(c - R_2 y) \\ y \end{bmatrix}$, $c = [I_r, 0]Q^T b$, where $y \in R^{m-r}$ is arbitrary.

A related problem is the *linear least squares problem* where a solution x of the system of equations $Ax = b$ is to be found that minimizes $\|b - Ax\|_2$. This is a more general problem than simply solving $Ax = b$, since solving it provides the "best" solution in the above sense, even when an exact solution does not exist. The least squares problem is discussed further in the next section.

A.3 SINGULAR VALUES AND SINGULAR-VALUE DECOMPOSITION

The singular values of a matrix and the Singular Value Decomposition Theorem play a significant role in a number of problems of interest in the area of systems and control, from the computation of solutions of linear systems of equations, to computations of the norm of transfer matrices at specified frequencies, to model reduction, and so forth. In the following we provide a brief description of some basic results and introduce some terminology.

Consider $A \in \mathscr{C}^{n\times n}$ and let $A^* = \bar{A}^T$, the complex conjugate transpose of A. $A \in \mathscr{C}^{n\times n}$ is said to be *Hermitian* if $A^* = A$. If $A \in R^{n\times n}$, then $A^* = A^T$, and if $A = A^T$, then A is *symmetric*. $A \in \mathscr{C}^{n\times n}$ is *unitary* if $A^* = A^{-1}$. In this case $A^*A = AA^* = I_n$. If $A \in R^{n\times n}$ then $A^* = A^T$ and if $A^T = A^{-1}$, i.e., if $A^T A = AA^T = I_n$, then A is *orthogonal* (refer to Section 6.2).

Singular values

Let $A \in \mathscr{C}^{m\times n}$ and consider $AA^* \in \mathscr{C}^{m\times m}$. Let $\lambda_i, i = 1, \ldots, m$, denote the eigenvalues of AA^*, and note that these are all real and nonnegative numbers. Assume that $\lambda_1 \geq \lambda_2 \geq \cdots \lambda_r \geq \cdots \geq \lambda_m$. Note that if $r = rank\, A = rank\,(AA^*)$, then $\lambda_1 \geq \lambda_2 \geq \cdots \geq \lambda_r > 0$ and $\lambda_{r+1} = \cdots = \lambda_m = 0$. *The singular values σ_i of A* are the positive square roots of $\lambda_i, i = 1, \ldots, \min(m, n)$. In fact, the nonzero singular values of A are

$$\sigma_i = (\lambda_i)^{1/2}, \qquad i = 1, \ldots, r, \tag{3.1}$$

where $r = rank\, A$, while the remaining $(\min(m, n) - r)$ of the singular values are zero. Note that $\sigma_1 \geq \sigma_2 \geq \cdots \geq \sigma_r > 0$, and $\sigma_{r+1} = \sigma_{r+2} = \cdots = \sigma_{\min(m,n)} = 0$. The singular values could have also been found as the square roots of the eigenvalues of $A^*A \in \mathscr{C}^{n\times n}$ (instead of $AA^* \in \mathscr{C}^{m\times m}$). To see this, consider the following result.

LEMMA 3.1. Let $m \geq n$. Then

$$|\lambda I_m - AA^*| = \lambda^{m-n}|\lambda I_n - A^*A|, \tag{3.2}$$

i.e., all eigenvalues of A^*A are eigenvalues of AA^* which also has $m - n$ additional eigenvalues at zero. Thus, $AA^* \in \mathscr{C}^{m\times m}$ and $A^*A \in \mathscr{C}^{n\times n}$ have precisely the same r nonzero eigenvalues ($r = \textit{rank}\, A$); their remaining eigenvalues, $(m - r)$ for AA^* and $(n - r)$ for A^*A, are all at zero. Therefore, either AA^* or A^*A can be used to determine the r nonzero singular values of A. All remaining singular values are zero.

Proof of the lemma. The proof is based on Schur's formula for determinants. In particular, we have

$$\begin{aligned} D(\lambda) &= \begin{vmatrix} \lambda^{1/2} I_m & A \\ A^* & \lambda^{1/2} I_n \end{vmatrix} = |\lambda^{1/2} I_m||\lambda^{1/2} I_n - A^*\lambda^{-1/2} I_m A| \\ &= |\lambda^{1/2} I_m||\lambda^{-1/2} I_n||\lambda I_n - A^*A| \\ &= \lambda^{(m-n)/2} \cdot |\lambda I_n - A^*A|, \end{aligned} \tag{3.3}$$

where Schur's formula was applied to the (1, 1) block of the matrix. If it is applied to the (2, 2) block, then

$$D(\lambda) = \lambda^{(n-m)/2} \cdot |\lambda I_m - AA^*|. \tag{3.4}$$

Equating (3.3) and (3.4), we obtain $|\lambda I_m - AA^*| = \lambda^{m-n}|\lambda I_n - A^*A|$, which is (3.2). ∎

EXAMPLE 3.1. $A = \begin{bmatrix} 2 & 1 & 0 \\ 0 & 0 & 0 \end{bmatrix} \in R^{2\times 3}$. Here $\textit{rank}\, A = r = 1$, $\lambda_i(AA^*) = \lambda_i\left(\begin{bmatrix} 2 & 1 & 0 \\ 0 & 0 & 0 \end{bmatrix}\begin{bmatrix} 2 & 0 \\ 1 & 0 \\ 0 & 0 \end{bmatrix}\right) = \lambda_i\left(\begin{bmatrix} 5 & 0 \\ 0 & 0 \end{bmatrix}\right) = \{5, 0\}$, and $\lambda_1 = 5$, $\lambda_2 = 0$. Also, $\lambda_i(A^*A) = \lambda_i\left(\begin{bmatrix} 2 & 0 \\ 1 & 0 \\ 0 & 0 \end{bmatrix}\begin{bmatrix} 2 & 1 & 0 \\ 0 & 0 & 0 \end{bmatrix}\right) = \lambda_i\left(\begin{bmatrix} 4 & 2 & 0 \\ 2 & 1 & 0 \\ 0 & 0 & 0 \end{bmatrix}\right)$, and $\lambda_1 = 5$, $\lambda_2 = 0$, and $\lambda_3 = 0$. The only nonzero singular value is $\sigma_1 = \sqrt{\lambda_1} = +\sqrt{5}$. The remaining singular values are zero. ∎

There is an important relation between the singular values of A and its induced Hilbert or 2-norm, also called the spectral norm $\|A\|_2 = \|A\|_s$. In particular,

$$\|A\|_2 (= \|A\|_s) = \sup_{\|x\|_2 = 1} \|Ax\|_2 = \max_i\{(\lambda_i(A^*A))^{1/2}\} = \bar{\sigma}(A), \tag{3.5}$$

where $\bar{\sigma}(A)$ denotes the largest singular value of A. Using the inequalities that are axiomatically true for induced norms (see Subsection 1.10B), it is possible to establish relations between singular values of various matrices that are useful in MIMO control design. The significance of the singular values of a gain matrix $A(j\omega)$ is discussed later in this section.

There is an interesting relation between the eigenvalues and the singular values of a (square) matrix. Let λ_i, $i = 1, \ldots, n$, denote the eigenvalues of $A \in R^{n\times n}$, let $\underline{\lambda}(A) = \min_i |\lambda_i|$, and let $\bar{\lambda}(A) = \max_i |\lambda_i|$. Then

$$\underline{\sigma}(A) \leq \underline{\lambda}(A) \leq \bar{\lambda}(A) \leq \bar{\sigma}(A). \tag{3.6}$$

Note that the ratio $\bar{\sigma}(A)/\underline{\sigma}(A)$, i.e., the ratio of the largest and smallest singular values of A, is called the *condition number of* A, and is denoted by *cond* (A). This is a very useful measure of how well conditioned a system of linear algebraic equations

$Ax = b$ is (refer to the discussion in previous section). The singular values provide a reliable way of determining how far a square matrix is from being singular, or a nonsquare matrix is from losing rank. This is accomplished by examining how close to zero $\underline{\sigma}(A)$ is. In contrast, the eigenvalues of a square matrix are not a good indicator of how far the matrix is from being singular, and a typical example in the literature to illustrate this point is an $n \times n$ lower triangular matrix A with -1's on the diagonal and $+1$'s everywhere else. In this case, $\underline{\sigma}(A)$ behaves as $1/2^n$ and the matrix is nearly singular for large n while all of its eigenvalues are at -1. In fact, it can be shown that adding $1/2^{n-1}$ to every element in the first column of A results in an exactly singular matrix (try it for $n = 2$).

Singular-value decomposition

Let $A \in \mathscr{C}^{m\times n}$ with $rank\, A = r \le \min(m, n)$. Let $A^* = \bar{A}^T$, the complex conjugate transpose of A.

THEOREM 3.1. There exist unitary matrices $U \in \mathscr{C}^{m\times m}$ and $V \in \mathscr{C}^{n\times n}$ such that

$$A = U\Sigma V^*, \tag{3.7}$$

where $\Sigma = \begin{bmatrix} \Sigma_r & \vdots & 0_{r\times(n-r)} \\ \cdots & \vdots & \cdots\cdots \\ 0_{(m-r)\times r} & \vdots & 0_{(m-r)\times(n-r)} \end{bmatrix}$ with $\Sigma_r = diag(\sigma_1, \sigma_2, \ldots, \sigma_r) \in R^{r\times r}$ selected so that $\sigma_1 \ge \sigma_2 \ge \cdots \ge \sigma_r > 0$.

Proof. For the proof, see for example, Golub and Van Loan [6] and Patel et al. [7]. ∎

Let $U = [U_1, U_2]$ with $U_1 \in \mathscr{C}^{m\times r}$, $U_2 \in \mathscr{C}^{m\times(m-r)}$ and $V = [V_1, V_2]$ with $V_1 \in \mathscr{C}^{n\times r}$, $V_2 \in \mathscr{C}^{n\times(n-r)}$. Then

$$A = U\Sigma V^* = U_1\Sigma_r V_1^*. \tag{3.8}$$

Since U and V are unitary, we have

$$U^*U = \begin{bmatrix} U_1^* \\ U_2^* \end{bmatrix}[U_1, U_2] = I_m,\ U_1^*U_1 = I_r \tag{3.9}$$

and

$$V^*V = \begin{bmatrix} V_1^* \\ V_2^* \end{bmatrix}[V_1, V_2] = I_n,\ V_1^*V_1 = I_r. \tag{3.10}$$

Note that the columns of U_1 and V_1 determine orthonormal bases for $\mathscr{R}(A)$ and $\mathscr{R}(A^*)$, respectively. Now

$$AA^* = (U_1\Sigma_r V_1^*)(V_1\Sigma_r U_1^*) = U_1\Sigma_r^2 U_1^*, \tag{3.11}$$

from which we have

$$AA^*U_1 = U_1\Sigma_r^2 U_1^* U_1 = U_1\Sigma_r^2. \tag{3.12}$$

If u_i, $i = 1, \ldots, r$, is the ith column of U_1, i.e., $U_1 = [u_1, u_2, \ldots, u_r]$, then

$$AA^*u_i = \sigma_i^2 u_i, \qquad i = 1, \ldots, r. \tag{3.13}$$

This shows that the σ_i^2 are the r nonzero eigenvalues of AA^*, i.e., σ_i, $i = 1, \ldots, r$, are the nonzero singular values of A. Furthermore, u_i, $i = 1, \ldots, r$, are the eigenvectors of AA^* corresponding to σ_i^2. They are the *left singular vectors of* A. Note that the u_i are orthonormal vectors (in view of $U_1^* U_1 = I_r$). Similarly,

$$A^*A = (V_1 \Sigma_r U_1^*)(U_1 \Sigma_r V_1^*) = V_1 \Sigma_r^2 V_1^*, \tag{3.14}$$

from which we obtain

$$A^*AV_1 = V_1 \Sigma_r^2 V_1^* V_1 = V_1 \Sigma_r^2. \tag{3.15}$$

If v_i, $i = 1, \ldots, r$, is the ith column of V_1, i.e., $V_1 = [v_1, v_2, \ldots, v_r]$, then

$$A^*Av_i = \sigma_i^2 v_i, \qquad i = 1, 2, \ldots, r. \tag{3.16}$$

The vectors v_i are the eigenvectors of A^*A corresponding to the eigenvalues σ_i^2. They are the *right singular vectors* of A. Note that the v_i are orthonormal vectors (in view of $V_1^* V_1 = I_r$).

The singular values are unique, while the singular vectors are *not*. To see this, consider

$$\hat{V}_1 = V_1 \, diag\,(e^{j\theta_i}) \qquad \text{and} \qquad \hat{U}_1 = U_1 \, diag\,(e^{-j\theta_i}).$$

Their columns are also singular vectors of A (show this).

Note also that $A = U_1 \Sigma_r V_1^*$ implies that

$$A = \sum_{i=1}^{r} \sigma_i u_i v_i^*. \tag{3.17}$$

The significance of the singular values of a gain matrix $A(j\omega)$ is now briefly discussed. This is useful in the control theory of MIMO systems. Consider the relation between signals y and v, given by $y = Av$. Then

$$\max_{\|v\|_2 \neq 0} \frac{\|y\|_2}{\|v\|_2} = \max_{\|v\|_2 \neq 0} \frac{\|Av\|_2}{\|v\|_2} = \bar{\sigma}(A)$$

or

$$\max_{\|v\|_2 = 1} \|y\|_2 = \max_{\|v\|_2 = 1} \|Av\|_2 = \bar{\sigma}(A). \tag{3.18}$$

Thus, $\bar{\sigma}(A)$ yields the maximum amplification, in energy terms (2-norm), when the transformation A operates on a signal v. Similarly,

$$\min_{\|v\|_2 = 1} \|y\|_2 = \min_{\|v\|_2 = 1} \|Av\|_2 = \underline{\sigma}(A). \tag{3.19}$$

Therefore,

$$\underline{\sigma}(A) \leq \frac{\|Av\|_2}{\|v\|_2} \leq \bar{\sigma}(A), \tag{3.20}$$

where $\|v\|_2 \neq 0$. Thus the gain (energy amplification) is bounded from above and below by $\bar{\sigma}(A)$ and $\underline{\sigma}(A)$, respectively. The exact value depends on the direction of v.

To determine the particular directions of vectors v for which these (max and min) gains are achieved, consider (3.17) and write

$$y = Av = \sum_{i=1}^{r} \sigma_i u_i v_i^* v. \tag{3.21}$$

Notice that $|v_i^* v| \leq \|v_i\| \|v\| = \|v\|$, since $\|v_i\| = 1$, with equality holding only when $v = \alpha v_i, \alpha \in \mathscr{C}$. Therefore, to maximize, consider v along the singular value directions v_i and let $v = \alpha v_i$ with $|\alpha| = 1$ so that $\|v\| = 1$. Then in view of $v_i^* v_j = 0, i \neq j$, and $v_i^* v_j = 1, i = j$, we have that $y = Av = \alpha A v_i = \alpha \sigma_i u_i$ and $\|y\|_2 = \|Av\|_2 = \sigma_i$, since $\|u_i\|_2 = 1$. Thus, the maximum possible gain is σ_1, i.e., $\max_{\|v\|_2 = 1} \|y\|_2 = \max_{\|v\|_2 = 1} \|Av\|_2 = \sigma_1 (= \bar{\sigma}(A))$, as was shown above. This maximum gain occurs when v is along the right singular vector v_1. Then $Av = Av_1 = \sigma_1 u_1 = y$ in view of (3.17), i.e., the projection is along the left singular vector u_1, also of the same singular value σ_1. Similarly, for the minimum gain we have $\sigma_r = \underline{\sigma}(A) = \min_{\|v\|_2 = 1} \|y\|_2 = \min_{\|v\|_2 = 1} \|Av\|_2$, in which case $Av = Av_r = \sigma_r u_r = y$.

Additional interesting properties include

$$\mathscr{R}(A) = \mathscr{R}(U_1) = span\{u_1, \ldots, u_r\} \tag{3.22}$$

$$\mathscr{N}(A) = \mathscr{R}(V_2) = span\{v_{r+1}, \ldots, v_n\}, \tag{3.23}$$

where $U = [u_1, \ldots, u_r, u_{r+1}, \ldots, u_m] = [U_1, U_2]$ and $V = [v_1, \ldots, v_r, v_{r+1}, \ldots, v_n] = [V_1, V_2]$.

Least squares problem

Consider now *the least squares problem* where a solution x to the system of linear equations $Ax = b$ is to be determined that minimizes $\|b - Ax\|_2$. Write $\min_x \|b - Ax\|_2^2 = \min_x (b - Ax)^T (b - Ax) = \min_x (x^T A^T Ax - 2b^T Ax + b^T b)$. Then $\nabla_x (x^T A^T Ax - 2b^T Ax + b^T b) = 2A^T Ax - 2A^T b = 0$ implies that the x that minimizes $\|b - Ax\|_2$ is a solution of

$$A^T Ax = A^T b. \tag{3.24}$$

Rewrite this as $V_1 \Sigma_r^2 V_1^T x = (U_1 \Sigma_r V_1^T)^T b = V_1 \Sigma_r U_1^T b$ in view of (3.14) and (3.8). Now $x = V_1 \Sigma_r^{-1} U_1^T b$ is a solution. To see this, substitute and note that $V_1^T V_1 = I_r$. In view of the fact that $\mathscr{N}(A^T A) = \mathscr{N}(A) = \mathscr{R}(V_2) = span\{v_{r+1}, \ldots, v_n\}$, the complete solution is given by

$$x_w = V_1 \Sigma_r^{-1} U_1^T b + V_2 w \tag{3.25}$$

for some $w \in R^{m-r}$. Since $V_1 \Sigma_r^{-1} U_1^T b$ is orthogonal to $V_2 w$ for all w,

$$x_0 = V_1 \Sigma_r^{-1} U_1^T b \tag{3.26}$$

is the optimal solution that minimizes $\|b - Ax\|_2$ (prove this).

The *Moore-Penrose pseudoinverse* of $A \in R^{m \times n}$ can be shown to be

$$A^+ = V_1 \Sigma_r^{-1} U_1^T. \tag{3.27}$$

It was seen that $x = A^+b$ is the solution to the least squares problem. Furthermore, it can be shown that this pseudoinverse minimizes $\|AA^+ - I_m\|_F$, where $\|A\|_F$ denotes the *Frobenius norm* of A that is equal to the square root of $trace[AA^T] = \sum_{i=1}^m \lambda_i(AA^T) = \sum_{i=1}^m \sigma_i^2(A)$. It is of interest to note that the Moore-Penrose pseudoinverse of A is defined as the unique matrix that satisfies the conditions (i) $AA^+A = A$, (ii) $A^+AA^+ = A^+$, (iii) $(AA^+)^T = AA^+$, and (iv) $(A^+A)^T = A^+A$.

Note that if *rank* $A = m \le n$, then it can be shown that $A^+ = A^T(AA^T)^{-1}$; this is in fact the *right inverse of* A, since $A(A^T(AA^T)^{-1}) = I_m$. Similarly, if *rank* $A = n \le m$, then $A^+ = (A^TA)^{-1}A^T$, the *left inverse of* A, since $((A^TA)^{-1}A^T)A = I_n$.

Singular values and singular-value decomposition are discussed in a number of references. See for example Golub and Van Loan [6], Patel et al. [7], Petkov et al. [8], and DeCarlo [4].

A.4 SOLVING POLYNOMIAL AND RATIONAL MATRIX EQUATIONS USING INTERPOLATION METHODS

Many system and control problems can be formulated in terms of matrix equations where polynomial or rational solutions with specific properties are of interest. It is known that equations involving just polynomials can be solved by either equating coefficients of equal power of the indeterminate s, or equivalently, by using the values obtained when appropriate values for s are substituted in the given polynomials. In the latter case one uses results from the classical theory of polynomial interpolation. Similarly, one may solve polynomial matrix equations using the theory of polynomial matrix interpolation. This approach has significant advantages. Full details can be found in Antsaklis and Gao [1] (see also Gao and Antsaklis [5]).

First, some required results from the theory of polynomial and rational matrix interpolation are briefly summarized. These are then used to determine solutions of polynomial and rational matrix equations.

Polynomial matrix interpolation

Consider first the polynomial case. The following is a fundamental result of the theory of polynomial interpolation: given l distinct complex scalars s_j, $j = 1, \ldots, l$, and l corresponding complex values b_j, there exists a unique polynomial $q(s)$ of degree $l - 1$ for which

$$q(s_j) = b_j, \qquad j = 1, \ldots, l. \tag{4.1}$$

Thus, an nth-degree polynomial $q(s)$ can be uniquely represented by the $l = n + 1$ interpolation (points or doublets or) pairs (s_j, b_j), $j = 1, \ldots, l$.

The polynomial matrix interpolation theory deals with interpolation in the matrix case. In the following we cite a basic result upon which the solution to the polynomial matrix interpolation problem rests.

Let $S(s) \triangleq$ *block diag* $([1, s, \ldots, s^{d_i}]^T)$, where the d_i, $i = 1, \ldots, m$, are nonnegative integers. Let $a_j \neq 0$ and b_j denote $m \times 1$ and $p \times 1$ complex vectors, respectively, and let s_j be complex scalars.

THEOREM 4.1. Given interpolation triplets (s_j, a_j, b_j), $j = 1, \ldots, l$ (i.e., interpolation points), and nonnegative integers d_i with $l = \Sigma_{i=1}^m d_i + m$ such that the $(\Sigma_{i=1}^m d_i + m) \times l$

matrix

$$S_l \triangleq [S(s_1)a_1, \ldots, S(s_l)a_l] \tag{4.2}$$

has full rank, there exists a unique $p \times m$ polynomial matrix $Q(s)$, with ith-column degree equal to d_i, $i = 1, \ldots, m$, for which

$$Q(s_j)a_j = b_j, \qquad j = 1, \ldots, l. \tag{4.3}$$

Proof. Since the column degrees of $Q(s)$ are d_i, $Q(s)$ can be written as

$$Q(s) = QS(s), \tag{4.4}$$

where the $p \times (\sum_{i=1}^{m} d_i + m)$ matrix Q contains the coefficients of the polynomial entries. Substituting into (4.3), Q must satisfy

$$QS_l = B_l, \tag{4.5}$$

where $B_l \triangleq [b_1, \ldots, b_l]$. Since S_l is nonsingular, Q, and therefore $Q(s)$, are uniquely determined. ■

It should be noted that when $p = m = 1$ and $d_1 = l-1 = n$, the above theorem reduces to the Polynomial Interpolation Theorem. In that case, for $a_j = 1$, S_l reduces to a Vandermonde Matrix that is nonsingular if and only if s_j, $j = 1, \ldots l$, are distinct (show this).

EXAMPLE 4.1. Let $Q(s)$ be a $1 \times 2 = p \times m$ polynomial matrix and let the following $l = 3$ interpolation points $\{(s_j, a_j, b_j), j = 1, 2, 3\}$ be specified: $\{(-1, [1, 0]^T, 0), (0, [-1, 1]^T, 0), (1, [0, 1]^T, 1)\}$. In view of Theorem 4.1, $Q(s)$ is uniquely specified when d_1 and d_2 are chosen so that $l = 3 = \Sigma d_i + m = (d_1 + d_2) + 2$, or $d_1 + d_2 = 1$, assuming that S_3 has full rank. Clearly, there is more than one choice for d_1 and d_2. The resulting $Q(s)$ depends on the particular choice for the column degrees d_i:

(i) Let $d_1 = 1$ and $d_2 = 0$. Then $S(s) = \textit{block diag}\,([1, s]^T, 1)$ and (4.5) becomes:

$$QS_3 = Q[S(s_1)a_1, S(s_2)a_2, S(s_3)a_3] = Q\begin{bmatrix} 1 & -1 & 0 \\ -1 & 0 & 0 \\ 0 & 1 & 1 \end{bmatrix}$$
$$= [0, 0, 1] = B_3,$$

from which we obtain $Q = [1, 1, 1]$ and $Q(s) = QS(s) = [s + 1, 1]$.

(ii) Let $d_1 = 0$ and $d_2 = 1$. Then $S(s) = \textit{block diag}\,(1, [1, s]^T)$ and (4.5) yields $Q = [0, 0, 1]$, from which we have $Q(s) = [0, s]$, which is clearly different from (i). ■

Rational matrix interpolation

Similar to the polynomial matrix case, the problem here is to represent a $p \times m$ rational matrix $H(s)$ by interpolation triplets or points (s_j, a_j, b_j), $j = 1, \ldots, l$, which satisfies

$$H(s_j)a_j = b_j, \qquad j = 1, \ldots, l, \tag{4.6}$$

where the s_j are complex scalars and the $a_j \neq 0$ and b_j are complex $m \times 1$ and $p \times 1$ vectors, respectively.

It can be shown that the rational matrix interpolation problem reduces to a special case of polynomial matrix interpolation. To see this, we write $H(s) = \tilde{D}^{-1}(s)\tilde{N}(s)$, where $\tilde{D}(s)$ and $\tilde{N}(s)$ are $p \times p$ and $p \times m$ polynomial matrices, respectively. Then (4.6) can be written as $\tilde{N}(s_j)a_j = \tilde{D}(s_j)b_j$, or as

$$[\tilde{N}(s_j), -\tilde{D}(s_j)]\begin{bmatrix} a_j \\ b_j \end{bmatrix} = Q(s_j)c_j = 0, \qquad j = 1, \ldots, l, \tag{4.7}$$

i.e., the rational matrix interpolation problem for a $p \times m$ rational matrix $H(s)$ can be viewed as a polynomial interpolation problem for a $p \times (p+m)$ polynomial matrix $Q(s) \triangleq [\tilde{N}(s), -\tilde{D}(s)]$ with interpolation points $(s_j, c_j, 0) = (s_j, [a_j^T, b_j^T]^T, 0)$, $j = 1, \ldots, l$. There is also the additional constraint that $\tilde{D}^{-1}(s)$ exists.

Solution of matrix equations

In this segment, polynomial matrix equations of the form $M(s)L(s) = Q(s)$ are considered. The main result is Theorem 4.2, which essentially states that all solutions $M(s)$ of degree r can be derived by solving Eq. (4.16). In this way, all solutions of degree r of the polynomial equation, if they exist, are parameterized. The Diophantine Equation is an important special case and is examined at length. It is also shown that Theorem 4.2 can be applied to solve rational matrix equations of the form $M(s)L(s) = Q(s)$.

Consider the equation

$$M(s)L(s) = Q(s), \tag{4.8}$$

where $L(s)$ and $Q(s)$ are given $t \times m$ and $k \times m$ polynomial matrices, respectively. We wish to determine the $k \times t$ polynomial matrix solutions $M(s)$ when they exist.

First consider the left-hand side of Eq. (4.8). Let

$$M(s) \triangleq M_0 + \cdots + M_r s^r \tag{4.9}$$

and let $d_i \triangleq deg_{ci}\,[L(s)]$, $i = 1, \ldots, m$, denote the column degrees of $L(s)$. If

$$\hat{Q}(s) \triangleq M(s)L(s), \tag{4.10}$$

then $deg_{ci}\,[\hat{Q}(s)] = d_i + r$ for $i = 1, \ldots, m$. According to the Polynomial Matrix Interpolation Theorem, Theorem 4.1, the matrix $\hat{Q}(s)$ can be uniquely specified, using $\Sigma_{i=1}^m (d_i + r) + m = \Sigma_{i=1}^m d_i + m(r+1)$ interpolation points. Therefore, consider l interpolation points (s_j, a_j, b_j), $j = 1, \ldots, l$, where

$$l = \Sigma_{i=1}^m d_i + m(r+1). \tag{4.11}$$

Let $S_r(s) \triangleq block\ diag\,([1, s, \ldots, s^{d_i+r}]^T)$ and assume that the $(\Sigma_{i=1}^m d_i + m(r+1)) \times l$ matrix

$$S_{rl} \triangleq [S_r(s_1)a_1, \ldots, S_r(s_l)a_l] \tag{4.12}$$

has full rank, i.e., the assumptions in Theorem 4.1 are satisfied. Note that for distinct s_j, S_{rl} will have full column rank for almost any set of nonzero a_j. Now in view of Theorem 4.1, the matrix $\hat{Q}(s)$ that satisfies

$$\hat{Q}(s_j)a_j = b_j, \qquad j = 1, \ldots, l, \tag{4.13}$$

is uniquely specified, given these l interpolation points (s_j, a_j, b_j). To solve (4.8), these interpolation points must be appropriately chosen so that the equation $\hat{Q}(s)\ (= M(s)L(s)) = Q(s)$ is satisfied.

We write (4.8) as

$$ML_r(s) = Q(s), \tag{4.14}$$

where $M \triangleq [M_0, \ldots, M_r]$ and $L_r(s) \triangleq [L^T(s), \ldots, s^r L^T(s)]^T$ are $k \times t(r+1)$ and $t(r+1) \times m$ matrices, respectively. Let $s = s_j$ and postmultiply by a_j, $j = 1, \ldots, l$.

Note that s_j and a_j, $j = 1, \ldots, l$, must be such that S_{rl} has full rank. Define

$$b_j \triangleq Q(s_j)a_j, \qquad j = 1, \ldots, l, \tag{4.15}$$

and combine the above equations to obtain

$$ML_{rl} = B_l, \tag{4.16}$$

where $L_{rl} \triangleq [L_r(s_1)a_1, \ldots, L_r(s_l)a_l]$ and $B_l \triangleq [b_1, \ldots, b_l]$ are $t(r+1) \times l$ and $k \times l$ matrices, respectively.

THEOREM 4.2. Given $L(s)$ and $Q(s)$ in (4.8), let $d_i \triangleq deg_{ci}[L(s)]$, $i = 1, \ldots, m$, and select r to satisfy

$$deg_{ci}[Q(s)] \leq d_i + r, \qquad i = 1, \ldots, m. \tag{4.17}$$

Then a solution $M(s) = M[I, sI, \ldots, s^r I]^T$ of degree r exists if and only if a solution M of (4.16) exists. ■

It is not difficult to show that solving (4.16) is equivalent to solving

$$M(s_j)c_j = b_j, \qquad j = 1, \ldots, l, \tag{4.18}$$

where

$$c_j \triangleq L(s_j)a_j, \qquad b_j \triangleq Q(s_j)a_j, \qquad j = 1, \ldots, l. \tag{4.19}$$

The $M(s)$ that satisfy (4.18) are obtained by solving

$$MS_{rl} = B_l, \tag{4.20}$$

where $S_{rl} \triangleq [S_r(s_1)c_1, \ldots, S_r(s_l)c_l]$ has dimensions $t(r+1) \times l$, and $S_r(s) \triangleq [I, sI, \ldots, s^r I]^T$ has dimensions $t(r+1) \times t$, and $B_l \triangleq [b_1, \ldots, b_l]$ has dimensions $k \times l$. Solving (4.20) is an alternative to solving (4.16).

Constraints on solutions. When there are more unknowns than equations [$t(r+1)$ and $l = \Sigma_{i=1}^m d_i + m(r+1)$, respectively] in (4.16), this freedom can be exploited so that $M(s)$ satisfies additional constraints. In particular, $k \triangleq t(r+1) - l$ additional linear constraints, expressed in terms of the coefficients of $M(s)$ (in M), can in general be satisfied. The equations describing the constraints can be used to augment Eqs. (4.16). In this case the equations to be solved become

$$M[L_{rl}, C] = [B_l, D], \tag{4.21}$$

where $MC = D$ represents the k linear constraints imposed on the coefficients M, and C and D are matrices (real or complex) with k columns each. ■

The Diophantine Equation

An important case of (4.8) is the *Diophantine Equation*

$$X(s)D(s) + Y(s)N(s) = Q(s), \tag{4.22}$$

where the polynomial matrices $D(s)$, $N(s)$, and $Q(s)$ are given and $X(s)$, $Y(s)$ are to be determined. Note that if

$$M(s) = [X(s), Y(s)], \qquad L(s) = \begin{bmatrix} D(s) \\ N(s) \end{bmatrix}, \tag{4.23}$$

then it is immediately clear that the Diophantine Equation is a polynomial equation of the form (4.8) and all previous results apply. Theorem 4.2 guarantees that all solutions of (4.22) of degree r are determined by solving (4.16). In systems and control theory, the Diophantine Equation that is used involves a matrix $L(s) = [D^T(s), N^T(s)]^T$, which has rather specific properties. These are exploited to solve the Diophantine Equation and to derive conditions for existence of solutions of (4.22) of degree r. It can be shown that the following result is true.

THEOREM 4.3. Let r satisfy

$$deg_{ci}[Q(s)] \le d_i + r, \qquad i = 1, \ldots, m, \text{ and } r \ge \nu - 1, \tag{4.24}$$

where ν is the observability index of the system $\{D, I, N, 0\}$. Then the Diophantine Equation (4.22) has solutions of degree r that can be determined by solving (4.16) [or (4.20)]. ■

EXAMPLE 4.2. Let

$$D(s) = \begin{bmatrix} s-2 & 0 \\ 0 & s+1 \end{bmatrix}, \qquad N(s) = \begin{bmatrix} s-1 & 0 \\ 1 & 1 \end{bmatrix}, \qquad \text{and} \qquad Q(s) = \begin{bmatrix} 1 & 0 \\ 0 & 1 \end{bmatrix}.$$

We have $d_1 = d_2 = 1$, $deg_{ci}\, Q(s) = 0$, $i = 1, 2$, and $l = 2 + 2(r+1)$.
For $r = 1$, $s_j = -2, -1, 0, 1, 2, 3$, and

$$a_j = \begin{bmatrix} 0 \\ 1 \end{bmatrix}, \begin{bmatrix} 1 \\ 3 \end{bmatrix}, \begin{bmatrix} 0 \\ -1 \end{bmatrix}, \begin{bmatrix} -1 \\ 3 \end{bmatrix}, \begin{bmatrix} -1 \\ 1 \end{bmatrix}, \begin{bmatrix} 1 \\ -1 \end{bmatrix}.$$

A solution is given by

$$M(s) = [X(s), Y(s)] = \begin{bmatrix} s & -1 & -s & s+1 \\ \frac{1}{3} & \frac{1}{3} & 0 & -\frac{1}{3}s + \frac{2}{3} \end{bmatrix}.$$ ■

Solving rational matrix equations

Now let us consider the rational matrix equation

$$M(s)L(s) = Q(s), \tag{4.25}$$

where $L(s)$ and $Q(s)$ are given $t \times m$ and $k \times m$ rational matrices, respectively. The polynomial matrix interpolation theory developed above can be used to solve this equation and determine the rational matrix solutions $M(s)$ of dimension $k \times t$. Let $M(s) = \tilde{D}^{-1}(s)\tilde{N}(s)$ be a polynomial fraction form of $M(s)$ that is to be determined. Then (4.25) can be written as

$$[\tilde{N}(s), -\tilde{D}(s)] \begin{bmatrix} L(s) \\ Q(s) \end{bmatrix} = 0. \tag{4.26}$$

Note that one could equivalently solve

$$[\tilde{N}(s), -\tilde{D}(s)] \begin{bmatrix} L_p(s) \\ Q_p(s) \end{bmatrix} = 0, \tag{4.27}$$

where $[L_p(s)^T, Q_p(s)^T]^T = [L(s)^T, Q(s)^T]^T \phi(s)$ is a polynomial matrix with $\phi(s)$ the least common denominator of all entries of $L(s)$ and $Q(s)$. In general, $\phi(s)$ may be any matrix denominator in a right fractional representation of $[L(s)^T, Q(s)^T]^T$. The problem to be solved is now of the form (4.8), a polynomial matrix equation, where $L(s) = [L_p(s)^T, Q_p(s)^T]^T$ and $Q(s) = 0$. Therefore, all solutions $[\tilde{N}(s), -\tilde{D}(s)]$ of degree r can be determined by solving (4.16) or (4.20). Let $s = s_j$ and postmultiply (4.27) by a_j, $j = 1, \ldots, l$, with a_j and l chosen properly. Define

$$c_j \triangleq \begin{bmatrix} L_p(s) \\ Q_p(s) \end{bmatrix} a_j, \qquad j = 1, \ldots, l. \tag{4.28}$$

The problem now is to determine a polynomial matrix $[\tilde{N}(s), -\tilde{D}(s)]$ that satisfies

$$[\tilde{N}(s_j), -\tilde{D}(s_j)]c_j = 0, \qquad j = 1, \ldots, l. \tag{4.29}$$

Note that restrictions on the solutions can easily be imposed to guarantee that $\tilde{D}^{-1}(s)$ exists and/or that $M(s) = \tilde{D}^{-1}(s)\tilde{N}(s)$ is proper. Additional constraints can be added so that the solution satisfies additional specifications [see (4.21)].

Pole placement

Output feedback. All proper output controllers of degree r (of order mr) that assign all the closed-loop eigenvalues to arbitrary locations are characterized in a convenient way using interpolation results.

Given $N(s)D^{-1}(s) = H(s)$, which is assumed to be proper, we are interested in solutions $[X(s), Y(s)]$ of dimensions $m \times (p+m)$ of the Diophantine Equation where only the roots of $|Q(s)|$ are specified. Furthermore, $X^{-1}(s)Y(s) = C(s)$ should exist and be proper since it represents the controller. Here the equation to be solved is

$$(X(s_j)D(s_j) + Y(s_j)N(s_j))a_j = 0, \qquad j = 1, \ldots, l, \tag{4.30}$$

or $ML_{rl} = 0\,(l = \Sigma_{i=1}^{m} d_i + mr)$. Thus, the $\Sigma_{i=1}^{m} d_i + mr$ roots of $|X(s)D(s) + Y(s)N(s)|$ are to be assigned the values s_j, $j = 1, \ldots l$. Note the difference between the problem studied earlier, where $Q(s)$ is known, and the problem studied here, where only the roots of $|Q(s)|$ (or of $|Q(s)|$ within multiplication by some nonzero real scalar) are given. In the present case, the vectors a_j can be viewed as design parameters and can be selected almost arbitrarily to satisfy requirements in addition to pole assignment. Note that this design approach is rather well known in the state feedback case, as is discussed later in this section (see also Chapter 4). The following result can be shown.

THEOREM 4.4. Let $r \geq \nu - 1$. Then $(X(s), Y(s))$ exists such that all the $n + mr$ zeros of $|X(s)D(s) + Y(s)N(s)|$ are arbitrarily assigned and $X^{-1}(s)Y(s)$ is proper. ■

EXAMPLE 4.3. Let $D(s) = \begin{bmatrix} s-2 & 0 \\ 0 & s+1 \end{bmatrix}$ and $N(s) = \begin{bmatrix} s-1 & 0 \\ 1 & 1 \end{bmatrix}$ with $n = deg\,|D(s)| = 2$. In this case there are $deg\,|X(s)D(s) + Y(s)N(s)| = n + mr = 2 + 2r$ closed-loop poles to be assigned. Note that $r \geq \nu - 1 = 1 - 1 = 0$.

(i) For $r = 0$ and $\{(s_j, a_j), j = 1, 2\} = \{(-1, [1, 0]^T), (-2, [0, 1]^T)\}$, a solution of $ML_{rl} = 0$ is

$$M = \begin{bmatrix} 2 & 0 & -3 & 0 \\ 0 & 2 & 1 & 2 \end{bmatrix}.$$

For this case, $M = M(s) = [X(s), Y(s)]$ and $C(s) = X^{-1}(s)Y(s) = \frac{1}{2}\begin{bmatrix} -3 & 0 \\ 1 & 2 \end{bmatrix}$, which is a static output controller.

(ii) For $r = 1$, and $\{(s_j, a_j), j = 1, \ldots, 4\} = \{(-1, [1, 0]^T), (-2, [0, 1]^T, (-3, [-1, 0]^T), (-4, [0, -1]^T)\}$, a solution of $ML_{rl} = 0$ is given by $[X(s), Y(s)] = \begin{bmatrix} s-7 & -1 & 12 & s+1 \\ 5 & s+4 & -6 & s+4 \end{bmatrix}$. Note that $C(s) = X(s)^{-1}Y(s)$ exists and is proper. ■

State feedback. Let A, B, and F be $n \times n$, $n \times m$, and $m \times n$ real matrices, respectively. Note that $|sI - (A + BF)| = |sI - A| \cdot |I_n - (sI - A)^{-1}BF| =$

$|sI - A| \cdot |I_m - F(sI - A)^{-1}B|$. Now if the desired closed-loop eigenvalues s_j are different from the eigenvalues of A, then F will assign all n desired closed-loop eigenvalues s_j if and only if

$$F[(s_jI - A)^{-1}Ba_j] = a_j, \qquad j = 1, \ldots, n. \tag{4.31}$$

The $m \times 1$ vectors a_j are selected so that $(s_jI - A)^{-1}Ba_j$, $j = 1, \ldots, n$, are linearly independent vectors. Alternatively, one could approach the problem as follows (see also Subsection 4.2B of Chapter 4). Let $M(s)$ and $D(s)$ be rc polynomial matrices of dimensions $n \times m$ and $m \times m$, respectively, such that $(sI - A)^{-1}B = M(s)D^{-1}(s)$. An internal representation equivalent to $\dot{x} = Ax + Bu$ in polynomial matrix form is $Dz = u$ with $x = Mz$ (see Subsection 7.3A of Chapter 7). The eigenvalue assignment problem now is to assign all the roots of $|D(s) - FM(s)|$, or to determine F so that

$$FM(s_j)a_j = D(s_j)a_j, \qquad j = 1, \ldots, n. \tag{4.32}$$

Note that this formulation does not require that s_j be different from the eigenvalues of A as in (4.31). The $m \times 1$ vectors a_j are selected so that $M(s_j)a_j$, $j = 1, \ldots, n$, are independent. Note that $M(s_j)$ has the same column rank as $S(s_j) = \textit{block diag}\,([1, s_j, \ldots, s_j^{d_i - 1}]^T)$, where d_i are the controllability indices of (A, B). Therefore, it is possible to select a_j so that $M(s_j)a_j$, $j = 1, \ldots, n$, are independent, even when the s_j are repeated. In general there is great flexibility in selecting the nonzero vectors a_j. For example when the s_j are distinct, which is a very common case, the a_j can be selected almost arbitrarily. For all the appropriate choices of a_j [$M(s_j)a_j$, $j = 1, \ldots, n$, linearly independent], the n eigenvalues of the closed-loop system will be at the desired locations s_j, $j = 1, \ldots, n$. Different a_j correspond to different F, which results in general in different closed-loop behavior.

The exact relation of the eigenvectors to the a_j can be determined by $[s_jI - (A+BF)]M(s_j)a_j = (s_jI - A)M(s_j)a_j - BFM(s_j)a_j = BD(s_j)a_j - BD(s_j)a_j = 0$. Therefore, $M(s_j)a_j = v_j$ are the closed-loop eigenvectors corresponding to s_j.

One may select a_j in (4.32) to impose constraints on the gains f_{ij} in F. For example, one may select a_j so that a column of F is zero (take the corresponding row of all a_j to be nonzero), or so that an element of F is zero. Alternatively, one may select a_j so that additional design goals are attained.

Note that this approach for eigenvalue/eigenvector assignment by state feedback has also been discussed in Subsection 4.2B of Chapter 4. For further details, see Antsaklis and Gao [1].

A.5 REFERENCES

1. P. J. Antsaklis and Z. Gao, "Polynomial and Rational Matrix Interpolation: Theory and Control Applications," *Int. J. of Control*, Vol. 58, No. 2, pp. 349–404, August 1993.
2. M. S. Bazaraa, H. D. Sherali, and C. M. Shetty, *Nonlinear Programming Theory and Algorithms*, 2d edition, Wiley, New York, 1993.
3. D. P. Bertsekas and J. N. Tsitsiklis, *Parallel and Distributed Computation—Numerical Methods,* Prentice-Hall, Englewood Cliffs, NJ, 1989.
4. R. A. DeCarlo, *Linear Systems. A State Variable Approach with Numerical Implementation,* Prentice-Hall, Englewood Cliffs, NJ, 1989.

5. Z. Gao and P. J. Antsaklis, "New Methods for Control System Design Using Matrix Interpolation," *Proc. of the IEEE Conference on Decision and Control,* Orlando, FL, December 1994.
6. G. H. Golub and C. F. Van Loan, *Matrix Computations,* The Johns Hopkins University Press, Baltimore, 1983.
7. R. V. Patel, A. J. Laub, and P. M. VanDooren, eds., *Numerical Linear Algebra Techniques for Systems and Control,* IEEE Press, Piscataway, NJ, 1993.
8. P. Hr. Petkov, N. D. Christov, and M. M. Konstantinov, *Computational Methods for Linear Control Systems,* Prentice-Hall International Series, 1991.
9. G. W. Stewart, *Introduction to Matrix Computations,* Academic Press, New York, 1973.

INDEX